P9-EED-156

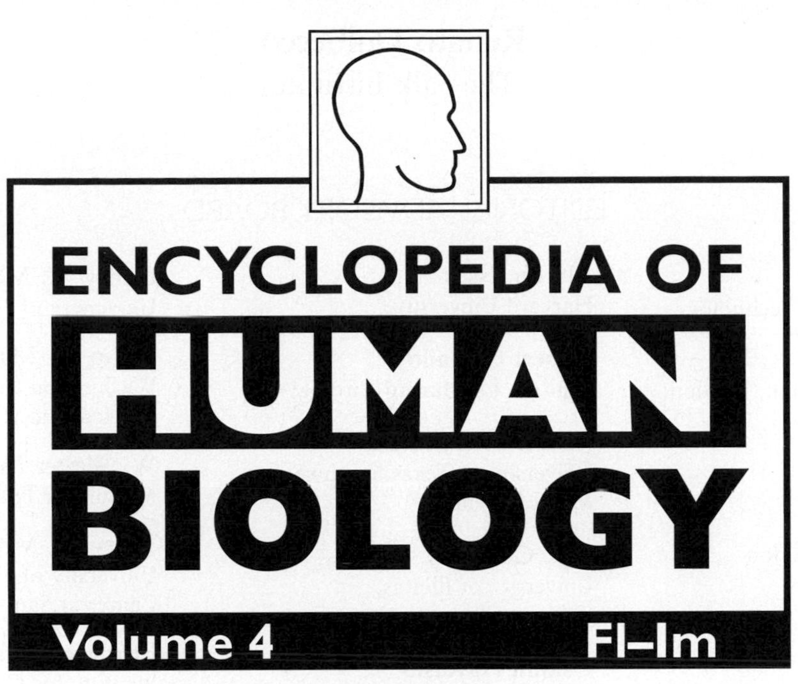

Second Edition

EDITOR-IN-CHIEF

Renato Dulbecco
The Salk Institute

R

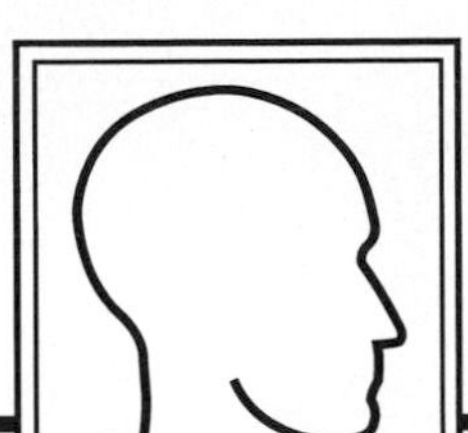

ENCYCLOPEDIA OF HUMAN BIOLOGY

Volume 4 Fl–Im

Second Edition

Editor-in-Chief

RENATO DULBECCO

The Salk Institute

La Jolla, California

ACADEMIC PRESS

San Diego London Boston New York Sydney Tokyo Toronto

This book is printed on acid-free paper. ∞

Academic Press
a division of Harcourt Brace & Company
525 B Street, Suite 1900, San Diego, California 92101-4495, USA
http://www.apnet.com

Academic Press Limited
24-28 Oval Road, London NW1 7DX, UK
http://www.hbuk.co.uk/ap/

Library of Congress Cataloging-in-Publication Data

Encyclopedia of human biology / edited by Renato Dulbecco. -- 2nd ed.
p. cm.
Includes bibliographical references and index.
ISBN 0-12-226970-5 (alk. paper: set). -- ISBN 0-12-226971-3 (alk. paper: v. 1). -- ISBN 0-12-226972-1 (alk. paper: v. 2). -- ISBN 0-12-226973-X (alk. paper: v. 3). -- ISBN 0-12-226974-8 (alk. paper: v. 4). -- ISBN 0-12-226975-6 (alk. paper: v. 5). -- ISBN 0-12-226976-4 (alk. paper: v. 6). -- ISBN 0-12-226977-2 (alk. paper : v. 7). -- ISBN 0-12-226978-0 (alk. paper: v. 8). -- ISBN 0-12-226979-9 (alk. paper: v. 9)
1. Human biology--Encyclopedias. I. Dulbecco, Renato, date.
[DNLM: 1. Biology--encyclopedias. 2. Physiology--encyclopedias.
QH 302.5 E56 1997]
QP11.E53 1997
612'.003-dc21
DNLM/DLC
for Library of Congress 97-8627
CIP

PRINTED IN THE UNITED STATES OF AMERICA
97 98 99 00 01 02 EB 9 8 7 6 5 4 3 2 1

CONTENTS OF VOLUME 4

Contents for each volume of the Encyclopedia appears in Volume 9.

H

PREFACE TO THE FIRST EDITION

We are in the midst of a period of tremendous progress in the field of human biology. New information appears daily at such an astounding rate that it is clearly impossible for any one person to absorb all this material. The *Encyclopedia of Human Biology* was conceived as a solution: an informative yet easy-to-use reference. The Encyclopedia strives to present a complete overview of the current state of knowledge of contemporary human biology, organized to serve as a solid base on which subsequent information can be readily integrated. The Encyclopedia is intended for a wide audience, from the general reader with a background in science to undergraduates, graduate students, practicing researchers, and scientists.

Why human biology? The study of biology began as a correlate of medicine with the human, therefore, as the object. During the Renaissance, the usefulness of studying the properties of simpler organisms began to be recognized and, in time, developed into the biology we know today, which is fundamentally experimental and mainly involves nonhuman subjects. In recent years, however, the identification of the human as an autonomous biological entity has emerged again—stronger than ever. Even in areas where humans and other animals share a certain number of characteristics, a large component is recognized only in humans. Such components include, for example, the complexity of the brain and its role in behavior or its pathology. Of course, even in these studies, humans and other animals share a certain number of characteristics. The biological properties shared with other species are reflected in the Encyclopedia in sections of articles where results obtained in nonhuman species are evaluated. Such experimentation with nonhuman organisms affords evidence that is much more difficult or impossible to obtain in humans but is clearly applicable to us.

Guidance in fields with which the reader has limited familiarity is supplied by the detailed index volume. The articles are written so as to make the material accessible to the uninitiated; special terminology either is avoided or, when used, is clearly explained in a glossary at the beginning of each article. Only a general knowledge of biology is expected of the reader; if specific information is needed, it is reviewed in the same section in simple terms. The amount of detail is kept within limits sufficient to convey background information. In many cases, the more sophisticated reader will want additional information; this will be found in the bibliography at the end of each article. To enhance the long-term validity of the material, untested issues have been avoided or are indicated as controversial.

The material presented in the Encyclopedia was produced by well-recognized specialists of experience and competence and chosen by a roster of outstanding scientists including ten Nobel laureates. The material was then carefully reviewed by outside experts. I have reviewed all the articles and evaluated their contents in my areas of competence, but my major effort has been to ensure uniformity in matters of presentation, organization of material, amount of detail, and degree of documentation, with the goal of presenting in each subject the most advanced information available in easily accessible form.

Renato Dulbecco

PREFACE TO THE SECOND EDITION

The first edition of the *Encyclopedia of Human Biology* has been very successful. It was well received and highly appreciated by those who used it. So one may ask: Why publish a second edition? In fact, the word "encyclopedia" conveys the meaning of an opus that contains immutable information, forever valid. But this depends on the subject. Information about historical subjects and about certain branches of science is essentially immutable. However, in a field such as human biology, great changes occur all the time. This is a field that progresses rapidly; what seemed to be true yesterday may not be true today. The new discoveries constantly being made open new horizons and have practical consequences that were not even considered previously. This change applies to all fields of human biology, from genetics to structural biology and from the intricate mechanisms that control the activation of genes to the biochemical and medical consequences of these processes.

These are the reasons for publishing a second edition. Although much of the first edition is still valid, it lacks the information gained in the six years since its preparation. This new edition updates the information to what we know today, so the reader can be confident of its full validity. All articles have been reread by their authors, who modified them when necessary to bring them up-to-date. Many new articles have also been added to include new information.

The principles followed in preparing the first edition also apply to the second edition. All new articles were contributed by specialists well known in their respective fields. Expositional clarity has been maintained without affecting the completeness of the information. I am convinced that anyone who needs the information presented in this encyclopedia will find it easily, will find it accessible, and, at the same time, will find it complete.

Renato Dulbecco

A GUIDE TO USING THE ENCYCLOPEDIA

The *Encyclopedia of Human Biology, Second Edition* is a complete source of information on the human organism, contained within the covers of a single unified work. It consists of nine volumes and includes 670 separate articles ranging from genetics and cell biology to public health, pediatrics, and gerontology. Each article provides a comprehensive overview of the selected topic to inform a broad spectrum of readers from research professionals to students to the interested general public.

In order that you, the reader, derive maximum benefit from your use of the Encyclopedia, we have provided this Guide. It explains how the Encyclopedia is organized and how the information within it can be located.

ORGANIZATION

The *Encyclopedia of Human Biology, Second Edition* is organized to provide the maximum ease of use for its readers. All of the articles are arranged in a single alphabetical sequence by title. Articles whose titles begin with the letters A to Bi are in volume 1, articles with titles from Bl to Com are in Volume 2, and so on through Volume 8, which contains the articles from Si to Z.

Volume 9 is a separate reference volume providing a Subject Index for the entire work. It also includes a complete Table of Contents for all nine volumes, an alphabetical list of contributors to the Encyclopedia, and an Index of Related Titles. Thus Volume 9 is the best starting point for a search for information on a given topic, via either the Subject Index or Table of Contents.

So that they can be easily located, article titles generally begin with the key word or phrase indicating the topic, with any descriptive terms following. For example, "Calcium, Biochemistry" is the article title rather than "Biochemistry of Calcium" because the specific term *calcium* is the key word rather than the more general term *biochemistry*. Similarly "Protein Targeting, Basic Concepts" is the article title rather than "Basic Concepts of Protein Targeting."

TABLE OF CONTENTS

A complete Table of Contents for the *Encyclopedia of Human Biology, Second Edition* appears in Volume 9. This list of article titles represents topics that have been carefully selected by the Editor-in-Chief, Dr. Renato Dulbecco, and the members of the Editorial Advisory Board (see p. ii for a list of the Board members). The Encyclopedia provides coverage of 35 specific subject areas within the overall field of human biology, ranging alphabetically from Behavior to Virology.

In addition to the complete Table of Contents found in Volume 9, the Encyclopedia also provides an individual table of contents at the front of each volume. This lists the articles included within that particular volume.

INDEX

The Subject Index in Volume 9 contains more than 4200 entries. The subjects are listed alphabetically and indicate the volume and page number where information on this topic can be found.

ARTICLE FORMAT

Articles in the *Encyclopedia of Human Biology, Second Edition* are arranged in a single alphabetical list by title. Each new article begins at the top of a right-hand page, so that it may be quickly located. The author's name and affiliation are displayed at the beginning of the article. The article is organized according to a standard format, as follows:

- Title and author
- Outline
- Glossary
- Defining statement
- Body of the article
- Bibliography

OUTLINE

Each article in the Encyclopedia begins with an outline that indicates the general content of the article. This outline serves two functions. First, it provides a brief preview of the article, so that the reader can get a sense of what is contained there without having to leaf through the pages. Second, it serves to highlight important subtopics that will be discussed within the article. For example, the article "Gene Mapping" includes the subtopic "DNA Sequence and the Human Genome Project."

The outline is intended as an overview and thus it lists only the major headings of the article. In addition, extensive second-level and third-level headings will be found within the article.

GLOSSARY

The Glossary contains terms that are important to an understanding of the article and that may be unfamiliar to the reader. Each term is defined in the context of the particular article in which it is used. Thus the same term may appear as a Glossary entry in two or more articles, with the details of the definition varying slightly from one article to another. The Encyclopedia includes approximately 5000 glossary entries.

DEFINING STATEMENT

The text of each article in the Encyclopedia begins with a single introductory paragraph that defines the topic under discussion and summarizes the content of the article. For example, the article "Free Radicals and Disease" begins with the following statement:

A FREE RADICAL is any species that has one or more unpaired electrons. The most important free radicals in a biological system are oxygen- and nitrogen-derived radicals. Free radicals are generally produced in cells by electron transfer reactions. The major sources of free radical production are inflammation, ischemia/reperfusion, and mitochondrial injury. These three sources constitute the basic components of a wide variety of diseases. . . .

CROSS-REFERENCES

Many of the articles in the Encyclopedia have cross-references to other articles. These cross-references appear within the text of the article, at the end of a paragraph containing relevant material. The cross-references indicate related articles that can be consulted for further information on the same topic, or for other information on a related topic. For example, the article "Brain Evolution" contains a cross reference to the article "Cerebral Specialization."

BIBLIOGRAPHY

The Bibliography appears as the last element in an article. It lists recent secondary sources to aid the reader in locating more detailed or technical information. Review articles and research papers that are important to an understanding of the topic are also listed.

The bibliographies in this Encyclopedia are for the benefit of the reader, to provide references for further reading or research on the given topic. Thus they typically consist of no more than ten to twelve entries. They are not intended to represent a complete listing of all materials consulted by the author or authors in preparing the article.

COMPANION WORKS

The *Encyclopedia of Human Biology, Second Edition* is one of an extensive series of multivolume reference works in the life sciences published by Academic Press. Other such works include the *Encyclopedia of Cancer, Encyclopedia of Virology, Encyclopedia of Immunology,* and *Encyclopedia of Microbiology,* as well as the forthcoming *Encyclopedia of Reproduction.*

Flow Cytometry

OLE DIDRIK LAERUM
University of Bergen

GLOSSARY

Analytical cytology Areas of cytology that deal with quantitative and qualitative analyses of cells

Aneuploidy Cellular DNA content different from a DNA index of 1 or 2

BrdU Bromodeoxyuridine; a synthetic analogue of a nucleotide base in DNA

DNA index Cellular DNA content as a quotient of DNA in diploid cells from the same species

FITC Fluoroisothiocyanate; a fluorescent dye that emits green/yellow light after excitation with blue light

Fluorescence Emission of light caused by a light source of different, usually shorter, wavelength

Light scatter Spread of light caused by a cell as it passes a narrow laser beam

FLOW CYTOMETRY IS A SERIES OF METHODS THAT enables automated quantitative and qualitative measurements on single cells in suspension. It is considered one of the major methodological breakthroughs in the field of analytical cytology during recent decades. Its methods have gained an important position in biological and medical research.

I. PRINCIPLES

Flow cytometry is a series of methods wherein single cells in suspension pass through a light beam with high velocity by which automated quantitative and qualitative measurements can be performed. They include both the deviation of light as a measure of size and optical properties of the cells (light scattering) and the absorption of light and emission of fluorescence from cell components that have been stained with a fluorescent dye. Multiple cell parameters can be measured simultaneously, and with computer technology these parameters can be analyzed in relation to each other. The flow of cell suspension can be broken up into droplets that are electrically charged, and cells containing specific signals can be sorted out by passing a static electrical field. There are two main principles for flow cytometry. One is where cells pass a narrow laser beam; the other is by use of microscope optics where cells pass a light field provided by the microscope lamp, and fluorescence is measured through the optical system. These are shown in Fig. 1.

II. DIFFERENT TYPES OF INSTRUMENTS IN CURRENT USE

Since the late 1960s, many laboratories have built their own machines based on the aforementioned principles. In addition, several commercial machines became available in the early 1970s. These were built both as analyzers and as cell sorters using electrostatic deflection of charged droplets as a cell-sorting mechanism. Today, several laser-based analyzers are available. Some are highly automated, in which cells, stained by simple methods, are automatically measured by use of multiple parameters. Most of the laser-based machines can also have cell-sorting devices. Their rate of measuring or sorting is usually between 1000 and 5000 single cells per second; however, for discriminating and sorting rare subpopulations, it takes a relatively long time to obtain a sufficient quan-

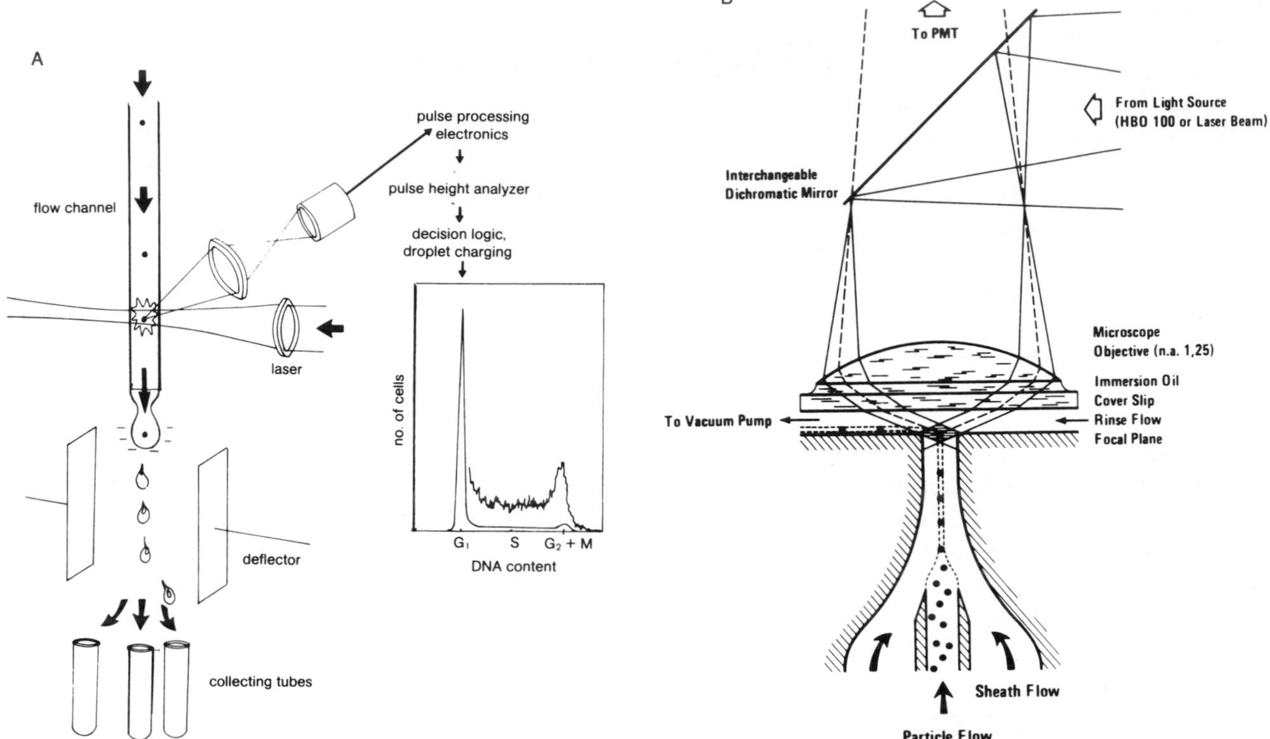

FIGURE 1 The two main principles of flow cytometry: (A) using a laser, where the measured cells can also be sorted in charged droplets passing a static electrical field, and (B) using microscope optics, where excitation and emitted light go through the same optical system as described by Göhde. [From M. R. Melamed *et al.* (1979). "Flow Cytometry and Cell Sorting." John Wiley & Sons, New York.]

tity of the desired cells. This can be overcome by performing flow analysis and sorting under high pressure, by which up to 20,000 single cells can be analyzed and sorted per second.

Several microscope-based systems are also on the market, either as closed-flow systems or as open-air systems where the cells pass over a coverslip in an illuminated field.

III. CELL PARAMETERS THAT CAN BE MEASURED

Although light scatter is not a reliable marker for cell size, it is useful for discriminating cells based on the combination of size and optical properties. If a small slit is placed between the recording light and the flow, the narrow light beam will scan each cell as it passes by. With a photomultiplier, the scattered light can be recorded and give a measure of the cell shape. The same can be done if mitotic cells are lysed so that single-chromosome suspensions are obtained. Using this procedure, DNA content and shape and also the centromeric index of the chromosomes can be measured.

Light absorption was used earlier to some extent but has entirely been replaced by fluorescence measurements because fluorescence is a far more sensitive marker than light absorption.

IV. COMMON PREPARATION AND STAINING TECHNIQUES

Staining can be performed on fresh cells, on cells fixed in suspension by ethanol or formaldehyde, and by enzymatic resuspension of cells that have been fixed and embedded in paraffin.

To measure different cell components by fluorochrome staining, the dye must be specifically bound to certain cell components. If it also binds to other components, these can be removed beforehand by enzyme treatment. An example is the staining of DNA with ethidium bromide. This dye will also stain RNA and can be unspecifically bound to cell protein. Cells in suspension are therefore first treated with pepsin to remove the cytoplasm, and thereafter with RNAse to remove single-stranded RNA. This provides a specific staining of the remaining DNA.

Several fluorochromes can be used at the same time, provided that they bind specifically to different cell components and fluoresce at different wavelengths. One example is the staining of DNA with propidium iodide, which emits red fluorescence, and protein with fluoroisothiocyanate (FITC), which emits green/yellow fluorescence.

To obtain staining of the cells, their membranes have to be opened to give access to nonpenetrable dyes, either by fixation in suspension (e.g., by use of ethanol) or they can be lysed by use of detergents, leaving naked nuclei accessible for staining. The fact that some stains do not enter through an intact cell membrane can be used for viability testing, by which fluorescent cells are detected and scored as nonviable. Staining and measurement can also be performed on cell samples that have been frozen and stored at low temperatures for prolonged periods. For some studies, fixed cells are not a necessary requirement.

For different fluorochromes that possess the same excitation wavelength spectrum, one single laser can be used. If they have different excitation spectra, lasers with different wavelengths can be mounted serially, and the fluoresence emitted from the same cell can be measured with a slight time difference. For systems using a mercury lamp, the desired wavelengths can be picked up by using specific optical filters.

A new area was opened when monoclonal antibodies conjugated to a fluorochrome were used for quantitative measurements of different antigens on single cells by flow cytometry. With this method, direct as well as indirect immunofluorescence are used, provided that the emitted light is sufficient for discriminating the antigens from the background of light and electronic noise. If specific antibodies are available, the types of cell components that can be measured quantitatively are almost unlimited. [*See* Monoclonal Antibody Technology.]

If cells take up a chromogenic substrate for a chemical or enzymatic reaction, the rate of this reaction can be measured quantitatively by use of the fluorescent product. In some cases, autofluorescence of naturally occurring substances can also be measured, although this is not as common as the other techniques.

V. APPLICATIONS IN BIOLOGY

Because there is almost no limitation of measurements that can be performed on single cells, including various cell components as well as the kinetics of different reactions, only main areas of applications will be mentioned.

A. Cell Kinetics

By specific staining of DNA and measurement of a cell population, its distribution around the cell cycle can be evaluated within a few minutes. Cells of the G_1 phase of the cell cycle show a fluorescence corresponding to two copies of each chromosome (DNA index = 1); cells in the DNA synthesis phase have an increasing amount of cellular DNA, whereas cells in the G_2 or mitotic phases double their DNA content (Figs. 2 and 3). Depending on the resolution of the technique, the fluorescence of the different phases shows some overlap, but with computer programs using parametric or nonparametric statistical methods, the percentage of cells in the different cell-cycle phases can be quantitated. Unless special preparation methods are used, G_1 and G_0 cells cannot be discriminated from each other, because they have the same DNA index = 1, neither can G_2 and mitotic cells. [*See* DNA Synthesis.]

If cells synthesizing DNA are offered the nucleoside analogue bromodeoxyuridine (BrdU), this will be incorporated into DNA instead of thymidine. Using fluorescent staining of the chromosomes or using FITC-labeled antibodies to BrdU, single-cell fluorescence will be dependent on how much BrdU it has taken up, that is, its position on the cell cycle can be identified. Whereas an ordinary cellular DNA distribution curve gives only a static picture of the cell cycle, the BrdU technique gives the actual rate of cell

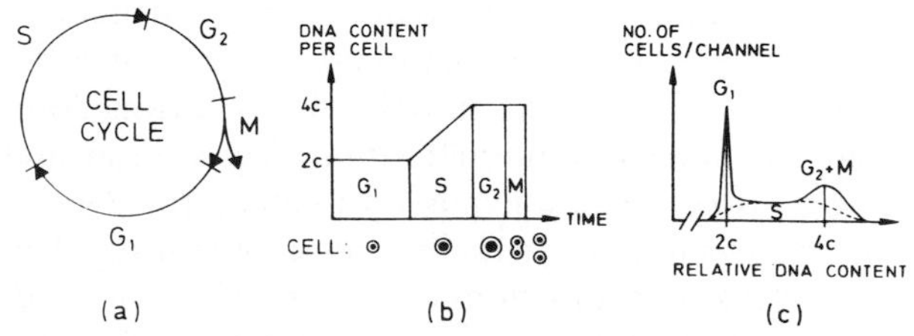

FIGURE 2 Schematic diagrams of (a) cell cycle, (b) cellular DNA content related to the cell cycle, and (c) the resulting cellular DNA distribution curve, where the areas of the different cell-cycle phases are indicated.

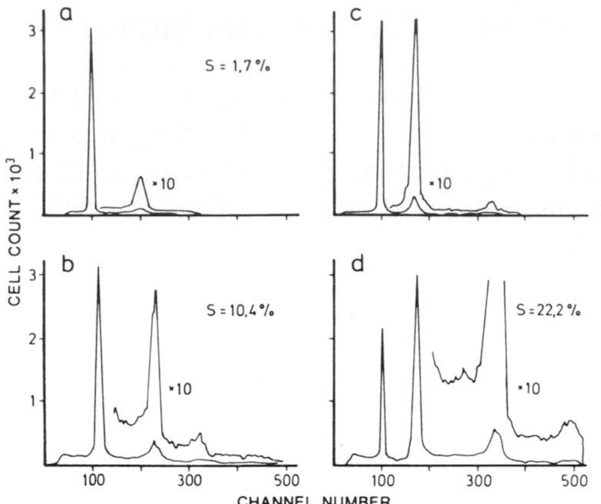

FIGURE 3 Example of DNA distribution curves in normal and malignant human cells obtained by aspiration from the uterus: (a) normal endometrium at the sixteenth day of the cycle, (b) a diploid endometrial carcinoma, (c) ascitic fluid from a patient with an aneuploid tumor, and (d) the primary aneuploid endometrial carcinoma (DNA index = 1.7). The proportions (%) of cells with S-phase DNA content are indicated for each curve. The areas of the curves corresponding to the S and G_2 phases are also shown with an amplification of 10×. [From O. E. Iversen (1986). *Am. J. Obstet. Gynecol.* **155**, 770–776.]

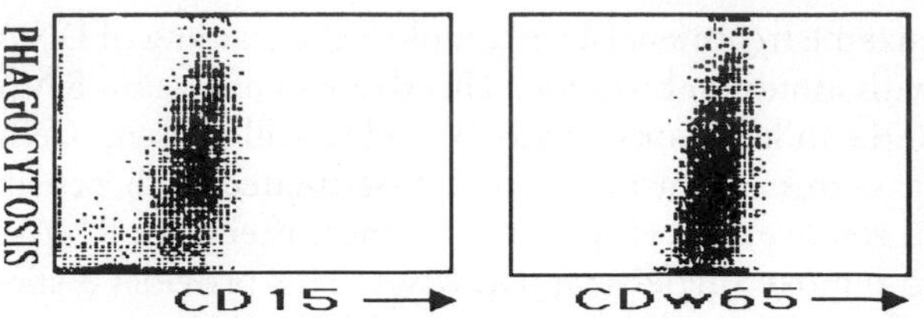

Combined measurement of cell surface immunofluorescence (x - axis) and phagocytosis of ethidium monoazide labeled fungi (y - axis). The histograms were gated from granulocytic cells as recognized by light scatter. A subpopulation of CD15 - weak cells with low phagocytic capacity was demonstrated

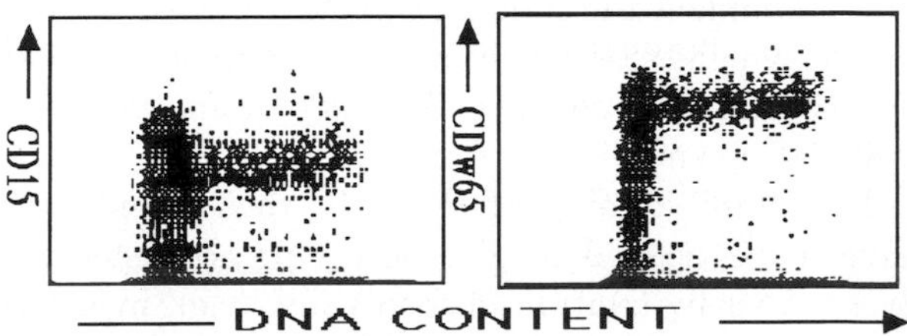

Combined measurement DNA content (x - axis) and cell surface immunofluorescence (y -axis) of bone marrow cells. CDw65 recognizes the total myeloid compartment, whereas CD15 is granulocyte specific. An increase in CDw65 expression is seen on cells in S and G2/M phases of the cell cycle.

FIGURE 4 Example of cluster analysis of two simultaneously measured parameters in human bone marrow cells. The percentages of each subpopulation can be quantitated and further characterized by electronic gating. (Courtesy of F. Lund-Johansen.)

cycling. Antibodies to different cell-cycle phases have also been used for their discrimination. Cell kinetics of, for example, human tumors are now being studied *in vivo* by using these labeling techniques and subsequent flow cytometry on biopsy materials.

B. Immunology and Leukocyte Functions

For many years, flow cytometry has been used for identification and quantitation of cells with different cell-surface antigens (Fig. 4). With monoclonal antibodies to such antigens, the numbers of different subclasses of lymphopoietic cells can be rapidly quantitated and separated in samples from both peripheral blood and lymphoid organs. The binding of complement and antibodies to immune cells can also be quantitated. In addition, the different aspects of phagocytosis can be identified and quantitated, by use of either fluorescent microorganisms or beads. By various combinations, adhesion, internalization, degradation, and exocytosis of bacteria exposed to human phagocytes can be measured sequentially and automatically. Defects in leukocyte functions can likewise be estimated. In addition, the killing of viable bacteria can be quantitated, as well as the degradation of both DNA and protein in bacteria exerted by phagocytes. Intracellular pH in granulocytes and macrophages degrading bacteria (phagosomal pH) can also be measured.

C. Hormone Research

By using fluorescent hormones, their binding and internalization can be studied and quantitated. The same applies to the binding of hormones to receptors and their internalization.

D. Measurements of Various Cell Components

Depending on the component that is stained, specific quantitative measurements can be performed of, for example, organelles, parts of the membrane, nuclear components, isolated chromosomes, RNA, lysosomes, and components of the cytoskeleton.

E. Use in Cancer Research

So far flow cytometry in cancer research has concentrated on the identification of abnormal distribution of DNA in tumors, including aneuploidy, altered cell-cycle parameters, and identification of abnormal chromosomes.

The slit-scan method and two-parameter measurements of, for example, cellular protein and DNA enable malignant cells to be discriminated from normal cells. In recent years, such parameters, and mainly DNA aberrations, have gained increasing attention for establishing prognosis in various tumor types.

F. Chromosome Analysis

Both normal and abnormal chromosomes can be measured and separated when chromosomes from mitotic cells are suspended. This applied to both their individual DNA content and their shape. So far, most success has been obtained by measuring cells from animals with few chromosomes (e.g., hamster, although human chromosome analysis is now performed routinely). Combined with different fluorescent DNA probes, the potential for identification of different genes is great. Thus, fluorescent *in situ* hybridization is being used extensively in flow cytometry.

G. Prokaryotes, Viruses, and Plant Cells

DNA, RNA, and protein can be measured in prokaryotic as in eukaryotic cells. This includes the distribution of DNA in growing bacterias, as well as their production of components with native fluorescence (e.g., chlorophyll). Also, viral DNA and RNA can be quantitated, and in some instances even single molecules can be detected with flow cytometry. Special devices are also available for the measurement of large cells, increasing the diameter of the sample flow. In addition, viral, bacterial, and plasmodial infections in cells can be identified and quantitated. Also, plant cells and algae can be analyzed with flow cytometric methods.

VI. CURRENT DEVELOPMENTS

Flow cytometry should be considered as a family of methods that combine high technology and automated measurements with advanced staining and preparation techniques for single cells. The flow cell and mechanical parts of the equipment have essentially remained the same during the last decade, although several practical improvements have been made. Developments have mainly been in preparation and staining techniques, by which almost all types of molecules and components can be identified and measured quantitatively in various cells. In addition, applications on different cell and organ types have rapidly expanded.

At present, one area of development is computer technology. Because large amounts of data are generated per second, the analysis is dependent on appropriate computer programs and large computers. This is especially the case when several different parameters are measured simultaneously and are analyzed in relation to each other. In recent years, great improvements have been achieved in multiparameter analysis and data handling. In this way multiple subpopulations within the same cell sample can be discriminated and measured for other parameters or sorted for further studies of biological or morphological parameters. This is facilitated by the use of a list mode computer function.

Simpler and less expensive machines are now available that automatically measure multiple parameters on cells under highly standardized conditions.

Some of the applications under rapid development are analyses of gene amplifications in mammalian cells, chromosome classification from both mammalian and plant cells, the measurement of chromatin in germ cells, identification of different gene products by *in situ* hybridization methods, and, not least, clinical applications that lead to diagnostic and prognostic improvements in several human diseases.

BIBLIOGRAPHY

Bjerknes, R., Bassoe, C.-F., Sjursen, H., Laerum, O. D., and Solberg, S. O. (1989). Flow cytometry for the study of phagocyte functions. *Rev. Infect. Dis.* **11**, 16.

Brown S. C., and Bergounioux, C. (1988). Plant flow cytometry. *In* "Flow Cytometry: Advanced Research and Clinical Applications" (A. Yen, ed.), Vol. II. CRC Press, Boca Raton, FL.

Gray, J. W. (ed.) (1989). "Flow Cytogenetics." Academic Press, London.

Laerum, O. D., and Bjerknes, R. (eds.) (1992). "Flow Cytometry in Hematology." Academic Press, London/San Diego.

La Via, M. F., Hurtubise, P. E., Hudson, J. L., and Stites, D. P. (eds.) (1988). "Clinical Applications of Cytometry." *Cytometry,* Suppl., **3**, 1.

Melamed, M. R., Mullaney, P. F., and Mendelsohn, M. L. (eds.) (1979). "Flow Cytometry and Cell Sorting." John Wiley & Sons, New York.

Owens, M., and Loken, M. (1994). "Flow Cytometry Principles for Clinical Laboratory Practice." Wiley–Liss, New York.

Robinson, J. P. (ed.) (1993). "Handbook of Flow Cytometry Methods." Wiley–Liss, New York.

Shapiro, H. M. (1994). "Practical Flow Cytometry," 3rd Ed. Wiley–Liss, New York.

Van Dilla, M. A., Dean, P. N., Laerum, O. D., and Melamed, M. R. (eds.) (1985). "Flow Cytometry: Instrumentation and Data Analysis." Academic Press, London.

Yen, A. (ed.) (1989). "Flow Cytometry: Advanced Research and Clinical Applications," Vols. I nad II. CRC Press, Boca Raton, FL.

Follicle Growth and Luteinization

COLIN M. HOWLES
Serono Laboratories

GLOSSARY

Aromatization Enzymatic conversion of androgens to estrogens, which occurs mainly in granulosa cells; this aromatase enzyme system present in granulosa cells is stimulated by follicle-stimulating hormone

Atresia Degeneration of the follicle and its cells leading to death of the oocyte; it occurs continuously at all stages of follicular development

Corpus luteum Tissue that develops after ovulation from the thecal and granulosa cells of the ruptured follicle; site of progesterone and estrogen synthesis; it regresses if implantation of an embryo does not occur

Graafian follicle Well-developed antral follicle, which is destined to ovulate (named after the anatomist Reiner de Graaf, 1641–1673)

Granulosa cells Cells that surround the oocyte throughout its development; they possess aromatase enzymes that stimulate the conversion of androgens to estrogens, and proliferate during follicular development

Luteinization Morphological and biochemical changes that occur in the follicle cells after ovulation

Menopause The last menses in women; it marks cessation of cyclical ovarian activity

Menstrual cycle In the ovary and female genital tract, the cyclical changes over a 28-day period governed by biphasic variations in ovarian and gonadotrophin hormone secretion; the cycle is divided into two discrete parts, the follicular and luteal phases. Menses (bleeding associated with the loss of the endometrium) marks the beginning of the next menstrual cycle

Recruitment Process in the late luteal phase of the previous menstrual cycle when a group, or cohort, of small antral follicles is recruited into the final growth phase by raised blood levels of follicle-stimulating hormone

Selection Process in which one of the antral follicles (the most well developed in terms of granulosa cell number and aromatase action) is selected to continue development and become a mature Graafian follicle

Thecal cells Cells that consist of two layers, the theca externa and interna cells (the latter possessing a rich blood supply); theca interna synthesize androgens under the influence of luteinizing hormone

THE FUNDAMENTAL UNIT OF THE OVARY IS THE follicle that consists of the female germ cell (oocyte) surrounded by a series of specialized cell layers, the granulosa and the thecal cells (Fig. 1). These cells are responsive to gonadotrophic hormone stimulation and cooperate to synthesize the most important ovarian steroid, estradiol-17β.

Every month during the woman's reproductive life, one oocyte is released (ovulation) from the single mature follicle that has completed development. If the oocyte is fertilized, the resultant pre-embryo is transported into the uterus, where, if hormonal conditions are appropriate, it will implant and develop into a new and unique individual.

The optimum hormonal environment for implantation is coordinated by the corpus luteum, the remains of the follicle once the oocyte has been released. After ovulation, the follicle cells undergo luteinization. This is the collective term for a number of important biochemical and morphological changes that occur in the cells.

The physiological mechanisms that control the

ENCYCLOPEDIA OF HUMAN BIOLOGY, Second Edition, VOLUME 4.

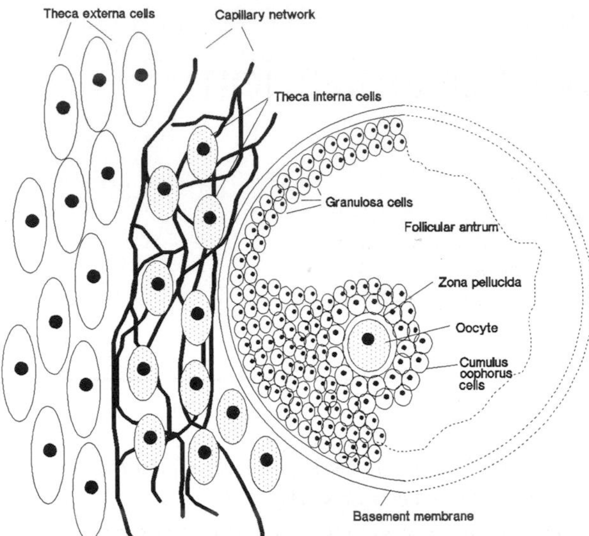

FIGURE 1 Diagram of a Graafian follicle showing the arrangement of follicle cells around the oocyte. The thecal cells consist of two layers: theca interna and theca externa. The theca interna cells possess a rich blood supply and are separated from the avascular granulosa cells by the basement membrane (follicle wall). (From Baird, 1984.)

growth of the follicle, ovulation, and fertilization are certainly complex, but they have evolved to maximize the reproductive potential of the species.

I. EARLY OOGENESIS

Unlike the male, the female germ cell (oogonia) undergoes mitotic proliferation in the ovary prior to the time of birth. All of these oogonia possess 46 chromosomes. From about the third month of gestation, increasing numbers of oogonia start to enter their first meiotic division, thereby becoming primary oocytes with a chromosome complement of 23. By the time of birth, or soon after, all female germ cells are primary oocytes. [*See* Oogenesis.]

At all stages of follicle development, however, degeneration, or atresia, occurs. In the human female, about 99% of oocytes are lost in this way. The consequence of atresia is that the number of oocytes is gradually reduced from birth until the time of menopause in women, when very few oocytes can be detected in histological sections of the ovary. Changes in germ cell numbers during the life of a human female are illustrated in Fig. 2.

Soon after formation, the primary oocyte becomes surrounded by a single layer of flattened epithelial (granulosa) cells. This group of cells constitutes the primordial follicle. Further meiotic division of the primary oocyte is halted at the dictyotene stage of prophase and it may remain in this state for up to 50 years. It is not known why the oocytes are stored in this protracted meiotic state.

The control of the next stage of folliculogenesis is not well understood but seems to be independent of gonadotrophin support. The oocyte undergoes its major growth phase marked by massive synthetic activity and morphological changes. The granulosa cells become cuboid in shape and increase in number, forming four or five layers around the oocyte, a preantral follicle.

However, during the next phase of follicular development, which occurs from puberty onward, growth is dependent on continuous secretion of hormones from the anterior pituitary. At the beginning of each

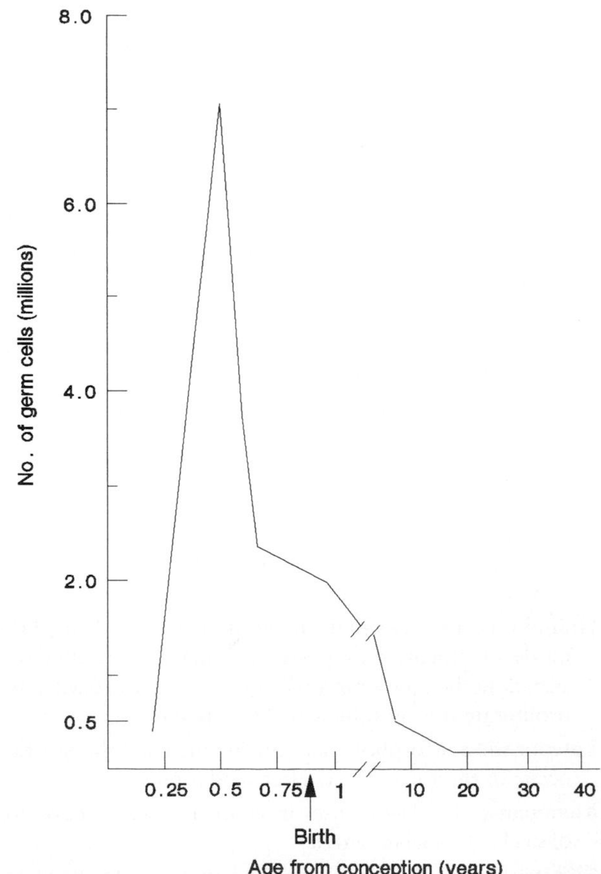

FIGURE 2 Total number of germ cells in the human female during life. Peak numbers occur before birth and then continuously decline until menopause. [From T. G. Baker (1971). *Am. J. Obstet. Gynecol.* 110, 746.]

menstrual cycle, a group, or cohort, of follicles enters the gonadotrophin-dependent phase of growth. In the next section, the role of these hormones in the human female will be discussed.

II. GONADOTROPHINS AND THE OVARY

Gonadotrophin-releasing hormone (GnRH) is secreted episodically into the portal blood system linking the hypothalamus to the anterior pituitary. GnRH release is affected by steroid feedback from the ovary as well as external environmental cues. GnRH stimulates specialized cells in the pituitary (the gonadotrope cells) to secrete the gonadotrophins—luteinizing hormone (LH) and follicle-stimulating hormone (FSH). [*See* Hypothalamus; Pituitary.]

The functioning of the adult ovary is dependent on the secretion of these gonadotrophic hormones. These interrelationships are diagrammatically represented in Fig. 3. The main properties of the two gonadotrophins, LH and FSH, and their action in the human female are shown in Table I.

The effect of LH and FSH on the ovary is to promote the synthesis of estradiol-17β, the predominant estrogen secreted by the ovary. In turn, estradiol, acting on the hypothalamic–pituitary axis, modifies LH and FSH secretion. During the luteal phase progesterone is the predominant ovarian steroid that modulates gonadotrophin secretion.

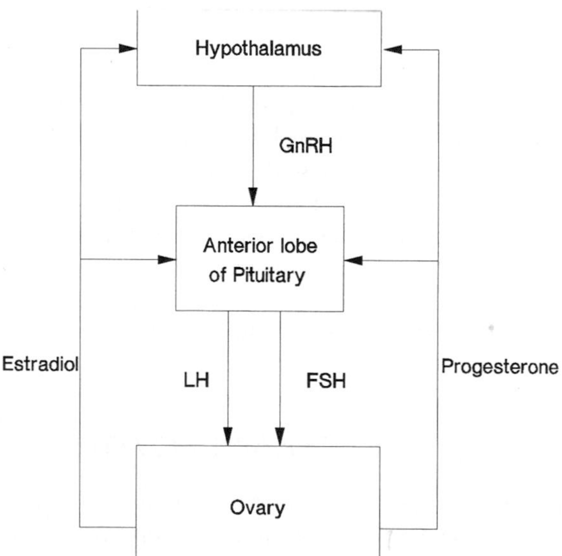

FIGURE 3 Hormone secretion between the hypothalamus, pituitary, and ovary. Estradiol and progesterone are the most important ovarian steroids that modulate gonadotrophin secretion in the follicular and luteal phases, respectively. Depending on the stage of the menstrual cycle, these steroids can have both positive and negative feedback effects upon gonadotrophin secretion (see text).

Before discussing in detail follicular growth, it is necessary to first consider how estradiol is synthesized by the ovary. Estradiol production is a cooperative venture involving both the theca interna and the granulosa cells of the follicle.

III. THE TWO-CELL THEORY OF FOLLICULAR STEROIDOGENESIS

The requirement for cell cooperation in the production of estradiol-17β was first demonstrated by B. Falck in 1959. In a series of classic experiments, he showed that estrogen formation by the rat follicle depends on the joint action of the theca interna and the granulosa cell layers.

In 1962, R. Short first proposed a two-cell theory to explain follicular steroidogenesis. This hypothesis was subsequently modified, as later experiments showed that, although thecal and granulosa cells can independently synthesize estrogens, the yield is greatly increased if both cell types are incubated together. This interaction of two cell types, independently stimulated by the two gonadotrophins, LH and FSH, led to the final realization of a "two-cell type, two-gonadotrophin" theory for estradiol synthesis in the ovary, as illustrated in Fig. 4.

Although this theory requires the involvement of both gonadotrophins for normal follicular steroidogenesis to occur, recent data suggest that normal folliculogenesis can occur independently of LH secretion. In these experiments, follicular development was induced and sustained by administering recombinant human FSH (r-hFSH), however, estradiol secretion was negligible.

The principal androgens, androstenedione and testosterone, are produced by the thecal cells, which are stimulated by LH. The androgens pass into the blood or are transported across the basement membrane of the follicle into the granulosa cell, where they are converted into estradiol by the aromatase enzyme stimulated by FSH. Estradiol is then secreted back into the blood or accumulates in the follicular fluid, keeping the intrafollicular environment highly estrogenic. For a mature follicle, the concentration of estradiol in the follicular fluid can be 1000 times that circulating in the blood. This could further potentiate

TABLE I
Properties and Action of the Gonadotrophic Hormones

	LH	FSH
Secreted from:	Anterior pituitary under the influence of GnRH release from the hypothalamus	
Composition:	Glycoprotein made up of an alpha and beta chain with 16.4% carbohydrate residues	Glycoprotein made up of an alpha (identical to that in LH) and beta chain with 25.9% carbohydrate residues
Molecular weight:	34,000	32,600
Cells acted on:	Thecal cells of antral follicles; granulosa cells of preovulatory follicles; luteal cells of the corpus luteum	Granulosa cells of preantral and antral follicles

LH and FSH action, as estradiol has been shown to increase the sensitivity of the follicle to gonadotrophin stimulation.

Evidence from numerous *in vitro* experiments has demonstrated the importance of paracrine factors [inhibin, activin, insulin-like growth factor-I and -II (IGF-I and IGF-II)] in the control of ovarian steroidogenesis. Of particular importance is inhibin, which is secreted from FSH-stimulated granulosa cells. Inhibin has been shown to promote LH-stimulated androgen synthesis *in vitro* by human thecal cells. These findings have important clinical implications as they help to explain why, in situations where there are very low LH levels, normal follicular growth and ovarian steroidogenesis can be driven by exogenous FSH administration alone.

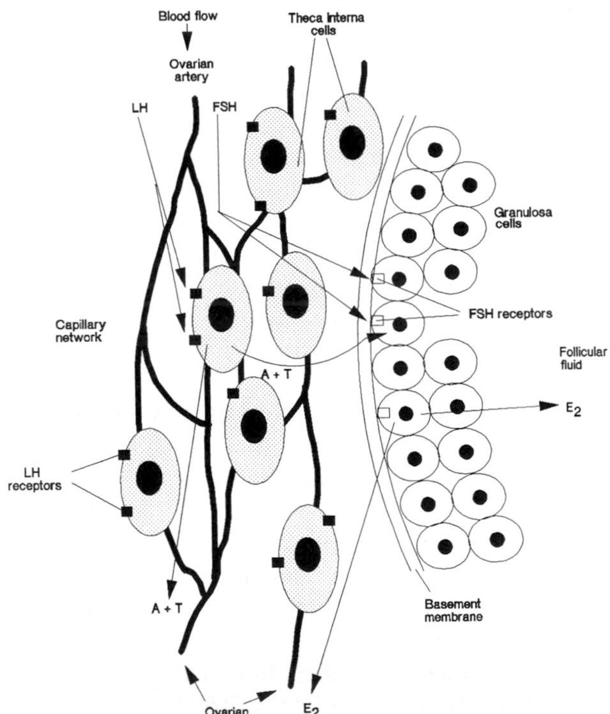

FIGURE 4 Schematic representation showing the action of the gonadotrophins on the follicle cells and the synthesis of estradiol. LH attaches to specific receptors (■) on theca interna cells. Under LH stimulation, these cells produce androgens (androstenedione, A; testosterone, T). These steroids are secreted into the blood or pass through the basement membrane of the follicle and enter the granulosa cells. FSH interacts with receptors (□) on the granulosa cells and activates the aromatase enzyme system, which converts androgens to estradiol (E_2). E_2 is either secreted out of the follicle and into the bloodstream or concentrated in follicular fluid. (From Baird, 1984.)

The importance of this two-cell type–two-gonadotrophin interaction for normal follicular development is aptly illustrated in the clinical condition known as polycystic ovarian disease (PCOD). Although the etiology of this condition is complex, it is sufficient to say that in 80% of subjects LH secretion is higher than normal, which is a contributing factor to excess production of androgens by the thecal cells. Patients with PCOD can suffer with chronic anovulation and are more likely to be obese. The ovaries of these women contain numerous antral follicles that are androgenic rather than estrogenic. This can be rectified by the administration of FSH, which initiates normal follicular growth and estradiol production.

IV. FOLLICULAR GROWTH

It takes approximately 85 days, or three ovarian cycles, for the development of a preovulatory, or Graafian, follicle from an early preantral follicle. At the onset of this development, a large number of primary follicles are capable of initiating growth. Those that do begin to grow are influenced by many factors, such as age of the woman as well as hormone activity and possibly nutritional status. Only one follicle will

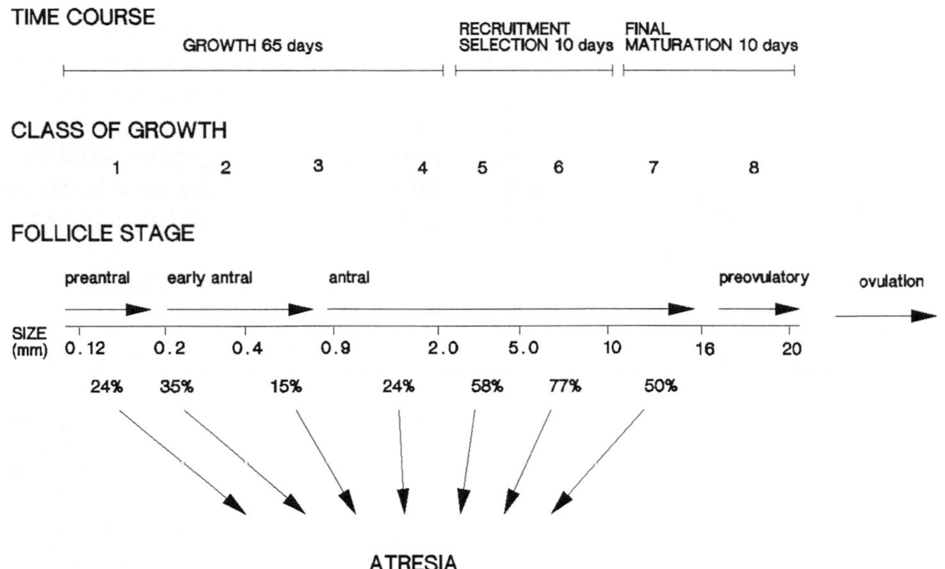

FIGURE 5 Time course of follicle growth. Class 1 is primary, and Classes 7 and 8 are Graafian follicles. At all stages of development, follicles degenerate (atresia). In the human female, only one follicle usually undergoes ovulation each month.

eventually become mature, ovulating and releasing the oocyte; the rest will undergo atresia.

The life cycle of a follicle during this 85-day period is shown in Fig. 5. There are eight classes of growing follicles, and at each stage many become atretic. Once the follicle reaches about 0.2 mm in diameter, fluid starts to collect between the granulosa cells. These pockets of fluid grow and merge together to form an antrum, hence the term antral follicle.

A. Follicular Recruitment

After about 65 days of growth, a final cohort of about 15–20 small antral follicles will enter the gonadotrophin-dependent phase of growth. Figure 6 depicts the gonadotrophin and steroid hormone changes that occur during the follicular phase of the human menstrual cycle. The length of the follicular phase is primarily determined by the rate at which the principal antral follicle matures. This follicle exerts its control by secreting hormones that orchestrate the pattern of gonadotrophin release from the anterior pituitary. This process will be addressed in the next section.

If pregnancy does not occur, the corpus luteum regresses, leading to a decline in progesterone levels. Concomitant with this reduction in corpus luteum activity, FSH levels increase. Only those small antral follicles that have acquired gonadotrophin receptors, coincident with this intercycle rise in FSH, will be recruited into the next growth phase.

The most crucial event for the further development of an antral follicle is the activation of the aromatase system by FSH. It is now widely accepted that each small antral follicle (about 4 mm in diameter) has a threshold requirement for stimulation by FSH. This was first postulated by Jim Brown in 1978. Brown's hypothesis was formulated from the treatment of infertile patients with exogenously administered gonadotrophins of pituitary origin. It was found that the ovary could detect and respond to changes in blood levels of gonadotrophins in the region of 10–30%. Thus, a modest rise in FSH levels is all that is required to initiate the recruitment of antral follicles.

Because follicular development in the human female is highly asynchronous, at the time of the intercycle rise in FSH the ovaries contain a cohort of follicles with varying sensitivities to FSH. The follicle with the lowest FSH threshold will be the first to undergo activation of the aromatase system and begin estradiol production.

B. Hormone Production from the Follicle

The production of steroids and the increase in follicular size are intimately linked. The proliferation of granulosa cells increases the ability of the follicle to aromatize androgens to estradiol. (The hormone–cell interactions and steroid production of developing antral follicles are summarized in Table II.) This action

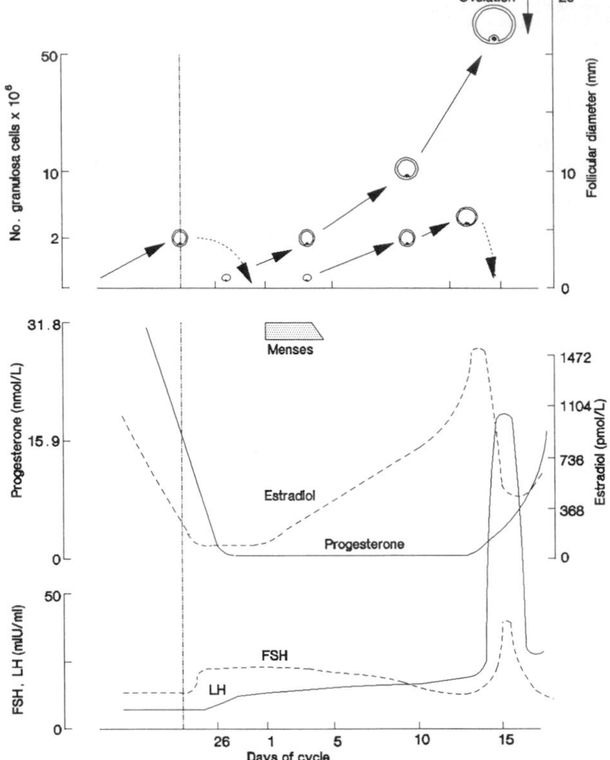

FIGURE 6 Top graph depicts the growth of antral follicles, and bottom graphs the plasma levels of gonadotrophins, estradiol, and progesterone during the follicular phase of a menstrual cycle. Only those follicles at the appropriate stage of development concomitant with the intercycle rise in FSH are recruited into the final growth stage. (From Baird, 1984.)

is a classic example of a positive feedback loop, and it ensures an increasing capacity of the follicle to convert androgens to estradiol, leading to a surge in estradiol production.

The increase in estradiol secretion from the ovary has been demonstrated to have a gonadotrophin-suppressive effect on the hypothalamic–pituitary axis. In particular, FSH secretion begins to decline prior to the midfollicular phase of the menstrual cycle. However, it has been hypothesized that a nonsteroidal factor (inhibin) may also play a role. It was not until the early 1980s that experiments proved the existence of inhibin. These studies involved the use of follicular fluid that contains high concentrations of inhibin-like activity. When injected *in vivo,* FSH secretion was selectively inhibited.

More recently, inhibin has been purified and shown to be a glycoprotein of about 32,000 molecular weight. It is composed of two subunits linked by disulfide bridges. Inhibin is produced by granulosa cells in response to FSH and androgen stimulation, secreted into the circulation, or concentrated in follicular fluid. However, recent evidence suggests that inhibin does not play a role in the modulation of FSH secretion during follicular development. In the spontaneous menstrual cycle, changes in inhibin concentrations are not related to those in FSH. As we shall see later, inhibin becomes important during the luteal phase.

TABLE II
Hormone–Cell Interaction and Subsequent Steroid Synthesis in Developing Antral Follicles

	Thecal cell	Granulosa cell
Hormone	LH	FSH (estradiol)
Steroid synthesis	Acetate/cholesterol to androgens	Androgens to estradiol

C. Selection of the Dominant Follicle

It seems, therefore, that FSH secretion in the follicular phase is inhibited by estradiol. The decline in blood FSH levels is in response to increased estradiol secretion from growing follicles. The result of this reduction in FSH during the spontaneous menstrual cycle is the "selection" of one antral follicle, which continues development and ultimately ovulates.

When the fall in FSH occurs, the selected follicle is less dependent on circulating levels of FSH. This is probably because it had the lowest FSH threshold at the onset of the intercycle FSH rise. The follicle will have had longer to activate its aromatase system, leading to higher estradiol production and greater granulosa cell proliferation than its rivals.

Once selected, the follicle is called a Graafian follicle. Selection occurs by about Day 7 of the menstrual cycle. The rest of the cohort of follicles become atretic as FSH is suppressed below their own threshold level.

It is possible to overcome the follicle selection procedure by the administration of exogenous gonadotrophins. After gonadotrophin injections, FSH levels are elevated for a longer time period, thus allowing other follicles to continue development. This principle is practiced in patients undergoing *in vitro* fertilization treatment. Exogenous gonadotrophins can be given daily throughout the follicular phase of the cycle to promote "superovulation."

Another physiological process concerning selection may also contribute to the emergence of a dominant follicle. Rising levels of estradiol, in conjunction with FSH, induce the appearance of LH receptors on the outer layer of granulosa cells. Thus, there is a gradual change in distribution for gonadotrophin receptors, which may be critical for further follicular development. *In vitro* studies have shown that granulosa cells possessing both FSH and LH receptors responded identically to both hormones in terms of aromatase activity and steroid production. This may mean that the presence of both LH and FSH receptors on granulosa cells may further protect the emerging dominant follicle from declining FSH concentrations.

V. THE LH SURGE AND OVULATION

Once LH receptors have been fully acquired by the granulosa cells, the Graafian follicle can enter the final or preovulatory phase of growth. This terminal growth phase is signaled by a surge in gonadotrophin output, the LH surge. As the large Graafian follicle reaches maturity, estradiol output reaches a peak. In such a highly estrogenic environment, the pulse frequency of GnRH is more rapid and the sensitivity of the pituitary gonadotrope cells to GnRH is greatly enhanced. These events lead to a massive discharge of LH.

The effect of the LH surge is twofold: it causes profound changes to the structure and function of the follicle and, second, stimulates the resumption of meiosis in the oocyte.

The follicle undergoes a final rapid growth phase, mainly due to an increase in the volume of follicular fluid. Also, major changes occur in the endocrinological activity of the follicle cells.

Of major importance is that the granulosa cells can no longer produce estradiol by aromatization and, thus, lose their FSH receptors. Instead, they start to synthesize progesterone through LH stimulation. These changes in hormone–cell interactions are summarized in Table III. This results in an increase in progesterone secretion from the follicle, concomitant with the rise in LH.

TABLE III
Hormone–Cell Interaction and Steroid Synthesis in a Mature Graafian Follicle

	Thecal cell	Granulosa cell
Hormone	LH	LH
Steroid synthesis	Acetate/cholesterol to androgens	Acetate/cholesterol to progesterone

The production of progesterone may also be important in facilitating the positive feedback effects of estradiol on LH release. Recently, women undergoing an *in vitro* fertilization attempt have shown that a single injection of progesterone can elicit an LH surge. Thus, LH secretion from the pituitary is maximized, ensuring that final follicular maturation is completed.

While progesterone is being synthesized from the follicle, the chromosomes of the oocyte progress through the first meiotic division. Although the mechanisms are not fully understood, it has been proposed that rising levels of LH either inhibit the action or block the synthesis of an oocyte maturation inhibitor factor, thus allowing terminal maturation to occur.

In addition, the dose-dependent effect of LH on granulosa cell proliferation has been demonstrated *in vitro*, where high-dose LH caused enhanced synthesis of progesterone, suppression of aromatase activity, and inhibition of cell growth. These observations have led to the concept of an "LH ceiling hypothesis" (proposed by Steve Hillier) to explain the clinical observations demonstrating that excessive LH stimulation can disrupt normal preovulatory follicle development.

This phenomenon has been highlighted in patients undergoing superovulation for *in vitro* fertilization. In one study, high levels of LH were associated with failure of implantation and early pregnancy loss, whereas low levels of LH were associated with the establishment of ongoing pregnancy. Table IV summarizes the results of this study. This effect of LH on pregnancy outcome has been confirmed by other studies, including one in women not undergoing any infertility treatment. It is now widely recognized that inappropriately raised LH concentrations are a significant cause of miscarriage. Further work is required to elucidate the mechanism of LH action on oocyte maturation.

Final meiotic division of the oocyte is peculiar in that the distribution of cellular material is grossly unequal. Only a very small amount of cytoplasm accompanies one-half of the divided chromosomes; this forms what is called the first polar body, which is extruded to one side of the maturing oocyte.

Furthermore, the follicle wall undergoes dramatic changes. The rapid expansion of the follicle at this time stretches the follicle wall and, probably through the action of collagenase enzymes, particularly plasmin, and prostaglandins, the wall starts to break down. Where this occurs an outward bulge appears, the stigma, which eventually ruptures, releasing the

TABLE IV
LH Secretion in the Late Follicular Phase of Treatment and Outcome of *in Vitro* Fertilization[a]

Outcome of treatment	Number of patients	Urinary LH secretion (IU/liter/hr ± SEM)
Ongoing pregnancy	88	0.17 ± 0.01
Nonpregnant	92	0.20 ± 0.01
Miscarried	29	0.26 ± 0.03

[a]Refer to Howles and Macnamee (1989).

follicle contents. By this time, the oocyte is connected to the mass of granulosa cells only by a very thin stalk of cells, which easily breaks, allowing the oocyte to be extruded in the flow of follicular fluid.

An oocyte, freshly harvested from a preovulatory follicle of a woman undergoing an *in vitro* fertilization treatment, is shown in Fig. 7. Note the sunburst arrangement of cumulus oophorus cells around the oocyte; this is highly characteristic of a mature human oocyte.

VI. LUTEINIZATION AND THE FORMATION OF THE CORPUS LUTEUM

In the final stages leading up to ovulation, the follicle cells lose their LH receptors and become desensitized to further LH stimulation. However, about 2–3 days after ovulation, they recover their ability to respond to LH. In rodents, this recovery of receptors is stimulated by prolactin. Progesterone becomes the main secretory product of the collapsed follicle, which is now called the corpus luteum. The granulosa cells no longer divide but increase in size and undergo internal structural changes. These include the production of a carotenoid pigment called lutein, which gives the corpus luteum its yellowish appearance in many species. Such cells are said to be luteinized.

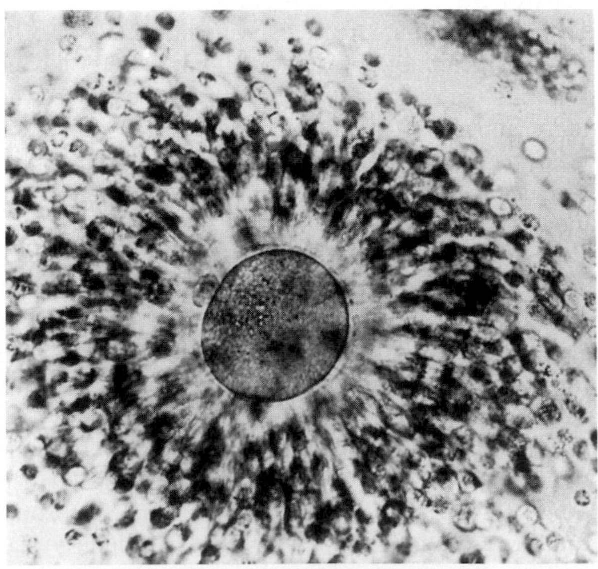

FIGURE 7 Photograph of a mature human oocyte, just after having been removed from the follicle for *in vitro* fertilization. The oocyte is surrounded by a mass of cumulus cells showing a dense corona radiata (sunburst appearance).

In addition to the secretion of progesterone, the corpus luteum of some higher primate species (including the human) produces estradiol. Both of these hormones are necessary to prepare the endometrium in the event of an embryo implanting.

During the luteal phase, gonadotrophin secretion is low, suppressed by the negative feedback effects of progesterone and estradiol. In addition to this negative feedback effect on gonadotrophin release, recent research has demonstrated that the corpus luteum also secretes large quantities of inhibin. Thus, inappropriate follicular development is completely inhibited during this phase of the cycle.

Normally, the corpus luteum starts to regress about 1 week after ovulation. However, if implantation occurs, regression of the corpus luteum and, hence, endometrial degeneration will not occur. It has been shown that the implanting embryo secretes a glycoprotein, human chorionic gonadotrophin, which extends the life of the corpus luteum. Human chorionic gonadotrophin has a similar action to LH and promotes hormone synthesis in the corpus luteum.

In the case of an embryo not implanting, the fall in progesterone and inhibin concentrations from the waning corpus luteum lead to a resumption of gonadotrophin secretion. Thus, the intercycle rise in FSH leads to a new phase of follicular recruitment and growth.

BIBLIOGRAPHY

Baird, D. T. (1984). The ovary. *In* "Reproduction in Mammals: Book 3" (C. R. Austin and R. V. Short, eds.). Cambridge Univ. Press, Cambridge, England.

Baird, D. T. (1987). A model for follicular selection and ovulation: Lessons from superovulation. *J. Steroid Biochem.* **27**, 15.

Balen, A. H., Tan, S. L., and Jacobs, H. S. (1993). Review. Hypersecretion of luteinising hormone: A significant cause of miscarriage. *Br. J. Obstet. Gynaecol.* **100**, 1082–1089.

De Cherney, A. H., Tarlatzis, B. C., and Laufer, N. (1985). Follicular development: Lessons learned from human *in vitro* fertilization. *Am. J. Obstet. Gynecol.* **153**, 911.

Glasier, A. F., Baird, D. T., and Hillier, S. G. (1989). FSH and the control of follicular growth. *J. Steroid Biochem.* **32,** 167.

Hillier, S. G. (1994). Review. Current concepts of the roles of follicle stimulating hormone and luteinizing hormone in folliculogenesis. *Hum. Repro.* **9,** 188.

Hillier, S. G., Harlow, C. R., Shaw, H. J., Wickings, E. J., Dixson, A. F., and Hodges, J. K. (1988). Cellular aspescts of pre-ovulatory folliculogenesis in primate ovaries. *Hum. Repro.* **3,** 507.

Howles, C. M. (1985). Follicle growth and luteinization. *In* "Implantation of the Human Embryo" (R. G. Edwards, J. M. Purdey, and P. C. Steptoe, eds.). Academic Press, London.

Howles, C. M., and Macnamee, M. C. (1989). "The Endocrinology of Superovulation; The Influence of Luteinizing Hormone." Excerpta Medica Asia Ltd., Hong Kong.

Howles, C. M., Macnamee, M. C., and Edwards, R. G. (1987). The effect of progesterone supplementation prior to the induction of ovulation in women treated for *in vitro* fertilisation. *Hum. Repro.* **2,** 91.

Food Acceptance: Sensory, Somatic, and Social Influences

D. A. BOOTH
University of Birmingham, England

GLOSSARY

Cognition Objective thought, that is, mental processes capable of relating to some reality (complex information processing). This includes intention, emotion, and imagination, as well as reasoning and perception. In current scientific usage, cognitive processes do not have to be conscious (nor in a human mind)

Food Chemical mixture in a form that is regarded by an organism as suitable for ingestion. Alternatively, an actually nutritive material. (Ingestates consisting primarily of water are generally called drinks, and in that usage food is solid nutriment.) The totality of an organism's foods and drinks is its diet

Physiological Refers to physical processes in any part of the body, including the gastrointestinal tract, liver, and brain, that (in this context) can have a normal causal effect on behavior toward food or drink. (Many other physiological processes in the central nervous system are also necessary for ingestion and all other behaviors or cognitions to occur and for sensory and social, as well as somatic, influences to be exerted.)

Sensory Perceptible aspect of the constitution of (in this context) a food or a drink. Sensory vocabulary is what can be used to describe physicochemical characteristics, but the use of a sensory descriptor score is not an objective measure of food constitution until shown to be such

Social Pertaining either to one or more other people or to cultural, economic, or institutional processes operative in a society

THE ACCEPTANCE OF FOOD IS A MOMENTARY ACT by which a person takes control of the ingestion of a food item, by purchasing it for eating, by preparing it for someone to eat, or by eating it. Food acceptance is in response to sensed characteristics of the material but also depends on bodily needs and on socioeconomic context. The interactions of these influences on individual consumers' food acceptances might be more or less stable over time within an adult consumer, and more or less similar in structure and strengths across consumers within a culture. Nevertheless, sales, frequencies, or averages of responses from a sample of consumers, and average or individual intakes of a food, are the result of many diversely structured pieces of acceptance behavior. Therefore, accurate measures of that behavior require the analysis of momentary situation-dependent influences on acceptance of particular foods in representative individuals.

I. COGNITIVE INTEGRATION

A. Varieties of Food Acceptance Behavior

A variety of human actions can coherently and usefully be regarded as accepting a food.

1. Purchase Acceptance

Taking a product off the shelf and to the checkout counter of a food store is the act that is crucial for commerce. To explain and predict sales adequately, we might need to understand more about the influences on food acceptance in the form of purchases.

Elaborate statistical modeling of preference data

ENCYCLOPEDIA OF HUMAN BIOLOGY, Second Edition, VOLUME 4.

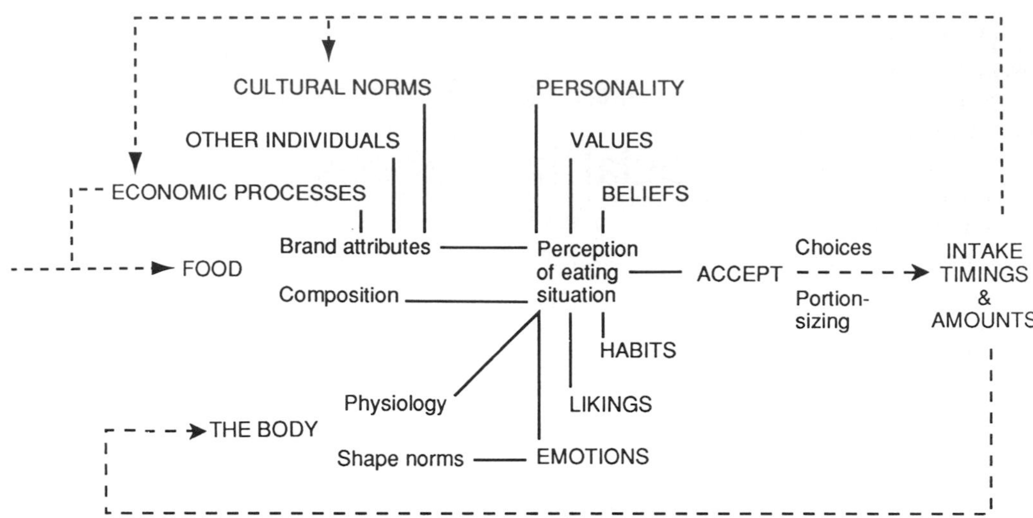

FIGURE 1 Immediate, mediating, and background sources of variation in people's food acceptance, with the classificatory (solid lines) or causal (dashed lines) links between sources within and outside the individual mind.

and/or sensory or other descriptive scores from consumer panels has increasingly been used in attempts to solve this problem. This is not, however, a mere data-analytic matter. The issues are the scientific ones of characterizing the causal processes within the behavior of consumers (Fig. 1). Research on the determinants of food acceptance therefore requires the science of cognitive psychology, which has also been applied in other technologically complex areas (e.g., human–computer interaction and ergonomics) more broadly.

2. Eating Acceptance

Food is almost always bought to be eaten. People selecting for their own consumption from a store, buffet counter, or restaurant menu are effectively choosing what to eat. Furthermore, eating quality is likely to be a major influence on subsequent repurchasing behavior. Also, foods cannot affect health without being eaten.

Therefore, the preparation, serving, and ingestion of food already purchased (or grown domestically) are further examples of acceptance. Arguably, eating acceptance is the foundational criterion, or at least the central type, of food acceptance.

B. Scientific Concepts Related to Food Acceptance

Food acceptance, both at ingestion and during the "search" for food in shops and restaurants, is a phenomenon that has long been of major interest within the biological, medical, behavioral, and anthropological/sociological sciences.

1. Terminology in Behavioral Disciplines

Eating habits and their social roles (e.g., food symbolism) are a major part of every human culture, both rural and industrialized. They have therefore been a focus of attention by many anthropologists and some sociologists.

The intake of materials resulting from food acceptance behavior is also a crucial influence on physiological processes (e.g., exchanges of water, sodium, energy, nitrogen, and other nutrients between the organism and the environment). However, studies of the intake of solids and fluids in regulatory physiology often lack the perceptual data to justify use of the behavioral concept of the sensory, physiological, or social control of food acceptance. As a result the physiologists and biopsychologists of food intake have often seriously distorted commonsense, but scientifically usable, concepts such as palatability, hunger, thirst, appetites, and satieties.

In the science of behavior itself, attention to food-oriented behavior has declined along with the dissolution of the specialism of "motivation" in psychology. Food intake has become a topic for neuroscientists. Food reinforcement is merely a tool for dissecting out learning mechanisms. Food attitudes only occasionally interest contemporary social psychologists. This lack of interest is paradoxical since eating and drinking are the most common cognitively rich human activities.

2. Terminology in Nutrition

Foods and beverages cannot affect health or disease unless they are in a diet (i.e., people eat and drink them). Dietary choices, sequences, and patterns, as perceived by the eater, are behavioral structures in the minds of individuals, equivalent to food acceptance. Dietary intakes are entirely the result of these cognitive–behavioral processes. That is, usual nutrient intakes result from decisions made in terms of cultural constructs of foodstuffs and eating practices.

There is concern over the effects of fermentables on the teeth, the timing of energy choices on the waistline, and preferences for foods high in saturates on the heart. Data on these dietary patterns are lost when weights of foods are totaled across the day. Clinical and community nutrition would advance by integrating soundly behavioral (and sociological) concepts equivalent to food acceptance into their research and practice.

3. Appetite: Cognitive Organization in Food Acceptance Behavior

Appetite is the disposition to eat and drink, including the approaches to food in a shop, restaurant, or refrigerator.

This basic psychological sense of terms such as "appetite," "hunger," "thirst," and "satiety" must be distinguished from *ad libitum* dietary intake. Food intake is relevant to appetite, but only as a measure of the cumulative effect of expressions of appetite in the changing circumstances of a series of food acceptances. Thus, without full specification of the sensory, physiological, and social contexts influencing each momentary acceptance that contributed to the recorded total intake, the physical disappearance tells us nothing about the organization of the behavior.

It also must be appreciated that it is logically incoherent to take the word "appetite," even in its everyday usage, to refer to subjectively conscious mental processes alone. The appetite is in the directedness of what we actually do, and only secondarily in the private experiences that go with that observable performance. [*See* Appetite.]

II. EMPIRICAL UNIT OF ACCEPTANCE BEHAVIOR

Food acceptance is a single momentary overt act of a person (or other organism) on a particular occasion. Thus, any acceptance of a food occurs in a particular context of alternative foods, bodily condition, physical and social environment, and other aspects of mental state. Also, any measure of behavior toward food that results from more than one act by one person toward one food in one situation might be confounded by variations over time in that person or by differences among people, foods, or situations.

A. Choice and Relative Acceptance

When two or more foods are present, the acceptance of one of them is an overt choice among the foods. In this context on this occasion, the chosen food has been preferred over the other food(s). Repetition of the choice by that person in similar contexts can establish that the choice was not random, or in some circumstances this can perhaps be assumed on other grounds.

Even if only one food is ever present at one time (i.e., monadic presentation), opportunities to accept can be provided in similar contexts for two or more foods in succession. Then, from the acceptance or refusal of the different foods, we can construct an observed relative acceptance.

This combination of acceptance and refusal on separate occasions can be represented by the same parameter as the combination of acceptances and refusals among the same set of foods presented simultaneously on repeated occasions (e.g., the proportion of acceptances of one food to the total acceptances from the set). Qualitative similarity of successive and simultaneous relative acceptances is evidence for the same underlying preference structure. Quantitative similarity is unlikely, however, since the strongest preference is freer to dominate behavior in the simultaneous test. The successive (i.e., monadic) test therefore has the advantage of sensitivity to less than the strongest disposition to accept a food. However, the real purchase or eating situation might be one of choice.

The issue that arises is always crucial: What are the objectives of the investigation? In particular, what situation of acceptance are we seeking to understand? In practice, using a familiar range of food products or variants of a food and monadic presentations within a fairly naturalistic situation provides sufficiently realistic and precise data to characterize and even to quantitate influences on either multi-item choice or single-item acceptance/rejection in the normal context.

B. Degrees of Acceptance and Graded Influences

The discussion so far has been couched in terms of acts of acceptance or rejection of particular foods in a particular situation. An acceptance is categorical—it occurs or it does not (except perhaps when two or more foods are put into the mouth or the shopping basket at once). Degrees of acceptance would be quantitatable as frequencies of acceptances (or refusals) in repeated tests.

Nevertheless, a single act can be used to generate an estimate of such frequencies over many yes/no tests. With people the simplest way to do this is to ask for a verbal judgment of the frequency with which they would accept a monadically presented item in some defined or assumed context. Alternatively, they can be asked to express a degree of liking or pleasantness; this appears to be more introspective, but that is exposed as an illusion of language when the hedonic rating is anchored on acts in specified situations. Then these ratings become the same as acceptance frequency ratings. In any case the qualitative validity and quantitative calibration of acceptance ratings of any sort are based on the actual frequencies of acceptance in the real-life decisions.

An influence on acceptance might also be inherently categorical (i.e., either present or absent), having no intermediate strengths. Brand name and food type are examples. However, these categories could be resolvable into sets of perceptible characteristics of the brand image or the food type's composition or uses; in other words, brand names or food types might scale onto several continua, whether sensory (e.g., sweet), somatic (e.g., quick filling), or social (e.g., breakfasttime or preschool age). The basis can be identified for either a categorical difference or a metric continuum by finding the observable characteristic(s) of an influence on acceptance that regresses linearly and precisely onto degrees of acceptance by the individual. This regression provides a measure of sensitivity to that influence, namely, the just discriminable difference in the influence by that response.

C. Generalizing

The usual concept of the *acceptability* of a food is highly abstract and often rather unrealistic. Food acceptability, as commonly conceived, generalizes across consumers, eating or purchasing contexts, and sets of foods among which choices are being made. In consequence, sensory acceptability or palatability is commonly misconstrued as an inherent characteristic of the food. Even if the sensory influences on acceptance of a food did not vary among people or situations (as they commonly do), palatability would still be a characteristic of behavior toward the food, not of the food itself—of a causal relationship between a person and a food, not of one of the terms in that relationship.

At a moment in a meal, whether or not more of a food will be eaten depends on how boring and filling it has so far become and on anticipatory decisions when serving onto the platter or when purchasing the items for the meal. Whether another food now becomes accepted could depend on any of these factors for other available foods and on the next food's suitability to the upcoming stage of the meal. In Western cuisines the acceptabilities of meat pie and apple pie reverse between the first course and the dessert, although, of course, they do not compete when shopping for an entire meal.

Acceptance of a food is often also contingent on the time of day (cf. breakfast foods, between-meal hot drinks, and snack items), company at the table (e.g., spouse, children, and guests), season of the year (e.g., hot or cold weather), health concerns, specialness of the occasion, and so on. Still further factors can be particularly influential at the point of purchase (e.g., price, advertising messages, and package design and information). Thus, a generalization about food acceptance must either specify the particular context assessed, in all its potentially influential aspects, or include the effect of an influence within the generalization.

Most challenging of all, people differ from one another in how they put all of these sensory, somatic, and social influences together into the habitual acceptance decisions. Until recently, all approaches to influences on food acceptance have assumed that there are simply additive combinations among influences, and that therefore the analysis of data lumped across people would not be qualitatively, or even quantitatively, too misleading. In fact, many influences simply do not add to each other. Instead, the distances of distinct influences from their best accepted level are combined in multi-dimensional space according to the Pythagorean Theorem (square root of sum of squares). Furthermore, qualitatively different structures of interactions among influences can be quite common. Indeed, there has long been evidence that the results of sensory tests with and without brand names sometimes differ greatly; that acceptabilities of food products after use often differ from, and can be even more diverse than, acceptabilities at first acquaintance and that the awareness parameter in a brand life-cycle model has nothing to do with the repurchase parameter.

D. Measurement of Influences within Acceptance Behavior

1. Disconfounding of Influences

To begin to understand the bases of food acceptance, the classes of influence on it must be distinguished and varied independently of each other. Acceptances can be categorized according to the influence(s) on them, as follows.

a. Palatability and Satiety

The palatability of a food can be assessed only by the difference that the purely sensory factors make to intake, or to some other measure of acceptance. The immediate sensory influences must not be confounded by other influences (e.g., secondary cognitive or physiological effects of palatibility or the effects of the swallowed food).

Conversely, meal size cannot be a valid measure of postingestional effects unless both sensory effects and changes in reactions to sensory factors induced by the stimulation from them (e.g., "sensory satiety") have been excluded.

Even the fine details of the temporal pattern of eating movements or food intake are insufficient to diagnose influences within acceptance behavior. Dissociations (i.e., zero correlations) between parameters of the micropattern over different meals (or different eaters) would be evidence that there are distinct influences. What these influences are, though, can only be determined by varying putative causes and getting a closely related variation specifically in one of the dissociated parameters. To have valid measures, we must show how the ingestive movements in response to a pang do differ qualitatively from those in response to a craving.

b. Nutrient-Specific Effects

Similarly, the choice of one nutrient preparation over another is not necessarily controlled by nutritional physiology. If the foods differing in nutrient composition also differ in sensory characteristics, with or without palatability differences, the relative acceptances of these foods might have nothing to do with nutrition.

For example, a wish to eat conventional snack foods, which are rich in carbohydrate and fat content and relatively low in protein content, should not be interpreted as a carbohydrate craving, or as being mediated by effects of low protein content on the brain, until an explanation in terms of a habit of eating this range of food types for their sensory characteristics has been excluded.

By the same criterion, differences in food acceptance (whether as actual intakes or in appetite ratings), induced by foods that obviously differ in character as well as in nutrient composition, are not evidence for differences in the satiating or appetizing power of these nutrients, or indeed of the sensory differences (e.g., sweetness). People acquire strong expectations from experiences of recognizable foods. These learned responses can be sufficient to affect appetite hours after eating a particular type of menu. The acceptance data need not reflect any physiological action of the food.

2. Ascribed Differences versus Observed Differences

People can ascribe different strengths to influences on their food acceptance (e.g., "Taste is more important to me than texture," "I never get hunger pangs," or "I rate healthiness higher than palatability"). This sort of data has often been used to assess the relative importance of categories of influence. However, what people are aware of, especially in retrospect, may have little to do with what is actually influencing them. It might reflect no more than their own positions on conventional theories and current topics of discussion.

That is to say, the wording of a rating cannot in itself make it a measure of what those words refer to, be it a subjective experience or an objective aspect of food composition, bodily state, or the context of the eating or purchasing occasion. So-called "direct" scaling is an illusion. Whether the rating uses a magnitude estimation procedure, visual analogs, or multiple categories, what the scores actually measure depends entirely on the precision of their relationship to an observed factor in the food, the body, or the context, or, indeed, to acceptance responses.

Furthermore, for a verbal test to identify influences on acceptance, its results would have to measure the relationship of differences in acceptance to differences in the putative influences, whether they be in the food's sensed characteristics, the physiological state affected, or the cultural role of the food.

3. Is Preference a Dimension?

This also means that it is not coherent to treat acceptance data (or preferences and likings) as quantitative values on a scale or dimension, as they are commonly treated in consumer behavior and economics. The real phenomenon, the causal process, or the theoretical entity is acceptance from within a perceived situation.

There is a psychological scale in preferences, but it is the relationship between the multidimensional combination of the influences operative in a person and the acceptance, or hedonic response. This relationship is latent in these data by themselves, hidden behind or underneath the observations. It is not recoverable from the preference data alone unless the latent structure is simple enough to be guessed and we have enough data to discriminate this causal model from other models.

There are several psychological reasons why the structure of influences on the acceptance of food is usually much more complicated than allowed by the assumption of a preference dimension.

Some consumers might put the influences together in ways qualitatively different from others' preference structures. Also, what influences the acceptance of a food might vary with the situation (e.g., whether other foods are present, expected, or recently experienced, or what the somatic or social context is).

Moreover, the preferred value of an influence sometimes is at an intermediate level. This is certainly the case for most of the salient sensory influences on acceptance. This creates a preference peak instead of the monotonic relationship between an influence and acceptance that standard modeling techniques assume. Different consumers have different peak values or ideal points for each influence. Most devastating of all to the grouped analysis of preference structures, people put together influences in idiosyncratic ways to generate overall acceptance, preference, and choice.

Therefore, we have only one general approach to the understanding and measurement of food acceptance behavior. This is to collect the individual's acceptance data together with data on the major influences relevant to the objectives of the investigation and to complete an analysis of each person's preference structure before attempting to generalize or aggregate.

4. Individual Measurement of Acceptance Structure

Each discrete influence on a person's food acceptance forms a subscale or psychophysical function on the acceptance responses. Even if this causal relationship is peaked, this peak is symmetrically linear whenever the consumer is allowed to express unbiased personal preference in a sufficiently familiar situation (Fig. 2). The basis of this theory is that people decide what they want on the basis of the differences of the alternatives

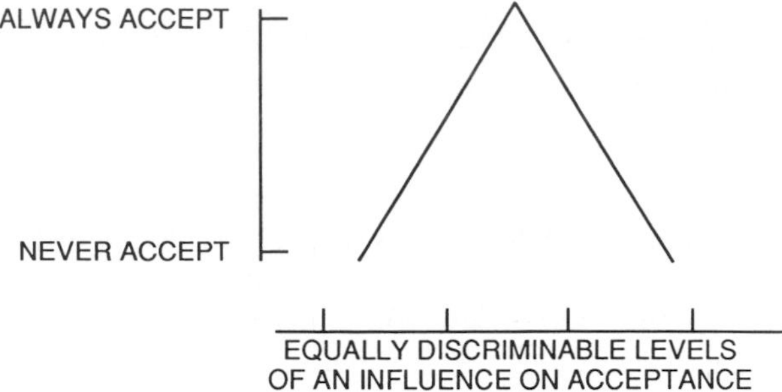

FIGURE 2 The acceptance triangle (also called the appetite triangle, the isosceles tolerance triangle, and the hedonic inverted V) represents the cognitive mechanism by which the excitation of an individual consumer's disposition to accept food declines proportionally to the discriminable distance of the presented level of a discrete or integrated influence on acceptance from the personally acquired most facilitatory level of that influence.

available from the most preferred levels of the operative influences, measured in units of equal discriminability.

Then, in the probably not uncommon case of a food acceptance behavior that is explicable by a single integration point (Fig. 3), acceptance has a "cognitive algebra" of Pythagorean combinations.

The personal acceptance structure can be expressed graphically as a response space. This maps the variation in acceptance by that individual against variations in all of the potential influences over their tolerated ranges. These personal response spaces can simply be summed across a representative sample of consumers, to provide highly precise, completely operational, and totally disaggregatable estimates of the

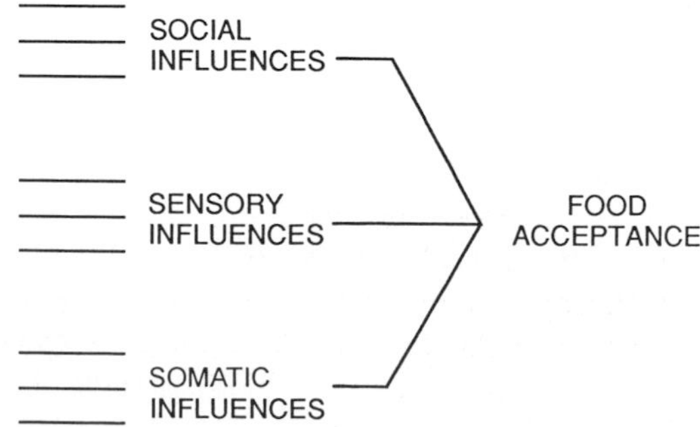

FIGURE 3 Different categories of immediate influence on an individual purchaser's or eater's acceptance of a food item, integrated through a single personal equation established by effects of past experience on inherited potentialities. This is the simplest type of causal model and such a multidimensional discrimination hierarchy should be refuted before multinodal attitudinal networks (such as in Fig. 1) are used in data analysis.

response to particular propositions in any segment of the market.

III. SENSORY INFLUENCES

A. Limits on Sensory Evaluation

The foregoing theoretical and methodological criteria for scientific understanding of influences on food acceptance have, until recently, appeared unachievably strict to those working on the sensory evaluation of foods, even when tackling relatively fundamental empirical or methodological issues. The lack of a psychologically scientific basis for sensory methods has not been noticed because the established atheoretical approaches to human data have helped food research and development personnel out of many problems that are beyond chemical or physical theory and measurement.

A technologically informed search for statistical differences or patterns in verbal responses to food samples has often sufficed to point to practical solutions to common problems, such as matching the current brand when the source of an ingredient changes, identifying and avoiding off-quality features (e.g., staling, spoiling, and taints), or working out the processes that yield the most popular version of a food for an acceptable level of expenditure by the manufacturer.

Moreover, the procedures for getting such pragmatic conclusions from descriptions of many characteristics of food samples by panels of expert, trained, or untrained assessors have been extended to more general issues. Careful use of quantitative descriptive analysis (i.e., "profiling") has illuminated fundamental problems in food perception, at least in cases in which a monotonic relationship exists in most or all assessors between the levels of each relevant physicochemical factor in the food and the rated intensity of a readily described sensation.

However, sensory interactions are more difficult to elucidate. This is because additive combinations of intensity ratings will not suffice, even if every panelist uses qualitatively the same rule for integrating the two or more physicochemical influences on the complex sensation. Furthermore, without a realistic formula for the interaction, quantitative estimates of the sensed effects of different formulations become unreliable. This could be of practical significance, as well as blocking the development of theory.

Most crucial of all for the extension of sensory evaluation to elucidate influences on food acceptance, there is typically a nonmonotonic relationship between the acceptance response and levels of a physicochemical factor or the relevant intensity scores. This means that the group average response contains little, if any, more information than the distribution of most preferred levels across the panel, obtainable from a preference test without any sensory data. When more than one sensory factor is influencing acceptance, the interactions that enter into the response are lost by constructing a response space from panel data.

Advanced statistical models have been developed to tackle these problems. These have sometimes revealed possibilities for product development not readily identified by conventional market research (or by experts or sensory panels). However, the power of such approaches to distinguish between qualitatively alternative consensus models of sensory integration into acceptance is usually low. The prevalence of the consensus is hard to estimate, and nonconsensus patterns (as well as their prevalences) are not identified.

The reliability of the estimates of the consensus optimum is low for all of these reasons and is much lower than those of a psychological approach at the same data-gathering cost. Even more serious, the validity of the estimates of response in the actual market is quite dubious. The polynomials or ideal points extracted are distorted by forcing everyone through the same set of samples without regard to personal ranges of tolerance. This also degrades the estimate of the individual's quantitative variant of the consensus preference structure. Worst of all from the practical point of view, as well as for growth in scientific understanding, the conclusions of a pure data-fitting operation apply only to the data set. In contrast, if the data had been collected and analyzed in a way that could identify ways in which the tested sensory variations were influencing each consumer's acceptances in a normal situation of choice, then some generalizations to other stimulus sets could be made, as well as a rather precise aggregate response surface constructed for whatever subset of the panel was of interest.

B. Sensory Integration in Individuals' Acceptances

Most appearances, flavors, and textures are themselves based on two or more physicochemical factors in the foodstuff. Insofar as these integrated percepts enter consciously or unconsciously into acceptance, each individual operates according to an algebraic formula for combining the inputs.

1. Intramodal Integration

a. Visual and Tactile Pattern

Psychological research on visual object recognition is, at the moment, grappling with highly abstract fundamental issues, often using verbal and pictorial materials. Visual integration phenomena directly relevant to food acceptance have yet to be examined.

The perception of mechanoreceptor patterns (i.e., tactile textures) has not been extensively studied. Some work on roughness and smoothness of touch to the skin of the arm should be extended to the feel of food in the mouth. Many of the textural sensations from homogeneous fluid foods are empirically related to various measures of viscosity, but the precise uses that the tongue and the palate make of viscous forces, and how they are combined with other physical actions of the fluid, remain to be elucidated. The geometry of the globules in cream is important to its smoothness, for example, over and above the effect of viscosity or thickness. Solid and inhomogeneous foods are not understood psychophysically at all, and progress is slow because of the physical transformations of the stimulus during mastication.

b. Aroma

Olfactory pattern recognition will probably be elucidated first by working "top-down" from the perceptual relationships among food aroma concepts, to complement the currently dominant "bottom-up" approach to odorant receptors and olfactory bulb neurophysiology. This would replace the traditional structure–activity approach of building statistical spaces out of supposed descriptions of subjective features corresponding to receptor types. Instead, personal summations and—more likely—interactions of molecular moieties in the recognition of familiar namable odor sources would be characterized by the discrimination analysis outlined earlier. [*See* Olfactory Information Processing.]

c. Taste

Gustatory integration is now gaining attention, although effective links to food acceptance have yet to be published.

Group analysis of raw ratings have been applied to mixtures of different sweeteners. The results have supported qualitative rules of combination; however, it has been pointed out that physical interactions could account for the observations and so these are not necessarily cognitive integration rules. Moreover, these designs are not strong enough to yield quantitative cognitive integration formulas; in any case this might require individual analysis.

There are many obvious hypotheses as to how tastant perceptions interact to contribute to well-known food flavors. Psychophysical data support the intuitive impression of generally subtractive relationships between sweetness and either sourness or bitterness. Individual combination of discriminable differences from ideal levels of sweetener and coffee solids in instant coffee provides further support.

The individualized discrimination approach to the tastants' contribution to likings for real foods could elucidate issues such as whether umami (the taste of, e.g., monosodium glutamate in tomato or chicken) is a fifth "basic" taste or only a mixture of the four classic tastes, as it certainly can be described subjectively and classified neurophysiologically. [*See* Tongue and Taste.]

2. Cross-modal Integration

An advantage of taking the cross-modal integration approach to sensory influences on acceptance is that work can progress in parallel at different levels of interaction, or "chunking." The natural stimuli for food aromas, tastes, textures, and visual appearances can be used to study the higher level of integration into the entire food percept and on into its most preferred form for a certain purpose.

C. Sensory Influences on Later Food Acceptance

If the only perceptible differences between the foods in the test situation are inherent features (rather than verbal or pictorial labels or packaging, e.g., or sight of the choices that others are making), then sensory influences must at least have contributed to the choice or relative acceptance of one food over the other(s) and indeed to any aftereffects of ingestion.

However, the perceived physical or chemical characteristics of a food can be heavily interpreted, and these attributions affect immediate or subsequent acceptance of foods as much as the "pure" sensations: for example, sweetness or creaminess might be sensually attractive, but feared for its usual health implications. The latter would be a clearly "cognitively" or conceptually mediated sensory influence.

Such effects are immediate, on the acceptance of the same food as is sensed, while it is being sensed. That is what is usually meant by palatability or sensory influences on acceptability. Also, however, the high acceptability of a food, or even the innately liked taste of sweetness itself, can elicit physiological responses that could affect subsequent acceptance of any foods. Proposed mediators of possible delayed effects of sweetness, for example, include neurally triggered changes in liver metabolism or the secretion of insulin from the pancreas or of endogenous opioids in the brain; there is, though, no satisfactory evidence for such effects in adults.

It is arbitrary to assume that a delayed sensory effect is mediated either conceptually or physiologically unless the alternative has been excluded by sensitive measurements. This cannot be done by a single verbal rating or just one biochemical assay, because a wide variety of mechanisms of conceptual and physiological mediation are possible, some quite complex. On the cognitive front, for example, sweetness from an artificial sweetener in a test food might be interpreted by someone who is trying to limit eating as indicating the consumption of a substantial amount of sugar calories, an idea that might so worry some people that they despair and lapse to unrestrained eating, instead of cutting back in compensation. We have evidence for this disparity between dieters and nondieters after eating soup with a "rich" aroma and texture. This might also happen with a low-calorie creamer.

IV. SOMATIC INFLUENCES

Naturally enough, it has mostly been biological psychologists and medical physiologists who have sought to characterize physiological influences on human appetite. As we have seen, however, this is not a biological issue in any sense that escapes thorough empirical analysis of the cognitive and psychosocial influences on the observed food acceptances (or on verbal data predictive of ingestion). On the other hand, progress on this set of issues cannot be made either by an exclusively psychological approach that lacks effective manipulations or by measurements of the hypothesized processes under the skin. As yet, expertise in both human physiology and cognitive psychology, with access to the necessary facilities, is rare in one person or in an effective collaboration. As a result, disappointingly little can presently be said that is definite about biological aspects of food acceptance in human consumers.

As was well recognized 20 years ago, although it has often been neglected more recently, the main methodological problem is that physiological influences on food acceptance are normally thoroughly confounded with cognitive influences. Emptying of the stomach is rather precisely related to awareness of the lapse of time since the last meal. Filling of the stomach during a meal is even more closely associated with the perceived amount eaten. Different postingestional effects of the nutrient composition of foods cannot be separated from differences in the "image" of those foodstuffs and their sensed constituents.

Unfortunately, most of those experiments that have exposed postingestional influences of food eaten at a meal, by disguising its amount or composition, have recorded only the amounts or temporal patterns of subsequent food intake and have not attempted to measure or control postingestional effects or their timing. Differences in acceptance pattern among conditions (or people) fail to identify the somatic influences that are operative.

A sufficiently high dose of glucooligosaccharides produces a suppression of actual and rated food acceptance after 20–30 minutes. At that time most of this carbohydrate has left the stomach and glucose is being absorbed rapidly. This is consistent with the suppression of appetite by energy production from the glucose (from animal experiments, thought to be in the liver). Action on receptors in the intestinal wall has not been excluded, though. This decrease in food acceptance is no greater than that related to guessing that caloric content of the meal is high; this shows the power of autosuggestion.

Infusion of fat into the small intestine has been shown to delay the rise of rated hunger sensations, whereas this sort of effect is not produced by intravenous infusion. This points more conclusively to the importance of gut wall stimulation among the physiological influences on human food acceptance.

V. SOCIAL INFLUENCES

Consumer survey data and sales trends collected by the food business provide material for elaborate discussion of socioeconomic, interpersonal, and cultural influences on food acceptance. The psychological content would, however, be largely intuitive and speculative. This is because, once again, little research has been done that tests how each consumer makes purchasing decisions about foods, let alone chooses when, what, and how much to eat. Thus, the important task is to put social influences on acceptance on the same scientific footing as sensory influences.

Interpersonal and cultural influences are not inherently any less objective than sensed food qualities or perceived physiological processes. The difference between social factors and the sensory and somatic factors is that many social influences are mediated linguistically; that is, they exist only because of communicable symbolic meaning. As a result they are not effectively describable in purely physicochemical terms. Nevertheless, the words or sentences applied to the food on the package or in the advertisement are entirely objective potential influences on speakers of the language who have been educated into the constructs being communicated. The same applies to nonverbal symbols in the pictorial "language" being used.

Some symbols have inherent potential for scaling on a dimension of influence (e.g., price, shelf-life date, and perhaps recommended storage temperature). These and essentially nominal labels must be scaled, however. Scaling consists of analyzing out the relative strengths of influence of the labels on one or more underlying dimensions accounting for a descriptive or acceptance response. This is most realistically and hence effectively achieved by mapping the labels onto differences in response, in the same way that influences are identified from inherently graded physical influences (e.g., food composition).

A neat illustration is provided by the interaction of the labeled nature of a food constituent with the taste of that constituent, in a manner that is rational to the meaning of the label. The label for a sweetener might be "low calorie" or "sugar," for example. The influence of the label should multiply by the level of sweetener to generate an estimate of calorie content. This is indeed observed in some consumers, so that the fattening potential of the food and its influence on food choice are rated to be higher for a strongly sweet-tasting food labeled "sugar-sweetened" than can be accounted for by the addition of the effects of the sugar label alone and sweetness alone. In other consumers, labeled calories and sweetness are treated as distinct dimensions or even added onto a single dimension.

Similar interactions of the effect of prices with other salient attributes of the range of alternatives can provide realistically scaled price–purchase elasticity functions. In recent years such trade-offs between brand attributes have been characterized by nonmetric methods of group data analysis. For much the same data-gathering cost the individualized discrimination approach to cognitive model testing gives precise quantitative diagnoses of attribute interactions in each consumer sampled. The results not only imply practical recommendations for food design, they also tell us something theoretical about the mechanisms of acceptance in the tested situation.

BIBLIOGRAPHY

Bennett, G. A. (1988). "Eating Matters. Why We Eat What We Eat." Heinemann Kingswood, London.

Boakes, R. A., Burton, M. J., and Popplewell, D. A. (eds.) (1987). "Eating Habits." John Wiley & Sons, Chichester, England/New York.

Booth, D. A. (1994). "Psychology of Nutrition." Taylor & Francis, London/New York.

Capaldi, E. D., and Powley, T. L. (eds.) (1990). "Taste, Experience, and Feeding." American Psychological Association, Washington, D.C.

Dobbing, J. (ed.) (1987). "Sweetness." Springer-Verlag, Berlin/New York.

Fieldhouse, P. (1995). "Food and Nutrition" 2nd Ed. Chapman & Hall, London/New York.

Friedman, M. I., and Kare, M. R. (eds.) (1990). "Chemical Senses: Appetite and Nutrition." Dekker, New York.

Logue, A. W. (1991). "The Psychology of Eating and Drinking," 2nd Ed. Freeman, New York.

Manley, C. H., and Morse, R. E. (eds.) (1988). "Healthy Eating—A Scientific Perspective." Allured, Wheaton, IL.

Mennell, S., Murcott, A., and van Otterloo, A. H. (1992). "The Sociology of Food: Eating Diet and Culture." Sage, London/New York.

Ramsay, D. J., and Booth, D. A. (eds.) (1991). "Thirst. Physiological and Psychological Aspects." Springer-Verlag, London/Berlin.

Ritson, C., Gofton, L., and McKenzie, J. (eds.) (1986). "The Food Consumer." John Wiley & Sons, Chichester, England/New York.

Shepherd, R. (ed.) (1989). "Handbook of the Psychophysiology of Human Eating." John Wiley & Sons, Chichester, England/New York.

Solms, J., Booth, D. A., Pangborn, R. M., and Raunhardt, O. (eds.) (1988). "Food Acceptance and Nutrition." Academic Press, London.

Thomson, D. M. H. (ed.) (1989). "Food Acceptability." Elsevier, London.

Food Allergies and Intolerances

STEVE L. TAYLOR
University of Nebraska

GLOSSARY

Anaphylactoid reaction Intolerance characterized by the nonimmunological release of mediators from mast cells and basophils

Basophil Blood cell that contains the mediators of allergic disease

Celiac disease Malabsorption syndrome apparently caused by a cell-mediated immunological response to proteins from wheat, rye, barley, triticale, and oats

Food allergy Form of food sensitivity characterized by an abnormal immunological response to food components, usually proteins

Food intolerance Form of food sensitivity that involves nonimmunological responses to food components

Idiosyncratic reaction Intolerance that occurs through an unknown mechanism

Immunoglobulin E (IgE) Class of antibodies that is involved in allergic responses

Mast cell Cell present in many tissues that contains the mediators of allergic disease

Metabolic food disorder Intolerance characterized by a defective capacity for the metabolism of a specific food component or by a heightened sensitivity to a food component caused by a defect in a specific metabolic pathway

FOOD ALLERGIES AND INTOLERANCES ARE THOSE adverse reactions to foods and food ingredients that affect only certain individuals in the population. Most individuals could ingest these foods with no adverse consequences. True food allergies involve an abnormal immunological response to some substance in the food, usually a naturally occurring protein. Food intolerances can occur through a variety of mechanisms, including (1) metabolic disorders, in which the individual's genetic makeup fails to provide sufficient quantities of an enzyme that is critical for the metabolism of a foodborne substance; (2) anaphylactoid reactions, which are allergy-like illnesses caused by the nonimmunological release of potent mediators such as histamine; and (3) idiosyncratic reactions that occur through unexplained mechanisms.

I. INTRODUCTION AND CLASSIFICATION

The Roman philosopher Lucretius was quoted as saying, "One man's food may be another man's poison." Although Lucretius was not referring to food allergies and intolerances, his quote fits these illnesses very well. Though many foodborne illnesses have the potential to affect everyone in the population equally, food allergies and intolerances are individualistic—they affect only a few individuals in the population.

Collectively, food allergies and intolerances can be called food sensitivity, which is defined as an abnormal physiological response to a particular food. This same food would be safe and nutritious for the vast majority of consumers. Only a small percentage of consumers are affected by food sensitivities. Food allergies involve an abnormal immunological response to a particular food or food component, usually a protein. Food intolerances occur through a variety of nonimmunological mechanisms.

II. FOOD ALLERGIES

True food allergies can occur through two different immunological mechanisms, the humoral or anti-

ENCYCLOPEDIA OF HUMAN BIOLOGY, Second Edition, VOLUME 4.

body-mediated mechanism and the cellular mechanism. The most common type of food allergies involves mediation by immunoglobulin E or IgE. [*See* Allergy.]

A. IgE-Mediated Food Allergies

IgE-mediated food allergies are sometimes called immediate hypersensitivity reactions because of the short onset time (minutes to a few hours) between the ingestion of the offending food and the onset of allergic symptoms. In IgE-mediated food allergies, the body produces allergen-specific IgE in response to exposure to a food allergen, usually a protein. The IgE antibodies are highly specific; the formation of IgE specific to one food will not confer allergy to another food unless the two foods are closely related and cross-reactive. Some food proteins are more likely to elicit IgE antibody formation than others. Although exposure to the food is critical to the development of allergen-specific IgE, exposure will not invariably result in the development of IgE antibodies even among susceptible people. Many factors are likely to influence the formation of allergen-specific IgE antibodies, including the susceptibility of the individual, the immunogenic nature of the food and its proteins, the age of exposure, and the dose, duration, and frequency of exposure.

The allergen-specific IgE is produced by plasma cells and attaches itself to the outer membrane surfaces of specialized cells called mast cells and basophils. Mast cells are present in many different tissues and basophils are found in the blood. In this process, known as sensitization, the mast cells and basophils become sensitized and ready to respond to subsequent exposure to that particular food allergen. However, sensitization is symptomless, and no adverse reactions will occur without subsequent exposure. Sensitization to a particular allergen distinguishes allergic individuals from nonallergic individuals.

Once the mast cells and basophils are sensitized, subsequent exposure to the allergen results in the release of a host of mediators of allergic disease that are either stored or formed by the mast cells and basophils. The allergen cross-links two IgE molecules on the surface of the mast cells and basophils, which elicits the mediator release. Mast cells and basophils contain large numbers of granules that store certain mediators that are released during this process. Several dozen different mediators have been identified. Histamine is one of the primary mediators of IgE-mediated allergies and is responsible for many of the early symptoms associated with allergies. Many of the other mediators, such as the prostaglandins and leukotrienes, are involved in the development of inflammation. The interaction of a small amount of allergen with the allergen-specific IgE antibodies results in the release of comparatively large quantities of mediators into the bloodstream and tissues. Thus, exposure to trace amounts of allergens can elicit symptoms. This mechanism of IgE-mediated reactions is involved in many different types of allergies to foods, pollens, mold spores, animal danders, bee venom, and pharmaceuticals. Only the source of the allergen is different.

TABLE I
Symptoms of IgE-Mediated Food Allergies

Gastrointestinal symptoms	Cutaneous symptoms
Vomiting	Urticaria (hives)
Diarrhea	Dermatitis
Nausea	Angioedema
Respiratory symptoms	Other symptoms
Rhinitis	Anaphylactic shock
Asthma	Laryngeal edema

A variety of symptoms can be associated with IgE-mediated food allergies (Table I). The nature of the symptoms will depend on the tissues affected by the various mediators. Fortunately, allergic individuals typically suffer from only a few of the many symptoms. With food allergies, symptoms associated with the gastrointestinal tract are quite common since this is the organ of insult after ingestion of food. Skin symptoms are also fairly common. Respiratory symptoms are less frequently involved with food allergies than with various inhaled allergens such as pollens and animal danders. However, asthma is a very serious, though uncommon, manifestation of food allergies.

Most of the symptoms of IgE-mediated food allergies are not particularly definitive. The gastrointestinal manifestations of food allergies can also be associated with many other causes. Though there are millions of asthmatics, relatively few are allergic to foods.

Anaphylactic shock is by far the most serious manifestation of food allergies. Thankfully, very few individuals with food allergies experience anaphylactic shock after ingestion of the offending food. It involves multiple organs and is often accompanied by gastrointestinal, cutaneous, and respiratory symptoms in combination with a dramatic fall in blood pressure and

TABLE II
Most Common Allergenic Foods Involved in IgE-Mediated Allergies

Infants	Adults
Cows' milk	Peanut
Eggs	Crustacea
Peanut	Tree nuts
Soybean	Eggs
Wheat	Cows' milk
Tree nuts	Fish
	Molluscs
	Soybean

cardiovascular complications. Death can ensue within minutes of the onset of anaphylactic shock.

The severity of an allergic reaction will be dependent to some extent on the amount of the offending food that is ingested. If an exquisitely sensitive individual inadvertently eats a large amount of the offending food, their reaction will be more serious than if the exposure was to a trace amount. Yet, trace quantities can elicit noticeable reactions owing to the large release of mediators.

The most common allergenic foods in the United States are listed in Table II. To some extent, the most common allergenic foods vary from one culture to another. Peanut allergy is more common in the United States than in most other countries, probably because peanuts are eaten more often in the United States and peanut butter is introduced at an early age. In Japan, soybean allergy is very common, probably because the Japanese eat large quantities of soybeans. Any food that contains protein has the potential to elicit an allergic reaction in someone. The most common allergenic foods tend to be foods with high protein content that are frequently consumed. The exceptions are beef, pork, chicken, and turkey, which are uncommon allergenic foods despite their frequent consumption and high protein content.

The prevalence of IgE-mediated food allergies is not precisely known. Infants have the highest prevalence of such allergies, estimated at 4–8%. Cows' milk is the most common allergenic food in infancy, followed by eggs and peanuts. The prevalence of IgE-mediated food allergies in older children and adults is even less certain. However, the prevalence in adults is thought to drop below 1% of the population. Thus, many infants and young children outgrow their IgE-mediated food allergies. The reasons for the development of tolerance to previously allergenic foods are not understood, but may involve the development of blocking antibodies of other types, especially IgG and IgA. Allergies to some foods, such as cows' milk and eggs, are more frequently outgrown than allergies to other foods, such as peanuts. Thus, peanut allergy is the most common food allergy among adults, at least in the United States.

The diagnosis of food allergies is typically a stepwise practice. A good medical diagnosis is important because parental diagnosis and self-diagnosis are often wrong, identifying incorrect foods and identifying too many foods. Most individuals with IgE-mediated food allergies are allergic to one or two foods, rarely to more than three foods. Occasionally, allergies can exist to a group of closely related foods, for example, the crustacea (shrimp, crab, lobster, and crayfish). Thus, the goal of medical diagnosis is to establish a cause-and-effect relationship between one or a few foods and the onset of allergic symptoms. Once that has been determined, the involvement of IgE must be ascertained.

Most physicians begin the diagnosis by compiling a careful history of the patient's adverse reactions. Critical information would include the identity of the foods eaten immediately before the onset of symptoms, the amount of various foods consumed, the type, severity, and consistency of symptoms, and the time intervals between eating and the onset of symptoms. In complicated situations, histories are often needed from several episodes to reach a probable diagnosis. Challenge tests with the suspected food(s) are often necessary to establish with certainty the role of a specific food in the reaction. The double-blind, placebo-controlled food challenge (DBPCFC) is considered the most reliable procedure. In the DBPCFC, neither the patient nor the medical personnel know when the food (in capsules or disguised in another food or beverage) is going to be administered and when the placebo is to be administered. Thus, the DBPCFC is free of bias. Single-blind and open challenge tests also have value in some situations. Sometimes, history alone is sufficient to make the diagnosis if the cause-and-effect relationship is particularly compelling. Challenge tests are seldom used on individuals who experience life-threatening allergic reactions.

The diagnosis of an IgE-mediated mechanism can be made with either the skin prick test (SPT) or the radioallergosorbent test (RAST). The simplest procedure is the SPT, in which a small amount of a food extract is applied to the patient's skin and the site is pricked with a needle to allow entry of the allergen.

A wheal-and-flare response (basically a hive) at the skin prick site demonstrates that IgE affixed to skin mast cells has reacted with some protein in the food extract. Histamine usually serves as the positive control. An alternative procedure is the RAST, which uses a small sample of the patient's blood serum. In the RAST, the binding of serum IgE to food protein bound to some solid matrix is assessed using radiolabeled or enzyme-linked antihuman IgE. The RAST is considerably more expensive than the SPT and is equally reliable, sensitive, and specific. The RAST is preferred for patients with extreme sensitivities. It should be emphasized that a positive SPT or RAST in the absence of a history of allergic reactions to that particular food is probably meaningless. The SPT and RAST are the most frequently used and reliable tests to assess the role of IgE in an adverse reaction.

The primary means of treatment for IgE-mediated food allergies is the specific avoidance diet, for example, if allergic to peanuts, don't eat peanuts. With IgE-mediated food allergies, the degree of tolerance for the offending food is often very low, so the construction of a safe and effective avoidance diet can be quite difficult. The contamination of one food with another from the use of shared processing equipment and preparation equipment (utensils, cooking surfaces, pots and pans, etc.) can be sufficient to elicit an allergic reaction. Also, patients must have considerable knowledge of food composition and recognize the sources of ingredients. For example, casein, whey, and lactose are ingredients derived from cows' milk and would be hazardous for individuals with cows' milk allergy. The ingredient must contain the protein component, which some foods (e.g., peanut oil and soybean oil) do not unless they have been contaminated during use. The careful scrutiny of food labels is critical to avoidance of allergic reactions. But undeclared uses of food can occur, especially through restaurant meals. As a result, many inadvertent exposures occur among allergic consumers who are attempting to avoid their offending food(s).

Cross-reactions are another perplexing issue for food-allergic consumers. Cross-reactions do not always occur between closely related foods, for example, many individuals are allergic to peanuts and most of these are allergic only to peanuts. A few of these individuals are cross-reactive with one or more other legumes, usually soybeans. Cross-reactions frequently occur among the crustacea but less commonly among the finfish. Cows' milk and goats' milk invariably cross-react, as do the eggs of various avian species. Cross-reactions can also occur between foods and other environmental allergens. The most common examples are the cross-reactions between some fresh fruits and vegetables and certain pollen allergies in some individuals and the cross-reaction between natural rubber latex and bananas, kiwis, and chestnuts. The basis for cross-reactions is frequently not known.

Attempts to prevent the development of food allergies in susceptible infants are often futile. Infants born to parents with histories of allergic disease are much more likely than other infants to develop food allergies. The avoidance of commonly allergenic foods such as cows' milk, eggs, and peanuts primarily through breast feeding appears to delay but not prevent the development of food allergies.

A few specialized hypoallergenic foods are available in the marketplace. These foods are intended for infants who have developed allergies to infant formula made with cows' milk. The most effective hypoallergenic infant formulae are based on extensively hydrolyzed casein. Although casein is a common milk allergen, the hydrolysis of its peptide bonds renders it safe for most cows' milk-allergic infants.

Other approaches to the treatment of food allergies are not prophylactic. Immunotherapy (e.g., allergy shots or sublingual food drops) is considered experimental and controversial therapy for food allergies. Other dietary approaches such as rotation diets are similarly controversial.

Pharmacological approaches can be used to treat the symptoms of allergic reactions. In particular, epinephrine (also known as adrenalin) is prescribed for individuals who experience life-threatening food allergies. Its early administration after inadvertent exposure to the offending food can be life-saving for such patients. Antihistamines are another group of drugs used frequently to counteract the symptoms of allergic reactions, yet they counteract only those symptoms associated with histamine and not those attributable to other mediators.

B. Cell-Mediated Food Allergies

Cell-mediated food allergies are sometimes called delayed hypersensitivity reactions. The symptoms of these reactions typically appear 6–24 hr after consumption of the offending food. Cell-mediated allergies involve the interaction of food allergens with sensitized lymphocytes. These reactions occur without the involvement of antibodies. The interaction between the allergen and the sensitized lymphocyte results in lymphokine production and release, lymphocyte proliferation, and the generation of cytotoxic T lymphocytes. Lymphokines are soluble proteins that exert potent effects on tissues and cells, which results in localized inflammation. Lymphocyte proliferation increases the number of reactive cells, which magnifies the inflammatory process. The generation of cytotoxic or killer T cells results in cell destruction.

The T lymphocytes responsible for cell-mediated allergies are prevalent in the gut-associated lymphoid tissue. Gut lymphocytes are likely to be very critical in food-related, delayed hypersensitivity. However, the inaccessibility of these lymphocytes has made the evaluation of the role of cell-mediated reactions in food allergies quite difficult. As a result, the prevalence and importance of cell-mediated food allergies remain unknown.

Celiac disease, also known as gluten-sensitive enteropathy, is the one illness that seems likely to involve a cell-mediated mechanism. Celiac disease occurs in certain individuals following the ingestion of wheat, rye, barley, triticale, and sometimes oats. Although the mechanism of celiac disease is not completely understood, it appears to involve an immunocytotoxic reaction mediated by intestinal lymphocytes. On ingestion of proteins from the offending grains, the absorptive cells of the small intestinal epithelium are damaged. The absorptive function of the small intestine is thus severely compromised, resulting in a malabsorption syndrome. The symptoms of celiac disease include diarrhea, bloating, weight loss, anemia from compromised iron absorption, bone pain from inadequate calcium absorption, chronic fatigue, weakness, muscle cramps, and, in children, failure to gain weight and growth retardation. Celiac disease is an inherited trait, but its inheritance is complex. The prevalence of this disease in the United States is about 1 in every 3000 individuals, although it occurs more frequently in Europe and Australia. Celiac disease rarely occurs in people of Chinese or African heritage. Its symptoms may begin at any age, and environmental factors, such as viral illness, may contribute to the onset of celiac disease in some cases. No evidence has been found of spontaneous recovery from the illness.

Celiac disease is triggered by the ingestion of the protein fractions, specifically the gluten proteins, of wheat and related grains. The gluten proteins are storage proteins in these grains that give them their characteristic baking functionality. Other grains, such as corn and rice, do not contain gluten proteins. As with IgE-mediated food allergies, most consumers do not react adversely to the ingestion of gluten. It is thought that the gluten proteins elicit a cell-mediated immune response in the intestinal lymphocytes.

A definitive diagnosis of celiac disease would necessitate a small bowel biopsy, in which the biopsy material is examined for evidence of flattened intestinal villi, which is characteristic of the disease. A normal appearance of subsequent biopsy material is restored after avoidance of wheat and related grains. Alternatively, a blood sample from the patient can be examined for the presence of antigluten antibodies. Many times, the diagnosis is made tentatively on the basis of symptomatic improvement after avoidance of wheat and related grains.

The treatment of celiac disease involves the total avoidance of wheat, rye, barley, triticale, and usually oats. Some, but not all, patients can tolerate oats. Ingredients that contain protein and are made from these grains must also be avoided, for small amounts of these foods can elicit adverse reactions. Unlike IgE-mediated allergies, the symptoms are usually delayed in onset by several hours to days. The normal absorptive function of the small intestine is restored within a few days with gluten avoidance.

III. FOOD INTOLERANCES

As noted earlier, food intolerances involve nonimmunological mechanisms. Food intolerances can be divided into three categories: metabolic disorders, anaphylactoid reactions, and idiosyncratic illnesses. Like the food allergies, food intolerances are individualistic adverse reactions to foods; most consumers tolerate the offending foods, or ingredients, very well. In contrast to food allergies, individuals suffering from food intolerances can usually tolerate small amounts of the offending food in their diet.

A. Metabolic Food Disorders

Metabolic food disorders result from inherited defects in the ability to metabolize some component of food. Lactose intolerance is the most common example.

1. Lactose Intolerance

Lactose intolerance results from a deficiency of the enzyme lactase or β-galactosidase in the intestinal mucosa. As a result, lactose, the principal sugar in milk, cannot be properly digested into its constituent monosaccharides, galactose and glucose. Normally, lactose is hydrolyzed to galactose and glucose, which are absorbed through the intestinal mucosa into the blood and used as sources of energy. The disaccharide lactose cannot be absorbed without hydrolysis to galactose and glucose. In individuals lacking β-galactosidase, undigested lactose passes from the small intestine into the large intestine. The large intestine is teeming with bacteria that metabolize the lactose to carbon dioxide and water. The result is abdominal cramping, flatulence, and frothy diarrhea, the primary symptoms of lactose intolerance.

The prevalence of lactose intolerance increases with advancing age. Lactose intolerance can have its onset as early as the teenage years, but often develops later in life. The intestinal activity of β-galactosidase tends to diminish with age so the symptoms of lactose intolerances often worsen with increasing age. The prevalence of lactose intolerance varies with different ethnic groups. Though the prevalence among Caucasian Americans is only 6–12%, lactose intolerance occurs in 60–90% of some ethnic groups, including Greeks, Jews, Arabs, African-Americans, and many Asians.

The diagnosis of lactose intolerance is established by use of the lactose tolerance test. In this test, a fasting person is given an oral dose of 50 g of lactose. In normal individuals, this oral lactose challenge would result in an increase in blood glucose of at least 25 mg/dl over a few hours. In those individuals with lactose intolerance, the rise in blood glucose is less than 25 mg/dl.

The usual treatment for lactose intolerance is the avoidance of dairy products containing lactose. However, many individuals with lactose intolerance can tolerate some lactose in their diets, often as much as an 8-oz. glass of milk. The 50-g lactose challenge used in the lactose tolerance test is equivalent to the amount of lactose in a quart of milk, so it is possible to be clinically lactose intolerant but still able to ingest a glass of milk with no symptoms. Lactose-intolerant individuals can often consume yogurt and acidophilus milk that contain active microbial cultures with β-galactosidase. Lactose-hydrolyzed milk is also sometimes available in the marketplace. The tolerance of lactose-intolerant individuals for lactose varies, so the development of a safe and effective avoidance diet must also be individual. Tolerance for lactose may decrease over time. [*See* Lactose Malabsorption and Intolerance.]

2. Favism

Some individuals experience acute hemolytic anemia following the ingestion of broad beans, also known as fava beans. This illness, known as favism, is manifested by pallor, fatigue, dyspnea, nausea, abdominal and/or back pain, fever, and chills. Occasionally, more severe symptoms occur, including hemoglobinuria, jaundice, and renal failure. Favism affects individuals with a deficiency of the enzyme glucose-6-phosphate dehydrogenase (G6PDH) in their red blood cells. The red blood cells of individuals with G6PDH deficiency are susceptible to oxidative damage by two substances, vicine and convicine, found in fava beans. G6PDH deficiency is the most common enzymatic defect affecting human populations worldwide, with perhaps 100 million affected individuals. The prevalence of this defect varies among ethnic groups and is highest among Asian Jews in Israel, Sardinians, Cypriot Greeks, African-Americans, and Africans. Favism would be extraordinarily common if fava beans were more widely consumed.

B. Anaphylactoid Reactions

Anaphylactoid reactions are caused by substances that can induce the spontaneous release of histamine and other mediators from mast cells and basophils without the intervention of IgE. The symptoms of an anaphylactoid reaction mimic those of IgE-mediated allergies, but it would not be possible to demonstrate the existence of allergen-specific IgE. The evidence for the involvement of anaphylactoid reactions in food sensitivities is circumstantial. No foodborne substances have been identified as mast cell-destabilizing agents, although several pharmaceuticals demonstrate such effects.

Strawberry "allergy" is considered the premier example of an anaphylactoid reaction, and they are well known to cause hives in some individuals. Strawberries contain little protein, no allergens have been found, and no evidence exists for strawberry-specific IgE. An anaphylactoid mechanism is a logical possibility, but more proof is required.

C. Idiosyncratic Illnesses

Idiosyncratic illnesses refer to adverse reactions to foods experienced by certain individuals for which the mechanism is unknown. Many such illnesses may exist, although the association with specific foods or food ingredients has often not been proven. Double-blind, placebo-controlled challenge procedures are useful in the establishment of a cause-and-effect relationship to a specific food or ingredient, but such diagnostic procedures have not been widely employed. Obviously, examples of idiosyncratic illnesses abound, and the symptoms involved in these reactions range from very minor to life-threatening. The mechanisms, though unknown, are also likely to be quite variable. The best example of an idiosyncratic illness is sulfite-induced asthma, because the association with sulfite ingredients is well established but the mechanism is unknown. This illness is discussed separately in another article [*see* Sulfites in Foods]. Other idiosyncratic illnesses, though poorly documented, are caused by tartrazine, aspartame, and monosodium glutamate.

1. Tartrazine

Tartrazine is an azo dye, also known as FD&C Yellow #5, that is allowed as a colorant in foods, beverages, and pharmaceuticals. Tartazine has been alleged to elicit urticaria (hives) and asthma in sensitive individuals. However, the association of tartrazine in elicitation of these symptoms remains unproven. This additive is likely involved very rarely in urticaria and asthma.

2. Aspartame

Aspartame is an approved food additive that is used as a nonnutritive sweetener and sold under the trade name Nutrasweet®. Aspartame has been implicated as a causative agent in urticaria and headache. However, well-designed clinical studies have failed to confirm this relationship. Aspartame is unlikely to be involved in idiosyncratic illnesses.

3. Monosodium Glutamate

Monosodium glutamate (MSG) is a widely used food ingredient that has flavor-enhancing properties. MSG ingestion has been linked primarily with Chinese restaurant syndrome (CRS) and asthma. However, the evidence supporting a cause-and-effect relationship between MSG and these or other symptoms is rather weak. The conclusion from several large trials of MSG and its role in CRS is that MSG may affect a very small segment of the population, but only after very high levels of exposure. CRS is characterized by headache, chest tightness, a burning sensation along the back of the neck, nausea, and diaphoresis occurring within minutes of the ingestion of a Chinese restaurant meal. CRS involves very subjective reactions, making it difficult to study objectively. CRS may be caused by some ingredient other than MSG. A few cases of MSG-induced asthma have been documented; high doses were necessary to provoke asthma in these few sensitive individuals. MSG-induced asthma appears to be, at worst, a rare phenomenon whose mechanism remains unknown.

IV. CONCLUSION

Food allergies and intolerances affect a small percentage of people. The reactions can range from mild reactions to life-threatening ones. With the exception of the idiosyncratic illnesses, food allergies and intolerances are mostly caused by naturally occurring substances in foods. Food allergies can be triggered by ingestion of very small amounts of the offending food, whereas food intolerances typically involve some degree of tolerance for the offending food.

BIBLIOGRAPHY

Brostoff, J., and Challacombe, S. J. (eds.) (1987). "Food Allergy and Intolerance." Bailliere Tindall, London.

Lemke, P. J., and Taylor, S. L. (1994). Allergic reactions and food intolerances. *In* "Nutritional Toxicology" (F. N. Kotsonis, M. Mackey, and J. Hjelle, eds.), p. 117. Raven, New York.

Metcalfe, D. D., Sampson, H. A., and Simon, R. A. (eds.) (1991). "Food Allergy-Adverse Reactions to Foods and Food Additives." Blackwell, Boston.

Perkin, J. E. (ed.) (1990). "Food Allergies and Adverse Reactions." Aspen Publishers, Gaithersburg, MD.

Schwartzstein, R. M. (1992). Pulmonary reactions to monosodium glutamate. *Pediatr. Allergy Immunol.* **3**, 228.

Simon, R. A. (1986). Adverse reactions to food additives. *N. Engl. Reg. Allergy Proc.* **7**, 533.

Strober, W. (1986). Gluten-sensitive enteropathy: A nonallergic immune hypersensitivity of the gastrointestinal tract. *J. Allergy Clin. Immunol.* **78**, 202.

Zeiger, R. S., and Heller, S. (1995). The development and prediction of atopy in high-risk children: Follow-up at age seven years in a prospective randomized study of combined maternal and infant food allergen avoidance. *J. Allergy Clin. Immunol.* **95**, 1179.

Food and Nutrition, Perception

RICHARD D. MATTES
KAREN L. TEFF
Monell Chemical Senses Center

GLOSSARY

Cephalic phase response Vagally mediated preabsorptive physiologic response to sensory stimulation

Chemical senses Taste and smell

Flavor Amalgam of input from the auditory, visual, somesthetic, olfactory, and gustatory systems

Gustatory Pertaining to taste

Hedonic Characterized by pleasure

Olfactory Pertaining to smell

Satiety Decreased hunger

Somesthetic Pertaining to touch, temperature, or pain

Trigeminal Fifth cranial nerve, which conveys information about touch, temperature, pain, and chemical irritation

AMONG THE NUMEROUS FACTORS REGULATING ingestive behavior are the flavors of foods. The perception of these flavors is determined by the interaction of the sensory properties of a given food and the sensory capabilities of the particular individual. Since these sensory parameters vary between and within both products and individuals over time, characterization of the role of sensory factors in food consumption has proven elusive. Nevertheless, evidence that sensory perception and food intake are related is available from clinical populations with disorders of food intake or sensory function as well as from healthy individuals. In addition to an influence on food selection, sensory stimulation, particularly chemosensory stimulation, may alter the utilization of nutrients derived from ingested foods. Finally, the use of these nutrients can, in turn, alter sensory function since sensory systems are composed of tissues with nutrient requirements.

I. SENSORY MEASURES

Sensory measurement as it relates to foods is complicated by the fact that there are two levels at which assessment is required. The first concerns the sensory properties of foods; the second concerns the sensory capacities of individuals. Sensory properties of foods consist of attributes such as texture, color, and odor as determined by each item's chemical composition. Whether these characteristics are detected and how they are interpreted are determined by the function of each individual's sensory systems. The interaction between the two determines the perceived flavor of foods. For example, humans are polymorphic with respect to their sensitivity to phenylthiocarbamide (PTC) and related compounds. Some individuals (approximately 70% of the Caucasian American population) find low concentrations of such compounds in the food supply to have a bitter taste, whereas others fail to perceive any bitterness from the same foods. This may influence the acceptability of these items since bitterness is generally a negative taste attribute.

Study of the sensory properties of foods has generally fallen into the domain of food science. This knowledge is used for applications such as new product development, product matching, product improvement, quality control, and product ratings. A variety of tests are used for each of these purposes. Difference and sensitivity tests are designed to determine whether particular product characteristics are discernible. Judgments about whether a stimulus is present in a given context (e.g., sucrose in ketchup) or what the quality of a stimulus may be (e.g., salty for NaCl in soup) are the two most common sensitivity measures. They are termed detection and recognition thresholds, respectively, and are only statistical concepts. They are merely the level of stimulus (e.g., spice) required so that the number of correct responses (e.g., present or not present) by observers in a defined testing situation will exceed chance performance with a preset level of probability. The difficulty of the task and, as a consequence, the absolute threshold value will be determined by testing conditions.

Descriptive sensory tests are used to identify and quantify product attributes. For example, judgments may be sought for the viscosity of an array of foods with varying gelatin levels. Once again, the absolute values of responses will be determined by the testing paradigm. Viscosity ratings for a particular food following sampling of a large array of either less viscous or more viscous samples will probably be higher in the first instance than in the second.

The third class of tests, affective tests, provide insights on the palatability and acceptability of products. Sensory responses may be obtained for levels of specific product attributes (e.g., sourness of lemonade) or for the product as a whole. Although the focus is on the product, wide intra- and interindividual variability in hedonic responses necessitates the use of a well-defined panel of judges tested under highly controlled conditions. An example of an intrasubject factor is the level of hunger or satiety of the respondent. Having subjects rate a rich dessert item when hungry or full will likely lead to discrepant responses. Differing previous experiences with foods can result in discrepant hedonic ratings between subjects. Whereas overtly bitter foods are generally regarded as distasteful in Western cultures, where exposure to such items is limited, they are considered more acceptable to groups whose customary diet includes bitter items.

Study designs for product evaluation generally involve obtaining sensory evaluations of foods or beverages with varying properties from observers with carefully controlled or at least well-defined characteristics (e.g., age, sex, educational level, economic status). This is particularly true of sensitivity and descriptive type tests. Instrumental analyses are also used, but in many instances the sensitivity of human sensory systems cannot be duplicated. Trained panelists are commonly used for these purposes. A trained panelist is an individual who, through practice and experience, is very familiar with a product or line of products and can reliably evaluate each of its attributes. It should be emphasized that these individuals are not selected on the basis of some innate supersensory capacity. Instead, they have learned to use their sensory capacities more fully. In contrast, untrained panelists are individuals with no particular expertise with a product. Often they are potential consumers of a product and are used primarily for tests of product acceptability.

Interest in the sensory capacities of humans has been particularly strong in the field of psychology as well as the basic physical sciences (e.g., physiology, chemistry, anatomy). Measurements are again generally made in the three domains of sensitivity, description, and affect. However, because of the difference in orientation (i.e., focus on the respondent instead of the stimulus), testing procedures often vary. Test stimuli are commonly held constant and individual variation in responses to these stimuli are the measures of interest.

Individual differences in sensory function may be attributable to both innate physiologic as well impinging environmental or external factors. An example of one innate physiologic influence is an individual's genetic constitution, which determines their sensitivity to the odors of different compounds. An undetermined number of individuals in the population have specific anosmias, the inability to perceive the odor of low concentrations of certain compounds (e.g., androstenone—the odorant that attracts swine to truffles). Since the perceived odor of a food is comprised a combination or profile of the volatile compounds that it releases, the absence of one constituent part will alter the overall quality of the food's odor.

External influences on sensory function include factors such as state of hunger, health status, and smoking habits. As these factors are in a constant state of flux, their influence and, as a consequence, the sensory function of an individual can vary widely over time. For example, nasal congestion may reduce olfactory (but not gustatory) sensitivity, smoking diminishes taste sensitivity acutely, and being hungry can enhance the pleasantness of tastes and smells.

Preference is the sensory attribute most strongly associated with food selection and ingestion. Preference responses are obviously based on discernment of product characteristics, but measures of sensitivity and intensity hold little predictive power with respect to ingestive behavior. They appear to be translated into an integrated hedonic message that is used to evaluate the acceptability of the item. Given the complexity of information used to derive such a judgment, it is not surprising that measurement of preferences is similarly complex. Responses to questions about the preferred frequency of ingestion of a food, preferred level of an attribute in the food, or preference for foods with a given profile of attributes can, and typically do, provide different impressions. For example, an affectively neutral food (e.g., bread, milk, butter) may be ingested frequently whereas a highly preferred item (e.g., pumpkin pie, lobster, caviar) may be ingested only on special occasions. A food with a high level of a given component (e.g., salt on potato chips) may be highly preferred, but other items with high levels of the same component can be disliked (e.g., anchovies). Only when multiple dimensions of taste preferences are assessed can predictions be made about consummatory behavior.

II. SENSORY INFLUENCES ON FOOD INTAKE

Human sensory systems evolved, in part, to aid in the identification, procurement, and ingestion of healthful foodstuffs. Each system contributes different information, but because the chemical senses serve as the final checkpoint before ingestion, research has focused primarily on these senses.

Whether taste is composed of four primary qualities (sweet, sour, salty, bitter) that in different combinations yield all gustatory sensations or whether there are multiple unique tastes remains controversial. Most researchers support the four-quality view and it has been proposed that sensitivity to each of the qualities has conveyed an adaptive protective benefit. Sweet taste may have aided in the identification of energy sources since many carbohydrates are sweet. Salt sensation could have helped ensure intake of adequate electrolytes as many possess a salty note. Sour sensitivity could have been used to avoid substances at pH extremes and it was useful to recognize and avoid bitter substances since a relatively high proportion of such compounds are toxic. The taste of monosodium glutamate (*Umami* in Japanese) may represent a fifth basic taste and could signal the presence of protein sources. [*See* Tongue and Taste.]

The prevailing view with regard to olfaction is that there are no olfactory primaries. This sensory system is more like the immune system in its ability to recognize a seemingly endless array of unique compounds. Whereas teleological arguments have been made with respect to "gustatory wisdom" and food intake (see earlier), no such speculation has been offered for the olfactory system. It should be noted that the oral cavity also receives trigeminal innervation (adding mouth feel, temperature, and pain). This system probably mediates the perception of fats in the diet as well as acids, bases, astringent compounds, and the textural properties of foods. [*See* Olfactory Information Processing.]

In part because of the currently greater interest in the role of other physiological systems (e.g., neural, endocrine) in the control of ingestive behavior, the contribution of sensory input under present living conditions is not well characterized. Perhaps the most compelling evidence stems from assessments of individuals with nutrient-related health disorders. There are case reports of individuals with salt-wasting pathologies who crave salt. Protein-calorie malnourished children have also shown a greater preference for stimuli containing an amino acid or protein source than nutritionally replete peers. There are few additional examples of such gustatory wisdom, although subtle sensory-based adjustments may occur without notice since the potential problems are mitigated. In contrast, the historical failure of individuals with scurvy, beriberi, and pellagra to correct their problem with available foods argues against a functional role of gustatory wisdom. Moreover, appropriate dietary responses to many chronic diseases, such as diabetes, hypertension, obesity, and dental caries, are not only absent, but efforts to promote more healthful eating are often met with strong opposition due to the perceived lower palatability of foods comprising the recommended diets.

Food cravings and food aversions, which are often based on the sensory qualities of foods, also hold nutritional implications. For example, cravings expressed by pregnant women, bulimic patients, and depressed patients treated with tricyclic antidepressants have generally involved sweet items. Such cravings can influence nutritional status if the sought after foods are ingested to the exclusion of items supplying needed nutrients, or if the craved items contain compounds that alter the bioavailability of nutrients provided by other foods.

An estimated one-third to two-thirds of the U.S. population have formed a food aversion at some point in their life. These aversions are marked decreases in the acceptability of particular foods after their ingestion has been associated with illness such as nausea or emesis. In clinical populations (e.g., cancer patients), aversions are typically specific (i.e., targeted to a few foods) and transient (i.e., durations of weeks to months). Aversions are specific, but often of longer duration among the general population. Because of the specificity, aversions rarely compromise diet quality and nutritional status, although they can be problematic in individual cases. High-protein items (e.g., red meat) are the most commonly affected foods, with sweets (especially chocolate), caffeinated beverages (e.g., coffee), and high-fat items being implicated often. The taste or smell of aversive foods is sufficient to elicit revulsion responses. [*See* Pregnancy, Dietary Cravings and Aversions.]

There are estimates that up to 20 million Americans suffer from taste and/or smell disorders. These range from slight diminutions of sensitivity to complete loss of function. Persistent tastes or smells and distorted sensations are also reported. These sensory abnormalities lead to decreased food enjoyment in a majority of sufferers, but alter dietary behavior to a clinically significant degree (i.e., changes of body weight exceeding 10% of predisorder weight) in only about 15–20% of affected individuals (based on findings from taste and smell research centers). Some patients report increasing intake in an attempt to achieve a sought after sensory experience or to mask an unpleasant sensation, whereas others reduce consumption in frustration with the lack of enjoyment with food or because food elicits an unpleasant sensation.

Diverse experimental evidence supports a view that sensory factors influence ingestive behavior in healthy individuals. Studies focusing on individual meals have shown that the palatability of foods influences eating behaviors such as rate of chewing and number of chews per bite. It is inferred that such effects impact on actual food intake. Other work has shown that sensory variety in a meal promotes greater total intake relative to when a single food is available. When chemosensory receptors are bypassed by delivering food directly into the stomach via a tube (intragastric feedings), patients often express dissatisfaction due to the lack of oral stimulation. Such patients can maintain adequate caloric intake via this intragastric feeding regimen. However, when oral intake is permitted, intragastric feeding does not induce a compensatory reduction in voluntary oral intake that offsets the extra calories derived from the intragastric feed. Thus, the desire for oral stimulation may override short-term satiety cues.

Common experience supports a role for sensory factors in the selection and ingestion of foods in a longer time frame. The sensory attributes of diets from different cultures are widely discrepant and are resistant to change, as evidenced by the reluctance of immigrants to modify their dietary practices. Efforts to supplement the diets of developing nations with various nutrients by introducing them into staple foods have frequently been unsuccessful due to some change they impart in a sensory property of the food. For example, efforts to add iron to sugar have not succeeded, in part because the sugar becomes discolored, rendering it unacceptable to consumers. Weak sales of "natural" or modified (e.g., low sodium, low fat) products that are purportedly more healthful, owing to their lower palatability, also attest to the importance of the sensory factors in food selection and ingestion.

There are numerous mechanisms by which sensory factors can influence dietary behavior. First, it has been argued that humans possess certain innate sensory preferences. The most commonly cited example is for the sweet taste. This quality is able to promote fetal drinking and increase sucking responses in neonates. Changes in mimetic reflexes as well as certain autonomic responses (e.g., heart rate) have also been recorded. In adults, there is evidence that 40–50% of calories are derived from primarily sweet-tasting foods. Preference for the salty taste may also be innately determined, but shows a maturational lag. It is manifest at approximately 6 months of age. Adults consume approximately 40% of calories from primarily salty-tasting foods. [*See* Salt Preference in Humans.]

Food intake may also be negatively influenced by sensory qualities. Deviations in any sensory attribute (e.g., color) will likely lead to rejection of a product with which an individual is familiar. In addition, some sensory properties may be innately undesirable and, in the absence of repeated exposures or cultural norms that promote acceptability, are avoided. Foods with predominantly sour and bitter tastes may reflect this since they constitute only about 10% of total energy intake. Often times when an unpreferred quality are consumed, but only after the undesirable sensory properties are modified (e.g., adding cream and sugar to a bitter beverage).

A third mechanism stems from the view that humans have an innate preference for sensory variety. One clear example of this mechanism is the premium placed on novel foods by populations forced to subsist on diets with restricted variety. Monotonous diets have actually been advocated as a means to achieve weight reduction and can be effective over a short period. However, frustration with this approach builds over time and long-term compliance is universally poor.

One final mechanism (though this list is not intended to be exhaustive) involves the role of sensory cues as predictors of the metabolic consequences of foods. Through experience, the satiety value of foods is learned and freely selected portion sizes are adjusted accordingly. Sensory cues, which have consistently been associated with each food, can supply the needed information. In studies where the energy values of meals have been surreptitiously manipulated, 25–50% of study participants were found to behave as though they were attending to the previously learned sensory cues rather than the actual metabolic implication of the meal. That is, they ingested less of a meal previously associated with a higher energy content and more of a meal previously associated with a low energy content. Humans are able to maintain adequate levels of energy intake in the absence of sensory input; however, when sensory information is available, it may be prepotent at least in some individuals over the short term.

III. SENSORY INFLUENCES ON NUTRIENT UTILIZATION

Sensory stimulation in the head and neck region exerts nutritional effects beyond its role in food selection via cephalic phase responses. These salivary, gastric, pancreatic exocrine and endocrine, thermogenic, cardiovascular, and renal responses to the thought, sight, sound, odor, feel, or taste of foods appear to prime the body to better absorb and utilize ingested nutrients. Though the existence of such reflexes has been documented in humans, their physiological role remains uncertain. This is due, in part, to the fact that they are short-lived (i.e., lasting several minutes), tend to be small relative to the comparable postingestive response, and are difficult to elicit reliably under experimental conditions. Nevertheless, evidence that postingestive responses are often shorter in duration, lower in magnitude, or in some way altered in the absence of any cephalic phase component supports a view that they serve to optimize energy and nutrient utilization. Generally, the more components of the normal ingestive process recruited, the larger will be the cephalic phase response. Thus, progressing from thoughts of food to combining this with the sight and then the smell followed by the taste of the item and finally swallowing, the magnitude of the response increases.

In some, but not all, studies the mere sight of food elicits a cephalic phase release of saliva. The response is strongest for items that evoke the largest release of saliva when ingested. Olfactory stimulation is a sufficient condition as odors alone can trigger a release of saliva. Gustatory stimulation, particularly with acidic solutions, elicits the largest cephalic phase salivary response. Whole-mouth salivary flow rates can rise from basal levels of approximately 0.3 ml/min to well over 1 ml/min. The nature of the response may be modified by an individual's state of hunger and the palatability of the stimulus. [*See* Salivary Glands and Saliva.]

To the extent that the cephalic phase salivary response influences the release of saliva during food ingestion, it may exert an impact on digestion via three mechanisms. First, the passage of a food through the gastrointestinal tract is facilitated by the addition of fluid and glycoproteins derived from saliva. Second, saliva contains digestive enzymes that can initiate the breakdown of starch and lipids. The importance of starch digestion in the oral cavity is typically minimal since amylase is rapidly inactivated in the acidic environment of the stomach. Lipid digestion can be more important under certain conditions, such as feeding of the preterm infant. In such cases, salivary lipase is the principal digestive enzyme for lipids. Finally, there is limited evidence that saliva entry into the stomach inhibits gastric emptying.

Thought, sight, smell, and taste are each singly capable of eliciting a release of gastric acid. Cognitive, visual, and olfactory stimulation can elicit a release that averages between one-fourth and two-thirds of that noted following mastication, but not swallowing, of food. Gustatory stimulation is the strongest cephalic phase trigger. Responses are also directly related to stimulus palatability. Stimulation with an unpleasant or tasteless stimulus can have no effect on gastric acid release, whereas provision of a preferred food will result in a marked release of acid. An exaggerated response has been reported in duodenal ulcer patients.

A nutritional impact of the cephalic phase gastric response can be presumed owing to the magnitude of the response and importance of gastric digestive functions. The release of gastric acid may serve to activate preenzymes (e.g., pepsinogen) that then promote the formation of numerous digestive enzymes. The acidic secretion, as well as gastrin, may also enhance pancreatic secretion.

The passage of gastric contents into the duodenum elicits a release of pancreatic secretions, but an independent effect of cephalic stimulation has also been documented. Responses are observed in individuals who are achlorhydric (i.e., absence of hydrochloric acid in gastric juice) or who have undergone procedures to block the passage of gut contents into the duodenum. Combined visual, olfactory, and trigeminal stimulation is a sufficient stimulus for a pancreatic exocrine response, but taste is more potent. More palatable taste stimuli also lead to larger responses. Sham feeding (ingestion of food coupled with diversion of the food out of the body via a gastric fistula) has been associated with elevations of trypsin, lipase, and chymotrypsin of 35–66%, 16–50%, and 25–60%, respectively. The total volume of the pancreatic secretion is enhanced, but there appears to be a particularly marked effect on release of digestive enzymes. Thus, an especially enzyme-rich secretion is produced. A cephalic phase release of pancreatic polypeptide has also been described. Pancreatic polypeptide is believed to inhibit gastric and pancreatic exocrine secretion and to relax the gallbladder and may thereby influence digestion.

The cephalic phase contribution to pancreatic exocrine responses to food ingestion may serve to augment the release of not only digestive enzymes but satiety hormones as well. For example, cholecystokinin, which can be released through a cephalic phase mechanism, is reported to possess satiety properties. In fact, the cephalic phase cholecystokinin response is attenuated in obese humans and perhaps could contribute to a lack of satiety in obese individuals.

The most extensively studied cephalic phase response is that of insulin. The thought, appearance, and odor of food have led to elevations of plasma insulin, though only a subset of test subjects display such a response in any given study. The poor reliability of the response may be due to its fragility, variations in experimental conditions, or the populations tested. For example, both attenuated and exaggerated responses have been reported in obese relative to normal-weight individuals and palatable stimuli are reported to either enhance or have no effect on the magnitude of cephalic phase insulin release.

Despite the reported variability, the presence of cephalic phase insulin release may be critical for the regulation of blood glucose. Animal studies indicate that the intragastric administration of glucose that bypasses the oropharyngeal receptors and therefore does not elicit cephalic phase reflexes results in hyperglycemia and hyperinsulinemia. In addition, although the role of insulin in the control of hunger and satiety has not been directly established, the cephalic phase insulin response may contribute to these sensations and their modification.

A larger thermogenic response to food ingested orally as compared to intragastrically suggests a role for sensory stimulation in postprandial thermogenesis. Mixed findings have been reported on an influence of food palatability on postprandial energy expenditure. Some work demonstrates enhanced thermogenesis following ingestion of a preferred meal, whereas other data reveal no effect. The extent to which sensory factors influence overall energy balance through modulation of the thermogenic response to foods is unknown.

Sham feeding also reportedly reduces cardiac output and may do so by increasing splanchnic blood flow. This could be a direct effect of oral stimulation on the sympathetic nervous system or one mediated by a secondary hormonal action. Increased blood flow in the splanchnic region could promote nutrient absorption.

Fluid balance may be influenced by sensory stimulation as well. Stimulation of the oral cavity with salt leads to antidiuresis in humans. Studies in rats indicate that this may be attributable to a neural effect on vasopressin release from the hypothalamus.

Direct evidence for the effect of the cephalic phase responses on nutrient digestion and utilization is still hypothetical. However, suggestive observations have been reported. First, in a number of independent studies of preterm infants, those infants provided with oral stimulation during nasogastric feeds displayed enhanced growth efficiency. That is, greater weight gain was achieved on a comparable level of energy intake. The stimulus in each case was a rubber pacifier. Similar results have been noted in infants provided nonoral tactile/kinesthetic stimulation so that mediation by cephalic phase responses cannot be assumed. Second, patients receiving total parenteral nutrition and thus receiving no oral stimulation have been reported to require unexpectedly high levels of energy to maintain body weight. However, recent studies also raise questions about this observation. Third, excursions of plasma glucose are greater when insulin is administered to diabetic individuals 15–30 minutes after a meal (thereby eliminating a priming signal) compared to when it is administered concurrently with food. Furthermore, the inhibition of early insulin release in normal subjects also results in hyperglycemia. Thus, under conditions of normal health and food intake, cephalic phase responses may serve more as the trigger or primer for responses that occur as food passes along the gastrointestinal tract than as an effective digestive response itself.

IV. NUTRITIONAL INFLUENCES ON SENSORY FUNCTION

All sensory systems are composed of metabolically active neural and supporting tissues. Normal functioning of these tissues requires the provision of appropriate levels of nutrients. Deficiencies or excesses of nutrients or food constituents can result in a derangement of sensory function. Examples of such effects are blindness due to vitamin A deficiency, deafness due to excessive alcohol or quinine ingestion, burning mouth syndrome due to B vitamin deficiency, and loss of taste due to zinc deficiency.

The mechanisms by which nutritional disorders can alter sensory function are multiple, as exemplified by the impact of nutrient deficiencies on the sense of taste. For a stimulus to be perceived, it must gain access to its appropriate receptor. In vitamin A deficiency, keratinous material plugs the taste pore and impedes the access of taste stimuli to receptors in this area. Second, once a stimulus reaches a receptor, the interaction must result in the generation of a signal to identify the presence and nature of the stimulus. The transduction processes for the sense of taste are only now being elucidated. Whether specific nutrients such as zinc are involved is still a matter of speculation. In the case of vision, however, the role of vitamin A in the transduction process is well documented. Third, sensory receptors must be supported by other types of tissues. Nutrient deficiencies or excesses can result in the death of these supporting tissues. Deficiencies of numerous nutrients lead to atrophy of taste papilla that contain the taste cells. Fourth, once an effective stimulus–receptor interaction occurs, information about the presence of this stimulus must be relayed to the central nervous system for decoding. The function of neural tissue that conveys this information is again dependent on the organism's nutritional status. Deficiencies (e.g., pyridoxine) or excesses of nutrients or food contaminants (e.g., lead) can lead to neuropathies and taste changes. Finally, incoming sensory information must be processed centrally. Central processing disorders such as Wernicke-Korsakoff syndrome, resulting from alcoholism, or schizophrenia, which may in some cases also have nutritional antecedents, have also been associated with abnormal responses to taste and smell stimuli and with olfactory and gustatory hallucinations.

Normal variations in nutritional status (i.e., shifts from states of mild depletion to repletion) can also alter sensory judgments. However, in this instance, evidence indicates that the changes are confined to affective responses with reports of alterations in sensitivity being attributable to response bias. Food may be less appealing following a meal, but the functionality of the gustatory and olfactory systems are not impaired. The influence of marginal nutrient deficiencies on sensory function remains an area of considerable speculation, but of little research.

In summary, food selection and perhaps nutrient utilization are influenced by the interplay between the sensory properties of foods and the sensory capabilities of consumers. Mechanisms by which sensory information may exert its impact on these two processes have been proposed, although the importance of each remains incompletely characterized. In any case, through an impact on food intake and digestion, sensory factors influence an individual's nutritional status and the status of the sensory systems themselves. Any disruption in this reciprocal relationship between sensory function and nutrition will likely lead to a compromised status of each.

BIBLIOGRAPHY

Amerine, M. A., Pangborn, R. M., and Roessler, E. B. (eds.) (1965). "Principles of Sensory Evaluation of Food." Academic Press, New York.

Carterette, E. C., and Friedman, M. P. (eds.) (1974). "Handbook of Perception: Psychophysical Judgement and Measurement." Academic Press, New York.

Friedman, M. I., and Mattes, R. D. (1991). Chemical senses and nutrition. *In* "Smell and Taste in Health and Disease" (T. V. Getchell, R. L. Doty, L. M. Bartoshuk, and J. B. Snow, eds.), pp. 391–404. Raven, New York.

Kare, M. R., and Brand, J. G. (1986). "Interaction of the Chemical Senses with Nutrition." Academic Press, New York.

Kare, M. R., and Maller, O. (eds.) (1977). "The Chemical Senses and Nutrition." Academic Press, New York.

Kramer, A., and Szczesniak, A. S. (eds.) (1973). "Texture Measurements of Foods." D. Reidel, Dordrecht, Holland.

Mattes, R. D. (1987). Sensory influences on food intake and utilization in humans. *Hum. Nutr.: Appl. Nutr.* **41A,** 77–95.

Mattes, R. D. (1995). Nutritional implications of taste and smell disorders. *In* "Handbook of Clinical Olfaction and Gustation" (R. Doty, ed.), pp. 731–744. Marcel Dekker, New York.

Mattes, R. D., and Kare, M. R. (1993). Nutrition and the chemical senses. *In* "Modern Nutrition in Health and Disease" (M. E. Shils, J. A. Olson, and M. Shike, eds.), 8th Ed., pp. 537–548. Lea & Febiger, Philadelphia.

Meiselman, H. L., and Rivlin, R. S. (eds.) (1986). "Clinical Measurement of Taste and Smell." Macmillan, New York.

Schiffman, S. S. (1983). Taste and smell in disease: Part I. *N. Engl. J. Med.* **308,** 1275–1279.

Schiffman, S. S. (1983). Taste and smell in disease: Part II. *N. Engl. J. Med.* **308**, 1337–1343.

Sensory Evaluation Division of the Institute of Food Technologists (1981). Sensory evaluation guide for testing food and beverage products. *Food Technol.*, November, pp. 50–59.

Solms, J., Booth, D. A., Pangborn, R. M., and Raunhardt, O. (eds.) (1987). "Food Acceptance and Nutrition." Academic Press, New York.

Teff, K. (1994). Cephalic phase insulin release in humans: Mechanism and function. *In* "Appetite and Body Weight Regulation: Sugar, Fat, and Macronutrient Substitutes" (J. D. Fernstron and G. D. Miller, eds.), pp. 37–50. CRC Press, Boca Raton, Florida.

Food Antioxidants

DENNIS D. MILLER
Cornell University

AARON EDWARDS
National Starch & Chemical Company

GLOSSARY

Antioxidant Substance capable of slowing, stopping, or preventing oxidation

Carotenoids Class of yellow, orange, and red pigments found in higher plants. Of the more than 600 known carotenoids, about 50 have vitamin A activity. Carotenoids are fat soluble and can scavenge free radicals and quench singlet oxygen

Free radicals Chemical species that contain one or more unpaired electrons. Most free radicals are unstable and highly reactive. They are commonly identified by a dot signifying an unpaired electron, for example, ·OH is the formula for hydroxyl radical

Homolysis Symmetrical cleavage of a covalent bond such that one electron of the pair forming the bond goes with each product. Homolysis of covalent bonds yields free radicals

Hydroperoxide Hydrogen-containing peroxide

Lipid hydroperoxide Hydroperoxide formed from lipids, usually polyunsaturated fatty acids

Oxidation Reaction in which a chemical species either loses electrons, gains oxygen, or loses hydrogen

Phytochemicals Chemicals that occur naturally in plants

Polyunsaturated fatty acids (PUFA) Fatty acids that contain more than one carbon–carbon double bond per molecule. PUFAs are very susceptible to oxidation owing to the presence of the double bonds. Fats and oils from plant sources (soybean, corn, sunflower, peanut) tend to be high in PUFAs

Reactive oxygen species (ROS) Oxygen-containing species that are more reactive than ground-state molecular oxygen (triplet oxygen). ROS include hydrogen peroxide, singlet oxygen, superoxide, hydroxyl radicals, peroxyl radicals, and alkoxyl radicals

Reduction Reaction in which a chemical species either gains electrons, loses oxygen, or gains hydrogen

Saturated fatty acids Fatty acids with no carbon–carbon double bonds. Saturated fatty acids are much less susceptible to oxidation than unsaturated fatty acids. Animal fats and coconut and palm oils are particularly rich in saturated fatty acids

Singlet oxygen ($^1O_2$) Excited (reactive) form of molecular oxygen. It contains an empty π^*2p orbital and thus is an electrophile, and reacts readily with carbon–carbon double bonds in PUFA to form lipid hydroperoxides

Superoxide ($O_2\cdot^-$) Free radical formed when molecular oxygen gains one electron. It is produced by phagocytic cells to kill bacteria and viruses, and also forms when electrons "leak" from the electron transport chain and combine with oxygen

Triplet oxygen ($^3O_2$) Ground-state molecular oxygen. It is the predominant form of oxygen in air and is relatively unreactive toward most organic molecules at physiological temperatures unless reactions are enzyme catalyzed. Not considered a ROS

Vitamin E Fat-soluble vitamin that functions as an antioxidant to protect cell membranes from oxidation. Eight naturally occurring compounds (four tocopherols and four tocotrienols) have vitamin E activity. α-Tocopherol has the highest vitamin activity

CELLS IN OUR BODIES ARE CONSTANTLY BOMbarded by highly reactive chemical species known as free radicals. Free radicals are products of normal metabolism but they also form when radiation strikes our bodies. They are present in high concentrations in cigarette smoke. Free radicals can cause damage to cellular components that may eventually lead to heart disease, cancer, arthritis, and cataracts. Biochemists have shown convincingly that substances called antioxidants can effectively eliminate (or scavenge) reactive free radicals in cells. This led to the hypothesis that diets high in antioxidants may protect us from developing these serious chronic diseases. Over the last few years, numerous epidemiological, biochemical, and nutritional studies have lent support to this hypothesis and prompted nutritionists and other health professionals to promote diets that are rich in antioxidants. One of the most widely publicized of these studies was authored by Gladys Block and her colleagues at the University of California-Berkeley. They reviewed about 200 epidemiological studies on the role of dietary fruits and vegetables in cancer risk. In almost all of the studies, high intakes of fruits and vegetables were associated with reduced risk for cancer. This fits with the foregoing hypothesis because fruits and vegetables are particularly good sources of antioxidants. It is important to keep in mind, however, that other compounds in fruits and vegetables could influence cancer without inhibiting oxidative reactions.

I. OXIDATION IN BIOLOGICAL SYSTEMS

To understand the significance of antioxidants in foods and how they work, we must first understand what oxidation is and why some forms of oxidation can be damaging to key cell components. This section provides a brief overview of the complex topic of oxidation as it occurs in foods and living biological systems. Please see the glossary for definitions of oxidation, reduction, reactive oxygen species, and free radicals.

A. Oxygen's Role in Energy Metabolism

Early life on planet earth evolved in an oxygen-free environment until photosynthetic organisms appeared on the scene and began generating molecular oxygen (O_2). Once oxygen was present in the environment, organisms evolved mechanisms for using oxygen to more efficiently generate ATP from fuel molecules (e.g., carbohydrates and lipids). ATP is a high-energy molecule used by cells and organisms as the primary energy source for driving muscle contraction, the synthesis of biomolecules (e.g., proteins, DNA, phospholipids, hormones), and the active transport of molecules and ions in and out of cells. Most of the ATP produced in the oxidation of fuel molecules is generated when electrons flow from NADH and $FADH_2$ to O_2 via a series of electron carriers that make up the electron transport chain in mitochondria. The process of generating ATP from electron transfer is called oxidative phosphorylation. NADH and $FADH_2$ are generated in glycolysis, fatty acid oxidation, and the citric acid cycle. When NADH is *oxidized* to NAD^+, two electrons are released for transfer to O_2. Clearly, oxidation in this case involves both a loss of electrons and a loss of hydrogen.

The electron transport system in cells is designed to transfer two electrons to each atom of oxygen so that the resulting reduced oxygen can combine with protons to yield harmless water:

$$NADH \rightarrow NAD^+ + H^+ + 2e^-,$$

$$1/2\ O_2 + 2e^- + 2H^+ \rightarrow H_2O.$$

B. Reactive Oxygen Species: Free Radicals

Reactive oxygen species (ROS) are oxygen-containing species that are more reactive than ground-state molecular oxygen. Biologically significant ROS include most oxygen free radicals, hydroperoxides, and singlet oxygen.

Enzymes in the electron transport chain have tight control over the transfer of electrons from NADH and $FADH_2$, ensuring complete reduction of O_2 to H_2O. However, the process is not 100% efficient and one-electron transfers may also occur, resulting in the formation of a free radical called superoxide anion:

$$O_2 + 1e^- \rightarrow O_2\cdot^-.$$

Another source of free radicals is the homolytic cleavage of an O–H bond in water. This may occur when ionizing radiation strikes the water in our bodies or in food (we are constantly exposed to low levels of radiation from naturally occurring radioisotopes):

$$H_2O \rightarrow H\cdot + \cdot OH.$$

Hydroxyl radicals may also form from the decomposition of hydrogen peroxide in a reaction catalyzed by iron ions:

$$Fe^{2+} + H_2O_2 \rightarrow Fe^{3+} + OH^- + \cdot OH.$$

Another type of free radical called alkoxyl radicals form when transition metals such as iron catalyze the decomposition of lipid hydroperoxides:

$$ROOH + Fe^{2+} \rightarrow RO\cdot + Fe^{3+} + OH^-.$$

Lipid hydroperoxides are products of lipid peroxidation (see Section II)

Most free radicals are highly reactive species. They may attack a variety of chemical compounds within the cell to obtain another electron to pair up with the unpaired electron, that is, they oxidize compounds by removing an electron or radical.

Molecular oxygen is also a free radical. It contains *two* unpaired electrons in separate orbitals and exists in the so-called triplet state. The chemical formula for molecular oxygen is sometimes written as $^3O_2$ to emphasize the fact that it is in the triplet state. Triplet oxygen needs to gain two electrons to achieve nonradical status. Surprisingly, $^3O_2$ does not easily react with most biological molecules without the aid of specific enzymes. This is because its two unpaired electrons have the same spin state, making it energetically unfavorable to accept a pair of electrons with opposite spins (most electrons in biological molecules exist in pairs with opposite spins). Triplet oxygen does react readily with other free radicals to form new radicals (see the following). Other free radicals include alkoxyl and peroxyl radicals (Table I).

C. Reactive Oxygen Species: Nonradicals

Other reactive oxygen species important in biology are the nonradicals singlet oxygen, hydrogen peroxide, and lipid hydroperoxides. Singlet oxygen ($^1O_2$) has the same chemical formula as ground-state molecular oxygen but a different arrangement of electrons (it has an empty π^*2p orbital but no unpaired electrons). Thus it is not a free radical but it is electrophilic because it seeks to fill its empty π^*2p orbital. It reacts readily with double bonds, especially double bonds in polyunsaturated fatty acids, to produce lipid hydroperoxides:

$$RH + {}^1O_2 \rightarrow ROOH,$$

where RH is a polyunsaturated fatty acid and ROOH is a lipid hydroperoxide. The ability of $^1O_2$ to form lipid hydroperoxide is important because, as mentioned earlier, lipid hydroperoxides can decompose to free radicals.

Singlet oxygen is at a higher energy level than triplet or ground-state oxygen. It may be formed in foods and other biological systems, (e.g., skin) when light energy is captured by pigments called sensitizers. The excess energy in the sensitizer is subsequently transferred from the sensitizer to $^3O_2$:

$$\text{Sen} \xrightarrow{\text{uv light}} \text{Sen}^*,$$

TABLE I
Peroxides and Free Radicals That May be Present in the Body, in Foods, or in Cigarette Smoke[a]

Chemical name	Chemical formula	Characteristics
Hydrogen peroxide	H_2O_2	Forms when protons (H^+) combine with superoxide in a reaction that is catalyzed by the enzyme superoxide dismutase
Lipid hydroperoxides	R(OOH)-COOH	Form when unsaturated fatty acids in foods or in the body are oxidized
Hydrogen atom	H·	A simple free radical
Superoxide radical	$O_2^-\cdot$	A side product of energy metabolism in the body; produced when one electron is transferred to O_2
Hydroxyl radical	·OH	Forms when transition metal ions react with hydrogen peroxide and when ionizing radiation strikes water, causing homolysis of a O–H bond
Alkyl radical	R·	A free radical in which a carbon atom possesses an unpaired electron; forms by homolysis of a C–H bond
Peroxyl radical	$RO_2\cdot$	Forms when oxygen combines with an alkyl radical
Alkoxyl radical	RO·	Forms when lipid peroxides decompose
Nitrogen oxide radicals	NO· and $NO_2\cdot$	Present in cigarette smoke, may form in the body

[a] After B. Halliwell *et al.* (1995).

$$\text{Sen}^* + {}^3O_2 \rightarrow \text{Sen} + {}^1O_2.$$

Singlet oxygen is also formed when two peroxyl radicals react:

$$\text{ROO}\cdot + \text{ROO}\cdot \rightarrow \text{R{=}O} + \text{ROH} + {}^1O_2.$$

Hydrogen peroxide (H_2O_2) may form in cells when an enzyme called superoxide dismutase (SOD) acts on superoxide anion:

$$2O_2^-\cdot + 2H^+ \xrightarrow{\text{SOD}} H_2O_2 + O_2.$$

D. Reactions of Free Radicals in the Body

Since free radicals are unstable and seek another electron to combine with their unpaired electron, they may attack other molecules in their vicinity. This can have serious consequences if the molecules that are attacked are altered in such a way that they no longer function normally. We now know that free radicals may attack nucleic acids, proteins, carbohydrates, and lipids, all molecules with vital roles in the cell (Fig. 1). It is widely believed that free radical damage to these molecules can lead to cancer, heart disease, cataracts, and other chronic diseases associated with aging (Table II). [*See* Free Radicals and Disease.]

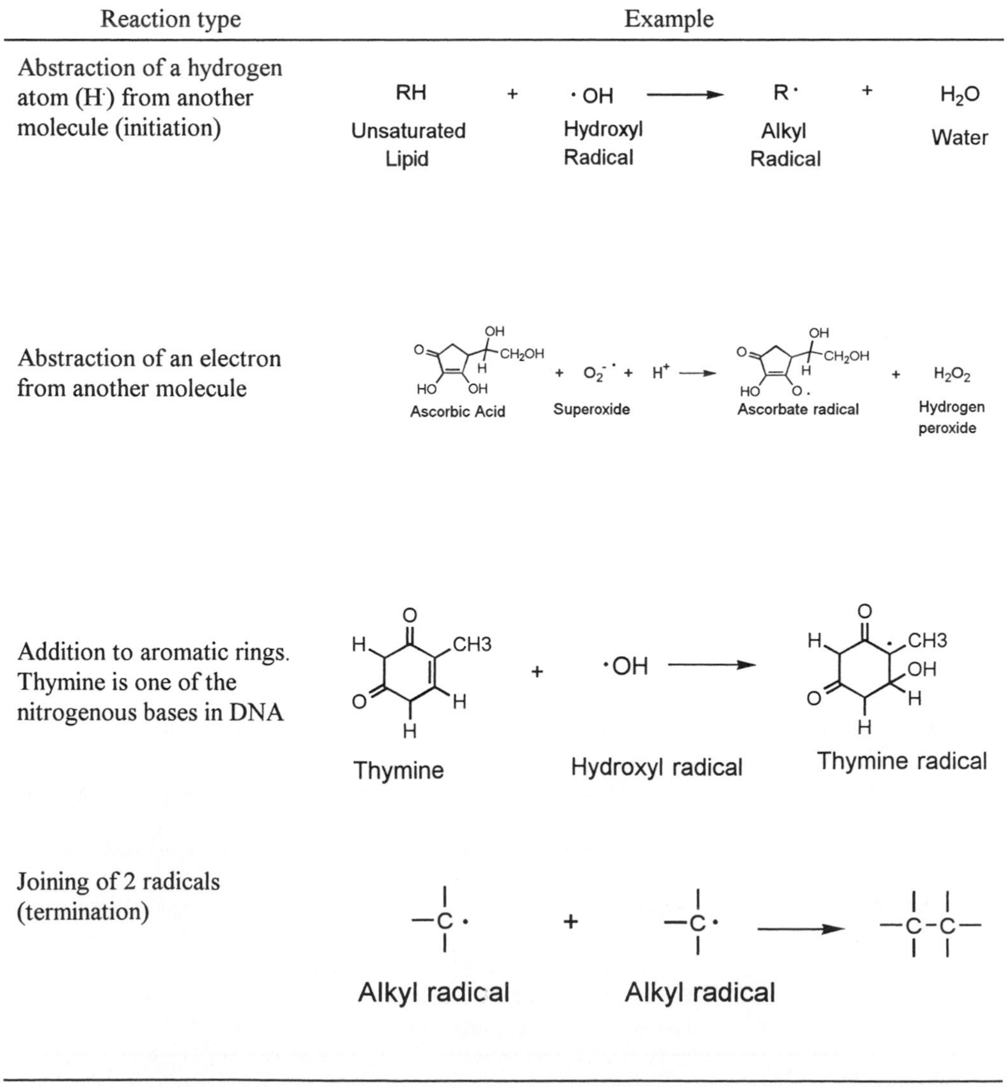

Reaction type	Example
Abstraction of a hydrogen atom (H·) from another molecule (initiation)	RH (Unsaturated Lipid) + ·OH (Hydroxyl Radical) → R· (Alkyl Radical) + H_2O (Water)
Abstraction of an electron from another molecule	Ascorbic Acid + $O_2^{-\cdot}$ (Superoxide) + H^+ → Ascorbate radical + H_2O_2 (Hydrogen peroxide)
Addition to aromatic rings. Thymine is one of the nitrogenous bases in DNA	Thymine + ·OH (Hydroxyl radical) → Thymine radical
Joining of 2 radicals (termination)	—C· (Alkyl radical) + —C· (Alkyl radical) → —C—C—

FIGURE 1 Free radical reactions that may occur in body cells and/or foods.

TABLE II
Consequences of Excessive Oxidation in the Body

- *Altered cell membrane function* caused by free radical attack on phospholipids, enzymes, and receptors in membranes
- *Enhanced atherogenicity of low density lipoproteins (LDL)* caused by changes in the surface charge on the LDL particles
- *Increased cancer risk* caused by DNA mutations and alterations in gene regulatory proteins
- *Increased risk of cataracts* caused by free radical attack on proteins in the lens of the eye
- *Arthritis*: free radical attack on proteins may turn them into antigens that stimulate the production of autoimmune antibodies

II. LIPID PEROXIDATION

To understand how antioxidants work, we must first understand how free radicals cause oxidation in the first place. As mentioned earlier, free radicals can damage DNA, proteins, carbohydrates, and lipids. We will use the example of lipid oxidation since it is a significant problem both in foods and in the body.

Lipid peroxidation (the oxidation of lipids) has been the subject of intense research for several decades because of its importance in foods and, more recently, its putative role in heart disease and cancer. Products of lipid peroxidation in foods include volatile aldehydes that often have unpleasant odors and flavors. Lipid peroxidation is a major cause of food deterioration and a great deal of effort has gone into developing strategies for reducing or preventing it. The process of lipid peroxidation is summarized in the following and in Fig. 2.

To begin the process of lipid peroxidation, a free radical must be generated. This step is often called the initiation step or simply *initiation*. As described in the foregoing, free radicals may be generated from the decomposition of lipid hydroperoxides (to form peroxyl or alkoxyl radicals), water (to form hydroxyl radicals), and hydrogen peroxide (to form hydroxyl radicals) or by the generation of superoxide.

Once a free radical is formed, it can abstract a hydrogen atom from chemical compounds containing hydrogens. This occurs fairly readily in unsaturated fatty acids because the hydrogens on carbons adjacent to a carbon–carbon double bond are susceptible to

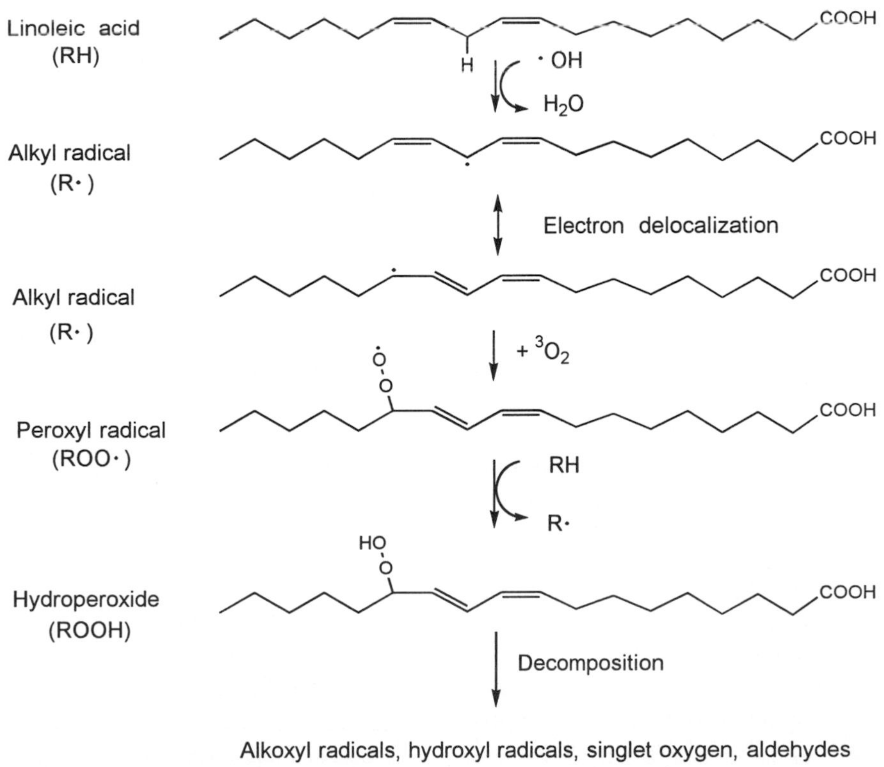

FIGURE 2 Lipid peroxidation of linoleic acid initiated by a hydroxyl radical. RH represents another unsaturated fatty acid.

abstraction. Hydrogens on saturated fatty acids are much harder to remove so peroxidation of saturated fatty acids does not occur to an appreciable extent. When a hydrogen atom (H·) is removed from a fatty acid, an alkyl radical (R·) forms. Alkyl radicals readily react with triplet oxygen to form peroxyl radicals, $RO_2\cdot$, which in turn can abstract a hydrogen from another unsaturated fatty acid to form a hydroperoxide and another alkyl radical. Thus, a chain reaction is set up because the initial reactant, R·, is regenerated. Moreover, additional radicals may be generated by the decomposition of the lipid hydroperoxides. Thus, the potential damage caused by free radicals is much greater than would be expected if chain reactions were not involved. Presumably, lipid peroxidation will continue indefinitely once it gets started as long as oxygen and unsaturated lipids are available and free radicals continue to be generated.

III. DEFENSES AGAINST OXIDATION

If oxygen generates free radicals in cells and free radicals are so damaging to cellular constituents, how can we survive as long as we do in our oxygen-containing atmosphere? How can we protect our foods against deterioration due to oxidation? Part of the answer is that our bodies and other biological systems have evolved defense mechanisms to remove or scavenge free radicals and to repair or excrete molecules damaged by free radical attack. It appears, however, that defenses intrinsic to body cells are not sufficient to adequately protect against excessive oxidation and we must supplement those defenses with dietary antioxidants.

We should not forget that even though peroxides and free radicals can cause serious damage to the body, they also play beneficial roles. For example, phagocytes (cells in the immune system that engulf and kill invading bacteria and viruses) purposefully generate hydrogen peroxide and superoxide, presumably to aid in the killing of the invading organism. There is also evidence that superoxide may play a role in intercellular signaling and growth regulation.

So there seems to be a delicate balance. Peroxides and free radicals are beneficial at low concentrations in the right places but may be toxic when concentrations are too high or when they are generated where they can cause harm. When this balance is tipped toward excess oxidation, cells and organisms are said to be under *oxidative stress*.

A. Enzymatic Defenses

Cells contains enzymes that can destroy free radicals and peroxides. These include superoxide dismutase, which converts superoxide to hydrogen peroxide, catalase, which converts hydrogen peroxide to water and molecular oxygen, and glutathione peroxidase, which catalyzes the decomposition of both hydrogen peroxide and lipid hydroperoxides (Table III). Glutathione is a tripeptide composed of glutamine, cysteine, and glycine residues. It is an important antioxidant in cells. Glutathione peroxidase requires selenium for activity. Thus, dietary selenium deficiency can impair the body's ability to defend against the damaging effects of peroxides.

B. Nonenzymatic Control of Oxidation

Apparently, the enzymatic defense systems just mentioned are not capable of completely protecting against damage by free radicals. Therefore, other mechanisms for removing free radicals are needed. This is where antioxidants come in. Antioxidants are chemical species that can prevent, slow, or eliminate oxidation. Since much of the damaging oxidation in cells occurs when free radicals attack various cell components, most of the antioxidants act by either scavenging free radicals or preventing their formation in the first place. Antioxidants in foods include α-tocopherol (vitamin E), ascorbic acid (vitamin C), β-carotene and other carotenoids, BHA, BHT, plant polyphenolics, and various metal chelating agents (EDTA, citric acid, phytic acid).

It should be clear from the preceding sections that oxidation of organic molecules requires oxygen and free radicals. In addition, transition metal ions may initiate and accelerate lipid peroxidation by catalyzing the formation of new free radicals. Light may initiate

TABLE III
Enzymes in the Body that Serve as Antioxidants

Superoxide dismutase (SOD):

$$2O_2^{-}\cdot + 2H^+ \rightarrow H_2O_2 + O_2$$

Catalase:

$$2H_2O_2 \rightarrow 2H_2O + O_2$$

Glutathione peroxidase:[a]

$$2GSH + H_2O_2 \rightarrow GSSG + 2H_2O$$

$$2GSH + ROOH \rightarrow GSSG + ROH + H_2O$$

[a]GSH and GSSG represent glutathione and oxidized glutathione, respectively.

peroxidation, transforming triplet oxygen into singlet oxygen and thereby allowing the formation of lipid hydroperoxides, which can decompose to form free radicals. How then might lipid peroxidation in foods and/or our bodies be controlled? Several strategies are available.

1. Elimination of Oxygen

Oxygen is required for many of the oxidation reactions in foods and the body. Therefore elimination of oxygen should prevent peroxidation. This strategy is used in many food products. It is accomplished by vacuum packaging or by flushing away oxygen with nitrogen gas and sealing the food in packages that are impermeable to oxygen. Obviously, this strategy cannot be used for controlling oxidation in the body because oxygen is required for energy metabolism.

2. Scavenging of Free Radicals

Given that many of the oxidative reactions that occur in foods and the body involve a free radical mechanism, it stands to reason that removal of free radicals would prevent or slow oxidation. In fact many of the most effective antioxidants in foods and the body are free radical scavengers. To understand how free radical scavengers work, we will take α-tocopherol (vitamin E) as an example (Fig. 3). Vitamin E is a phenolic compound, that is, it contains an aromatic ring with an attached hydroxyl group. The hydrogen in the hydroxyl group of a phenolic compound is relatively easily abstracted. Thus phenolic compounds will donate a hydrogen atom to free radicals converting them to nonradicals:

$$\alpha\text{-TH} + \text{R}\cdot \rightarrow \alpha\text{-T}\cdot + \text{RH}.$$

We still have a free radical, α-T·. However, phenolic free radicals are relatively stable and are incapable of abstracting a hydrogen from another unsaturated fatty acid. Thus, the chain is broken and lipid peroxidation is slowed. Phenolic radicals are stable because of resonance delocalization around the aromatic ring:

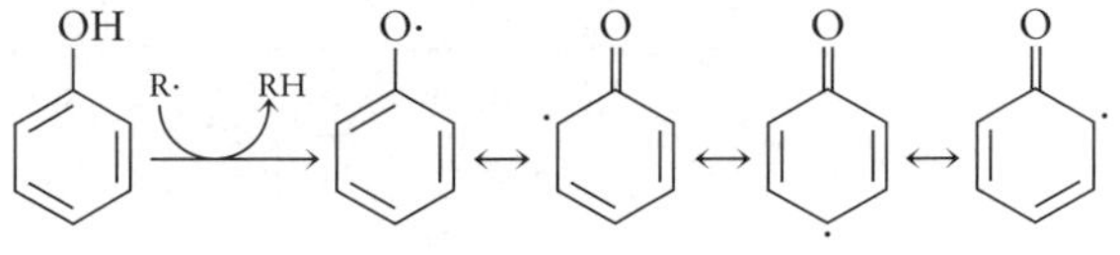

OH, R·, RH, O·, O, O, O

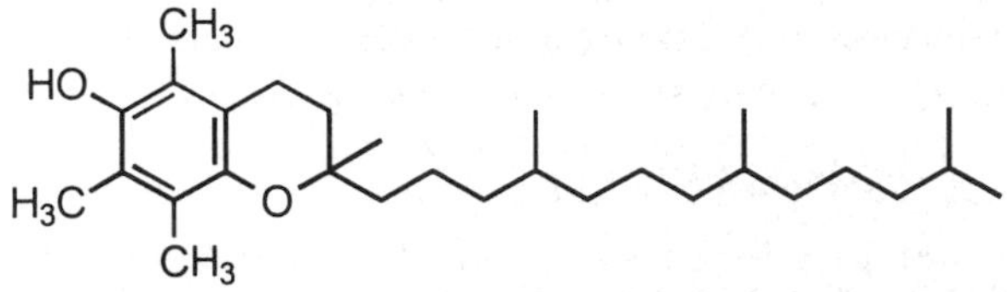

FIGURE 3 The structure of α-tocopherol showing the hydroxylated aromatic ring and the long hydrocarbon tail. α-Tocopherol is the most active form of vitamin E.

3. Chelation of Metal Ions

Since transition metal ions catalyze the formation of free radicals, we might expect that removal of metal ions would reduce oxidation. However, it is not possible or practical to remove them from foods or the body. Iron and copper are both essential nutrients, moreover, iron deficiency is a widespread nutritional problem. Therefore, even if it were possible to remove these metals from foods, it would be unwise to do so. In fact, iron is added to many foods to ensure adequate intakes by individuals in the population.

Fortunately, chelation of metal ions by chelating agents often reduces their effect on oxidation. Recall that metal ions can form complexes (coordination compounds) by attaching to various ligands through coordinate covalent bonds:

$$M^+ + :L \rightleftarrows [M\text{-}L]^+.$$

In this reaction, the metal ion is acting as a Lewis acid (an electron pair acceptor) and the ligand is acting as a Lewis base (an electron pair donor). Many ligands are organic molecules that contain atoms (usually oxygen or nitrogen) capable of donating a pair of electrons to form the coordinate covalent bond with metal ions. Often these molecules contain more than one donor atom, allowing the formation of multiple coordinate bonds between the metal ion and the ligand. When this is the case, the complex that forms is called a chelate. Structures of EDTA, a strong chelating agent capable of forming six bonds with metal ions, and iron-EDTA chelate are shown in Fig. 4.

When metals combine with ligands to form chelates, their electronic structures are altered and this affects their ability to participate in other reactions. As mentioned earlier, chelated metals may not catalyze oxidation reactions in the same way as free metal ions. Thus oxidation can often be controlled by adding a chelating agent to a food. The cells in plants and animals synthesize a variety of chelating agents that apparently function to prevent oxidation, including the proteins transferrin and ferritin. Transferrin is a protein that circulates in the blood serum. It binds

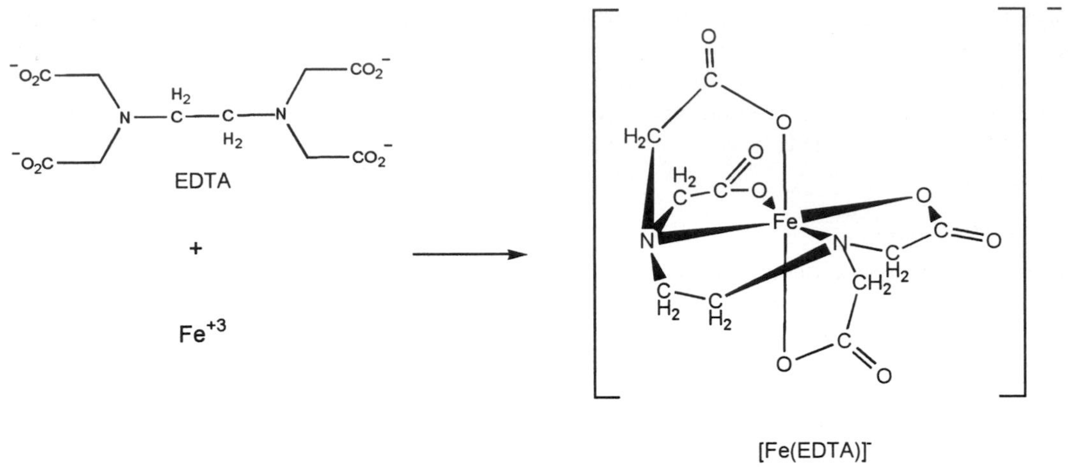

FIGURE 4 Ethylenediaminetetraacetate (EDTA) and its chelate with Fe^{3+}.

iron extremely tightly and serves to transport iron from the intestine, where it is absorbed, to body cells, where it is needed. Ferritin is a protein that sequesters iron in storage sites. It is present in greater concentration in liver, bone marrow, and spleen, important storage sites for iron.

C. Some Common Food Antioxidants

1. Radical Scavenging Antioxidants

a. Vitamin E

The term "vitamin E" refers to a family of compounds that contain a hydroxylated aromatic ring. These compounds are known as tocopherols and tocotrienols. The hydroxylated aromatic ring makes them members of the class of compounds known as phenols or phenolic compounds. The most active form of vitamin E in animals is α-tocopherol (Fig. 3). Vitamin E is highly lipophilic and resides primarily in cell membranes and lipoprotein particles in the blood (e.g., low density lipoprotein and very low density lipoprotein). Vitamin E protects against oxidative damage to polyunsaturated fatty acids by scavenging free radicals. In the process, a tocopherol radical is generated. As mentioned earlier, phenolic free radicals are relatively stable and will not abstract a hydrogen atom from another fatty acid, thereby breaking the chain reaction. Vitamin E may be regenerated from the tocopherol radical by the action of ascorbic acid (vitamin C).

Tocopherols and tocotrienols are widely distributed in plant foods. Vegetable oils, whole grain cereals, nuts, and green leafy vegetables are particularly good sources. The vitamin E content of animals foods is generally very low. [*See* Vitamin E.]

b. Vitamin C (Ascorbic Acid)

Ascorbic acid is a water-soluble antioxidant and is also a free radical scavenger. It regenerates vitamin E from tocopherol radicals and can also scavenge other free radicals, including superoxide, peroxyl, thiyl, and hydroxyl radicals. In addition, ascorbic acid can quench singlet oxygen in aqueous solution. Some nutritionists and promoters of nutrient supplements advocate vitamin C intakes far in excess of the RDA. The use of high-dose vitamin C supplements is controversial because of the possibility that vitamin C can act as a *pro-oxidant* under some conditions. Recall that ferrous iron (Fe^{2+}) can catalyze the formation of hydroxyl radical from hydrogen peroxide and that ascorbic acid is capable of reducing ferric iron (Fe^{3+}) to ferrous iron. *In vitro* experiments have clearly shown that under conditions where ferric iron and unsaturated fatty acids are present, ascorbic acid actually promotes lipid oxidation. Whether this also occurs *in vivo* is not known.

Citrus fruits, green peppers, cauliflower, broccoli, cabbage, and strawberries are particularly good sources of vitamin C. [*See* Ascorbic Acid.]

c. Carotenoids

Carotenoids are plant pigments composed of isoprene units covalently linked together giving them multiple conjugated double bonds (Fig. 5). Colors of carotenoids range from yellow to orange to red. Although

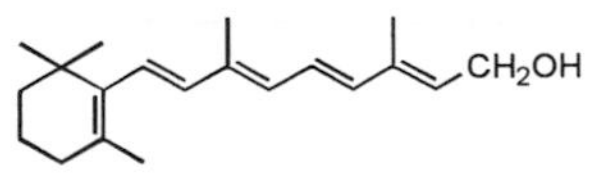

All-trans retinol (vitamin A) (liver, eggs, dairy products)

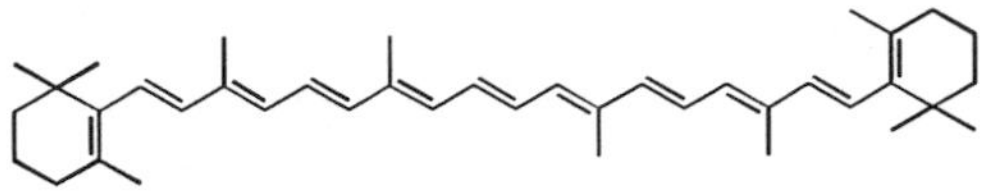

beta-carotene (carrots, squash, tomato,most fruits & vegetables)

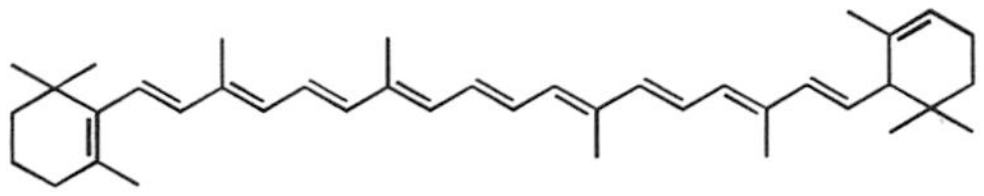

alpha-carotene (carrots, squash, tomato, most fruits & vegetables)

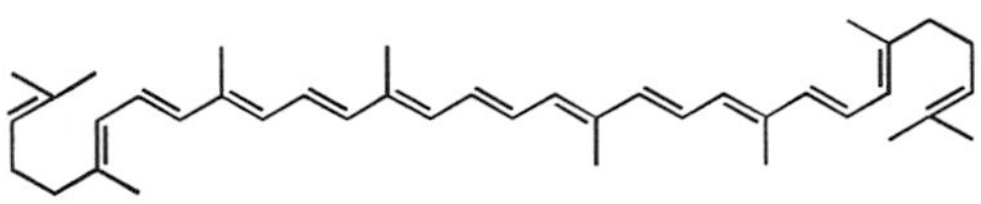

lycopene (tomatoes, watermelons, pink citrus, apricots)

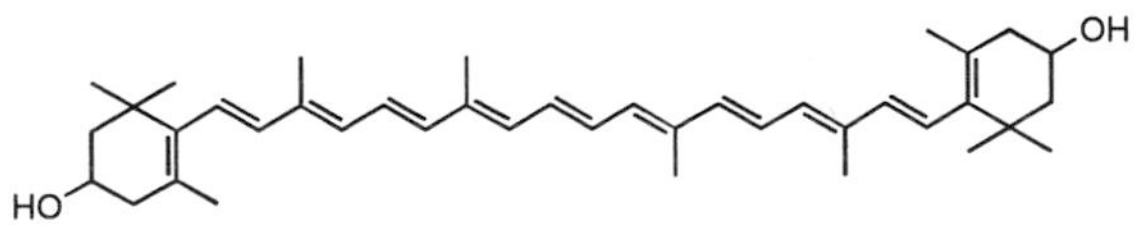

zeaxanthin (spinach, corn, green pepper)

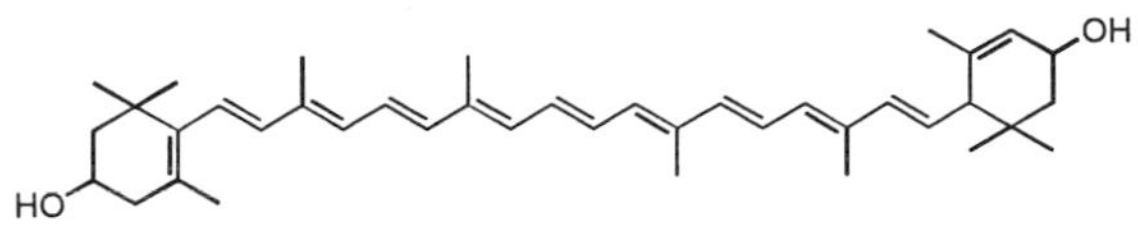

lutein (spinach, egg yolk)

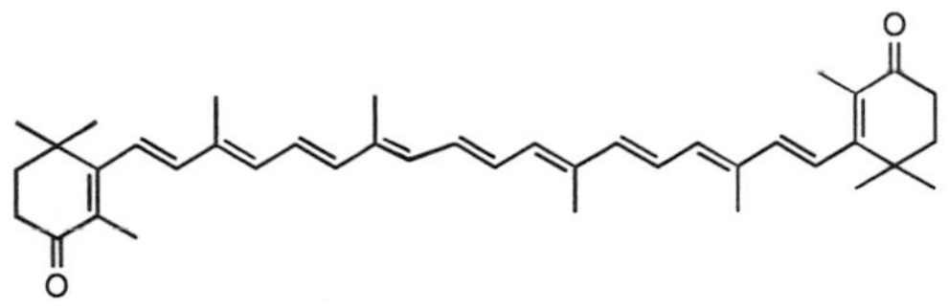

canthaxanthin (food colorant, mushrooms, shrimp, algae)

FIGURE 5 Structures and food sources of some selected carotenoids. Of the carotenoids listed, only retinol and α- and β-carotene have vitamin A activity.

animals cannot synthesize them, some animal foods contain carotenoids because animals absorb, modify, and deposit dietary carotenoids in tissues. The yellow in egg yolk, for example, is due to carotenoids. Some carotenoids may be converted to retinol (vitamin A) in the body, thus these carotenoids have vitamin A activity. There are two main groups of carotenoids, the carotenes and the xanthophylls. The carotenes are hydrocarbons (they are composed of only carbon and hydrogen) and the xanthophylls contain oxygen in their structures. Carotenoids may function as free radical scavengers; they also quench singlet oxygen.

Carotenoids with vitamin A activity are found in yellow and orange vegetables and fruits and in many dark green vegetables. Carrots, squash, sweet potatoes, spinach, broccoli, papayas, and apricots are good sources.

d. Plant Phenolic Compounds

In addition to vitamin E, plants contain many other phenolic compounds (Fig. 6). Many of these compounds contain aromatic rings with more than one hydroxyl group. Thus they are referred to as polyphenols or polyphenolic compounds. The general term "plant phenolics" includes compounds ranging from relatively low-molecular-weight phenolic acids such as caffeic acid to high-molecular-weight polymers known as tannins. Many plant phenolics are pigments. For example, the reds and blues in grapes, plums, cherries, and strawberries are due to the presence of a subclass of phenolic compounds called anthocyanins. Tea and coffee contain substantial concentrations of catechins and their derivatives. Catechins may be converted to tannins by oxidative polymerization reactions, and tannins impart an astringent taste to foods. Many spices are also rich sources of phenolic compounds and are added to foods to add flavor but also to prevent oxidation.

As we would expect, based on their chemical structures, plant phenolics are antioxidants. They exert their antioxidant effects by scavenging free radicals, quenching singlet oxygen, and, in some cases, chelating metal ions.

e. Synthetic Antioxidants

Many foods that contain polyunsaturated fatty acids are susceptible to oxidation. Thus food scientists have developed a variety of compounds that can prevent or retard lipid oxidation in foods.

The primary synthetic antioxidants used by the food industry are butylated hydroxyanisole (BHA), butylated hydroxytoluene (BHT), tertiary-butyl hydroquinone (TBHQ), and short-chain esters of gallic acid (propyl gallate) (Fig. 7). Antioxidant activity and lipid solubility of these compounds are influenced by the number of hydroxyl groups on the aromatic ring and the number and size of substituted alkyl groups. When choosing which synthetic antioxidant to use, food manufacturers must determine what qualities the antioxidant must have in order to be effective.

BHT is highly nonpolar because the two tertiary butyl groups on the molecule make it very lipid solu-

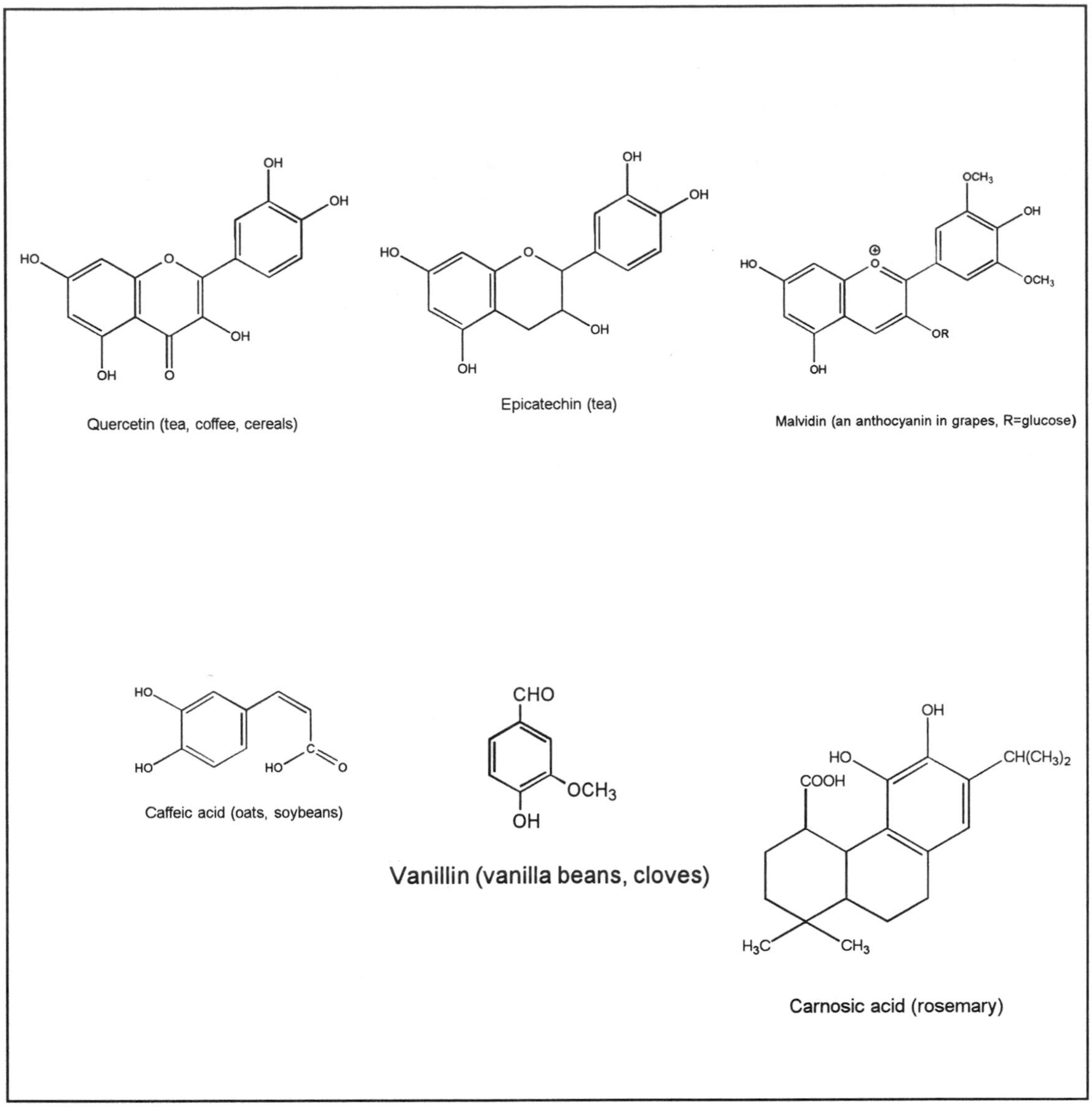

Quercetin (tea, coffee, cereals)

Epicatechin (tea)

Malvidin (an anthocyanin in grapes, R=glucose)

Caffeic acid (oats, soybeans)

Vanillin (vanilla beans, cloves)

Carnosic acid (rosemary)

FIGURE 6 Selected examples of naturally occurring phenolic antioxidants (sources of the compounds are shown in parentheses).

ble. However, the bulky tertiary butyl groups adjacent to the hydroxyl group limit its antioxidant activity.

BHA is a mixture of two isomers: 3-tertiary-butyl-4-hydroxyanisole and 2-tertiary-butyl-4-hydroxyanisole. The 3-isomer predominates, making up approximately 90% of the total. BHA, like BHT, is a hindered phenol but less so. However, the hindrance of the phenol does provide some favorable properties to both BHA and BHT. Because the phenol is somewhat protected, BHA and BHT exhibit "carry-through" properties in baked and fried foods. This means that they do not fully degrade during heating processes, and therefore continue to protect the food from oxidizing after it has been processed. BHA is sold as white waxy tablets and is often used in combination with BHT.

TBHQ has two hydroxyl groups that give it greater antioxidant activity than BHA or BHT. TBHQ has good carry-through properties in baked and fried foods, and does not cause discoloration upon reaction

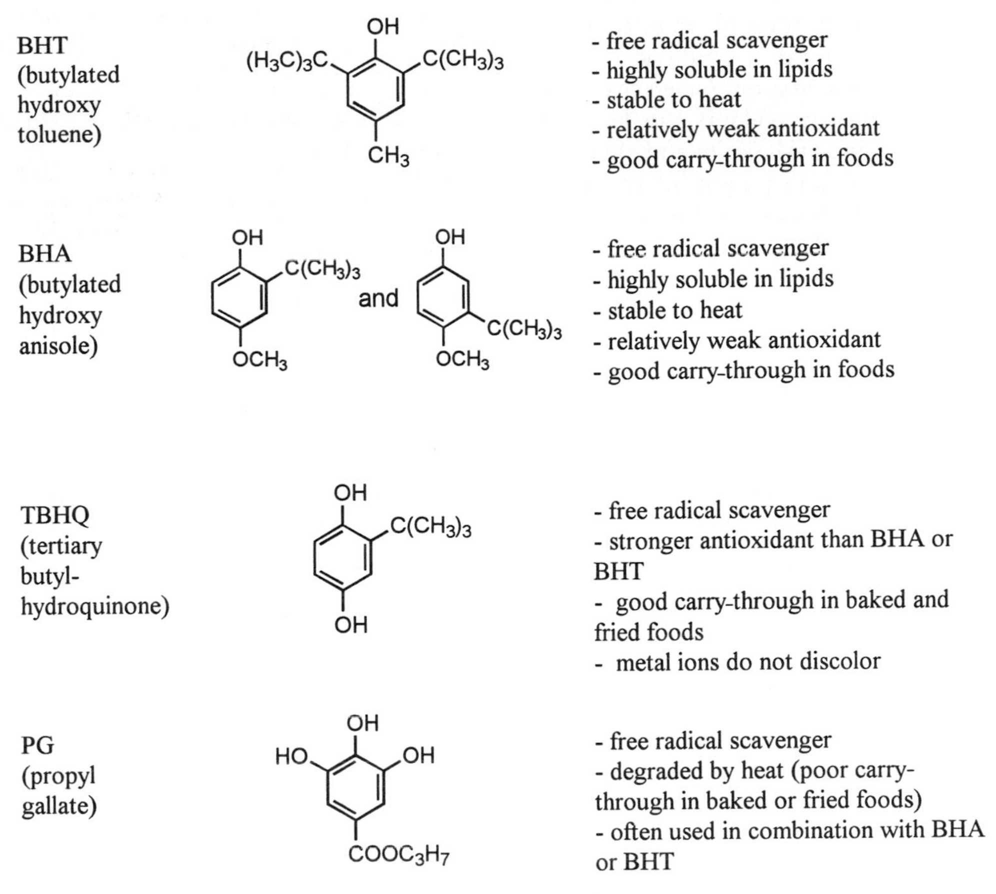

BHT
(butylated
hydroxy
toluene)
(H3C)3C
OH
C(CH3)3
CH3
- free radical scavenger
- highly soluble in lipids
- stable to heat
- relatively weak antioxidant
- good carry-through in foods
BHA
(butylated
hydroxy
anisole)
OH
C(CH3)3
OCH3
and
OH
C(CH3)3
OCH3
- free radical scavenger
- highly soluble in lipids
- stable to heat
- relatively weak antioxidant
- good carry-through in foods
TBHQ
(tertiary
butyl-
hydroquinone)
OH
C(CH3)3
OH
- free radical scavenger
- stronger antioxidant than BHA or BHT
- good carry-through in baked and fried foods
- metal ions do not discolor
PG
(propyl
gallate)
OH
HO
OH
COOC3H7
- free radical scavenger
- degraded by heat (poor carry-through in baked or fried foods)
- often used in combination with BHA or BHT

FIGURE 7 Synthetic antioxidants approved as food additives.

with metals such as iron. It is often added to polyunsaturated oils used for deep fat frying.

Propyl gallate is the *n*-propyl ester of gallic acid. Because of the presence of the three hydroxyls, gallic acid esters make very effective antioxidants. Free gallic acid would presumably make an effective antioxidant but its solubility in lipids is low. Thus, it is esterified with propionic acid to increase lipid solubility. Propyl gallate is degraded by heat and therefore provides poor carry-through properties in foods that are baked or fried. BHA or BHT is often used in combination with propyl gallate to provide carry-through properties to baked goods.

2. Chelating Agents

Chelating agents are frequently added to food to retard lipid oxidation. The most common chelating agents added to foods are ethylenediaminetetraacetic acid (EDTA) and citric acid. Many of the phenolic compounds in foods also function as chelating agents in addition to their radical scavenging actions.

IV. SUMMARY

Uncontrolled oxidation of lipids and other substances leads to serious diseases in humans and quality deterioration in foods. There is growing evidence that antioxidants in foods can prevent or slow excessive oxidation in body cells and in the foods themselves. Foods contain complex mixtures of antioxidants. These include the antioxidant vitamins (E, C, and β-carotene), carotenoids, and phenolic compounds. Most foods contain antioxidants, but fruits and vegetables are particularly rich sources. Antioxidants are also available in health food and drug stores in pill form. Food manufacturers may add natural or synthetic antioxidants to foods to control oxidation. Given that concentrated forms of some antioxidants can be toxic and that our understanding of the roles and effectiveness of specific antioxidants in preventing disease is still very limited, the best advice is to choose a varied diet that contains an abundance of fruits and vegetables.

BIBLIOGRAPHY

Aruoma, O. I. and Halliwell, B. (1991). "Free Radicals and Food Additives." Taylor & Francis, London.

Block, G., and Langseth, L. (1994). Antioxidant vitamins and disease prevention. *Food Technol.* **48**(7), 80–84.

Block, G., Patterson, B., and Subar, A. (1992). Fruit, vegetables, and cancer prevention: A review of the epidemiological evidence. *Nutr. Cancer* **18**, 3–4.

Combs, G. F. (1992). "The Vitamins: Fundamental Aspects in Nutrition and Health." Academic Press, San Diego.

Decker, E. A. (1995). The role of phenolics, conjugated linoleic acid, carnosine, and pyrroloquinoline quinone as nonessential dietary antioxidants. *Nutr. Rev.* **53**(3), 49–58.

Fennema, O. R. (1985). "Food Chemistry," 2nd Ed. Marcel Dekker, New York.

Halliwell, B., Murcia, M. A., Chirico, S., and Aruoma, O. I. (1995). Free radicals and antioxidants in food and *in vivo*: What they do and how they work. *Crit. Rev. Food Sci. Nutr.* **35**(1,2), 7–20.

Ho, C. T., Lee, C. Y., and Huang, M. T. (1992). "Phenolic Compounds and Their Effects on Health II: Antioxidants and Cancer Prevention." American Chemical Society, Washington, D.C.

Kinsella, J. E., Frankel, E., German, B., and Kanner, J. (1993). Possible mechanisms for the protective role of antioxidants in wine and plant foods. *Food Technol.* **47**(4), 85–89.

Thomas, J. A. (1994). Oxidative stress, oxidant defense, and dietary constituents. *In* "Modern Nutrition in Health and Disease" (M. E. Shils, J. A. Olson, and M. Shike, eds.), 8th Ed., Vol. 1, pp. 501–512. Lea & Febiger, Philadelphia.

Wong, D. W. S. (1989). "Mechanism and Theory in Food Chemistry." Van Nostrand–Reinhold, New York.

Food Groups

ELAINE B. FELDMAN
JANE M. GREENE
Medical College of Georgia

GLOSSARY

Anthropometric Refers to measurements of individuals that reflect growth, development, and nutritional status. Common measurements are height, weight, skinfold thickness at various sites, and arm muscle circumference. Norms for healthy persons are published

Collagen Characteristic protein of connective tissue and the most common protein in the body

Complex carbohydrate Primarily from plant sources, they are made of polysaccharides (long chains of sugars), mainly starch and dextrins, that break down to yield intermediate polymers of sugars or simple sugars (mono- or disaccharides)

Cruciferous Vegetables of the mustard family—broccoli, cabbage, cauliflower, and others—with a cross-shaped flower

Dietary fiber Nondigestible or partially digestible components of plants (fruits and vegetables) that may be complex carbohydrates, including insoluble cellulose and lignin, and soluble pectin and gums

Fat Dietary fat, of animal or vegetable origin, includes triglycerides (fats and oils) that are esters of glycerol with three long-chain fatty acids. Fatty acids are saturated (hard), monounsaturated, or polyunsaturated ($\omega 3$ and $\omega 6$ depending on the location of the terminal double bond), and of animal or vegetable origin. Cholesterol is a waxy steroid alcohol that is synthesized only by animal cells and therefore found only in animal foods, especially organ meat and egg yolks

Nutrient Substance in the diet, contained in food, that is used by the body in varying amounts for growth, maintenance, and repair. They may be essential (the body cannot synthesize them) and include water, energy, carbohydrates, protein, fat, vitamins, and minerals. Micronutrients are required in the diet in amounts less than 1 g/day

Protein Compound composed of a variety of amino acids, at least eight of which are essential. These nitrogen-containing compounds form the structure of the body (muscle) and perform vital functions (transport proteins, enzymes). The protein must be complete to sustain and maintain growth, that is, the proportion of essential amino acids must be at least 20%, and all individual essential amino acids must be present in appropriate proportions. The greater the percentage of nitrogen from essential amino acids, the greater the biological value of the protein

Vegetarian Person who eats no animal products (vegan), or no animal products other than milk (lacto-vegetarian), eggs (ovo-vegetarian), or milk and eggs (lacto-ovo-vegetarian)

TO PROMOTE HEALTH AND PREVENT DISEASE, THE daily diet must contain the 40+ essential nutrients. These are provided by a variety of foods, consumed as meals and snacks. The goal is to meet the nutrient needs for the vast majority of the healthy population, as determined by periodic review and recommendation by nutrition scientists (recommended dietary allowances). For convenience, foods are divided into groups so that a balanced diet, selected daily from a variety of foods from all the groups, in appropriate amounts, will provide all essential nutrients.

I. SOURCES OF POPULATION NUTRIENT DATA

Information on the nutritional status of Americans is obtained from data on food production; imports and exports; marketing, distribution, and storage of food; patterns of food consumption by ethnic groups, families, and individuals; clinical nutrition surveys; studies of physical development; laboratory tests of nutrient levels; vital statistics on morbidity and mortality; and epidemiological information relating diet to disease. The American diet has changed markedly since 1900, with increased consumption of meat, poultry, fish, dairy products, sugar and other sweeteners, fats and oils, and processed fruits and vegetables and decreased consumption of grain products, potatoes, sweet potatoes, fresh fruit, vegetables, and eggs. [*See* Diet.]

A. Ten-State Nutrition Survey and Health and Nutrition Examination Survey (HANES)

The Ten-State Nutrition Survey from 1968 to 1970 focused on low-income groups and evaluated the nutritional status and dietary practices of 40,000 persons. In 1971, HANES evaluated a sample of 28,943 persons aged 1 to 74 from 65 locations in the 48 contiguous U.S. states. HANES provided information on dietary intake, clinical and biochemical findings, anthropometric data, hemoglobin, serum iron, transferrin saturation, and serum cholesterol levels. In 1974 a follow-up survey, HANES 2, was carried out. HANES 3 was scheduled over two 3-year periods, from 1988 to 1991 and 1991 to 1994. Data from the first 3 years from a sample of 20,277 persons were made available in 1994 and reported consumption of macro- and micronutrients for ages 2 months to over 80 years and among gender and race-ethnic groups. Thirty-four percent of calories were derived from fat; fiber intake averaged 15 g/day (low). There were concerns over low iron intake in women, low calcium in women and adolescents, and increased sodium.

B. Continuing Survey of Food Intakes by Individuals

In 1985 and 1986 the U.S. Department of Agriculture (USDA) conducted the yearly Continuing Survey of Food Intakes by Individuals (CSFII) using a 1-day dietary recall and 5 days of dietary data obtained by telephone. The CSFII is a component of the National Nutrition Monitoring System, a set of federal activities that provides regular information on the nutritional status of the U.S. population. The CSFII sampled households of women 19–50 years old and their children 1–5 years old. Data were collected for 1500 women, 1100 men, and 500 children. In 1985, compared to 1977, men ate less meat (principally beef), whole milk, and eggs. They ate more fish, low-fat or skim milk, legumes, nuts and seeds, and carbonated soft drinks (regular and low calorie). The percentage of calories decreased from fat and increased from carbohydrate.

In 1986 the survey reported that one-third of meals were consumed away from home and only half the people ate breakfast. Women consumed more skim and low-fat milk than whole milk. Children consumed equal amounts of these products and drank more carbonated beverages than fruit drinks or "ades."

The intake for women was reported as 37% of calories from fat (13% saturated, 14% monounsaturated, and 7% polyunsaturated), 46% from carbohydrate, and 17% from protein. For children the figures were 35% of calories from fat (14% saturated, 13% monounsaturated, and 6% polyunsaturated), 51% from carbohydrate, and 15% from protein.

The CSFII also was conducted in 1989 to 1991. An initial report published in September, 1995 indicated that good sources of calcium and iron may be in short supply in the diets of young children, teenage girls, and women of childbearing age.

As of this writing, the most recent CSFII dates from 1994 to 1996.

II. DETERMINANTS OF FOOD INTAKE

A. Food Choices

People choose foods—not nutrients—depending on cultural, social, personal, and situational factors, including ethnicity and family tradition. Fads affect food choices, and associations with rewards or punishment may explain some selections. The average American diet contains nearly 200 different foods. People avoid foods that cause unpleasant symptoms and select those that are well tolerated. Food choices may be restricted or influenced by poverty, lack of transportation, limited availability of foods in stores, poor food storage facilities, lack of cooking facilities or skills, or limited time for food preparation. Advertising and food labeling strongly influence choices.

In the United States, economic factors tend to limit the variety of food intake rather than directly determine an inadequate diet. Low-income populations may be poorly educated and thereby less understanding of the food group classifications; their choices are determined by their likes and dislikes, influenced by advertising, and the appearance of meal items.

B. Ethnic Preferences

Diets of ethnic minority groups in the United States may be influenced by the nature of their traditional diets, and the extent to which the diet has been adapted to the typical American diet. Dietary patterns of African-Americans resemble the traditional diet more than does the diet of other minorities. The African-American diet shows a preference for "soul food" and contains yellow and dark-green leafy vegetables, pork, fish, and poultry, which provide vitamins A and C, thiamin, and protein. The extensive use of frying, overcooking of vegetables, high consumption of sodium, and a low intake of milk and other dairy products are unhealthy aspects of the African-American diet.

Hispanic American eat diets high in carbohydrates in the form of tortillas and rice. Corn, onions, tomatoes, and sweet potatoes are the dominant vegetables, with few leafy green vegetables and fruits. Dried beans are the dominant protein source, but Americanization of their diet is causing an increase in animal protein. In general the fiber content is high, and the proportion of animal fat to total calories is less than that in typical American diets. The diet may contain too much sodium and energy and may be deficient in calcium, iron, and vitamins A and C.

The primary source of energy in Asian/Pacific American diets is rice. Other carbohydrate foods such as wheat, noodles, and tubers are prominent. Compared to typical American diets, these diets are higher in vegetables, fruits, fish, and shellfish, but lower in meat and dairy foods. Native Americans have varying dietary patterns depending on their heritage and geographic location. The traditional diets consist of a combination of foods such as mutton, game, fish, tortillas, fried bread, fruits, roots, corn, and wild greens. The diet has been altered by the addition of processed foods such as bologna, potato chips, carbonated beverages, and refined sugar products. The typical diet consumed today is high in carbohydrate, saturated fat, sodium, and sugar. Calcium intake is low.

III. COMPONENTS OF A HEALTHFUL DIET

Guidelines have been published that advise Americans on the components of a healthful diet. In 1977 the Senate Select Committee on Nutrition and Human Needs issued "Dietary Goals for the United States," based on concerns that overnutrition may cause obesity, coronary heart disease, cancer, and stroke. This report recommended that Americans should consume less food, fat (especially saturated fat), cholesterol, refined sugar, and salt, and should increase consumption of fruits, vegetables, grain products, and unsaturated oils.

A. Nutrition Objectives for the Nation

National nutrition goals and objectives in relation to health promotion and disease prevention were first published in 1979 in the Surgeon General's Report on Health Promotion and Disease Prevention, titled *"Healthy People."* The 17 specific nutrition objectives to be attained by 1990 included reducing iron deficiency anemia in pregnant women; eliminating diet-induced growth retardation of infants and children; decreasing significant overweight by weight-loss regimens; reducing serum cholesterol and sodium ingestion; increasing breastfeeding; educating the population on diet-related diseases and food composition; providing nutrition information in labels, via employee and school cafeterias, in state school systems, and in health contacts; and improving the national nutrition status monitoring system. Only the objective of reducing sodium was achieved by 1990; there was no progress in decreasing overweight, which actually increased, and the remaining objectives were partially achieved.

The 21 nutrition objectives for the year 2000 are measurable and include intervention strategies. They are: reduce coronary heart disease deaths; reverse the rise in cancer deaths; reduce overweight; reduce growth retardation among children; reduce dietary fat intake; increase complex carbohydrate and fiber-containing foods (5+ servings of vegetables and fruits, 6+ grain products); increase the proportion of overweight who exercise; increase calcium intake; decrease salt and sodium intake; reduce iron deficiency; increase breastfeeding; prevent baby bottle tooth decay; use food labels to make nutritious food selections; achieve useful and informative nutrition labeling for processed, fresh, and ready-to-eat foods; increase the number of brand items reduced in fat and saturated fat and the proportion of restaurants and institutions offering low-fat, low-calorie food choices consistent with the Dietary Guidelines (see the following), children's food services consistent with the Dietary Guidelines, and home food services to the elderly; increase nutrition education in schools and worksites; and increase the proportion of primary care providers who provide nutrition services and referral. Annual tracking data will be provided throughout the decade.

B. USDA/HEW Dietary Guidelines

In 1980, the USDA and the Department of Health, Education, and Welfare (HEW) published *"Dietary Guidelines"* (updated in 1985, 1990, and 1995). These recommended that we: (1) eat a variety of foods to provide energy (calories), protein, vitamins, minerals, and fiber needed for good health; (2) balance the food you eat with physical activity; maintain or improve your weight to reduce incidence of high blood pressure, heart disease, stroke, certain cancers and the most common kind of diabetes; (3) choose a diet with plenty of grain products, vegetables, and fruits, that provide needed vitamins, minerals, fiber and complex carbohydrates; (4) choose a diet low in fat, saturated fat, and cholesterol to reduce risk of heart disease and certain types of cancer and to help maintain a healthy weight; (5) choose a diet moderate in sugars to help reduce caloric intake and tooth decay; (6) choose a diet moderate in salt and sodium to help reduce risk of high blood pressure; and (7) if you drink alcoholic beverages, do so in moderation as alcoholic beverages supply empty calories and can cause many health problems and accidents and lead to addiction.

C. Diet and Cancer Report

The National Research Council and the American Cancer Society have presented guidelines aimed at cancer prevention that are similar to those just listed. Recommendations that are specific for cancer prevention are: reduce fat intake to 30% or less of calories; increase fiber intake to 20–30 g/day; eat five or more servings of vegetables and fruits daily; and minimize consumption of foods preserved by salt curing (including salt pickling) or smoking. Additional cautions are to minimize contamination of foods with carcinogens from any source; identify mutagens in food; and remove or minimize their concentration unless the nutritive value of foods is jeopardized or other potential hazard is introduced.

D. Surgeon General's *"Report on Nutrition and Health"*

In 1988, dietary changes were recommended in the Surgeon General's *"Report on Nutrition and Health"* in order to improve the health of Americans. The highest priority is to reduce intake of foods high in fats and increase intake of foods high in complex carbohydrates and fiber. The specific recommendations for most people (Table I) deal with fats and cholesterol, energy and weight control, complex carbohydrates and fiber, sodium and alcohol. Other issues for some people are fluoride, sugars, calcium, and iron. As of this writing, the next *report* was due in 1996. The new format will discuss specific topics, for example, dietary fat and health, and new reports will be issued every two years.

IV. THE FOOD GROUPS

A. The Food Pyramid

In 1956, simple, specific guidelines enabling an American household to plan and consume meals that meet the recommended dietary allowances were established by the USDA in the form of four major food groups—*the* "basic four" plan. The plan recommended a specific number of daily servings from each group: milk and milk products, meat and meat substitutes, fruits and vegetables, and breads and cereals. In April 1992, the USDA released the Food Guide Pyramid to replace the basic four food groups as a nutrition education tool. The pyramid provides dietary guidance for healthy American two years of age and older. The predominant message is to eat a variety of foods from each food group in moderate amounts and in proportion to individual nutrient and energy needs. The pyramid also recommends less sugar and fat in the diet and increased fiber. The five food groups depicted in Fig. 1 are:

- bread, cereal, rice, and pasta group (6–11 servings/day);
- vegetable group (3–5 servings/day);
- fruit group (2–4 servings/day);
- meat, poultry, fish, dry beans, eggs, and nuts group (2–3 servings/day);
- milk, yogurt, and cheese group (2–3 servings/day).

TABLE I
Recommendations from the Surgeon General's "Report on Nutrition and Health" (1988)[a]

Fats and cholesterol	Reduce consumption of fat (especially saturated fat) and cholesterol. Choose foods relatively low in these substances, such as vegetables, fruits, whole-grain foods, fish, poultry, lean meats, and low-fat dairy products. Use food preparation methods that add little or no fat.
Energy and weight control	Achieve and maintain a desirable body weight. To do so, choose a dietary pattern in which energy (caloric) intake is consistent with energy expenditure. To reduce energy intake, limit consumption of foods relatively high in calories, fats, and sugars, and minimize alcohol consumption. Increase energy expenditure through regular and sustained physical activity.
Complex carbohydrates and fiber	Increase consumption of whole-grain foods and cereal products, vegetables (including dried beans and peas), and fruit.
Sodium	Reduce intake of sodium by choosing foods relatively low in sodium and limiting the amount of salt added in food preparation and at the table.
Alcohol	To reduce the risk for chronic disease, drink alcohol only in moderation (no more than two drinks a day), if at all. Avoid drinking any alcohol before or while driving, operating machinery, taking medications, or engaging in any other activity requiring judgment. Avoid drinking alcohol while pregnant.
Fluoride	Community water systems should contain fluoride at optimal levels for prevention of tooth decay. If available, use other appropriate sources of fluoride.
Sugars	Those who are particularly vulnerable to dental caries (cavities), especially children, should limit their consumption and frequency of use of foods high in sugars.
Calcium	Adolescent girls and adult women should increase consumption of foods high in calcium, including low-fat dairy products.
Iron	Children, adolescents, and women of childbearing age should be sure to consume foods that are good sources of iron, such as lean meats, fish, certain beans, and iron-enriched cereals and whole-grain products. This issue is of special concern for low-income families.

[a]USDHHS, PHS Publication No. 88-50210.

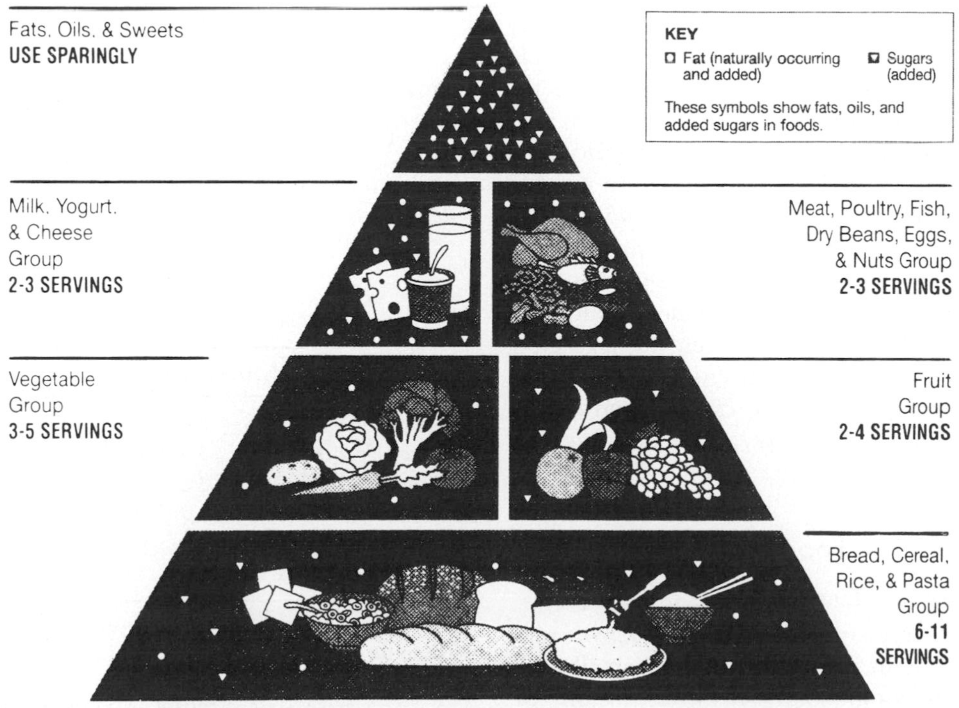

FIGURE 1 The Food Pyramid: A Guide to Daily Food Choices.

TABLE II
Food Groups and Some Major Nutrients

	Nutrient			Vitamins			Minerals	
Food group	Protein	CHO	Fiber	A	B	C	Ca	Fe
Milk, yogurt, cheese	•			•	Riboflavin		•	
Meat, poultry, fish, dry beans, eggs, nuts	•				•			•
Fruits			•	•		•		
Vegetables			•	•	Folate	•		
Bread, cereal, rice, pasta		•			•			•

The pyramid is completed with fats, oils, and sweets, to be used sparingly. It is important that people include beverages when reporting food consumption. Among the other educational messages of this pyramid are that some food from each of the major food groups should be eaten every day but that the range of servings may vary, that no one food group is more or less important, and that fat and sugars may be naturally occurring in specific foods or added as such. The pyramid does not tackle the problem of defining portion size. The major nutrients provided by each group are shown in Table II.

Individuals must select meals and snacks from each group to obtain all essential nutrients that are not distributed evenly within each group. Thus, servings of fruit should include one vitamin C-rich food daily and servings of vegetables should include one vitamin A-rich food every other day. While adhering to the food pyramid plan, *total nutrient intake* can vary widely depending on age, sex, size, and activity level of an individual. At the least, the minimum number of servings in each range should be consumed daily. The lowest range of servings provides about 1600 calories and is adequate for elderly persons and sedentary women. The middle range provides about 2200 calories and is adequate for children, teenage girls, active women, and inactive men. The maximum serving range provides about 2800 calories and can be used as a guide for teenage boys, active men, and very active women (Table III).

B. The Food Groups

1. Milk, Yogurt, and Cheese

Milk is a good source of many nutrients. Cow's milk protein is 80% casein. The whey includes lactalbumin and various immunoglobulins. Milk fat is easily digested and varies from 4% in whole milk to 2% in low-fat milk to less than 0.5% in skim milk. The carbohydrate in milk, lactose, is less sweet than sucrose and not readily digested by some ethnic groups and sick people. Calcium, present in large quantities, is generally absorbed more readily than the calcium in other foods (Table IV). Milk contains minimal iron and is a useful source of riboflavin and nicotinic acid. Its low ascorbic acid content is destroyed by pasteurization. Vitamin D is generally added to milk.

The fat content of cream varies from 10% in half-and-half to 35% in whipping cream. The fat content of evaporated milk may vary, while condensed milk has added sugar. Skim milk contains the protein, calcium, and B vitamins in the original milk without fat and with less cholesterol. Yogurt is a nutritious and convenient food with variable fat content.

Most cheese is made from milk clotted using rennet, and contains the same protein and fat as milk and many of the other nutrients. Most cheeses contain 25–35% protein and 16–40% fat, and are rich in calcium, vitamin A, and riboflavin.

TABLE III
How Many Servings Do You Need Each Day?

Calorie level:[a]	About 1600	About 2200	About 2800
Bread group	6	9	11
Vegetable group	3	4	5
Fruit group	2	3	4
Milk group	2–3[b]	2–3[b]	2–3[b]
Meat group	2, for a total of 5 oz.	2, for a total of 6 oz.	3, for a total of 7 oz.

[a]These are the calorie levels if you choose low-fat, lean foods from the five major food groups and use foods from the fats, oils, and sweets group sparingly.

[b]Women who are pregnant or breastfeeding, teenagers, and young adults to age 24 need 3 servings.

TABLE IV
Equivalent Providers of Calcium in the Diet[a]

Milk	1 cup
Yogurt	1 cup
Pudding	1 cup
Custard	1 cup
Nonfat dry milk	1/3 cup
Cottage cheese	1 1/3 cup
Processed cheese	1 1/3 oz.
Cheddar cheese	1 1/3 oz.
Ice cream	1 1/2 cup

[a]From E. B. Feldman (1988). "Essentials of Clinical Nutrition," Table 4-6, p. 92. F.A. Davis Co., Philadelphia. Used with permission.

2. Meat, Poultry, Fish, Dry Beans, Eggs, and Nuts

Meat is an excellent source of protein. Its digestibility relates to the amount of muscle protein versus connective tissue, collagen, and fat. The collagen may vary from 2.5 to 23.6% and fat from 5 to 50%. Tenderness is associated with fat (marbling). Lean meat contains about 20% protein and 5–10% fat; the protein is of high biological value. Pork and chicken have a higher protein-to-fat ratio than either beef or lamb. Meats are usually rich in iron and zinc, contain little calcium, and are important sources of nicotinic acid and riboflavin. Muscle provides moderate amounts of vitamin B_{12}, but little vitamin A or ascorbic acid. Current recommendations limit meat to no more than 6 ounces per day.

Fish is an important source of protein of high biological value. Lean fish, such as cod, haddock, and sole, contain less than 1% fat and about 10% protein. They are relatively low in calories and are easily digested. Fatty fish, such as herring, salmon, and sardines, contain 8–15% fish oil, doubling the calories. Halibut, mackerel, and trout have an intermediate fat content. Fish roe contains 20–30% protein and 20% fat. Fish oils are rich sources of vitamins A and D and long-chain, polyunsaturated $\omega 3$ fatty acids. Iodine and fluoride are ample in marine fish, and small whole fish are high in calcium.

Shellfish have little fat and calories. The protein content of oysters, mussels, and other molluscs is about 15%. Oysters are the richest food source of zinc.

The average egg (60 g) contains 6 g protein and 6 g fat and yields 80 calories. Egg proteins are mostly albumin, with the highest biological value of all food proteins for human adults. The yolk is a fair source of vitamin A and contains significant amounts of B vitamins. The average large egg contains about 215 mg of cholesterol.

Textured vegetable protein derived from soybeans is flavored to resemble meat. The natural ingredients contain no vitamin B_{12}. Vegetable proteins have less of the amino acid methionine than do animal proteins (Table V) and may be lower in iron, thiamin, and riboflavin than meat.

Legumes are seeds of the family that includes peas, beans, peanuts, and lentils. Their high protein content (20 g per 100 g dry weight) qualifies legumes as meat substitutes. Their low content of sulfur-containing amino acids reduces the biological value of the protein; they are rich in lysine, which is deficient in many cereals (Table V). A combination of legumes and cereal proteins may have a nutritive value as good as animal proteins and is an excellent source of fiber.

TABLE V
Food Sources Providing Complementary Plant Proteins[a]

Food	Amino acids deficient	Complementary protein
Grains	Isoleucine Lysine	Rice, corn, wheat, rice + legumes Wheat + peanut + milk Wheat + sesame + soybean Rice + sesame Rice + brewer's yeast
Legumes	Tryptophan Methionine	Legumes + rice Beans + wheat Beans + corn Soybeans + rice + wheat Soybeans + corn + milk Soybeans + wheat + sesame Soybeans + peanuts + sesame Soybeans + peanuts + wheat + rice
Nuts and seeds	Isoleucine Lysine	Peanuts + sesame + soybeans Sesame + beans Sesame + soybeans + wheat Peanuts + sunflower seeds
Vegetables	Isoleucine Methionine	Broccoli Brussels sprouts + sesame seeds Cauliflower — Brazil nuts or Green peas — mushrooms Lima beans Greens + millet or converted rice

[a]From E. B. Feldman (1988). "Essentials of Clinical Nutrition," Table 4-11, p. 103. F.A. Davis Co., Philadelphia. Used with permission.

Legumes are a good source of B vitamins, except for riboflavin. Legumes lack ascorbic acid but sprouted legumes will prevent scurvy. Soybeans are high in protein; the whole dry grain contains 40% protein and up to 20% fat. Soya also provides B vitamins. Peanuts contain about 20% fat.Other legumes include a variety of beans, and although their digestion and absorption are virtually complete, flatulence may be a by-product.

3. Fruits

Fruits have many pleasing flavors and can serve as desserts without excessive calories. Ascorbic acid is an essential nutrient abundant in fruits. Fruits are sources of dietary fiber and contain carotene and small quantities of B vitamins. Most fruits have little or no protein or fat and contain 5–20% carbohydrate. Fructose and glucose are the major sugars in ripe fruits. Bananas may serve as a useful energy source, but provide no protein. Fruits are high in potassium.

4. Vegetables

Vegetables include leaves, roots, flowers, stalks, and gourds. Their chief nutritional value is for carotene, ascorbic acid, folate, and dietary fiber. Calcium and iron may be present in significant amounts but absorption is variable. Leafy vegetables may provide some B vitamins (riboflavin), but vegetables are poor sources of energy and protein. Potatoes are the inexpensive food that is best capable of supporting life as the sole diet, with starch supplying most of the calories. Although low in protein, it is of relatively high biological value. Potatoes are high in fiber and a good source of potassium; they are easily digested and well absorbed. The quality and nutritional value of canned and frozen vegetables compare favorably with those of fresh produce.

A single serving of fruit or vegetable is one-half cup, or one medium-size piece. The National Research Council report on *"Diet and Health"* (1989) recommends five or more servings every day of vegetables and fruits, especially green and yellow vegetables and citrus fruits.

5. Breads, Cereals, Rice, and Pasta

Cereal grains are the most important single food in many countries and are consumed as bread and in flour products. Corn, wheat, barley, oats, and rye are the principal cereals of North America.

Whole-grain cereals provide energy, good-quality protein, and appreciable amounts of calcium and iron. Cereals contain no ascorbic acid and practically no vitamin A; yellow corn contains significant amounts of carotene. Whole-grain cereals contain adequate amounts of B vitamins except for corn, in which the bound nicotinic acid is not biologically available. Milling and discarding the outer portion of the seed diminishes the B vitamin content of wheat and rice. Wheat may contain 10–20 g protein per 100 g, with lysine the limiting amino acid. In the average flour used to make white bread, protein provides about 13% of the energy. Whole-what flour contains three times as much dietary fiber as white flour. Whole wheat is also high in phytate, which binds minerals, making them unavailable. Thus, although whole-wheat flour contains appreciable calcium, iron, and zinc, absorption may be limited. In the United States, 100 g of white flour is enriched with up to 0.44 mg thiamin, 0.26 mg riboflavin, 3.5 mg nicotinamide, and 2.9 mg iron. In some states, calcium and vitamin D may also be added. Addition of folic acid has been approved. Pasta is made from a hard variety of wheat but utilizes that portion relatively poor in B vitamins; it is frequently enriched. Highly refined rice is almost devoid of vitamins. Parboiling fixes the vitamins so that they are not removed with milling and is the simplest preventive measure against beriberi (thiamin deficiency). Most rice contains 6.5–8.0 g of good-quality protein per 100 g. The principal protein in corn is incomplete, lacking the amino acids lysine and tryptophan. The preparation of tortillas makes nicotinic acid biologically available from corn by heating the grains in lime water to soften them. Oatmeal contains more protein (12 g per 100 g) and more oil (8.5 g per 100 g) than other common cereals and is rich in soluble fiber. Barley produces the malt for brewers and is the basis of the best beers and some whiskey. Bread made from rye flour is rich in B vitamins and also contains fiber. The chief nutritive value of most dry breakfast cereals is derived from the addition of milk. Some cereals are fortified with B vitamins, iron, and, most recently, calcium.

6. Fats, Oils, and Sweets

The small tip of the pyramid shows the additional category of fats, oils, and sweets. Enough fat is normally consumed in the usual foods found in the other groups of the pyramid. Additional fats from vegetable oils and salad dressings, lard, butter, margarine, and cream, and from sugars, soft drinks, candies, and sweet desserts are high in calories compared to the nutrients they contain (i.e., they are "calorie dense" and provide relatively "empty" calories). They should be eaten sparingly by most people.

7. Beverages

In addition to the water naturally occurring in foods from each of the groups, one must consume adequate fluids each day as either water or other liquids. Natural bottled water can include minerals and carbon dioxide. Seltzer can be low calorie or be sweetened with sugar (sucrose), corn syrup, or other fruits or syrups (fructose). It is important that the major portion of the added fluid be noncaloric, as many empty calories can be added to the diet in the form of soft drinks and other sugar-sweetened beverages.

V. IMPROVING MEALS AS SOURCES OF NUTRIENTS

A. Nutritional Quality

Smaller portions of a wider variety of foods can improve the nutritional quality of meals without increasing the calories (Table VI). For example, 6 ounces of steak with a baked potato and a tomato contain fewer calories (480 kcal) than 10 ounces of steak and a tomato (655 kcal). Substituting a roll for a high-fat salad dressing transforms a 740-kcal chef's salad to a more healthful 590-kcal meal, with room for a fruit dessert. Decreasing the intake of fatty and protein-heavy foods and increasing the proportion of starchy foods and fruit will provide more satisfying meals that offer more food, fiber, vitamins, and minerals.

Simple changes in food preparation methods and seasoning (Table VII) lower calories without altering vitamins, minerals, or protein. More calories and nutrients can be obtained from healthful between-meal snacks such as milk shakes or peanut butter or other nut butter.

B. Determining Composition of Food

Food composition tables that provide average nutrient values based on quantitative analyses have been available for about 100 years. Most tables include data for five vitamins (vitamin A, thiamin, riboflavin, niacin, ascorbic acid), calcium, iron, energy, protein, carbohydrate, and fat. The USDA periodically publishes updated food composition tables (e.g., Handbook No. 8, "*Composition of Foods*") in various sections. USDA Handbook No. 456 presents values for foods

TABLE VI
How to Improve Your Meals[a]

	Common choice		Wiser choice	
	Menu	kcal	Menu	kcal
Breakfast	Orange juice, 1/2 cup Black coffee Fruit-flavored yogurt	320	Orange juice, 1/2 cup Skim milk, 1/2 cup Whole-grain cereal, 1/2 cup Toast, 2 slices Butter/margarine, 1 tsp	325
Lunch	Tuna salad Coleslaw, potato salad Sliced tomato Crackers Mineral water	930	Sliced turkey Carrot sticks Tomato Whole-wheat bread Cantaloupe, 1/2 Milk, 1 cup	540
Dinner	Broiled chicken, 1/2 Tossed salad French dressing, 2 Tbs Green beans	410	Broiled chicken leg Tossed salad Dressing, 1 Tbs Peas and onions, 1 cup Roll Grapes, small bunch	575
Snack	Omitted		Banana, popcorn (3 cups) 11 saltines, fruit or fruit juice and mineral water	150

[a]Adapted from E. B. Feldman (1988). "Essentials of Clinical Nutrition," Table 4-8, p. 99. F.A. Davis Co., Philadelphia. Used with permission.

TABLE VII
Suggestions to Decrease Calories[a]

Suggestion	Examples	Amount	Calorie content
Use skim or low-fat dairy products	Whole milk	8 fl. oz.	160
	Skim milk	8 fl. oz.	90
Try seasonings or lemon juice instead of butter or margarine to bring out natural vegetable flavors	Butter	1 tsp	35
	Margarine	1 tsp	35
	Lemon juice	1 tsp	1
Broil, bake, stew, or roast trimmed meat instead of frying. Use a rack to hold roast out of drippings. Skim excess fat from stew and soup.	Vegetable oil	1 Tbs	125
	Lard	1 Tbs	115
	Vegetable shortening	1 Tbs	110
Poach or boil eggs rather than frying or scrambling	Poached egg	1	80
	Boiled egg	1	80
	Scrambled egg	1	110
	Fried egg	1	110

[a]From E. B. Feldman (1988). "Essentials of Clinical Nutrition," Table 4-9, p. 100. F.A. Davis Co., Philadelphia. Used with permission.

using household measures rather than the standard 100-g portions. [*See* Nutritional Quality of Foods and Diets.]

C. Labeling

The federal government oversees the labeling of most foods through the USDA and the Food and Drug Administration (FDA). The FDA requires listing ingredients of all foods with more than one ingredient. Ingredients are listed in descending order by weight. The Nutrition Labeling and Education Act of 1990, the first major legislation related to food labeling in over 20 years, went into effect in May 1994. The new labels, identified by the heading "Nutrition Facts," must include serving size in standard household measure; number of servings per container; total calories; total fat and saturated fat; cholesterol; sodium; total carbohydrates, complex carbohydrates, and sugar; total protein; and dietary fiber. The information is related as percentage daily value of one serving to a normal daily diet that contains 2000 or 2500 calories, 65 or 80 g fat (30% of calories), 20–25 g saturated fat, 300 mg cholesterol, 2400 mg sodium, 300–375 g total carbohydrates, and 25–30 g of fiber.

The law also gives specific definitions of descriptors used on labels (Table VIII). For the first time, specific health claims are allowed on labels that describe the link between a food or nutrient and a disease. The eight connections supported by scientific evidence include: a link between calcium and a lower risk of osteoporosis; fat and a greater risk of cancer; saturated fat and cholesterol and a reduced risk of coronary heart disease; fiber-containing grain products, fruits, and vegetables and a reduced risk of cancer; fruits, vegetables, and grain products that contain fiber and a reduced risk of coronary artery disease; sodium and a greater risk of high blood pressure;

TABLE VIII
Commonly Used Label Terms[a]

Term	Definition
Free	Contains no or trivial amounts of one or more of these components: fat, saturated fat, cholesterol, sodium, sugars, or calories
Reduced	A nutritionally altered product contains at least 25% less of a nutrient or calories than the regular or referenced product
Lean	Less than 10 g fat, 4.5 g or less saturated fat, and less than 95 mg cholesterol per 100-g serving
Less	A food whether altered or not that contains 25% less of a nutrient than the referenced food
Light	A nutritionally altered product contains one-third fewer calories or half the fat of the referenced food or the sodium content of a low-calorie, low-fat food has been reduced by 50%; light in sodium may be used on food in which the sodium content has been reduced by at least 50%
Extra lean	Less than 5 g fat, less than 2 g saturated fat, and less than 95 mg cholesterol per 100-g serving
Low	Low fat: 3 g or less per serving Low saturated fat: 1 g or less per serving Low sodium: 140 mg or less per serving Very low sodium: 35 mg or less per serving Low cholesterol: 20 mg or less and 2 g or less saturated fat per serving Low calorie: 40 calories or less per serving
Fewer	Synonym for low
High	The food contains 20% or more of the Daily Value for a particular nutrient
More	A serving of a food whether altered or not contains a nutrient that is at least 10% of the Daily Value more than the referenced food
Good source	One serving of a food contains 10 to 19% of the Daily Value for a particular nutrient

[a]Nutrient content descriptors as defined by federal law.

fruits and vegetables and a reduced risk of cancer; and folic acid and a reduced risk of neural tube defects.

VI. SPECIAL DIETARY NEEDS

A. Vegetarian Diets

Some people adopt vegetarian diets, often out of philosophical or religious conviction. Forms of vegetarianism include lacto-ovo-vegetarian, lacto-vegetarian, ovo-vegetarian, and strict or pure vegetarian (vegan) (see Glossary). Vegetarians usually consume significantly less total fat, saturated fat, and cholesterol, and more polyunsaturated fat, carbohydrate, and dietary fiber than in the usual American diet. Meeting protein needs and consuming all the essential amino acids is increasingly difficult in the more restricted forms of vegetarianism. The amino acids missing in any one grain or vegetable can be replaced by consuming several together in the same meal or adding dairy products (Table V). Vegan diets may be nutritionally inadequate if there is undue reliance on a single plant food source. Infants and children on vegan diets are particularly likely to develop symptomatic clinical nutritional deficiencies, with retarded growth and development. The four groups of plant foods relied on by vegans are legumes, cereal grains, fruits and vegetables, and nuts and seeds.

Other nutrients that may be deficient in vegan diets are vitamin B_{12}, vitamin D (for children not exposed to sunlight), riboflavin, calcium (especially for children and women), and iron (for women of childbearing age). Generous servings of green leafy vegetables, dried beans, sesame seeds, onions, and soybean milk will supply riboflavin and calcium, whereas beans, seeds, nuts, green leafy vegetables, dried fruits, and grains will supply iron.

B. The Athlete

Following the recommended dietary allowances will provide all necessary nutrients for a physical conditioning program. The diet that contributes to the best performance by the recreational or world-class athlete is that which is best for the nonathlete: a nutritionally balanced diet supplying appropriate quantities of water, energy, protein, fat, carbohydrate, vitamins, and minerals. Commercially promoted food supplements and drugs offer nothing to the healthy, well-nourished athlete and should be rejected.

VII. MODIFICATIONS FOR THE LIFE CYCLE

A. Pregnancy and Lactation

The critical period in which the developing fetus is influenced most by diet is between 17 and 56 days after conception. Many women do not know that they are pregnant until beyond this period; therefore the diet should be optimized prior to pregnancy. Any unhealthy eating practices or nutritional deficiencies should be corrected prior to conception. Guidelines for nutrition during pregnancy and lactation from the Institute of Medicine, National Academy of Science, were updated in 1992. The pregnant woman should increase her food intake by about 300 calories per day, increase protein 20% and calcium 50%, and double the folic acid and iron intakes for optimal outcome. The normal-weight pregnant woman should gain 25–35 pounds, underweight and young adolescents up to 40 pounds, whereas the overweight woman should not gain more than 15–25 pounds. Weight gain should begin in the second trimester. Additional calories are usually provided by complex carbohydrate (bread, vegetable, and fruit groups). The diet should include increased servings of the milk and meat group to provide extra protein. The calcium requirements are met by a higher intake of the milk group and foods rich in calcium. Iron supplementation is recommended, and many nutritionists recommend folate supplementation in order to achieve folate status that is adequate to prevent neural tube defects.

The lactating woman's major increased needs are for calories, protein, and calcium, especially if lactation exceeds 3 months. These needs can be met by including 1 quart of whole milk per day in the diet.

B. Pediatric Needs

Infancy is the only time in life when a single food comprises the entire diet. Though commercial infant formulas meet nutritional needs, breastfeeding is recommended to nourish the healthy term infant, particularly for the first 6 months. Formula may be replaced with up to 1 quart per day of cow's milk at about the first birthday. At that time the child will be consuming iron-rich foods such as fortified infant cereals or meats. Solid foods are added to the baby's diet during the first half of the first year (4–5 months), with the first food usually being a single-grain cereal. Strained fruits and yellow vegetables are then added, followed by green vegetables, meats, and egg yolks. Citrus, seafood, nut butters, chocolate, nitrate-containing vegetables, and egg whites are not introduced until the end of the first year. The older child should eat a variety of foods distributed according to the food pyramid. The adolescent requires increased calories, calcium, and iron. These needs are not met (except for calories) by a diet of soft drinks, French fries, candy, and potato chips.

C. Geriatric Needs

Caloric needs are less, depending on the level of activity, for the elderly. Inactive elderly persons should consume foods of increased nutrient density, such as lean meat, fish, eggs, milk, and vegetables, and curtail fats and carbohydrates. Intakes of calcium, iron, zinc, selenium, and vitamins D, B_6, folate, B_{12}, and C may be inadequate or marginal, especially in low-income elderly. Good sources of these nutrients include dark green vegetables, whole-grain and enriched cereals, pasta and bread, meat, fish, poultry, dried beans and peas, milk, and milk products. A multivitamin supplement that is formulated for the elderly may be appropriate when energy intake is less than 1500 kcal/day or when the diet is restricted.

BIBLIOGRAPHY

Agricultural Research Service (1975). "Composition of Foods," Agricultural Handbook No. 8. U.S. Department of Agriculture, Washington, D.C.

Feldman, E. B. (1988). "Essentials of Clinical Nutrition." F. A. Davis Co., Philadelphia.

National Academy Press (1989). "Diet and Health." National Academy Press, Washington, D.C.

National Academy of Sciences (1989). "Recommended Dietary Allowances," 10th Ed. National Academy of Sciences, Washington, D.C.

Pennington, J. A. T. (1994). "Bowes and Church's Food Values of Portions Commonly Used," 16th Ed. Lippincott, Philadelphia.

Food Microbiology

CARL A. BATT
Cornell University

IN THE FIELD OF FOOD MICROBIOLOGY, WE SEEK to understand the role of microorganisms in the production, spoilage, and safety of foods. As food microbiologists, we attempt to not only understand their role in, for example, the spoilage of foods, but in the case of fermentations we try to enhance or accelerate these processes. In contrast, where microorganisms spoil the food or their presence in food represents a potential health hazard, we would attempt to control their initial numbers and proliferation. Food microbiologists can be found in industry, academia, and government. Scientists from a variety of disciplines may find their way into the field and their training in microbiology, biochemistry, and genetics is applicable to problems in food microbiology. Solutions to problems in food microbiology may be formulated by extending observations made in the more basic sciences. Typically, however, few solutions are realized owing to the complexity of the food matrix in which the microorganisms grow. Although empirical discovery is the primary route to understanding, there are basic fundamentals that have been established and some broad-based tenets have been constructed.

I. HISTORY

A. Ancient Fermentations

The microbiology of foods dates back in history to well before even the initial discovery of microorganisms and the eventual appreciation of their positive and negative effects on foods. The examination of foods recovered from archaeological studies has been illuminating in terms of our understanding of ancient food-handling practices. Chemical analyses and electron microscopic examination have suggested how microorganisms might have played a role in fermentation processes. In ancient times, fermentation processes were spontaneous. The microbial flora that initially populated the food, as well as that flora that became dominant after the "fermentation" process was complete, were largely dictated by the type of food and the environmental conditions. Nevertheless, ancient humans enjoyed a number of different fermented foods, including beer, wine, and cheese. Whether their attraction to the products of alcoholic fermentations was more for the physiological effects as compared to their organoleptic qualities is difficult to determine retrospectively. The prescribed health benefits from the consumption of beer by the ancient Egyptians may have simply been an attempt to justify the means.

The first food fermentations were clearly accidental, arising from the contamination of foodstuffs collected from the surrounding environment. Perhaps because of their initial role in religious life, alcoholic beverages appear to be the first examples of products of food microbiology. A variety of evidence traces the origins of beer fermentation to the Sumerians in 7000 BC. Records confirm that beer was already being drunk in Mesopotamia in approximately 4000 BC. The process as it was refined for the next 4000 years appears to be a by-product of bread making. Ancient Egyptian brewesses discovered a key element of beer fermentation when they used sprouted dry grain in preparing bread, which provided the enzymes necessary for the breakdown of starch. Since the endogenous yeast flora that would be responsible for the production of ethanol do not normally break down complex carbohydrates, adjunct ingredients are needed to initiate the fermentation process. Eventually a number of events led to the discovery of the microorganisms that were responsible for the conversion of carbohydrate to ethanol. That coupled to the elucidation of the biochemical steps in the pathway formed the knowledge base for this fermentation.

The eventual success of food fermentations was dependent on making these processes reproducible and predictable. Knowledge gained by examining ancient fermentations can be productive and illuminating. A brilliant example of this retrospective examination was reported by Delwen Samuel of the University of Cambridge and Peter Bolt of Scottish Courage Breweries (Edinburgh, Scotland). By examining residues collected from the time of Tutankhamun and the historical record of the agricultural practices some 3500 years ago, a hypothetical process was formulated and implemented on a small-scale commercial basis. Its reduction to practice and the splendid results suggest that the beer-making process used by the ancient Egyptians had indeed been duplicated. Few scientific discoveries usually carry such organoleptically pleasing results.

B. Death

A knowledge of food safety began to accumulate as humans attempted to understand their own mortality and the reasons for death. Massive food poisoning outbreaks probably occurred as ancient societies became more organized and began dining en masse. An ignorance of safe food-handling practices coupled with a lack of refrigeration probably provided a robust environment for the growth of pathogens.

The earliest documented accounts of food poisoning outbreaks were not bacterial in nature, but due to ergot alkaloids, a secondary metabolite produced by *Claviceps*. This fungus is known to infect grains, including rye, wheat, barley, and oats. Consumption of these compounds led to mass epidemics during the 1200s, primarily in central Europe, but the last major outbreak was recorded as recently as the 1950s.

Death being the ultimate form of illness, one of the first discoveries of food-borne microbial disease was botulism. Botulism is derived from the Latin *botulus,* meaning sausage. The first outbreak was recognized in 1793 but the bacterium was not isolated for almost another 100 years. It was Emperor Leo who outlawed the consumption of blood sausage, presumably after he realized that deaths in the population were associated with this sausage. Few food safety problems have been handled in so unequivocal a manner by direct edict.

C. Key Personnel

Microbiology is one of the more recent scientific disciplines to be established, and considerable skepticism once existed that acted as an impediment to the discovery of biological processes. Perhaps no individual scientist, in virtually any field of science, has had as large an impact as Louis Pasteur. Originally trained as a chemist, Pasteur showed tremendous insight and contributed to the fields of both food fermentation and food safety. His accomplishments spanned fields as diverse as wine production and meat spoilage.

As detailed previously, the first fermentations were accidental, the result of the proliferation of endogenous microorganisms present in the starting substrate. Modern fermentations required that these endogenous microorganisms be identified and cultivated separately. The development of methods for the propagation of microorganisms was aided by the discovery of solid microbial cultivation medium that included agar, one of many accomplishments by Robert Koch and his colleagues in the late 1800s. One of the first fermented products for which the microbial flora were dissected was beer—an accomplishment of Christian Hansen, who would also make significant contributions to dairy fermentations. One of the major dairy starter culture houses bears his name in tribute to his efforts.

The preservation of food was an important milestone as it permitted the expansion of markets for a given food. Although fermentation is a form of food preservation, the fermented product does not have the same taste and texture as the starting materials. Thermal processing was the invention of Francois (Nicholas) Appert, a confectioner. In 1809, he won a prize from the French government for developing the process to preserve meat, which involved heating filled glass jars in boiling water. The true nature of the process whereby the microorganisms were killed by heat would, however, not be revealed until the work of Louis Pasteur some 50 years later.

II. FOOD FERMENTATIONS

The discovery that microorganisms could positively affect the organoleptic quality and storage life of foods, a process that includes numerous chemical and physical changes, was accidental. In general, the fermentation process involves the breakdown of complex macromolecules and the conversion of the resulting substrate into other metabolic end products. In most cases this specifically involves the breakdown of a complex carbohydrate and the production of an acid or alcohol as the end product. The development of a particular acidogenic or alcoholic fermentation process may be more a function of the interest in and exploitation of these types of fermentation rather than a fundamental thermodynamic driving force. Fermentative microorganisms carry out this process to alter their environment by reducing the available substrate, as well as to generate a concentration of metabolite that would discourage the growth of competitors. Fermentations may also be a combination of enzymatic and microbial processes, with the former contributed by the raw ingredients and the latter by endogenous microflora. Modern adaptations of these natural fermentations have mastered these previously haphazard events to ensure the uniformity and predictability of the processes.

A. Beer

As mentioned previously, beer fermentation owes its origins to the ancient Egyptians, who first brewed beer using bread as a substrate. Throughout the intervening years, beer brewing has evolved through a constant alteration in the substrates used in the process and an understanding of the microbiology of the fermentation. Most modern brewing practices are derived from the sixteenth-century Bavarians, who discovered the value in using "bottom-fermenting" yeast and hops as key ingredients. To preserve their product, the Bavarian Purity Law was established, which mandated that only a product composed solely of malted cereals, water, yeast, and hops could be called beer. Today that law is still enforced in Bavaria, although liberal interpretations of the ingredients list are practiced elsewhere.

The brewing process begins with the preparation of the malt. Malt is composed of sprouted grains, which provide amylotic enzymes that assist in the breakdown of starch. The grain is sprouted by soaking in water and allowed to proceed for 4–7 days. After sufficient time is allowed for the sprouting process, it is then halted by heating. The wort is then prepared using a starch base such as corn, wheat, or rice and hydrolysis is carried out by the amylolytic enzymes in the malt. The hydrolyzed starch can then be converted to ethanol upon the addition of the starter culture yeast. Two different types of yeast can be used in the production of beer, a top-fermenting *Saccharomyces cerevisiae* and a bottom-fermenting *S. carlsbergensis*. It is the latter bottom-fermenting yeast that proved to be one of the key improvements in the brewing process introduced during the Middle Ages by Bavarian brewers.

Once the yeast is "pitched" into the wort to start the fermentation process, both ethanol and other minor fermentation end products accumulate over the course of several days. Glucose and larger-molecular-weight dextrans are metabolized by the yeast. The process ends with the depletion of substrate and the accumulation of ethanol, and in most processes the yeast is removed. Some additional finishing is required and the beer may be aged before being packaged.

B. Wine

Wine is the result of a fermentation similar to that of beer except that there are no deliberate substrate processing steps similar to malting. In addition, owing to the absence of any heat treatment prior to fermentation, the flora in the fermentation phase of wine are likely to be more complex. The different varieties of grapes and the yearly differences in the sugar and tannin content of grapes also contribute to the vintage variations. Yeast strains used in the production of wine vary and, as in beer brewing, these different strains are one of the closely guarded secrets of the vintner. The yeast serves to convert the sugar in the grapes to ethanol, but the flavor and aroma are the result of a complex and only marginally defined set of metabolic reactions. Some ancillary reactions are carried out not by the yeast but by the endogenous microflora. The malolactic fermentation carried out by *Leuconostoc* is of concern as it results in wine spoilage. L-Malic acid is decarboxylated to form lactic acid and carbon dioxide.

C. Cheese

The discovery of fermented dairy products were undoubtedly accidental, the serendipitous result of leaving milk out until it coagulated. The primary flora that would have been responsible are lactic acid bacteria, so called because of their propensity to produce lactic acid as a nearly exclusive metabolite end product. Lactic acid bacteria include members of the genera *Lactococcus, Lactobacillus,* and *Pediococcus.* These are gram-positive, non-spore-forming cocci or rods.

Modern dairy fermentations and specifically cheese begin with milk, usually bovine in origin. Other milks from sheep, goats, and buffalo can also be used and products from these milks are common in less developed countries. Depending on the process, the milk may or may not be pasteurized, which has an obvious impact on the microflora in the fermentation. During the fermentation process, lactic acid bacteria (which may be added as a deliberate starter culture and/or be present in the endogenous flora) convert lactose to lactic acid. With these bacteria, the metabolism of lactose is virtually exclusive with greater than 90% substrate to product conversion. In addition to the production of lactic acid, other metabolic products, including diacetyl, contribute to the flavor of cheese. Other reactions that take place include the hydrolysis of triglycerides to free fatty acids and the breakdown of proteins into peptides and amino acids.

The lactic acid bacterial fermentation results in the formation of a curd, which is the coagulated aggregation of the proteins found in milk. The production of lactic acid as well as, in certain cases, the addition of processing aids, including chymosin, denature the caseins in milk leading to the formation of the curd. The nuances in the flavor of cheeses are a function of the starter culture and nonstarter bacteria. For certain cheeses, including blue and Roquefort, fungi are added that carry out secondary reactions usually involving proteolysis or lipolysis. Because of their aerobic nature, these adjunct microorganisms grow on the surface.

III. FOOD SAFETY

Food poisoning can be classified by two distinct descriptors, intoxications and infections. The former is the result of the ingestion of toxins that may be the product of foodborne microorganisms. As such, the microorganisms need not be consumed nor vestiges of their presence in the food even detected. In a broader sense, the ingestion of any chemical toxicant that results in disease is an intoxication, with the exception of those reactions that are due to an allergic response or are mediated by an inherited metabolic disorder. Infections, on the other hand, are the result of the consumption of the pathogenic organisms. The nature of the poisoning is expressly a function of the etiologic agent. Certain microorganisms cause infections, including *Salmonella, Escherichia coli, Clostridium perfringens, Campylobacter, Yersinia,* and *Listeria.* Other, such as *Staphylococcus, Clostridium botulinum,* and *Bacillus cereus,* cause intoxications. The severity of the disease varies with both infections and intoxications, with some of each class being fatal. Other food poisoning agents include viruses and parasites, whereas mycotoxins can cause illness but are more of concern due to their long-term health consequences.

A. Intoxications

1. *Staphylococcus gastroenteritis*

Staphylococcus aureus is a gram-positive, non-spore-forming coccus that grows in chains. It is a normal inhabitant of a number of environments and is a common resident of human skin. It grows under a wide variety of conditions and can grow at lower water activities than most other bacteria ($A_w = 0.85$). As a consequence, *S. aureus* food intoxications are frequently associated with foods that have low water activity.

Although *S. aureus* is able to cause infection, most foodborne illness is due to intoxication. This microorganism produces a powerful enterotoxin that causes gastrointestinal distress with symptoms including nausea, retching, abdominal cramps, and diarrhea. *Staphylococcus aureus* enterotoxin is notable because it is heat stable, able to withstand heating at, for example, 110°C for >60 min. Therefore, in certain processed foods, the enterotoxin can be present in the absence of any viable *S. aureus.* The term *preformed enterotoxin* is used to define illness from enterotoxin where no evidence for *S. aureus* cells is found.

2. Gram-Positive Spore Formers

The gram-positive spore-forming bacteria include the genera *Bacillus* and *Clostridium.* These organisms are different in that the former is an aerobe whereas the latter is an anaerobe. They are notable not only because they include one of the more deadly foodborne microorganisms, *Colstridium botulinum,* but also because of their extreme heat resistance. The current regulations regarding the proper thermal processing of canned foods is largely based on the heat resistance of these spore formers (see Section IV).

Bacillus cereus is a frequent contaminant of dried dairy ingredients largely due to its ability to survive spray drying. It produces two types of toxins, an enterotoxin and an emetic toxin. The former causes gastroenteritis and the latter causes vomiting. It is not considered to be a life-threatening foodborne pathogen, although because of its potential contamination of infant formula (which may include milk powder as a major ingredient), specific regulations on maximum levels of *B. cereus* in infant formula do exist.

Clostridium perfringens causes gastroenteritis due to the production of an enterotoxin. Because the enterotoxin can be produced in the intestinal tract, this organism is typically an infective disease. The enterotoxin is spore specific, being produced during sporulation of the vegetative cell. As an anaerobe, *C. perfringens* proliferates under low-oxygen conditions and requires a relatively rich medium for growth. An increased incidence of illness due to *C. perfringens* has been reported in slow-cooked meat products including roast beef. Reaching proper cooking temperatures in excess of 165°C and then rapidly cooling the food product after cooking helps to reduce the likelihood of *C. pefringens* contamination.

B. Infections

1. *Listeria monocytogenes*

The gram-positive bacterium *Listeria monocytogenes* causes a variety of diseases in humans as well as in animals. Clinically, infection with *L. monocytogenes* can lead to septicemia, encephalitis, or abortion depending on the infected host and perhaps the strain present. The organism, similar to other bacteria such as *Salmonella* and *Yersinia,* is capable of intracellular growth. It has the unique ability to evade the normal host defense mechanisms that include the ability of macrophages to kill bacteria. *Listeria monocytogenes* possesses a number of virulence factors that encode the ability to lyse blood cells (hemolysins) and to transit both within and between host cells.

Listeria monocytogenes was recognized as a pathogen almost 50 years ago, but only in the last 15 years did it come to prominence as a food pathogen. It has the ability to grow at refrigerator temperatures (4°C), making it problematic in foods normally assumed to be safe under those storage conditions. Because it can grow at refrigeration temperatures, the normal warning systems that protect the consumer are not effective. A typical consumer would not ordinarily consume spoiled food, which is characterized by an off-odor or appearance. Spoilage organisms are usually not pathogenic and they are useful *indicators* that a food has exceeded its effective shelf-life. Under refrigeration, these spoilage organisms do not grow and therefore do not spoil the food, while *L. monocytogenes* is still capable of proliferating.

2. *Salmonella*

Salmonella is a gram-negative bacterium that is closely related to *Escherichia coli.* Speciation of *Salmonella* is based on serology and there are greater than 2200 different serotypes. *Salmonella* is responsible for the most reported foodborne disease outbreaks involving approximately 2000–3000 cases per year in the United States. This is probably a vast underestimate, as many cases of foodborne poisoning go unreported due to their relatively mild consequences or the uncertain nature of their origin. Some reported single outbreaks of salmonellosis have been estimated to involve more than 200,000 people on the basis of the consumption of contaminated product. Most of these outbreaks involve either water or foods of animal origin. In general this is due to the normal habitat of *Salmonella,* which is the intestinal tract of animals. The ability of certain animals to carry this organism yet remain virtually asymptomatic is a function of the particular host range of the *Salmonella* strain. Among the wide variety of *Salmonella* species, some are particularly linked to a given host whereas others are not "host adapted."

The prevalence of *Salmonella* contamination of poultry is of primary concern as it affects human health. Estimates ranging as high as 100% have been reported for its contamination of poultry. As a consequence, significant efforts have been put forward to control contamination in flocks. One promising avenue is the administration of competitive microflora that exclude the colonization of the bird's intestinal tract by *Salmonella.* However, the most effective measure for reducing salmonellosis in humans is educational, that is, making the consumer aware of proper handling techniques and the need to thoroughly cook poultry products.

3. *Escherichia coli*

Escherichia coli, similar to *Salmonella,* is a gram-negative bacterium. It can cause a number of different diseases depending on its complement of virulence factors and the immune status of the host that it infects. The most common types of disease caused by *E. coli* include enteropathogenic, enteroinvasive, enterotoxigenic, and enterohemorrhagic. In general the most common symptom is diarrhea, which when accompanied by blood is a vivid indication of *E. coli.* Similar to *Salmonella, E. coli* is also a normal inhabitant of the intestinal tract of animals making it ubiquitous in situations where fecal material might contaminate the environment. Few strains of *E. coli* are pathogenic and hence its mere detection in a food product is not necessarily unusual. However, it is a good indication of fecal contamination and *E. coli* along with coliforms and, more narrowly, fecal coliforms are common indicators of food safety.

Of recent note is the attention given to one serotype of *E. coli,* O157:H7. Like *Salmonella, E. coli* can be classified by serotyping, which is dependent on the O and H antigens carried on the cell surface. This particular serotype has been associated with a number of recent food-poisoning outbreaks involving ground beef and apple cider. First recognized in 1981 when it was the causative agent in a foodborne outbreak involving ground beef, *E. coli* O157:H7 has been the object of considerable attention in the United States. One unexpected source of *E. coli* O157:H7 has been unpasteurized apple cider and products formulated with unpasteurized apple juice. This food product was not assumed to be a likely source of enteric contamination due to its relatively low pH. Recent evidence has suggested that *E. coli* O157:H7 may be more tolerant to low pH relative to other *E. coli* strains. [*See* Bacterial Infections, Detection.]

C. Others

1. Parasites

Parasites are defined as animals that require a second animal host for their livelihood. They grow and replicate only within the host, although they can survive for varying lengths of time outside of the host. In foods, they do not replicate, nor can they be grown in standard microbiological media and conditions. As such their detection usually employs a direct inspection of the food, frequently using a microscopic-based format. Examples of parasites include *Trichinella, Giardia, Cryptosporidium,* and more recently *Cyclospora.* Historically *Trichinella* contamination of pork was an issue, though more recently attention has been focused on the contamination of water supplies with *Cryptosporidium,* perhaps the result of fecal material originating from domestic or wild animals.

2. Viruses

Viruses are similar to parasites in that they require a host for their replication. They are simple in their structure and biology, possessing only the necessary nucleic acid (either DNA or RNA) to encode their structural proteins and perhaps a few proteins for replication. Foods are just one means through which viruses are transmitted. The viruses that are typically transmitted through foods are not usually life-threatening, hence an accurate survey of their significance is not available. The symptoms of food-borne viral infections usually include gastrointestinal distress.

Where food is a major vehicle for transmission to humans at least two viruses are of significance. The rotaviruses cause gastrointestinal illness and are frequently associated with day care facilities. They are approximately 75 nm in diameter and have a genome that is double-stranded RNA. Another virus of importance in foods is the Norwalk-like virus, named after the city where it was isolated. As with many viruses, Norwalk-like are a collection of very similar viruses that can, however, be distinguished immunologically. This virus causes a gastroenteritis similar to that produced by bacteria, including *Staphylococcus.* It is 30–35 nm in size and contains single-stranded RNA.

3. Mycotoxins

Mycotoxins are metabolic end products of fungi and are typically secondary metabolites produced during the sporulation stages of growth. Their toxicity has not always been accurately determined owing to differences in sensitivity of animal models to a particular mycotoxin. Many mycotoxins are carcinogens but typically require activation by one or more mammalian enzymes. Their primary target in animals is therefore the liver, where they are activated in, ironically, the body's attempt to detoxify them. Their true food safety impact is difficult to assess because of long-term potential health problems. [*See* Mycotoxins.]

The most studied mycotoxin is aflatoxin, which is produced by certain strains of the fungus *Aspergillus flavus.* It is the most potent hepatocarcinogen known to humans, although exacting measurements of its carcinogenicity have not been carried out. Four different aflatoxins have been reported, designated aflatoxin B1, B2, G1, and G2. The biosynthetic pathway for the aflatoxin has not been completely elucidated but its precursors originate from the polyketide synthesis.

Aspergillus flavus is ubiquitous in the environment but tends to proliferate under conditions that are not favorable to bacterial growth. Growth and aflatoxin production are most common on dried grains, including corn, especially when the drying process is not adequate to reduce the water activity below the requirements for fungal growth. Aflatoxin is therefore a problem not only because of the contamination of food destined for human consumption but because of its presence in animal feed. Losses from aflatoxin contamination of animal feed are again difficult to estimate as they are sometimes manifested in reduced growth rates and feed conversion efficiencies. There is also a secondary problem in the transmission of aflatoxin through animals into foods destined for human consumption. Specifically, aflatoxin M1 is a metabolic derivative of aflatoxin B1, which is found in milk. It is still toxic and carcinogenic, although the extent of its actual occurrence in milk is not known.

IV. CONTROL OF MICROBIAL GROWTH IN FOODS

In virtually every food, there is a need to control the growth of microorganisms to ensure both safety and quality. Even in fermented foods, conditions must be established during the fermentation process to enhance the growth of the starter culture, but more importantly to retard the growth of contaminating flora. In foods, the shelf life of a product is defined by the maximum time that a product can be stored under the prescribed conditions before either its quality and/or its safety falls below a set limit. Shelf life can vary from days to years; it has been suggested that the shelf life of properly processed canned foods is virtually indefinte.

There are two basic factors that control the growth of microorganisms in foods, the extrinsic and the intrinsic factors. The *extrinsic* factors include the environment surrounding the food, (e.g., the oxygen tension), whereas the *intrinsic* factors are the properties of the food itself (e.g., the pH or the water activity). Microorganisms as a collective population are capable of growth over a wide range of both intrinsic and extrinsic factors. There are, for example, microorganisms that can grow at refrigerated temperatures (4°C) and others that can grow at extremely high temperatures exceeding 100°C.

Although a given microbial population can be controlled through the judicious selection of intrinsic and extrinsic factors, there is virtually no food system that can be formulated that will totally exclude the growth of all microorganisms. The microbial flora will change depending on the extrinsic and intrinsic factors. For example, at low water activity the molds may predominate. They would normally be excluded by the more rapidly growing bacterial flora except that the growth of the bacterial flora is usually suppressed at water activities of less than 0.8. Similarly, at low pH, certain bacteria, including *Lactobacillus,* may predominate to the exclusion of other organisms. Some microorganisms can alter their environment or the intrinsic properties of the food to exclude the growth of other competitive microflora. This is the basis for fermentation processes as described in a previous section.

The shelf life of a product is also determined by the initial microbial load in the product, therefore one of the most fundamental steps in food product manufacture is processing. Processing can be accomplished using a variety of technologies, including heating, drying, and freezing. Heating or thermal processing was perhaps the single greatest advance in food manufacturing by providing a significant increase in the shelf-life and hence the widespread distribution of the product. Prior to commercial thermal processing, the shelf life of most products was limited, perhaps with the exception of fermented foods.

Thermal processing can range from mild heating through pasteurization to heating regimens known as *12-D kills.* This latter term was specifically designed for thermal processing of canned foods, where the concern was to eliminate the potential for botulism. Two key terms in thermal processing are the D-value and Z-value. The *D-value* is the time necessary to reduce the microbial population 10-fold at a prescribed temperature. For example, the D-value for *C. botulinum* can be as high as 30 min at 100°C. D-values can vary several 100-fold, with the spore formers being the most thermal resistant. A 12-D process is the time and temperature necessary to reduce the population of a standard spore-forming test organism (e.g., FS617) by 12 logarithms. FS is a term for a microorganism whose growth in a canned food leads to a flat (F) and sour (S) condition. The flat refers to the lack of bulge in the can, due to a lack of gas production. The sour reflects the low pH of the spoiled product, hence acid production. In theory, a 12-D process should reduce the likelihood of *C. botulinum* to less than one in 10^{12} cans.

The *Z-value* refers to the change in temperature necessary to effect a 10-fold difference in the D-value. Heating, in addition to destroying microorganisms, causes a deterioration in the organoleptic properties (i.e., most notably the texture) and the nutrient content of foods. D-values and Z-values can be calculated for these nonmicrobial destruction processes as well. However, the Z-values do differ dramatically and therefore can be used to optimize the destruction of the microorganism with only a limited loss in texture or vitamin content. In general, as the temperature is increased, the rate of vitamin or texture loss is lower in comparison to the inactivation of microorganisms.

Therefore "high-temperature, short-time" (HTST) process such as that used in certain milk products is preferred. Not all food products can be processed using HTST regimens, and this is particularly true of particulate foods or any food that has a low thermal conductivity.

V. HAZARD ANALYSIS CRITICAL CONTROL POINT

The absolute safety of a food can be ensured only by testing each and every bite. In medieval times, the king would employ a taster whose job it was to sample the food prior to the king's consumption. If and only if the taster survived was the food then fit for a king. Obviously this would be a suitable assay for acutely toxic material, but none of the bacterial pathogens or their toxins would work in such a rapid fashion. In lieu of this system, end product sampling and testing has certain merits, but given the low statistical probability of an outbreak, the sampling frequency must be high. Furthermore, an assay conducted immediately after processing does not ensure that the product will be safe throughout the projected shelf life, especially if the product is abused or not properly cooked by the consumer. [*See* Food Toxicology.]

Hazard Analysis Critical Control Point (HACCP) is an approach to food safety that attempts to predict and then control the specific steps in a given process that, should they fail, will lead to a potentially unsafe product. Originally developed in the 1970s in a joint effort by the National Aeronautics and Space Administration (NASA) and Pillsbury, it applies not only to food processes but to any process where failure can occur due to a problem in one or more steps. In HACCP, the entire process is examined and specific steps in the process or key ingredients are examined to determine which, should they fail to meet a certain specification, might lead to an overall failure in the process. These steps are then termed critical control points, of which there may be one or more for any given food process. For example, in the processing of fluid milk, the pasteurization step and the need to heat the milk at 145°F for 30 min would be a critical control point. Monitoring of that critical control point might consist of having a thermocouple in the heating section and a flow gauge to determine residence time in the heating region of the pasteurizer. In fact, for several food processes, many critical control points are not only common sense but also standard practice. In other situations, HACCP is only now being considered, and for a few, HACCP is being mandated by regulatory agencies. Some critical control points may incorporate microbiological testing, for example, with ingredients or between shift cleaning of equipment. Specifically, in a food process where the processing temperatures may not be very severe, ingredients need to be tested to ensure that they are free of *Salmonella*. Specifications for ingredients are useful because it allows the food processor to produce a food that employs a process that may not be rigorous enough to kill a particular pathogen. They ensure the safety by not allowing a pathogen to contaminate the ingredients.

VI. FUTURE DIRECTIONS

There will be constant demands to improve the safety of the food supply, which invariably must be balanced against the economic premium that their implementation imposes. The field of *risk assessment* has been established to weigh various factors in estimating the impact of food safety in terms of potential consequences and likelihood of reducing potential safety problems. A key element in the process of ensuring food safety is the detection of pathogens in the food. Traditional methods for detection rely on growth of the targeted organism under conditions that also aim to suppress the growth of other organisms. Despite their labor intensity and relatively long time to completion, growth-based microbiological methods are still the standard methods for detection.

Modern technology for ensuring the safety of foods is being developed, driven principally by the need to develop quick methods for diagnosis of human disease. One of the more fundamental differences in opinions among food microbiologists is the use of indicator organisms as a sentinel for the presence of pathogens. The use of indicators is not a new concept, in terms of both qualitative and quantitative specifications. One global indicator that is attracting increased attention is the quantification of adenosine triphosphate (ATP) levels to determine relative biological loads. On processing equipment that contacts food, cleaning practices should reduce the biological load to negligible levels. ATP levels can be rapidly assessed using biological reagents that were originally characterized in fireflies. The enzyme luciferase in the presence of ATP and luciferin emits a photon of light that can be quantified using a luminometer. The process is very rapid and typically results can be obtained within 5 min. The development of low-cost luminometers and the accessory reagents have spurred the use of ATP as a measure of total microbial load and hence cleanliness. Yet how this measure relates to the potential for pathogen contamination is difficult to predict. The primary value of ATP measurements may lie in their integration into a HACCP plan. It can then provide a measure of assurance that cleaning and anticipated reductions in the microbial load are achieved on a regular basis.

Specific tests for pathogens will continue to be improved in terms of specificity and speed.One likely assay format to be incorporated into food pathogen tests is the polymerase chain reaction (PCR). This method for amplifying nucleic acid sequences as delineated by a set of oligonucleotide primers is a very powerful means to detect target sequences that uniquely define, for example, a bacterial pathogen. The speed at which these sequences can be amplified opens new avenues for the development of unique tests to look for virtually any pathogen in any food system. However, there are obstacles to be overcome, which in part involve the purification of the targeted DNA sequence from the food matrix with an efficiency that is consistent with a need to detect very small numbers of bacteria in relatively large food samples. [*See* Polymerase Chain Reaction.]

BIBLIOGRAPHY

Cliver, D. O. (ed.) (1990). "Foodborne Diseases." Academic Press, San Diego.

Jay, J. M. (ed.) (1996). "Modern Food Microbiology," 5th Ed. Chapman Hall, New York.

Food Toxicology

GILBERT S. STOEWSAND
Cornell University, Geneva Campus

GLOSSARY

Ames assay Rapid screening test (developed by Bruce Ames) for genotoxic food components/chemicals using mutant *Salmonella* bacteria

Ciguatoxin Common toxin of snappers and sea bass as well as many other fish

Decision tree protocol Protocol for risk decisions after testing acute and chronic toxicities of food components

Phytoalexin Toxic compound produced in plants under environmental and pathogenic stress for plant defense

Saxitoxin Potent toxin of shellfish

Solanine Toxic glycoalkaloid of potato and other Solanaceae plants

Trace element species Element with a specific valence state or in a specific complex/binding state

FOOD TOXICOLOGY CAN BE DEFINED AS THE study of adverse chemicals produced, incorporated, or carried by plant or animal material normally ingested and assimilated by a living organism. These three types of adverse food chemicals are usually described as natural toxicants, intentional additives, and environmental contaminants. Food poisons usually infer chemicals that are highly toxic, for example, the highly fatal neurotoxins from the bacillus *Clostridium botulinum,* found as natural contaminant endospores from soil and aquatic environments in many foods that have undergone insufficient heat processing. Food allergy represents immunological toxic reactions to natural substances in foods, as specific proteins or intentional additives (e.g., various food colors). This article will not consider food allergies, food poisons, or toxicants produced by bacteria, viruses, parasites, or fungi.

I. BACKGROUND

A. Historical

Toxicology dates to the earliest humans who used animal venoms and plant extracts for hunting, waging war, and assassinations. Food toxicology also dates back to those times when hunters and gatherers observed that eating certain plants and animals made them ill or caused death and should be avoided as a source of food. Later, farmers observed toxic symptoms and death of livestock when fed or grazed on certain plants. The Bible mentions that the Hebrews were punished with "quail meat between their teeth." This gives the impression of a rather acute toxicity. It has now been observed that hemlock seed is practically harmless to quail, yet dogs fed this quail meat quickly developed symptoms with fatal results similar to humans poisoned with hemlock. In the first century AD, Pliny the Elder wrote: "So many poisons are employed to force wine to suit our taste—and we are surprised that it is not wholesome!"

The father of modern toxicology was M. J. B. Orfila, a Spaniard born on the island of Minorca in 1787. He studied chemistry and mathematics and subsequently medicine in Paris. Although his interests centered on the harmful effects of chemicals as well as therapy, he was the first to introduce quantitative methodology into the study of the actions of chemicals on animals. He was the author, in 1815, of the first book devoted entirely to studies of the harmful effects of chemicals. He could also be consid-

ENCYCLOPEDIA OF HUMAN BIOLOGY, Second Edition, VOLUME 4.

ered the father of modern food toxicology as his research concerned primarily naturally occurring substances in plants. Many present-day hybrid food plants that are resistant to insects and disease have been bred from nonfood plants, within similar botanical families, containing naturally occurring toxicants (see Section IV,C). This may result or be part of the development of the plant's resistance. The potential harmfulness and safety of many of these natural compounds in plant foods have yet to be completely identified and studied.

B. Multidisciplinary Field

Food toxicology is not only an applied science but also an art that demands scientific adaptability and flexibility. It is usually taught in departments of food science, nutrition, animal science, veterinary science, or entomology in a somewhat fragmented fashion. In some cases, food toxicology is taught as an outgrowth of traditional toxicology normally presented in departments of pharmacology within schools of medicine or pharmacy. In other cases, food toxicology evolved from agricultural college laboratory research on pesticide chemistry and was taught as a part of food chemistry or applied biochemistry. Regardless of its origins or place of study, modern food toxicology borrows freely from several of the basic sciences. A knowledge of, and an ability to study, the interaction between chemicals in a food matrix and biological mechanisms is founded on a background from all of the basic physical, chemical, and biological subjects. A broad training in chemistry, biochemistry, biology, microbiology, statistics, physiology, pharmacognosy, nutrition, and public health affords individuals to have the ability and foresight to observe, investigate, and solve problems of toxicants in our foods as well as their mechanisms of action.

C. Food Safety Regulations

Early regulatory authorities focused much of their attention on adulterated food and drugs. Protective laws date back to India in 300 BC, where the first rules were written against the adulteration of grain, scents, and medicine. The first European law regulating some foods appeared in England in 1215 and in regulating the sale of bread in 1266. General food protection laws were established in England, Germany, and Sweden between 1860 and 1870. However, a more comprehensive safeguard of public health was written into the Sale of Foods and Drugs Act in England in 1875, which was used as somewhat of a guide for the first food and drug laws passed by the United States Congress 31 years later.

National concern for food safety and protection of the public in the United States against misbranded and adulterated food resulted in the passage of the Pure Food and Drug Act in 1906. Research on food safety and food analysis prior to passage of this first food and drug law was published by Harvey W. Wiley, head of the chemistry division of the U.S. Department of Agriculture (USDA). He used employees of the USDA as human subjects (the "poison squad") to show widespread use of poisonous preservatives and dyes in foods. These results, together with much publicity regarding filthy conditions in meat-packing plants and worthless or dangerous patent medicines, induced President Theodore Roosevelt and the Congress to pass these first food and drug laws. This Act has been amended, revised, or added to several times (i.e., in 1938, 1954, 1958, 1960, 1962, 1968, 1976, 1994, and 1996). The modern U.S. Food and Drug Administration (FDA) was established in 1931. Oversight for the safety of foods is provided by the FDA.

Tolerance limits (maximum permissible quantity) for pesticide residues, as well as approved pesticide use, on raw agricultural commodities and processed foods are established by the Environmental Protection Agency (EPA). The FDA has the authority to enforce, by seizure of the food products if necessary, the limits established by the EPA. The Food Quality Protection Act of 1996 replaced the 1958 Delaney Clause's zero tolerance for cancer-causing pesticide or additive residues in foods. There is now a "reasonable certainty of no harm" standard.

Toxic air and water pollutants can be sources of numerous environmental contaminants in foods. The major federal law to combat air pollution is the Clean Air Act of 1970. Control of toxic effluents into the nation's waters is addressed by the 1978 amended Federal Water Pollution Control Act of 1972. EPA enforces all environmental laws.

II. FOOD SAFETY ASSESSMENT

A. Determination of Toxic Materials

When a whole food, component of a food, or a toxicant known to be in a food has to be analyzed and determined for safety, the food toxicologist must proceed with a number of steps. The questions of what known toxicant(s) is present, its concentration, population exposure, acute and chronic toxicity, and possibly a way to eliminate or detoxify the toxicant(s) must all be addressed. Food toxicants are usually always present in very low amounts because chemicals with any significant level of toxicity are rejected as foods. People develop a distaste for food after it is associated with any episodes of illness.

A decision tree protocol, used to determine whether there exists an unacceptable or acceptable risk after each testing step of food components, has been proposed by a Scientific Committee of the U.S. Food Safety Council. A summary of this protocol is presented in Fig. 1. Other similar kinds of decision trees regarding (1) the safety evaluation of whole foods, (2) single chemical entities, and (3) food ingredients derived from genetically modified microorganisms have been made by the International Food Biotechnology Council. Numerous modifications have and will continue to occur, but these decision tree concepts have strong support from the scientific community.

The analysis of toxicants requires both an assay for detecting the toxic material and a method for separating it from the rest of the complex food matrix. Identification of unknown chemicals has dramatically improved since the development of analytical instruments such as ultraviolet, infrared, nuclear resonance, and mass spectroscopy. Additional analytical techniques specific for trace element toxicants are discussed in Section III. Food toxicant analysis essentially goes through four steps: (1) proper sampling; (2) extraction; (3) cleanup; and (4) chromatography.

B. Toxicity Testing

1. Acute toxicity is defined as the adverse effects occurring within a short time of intake of a single dose or multiple doses within 24 hr. Groups of laboratory animals, usually rats or mice, can be given one dose to observe the quantal or "all or none" response, or multiple doses to study the "graded response." Mortality and overt toxicity signs are examples of quantal data, whereas enzyme activity, hematology data, and body weight are quantitative parameters observed in a short-term graded response. The LD_{50} is a statistically derived single dose of a substance that can be expected to cause death in 50% of a group of animals. It is actually not a biological constant but a statistical term designed to describe the lethal response of a compound in a particular population under a specific set of experimental conditions. In essence, the dose–response curve, the time to death, toxicity symptoms, and pathologic findings are all vital and perhaps even more critical than the LD_{50} in the evaluation of acute toxicity.

2. Subchronic toxicity tests are studies of food toxicants designed to last usually for 90 days but can extend to 1 year. These tests are designed to determine responses produced by low-dose repeated exposure

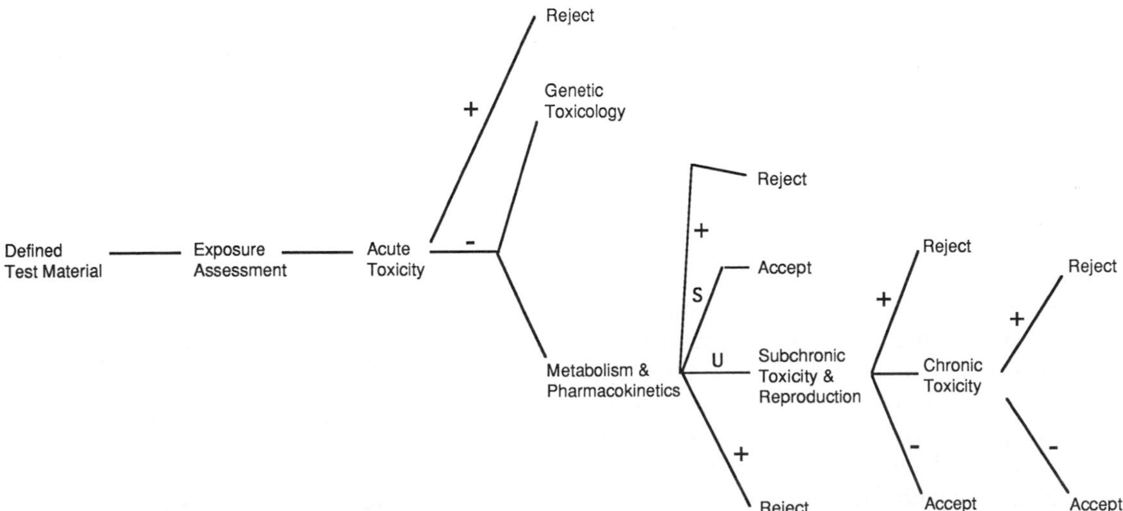

FIGURE 1 Decision tree protocol. The defined test material is a specific food chemical or component. The component may be rejected, because of unacceptable risks, after each testing stage. +, socially unacceptable risk; −, does not present a socially unacceptable risk; S, metabolites known and safe; U, metabolites unknown or of doubtful safety.

of a substance to a test animal. The substance can be administered to groups of animals by gavage, drinking water, or directly mixed into a defined diet. Three to four doses of the material should be given to 10–20 rodents per each sex per dose group. In general, not more than 10% of the animals in the high-dose group should die within a 90-day study, and no deaths should occur within the lower-dose groups. The usual observations and evaluations are: daily observations, periodic physical examinations, monitoring body weight and food consumption, and analyses of hematology, biochemistry, and urinary parameters. When the animals are killed at the end of the study, organ weights are recorded and a histopathologic evaluation is performed.

3. Chronic tests are long-term studies that are designed to test relatively low-level exposure of a food toxicant. The studies are designed to detect the origin of any adverse response regarding the animals' structural and functional entities. They determine the margin of safety between the substance's use or presence in a food and toxicity. Chronic studies, usually measured in years depending on the experimental animal, test the tumor induction or carcinogenicity potential of a substance. Generally the food toxicologist needs the assistance of an experienced animal pathologist for both routine and special pathology in differentiating between degenerative or atrophic changes of tissues and normal variations.

To learn why a chemical is carcinogenic and to provide an improved basis for estimating human risk, it is required to perform absorption, distribution, and excretion studies of the substance and its metabolites. Specific protocol directions and suggestions for carcinogenicity testing and monitoring continue to be the subject of numerous reviews by individuals, expert panels, and government agencies. [*See* Chemical Carcinogenesis.]

4. Teratogenesis testing is done to observe possible embryonic developmental abnormalities from a food toxicant. The human fetus is most susceptible to anatomical defects at about 30 days of gestation, that is, during organogenesis. Thus, teratogenesis studies in animal models should include 1- to 2-day treatments during their particular organogenesis time frame, as well as continuous treatment during gestation. In addition, tests of reproductive toxicity may include treatment of both males and females prior to mating, treatment of females through lactation, and continued testing of the offspring.

5. Genetic testing is done to determine if the substance induces mutation or inheritable changes in the genetic information of the cell. The decision tree approach (see Fig. 1) proposes a battery of genetic tests early in the testing scheme. Certain mutagenic substances are also carcinogenic. Point mutations (localized changes in DNA) and frame shift mutations (DNA base pair additions or deletions) of substances have been tested for many years by the Ames assay. This assay is accomplished by a specially constructed strain of *Salmonella typhimirium* bacteria. There are many other microbial organisms as well as mammalian cell lines used in various kinds of mutagen tests. Macrolesions, that is, structural and/or numerical changes in chromosomes, are also studied using cytological analysis. In the future, genetic testing will probably be done with the use of cultured human somatic as well as germ cells. Damage to germ cells has the important potential for transmission to the next generation. [*See* Food Microbiology; Genetic Testing.]

III. TRACE ELEMENT TOXICANTS

A. Scope

Trace elements can be found in foods as natural constituents, intentional food additives, or environmental contaminants. Although Table I lists trace elements as essential or nonessential for humans, all of them can produce adverse effects if their intakes are suffi-

TABLE I
Elements in Human Diets

Essential	Nonessential
Boron	Aluminum
Calcium	Arsenic
Chlorine	Barium
Chromium	Bromine
Copper	Cadmium
Fluoride	Germanium
Iodine	Lead
Iron	Lithium
Magnesium	Mercury
Manganese	Nickel
Molybdenum	Rubidium
Phosphorus	Silicon
Potassium	Tin
Selenium	Vanadium
Zinc	

ciently high. The statement by the alchemist and physician Paracelsus almost 500 years ago—"All substances are poisons. . . . The right dose differentiates a poison from a remedy"—is very appropriate for essential elements. Whatever method that is used for the calculation of safety limits should be based on the highest amount of the element that had been widely consumed by enough people for a long enough period of time to provide assurance that no adverse effects occur. Unfortunately, there are usually not enough clinical data for this basis, so numerous types of formulas have been used to calculate safety limits (e.g., Hathcock's Nutrient Safety Limit, or NSL).

Many of the nonessential trace elements for humans have been shown to be essential to animals, including, according to one laboratory, even toxic metals such as cadmium and tin. However, most of these data are limited, difficult to verify, and very controversial.

The most significant toxic effects of trace elements are mutagenicity, teratogenicity, and carcinogenicity. Numerous countries follow the FAO/WHO Codex Alimentarius Commission on tolerance limits of some of the toxic trace elements in foodstuffs, additives, and drinking water (e.g., that the daily intake of lead from all sources should not exceed 0.43 mg). Examples of the concentration ranges of three potentially toxic elements—mercury, lead, and tin—in major food products are presented in Table II.

B. Toxicological Evaluation

The amounts of trace elements in foods permit an estimate of dietary intake but do not give any information as to the element's toxicity. This depends to a great extent on the physical and biochemical properties of the trace element species. For example, the trace element chromium (Cr) when inhaled in air as Cr(VI) is highly carcinogenic, but Cr(III) usually present in foods in very small amounts (1 ng. or 10^{-9} g) is an essential element (see Table I). Indeed, future legislation will refer to the maximum content for each element species rather than simply list the total amount of each element in foods. [*See* Nutrition, Trace Elements.]

The problem is that very little is known about the species of most trace elements in food or how processing or cooking affects or changes species; furthermore, most of the information is qualitative. In addition, almost nothing is known about the interaction of other food components with trace metal toxicity (e.g., protective factors of selenium toxicity in tuna fish).

To make an evaluation of the toxicity of a trace element, it is necessary to know its bioavailability, that is, how much is absorbed and utilized *in vivo*. Usually an element is not absorbed completely, but the fraction that is absorbed is dependent on the amount in the diet, the oxidation state of the element, the chemical species, and the presence of interfering or enhancing factors in the food. Unfortunately, an experimental study of trace element speciation in a complex matrix as food is a costly and difficult undertaking. In addition to detection sensitivity, there is the important need to avoid altering the speciation of the element during the various analytical steps.

TABLE II

Concentration Ranges of Three Potentially Toxic Elements in Major Food Products

Food	Mercury (μg/g)	Lead (μg/g)	Tin (μ/g)
Meat and poultry	0.001–0.136	0.004–15.7	<0.2–44
Raw fish and shellfish	0.06–10	0.008–25	<0.2–14
Vegetables and fruits	0.010–0.040	<0.0006–16.5	<0.2–59
Bakery goods and cereals	0.010	0.002–3.2	
Milk and dairy products	0.005	0.0003–0.7	<0.1–40

C. Analytical Techniques

It is possible now to detect elements, as well as other food chemicals, at levels of picograms (10^{-12}) per kilogram amounts by means of mass spectrometry. Concentration of the element with different oxidation states can be measured by the use of (1) electrothermal atomization–laser excited atomic fluorescence spectrometry, (2) laser-enhanced ionization spectrometry, and (3) inductively coupled plasma–mass spectrometry.

Multielement analysis is desirable to provide a scan of possible toxic levels of elements that may be present in one specific food. This analysis can enable more than one determinant to be measured in a single sample digest and greatly improves the speed of analysis. Modern instrumentation includes X-ray fluorescence, neutron activation, potentiometric stripping analyzer, proton-induced X-ray emission, and simultaneous multielement atomic absorption spectrometry with a continuum source.

IV. NATURAL TOXICANTS

A. Animal Tissue

Outside of reports of toxic quail meat, as described in Section I, and vasoactive amines produced by bacteria on putrefying meat, there is not a great deal of evidence of natural toxicants in mammalian or avian species. However, there are toxins in animal liver that have produced toxic responses when consumed to an excess by animals and humans. For centuries Asiatic people have used dried bear liver as a folk medicine due to its tranquilizing and pain-killing effects. Bile acids, mainly cholic acid, synthesized in the liver can act as a suppressant on the central nervous system. Beef and other animal livers eaten in the United States do not contain sufficient quantities of bile acids to produce such effects.

Vitamin A is an essential vitamin necessary for normal growth and vision. However, it is toxic to humans when ingested in excess of 2 million International Units (IU). Polar bear liver contains about 1.8 million IU of vitamin A per 100 g of fresh liver. Early in the twentieth century, explorers in the Arctic and their dogs experienced various symptoms of joint pains, swelling, bleeding lips, and some fatalities. This was later shown to be a toxic syndrome from extreme levels of vitamin A intake when polar bear liver was consumed.

B. Marine Animal Tissue

Fish toxins are of two types, oral biotoxins when the fish is eaten and the large molecular venoms that are injected by specialized venom apparatus. Food toxicologists are concerned with toxic marine flesh from fish termed ichthyosarcotoxic fish. Widely consumed fish that can exhibit various types of poisons, depending on numerous but not completely known factors including environmental changes in their food chain, are listed in Table III.

TABLE III
Seafood Toxins

Fish	Type of poison
Shellfish	Saxitoxin
Snapper, sea bass	Ciguatoxin
Tuna, mackerel	Scombrotoxin
Sardines, anchovies	Clupeotoxin
Pufferfish	Tetrodotoxin
Mullet, sea chub	Hallucinogenic poison

1. Saxitoxin

Clams, mussels, and other shellfish feeding on certain species of marine algae or dinoflagellates, especially *Gonyaulax* spp., accumulate a potent toxin known as saxitoxin (Fig. 2). This compound was first isolated in pure form in 1954 from California mussels and the Alaska butter clam. At certain times, depending on water temperature, pH, and other factors, the dinoflagellates grow rapidly. At a concentration of about 20,000 cells per milliliter of seawater, the water appears brownish-red in color and is called red tide. The shellfish do not appear to be harmed by consuming the poisoned algae and slowly excrete the toxin. However, people eating such shellfish develop a serious condition of paralytic shellfish poisoning. The amount of this neurotoxin required for sickness and death in humans varies considerably and seems to indicate a tolerance to the poison from continuous intake of small doses. Symptoms of paralytic shellfish poisoning begin with a numbness in the lips, tongue, and fingertips (paresthesia) that is apparent within a few minutes of eating the shellfish. This is followed by numbness in the legs, arms, and neck, with muscular incoordination (ataxia). Death results from respiratory paralysis within 2–12 hr after the dose. If one survives 24 hr, the prognosis is good with no lasting effects. FDA has

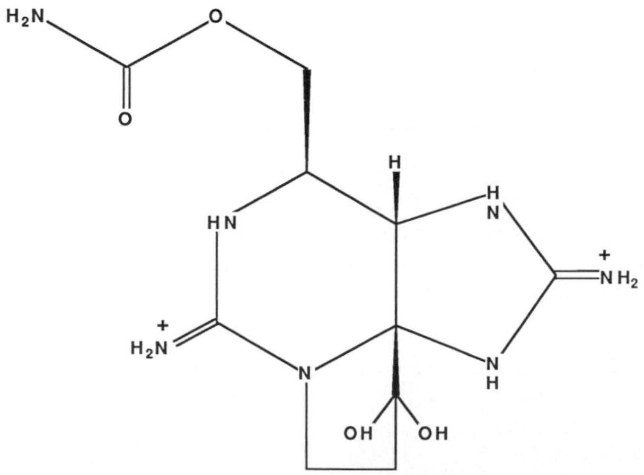

FIGURE 2 Structure of saxitoxin.

regulated the poison to a maximum allowable level in all seafood products of 80 μg/100 g.

2. Ciguatoxin

Besides snappers and sea bass, as indicated in Table III, there are upward of 400 fish species that possess ciguatoxin. It is also associated with fish consuming dinoflagellates, especially *Porocentrum* spp. in tropical or warm, temperate-zone reefs. Evidence indicates that the primary cause of this toxicity is bacteria living on these and other dinoflagellates, but ciguatera poisoning is not a form of typical bacterial food poisoning. It is usually never fatal, but neurological symptoms of extremity paresthesias, ataxia, muscle pain, and weakness are almost always observed. There are no accurate statistics worldwide as to the incidence rate of ciguatoxin poisoning. Recovery can take from 2 days to years. There is no way to detect a ciguatoxic fish by appearance, since freshness has no bearing on toxicity, and cooking does not destroy the toxin. Large predatory reef fish should be eaten with caution.

3. Scombrotoxism

Scombrotoxism is caused by the improper preservation of scombroid fish, such as tuna and mackerel, whereby various bacteria act on the fish muscle histidine to form histamine. It has also been called histamine poisoning and is detected by a very sharp or peppery taste. Symptoms usually last for a few hours and include dizziness, cardiac palpitation, rapid pulse, and abdominal pain. There is danger of shock. To prevent poisoning, prompt fish refrigeration or consuming the fish soon after capture is necessary.

4. Clupeotoxism

Clupeiform fish (i.e., sardines, herring, and anchovies) in tropical island areas feeding on toxic dinoflagellate blooms produce clupeotoxin, especially in the viscera. The actual source and nature of the poison have never been identified. Symptoms of the poisoning are usually violent, with the indication of a sharp metallic taste on the first taste of the fish. Tachycardia, cold and clammy skin, and a drop in blood pressure, following by various neurological disturbances, rapidly ensue. Death may occur in 15 min. The fatality rate is very high. Pruritus and various types of skin eruptions have been reported in victims who have survived. Clupeotoxism has been related to ciguatera poisoning, because of initial similar symptoms, but this has never been documented.

5. Tetrodotoxin

Tetrodotoxin from pufferfish is the major cause of food intoxications in Japan. The sale of toxic puffers is regulated carefully, as this toxin is one of the most violent forms of marine biotoxications. There has been a great deal of research accomplished on this toxin and the mechanism of action is known. No cooking or drying procedure controls the poison.

6. Hallucinogenic Poison

Certain types of reef fish in the tropical Pacific and Indian oceans, such as mullet and sea chub, can produce hallucinogenic symptoms within minutes to 2 hr after ingestion. The poison primarily effects the central nervous system, but there have been no fatalities reported and the form of poisoning generally is mild. The nature of the poison is unknown and toxicity is very sporadic, somewhat uncommon yet unpredictable. The poison is not destroyed by cooking, and hallucinogenic fish cannot be detected by their appearance.

C. Plant Foods

Plant foods are essentially a combination of chemicals. Some of these chemicals are required nutrients, or nonnutrients that may inhibit diseases such as coronary thrombosis or cancer, but some may be toxicants. Many toxicants are present in either a free or a bound state (e.g., glycosides). There are still many natural chemicals in every plant that have not been completely identified. Indeed, there are probably one or more toxicants naturally occurring, albeit at low levels, in every plant food depending on various environmental, genetic, and as yet unknown factors.

A common vegetable may have a level of toxicant so small that it is normally metabolized and excreted by the consumer without any effect, or the food is never consumed in continuous or high enough levels to observe either acute or chronic toxic symptoms. In addition, a known toxicant that is in specific strains of a common vegetable can be bred out by either traditional plant breeding or by genetic engineering. For example, a black Puerto Rican variety of lima bean yields about 300 mg of cynanide per 100 g of seed. This is fatal for an adult human who consumes 100 g of these lima beans. Serious incidences of cyanide poisoning occurred in Europe in the early part of the twentieth century from these types of tropical imported lima beans. Present-day strains of lima bean contain under 10 mg of HCN per 100 g of seed, which permits a rather large margin of safety for all consumers.

Important, common food plants may be related to very toxic poisonous plants. For example, the common tomato and potato are both members of the plant family Solanaceae. This family includes belladonna (deadly nightshade), which produces atropine; jimson weed, which produces scopolamine; henbane and mandrake, which produce atropine and other tropane alkaloids; tobacco, which produces nicotine; ground cherry, which produces solanine (a toxic glycoalkaloid, Fig. 3c); and trumpet flowers and hairy nightshade, which produce solanaceous alkaloids. All of these chemicals are extremely toxic, but some (i.e., atropine and scopolamine) are useful medicinals. The vine and unripe fruit of the tomato and potato leaves, green tubers, or tuber sprouts can have acutely toxic levels of solanine (see Fig. 3c), similar to the level (in excess of 1000 mg/kg of fresh plant material) occurring in nonfood, poisonous plant members of Solanaceae. It should be pointed out that a number of reported potato glycoalkaloid poisonings of people, especially in Europe, occurred in the early part of the twentieth century. The United States has not documented a single poisoning in the past 50 years, even though the consumption of potatoes and potato products averages about 125 pounds per year. Continued analyses of potatoes, especially new varieties, in the United States and Europe indicate that tubers have much less than 200 mg/kg fresh tuber, the maximum level of safety.

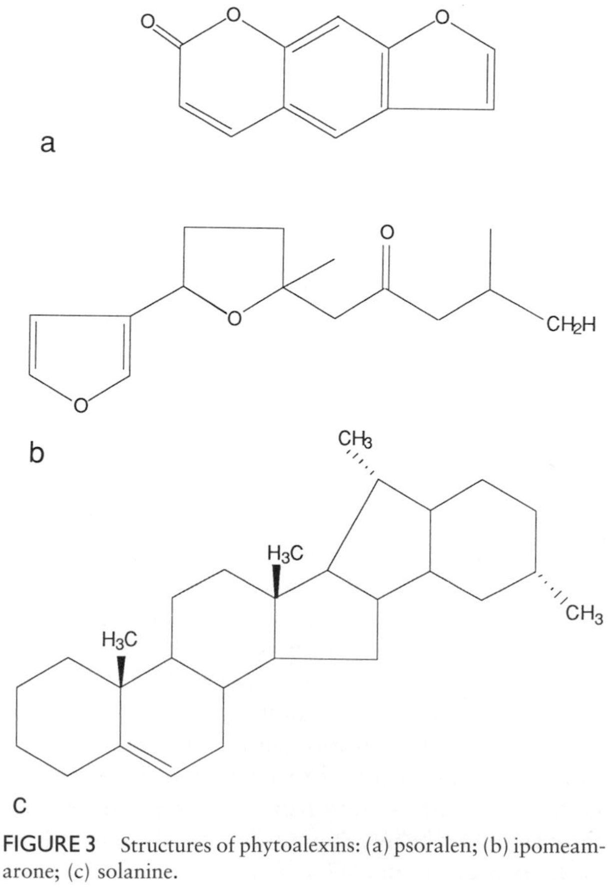

a

b

CH_2H

CH_3

H_3C

H_3C

CH_3

c

FIGURE 3 Structures of phytoalexins: (a) psoralen; (b) ipomeamarone; (c) solanine.

Phytoalexins, called stress metabolites, are synthesized toxic compounds that are thought to be needed for the plant's defense against pathogens and physical damages. These compounds exhibit toxicity not just to pathogenic microorganisms but also to insects, animals and humans, dependent, of course, on the amount present. Various stimuli, such as bacterial or viral infection, cold, ultraviolet light, heavy metals, fungicides, herbicides, or nematode attack, can contribute to pytoalexin production. Thus, a number of these toxic phytoalexins can be described as "natural pesticides." The toxicologic aspects of most phytoalexins have received little attention. Examples of three phytoalexins (Fig. 3) are: furanocoumarins, such as psoralen (celery and parsley), which causes phototoxic effects on the skin of harvesters and handlers; ipomeamarone (sweet potato), which causes liver degeneration and pulmonary edema in animals; and solanine (potato), which is a human plasma cholinesterase inhibitor and animal teratogen. Presently, there are more than 20 food plants known that can synthesize one or more phytoalexins during times of stress.

One or more plant genes can express naturally occurring toxicants, an example of which was observed in Australia during the early 1980s. Many individuals, within a few weeks, reported stomach cramps, diarrhea, vomiting, and headaches after consuming zucchini squash, which was also described as unusually bitter. Purgative bitter principals have been previously known, for at least 30 years, in various nonfood plants within the family Cucurbitaceae, the same family of zucchini squash. Tetracycline triterpenes, known as cucurbitacins (Fig. 4, cucurbitacin E) and studied some years earlier as a possible drug with purgative, emetic, and narcotic properties but abandoned owing to their high acute toxicities, were present in the toxic squash. Bitter zucchini squash was later found in a few plants growing in Alabama gardens and California fields. Production of this squash was concluded to be the result of accidental, aberrant outcrossings in seed-producing fields. Great care is now taken in marketing commercial squash seeds.

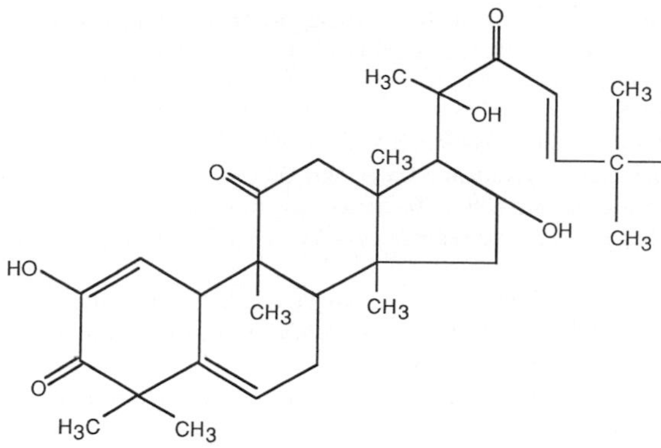

FIGURE 4 Structure of cucurbitacin E.

Plant breeders, either by traditional breeding methods or by modern recombinant DNA methodologies, purposely develop disease- and insect-resistant plant foods. These crops then require reduced, or even no, chemical pesticide applications. However, in many cases, there is a lack of knowledge of potential toxicants present in the newly developed hybrid food plant. One recent example is a virus-resistant hybrid lettuce developed from crosses with wild, nonfood lettuce strains, *Lactuca saligna* and *L. virosa*. This development was extremely important since there are no chemical pesticides efficacious against plant viruses. The newly developed hybrid lettuce was grown, marketed, and consumed without any apparent ill effects. However, the wild-type lettuce strains that were used in the traditional breeding project to produce the hybrid lettuce have been known for many years to contain high levels of sesquiterpene lactones, known as lactucins (Fig. 5), which have antibiotic, cytotoxic, and allergenic properties. Fortunately, later analyses of the hybrid lettuce strains showed quite small amounts of sesquiterpene lactones. Although lactucins may or may not be the effective natural compound that causes viral resistance, the levels of these compounds, with known detrimental biological effects, should have been of major concern and, at least, quantitated prior to commercial, public release.

An "Integrated Breeding and Environmental Strategy" program was suggested some years ago for breeders and developers of vegetables, cereals, fruit, and oil-bearing crop plants that could or may possess possible natural toxic compounds. An evaluation of these recognized toxic substances, including safety assessment, would be accomplished prior to market release. For the most part, with some exceptions, this has never been done, probably because of the costs and lack of specific enforced regulations.

Some natural food toxicants can show numerous toxicologic effects and yet may have potential health values. Glucosinolates (Fig. 6) are natural plant bioactive organosulfur compounds in Cruciferae vegetables (e.g., cabbage, broccoli, brussels sprouts). There are about 100 known glucosinolates in the plant kingdom, 10–12 distinct vegetable types, that are enzymatically metabolized to various compounds. These metabolites, studied over the past 50 years, have shown diverse biological effects from goiter development to anticarcinogenic properties in experimental animals. One metabolite, indole-3-carbinol from the

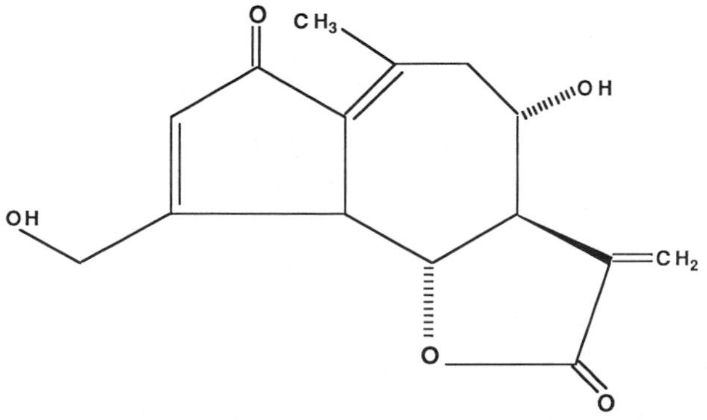

FIGURE 5 Structure of lactucin.

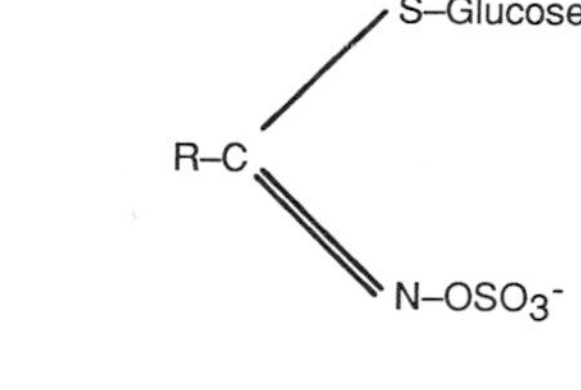

a

b

FIGURE 6 Glucosinolates: (a) general structure and (b) structure of 3-indolymethyl glucosinolate.

breakdown of 3-indolylmethyl glucosinolate (Fig. 6b), seems promising as a potential preventive agent against human breast and uterine cancers.

BIBLIOGRAPHY

Bier, R. C., and Nigg, H. N. (1994). Toxicology of naturally occurring chemicals in food. *In* "Foodborne Disease Handbook," (Y. H. Hui, J. R. Gorham, K. D. Murrell, and D. O. Cliver, eds.), Vol. 3. Marcel Dekker, New York.

Halstead, B. W. (1994). Fish toxins. *In* "Foodborne Disease Handbook" (Y. H. Hui, J. R. Gorham, K. D. Murrell, and D. O. Cliver, eds.), Vol. 3. Marcel Dekker, New York.

Hathcock, J. N. (1996). Safety limits for nutrients. *J. Nutr.* **126,** 2386S.

Shibamoto, T., and Bjeldanes, L. F. (1993). "Introduction to Food Toxicology." Academic Press, San Diego.

Stoewsand, G. S. (1995). Bioactive organosulfur phytochemicals in *Brassica oleracea* vegetables—A review. *Food Chem. Toxicol.* **33,** 537.

Tamaki, H., Robinson, R. W., Anderson, J. L., and Stoewsand, G. S. (1995). Sesquiterpene lactones in virus-resistant lettuce. *J. Agric. Food Chem.* **43,** 6.

Ybanez, N., and Montoro, R. (1996). Trace element food toxicology: An old and ever-growing discipline. *Crit. Rev. Food Sci. Nutr.* **36,** 299.

Forearm, Coordination of Movements

C. C. A. M. GIELEN
University of Nijmegen

GLOSSARY

Biceps Muscle in the upper arm that contributes to flexion and supination of the arm

Electromyographic activity (EMG) Electrical activity of muscle related to activation

Force–velocity relation Relation that describes muscle force as a function of the rate of change in muscle length at a constant activation of muscle

Gastrocnemius Muscle in the lower leg that contributes to extension of the ankle

Intramuscular EMG EMG activity recorded with intramuscular wire or needle electrodes

Isometric contraction Contraction of a muscle when it is held at a constant length by keeping the limb at a fixed position

Isotonic contraction Contraction of muscle at constant force

Motor unit (MU) Ensemble of motoneuron in the spinal cord and its axon and all muscle fibers innnervated by that axon

Pronation Rotation of the hand along an axis through the forearm, counterclockwise to supination

Recruitment Phenomenon in which more motor units become active with increase of muscle force in a fixed order

Recruitment threshold Muscle force at recruitment of a motor unit

Supination Rotation of the hand along an axis through the forearm such that the backside of the hand turns downward with the (right hand) thumb to the right

Triceps Muscle in the upper arm that contributes to elbow extension

Twitch Muscle force as a function of time after a pulse-like activation of muscle (fiber) or MU

THE COORDINATION OF MOVEMENTS REQUIRES that a number of muscles are activated each with a precise intensity, duration, and timing between the activation of various muscles. Usually a distinction is made between coding of amplitude and duration of the movement and direction of the movement. Amplitude and duration of movements are varied by changing the intensity and duration of the muscle activation pattern in a specific way. Direction is coded by selection of a specific activation for (parts of) the relevant muscles. The muscle activation pattern is the result of contributions of the central nervous system and (afferent) signals reaching it from the periphery, depending on the type of movement and the phase of the movement.

I. COORDINATION OF MOVEMENT AMPLITUDE AND DURATION

In general, goal-directed movements have one accelerating phase and one declerating phase if their trajectory is approximately along a straight line. This is not true any longer if an obstacle requires that the movement is curved or if the movement requires a specific curved trajectory. In these conditions the movement is segmented and consists of a sequence of trajectories, each with an accelerating and declerating phase. [*See* Movement.]

Corresponding to the accelerating and decelerating phase, the activation of muscles has two components:

ENCYCLOPEDIA OF HUMAN BIOLOGY, Second Edition, VOLUME 4.

one burst of electromyographic (EMG) activity for the muscles that start the movement (agonist) and a burst of EMG activity in the muscles that brake the movement (antagonist). For movements made as fast as possible, a third burst of EMG activity may be observed in the agonist muscle. It is generally accepted that this third burst contributes to stabilization of the limb at the final position. The few data available suggest that the third burst in the triphasic activation is not simply due to reflex components, but that it has, at least partially, a central origin. Its presence and amplitude are to some extent under voluntary control and can be changed independent of the amplitude and duration of the first and second bursts. [*See* Muscle Dynamics.]

Remarkably, all goal-directed movements have a similar shape more or less independent of the amplitude and duration. Each velocity profile is bell-shaped; velocity profiles of movements with different amplitude and duration can be superimposed after proper scaling in time and/or amplitude. If the amplitude is varied, keeping the movement duration constant, the velocity profiles are scaled in amplitude proportionally to movement amplitude. After appropriate scaling, all corresponding EMG bursts in the agonist muscles superimpose. This is not true for the EMG burst in the antagonist muscles, indicating a more complex activation of the antagonist muscles for braking movements as a function of movement duration and amplitude.

When movement duration is varied, keeping movement amplitude constant, the duration of the accelerating and decelerating phase of the movement is scaled by the same amount. However, the duration of the two phases varies with the inverse of the square of that scaling factor. This is not true for EMG in the antagonist muscle, presumably owing to muscle nonlinearities. These observations can be generalized to both single- and multijoint movements.

The fact that the shape of the velocity profiles is more or less the same for single- and multijoint movements despite the large variability in duration and amplitude has led to the idea of the generalized motor program. This idea implies that the activation pattern of the muscles is based on a general and abstract representation of a motor program, which acts as an elementary unit for any type of action and which can adjust movements for different task conditions by simply scaling a basic velocity profile. However, in movements like grasping and for movements aiming at targets of variable size, the velocity profiles were distinctly different from velocity profiles of fast movements to large targets. These data clearly indicate that planning and control of trajectories are task dependent and cannot always be reduced to a generalized motor program. A coherent model, which describes and explains the movement trajectories for different conditions, is still lacking. [*See* Motor Control.]

II. COORDINATION OF MOVEMENT DIRECTION

One of the main problems concerning the coordination of muscle activation for movements or torques in a particular direction is that for some motor tasks, the number of muscles acting across a particular joint is larger than strictly necessary. For example, there are at least seven muscles acting across the elbow joint, contributing to flexion/extension and supination/pronation (i.e., rotation along the longitudinal axis of the forearm). This gives rise to the possibility that a torque or movement in a particular direction can be realized by a large variety of different muscle activation patterns. Despite this redundancy, a unique activation pattern of muscles is observed in various subjects for isometric contraction in a particular direction. Note that we do not imply that the motor system is redundant in general, because different muscles may have muscle fibers with different histochemical properties and with a different fiber architecture, which gives each muscle particular functional properties.

A. Muscle Activation in Isometric Contractions

To investigate the activation of muscles in detail, the activation of single motor units in human arm muscles has been measured. Muscle force is graded by two mechanisms: recruitment (i.e., the increase of the number of active motor units) and firing rate of motor units. The relative contribution of each of these mechanisms is different for proximal (i.e., closer to the center of the body) and distal (i.e., farther away from the center of the body) muscles, with a more prominent role for recruitment in proximal muscles and a more prominent role for rate modulation in distal muscles.

The motor units in a muscle all have a different recruitment threshold in the range from very small to very large forces. Because a motor unit is thought to be recruited every time the input to that motoneuron exceeds a particular level, the recruitment threshold

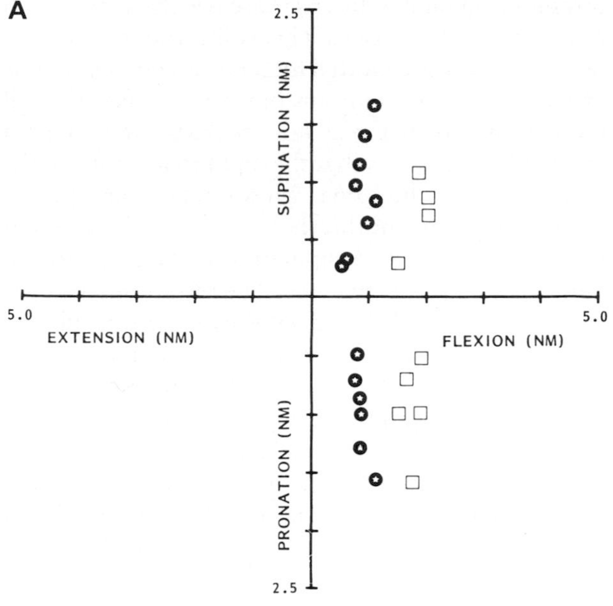

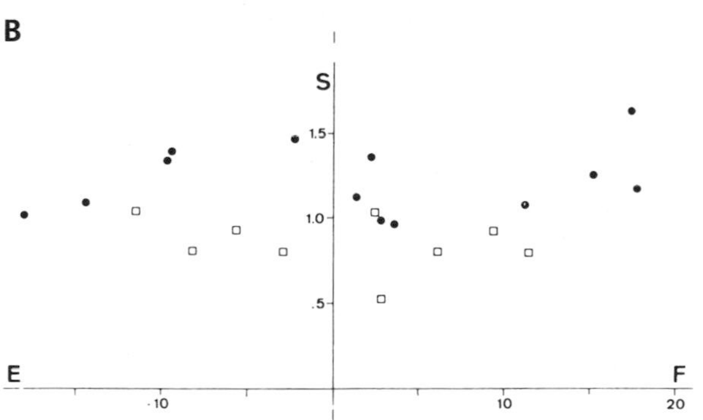

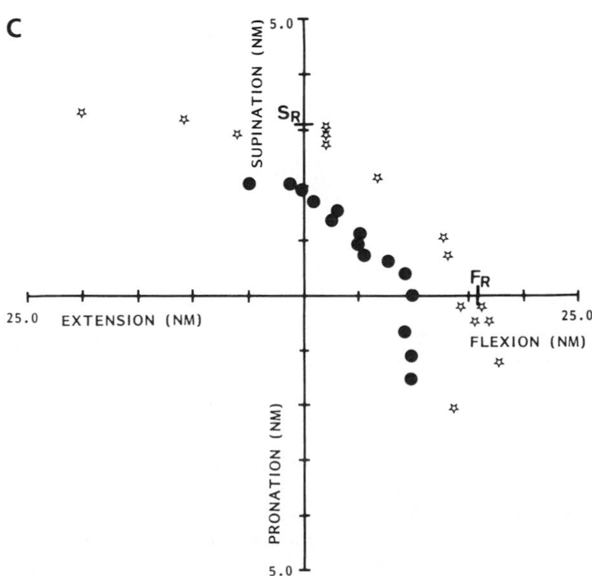

of a motor unit may be considered as a condition in which the total synaptic input to that motor unit is constant. This phenomenon was used to study the input to motoneurons in different experimental conditions.

The human biceps brachii is frequently called a multifunctional muscle because m. biceps contributes to torques in flexion and supination direction. This multifunctional character is reflected in the activation of motor units. Figure 1 shows the recruitment threshold of several motor units in m. biceps brachii, caput longum. For the motor units in Fig. 1A, the recruitment threshold depends on the torque in flexion direction only. Whatever the torque in supination direction, the recruitment threshold for flexion remains the same. This indicates that motor units of this type are activated for flexion torques in the elbow.

Figure 1B shows the recruitment behavior of another type of motor units in m. biceps, caput longum. The recruitment threshold of motor units of this type depends on the torque in supination direction, whatever the torque in flexion direction.

The majority of motor units in m. biceps, caput longum, are recruited for torques in both flexion and supination direction. An example of this type of behavior is shown in Fig. 1C. All recruitment thresholds fall along a straight line in the flexion/supination plane, which can be described by the equation

$$\frac{F}{F_r} + \frac{S}{S_r} = 1.$$

In this formula, F and S refer to the torque in flexion and supination direction, respectively, and F_r and S_r refer to the recruitment threshold in flexion and supination direction, respectively. For extension, the recruitment threshold of these motor units depends on

FIGURE 1 Example of motor-unit behavior in m. biceps for three motor-unit subpopulations. With different symbols (squares, circles, and stars), recruitment behavior is illustrated for two units. Each symbol indicates the combination of torques at which a motor unit is recruited. Recruitment threshold is plotted in Newton meters (Nm) in flexion (F), extension (E), supination (S), and pronation. For one motor unit (C), the recruitment threshold for flexion and supination direction is indicated along the horizontal and vertical axis, respectively. (A, B) Recruitment behavior of motor units, the recruitment threshold of which depends only on flexion torque or supination torque, respectively. (C) Data are shown for two motor units for which the recruitment threshold depends on torques in both flexion and supination direction. [Reprinted with permission from E. J. van Zuylen *et al.* (1988).]

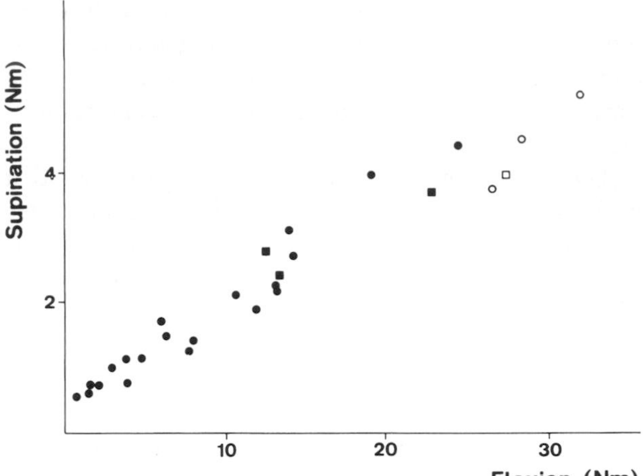

FIGURE 2 Recruitment threshold [in Newton meters (Nm)] for flexion and supination for motor units of the type shown in Fig. 1C. Circles and squares refer to data of motor units in the long head and short head, respectively, of m. biceps. The fact that all data points tend to fall along a straight line suggests that the ratio of recruitment thresholds in flexion and supination direction is constant. [Reprinted with permission from B. M. ter Haar Romeny *et al.* (1984).]

the torque in supination direction only, and for pronation, the recruitment threshold depends on the torque in flexion direction only.

The combinations of torques in Fig. 1C, where the two motor units are recruited, fall along parallel lines. This is related to the fact that the ratio of recruitment thresholds in human biceps muscle is the same for all motor units of this type. This is illustrated in Fig. 2, where F_r and S_r are plotted for a representative sample of motor units in m. biceps brachii. The fact that all the data points in Fig. 2 fall approximately along a straight line indicates that the input to all motor units of this type is the same. Motor units may have a different recruitment threshold, but the relative input related to flexion and supination is the same for all motor units of the same type. The results in Fig. 1 are very consistent and reproducible over various subjects and indicate that the coordination of muscles for torque in a particular direction is unique except for a cocontraction term of antagonistic muscles.

The motor units of a particular subtype are not randomly distributed in the muscle but are clustered in a specific location of muscle. In fact, there is a specific relation between location of a motor unit in muscle and its recruitment behavior. This is illustrated in Fig. 3, which shows the intramuscular EMG activity recorded with intramuscular wire electrodes in the lateral and medial side of m. biceps brachii. When flexion torque is increased, activity at the lateral side increases, and when flexion torque decreases and supination torque increases, the center of activity shifts from the lateral to the medial side of m. biceps brachii.

These results indicate that the activation of the motoneuron population of m. biceps brachii is inhomogeneous. However, for each subpopulation the results are compatible with the notion of a homogeneous

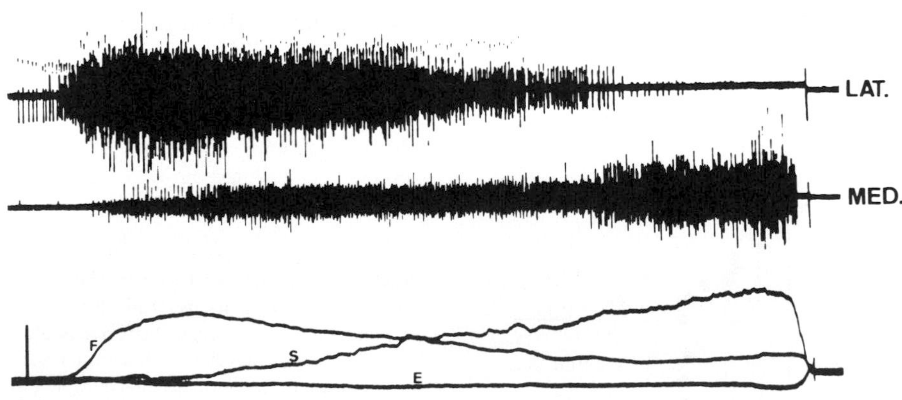

FIGURE 3 Intramuscular recordings of electromyographic activity obtained at the lateral side (top trace) and medial side (middle trace) in m. biceps, caput longum, when isometric flexion torque is initially increased and gradually decreased while supination torque is gradually increased. Lower trace shows torques in flexion (F), supination (S), and exorotation (E). Exorotation is a rotation in the shoulder joint along a longitudinal axis along the upper arm. [Reprinted with permission from B. M. ter Haar Romeny *et al.* (1984).]

activation of a particular pool of motoneurons. The idea of homogeneous activation is related to the notion of the size principle. According to this principle, the same input is received by each of a set of motoneurons, and the recruitment threshold of each motoneuron is related to the size of the motoneurons cell body. This idea predicts that the recruitment order of motoneurons is the same in all conditions. Without further arguments, the size principle has been generalized in the literature to the whole motoneuron pool of a muscle, although this is not implied by the strict definition of the size principle. The data in Fig. 1 clearly reveal an inhomogeneous activation of the motoneuron pool of m. biceps, and similar data have been presented for almost all muscles. However, these data do not argue against the size principle. Rather the data in Fig. 1 suggest that the motoneuron pool of a muscle has several groups and that each group receives a distinct but homogeneous activation. Recent experiments in animals have confirmed the compartmentalization of the motor-unit pool in a single muscle in several subpopulations and have shown that the motoneurons of motor units of the same subtype are clustered together in the spinal cord. This observation suggests that for movements or torques in different directions, different locations in the spinal cord may be activated.

The compartmentalization of motor units of a particular type in a specific part of muscle has important methodological implications because numerous studies have used surface EMG activity as a measure of muscle activation. Because EMG activity recorded with surface electrodes reflects the weighted activity of some part of the muscle fibers, the EMG activity strongly depends on the position of the surface electrodes relative to the muscle. Therefore, EMG activity recorded with surface electrodes cannot provide a complete picture of muscle activation.

Because motor units that were active during flexion and supination were localized in different parts of m. biceps, the question arose whether motor units in different subpopulations had a different mechanical advantage. The mechanical advantage of a muscle is the mechanical effect of muscle force to the force exerted by the limb. Evidently, this mechanical advantage is optimal when the pulling direction of muscle is orthogonal to the bone of insertion. To investigate this issue, the contribution of motor-unit twitches in supination and flexion direction was determined. Because motor units may have a different twitch amplitude, the argument was that if all motor units contribute equally to flexion and supination torques, the ratio of twitches in flexion and supination direction should be the same for all motor units. Any systematic violation of a constant ratio for motor units in different subpopulations would suggest a relation between recruitment behavior and mechanical advantage. It appeared that no relation whatsoever exists between the ratio of twitches in flexion and supination directions for motor units in human arm muscles. This indicates that the compartmentalization in m. biceps reflects a neural organization rather than a difference in mechanical advantage.

An inhomogeneous activation of the population motor units has been found in almost all muscles involved with flexion/extension and supination/pronation. It has also been found in muscles that were not considered multifunctional, such as m. triceps, which generates a torque in extension only. For a small sample of motor units in m. triceps, the recruitment threshold depends on torque in extension direction only. However, there is a larger group of motor units in m. triceps that is also activated for torques in supination and pronation direction.

The activation for torques in supination direction can be understood from the fact that m. biceps is activated for torques in supination direction. Because m. biceps has a mechanical advantage with a component in flexion direction, a torque in supination direction can be obtained only if m. triceps compensates for the torque component in flexion direction by m. biceps. The activation of motor units in m. triceps for pronation can be understood from the contribution of m. pronator teres to torque in pronation and flexion direction.

The mechanical advantage of muscle changes as a function of joint angle. This is illustrated in Fig. 4, which shows the combinations of torque where a single motor unit in m. biceps is recruited at different flexion angles in the elbow joint. Clearly the ratio of recruitment thresholds in flexion and supination direction, F_r/S_r, changes as a function of joint angle. Similar changes in recruitment behavior have been observed in nearly all human arm muscles. These results indicate that the coordination of muscles and the precise distribution of muscle activations are functions of joint angle.

To provide an explanation for the inhomogeneous activation of the motor-unit pool of muscle, the concept of task group has been proposed, in which functional groups of alpha and gamma motoneurons and spindle afferents are programmed to achieve optimal control of movements. Alpha motoneurons are the motoneurons that activate muscle fibers and thereby

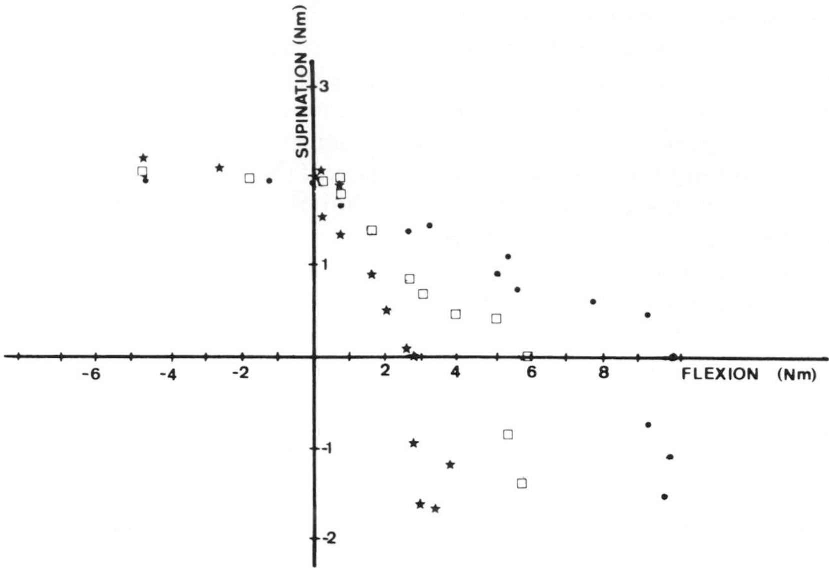

FIGURE 4 Recruitment threshold in Newton meters (Nm) for a single motor unit in m. biceps brachii (caput longum). Each data point indicates the recruitment threshold for a combination of forces in flexion/extension and supination/pronation. Different symbols refer to data obtained at different angles between forearm and upper arm: 100° flexion (circles), 145° flexion (squares), and 175° flexion (stars).

produce force. The gamma motoneurons activate the muscle spindles, thereby regulating their sensitivity. The problem with the notion of task group is that the criterion for "optimal" is not always properly defined. However, several suggestions have been made:

- Compartmentalization due to a different mechanical advantage. Some muscles have a broad attachment at the tendon such that different muscle fibers may have a different mechanical advantage (e.g., in cat sartorius muscle).
- Compartmentalization due to a different functional role. For example, some biarticular muscles are activated during both shortening and lengthening. In these two conditions, different motor-unit groups may be activated.
- Compartmentalization due to different mechanical properties of muscle fibers. For some muscles (e.g., gastrocnemius), muscle fibers in different muscle locations have different properties with respect to twitch contraction time, fatigability, and twitch amplitude.
- Compartmentalization reflecting a neural organization, such as in the human m. biceps brachii.

Presumably, all of these mechanisms may underlie the compartmentalization of muscle in general.

B. Muscle Activation in Reflex-Induced Contractions

In general, it can be stated that the same recruitment order is found in isometric contractions and in contractions that result from the reflex-induced activation of muscle. However, for a proper description of the reflex-induced activation of motor units, a distinction has to be made for the different reflex components.

It is well known that reflex activity is segmented into several reflex components, each with a different functional role and presumably mediated by different neural pathways. The precise pathways involved in all reflex components are not yet known. For the first reflex component (also called tendon reflex, M-1 reflex, or short-latency reflex), it is generally accepted that it is predominantly caused by a short route involving a single neuron directly activated by stretch-sensitive muscle spindles (presumably I_a afferents); however, other pathways may also contribute to it. The main function of the shortlatency reflex component is thought to be to compensate for the sudden decrease in muscle force caused by the break of actin–

myosin bonds, which occurs when muscle fibers are stretched by more than 0.2% of their rest length. The gain of this reflex component is not constant, but is modulated in various phases of movements and for different motor tasks by presynaptic mechanisms.

The second reflex component is frequently called long-latency reflex component or M-2 reflex. Presumably more than a single type of sensor is involved. The experimental results so far have demonstrated a role for muscle spindle afferents (both I_a and type II endings) and cutaneous afferents. The long-latency reflex component is thought to be mediated by pathways, involving many neurons in the spinal cord and presumably also a long-loop pathway involving the motor cortex.

The long-latency reflex component serves a role in the coordination of an adequate response. As explained before with regard to the activation of motor units in m. triceps, for torques in supination or pronation direction some muscles may be activated in isometric contractions even if the isometric torque is orthogonal to the torque that results from activation of that muscle. Similarly, m. triceps is activated in a long-latency reflex by torque perturbations in pronation direction, which do not stretch or shorten the length of m. triceps. The time interval of activation of motor units in m. triceps after a torque perturbation in pronation direction corresponds to that of the long-latency reflex. This shows that muscles that are not stretched may reveal long-latency reflex activity if the adequate coordination of movements requires so. This reflex activity may be excitatory, as in the case of m. triceps in response to perturbations in pronation direction, or inhibitory, such as observed for motor units in m. brachialis in response to perturbations in pronation direction (Fig. 5).

The amplitude of short- and long-latency reflex components may be different in different muscles; however, within each reflex component, the recruitment order appeared to be the same, equal to that in isometric contractions.

C. Muscle Activation in Voluntary Movements

To study in detail the activation of motor units in voluntary contractions, single motor-unit activity has been studied in movements at constant velocity against a constant force acting in flexion or extension direction at the wrist. Because the relative activation of muscle changes as a function of joint angle, the recruitment behavior and firing rate were studied in a small range of positions centered around the same fixed elbow angle. This procedure also eliminates difficulties in interpretation due to the variable mechanical advantage of muscle at various joint angles.

The activation of motor units in voluntary movements is quite different from that in isometric contractions. Already at very slow shortening velocities (as low as 2 deg/sec flexion in the elbow joint), the recruitment threshold of motor units in m. biceps is considerably decreased (about 30%) with respect to the isometric recruitment threshold. This decrease was found at the lowest contraction velocities that could be tested. For larger shortening velocities the recruitment threshold decreases more gradually, in quantitative agreement with the well-known force–velocity relation. The difference in recruitment threshold in isometric and isotonic conditions increases proportionally to the isometric recruitment threshold, and the ratio of recruitment thresholds in isometric and iso-

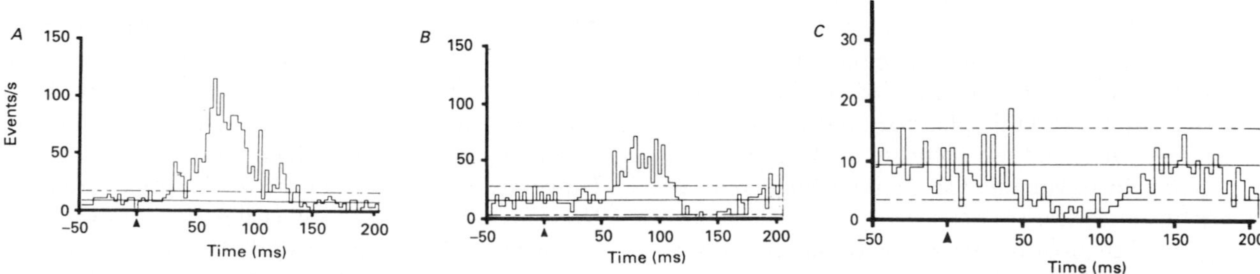

FIGURE 5 Motor-unit activity in m. triceps (A, B) and brachialis (C) elicted by torque perturbations in flexion direction (A) and pronation direction (B, C). Motor-unit responses, plotted as the number of action potentials per unit of times in m. triceps, were tested for a preload in flexion direction. Motor units in brachialis were tested by torque perturbation superimposed on an extension preload. Note the excitation in m. triceps in the time interval between 50 and 100 msec and the decreased activity for the motor unit in m. brachialis in the same period. [Reprinted with permission from Gielen *et al.* (1988).]

tonic contractions at 2 deg/sec is approximately constant for all motor units in m. biceps.

In other muscles the recruitment behavior in isometric and isotonic contractions may differ from that observed in m. biceps. For example, in m. brachialis, the recruitment threshold in isometric and isotonic conditions was the same. If the angular velocity in the elbow joint is used to calculate the shortening velocity of muscle fibers in m. biceps and m. brachialis, the shortening velocity of fibers is approximately the same in the two muscles. This fact and the observation of a difference in recruitment behavior in isometric contractions and in movements in biceps and brachialis indicate that the relative activation of human arm muscles is different in isometric contractions and in movements.

At a higher level, the coordination of all muscles involved in a multijoint movement reflects a rather stereotyped behavior. There is good evidence that the organization of limb movements involves a series of sensorimotor transformations between intrinsic and extrinsic coordinates. The central nervous system has at its disposal extra degrees of freedom in the performance of any motor task. Thus a given position of the hand in space, relative to the trunk, can be achieved with an infinite variety of postural orientations of the arm because the arm has seven degrees of freedom (three at the shoulder, one at the elbow, and three at the wrist), whereas only three parameters are required to describe hand location. Nevertheless, in the case of arm movements at the shoulder and elbow, it has been found that any given task is performed in a stereotyped manner with a low variability within and between subjects.

This observation is reflected in invariances in movements that hold true for entire classes of limb movements. For example, in drawing or handwriting, the magnitude of shoulder and elbow angular motions scales roughly with the size of the figure drawn. Furthermore, although shoulder and elbow movements are tightly coupled (constant phase relation), motions at distal joints are loosely coupled to those at the proximal joints (variable phase relation). Motion at distal joints increases the accuracy of the movement as indicated by the smaller variability of finger relative to wrist trajectories.

In general the relative activation of muscles in single- and multijoint movements is very much constant. However, an exception has to be made for very fast movements, when muscles that have predominantly "fast" muscle fibers may be activated preferably at the expense of muscles with predominantly "slow" muscle fibers. An example has been found in rapid movements in the cat ankle, where gastrocnemius seems to be activated almost exclusively, with m. soleus almost silent.

D. From Joint Rotations to Limb Displacement

One of the interesting problems with joint rotations is that rotations about noncolinear axes in space do not commute. If you take a die and rotate it along two orthogonal axes in space, you will notice that the final orientation of the die depends on the order of the rotations.

This raises the question as to whether the orientation of limbs in joints with three degrees of freedom, such as the eyes, head, and shoulder, depends on previous rotations. The answer is they do not. The orientations of the eye, head, and upper arm are uniquely defined for each pointing direction of the arm or gaze direction. This result is obtained by a specific reduction in degrees of freedom from three to two: all positions of the upper arm (both during movements and during stationary pointing) can be obtained by a rotation vector from a fixed reference position in such a way that all rotation vectors lie in a two-dimensional plane. Notice that this implies a reduction in degrees of freedom from three to two. To achieve this result, the angular velocity vectors that describe a movement have a fixed axis for each movement. The angular *velocity* rotation vectors do not fall in a plane, however, but they can have any orientation in 3D space, contrary to the rotation vectors, which describe the *position* of the eye, head, and arm.

To make straight movement trajectories with joint rotations requires a tight coordination in multiple joints. However, it also illustrates the necessity of biarticular muscles. Consider, for example, Fig. 6, which gives a schematic representation of a human arm that exerts an external force F. In Fig. 6, the lever arm relative to the external force F has the same orientation for the elbow and shoulder. Therefore, the torque in elbow and shoulder contributes to flexion. However, a movement in the direction of F requires elbow extension and shoulder flexion. Because of the opposite sign of torque T and change in joint angle $\Delta\phi$, the work $W = T \cdot \Delta\phi$ is negative. If only monoarticular muscles would be available, they would dissipate work, rather than contribute positive work, as the monoarticular shoulder muscles do. This dissipa-

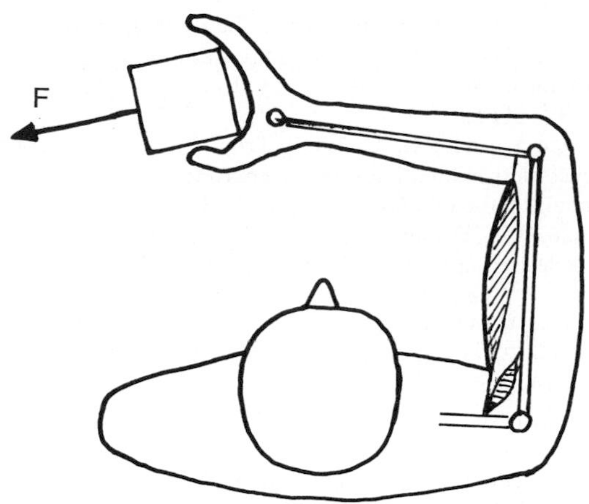

FIGURE 6 Schematic representation of the human arm. Notice that a movement in the direction *F* requires flexion of the shoulder and extension of the elbow. Also notice that a force in direction *F* requires a flexion torque in both elbow and shoulder. The most efficient way for making a movement in the direction *F* while exerting a force F is to activate the biarticular biceps muscle, assisted by additional activation of the monoarticular shoulder muscle to generate the required torque in the shoulder.

tion of energy, which is economically very unfavorable, can be partly relieved by using biarticular muscles. A biarticular muscle like m. biceps contributes to torque in both elbow and shoulder, but does not change its length as much as a monoarticular elbow muscle would, since the changes in elbow and shoulder angles give rise to opposite changes in length for m. biceps.

Similar apparent conflicts between torque and change in joint angle occur in daily motor acts such as walking, running, and cycling. Thanks to the presence of biarticular muscles, these motor acts can be performed with the efficiency and elegance that are observed in animal and humans.

III. ROLE OF CENTRAL AND PROPRIOCEPTIVE INPUTS IN MUSCLE ACTIVATION

Presumably because of the relatively large number of sensor types and their complex behavior, the precise role of muscle receptors and their contribution to movement control are unknown. It is generally accepted that the initial accelerating part of fast goal-directed movements is made without the use of afferent feedback. Presumably the first effect of proprioceptive information on the motor program coming from the muscles themselves or the surrounding skin becomes evident about 115 msec after a perturbation or a deviation from the intended movement trajectory. Reflex mechanisms may affect the muscle activation pattern at shorter delay times, and they give rise to the same relative activation of muscles as observed in isometric contractions. Because it has been demonstrated that variability in the accelerating phase is initially high and that it decreases during the first 100 msec of the movement (i.e., before sensory information can become effective), this has been interpreted as evidence for the use of an internal feedback loop.

After the initial accelerating phase, sensory information may affect the activation pattern of muscles. On the basis of all available evidence, virtually everybody now agrees that muscle sensors contribute to position sense and to the control of movements. Recent experiments have demonstrated that the type of information used depends on the motor task. For example, it has been shown that an observer who is instructed to maintain his arm at a given target position while his biceps is vibrated by pulses with a repetition rate of about 100 Hz flexes the arm. (The subject cannot see the arm!) When he is instructed to move the arm fast to the target position, he will be surprised by the instruction because he thinks he is already at the target. However, when he is pushed to obey the instruction, he makes a movement and brings the arm accurately to the target. This suggests that certain motor acts employ a quite different map of limb position from that perceived. Electromyographic recordings ruled out a simple mass-spring strategy and demonstrated that subjects behaved as if the motor system really knew the correct position of the arm even though it was not perceived.

With regard to other types of sensors, the role of joint receptors in motor control is still unknown. Most experimental observations indicate that joint receptors are effectively excited only when the limb is moved to one of its extremes. In the middle of the range, most, if not all, receptors are normally silent. However, there is evidence that joint receptors may modulate the reflex gain not only in the extremes of the physiological range, but throughout the full range of joint action.

Recently, more and more information suggests that cutaneous receptors do contribute to normal movements and to reflex activity. However, there is no consensus yet about their precise role. The same is true for the Golgi-tendon organs, although several hypotheses have been put forward as to their role, for example, for stiffness regulation.

BIBLIOGRAPHY

Gielen, C. C. A. M., Ramaekers, L., and van Zuylen, E. J. (1988). Long-latency reflexes as coordinated functional responses. *J. Physiol.* **407**, 275–292.

Haar Romeny, B. M. ter, Denier van der Gon, J. J., and Gielen, C. C. A. M. (1984). Relation between location of a motor unit in the human biceps brachii and its critical firing levels for different tasks. *Exp. Neurol.* **85**, 631–650.

Jeannerod, M. (1988). "The Neural and Behavioural Organization of Goal-Directed Movements." Clarendon Press, Oxford.

Sittig, A. C., Denier van der Gon, J. J., and Gielen, C. C. A. M. (1985). The attainment of target position during step-tracking movements despite a shift of initial position. *Exp. Brain Res.* **60**, 407–410.

van Ingen Schenau, G. J. (1989). From rotation to translation: Constraints in multijoint movements and the unique role of bi-articular muscles. *Hum. Mov. Sci.* **8**, 301–337.

Wadman, W. J., Denier van der Gon, J. J., and Derksen, R. (1980). Muscle activation patterns for fast goal-directed arm movements. *J. Hum. Mov. Stud.* **6**, 19–37.

Windhorst, U., Hamm, T. M., and Stuart, D. G. (1989). On the function of muscle and reflex partitioning. *Behav. Brain Sci.* **12**, 629–681.

Zuylen, E. J. van, Gielen, C. C. A. M., and Denier van der Gon, J. J. (1988). Coordination and inhomogeneous activation of human arm muscles during isometric torques. *J. Neurophysiol.* **60**, 1523–1548.

Fragile X Syndromes

RANDI J. HAGERMAN
University of Colorado Health Sciences Center and The Children's Hospital

GLOSSARY

Autism Syndrome characterized by impairment in reciprocal social interaction and in verbal and nonverbal communication, such that individuals appear isolated and are disinterested in social interactions

CGG repeat Three nucleotides—cytosine, guanine, and guanine—that occur in a repeated fashion within the FMR1 gene

Full mutation Large expansion of the CGG repeat number to greater than 200 repeats and methylation of the FMR1 gene. This finding is associated with fragile X syndrome because the FMR1 gene is not active and does not produce the FMR1 protein (FMRP)

Heterozygote (or carrier female) Female who has one normal X chromosome and one X chromosome with the FMR1 mutation. She may be unaffected with a premutation or affected with the full mutation

Macroorchidism Large testicles; in the adult, a testicular volume of ≥30 ml represents macroorchidism

Mental retardation (MR) IQ less than 70

Nonpenetrant male Male who carries the fragile X premutation but is not affected by the syndrome and usually does not demonstrate the fragile X chromosome on cytogenetic testing

Premutation Small expansion of the CGG repeat number to the range of 50 to 200 repeats within the FMR1 gene. Individuals who carry the premutation are usually unaffected by the fragile X syndrome

X chromosome One of the chromosomes that determines the sex of an individual. Females have two X chromosomes and males have one X and one Y chromosome

X-linked MR Cause of MR that is characterized by an abnormal gene on the X chromosome. Therefore, these disorders affect more males in a family tree than females and cannot be passed from father to son because the father gives his Y chromosome (not his X chromosome) to his sons

THE FRAGILE X SYNDROME IS A GENETIC DISorder associated with a fragile site or a partial break on the long arm of the X chromosome at the q27.3 region. In 1991 the mutated gene that causes fragile X syndrome was identified and sequenced. It is called the Fragile X Mental Retardation 1 gene or FMR1 and is located at Xq27.3. A trinucleotide repetitive sequence (CGG) was found within FMR1, such that normal individuals have approximately 6 to 50 repeats, carrier individuals have 50 to 200 repeats (premutation), and individuals who are affected by fragile X syndrome have >200 CGG repeats (full mutation). The full mutation is associated with methylation of the FMR1 gene, which turns off transcription and translation of the FMR1 gene into protein. It is the absence of the FMR1 gene protein (FMRP) that causes fragile X syndrome.

The clinical picture of fragile X syndrome includes mental retardation, often with autistic-like characteristics such as poor eye contact, hand-flapping, and hand-biting. Physical features are also associated with this syndrome, including macroorchidism or large testicles, a long narrow face, and large or prominent ears. Hyperextensibility of the finger joints, flat feet, and mitral valve prolapse are also commonly seen clinically. Males who are affected with fragile X syndrome are usually retarded, although the spectrum of involvement ranges from learning disabilities with a normal IQ to profound retardation. All males with the full mutation are clinically affected by the syndrome, that is, they usually demonstrate mental retardation and associated physical features. Approximately one-half of females with the full mutation are cognitively impaired with a borderline IQ or with mental retardation. Individuals who carry the premutation (50 to 200 CGG repeats) are usually unaffected intellectually.

Because this disorder is carried on the X chromosome, it is inherited in an X-linked fashion, that is, it is passed on from mother to son and cannot be passed on from father to son. The full mutation occurs only when the mutation is passed on by a mother who has the premutation or a full mutation. Carrier males with the premutation will pass on the premutation only to all of their daughters but to none of their sons, who receive the Y chromosome. The overall incidence of individuals affected with this syndrome in the general population is approximately 1 in 2,000.

I. EPIDEMIOLOGY, CYTOGENETIC, AND MOLECULAR ASPECTS

The fragile X chromosome (Fig. 1) was first visualized in 1969 in a family with four retarded males over three generations. Work in Australia led to the association of the fragile X chromosome with the physical phenotype in the mid-1970s, however, individuals were rarely diagnosed with this disorder during the 1970s in the United States. This was because the demonstration of the fragile X chromosome is dependent on the use of a specific type of tissue culture media that was described by G. Sutherland in the late 1970s. The cytogenetic testing done on a peripheral blood sample requires that the cells be grown in tissue culture media that is deficient in folic acid and/or thymidine. Expression of the fragile X chromosome can also be induced by inhibitors of folic acid, such as methotrexate, or by inhibitors of thymidylate synthetase, such as 2-deoxy-5-fluorouridine (FUdR). Under these conditions, the fragile X chromosome will be seen in 1 to 50% of the cells analyzed. Although all the cells carry a mutation at the fragile X site, the visualization of the fragile site at Xq27.3 is not seen in every cell for unknown reasons.

Usually only individuals with the full mutation demonstrate the fragile site. Individuals with the premutation are usually negative on cytogenetic studies but they are still at high risk to have children with fragile X syndrome and the full mutation (Fig. 2). An analysis of the FMR1 gene can be done with DNA testing of a blood sample. DNA testing is less expensive and more accurate than cytogenetic testing for fragile X syndrome or carrier status. However, DNA testing for FMR1 mutation does not include testing of other genes or structural defects throughout the genome, which is done in cytogenetic testing. Therefore, in the workup of mental retardation or autism of unknown etiology, both DNA and chromosome testing are usually carried out, whereas in the workup of a known fragile X family, only the DNA testing is necessary to identify carriers with the premutation and individuals more affected with the full mutation.

II. PHYSICAL FEATURES

A. Males

Males who are affected by the fragile X syndrome usually demonstrate a triad of physical features, including macroorchidism, large or prominent ears, and a long narrow face.

Large or prominent ears are seen in approximately 80% of fragile X males at all ages. The ears are almost never malformed. However, they may demonstrate a cupping of the upper part of the pinna. In the prepubertal male, prominent ears are the most common feature that can be detected on physical examination (Fig. 3). Prepubertal individuals only occasionally demonstrate a long narrow face, but this finding is more common in postpubertal males. Prognathism or a prominence of the lower jaw is a frequent finding in older fragile X males. A high-arched palate is also commonly seen in males and this may be associated with crowding of the teeth or malocclusion.

Macroorchidism is seen in approximately 80% of adult fragile X males and it is quantified by a testicular volume of 30 ml or larger. Typically adult fragile X males have a testicular volume of 50 to 60 ml, although individuals with a volume as large as 100 ml have been reported. Macroorchidism is much more difficult to diagnose in the prepubertal male and only one-third of prepubertal individuals have a testicular volume that is larger than normal. The testicle increases most dramatically in size during the early pubertal years, perhaps related to gonadotrophin stimulation.

Approximately 60% of prepubertal fragile X males demonstrate hyperextensibility of the finger joints,

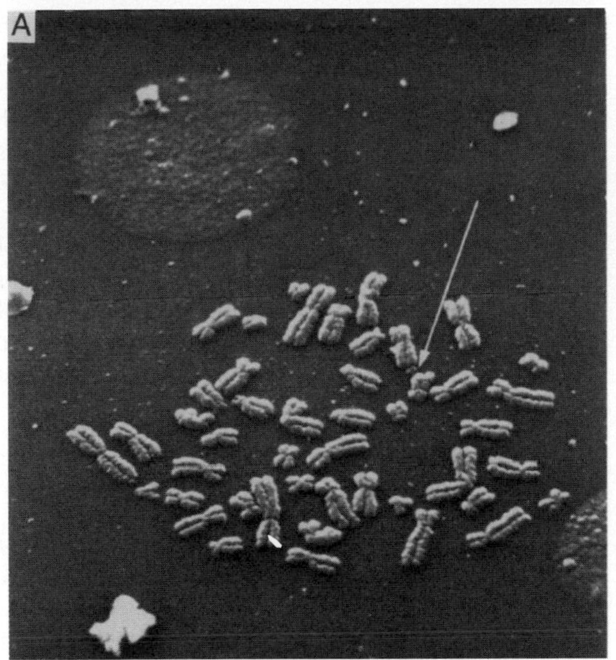

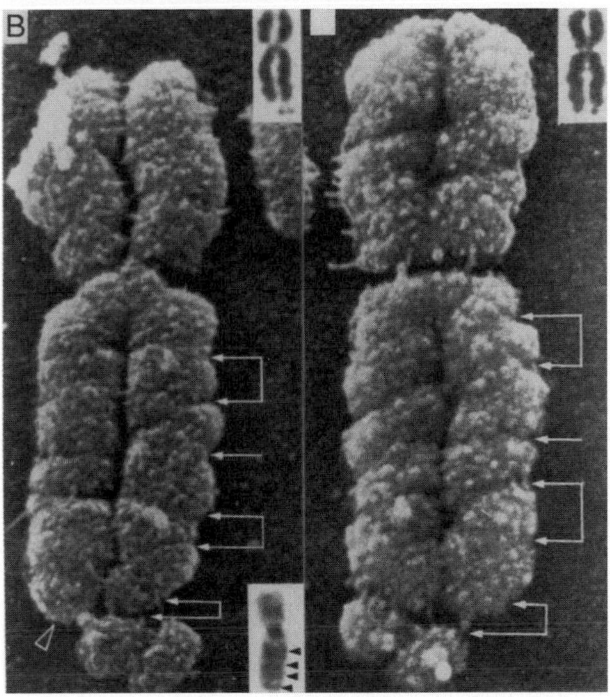

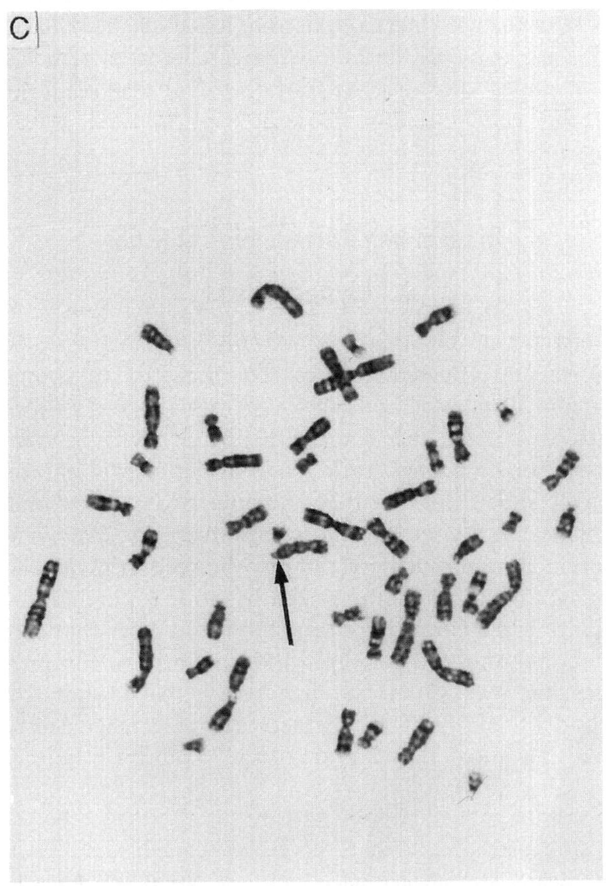

FIGURE 1 (A) A scanning electron micrograph of chromosomes in metaphase with arrow pointing to the fragile site on the X chromosome. (B) Closeup of two fragile X chromosomes with the lowest arrow pointing to the fragile site at Xq27.3. [Reprinted with permission from C. J. Harrison *et al.* (1983). The fragile X: A scanning electron microscope study. *J. Med. Genet.* **20**, 280–285.] (C) Metaphase spread viewed from light microscopy. The arrow is pointing to the fragile site at q27.3 on the X chromosome.

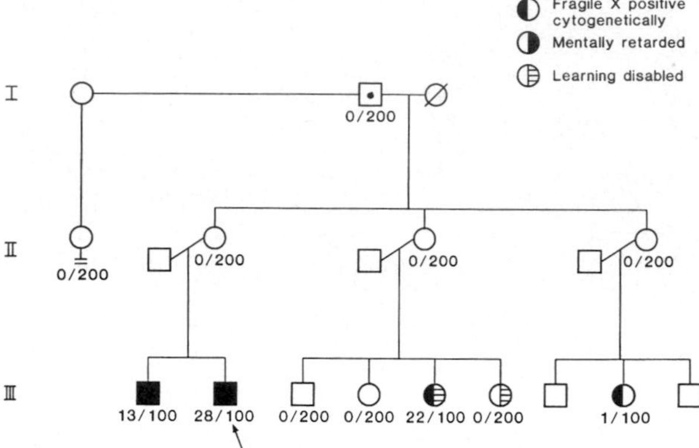

FIGURE 2 A fragile X pedigree. Cytogenetic results (number of fragile X positive lymphocytes/total number of lymphocytes evaluated) are noted below each patient. Generation I has a nonpenetrant male (grandfather of affected males) who has been documented to carry the fragile X gene by DNA studies.

most easily detected by bending the fingers back at the knuckles. Hyperextensibility is defined by having the fingers bend to a 90° or greater extension at the knuckle (Fig. 4). Double-jointed thumbs and looseness of other joints, such as the elbow or ankle, are also commonly seen. Flat feet often with pronation are seen in over 50% of fragile X children and adults. Other features that occur in 25 to 50% of fragile X males include a pectus deformity of the chest (a pushed-in appearance to the chest), a weak eye muscle causing an in-turning or out-turning of one of the eyes, a single palmar or simian crease, and a large head circumference particularly in early childhood. CT scans are usually normal in fragile X males; however, more recent studies of brain structure using MRI scans have documented a small size of the cerebellar vermis, particularly the posterior region, and enlarged lateral ventricles in fragile X males and significantly affected females. The cerebellar finding appears to be related to the motor and sensory processing problems in fragile X syndrome.

Individuals who are carriers of the premutation usually do not have typical physical features of fragile X syndrome, although some may have prominent ears. It has been postulated that a connective tissue abnormality or dysplasia is associated with the fragile X syndrome because of the physical findings of flat feet, loose joints, and large or prominent ears. Other disorders of connective tissue such as Marfan syndrome and Ehlers-Danlos syndrome also demonstrate similar physical features. Because these other disorders commonly have cardiac problems associated with connective tissue abnormalities, fragile X patients have also been studied for cardiac abnormalities. Approximately 50% of adult fragile X males demonstrate mitral valve prolapse, which is a floppiness of the mitral valve, sometimes leading to regurgitation of blood through the mitral valve during systole. Mitral regurgitation, however, is only occasionally seen in males with fragile X syndrome. [*See* Connective Tissue.]

B. Females

Females who are intellectually impaired because of the fragile X syndrome are most likely to demonstrate the typical physical features. A long narrow face and prominent ears are commonly seen in retarded or learning disabled females (Fig. 3D). Other findings, such as hyperextensible finger joints, a high-arched palate, and mitral valve prolapse, have also been documented in fragile X females. An occasional female is reported with large ovaries but it is unclear whether this is a common finding. Unaffected carrier females sometimes demonstrate physical features of fragile X.

III. BEHAVIORAL FEATURES

A. Hyperactivity

Behavior problems are commonly seen in fragile X males and affected females. The majority of young fragile X males demonstrate hyperactivity and a short attention span; they have difficulty in focusing their attention on a single task. Their attention will impulsively shift from one topic or object to another and this is usually associated with an increased activity level. Their impulsivity can also be seen in their language, which is characterized by poor topic maintenance and a cluttering of thoughts and ideas that can be communicated in a rapid and sometimes incomprehensible way. Often the diagnosis is made when the child is evaluated for hyperactivity or language delay, which is noted after 2 years of age. Younger fragile X

FIGURE 3 (A) Young fragile X boy, note cupping of the ears. (B) Young fragile X boy with only mildly prominent ears. (C) Adult fragile X male, note long face. (D) Two young fragile X boys and their sister, who is also fragile X positive and learning disabled.

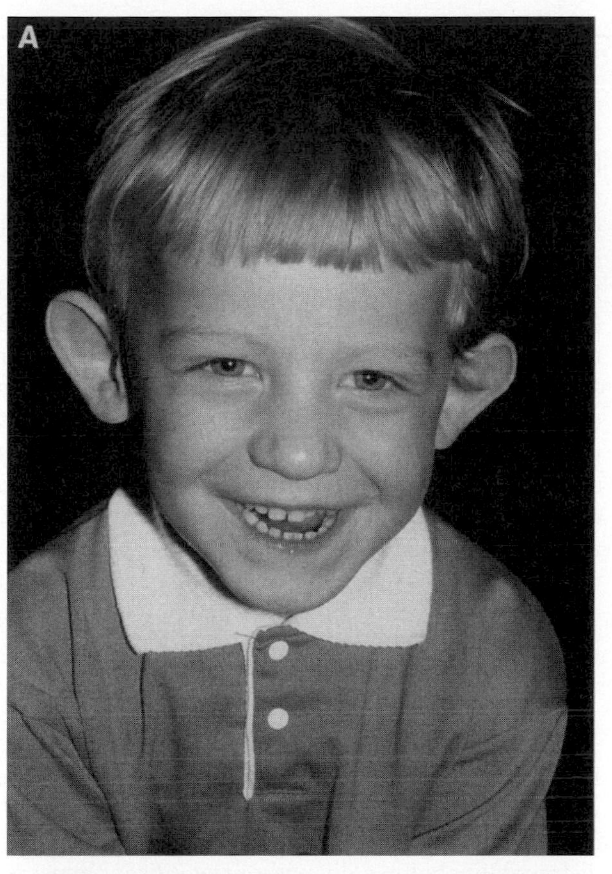
A

B
PANAMA
JACK

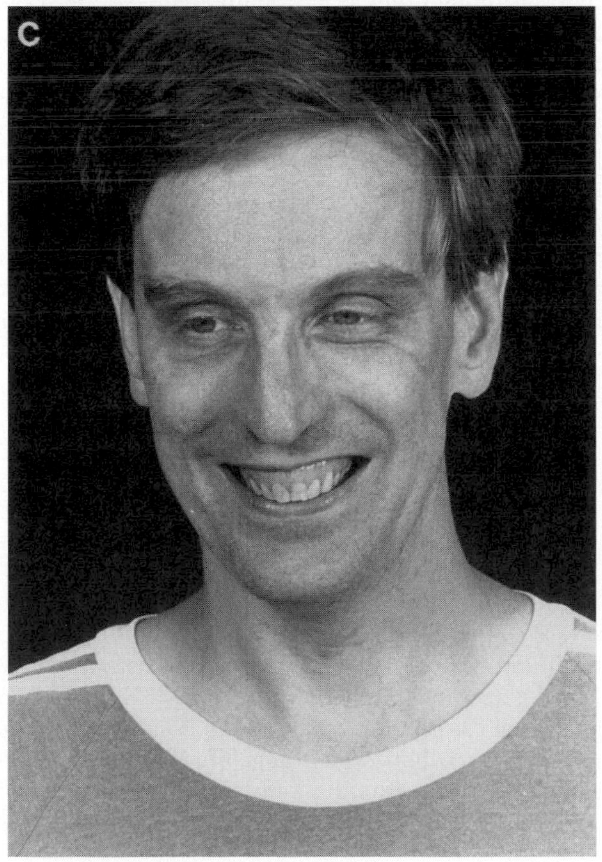
C

D
NIKE

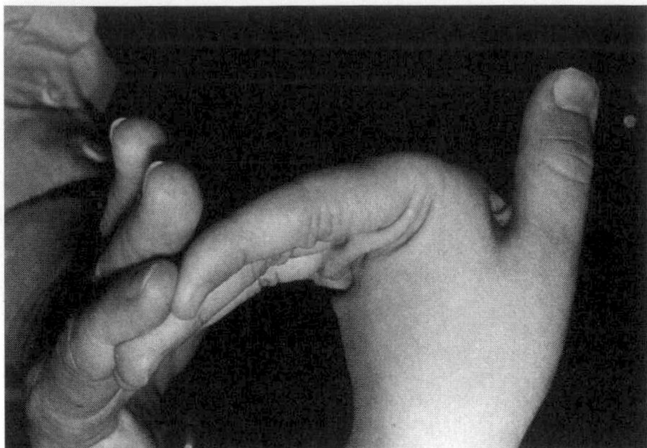

FIGURE 4 Hyperextensible finger joints. Note that the fingers extend >90° from the neutral position.

boys may also demonstrate hypotonia or poor muscle tone in addition to their hyperextensible joints.

B. Autism

Other unusual features are commonly seen with hyperactivity and have been described as autistic-like. These features include poor eye contact, shyness or social interactional problems, hand-flapping when they are excited or angry, and other hand stereotypies or unusual hand movements. Obsessive and compulsive behavior, sometimes ritualistic behavior, can be seen and repetition or perseveration in both behavior and language are very common. Often a child will do an activity over and over again in a compulsive fashion or will speak in repetitive phrases. Other autistic features include tactile defensiveness or sensitivity or aversion to touch, overreactions to minimal stimuli, and frequent tantrums. Hand-biting is seen in the majority of young fragile X males and may persist into adulthood, causing calluses on the finger where repetitive biting occurs.

Although the majority of young fragile X males demonstrate autistic features, a full diagnosis of infantile autism or autistic disorder is seen only in approximately 16% of fragile X males. This diagnosis is made only when a pervasive lack of relatedness and lack of interest in social interactions exist. The majority of fragile X boys are friendly and yet autistic-like features interfere with normal relatedness.

Populations of autistic males have been screened for fragile X syndrome and it has been detected by cytogenetic studies in approximately 7% of autistic males. Fragile X syndrome therefore represents the most common inherited cause of autism that can be identified. The fragile X syndrome is also responsible for 30 to 50% of all forms of X-linked mental retardation. In institutional screening studies of retarded individuals, fragile X syndrome has been identified in 1 to 10% of various populations that have been studied. If one evaluates children or adults who are retarded for unknown reasons, fragile X can be detected in 2 to 10% of these individuals. Therefore, any child or adult who presents with mental retardation or autism of unknown etiology should be tested for fragile X by performing DNA studies or cytogenetic studies on a peripheral blood sample using appropriate tissue culture conditions. Although the yield of fragile X positive individuals may be approximately 2 to 10%, this is a very significant yield in the evaluation of children or adults with developmental disabilities. [*See* Autism.]

C. Females

Females who are affected with fragile X syndrome with either learning disabilities or mild mental retardation often demonstrate attentional problems with distractibility but usually without significant hyperactivity. Although they occasionally demonstrate some autistic features such as hand-flapping or hand-biting, they more commonly demonstrate shyness and social withdrawal. The shyness can be profound, particularly in adolescence, and feelings of social isolation leading to depression are quite common. Depression is also a finding that is frequently seen in mothers of fragile X children. It is unclear whether this finding is related to the environmental stresses of raising a difficult child with fragile X syndrome, or whether the fragile X mutation when present just in the carrier state can predispose an individual to depression. [*See* Depression.]

IV. COGNITIVE FEATURES

A. Males

Intellectual abilities in fragile X males can range from severe learning disabilities in the normal IQ range to profound retardation. The majority of adult fragile X males are moderately retarded. Fragile X boys often present in the preschool years with IQ abilities in the low normal to borderline range. These higher cognitive abilities are usually a reflection of their areas of cognitive strength, which include single word expressive vocabulary abilities, isolated visual perceptual abilities such as visual matching tasks, and imitation. Almost all fragile X males, however, demonstrate significant weaknesses in higher linguistic abilities such as abstract reasoning skills, ability to generalize and solve novel problems, and memory deficits. These abilities, particularly the reasoning abilities, become more important in middle and later childhood and are reflected in the score on IQ testing. Many fragile X children will demonstrate a decrease in their IQ score through middle and later childhood and adolescence. Fragile X children do not lose cognitive abilities and, in fact, they continue to learn and increase their ability through time. However, the emphasis of cognitive testing in later childhood and adolescence stresses their weak areas, such as abstract reasoning. Math is a consistent weakness for almost all fragile X males and it usually is their most profound deficit in the academic area. Many higher-functioning fragile X males will learn to read and they may even maintain reading and spelling abilities at their grade level.

B. Females

Approximately 50% of females with the full mutation have cognitive deficits, including a borderline IQ or mental retardation. Of females with the full mutation and a normal IQ, disabilities are common. Females who are cognitively impaired with learning disabilities or mental retardation usually demonstrate their greatest deficit in mathematics. Cognitive testing using the Wechsler Intelligence Scale has demonstrated a profile of strengths and weaknesses in the subtest scores. A pattern of weakness can be seen in the subtests of Arithmetic, Digit Span, and Block Design. Areas of strength include the Vocabulary and Digit Symbol subtests. These profiles are seen in the analysis of group data but may be inconsistently present when looking at a particular individual.

V. INHERITANCE PATTERN

The discovery of a new form of mutation in fragile X syndrome, specifically a trinucleotide repetitive sequence that grows from generation to generation, has led to similar discoveries in other disorders that become worse (start earlier or have a more severe form) from generation to generation. This phenomenon is called anticipation. Myotonic dystrophy, Huntington's chorea, and spinocerebellar atrophy all demonstrate anticipation and all have been found to have a similar mutation pattern as seen in fragile X, that is, a trinucleotide repeat sequence that increases in size from generation to generation.

VI. TREATMENT

A. Medical Follow-up

Medical follow-up of the fragile X syndrome is important for a variety of reasons. Medical problems such as hernias and joint dislocations are often associated with the connective tissue abnormalities in fragile X syndrome and treatment may be required. Medical intervention or medication can be helpful for behavioral problems, particularly hyperactivity. The physician can also be important in organizing and coordinating an overall treatment program that requires special education and individual therapies as described in following sections. There is no cure for the fragile X syndrome, but an overall treatment plan can help fragile X children reach their maximal potential.

Fragile X children often suffer from frequent and recurrent middle ear infections in early childhood. This problem requires vigorous medical intervention, including antibiotics and often the placement of polyethylene tubes through the tympanic membrane to ventilate the middle ear space and to normalize hearing. If hearing is not normalized, the sequelae of language problems associated with a conductive hearing loss will further compound the language deficits that are associated with the fragile X syndrome.

The looseness of connective tissue, which is a frequent problem in fragile X, can cause significant flat feet that may require orthopedic intervention, a high narrow palate, and mitral valve prolapse, which requires antibiotic prophylaxis to prevent subacute bacterial endocarditis. This prophylaxis is required when children undergo dental procedures or surgical procedures that could contaminate the blood with bacteria.

Strabismus or a weak eye muscle may be seen in approximately 40% of fragile X children and strabismus requires ophthalmological treatment, such as surgery or patching to strengthen the weak eye muscle. All fragile X males and cognitively impaired females should be evaluated by an ophthalmologist before 4 years of age to detect visual problems at an early stage.

A seizure disorder can be seen in up to 20% of fragile X children and this requires anticonvulsant medication. EEG abnormalities, including a slowing of background activity and spike wave discharges similar to benign rolandic spikes, can be seen in approximately 50% of children with fragile X syndrome. When seizures are clinically apparent, they are usually infrequent and easily controlled with anticonvulsants. The seizures may be partial motor seizures, grand mal, petit mal, or temporal lobe seizures. The clinical seizures usually disappear in adolescence, as do benign, rolandic seizures. [*See* Seizure Generation, Subcortical Mechanisms.]

Medical intervention for behavior problems, specifically hyperactivity, is often warranted. Central nervous system stimulant medication, such as methylphenidate, dextroamphetamine, or pemoline, can help young children with fragile X syndrome to decrease hyperactivity and to improve their attention span. However, medication should be only one aspect of a total treatment program for fragile X children, which should also include special education help, speech and language therapy, and occupational therapy.

B. Speech and Language Therapy

Often fragile X children present with very significant language delays that are first identified between 2 and 3 years of age. Speech and language therapy can be helpful for the articulation deficits that are commonly seen, but perhaps, most importantly, this therapy can enhance both expressive and receptive language abilities. The language therapist can work individually with children to improve auditory processing and attention and can also address early deficits in abstract reasoning skills and in the ability to generalize.

Pragmatic aspects of communication are also areas of deficit for fragile X children, perhaps because of the autistic features that are common. Group language therapy can enhance social communication abilities and can also improve pragmatic aspects of communication.

C. Occupational Therapy

Occupational therapy is helpful for improving both fine and gross motor coordination even in very young fragile X children. Many of the behavioral difficulties associated with the fragile X syndrome have been ascribed to sensorimotor integration deficits. Fragile X children have difficulty in processing a variety of sensory input, including tactile, auditory, and visual input. They frequently become overwhelmed with stimuli and this may precipitate behavior problems such as tantrums. Sensorimotor integration therapy is helpful in decreasing behavior difficulties and in helping a child feel comfortable and less anxious with a variety of stimuli. Hypotonia, motor incoordination, joint stability, motor planning, and calming techniques can all be addressed in occupational therapy.

D. Special Education

All children who are affected by the fragile X syndrome will require special education help. For higher-functioning individuals, this may involve only pull-out remediation from a regular classroom setting. Many children with fragile X syndrome can be included into the regular classroom with the help of an aide who can modify assignments. For more significantly retarded individuals, a self-contained placement may be necessary, particularly when severe hyperactivity and a short attention span are complicating features. Fragile X children often respond well to mainstreaming because they usually mimic the behavior of children who surround them. If the classmates have normal behavior, then the fragile X child is more likely to develop normal behavior patterns. If fragile X children are placed in a group of children with significantly deviant behavior, then fragile X children often mimic more deviant behavioral characteristics.

Special education help is also essential for learning disabled individuals, particularly females and higher-functioning males. Math deficits are the most common academic problem and individualized help is usually necessary. Many learning disabled individuals demonstrate high achievement in reading and spelling, although deficits can also be seen that require remediation. Attentional problems may be present in normal IQ individuals and stimulant medication, in conjunction with special education, is useful.

Counseling or psychotherapy can be helpful on an individual basis and in family work. Behavior modification techniques can be taught to parents to guide them in improving their child's behavior. Individual work can be useful, particularly for the adolescent or adult male with fragile X syndrome, to help him deal with sexuality issues and aggression.

VII. GENETIC COUNSELING

Genetic counseling is probably the most important reason for identifying children and adults with the fragile X syndrome. Once a single individual has been identified, then the whole family pedigree must be evaluated so that carrier females can be identified and counseled appropriately concerning their risk of having subsequent children with the fragile X syndrome. DNA FMR1 testing should be carried out on all individuals in a family tree who are at risk for carrying the premutation or who may be affected by the syndrome; this testing is available at most major medical centers. Although the fragile X syndrome has a similar incidence to Down syndrome, the recurrence risk in fragile X is far higher. A carrier female has a 50% risk of passing on the FMR1 mutation with each child. If the mother carries the full mutation, her risk of having a significantly affected child is increased compared to a mother who carries the premutation. However, a woman with >90 CGG repeats will pass on the full mutation to her children every time that X chromosome is transmitted. Expansion to a full mutation occurs only when the mutation is passed on by a female. Males who carry the premutation will pass it on only to their daughters. Their sons inherit the Y chromosome, not the X, so sons of carrier fathers are spared from the fragile X syndrome (Fig. 5).

FIGURE 5 A family with three affected fragile X boys. The mother is an unaffected carrier female.

Once a child is diagnosed with fragile X syndrome it is helpful to have the family contact other parents who have children with similar problems related to fragile X. Parent support groups have been established in almost all states. To find a parent support group in your area, contact the National Fragile X Foundation at 1-800-688-8765. In addition, parent educational materials can be obtained through this foundation. It is essential for parents to understand basic issues regarding inheritance of fragile X syndrome.

Males with the fragile X syndrome have the full mutation in their blood and other tissues with the exception of their sperm. The sperm always carries the premutation only. This recent finding suggests that the expansion to the full mutation does not occur in the egg, but instead expansion occurs early on in embryonic development after the cells that are destined to be gametes (sperm or eggs) are separated from the rest of the embryo and protected against the CGG expansion to the full mutation. If males with fragile X syndrome reproduce they would have unaffected sons and all of their daughters would have the premutation, so they would not be cognitively affected by fragile X syndrome. Because of the overall high recurrence risk, it is absolutely essential for medical professionals to identify this disorder, which at the present time is significantly underdiagnosed. [*See* Genetic Counseling.]

BIBLIOGRAPHY

Ashley, C. T., Jr., and Warren, S. T. (1995). Trinucleotide repeat expansion and human disease. *Annu. Rev. Genet.* **29,** 703–728.

Hagerman, R. J., and Cronister, A. (eds.) (1996). "Fragile X Syndrome: Diagnosis, Treatment and Research," 2nd ed. Johns Hopkins Univ. Press, Baltimore.

Hagerman, R., Staley, L., Brown, W. T., Taylor, A., Meadows, K., Dorn, M., Stoorman, S., Neri, G., Chiurazzi, P., Levitas, A., Spiridigliozzi, G. A., O'Connor, R., Weber, J. D., Braden, M., and Sudhalter, V. (1995). Conference summary: Fourth International Conference on Fragile X and X-Linked Mental Retardation. Sponsored by the National Fragile X Foundation. *Dev. Brain Dysfunction,* **8,** 167–184.

Rousseau, F., Heitz, D., Biancalana, V., Blumenfeld, S., Kretz, C., Boue, J., Tommerup, N., Van Der Hagen, C., DeLozier-Blanchet, C., Croquette, M-F., Gilgenkrantz, S., Jalbert, P., Voelckel, M-A., Oberle, I., and Mandel, J-L. (1991). Direct diagnosis by DNA analysis of the fragile X syndrome of mental retardation. *New Eng. J. Med.* **325,** 1673–1681.

Rousseau, F., Heitz, D., Tarleton, J., MacPherson, J., Malmgren, H., Dahl, N., Barnicoat, A., Matthew, C., Mornet, E., Tejadu, I., Maddaline, A., Spiegel, R., Schinzel, A., Marcos, J. A. G., Schorderet, D. F., Schaap, T., Maccioni, L., Russo, S., Jacobs,

P. A., Schubart, A. C., and Mandel, J. L. (1994). A multicenter study on genotype–phenotype correlations in the fragile X syndrome using direct diagnosis with probe StB12.3: The first 2253 cases. *Am. J. Hum. Genet.* **55**, 225–237.

Verkerk, A. J., Pieretti, M., Sutcliffe, J. S., Fu, Y. H., Kuhl, D. P., Pizzuti, A., Reiner, O., Richards, S., Victoria, M. F., Zhang, F., Eussen, B. E., van Ommen, G. J., Blonden, L. A. J., Riggins, G. J., Chastain, J. L., Kunst, C. B., Galjaard, H., Caskey, C. T., Nelson, D. L., Oostra, B. A., and Warren, S. T. (1991). Identification of a gene (FMR-1) containing a CGG repeat coincident with a breakpoint cluster region exhibiting length variation in fragile X syndrome. *Cell* **65**, 905–914.

Free Radicals and Disease

MAHESH S. JOSHI
JACK R. LANCASTER, JR.
Louisiana State University Medical Center

GLOSSARY

Cytokines Group of soluble proteins that are immune hormones and act to transmit messages from one immune cell to another

Free radical Any species that has one or more unpaired electrons. Because of their rapid reactive nature, free radicals are transient

Lipid peroxidation Oxidative destruction of polyunsaturated fatty acids in the cell membrane to form lipid hydroperoxides

Oxidative stress Imbalance between levels of reactive oxygen species and the antioxidants, with the reactive oxygen species being in excess

Phagocytes Cells such as leukocytes that characteristically engulf foreign material and consume debris and foreign bodies. They constitute the primary defense against invading pathogens

Respiratory burst Interaction of an external stimuli with cells, resulting in a burst of oxidative activity. When phagocytes are exposed to a stimulus, their oxygen uptake rises rapidly, on the order of 50-fold over that of the resting cells

Sepsis Toxic condition resulting from the spread of bacteria or their products from the site of infection. It is the host's disproportionate inflammatory response that mediates the pathophysiology of this disease

A FREE RADICAL IS ANY SPECIES THAT HAS ONE OR more unpaired electrons. The most important free radicals in a biological system are oxygen- and nitrogen-derived radicals. Free radicals are generally produced in cells by electron transfer reactions. The major sources of free radical production are inflammation, ischemia/reperfusion, and mitochondrial injury. These three sources constitute the basic components of a wide variety of diseases. Free radicals are produced *in vivo* under normal metabolic conditions and are also generated during exposure to radiation. Radicals are an important part of host defense against foreign organisms. When produced in excess, they can be lethal to the host owing to their ability to modify a spectrum of biomolecules.

I. INTRODUCTION

Free radicals and related oxidants are implicated as causative agents in a vast array of disease states. These include aging, cancer, hemorrhagic shock, arthritis, various pulmonary disorders, and acquired immunodeficiency syndrome (AIDS). Examples of free radicals are superoxide (O_2^-), thiyl ($RS^\cdot$), trichloromethyl ($CCl_3^\cdot$), hydroxyl ($^\cdot OH$), nitrogen dioxide ($NO_2^\cdot$), and nitric oxide ($NO^\cdot$). Radicals can form other species that, although not radicals, can be classified as reactive oxygen species and thus also contribute to oxidative injury. An example is hydrogen peroxide (H_2O_2). Radicals and oxidants vary widely in their reactivity toward different tissues. Hydroxyl radical is the most potent of all radical species. The cellular damage is the result of DNA lesions, protein denaturation, and peroxidation of unsaturated lipids. Among the biomolecules, lipids are most prone to radical

ENCYCLOPEDIA OF HUMAN BIOLOGY, Second Edition, VOLUME 4.

attack since the oxidation of lipids follows a self-perpetuating chain reaction. The extent of tissue damage taking place upon production of radical species depends on the site of radical formation, the type of radical formed, and proximity of molecules sensitive to free radical damage. Leukocytes have the potential to release large amounts of radical species and cause tissue and cellular destruction. Leukocytes act as a primary cellular defense against invading microbes and are an important component of inflammatory responses. Because of their short half-life, the detection and measurement of free radical generation have relied largely on the determination of products of radical reaction. Cells have devised various lines of defenses, including enzymatic and nonenzymatic antioxidants, against the injurious effects of radical species. The imbalance between the production of free radicals and cellular defenses leads to the deleterious consequences.

II. BASIS OF FREE RADICAL INJURY

A. Biological Sources of Free Radicals

1. Respiratory Burst

Upon exposure to a stimulus, phagocytes start consuming large amounts of oxygen, an event termed "oxidative burst" that results in the generation of reactive oxygen species. Whereas resting phagocytes produce minimal levels of superoxide, those exposed to a stimulus may produce 100 million superoxide molecules per second per cell. Superoxide and H_2O_2 are two main oxygen species produced during this process. These two react in the presence of metal ions to generate more potent radicals like hydroxyl radical, which is so reactive that it reacts with everything that is close to it. Phagocytes are capable of producing large levels of H_2O_2 to oxidize intracellular components in erythrocyte targets, to destroy tumor cells, to injure endothelial cells, and to damage leukocyte functions. The H_2O_2 also participates in myeloperoxidase-catalyzed oxidation of halides, resulting in the formation of hypohalous acid, most notably hypochlorous acid (HOCl). The products thus formed have antimicrobial activity. The multicomponent NADPH oxidase located in the plasma membrane of neutrophil catalyzes the formation of superoxide. The superoxide then dismutates to H_2O_2. In an inherited condition called chronic granulomatous disease, granulocytes fail to produce superoxide, which is the result of diminished levels of NADPH oxidase. [*See* Inflammation; Phagocytes.]

In addition to their antimicrobial activity, the reactive oxygen species generated during an oxidative burst are also found to be responsible for host tissue injury and inflammation. The symptoms of tissue damage are redness, swelling, pain, and loss of function. The synovial fluid in the swollen knee joints of rheumatoid arthritis patients contains large number of activated neutrophils. The reactive oxygen species produced by these neutrophils may contribute to the joint injury. The HOCl produced during an oxidative burst by phagocytes can react rapidly with amino acids, sulfhydryl groups, thioesters, amines, and other unsaturated carbon centers. HOCl also inactivates α_1-antiproteinase, the major circulating inhibitor of serine proteinases in body fluids, resulting in the uncontrolled action of several proteinases.

Nitric oxide is another free radical produced by several types of cells including macrophages. This radical molecule plays diverse roles as a vasodilator, neurotransmitter, and immunomodulator. $NO^{\cdot}$ is synthesized as a result of five-electron oxidation of arginine catalyzed by nitric oxide synthase (NOS). The macrophages have a form of NOS that is inducible by a stimulus, and upon activation, for example, exposure to microbes, they are shown to produce high levels of $NO^{\cdot}$. Murine macrophages produce nitrite (NO_2^-) and nitrate (NO_3^-), the stable end products of $NO^{\cdot}$ oxidation, in response to *in vitro* treatment with lipopolysaccharide and cytokines. $NO^{\cdot}$ synthesis is increased during inflammation and the symptoms of inflammation can be treated with NOS inhibitors. [*See* Respiratory Burst.]

2. Ischemia-Reperfusion

Ischemia-related pathological processes account for the majority of deaths in the United States and a burst of oxygen-derived free radical formation is at least partially responsible for the damage that results. Although the mechanisms for cellular and tissue injury during ischemia-reperfusion are complex and variable, a representative scenario begins with cellular energy depletion as a result of ATP breakdown. This energy depletion results in calcium influx, which in turn activates endogenous proteases leading to the conversion of xanthine dehydrogenase to xanthine oxidase. These enzymes are two forms of the same gene product. Whereas xanthine dehydrogenase utilizes NAD^+ as an electron acceptor, xanthine oxidase uses O_2 as an electron acceptor. Upon reperfusion,

O_2 reenters the tissue and gets reduced to large amounts of O_2^- by xanthine oxidase.

Xanthine oxidase:

$$\text{hypoxanthine} + H_2O + 2O_2 \rightarrow \text{xanthine} + 2O_2^- + 2H^+$$

$$\text{xanthine} + H_2O + 2O_2 \rightarrow \text{uric acid} + 2O_2^- + 2H^+$$

Xanthine dehydrogenase:

$$\text{hypoxanthine} + H_2O + NAD^+ \rightarrow \text{xanthine} + NADH + H^+$$

Xanthine oxidase was the first documented biological source of superoxide. Superoxide thus produced, in collaboration with H_2O_2 and transition metal ions, initiates the formation of more potent secondary radicals that cause the tissue damage. Thus ischemia itself does not bring about this tissue injury, rather the injury takes place during the reoxygenation process. Evidence has also been accumulated to show that neutrophils are another source of tissue injury during ischemia-reperfusion. Administration of free radical scavengers such as superoxide dismutase (SOD), catalase, and dimethylthiourea has been found to inhibit the infiltration of neutrophils into reoxygenated tissues, suggesting that superoxide radical directly or indirectly initiates the infiltration of neutrophils into tissues during the postischemic period. These neutrophils, which contain the free radical-producing enzymes NADPH oxidase and myeloperoxidase, trigger the tissue damage already initiated by xanthine oxidase. Depletion of neutrophils has been shown to significantly attenuate postischemic parenchymal and microvasculature dysfunction. The injury resulting from free radical generation will be more pronounced because of the observed drop in antioxidant defenses during ischemia-reperfusion. [*See* Superoxide Dismutase.]

3. Mitochondria

The mitochondrion is another source of production of free radicals. The major cause of radical production in mitochondria is the partial reduction of O_2 by electrons leaking from the electron transport chain. Superoxide is the main oxygen species generated by mitochondria. The rate of superoxide radical production is highest when the components of the electron transport chain are in the reduced state. More than 90% of O_2 consumed by the human body is used by mitochondrial cytochrome oxidase, of which only 2–5% goes toward the production of reactive oxygen species. The reactive oxygen species production takes place mainly at complexes I, II, and III of the respiratory chain as a result of one-electron reduction of O_2. Superoxide thus generated undergoes dismutation to H_2O_2 by manganese superoxide dismutase (MnSOD). It has been demonstrated that the rate of oxygen species generation rises with an increase in O_2 concentration. Factors that enhance the leakage of electrons from the respiratory chain include several pathological events such as ischemia-reperfusion and sepsis. [*See* Mitochondrial Respiratory Chain.]

III. REACTIONS OF FREE RADICALS AND THEIR TARGETS

The free radicals vary widely in their reactivity toward the tissue target molecules. Superoxide, although a radical, can cause little damage by itself. Rather, it is the reaction of superoxide with other radical species that leads to the formation of more potent oxidizing agents. The dismutation of superoxide either spontaneously or enzymatically catalyzed by SOD generates H_2O_2:

$$O_2^- + O_2^- + 2H^+ \rightarrow H_2O_2 + O_2$$

H_2O_2 is also produced by several other enzymes such as L-amino acid oxidase, glucose oxidase, and monoamine oxidase. H_2O_2 can readily cross the cell membrane and thus take part in reactions away from its site of production. Like superoxide, H_2O_2 also has limited reactivity. Simultaneous generation of superoxide and H_2O_2 inside the cell has been shown to cause DNA damage, but neither of these two reacts directly with DNA. Both O_2^- and H_2O_2 produced during phagocytosis or ischemia-reperfusion participate in a reaction catalyzed by transition metals to generate hydroxyl radical. Copper also catalyzes the generation of $OH^\bullet$. Both ferrous and ferric salts react with H_2O_2, the ferrous compounds reacting faster. The importance of iron in this type of oxidative damage raises the question of the source of intracellular free iron. The source of iron is not clearly elucidated in most of the circumstances. Almost all of the iron pool within the cell is not in free form as it is sequestered mainly by transferrin, a transport protein, and ferritin, a storage protein, thus acting as antioxidants. The storage capacity of ferritin is quite high, namely, 4500 mol of iron per mole of protein. Thus, in the absence of free iron, O_2^- and H_2O_2 at physiological concentrations may have limited damaging effects. The blood plasma of healthy humans is found to contain no free iron and copper. Large amounts of O_2^- formed during an oxidative stress mobilize iron from ferritin. Intracellular degradation of ferritin also leads to the liberation of iron. Hydrogen peroxide can degrade heme proteins to release iron and thereby promote cell injury.

The major biological targets of free radicals are deoxyribonucleic acids, unsaturated lipids, and proteins. Neither H_2O_2 nor O_2^- can cause a DNA strand break, but hydroxyl radicals can cause cellular damage by nucleic acid base modification and DNA strand scission. The DNA strand scission takes place as a result of hydroxyl radical interaction with the sugar–phosphate backbone. Reaction with thymidine also generates lesions, producing single-stranded breaks. Thus the appearance of single-stranded DNA breaks is an indication of the interaction with free radicals. This strand scission can be prevented using hydroxyl radical scavengers. Another mechanism of DNA damage during an oxidative stress is due to rises in intracellular free Ca^{2+}, which in turn activates nucleases, leading to the fragmentation of DNA.

Membrane lipids are another class of biomolecules that are susceptible to free radical attack. The peroxidation of lipids by radical species has been implicated in a variety of tissue injury and disease states. Cell membranes are rich sources of polyunsaturated fatty acids, and lipid peroxidation is defined as the oxidative destruction of polyunsaturated fatty acids present in cell membranes, leading to the formation of lipid hydroperoxides. When a free radical reacts with a nonradical, another free radical is produced, thus initiating a chain reaction. This forms the basis of oxidation of lipid molecules, for example, polyunsaturated fatty acids. The following scheme defines the process of lipid peroxidation:

$$LH + R^{\cdot} \rightarrow L^{\cdot} + RH$$

$$L^{\cdot} + O_2 \rightarrow LOO^{\cdot}$$

$$LOO^{\cdot} + L'H \rightarrow LOOH + L'^{\cdot}$$

where LH is the lipid molecule, $R^{\cdot}$ is the radical species, and LOOH is the lipid hydroperoxide. The radicals react with a lipid molecule by abstracting a hydrogen atom and forming a radical. This radical, upon addition of oxygen, generates lipid peroxyl radical, which abstracts a hydrogen atom from another lipid to form a lipid hydroperoxide and another lipid radical. Transition metals like iron and copper participate in lipid peroxidation in two ways. First, they catalyze the formation of initiating species and then they promote peroxidation by reacting with lipid hydroperoxides and decomposing them to peroxyl radicals:

$$LOOH + Fe^{2+} \rightarrow LO^{\cdot} + Fe^{3+} + OH^-$$

$$LOOH + Fe^{3+} \rightarrow LOO^{\cdot} + Fe^{2+} + H^+$$

The cascade of free radical formation amplifies the damage done to membrane proteins and DNA. The formation of lipid peroxides within the membrane disrupts its functioning, altering the fluidity and letting the ions such as Ca^{2+} leak through the membrane. Owing to their hydrophobic nature, lipid radicals almost always react with membrane-associated components. Peroxyl radicals may react with membrane proteins and thus inactivating enzymes and receptors critical to cell function. Examples of this are losses of cytochromes P-450 and glucose-6-phosphatase activity in microsomes. The cell membrane proteins can also be damaged by aldehydes like malondialdehyde and 4-hydroxy-2,3-*trans*-nonenal, the byproducts of lipid peroxidation. These products diffuse away from their site of formation and cause tissue injury at a distant site.

Protein denaturation or damage due to free radical attack is dependent on their amino acid content and the importance and location of susceptible amino acids. Proteins consisting of sulfhydryl groups and unsaturated amino acids are more susceptible to free radical damage and hence proteins containing cysteine, methionine, histidine, and phenylalanine are subject to amino acid modification. This modification in amino acids may then lead to disruption in secondary, tertiary, and quaternary structures of proteins. The requirement of a particular amino acid for the activity of an enzyme or location of an amino acid for the structural integrity determines the susceptibility of proteins to oxidative damage. Hydrogen peroxide and other peroxides oxidize glutathione through the catalytic action of glutathione peroxidase, thus depleting an important antioxidant that is essential for maintenance of the intracellular reducing pool:

$$2GSH + ROOH \rightarrow GSSG + ROH + H_2O$$

The brain enzyme glutamate synthetase, whose function is glutamate removal, is susceptible to radical damage. Decreased glutamate synthetase activity and increased appearance of protein degraded end products have been observed in brains of older humans.

Nitric oxide and its reaction product with superoxide, peroxynitrite ($ONOO^-$), are responsible for causing inactivation of a number of physiologically important proteins:

$$NO^{\cdot} + O_2^- \rightarrow ONOO^- + H^+ \rightarrow ONOOH$$

The most widely studied reaction of $NO^{\cdot}$ is its interaction with the iron of heme- and nonheme- containing proteins to form dinitrosyl iron complexes, thereby inactivating the proteins. Thus $NO^{\cdot}$ inactivates aconitase, an enzyme of the tricarboxylic acid cycle, and cytochromes P-450, a class of proteins involved in drug metabolism. $NO^{\cdot}$ also inactivates another heme-containing enzyme, catalase, which provides protection against oxidative damage by breakdown of H_2O_2. Inactivation of critical iron–sulfur (Fe–S)-containing enzymes of mitochondrial respiration takes place due to their reactivity with $NO^{\cdot}$. In contrast, the interaction of $NO^{\cdot}$ with the heme of guanylate cyclase leads to the stimulation of its enzyme activity. This results in elevation in cGMP levels and vasorelaxation. One of the widely studied reactions of $NO^{\cdot}$ is the formation of NO_3^- and methemoglobin as a result of interaction between $NO^{\cdot}$ and oxyhemoglobin:

$$\text{heme} \cdot Fe^{2+}{-}O_2 + NO^{\cdot} \rightarrow NO_3^- + \text{heme} \cdot Fe^{3+}$$

$$\text{heme} \cdot Fe^{2+}{-}NO + O_2 \rightarrow \text{heme} \cdot Fe^{3+} + NO_3^-$$

The formation of this nitrosylhemoglobin has been demonstrated during allogenic organ transplantation and vasodilator administration in humans.

The peroxynitrite is not a free radical but an anion formed as a result of interaction between two free radicals. The formation of peroxynitrite thus dramatically increases the toxicity associated with the presence of either O_2^- or NO alone. The protonated form of $ONOO^-$, peroxynitrous acid (ONOOH), decomposes to form NO_3^-. It has been proposed that an activated isomer of peroxynitrous acid, HOONO*, can be formed in the steady state during the decomposition of HOONO and this species has "hydroxyl radical-like reactivity." The toxic reactivity of peroxynitrite is varied as a potent oxidant as well as a nitrating and hydroxylating agent. It can directly oxidize low-molecular-weight-SH groups such as glutathione and -SH groups in proteins and thus deplete intra- and extracellular antioxidant levels. Reaction of $ONOO^-$ with metal ions has been shown to produce a powerful nitrating agent that resembles the nitronium ion (NO_2^+). Thus, the tendency of $ONOO^-$ to react with sulfhydryls and metals significantly contributes to the *in vivo* toxicity of $ONOO^-$ by inactivating vital enzymes in the energetic cellular metabolism. Peroxynitrite reacts with phenolic groups such as tyrosine residues in proteins to form nitrotyrosine. The determination of nitrotyrosine has been used as a means of detecting the formation of $ONOO^-$.

Tyrosine + ONOOH → 3-Nitrotyrosine

OH; $H_2N{-}CH{-}COOH$ with CH_2 — Tyrosine

OH, O_2N; $H_2N{-}CH{-}COOH$ with CH_2 — 3-Nitrotyrosine

Recently, the reaction of peroxynitrite with carbon dioxide has been brought to attention. This results in the formation of nitroperoxycarbonate anion ($O{=}N{-}OOCO_2^-$), which in turn rearranges to nitrocarbonate anion ($O_2N{-}CO_2^-$). The nitrocarbonate anion thus formed has been proposed to act as an oxidant and nitrating species in biological systems, thus generating peroxynitrite. These observations underline the importance of CO_2 in the cell culture studies of peroxynitrite. Peroxynitrite has been shown to inactivate a number of enzymes, including superoxide dismutase, $\propto_1$-antiproteinase, and prostacyclin synthase. Extensive nitration of tyrosine occurs around alveolar inflammatory cells in patients with respiratory distress syndrome, pneumonia, and sepsis.

IV. DEFENSES AGAINST FREE RADICAL TOXICITY

Free radical species are constantly produced *in vivo* either accidently or by design. Our body has devised various lines of defenses to guard against the deleterious effects of a plethora of radical species. These defenses (antioxidants) include thiols, enzymes, and transition metal chelators. These antioxidants act to lower the effective levels of intracellular oxidant species, which may otherwise cause cellular and tissue injury.

A. Low-Molecular-Weight Antioxidants

Glutathione (GSH) is the major low-molecular-weight thiol that serves the role as antioxidant. GSH is present at millimolar quantities in most mammalian cells. GSH in cytosol serves as a ready source of reducing equivalents and is released from cells during oxidative stress. It acts by reducing H_2O_2, the precursor of hydroxyl radical, to water through the catalytic action of glutathione peroxidase. In this process, glutathione gets oxidized to glutathione disulfide (GSSG). The disulfide is reduced to GSH at the expense of NADPH catalyzed by glutathione reductase:

$$2GSH + H_2O_2 \rightarrow GSSG + 2H_2O$$

$$GSSG + NADPH + H^+ \rightarrow 2GSH + NADP^+$$

The intracellular ratio of GSH/GSSG must be maintained at significantly higher levels to overcome the damaging effects of peroxides. GSH may also play a role in the protection process by reducing disulfide linkages, by serving as a storage form of cysteine, and by acting as a cofactor for many enzymes. GSH has been suggested to be a protective agent during cardiopulmonary bypass surgery.

α-Tocopherol (vitamin E) is the primary chain-breaking antioxidant, which acts by trapping peroxyl radical in the hydrophobic milieu. The vitamin E radical formed as a result is fairly stable and hence vitamin E is considered a good antioxidant. Other examples of low-molecular-weight antioxidants include dimethyl sulfoxide, dimethyl thiourea, ascorbic acid, lazaroids, β-carotene, and allopurinol. [*See* Vitamin E.]

B. Enzymatic Antioxidants

Among the enzymatic antioxidants, SODs have been most widely studied and used to treat the injury caused by oxygen radicals. SODs are present in every type of aerobic cell. Their function is to catalyze the dismutation of O_2^- to H_2O_2 and O_2:

$$2O_2^- + 2H^+ \rightarrow H_2O_2 + O_2$$

These enzymes have a transition metal in the active site, which is copper (Cu^{2+}, Cu^+), nonheme iron (Fe^{3+}, Fe^{2+}), or manganese (Mn^{3+}, Mn^{2+}). Eukaryotic cells contain two isozymes of SOD, of which Cu/Zn-SOD is localized mainly in the cytosol and Mn-SOD is synthesized in mitochondria. SOD is widely used to prevent the injury to tissues during ischemia-reperfusion and is also used to treat injury associated with inflammation. It should also be noted that SOD can act as a prooxidant at higher concentrations. Low-molecular-weight SOD mimics have been synthesized to overcome the problem associated with SOD permeability through cell membrane and its high cost of production.

Catalase is another radical-scavenging enzyme that acts in concert with SOD to combat the lethal effects of O_2^-. The H_2O_2 produced by SOD becomes a substrate for catalase and gets degraded to H_2O:

$$2H_2O_2 \rightarrow 2H_2O + O_2$$

Thus, catalase plays an important role in detoxification, since SOD alone might elevate the levels of H_2O_2 and generate hydroxyl radicals through a Haber–Weiss reaction. Many animal studies have been carried out with the combined administration of catalase and SOD to treat ischemia-reperfusion injury. Catalase is present in all kinds of mammalian cells and is localized in the peroxisomes and microperoxisomes.

Glutathione peroxidase reduces not only H_2O_2 but also other hydroperoxides by using GSH as a reducing agent and forming glutathione disulfide. The disulfide is then reduced back to GSH through glutathione reductase at the expense of NADPH, thus replenishing the levels of GSH lost during the degradation of peroxides:

$$ROOH + GSH \rightarrow GSSG + ROH + H_2O$$

$$GSSG + NADPH + H^+ \rightarrow 2GSH + NADP^+$$

There are two isoforms of glutathione peroxidase, one of which requires selenium as a cofactor. The selenium-dependent isoform reduces both lipid hydroperoxides and H_2O_2, whereas the selenium-independent isoform reduces only H_2O_2.

C. Transition Metal Chelators

Transition metal chelators do not directly interact with free radicals, instead they provide protection by chelating the metal ions that catalyze the formation of free radicals. Iron is the major transition metal that participates in the free radical formation. Deferoxamine is the most commonly used iron chelator to scavenge the oxygen radicals. Other iron chelators like ethylenediaminetetraacetic acid (EDTA) and nitrilotriacetic acid (NTA) potentiate the toxicity. Deferoxamine is a specific chelator of Fe^{3+} and can be injected into humans and experimental animals. It is a bacterial siderophore isolated from the species *Streptomyces pilosus*. Deferoxamine administration has been widely used in treatment of diseases associated with iron overload like thalassemia and to limit the injury during ischemia-reperfusion. The sensitivity of isolated rat hepatocytes to H_2O_2 killing has been reduced by pretreating the cells with deferoxamine. In addition to its role as a metal chelator, deferoxamine has been recently shown to scavenge $ONOO^-$. Hence its protective action during oxidative injury does not necessarily imply that it is acting by chelating iron.

V. DISEASES ASSOCIATED WITH FREE RADICAL INJURY

Although increased accumulation of radical species has been observed in a number of diseases (Table I), the exact role played by these species is not yet clearly understood.

A. Rheumatoid Arthritis

In rheumatoid arthritis, thickening of synovial lining takes place, resulting in folds and infiltration by blood vessels and inflammatory cells. The synovial fluid from the knee joints of rheumatoid arthritis patients has been found to contain higher levels of lipid peroxidation products. The hyaluronic acid synthesized by rheumatoid synovium is responsible for the viscosity of synovial fluid and is susceptible to breakdown by interaction with free radicals. It has been demonstrated that synovial fluid from rheumatoid arthritis patients contains degraded hyaluronic acid, implicating that oxygen species are responsible for this disease. How are oxygen species generated in rheumatoid arthritis? Radical species may be produced by activated macrophages in the inflamed synovial membrane and by activated neutrophils in the synovial cavity.

TABLE I
Some Diseases Associated with Free Radicals

Reperfusion injury
Adult respiratory distress syndrome
Rheumatoid arthritis
Atherosclerosis
Neurodegenerative diseases
Inflammatory bowel disease
Cancer
Sepsis
Myocardial infarction
Aging

B. Adult Respiratory Distress Syndrome

Adult respiratory distress syndrome (ARDS) is a form of acute lung injury, causes of which are varied, such as hypoxia, sepsis, and polytrauma. The mortality rate of ARDS is more than 50%. This syndrome has been shown to be mediated by inflammation as a result of the presence of neutrophils, blood platelets, and other inflammatory mediators in the biopsy specimens of ARDS patients. Increased levels of myeloperoxidase and oxidized α_1-antiproteinase in bronchoalveolar lavage fluids and exhalation of excess H_2O_2 in the breath of ARDS patients are indicative of involvement of radical species. No treatment is currently available that is directly aimed at the pathogenic factors of this disease. All attempts at present are directed toward developing strategies to prevent early activator processes, such as injecting antibodies directed against bacterial lipopolysaccharide and tumor necrosis factor.

C. Sepsis

Septic shock is manifested by a dramatic drop in blood pressure as a result of severe infection with gram-negative bacteria. This is a deadly disease that results in failure of various organs such as liver, kidney, and heart, ultimately leading to death. According to one estimation, about 200,000 patients are prone to septic shock annually in the United States alone. It has been demonstrated in animals that endotoxin and cytokines such as tumor necrosis factor and interferon reproduce many of the cardiovascular features of sepsis. The severe hypotension is caused by the excessive production of $NO^\cdot$ induced by the bacterial endotoxin. The bacterial endotoxin or lipopolysaccharide triggers the release of various cytokines, which in turn up-regulate the inducible form of $NO^\cdot$ synthase. These cytokines are also responsible for liver failure and cirrhosis. The endotoxin-mediated hypotension is associated with increased $NO^\cdot$ synthesis by vascular endothelial cells. Elevated levels of circulating NO_2^- and NO_3^-, the stable products of $NO^\cdot$ oxidation, have been documented in septic shock patients.

L-NMMA (N-monomethyl-L-arginine) an inhibitor of NO˙ synthase, is shown to prevent the endotoxin shock in animal models and has been used successfully to restore normal arterial tension in septic shock patients. Further investigations are directed at the usefulness of NO˙ synthase inhibitors in treating sepsis because of the serious side effects associated with them.

VI. SUMMARY

Free radicals are produced in normal human metabolism. When produced in excess, they prove to be deleterious to the system. Free radicals, once produced, may react with other free radicals to cause more damage. Hydroxyl radical and peroxynitrite are the two reactive oxygen species capable of causing severe damage. Important biomolecules are susceptible to free radical damage, which leads to the development of various disease states. The human body has devised effective defense mechanisms to combat the lethality of free radicals. In conditions of severe radical production, the protective mechanisms may not be adequate to control the injury.

BIBLIOGRAPHY

Grisham, M. B. (1994). "Reactive Metabolites of Oxygen and Nitrogen in Biology and Medicine." R. G. Landes Company, Austin, Texas.

Halliwell, B., Gutteridge, J. M. C., and Cross, C. E. (1992). Free radicals, antioxidants, and human disease: Where are we now? *J. Lab. Clin. Med.* **119**, 598.

Ignarro, L., and Murad, F. (eds.) (1995). "Nitric Oxide: Biochemistry, Molecular Biology, and Therapeutic Implications," *Advances in Pharmacology* **34**. Academic Press, San Diego.

Janssen, Y. M. W., Houten, B. V., Borm, P. J. A., and Mossman, B. T. (1993). Biology of disease: Cell and tissue responses to oxidative damage. *Lab. Invest.* **69**, 261.

Kehrer, J. P. (1993). Free radicals as mediators of tissue injury and disease. *Crit. Rev. Toxicol.* **23**, 21.

Kerrigan, S. L., and Stotland, M. A. (1993). Ischemia reperfusion injury: A review. *Microsurgery* **14**, 165.

Kerwin, J. F., Lancaster, J. R., Jr., and Feldman, P. L. (1995). Nitric oxide: A new paradigm for second messengers. *J. Med. Chem.* **38**, 4343.

Lancaster, J. R., Jr. (1992). Nitric oxide in cells. *Am. Sci.* **80**, 248.

McCord, J. M. (1993). Human disease, free radicals, and the oxidant/antioxidant balance. *Clin. Biochem.* **26**, 351.

Packer, L. (ed.) (1996). Nitric oxide: Physiological and pathological processes. *Meth. Enzymol.* (*Part B*) **269**.

Pryor, W. A., and Squadrito, G. L. (1995). The chemistry of peroxynitrite: A product from the reaction of nitric oxide with superoxide. *Am. J. Physiol.* **268**, L699–L722.

Fusimotor System

DAVID BURKE
Prince of Wales Medical Research Institute and University of New South Wales

GLOSSARY

Alpha Skeletomotor

Beta Skeletofusimotor

Extrafusal fibers Force- and movement-generating muscle fibers that compose the bulk of the muscle, excluding the spindle

Fusimotor Motor to intrafusal muscle fibers in the muscle spindle

Gamma Fusimotor

Intrafusal fibers Muscle fibers in the muscle spindle

Skeletofusimotor Motor to both intrafusal and extrafusal muscle fibers

Skeletomotor Motor to extrafusal muscle fibers

Spinal monosynaptic pathway Simplest component of the stretch reflex, in which sensory fibers from muscle spindles directly excite motoneurons in the spinal cord

Stretch reflex Involuntary contraction of muscle, produced by activation of stretch-sensitive receptors

MUSCLE IS MORE THAN JUST A FORCE-GENERATING machine: it contains many receptors, some of which are encapsulated (e.g., the muscle spindle and tendon organ) but most of which are free nerve endings, responsive to injury and to metabolic, thermal, and mechanical stimuli within the muscle. The muscle spindle is a unique mechanoreceptor. It is innervated by motor nerve fibers through which the central nervous system can alter the response of spindle endings and thereby modify the messages it receives from the spindle endings. Muscle spindles are elongated sensory endorgans, lying in parallel with the contractile elements of muscle (Fig. 1). They are mechanoreceptors, primarily responsive to changes in length, whether produced by stretch or shortening of the whole muscle or by contraction of nearby muscle fibers. The Golgi tendon organ is the second major encapsulated mechanoreceptor in muscle, located at musculotendinous and musculofascial junctions (Fig. 1). Muscle fibers insert into the capsule of this receptor, which is therefore "in-series" with muscle fibers and responsive to their contraction.

The information from spindle endings is involved in a variety of reflex actions in the spinal cord and at higher levels in the central nervous system, apart from the classic monosynaptic pathway to spinal cord motoneurons. In addition, muscle spindle activity constitutes one of the major sensory cues for kinesthesia. This article addresses the anatomy of the motor innervation of the muscle spindle and current concepts about how it functions in human subjects. [*See* Proprioceptors and Proprioception.]

I. ANATOMICAL CONSIDERATIONS

A. General

The muscle spindle has a fusiform (spindle-like) appearance and contains a number of modified muscle fibers (*intrafusal* muscle fibers), which produce negligible force at the muscle tendon when they contract. The motor innervation of the spindle is directed to these intrafusal muscle fibers and is termed *fusimotor*. Because the contractile bulk of the muscle lies outside the spindle, the term *extra-fusal* muscle is used when referring to the force-generating muscle fibers that constitute the bulk of muscle (Fig. 1). These muscle

ENCYCLOPEDIA OF HUMAN BIOLOGY, Second Edition, VOLUME 4.

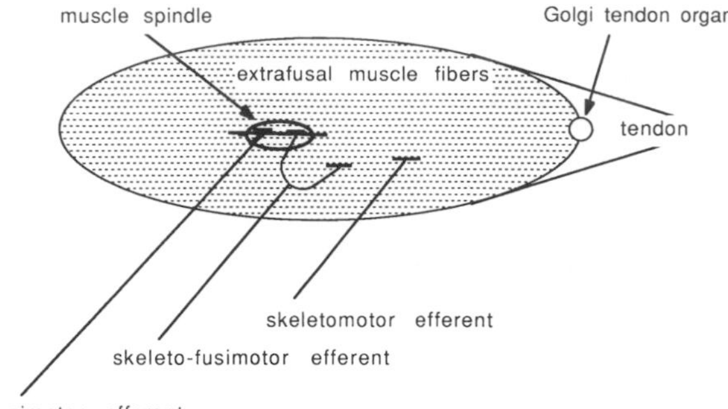

FIGURE 1 Simplified diagram of a muscle containing a single muscle spindle lying in parallel with "extrafusal" muscle fibers but considerably shorter than the muscle and a single Golgi tendon organ at the musculotendinous junction, in series with the extrafusal muscle. Motor innervation is by alpha motor axons (exclusively "skeletomotor"), gamma motor axons (exclusively "fusimotor"), or beta motor axons (directed to extra- and intra-fusal muscle, hence "skeletofusimotor"). Afferent innervation is detailed in Fig. 2.

fibers are innervated by *skeletomotor* axons. Motor axons directed only to intrafusal muscle are of small diameter and relatively slow conduction velocity and are often referred to as *gamma* efferent axons from "gamma" motoneurons. Similarly, skeletomotor axons are often called *alpha* and arise from "alpha" motoneurons. They are of larger diameter and have conduction velocities of some 30–60 m/sec. Some motor axons innervating extrafusal muscle send branches to nearby muscle spindles, thereby have a mixed function, and are termed *skeletofusimotor* or *beta*. [*See* Muscle Dynamics; Muscle, Physiology and Biochemistry.]

B. Muscle Spindles

Except for the facial and jaw-opening muscles, all skeletal muscles contain a number of muscle spindles distributed throughout the bulk of the muscle, the absolute numbers varying from fewer than 50 for the intrinsic muscles of the hand to more than 1000 for the thigh muscles. However, given the small size of the individual intrinsic muscles of the hand, the density (number of spindles per gram of muscle) is significantly higher for these muscles than for the much larger thigh muscles. It is generally believed that the greater density of spindles in the more distal muscles that operate the hand contributes to the precision with which the hand can be used, but attempts to demonstrate that proprioceptive sensations are more acute distally in the human upper limb have failed to do so. Presumably much of the precision with which the hand can be used resides not so much on the sensory side but on the motor side: in the greater access of corticospinal pathways transmitting the volitional drive from motor cortex to motoneurons innervating the more distal muscles. [*See* Motor Control.]

Muscle spindles are commonly about 10 mm in length, although their size varies with the muscle in which they are located. Human spindles are slightly larger in absolute terms than those of other animals (e.g., the cat), but they are relatively small when compared with the size of human muscles and, in particular, with the length of the extrafusal muscle fibers, with which they lie in parallel. As a result, human spindles commonly attach directly or indirectly to the adjacent extrafusal muscle fibers and are likely to be disturbed by the contraction of these muscle fibers. Hence, muscle spindles can be stretched or shortened not only by changes in overall muscle length, such as occur when a joint is rotated, but also by disturbances created when nearby extrafusal muscle fibers contract, even if that contraction does not produce joint movement and a change in overall muscle length.

Human spindles contain 2–14 intrafusal muscle fibers, which can be divided into three types on anatomical and physiological grounds. The more plentiful type is the "nuclear chain fiber" (commonly 3–10 per spindle), in which 20–50 nuclei lie side by side in a chain down the center of the fiber (Fig. 2). There are two types of intrafusal muscle fiber in which the nuclei are congregated at the equator to form a swelling of the fiber, with some 5–6 nuclei abreast. These "nu-

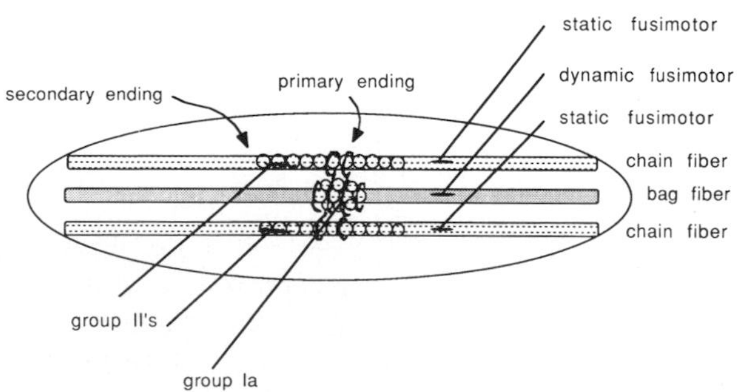

FIGURE 2 Simplified diagram of a muscle spindle containing a single nuclear bag (bag_1) and two nuclear chain fibers, with one primary ending (giving rise to a group Ia afferent) and two secondary endings (each giving rise to a group II afferent). Fusimotor innervation is directed to the bag_1 fiber (dynamic) and to the chain fibers (static).

clear bag fibers" can be differentiated anatomically into "bag_1" and "bag_2," the latter being thicker and longer and having a lot of surrounding elastic tissue. Each spindle usually has one or more bag_1 fibers and one bag_2 fiber. For simplicity, the diagrammatic spindle in Fig. 2 contains only one nuclear bag fiber (a bag_1 fiber) innervated by a dynamic fusimotor axon. However, some spindles lack bag_1 fibers completely. The response to passive stretch of such spindles does not differ appreciably from those with bag_1 fibers; that is, spindles that lack bag_1 fibers do not lack a dynamic response to passive stretch. This evidence supports the view that the dynamic response to stretch is generated by the bag_2 fiber not, as was previously thought, the bag_1 fiber. In the passive spindle not under fusimotor drive, the resting discharge and the response to stretch arise from terminals on the bag_2 and chain fibers; those on the bag_1 fiber contribute little. However, as discussed later, this is not the case when the spindle is subjected to fusimotor activity: dynamic fusimotor axons innervate the bag_1 fiber almost exclusively and enhance the dynamic response to stretch through actions on this intrafusal fiber.

There are two types of sensory ending on the intrafusal fibers (Fig. 2). Each spindle has one primary ending, which commonly produces spirals around the equator of the intrafusal fibers. The primary ending gives rise to a single large myelinated afferent axon, the group Ia afferent, which in human subjects conducts at up to 60–70 m/sec in the upper limb and up to 50–60 m/sec in the lower limb. These velocities are much slower than the equivalent values in the cat (up to 120 m/sec). Given the greater length of nerve pathways in humans than in cats, sensory feedback from muscle will take much longer to reach the central nervous system in human subjects. It is likely then that the way in which the nervous system uses feedback from muscle spindles differs in different animal species. Each spindle may have a number of secondary endings, lying predominantly on the nuclear chain fiber, on one or both sides of the equator. There are one to five secondary endings per spindle, each giving rise to a small myelinated afferent axon (group II afferent). The conduction velocity for human group II afferents has not been determined, but based on their size (and by analogy with other animals) it is likely to be significantly slower than the group Ia velocity. Because of the intrafusal muscle fibers on which their respective sensory endings lie, group Ia afferents transmit a signal proportional to the velocity *and* the extent of stretch, whereas group II afferents encode predominantly the extent of stretch.

C. Fusimotor Innervation

The poles of intrafusal muscle fibers are innervated by a number of small myelinated (gamma) efferent axons, which come from gamma motoneurons in the spinal cord. In addition, some spindles (perhaps 20–30%) receive branches of large myelinated efferent axons, which are directed to extrafusal muscles, and thereby innervate both extrafusal and intrafusal muscle fibers (Fig. 1). This skeletofusimotor (or beta) innervation is the sole motor innervation of spindles in amphibia but was thought to be uncommon in mammals and an evolutionary relic in humans. However, anatomical studies of primate spindles have shown it to be a significant form of spindle innervation, and physiological studies in the cat indicate that it may be an important modulator of spindle discharge in that species. Of necessity, beta innervation implies that any effect on the spindle is compulsorily tied to a muscle contraction. The evolutionary "advantage" of the gamma system is that it confers the potential for control of spindle output independent of the contractile state of the muscle.

Contraction of intrafusal muscle fibers caused by the activity of gamma or beta efferent axons can change both the resting deformation of the sensory terminals on those fibers and the additional deformation produced when the spindle is stretched (Fig. 2). Efferents innervating the bag_1 intrafusal fiber, be they gamma or beta, increase the sensitivity of the primary ending to the velocity of stretch and are therefore known as "dynamic" fusimotor (or skeletofusimotor) efferents. Efferents innervating the bag_2 and chain fibers have a number of actions, including reducing the velocity sensitivity of the primary ending while increasing its overall discharge. They also increase the sensitivity of secondary endings to the extent of stretch. These efferents are known as "static" fusimotor (or skeletofusimotor) efferents. In the cat, it has been shown that stimulation within the central nervous system can affect dynamic fusimotor efferents independently from the static, and even the static efferents directed to chain fibers separately from those directed to bag_2 fibers. Comparable studies have not been performed in awake behaving animals, and methodological (and ethical) constraints have prevented such studies in human subjects. The extent to which the human nervous system ever accesses these subdivisions of the fusimotor system selectively and the conditions under which it normally does so have been the subject of speculation. Suffice it to say that the anatomical substrates for independent control exist in the human spindle.

In summary, there are three ways in which the discharge from the muscle spindle can be altered: by a change in length of the muscle; by the mechanical deformation produced by the contraction of nearby extrafusal muscle fibers; and by the activity of fusimotor efferents.

II. THEORIES OF FUSIMOTOR FUNCTION

With the gamma efferent system, attention has focused on its potential to adjust the discharge of spindles independently of the degree of contraction, with theories based on situations in which selective activation of gamma efferents might be functionally advantageous. Although there has been a lot of theorizing, the experimental evidence has been mainly negative. To date, there is only one circumstance under which activation of spindle endings has been demonstrated to occur in the absence of muscle stretch or contraction: reflex activation of fusimotor neurons by cutaneous afferent volleys. Stimulation of cutaneous afferents from the foot can reflexly alter the discharge of spindle endings in muscles that act on the ankle in human subjects engaged in the task of standing without support. In addition, cutaneous afferents from the hand can alter the activity of spindles in forearm muscles, but this may occur in subjects who are not engaged in any specific task. Whether this difference indicates that cutaneous afferents more readily access fusimotor neurons innervating upper rather than lower limb muscles remains to be established, but it would be consistent with a greater role of cutaneous afferent feedback in the control of the hand than in the control of the foot.

The ability of cutaneous afferent volleys to access fusimotor neurons selectively contrasts with the negative findings that have come from studies of the preparation for and initiation of voluntary movement (see Sections II,A and II,B) and studies of voluntary maneuvers that enhance proprioceptive reflexes (see Section II,D). These latter studies have a common feature: volition. It is possible that the descending pathways activated by volition or intent always activate skeletomotor and fusimotor neurons together. This would not be expected from animal data, because stimulation of different regions of the brain can activate fusimotor neurons selectively if the stimulus parameters are carefully chosen. However, this may not be the way that such pathways are activated in natural movements. Whatever the explanation, at the moment, human fusimotor neurons have been activated selectively by cutaneous afferent volleys through a spinal reflex pathway, but not by various forms of voluntary activity.

One purpose of the beta efferent system could be to adjust the sensitivity of spindle endings in a contracting muscle in relation to the contraction, perhaps to compensate for the unloading of the spindle when the contracting muscle shortens. However, those spindles not receiving beta innervation would still be unloaded unless their gamma innnervation was simultaneously activated. There is some experimental evidence for this possibility, as discussed in Section II,D.

A. Preparation for Movement

Changes in brain potentials precede any voluntary movement by hundreds of milliseconds, and reflex pathways are sensitized in advance of and in preparation for the movement. It was thought that this sensitization involved adjusting the responsiveness of muscle spindle endings so that their heightened discharge would tone up reflex pathways. In addition, sensitized receptors would be better able to detect any disturbance that might compromise the intended movement. In reality, however, muscle spindle endings increase their discharge only after the muscle has started to contract. If the subject involuntarily tenses up in anticipation of the need to act, spindle endings will be activated in the tense muscles, but there is no evidence that motor preparation, even when given a warning, results in selective activation of the gamma efferent system. Attempt to train subjects to activate spindle endings using biofeedback techniques have been unsuccessful.

B. Initiation of Movement

The control of movement is often likened to a servo system as used by engineers. An influential theory in motor control postulated that some movements, particularly slow ones, were initiated indirectly using the gamma system to activate spindle endings. The resultant discharge in their afferents would reflexly activate the alpha motoneurons of the desired muscle (using a pathway in the spinal cord). The final nail in the coffin of this theory was the demonstration that when a subject voluntarily contracted a muscle, the increase in spindle discharge occurred after rather than before the onset of the contraction. Perhaps the most popular current theory of the role of the fusimotor/muscle spindle system is that through its reflex pathways, the muscle spindle activity modulates contraction rather than initiating it. This possibility is considered further in the next section. [*See* Movement.]

C. Maintenance of Spindle Discharge during Movement

If the purpose of the increase in fusimotor drive that accompanies a voluntary contraction is to maintain the level of spindle feedback and thereby reflex assistance to the movement, one would expect the discharge of spindle endings to change little during voluntary movements that produce shortening of the active muscles. However, they do this only under specific circumstances. In free and relatively fast shortening movements, spindle endings fall silent. Their responsiveness to perturbations delivered during such movements is insufficient to produce much reflex assistance to the movement. It appears that free fast movements are performed "open-loop," without reflex assistance from muscle spindles. Indeed, given the delays that must occur in human subjects because of their longer pathways and slower conduction velocities (see Section I,B), powerful reflex compensation for a transient disturbance could occur too late to be of benefit and might instead disrupt subsequent stages of the movement. This is likely to be the situation in patients suffering from spasticity, with gross exaggeration of spinal reflex excitability.

If contractions are "isometric" (i.e., there is no external change in muscle length) or if movements proceed slowly against a load, the fusimotor activation that accompanies the contraction is capable of maintaining and even increasing the discharge of spindle endings and thereby of modulating the contraction. In such contractions, the reflex assistance improves the subject's ability to control the output of the active muscle. These are the types of contractions used when performing a delicate motor skill or when learning a totally new motor task. Movement is then slow and cautious, with the joint braced by simultaneous contraction of a muscle and its antagonist, and movement is allowed to proceed by adjusting this co-contraction. Under such circumstances the fusimotor-induced spindle discharge could make a significant contribution to the control of movement.

D. Regulation of Intensity of Stretch Reflexes

The afferent input from muscle spindles can excite the motoneurons of their own muscle through a variety of pathways, the best known of which is the spinal monosynaptic pathway. The intensity of stretch reflexes can therefore be modified at three levels: by altering the muscle spindle input to the spinal cord; by altering the transmission of the afferent information to the motoneurons in the spinal cord; and by altering the excitability of the motoneurons. It cannot be denied that if other factors remain constant, a bigger afferent input into the spinal cord will evoke a larger reflex response, but it does not appear that a fusimotor mechanism is the major mechanism for altering the intensity of stretch reflexes. Powerful mechanisms exist within spinal circuitry for altering the excitability of all reflex pathways, even the simplest, the monosynaptic reflex. The fusimotor system does not appear to be called into action as a mechanism of altering reflex strength. For example, it is easy to demonstrate that tendon jerks in noncontracting muscles can be accentuated by the deliberate contraction of remote muscles. Potentiation of the knee and ankle jerks by fist clenching is commonly used by clinicians to determine whether a depressed reflex is absent or not. The process of potentiation (known as "reflex reinforcement") by the performance of this maneuver (known as the Jendrassik maneuver) was thought to be due to activation of the gamma efferent system, such that the spindle response to tendon percussion was enhanced. This view has proved erroneous: the reflex enhancement occurs centrally within spinal cord circuits, not peripherally at the spindle level.

The conclusion that the fusimotor system is not the main mechanism for controlling the strength of proprioceptive reflexes does not contradict the view that muscle spindle feedback provides support to active skeletomotor neurons through proprioceptive reflexes (see Section II,C). Indeed, when the nerve to a muscle is blocked completely (by local anesthetic) so that feedback from the muscle cannot reach the motoneuron pool, subjects attempting to contract the paralyzed muscle lose the ability to drive motoneurons as intensely as normal. The discharge rates of active motoneurons are then roughly 30% less than normal, and this occurs across the range of contraction strengths. Clearly, the motoneuron pool is deprived of a source of supportive excitation; at least some of this support comes from spindle endings. [*See* Proprioceptive Reflexes.]

The view that the fusimotor system was not designed primarily to control reflex strength has implications for patients with neurological diseases that produce changes in the strength of reflexes. Based on the finding that excessive fusimotor drive occurred in, and was necessary for the development of, decerebrate rigidity in the cat after transection of the midbrain, the reflex abnormalities that occur in patients with spasticity and rigidity have been attributed to excessive fusimotor drive. Selective overactivity of dynamic fusimotor neurons was postulated to account for the exaggeration of tendon jerks and the phasic, velocity-dependent nature of muscle tone in spastic patients. Selective overactivity of static fusimotor neurons was postulated to account for the rigidity that occurs in Parkinson's disease. It is now known, however, that there are abnormalities of reflex transmission within the central nervous system in both disorders and that an excessive muscle spindle input is insufficient by itself to produce either condition. Whether overactivity of the fusimotor system ever occurs in these disorders is not known, but at most it could only be a minor factor in the reflex disturbance.

III. FUSIMOTOR SYSTEM AS A MOTOR SYSTEM

When normal human subjects relax a muscle completely, the level of fusimotor drive directed to that muscle is low, insufficient to affect spindle discharge. This finding is perhaps not unexpected if the purpose of the fusimotor system is the control of movement. To be able to mobilize sensory feedback from muscle would be of greatest value during a motor act but would serve little purpose when the motor system was quiescent. To explain why a totally independent control of spindle discharge has evolved remains an elusive problem: perhaps it lies in a variety of complex motor tasks not yet studied experimentally (e.g., in tasks involving inactive or relatively inactive synergists or antagonists to the active muscle group). In this respect, it may be relevant that cutaneous sensory inputs from the sole of the foot can preferentially affect gamma fusimotor neurons directed to the tibialis anterior of standing human subjects. During normal quiet stance, balance is maintained by continuous activity in the calf muscle group, triceps surae, but there is little activity, if any, in the antagonist, tibialis anterior, unless the subject inadvertently sways backward. Under these circumstances, cutaneous inputs from the sole of the foot appear capable of adjusting the spindle feedback from the inactive tibialis anterior. Feedback from peripheral sources such as this is integrated with visual and vestibular inputs to produce the coordinated body movements that maintain balance.

BIBLIOGRAPHY

Aniss, A. M., Diener, H.-C., Hore, J., Burke, D., and Gandevia, S. C. (1990). Reflex activation of muscle spindles in human pretibial muscles during standing. *J. Neurophysiol.* **64**, 671–679.

Boyd, I. A., and Gladden, M. H. (eds.) (1985). "The Muscle Spindle." Stockton, New York.

Burke, D. (1981). The activity of human muscle spindle endings during normal motor behavior. *In* "International Review of Physiology. Neurophysiology IV" (R. Porter, ed.), pp. 91–126. University Park Press, Baltimore.

Burke, D. (1983). Critical examination of the case for or against fusimotor involvement in disorders of muscle tone. *In* "Motor Control Mechanisms in Health and Disease, Advances in Neurology" (J. E. Desmedt, ed.), Vol. 39, pp. 133–150. Raven, New York.

Burke, D., and Gandevia, S. C. (1989). The peripheral motor system. *In* "The Human Nervous System" (G. Paxinos, ed.), pp. 125–145. Academic Press, New York.

Hulliger, M. (1984). The mammalian muscle spindle and its central control. *Rev. Physiol. Biochem. Pharmacol.* **101**, 1–110.

Macefield, V. A. G., Gandevia, S. C., Bigland-Ritchie, B., Gorman, R. B., and Burke, D. (1993). The firing rates of human motoneurones voluntarily activated in the absence of muscle afferent feedback. *J. Physiol (Lond.)* **471**, 429–443.

Matthews, P. B. C. (1981). Evolving views on the internal operation and functional role of the muscle spindle. *J. Physiol. (Lond.)* **320**, 1–30.

Vallbo, Å. B., Hagbarth, K.-E., Torebjörk, H. E., and Wallin, B. G. (1979). Somatosensory, proprioceptive, and sympathetic activity in human peripheral nerves. *Physiol. Rev.* **59**, 919–957.

Gangliosides

A. SUZUKI
T. YAMAKAWA
Tokyo Metropolitan Institute of Medical Science

GLOSSARY

Ceramide Hydrophobic portion of glycosphingolipids, consisting of sphingosine (a long-chain base; D-*erythro*-1,3-dihydroxy-2-amino-4,5-*trans*-octadecene and its analogs) and a fatty acid, which is attached to the sphingosine through an acid–amide linkage

Glycolipids Compounds consisting of a carbohydrate chain and a lipid. Glycosphingolipids and glycoglycerolipids differ in the presence of sphingosine and glycerol, respectively. Major glycolipids in mammalian cells are usually glycosphinogolipids, and the term "glycolipids" is frequently used instead of "glycosphingolipids"

Hydrophobic interaction Interaction contributing to the aggregation of hydrophobic molecules or nonpolar portions of molecules in water in order to minimize their disruptive effect on the hydrogen-bonded network of water molecules. The aggregates are called micelles or liposomes

Sialic acids Family of derivatives of neuraminic acid (5-amino-3,5-dideoxy-D-glycero-D-galactononulosonic acid), which is a nine-carbon sugar compound. Glycoconjugates containing sialic acid(s) are called sialoglycoproteins in the case of proteins and gangliosides in the case of glycolipids

GANGLIOSIDES ARE A FAMILY OF GLYCOLIPIDS composed of a ceramide and a carbohydrate chain containing sialic acid(s). Ceramide is the hydrophobic part, which anchors the carbohydrate chain to membranes. More than 100 gangliosides have been reported, which differ mainly in the carbohydrate structure. Because of these features, gangliosides can play roles in the recognition between cells or between cells and macromolecules. Recently, gangliosides have attracted attention as physiologically active substances, modulators for membrane receptors, receptors for toxins and viruses, and cancer-associated antigens.

I. STRUCTURAL CHARACTERISTICS

A. Discovery

The first indications of gangliosides were observed by K. Landsteiner and P. A. Levene of the Rockefeller Institute for Medical Research (New York) in 1925–1927 and by E. Walz of the University of Tübingen (Germany) in 1927. They reported lipid fractions that showed a chemical reaction involving color development for the detection of sialic acid, but they did not realize that this is due to a family of biologically important lipid compounds. In 1935–1938, E. Klenk of the University of Cologne (Germany) published a series of papers on the glycolipid that accumulated in the brains of patients with Tay–Sachs disease, and his work is regarded as being definitive as to the description and characterization of gangliosides.

Prior to elucidation of the chemical structure of gangliosides, the structure of sialic acid, which is a component of gangliosides, was debated by E. Klenk and G. Blix of the University of Uppsala (Sweden). Klenk isolated a new nitrogen-containing organic acid

ENCYCLOPEDIA OF HUMAN BIOLOGY, Second Edition, VOLUME 4.

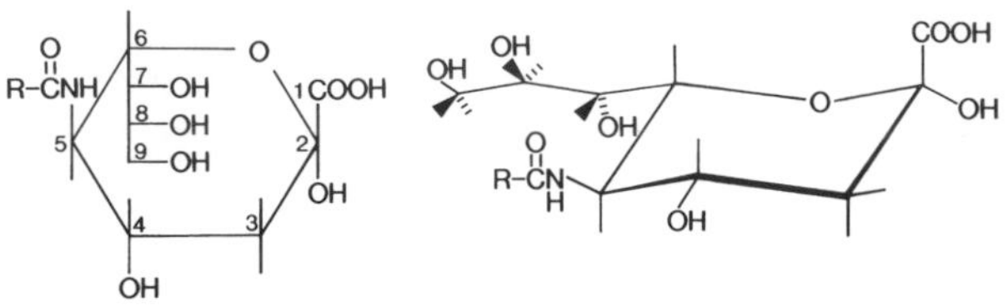

FIGURE 1 Structures of sialic acids. R : CH_3, *N*-acetyl-neuraminic acid; CH_2OH, *N*-glycolylneuraminic acid.

in a crystalline state from a ganglioside fraction obtained from bovine brain and named it "Neuraminsäure" in 1941, while Blix isolated the same acid from bovine submaxillary mucin, which is a glycoprotein, and proposed the name "sialic acid" in 1952. They are the same substance and the correct structure of neuraminic acid or sialic acid was finally proposed by A. Gottschalk of The Walter and Eliza Hall Institute (Melbourne, Australia) in 1955. S. Roseman established the skeleton of sialic acid in 1958, which is composed of *N*-acetylmannosamine and pyruvate. Thus, the structure of sialic acid is that shown in Fig. 1. The first report on the occurrence of a ganglioside in an extraneural tissue and in the cell membrane was made by T. Yamakawa in 1951, with the isolation of a sialyllactosylceramide or G_{M3} from horse erythrocytes, which he named "hematoside."

A variety of structural modifications of neuraminic acid have been reported so far, two major modifications being acylation of the amino group and acetylation of hydroxyl groups. Concerning the modification of the amino group, the gangliosides in normal human tissues contain only *N*-acetylneuraminic acid (NeuAc) and ones containing *N*-glycolylneuraminic acid (NeuGc) are found only in cancer tissues. Interestingly, mammalian species other than humans contain NeuGc in normal tissues. Human tissues also contain some gangliosides, which are acetylated at the C-9 position of *N*-acetylneuraminic acid. We use "sialic acid" as the generic term for neuraminic acids with various modifications. Sialic acids of gangliosides include NeuAc, NeuGc, 4-*O*-acetyl-NeuGc, and 9-*O*-acetyl-NeuAc.

The structural elucidation of gangliosides was attempted by several laboratories, and finally the correct structures of a series of brain gangliosides were proposed in 1963 by R. Kuhn and H. Wiegandt of the Max-Planck Institute (Heidelberg, Germany). Since then, the isolation and structural characterization of a variety of gangliosides have been reported. Table I lists gangliosides, which are relatively abundant in human tissues, with notes on their biological functions. Figure 2 shows the structure of one ganglioside, G_{M1}.

A nomenclature widely used for gangliosides is based on the system proposed by L. Svennerholm, and another, used here, follows recommendations by the Nomenclature Committee of the International Union of Biochemistry. The latter nomenclature is based on that of neutral glycosphingolipids, which always comprise the backbone structures of gangliosides.

The structures of neutral glycosphingolipids are divided into five major types on the basis of their core oligosaccharide structures. The five groups are named gala-, ganglio-, lacto-, neolacto-, and globo-series glycolipids, according to the trivial names for their oligosaccharide structures. The core structures and their names are shown in Fig. 3. The gangliosides are therefore named sialo- or sialyl-R-osylceramide, where R is derived from the names of the neutral oligosaccharides (e.g., sialyllactosylceramide, sialylgangliotriaosylceramide, sialylneolactotetraosylceramide). The position of the monosaccharide and the carbon atom of the monosaccharide to which sialic acid is attached should be determined because there are positional isomers. Thus, the position and the carbon number are indicated by a Roman numeral and a superscript Arabic numeral, respectively. The configuration of the ketosidic bond of sialic acids in gangliosides is also quite important and it was demonstrated to be α by R. Ledeen and R. K. Yu. Gangliosides reported so far contain exclusively sialic acids with α configuration. Table I lists the structures and names of gangliosides according to both nomenclatures.

B. Characteristics as Membrane Components

Gangliosides are glycosphingolipids that contain one mole or more of sialic acids in their molecules and thus they share characteristics with glycosphingolipids, as amphipathic molecules, and with sialic acids, as acidic sugars. Glycosphingolipids contain both hydrophobic (or lipophilic) and hydrophilic structures. The hydrophobic structure is named ceramide, which is composed of sphingosine (a long-chain base) and a fatty acid. The name "sphingosine" was given by J. L. W. Thudichum to a component of the brain lipids, cerebroside and sphingomyelin, the function of which was unclear. [*See* Sphingolipid Metabolism and Biology.]

Lipid bilayers made of phospholipids and cholesterol are the basis of biological membranes, and gan-

TABLE I

Structures of Major Gangliosides Found in Human Tissue

Gala series			
G_{M4}[a]	I^{3}NeuAc-GalCer[b]	NeuAcα2–3Galβ1-Cer	A myelin marker
Ganglio series			
a series			
G_{M3}	II^{3}NeuAc-LacCer	NeuAcα2–3Galβ1–4Glcβ1-Cer	Modulator of epidermal or fibroblast growth factor receptors, causes the differentiation of HL-60 cells into macrophage-like cells, hematoside[c]
G_{M2}	II^{3}NeuAc-Gg$_3$Cer	GalNAcβ1–4(NeuAcα2–3)Galβ1–4Glcβ1-Cer	Accumulated in the brains of patients with Tay–Sachs disease
G_{M1}	II^{3}NeuAc-Gg$_4$Cer	Galβ1–3GalNAcβ1–4(NeuAcα2–3)Galβ1–4Glcβ1-Cer	Receptor for cholera toxin, promotes neurite outgrowth
G_{D1a}	IV^{3}NeuAc-, II^{3}NeuAc-Gg$_4$Cer	NeuAcα2–3Galβ1–3GalNAcβ1–4(NeuAcα2–3)Galβ1–4Glcβ1-Cer	
G_{T1a}	IV^3(NeuAc)$_2$-, II^{3}NeuAc-Gg$_4$Cer[d]	NeuAcα2–8NeuAcα2–3Galβ1–3GalNAcβ1–4(NeuAcα2–3)Galβ1–4Glcβ1-Cer	
	IV$^2_\alpha$Fuc-, II^{3}NeuAc-Gg$_4$Cer	Fucα1–2Galβ1–3GalNAcβ1–4(NeuAcα2–3)Galβ1–4Glcβ1-Cer	
	II^{3}NeuAc-Gg$_5$Cer	GalNAcβ1–4Galβ1–3GalNAcβ1–4(NeuAcα2–3)Galβ1–4Glcβ1-Cer	
b series			
G_{D3}	II^3(NeuAc)$_2$-LacCer	NeuAcα2–8NeuAcα2–3Galβ1–4Glcβ1-Cer	Melanoma antigen
G_{D2}	II^3(NeuAc)$_2$-Gg$_3$Cer	GalNAcβ1–4(NeuAcα2–8NeuAcα2–3)Galβ1–4Galβ1-Cer	Melanoma antigen
G_{D1b}	II^3(NeuAc)$_2$-Gg$_4$Cer	Galβ1–3GalNAcβ1–4(NeuAcα2–8NeuAcα2–3)Galβ1–4Glcβ1-Cer	Tetanus toxin receptor
G_{T1b}	IV^{3}NeuAc-, II^3(NeuAc)$_2$-Gg$_4$Cer	NeuAcα2–3Galβ1–3GalNAcβ1–4(NeuAcα2–8NeuAcα2–3)Galβ1–4Glcβ1-Cer	Tetanus and botulinus toxin receptor
G_{Q1b}	IV^3(NeuAc)$_2$-, II^3(NeuAc)$_2$-Gg$_4$Cer	NeuAcα2–8NeuAcα2–3Galβ1–3GalNAcβ1–4(NeuAcα2–8NeuAcα2–3)Galβ1–4Glcβ1-Cer	Neurite outgrowth activity, botulinus toxin receptor
c series			
G_{T3}	II^3(NeuAc)$_3$-LacCer	NeuAcα2–8NeuAcα2–8NeuAcα2–3Galβ1–4Glcβ1-Cer	
G_{T2}	II^3(NeuAc)$_3$-Gg$_3$Cer	GalNAcβ1–4(NeuAcα2–8NeuAcα2–8NeuAcα2–3)Galβ1–4Glcβ1-Cer	
G_{T1c}	II^3(NeuAc)$_3$-Gg$_4$Cer	Galβ1–3GalNAcβ1–4(NeuAcα2–8NeuAcα2–8NeuAcα2–3)Galβ1–4Glcβ1-Cer	
G_{Q1c}	IV^{3}NeuAc-, II^3(NeuAc)$_3$-Gg$_4$Cer	NeuAcα2–3Galβ1–3GalNAcβ1–4(NeuAcα2–8NeuAcα2–8NeuAcα2–3)Galβ1–4Glcβ1-Cer	
α series			
$G_{D1\alpha}$	IV^{3}NeuAc-, III^{6}NeuAc-Gg$_4$Cer	NeuAcα2–3Galβ1–3(NeuAcα2–6)GalNAcβ1–4Galβ1–4Glcβ1-Cer	Cholinergic synapse
$G_{T1a\alpha}$	IV^{3}NeuAc-, III^{6}NeuAc-, II^{3}NeuAc-Gg$_4$Cer	NeuAcα2–3Galβ1–3(NeuAcα2–6)GalNAcβ1–4(NeuAcα2–3)Galβ1–4Glcβ1-Cer	Cholinergic synapse
Ganglio series others			
G_{M1b}	IV^{3}NeuAc-Gg$_4$Cer	NeuAcα2–3Galβ1–3GalNAcβ1–4Galβ1–4Glcβ1-Cer	Erythrocytes
Lacto series			
	IV^{3}NeuAc-Lc$_4$Cer	NeuAcα2–3Galβ1–3GlcNAcβ1–3Galβ1–4Glcβ1-Cer	
	IV^{3}NeuAc, III^{4}αFuc-Lc$_4$Cer	NeuAcα2–3Galβ1–3(Fucα1–4)GlcNAcβ1–3Galβ1–4Glcβ1-Cer	Sialyl-Le^a
	IV^3(NeuAc)$_2$-Lc$_4$Cer	NeuAcα2–8NeuAcα2–3Galβ1–3GlcNAcβ1–3Galβ1–4Glcβ1-Cer	
Neolacto series			
	IV^{3}NeuAc-nLc$_4$Cer	NeuAcα2–3Galβ1–4GlcNAcβ1–3Galβ1–4Glcβ1-Cer	Sialylparagloboside[c] erythrocytes
	IV^{6}NeuAc-nLc$_4$Cer	NeuAcα2–6Galβ1–4GlcNAcβ1–3Galβ1–4Glcβ1-Cer	
	IV^{3}NeuAc-, III^{3}αFuc-nLc$_4$Cer	NeuAcα2–3Galβ1–4(Fucα1–3)GlcNAcβ1–3Galβ1–4Glcβ1-Cer	Sialyl-Le^x
	IV^3(NeuAc)$_2$-nLc$_4$Cer	NeuAcα2–8NeuAcα2–3Galβ1–4GlcNAcβ1–3Galβ1–4Glcβ1-Cer	
	IV^{3}β(NeuAc-3Gal)-nLc$_4$Cer	NeuAcα2–3Galβ1–3Galβ1–4GlcNAcβ1–3Galβ1–4Glcβ1-Cer	
	IV^{3}β(NeuAc-3GalNAc)-, II^{2}αFuc-nLc$_4$Cer	NeuAcα2–3GalNAcβ1–3Galβ1–4GlcNAcβ1–3(Fucα1–2)Galβ1–4Glcβ1-Cer	
	VI^{3}NeuAc-nLc$_6$Cer	NeuAcα2–3Galβ1–4GlcNAcβ1–3Galβ1–4GlcNAcβ1–3Galβ1–4Glcβ1-Cer	
	VI^{3}NeuAc-, V^{3}αFuc-nLc$_6$Cer	NeuAcα2–3Galβ1–4(Fucα1–3)GlcNAcβ1–3Galβ1–4GlcNAcβ1–3Galβ1–4Glcβ1-Cer	
	VI^{3}NeuAc-, IV^{6}β(Galβ-4GlcNAc)-nLc$_6$Cer	NeuAcα2–3Galβ1–4GlcNAcβ1–3(Galβ1–4GlcNAcβ1–6)Galβ1–4GlcNAcβ1–3Galβ1–4Glcβ1-Cer	
	VI^{3}NeuAc-, IV^{6}β(Fucα-2Galβ-4GlcNAc)-nLc$_6$Cer	NeuAcα2–3Galβ1–4GlcNAcβ1–3(Fucα1–2Galβ1–4GlcNAcβ1–6)Galβ1–4GlcNAcβ1–3Galβ1–4Glcβ1-Cer	
	VI^{3}NeuAc-, IV^{6}β(NeuAc-3Galβ-4GlcNAc)-nLc$_6$Cer	NeuAcα2–3Galβ1–4GlcNAcβ1–3(NeuAcα2–3Galβ1–4GlcNAcβ1–6)Galβ1–4GlcNAcβ1–3Galβ1–4Glcβ1-Cer	
Globo series			
	V^{3}NeuAc-Gb$_5$Cer	NeuAcα2–3Galβ1–3GalNAcβ1–3Galα1–4Galβ1–4Glcβ1-Cer	SSEA-4 in mouse, human teratocarcinoma
	V^3(NeuAc)$_2$-Gb$_5$Cer	NeuAcα2–8NeuAcα2–3Galβ1–3GalNAcβ1–3Galα1–4Galβ1–4Glcβ1-Cer	
	V^{3}NeuAc-, V^{6}NeuAc-Gb$_5$Cer	NeuAcα2–6(NeuAcα2–3)Galβ1–3GalNAcβ1–3Galα1–4Galβ1–4Glcβ1-Cer	Erythrocytes

[a]Nomenclature according to Svennerholm's system.
[b]Nomenclature recommended by the Nomenclature Committee of the International Union of Biochemistry.
[c]Trivial names.
[d](NeuAc)$_n$: (NeuAcα2–8)$_{n-1}$ (NeuAcα2–3). The anomeric linkage of NeuAc- is of the α configuration. The β configuration is not found, and α is omitted.

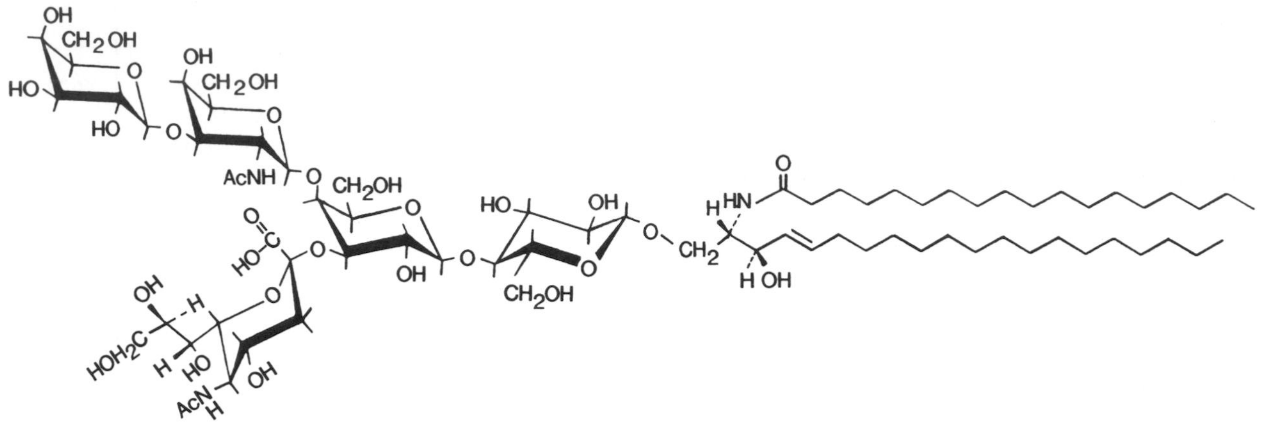

FIGURE 2 Structure of G_{M1} ganglioside. The configuration of each monosaccharide is arbitrary, and the conformation of the molecule anchored into the cell membrane is a matter requiring further investigation.

gliosides are able to act as membrane components because their ceramides can interact with these membrane components through hydrophobic interactions. [*See* Cholesterol; Lipids; Phospholipid Metabolism.] In the plasma membrane of cells, however, ganglioside–carbohydrate chains are considered to be anchored to the outer leaflet of the lipid bilayer. This is due to the biosynthesis of their carbohydrate chains in cells. The carbohydrate chains of gangliosides are biosynthesized in the Golgi apparatus, which is a cellular organelle consisting of multilayers of membrane cisternae. Cells biosynthesize most carbohydrate chains of their constituents, including gangliosides, and of secetion products on the inner surface of Golgi cisternae but not on the outer surface, which faces the inside of cells, or the cytosol. The inner leaflet of the Golgi membrane fuses with the outer leaflet of plasma membranes. Because of this mechanism of membrane biogenesis, the carbohydrate chains of gangliosides or other carbohydrate-containing molecules (e.g., membrane-bound glycoproteins) are extended into the space outside the cells.

II. REGULATION OF EXPRESSION

A. Biosynthesis

Table I indicates that gangliosides differ in their carbohydrate structures. The machineries producing the

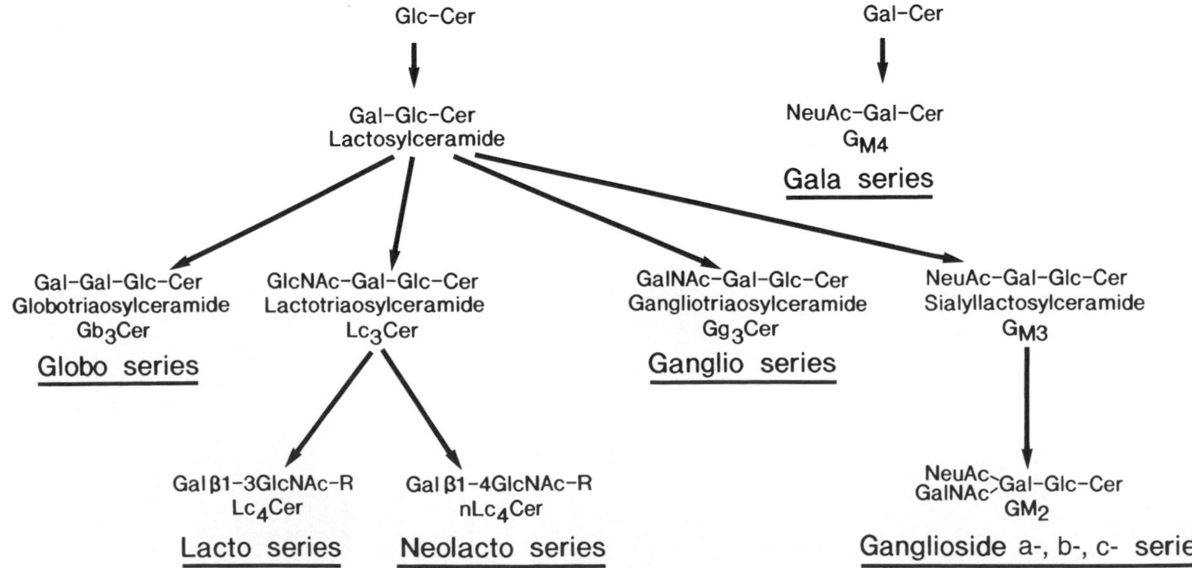

FIGURE 3 The five core structures of neutral glycosphingolipids for gangliosides.

various kinds of carbohydrate structures are located in the Golgi apparatus, and glycosyltransferases are primarily responsible for the production of the various structures. Figure 4 shows the major biosynthetic pathways for gangliosides. Every step in the elongation of carbohydrate chains requires a glycosyltransferase, which sometimes has a broad, sometimes a strict, substrate specificity. At present it is unknown how many glycosyltransferases in the Golgi apparatus are involved in ganglioside biosynthesis, but it is believed to be more than a few, at least several tens. Without this characteristic, variations in ganglioside structures cannot be produced in the tissues or cells of humans and other mammals.

Factors other than glycosyltransferases are also required for the production of ganglioside carbohydrates. These include sugar nucleotides as donors of sugars, which are transferred to precursor gangliosides or neutral glycolipids; enzymes involved in sugar nucleotide production; and translocators for the transport of sugar nucleotides into the internal space of the Golgi apparatus from the cytosol. Sugar nucleotides containing one or two high-energy phosphate bonds are synthesized in the cytosol, but CMP-NeuAc is exceptional, being synthesized in the nucleus and then transported to the cytosol.

It is also interesting that galactose (Gal), glucose (Glc), *N*-acetylgalactosamine (GalNAc), and *N*-acetylglucosamine (GlcNAc) form UDP derivatives, fucose (Fuc) and mannose (Man) GDP derivatives, and

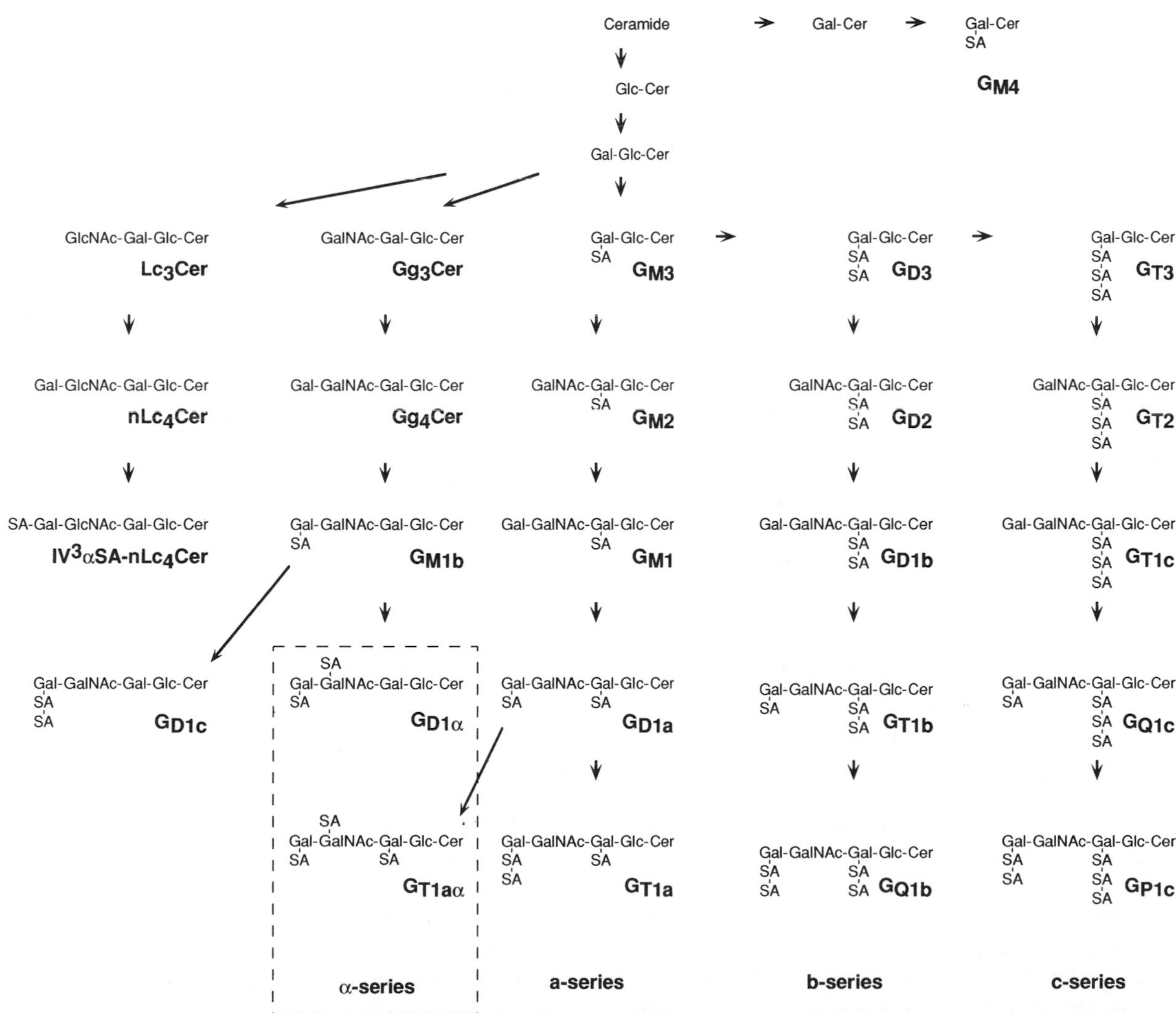

FIGURE 4 The major biosynthetic pathways in human tissues. The a, b, and c pathways and G_{M4} synthesis occur in neural tissues. The neolacto series pathway occurs in extra-neural tissues. SA, sialic acid.

NeuAc, a CMP derivative. Two isomers, α and β, are formed when a sugar attaches to any other molecules through the C-1 carbon, an anomeric carbon atom. The isomers of all of these nucleotide sugars except for CMP-NeuAc and GDP-Fuc are of the α configuration, but in gangliosides those of Gal and GalNAc are either of the α or β configuration, those of Fuc and NeuAc are of the α configuration, and that of Glc is of the β configuration, indicating that the anomeric linkages in biosynthesized gangliosides are determined by the nature of the glycosyltransferases. Glycosyltransferases also determined which of the four hydroxyl groups in hexoses and which of the three in hexosamines are used as the linkage position for a newly added sugar.

An interesting feature is that gangliosides found in various tissues of mammals, including humans, differ among tissues, but are constant for each tissue from the same species, matched as to age and sex, indicating that the mechanisms producing these characteristics are genetically determined. For example, in mammals the tissues that contain gangliosides at the highest concentration are neural tissues. The major biosynthetic pathways for a-, b-, and c-series gangliosides and G_{M4} are shown in Fig. 4. For the production of a large amount of gangliosides, neural tissues must activate a mechanism that produces the gangliosides much more strongly than that in other tissues. At the same time, extraneural tissues in humans contain sialyllactosylceramide and neolacto-series gangliosides, as the major gangliosides, indicating the activation of a different ganglioside-producing mechanism.

Another reason for the heterogeneity of gangliosides is species differences. Brain gangliosides in vertebrates are well conserved and are of the a-, b-, and c-series, but gangliosides in visceral organs differ among species. In erythrocytes, sialyllactosylceramide or G_{M3} is the major ganglioside in horses, dogs, pigs, and humans, disialyllactosylceramide is found in cats, pandas, and bears, and a-series gangliosides are predominant in rats and mice. This heterogeneity is also considered to be genetically determined.

Individual differences in gangliosides were also reported for sialyl-Le^a (a derivative of Lewis antigens) in humans, sialic acid species of hematoside or G_{M3} in dog erythrocytes, and a-series gangliosides in mouse liver. These three cases of heterogeneity (i.e., tissue-, species-, and individual-specific differences) should reflect how the biosynthetic mechanisms are activated. We have just started to understand the mechanisms that produce this heterogeneity by the introduction of molecular biology for glycosyltransferases and tissue-specific response elements acting on the transferase genes.

B. Changes in Ganglioside Expression

The tissue-specific composition of gangliosides is the result of changes in ganglioside expression during early embryogenesis, differentiation, and organogenesis. These changes are quite interesting because of the possibility that they play roles as signals for determination of the differentiation direction of cells in embryos and fetuses.

During early embryogenesis in mice, dramatic changes in carbohydrate expression have been demonstrated by immunostaining with monoclonal antibodies. Carbohydrate antigens are expressed at particular stages in mouse early embryogenesis, called stage-specific embryonic antigens (SSEAs). The epitopes of SSEA-1, -3, and -4 are Galβ1–4(Fucα1–3)GlcNAcβ-, Galβ1–3GalNAcβ1–3Galα1–4Galβ1-, and NeuAcα2–3Galβ1–3GalNAcβ1–3Galα1–4Galβ1-, respectively. Anti-SSEA-1 antibody and SSEA-1 oligosaccharide can arrest further embryogenesis after the morula stage. SSEA-4 is the only ganglioside determined so far, but other gangliosides might also be molecules involved in early mouse embryogenesis. In the human fetus, the presence of SSEAs has also been demonstrated.

In neural tissues, striking changes in ganglioside expression were demonstrated during development in the late stage of gestation and after birth. The main period of ganglioside accretion coincides with the period of accelerated neuronal membrane formation, just before and during the period of intensive synaptogenesis. The proliferative state of neuronal and glial precursor cells is characterized by the predominant expression of G_{D3}. The developmental period between neurogenesis and synaptogenesis seems to be characterized by predominant synthesis of b-series gangliosides, mainly G_{Q1b}, the synthesis of a-series gangliosides being triggered last during development. An understanding of the regulation of these changes is quite important in relation to the developments of the network formation and neural functions.

Every mammal carries genetic information for the expression of gangliosides in its DNA sequences. However, another important control mechanism for ganglioside expression is epigenetic control, which includes the following. Cells can change gangliosides when they are cultured with ground substances or different types of cells. Changes in ganglioside compositions are found in the brains of fish fed at low temperatures. Changes in gangliosides were also observed in the brains of experimental animals as a result of learning. These changes in ganglioside contents and compositions are under epigenetic control and possibly are one of the biochemical bases for neuronal plasticity. [*See* Nervous System, Plasticity.]

The tissue-specific expression of gangliosides, as just described, can be explained as the summation of cell type-specific expressions of gangliosides. It can be said, therefore, that the expression of gangliosides in individual cells is controlled or determined by genetic and epigenetic regulation. This concept can be extended to the idea that particular ganglioside composition is a marker for a particular cell, depending on its particular stage of differentiation.

C. Three-Dimensional Structures

Gangliosides are transported onto the surfaces of cells and their carbohydrate portions extend out from the plasma membrane. The carbohydrates are recognized by lectins, toxins, antibodies, viruses, bacteria, and other recognition molecules or organisms, and this recognition process must be explained in terms of the three-dimensional structures of gangliosides anchored in membranes. A ganglioside has multiple binding sites, as shown using monoclonal antibodies. For one particular ganglioside, several monoclonal antibodies with different binding specificities can be produced. Three-dimensional structures of gangliosides are unknown because it is difficult to obtain crystals needed for X-ray crystallography. Recent advances in nuclear magnetic resonance (NMR) spectroscopy, however, give hope that it will be possible to analyze the structures of gangliosides anchored in membranes. The physiological functions of gangliosides are assumed to be mediated by structural changes in ganglioside-recognizing molecules, following the binding of gangliosides. Therefore, an understanding of the three-dimensional structures of gangliosides anchored in membranes will be the basis for an understanding of their biological functions.

The possibility was proposed that the carbohydrate chains of gangliosides are not straight on the cell-surface membrane, but may be bent and thus lay on the membrane. One side of the surface of carbohydrate chains is rich in hydrophobic atoms or functional groups like -CH, CH_3CONH-, and -CH_3, and the other side is rich in hydrophilic atoms like -OH and -COO^-. The hydrophilic domain faces the membrane and may interact with polar groups of phospholipids. The hydrophobic face can be recognized by ganglioside carbohydrate-recognition molecules from the outside of cells. This hypothesis will be verified by NMR-spectroscopic studies of gangliosides incorporated into micelles or the membranes themselves.

In the ceramide portion of gangliosides, there are heterogeneities in both sphingosines and fatty acids. Regarding human brain gangliosides, the major molecular species of sphingosines is 1,3-dihydroxy-2-amino-*trans*-4-octadecene (C18-sphingenine or C18 : 1-sphingosine). With aging or an increase in the molar composition of sialic acids, the proportion of C20-sphingenine (C20 : 1-sphingosine) to C18-sphingenine increases. The major fatty acid of brain gangliosides, other than G_{M4}, is stearic acid (C18 : 0 fatty acid), followed by palmitic acid. Regarding the gangliosides of visceral organs, the fatty acids are more heterogeneous than those of the brain, palmitic acid and longer-chain fatty acids (C22 : 0, C24 : 0, and C24 : 1) being predominant. These heterogeneities of the ceramide portion were reported to affect the interaction between gangliosides and the molecules that recognize them. It is assumed that the three-dimensional structures of the carbohydrates of gangliosides in membranes are somehow modified when the ceramide portions are different.

D. Degradation

The degradation of gangliosides mainly occurs in lysosomes, in which a part of the plasma membrane is pinocytosed (i.e., engulfed into the cell) and the pinocytic vesicles fuse with primary lysosomes, which contain various carbohydrate hydrolases as well as other degradative enzymes. If one of the carbohydrate hydrolases is deficient, the cell cannot break down the carbohydrate structures that are susceptible to that enzyme's action, and carbohydrate-containing substances remain undergraded, thus accumulating in lysosomes. The damage to cells caused by such accumulation leads to dysfunctions or cell death, especially in tissues or cells exhibiting high activity of biosynthesis of carbohydrate-containing substances. Such metabolic disorders are the basis of a group of inherited diseases, and Tay–Sachs disease is a well-known ganglioside storage disease.

An understanding of the pathogenesis of Tay–Sachs disease was not reached until the ganglioside research by Klenk. He noticed and isolated a glycolipid that is highly accumulated in patients' brains. He coined the term "ganglioside" because glycolipids exhibiting similar characteristics to the patients' glycolipid were abundant in gray matter and ganglions, which are masses of neuronal cells, in normal brains. The ganglioside accumulated in the brains of Tay–Sachs disease cases is G_{M2} ganglioside, which contains a βGalNAc linkage at its terminus. This linkage is normally degraded by β-hexosaminidase in lysosomes. In the lysosomes of patients with Tay–Sachs disease, the activity of β-hexosaminidase is deficient. β-Hexosaminidases are dimers, consisting of α and β subunits, the dimers formed being α–α, α–β, and β–β; they require an activator protein for their activity. A defect of either one of these three components causes Tay–Sachs disease. The clinical course of the disease is variable and dependent on which component is damaged. Recent molecular biological studies have revealed that various types of mutations occur in the genes of enzyme subunits and the activator protein and that the clinical manifestations differ slightly among mutations.

Another inherited disease concerning gangliosides is G_{M1} gangliosidosis, in which a deficiency of β-galactosidase activity causes accumulation of the G_{M1} ganglioside and oligosaccharides containing β-galactoside linkages at their termini. In this case, mutations in the gene encoding β-galactosidase are responsible. Galactosialidosis is a disease in which the activities of β-galactosidase and α-sialidase are deficient. There is a protein that protects both enzymes from rapid degradation, and functional loss of this protective protein is responsible for this disease.

The activities of some carbohydrate hydrolases have also been demonstrated in membranes or the cytosol, membrane-bound hydrolases having attracted particular attention because they can modify the carbohydrate structures of gangliosides in membranes *in situ*. The degraded components or partially degraded gangliosides are recycled to the Golgi apparatus for biosynthesis.

The linkages of sialic acids, which are modified at their hydroxyl groups through acetylation, are resistant to hydrolysis by sialidases; thus gangliosides having these sialic acids cannot be degraded further. Their degradation is controlled by a deacetylase. A transgenic experiment of bacterial deacetylase cDNA into mice demonstrated that tissue construction of the retina and kidney of adult mice is disturbed. It is supposed that O-acetylated sialic acid-containing gangliosides would be recognized by molecules in a specific manner, though they have not been identified yet, and play roles in the formation of normal tissue construction.

E. Genes Controlling Ganglioside Expression

An example of genes controlling glycoconjugates is the ABO blood group genes, which are located on a single locus of human chromosome 9. On the ABO gene locus, three alleles are present. The A allele encodes α-*N*-acetylgalactosaminyltransferase producing the A antigen, GalNAcα1–3(Fucα1–2)Galβ1-; the B allele encodes α-galactosyltransferase producing the B antigen, Galα1–3(Fucα1–2)Galβ1-; and the O allele encodes no active transferases, resulting in the expression of no A and B antigens but the H antigen, Fucα1–2Galβ1-, as their precursor carbohydrate chain. This is the biochemical basis for the polymorphic expression of the ABH blood group antigens. Recently, several single-base mutations in open reading frame of the genes have been reported to be the basis of the variations. This is one model of regulation mechanism for carbohydrate chain expression. Genes controlling the expression of gangliosides are also a matter of interest, because, if gangliosides have critical functions for embryonic and fetal developments and neural tissue construction as described in Section IV of this article, these genes and their regulations are the basis for ganglioside expression and thus for ganglioside functions. Glycosyltransferase genes responsible for gangliosides provide important tools for us to manipulate ganglioside expression and functions.

Thus, cloning of glycosyltransferases has opened a new aspect. cDNA of β-1,3-GalNAc transferase, which transfers GalNAc from UDP-GalNAc to G_{M3} and synthesizes G_{M2}, has been cloned by an expression cloning method and sequenced. The enzyme has the same construct with other glycosyltransferases involved in biosynthesis of glycoprotein carbohydrate chains, that is, the enzyme has the membrane-spanning domain near the NH_2 terminus and the catalytic domain in the half of the molecule at the COOH-terminal side. The proteins with these features are called type 2 membrane proteins. Transfection of the enzyme cDNA to cultured cells changed cellular gangliosides, increased G_{M2} content in most cases, and increased G_{D2} and Gg_3Cer in some cases, suggesting that the enzyme can synthesize G_{D2} from G_{D3} and Gg_3Cer from Lac-Cer as well. *In situ* hybridization with the GalNAc transferase cDNA indicated that the message level is regulated in a development-dependent and a space-dependent manner during the formation of nervous tissues in mice. A targeting experiment for this GalNAc transferase gene is now under way and results will give us interesting information.

Three sialytransferases for the sialylation of glycoprotein carbohydrate chains were purified and their cDNAs were cloned. The consensus sequences called the sialyl motif were discovered, and then more than 10 sialytransferase cDNAs were cloned on the basis of this sialyl motif. One of them is α-2,8-sialytransferase, responsible for the synthesis of G_{D3}. The same cDNA has also been cloned by an expression cloning method with anti-G_{D3} antibody. G_{D3} is the key ganglioside for the biosynthesis of b-series gangliosides, and the expression of G_{D3} is closely related to neuronal and glial cell development in the fetus. Therefore, further studies on the sialyltransferase and its gene will make clear the functions of G_{D3} in neural tissues.

Normal human tissues do not express NeuGc-containing glycoconjugates, as already mentioned. The enzymatic reaction regulating NeuGc expression is hydroxylation of CMP-NeuAc, which converts CMP-NeuAc to CMP-NeuGc. NeuGc, then, is incorporated into glycoconjugates by the action of sialyltransferases. The hydroxylation reaction is carried out by a complex of reactions, including an electron transport system composed of NADH cytochrome b5 reductase, cytochrome b5, and a terminal hydroxylase. The hydroxylase was purified from mouse liver and its cDNA was cloned. Southern analysis indicated that human genome contains base sequences that are cross-hybridized with mouse hydroxylase cDNA. It would be quite interesting to know whether humans retain the ability to express this hydroxylase and thus NeuGc in carbohydrate chains of gangliosides and glycoproteins. Further studies with this cDNA will make clear how normal human tissues suppress transcription of the hydroxylase gene and whether NeuGc is an ideal cancer-associated antigen or not.

We have just obtained cDNAs of about 20 glycosyltransferases and these cDNA are tools for changing carbohydrate chains in cultured cells and experimental animals. Other human genes related to ganglioside expression and cloned and mapped on human chromosomes so far are β-galactosyltransferase, involved in the synthesis of lactose in mammary gland and Galβ1–4GlcNAcβ1 structure in other tissues on chromosome 9; fucosyltransferase-III (FucT-III), FucT-V, and FucT-VI on chromosome 19 in a tandem manner (FucT-III is the gene for the Lewis blood group); and FucT-IV and VII on chromosomes 11 and 9, respectively. These fucosyltransferases have different substrate specificity and attract attention in relation to the expression of sialyl-Lewis x (sialyl-Le^x) and sialyl-Lewis a (sialyl-Le^a) antigens, which are tumor markers and ligands for E-selectin on endothelial cells. Gene loci of other molecules regulating gangliosides are activator proteins for β-hexosaminidase on chromosome 5 and the β-hexosaminidase α and β subunits on chromosomes 15 and 5, respectively. In mice, four genes regulating the polymorphic expression of gangliosides were mapped on specific chromosomes, *Ggm1* for mouse liver G_{M1} at 1 centimorgan (cM) centromeric to *H-2K* on chromosome 17, *Ggm2* for mouse liver G_{M2} on chromosome 10, *Gs14* for erythrocyte G_{M4} on chromosome 3, and *Gs16* for kidney V_3NeuGc-Gb_4Cer on chromosome 19. *Ggm2* may regulate transcription of the GalNAc transferase gene for G_{M2} synthesis in a liver-specific manner.

Studies of genes controlling the expression of carbohydrate chains of glycoconjugates will establish a paradigm for a biological system using remarkably heterogeneous structures, which have been produced during evolution. And transgenic and gene targeting experiments will give us indications for the functions of gangliosides *in vivo*.

III. GANGLIOSIDES IN CANCER

The changes in the ganglioside composition in cancer cells were demonstrated by S. Hakomori and coworkers in 1968, using baby hamster kidney fibroblasts and their polyoma virus transformants. Extensive studies of gangliosides in cancer tissues have also been done and the changes in gangliosides in cancer tissues were summarized: (1) incomplete synthesis with or without accumulation of precursor glycolipids; (2) induction of synthesis of new glycolipids; and (3) organizational changes of glycolipids in membranes.

Recent research involving introduction of genes into cells has demonstrated that the introduction of oncogenes into normal fibroblasts induces changes in ganglioside expression. In several cases, the changes were mediated by increases in glycosyltransferase activities. The precise mechanisms underlying the changes produced by malignant transformation, however, remain unknown. [*See* Oncogene Amplification in Human Cancer.]

Extensive research on cancer-associated antigens and cancer immunity has been performed with the hope of clinical application to cancer diagnosis and therapy. The detection of gangliosides as cancer-associated antigens and immunotherapy with monoclonal antibodies against gangliosides are examples. The detection of sialyl-Le^a antigen in sera of patients is applied to the diagnosis of pancreatic cancers, unfortunately however, preferentially for the late stages. In mice, some tumors were successfully treated by the injection of monoclonal antibodies against a glycolipid expressed in abundance on the cancer cells. Human and murine monoclonal antibodies against human melanoma cells recognizing the gangliosides G_{M2}, G_{D2}, and G_{D3} were used in trials for the treatment of melanoma patients. Local and systemic injections of the monoclonal antibodies were successful in terms of regression of the tumor size or suppression of metastatic invasions. At the same time, antibodies against these gangliosides were detected in sera of patients with melanomas, and a relationship between the prognosis of the cancer and the level of the antibodies was recognized. Therefore, the possibility of active immunization for elevating the titer of these antibodies is being investigated.

As mentioned in Section II,B, the altered carbohydrate structures in cancer are known to mimic those in the stages of embryogenesis and the differentiation of fetuses. Antigens of this category are called oncofetal antigens. At present, we do not know what biochemical events are the basis for the production of oncogenic and fetal antigens, and how to approach this problem is an important and fundamental subject.

IV. BIOLOGICAL FUNCTIONS

A. Neural Functions

Among neurons of the brains of Tay–Sachs disease patients, some have swollen or giant neurites (i.e., meganeurites). These are packed with numerous membranous cytoplasmic bodies, composed of accumulating substances, including the G_{M2} ganglioside. Many fine neurites originate from the meganeurites. In addition, the administration of brain gangliosides promotes the regeneration and reinnervation processes of both cholinergic and adrenergic nerve fibers in animals. These two observations suggest that brain gangliosides might induce neurite growth and axonal elongation. Many investigators have been interested in these effects and have tried to demonstrate them *in vitro*, using primary and established neural cell cultures. It was shown that mixtures of brain gangliosides or purified G_{M1}, G_{D1a}, and G_{Q1b} increase the number of neurites and their length. A few-nanomolar concentration of G_{Q1b} produced these effects in human neuroblastoma cell lines. The carbohydrate structure of G_{Q1b} is strictly required for its effect: G_{D1a}, G_{D1b}, G_{T1a}, G_{T1b}, and G_{Q1c} were not active in this experimental system. On the basis of these results, it was proposed that neurons have receptors recognizing G_{Q1b}, which could play an important role in the formation of neuron networks. However, it is not answered yet how signals are triggered and transmitted through cell-surface membranes after the binding of gangliosides to the receptors, how the signals reach nuclei, and how the biosynthesis of proteins and other components needed for neurite outgrowth and elongation is triggered. Research to answer these questions is in progress.

Production of a series of monoclonal antibodies against brain gangliosides enabled us to visualize cell-type-specific distributions of gangliosides in the brain. In rat cerebellum, G_{M1} distributes in myelin and glial cells, G_{D1a} is in the molecular layer, and G_{D1b} and G_{Q1b} are in the granular layer. The G_{M1} distribution is a controversial result, compared with that of biochemical analysis, however, histological results suggest that establishment and maintenance of the nervous system network require differential expression of gangliosides. Interestingly, patients with myeloma secrete monoclonal antibodies and some of them develop peripheral neuropathy. There is a relationship between the production of antibodies and neuropathy. Some of these monoclonal antibodies react with gangliosides G_{M2}, G_{M1}, G_{D1a}, or G_{D1b}. Thus, it is assumed that antiganglioside monoclonal antibodies bind nerve gangliosides and this is the pathology for this type of peripheral neuropathy. Some patients with Fisher syndrome and Guillain-Barré syndrome develop polychronal antiganglioside antibodies, which are anti-G_{Q1b} and anti-G_{M1} antibodies, respectively. Changes of antibody titers in sera correlate with severeness of clinical symptoms. Thus, the antibody production caused by bacterial infection seems to be the pathology of these two syndromes. These lines of evidence support the idea that the neuronal network is actually supported by gangliosides and their recognition molecules. Recognition molecules may play key roles for ganglioside functions.

Studies of subcellular distributions of gangliosides in neuronal cells have shown that they occur in synaptic membranes. Presynaptic terminals contain synaptic vesicles, which store chemical transmitters (e.g., acetylcholine). Neurotransmission takes place as the result of membrane fusion between synaptic vesicles and presynaptic membranes and the release of the chemical transmitter into the synaptic cleft, the space between pre- and postsynaptic membranes. The efficiency of the vesicle fusion is affected by membrane components such as phospholipids, cholesterol, and gangliosides. Ca^{2+} ions play important roles in the release of neurotransmitters. It was proposed that the interaction between gangliosides in presynaptic membranes and Ca^{2+} ions can modulate the process of vesicle fusion. Recently, the epitopes of the polychronal antibodies affecting cholinergic transmission have been determined and a new type of gangliosides named α-series has been identified. α-Series gangliosides contain NeuAcα2-6GalNAcβ structure (Table I and Fig. 4). This evidence suggests that α-series gangliosides have specific interaction with functional molecules involved in cholinergic synapses.

The role of gangliosides in neuronal functions is one of the major topics of ganglioside research. This is quite reasonable because the brain contains the highest concentration of gangliosides among all tissues or organs.

B. As Receptor Ligands for Toxins, Virus, and Mammalian Lectins

A brain lipid fraction containing gangliosides can neutralize the toxic activity of tetanus toxin, which is produced and excreted by *Clostridium tetani*. The toxin produces characteristic tonic spasms of voluntary muscles in humans, called opisthotonos, suggesting that it acts in the spinal cord to suppress inhibitory neurons, the action of which relaxes extended muscles during movement caused by muscle contraction. *In vitro*, the toxin binds more strongly to G_{D1b} and G_{T1b} than to G_{M2}, G_{M1}, or G_{D1a}, suggesting that synapses with different functions could have different and specific ganglioside compositions.

In 1972, W. E. van Heyningen and collaborators reported that mixture of gangliosides could abolish the toxic activity of cholera toxin. Extensive further research established that G_{M1} is specifically recognized by and binds to the B subunit of the cholera toxin. A cholera toxin molecule consists of five B subunits and one A subunit. The B subunits exhibit binding activity for G_{M1}, and the binding of the B subunits to G_{M1} on cell-surface membranes can cause insertion of the A subunit into the membranes through an unknown mechanism, possibly involving conformational changes of both the A and B subunits. The inserted A subunit can activate adenylate cyclase by ADP ribosylation. The elevated concentration of cAMP in the cells results in hypersecretion of chloride and water and impaired absorption of sodium in the small intestine. However, the biochemical details of the processes from the elevation of cAMP to water loss are not known yet. The specific interaction between G_{M1} and the B subunit is used as a reagent for a sensitive detection of G_{M1}.

The toxin produced by *Clostridium botulinum* is also neutralized by gangliosides, the G_{T1b} and G_{Q1b} gangliosides being the most effective.

Sialidase treatment of mammalian erythrocytes can destroy the receptor activity for the influenza, Sendai, and Newcastle disease viruses. Sialylated compounds are the virus receptors on epithelial cells of the respiratory tract as well as erythrocytes. G_{T1a}, G_{Q1b}, and G_{P1c}, which have the terminal structure NeuAcα2–8NeuAcα2–3Galβ1–3GalNAc, are possibly receptors for the Sendai virus. The question of whether or not the terminal structure of glycoproteins or gangliosides is actually responsible for the virus infection awaits further research.

Carbohydrate chain recognition molecules in animal tissues are important for glycoconjugates to achieve physiological functions *in vivo* since we can suppose that the recognition molecules specific to particular carbohydrate chains can transmit signals to the cells and trigger following biological responses. Recognition molecules that are specific to only gangliosides have not been reported, but the molecules recognizing carbohydrate chains carried by glycoproteins and gangliosides are known. E-Selectin expressed on endothelial cells recognizes sialyl-Le^x and sialyl-Le^a structures of glycoproteins and gangliosides on leukocytes, and plays a role for migration of leukocytes from blood vessels to tissues when tissues have inflammation. Colon cancer patients, if their tumors express sialyl-Le^x structure, have rather bad prognosis because of metastasis, and this is considered to be due to the enhanced attachment of cancer cells in blood to endothelial cells and resulting metastasis. Sialoadhesin expressed on mouse marginal zone macrophages recognizes NeuAcα2–3Galβ1- of GD1a, GT1b, and glycoproteins, and myelin-associated glycoprotein recognizes NeuAcα2–3Galβ1-3GalNAc of GT1b. Thus, one of the major functions of gangliosides is to be ligands for recognition molecules and to contribute to cell–cell interactions. Further studies will make these roles of gangliosides clear.

C. Regulation of Cell Growth and Differentiation

The exogenous addition of G_{M3} inhibits the growth of epidermal cells cultured in chemically defined media containing epidermal growth factor (EGF). This effect might be produced through the modulation of growth factor receptor functions. The binding of growth factors to growth factor receptors and sequentially induced phosphorylation of the receptors are the initial events. It is thought that the exogenously added gangliosides become inserted into the plasma membranes and reach receptors, modifying their three-dimensional structures. In this way the gangliosides could modify the rate of receptor phosphorylation, which is triggered by the binding of growth factors to the receptors. The growth of fibroblasts appears to be controlled by the content of G_{M3} in the plasma membrane, where interaction between fibroblast growth factor (FGF) and the FGF receptors is modified by G_{M3}. A membrane-bound sialidase seems to control the content of G_{M3} in the membranes, thus regulating cell growth. It is interesting that membrane-bound hydrolases, but not those in lysosomes, are involved in the turnover of membrane gangliosides. The modulation of cell growth by gangliosides seems to be a secondary control mechanism, which does not operate in an all-or-nothing fashion. This type of regulation or modulation would be physiologically important, but its mechanism is unknown.

HL-60 cells, a human myelogenous leukemia cell line, are induced to differentiate into macrophage-like cells by the exogenous addition of the G_{M3} ganglioside, but they differentiate into granulocyte-like cells upon the addition of sialylparagloboside ($IV^3NeuAc\text{-}nLc_4$-Cer). These observations suggest that gangliosides could regulate or determine the direction of differentiation of immature cells. Little information is available on biochemical events triggered by the addition of gangliosides at present, but studies are in progress.

Extensive further studies on the role of gangliosides in physiological cellular function and in formation of normal tissue organization are required. For this purpose, various approaches (e.g., enzymological, cell biological, and molecular biological) are needed. Given the limited space of this article, the authors have not been able to cover all aspects of ganglioside research. For further detailed information and discussion, readers are referred to the excellent review articles listed in the Bibliography.

BIBLIOGRAPHY

Hakomori, S.-I. (1981). Glycosphingolipids in cellular interaction, differentiation, and oncogenesis. *Annu. Rev. Biochem.* **50**, 733.

Kanfer, J. N., and Hakomori, S.-I. (1983). "Handbook of Lipid Research. Volume 3: Sphingolipid Biochemistry" (D. J. Hanahan, ed.). Plenum, New York.

Kotani, M., Kawashima, I., Ozawa, H., Terashima, T., and Tai, T. (1993). Differential distribution of major gangliosides in rat central nervous system detected by specific monoclonal antibodies. *Glycobiology* **3**, 137.

Marth, J. D. (1994). Will the transgenic mouse serve as a Rosetta Stone to glycoconjugate function? *Glycoconjugate J.* **11**, 3.

Varki, A. (1993). Biological roles of oligosaccharides: All of the theories are correct. *Glycobiology* **3**, 97.

Wiegandt, H. (1985). "Glycolipids. Volume 10: New Comprehensive Biochemistry." Elsevier, Amsterdam.

Ganglioside Transport

ROBERT W. LEDEEN
Albert Einstein College of Medicine

GLOSSARY

Anterograde axonal transport Transport of substances in the axon away from the cell body, toward the nerve ending

Ganglio series Family of gangliosides based on G_{M1}, most of which have the gangliotetraose (or a closely related) structure

Ganglioside Glycosphingolipid that possesses at least one sialic acid on the oligosaccharide chain

Gangliotetraose structure G_{M1} ganglioside or an oligosialo derivative of it

G_{M1} ganglioside Galactosyl-*N*-acetylgalactosaminyl-(*N*-acetylneuraminyl)-galactosyl-glucosyl-ceramide

Oligosialogangliosides Gangliosides having two to five sialic acids

Retrograde axonal transport Transport of substances in the axon back toward the cell body

GANGLIOSIDES FORM PART OF THE CARBOHYDRATE-rich layer that surrounds mammalian cells and determines their surface properties. In neurons they comprise a much larger proportion of this membrane-associated coating than in most other cells. Gangliosides were once thought to be localized primarily at the synapse. Although their precise localization in the neuron is still somewhat controversial, increasing evidence suggests they are distributed over a large portion of the neuronal surface, including cell body, synaptic region, and the extensive network of dendrites and axons. These latter, which are often several thousand times greater in length than the cell body, where synthesis occurs, require efficient transport/transfer mechanisms to convey gangliosides to the far reaches of this complex cell. Research over many years has elucidated several aspects of this process as well as that of "reverse flow" involving return to the cell body. These studies quite naturally employed animal models, as in most other areas of ganglioside research, but it is reasonable to assume that processes that are common to a variety of vertebrates, from fish to rabbits, also apply to the human. Axonal transport of gangliosides has been studied in both the central (CNS) and peripheral nervous systems (PNS), with generally similar results. In the course of this work, several ancillary findings relating to ganglioside distribution and metabolic behavior have come to light. The recent discovery of glycolipid transfer proteins has expanded our general concept of ganglioside movement. The subject of dendritic transport unfortunately has not yet been investigated in relation to gangliosides, perhaps owing to the lack of a suitable anatomical model. Experimental ingenuity will be needed in the future to overcome this rather large gap in our understanding of ganglioside transport. [*See* Gangliosides.]

I. AXONAL TRANSPORT IN THE CNS

The general procedure for axonal transport studies has been to inject a radiolabeled precursor into the tissue of a living animal in the vicinity of neuronal cell bodies, followed by detection of radiolabeled substances that have migrated a measured distance along the axon during a specified period of time. The optic system lends itself well to this approach because of the clear anatomical separation of cell bodies from axons (optic nerve/tract) and nerve endings (optic tectum, superior colliculus, lateral geniculate nucleus). This involves one cell type, the ganglion cells, whose cell bodies reside in the retina. Beginning with the goldfish, it was shown that intraocular injection of ganglioside precursors (e.g., glucosamine and *N*-acetylmannosamine) gives rise to radiolabeled gangliosides, which are translocated from the retina to the optic tectum. Because the optic system is generally crossed, axonal migration occurs to the tectum of the opposite side; the tectum of the same side is then used as control tissue to indicate background labeling that occurs systemically via the circulation. Subtraction of the latter from the former gives the amount of axonally transported radiolabel. Similar studies were subsequently carried out with the optic systems of chick, rat, rabbit, and other kinds of fish. [*See* Retina; Visual System.]

In such studies it is necessary to rule out the possibility that the precursor undergoes transport and is subsequently assembled in the axon or nerve terminal, but this possibility could be excluded. One might argue intuitively against such a mechanism on the basis of the large number of enzymes, substrates, and relevant organelles that would have to be transported. Direct experimental evidence against it often consists in the failure to observe movement of radiolabeled precursors into the soluble compartment of axons—nerve endings. In most of those cases where such movement was thought to occur, it subsequently turned out to be attributable to extraaxonal diffusion of administered radiolabel.

The velocity of anterograde ganglioside transport away from the cell body corresponds to the fast phase of protein transport, estimated at 70–100 mm/day in the goldfish optic system and approximately two to four times as fast in the rat or rabbit. This wave, now recognized as also characteristic of phospholipids, glycoproteins, and membrane components in general, stands in contrast to the slower phases involving cytoskeletal elements and other classes of proteins.

Because the gangliosides of the nervous system are a complex mixture of structural types, it was of some interest to determine the molecular pattern undergoing axonal transport. Using the rabbit optic system, it was shown that all detectable gangliosides migrate out of the retina at the same speed. That study further revealed that the types of ganglioside being transported in this single neuron class (retinal ganglion cells) are qualitatively similar to those of whole rabbit brain. From this it can be inferred that the complex molecular pattern of gangliosides in brain arises primarily from the multiplicity of structures in individual neurons, rather than summation of simple but variable patterns in different neuronal types. However, at higher sensitivity it might be possible to detect specificity in regard to one or more ganglioside structures, as shown for a group of minor gangliosides that appear to be unique to a particular class of neuron. There also may be differences in regard to the relative proportions of individual gangliosides, which undoubtedly do exist.

Some interesting conclusions have also been drawn from such studies in regard to ganglioside localization in the neuron. One is that axons do not simply serve as conduits for the flow of gangliosides to nerve endings, but themselves receive a fraction of those transported. Axonal and nerve ending membranes thus appear to function as a unit in the uptake and turnover of these molecules. This behavior has been contrasted with that of glycoproteins, which are targeted primarily for the nerve ending, a much smaller portion entering the axon. Another important difference is reflected in the fact that 30% or so of the gangliosides synthesized in the cell body region (e.g., retina) are committed to transport, compared with about 12% for glycoproteins and 1–2% for proteoglycans. These findings are consistent with other studies showing that although all classes of glycoconjugate are represented in membranes of the axon and nerve ending, gangliosides predominate. They lead to the prediction that analysis of the axon membrane will reveal an especially heavy preponderance of these substances.

The form in which gangliosides migrate down the axon has not been well defined. There are indications that they are part of prepackaged lipid–protein vesicles, of which there may be more than one type. The importance of protein was emphasized by the observation that injection of protein-synthesis inhibitors along with ganglioside precursors effectively blocked axonal transport of the newly synthesized gangliosides. Because gangliosides have been shown to occur in coated vesicles and some types of synaptic vesicles (organelles known to undergo fast transport away from the cell body), these would appear to constitute important transport vectors. Additional forms designed especially for membrane renewal may also exist. Axonal transport has been considered an example of endomembrane flow involving mechanisms of plasma membrane renewal common to all cells. In this scheme, a key role has been ascribed to the Golgi apparatus present in the cell body for the packaging of such vesicles as well as for biosynthesis of the gangliosides themselves.

Developmental questions have been addressed indirectly through analysis of neuronal growth cones, structures that constitute the neurite's leading edge during neuronal differentiation. Using newly developed methods for isolating these particles from embryonic rat brain, the plasma membranes obtained after osmotic disruption were shown to contain appreciable levels of ganglioside. This indicated that neuritic transport begins early in the life of a neuron, even before the process of differentiation has been completed. It is of some interest that glycolipids were shown by microscopic studies to be the predominant form of cell-surface glycoconjugate at this early stage of development; although some glycoproteins are also present at that time, the majority of these appear to enter the neuronal membrane at a later stage. [*See* Gangliosides and Neuronal Differentiation.]

II. AXONAL TRANSPORT IN THE PNS

Ganglioside transport has been studied more recently in the peripheral nervous system, with results that are generally similar to those obtained with the optic system. Thus, using rat sciatic nerve, gangliosides were shown to undergo fast transport away from the cell body at a rate of approximately 300–400 mm/day. As with the CNS, all molecular species appeared to migrate simultaneously.

Use of the sciatic nerve has permitted comparison of transported gangliosides in two different neuronal populations, motor and sensory. In both cases, fast transport was observed with generally similar molecular patterns in which species of the gangliotetraose type predominated. However, sensory neurons also had structures characteristic of other ganglioside families, along with neutral glycosphingolipids that were lacking in motoneurons. Despite the differences, these and related observations suggest that a preponderance of ganglio-series structures is an intrinsic feature of neurons in general.

An interesting difference in the outflow patterns of gangliosides and glycoproteins, after labeling of both by glucosamine injection into the dorsal root ganglia, was that gangliosides did not show the well-defined crest of radioactivity characteristic of glycoproteins, but rather a pattern with an attenuated crest in a series of rather flat curves representing different times. This result can be interpreted as indicating extensive exchange of gangliosides between mobile and stationary axonal structures, in contrast to glycoproteins, which are targeted primarily to the nerve ending. This is similar to the results obtained with the optic system discussed earlier.

A major advantage in using the PNS is that longer nerves (e.g., the sciatic) facilitate study of transport in both directions. The method of choice has been the double-ligation model, which permits observation of both flows simultaneously. The procedure is to inject radiolabeled precursor into the vicinity of the cell bodies (e.g., dorsal root ganglia, lumbosacral spinal cord) and then after variable periods of time tie two ligatures 1 cm apart on a "downstream" segment of the exposed sciatic nerve. Depending on the direction of flow, the radioactive molecules accumulate on one or the other side of each ligature. After a few hours, the tissue is dissected for analysis. By this means, gangliosides were shown to undergo bidirectional transport in both motoneurons and sensory neurons of rat sciatic nerve. A typical experiment of this kind is shown in Fig. 1. Essentially similar results were obtained for glycoproteins. The velocity of transport toward the cell body could not be calculated directly, but the early return of labeled gangliosides appeared consistent with the relatively rapid velocities (equivalent to one-half anterograde flow) previously estimated for proteins.

Use of the double-ligation model has facilitated comparison of the molecular species migrating in the two directions. Such experiments would help to answer long-standing questions concerning possible metabolic changes that might occur during the transport of gangliosides to and return from axonal and neve-ending membranes. It has been conjectured, for example, that because the enzyme neuroaminidase is present in synaptic plasma membranes, a portion of the oligosialogangliosides residing there might be converted to G_{M1}. Using the model depicted in Fig. 1, the procedure is to compare the gangliosides extracted from segment 2, which is toward the cell body in respect to the first ligature (representing transport away from the cell body), with those extracted from segment 4, which is on the opposite side in respect to the second ligature (representing transport toward the cell body). The general pattern distributions in relation to sialic acid number turned out to be similar (Fig. 2), indicating no apparent change in the relative proportions of mono-, di-, tri-, or tetrosialogangliosides; this suggested no major metabolic alterations due to neuraminidase in axons or nerve endings dur-

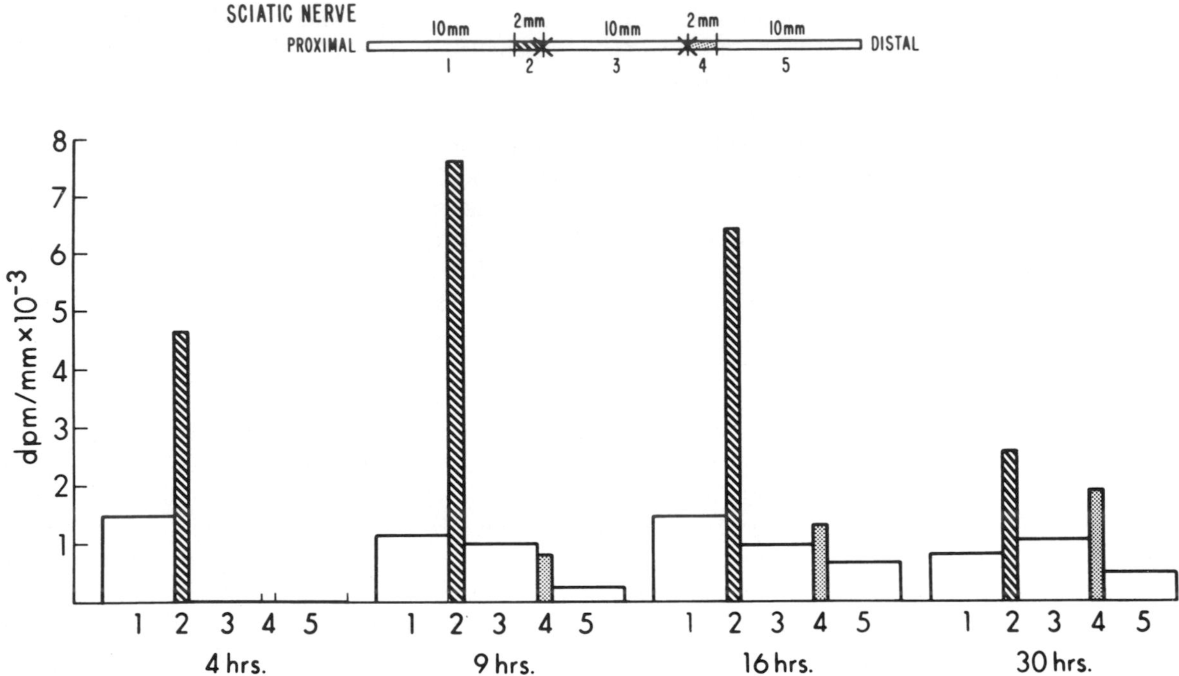

FIGURE 1 Profiles of ganglioside ^{3}H radioactivity in sciatic nerve segments (lower diagram). Ligatures were tied (upper diagram) at the times indicated after [^{3}H] glucosamine injection into motoneurons, and accumulations progressed for 2–3 hr. Gangliosides were isolated in pure form from each segment and counted. Radioactivity accumulation at segment 2 represented flow away from cell bodies, and that at segment 4, flow in the opposite direction. The latter was first discernible 16 hr after injection. [From D. A. Aquino *et al.* (1985). *J. Neurochem.* **45**, 1262, with permission.]

ing the course of the round trip. However, the possibility of metabolic change(s) of a minor nature remains open until the compositions of these two pools can be examined in more detail.

Observation of retrograde axonal transport provides at least a partial answer to the question of the mechanism of ganglioside turnover. The phenomenon of bidirectional transport indicates that at least a portion of the gangliosides situated in axonal and nerve-ending membranes return to the cell body for catabolism and/or recycling. There are suggestions that other mechanisms may also operate because the amount returning is possibly less than the amount leaving the cell body. One such mechanism is shedding from the surface; this was previously shown to occur for glycoproteins of certain animal cells and more recently for gangliosides as well. It could account for at least part of the gangliosides detected in the soluble compartment of brain. Another possibility is *in situ* degradation within the axon and nerve ending. However, although selective protein degradation has been reported in these structures, there is as yet no direct evidence pertaining to gangliosides.

III. GANGLIOSIDE TRANSFER PROTEINS

Although anterograde and retrograde axonal transport can account for some of the movement of gangliosides within the neuron, it is likely that additional mechanisms are needed for distribution to and from their sites of use. Transfer of gangliosides between membranes, analogous to the well-characterized transfer of phospholipids, may be one such mechanism because proteins capable of catalyzing such transfer and/or exchange have been isolated from the soluble compartment of brain. This activity was displayed toward all the major brain gangliosides and some neutral glycolipids as well, but not toward phosphatidylcholine. Transfer proteins possess a molecu-

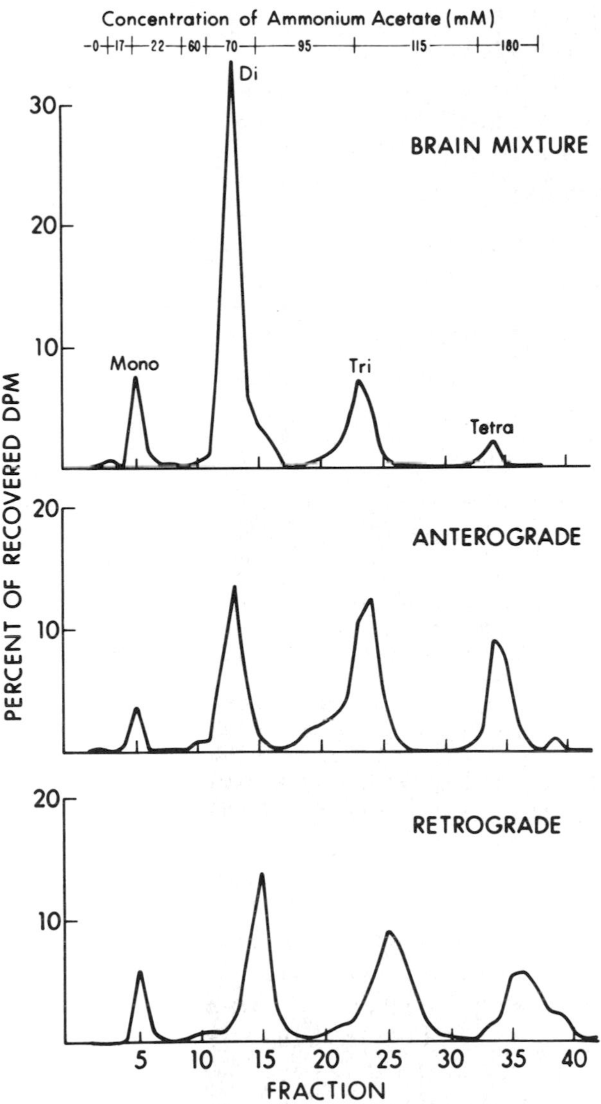

FIGURE 2 Fractionation of gangliosides on DEAE-Sephadex. A brain mixture is compared with gangliosides undergoing transport away from the cell body (anterograde) or toward the cell body (retrograde) transport in motoneurons of sciatic nerve. Gangliosides were labeled before transport by injection of [^{3}H]glucosamine into the lumbosacral spinal cord. [From D. A. Aquino *et al.* (1985). *J. Neurochem.* **45**, 1262, with permission.]

lar weight of approximately 20,000, similar to activator proteins for certain glycohydrolase enzymes, which also possess glycolipid transfer activity; however, they are clearly different proteins.

The physiological function of these transfer proteins is still unknown, as is their cellular locus. If they are cytosolic proteins, the question arises as to how they would interact with intracellular gangliosides, most of which are believed to be sequestered within the luminal portion of organelles (Fig. 3). Because transmembrane movement ("flip-flop") of glycolipids is considered unlikely, they would not readily become accessible to transfer. At this stage we cannot preclude the existence of vesicles or organelles with gangliosides on the cytoplasmic surface nor the possibility of a role within organelles (e.g., the Golgi matrix). Such possibilities can be better assessed when the localization of ganglioside transfer proteins has been determined. Should they turn out to be extracellular, a catalytic role in intercellular transfer would need to be considered (Fig. 3).

IV. SUMMARY

The predominant—perhaps sole—site of ganglioside synthesis in the neuron is the endoplasmic reticulum/Golgi complex, from which they enter the transport machinery of the axon (and persumably dendrites). Fast transport away from the cell body is the primary form of movement into the axons and nerve endings of both the CNS and PNS. Axonal transport back to the cell body, revealed by the double-ligation method applied to sciatic nerve, accounts for the turnover of at least some of the membrane-bound gangliosides of the axon and nerve ending. Similarities in the gross patterns of gangliosides moving in the two directions suggest there are few if any metabolic alterations of these molecules during transport and use. Comparison studies have revealed that transported gangliosides are targeted in both axonal and nerve ending membranes, in contrast to glycoproteins, which are destined primarily for the nerve ending. Ancillary findings were that virtually all gangliosides undergo transport simultaneously and that the molecular diversity detected in a single class of neurons (retinal ganglion cells) is qualitatively similar to that of whole brain. Hence the complexity of ganglioside structural types arises from the pattern within individual neurons. It would appear that ganglio-series structures are an intrinsic feature of neurons in general. Ganglioside transfer proteins, which catalyze exchange and/or net transfer of glycolipids between membranes, were shown to occur in the soluble fraction of brain, suggesting another mode of intra- or intercellular ganglioside movement. Current concepts of ganglioside transport after their synthesis in the cell body are summarized in Fig. 3.

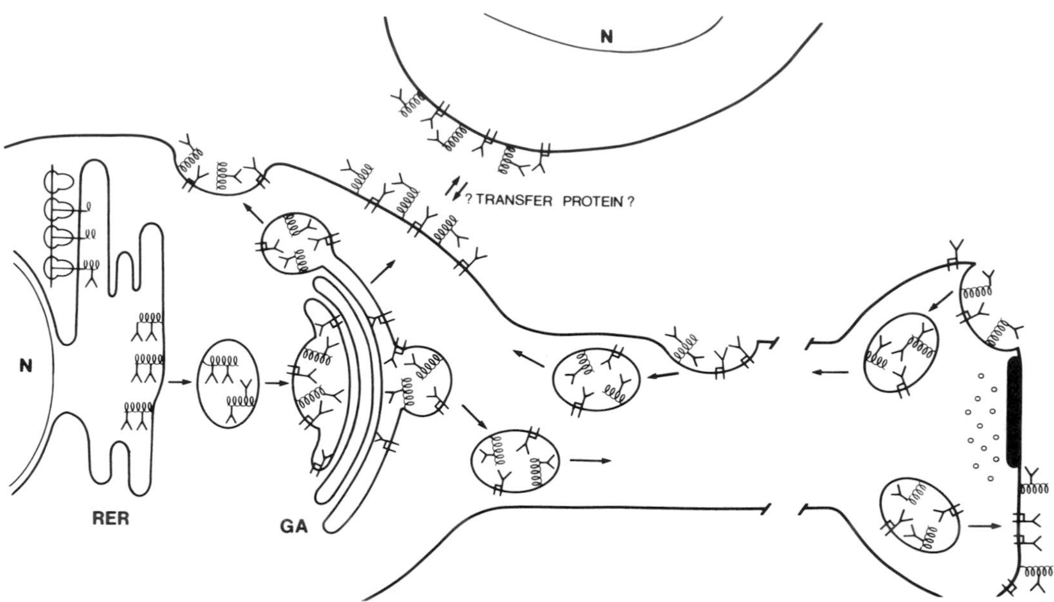

FIGURE 3 Aspects of ganglioside and glycoprotein movement in the neuron, as currently visualized. Asparagine-linked glycoprotein (ᘐᘐᘐ with the carbohydrate ≺ attached) commences synthesis in the rough endoplasmic reticulum (RER) and passes through the Golgi apparatus (GA). Gangliosides (⊐≺) are also synthesized at least partially in the GA, but the role of the RER is not known. From the GA, glycoconjugates migrate to the plasma membrane and to axonal and nerve-ending membranes via fast axonal transport. Return to the cell body is also depicted. Intracellular migration, including axonal transport, is thought to occur in vesicles, the carbohydrate portions being sequestered in the luminal compartments. Intercellular movement of gangliosides, catalyzed by ganglioside transfer protein, is presented as a hypothetical process. N, nucleus. [From R. W. Ledeen *et al.* (1987). *In* "Gangliosides and Modulation of Neuronal Functions" (H. Rahmann, ed.), pp. 259–274. Springer-Verlag, Berlin/Heidelberg, with permission.]

BIBLIOGRAPHY

Aquino, D. A., Bisby, M. A., and Ledeen, R. W. (1985). Retrograde axonal transport of gangliosides and glycoproteins in the motoneurons of rat sciatic nerve. *J. Neurochem.* **45**, 1262–1267.

Aquino, D. A., Bisby, M. A., and Ledeen, R. W. (1987). Bidirectional transport of gangliosides, glycoproteins and neutral glycosphingolipids in the sensory neurons of rat sciatic nerve. *Neuroscience* **20**, 1023–1029.

Gammon, C. M., Vaswani, K. K., and Ledeen, R. W. (1987). Isolation of two glycolipid transfer proteins from bovine brain: Reactivity toward gangliosides and neutral glycosphingolipids. *Biochemistry* **26**, 6239–6243.

Goodrum, J. F., Stone, G. C., and Morell, P. (1989). Axonal transport and intracellular sorting of glycoconjugates. *In* "Neurobiology of Glycoconjugates" (R. U. Margolis and R. K. Margolis, eds.), pp. 277–308. Plenum, New York.

Harry, G. J., Goodrum, J. F., Toews, A. D., and Morell, P. (1987). Axonal transport characteristics of gangliosides in sensory axons of rat sciatic nerve. *J. Neurochem.* **48**, 1529–1536.

Landa, C. A., Defilpo, S. S., Maccioni, H. J. F., and Caputto, R. (1979). Deposition of gangliosides and sialoglycoproteins in neuronal membranes. *J. Neurochem.* **37**, 813–823.

Ledeen, R. W., Skrivanek, J. A., Nunez, J., Sclafani, J. R., Norton, W. T., and Farooq, M. (1981). Implications of the distribution and transport of gangliosides in the nervous system. *In* "Gangliosides in Neurological and Neuromuscular Function, Development, and Repair" (M. M. Rapport and A. Gorio, eds.), pp. 211–223. Raven, New York.

Ledeen, R. W., Aquino, D. A., Sbaschnig-Agler, M., Gammon, C. M., and Vaswani, K. K. (1987). Fundamentals of neuronal transport of gangliosides. Functional implications. *In* "Gangliosides and Modulation of Neuronal Functions" (H. Rahmann, ed.), Nato ASI Series, Vol. H7, pp. 259–274. Springer-Verlag, Berlin, Heidelberg.

Rosner, H., Wiegandt, H., and Rahmann, H. (1973). Sialic acid incorporation into gangliosides and glycoproteins of the fish brain. *J. Neurochem.* **21**, 655–665.

Tettamanti, G., and Riboni, L. (1993). Gangliosides & Modulation of the Function of Neural Cells in Advances in Lipid Research (R. M. Bell, Y. Hannum, and A. Merrill, Jr. eds.), pp. 235–260. Academic Press, San Diego.

Yamada, K., Abe, A., and Sasaki, T. (1986). Glycolipid transfer protein from pig brain transfers glycolipids with β-linked sugars but not with α-linked sugars at the sugar–lipid linkage. *Biochem. Biophys. Acta* **879**, 345–349.

Yates, A. J., Tipnis, U. R., Hofteig, J. H., and Warner, J. K. (1984). Biosynthesis and transport of gangliosides in peripheral nerve. *In* "Ganglioside Structure, Function and Biomedical Potential" (R. Ledeen, R. K. Yu, M. M. Rapport, and K. Suzuki, eds.), pp. 155–168. Plenum, New York.

Gastric Circulation

EDWARD H. LIVINGSTON
PAUL H. GUTH
Veterans Administration Medical Center, University of California, Los Angeles, and Center for Ulcer Research and Education, Los Angeles

GLOSSARY

Acid secretion Metabolically active process of the stomach that secretes strong acid into the lumen, assisting in the digestion of food

Artery Vessel supplying oxygenated blood from the heart to a tissue

Capillary Small thin-walled vessel located between the arterial and venous sides of the circulation; where the exchange of oxygen and nutrients occurs

Erosion Damage confined to the mucosa and not extending through the muscularis mucosa

Mast cells Cells within the gastric mucosa that contain histamine. When histamine is released, it stimulates gastric acid secretion

Microcirculation Network of small arterioles, venules, and capillaries that are closely interfaced with the body cells. It is at this level of the circulation that nutrients and oxygen are exchanged and blood flow is controlled

Mucosa Mucous membrane consisting of a specialized epithelial layer on the luminal side of the gastric wall

Muscularis externa Layer of smooth muscle on the outer portion of the stomach wall that is invested with a thin layer of epithelium known as the serosa

Muscularis mucosa Thin layer of smooth muscle at the base of the mucosa (between the mucosa and the submucosa)

Portal vein Vein draining blood from the stomach and the intestines into the liver. From the liver the blood enters the vena cava and then travels to the heart

Sphincter Circularly oriented groups of smooth muscles surrounding various portions of the gastrointestinal tract (e.g., the pylorus) that contract or relax to regulate the diameter of the structure at that location. These serve to regulate the passage of contents at their locations

Submucosa Loose connective tissue just beneath the mucosa

Ulceration Local area of tissue destruction in the stomach extending through the entire mucosa, including the muscularis mucosa. When perforation occurs, the damage extends through the entire wall of the stomach

Vein Vessel draining blood from a tissue and connected to increasingly larger veins that ultimately enter the heart

GASTRIC CIRCULATION IS THE SYSTEM OF ARTERIES, veins, and capillaries that provides for the flow of blood to, through, and out of the stomach. The stomach is a major digestive organ in the gastrointestinal tract, lying between the esophagus and the duodenum in the abdominal cavity. The proximal stomach serves two major functions. The fundus is the temporary storage portion and stretches to accommodate incoming food. The relaxation of this portion with feeding is known as receptive relaxation and is mediated by the vagus nerve. As the stomach empties, the fundus gradually contracts to its original size. Digestion of food into particles of a size that can be absorbed by the intestine first begins with the act of chewing, breaking up the largest of particles. Food particles entering the stomach are large and must be broken down further before they enter the small bowel. The fundus and the midportion of the stomach, known as the corpus, secrete large amounts of highly concentrated (i.e., up to 0.15 *N* HCl) acid from their parietal cells. Another population of cells, the chief cells, secrete the proteolytic enzyme pepsin, which, in combination with the acid, breaks down the food particles chemically. The antrum is the muscular distal portion

ENCYCLOPEDIA OF HUMAN BIOLOGY, Second Edition, VOLUME 4.

of the stomach, which is capable of generating powerful contractions that break up the food particles by mechanical means. When the particles are small enough they are "pumped" by the antral contractions from the stomach into the duodenum.

With all of these functions to perform, the stomach necessarily has a high metabolic demand and therefore is richly vascularized. Blood supplies the necessary oxygen and nutrients to support the metabolic activity required for the secretory and contractile functions. Acid secretion requires the delivery of chloride and carbon dioxide to generate the hydrochloric acid that is secreted. Blood also carries away the metabolites accumulated from these processes. On average the stomach secretes about 1 liter of fluid per day, all of which is derived from the blood. This secretion is needed to facilitate the partial digestion and passage of ingested food distally into the gastrointestinal tract.

I. ANATOMY

Soon after the esophagus traverses the diaphragmatic hiatus, it enters the abdomen and terminates at the cardia of the stomach (Fig. 1). The stomach lies in the left upper quadrant of the abdomen and extends from left to right. The lower side of the stomach is called the greater curvature and the upper side is the lesser curvature. It fans out to the corpus, then tapers to the antrum and the pylorus. The pylorus is fixed in position on the right side of the abdomen. The fundus, corpus, and antrum do not have attachments anteriorly or posteriorly and are therefore relatively free to move as the stomach fills with food. [*See* Digestive System, Anatomy.]

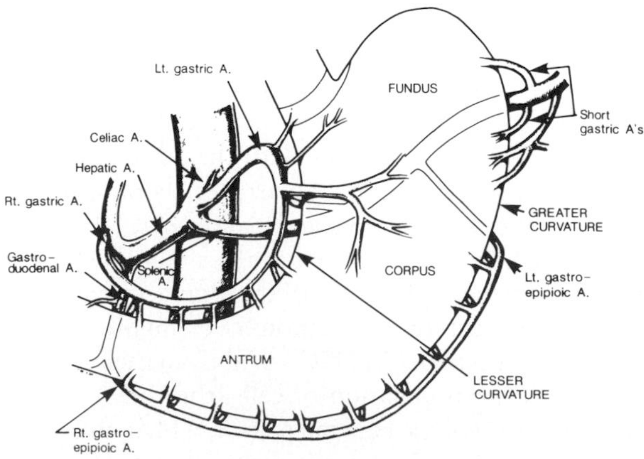

FIGURE 1 Anatomy of the stomach and its vasculature. A, artery.

The first major branch of the aorta after it enters the abdominal cavity is the celiac axis (Fig. 1). Branches of this artery provide the major blood supply to the stomach. The first branch of the celiac axis is the left gastric artery, which courses from the cardia to the pylorus along the lesser curvature. The left gastric artery supplies the distal esophagus, cardia, and the superior aspect of the corpus, antrum, and pylorus. The right gastric artery originates from the common hepatic artery, a branch of the celiac axis. This artery is found on the superior surface of the pylorus, where it anastomoses with the left gastric artery and supplies the superior pylorus. Both left and right gastric arteries send branches to the anterior and posterior walls of the stomach. The inferior portion of the stomach is supplied by the gastroepiploic arteries. The left gastroepiploic artery branches from the splenic artery, the second branch of the celial artery. This artery follows the greater curvature, anastomosing with the right gastroepiploic artery, which courses along the right side of the greater curvature, originating from the gastroduodenal artery, a branch of the common hepatic artery. The fundus is supplied by the short gastric arteries that branch from the splenic artery in the region of the splenic hilum.

The vascular supply of the stomach has a large number of interconnections, assuring that occlusion of one or several arteries will not lead to gastric ischemia. The right and left gastric arteries anastomose at the antropyloric area, the right and left epiploic arteries anastomose along the greater curvature, and the branches of the gastric and gastroepiploic arteries anastomose with each other along the anterior and posterior aspects of the stomach. Within the stomach, all of these vessels terminate in the submucosal arteriolar plexus, which connects all of the supplying vessels together. The gastroepiploic arteries anastomose with branches of the superior mesenteric artery, assuring vascular supply even if the celiac axis blood flow were to become compromised.

The veins of the stomach follow the course of the arteries, but terminate in various branches of the portal vein. The proximal stomach drains into the splenic vein, and the distal stomach empties into the superior mesenteric vein.

The interstitial spaces are drained by the lymphatics, which follow the course of the arteries, coalesce, and enter the lymph nodes (Fig. 2). The majority of the stomach is drained by lymphatics, which cluster about the left gastric artery, known as the superior

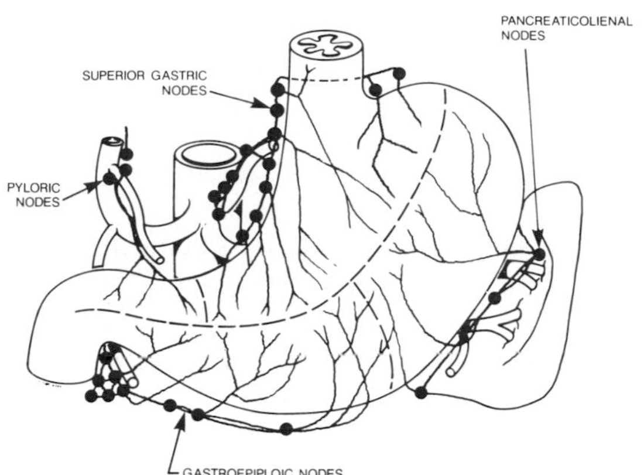

FIGURE 2 Lymph drainage of the stomach.

gastric lymph nodes. The second largest area is drained by the right gastroepiploic nodes and the pyloric nodes. The fundic region drains into the left gastroepiploic nodes and the pancreaticosplenic lymph nodes. A small portion of the pylorus and the antrum is drained by the right gastric lymph nodes.

II. MICROVASCULAR ANATOMY

The arteries supplying the stomach course along the external surface of the stomach and eventually branch into small vessels that perforate the muscularis externa (Fig. 3). As these small arteries travel through the muscle layer, they branch into smaller arteries, which supply the muscularis externa. The arteries then terminate in the submucosal arteriolar plexus. The submucosal plexus of arterioles (i.e., very small arteries) is a network of highly interconnected arcades of vessels. Vessels branch from this network, enter the mucosa, and terminate at the base of the mucosa in capillaries that traverse the mucosal layer and enter collecting venules in the luminal one-third of the mucosa. These venules traverse the mucosa and enter a plexus of venules (i.e., very small veins), which architecturally resemble the arterial plexus. Veins from the plexus join draining veins from the muscle layer, traverse the muscularis externa, and coalesce to form the large veins of the stomach.

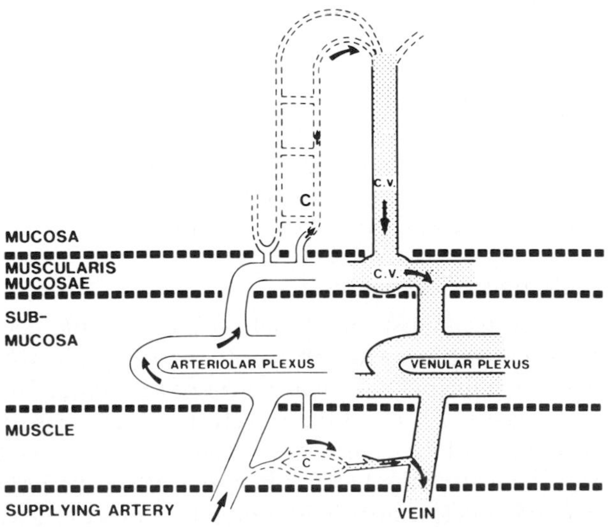

FIGURE 3 Gastric microcirculation. C, capillaries; C.V., collecting vein.

The capillaries of the corpus mucosa originate from arterioles at the base of the mucosa and extend to the luminal portion of the mucosa, entering draining venules. The capillaries form a network that surrounds the acid- and pepsin-secreting glands. Characteristic of the gastric mucosal capillaries is that they originate from the submucosal arterioles, enter the mucosa, travel the entire thickness of the mucosa, and then turn at the mucosal surface to enter the collecting venules in the upper one-third of the mucosa.

The blood supply to the stomach is highly redundant, not only because of the multiple interconnecting arteries supplying it, but also because of the rich network of anastomosing arterioles in the submucosa. This ensures adequate blood supply for the stomach even if one or more vessels should become damaged. One can envision an evolutionary advantage for the redundant blood supply: blood supply is critically important for mucosal resistance against the corrosive effects of luminal acid. Should the blood supply be interfered with, severe mucosal damage and even life-threatening hemorrhage could occur. The highly redundant blood supply guards against this possibility.

III. PHYSIOLOGY

A. Methods of Study

To assess the contribution of blood flow to various physiological and pathological conditions, it is necessary to accurately measure it. In experimental animals the amount of blood passing through some unit volume of tissue per unit time can be determined by measuring all of the venous drainage. This would give the total volume flow for all layers of the tissue. Flow in different tissue compartments (i.e., the mucosa, submucosa, or the muscularis externa) differs, depending on their individual metabolic demands. The majority of flow is distributed to the mucosa, which has the highest metabolic activity. Flow alterations in the mucosa might lead to pathological conditions such as ulcers. The available methods applicable to blood flow measurement in the human gastric mucosa are based on the clearance or extraction of a tracer, determination of flow velocity through capillaries, or measurement of the amount of hemoglobin in the tissue's vessels.

1. Clearance Techniques

Clearance techniques are based on the clearance of a tracer from the blood into the tissue or vice versa, whose concentration in the tissue can be measured. For a freely diffusible tracer (i.e., one that is soluble in blood and diffuses out of the blood into the tissue through the capillaries such that it rapidly achieves equilibrium between tissue and blood), the amount of tracer extracted from the blood by one pass through the capillaries is proportional to the flow (volume of blood per unit time) and the concentration difference:

$$Q_t = F(C_a - C_v)\ dt,$$

where Q_t is the quantity, in moles, of tracer in the tissue that changes over time; F is the flow, in milliliters per second; C_a is the concentration of tracer in the supplying arterial blood, in moles per liter; C_v is the concentration of tracer in the venous blood; and dt is the time interval, in seconds, over which the experiment is performed. The quantity of a material per unit volume in which it is dissolved is its concentration:

$$C_t = Q_t/V,$$

where C_t is the concentration of the tracer in the tissue, in moles per liter, and Q_t is the total quantity of substance, in moles, in the volume of tissue V, in liters. Combining these two equations yields

$$C_tV = F(C_a - C_v)\ dt$$

and

$$C_t = F(C_a - C_v)\ dt/V.$$

At equilibrium the ratio of concentration of the tracer in venous blood and in tissue is defined by the partition coefficient λ:

$$\lambda = C_v/C_t.$$

Substituting

$$C_t = F(C_a - C_t\lambda)\ dt/V$$

for the clearance of a tracer from previously saturated tissue (after the gas is removed from the inspired air), the C_a falls to 0:

$$C_t = F(-C_t\lambda)\ dt/V.$$

The solution of this differential equation is

$$C_t = C_o e^{-\kappa t},$$

where C_o is the tissue concentration of tracer at complete tissue saturation, and κ is $\lambda F/V$.

Thus, by measuring the rate constant for tissue clearance of the tracer (or uptake), the blood flow per unit volume can be determined. For determination of the constant, the tracer must be measured continuously over time. The presence of a tracer can be determined by radioactive decay or electrochemical reaction.

^{85}Kr has been used for the measurement of human gastric blood flow by injecting the tracer intraarterially until the stomach tissue was saturated with tracer. Upon reaching tissue saturation, the injection was stopped and the disappearance of tracer was determined by observation of radioactive decay.

Hydrogen gas concentration can be measured by the electrolytic oxidation of molecular hydrogen at the surface of a platinum electrode. The hydrogen gas clearance technique is a polarographic technique in which current (due to electrons released as the hydrogen is oxidized) passing through an electrode held at constant voltage is proportional to the concentration of the hydrogen being oxidized at the electrode surface. This technique measures flow in a small area of the mucosa around the electrode surface.

2. Partition Techniques

Partition techniques are based on the principle that weak bases (i.e., with pKa's of 5–10) remain uncharged in the pH 7.4 environment of the blood. The uncharged molecules diffuse freely across the plasma membranes of the gastric cells. Upon reaching the acid environment of the lumen, they ionize (i.e., become charged) and lose their lipid solubility. Consequently, they partition in the gastric lumen. The rate of luminal accumulation in the face of a constant plasma concentration of the tracer is proportional to the gastric blood flow. Aminopyrine, a weak base measurable by fluorescence or labeling with ^{14}C, was used extensively for many years for gastric blood flow determination. However, it was found to be actively secreted by parietal cells, leading to overestimation of blood flow in the presence of stimulated acid secretion. Partition techniques have been largely replaced by inert gas clearance techniques.

3. Laser Doppler Technique

The laser Doppler technique is based on the Doppler principle. If the tissue is illuminated with light of a uniform frequency, the reflected light will have a shift in frequency that is proportional to the velocity of cells moving in the field of observation. The velocity can be determined from the frequency (i.e., color) shift, and if the volume of tissue under study is known, the blood flow can be derived. This technique has the advantage that the incident and reflected light can be carried within fiber-optic light guides. The fiber-optic cable can be narrow and passed through an endoscope to measure gastric mucosal blood flow in humans. The measurement can be continuous over time. The major difficulties of the technique are failure to maintain adequate optical coupling between the tissue and the probe, resulting in motion artifact and, once coupling is established, the possibility of excessive pressure, causing mucosal ischemia (i.e., stoppage of blood flow). In addition, the resulting blood flow measurement is in volts, and as yet there is no valid way to convert this into blood flow units (i.e., milliliters per minute per gram of tissue).

4. Reflectance Spectrophotometry

In reflectance spectrophotometry, white light illuminates the tissue under study, and the reflected light is analyzed spectrophotometrically. By measuring the intensity of the reflected light at the wavelengths characteristic for hemoglobin and oxygenated hemoglobin, the relative concentration of hemoglobin and the fraction of it that is oxygenated can be measured. These measurements are not absolute and provide only indices of the tissue's hemoglobin content and hemoglobin oxygen saturation. However, the hemoglobin concentration is reduced with diminished blood flow and increases when blood flow increases. Similarly, the proportion of hemoglobin that is oxygenated is reduced with reduced blood flow. As with the laser Doppler technique, the incident and reflected light can be carried along fiber-optic light guides and used for measurement of human gastric mucosal blood flow through the endoscope. The same limitations apply: the optical coupling is difficult to maintain (although motion artifacts are not so great as with the laser Doppler), and if too much pressure is applied to the end of the fiber-optic cable, blood flow in the tissue is stopped.

B. Intrinsic Regulation

Blood flow to the stomach must supply needed oxygen and nutrients to the cells as well as carry away metabolic waste and back-diffusing acid. Therefore, systems regulating the amount of blood flow entering the tissue must be sensitive to the metabolic needs of the tissue or to the presence of acids or waste. Intrinsic regulation of blood flow involves the mechanisms by which changes in tissue perfusion are triggered by locally occurring events. This is different from extrinsic regulation, which is the control of blood flow by mechanisms that function primarily outside the stomach.

The fundamental mechanism for the regulation of blood flow entering a capillary bed is the control of arteriolar diameter. It is at the arteriole that the major resistance to flow is encountered. Arterioles have the capability of completely closing or dilating widely. These vessels are sensitive to the effects of the vasoactive agents discussed in Sections III,C and IV. The importance of these vessels in the regulation of flow entering the capillary bed can be inferred from Poiseuille's law,

$$Q = \frac{\pi \Delta P r^4}{8 v l},$$

which states that the flow through a vessel (Q, in milliliters per second) is equal to the constant π times the change in pressure (ΔP) across the vessel times the radius of the vessel raised to the fourth power (r^4), all divided by the product of eight times the viscosity constant for blood (η) times the length of the vessel (l). From this equation it is readily apparent that the amount of blood flow through a vessel is exquisitely sensitive to the diameter of the vessel. Thus, the arterioles, which are capable of large diameter changes, are the single most important regulator of the amount of blood entering the tissue.

A second potential mechanism for the control of blood flow is change in the number of capillaries receiving blood flow. In most tissues this is accomplished by opening or closing of the precapillary sphincters. At any given time a certain fraction of the capillaries are perfused. To increase the delivery of oxygen and nutrients, more sphincters are opened, perfusing a greater number of capillaries, thereby increasing the surface area available for capillary exchange. However, no evidence for this type of mechanism exists for gastric microcirculation. In the stomach, submucosal arteriolar diameter changes have the most significant regulatory influence on blood flow.

The blood flow into the gastric tissue, and therefore arteriolar diameter, must be sensitive to a variety of factors. An important role of blood flow is to provide oxygen to the tissues. It has been hypothesized that one mechanism for the control of blood flow is the effect of oxygen on the arterioles. Contraction of the arterioles is an active metabolic phenomenon dependent on the presence of oxygen. If tissue levels of oxygen decrease, less would be available to support the arteriolar contraction. In this setting the arterioles would dilate, increasing the perfusion of the tissue. This sort of mechanism probably does not exist in the stomach, because of the anatomic arrangement of the vessels. In the stomach the arterioles that regulate mucosal blood flow are in the most basal portion of the mucosa, up to 1 mm away from the superficial mucosa. The likelihood that precapillary arterioles can detect the oxygen demands of tissue that far away is small. Similarly, no evidence exists that deficits of specific nutrients required for tissue metabolism can trigger increases in blood flow. Most likely, the response to gastric mucosal needs for blood flow results from the accumulation of products, rather than the substrates of metabolism.

One of the more likely regulators of blood flow is adenosine. Adenosine is a nucleotide that transports energy within cells by carrying phosphate groups. Utilization of energy occurs by phosphorylated adenosine releasing its phosphate groups. As a consequence, when cells are metabolically active, adenosine accumulates inside and outside the cell. Most studies of the relationship between adenosine and blood flow regulation have been performed in the heart. It appears that within the cardiac circulation, adenosine serves as the dominant regulator of blood flow. During conditions of low oxygenation or increased cardiac work, adenosine accumulates in the interstitial space, resulting in vasodilatation and increased blood flow.

Adenosine appears to be a potent arteriolar dilator in all tissues studied. Studies of gastric circulation confirm that adenosine probably contributes to the regulation of gastric mucosal blood flow. Topical application of adenosine to gastric microvessels results in gastric mucosal hyperemia (i.e., an increased amount of blood). Increases in acid secretion lead to large increases in metabolic demand, concomitant with large increases in gastric mucosal blood flow. Treatment with the specific adenosine inhibitor 8-phenyltheophylline inhibits the blood flow response to stimulated acid secretion.

Histamine might also participate in the regulation of gastric mucosal blood flow. In the presence of injurious agents, histamine is released by mucosal mast cells. Histamine applied locally to the stomach results in increased mucosal blood flow and increases capillary permeability. It has been hypothesized that when the mucosa is damaged, histamine is released and signals the need for increased blood flow. The increased blood flow is needed to neutralize back-diffusing acid and carry away toxic substances.

The eicosanoids are a group of chemicals derived from arachidonic acid. They are divided into two major classes: the prostaglandins, which are synthesized by cyclooxygenase, and the leukotrienes, synthesized by lipoxygenase. Prostaglandins and leukotrienes mediate a variety of physiological responses and are released by nearly all tissues. The different prostaglandins and leukotrienes differ in their side-chain substitutions and have widely varying properties. Some are vasodilatory; others produce vasoconstriction. Capillary permeability can be altered by these compounds. Both prostaglandins and leukotrienes are released by cells within the gastric mucosa and contribute to the regulation of mucosal blood flow.

Diameter changes of the submucosal arterioles control mucosal blood flow. How the mucosa signals the submucosal arterioles to change their diameter in response to mucosal blood flow demands is not clear. A number of substances found within the mucosa are released either as a result of metabolic activity or to initiate processes such as acid secretion that result in augmented metabolic activity. Many of these agents (e.g., histamine, prostaglandins, and some gut peptides) are vasoactive. Theoretically, they could diffuse from the mucosa to the submucosa and, in turn, regulate mucosal flow by their action on the submucosal vessels. However, the probability that any of these agents could reach the submucosa is low. The interstitial space contains a number of enzymes that deactivate the mediators before they can travel far from where they are released. By rapidly deactivating these highly potent substances, distant unwanted effects of these agents are prevented. One hypothesis concerning the regulation of mucosal blood flow is based on the observation that nerves traverse the mucosa and could provide pathways for reflex control of the submucosal vessels. In this way the mediators found within the mucosa could stimulate mucosal sensory nerves, which, in turn, transmit a signal to other nerves that control submucosal arteriolar diameter. There is evidence that this sort of pathway exists for the hyperemic response to injurious agents penetrating into the mucosa.

Aside from regulating the amount of blood flow entering the tissue in response to metabolic needs, the local circulatory system must maintain adequate perfusion in the face of changes that might occur in the systemic circulation. Two intrinsic regulatory phenomena exist to perform this task: autoregulation and reactive hyperemia.

Autoregulation is the ability of the tissue to maintain nearly constant levels of perfusion despite wide variations in the pressure of the supplying arteries. In most tissues the blood flow changes little, despite wide variations in perfusion pressure. It is thought that the smooth muscle of arterioles has the ability to contract in proportion to the pressure exerted against it. Thus, when the perfusion pressure increases, the arteriolar smooth muscle contracts to hold constant the blood flow to the tissue. Another possible mechanism is that concentrations of oxygen or metabolites (e.g., adenosine or carbon dioxide) adjacent to the arterioles regulate their diameter. If perfusion pressure increased and flow through the tissue increased, these metabolites would be washed away at higher rates than during basal conditions, reducing their concentrations and leading to vasoconstriction and a return of flow to basal levels. Conversely, a decrease in perfusion pressure would lead to lower flow through the capillary bed, a rise in the concentrations of the metabolites, and a consequent vasodilatation and return of flow to basal levels.

In the stomach, autoregulation has been difficult to establish. Rather than autoregulate, blood flow tends to parallel perfusion pressure in the stomach in intact animals. With sympathetic denervation, however, autoregulation of the stomach approximates that of other tissues. Study of the oxygen extraction of the tissue demonstrates that this parameter is held nearly constant, despite wide variations in perfusion pressure or blood flow. Thus, autoregulation of the stomach occurs not so much as has been described in other tissues with constancy of blood flow in the face of changes in perfusion pressure, but rather as constancy of oxygen delivery to the tissue, despite variations in perfusion pressure.

Reactive hyperemia is the rebound increase in blood flow observed following a period of ischemia. Ischemia can be defined as a deficit of blood supply in relation to the metabolic needs of the tissue. When the restriction of blood supply is removed, there is a short-lived increase in blood flow above basal values. During the period of ischemia, tissue oxygen levels fall and metabolites such as carbon dioxide and adenosine accumulate. These effects serve to dilate the arterioles so that when the restriction of blood flow is removed, the flow is high. As the metabolites are washed away and tissue oxygen content is restored, the arterioles again constrict and flow returns to the basal level.

The relationship between gastric blood flow and acid secretion has been extensively studied. Studies using techniques not affected by the process of acid secretion (e.g., the hydrogen gas clearance technique) have indicated that there is most likely a linear relationship between gastric mucosal blood flow and acid secretion.

One difficulty that arises in the study of the acid secretion–blood flow relationship is that some of the agents used to control acid secretion have direct vasoactive effects. For example, histamine is a potent vasodilator, in addition to its effect on acid secretion. Histamine can potentially overestimate the positive relationship between blood flow increases resulting from stimulated acid secretion, because in itself it increases blood flow by producing vasodilatation. When used in too high a dose, histamine obscures the relationship by causing vasodilatation throughout the body, resulting in systemic hypotension, reducing gastric blood flow, while acid secretion remains elevated. Thus, in studying the relationship it is important to account for potential primary vasoactive effects of the agents being studied.

C. Extrinsic Regulation

The distribution of blood flow to different organs is regulated by the nervous system, which is important for the maintenance of homeostasis of the entire organism. The brain requires a constant supply of well-oxygenated blood for normal function, whereas the other organs of the body can tolerate temporary fluctuations in their blood supply. Nervous system control of the blood supply coordinates the blood supply to all of the organs with regard to the overall needs of the entire organism.

A variety of nuclei in the brain and the brain stem serve to control gastric blood flow. These receive inputs from sensory nerves distributed throughout the body and send out fibers that reach the vessels of the stomach. The most important nuclei are those contained within the hypothalamus, which mediate sympathetic nervous system function, and those of the medulla, which regulate the parasympathetic nervous system.

The sympathetic nerve supply to the stomach is derived from the 6th to the 10th spinal nerves and terminates in the celiac ganglion. From the ganglion, postganglionic fibers reach the stomach as discrete nerves, as nerves mixed with vagal fibers, or as nerves accompanying the arteries. The arterioles of the submucosal plexus are richly innervated with terminals from these nerves. There is some evidence that the capillaries of the mucosa are also innervated with sympathetic fibers. The sympathetic nervous system is activated to enhance an animal's response in a "fight-or-flight" situation. Therefore, pulse rate and blood pressure rise, blood flow to the brain and muscles increases, and blood flow to the viscera decreases. It is thought that the reduction in visceral blood flow provides more blood supply to organs for response to environmental threats.

Sympathetic nerves release norepinephrine as their transmitter. Stimulation of the sympathetic nervous system also activates the adrenal medulla, causing it to release epinephrine.These neurotransmittors act on receptors located on the arteriolar smooth muscle. There are two major classes of receptors: α-adrenergic receptors are vasoconstrictive, and β receptors are vasodilatory. Norepinephrine reduces gastric mucosal blood flow by activatig the α receptors of the arterioles of the submucosal plexus. Epinephrine can produce either vasodilatation, by activating the β receptors, or vasoconstriction, by its action on α receptors. The activity of these neurotransmitters on the gastric circulation is complex, and study is complicated by the systemic cardiovascular effects of both agents. To reliably identify their actions on the gastric circulation, they must be administered locally by close intraarterial infusion. Studies of both epinephrine and norepinephrine intraarterial infusions have revealed that they initially decrease gastric mucosal blood flow, but then, after several minutes, mucosal blood flow increases. This phenomenon is known as autoregulatory escape. When the sympathetic nerves supplying the stomach are stimulated, the autoregulatory escape is not complete, and the overall effect is the reduction of gastric mucosal blood flow. Teleologically, one can imagine that the purpose of reduced gastric blood flow in the fight-or-flight state is to make the blood available to the brain when the animal is threatened.

The parasympathetic nerve supply to the stomach is derived from the vagus nerve. Stimulation of the vagus releases acetylcholine, which serves to increase gastric motility, acid secretion, and mucosal blood flow. Similar to the adrenergic receptors, cholinergic receptors are located on the smooth muscle cells of the submucosal arterioles. Activation of the receptors with acetylcholine results in vasodilatation. Stimulation of the vagus results in increased gastric acid secretion, which could result secondarily in increased blood flow. However, studies in which the acid secretory response has been blocked have revealed that blood flow increases still occur, indicating a direct neurally mediated vasodilatory response. Furthermore, stimulation of the vagus nerve results in rapid vasodilatation of the submucosal arterioles, too rapid to be secondary to the increased metabolic demand of acid secretion.

In the vagus there are many populations of fibers. Although many of the actions of vagal nerve stimulation can be explained by the action of acetylcholine, only a small percentage of the vagal fibers are cholinergic. Many are adrenergic and others contain peptide neurotransmitters. Stimulation with low-frequency electrical impulses activate low-threshold (i.e., activated with impulse durations of less than 0.5 msec) noncholinergic fibers, and this type of stimulation results in gastric vasodilatation not mediated by acetylcholine. The cholinergic high-threshold fibers, when stimulated, increase gastric mucosal blood flow by a combination of direct neurally mediated vasodilatation as well as by increasing gastric acid secretion.

The visceral afferent nerves are a group of small fibers found in the gastric mucosa that are sensory. They detect changes in the mucosal environment and transmit information to the central nervous system. These fibers contain neuropeptides such as calcitonin gene-related peptide that are known vasodilators. Recently, evidence has been obtained indicating that when these transmitters into the tissue, resulting in increased mucosal blood flow.

IV. PATHOPHYSIOLOGY

The stomach is continuously exposed to the harshest environment of all the tissues of the body. The mechanisms involved in the protection against injury appear to hinge on adequate blood flow. Most laboratory studies of gastric injury use agents known to produce gastric mucosal injury in humans (i.e., aspirin, ethanol, and bile acids). When the stomach is exposed to these agents in the presence of hypotension and, therefore, reduced gastric mucosal blood flow, injury is much worse than in animals with normal blood flow. Conversely, if blood flow is increased, the injury is less. Blood flow is needed to supply the necessary oxygen and metabolites used in mounting a protective response.

One important factor in mucosal damage is acid. Agents that damage the mucosa produce more injury in the presence of gastric acid than when administered alone. Acid alone can injure the tissue. Therefore, one important component of mucosal defense is acid neutralization. Blood contains bicarbonate; thus, one function of blood flow in gastric mucosal protection is the delivery of bicarbonate to neutralize and carry away the acid.

It has been hypothesized that the way alcohol, aspirin, or bile salts damage the tissue is by breaking the gastric–mucosal barrier to acid back-diffusion. The barrier is the surface mucosal epithelium, which prevents the entry of luminal substances into the gastric tissue. Once the barrier is broken, acid from the gastric lumen enters the underlying tissues, damaging them. When this occurs, mucosal blood flow increases to neutralize and carry away the incoming acid and thus reduce the amount of damage. Recent studies have indicated that the hyperemia observed in response to barrier-breaking is mediated by sensory nerves. Inhibition of the function of these nerves eliminates the mucosal hyperemic response to barrier-breaking and greatly exacerbates the injury. Thus, adequate blood flow is necessary not only for maintenance of tissue integrity under normal circumstances, but also for protecting the mucosa against exogenous injurious agents.

It has been hypothesized for over 100 years that a blood flow deficit might be the cause of gastric ulceration. However, though vascular disease resulting in reduced gastric blood flow in humans is rare, gastric ulceration is common. Furthermore, as pointed out in Section III, the gastric vascular supply is highly redundant and a reduction in blood flow on a structural basis would require the simultaneous loss of many supplying vessels. If gastric ulceration results from a blood flow deficit, the deficit probably results from a disordered physiological response to the need for increased blood flow. The abnormality probably occurs in a local area, and it is in this area where the ulceration begins.

BIBLIOGRAPHY

Gannon, B., Browning, J., O'Brien, P., and Rogers, P. (1986). Mucosal microvascular architecture of the fundus and body of human stomach. *Gastroenterology* **86**, 866–875.

Granger, D. N., and Bulkley, G. (eds.) (1981). "Measurement of Blood Flow: Applications to the Splanchnic Circulation." Williams & Wilkins, Baltimore.

Granger, D. N., and Kvietys, P. R. (1981). The splanchnic circulation: Intrinsic regulation. *Annu. Rev. Physiol.* **43**, 409–418.

Guth, P. H., Leung, F. W., and Kauffman, G. L. (1989). Physiology of gastric circulation. *In* "Handbook of Physiology—The Gastrointestinal System I" (J. Wood, ed.). Am. Physiol. Soc., Washington, D.C.

Miller, T. A. (1988). Gastroduodenal mucosal defense: Factors responsible for the ability of the stomach and duodenum to resist injury. *Surgery* **103**, 389–397.

Gastric Microbiology and Gastroduodenal Diseases

ERNST J. KUIPERS
MARTIN J. BLASER
Vanderbilt University School of Medicine

GLOSSARY

Dyspepsia Complaints of upper abdominal discomfort

Gastric cancer Epithelial tumor of the stomach, mostly adenocarcinoma of the intestinal or diffuse type

Gastric lymphoma Lymphoproliferative disorder of the stomach, mostly of mucosa-associated lymphoid B-cell non-Hodgkin's type (MALT B lymphoma)

Gastritis Inflammation of the gastric mucosa

Helicobacter pylori A spiral or slightly curved gram-negative bacterium with two to six characteristic unipolar flagella

Peptic ulcer Defect in the integrity of the mucosa of the stomach or duodenum. The usual definition comprises lesions with a diameter of at least 0.5 cm that penetrate through the muscular layer of the mucosa

COLONIZATION OF THE STOMACH BY *Helicobacter pylori* can cause chronic inflammation of the mucosa. This bacterium has unique defense mechanisms, permitting long-term residence in the acidic gastric milieu. The chronic mucosal inflammation can lead to various disorders, in particular peptic ulcer disease and gastric cancer. Treatment requires a combination of antimicrobial agents, which can be enhanced by acid suppressive drugs. Such treatment has been shown to cure ulcer disease, but has yet to be shown to prevent gastric cancer.

I. *Helicobacter pylori*

Natural infection with *H. pylori* has been demonstrated only in humans and in nonhuman primates. This bacterium specifically colonizes the mucosa of the stomach and part of the duodenum (the small bowel immediately following the stomach). It is only rarely observed elsewhere in the gastrointestinal tract. *H. pylori* bacteria live in the mucus layer lining the stomach and can attach to the mucosal surface. Occasional bacteria may invade between mucosal cells. The bacterium has strong urease activity. This enzyme catalysis the hydrolysis of urea to yield ammonia and carbonic acid. Equilibration of ammonia with water raises the local pH, which aids survival in the acidic environment of the stomach. Survival in this environment is further made possible by the presence of flagella, which enable bacterial motility. *H. pylori* are highly diverse at the genetic level. This diversity supports the hypothesis that humans and prehumans were colonized since time immemorial. For three different genes—*cagA*, *vacA*, and *iceA*—variations are associated with different manifestations of infection. Approximately 50% of *H. pylori* strains contain *cagA* and adjacent genes. Persons infected with *cagA*$^+$ strains experience more severe gastritis and have a higher risk for ulcer disease and cancer than do those infected by strains lacking *cagA*. The *vacA* and *iceA* genes are present in each strain, but in different forms.

H. pylori infection is very common; an estimated 50% of the world population is infected. The prevalence of infection in developing countries exceeds 70%, whereas in the United States and other developed countries it is approximately 30%. This prevalence increases with age and is higher among subjects of lower socioeconomic status. The incidence of infection has decreased in the United States, western Europe, and Japan in recent decades with improvement of socioeconomic standards, a phenomenon that partly explains the higher infection prevalence among the elderly. It is believed that most infections occur during childhood, probably via direct transmission from people who harbor the organism. The exact route of transmission, however, is still unknown, but poor sanitation and crowding clearly facilitate the process. Once acquired, the bacterium usually is carried for life, which in most persons does not result in symptoms. Some persons are infected with more than one *H. pylori* strain, and occasionally infection with *Helicobacter heilmannii,* another *Helicobacter* species, is diagnosed. The presence of *H. pylori* can reliably be diagnosed by noninvasive methods such as determination of anti-*H. pylori* IgG antibodies in serum by enzyme-linked immunosorbent assay (ELISA), as well as by breath testing. With the latter method, subjects drink a small amount of ^{13}C- or ^{14}C-labeled urea. In persons carrying *H. pylori,* urea is transformed into labeled carbon dioxide, which enters the bloodstream, is exchanged in the lungs, and can be measured in an expiration sample. Invasive methods to diagnose *H. pylori* presence require upper gastrointestinal endoscopy. Biopsy specimens sampled during this procedure can be cultured for *H. pylori* or stained and studied under a microscope for the presence of spiral bacteria or the presence of urease activity.

The bacterium is sensitive to numerous antibiotics *in vitro;* however, monotherapy with any of these antibiotics only rarely eradicates the infection. The difficulty in treatment may relate to the ecological niche of *H. pylori* in the gastric mucus layer, where antibiotics may penetrate more slowly and may be inactivated more quickly due to the acidic pH. Therefore, current treatment regimens generally include an acid suppressive drug and at least two antimicrobial agents, with or without an additional bismuth-containing drug. With such combinations of three or four drugs, given for 7 to 10 days, the organism is effectively removed in approximately 90–95% of those infected persons that require treatment. Failures are mostly due to antibiotic resistance and to reduced patient compliance, particularly resulting from side effects. The prevalence of resistance, particularly to nitroimidazoles and macrolide compounds, is increasing and worrisome. After successful eradication, the chance of recurrent infection in adults in developed countries is less than 1% per year. Vaccinations could potentially prevent or treat this infection. Successful vaccination against *Helicobacter* infections has been achieved in laboratory mice. Although studies in humans are being performed, an approved vaccine is not expected within the coming years. [*See* Antimicrobial Drugs.]

II. GASTRITIS

Gastritis, inflammation of the stomach mucosa, can be caused by various disorders, including bacterial, fungal, viral, and parasitic infections, granulomatous disorders such as Crohn's disease and sarcoidosis, and by autoimmunity against the acid-producing parietal cells of the gastric mucosa, such as occurs in pernicious anemia. However, by far the most common cause of chronic gastritis in *H. pylori* infection. Virtually all *H. pylori*-infected persons develop gastritis at the time that they become infected and this inflammation persists throughout the infection. Gastritis is usually without symptoms. Importantly, resolution of gastritis after treatment of *H. pylori* infection has not unequivocally been shown to improve dyspeptic symptoms, unless ulceration had been present.

Because the bacteria live on the mucosal surface, gastritis is not caused by bacterial invasion of the mucosa, but by substances secreted by the bacteria that either directly damage mucosal cells or attract inflammatory cells. These substances include ammonia resulting from the urease activity and a variety of bacterial proteins. Signs of *H. pylori* gastritis are usually present throughout the entire stomach, yet are most pronounced in the lowest one-third of the stomach away from the acid-producing parietal cells that are present in the proximal two-thirds. However, in patients treated with drugs that profoundly suppress gastric acidity (e.g., proton pump inhibitors), the inflammation becomes more pronounced in the upper two-thirds of the stomach. [*See* Drug Therapy; Inflammation.]

Chronic gastritis can only be reliably diagnosed by the histological study of gastric mucosal biopsy specimens. The mucosa in the stomach of an uninfected person contains no or very few inflammatory cells. A diagnosis of gastritis is made if more than a few inflammatory cells are present in a gastric biopsy specimen when studied microscopically. Eradication of the infection leads to healing of gastritis.

III. PEPTIC ULCER DISEASE

Five to 10% of the population in developed countries are at some point of their lifetime affected by peptic ulcer disease. In the developed world, most ulcers occur in the duodenum, whereas in the developing world, ulcers in the stomach are more common. Peptic ulcer disease is a chronic condition; patients affected by ulcer disease often have repeated attacks. Ulcer symptoms include nausea, upper abdominal pain, and vomiting. In up to 10% of affected subjects the ulcers may be complicated by bleeding or perforation, conditions that require immediate medical treatment. These complications are life-threatening, particularly in the elderly.

A common cause of ulcer disease is the intake of aspirin or nonsteroidal anti-inflammatory drugs (NSAIDs). These drugs are very frequently used as pain relievers. However, research done since the early 1980s has shown that in the absence of these drugs, the vast majority of ulcers (>90%) occur in the presence of *H. pylori* infections. In developed countries, approximately 10 to 20% of persons carrying *H. pylori* are afflicted by ulcer disease as a result of the infection. The issue of why infection with *H. pylori* leads to repeated peptic ulceration in some subjects, but not in others, has not yet been solved. Possible explanations include differences in infecting strains, characteristics of the infected host, environmental cofactors, and differences in age at which infection occurred. Ulcer disease occurs more often in persons infected with *cagA*$^+$ strains than in those infected with a strain lacking the *cagA* gene. In addition, particular genotypes of the *vacA* and *iceA* genes are also associated with a higher risk for ulcer disease.

Host conditions that influence the development of peptic ulceration include sociobehavioral and genetic factors. Ulcer disease used to be considered a disease primarily affecting smokers and persons under stress. It now has become clear that these conditions do not lead to ulcer disease if a person does not carry *H. pylori*. In those infected persons who suffered from ulcer disease, ulcers do not recur after bacterial eradication even if stress and smoking behavior remain unchanged. A genetic factor that influences the risk for ulcer disease in *H. pylori*-infected subjects is blood group. For unknown reasons, ulcers occur somewhat more often in persons who are nonsecretors of ABO blood group antigens.

Epidemiological research has suggested that gastric ulcer disease is more common among persons who became infected close to birth, whereas duodenal ulcer disease would be more common among persons infected later than early childhood. This hypothesis facilitates explanation of the geographical differences and changes in the epidemiology of gastric and duodenal ulcer disease in the last century.

Treatment with acid suppressive drugs accelerates the healing of an ulcer, but mostly cannot prevent recurrent attacks, even when taken as maintenance therapy. Since the recognition of the role of *H. pylori* in ulcer disease, many studies worldwide have reported that eradication of the infection can cure ulcer disease. More than 50% of ulcer patients suffer from recurrent attacks within the first presentation of disease if the *H. pylori* infection remains untreated. However, after successful eradication of *H. pylori* infection, only few patients have recurrent attacks. If recurrent ulcers occur, they are usually caused by the intake of NSAIDs. Patients who are diagnosed with peptic ulcer disease therefore require testing for *H. pylori* infection. If they are infected, eradication therapy needs to be prescribed to prevent recurrent attacks. Ulcer disease is so far the only generally accepted indication for the treatment of *H. pylori* infections.

IV. GASTRIC CANCER

Chronic *H. pylori* gastritis can, after many years, lead to a loss of the glands of the gastric mucosa. These glands contain specialized parietal and chief cells, which produce acid and pepsinogens (proteolytic enzymes), respectively. A mucosa with a reduced number or total lack of glands is called atrophic. The surface cells in an atrophic mucosa often become partially replaced with intestinal type epithelium, resulting in so-called intestinal metaplasia. Approximately 50% of *H. pylori*-infected subjects ultimately develop such atrophy during their life. The chance of development of atrophy is higher in those with more severe gastritis. Subjects with more severe gastritis include those infected with *cagA*$^+$ strains. Drugs and surgical procedures that suppress gastric acidity also increase the severity of gastritis. This occurs by changes in the distribution of *H. pylori* bacteria in the stomach. Maintenance therapy with these drugs, as often prescribed for the relief of severe heartburn, also has been shown to increase the rate of development of atrophic gastritis, as would be predicted by this model.

The sequence of events leading from chronic gastritis to atrophy and intestinal metaplasia increases the risk for gastric cancer. Most cancers occur in subjects with gastric mucosa that shows atrophic and metaplastic changes. Evidence that *H. pylori* can cause gastric cancer via the sequence of atrophy and metaplasia came from various studies, most notably from a number of cohort follow-up studies. These studies showed that *H. pylori*-infected subjects more often develop gastric cancers involving the lower stomach at about a 10-fold higher rate during long-term follow-up than uninfected controls. In 1994, the International Agency for Research on Cancer, an arm of the World Health Organization, declared that *H. pylori* is a class I carcinogen, which is the highest rank given by this agency to cancer-causing agents. It is estimated that approximately 0.5 to 1% of infected subjects ultimately develop gastric cancer and that at least 60–70% of gastric cancers occurring in the developed world are the result of life-long infection with *H. pylori*. For that reason, large studies have begun to evaluate whether early *H. pylori* eradication can prevent the development of atrophy and cancer. Such studies are in particular being performed in localities with high incidences of gastric cancer, such as Japan and certain areas in South America. Epidemiological calculations suggest that population screening and treatment in the United States also could be cost-effective with respect to the prevention of gastric cancer in high-risk populations, but this has yet to be corroborated by clinical trials. For the time being, prevention of atrophy and cancer is not yet an accepted indication for the treatment of *H. pylori* infections. A possible exception is the prevention of atrophy in subjects chronically taking drugs that profoundly suppress gastric acidity.

The incidence of gastric cancer has significantly decreased in the United States from approximately 30 cancers per 100,000 inhabitants per year in 1930 to little over 5 cancers per year in the 1990s. This decrease parallels the decrease in the proportion of the population that is infected with *H. pylori* due to socioeconomic changes. As cancers of the distal stomach have decreased, cancers of the proximal stomach and esophagus have increased and preliminary data relate these phenomena to the absence of *H. pylori*. [*See* Gastrointestinal Cancer.]

V. GASTRIC LYMPHOMA

A very rare malignant tumor of the stomach is caused by the proliferation of B-type lymphoid cells within the mucosa. This low-grade tumor is therefore called mucosa-associated lymphoid tissue (MALT) lymphoma. It used to be that patients with this disease could only be treated with radiation therapy or stomach resection. With the recognition that the normal stomach mucosa of an uninfected person contains very few lymphoid cells, but that these cells are present in the mucosa of nearly all persons carrying *H. pylori*, research has focused on the relation between *H. pylori* and MALT lymphoma. It has become apparent that more than 95% of patients with this rare tumour were infected with *H. pylori*. When treated with *H. pylori* eradication therapy, 70% of patients showed total regression of this malignancy. This remarkable result has radically changed the treatment of this tumor. The therapy of choice has now become of a 1-week course of *H. pylori* eradication therapy.

BIBLIOGRAPHY

Blaser, M. J. (1992). Hypotheses on the pathogenesis and natural history of *Helicobacter pylori*-induced inflammation. *Gastroenterology* **102**, 720–727.

Blaser, M. J., Chyou, P. H., and Nomura, A. (1995). Age at establishment of *Helicobacter pylori* infection and gastric carcinoma, gastric ulcer, and duodenal ulcer risk. *Cancer Res* **55**, 562–565.

Blaser, M. J., Pérez-Pérez, G. I., Kleanthous, H., Cover, T. L., Peek, R. M., Chyou, P. H., Stemmerman, G. N., and Nomura, A. (1995). Infection with *Helicobacter pylori* strains possessing *cagA* is associated with an increased risk of developing adenocarcinoma of the stomach. *Cancer Res.* **55**, 2111–2115.

Forbes, G. M., Glaser, M. E., Cullen, D. J. E., Warren, J. R., Christiansen, K. J., Marshall, B. J., and Collins, B. J. (1994). Duodenal ulcer treated with *Helicobacter pylori* eradication: Seven-year follow-up. *Lancet* **343**, 258–260.

Forman, D., Webb, P., and Parsonnet, J. (1994). *Helicobacter pylori* and gastric cancer. *Lancet* **343**, 243–244.

Hansson, L. E., Nyrén, O., Hsing, A. W., Bergström, R., Josefsson, S., Chow, W. H., Fraumeni, J. F., and Adami, H. O. (1996). The risk of stomach cancer in patients with gastric or duodenal ulcer disease. *N. Engl. J. Med.* **335**, 242–249.

International Agency for Research on Cancer. (1994). "IARC Monographs on the Evaluation of Carcinogenic Risks to Humans," Vol. 61. Lyon.

Kuipers, E. J., Lundell, L., Klinkenberg-Knol, E. C., Havu, N., Festen, H. P. M., Liedman, B., Lamers, C. B. H. W., Jansen, J. B. M. J., Dalenbäck, J., Snel, P., Nelis, G. F., and Meuwissen, S. G. M. (1996). Atrophic gastritis and *Helicobacter pylori* infection in patients with reflux esophagitis treated with omeprazole or fundoplication. *N. Engl. J. Med.* **334**, 1018–1022.

Kuipers, E. J., Uyterlinde, A. M., Peña, A. S., Roosendaal, R., Pals, G., Nelis, G. F., Festen, H. P. M., and Meuwissen, S. G. M. (1995). Long term sequelae of *Helicobacter pylori* gastritis. *Lancet* **345**, 1525–1528.

Parsonnet, J., Harris, R. A., Hack, H. M., and Owens, D. K. (1996). Modelling cost-effectiveness of *Helicobacter pylori* screening to prevent gastric cancer: A mandate for clinical trials. *Lancet* **348**, 150–154.

Gastrointestinal Cancer

PELAYO CORREA
Louisiana State University Medical Center

GLOSSARY

Dietary fiber Undigestible food residue

Dysplasia Histopathologic pattern of abnormal cell proliferation characterized by loss of polarity and nuclear abnormalities, not invasive but considered precancerous

Foveolar cells Mucus-secreting epithelial cells lining the surface and the pits (foveolae) of the gastric mucosa

Goblet cells Specialized epithelial cells secreting acid mucins, which accumulate in the cytoplasm as a globus, later to be discharged

Hamartoma Overgrowth of mature tissues and cells that normally occur in the affected organs

Leiomyoma Benign tumor of smooth muscle

Leiomyosarcoma Malignant tumor of smooth muscle

Oxyntic Specialized epithelial cells, which secrete hydrochloric acid and are located in the corpus and fundus of the gastric mucosa

Peristalsis Wave contractions of the smooth muscle layers of the digestive organs

THIS ARTICLE PROVIDES A NARRATIVE ACCOUNT of the salient features of the biology of gastrointestinal cancer in humans. It emphasizes the precancerous process as well as the epidemiology and etiology of cancer in the context of the interaction between genetic and environmental influences. It highlights human ecology as the main determinant of risk and elaborates on the role of secular changes that have taken place in the human environment as a consequence of urbanization. The histologic features of the digestive system that are relevant to gastrointestinal cancer are summarized. The natural history and the clinical features of the tumors are briefly described. Finally, the available biologic information on the cancerous and precancerous process is used to provide guidelines of primary and secondary prevention.

I. GASTROINTESTINAL CANCER: A REFLECTION OF HUMAN ECOLOGY

Approximately one-third of all cancers in humans originate in the gastrointestinal mucosa. This proportion is similar in most populations, but there are marked interpopulation differences in the specific location and type of tumors. Marked shifts in this frequency with time have been observed in several populations, the most prominent of which is depicted in Fig. 1, which shows the time trends of mortality rates from cancers of the stomach and large bowel in U.S. populations. These shifts in the frequency of digestive cancer in the relatively short lapse of a few decades are a reflection of the rather dramatic changes in human ecology, particularly evident in new dietary habits in response to changes in life-style.

Although the reasons for such shifts are still somewhat controversial, there is general agreement that they mostly represent changes in dietary practices, most probably driven by the complex phenomenon of "urbanization." The change in gastrointestinal cancer frequency observed so dramatically in the United States has also occurred in other countries, although

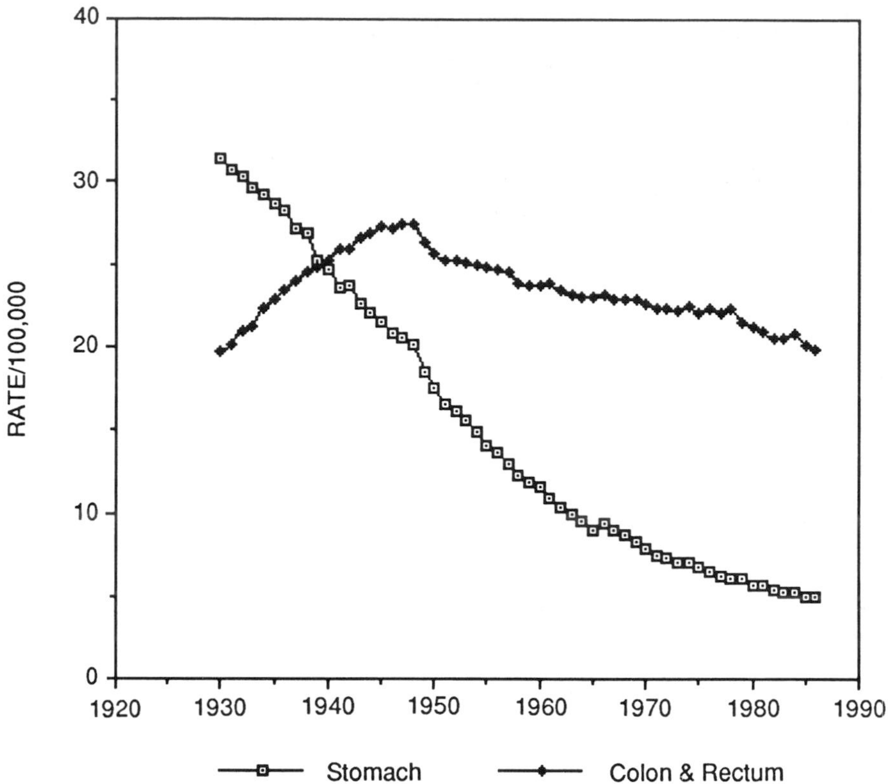

FIGURE 1 Trends of mortality rates for gastric and colorectal cancer. (Data compiled by E. Silverberg, American Cancer Society.)

it seems to have started somewhat later and so far reached less contrasting proportions. Many populations of the world still display the 1930 U.S. pattern of high gastric and low colorectal cancer rates. This is especially true of the Andean and mountainous parts of Central and South America and central Europe, as well as some populations that have kept a more "traditional" diet (e.g., the American Indians). The diet of such populations has several common characteristics. It consists mostly of locally grown agricultural products rich in starchy grains and roots and of other vegetable foods rich in complex carbohydrates. Such food items are rarely consumed fresh (unprocessed). They are usually ingredients of complex soups and other items extensively cooked at high temperatures. Extensive cooking may have resulted from empirical observations that diarrheic diseases followed the consumption of raw vegetables, owing to contamination with human excreta containing infective bacteria and parasites (e.g., cysts of *Entamoeba histolytica*). The practice of high-temperature cooking prevents infectious and parasitic diseases but also deprives the consumer of "protective" factors, which are present in fresh fruits and vegetables.

High-gastric/low-colorectal cancer risk diets are usually low in animal foods rich in fats and good-quality proteins, which in present-day traditional societies are expensive. These items (i.e., beef, eggs, chicken, and milk) have become more affordable in societies that have "benefited" from technological advances in agroindustry.

Another frequent habit in societies with high gastric/low colorectal cancer rates is excessive salt intake. Most of the world consumes amounts of salt that are at least one order of magnitude greater than the presumed dietary requirement. The reasons for this human preference for excessive salt intake are not clear. Newborn infants show no preference for salty fluids versus water, but by the age of 2–3 years they prefer salty food over its unsalted counterpart. The primate ancestors of humans were likely to have existed on a vegetarian diet poor in salt, which makes animals search out sources of salt in their environment. Many traditional indigenous diets are predominately vegetarian and need to be supplemented with salt intake. Salt was historically a highly valued commodity, a marked contrast with today's industrialized societies in which salt is cheap and plentiful. The taste for salty foods is largely acquired and can be modified: subjects accustomed to high-salt intake prefer salty foods but, when they change to a low-salt diet, dislike salty foods. The salt consumption differs greatly between populations: from higher than 25 g/day/subject to lower than 4 g in some isolated Indian tribes. The etiologic role of excessive salt in cardiovascular diseases and in cancer has been widely recognized. The Japanese government has sponsored an extensive education campaign, which in two decades has brought the average per capita daily intake of salt from approximately 24 g to 11 g. The rate of gastric cancer in Japan, the highest in the world, did not change for about 10 years after the salt intake reduction, but at the present time is falling rapidly. [*See* Salt Preference in Humans.]

Most traditional societies have a bulky diet with abundant undigestible residues of cellulose and similar compounds (so-called dietary "fiber"). This habit requires frequent defecation, which becomes a problem in urbanized societies. Dietary changes from traditional to modern and from rural to urban societies have brought marked reduction in undigestible dietary residues, which require less frequent defecation and increase the bowel transit time. This change has been accomplished gradually over the centuries, mostly by increasing the consumption of animal foods (low in residue) and reducing the undigestible residues from vegetables by decreasing their intake. For that purpose the technology in "refining" vegetables has been used, as in the case of highly refined wheat flour, which has much less residue than the original whole wheat flour. These changes have no apparent effect on gastric cancer frequency but do seem to be mainly responsible for the increase in colorectal cancer frequency. Societies that are heavily dependent on animal meats (e.g., Scotland, Australia, and Argentina) have the highest rates of colon cancer. When the consumption of animal products emphasizes dairy products such as eggs, milk, and cheese, the colorectal cancer rates are reduced, but the rates of coronary heart disease are increased. [*See* Dietary Fiber, Chemistry and Properties.]

The rates of colorectal cancer in the United States probably started to climb before 1930, as suggested by Fig. 1. Vital statistics before 1930 are unreliable and preclude inferences concerning the actual time of initiation of this "epidemic." Most traditional societies have rates that are lower than 10 deaths per 100,000 inhabitants per year.

Not all countries belong to one of the two patterns depicted in the United States in 1930 versus 1980. Some have excesses of both gastric and colorectal cancer, as in the case of France and Uruguay, whereas others have low rates of both tumors, as in the case of most of Africa and in northern Brazil. Such populations offer opportunities to investigate specific dietary practices linked to specific tumors. It appears that excessive residue in the diet does not necessarily contribute to gastric cancer rates if irritants such as excessive salt are avoided and protectors such as fresh fruits and vegetables are provided. Cancer protectors may also be found in fish, with a high content of omega 3 fatty acids (a special kind of unsaturated acids), which may exert a protective role similar to that of fresh fruits and vegetables.

Not all gastric and colorectal cancers are so markedly influenced by dietary factors. Infections and other nondietary related factors play a prominent role in some cases. Genetic factors determine if specific individuals are susceptible or resistant to the carcinogenic influences in the environment.

II. DIGESTIVE ORGANS: TISSUES AND CELLS

The gastrointestinal tract is a series of interconnected hollow organs specialized in digestive functions responsible for the breakdown of food, the absorption of its components, and the mobilization of the undigestible residue. The breakdown and absorption of food are accomplished by the inner glandular layer of these tubular organs, called the mucosa. The mobilization of residues is the function of a series of juxtaposed layers of smooth muscle whose zonal contractions and relaxation (peristalsis) move the food bolus forward. The muscular and mucosal layers are separated by the submucosa, a layer made of loose connective tissue that allows sliding movements between the other layers. The external surface of these tubular organs is covered by a thin layer of connective tissue lined by mesothelial (peritoneum) cells, which allow minimal friction from the peristaltic movements. A rich network of nerve filaments and nerve cells is present and regulates the digestive and peristaltic functions. Such regulation also has a hormonal component represented by endocrine cells that secrete specialized polypeptide hormones. The specialized cells and tissues described intermix with "support" cells and tissues, mostly the connective tissue composed of fibroblasts and their products (reticulum and collagen fibers), the vascular tissue through which blood and lymph flows, and the lymphoid tissue whose cells take care of the immunologic functions. [*See* Digestive System, Physiology and Biochemistry; Peptide Hormones of the Gut.]

All of these cells can give rise to neoplasia, but the great majority of them (adenomas and adenocarcinomas) originate in the mucosal glands. Somewhat less frequent are tumors originating in the smooth muscle cells (leiomyomas and leiomyosarcomas) and the lymphoid organs (lymphomas). Tumors originating in endocrine cells are infrequent, and tumors originating in other cells (nerve cells, mesothelial cells) are even more rare. Our discussion will concentrate on glandular tumors because of their frequency and because more is known about their etiology.

The gastric mucosa has two main components with specialized functions: the antrum (pyloric) and the corpus-fundus (oxyntic). Both are lined by foveolar cells secreting specialized neutral mucin, which serves as a lubricant and probably as a protector against acid secretions, apparently with help from bicarbonate, which is also secreted by foveolar cells. Blood group substances are also secreted along with the mucin. The glands of the oxyntic mucosa contain three different cell types: (1) the parietal cells (also called oxyntic cells) secrete hydrochloric acid and intrinsic factor; (2) the chief cells secrete pepsinogens (I and II), which on release into the acid gastric secretions are activated to form the proteolytic enzyme pepsin (this enzyme starts the process of digestion of ingested proteins); and (3) the endocrine cells, which in the oxyntic mucosa secrete some known polypeptide hormones such as serotonin and somatostatin as well as some hormones whose classification and function are still unknown.

The cells of the antral glands secrete mostly mucins and pepsinogen II. The endocrine cells of the antrum secrete gastrin, a stimulator of acid-pepsin secretion. Somatostatin, bombesin, and serotonin are also secreted by antral cells. The so-called gland necks are composed of special "neck" cells, which are the only ones capable of replication in the normal stomach. The loss of cells at the surface or in deeper glands leads to replication of neck cells, which force migration and specialization of the displaced cells. Neck cells secrete both mucins and pepsinogens, but they in actuality represent "stem" cells, which contain the genetic information to synthesize (after migration) the more specialized secretions of the glands and surface epithelial cells.

The small intestinal mucosa is formed by villi that increase the absorptive capacity of the organ. Such villi are taller and presumably more efficient in individuals living in temperate zones than in tropical climates. Migration from tropical to temperate zones results in modification of overall bowel morphology. At the base of the villi are the crypts, where replication of stem cells takes place. Such cells migrate to the surface and are shed at a rapid rate: the small intestinal epithelium renovates itself faster than any other tissue. This may be one of the explanations for the rarity of epithelial tumors in this organ: cells with abnormal mitosis or other DNA abnormalities are discarded before they have a chance to form new abnormal viable clones. The villi are mostly covered by enterocytes, columnar cells specialized in absorption and digestion. They contain α-1 antitrypsin, apolipoprotein, and lysozyme. Their surface is covered by tall and crowded microvilli (identified as a "brush border" in light microscopy) rich in disaccharidases and peptidases. Alternating with enterocytes are goblet cells, which secrete sialomucins. The endocrine cells of the small intestine contain several polypeptide hormones, including cholecystokinin, glucagon, gastrin, motilin, secretin, and somatostatin. Such cells give rise to carcinoids, also called APUD tumors.

The large intestinal mucosa (cecum, ascending, transverse, descending, sigmoid colon, and rectum) is lined by crypts arranged perpendicular to the surface, lined predominantly by mucus-secreting columnar and goblet cells. They secrete a mixture of two acid mucins: sialomucins predominate in the proximal colon, and sulfomucins predominate in the distal colon and rectum. The base of the crypts of the small and large intestine contains Paneth cells with large eosinophilic granules of unknown function. The granules are rich in lysozyme and immunoglobulins A and G, suggesting a role in the control of the luminal bacterial flora.

III. HETEROGENEITY OF GASTROINTESTINAL CANCER

The great majority of gastrointestinal tumors originate from the glandular epithelium and are therefore appropriately called *adenomas* (if benign) and *adenocarcinomas* (if malignant). Beyond this general affinity, subtypes of adenocarcinomas are determined by the organ of origin, the type of epithelial cells they mimic, and their biologic behavior.

Most of the gastric malignant tumors belong to one of two types, determined mostly by the cohesion of their cells. If the tumor cells adhere to each other by desmosomes and tight junctions, they form neoplastic glandular structures that resemble intestinal glands. This characteristic has given these tumors their name: they are called *intestinal-type* or *expansive tumors*. If the intercellular junctions are absent, the tumor cells diffusely infiltrate the gastric wall and do not form well-defined structures. For such reason they are called *diffuse-type* or *infiltrating carcinomas*. As a rule, the intestinal type of gastric carcinoma grows less rapidly, is more frequent in males and older persons, and more frequently is ulcerated than their diffuse counterparts.

Despite such obvious differences in biologic behavior, no clear differences in the characteristic of the cells forming intestinal or diffuse types of tumors have been documented. They both produce mucins in an irregular and anarchic pattern: gastric-type (neutral) mucins, small intestinal-type (sialic) mucins, and colorectal (sulfated) mucins have been reported in both types of tumors. Abnormal pepsinogens have also been reported. Both types of gastric adenocarcinomas may display patterns that resemble more primitive cells (anaplasia).

Adenocarcinomas of the small intestine are rather rare. The ampulla of Vater, where bile meets the intestinal content, is the most frequent site of origin. Other carcinomas originate in previously existing "benign" polyps such as those seen in Peutz-Jeghers syndrome or in familial polyposis.

Adenocarcinomas of the large intestine are usually mucin-producing tumors that vary in their degree of differentiation. They are especially frequent in the sigmoid and rectosigmoid portions. Most are believed to originate from preexisting adenomas.

Endocrine tumors are somewhat less frequent in the gastrointestinal tract. In some organs, such as the vermiform appendix and the rectoanal junction, they tend to remain small and limited to their site of origin. Those that invade and metastasize originate more frequently in the small intestine and less frequently in the stomach and colon. Those tumors secreting 5-hydroxytryptamine (5HT) are frequently called *carcinoids*, and those secreting one or more of the polypeptide hormones are named after their main secretion (i.e., glucagonoma). 5HT is metabolized to 5-hydroxyindole acetic acid (5-HIAA) by monoamine oxidase in the liver; 5-HIAA is excreted in urine and is useful in the diagnosis of the carcinoid syndrome. The latter is characterized by episodes of facial flushing (which can be induced with alcohol), cardiac lesions, and episodic diarrhea. The cardiac lesions consist of right-sided fibrosis, which may lead to stenosis of the tricuspid and pulmonary valves. Although several biologically active substances may be involved in the pathogenesis of the syndrome, serotonin is probably the major contributor because of its abundance in midgut carcinoids with which the syndrome is almost exclusively associated.

Gastrointestinal lymphomas as a rule originate in lymphoid cell populations that specifically home to these digestive organs. Contrary to other lymphomas, they tend to remain localized to their site of origin for a long time and disseminate to other organs only rarely. In industrialized Western countries, these tumors are seen predominantly in the stomach, followed by the small intestine, and less frequently the large intestine. The mucosal infiltrate by lymphocytes and plasma cells has received the name of MALT (mucosa-associated lymphoid tissue). MALT is not normally present in the gastric mucosa but it is regularly observed in inflammatory conditions, especially after infection with *Helicobacter pylori*. Well-differentiated gastric MALTomas are frequently associated with *H. pylori* infection and have been reported to disappear after successful anti-*Helicobacter* therapy. The tumors apparently originate from perifollicular B cells. They respond well to locally directed therapy. In the Middle East, a special type called the Mediterranean lymphoma predominates; it is principally localized in the small intestine. This tumor is part of a disease complex called immunoproliferative small intestinal disease (IPSID), first described in the tropics and studied in detail in the Middle East. IPSID is clinically manifested as a malabsorption syndrome and is seen mostly in young individuals. In its initial stages it behaves as an infectious disease and can be cured with broad-spectrum antibiotics. In more advanced stages it behaves as a frank lymphoma, involving extensively the small intestine. Histologically the intestinal villi are expanded by a dense infiltrate of mononuclear leukocytes that in advanced stages extends to the muscular layer. Lymph node involvement is a late and rare event. The tumor cells are B lymphocytes and plasma cells that synthesize an abnormal α_1 heavy-chain protein (it is sometimes called α_1 heavy-chain disease). A nonspecific antigenic drive originated by luminal bacteria is suspected in its pathogenesis. In north Africa and the Middle East, another peculiar lymphoma is observed in the gastrointestinal tract of children—Burkitt's lymphoma. These are rapidly spreading tumors made of dense agglomerations of primitive stem cells and occasional phagocytic macrophages, giving a microscopic pattern that has been compared with a starry sky. Another variant of gastrointestinal lymphoma is made of T cells, mostly associated with gluten intolerance and celiac disease. The tumor is derived from the intraepithelial T lymphocytes. [*See* Lymphoma.]

The stromal tissues of the gastrointestinal tract give rise to tumors that have most frequently been classified as leiomyomas (benign) or leiomyosarcomas (malignant), even though distinct features of smooth muscle differentiation are often lacking. Small nodular proliferations of smooth muscle, so-called seedling leiomyomas, are frequent autopsy findings, especially in the esophagus, stomach, and small intestine; they have no clinical significance. Malignant tumors form spherical masses that compress and ulcerate the mucosa and may produce blood-borne metastasis.

IV. PRECANCEROUS STATE

Most malignant gastrointestinal tumors are preceded by mucosal lesions that can be identified by morphologic techniques. The better-known cancer precursors have been described in the stomach and in the large bowel.

The intestinal (or expansive) type of gastric cancer is preceded by the following apparently sequential lesions: superficial gastritis (SG), chronic atrophic gastritis (CAG), intestinal metaplasia (IM), and dysplasia. The first two lesions involve cell damage and repair processes, whereas the last two probably represent cell mutational events. Cell products are gradually altered as the process advances, particularly the mucins, digestive enzymes, and the so-called fetal antigens. The normal neutral gastric mucin is altered in amount but not in quality during SG and CAG. When the atrophic foci express abnormal intestinal phenotypes (IM), neutral mucins are gradually replaced by acid mucins. These are mostly sialomucins in mature metaplasia, also called *type I-II, complete,* or *small intestinal type.* More advanced metaplastic changes (type III, incomplete, or colonic) express sulfomucins characteristic of the distal large bowel; they are frequently observed in the vicinity of dysplastic and early neoplastic foci.

The normal gastric pepsinogen I (precursor of pepsin, the only gastric digestive enzyme) gradually decreases as atrophy advances, and the normal chief cells are replaced by intestinal cells. Pepsinogen I can be measured in the blood, and its low level is a good indicator of advanced gastric atrophy (loss of chief cells) and indirectly of high gastric cancer risk. Intestinal digestive enzymes make their abnormal appearance in the gastric mucosa when it becomes intestinalized (IM). A full set of intestinal enzymes (sucrase, trehalase, leucine aminopeptidase, and alkaline phosphatase) is present in small intestinal metaplasia, which for that reason has been labeled *complete metaplasia.* Such enzymes gradually disappear when the metaplasia becomes "colonic" or "incomplete."

Gastric dysplasia is characterized by proliferation of abnormal closely packed tubular glands, which secrete little or no mucin and are lined by cells with large, hyperchromatic, and crowded nuclei. The cells' phenotype is neoplastic, but they lack the invasive capacity and remain within the bounds of the basal membrane that surrounds each gland. Gastric dysplastic lesions are mostly flat or slightly depressed, but occasionally the proliferating glands form a small mass that protrudes into the lumen and is called *adenomatous polyp.* Not all dysplastic foci progress to invasive cancer, but they are indicators of high risk and frequently coincide with small foci of invasive carcinoma present in other areas of the mucosa.

In the large intestine, a process of abnormal (benign) gland proliferation leads to the formation of tubular adenomas (also called adenomatous polyps), which are dragged distally by the fecal stream, forming a pedicle that separates the polyp proper from the mucosa from which it originated. Colonic adenomas increase in size and prevalence with age and are present in more than 50% of subjects older than 60 years in populations with high colon cancer rates (e.g., the United States, western Europe, Australia, and Argentina). In populations with low colon cancer rates (e.g., Africa), colonic adenomas are practically nonexistent. The high prevalence of colonic adenomas in Western societies obviously indicates that most polyps never progress to malignancy: It has been estimated that less than 5% of them do. Most colon carcinomas, however, are thought to originate from adenomatous polyps.

Several genetic alterations were detected in different stages of the colorectal precancerous process discussed in this section. In the hyperproliferative stage that precedes polyp formation, gene mutations or losses were observed in chromosome 5q. Clonal expansion of one of the hyperproliferating cells may lead to the formation of small adenomas in which the genome is hypomethylated. Hypomethylation may lead to aneuploidy and the loss of suppressor gene alleles. K-*ras,* which appears to occur in one cell of the small adenomas and throughout clonal expansion, results in larger and more dysplastic tumors. A key event is the loss of a large portion of chromosome 17p, associated with the progression of adenomas to carcinomas. Allelic losses in chromosome 17p lead to the removal of the wild-type p53 gene, which functions as a tumor suppressor. Losses of tumor suppressor genes in other chromosomes (18q) may also occur. The accumulated losses of suppressor genes correlate with tumor aggressiveness and capacity to metastasize. Although the order in which these genetic alterations appear may be of some importance, it seems that the accumulation of the alterations is the major determinant of progression of precancerous changes to ultimate invasion.

V. CAUSES OF GASTROINTESTINAL CANCER

Most malignant neoplasms in humans are the end result of the interplay of forces that induce or inhibit anarchic cell replication. Such interplay is especially prominent and complex in the gastrointestinal organs, in which it usually operates for many years before a final cancer outcome. Schematic models of the carcinogenic process have been proposed and usually described in epidemiologic terms as a "web of causation." The etiologic forces fall into two main categories: genetic and environmental. The latter term taken in its broadest sense refers to all circumstances surrounding the individual, including factors under his or her control, known as the voluntary environment (i.e., smoking, diet) as well the involuntary environment (i.e., industrial pollution). Within each category of etiologic forces are factors that favor and factors that impede the progress of carcinogenesis. In the genetic categories, such forces confer either susceptibility or resistance to the host. In the environmental category, such forces are usually known as carcinogens (initiators and promoters) or inhibitors. The entire process is the modulation of the carcinogenic process.

Causal associations in humans are usually established with epidemiologic methodology, which has been applied more extensively to adenocarcinomas of the stomach and the large bowel as summarized in the following paragraphs. With some exceptions, the etiology of other gastrointestinal tumors is largely unknown.

The genetic influences in gastrointestinal cancer are best exemplified by the familial polyposis syndromes, the most frequent of which is familial polyposis coli (FPC). This condition, observed in approximately one of every 7000 births, is characterized by the development of increasing numbers of adenomas in the large bowel mucosa until it is literally covered with such tumors. In resected specimens, the large bowel contains more than 1000 grossly recognizable adenomas, but more are identified histologically, making them literally innumerable. The disease is inherited in autosomal dominant fashion, and therefore the children of an affected parent have a 50% chance of developing the disease. As a rule the polyps can be seen in the third decade of life and become symptomatic in the fourth. If not resected, one or more of the polyps will inevitably develop into carcinoma; the mean age of diagnosis of carcinoma is 39 years. The susceptibility to carcinogenesis is more evident in the large bowel epithelial cells, but it is also evident in other gastrointestinal organs; even skin fibroblasts have an increased susceptibility to malignant transformation by oncogenic viruses and chemical carcinogens. Although the genetic influence in this disease is overriding, environmental factors are suspected to play a role. The clearest example of such influences is the occurrence of gastric adenomas in most FPC patients living in areas of high gastric cancer risk such as Japan but less frequent occurrence in FPC patients in Western populations with low gastric cancer risk. Other familial polyposis syndromes resulting in multiple colon adenomas are Gardner's syndrome and the Turcot syndrome, whose patients develop additional tumors of other organs, especially the bones and the central nervous system. Polyps somewhat different from the adenomas, in which tubular glands predominate, are the hamartomas or hamartomatous polyp, in which other tissues are prominent (e.g., smooth muscle, fibroblasts, and blood vessels). Hamartomatous polyps are seen in Peutz-Jeghers syndrome and in the juvenile polyposis syndrome. The former displays polyps mostly in the small intestine and less frequently in the colon and stomach; it is associated with patchy pigmentations around the lips. The malignant transformation of hamartomatous polyps is much less frequent than that observed in FPC.

Strong genetic influences are suspected in the pathogenesis of diffuse gastric carcinoma because of its association with blood group A, its limited interpopulation variability, and its stability in migrant populations. Our knowledge of the etiology and pathogenesis of diffuse gastric carcinoma is too limited to draw meaningful conclusions.

The intestinal type of gastric carcinoma, however, has been the subject of extensive interdisciplinary research, which has led to the postulation of an etiologic model outlined in Fig. 2. The key event in the process is a transformation of the genotype of normal cells into that of neoplastic cells whose replication is no longer under the control of physiologic forces. Such a sine qua non event is probably a mutation or a series of sequential mutations, which call for a genotoxic etiologic factor. This factor has not been identified, but it has been postulated to be an *N*-nitroso compound synthesized in the gastric cavity by the action of nitrite on nitrogen-containing substances. The nitrite is produced in the gastric lumen by anaerobic bacteria containing reductases, which transform ingested nitrate. The anaerobic bacteria colonize the stomach only when its normally acid pH (around 2.0) is elevated to levels around 5.0 or higher. These come about by focal loss of parietal cells as a result of

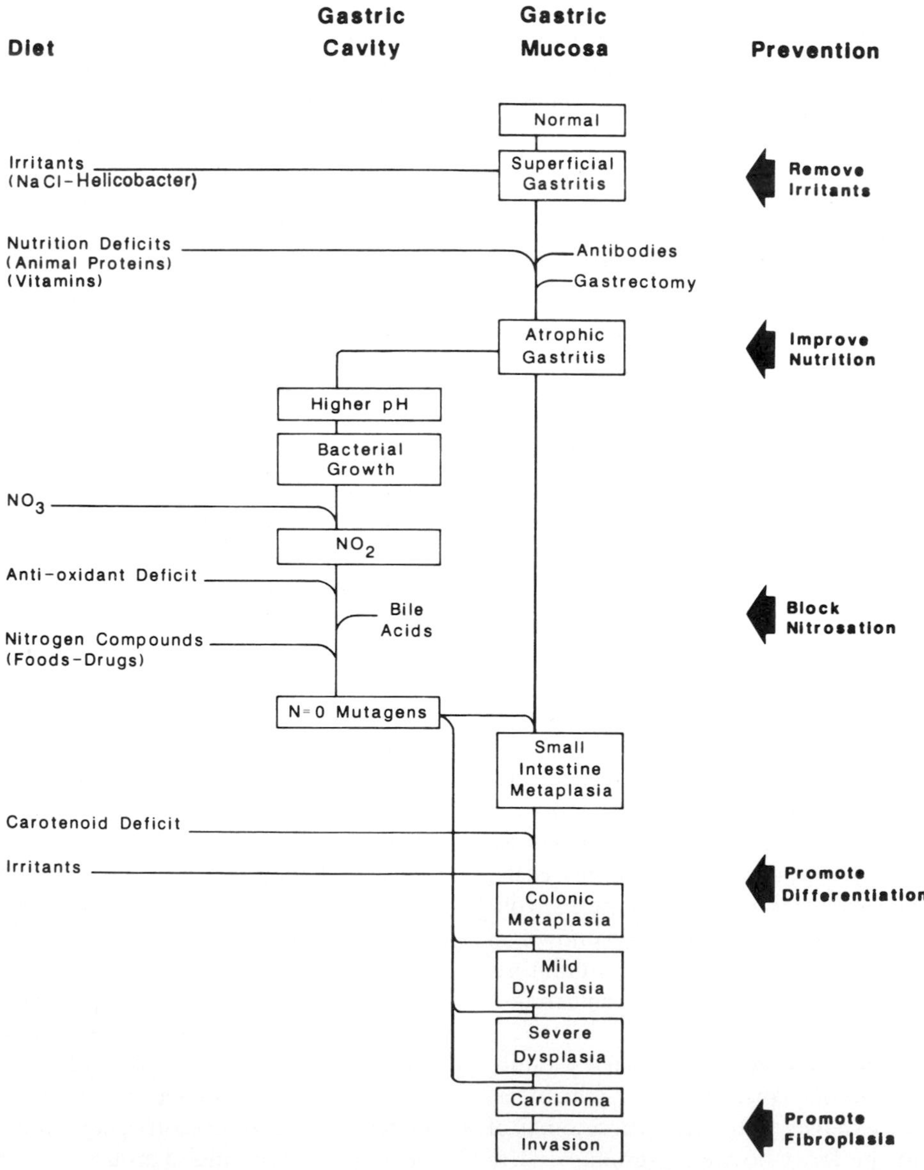

FIGURE 2 Etiologic model of gastric carcinogenesis in humans.

multifocal chronic atrophic gastritis, which is frequently complicated by intestinal metaplasia as explained earlier. The genotoxic event is therefore dependent on a long chain of previous (not genotoxic) events having to do with loss of cells and their inadequate replacement. Although the forces leading to cell loss and epithelial hyperplasia are not entirely understood, at least one factor [i.e., excessive salt (NaCl) intake] has been identified in epidemiologic and experimental studies.

Infection with *Helicobacter pylori* has been identified as an important risk factor in gastric carcinoma. Nested case–control studies in which blood serum was drawn and stored have been conducted in cohorts of individuals followed for a number of years in Hawaii, California, and England. The serum samples of subjects who developed gastric carcinoma during the period of observation (frequently 10 or more years) were tested for antibodies against *H. pylori* and compared with those of subjects who remained free of such tumors. The relative risks of developing gastric carcinoma were approximately three to six times higher in positive (infected) than in negative (not infected) individuals. The mechanisms by which *H. pylori* infection increases the risk of gastric carcinoma are being explored and appear to involve increased cell replication and impairment of ascorbic acid secretion into the gastric lumen. A locally formed carcinogen may be associated with increased synthesis of nitric oxide and related oxygen radicals by macrophages and polymorphonuclear leukocytes attracted to the gastric mucosa by *Helicobacter* bacteria present in the gastric lumen.

Even if the genotoxic event takes place, its effects can be inhibited or can be retarded by "protective" forces such as antioxidant micronutrients in the diet. Prominent among these protectors is beta-carotene, which has been found deficient in subjects with gastric dysplasia and has shown an inhibitor role in experimental gastric carcinogenesis. Ascorbic acid further plays a role in retarding the progression of cancer by increasing the fibroplastic barrier, which stands in the way of invasion. The complexity of the gastric cancer etiologic model contrasts with the simplicity of experiments in which large doses of carcinogens are administered to inbred strains of susceptible experimental animals. The proposed model is probably closer to human reality than the experiments, but such reality may be even more complex, as suggested by human genetic studies. One such study is the segregation analysis of CAG, a cancer precursor, the results of which are expressed in the model depicted in Fig. 3. This model proposes the existence of an autosomally transmitted major recessive susceptibility gene that is prevalent in the population. The expression of such a gene results in CAG, and its penetrance is modulated by age and by having an affected mother, which probably represent environmental forces. The mother's influence may be related to preparation of food for the family. The prevalence of CAG reaches approximately 80% at age 40 in homozygous individuals whose mother is affected by CAG; homozygous subjects whose mother is not affected reach such a high prevalence of CAG years later. Carriers of the gene and subjects presumably without it may develop gastritis much later because of excessive environmental influences, but they never reach high prevalence levels because competing disease risks decimate the population. The proposed model takes into account genetic and environmental interactions and may be useful in exploring neoplastic diseases other than gastric cancer.

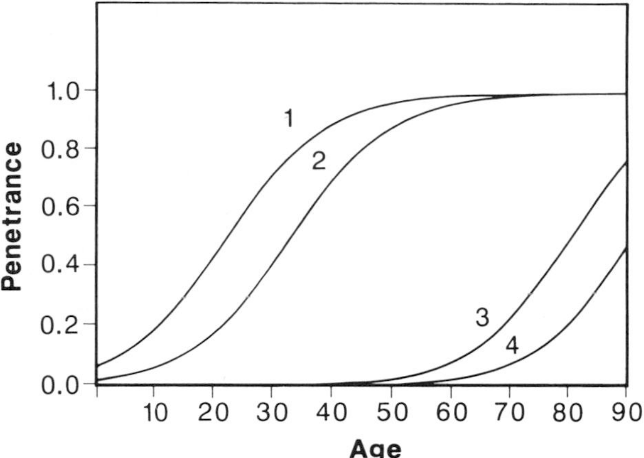

FIGURE 3 Theoretical model of the genetic etiology of chronic atrophic gastritis derived from segregation analysis of family clusters. Estimated penetrance fractions (CAG prevalence): (1) homozygous recessive (AA), mother affected; (2) homozygous recessive (AA), mother unaffected; (3) carrier (AB) or noncarrier (BB), mother affected; (4) carrier (AB) or noncarrier (BB), mother unaffected. [From G. E. Bonney *et al.* (1986). *Genet. Epidemiol.* 3, 213–224, with permission.]

The etiology of colorectal cancer offers considerable contrasts with stomach cancer; the demographic distributions approximate mirror images of each other. Colorectal cancer incidence rates are low in Africa and high in Western societies known for their high consumption of meat. After living in England and then in Africa for many years, Burkitt postulated an etiologic model outlined in Fig. 4. The model basi-

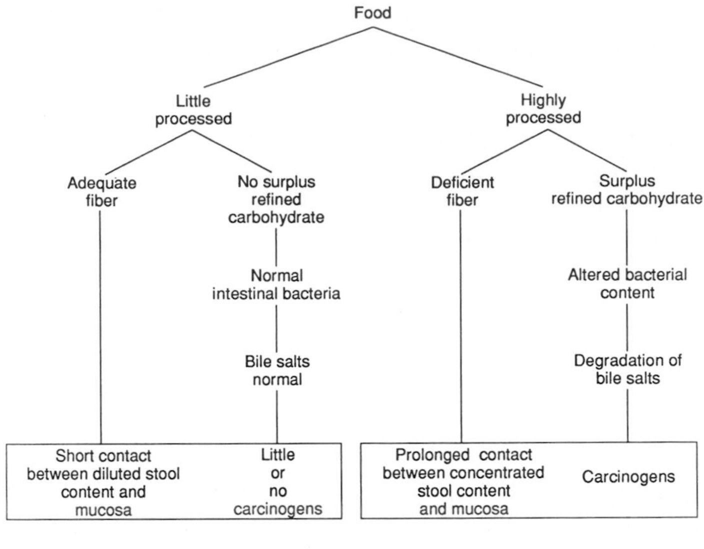

FIGURE 4 Diagrammatic representation of the possible relation between diet and cancer of the large bowel. [From D. Burkitt (1971). *Cancer* 28, 3–13, with permission.]

cally proposes that the high-residue diet of traditional populations prevents colon cancer by inducing rapid transit time and dilution of the bile acids, which otherwise may be carcinogenic. However, the mechanism by which prevention is accomplished is still unknown. Recent research on the causes of colon cancer has centered on the etiologic factors discussed next.

The search for genotoxic agents using mutagenesis tests with bacteria *in vitro* has centered on compounds found in fried foods and in feces. Several compounds formed by pyrolysis during food preparations at high temperature are highly mutagenic and induce nuclear aberrations in colonic cells. Although these compounds have not been identified in the human fecal stream, one of them [i.e., 2-amino-3-methylimid-azo-(4-5f)-quinoline (IQ)] has been found to be carcinogenic to the liver in subhuman primates. Whether any of the pyrolytic compounds plays a role in human colon carcinogenesis is still undetermined. Fecal mutagens are more abundant in Americans on a Western diet than in African traditional societies. A sizable portion of that mutagenicity is due to fecapentenes formed by fecal bacteria; considerable epidemiologic and experimental work, however, has failed to support fecapentenes as human carcinogenes. Another group of compounds suspected to play a genotoxic role are products of cholesterol metabolism, especially 3-ketosteroids, which induce chromosomal aberrations in colonic cells and which could be formed from cholesterol either in the colon or in the preparation or storage of food, and induce chromosomal aberration in colonic cells in culture. Proof of their relevance to humans is still missing. [*See* Cholesterol; Food Microbiology and Hygiene.]

Other dietary factors of possible relevance in human colorectal carcinogenesis are calcium deficiency and fecal pH. Calcium salts significantly reduce the toxic effects of bile acids, which are well-established cancer promoters on the colon. High-fat diets both increase the level of bile acids in the fecal stream and create a relative deficiency of calcium in the intestine. However, evidence implicating suboptimal calcium intake in colon carcinogenesis in humans is insufficient. The importance of fecal pH has recently been emphasized, because it is several units lower (more acidic) in populations at low colon cancer risks than in their high-risk counterparts. At high pH (about 9) most fecal bile acids are in solution. As the pH decreases, more of the acids are in their protonated form and drop out of solution, so that at pH 6, their concentration is very low. Diets low in fat and high in fiber and starch may prevent cancer at least in part by lowering fecal pH. [*See* Bile Acids.]

VI. NATURAL HISTORY

As described earlier, the precancerous process for most gastrointestinal tumors is so prolonged that it takes many years to reach the stage of invasive neoplasia. After it reaches such a stage, two phases have been observed. In the first one, the malignant tumor remains small and localized for years, but when the localized stage is surpassed, the tumor cells invade neighboring structure and metastasize to other organs at an accelerated pace. [*See* Metastasis.]

In the stomach, the localized stage is characterized by confinement of the tumor cells to the mucosa and later to the submucosa. For them the term *early cancer* has been coined in Japan. Cell kinetics studies conducted in Japan have estimated that the localized stage lasts longer than 15 years on average. These tumors are classified according to their architecture: type I is elevated, type II is approximately flat, and type III is ulcerated. They all have an extensive free surface through which malignant cells are exfoliated. When localized ("early") carcinomas are resected, the cure rate is high: more than 95% 5-year survival in Japan and somewhat lower rates in other countries. In Japan, screening campaigns using mobile X-ray units monitor apparently normal individuals. There are some indications that Japanese have a more favorable prognosis than other races after a diagnosis of cancer in the gastrointestinal as well as in other organs. When the localized phase is surpassed and the tumor invades the muscularis, the prognosis is irreversibly bad: less than 10% survival at 5 years. The tumor cells invade the perigastric lymph nodes and then spread through lymphatic and blood vessels to produce metastasis in distant organs, especially the liver, the peritoneum, and the cervical lymph nodes.

The localized stage in the large bowel is characterized by the presence of microcarcinomas in adenomatous polyps. Such foci are practically absent in small polyps (<1 cm in diameter) but increase in prevalence as the size of polyp increases. The occurrence of lymph node metastasis from resected polyps containing small superficial carcinoma is low: 1–9%. Factors that increase the probability of metastasis are poor differentiation of the carcinoma, extension to the surgical margin, and short stalk. Adenocarcinomas most frequently overcome the localized stage by invading the submucosa and penetrating the lymphatic and blood vessels. The metastasis to the mesenteric lymph nodes then extends to the peritoneal organs, the liver, and other distant organs. The local invasion of the intestinal wall is of considerable prognostic importance. It is the basis of the Duke's classification, which considers three stages: (A) when the invasion is limited to the mucosa, submucosa, and inner part of the muscularis; (B) when it reaches the serosa; (C) when regional lymph nodes have metastasis; and (D) distant metastasis. The 5-year survival rates after treatment are 99% for Duke's (A), 85% for (B), 67% for (C), and 14% for (D). These rates vary in different series and are considerably lower in African-Americans and lower socioeconomic groups.

VII. PREVENTION

Present knowledge about etiology as well as about precancerous and early cancerous lesions makes it possible to plan prevention strategies that cover two fields: primary prevention to avoid the development of the disease, and secondary prevention to avoid death from the disease.

Primary prevention of stomach cancer apparently has been taking place in most Western communities, as evidenced by declining incidence rates. It is not clear which of the many recent changes is responsible for the decline, but the invention and widespread use of refrigerators appears to have been a major factor. This has allowed fresh fruits and vegetables to be available year-round and decreased dependency on food preservation methods based on excessive use of salt and nitrites. Most communities with above average consumption of salt and below average consumption of fresh fruits and vegetables have high gastric cancer rates and should benefit from changes in such dietary practices.

Secondary prevention of gastric cancer depends on the identification of high-risk populations and individuals who carry precancerous or early cancerous lesions. This can be accomplished with markers of gastric atrophy (e.g., low pepsinogen I blood levels). Recent advances in endoscopic techniques have made it possible to screen high-risk individuals and take multiple gastric biopsies with a minimum of risk. If a small (localized or early) carcinoma is found, gastrectomy offers a high probability of cure. If dysplastic changes are found, close surveillance with frequently repeated gastroscopies is indicated. Research is being conducted on induction of regression of premalignant lesions with antioxidants such as retinoids and carotenoids. Retinoids accomplish such regression in experimental systems but are toxic to humans. Carotenoids are practically nontoxic but have been less explored. It remains to be determined if chemoprevention can be of practical value in humans. The prevalence of intestinal metaplasia of the gastric mucosa is high in many communities, and dietary or chemical prevention of its progression to cancer is an important and interesting challenge.

Primary prevention of colorectal cancer is a more difficult matter because there is no general consensus as to which of the etiologic factors is relevant and amenable to intervention. Increase in the intake of undigestible components of food ("roughage" or "fiber"), as well as decrease in fat intake, is generally recommended but has not been tested in large populations.

Secondary prevention of colorectal cancer again depends on the identification of high-risk groups. In two groups of individuals, the risk is so high (close to 100%) that total colectomy is indicated. These are patients with the familial colon polyposis syndrome and individuals with chronic (more than 10 years) active ulcerative colitis whose colonic mucosa shows unequivocal signs of dysplasia.

Otherwise, secondary prevention addresses the management of polyps of the large bowel, a common condition. One way to identify colonic polyps or early carcinoma is based on the search for occult blood in the stools. This is accomplished with commercially available kits for home use based on guaiac or similar substances. These kits are useful and may discover carcinomas with high probability of curable resection. However, both false-positive and false-negative results are frequent. The second stage in the identification of premalignant or early malignant lesions is endoscopy and biopsy of suspected lesions. These methods are generally available but are not yet being used with a frequency that makes a real impact on the incidence and mortality rates for colorectal cancer.

BIBLIOGRAPHY

Beauchamp, G. (1987). The human preference for excess salt. *Am. Scientist* **75**, 27–33.

Bruce, W. R. (1987). Recent hypothesis for the origin of colon cancer. *Cancer Res.* **47**, 4237–4242.

Correa, P. (1988). A human model of gastric carcinogenesis. *Cancer Res.* **48**, 3554–3560.

Fearon, E. R., and Vogelstein, B. (1990). A genetic model of colon tumorigenesis. *Cell* **61**, 759–767.

Joossens, J. V., Hill, M. J., and Geboers, J. (1985). "Diet and Human Carcinogenesis." Excerpta Medica, Amsterdam.

National Research Council (1982). "Diet, Nutrition and Cancer." National Academy Press, Washington, D.C.

Schottenfeld, D., and Fraumeni, J. (1982). "Cancer Epidemiology and Prevention." Saunders, Philadelphia.

Whitehead, R. "Gastrointestinal and Oesophageal Pathology." Churchill Livingston, New York. 2nd Ed., 1994.

Gaucher Disease, Molecular Genetics[1]

ARI ZIMRAN
Shaare Zedek Medical Center, Jerusalem

JOSEPH SORGE
Stratagene

ERNEST BEUTLER
The Scripps Research Institute

GAUCHER DISEASE IS THE MOST PREVALENT GLYcolipid storage disease. It is characterized by an accumulation mainly in macrophages of the sphingoglycolipid glucocerebroside, due to an inherited deficiency of the lysosomal enzyme glucocerebrosidase. Classically, Gaucher disease has been subdivided into three types based on the patient's clinical manifestations. Type I (adult or chronic) disease is by far the most common form, with involvement largely limited to the spleen, liver, and bone marrow; although other organ systems may be affected, the central nervous system is, by definition, spared. Type I is especially prevalent among the Ashkenazi Jewish population, in which the frequency of the disease genotype is approximately 1:850, although it is seen in other ethnic groups as well. There is tremendous heterogeneity in the severity of the clinical manifestations of type I disease, ranging from patients who are totally asymptomatic to patients who experience life-long debilitating disease. Most of the symptoms are related to the enlargement of the spleen and liver; decreased platelet count as well as anemia and decreased white blood cell count are among the most frequent signs. There is often a bleeding tendency, especially overt after superficial trauma, and fatigue is a common complaint. Bone pain, occasionally fractures, or destruction of the head of the femur, are also prominent features of symptomatic type I disease. Types II and III (both neuronopathic forms) are relatively rare; the neurologic component common to both types is abnormal eye movements in the horizontal plane. Type II (infantile or acute) disease is characterized by the appearance of several neurologic features in addition to the severe spleen and liver manifestations. The onset of symptoms is usually during the first six months of life, with progressive deterioration of neurologic function until death intervenes, often due to respiratory failure or infection, before the second year of life. This form has no ethnic predilection. Type III (juvenile or subacute) disease is marked by a less aggressive acceleration of the neurologic and visceral aspects seen in type II disease. Generally, the onset and course of both components are attenuated so that advent of symptoms is usually in the first decade and survival may extend into mid-adulthood. One form of this variant is relatively prevalent in the Norbottnian and Vesterbottnian territories of northern Sweden.

[1]This is manuscript 9240-MEM from The Scripps Research Institute. Supported by National Institutes of Health Grants DK36639 and RR00833 and The Stein Endowment Fund.

ENCYCLOPEDIA OF HUMAN BIOLOGY, Second Edition, VOLUME 4.

I. HISTORY

Gaucher disease was first described in a 32-year-old woman in 1882 by the French medical student Philippe Gaucher, who assumed that the large splenic cells that today bear his name were a manifestation of a primary neoplasm of the spleen. Twenty-five years later, F. Marchand noted the pathologic storage of material in reticuloendothelial cells, a lipid that was defined by H. Aghion in 1934 as glucocerebroside. The etiology of Gaucher disease was defined in 1965 by R. O. Brady in the United States and by A. D. Patrick in England as an inherited deficiency of the enzyme glucocerebrosidase. The first practical biochemical diagnostic test was developed in 1970 by E. Beutler and W. Kuhl. The cDNA was cloned and sequenced in 1985 by J. Sorge and coworkers at The Scripps Research Institute and independently by E. Ginns and coworkers at the National Institutes of Health. The first effective enzyme therapy was introduced by N. Barton and coworkers in the late 1980s and was approved by the Food and Drug Administration in April 1991.

II. CLONING AND CHARACTERIZATION

The cloning of the full-length cDNA of human glucocerebrosidase provided important insights into the biology of the enzyme at the levels of the DNA, mRNA, and protein, and made possible the identification and characterization of specific lesions causing Gaucher disease. The glucocerebrosidase cDNA was unusual in that there were two functional ATG initiating codons (Fig. 1): the upstream, inframe ATG initiates the synthesis of a protein containing a 39-amino acid leader sequence, whereas the downstream, inframe ATG initiates a 19-amino acid leader sequence. The leader sequence is presumably required for transport across the membranes of the rough endoplasmic reticulum, where it is cleaved from the mature protein and secreted into the lysosome. Coding for the amino-terminal sequence of the mature enzyme begins at nucleotide 211. Experiments using site-directed mutagenesis to remove either of the ATG codons have demonstrated that each ATG can function independently to produce an active glucocerebrosidase enzyme in cultured fibroblasts. Interestingly, both translational products are capable of being transported into the lysosomes. The significance of the presence of two ATG codons remains to be elucidated. In mice, only the downstream ATG codon exists; therefore it has been speculated, by analogy, that human tissues may preferentially employ one ATG codon or, alternately, that different signal peptides are processed selectively in various human tissues.

```
TTC TCT TCA TCT AAT GAC CCT GAG GGG ATG GAG TTT TCA AGT CCT
Phe Ser Ser Ser Asn Asp Pro Glu Gly Met Glu Phe Ser Ser Pro

TCC AGA GAG GAA TGT CCC AAG CCT TTG AGT AGG GTA AGC ATC ATG
Ser Arg Glu Glu Cys Pro Lys Pro Leu Ser Arg Val Ser Ile Met

GCT GGC AGC CTC ACA GGT TTG CTT CTA CTT CAG GCA GTG TCG TGG
Ala Gly Ser Leu Thr Gly Leu Leu Leu Leu Gln Ala Val Ser Trp

GCA TCA GGT|GCC CGC CCC TGC ATC CCT
Ala Ser Gly|Ala Arg Pro Cys Ile Pro
           ^
```

FIGURE 1 Sequence of 5′ portion of coding region of glucocerebrosidase cDNA including the initiator ATGs (underlined). The vertical line represents the known cleavage site for the mature placental enzyme. [From J. A. Sorge, C. West, W. Kuhl, L. Treger, and E. Beutler (1987). The human glucocerebrosidase gene has two functional ATG initiator codons. *Am. J. Hum. Genet.* **41**, 1016–1024. Reprinted with permission of the University of Chicago Press.]

III. STRUCTURE AND FUNCTION

The glucocerebrosidase gene is located on chromosome 1 at band q21; its total length is nearly 7 kilobases. It is tightly linked to the *PKLR* gene, mutations of which cause a deficiency of the liver/red cell type of pyruvate kinase. Approximately 16 kilobases downstream from the glucocerebrosidase gene is a highly homologous pseudogene. Both the active gene and the pseudogene have been cloned and sequenced. The active gene contains 11 exons, extending from base pair 355 to base pair 7232. The pseudogene is shorter than the active gene; its length is only 5769 base pairs, yet it is nearly 96% homologous to the functional gene. The pseudogene is actively transcribed, but no functional product is formed. There is also a 55-base pair deletion from part of exon 9 flanked by short inverted repeats and there are base pair changes scattered throughout. Alu sequences of 313, 626, 320, and 277 base pairs are present in introns 2, 4, 6, and 7, respectively, of the active gene but not of the pseudogene, accounting, in part, for the smaller size of the latter. Presumably these Alu sequences have been inserted into the active gene after the duplication of the ancestral gene that resulted in the presence of the pseudogene. A restriction map of both the gene and the pseudogene is shown in Fig. 2. A gene promoter has been identified containing two TATA boxes and two CAT-like boxes upstream from

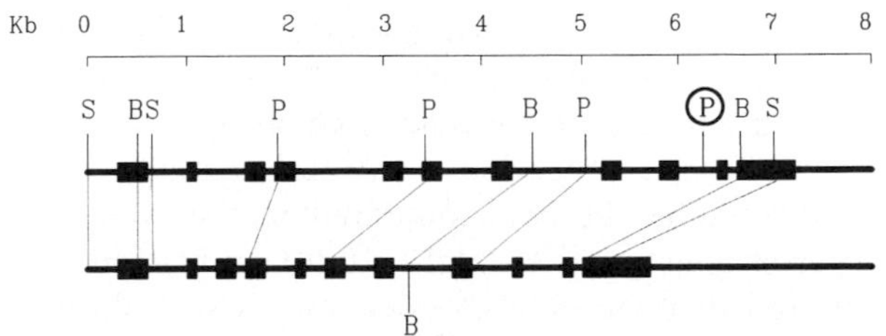

FIGURE 2 Restriction map of the glucocerebrosidase gene (top) and pseudogene (bottom) based on a published sequence. B, BamH1; P, PstI; S, SacI. The PstI site that is circled is unique to the functional gene and is one of the features that has been found to be useful in separating functional from pseudogene for the purposes of diagnosis.

the major RNA transcription initiation site. The functionality of the promoter has been confirmed by coupling it to a bacterial gene coding for the enzyme chloramphenicol acetyltransferase and assaying enzyme activity in transfected cells. Unlike the active gene promoter, enzyme activity directed by the pseudogene promoter was found to be very low; however, pseudogene mRNA has been detected. The availability of the complete sequences of both the active glucocerebrosidase gene and the pseudogene has been crucial to the development of strategies for molecular diagnosis of Gaucher disease.

IV. MUTATIONS

Over 50 mutations in the glucocerebrosidase gene have been discovered to date. Of these mutations, five account for approximately 97% of the mutations in Jewish patients with Gaucher disease. Most of the known mutations are point mutations; however, other mechanisms are also known to give rise to mutations in the glucocerebrosidase gene, including insertions, deletions, splicing, and crossing-over between the functional gene and the pseudogene. "Null" alleles, that is, those that cannot produce an active product, are apparently not compatible with life in the homozygous state. Mice whose glucocerebrosidase gene has been destroyed by targeted disruption do not survive past birth. Mutations that are never observed in patients with neuronopathic disease are designated as "mild," whereas those that are found in patients with neuronopathic disease are designated as severe. Table I summarizes the effect of combining the three classes of mutations that occur in Gaucher disease.

TABLE I
The Effect of Combining Null, Severe, and Mild Gaucher Disease Alleles on the Disease Genotype

		One Allele		
		Null	Severe	Mild
Other Allele	Null	Nonviable	Type II/III	Type I
	Severe	Type II/III	Type II/III	Type I
	Mild	Type I	Type I	Type I

A. Common Mutations

1. The 1226G Mutation: A→G (370Asn→Ser)

The most frequent mutation in Gaucher disease is a point mutation in exon 9 that is found exclusively in patients with type I disease and accounts for more than 75% of Gaucher alleles in Ashkenazi Jews and 36% of alleles among non-Jews. This mutation is the prototype of mild mutations (see Table I), and studies correlating the mutations detected in the gene with the severity of the clinical manifestations have shown an apparent association of the 1226G mutation with mild phenotypic expression of the disorder, as well as rather late age of onset of symptoms. Some patients are totally asymptomatic and are diagnosed only by chance following a family study or as an incidental finding when seeking medical attention for unrelated symptoms; the remainder generally present symptoms related to enlargement of the spleen and the decreased platelet count, for example, easy bruising, nosebleeds, and abdominal discomfort. Although many patients have the 1226G mutation on both alleles, a finding that is especially common among Ashkenazi Jewish patients, some are compound heterozygotes and the clinical course of the disease is influenced by the severity of the mutation on the other allele. However, it is apparent that the 1226G mutation on one allele is protective against the development of central nervous system involvement.

2. The 1448C Mutation: T→C (444Leu→Pro)

This point mutation in exon 10 has been identified in all three phenotypic forms of the disease. It is the second most common mutation in type I Gaucher disease but it is the most frequent mutation in patients with types II and III disease, including the Norbottnian variant. Most patients with this mutation are not Jewish, hence making this mutation the most common among non-Jewish Gaucher disease patients. It may be considered the prototype of the "severe" Gaucher disease mutations. Patients with type I Gaucher disease who are compound heterozygotes for this mutation usually have moderate to severe disease. The normal sequence of the pseudogene matches the sequence of the active gene with this mutation, a finding that suggests that its occurrence in Gaucher disease may represent the result of recombination events between the active gene and the pseudogene, such as gene conversion or unequal crossing-over. This situation also underscores the need for separation of active gene from pseudogene during molecular analysis. In addition, this mutation introduces a new site on the gene that can be cleaved by the restriction enzymes NciI and MspI. A unique DNA product is produced that is easily detected and can be used to identify the presence of this mutation for the purpose of genotyping.

3. The 84GG Mutation

In this mutation an extra guanine has been inserted at cDNA position 84. This is a null mutation, and the homozygous state is probably not compatible with life. It is the second most common mutation in the Ashkenazi Jewish population, accounting for 10–13% of mutated alleles. In the heterozygous state along with a severe mutation such as 1448C, the phenotypic expression may be associated with a neuronopathic form. In compound heterozygotes with the 1226G mutation symptoms usually appear before adulthood and are generally moderate to severe.

4. The IVS2(+1) Mutation

This mutation, in which adenine is substituted for a guanine at the first position of intron 2, prevents normal splicing of the mRNA and results in a loss of exon 2. It is a null mutation, and therefore the homozygous state is unknown. It represents 1–3% of the mutations encountered in Jewish patients with type I Gaucher disease and has been found in non-Jewish patients as well. Presentation of symptoms is in childhood and bone involvement is common. In general, these compound heterozygotes suffer from moderate to severe clinical disease manifestations.

5. The 1604A Mutation: G→A (496Arg→His)

To date only a few compound heterozygotes have been described with this mutation, all of whom had rather mild disease manifestations. However, the population frequency of the mutation approaches that of the 84GG mutation. Most Gaucher disease patients with this mutation who come to medical attention have inherited it together with a null mutation; occasionally it has been encountered in the compound heterozygous state with the 1226G mutation. It is inferred that the homozygous state is so mild as to preclude complaints specific to Gaucher disease.

B. Less Common Mutations

1. The 1342T Mutation: G→T (409 Asp→His)

Since this mutation, like the 1448C mutation, exists in the pseudogene, it may be due to gene conversion or recombination. Although rare in non-Jews, comprising only about 5% of mutated alleles, this mutation has recently been identified in an Arab enclave in Israel. In the homozygous state, patients present with eye movement disorders and a progressive heart valve disease, the former symptom defining type III disease. Onset of the ocular component is in early childhood and the eventual sclerosis of the heart valves is so serious as to often require surgical intervention before adulthood.

2. The 1504T Mutation: C→T (463Arg→Cys)

This mutation has so far been identified only in non-Jewish Gaucher disease patients, and may appear in all three types of Gaucher disease. It is therefore classified as a severe mutation.

C. Complex Mutations

A different kind of mutation has been identified in a number of Gaucher disease patients: cDNA has been found to represent a transcript from a fusion gene in which the 5′ end is the active gene and the 3′ end is the pseudogene. This fusion can apparently occur as a result of either molecular recombination events of unequal crossing-over or gene conversion. Although rare, this fusion gene provides information regarding the molecular anatomy of both the active gene and the pseudogene, and has significant implications with respect to the molecular pathogenesis and diagnosis of Gaucher disease.

1. The "NciI" Recombination: Rec NciI

Rec NciI is the designation given to three discrete point mutations: 1448T→C (444Leu→Pro),

1483G→C (456Ala→Pro), and 1497G→C (460Val→Val). This complex allele is generally associated with moderate to severe type I disease and, when appearing with the 1448C mutation, can cause type II disease.

2. The "TL" Recombination: Rec TL

A fourth 1342G→C transversion at amino acid 409 (409Asp→His) in addition to preceding three point mutations (Rec NciI) has been designated as rec TL. Parts of exon 9, intron 9, and exon 10 have the pseudogene sequence; hence, it has been hypothesized that these mutations are due to unequal recombination or gene conversion. Interestingly, when seen in the homozygous state, this point mutation can cause type III Gaucher disease: a unique variant with oculomotor apraxia and a progressive heart valve defect but only minimal other organ involvement. This mutation has been documented in the compound heterozygous state with the 1226G mutation, where it is associated with asymptomatic to moderate disease.

D. Other Point Mutations

Many other single point mutations have been described; they are generally considered private or rare, that is, appear in only one individual or a single family. Some of these mutations are associated with neuronopathic disease, whereas others are consistent with type I disease. Many more rare mutations have been encountered among non-Jews than among the Ashkenazi Jewish Gaucher disease patients.

E. Activator Saposin C

Saposin C, a small, heat-stable protein consisting of 80 amino acids, serves as an activator of glucocerebrosidase. A prosaposin product is translated and processed into four discrete activators of four related lysosomal enzymes. Mutations in the region of the saposin C activator of the prosaposin molecule resulted in Gaucher disease in two known cases.

V. DNA POLYMORPHISM

Twelve polymorphic sites have been documented in the introns and flanking sequences of the glucocerebrosidase gene. With a few minor exceptions, these polymorphic sites are all in linkage disequilibrium and therefore constitute two major haplotypes. Additional haplotypes are defined by two polymorphic sites in the tightly linked *PKLR* (liver/red cell pyruvate kinase) gene.

A restriction polymorphism involving a *Pvu*II site (intron 6, constituting a G→A single-base substitution at position 3931 of the genomic sequence) is characterized by the presence or absence of a 1.1-kilobase fragment after digestion with the restriction enzyme *Pvu*II: Pv1.1$^+$ and Pv1.1$^-$, respectively (Fig. 3). First of the polymorphic sites to have been identified, this polymorphism serves as a useful surrogate for the other sites, since it is in complete linkage disequilibrium with them. Analysis of the DNA of unrelated Gaucher disease patients showed that the Pv1.1$^-$ genotype was invariably present when the "common Jewish" mutation 1226G was found. Similarly, the 84GG mutation is always found in context of the Pv1.1$^+$ allele and the IVS2(+1) mutation with the Pv1.1$^-$ allele. In contrast, it appears that the 1448C mutation arose more than once, since it is commonly found with either type of allele. The fact that the 1226G mutation and the IVS2(+1) mutation are always associated with the Pv1.1$^-$ haplotype, and the 84GG mutation is associated with the Pv1.1$^+$ haplo-

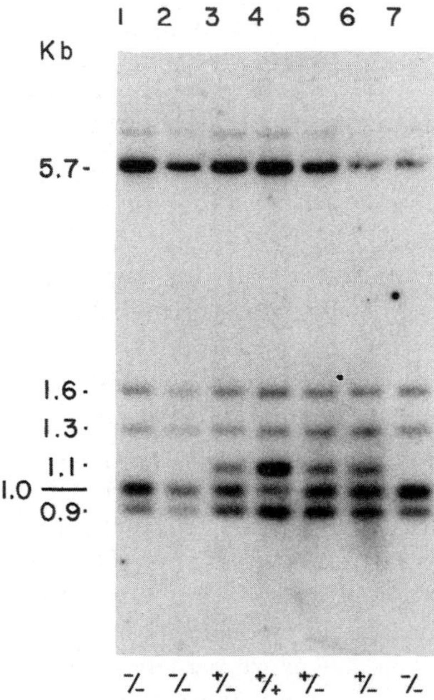

FIGURE 3 Southern blots of DNA digested with PvuII. When developed with a glucocerebrosidase cDNA probe, it is apparent that the DNA from some individuals has a 1.1-kilobase fragment, which is missing from others. The genotype is given at the bottom of each lane. [From E. Beutler (1988). Gaucher disease: New developments. *Curr. Hematol. Oncol.* 6, 1–26. Reprinted with permission from Yearbook Medical Publishers, Inc.]

type, is compatible with a single source for each of these mutations, which are most common in the Ashkenazi Jewish population. The premise that the evolutionary process allowed for expansion of these mutations despite the obvious deficiencies in the homozygote state speaks to the argument for a selective advantage within this ethnic group. To date, resistance to infection (e.g., tuberculosis) increased fertility, and enhanced intellectual capacity have all been tested and rejected as possible explanations for the selective advantage in carriers of the defective allele.

VI. DIAGNOSIS

Classically, the diagnosis of Gaucher disease had been established on the basis of histologic examination of bone marrow, spleen, or liver with identification of Gaucher cells. The development of a reliable enzymatic assay using peripheral blood lymphocytes or cultured skin fibroblasts obviates the need for an invasive procedure and has proven to be more accurate in that many other disorders, for example, leukemias and lymphomas, can cause the appearance of Gaucher-like cells. Low β-glucocerebrosidase activity in an assay is pathognomonic for Gaucher disease; even cells from amnionic fluid can be assayed for a definitive diagnosis. However, there are limitations to the ability of this assay to be predictive in that some heterozygotes are represented as false negatives and hence gene frequencies cannot be determined accurately. Moreover, the absolute amount of enzyme does not correlate with the severity of the disease, and the three types cannot be uniquely identified. The diagnosis of Gaucher disease can be achieved by identification of the mutation at the DNA level. Mutation analysis can establish complete and accurate diagnosis when the mutations on both alleles can be ascertained, and can be helpful even when only one mutation is determined. Ascertaining that a patient has the 1226G/1226G genotype may be encouraging in that it is generally predictive of a mild clinical course; whereas the 1448C/1448C genotype predicts a neuronopathic form of Gaucher disease. Thus, although only broad guidelines are available, genetic counseling based on molecular genotyping may be attempted. All methods used today for analysis of mutations causing Gaucher disease are dependent on the use of the polymerase chain reaction (PCR). [*See* Polymerase Chain Reaction.] Separation of the active gene from the pseudogene is critical and is accomplished by several methodologies, including allele-specific oligonucleotide amplification, allele-specific oligonucleotide hybridization, and the amplification refractory mutation system (ARMS).

VII. THERAPY FOR GAUCHER DISEASE

Until recently, treatment for Gaucher disease patients was largely symptomatic: splenectomy for relief of symptoms associated with the enlarged organ and/or the abnormally low blood counts; orthopedic surgery, including total joint replacement, to improve limb function and control pain; and regimens of antibiotics to fight infections. In a few cases, bone marrow transplantation has been curative, but this procedure, even when an HLA-matched donor is available, has many attendant risks that may not be justified in a disease in which an alternative, safer treatment modality, namely, enzyme replacement therapy, is available. Enzyme replacement with a human placental product, alglucerase (Genzyme Co., Cambridge MA), has proven to be both safe and effective in ameliorating the signs and symptoms of type I Gaucher disease. This approach is predicated on the targeting of glucocerebrosidase to macrophages by attaching to mannose-specific receptors. Although it has been shown that most of the enzyme does not attach to the macrophage-restricted classic high-affinity mannose receptor but to a lower-affinity promiscuous receptor, the preparation is effective. Clinical trials of the original high-dose (60 U/kg body weight) protocol of infusions every 2 weeks resulted in successful reduction of spleen and liver volumes as well as increased hemoglobin and platelet counts to within the normal ranges. Subsequently, treatment protocols of low dose (2.3 U/kg body weight) but infused three times a week have been found to be equally effective in achieving therapeutic results. Hence, in the hiatus before successful gene therapy, enzyme replacement therapy is the current treatment of choice for symptomatic type I patients. A recombinant form of the placental product, imiglucerase (Genzyme Co.), appears to be as or more effective on a unit-per-unit basis as the placental derivative.

In the foreseeable future, cure of the disease, not just palliation, may be achieved by gene transfer. Glucocerebrosidase cDNA has been successfully inserted into cultured fibroblasts and lymphoblasts from Gaucher disease patients, and the enzymatic defect was corrected (Fig. 4). Gaucher disease is a suitable

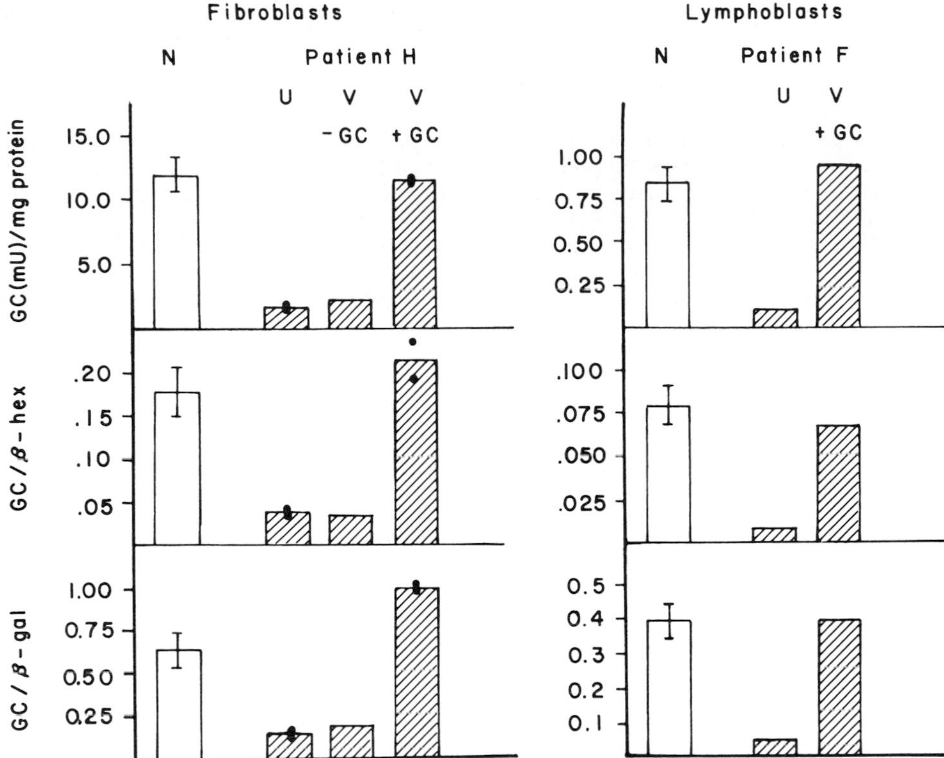

FIGURE 4 The restoration of glucocerebrosidase activity in fibroblasts of two type I Gaucher disease patients. N represents normal control cells. Enzyme assays were carried out on uninfected cells (U), cells infected with a retrovirus containing the neo gene but not the glucocerebrosidase cDNA (V − GC), and a retrovirus containing both the neo gene and the glucocerebrosidase cDNA (V + GC). [From J. Sorge, W. Kuhl, C. West, *et al.* (1987). Complete correction of the enzymatic defect of type I Gaucher disease fibroblasts by retroviral-mediated gene transfer. *Proc. Natl. Acad. Sci. USA* **84**, 906–909.]

paradigm for gene transfer to hematopoietic stem cells, since the cells expressing the enzymatic deficiency are progeny of these cells, which are accessible and easily manipulated *in vitro*. Infection of these hematopoietic stem cells from Gaucher disease patients with glucocerebrosidase cDNA in a recombinant retrovirus, and subsequent autotransplantation of the transformed cells, could cure the molecular defect without exposing the patient to the considerable risks of allogenic bone marrow transplantation. Unfortunately, the transduction of bone marrow stem cells by retroviruses has generally been inefficient, and the expression of the retrovirus genes in primitive cells, unlike the mature cultured skin fibroblasts, is very limited and is actively suppressed. Promising modes of gene transfer currently being explored as alternatives to retroviral transfer include the use of liposomes as carriers and the adenoassociated viruses as vectors. Once the barriers of efficient transduction on maintained expression are overcome, gene replacement therapy may become a feasible form of treatment. [*See* Gene Therapy.]

VIII. CONCLUSIONS

Gaucher disease is among the human genetic disorders that have been studied at the molecular level in considerable detail, resulting in improved diagnosis and ability to correlate clinical course with genetic mutation. Gaucher disease is also a model for the implementation of gene transfer therapy. Although the current mode of enzyme replacement therapy has vastly improved the quality of life of symptomatic type I patients, the hope for the future is a curative approach to genetic mutations.

BIBLIOGRAPHY

Beutler, E. (1991). Gaucher's disease. *N. Eng. J. Med.* **325**, 1354–1360.

Beutler, E. (1993). Gaucher disease as a paradigm of current issues regarding single gene mutations of humans. *Proc. Natl. Acad. Sci. USA* **90,** 5384–5390.

Beutler, E. (1993). Modern diagnosis and treatment of Gaucher's disease. *Am. J. Dis. Child* **147,** 1175–1183.

Beutler, E. (1995). Gaucher disease. *In* "Advances in Genetics" (T. Friedmann, ed.), pp. 17–49. Academic Press, San Diego.

Beutler, E., and Grabowski, G. (1995). Gaucher disease. *In* "The Metabolic Basis of Inherited Disease" (C. R. Scriver, A. L. Beudet, W. S. Sly, and D. Valle, eds.), pp. 2641–2670. McGraw–Hill, New York.

Correll, P. H., and Karlsson, S. (1994). Towards therapy of Gaucher's disease by gene transfer into hematopoietic cells. *Eur. J. Haematol.* **53,** 253–264.

Grabowski, G. A. (1993). Gaucher disease: Enzymology, genetics, and treatment. *Adv. Hum. Genet.* **21,** 377–441.

Grabowski, G. A., Gatt, S., and Horowitz, M. (1990). Acid β-glucosidase: Enzymology and molecular biology of Gaucher disease. *CRC Crit. Rev. Biochem. Mol. Biol.* **25,** 385–414.

Horowitz, M., and Zimran, A. (1994). Mutations causing Gaucher disease. *Hum Mutat* **3,** 1–11.

Karlsson, S., Correll, P. H., and Xu, L. (1993). Gene transfer and bone marrow transplantation with special reference to Gaucher's disease. *Bone Marrow Transplant* **11**(Suppl. 1), 124–127.

Gene Amplification

GEORGE R. STARK
The Cleveland Clinic Foundation

GLOSSARY

Bridge–breakage–fusion cycles Dicentric chromatids bridge as the centromeres are pulled apart at anaphase. Eventually they break between the centromeres, one part going to each daughter cell. After replication, the two broken chromatids fuse, generating a new dicentric chromatid and beginning a new cycle

Cellular oncogenes Normally encode proteins that carry out important regulatory functions in cells; when altered by mutation or overexpressed, they can contribute to tumorigenesis

Double minutes Form of amplified DNA visible under the light microscope in metaphase cells; these elements are often paired and have the appearance of normal chromatin; however, they lack centromeres and, therefore, segregate randomly upon cell division

Episomes Submicroscopic extrachromosomal DNA; in mammalian cells, they are probably precursors of double minutes

Inversion and translocation Processes in which the sequence of DNA is rearranged; in inversion, the orientation of part of the DNA within a chromosome is reversed; in a reciprocal translocation, segments of two chromosomes are exchanged

Recombination Reciprocal crossing-over in which two double-stranded DNAs exchange parts of their sequences; if the sequences at the site of the exchange are different, the exchange is nonhomologous (e.g., as in translocation)

Telomeres Structures at each end of each chromosome which prevent end-to-end joining (fusion)

THE PHENOMENON OF DNA AMPLIFICATION CONcerns an increase in a part of the genome of a cell, usually less than the amount contained in a full chromosome.

I. INTRODUCTION

In some lower organisms, amplification is a normal process involving developmentally programmed increases in DNA copy number in specific cells or tissues at specific times. The purpose is to achieve a rate of gene expression too rapid to be met by maximum expression of a single gene. In humans and other mammals, amplification is not known to occur during normal development; however, it is observed readily in tumor cells, where amplification of cellular oncogenes contributes in important ways to tumorigenesis and metastasis, and in cells in culture, where amplification of genes whose overexpression confers resistance to cytotoxic drugs is seen readily.

II. GENE AMPLIFICATION AND DEVELOPMENT

Developmentally regulated gene amplification has been best documented in two types of cells that need to make an unusually large amount of one gene product relatively quickly. One is the developing egg (i.e., oocyte) of many animals, which must make and store huge numbers of ribosomes. The other is the follicle cell of insects. In young *Xenopus* oocytes, ribosomal RNA genes are amplified about 1000-fold, producing many free ribosomal DNA molecules that cluster together to form about 1000 extra nucleoli. The follicle cells of insects synthesize and secrete the proteins that form the hard coat, or chorion, of insect eggs. In *Drosophila,* just before the chorion proteins are needed, the DNA sequences coding for them are amplified 20-fold in the sex-linked chorion cluster and 60- to 80-fold in the chorion cluster of the third chromosome. Although the problem of rapid synthesis of chorion genes is solved in developmental time in *Drosophila,* silk moths provide an interesting instance in which this has been done in evolutionary time, such that the present-day silk moths come endowed with approximately 25 copies of these genes in their germ line for developmentally regulated expression during choriogenesis. [*See* Genes.]

III. AMPLIFICATION OF GENES MEDIATING DRUG RESISTANCE

Most information about what happens during DNA amplification comes from studies of drug resistance in cell culture. Any drug whose toxic effect can be overcome by increased expression of a particular protein can be used to select resistant cells that have amplified the gene encoding that protein. Two examples are methotrexate, which selects for amplification of the dihydrofolate reductase gene, and hydroxyurea, which selects for amplification of ribonucleotide reductase. There are now 20–30 different genes whose amplification has been selected in this way. Multidrug resistance is a particularly interesting case. In this phenomenon, a cell selected for resistance to any one of a set of unrelated drugs (e.g., doxorubicin, vinblastine, colchicine) is resistant to all members of the set. The cause is amplification of a single gene, now called mdr, encoding a protein that functions as an efflux pump in the plasma membrane. The pump is relatively unspecific and uses the energy of adenosine triphosphate to push many different unwanted compounds out of cells. Because multidrug resistance involves several drugs routinely used in the chemotherapy of cancer, understanding and controlling this phenomenon is particularly relevant to future success in treating cancer.

IV. AMPLIFICATION OF CELLULAR ONCOGENES IN TUMORS

Amplification has yet to be detected in normal mammalian cells. It follows that the frequent amplification readily observed in drug-resistant cell lines and in tumors has somehow been activated abnormally in these cells or that amplifications lead to the death of normal cells but not tumor cells. Furthermore, the probability of amplifying DNA in abnormal cells increases with increasing tumor-forming potential. Because cellular oncogenes are often amplified in tumors, especially in advanced metastatic tumors, it is virtually certain that their overexpression contributes to tumor progression. Specific cellular oncogenes are amplified frequently in specific types of tumors. Some of the best correlations are amplification of N-myc in neuroblastoma, c-myc in breast and lung cancer, neu in breast and stomach cancer, and int-2 and hst in breast cancer and in squamous carcinoma. [*See* Oncogene, Amplification in Human Cancer.]

V. STIMULATION OF AMPLIFICATION

A wide variety of agents or treatments that damage DNA (ultraviolet or ionizing radiation, or carcinogens) or arrest DNA synthesis (hydroxyurea, aphidicolin, or hypoxia) can induce DNA amplification. A probable common denominator is the formation of broken DNA, which can participate in several amplification mechanisms. Also, severe cellular stress leads to the short-term expression of otherwise silent genes whose products, intended to deal with the emergency, may also facilitate amplification. Furthermore, a single abnormal process can lead to an unusual transient DNA structure, which can then rearrange in more than one way, leading to amplification, deletion, other chromosomal abnormalities such as inversion or translocation, or even loss of the affected chromosome. Detailed consideration of mechanisms and structures that might be involved in DNA amplification provides possible explanations for such linked responses. [*See* Chromosome Anomalies.]

VI. FORMS OF AMPLIFIED DNA

When the degree of amplification is high, the extra DNA can often be seen under the light microscope, either as expanded regions of chromosomes or separate from chromosomes as paired, acentric elements called double minutes (Fig. 1). There are also episomes, extrachromosomal forms too small to be seen in the microscope. Double minutes are observed commonly in human tumors analyzed immediately after biopsy; therefore, they are likely to be important in the formation or evolution of amplified DNA in tumors. They have a nucleoprotein structure similar to that of chromosomes and each contains about 5 million bases of DNA. The extrachromosomal forms of amplified DNA replicate only once per cell cycle. Because they lack centromeres, they segregate randomly during cell division and can either accumulate or become lost in a population of cells, depending on whether the selection used requires gain or loss of a particular gene for the cells to survive.

In intrachromosomal amplified DNA, a very large region of flanking sequences, as large as 10 million bases, is often coamplified together with each copy of the gene that is the target of selection. This is especially true in the initial event, when a single-copy sequence becomes amplified for the first time (Fig. 2a). If a population of cells already containing an amplification is exposed to a higher concentration of drug, further amplification will allow some cells to survive. In extreme cases, cells containing thousands of copies of a particular amplified gene have been obtained after several steps of selection. If the amplified DNA is

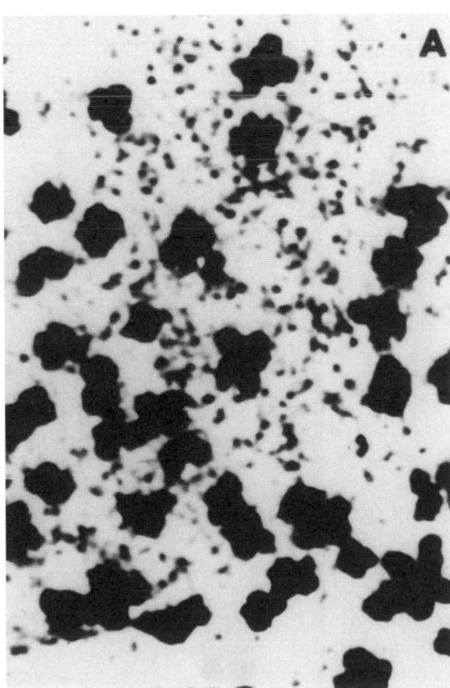

FIGURE 1 Double minute chromosomes. A partial metaphase chromosome spread from a human neuroblastoma cell line is shown.

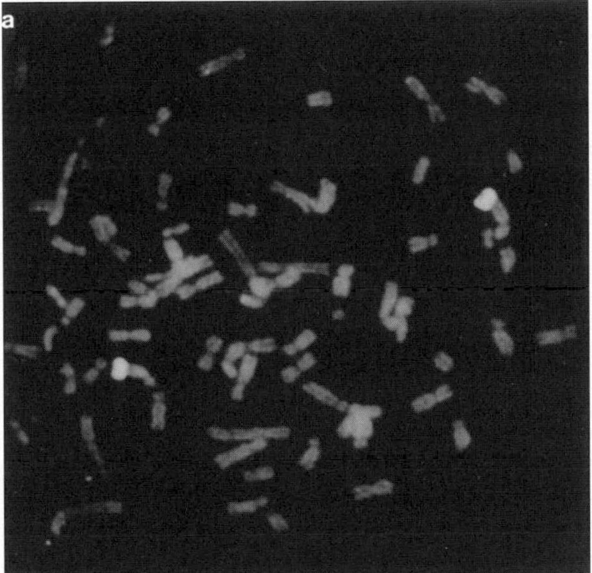

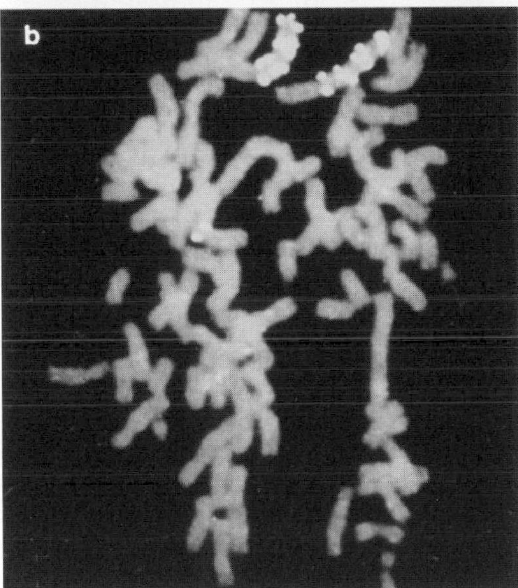

FIGURE 2 *In situ* hybridization to Syrian hamster cells containing amplification of a gene mediating resistance to an inhibitor of uridine monophosphate synthesis. Detection of a biotin-labeled DNA probe employed fluorescein-labeled avidin and antibodies directed against avidin. (a) The first step of amplification includes only a few extra copies of the gene. These are widely spaced and can be seen as pairs of spots on the two sister chromatids. (b) After the third step of amplification, 25 copies of the gene are present in a more condensed form.

extrachromosomal, increased resistance can be achieved by unequal segregation of the element at each mitosis without a new amplification event. However, if the amplified DNA is chromosomal, new amplification events are required, in which the genes conferring resistance become more concentrated (Fig. 2b) and much of the extraneous coamplified DNA is lost.

VII. MECHANISMS OF AMPLIFICATION

More than one mechanism is involved, based on the different structures and locations of the amplified DNA. The mechanisms proposed fall into two major classes. In each, unequal distribution of DNA sequences at division allows one daughter cell to accumulate DNA at the expense of the other. The first mechanism proposes that an extrachromosomal element arises by circularization and deletion of an intrachromosomal segment. This episome then replicates under normal control, once per cell cycle. However, lacking a centromere, it distributes unequally into the daughter cells at each cell division, so that some cells accumulate more copies than others.

In the second class of mechanisms, amplification follows the formation of dicentric chromosomes, which participate in bridge–breakage–fusion (BBF) cycles. As shown in Color Plate 1, asymmetric breakage of a dicentric chromosome during cell division leads to unequal distribution of some genes between the two daughter cells. The new dicentric chromosome formed at the end of each cycle initiates a subsequent cycle and, until the chain is broken, many copies of a particular gene can accumulate. Note that the process is pyramidal, with an increasing number of cells involved after each cycle. Note too that deletions are generated by the same process. The initiating event is likely to be breakage in many cases, accounting for the stimulation of amplification by agents that damage DNA. It is also likely that the loss of telomeres can stimulate fusion-first entry into BBF cycles. Telomeres protect the ends of chromosomes from fusing and, when they are lost or reduced in size, a dramatic increase in the number of dicentric chromosomes follows.

To test these and other mechanistic possibilities critically, it is important to examine individual cells soon after the initial event. New methods now allow detection of single copies of any DNA sequence in single cells with high resolution, using fluorescence to detect the site of hybridization of a cloned DNA probe (Fig. 2). The structure of newly amplified DNA is thus open to a detailed analysis. As the mechanisms of DNA amplification become better understood, so too will the connections between amplification and the other chromosomal abnormalities of tumor cells.

VIII. REGULATION OF AMPLIFICATION: p53

The protein p53, the "guardian of the genome," is lost or mutated in the majority of human tumors. A major function of p53 is to prevent the proliferation of cells that have suffered DNA damage, causing them to stop growing or to undergo programmed cell death. Normal cells, which contain wild-type p53, are not permissive for gene amplification but become so when p53 is lost or mutated. This process is now understood in broad outline, as it is known that the processes of amplification generate broken DNA and that broken DNA stimulates p53 to act. Therefore, tumor cells that lose p53 itself, or other components of the pathway through which p53 acts, become permissive not only for amplification but also for other chromosomal abnormalities that contribute to increased genome plasticity, a property of great advantage to tumor cells. A major challenge for the future is to understand the detailed processes of p53 activation and function, which will in turn explain how amplification is regulated.

BIBLIOGRAPHY

Livingstone, L. R., White, A. A., Sprouse, J., Livanos, E., Jacks, T., and Tlsty, T. D. (1992). Altered cell cycle arrest and gene amplification potential accompany loss of wild-type p53. *Cell* **70**, 923.

Nelson, W. G., and Kastan, M. B. (1994). DNA strand breaks: The DNA template alterations that trigger p53-dependent DNA damage response pathways. *Mol. Cell. Biol.* **14**, 1815.

Poupon, M.-F., Smith, K. A., Chernova, O. B., Gilbert, C., and Stark, G. R. (1996). Inefficient growth arrest in response to dNTP starvation stimulates gene amplification through bridge-breakage-fusion cycles. *Mol. Biol. Cell.* **7**, 345.

Saito, I., Groves, R., Giulotto, E., Rolfe, M., and Stark, G. R. (1989). Evolution and stability of chromosomal DNA coamplified with the CAD gene. *Mol. Cell. Biol.* **9**, 2445.

Smith, K. A., Gorman, P. A., Stark, M. B., Groves, R. P., and Stark, G. R. (1990). Distinctive chromosomal structures are formed very early in the amplification of CAD genes in Syrian hamster cells. *Cell* **63**, 1219.

Smith, K. A., Stark, M. B., Gorman, P. A., and Stark, G. R. (1992). Fusions near telomeres occur very early in the amplification of

CAD genes in Syrian hamster cells. *Proc. Natl. Acad. Sci. USA* **89**, 5427.

Stark, G. R. (1993). Regulation and mechanisms of mammalian gene amplification. *Adv. Cancer Res.* **61**, 87.

Stark, G. R., Debatisse, M., Giulotto, E., and Wahl, G. M. (1989). Recent progress in understanding mechanisms of mammalian DNA amplification. *Cell* **57**, 901.

Stark, G. R., Debatisse, M., Wahl, G. M., and Glover, D. M. (1990). DNA amplification in eucaryotes. *In* "Genome Rearrangement and Amplification: Frontiers in Molecular Biology" (D. Hames and D. M. Glover, eds.). IRL Press, Oxford.

Toledo, F., Buttin, G., and Debatisse, M. (1993). The origin of chromosome rearrangements at early stages of *AMPD2* gene amplification in Chinese hamster cells. *Curr. Biol.* **3**, 255.

Trask, B. J., and Hamlin, J. L. (1989). Early dihydrofolate gene amplification events in CHO cells usually occur on the same chromosome arm as the original locus. *Genes Dev.* **3**, 1913.

Wright, J. A., Smith, H. S., Watt, F. M., Hancock, M. C., Hudson, D. L., and Stark, G. R. (1990). DNA amplification is rare in normal human cells. *Proc. Natl. Acad. Sci. USA* **87**, 1791.

Yin, Y., Tainsky, M. A., Bischoff, F. Z., Strong, L. C., and Wahl, G. M. (1992). Wild-type p53 restores cell cycle control and inhibits gene amplification in cells with mutant p53 alleles. *Cell* **70**, 937.

Gene Mapping

GLEN A. EVANS
University of Texas Southwestern Medical School at Dallas

GLOSSARY

Genetic linkage Technique of constructing genetic maps by determining the amount of recombination occurring between genetic markers or polymorphisms

Haplotype Specific alleles of a group of linked markers on a chromosome

Heterozygosity (H) Quantitative measure of polymorphism that measures the probability that a random individual in a population is a heterozygote

Linkage disequilibrium Nonrandom association of alleles at linked loci

Polymorphism information content (PIC) Quantitative measure of polymorphism that measures the probability that one of two random repeats is a heterozygote and the other is a different genotype

Restriction fragment-length polymorphism (RFLP) Polymorphism in a DNA sequence that occurs within a restriction enzyme recognition site and may be detected by an alteration of bands on a Southern blot

Sequence tagged site (STS) Mapping object defined by a unique DNA sequence and a map location

Simple sequence repeat (SSR) polymorhpism Polymorphism defined by a series of simple sequences such as $(CA/GT)_n$, where the repeat number differs between individuals

Single-base-pair (SBP) polymorphism Single-base difference between two different individuals

Variable number tandem repeat (VNTR) polymorphism Polymorphism in the number of repeat units, where the unit is a sequence of 20 to 200 base pairs in length

Yeast artificial chromosome (YAC) Cloning technique and vector for constructing artificial chromosomes to replicate in yeast but containing a large human (or other species) fragment of DNA as an insert

GENE MAPPING REFERS TO THE GROUP OF TECHniques that generate a physical or conceptual interrelationship between genes as they are arrayed or organized on chromosomes. Gene mapping allows genes to be located at specific locations on chromosomes, allows diagrams or maps defining the interrelationship between genes and other markers to be constructed, and allows genes defined by their phenotype, such as inherited diseases, genetic traits, or other visible attributes, to be located relative to other markers on genetic maps. In biochemical terms, genes are sequences of DNA elaborating a protein and/or resulting in a detectable trait, disease, or phenotype. In theoretical terms, genes are units of inheritance contributing a single characteristic or phenotype to the organism. [*See* Genes.]

I. UNDERLYING THEORY

Since DNA is a linear molecule, gene maps are inherently one dimensional. Gene maps relate map objects, such as genes, polymorphic markers, sequence tagged sites (STS), repetitive sequences, or other structures encoded in the genome, by their order on the chromosome and display a distance metric separating the objects. Different types of gene maps differ in the underlying metric or distance measure separating the genes as well as in the method of measurement used to determine the order and distance. Genetic maps are constructed through the analysis of inheritance and the frequency of recombination between genes or markers that are located close together but separated through meiosis. The markers must be polymorphic, or different between different individuals, and the distance separating them must be measured in frequency of recombination or centiMorgans (cM).

Physical maps refer to maps that define the distance between genes or other markers in actual physical distance along the DNA molecule, defined in base pairs (bp), kilobases (1000 bp), or megabases (1,000,000 bp). The basic units of physical mapping, by convention, are unique mapping markers defined by DNA sequence known as sequence tagged sites. A specialized type of physical map, constructed using sets of somatic cell hybrid cell lines grown in culture, is called radiation hybrid (RH) mapping. RH mapping defines the physical distance between markers by the frequency of radiation-induced breakage. The distance separating markers is measured in centiRays (cR), referring to the frequency of chromosome breakage at a known does of ionizing radiation. Table I lists types of maps and the associated metrics.

II. TYPES OF MAPS AND TECHNIQUES FOR MAPPING

A variety of techniques have been used for gene mapping that generate maps of differing resolution. Depending on the use or requirement, techniques may be applied for determining the chromosome that a gene is located on, the location of a gene on a specific chromosome, the location of a gene or marker within a genetic or physical map, or the location of an inherited phenotype within known polymorphic markers. The resolution is the minimum size or region of DNA to which a gene or marker can be mapped and is defined in base pairs. Table II lists commonly used techniques for gene mapping and the approximate resolution that can be achieved with each technique.

TABLE I
Types of Genetic Maps

Type	Metric	Unit
Genetic	Recombination	cM
Radiation hybrid	Radiation breakage	cR
Physical	DNA sequence	bp, kb, mb

TABLE II
Genetic Mapping Methods and Resolution

Mapping method	Approximate resolution
Genetic mapping	
Genetic linkage	2 to 10 cM
Linkage disequilibrium	0.1 to 1 cM
Physical mapping	
Cytogenetic	50 mb
Somatic cell hybrid	50 to 250 mb
Hybrid panel	50 to 250 mb
Breakpoint panel	10 to 50 mb
Radiation hybrid	50 to 500 kb
FISH (metaphase chromosomes)	2 mb
FISH (interphase chromosomes)	10 to 100 kb
DIRVISH (extended DNA)	5 to 100 kb
STS content mapping	200 kb to 1 mb
YAC contig mapping	200 kb to 1 mb
Cosmid contig mapping	10 to 100 kb
Sequence sampled mapping	1 to 5 kb
DNA sequence	1 bp

A technique with a mapping resolution of 2 mb means that two markers that map to the same location cannot be ordered if they are less than 2 mb apart. Markers that map to the same region within the mapping resolution are said to map to the same "bin."

Techniques for genetic mapping include genetic linkage, two-point genetic linkage mapping, multipoint genetic linkage mapping, and linkage disequilibrium mapping. Techniques for physical mapping include fluorescence *in situ* hybridization (FISH), somatic cell hybrid mapping, radiation hybrid mapping, STS content mapping, yeast artificial chromosome (YAC) contig mapping, and DNA sequencing. [*See* Genetic Maps.]

III. GENETIC MAPPING

Genetic mapping refers to the analysis of the inheritance of polymorphic markers through families or pedigrees and determination of the degree of genetic linkage. Linkage occurs when two or more markers are inherited together and separated only through rare recombination. The closer two markers are on a chromosome, the higher the degree of linkage and the lower the frequency of recombination. Two markers that segregate at random are said to be unlinked.

Techniques for genetic mapping depend on the detection and analysis of DNA polymorphisms. Polymorphisms are DNA sequences that differ between two or more different individuals in a population and the sequence variations can be traced through pedigrees with Mendelian inheritance. Several types of polymorphisms are commonly used in genetic mapping and are listed in Table III. Restriction fragment-length polymorphisms (RFLP) are variations in sequence length due to single-base changes in restriction enzyme recognition sizes. RFLPs usually demonstrate two alleles and are generally detected by a procedure known as Southern blotting and DNA hybridization. A second type of DNA polymorphism is a variable number tandem repeat (VNTR) polymorphism, which are variations in the number of a single tandem repeating units. Commonly used VNTRs are repeats of a 15- to 200-bp motif repeated 10 to 200 times. VNTR polymorphisms generally have between two and several hundred alleles. They are detected by Southern blotting and DNA hybridization, or by polymerase chain reaction (PCR) amplification followed by gel electrophoresis. A third type of polymorphism is a simple sequence repeat (SSR) polymorphism, which is a variation in the number of repeat units of a simple sequence between individuals. The most commonly used SSR for genetic mapping is the $(CA/GT)_n$ repeat, also known as a microsatellite sequence. SSR polymorphisms generally have two to several hundred alleles and are detected and analyzed by PCR amplification and gel electrophoresis. Finally, a fourth type of polymorphism is defined by any single-base DNA sequence change, a single-base pair (SBP) polymorphism. SBPs occur frequently in the human genome with 1/500 to 1/1000 base pairs being polymorphic. SBPs have from two (biallelic SBP) to four alleles; they are detected and analyzed using allele-specific oligonucleotide hybridization. [*See* Genetic Polymorphism.]

Markers that are more polymorphic are in general more useful for genetic mapping. Quantitative measures of the amount of polymorphism inherent in a marker are thus important in mapping studies. Quantitative measures of polymorphism are heterozygosity (H) and polymorphism information content (PIC). H is the probability that a random individual in a population is a heterozygote at a particular polymorphic locus and is calculated as

$$H = 1 - \sum p_i^2$$

where p_i is the population frequency of the ith allele. H is the measure of polymorphisms that is most useful in genetic linkage studies.

A second, more complex, measure of genetic polymorphism is PIC, which is calculated as

$$PIC = 2 \sum \sum p_i p_j (1 - p_i p_j)$$

where p_i and p_j are allele population frequencies. PIC is a measure of the probability that one of two random individuals (e.g., parents in a mating) is a heterozygote and the other has a different genotype, or combination of alleles at that marker. The maximum PIC occurs when all alleles have an equal frequency. PIC is most useful in genetic counseling and gene mapping through the use of genetic markers. Both H and PIC values range from 0 to 1, and the higher the value, the more polymorphic the locus. For instance, the H of the ABO blood group is about 0.48, whereas the H of a VNTR with five alleles may be more than 0.8. For polymorphisms with a large number of alleles, H and PIC are approximately equal.

Genetic mapping is carried out by detecting the segregation of a polymorphic DNA sequence and a visible trait or disease through a pedigree by inheritance. If the polymorphism and trait are genetically linked, they will, in general, be inherited together except where recombination separates them. The simplest measure of recombination frequency is the recombination fraction, or the number of recombinations that occur divided by the number of meiosis events. The unit of recombination is the centiMorgan, where 1 recombination in 100 meiosis events represents 1 cM or 1% recombination. The human genome is approximately 3300 cM. [*See* Human Genome.]

A more complex measure of recombination that is widely used for genetic mapping is the LOD score:

$$LOD = \log_{10} [P\text{ (family}|\text{linkage)}] / [P\text{ (family}|\text{independent assortment)}]$$

The LOD score is a measure of the likelihood of the odds that two markers are linked as opposed to un-

TABLE III
Types of Human DNA Polymorphisms

Type	Description
RFLP	Restriction fragment-length polymorphism
VNTR	Variable number of tandem repeat
SSR	Simple sequence tandem repeat [microsatellite, $(CA/GT)_n$]
SBP	Biallelic single-base pair repeat

linked and are randomly assorted. The LOD method is the most widely used method for calculating the probability of genetic linkage. Since LOD is a $\log_{10}$, an LOD score of 4 means odds of 10,000 : 1 in favor of linkage. LOD scores may be added to allow the combination of data from different pedigrees and families and LOD scores are usually plotted for a range of recombination fraction. Genetic heterozygosity is revealed when LOD scores from one or more families are inconsistent. By convention, an LOD score greater than 3 (odds of 1000 : 1 in favor of linkage) is generally accepted as evidence for linkage of two genetic markers and a score less than −2 is generally accepted as excluding linkage.

A genetic map may be constructed by analyzing a series of linked polymorphisms in a pedigree and calculating an LOD score for each pair of markers. The LOD scores indicate linkage and the recombination fraction at which the maximum LOD score is achieved indicates the most likely distance separating the markers. This method is known as two-point genetic mapping since each pair of markers is independently compared. Genetic maps may also be constructed using multipoint mapping, in which complex algorithms are used to carry out simultaneous comparisons of multiple polymorphic markers and the most probable order and distances are determined.

A related technique for genetic mapping utilizes linkage disequilibrium to locate genes or polymorphism markers within a known genetic map. Linkage disequilibrium is the nonrandom association of alleles at linked loci. In any genetic map, the specific collection of alleles on each chromosome is known as the haplotype. Genes linked to these markers may be located in the map by determining the region that is identical by descent or identical by inheritance. Linkage disequilibrium diminishes through time as a result of recombination and is a widely used mapping method for locating disease genes or phenotypic traits within a genetic map.

IV. PHYSICAL MAPPING

Physical mapping refers to maps that use actual DNA length in base pairs as a measure of mapping distance. Many different techniques for physical mapping have been derived that differ in the resolution of the map produced. Some of the techniques used in gene mapping are listed in Table II, along with the approximate resolution of the mapping method.

Fluorescence *in situ* hybridization is a mapping method in which a DNA probe is labeled with fluorescent compounds and DNA hybridization carried out to metaphase chromosomes attached to a microscope slide. The location of the gene or DNA probe can be determined using epifluorescence or confocal microscopy and a position determined relative to cytogenetic band position. Chromosomal DNA is in an extended form during interphase and FISH carried out on interphase nuclei has also been used as a mapping method. Interphase FISH allows the mapping resolution to be extended from 2 mb to 50 kb. Using this technique, detailed maps of chromosomal regions have been constructed. Direct visualization FISH (DIRVISH) using *in situ* hybridization is used with stretched prepared DNA. In this process, cellular lysates are fixed to microscope slides, DNA is extended, and FISH is carried out with a possible resolution of 5 kb. Using FISH, several detailed physical chromosomal maps have already been constructed.

Mapping techniques that utilize somatic cell hybrids depend on the construction of specialized cell lines growing in tissue culture. Hybrid cell lines are formed when a mouse or hamster cell is fused with a normal human cell and cell lines are derived that contain the normal complement of mouse or hamster chromosomes but also carry one or more human chromosomes. Hybrid panels are collections of cell lines where each individual line carries a different complement of human chromosomes. Analyzing each cell line in the panel for the presence of a specific human gene can be used to localize a gene to a specific human chromosome. Monochromosomal somatic cell hybrids are cell lines carrying only a single human chromosome and are very valuable for mapping purposes. Hybrid cell line panels may also be constructed with fragments of human chromosomes derived from translocated or deleted chromosomes and can be used to regionally localize genes to bands or cytogenetically defined regions of chromosomes. A specialized physical mapping technique, radiation reduction somatic cell hybrid mapping or radiation hybrid mapping, uses ionizing radiation to fragment human chromosomes carried in a somatic cell hybrid cell line, generating hundreds of cell lines carrying small chromosomal fragments. The size of the fragments, and thus the mapping resolution, is dependent on the dose of radiation used for the chromosomal breakage. Analysis of a collection of cell lines with a set of specific probes can result in a radiation hybrid map. An RH map is similar to a physical map but the distances separating map objects are measured by the frequency of radiation breakage or cR. On average, cR is approximately 50 kb, though it varies enormously between chromosomal regions.

For the purposes of most human gene mapping and for applications in the Human Genome Project, physical mapping techniques now utilize a standardized map object, the STS. A sequence tagged site is a unique sequence in the genome that can be detected by PCR amplification and has a defined map location. The implication of the use of STSs as standardized mapping tools is that map tools can be transmitted by information only, instead of requiring the exchange of an actual DNA probe. Large collections of STSs have been produced and define detailed physical maps of all human chromosomes. [*See* Chromosomes, Molecular Studies.]

The highest-resolution human chromosomal maps have been constructed through clone contig building using STS analysis. One valuable approach is known as yeast artificial chromosome contig mapping by STS content. Yeast artificial chromosomes are molecular clones in which 100-kb to 5-mb fragments of human DNA are cloned in a yeast vector containing a yeast centromere, telomere, and origins of replication, as well as drug resistance genes for selection. Large libraries of YAC clones representing the human genome have been produced and analyzed with equally large panels of STSs. Where one YAC clone is found to contain two STSs, these STSs are thus physically linked. The data from screening YAC clones with large numbers of STSs can be assembled into an STS content map linking clones and markers covering large regions of human chromosomes. Sets of overlapping clones linked by STSs are known as contigs (for contiguously overlapping clones), and not only provide order and distance but also supply the physical material (cloned DNA) for further analysis and for DNA sequencing. STS content maps at approximately 100-kb resolution have now been constructed for the entire human genome through the Human Genome Project.

V. DNA SEQUENCE AND THE HUMAN GENOME PROJECT

The Human Genome Project was established in 1990 as an organized large-scale biology project, funded by the National Institutes of Health and the Department of Energy, with the goal of constructing detailed genetic and physical maps and determining the complete DNA sequence of the human genome, as well as the DNA sequences of model organisms such as *Escherichia coli, Saccharomyces cerevisiae, Caenorhabditis elegans, Drosophila melanogaster,* and possibly *Mus musculus.* The mapping phase of the Human Genome Project was completed in 1996 with a complete 0.5-cM resolution genetic map of the human genome, 100-kb resolution physical maps, and STS content YAC contig maps of all human chromosomes. The DNA sequencing of the human genome is expected to be completed between the years 2003 and 2005.

The complete DNA sequence represents the ultimate and most detailed physical map of the human genome, which consists of 3 × 109 bp (haploid), organized as 23 chromosomes and a 16,600-bp mitochondrial genome. The genome contains an estimated 100,000 genes. Two to 3% of the genome represents protein-encoding regions, approximately 5% highly repetitive satellite sequences, 30% moderately repetitive sequences, and 60% DNA in intragenic and intergenic regions that are essentially of unknown function. About 1/500 to 1/1000 bases are polymorphic between any two individuals. Based on the human genome sequence, a detailed map will be assembled that will comprise the basic blueprint for the construction of a human being.

BIBLIOGRAPHY

Chumakov, I. M., *et al.* (1995). A YAC contig map of the human genome. *Nature* **377,** 175–298.

Cooper, N. C. (ed.) (1994). "The Human Genome Project: Deciphering the Blueprint of Heredity." University Science Books, Mill Valley, California.

Davies, K. E., and Tilgman, S. M. (eds.) (1992). "Strategies for Physical Mapping." Cold Spring Harbor Laboratory Press, Cold Spring Harbor, New York.

McKusick, V. A. (1994). "Mendelian Inheritance in Man: A Catalog of Human Genes and Genetic Disorders," 11th Ed. Johns Hopkins Univ. Press, Baltimore.

Schuler, G. D., *et al.* (1996). A gene map of the human genome. *Science* **274,** 540–546.

Weiss, K. M. (1993). "Genetic Variation and Human Disease: Principles and Evolutionary Approaches." Cambridge Univ. Press, Cambridge, England.

Genes

H. ELDON SUTTON
University of Texas at Austin

GLOSSARY

Allele Alternate form of a gene

Chromosome Structure found in the nuclei of eukaryotes that contains DNA and proteins; the means by which a cell distributes genes during cell division

Complementation Formation of a normal phenotype when two "alleles" are on different members of a pair of homologous chromosomes; ordinarily this indicates that the mutations are in different loci

DNA Deoxyribonucleic acid; the molecule that serves for the primary storage of genetic information; a double-stranded structure composed of nucleotides

Eukaryotic Having cells with nuclei; prokaryotes, such as bacteria, do not have nuclei

Exon Segments of a gene that appear in messenger RNA after processing of the primary RNA transcript

Gene Basic unit of heredity

Intron Segments of a gene that are removed during processing of the primary RNA transcript

Mendelian Heredity characterized by segregation of either of two alleles into gametes and recombination in zygotes

Mutation Any abrupt heritable change in the genetic material

Protein Large linear molecules composed of amino acid subunits; the sequence of amino acids is coded by genes

Transcription Process by which genetic information encoded in DNA is copied into RNA

GENES ARE THE BASIC UNITS OF BIOLOGICAL heredity. They are the elements of the blueprint, the genotype, received by a cell from its parents from which it builds a complete organism. The characteristics of that organism, the phenotype, are a product of the interaction of the genotype with the environment. Genetic information is encoded in the nucleotide sequences of DNA in a linear, quaternary code. This information eventually is translated into the kinds and amounts of proteins that an organism can synthesize. Variations in the coded information account for the inherited variations that exist among individuals in a population, among populations within a species, and among different species. Since present-day genes in all species evolved from a few primordial genes by duplication and diversification, the evolutionary relationships among genes and among species can often be noted in the similarities of gene structure.

I. DEFINITIONS OF A GENE

A. Genes as Units of Variation

The observations of Mendel and the early Mendelians were concerned with inherited variation. The physical basis for the variations was, of course, unknown. Only by observing variation could one deduce the existence of a unit that, in its various forms (allelomorphs or, more commonly, alleles), accounted for the inheritance of different phenotypes. This unit was given the name gene by W. Johannsen in 1909.

The gene defined as a unit of variation is still often used. As methods have been developed to examine the molecular details of gene structure, many examples have been discovered in which the variation from wild type actually involves elimination of clusters of genes (defined as functional units; see the following section). Such deletions may be transmitted strictly by Mendelian laws, and for many purposes the molecular basis of the variation is not important.

B. Genes as Units of Function

A major shift in the concept of a gene occurred with studies in the 1930s and 1940s on gene function. Each mutant allele was associated with loss of function of a single specific enzyme. The normal ("wild-type") allele was therefore responsible for production of a functional enzyme. These results were anticipated by studies on metabolic diseases in humans, but such studies had little impact because they represented too great a jump from established knowledge and, because the observations were on humans, experimental studies were limited.

Enzymes are proteins, the only important exceptions being some RNA molecules that have recently been found to catalyze certain reactions. The connection between genes and proteins was established in 1956 with the demonstration that the amino acid sequence of human sickle-cell hemoglobin is altered as compared with the sequence in normal hemoglobin. Therefore, one function of genes is to specify the amino acid sequence (primary structure) of proteins. Other studies showed that mutations in the tryptophan synthetase gene of *Escherichia coli* can be arranged by genetic analysis into a linear sequence and that the corresponding amino acid changes in the protein product have the same linear sequence; that is, the two structures are colinear. Therefore, one could consider a gene as a DNA sequence that codes for a corresponding amino acid sequence in a protein.

C. Genes as Units of Transcription

As knowledge of the molecular events of DNA function and protein synthesis has expanded, the definition of a gene has also evolved. It is now accepted that the function of a gene is largely described by the amount and structure of the RNA that is transcribed. Furthermore, some genes, such as ribosomal RNA or transfer RNA genes, have RNA as their final product rather than protein. Therefore, a gene can be more generally defined as a unit of transcription. Because the amount of transcribed RNA is regulated by adjacent DNA nucleotide sequences, these can also be included within the boundaries of the gene, even though they are not transcribed. These contiguous regulatory sequences are to be distinguished from more remote DNA segments (enhancer regions) that also influence transcription of particular genes. [*See* DNA and Gene Transcription.]

D. Pseudogenes

DNA probes have allowed the detection of specific nucleotide sequences and, hence, of any gene or other DNA region that contains sequences complementary to the probe. Although many probes detect a single nucleotide sequence in the genome, others detect more. In some instances where several complementary copies exist, only one is ever transcribed; the remainder are pseudogenes, that is, genes whose structures resemble those of functional genes but are not functional. Pseudogenes are thought to have arisen primarily by gene duplication with subsequent changes that caused loss of function. Some ppear to have resulted from insertion of cDNA; that is, a DNA copy is made from transcribed and processed RNA and is then inserted back into the DNA but without the essential regulatory regions.

E. Alleles

An allele is defined as an alternate form of a gene (alternate being understood in the modern sense as a variant DNA sequence). Until it became possible to analyze gene structure and function at the molecular level, the demonstration that two genetic variants were modifications of the same gene was formally impossible; rather, arguments were based on the absence of crossing-over (which produces recombinant products) when the two variants were in heterozygous combination in the same diploid cells. Supporting evidence was provided by the inability of the two variants to complement each other, suggesting that the defects were in the same functional group. We now know that crossing-over can occur within genes, albeit extremely rarely, providing false evidence of nonallelism. Also, there are interpretations of complementation other than nonallelism.

Nonallelism of distant loci can readily be established by the frequency of recombination between two variants; however, if two variants map very close to each other (i.e., there is a very low frequency of recombination between them), convincing evidence of allelism or nonallelism can be provided only by knowledge of the DNA changes associated with the variants.

Because any change in nucleotide sequence can be transmitted as a stable variant, the number of possible alleles at any locus is very large. Many will be lethal, but many others, perhaps most, are likely to be viable,

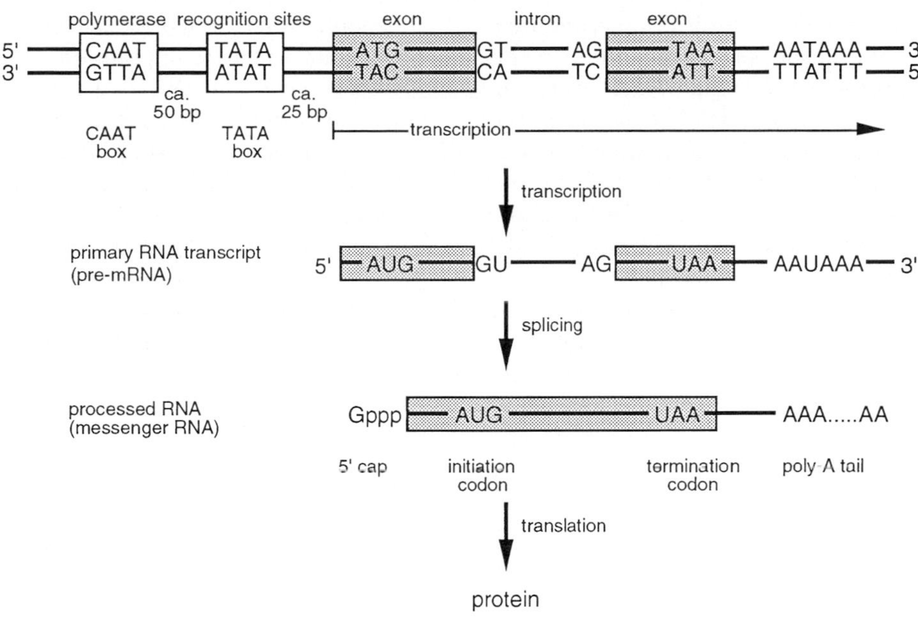

FIGURE 1 Diagram of a typical eukaryotic gene. The sense strand is at the top, and the antisense strand, which serves as the template for transcription, is on the bottom. Only one intron is shown, but the coding regions (exons) may be interrupted by dozens of introns. The primary RNA transcript is processed to remove introns (splicing) and the 5′ cap and poly-A tail are added to form messenger RNA (mRNA). The information coded in mRNA is used in translation to determine the sequence of amino acids in the corresponding protein. [Adapted, with permission, from H. E. Sutton (1988). "An Introduction to Human Genetics," 4th Ed., p. 228. Saunders College, Philadelphia.]

at least in heterozygous combination with a "normal" allele.

II. STRUCTURE AND FUNCTION OF EUKARYOTIC GENES

A. DNA Structure of Genes

Typical eukaryotic genes that code for proteins have the structure shown at the top of Fig. 1. The top strand is the sense strand of a DNA double helix, the 5′ to 3′ direction conventionally written from left to right. The complementary 3′ to 5′ strand, the antisense strand, serves as the template for RNA transcription. The 5′ to 3′ direction of the sense strand and of the transcribed RNA corresponds to the N-terminal to C-terminal direction of the corresponding polypeptide chain.

The protein-coding regions, present in exons, are separated by regions that do not code. These introns or intervening sequences are transcribed into RNA but must be removed before protein synthesis. The number of exons varies greatly. For example, the globin genes have three; other genes are known to have several dozen. The relative sizes of the exons and introns are quite variable, with many introns being larger than the entire coding region of a gene.

B. Transcription of Genes

RNA transcription refers to copying the nucleotide sequence of a segment of DNA into a complementary sequence of RNA (Fig. 1). Transcription is initiated some dozens of nucleotides in the 5′ direction (upstream) from the coding region and continues in the 3′ direction (downstream) past the coding region. The exact point of initiation is determined by the location of the sites at which the RNA polymerase that catalyzes transcription is bound. The point of termination of transcription is not well defined. The nucleotide sequence AATAAA occurs at some distance downstream from the coding region, and transcription extends beyond that, presumably until some nucleotide sequence signals the termination.

The amount of transcription, indeed, whether or not transcription occurs at all, is dependent on regulatory elements (nucleotide sequences) in the 5′ direction from the point at which transcription is initiated. Two important nucleotide sequences often occur: TATA or some close variation (the TATA box) commonly occurs some 25 nucleotides 5′ from the initiation point, and CAAT (the CAAT box) some 50 nucleotides from the initiation point. Various combinations of proteins (transcription factors) bind to these and other sequences in the 5′ direction, resulting in the binding of RNA polymerase, which moves in the 3′ direction with respect to the sense strand, catalyzing the formation of a strand of RNA that is complementary to the template strand. Mutations in other upstream regions of specific genes may alter the rate of transcription, although attempts to develop general rules that relate DNA structure to transcription have been only partially successful.

A number of mutations downstream from the coding regions are known to reduce protein synthesis. Many are thought to affect translation rather than transcription.

The coding region and its contiguous regulatory regions were noted earlier as constituting a gene when defined as a transcription unit. Enhancer regions introduce some ambiguity into this definition. An enhancer is defined as a DNA segment that enhances transcription of a specific gene that is located on the same chromosome but that is ordinarily noncontiguous. The enhancer may be in either the 5′ or 3′ direction from the coding region and may be thousands of nucleotides distant. Or it may even be within an intron. Yet, it affects only the gene on the same chromosome and not the same gene on a homologous chromosome. Therefore, there must be a physical interaction between the enhancer region and the target gene and not interaction by means of a diffusible gene product.

Transcription is also regulated by the protein products of other genes that produce transcription factors. Such trans regulation is apparently responsible for some of the events of differentiation and for regulation of physiological functions. Though it is useful to recognize the function of certain genes as regulatory, the structures of regulatory genes themselves are not known to be different from other genes. [*See* DNA and Gene Transcription.]

C. Comparison of Eukaryote Genes with Other Genes

The basic patterns of genetic coding, transcription, and translation are remarkably similar throughout the biological world. Nevertheless, there are variations in detail, some of which are instructive about the evolutionary relationships among different major biological groups. The gene structure of eukaryotes (i.e., organisms whose cells have nuclei and chromosomes) has been described earlier. These patterns apply to such diverse eukaryotes as humans, plants, and yeast.

The genes of prokaryotes (organisms whose cells lack nuclei and chromosomes) have some characteristic differences from those of eukaryotes. One difference is the absence of introns in most prokaryotes (bacteria). Prokaryote genes code without interruption from the initiation codon to the termination codon. Another difference is the organization of some prokaryote genes into clusters that are transcribed as a unit.

Mitochondrial genes are similar to prokaryote genes rather than eukaryote nuclear genes in some respects, supporting the idea that mitochondria arose as prokaryote symbionts. For example, mitochondrial genes have few introns. Several of the mitochondrial codons vary slightly from those of eukaryote nuclear genes. Because all mitochondrial genes are functional at all times, mitochondrial genes should not need the more complex regulatory systems of nuclear genes. But in most respects, mitochondrial genes conform to the general structure of genes found in all biological systems.

D. Gene Rearrangement

All the cells of multicellular organisms typically contain the same complement of genes and chromosomes that were present in the fertilized egg. An obvious exception are the products of meiosis—eggs and sperm—with half the chromosome complement (and half the genes). Another exception in mammals is red blood cells. As red cells mature, their nuclei disintegrate, producing a cell that is anuclear. Other specialized cells may be polyploid (liver) or multinuclear (muscle). In each of these variations, the structure of individual genes is unchanged. [*See* Meiosis.]

There are now several well-studied examples in eukaryotes in which DNA rearrangement occurs as a part of normal differentiation. The only example in mammals involves certain genes necessary for immunity. The great variety of potential antibody responses to antigenic stimulation are possible largely because of the ability of the antibody genes to rearrange during the development of the lymphocytes. An example of this rearrangement is shown in Fig. 2. Each gene is actually a complex of gene parts. In the case of anti-

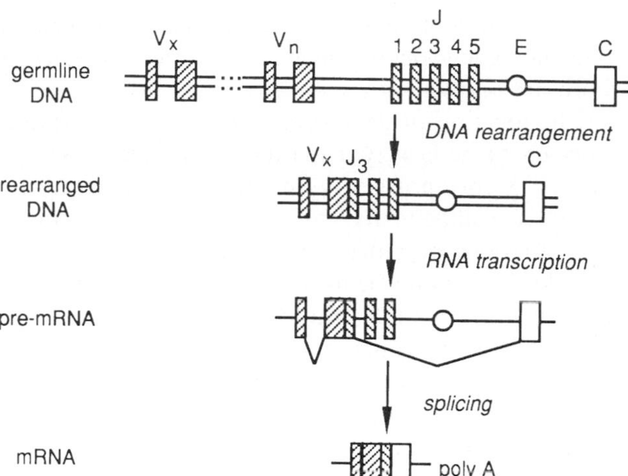

FIGURE 2 Genetic structure of the κ chain immunoglobulin complex. During maturation of a B lymphocyte, a V region is joined directly to a J region by deletion of the intervening DNA. [Reproduced, with permission, from H. E. Sutton (1988). "An Introduction to Human Genetics," 4th Ed., p. 405. Saunders College, Philadelphia.]

bodies produced by B cells, the genes that code for light chains (L chains) consist of three components: V regions (which code for the N-terminal "variable" part of the polypeptide chain), J regions (which code for a joining segment), and C regions (which code for the C-terminal "constant" segment). Heavy-chain genes have an additional component—the D region—that is located between the J and C regions and adds diversity to the final product. During development of a stem cell destined to become a B cell, stretches of DNA are eliminated so that the mature B cell has a contiguous array of a V segment and a J segment (Fig. 2).

This new array serves as the transcription unit for that gene complex. Since the joining of DNA segments is random and varies from one chromosome to another, not every rearrangement produces a functional transcription unit. If the first attempt is successful, no rearrangement occurs on the homologous chromosome. If the first attempt is not successful, rearrangement on the homologous chromosome occurs. That also may or may not be successful. In the case of the L chains, there are two such complexes, on chromosomes 8 and 22. The H chain is represented by a single complex on chromosome 14. A cell that successfully rearranges to produce an L chain and an H chain is retained in the immune repertory to produce that particular antibody. Many cells are not successful and are lost.

The random rearrangement of the DNA has the potential to produce an enormous number of different successful combinations, thus generating the required antibody diversity. When an antibody-producing cell is stimulated to grow by presence of the corresponding antigen (or rarely by malignant transformation), the rearranged DNA is faithfully replicated as any other DNA would be, expanding the number of cells that code for that particular antibody. Because the rearrangements occur only in the cells of the immune system, all other somatic cells continue to have the original DNA, sometimes referred to as germline DNA.

Rearrangement of the genes that code for T-cell receptors also occurs in a similar manner. The antibody genes and the T-cell receptor genes are evolutionarily related and are part of a gene family (see Section IV). Not all members of the family can rearrange, and rearrangement is not known to occur in any other mammalian gene family. [*See* T-Cell Receptors.]

E. Imprinting

According to traditional Mendelian rules, the activity of a gene is independent of the parent from which it was received. Although this is true of the vast majority of genes studied, there are now a few well-documented examples of differential activity of an allele, depending on whether it was received from the father or the mother. The basis for imprinting is not certain, but the most likely explanation is transmission of the state of methylation of the DNA. Genes that are highly methylated are typically not very active. Traditional models of DNA replication do not take the methylation pattern into consideration and, indeed, for most genes it must not be important, but for some it may be. Just as the inactive X chromosome remains inactive through numerous mitotic replications, some genes may be transmitted in an inactive form, even through meiosis. Activation would appear to be possible in some instances. [*See* Genetic Imprinting.]

III. ORGANIZATION OF HUMAN GENES

A. Number of Genes

The number of nucleotide pairs in the DNA of a haploid set of human or other mammalian chromosomes is approximately 3 billion. If genes were closely packed and a typical gene required 3000 nucleotides for the coding and regulatory regions and for intron sequences, there would be room for 1.5 million genes.

Both genetic arguments and experimental observations indicate this number of genes to be much too high. The genetic arguments hinge on the mutation rate of genes. If detrimental mutations occurred at a rate of one per million loci per gamete, a figure that is lower than the observed rates, each zygote would have on the average three new mutations plus all the mutations that were transmitted from prior generations. Various models suggest that such a number of mutations would overwhelm the ability of natural selection to eliminate them.

Two kinds of observations also support a lower number of genes. One is the direct analysis of nucleotide sequences. There are substantial stretches of noncoding DNA between functional genes, DNA for which there is no obvious or demonstrated function. Another consideration is the great variation in the total amount of DNA among various species. Mammals are relatively similar to each other in having some 3 billion nucleotide pairs per haploid genome; however, closely related vertebrates such as toads (*Bufo bufo*) have twice as much DNA as humans, and the amphibian *Amphiuma* has 76 billion nucleotides. *Tradescantia,* a plant, has 53 billion nucleotides. We humans are reluctant to concede that amphibians and plants are genetically that much more complex than we are. The alternate hypothesis is that most of the DNA in those species does not code for genes, and the same is probably true for *Homo sapiens.*

The conventional estimate of the number of human genes is 50,000–100,000. This estimate is not based on secure analytical arguments but rather on speculation that we need at least 50,000 loci to code for all the required mammalian functions, and much more than 100,000 would create problems with the mutation rate.

B. Genes and Chromosomes

Each chromosome appears to have a single molecule of DNA stretching from one end of the chromosome to the other. Therefore, the thousands of genes on a single chromosome must be arrayed along this molecule. How the genes are organized with respect to each other is largely unknown. There are a number of clusters of genes with related functions and structures, and such clusters could have functional significance. Or they could be a reflection of evolutionary origins by tandem duplication from a single ancestral gene, with chance not yet having separated them through chromosome rearrangement. Probable examples of both are known.

The location of genes on particular chromosomes appears to be largely a matter of evolutionary accident. The chromosomes of mammals have undergone many rearrangements, with new, highly functional combinations of genes being generated many times. In addition, many healthy persons are known who have inversions or balanced translocations. So long as the genomic complement is complete, there are many arrangements that are functional.

This does not mean that all genes are oblivious to their neighbors. There appear to be clusters of genes that are regulated as a unit, and presumably any breaking up the cluster by chromosome rearrangement would be detrimental. Evidence is lacking at present on the number of such clusters and their typical sizes. Nor do we have any idea how tolerant genes or gene clusters are of their location; we can only be sure of the successful rearrangements and not the unsuccessful.

If one divides the number of nucleotides by the approximate number of genes in a mammalian genome, the nucleotides per gene is very much larger than would be accounted for by the known requirements of structural genes and their regulatory regions. This excess amount of DNA appears to be distributed along the chromosomes between coding regions, and some also occurs as introns. Much of this excess DNA is unique sequence DNA; that is, the particular sequence of nucleotides occurs only once in the genome. Coding regions of genes are other examples of unique sequence DNA.

Some 30% of the total DNA in mammals is repetitive DNA. In the case of highly repetitive DNA, there are millions of copies of short nucleotide sequences. Middle repetitive DNA consists of thousands to hundreds of relatively longer sequences. The *Alu* sequence in humans, an example of middle repetitive DNA, is approximately 300 nucleotides in length, with some 500,000 copies dispersed throughout the genome. The function of repetitive DNA is unknown. It may have some as yet undescribed function, or it may be "junk" DNA—DNA that has no function other than to replicate itself.

Chromosomal regions are classified as euchromatin and heterochromatin on the basis of their staining behavior, the heterochromatin being more easily stained by nuclear stains. In general, euchromatin consists of unique sequence DNA and is the location of functional genes. Heterochromatin consists of repetitive DNA and lacks functional genes. As more is learned about the fine structure of DNA and chromosomes, there are likely to be exceptions to these general statements, but they are well documented as rules in many species. Heterochromatic regions may occur in any part of a chromosome, but the regions around the centromere are likely to be heterochromatic.

IV. EVOLUTION OF GENES

A. Universality of the Genetic Code

The fact that all genes in all species use virtually the same genetic code and have most of the same organizational features can most readily be interpreted as indicating that all genes are descendants of one primordial gene, or at least one form of primordial life. In addition to this common heritage, however, similarities in gene structure among groups of genes suggest a more recent divergence from each other and inform us about the evolutionary relationships not only of the genes but also of various taxa.

B. Mutations in Genes

A mutation can be defined as any change in DNA sequence. Mutations can occur in the germ line, in which case they may be transmitted to offspring. Or they can occur in somatic cells (all cells except germ cells), in which case they cannot be transmitted to offspring. Mutations are rare, but they nevertheless account for the great genetic diversity both within each species and among species. Mutations may be nucleotide substitutions, additions, or deletions. The consequences of such small changes—small in the physical sense—may be inconsequential for gene function or they may be very large, particularly if they occur in critical regions, such as the DNA sequences that regulate transcription. Most mutations are neutral or detrimental and will ultimately be eliminated by natural selection, but occasionally these random changes in the structure of a gene may result in a more functional locus that will be favored by selection.

Mutation may also result in duplication or deletion of entire genes. Deletion of a critical gene is likely to be detrimental to the organism, and the mutant form will be eliminated. Duplications may be tolerated and allows mutations to occur and new gene functions to evolve without losing the original gene functions. Such repeated gene duplication presumably accounts for the large number of loci in the repertory of higher organisms.

In addition to duplication and divergence of genes, other rearrangements among genes may be adaptive. For example, the selective advantage of introns within the coding regions of genes has not been proved. One attractive hypothesis is that each exon codes for a separate protein domain. Indeed, there are many illustrations of this principle, but other genes appear to violate it. In situations in which each exon corresponds to a functional domain, one could imagine that entire domains might be translocated to other genes, thereby providing new functional combinations. Such seems to be the case, for example, in several transmembrane proteins, in which the exons that code for the transmembrane domain are more closely related structurally than would be expected by parallel evolution. It is thought that the different domains may have evolved separately and then been joined by chromosome rearrangements to produce new functional units.

Such "exon shuffling" must be understood in the traditional concepts of evolutionary changes. There is no reason to suppose that exons are traded about other than by rare, chance chromosome rearrangements. On the other hand, the occasional finding of repetitive sequences such as *Alu* within introns might be expected to promote mispairing of chromosomes. If this were followed by crossing-over or translocation, new genes would be generated.

C. Gene Families

A number of groups of proteins—and their corresponding genes—share structural features and functions. Such groups are referred to as gene families. One example would be the globins that form hemoglobins and myoglobulins. Each human haploid genome has several α-globin-like loci on chromosome 16, several β-globin-like loci on chromosome 11, and a myoglobin locus on chromosome 22; however, examination of the nucleotide sequences of these genes, the organization of introns and exons, the amino acid sequences, the three-dimensional structures of the protein products, and the functions provides convincing evidence that all evolved from a single, ancient globin-like gene. The detection of globin-like DNA sequences in such remote relatives as yeast suggests that the globin family may be quite old and quite extensive.

Other gene families that have been identified include the serine proteases and the immunoglobins.

Serine proteases are enzymes that catalyze the breakdown of peptide bonds, with serine at the active site of the enzyme. Examples are trypsin, chymotrypsin, and pepsin. Other members of the family include haptoglobin, a plasma protein not known to have proteolytic activity. Even though the serine proteases have diverged greatly in the types of peptide bonds attacked and the conditions under which maximal activity occurs, their shared structures support a common origin.

The genes that code for immunoglobulins (Ig) and T-cell receptors (TCR) are part of an especially interesting superfamily of genes. Among the properties shared by Ig and TCR genes and no other known mammalian genes is the rearrangement of DNA during maturation of lymphocytes. Such rearrangement in somatic cells must be extremely rare, although DNA rearrangement also occurs in germline genes that code for trypanosome antigens. Other members of the Ig–TCR superfamily do not undergo rearrangement, nor do the Ig and TCR genes in any tissue other than lymphocytes. Thus, programmed DNA rearrangement can occur but is exceedingly rare. [*See* Lymphocytes.]

D. Phylogenetic Comparisons of Genes

As the ultimate repository of biological heritage, DNA has the potential to tell us how closely related all existing species are. From homologous nucleotide sequences, it is possible to construct phylogenetic trees that account for the evolutionary origins of all species for which data are available.

As yet, no remarkable revisions of phylogeny have been required by the DNA sequence data. On the other hand, the consistency of DNA structure and fossil or other phylogenetic evidence support the value of DNA as a means of reconstructing phylogeny. DNA has proved valuable in understanding the relationships among closely related species, subspecies, or even local populations.

V. GENETIC VARIATION IN HUMANS

A. Mutation and Genetic Variation

Mutation is a fundamental property of genes. The changes in genetic information, which we call mutation, may have little or no effect on the function of a gene, or they may interfere profoundly with function. A single nucleotide substitution may have no effect on the gene product or it may make the gene unable to produce any product at all. Therefore, many mutant alleles are normal variants, whereas others interfere with normal development.

Mutations are rare events. The measurement of mutation rates is difficult in humans but is on the order of one per 100,000 gametes per locus for many mutations that cause dominant disease. The rate of production of normal variant alleles, essentially nucleotide substitutions, is probably the same order of magnitude. The most likely fate of a new mutation is elimination, either by natural selection or by chance, even for favorable mutations. A few will attain substantial frequencies, and rarely one may displace the original wild-type allele, again either by natural selection or by chance. [*See* Mutation Rates.]

B. Normal Genetic Variation in Humans

The existence of alternate alleles in a population allows each member of the population to be genetically unique. With the exception of identical twins, each person possesses a combination of alleles that has never existed before, nor will it exist again. This is assured by the variety of choices available at tens of thousands of loci.

In discussing genetic variability, distinguishing between polymorphic loci and monomorphic loci is useful. Monomorphic loci are those at which there is only one common allele. Polymorphic loci are those with two or more common alleles. An arbitrary definition of "common" is 1% or greater. Monomorphic loci may have rare variant alleles, not one of which has a frequency as large as 1%. In the case of polymorphic loci such as ABO and Rh blood groups, two or more alleles have frequencies greater than 1%. The genes that account for most of the individual variations are polymorphic.

There are various estimates of the proportion of loci that are polymorphic. Such estimates are based on variations in the protein products of genes, usually detected by electrical charge variations that lead to variations in electrophoretic mobility. Studies of enzymes in *Drosophila* and in humans indicate at least one-third of loci to be polymorphic. On the other hand, studies that include structural proteins give a lower frequency, on the order of 6% in humans. In either case, there is ample variation to account for the many genetic differences that distinguish each person. No population-based studies have yet been reported on DNA variations that might influence the function of loci without altering the structure of the gene product. We may assume that additional polymorphisms occur at this level. [*See* Genetic Polymorphism.]

In addition to polymorphic alleles, there are rare normal variants at many loci. Because of their rarity, few have been studied in detail. The fact that they are found in normal persons does not mean that there is no effect on the phenotype. Usually we have no way to relate the individuality of the phenotype to variation at particular loci. We suspect, however, that the phenotypic difference that we observe among individuals are often due to the cumulative effects of variations at many loci. As information has accrued on the extent of genetic variation, the view has changed from one in which occasional variation occurs, rigidly limited by natural selection, to one in which variation is very widespread.

C. Genetic Variation and Disease

We assume that mutations occur at all 50,000–100,000 human genes, and many of the mutant alleles cannot function normally. Some will be eliminated immediately as dominant lethals, and we may never observe the mutant form. Others may cause disease. The number of loci at which mutant alleles are known to cause disease exceeds several thousand and the true number is probably considerably higher. Some inherited diseases such as cystic fibrosis, Tay-Sachs disease, and sickle-cell anemia are relatively common; others may be very rare, affecting one person in tens of thousands.

The impact of genetic variation on health and health care has been estimated in various ways, none very satisfactorily. Some 1% of newborns have some genetic defect, often trivial but sometimes fatal. Even diseases more characteristic of older ages such as heart disease and cancer often have a large genetic component. Understanding the full extent of the genetic contribution to disease and the specific genes involved should enable us to reduce the burden somewhat, though perhaps never to zero.

BIBLIOGRAPHY

Alberts, B., Bray, D., Lewis, J., Raff, M., Roberts, K., and Watson, J. D. (1994). "Molecular Biology of the Cell," 3rd Ed. Garland Publishing, New York.

Darnell, J., Lodish, H., and Baltimore, D. (1990). "Molecular Cell Biology," 2nd Ed. Scientific American Books, New York.

Lewin, B. (1994). "Genes V." Oxford Univ. Press, New York.

Sutton, H. E. (1988). "An Introduction to Human Genetics," 4th Ed. Harcourt Brace Jovanovich, San Diego.

Watson, J. D., Hopkins, N. H., Roberts, J. W., Steitz, J. A., and Weiner, A. M. (1987). "Molecular Biology of the Gene," Vol. 1, 4th Ed. Benjamin/Cummings, Menlo Park, CA.

Gene Targeting Techniques

LYNN M. CORCORAN
The Walter and Eliza Hall Institute of Medical Research, Australia

GLOSSARY

Allele One of the alternative versions of a gene, occupying a given region on a chromosome

Base pair The basic unit of DNA, comprising the two bases (A + T or C + G) that face one another in the DNA double helix

Chimera Here this term is used for an animal that has a mixed derivation, composed in part from both embryonic stem (ES) cells and from the resident embryonic cells of the blastocyst into which the ES cells were injected

Clone Large number of cells (or molecules) that are identical and are derived from a single ancestral cell (or molecule)

Genome Sum total of all genetic material (DNA) carried by an organism

Germline Genetic material that an organism passes to its offspring

Heterozygous Having two different alleles (versions) of a particular gene

Homozygous Having two identical alleles (versions) of a particular gene

In vivo In the context of a whole, living organism

Locus Position on a chromosome at which a particular gene lies

Marker Gene that confers a unique trait to the cell carrying it

PCR primer Short piece of DNA (~20 bases) that matches sequences in the chromosome. It is used to initiate (or "prime") the synthesis of a longer piece of DNA by enzymatic extension

Promoter Region of DNA at the front of a gene that controls the transcription of that gene

Selection Use of particular conditions to allow the survival of only those cells that carry a particular gene (the selectable marker). For example, a drug can be used to select for the presence of a drug-resistance gene

Stem cell Primitive cell that is the precursor of a wide range of differentiated cell types, and that also renews itself

Totipotent Having the potential to differentiate to all cell types. Not restricted in differentiation potential

Transcription Process of copying a gene's DNA into RNA

Translation Process of synthesizing a protein from the information held in the sequence of an RNA molecule

COMPLEX ORGANISMS SUCH AS MAMMALS DEvelop through the carefully orchestrated actions of a large number (~100,000) of genes. Each gene makes a particular contribution to the development and the physiology of the animal. Most of our current knowledge about gene function is incomplete, coming from experiments using cells in culture or molecules in test tubes. Using the recently developed technique of gene targeting, it is possible to measure a single gene's contribution to the making of a whole animal (a mouse in this instance). Gene targeting starts with a cloned version of a specific gene, proceeds with its precise alteration through genetic engineering, and culminates in the generation of an animal that bears a single predetermined mutation. Careful observation of the "mutant" mouse can illuminate those processes that depend on the product of the targeted gene, as these fail to occur when the gene is mutated. Thus, gene targeting is a tool that will lead to a greater understanding of the function of individual mammalian genes, providing mouse models for normal human biology and for human genetic disease.

I. GENERAL PRINCIPLES OF THE PROCEDURE

A. Homologous Recombination in Mammalian Cells

Two recent observations catalyzed the development of gene targeting technology. One came through a recognition of what happens to DNA when it is introduced into a mammalian cell. Most commonly, when exogenous or "foreign" DNA enters the cell, it is simply lost. Cells are not inclined to allow the intrusion of foreign DNA, probably as a defense against viral infection. However, at a low frequency (in approximately 1 in 1000 treated cells), the incoming DNA will invade the cell's DNA and insert itself into one of the chromosomes in a process called random integration (Fig. 1A). During this process, there need be no relationship between the incoming DNA and the site at which it integrates: random integration can occur anywhere in the cell's chromosomal DNA. Once integrated, the exogenous DNA is stably inherited just as any other part of the cell's own DNA.

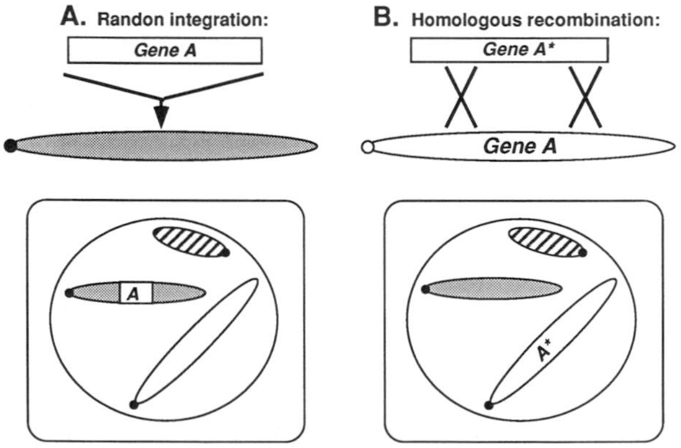

If the incoming DNA is closely related to the cell's DNA (e.g., mouse DNA entering a mouse cell), then a very small proportion of cells that stably take up the introduced DNA may acquire it by a different mechanism: homologous recombination (Fig. 1B). For this to occur, the incoming DNA must find matching sequences on one of the chromosomes. The matching or "homologous" sequences of the cell and the incoming DNA will pair up, and an exchange reaction occurs in which the chromosomal sequences are *replaced* by the introduced DNA. Once again, the incoming DNA will be stably inherited after joining the cell's chromosome. By comparison to random integration, homologous recombination is rare, occurring ~1000 times less frequently. Nevertheless, the knowledge that homologous recombination *does* occur in mammalian cells became a starting point for gene targeting, by showing that normal cellular genes could be replaced by DNA that originated outside the cell. (For simplicity and brevity, henceforth the terms "homologous recombination" and "homologous recombinant" will be abbreviated "HR".)

A. Randon integration: Gene A
B. Homologous recombination: Gene A* Gene A

FIGURE 1 Random integration versus homologous recombination (HR). The bar designated "Gene A" represents a piece of DNA that has been introduced into a cell. The cell is depicted as a rectangle, with an internal circle representing the nucleus. The shading and stripes on the chromosomes indicate that they are unrelated to the incoming DNA. (A) Random integration into an unrelated region on a chromosome. (B) The "X" symbols indicate crossing-over between the incoming DNA and a homologous region in the DNA of the chromosome. The asterisk is used to indicate that an exchange has occurred, replacing the chromosomal gene A with the version (A*) carried by the incoming DNA.

B. Embryonic Stem Cells

The second critical component of gene targeting technology came with the availability of embryonic stem (ES) cells, first cultured in 1981. These cells are derived from the inner cell mass of very early (3.5-day-old) mouse embryos, called blastocysts (Fig. 2). ES cells are small, undifferentiated, rapidly dividing cells that are amenable to many of the manipulations usually employed for molecular and cellular biology, such as continuous culture, single-cell cloning, drug selection, and introduction of foreign DNA. Most importantly, they are totipotent: ES cells can be returned to an embryonic environment by injection into a blastocyst, where they will resume normal development and can contribute to all tissues of the growing mouse embryo. As ES cells can particpate in formation of the reproductive tissues, genes carried in the ES cells can be passed on to the mouse's offspring. Gene targeting was born through the combined use of HR technology to modify specifically and precisely a single genetic locus, and ES cell technology to deliver the mutation back to the mouse. Figure 3 provides an outline of the basic steps in a conventional gene targeting experi-

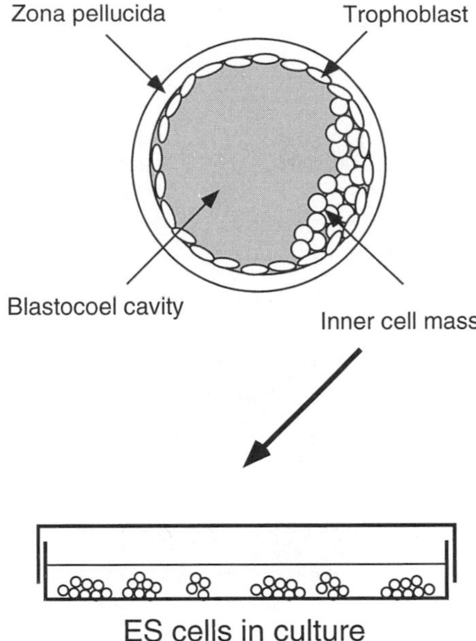

FIGURE 2 The origin of embryonic stem (ES) cells. At the top of the figure is a representation of a mouse blastocyst. ES cells are derived from the inner cell mass of the blastocyst and can be grown in culture (bottom).

ment. The components and procedures will be discussed in detail.

C. Contrast to Transgenic Mice

Another means of studying gene function *in vivo* is through the production of transgenic mice. These differ from gene targeted mice in several important respects. First, transgenic mice are made from fertilized eggs into which naked DNA has been injected. Second, the foreign DNA, or "transgene," integrates randomly and unpredictably, in contrast to gene targeting, where integration (by HR) occurs only at the target gene. The site of integration of the transgene and the number of copies that integrate into the chromosome are out of the experimenter's control, and yet both factors can strongly influence the behavior of the transgene. Furthermore, integration of the transgene can interfere with the function of a critical gene at or near the integration site. Finally, by the nature of the process, transgenic mice have usually *gained* new functions. In contrast, most gene targeting experiments to date have resulted in a specific *loss* or alteration of gene function.

II. THE NECESSARY COMPONENTS

A. Mice

C57BL mice are most commonly used as a source of "host" blastocysts (those that will be injected with genetically altered ES cells). Blastocysts of this pure, inbred mouse strain have been shown to be particularly receptive to injected ES cells, yielding chimeric mice (animals of mixed host and ES-cell derivation) at a high frequency. Furthermore, C57BL mice have black fur, which provides a convenient visual marker (see Section II,B). Blastocysts are harvested from the lumen of the uterus 3.5 days after mating (Fig. 4A).

After the blastocysts have been injected with ES cells, they must be returned to the uterus of a foster mother to complete development (Fig. 4B). The foster mothers are pseudopregnant female mice. These are females that have been mated (2.5 days earlier) with vasectomized males and are therefore hormonally prepared to receive and gestate the embryos (they bear no embryos of their own). The foster mother and the embryos must be at a compatible stage for the pregnancy to go to term. Foster mothers can be derived from any mouse strain, but genetically mixed strains are superior to inbred strains.

B. Cells

There are a large number of ES cell lines available, and most have been cultured over many years. All ES cell lines are derived from male embryos, which appear to be more genetically stable in culture than female cell lines. Most lines are derived from the 129 strain of mouse, as these have been shown to work well in chimera formation. Thus, 129-derived ES cells are most commonly injected into C57BL blastocysts to generate chimeras (Fig. 5A). The 129 mouse has a pale yellow to light brown coat (determined in large part by the "agouti" coat color gene) that can be visually distinguished from the C57BL-derived coat in a chimeric mouse (Fig. 5B). In some cases, it may be desirable to have a pure, rather than mixed, genetic background, so ES cell lines have also been derived from the C57BL mouse strain.

The quality of the ES cell line is probably the major determining factor of success in a gene targeting experiment. One should start with a recently proven source of cells and tend them carefully to ensure that the cells do not differentiate during the culture period. Standard conditions include medium made with specially tested components that promote

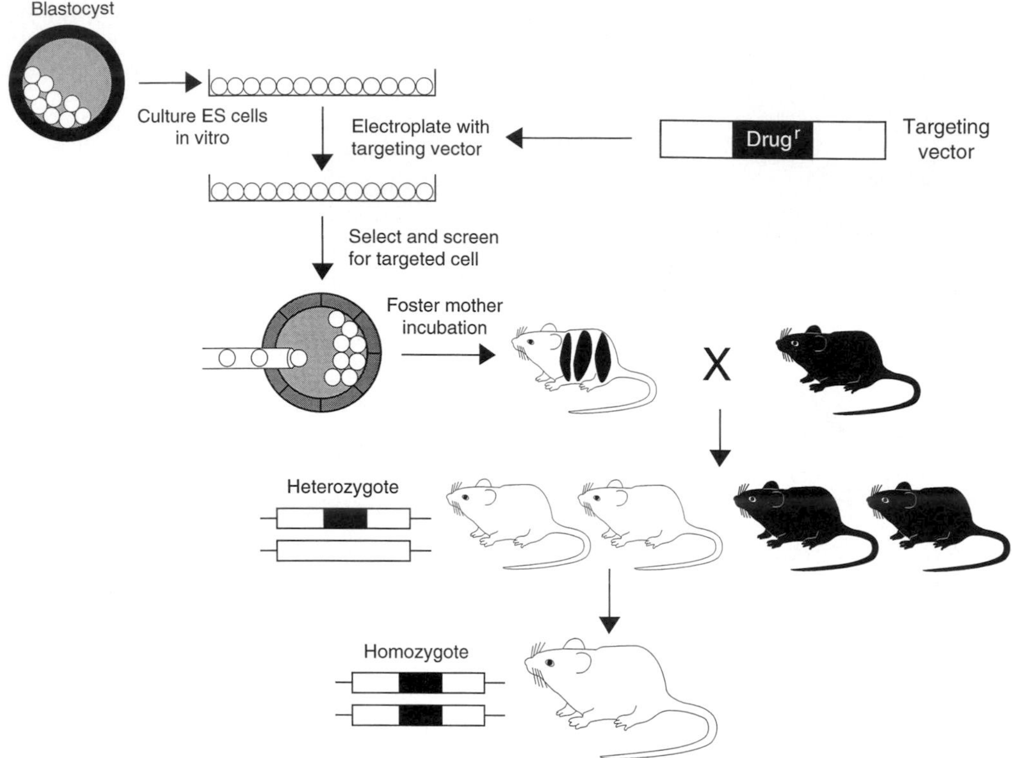

FIGURE 3 A conventional gene targeting protocol. Shown are the derivation of ES cells, introduction of the targeting vector, drug selection and screening, injection of targeted cells into a recipient blastocyst, and the generation of a chimeric mouse (mottled). The mating strategy used to generate homozygous mutant mice is shown, with black indicating inheritance of recipient blastocyst (C57BL)-derived genes and white indicating ES cell (129)-derived gene inheritance, or germline transmission.

growth without differentiation and the presence of an underlayer of support cells upon which the ES cells can grow. Embryonic mouse fibroblasts are used for this purpose. In addition, a hormone called LIF/DIA (leukemia inhibitory factor/differentiation inhibiting activity) is included to further ensure that the ES cells maintain their undifferentiated, totipotent state until they are returned to the embryo. Some workers use LIF as a substitute for the embryonic fibroblasts.

C. The Targeting Vector

The DNA that will deliver the mutation to the gene in the ES cell is called the targeting vector. It comprises a stretch of DNA [~6000 to 10,000 base pairs (bp)] matching the gene to be targeted and flanking the region of the gene that is to be mutated. This DNA is cloned into a bacterial plasmid for easy manipulation. As stable integration of DNA into cells is a rare event (~1 in 1000 treated cells), a selectable marker is included in the targeting vector. This is usually a gene that confers drug resistance to the cell that has stably taken up the DNA, so that only these cells will survive in culture medium containing the drug. (Because this type of marker gives a selective *advantage* to the cells that bear it, it is called a *positive* selectable marker. Later, an example of *negative* selection will be given.)

There are two general types of targeting vector, the insertion and the replacement vectors (Fig. 6). Both participate in HR at similar frequencies. A single crossover between the chromosomal gene and an insertion-type vector generates a disrupted gene that contains all chromosomal sequences, the drug resistance marker, and additional sequences from the targeting vector plasmid. HR with this type of vector generates a duplication of part of the gene. Normally such a duplication would be undesirable, as duplicated sequences can participate in spontaneous *in-*

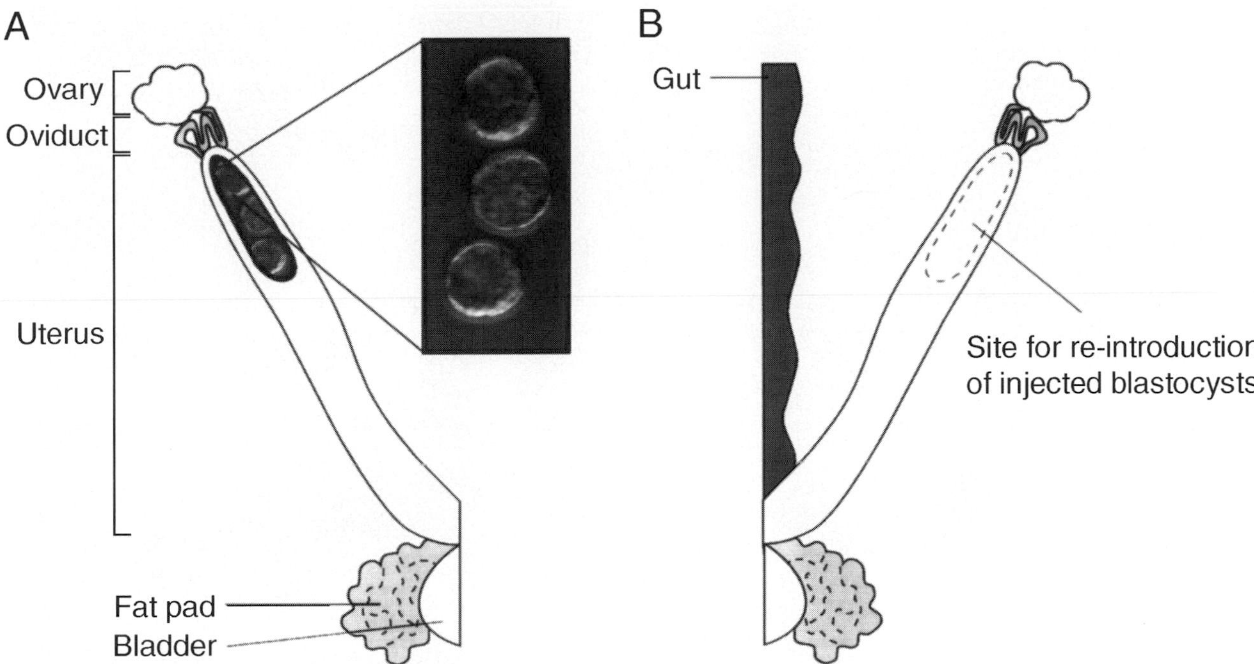

FIGURE 4 The female mouse reproductive system. (A) Blastocysts are located in the lumen of the uterus 3.5 days after fertilization. (B) Blastocysts that have been injected with ES cells are returned to the indicated site in the uterus of a pseudopregnant foster mother for gestation.

*tra*chromosomal HR, but one application discussed later makes use of this feature.

A more commonly used targeting vector is the replacement type (Fig. 6B). In this situation, a double crossover occurs between the vector and the chromosome, and a segment of the chromosomal gene is replaced by the corresponding part of the targeting vector. The drug resistance gene is incorporated, and some of the target gene may be intentionally deleted. Such a deletion is a simple way to functionally silence the target gene, if gene ablation or "knockout" is the desired goal. Whether an insertion or replacement vector is used, foreknowledge of the gene's structure and its functional domains is essential in designing the mutation, and in planning the means of recognizing rare HR events (the screening protocol, discussed in Section IV).

III. STRATEGIES TO IMPROVE THE EFFICIENCY OF "GENE TARGETING" IN ES CELLS

As stated earlier, HR is a relatively rare event in mammalian cells. It is thought to be catalyzed by DNA repair enzymes and peaks in cells at the time of DNA replication prior to cell division. The frequency of HR varies enormously (~1000-fold) from one gene locus to another, but the reasons for this are not known. Several strategies have evolved that either increase the frequency of HR or improve the discrimination between ES cell clones that have become drug resistant through HR rather than through random integration.

The frequency of HR is generally improved by increasing the length of homology between the targeting vector and the chromosomal gene, up to a limit of a few thousand base pairs on either side of the selectable marker, when the targeting vector becomes too large to be manipulated easily. The transcriptional activity of the target gene in ES cells has surprisingly little influence on HR frequency. In some cases, very significant improvement (~20-fold) of HR frequency can be achieved when the targeting vector is made from DNA of the same mouse strain (isogenic) as the ES cell line with which it is to be used (129 in most cases). It was found that the DNA of two different mouse strains differs at least once (by base insertion, deletion, or replacement) every 160 bp, and such mismatches can probably inhibit HR.

Improved discrimination between random integrants and HRs can be achieved by using a combination

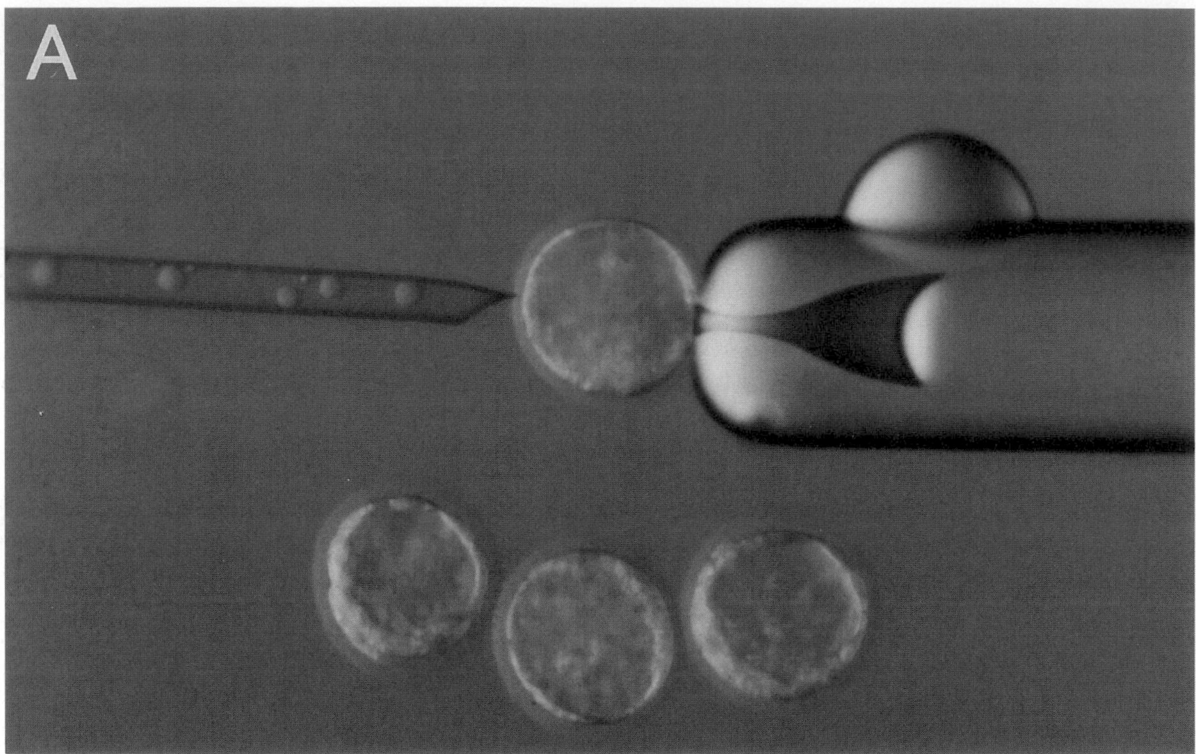

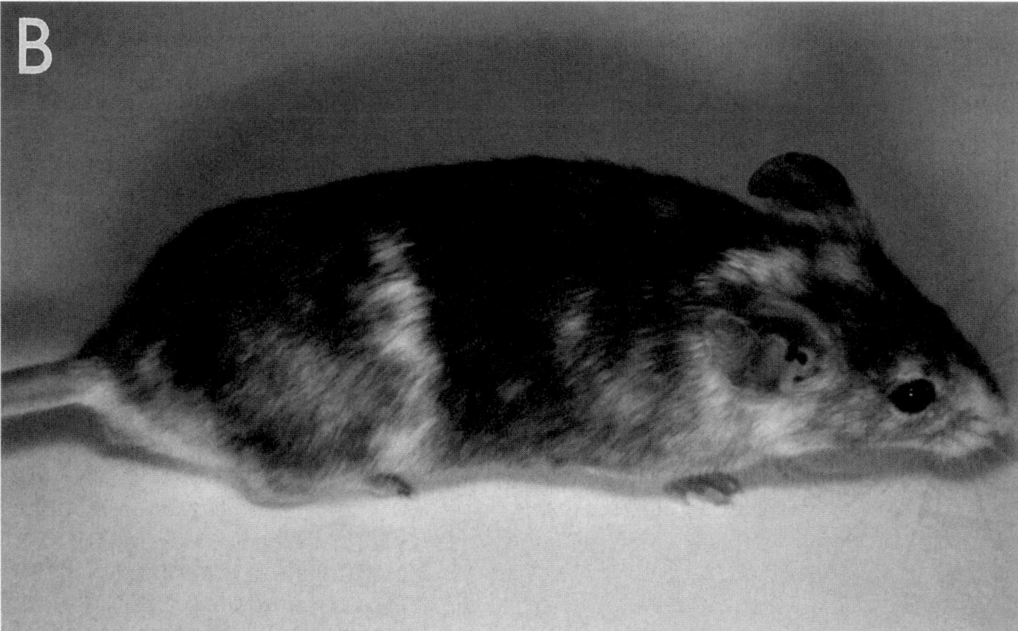

FIGURE 5 Blastocyst injection and a chimeric mouse. (A) The pointed needle, containing ES cells, will pierce the wall of the blastocyst, and a small number of ES cells will be deposited inside. The inner cell mass of each blastocyst is discernible as a thickening along the wall of the embryo. The large, blunt needle is a holding pipette that secures the blastocyst (with suction) during the injection. (B) A chimeric mouse, showing the characteristic mixed coat color.

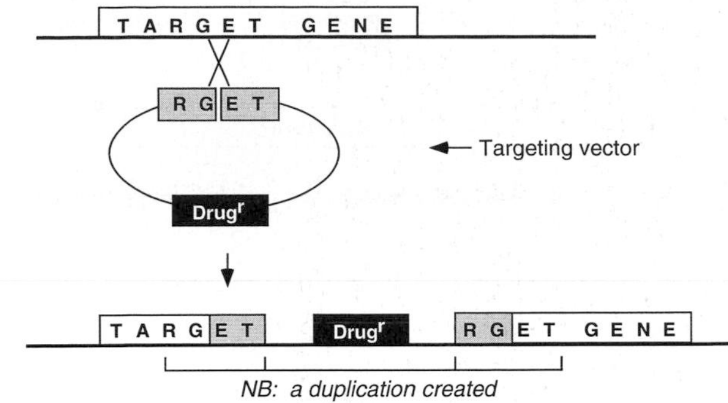

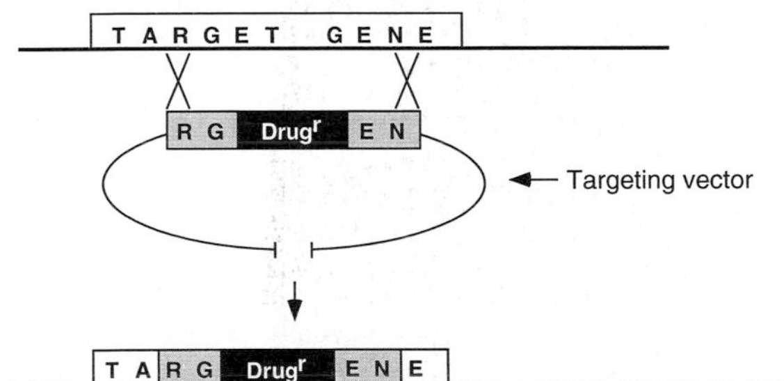

FIGURE 6 Two types of targeting vector. In both panels, the target gene (open box) is shown as a specific region on a linear chromosome (thin line). The targeting vectors contain a region of homology (shaded) to part of the target gene and a drug resistance gene. (Gaps in the circular structures of the two targeting vectors indicate that these plasmids are introduced into ES cells in a linear form in order to increase the efficiency of their integration into the ES cell genome.) The "X" symbols indicate the position of crossing-over between the targeting vector and the target gene on the chromosome. The structure shown at the bottom of each panel depicts the product of HR at the target gene.

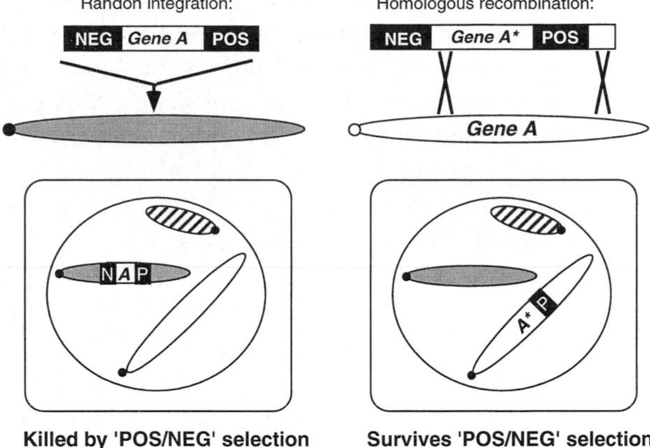

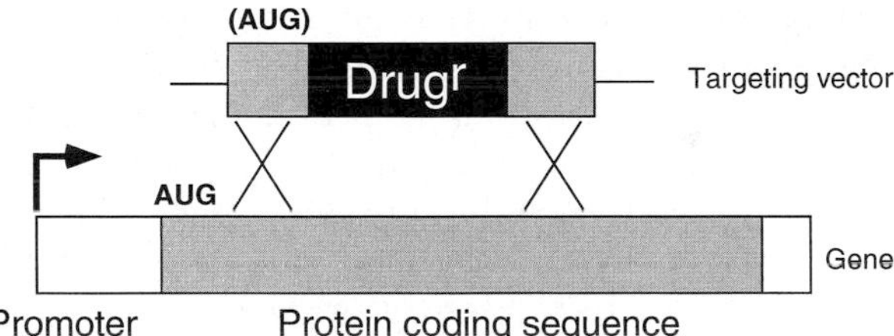

FIGURE 7 Strategies to improve the yield of ES cells with HRs. (A) Positive/negative selection. Symbols used are the same as in Fig. 1. In addition, NEG and POS are negative and positive selectable markers, respectively. (B) Promoter fusions. The shaded region is the part of the target gene that encodes its protein product. "AUG" is the beginning of the protein-coding information.

of positive and negative selection. Positive selection requires the acquisition of a drug resistance gene, as described earlier. Negative selection imposes a strong *disadvantage* to cells bearing the negative gene marker, usually a gene encoding a toxin or creating a toxic product. Here the targeting vector carries both a positive and a negative selectable marker (Fig. 7A). The positive/negative strategy works because random integration and HR occur through different mechanisms. During random integration, the incoming DNA integrates into the chromosome via its ends, so the whole length of the DNA integrates. As a result, both the positive and the negative markers are incorporated. Such ES cells would survive positive selection, but they would be killed by simultaneous negative selection. During HR, however, crossovers will occur only where the targeting vector and the chromosomal site are homologous. If the negative marker is placed at one end of the vector, it will be eliminated during the exchange process. ES cells that have undergone HR in this way survive both positive and negative selection, resulting in enrichments of ~2- to 100-fold over positive selection alone.

A very effective way of increasing the yield of ES cell clones that have undergone HR can be employed if the gene to be targeted is active (transcribed and translated) in ES cells. Here the targeting vector has target gene sequences flanking an incomplete drug resistance gene, one that lacks the necessary signals required for its expression (a start site for either mRNA or protein synthesis; see Fig. 7B). Random integration is unlikely to provide these signals, and such ES cells will not survive positive selection with the drug. However, HR will bring the selectable marker into an active site (the target gene), thus providing the signals needed to turn on the drug resistance gene. Such a "promoter or gene fusion" strategy can engender a 100-fold improvement in the yield of correctly targeted ES cell clones.

IV. SCREENING FOR HOMOLOGOUS RECOMBINANTS

Once the targeting vector has been constructed, it is introduced into ES cells, generally by electroporation (subjection of the cell/DNA suspension to a brief high-voltage pulse). The ES cells are returned to culture, and positive selection is applied (drug is added to the medium). After 8 to 10 days in culture, cells that have not acquired the targeting vector (carrying the drug resistance gene) have died. Those cells in which the targeting vector has integrated will have multiplied into small colonies (each of approximately several hundreds of cells). As ES cells grow in tight clumps, each discrete colony is likely to represent a clone of cells, all derived from a single parental cell. It is necessary to screen each drug-resistant clone to determine whether it acquired the targeting vector through random integration or through HR. Two methods are commonly used, and both rely on the fact that HR will alter the target gene's structure, whereas random integration will not. Therefore, the structure of the target gene is examined, looking for the predicted change.

If the HR frequency is low (say, <1 in 100 drug-resistant clones), screening can be done on pools of colonies, using the polymerase chain reaction (PCR). Primers are designed to detect a novel DNA junction, generated when the targeting vector recombines with the target gene (Fig. 8A). In the absence of HR, the targeting vector would not be found near the target gene, so no PCR product would be made. [*See* Polymerase Chain Reaction.]

If HR frequency is high, clones can be screened individually by genomic Southern analysis (Fig. 8B). In this procedure, each clone's DNA is fragmented using an enzyme that breaks, or cleaves, the DNA very specifically at particular sites, generating a collection of fragments. Fragments from a particular chromosomal region are selectively identified using a short DNA probe from the region (Fig. 8B). Integration of the targeting vector can introduce a new enzyme cleavage site, as in Fig. 8B, thus altering the size of the fragment detected. If HR has occurred, a novel fragment of a predictable size will be seen. If random integration has occurred, the structure of the target gene will not be altered, so the cleavage pattern will not change. It is not possible to know in advance what the targeting frequency at a given target gene will be. One usually starts by screening individual clones by the Southern method, and if no HRs are detected among the first hundred or so clones tested, a larger screen employing PCR is

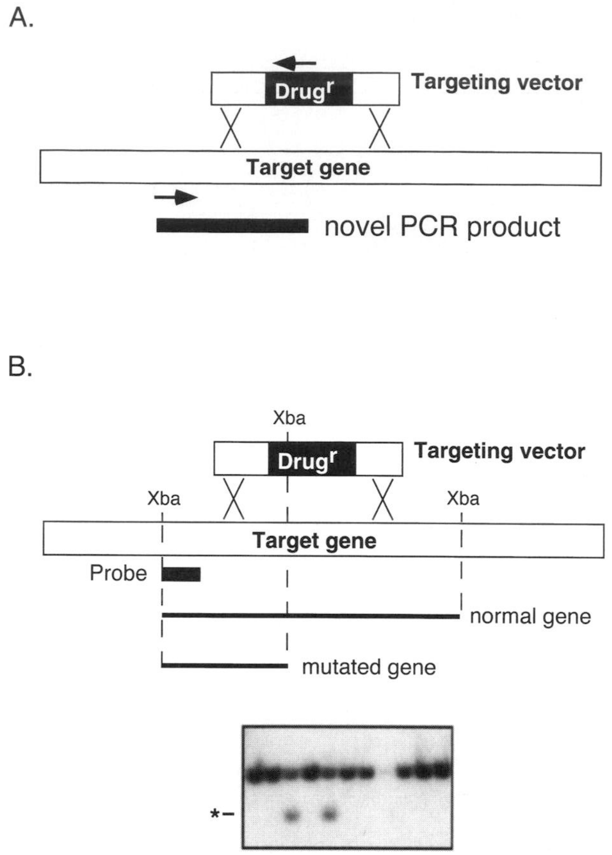

FIGURE 8 Screening for targeted ES cell clones. (A) A PCR strategy. Arrows mark the position of primers, and the solid bar reflects the PCR product expected if HR has occurred. (B) A genomic Southern screening strategy. When the enzyme XbaI is used to cleave the DNA, the fragments generated from the normal gene and from the mutated (targeted) gene are easily distinguishable (thick lines). At the bottom, actual Southern data are shown. The common (upper) band is derived from the normal allele, whereas the band marked with an asterisk comes from an allele targeted by HR.

undertaken. Once one to a few clones bearing HRs (and therefore the desired mutation) are isolated, they are prepared for reintroduction to a blastocyst (see Figs. 3 and 5A).

V. BREEDING FOR HOMOZYGOUS MUTANTS

Once ES cell clones have been isolated that carry the correctly mutated gene, a small number (5–15) of such cells are returned to the interior of a blastocyst by microinjection (Figs. 3 and 5A). The injected blastocysts are then returned to the uterus of a pseudopregnant female (Fig. 4B). Seventeen to 18 days later, a litter should be born, and when the pups are a week old (and their coats start to appear), it will be possible to determine whether any chimeras have been generated (as in Fig. 5B). The chimeras must be mated, to test whether their ES cell-derived genes can be inherited or "passed through the germline." Coat color is used as a convenient test for such "germline transmission." Male chimeras (combined agouti and black) are mated with C57BL (black) females (Fig. 9). Agouti pups indicate that germline transmission has occurred, as the agouti gene comes from the ES cell (129-derived) and is dominant. Because the injected ES cell was homozygous for agouti (two copies of the agouti gene), but heterozygous for the target gene (one normal allele of the gene and one mutated allele), each agouti mouse must be tested for the presence of the mutated gene (Fig. 9A). This can be done by the PCR or genomic Southern strategies (see Fig. 8). Mice that carry a copy of the mutated gene must then be mated to one another in order to generate progeny that are homozygous for the mutated allele (Fig. 9B). These are the mice that should be affected by the mutation, as they lack a normal copy of the gene.

A

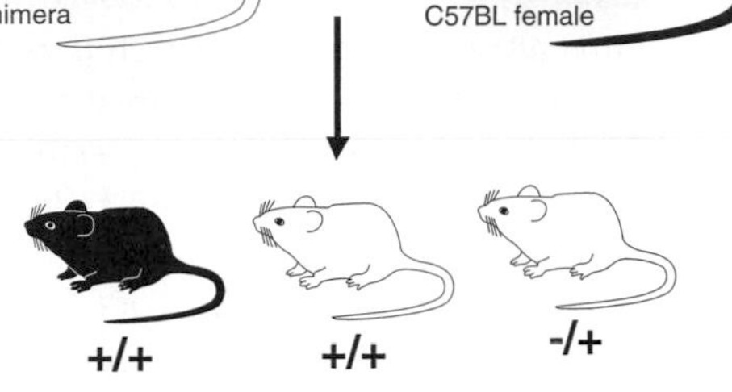

B

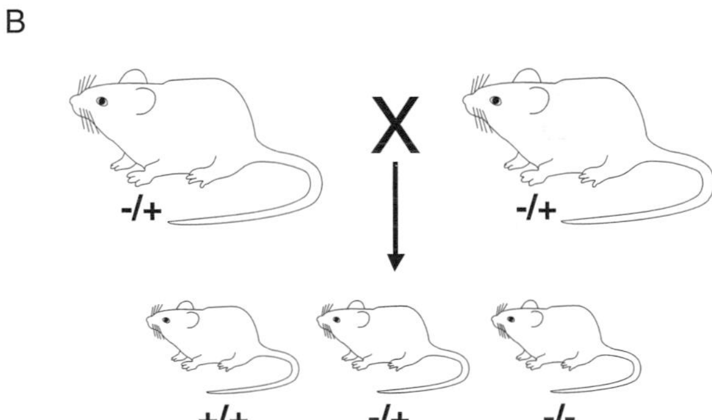

FIGURE 9 Breeding for homozygous mutant mice. In this figure, the color coding of the mice is the same as that used in Fig. 3. Here "X" means a mating. (A) Germline transmission, ending with a mouse that is heterozygous for the desired gene mutation (−/+). (B) Intercross between heterozygotes to create a wholly mutant mouse (homozygous mutant; −/−).

VI. TYPES OF MUTATION POSSIBLE

A. Gene Ablation

In most of the several hundred gene targeting experiments performed to date, the gene of interest was ablated to create an animal with a complete gene deficiency. Gene ablation experiments have been extremely informative in defining a "sphere of influence" for each of the genes targeted. Some genes, however, play such critical roles during development that their loss cannot be tolerated; ablation of such genes is lethal. For this and other reasons, more sophisticated strategies have evolved that allow the generation of subtle mutations (in contrast to gene ablation) or mutations that are restricted to particular tissues or to particular times of development.

B. Subtle Mutations

A conventional gene targeting vector mutates the gene of interest but leaves behind a piece of foreign DNA: the selectable marker (Fig. 6). In a gene ablation experiment, this may be of little consequence, but if creation of a subtle mutation is the aim, then the selectable marker must be eliminated from the targeted gene. Otherwise, foreign DNA sequences contained in the selectable marker might interfere with the behavior of the target gene in an unpredictable way. The "hit-and-run" (also called "in-and-out") strategy starts with a conventional gene targeting step using an insertion-type vector carrying both a positive and a negative selectable marker, in addition to a subtle mutation in the target gene (Fig. 10). After HR between the vector and the target gene, a duplication is created. At a low frequency (less than ~1 in 1000), *spontaneous, intrachromosomal* HR occurs between the duplicated sites, and the intervening selectable markers are deleted. By applying negative selection, clones in which this has occurred can be recovered. The result is a fine rather than a gross alteration to the target gene. A novel alternative for generating subtle mutations is described in the next section.

C. Mutation via Recombinase-Mediated HR

Certain bacterial viruses and strains of yeast carry enzymes that perform very simple and specific HR reactions. Only two components are necessary: the enzyme and a short piece of DNA carrying the recognition sequence of the enzyme (each 34 bp long). The P1 bacterial virus uses the Cre (catalyzes recombination) enzyme, and the yeast protein is called FLP, or "flip." Their recognition sequences are called lox (locus of crossing-over) sites and FRTs (FLP recognition targets), respectively. Despite their simple origins, both recombination systems have been shown to work well in mammalian cells, catalyzing the recombination reaction depicted in Fig. 11A. Although the Cre/lox system will be discussed here primarily, the FLP/FRT system also holds promise.

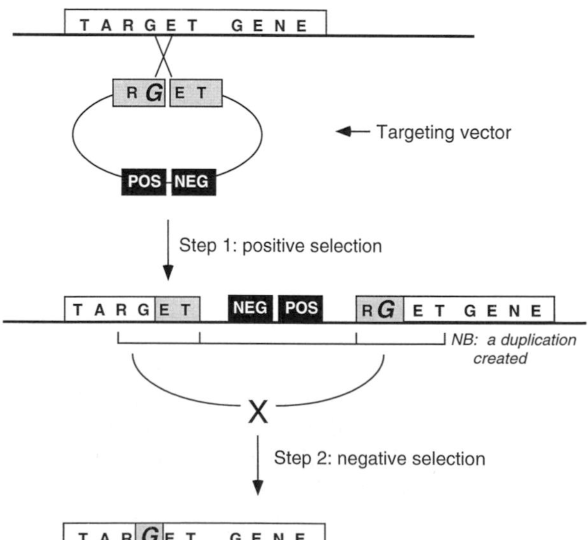

FIGURE 10 The "hit-and-run" or "in-and-out" strategy for creating subtle mutations. Symbols are as in Figs. 6 and 7. In addition, the large, italicized *G* represents a subtle mutation of the target gene sequence, and the curved line and "X" at step 2 indicate an *intra*chromosomal crossover.

1. Subtle Mutations

One application of the Cre/lox system is the generation of subtle mutations. A conventional gene targeting vector delivers the subtle mutation and the selectable marker, *flanked by lox sites,* to the target gene locus in the ES cell (Fig. 11B). A second, Cre-producing plasmid is then introduced to the cell in a form that is unlikely to integrate into the cell's DNA, and a transient burst of Cre activity occurs. During this short time, the Cre protein efficiently finds its recognition sites (the lox sites) and catalyzes a recombination between them, resulting in the deletion of the intervening DNA. Only the subtle mutation and one lox site remain at the target gene.

2. Mutations Restricted to Certain Cells of the Mouse

The Cre/lox system can also be used to generate mice with targeted mutations that are manifest only in particular tissues, or at particular times during mouse development. In this way, genes that are essential for development or for survival can be studied in a restricted group of cells, without compromising the entire animal. In the example described next and shown in Fig. 11C, the target gene would be mutated only in neurons in the brain.

The first targeting step is again a conventional one, placing the lox sites and selectable marker at the target gene in ES cells. However, the targeting vector is designed so that the targeted locus still functions normally. A mouse is generated, and as the "mutated" allele behaves normally, this mouse should not be affected by the change. A second mouse strain is now used. This strain carries a Cre transgene that is active only in neurons (Fig. 11C). The mice are mated to bring together the Cre transgene and the targeted gene with the lox sites. In neurons of the progeny, Cre protein will catalyze recombination between the lox sites, creating a debilitating mutation at the target

A. Site-specific recombination mediated by Cre/lox

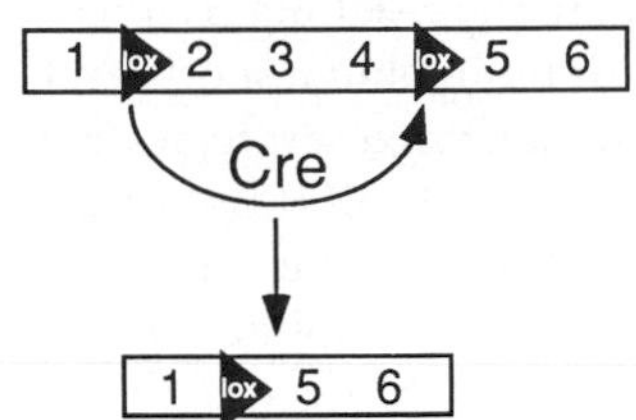

B. Subtle mutation

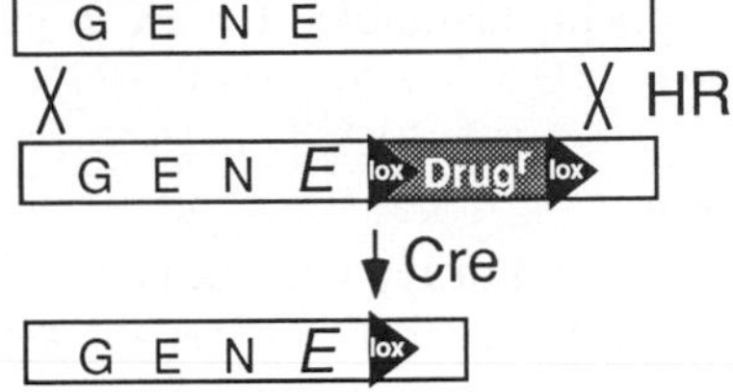

C. Tissue- or stage-specific mutation

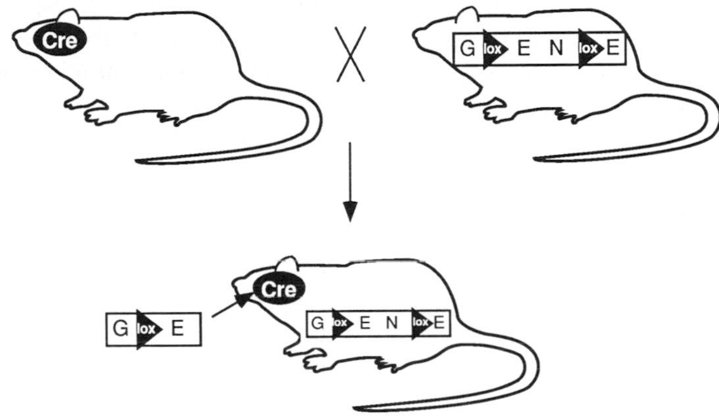

D. Gene replacement

FIGURE 11 Mutation strategies available using recombinase-mediated HR. In all panels, triangles indicate the position and orientation of a lox site. "Cre" indicates that the Cre recombinase is active and will catalyze the recombination of two lox sites. (A) The generic Cre-mediated reaction. (B) Two steps to a subtle mutation using Cre/lox. The first is a conventional HR step, placing the subtle mutation (*E*), lox sites, and Drug^r gene at the target locus, and the second is a deletion mediated by Cre. (C) The mouse on the top left is a transgenic mouse expressing Cre only in neurons (black oval represents the brain). This mouse is mated with a mouse with a targeted gene bearing lox sites. Progeny of this mating will have a Cre-mediated gene deletion in the brain (arrow), but not in other tissues. (D) Gene replacement. The hatched regions represent DNA adjacent to the target gene, which are included in the targeting vector to catalyze HR.

gene. No other cell types make Cre protein, so the mutation will be restricted to neurons. In one such experiment, the Cre-mediated deletion was shown to occur in most but not all cells where Cre was present, so the target tissue was a mosaic of normal and mutant cells. (This may simply reflect the level of Cre protein achieved in each cell, which is in turn determined by the activity of the Cre transgene.) Despite this limitation, the system represents a valuable tool for studying gene function in restricted cell populations. A collection of transgenic mice expressing Cre protein in particular tissues is being assembled for use in experiments of this type.

3. Gene Replacement

Another valuable adaptation of the Cre/lox system is the gene replacement strategy. In this case, the targeting vector contains two regions of homology to the target gene, flanking the lox-bound selectable marker genes. An additional piece of DNA is included: the new DNA that is to replace the resident gene at the target locus (Fig. 11D). In this way, a gene replacement vector is conceptually the same as a vector for generating a subtle mutation in the Cre/lox system. (Compare the organization of the targeting vectors in Figs. 11B and 11D.) After the gene is targeted in ES cells by the initial HR (whereby the target gene is replaced by the vector DNA), transient exposure to Cre protein, as described earlier, will catalyze the deletion of the selectable marker, leaving behind the new gene.

Gene replacement can be used to study the expression pattern of a gene *in vivo*, that is, to find the cells in which the gene of interest is active. Here, the target gene is replaced by a marker gene, most commonly one for the enzyme β-galactosidase. This enzyme catalyzes a reaction, generating a bright blue product. The gene replacement procedure places the β-galactosidase gene under the same control as the target gene, therefore mimicking its expression. Cells in which the target gene is active (and *only* those cells) will turn blue when stained appropriately. Such a gene replacement can be designed simultaneously to mutate the target gene, so an animal generated by this strategy can also be used to study the consequences of the mutation. Tagging a gene with the β-galactosidase marker can be done either by the Cre/lox system or by conventional gene targeting, using a hit-and-run strategy.

VII. CURRENT AND POTENTIAL APPLICATIONS OF GENE TARGETING

Some current applications of gene targeting have already been mentioned: gene ablation, subtle mutation, tissue- or stage-specific mutation, and gene replacement, with the particular application of expression tagging. Each of these strategies could be applied equally to the body of a target gene, or to the nearby sequences that determine the gene's activity, thus eventually teaching us why certain genes are active in some tissues but not others. Conventional or recombinase-mediated strategies can be used to delete large regions (>15,000 bp) of a chromosome for application to large genes or to gene clusters.

A practical application of gene replacement is the generation of mice with humanized antibody genes. Specialized mouse antibodies (monoclonal antibodies) are currently used therapeutically and are prized for the specificity of their action. However, a serious limitation is that a human immune system recognizes mouse antibodies as foreign and learns to reject them. At this point, the therapy fails. Humanized monoclonal antibodies should not suffer from this limitation. Another potential application of gene targeting to human disease is in genetic diseases of the blood (immune deficiencies, anemias). Blood cells are all derived from blood stem cells found in the bone marrow. If gene targeting could be applied to blood stem cells from a patient's bone marrow, it may be possible to specifically correct the mutated gene. Blood cells derived from the corrected stem cells would behave normally when returned to the patient and could permanently cure the disorder. [*See* Hemopoietic System.]

Already gene targeting has generated several mouse models of human genetic diseases, such as cystic fibrosis, bowel cancer, atherosclerosis, and thalassemia. We have learned that single genes control the development of certain regions of the brain, of whole organs, and of red blood cells. We have learned that many protooncogenes (cancer-causing in a mutant form) are essential for life. The current versatility of gene targeting technology, and the rate of evolution of new methods, should significantly advance our understanding of what each of our genes does for us. [*See* Genetic Diseases; Human Genetics.]

BIBLIOGRAPHY

Bronson, S. K., and Smithies, O. (1994). Altering mice by homologous recombination using embryonic stem cells. *J. Biol. Chem.* **269,** 27155–27158.

Chambers, C. A. (1994). TKO'ed: Lox, stock and barrel. *BioEssays* **16,** 865–867.

Fung-Leung, W.-P., and Mak, T. W. (1992). Embryonic stem cells and homologous recombination. *Curr. Opin. Immunol.* **4,** 189–194.

Joyner, A. L. (1991). Gene targeting and gene trap screens using embryonic stem cells: New approaches to mammalian development. *BioEssays* **13**, 649–656.

Kilby, M. N., Snaith, M. R., and Murray, J. A. H. (1993). Site-specific recombinases: Tools for genome engineering. *Trends Genet.* **9**, 413–421.

Koller, B. H. (1992). Altering genes in animals by gene targeting. *Annu. Rev. Immunol.* **10**, 705–730.

Melton, D. W. (1994). Gene targeting in the mouse. *BioEssays* **16**, 633–638.

Smithies, O. (1993). Animal models of human genetic diseases. *Trends Genet.* **9**, 112–116.

Yong-Rui, Z., Muller, W., Gu, H., and Rajewsky, K. (1994). Cre-loxP-mediated gene replacement: A mouse strain producing humanized antibodies. *Curr. Biol.* **4**, 1099–1103.

Gene Therapy

JAMES A. ZWIEBEL
National Cancer Institute

GLOSSARY

Allogeneic Cells or tissues that are derived from a genetically nonidentical individual

Angiogenesis Formation of new blood vessels

Antigen A molecule that induces an immune response resulting in recognition of and reactivity with the antigen

Autologous Cells or tissues that are derived from the same individual

Cytokine A protein produced by one cell population that activates responses in the same or another cell population

Hematopoietic cells Cells arising from the bone marrow, including red blood cells, white blood cells, and platelets

Major histocompatibility complex These cell surface proteins allow the immune system to recognize foreign cells and antigens

Plasmid A circular form of DNA that is used for cloning genes in bacterial cells

Replication-competent virus Virus that is capable of reproducing itself and of spreading infection (also known as helper virus).

Somatic cells Those cells of the body other than the reproductive (germ) cells

Stem cells Cells that are capable of giving rise to differentiated cells of various cell lineages throughout the lifetime of the individual

Transduction Stable insertion of a transgene into a cell

Transgene A recombinant gene that has been inserted into a cell using gene transfer

Vector An agent that is capable of inserting new genetic information into a cell

Viral titer A measure of the concentration of infectious particles

I. INTRODUCTION

A. Definition and Background

Gene therapy can be defined as the insertion of genetic material into cells for the treatment of disease. Although gene therapy was envisaged initially as a means of treating inherited genetic diseases, disciplines as diverse as oncology, cardiology, endocrinology, and infectious diseases are currently developing novel therapeutic strategies that have in common gene transfer into cells. The result of the burgeoning interest in gene therapy has been a rapid growth in the number of clinical trials throughout the United States, Europe, and Israel that utilize gene transfer, either for cell marking or for therapeutic intent. Although the usefulness of gene transfer for patient care is not yet established, a profusion of different strategies and applications of clinical gene transfer continues to emerge. Its potential impact is likened to that of the introduction of antibiotics—gene therapy eventually may become part of the standard clinical armamentarium. For now, limitations to gene transfer and expression have impeded the clinical development of this treatment.

What are the elements necessary for carrying out gene therapy? What diseases are being targeted? What has been learned from the initial clinical trials? This article addresses these questions to provide a perspective upon the current status of gene therapy.

ENCYCLOPEDIA OF HUMAN BIOLOGY, Second Edition, VOLUME 4.

In 1982, just a few years after the introduction of recombinant DNA technology, an attempt was made to correct β-thalassemia by inserting the correct gene into the bone marrow cells of several affected individuals. That unauthorized and ineffective experiment focused the attention of both scientific and lay communities on the possibility of modifying the human genome as well as the potential for abuse. A number of principles emerged from the ensuing debate. First, the modification of somatic cells to treat a disease in a *particular* individual was felt to be acceptable, although germ line gene transfer, resulting in the introduction of a heritable trait, was not, nor was gene transfer for the purpose of enhancing a trait, such as height. Second, in order to ensure that gene therapy procedures would be based on sound scientific grounds and adhere to acceptable ethical and social norms, approval to carry out a clinical protocol required an extensive review process designed to ensure the safety of the patient and others. This review involved committees from the local institution, the National Institutes of Health (NIH), and the Food and Drug Administration (FDA) (discussed further below). "Points to Consider" guidelines for treating patients that were developed by both the NIH and the FDA provided a framework and a uniform standard for the review process. The initial approval of a gene marking study in 1988 and for a gene therapy study in 1990 opened the door to a flood of proposals. Currently there are approximately a thousand patients enrolled in nearly 200 approved gene transfer protocols in the United States (primarily) and throughout the world.

B. Gene Therapy: Desirable Attributes

Optimally, gene therapy would be easily administered and would allow targeting of specific cells in an efficient manner. In addition, the inserted genes would be expressed only when required. For example, regulated expression of an insulin gene would be necessary to avoid the deleterious extremes of hyper- or hypoglycemia. As yet, efficient gene transfer, coupled with the temporal and spatial control of transgene expression, has not been accomplished. Currently available gene transfer vectors suffer from problems of inefficient gene transfer, loss of gene expression, or adverse effects. The immediate goal is to achieve constitutive transgene expression in gene-modified cells in order to produce therapeutically relevant amounts of a protein. For now, targeted gene delivery and regulated gene expression remain elusive aims. However, newer gene transfer techniques and an emerging understanding of gene regulation undoubtedly will bring us closer to accomplishing these goals.

C. *Ex Vivo* vs Direct *in Vivo* Gene Therapy

The inefficiency of early attempts at introducing new genetic information into cells made it necessary to cultivate cells *in vitro* in order to carry out gene transfer successfully. In addition, the ease of accessing and reimplanting bone marrow cells, the initial targets for gene therapy, was another reason for pursuing the *ex vivo* approach to gene therapy. Carrying out the gene transfer procedure *in vitro* does have additional advantages. First, by first isolating the target cell, it is possible to achieve tissue specificity of transgene expression. Second, it may be possible to optimize production of the therapeutic agent by selecting for its expression in the gene-modified cell population. Third, one may be better able to ensure the safety of the gene transfer procedure by assaying, for example, for the presence of replication-competent virus before administering the gene-modified cell to the patient. However, the *ex vivo* gene transfer method requires sophisticated tissue culture expertise that is both costly and labor-intensive. Moreover, the introduction of gene-modified cells may be complicated by the inability of those cells to become efficiently reincorporated into the tissues of the recipient.

Direct *in vivo* gene transfer is much simpler because it does not require cell isolation and cultivation. A greater ease of administration eventually might facilitate the availability of gene therapy to the local community setting. However, direct *in vivo* gene transfer in general still suffers from a poor gene transfer efficiency and the inability to target a specific tissue or cell type. Hence, transgene expression by nontarget tissues, including the germ line, may be an undesired consequence of the *in vivo* gene transfer procedure.

Currently, both *ex vivo* and *in vivo* gene transfer methods are being used for gene therapy. For example, cystic fibrosis is being treated using an adenoviral vector to deliver the cystic fibrosis transmembrane regulator gene directly to bronchial epithelial cells *in situ*. In contrast, for the treatment of severe combined immunodeficiency due to adenosine deaminase (ADA) deficiency, lymphocytes or bone marrow progenitor cells are removed from the patient, transduced with a retroviral vector containing the ADA gene, and then reinfused. The diseases currently under study with approved clinical protocols are listed in Tables I and II.

TABLE I
NIH-Approved Cancer Gene Therapy Protocols[a]

Disorder	Cellular target	Category/gene	Method/vector
Malignancy	Tumor cells, fibroblasts, tumor-infiltrating lymphocytes, muscle	Immunotherapy: interleukin(IL)-2, IL-4, IL-12, HLA-B7, interferon-γ, tumor necrosis factor, granulocyte–macrophage colony-stimulating factor, B7 immune accessory molecule, β_2-microglobulin, MART-1 melanoma antigen, prostate-specific antigen, carcinoembryonic antigen, HER-2/Neu, chimeric T-cell receptor reactive with protein expressed on tumors Tumor suppressor molecules: p53, retinoblastoma protein, BRCA-1 Oncogene downregulation: anti-erbB-2 Chemoprotection: multidrug resistance gene Pro-drug activation: herpes simplex virus thymidine kinase/ganciclovir, cytidine deaminase/5-fluorocytosine Antisense: K-*ras*, c-*myc*, insulin-like growth factor I, TGF-β2	*Ex vivo, in vivo*/retrovirus, adenovirus, liposomes

[a]Source: National Institutes of Health, Office of Recombinant DNA Activities.

II. HOW GENES ARE INSERTED INTO CELLS

Although great strides have been made over the last decade in vector development, inefficient gene transfer and/or unstable gene expression remains the greatest limitation to achieving successful gene therapy. To address these issues, efforts are ongoing toward developing novel vectors as well as modifying the design of existing ones.

A. Viral Vectors

A list of currently available methods for introducing new genetic information into cells is listed in Table III. Until recently, *retroviruses* were the most popular means of transducing cells because of their ability to integrate into the cellular genome in a stable fashion and to permit efficient gene expression. However, they suffer from a number of shortcomings, including the need for cell division in order for gene integration to occur, a limit in the size of the gene that can be transferred, and the potential risk of causing neoplastic transformation by inserting nearby and activating a cellular protooncogene. [*See* Retroviruses as Vectors for Gene Transfer.]

Due to their stability, *adenoviral vectors* can be concentrated to titers that are several logs higher than those achievable with retroviral vectors. The high viral titer allows efficient *in vivo* gene transfer to occur. Moreover, adenoviruses can infect nondividing cells. However, following adenoviral infection, the transgene exists as an episome (an unintegrated gene) that is eventually lost from the cell. The expression of adenoviral proteins by infected cells also results in a cellular and humoral immune response that abrogates transgene expression, may result in toxicity, and interferes with subsequent attempts to transduce cells. Adenoviral vectors may also be directly cytopathic and are fatal to animals when large quantities of virus are administered.

Adeno-associated viruses (AAV) can also be concentrated to high titer. Although wild-type AAV integrates into host DNA at a site on chromosome 19, recombinant vectors appear to have lost site-specific integration and may be prone to gene rearrangement.

Pox viruses, including *vaccinia* and *fowl pox,* are among a number of different vectors being used for antitumor vaccination strategies, whether to deliver recombinant cytokines or tumor antigens, in order to elicit host antitumor immune responses.

Modified *herpes viruses* are being studied for their ability to target neuronal cells with recombinant genes. Although the expression of foreign genes in neuronal cells has been successful with herpes virus vectors, the vectors used for gene transfer have been contaminated with replication-competent virus, resulting in significant neuronal toxicity. Additional

TABLE II
NIH-Approved Noncancer Gene Therapy Protocols[a]

Disorder	Cellular target	Gene	Method/vector	Comments
Severe combined immunodeficiency				
1. Adenosine deaminase (ADA) deficiency	Lymphocytes, hematopoietic progenitor cells from bone marrow and umbilical cord blood	ADA	*Ex vivo*/retrovirus	Severe combined immunodeficiency; first approved human gene therapy study (see text)
2. X linked	Hematopoietic progenitors from bone marrow and umbilical cord blood	Chain of multiple cytokine receptors	*Ex vivo*/retrovirus	
Hypercholesterolemia	Hepatocytes	Low-density lipoprotein receptor	*Ex vivo*/retrovirus	Improvement in cholesterol level seen
Cystic fibrosis	Bronchial epithelium	CFTR	*In vivo*/adenovirus, adeno-associated virus, liposomes	See text
Gaucher disease	Hematopoietic progenitor cells	Glucocerebrosidase	*Ex vivo*/retrovirus	
Hunter syndrome (mucopolysaccharidosis type II)	Lymphocytes	Iduronate-2-sulfatase	*Ex vivo*/retrovirus	
AIDS	Hematopoietic progenitor cells, lymphocytes	Herpes simplex thymidine kinase gene, transdominan *Rev* protein, antisense TAR, TAR and RRE decoys. TAT and Rev hammerhead ribozyme, envelope protein vaccine, intracellular antibody against HIV-1 envelope protein, chimeric T-cell receptor	*Ex vivo*/retrovirus	See text
α_1-Antitrypsin deficiency	Respiratory epithelium	α_1-Antitrypsin	*In vivo*/liposome	
Rheumatoid arthritis	Synovium	Interleukin-1 receptor antagonist	*Ex vivo*/retrovirus	See text
Fanconi's anemia	Hematopoietic progenitor cells	Complementation group C	*Ex vivo*/retrovirus	
Peripheral artery disease; coronary artery restenosis	Vascular endothelial cells, smooth muscle cells	Vascular endothelial growth factor	*In vivo*/plasmid DNA	Intravascular administration by gel-coated balloon catheter; see text
Chronic granulomatous disease	Hematopoietic progenitor cells	$p47^{phox}$	*Ex vivo*/retrovirus	Transient appearance of gene-corrected neutrophils in a patient
Purine nucleoside phosphorylase deficiency	Lymphocytes	Purine nucleoside phosphorylase	*Ex vivo*/retrovirus	T-cell immune deficiency
Ornithine transcarbamylase deficiency	Hepatocytes	Ornithine transcarbamylase	*In vivo*/adenovirus	Urea cycle disorder

[a]Source: National Institutes of Health, Office of Recombinant DNA Activities.

modifications of the viral genome should improve the safety of herpes vectors. [*See* Herpesviruses.]

Of the nonviral methods of gene transfer, perhaps the simplest is the injection of an uncoated *plasmid DNA* vector directly into muscle. Despite the very low efficiency of gene transfer that has been achieved *in vivo* with plasmid alone, intramuscular injection has led to surprising, and effective, levels of transgene expression in animal models. Long-term therapy could be accomplished with repetitive injections of DNA. However, even short-term gene expression could be effective as a vaccination strategy for the treatment of both infectious diseases and cancer. Even a transient expression of microbial or tumor antigens to immune cells may lead to potent immunological responses.

TABLE III
Gene Transfer Methods

Method	Stable integration?
Viral	
Murine retrovirus	Yes
Adenovirus	No
Adeno-associated virus	Yes
Herpes virus	No
Pox virus	No
Nonviral	
Calcium phosphate	Possibly
Liposomes	Possibly
DNA–ligand complexes	Possibly
Bioballistics	Possibly

By permitting selective targeting of cells that display a particular cell surface receptor, *receptor-mediated gene uptake* by cells, which utilizes a complex of DNA conjugated to a protein ligand (a molecule that specifically interacts with a cell receptor), is also a promising method for clinical gene transfer. The addition of adenoviral components to the DNA–protein complex improves gene transfer by facilitating the release of the internalized transgene that ordinarily becomes trapped within cellular endosome. Nonetheless, because DNA integration is unlikely to occur, the transgene eventually will be lost from the cell. However, the simplicity and the potential for repeated direct *in vivo* injection account for the appeal of receptor-mediated DNA gene transfer.

Liposomes are another potentially useful vehicle, with impressive reports of efficient gene transfer *in vivo* into diverse cells such as lung epithelium, vascular cells, and myoblasts in experimental models. Liposome-mediated gene transfer trials in cancer and cystic fibrosis are currently underway.

Bioballistics, utilizing the "gene gun," uses DNA-coated metal particles propelled under high velocity to introduce genes into cells. This method may prove to be useful as a vaccination method.

III. GENE TRANSFER INTO HEMATOPOIETIC STEM CELLS

Bone marrow cells were viewed as ideal targets for gene transfer for a number of reasons. Bone marrow is readily obtained and manipulated *ex vivo*. Moreover, by transducing hematopoietic stem cells (HSC), one would be able to repopulate the bone marrow and provide a lifetime supply of gene-corrected cells. Indeed, early studies in mice, using retroviral-mediated gene transfer (see below), demonstrated the feasibility of this approach, with an efficient transfer of marker genes into cells that were capable of repopulating the bone marrow of lethally irradiated animals. Long-term expression of transgenes was seen occasionally as well, indicating that stem cells had been transduced. However, in subsequent monkey studies there was a loss of expression that was associated with the disappearance of the transgene in those animals that received genetically modified bone marrow cells. These results suggested that, unlike with mice, the transduction of primate bone marrow stem cells was very inefficient.

Subsequently, a number of groups have demonstrated prolonged *in vivo* recombinant gene expression in human bone marrow, albeit at a low frequency. In a gene-marking study carried out at St. Jude, leukemia patients who were undergoing high-dose chemotherapy received bone marrow cells of their own that had been exposed to a retroviral vector containing a marker gene. There was no evidence of leukemia contamination of the bone marrow at the time of transplantation, even by sensitive polymerase chain reaction (PCR) analysis. However, the finding of gene-marked cells at the time of subsequent relapse indicated that leukemic cells, in fact, had contaminated the administered bone marrow. Moreover, despite the passage of many months since the transplant, even normal bone marrow cells were found to carry the transgene, indicating that HSC had been transduced by the retroviral vector. In other studies, patients with severe combined immunodeficiency due to ADA who received HSC transduced with vectors containing the ADA gene had persistent transgene expression in cells from the blood and bone marrow. These encouraging results, suggesting that hemopoietic stem cell transduction has occurred, appear to be due to two factors. First, retroviral integration may have been promoted by stimulating bone marrow stem cells to cycle with recombinant cytokine molecules [such as the combination of interleukin (IL)-3, IL-6, and stem cell factor] that were added at the time of retroviral infection. Second, the practice of enriching for bone marrow pluripotential cells using antibodies to CD34, a molecule expressed on stem cells, has an increased likelihood of infecting the desired target cell by increasing the number of viral particles per cell. It also has been suggested that purifying the small population of stem cells that are capable of being transduced may also facilitate their engraftment.

Similarly, correction of hemophilia may also be possible with HSC gene therapy, despite a low transduction efficiency, as only a small increase in factor VIII or IX levels may significantly ameliorate the frequency and severity of bleeding complications. Unfortunately, this is not true for some of the other hematologic diseases that are potentially curable with allogeneic bone marrow transplantation. For example, although thalassemia theoretically could be treated by inserting a normal globin gene into erythroid progenitor cells, balanced synthesis of the α- and β-globin chains is necessary to correct the cellular abnormality. Such tightly regulated gene expression is not yet possible.

Greater attention is now being focused on the possibility of correcting genetic abnormalities in the fetus. *In utero* gene transfer is attractive both as a means of preempting irreparable injury as well as affording a period of immune tolerance that may permit allogeneic implants of gene-corrected cells. An *in utero* sheep bone marrow transplant–gene transfer model has provided encouraging results and surgical manipulation of the human fetus is now possible.

Current applications of hematopoietic cell gene transfer include both gene marking, the insertion of gene sequences to permit tracking the fate of cells *in vivo,* and gene therapy. Some examples of therapeutic gene transfer are listed in Table II.

IV. GENE TRANSFER INTO DIFFERENT CELL TYPES

The past several years has seen a rapid increase in gene therapy strategies under development in nearly all medical specialties. Gene augmentation therapy, the insertion of genes into cells to augment an existing function or, alternatively, to confer upon the cell a new function, allows virtually every cell and tissue to be a potential target for gene transfer. In addition, exploiting the biological characteristics of particular cell types may facilitate local or systemic gene delivery. Some of the cell types under study warrant separate consideration and are highlighted below.

A. Lymphocytes

Lymphocytes have a number of attributes which make them attractive vehicles for delivery of recombinant molecules. First, these cells will respond to antigen or cytokine stimulation by expanding their numbers by many orders of magnitude. Second, because lymphocytes exist within the circulation and migrate throughout the body, they are attractive vehicles for systemic drug delivery. Third, because lymphocytes are sequestered within the circulatory system, they do not require the provision of a vascular supply that is necessary for implants of nonhematopoietic cells. Despite these appealing characteristics, coupled with their ability to express foreign genes, lymphocytes are not efficiently transduced with available vectors. [*See* Lymphocytes.]

B. Fibroblasts

Despite their ease of cultivation and promising *in vitro* transgene expression, genetically modified fibroblasts have repeatedly failed to sustain transgene expression *in vivo*. Studies to investigate this problem have revealed that there is a shut off of gene expression of implanted cells. However, these cells may still be useful where only short-term gene expression is desired, as for the treatment of cancer. For example, investigators in Pittsburgh and in San Diego are innoculating cancer patients with cytokine-expressing autologous fibroblasts, plus irradiated autologous tumor cells, in order to activate an antitumor immune response. In this case, short-term cytokine expression may be sufficient for immune activation and, in fact, the eventual loss of cytokine gene expression may avoid toxicity from sustained cytokine production.

C. Hepatocytes

Because of their importance for both synthetic and metabolic pathways, hepatocytes are an important target for gene replacement therapy. Using the *ex vivo* gene transfer approach, investigators at the University of Pennsylvania successfully expressed the low-density lipoprotein receptor (LDLR) *in vivo* in deficient hepatocytes of the animal model for familial hypercholesterolemia, the Watanabe rabbit. The results of the preclinical studies enabled the investigators to carry out a human gene transfer protocol (see below).

While promising, the considerable tissue culture support that is required for *ex vivo* hepatocyte gene transfer restricts the application of this procedure to properly equipped centers. Seeking to overcome this limitation, investigators at Baylor have had success using both retroviral and adenoviral vectors for direct *in vivo* hepatocyte transduction. Phenylalanine hydroxylase (phenylketonuria), α_1-antitrypsin (α_1-antitrypsin deficiency), and factor IX genes (hemophilia

B) have been successfully expressed in animals, although the level of gene expression may be insufficient to be clinically effective.

D. Respiratory Epithelial Cells

The respiratory epithelium is easily accessible and has been genetically modified *ex vivo,* using either retroviral-mediated gene transfer or receptor-mediated endocytosis of transferrin–polylysine–DNA complexes, and directly *in vivo,* using liposomes, retroviral vectors, or adenoviral vectors. Several investigators received NIH and FDA approval to treat cystic fibrosis with adenoviral vectors or liposomes that contain the cystic fibrosis transmembrane regulator gene. Introduction of vectors directly into the patient's airway holds promise for the treatment of other respiratory tract disorders, including α_1-antitrypsin deficiency and lung cancer (see Section V).

E. Vascular Cells

The cells of the blood vessel wall, which include endothelial cells and vascular smooth muscle cells, are attractive targets for gene transfer for a number of reasons. First, endothelial cells are strategically located for secretion of therapeutic products into the bloodstream. Second, they have already been shown to be capable of expressing recombinant proteins such as rat growth hormone, tissue plasminogen activator, and factor IX in a stable and efficient manner *in vitro.* Third, genetic modification of vascular smooth muscle would be useful for the study and treatment of conditions that affect the vessel wall, including thrombosis and atherosclerotic disorders.

Both *ex vivo* and direct *in vivo* gene transfer methods have been used experimentally to introduce recombinant genes into the vascular compartment. These methods include (i) balloon catheters to seed blood vessels with endothelial cells previously genetically modified *ex vivo,* (ii) *in situ* transduction of vascular cells using specially designed balloon catheters to infuse the vessel with various vectors (see Section V,F), and (iii) seeding of implantable vascular prostheses, such as grafts and stents, with genetically modified endothelial cells. All these procedures share the common goal of delivering agents to the vicinity of the vascular wall in high local concentrations.

Endothelial cells also exist in the microvasculature that is found in all tissues, where capillary beds form vascular networks with a surface : volume ratio that is much greater than in larger, smooth muscle-walled vessels. Ordinarily, following prenatal development, angiogenesis may occur during would healing, placental growth, and, pathologically, in association with cancer and diabetes. One possibility being studied is to incorporate genetically modified endothelial cells (GMEC) into microvessels generated by an angiogenesis growth factor such as fibroblast growth factor. Thus, the possibility exists of targeting GMEC into angiogenically active sites, such as tumor deposits or "neo-organoids" that have been generated using artificial tissue matrix implants. Once in place in these sites, GMEC could express transgenes that, in one instance, would interfere with tumor growth (e.g., the local production of a cytokine to induce an antitumor immune response) or, alternatively, manufacture a missing protein, such as factor VIII.

F. Muscle Cells

Myogenesis (new muscle formation) results from the fusion of individual precursor cells called myoblasts into multinucleated muscle fibers. Myoblasts, which exist in muscle tissue as satellite cells along muscle fibers, are amenable to *ex vivo* gene transfer, as has been shown experimentally with dystrophin, as well as with other genes ordinarily not expressed in muscle cells, such as factor IX and growth hormone. These gene-modified cells can be expanded *in vitro* and then reinjected into muscle tissue where they may fuse with existing muscle fibers.

Direct *in vivo* gene transfer with viral vectors, DNA liposomes, or even naked plasmid DNA also has been accomplished in animals by injecting directly into striated muscle. Thus, myoblast gene transfer holds promise for either gene replacement or gene augmentation in the treatment of a variety of disorders. However, a recent study of myoblast transfer for the treatment of Duchenne's muscular dystrophy demonstrated that there was hardly any persistence of injected myoblasts over 6 months, nor was there improvement in muscle strength. This trial points out the problem of carrying over into the clinic a procedure that, while shown to be effective in animals, may fail to work in patients. [*See* Muscle Development, Molecular and Cellular Control.]

G. Cells of the Central Nervous System

Gene therapy may eventually have a role in the treatment of neurological disorders, including epilepsy, Alzheimer's disease, trauma, and Parkinson's disease. Although dividing embryonal neuronal cells in animals have been transduced with retroviral vectors, either by direct *in vivo* gene transfer or by *ex vivo* transduction and cell implantation, their nondividing adult counterparts are not amenable to retroviral-mediated gene transfer. As discussed earlier, modified herpes virus vectors have been used experimentally to introduce foreign genes into neuronal cells. In addition, retroviral-transduced fibroblasts secreting nerve growth factor have been engrafted successfully into the brains of animals.

V. GENE THERAPY OF SPECIFIC DISEASES

A. Adenosine Deaminase Deficiency

In a clinical test of this approach in patients, investigators in Italy, Holland, at Children's Hospital in Los Angeles, and at the NIH treated 10 children with severe combined immunodeficiency (SCID) secondary to ADA deficiency (Table IV). In the absence of ADA, an enzyme necessary for cellular purine metabolism, there is an accumulation of toxic metabolites that leads to the destruction of lymphocytes, particularly T cells. The loss of lymphocytes results in defective cellular and humoral immunity and in life-threatening infections. ADA deficiency has a number of features that made it an attractive gene therapy candidate. First, normal ADA gene expression is under simple regulation (i.e., there is no complex feedback). Second, a wide range of expression is associated with normal cellular function. Third, gene-corrected lymphocytes theoretically would possess a survival advantage over unmodified cells, permitting the ADA-expressing cells to expand their numbers *in vivo*. Consequently, there would be no need to administer chemotherapy to the patient "make room" for the gene-corrected cells in the blood and bone marrow. Finally, hematopoietic cells are easily obtained and manipulated *ex vivo*.

In the NIH and Milan studies, peripheral blood lymphocytes (PBL) were chosen for therapy for a number of reasons: (1) PBL are long-lived, attested by the persistence of antimicrobial immunity for many years; (2) these cells are easily obtained by a procedure called apheresis; and (3) their numbers can be expanded rapidly using a combination of an antibody (OKT-3) and a cytokine (IL-2). However, to provide a lifelong source of gene-corrected cells, hematopoietic stem cells, the cells that give rise to the cellular components of blood and the lymphoid system, would have to be transduced by the retroviral vector.

Three of the groups (Milan, Los Angeles, and Holland) attempted to introduce the ADA gene into repopulating stem cells using retroviral vectors. In Los

TABLE IV
ADA Gene Therapy Trials

Institution	Number/age of patients	Cells transduced	Total number of treatments	Percentage of cells positive for the ADA gene more than 1 year after therapy completed	ADA activity after gene therapy treatment
NCI	Two/4, 9 years	Peripheral blood lymphocytes (PBL)	11 over 2 years 12 over 1.5 years	≤1% (PBL)	Increase in circulating T lymphocytes in one of two patients
USC	Three/4 days	CD34$^+$ cells[a] from umbilical cord blood	Single infusion	~1% of CD34$^+$ bone marrow cells	Increased only in selected[b] bone marrow cells
Milan	Two/2 years	PBL and unfractionated bone marrow hematopoietic cells	9 over 2 years 5 over 10 months	2–5% of PBL; 17–25% of bone marrow progenitor cells	Increased in both selected[a] and unselected blood and bone marrow cells
Holland	Three/ages not reported	CD34$^+$ cells from bone marrow	Single infusion	Not reported[c]	Not reported

[a]CD34$^+$ cells are an enriched population of blood and bone marrow cells and include the stem cells that give rise to all the hematopoietic lineages (i.e., myeloid, erythroid, and lymphoid cells).

[b]Cells were selected using the toxic neomycin analog G418. Expression of the *neo*^{R} gene contained in the retroviral vector allows transduced cells to survive in G418.

[c]At 6 months, bone marrow cells in one of three patients were positive for the transferred gene using PCR analysis.

Angeles, ADA-deficient infants received gene-corrected $CD34^+$ cells that had been obtained from their umbilical cord blood, a rich source of these cells. As discussed earlier, the stem cell-containing $CD34^+$ cells are named for the molecule that is expressed on their surface. By concentrating these cells using an anti-CD34 antibody, it is possible to achieve a higher ratio of viral particles per stem cell and, hence, more efficient gene transfer.

Before undergoing the gene transfer procedure, the ADA-deficient children began to receive ADA enzyme replacement therapy consisting of the ADA protein stabilized with polyethylene glycol. This PEG–ADA preparation had been shown to be effective in increasing the number of circulating lymphocytes and of improving the immune function of these patients. PEG–ADA therapy also facilitated the collection of sufficient numbers of T lymphocytes for gene transfer. Investigators at the NIH and in Milan administered ADA gene-modified cells over the course of 1 to 2 years. The procedure was found to be safe and resulted in the stable transfer of the ADA gene, albeit at low levels (Table IV). Moreover, patients who underwent the gene transfer procedure had a significant improvement in their immune function that was associated in most cases with a sustained rise in the level of ADA enzyme activity in the blood. Although encouraging, these results must be interpreted with caution, as the patients have continued to receive PEG–ADA throughout the period following the administration of gene-modified lymphocytes.

Blood and bone marrow cells from neonates who were treated with gene-modified umbilical cord blood cells were found to contain and to express the introduced ADA gene for at least 1.5 years (Table IV). So far, evidence of clinical benefit has been lacking. While much remains to be done to improve the level of efficiency of gene transduction, these studies have demonstrated the feasibility and safety of carrying out retroviral-mediated gene transfer with both lymphocytes and hematopoietic stem cells.

B. Familial Hypercholesterolemia

Familial hypercholesterolemia is a disorder resulting from a mutation in the gene that encodes for the LDLR in hepatocytes. This results in an ability of the liver to clear cholesterol from the blood and an accelerated form of coronary artery disease. Investigators at the University of Pennsylvania carried out an *ex vivo* gene transfer procedure whereby hepatocytes obtained by a partial hepatectomy were cultivated and transduced with a retroviral vector encoding the LDLR gene. Following the administration of the gene-modified cells, several patients exhibited a lowering of serum cholesterol levels. While demonstrating the feasibility of the *ex vivo* gene transfer procedure, this study has also pointed to the need for a more efficient and less invasive method of liver cell gene transduction.

C. Cystic Fibrosis

Just 5 years after the cloning of the *cystic fibrosis transmembrane regulator* (CFTR) gene, whose inactivation by genetic mutation results in the disease, six groups of investigators were already carrying out clinical trials attempting to restore a functional CFTR gene to the respiratory epithelial cells of cystic fibrosis (CF) patients. The CFTR gene encodes for an adenosine 3′,5′-monophosphate (cAMP)-regulated chloride channel that is necessary for maintaining both the hydration and the clearing of lung secretions. CFTR dysfunction leads to chronic airway infection that eventually results in respiratory failure and death. Following extensive studies in animals that demonstrated the utility of adenoviral-mediated gene transfer directly into the lung, a number of protocols were initiated to determine the feasibility of using adenoviral vectors to introduce the CFTR directly to the cells lining the nasal or lung passages of CF patients. Preliminary results from these clinical trials have indicated that limited gene transfer did occur, but was short-lived, apparently as a result of inflammatory responses to the adenoviral vector. The focus of current studies is on strategies to evade the host antiviral immune responses that have limited both gene transfer and prolonged expression of the CFTR gene. [*See* Cystic Fibrosis, Molecular Genetics.]

D. AIDS

More than a dozen gene transfer trials have been approved in the United States for the treatment of patients infected by the human immunodeficiency virus-1 (HIV-1). While several of these studies involved genetically marked cells to track lymphocytes that are infused as a form of immunotherapy, the majority of clinical protocols are intended to test a number of novel strategies for the therapy of HIV infection. Introduction into cells of a *transdominant negative* gene that codes for a protein that interferes with viral replication has been shown experimentally to inhibit HIV infection. Other approaches have in common a means of interfering with the expression of essential HIV genes and include (a) either *antisense* or *ribozyme* RNA capable of inactivating viral genes; (b) RNA molecules that function as *decoys*, or intracellular sinks, for key viral proteins; and (c) a gene coding for an *intracellular antibody* against a viral protein. [*See* Acquired Immune Deficiency Syndrome, Virology.]

E. Cancer

Cancer gene therapy protocols may be divided broadly into the categories of *gene marking* and *gene therapy*. Gene-marking studies are intended to facilitate the evaluation of cellular therapies for both cancer and AIDS. With the PCR assay's sensitivity of detecting as few as 1 gene-marked cell in 100,000, the fate of cells, whether circulating in the blood or having settled in tissues, can be determined for many months following their administration. Investigators at the NIH first demonstrated the feasibility of gene marking by tracking the fate of lymphocytes administered as a form of immunotherapy for the treatment of melanoma. As discussed earlier, a study in children demonstrated that leukemia-contaminating transplanted bone marrow contributed to disease relapse.

Gene transfer with therapeutic intent comprises the majority of protocols for patients with cancer. The different therapeutic approaches are summarized in Table III.

By borrowing a strategy used by many tumor cells, gene transfer may be used to prevent bone marrow toxicity caused by chemotherapeutic agents. The *multidrug resistance* (MDR) *gene* encodes for a membrane protein that ordinarily protects normal cells by pumping toxins out of them. However, overexpression of the gene in tumor cells results in the acquisition of resistance to a variety of commonly used anticancer drugs. Transduction of normal bone marrow cells with the MDR gene as a means of *chemoprotection* is currently being investigated in clinical trials for a number of tumors.

Transgenic *immunotherapy* involves gene transfer for the purpose of inducing antitumor immune responses. This approach is based on the observation that tumor cells, often expressing potentially immunogenic proteins, appear to induce a state of immune tolerance that enables the tumor to evade immune destruction. The underlying hypothesis is that local cytokine delivery may improve tumor antigen presentation and activation of immune effector cells, including natural killer cells, cytotoxic T-lymphocytes, neutrophils, and macrophages. In transgenic immunotherapy, this is achieved by inserting immune-activating genes into tumor cells themselves or into adjacent fibroblasts. Numerous studies in animals have demonstrated the feasibility of this strategy, with the generation of local and/or systemic antitumor immunity. Tumor antigenicity may be enhanced, for example, by introducing a foreign major histocompatibility complex (MHC) gene to activate the immune system or by introducing the interferon-γ gene for improving tumor antigen presentation to the immune system. Alternatively, the genes for cytokines such as IL-2 and IL-4 may be inserted into tumor cells in order to activate immune effector cells.

Because rapidly growing or bulky tumor may overcome the ability of the immune system to mount an effective antitumor response, immunologic gene therapy may be most useful (i) in an adjuvant setting following cytoreduction with surgery, radiation, and chemotherapy, and (ii) for immunizing individuals at risk for cancer.

Sensitization of tumor cells to chemotherapy using a method called *pro-drug activation* is another kind of cancer gene therapy. Herpes simplex virus thymidine kinase (HSVTK) renders mammalian cells susceptible to killing by the antiviral agent ganciclovir due to the ability of HSVTK to phosphorylate the drug to the active, toxic compound. Subsequently, a number of investigators discovered that a "bystander effect" existed, whereby cells that were not expressing the HSVTK gene themselves were also killed by ganciclovir. Although the mechanisms of bystander killing largely remain to be elucidated, they appear to include the direct transfer of activated ganciclovir from one cell to another, activation of the immune system by dying cells, and damage to the tumor blood supply caused by the release of tumor necrosis factor-α. The bystander phenomenon has important implications for cancer therapy, suggesting that transduction of all the cells in a tumor, not achievable with currently available gene transfer methods, may not be necessary for tumor eradication. This approach is currently being tested in numerous protocols for a variety of different tumors.

The correction of commonly occurring genetic abnormalities, involving either the activation of oncogenes or the loss or mutation of tumor suppressor genes, is being investigated using *antisense, intracellu-*

lar antibody, or *suppressor gene replacement* therapy, respectively, as summarized in Table II.

F. Gene Therapy of Cardiovascular Disease

Investigators in Boston are using gene transfer into the vascular wall to treat two common cardiovascular disorders: restenosis following coronary artery angioplasty and peripheral vascular disease. In both instances, a plasmid containing a gene encoding an angiogenesis factor, *vascular endothelial growth factor* (VEGF), is applied to the wall of the diseased blood vessel using an angioplasty balloon catheter coated with a hydrogel polymer containing the angiogenic substance. The expression of the VEGF molecule in the vessel following angioplasty is intended to foster repair of the damaged endothelial cell monolayer, thereby averting the inflammation that precipitates reobstruction. In the case of peripheral leg ischemia due to peripheral vascular occlusion, VEGF is used to stimulate angiogenesis to improve the blood supply and to restore oxygen delivery to the tissues. One patient suffering from leg ischemia treated in this fashion was reported to have experienced an improvement in blood flow that was associated with an increase in the number of collateral blood vessels in the affected extremity. These elegant studies, incorporating site-directed *in vivo* gene transfer, hold the promise for nonsurgical remedies to a variety of cardiovascular disorders.

G. Other Diseases

In addition to cystic fibrosis, other single gene diseases are being treated using gene replacement therapy (Table II). For example, in *Gaucher disease,* where the accumulation of lipid in macrophages leads to organ enlargement, bone damage, and, in the juvenile form of the disease, neurologic impairment and mental retardation, a number of trials are underway with the intention of restoring glucocerebrosidase enzyme activity in hematopoietic cells. [*See* Gaucher Disease, Molecular Genetics.]

The treatment of multifactorial diseases, such as cancer, requires strategies that go beyond simple gene replacement. *Rheumatoid arthritis,* for example, is being treated by attempting to interfere with a molecule that plays a key role in the inflammatory process that destroys joints, interleukin-1. By inserting the *interleukin-1 receptor antagonist* gene into the synovial cells that line affected joints, it is hoped that inhibition of the pro-inflammatory protein, interleukin-1, will ameliorate the destructive process.

VI. SAFETY AND APPROVAL OF CLINICAL GENE TRANSFER PROCEDURES

Concerns about retroviral vector safety center primarily on (a) the possibility of activation of cellular protooncogenes by the nearby insertion of the retrovirus, resulting in uncontrolled cell proliferation and tumor formation; and (b) transduction of the reproductive or germ line cells, with transmission of the transgene as an inherited trait in all subsequent generations. The likelihood of these untoward events occurring is very small provided continued retroviral infection due to the presence replication-competent virus ("helper virus") does not occur. This problem has been ameliorated through the use of carefully designed retroviral packaging cells and stringent surveillance for the presence of helper virus. However, the concern over the potential for malignant transformation was borne out in a study carried out by a group of NIH investigators. They discovered a high frequency of lymphomas in primates that had undergone bone marrow transplantation with cells that had been exposed to a helper virus-containing retroviral vector preparation. Helper virus was found to have integrated in the DNA of the tumor cells. Careful screening of vector stocks for the presence of helper virus should prevent this complication from occurring in patients.

As mentioned earlier, adenoviruses may lead to the death of infected cells due to the inflammation that develops in response to viral protein expression. In addition, "first generation" adenoviral vectors, those with deletions of the viral E1 and E3 genes, have been shown to elicit responses in both the cellular and the humoral arms of the immune system. The result is the production of neutralizing antibodies that clear the virus before gene transfer can occur as well as the elimination of vector-infected cells. Further modifications to adenoviral vectors, particularly the deletion of the adenoviral E4 or E2 region, may succeed in reducing or eliminating these problems. Although no significant toxicity has been observed so far with either liposomal or receptor-mediated gene transfer, experience with these vectors is limited. It is certain that the safety of gene transfer will continue to be carefully studied by investigators and federal regulators alike.

Until recently, all gene therapy protocols were required to undergo an extensive review process that is intended to protect the rights and ensure the safety of the patient. Beginning with local institutional review boards, the review process has included a thorough review by both the FDA and the Recombinant DNA Advisory Committee (RAC) of the NIH, which passed on its recommendations to the NIH director for final approval. In order to streamline a process that often took as much as 2 years to complete, and since most protocols utilize similar gene transfer technologies, heretofore the RAC will review only proposals that contain truly novel gene transfer procedures or vectors. The FDA, as part of its normal regulatory function, now provides sole review for most gene therapy proposals. These changes should expedite the implementation of clinical trials and avoid the lengthy delays that frequently existed before patients could receive these therapies.

VII. FUTURE DIRECTIONS

In 1995, a special panel was convened at the request of the director of the NIH to assess the status of gene therapy and to provide recommendations for future research. Among the panel's findings was the conclusion that, despite the great promise of gene therapy and the proliferation of numerous protocols, there has been no definite evidence of clinical efficacy, with significant shortcomings in the currently available vectors. In addition, there was felt to be a lack of sufficient understanding of the basic pathophysiology of diseases, as well as a failure of many of the clinical gene therapy protocols to assess adequately the efficacy of the gene transfer procedure. However, due to the limitations of many animal models of disease, clinical studies were felt to be a legitimate and necessary means of moving the field forward. Finally, the panel found that too often the results of gene therapy studies have been exaggerated, resulting in unrealistic expectations and disappointments that threaten to undermine further support for gene therapy development.

Consequently, the panel recommended that efforts should be focused on optimizing gene transfer and expression, including developing the means of expressing genes in specific tissues and in a regulated fashion. The panel also felt it necessary to place a greater emphasis on research into disease pathogenesis, i.e., the mechanisms of diseases, in order to formulate rational gene therapy strategies. Furthermore, the panel urged that higher standards should be imposed on clinical protocols and that investigators and journalists alike exercise restraint in reporting new findings. Finally, the panel concluded that the current investment in gene therapy research is adequate and recommended that gene therapy research proposals be subjected to the same high standards of peer review that are demanded of other forms of biomedical research.

The panel's report fairly reflects what has been achieved since the first clinical gene transfer protocol was initiated at the NIH. Although the premise underlying gene therapy remains sound, i.e., the use of genetic engineering for the treatment of diseases, the success of gene transfer in preclinical studies has not yet carried over into the clinic. Despite the absence of dramatic cures, much has been learned from the successes and the failures of the initial clinical studies. Based on data from these early trials, investigators are now developing new vectors that will be tested in both preclinical and clinical gene therapy studies. Given the complexity of the diseases being targeted with gene transfer, it is hardly surprising that a greater understanding of disease pathophysiology and better methods of introducing and expressing recombinant genes in the body are required. Through the application of well-established principles of scientific investigation, gene therapy should, in the words of the panel, fully realize its "extraordinary potential, in the long term, for the management and correction of human disease."

REFERENCES

Brenner, M. K. (1996). Gene transfer to hematopoietic cells. *N. Engl. J. Med.* **335**, 337.

Brenner, M. K., and Moen, R. C. (eds.) (1996). "Gene Therapy in Cancer." Dekker, New York.

Crystal, R. G. (1995). Transfer of genes to humans: Early lessons and obstacles to success. *Science* **270**, 404.

Friedmann, T. (1996). Human gene therapy: An immature genie, but certainly out of the bottle. *Nature Med.* **2**, 144.

Ledley, F. D. (1995). Nonviral gene therapy: The promise of genes as pharmaceutical products. *Hum. Gene Ther.* **6**, 1129.

Mulligan, R. C. (1993). The basic science of gene therapy. *Science* **260**, 929.

Orkin, S. H., and Motulsky, A. G. (1995). "Report and Recommendations of the Panel to Assess the NIH Investment in Research on Gene Therapy." NIH, Bethesda, MD (the report can be found on the worldwide web at http://www.nih.gov/news/panelrep.html).

Rosenfeld, M. A., and Collins, F. S. (1996). Gene therapy for cystic fibrosis. *Chest* **109**, 241–252.

Wilson, J. M. (1996). Adenoviruses as gene-delivery vehicles. *N. Engl. J. Med.* **334**, 1185.

Genetically Engineered Antibody Molecules

SHERIE L. MORRISON
University of California, Los Angeles

GLOSSARY

Antibody-dependent cellular cytotoxicity (ADCC) Cell-killing reaction in which Fc receptor bearing killer cells recognize target cells via specific antibodies

Bacteriophage Viruses that infect bacteria. Bacteriophage λ is a temperate phage that can grow lytically where it lyses the bacteria and forms a clear plaque on a lawn of bacteria. Infection with filamentous phages such as M13 is not lethal and instead of lysing, the growth rate of the host bacteria slows and they form turbid plaques on the bacterial lawn

Chimeric Assembled from diverse sources not normally found associated

Complement Group of serum proteins important for the lysis of foreign cells and pathogens; they also play an important role in phagocytosis

Complement dependent cytotoxicity (CDC) Cell killing mediated using complement

Constant region Portion of the antibody molecule exhibiting little variation and determining the isotype of the antibody

Drug resistance Ability to grow in the presence of drugs that are normally toxic

Fc Portion of the antibody responsible for binding the antibody receptors on cells and activating complement

Glycosylation Attachment of carbohydrate (sugar) residues; proteins containing carbohydrate residues are glycoproteins

Hybridoma Cell derived by fusion of a normal cell, usually a lymphocyte, with a tumor cell

Isotype Class of the antibody

Monoclonal antibody Homogenous antibody derived from a single clone of antibody-producing cells

Plasmid Circular DNA segment that replicates extrachromosomally in bacteria

Spheroplasts Bacteria whose cell walls have been removed by treatment with lysozyme

Transfection Introduction of foreign DNA into a cell; the foreign DNA may be transiently expressed or integrated into the chromosome and stably replicated and expressed

Variable region Variable portion of the antibody molecule that is responsible for antigen binding

Vector Piece of DNA, usually from a plasmid or virus, used to deliver genetic information to a cell by transfection

ANTIBODIES HAVE LONG BEEN APPRECIATED FOR their exquisite specificity in binding the antigenic determinant they recognize. Because of this specificity, antibodies seemed the ideal "magic bullet" for targeting diagnosis or therapy to specific cells. However, even the monoclonal antibodies produced by hybridoma cell lines have some properties that make them less than ideal for this purpose: (1) It has proven difficult to produce monoclonal human antibodies, and (2) the isotype of the resulting antibodies often is inappropriate for the desired biologic properties. Genetically engineered antibodies produced by expressing cloned genes in the appropriate cell type provide an approach for producing antibodies with superior properties. It is now possible to produce chimeric antibodies with gene segments derived from diverse sources. Variable regions can be expressed joined to constant regions from either the same or different species; constant regions with improved biologic properties can be produced. Bacteriophage expression systems hold the promise of being able to produce specific human antibodies in the absence of animal immunization. Genetically engineered mice are also available which produce specific antibodies that are totally human in sequence. In addition, molecules that bind antigen but are never found in nature (e.g., fusions of antibodies with nonantibody proteins) can be manufactured, as can antibodies that function as enzymes.

I. PROPERTIES OF ANTIBODY MOLECULES

Antibodies are among the most versatile of proteins. They are designed to achieve specificity of binding so that they can distinguish foreign substances from the naturally occurring components of the body. They are also capable of inducing biologic activities that can destroy and eliminate undesirable substances such as bacteria.

All antibodies have a similar structural organization and are formed by polypeptide chains held together by noncovalent forces and disulfide bridges. In the basic structure, two pairs of identical heavy (H) and light (L) chains form a bilaterally symmetric structure (Fig. 1). The polypeptide chains fold into globular domains separated by short peptide segments. The H chain has four or five domains depending on the isotype. The L chain has two domains. The N-terminal domains of each chain constitute the variable region that carries the antigen-combining site and determines the specificity of the antibody. The remainder of the antibody constitutes the constant region, which is responsible for the effect of the antibody in the body.

Antibodies of the same specificity can have different H-chain constant regions and, therefore, exhibit different effects. In humans the different constant regions produce antibodies of different isotypes: IgM, IgD, IgG1–4, IgA1, IgA2, and IgE. IgM, IgA1, and IgA2 differ from the other isotypes in that they are constituted of multiples of the basic H_2L_2 structure and contain an additional polypeptide chain, the J chain. They can also be secreted onto the mucosal surfaces associated with secretory component (SC), a product of the epithelial cell. There are also two different isotypes of light chains, κ and λ. The L-chain isotypes do not appear to influence the effect of the antibody molecule. All antibodies are glycoproteins with the carbohydrate content varying among different isotypes. The carbohydrate present in the constant region of IgG antibodies has been shown to be essential for many of its effects.

Antibodies are unusual proteins in that the genes that encode them must be assembled after birth to produce a functional protein. Hundreds of different H- and L-chain variable region genes are present in the genome. For a variable region to be expressed, it must be positioned next to a J segment for the L chain or a DJ segment for the H chain by a DNA rearrangement. The variable regions of the chains are expressed first with an IgM constant region, but as the immune response matures, further DNA rearrangements occur at the H-chain locus, and the variable regions are expressed with different constant regions (Fig. 2). In the case of the L chains, the variable region of the κ chain is rearranged first. As soon as a functional κ-chain protein is produced, rearrangement stops. If both κ-chain alleles have rearranged but failed to produce a functional protein, rearrangement at the λ locus occurs. Each antibody-producing cell produces only one functional H chain and one functional L chain, a phenomenon called *allelic exclusion*. It should be noted that the domain structure seen in the antibody molecule is reflected in the structure of the genes that encode it (Fig. 1).

During the normal immune response, a wide variety of antibodies are produced. These include antibodies with different variable regions, which recognize the same antigen, and antibodies with the same variable region associated with different constant regions. Different individuals will make different immune responses. This heterogeneity in the immune response has made it difficult to use antisera for many applications. [*See* Immune Surveillance.]

A significant breakthrough was made when it became possible to produce *hybridoma antibodies*. Hybridoma antibodies are *monoclonal* (i.e., they are the product of a single antibody-producing cell) and are therefore homogeneous, with a single variable region associated with only one constant region. Although these homogeneous antibodies have many advantages, they still have some inherent limitations, as

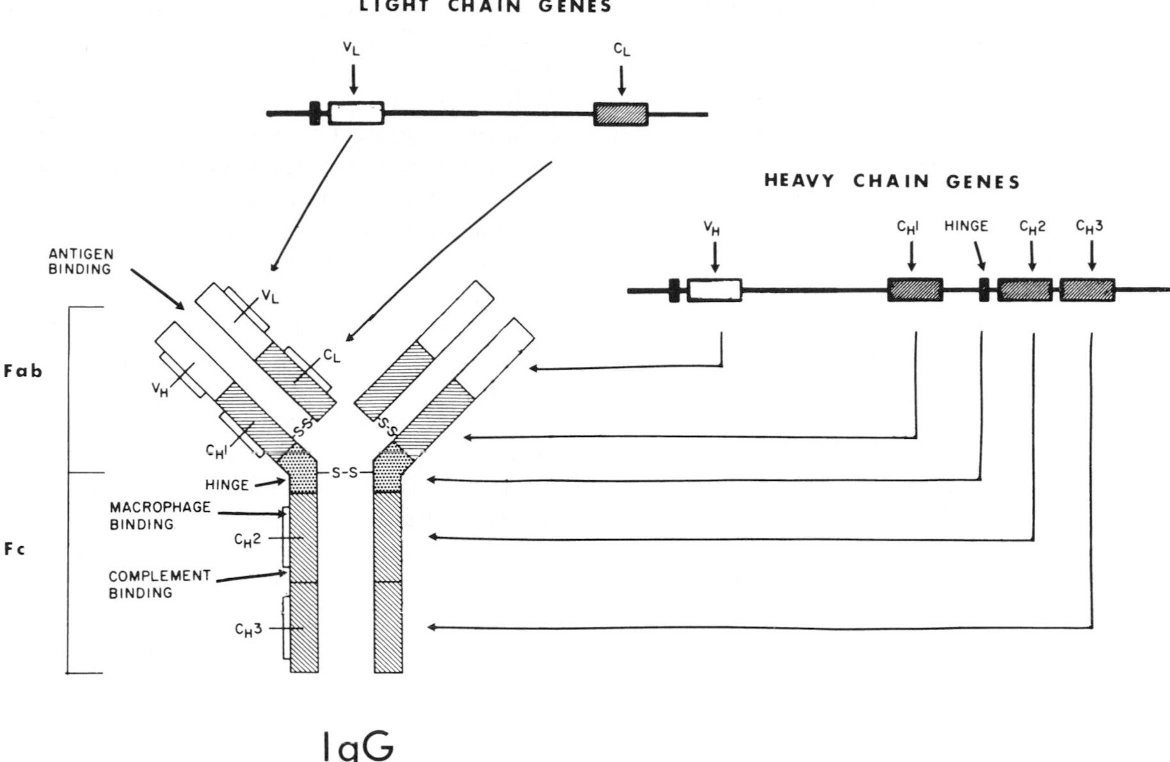

FIGURE 1 Diagram of an antibody molecule and of the genes that encode the heavy and light chains. The antibody molecule is divided into discrete functional domains: two domains (V_L and C_L) constitute the light chains, whereas five domains (V_H, C_H^1, hinge, C_H^2, and C_H^3) make up the heavy chain. The variable region domains (V_L and V_H) make the antibody-combining sites and are designated the Fv region. The effector functions of the molecule (e.g., the ability to activate complement or bind to cellular receptors) are properties of the constant region domains. The hinge provides flexibility in the antibody molecule, facilitating antigen binding and some effector functions. In the genes, each domain is encoded by a discrete exon (indicated by boxes), separated by intervening sequences (introns) indicated by the line; the intervening sequences are present in the primary transcript but are removed from the mature mRNA by splicing. The heavy and light chains both contain hydrophobic leader sequences (indicated by the black exon) necessary for their secretion. This leader sequence is present in the newly synthesized heavy and light chains but is cleaved from them after they enter the endoplasmic reticulum and therefore is not present in the mature antibody molecule. The enzyme papain cleaves the molecule into an Fab fragment containing the antibody-binding site and an Fc fragment. [Modified from S. L. Morrison (1985). *Science* **229**, 1202–1207, with permission.]

most are of rat or mouse origin. It has been proven more difficult to produce monoclonal human antibodies useful for *in vivo* immunotherapeutic applications. In addition, hybridoma antibodies are homogeneous not only with respect to their variable regions, but also with respect to their constant regions and hence might not have the desired effects. In addition, they may not possess the exact desired binding specificity or affinity. [*See* Monoclonal Antibody Technology.]

One approach to producing improved antibody molecules is to use recombinant DNA techniques to produce antibodies with improved antigen-binding specificities and effects. An advantage of this approach is that limitations are not placed on producing antibodies as they exist in nature. Instead, antibody molecules with improved properties such as binding specificities, pharmacokinetics, and effector functions such as complement activation and Fc receptor binding can be produced. Additionally, novel functions can be introduced into the antibody molecule not normally found there. The availability of these antibody molecules promises to revolutionize our ability to use antibody molecules for diverse applications.

Genetically engineered antibody molecules can be produced in bacteria, yeast, insect cells, plants, and mammalian cells. To produce a functional antibody, proper assembly and proper glycosylation must take place. These requirements and the need for the anti-

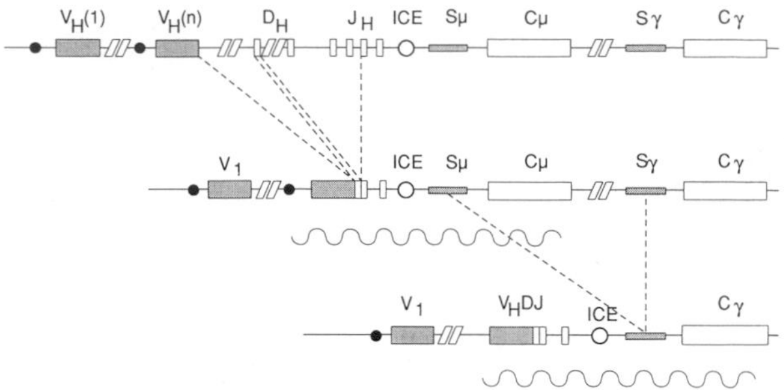

FIGURE 2 Structure of the heavy-chain locus before expression (top), during expression of an IgM protein (middle), and after switching to production of an IgG antibody (bottom). The locus has been simplified to show only one IgG gene. ICE designates the enhancer that is found in the intervening sequence. Before expression the V_H, D_H, and J_H exist as discrete segments of DNA (top). To produce a functional heavy-chain gene, a V must be joined to a D and J_H segment of DNA; the DNA segments between the joined elements are usually deleted from the genome. The first heavy chain produced during B-cell development is of the IgM isotype (middle). As B-cell development proceeds, class switching occurs by DNA, rearrangement, and antibodies are produced with a different constant region joined to the same variable region. In the example, an antibody of the IgG isotype is produced (bottom).

body to be secreted determine the choice of the cell type used to produce the antibody molecule.

II. ANTIBODY PRODUCTION IN BACTERIA AND YEAST

Expression of gene products in bacteria has a great deal of appeal. Bacteria can be grown in large quantities at relatively little cost and can be used to produce large quantities of proteins efficiently. However, antibodies are glycoproteins, and because bacteria do not glycosylate proteins, any function of the antibody molecule that depends on glycosylation will not be exhibited by antibodies synthesized in bacteria. In addition, production of a functional antibody molecule requires correct assembly and disulfide bond formation between the H and L chains, which is frequently difficult to obtain in bacteria.

Attempts to generate intact functional antibodies by expression in bacteria have met with limited success. The majority of the H and L chains produced in bacteria end up as insoluble material that accumulates in inclusion bodies. Attempts to renature these insoluble products resulted in antibodies exhibiting minimal function.

Bacterial expression of antibody fragments has been more successful. Although cytoplasmic expression of antibody in *Escherichia coli* results in the formation of insoluble and inactive protein aggregates (inclusion bodies), which require refolding and renaturation to be active, antibody fragments supplied with leader sequences can be expressed and secreted in fully functional forms. When two chains are made in the same cell, they can assemble into functional binding sites. Clearly, antibody fragments produced in bacteria cannot at this time be used for applications that require an intact antibody molecule or any of the functional properties associated with its glycosylation.

Antibodies and antibody fragments have also been produced in yeast, which glycosylate their proteins. However, the added carbohydrate differs in structure from the carbohydrate added by mammalian cells. Antibody-dependent cellular cytotoxicity (ADCC) and complement-dependent cytotoxicity (CDC) are two effector functions that are not exhibited by antibodies lacking carbohydrate. Antibodies possessing the yeast carbohydrate were able to mediate ADCC but were unable to mediate CDC. Therefore, antibodies require carbohydrates similar in structure to those added by mammalian cells to exhibit their full range of biologic functions.

III. GENE TRANSFECTION OF MAMMALIAN CELLS

Expression of genes in mammalian cells such as myeloma and hybridoma cell lines and Chinese hamster ovary cells, which efficiently express, glycosylate, assemble, and secrete functional antibody molecules, provides an alternative to expression in bacteria and yeast. The introduced genes, however, are not usually expressed at a high level.

Several methods are available for introducing foreign DNA into mammalian cells. $CaPO_4$-precipitated DNA is effective in fibroblasts but not in lymphoid cells. Protoplast fusion in which bacteria containing the genes of interest are converted to spheroplasts using lysozyme and fused to the cell to be transfected provides one approach. However, protoplast fusion is technically difficult. Electroporation and liposome-mediated transfection provide two more tractable methods for transfection. For electroporation, the cells are suspended in a solution of DNA containing the genes to be expressed and are then subjected to an electrical pulse which makes pores in the cells and enables them to take up and express the DNA. For liposome-mediated transfection, DNA complexed with a liposome suspension is added directly to cells which are able to take up and express the DNA. Using these approaches, transfection frequencies between 10^{-3} and 10^{-6} can be routinely achieved and are sufficient in isolating the desired transfectant cell line. Myeloma cell lines that do not produce antibodies are routinely used.

IV. VECTORS USED FOR TRANSFECTION

Even under optimal conditions, gene transfection into eukaryotic cells is an inefficient procedure. Vectors containing selectable markers are therefore used to select rare, stably transfected cell lines. The most commonly used vectors to date are based on the pSV2 vectors shown in Fig. 3 and contain three essential elements: a plasmid origin of replication, a gene encoding a biochemically selectable phenotype in bacteria, and a gene encoding a biochemically selectable

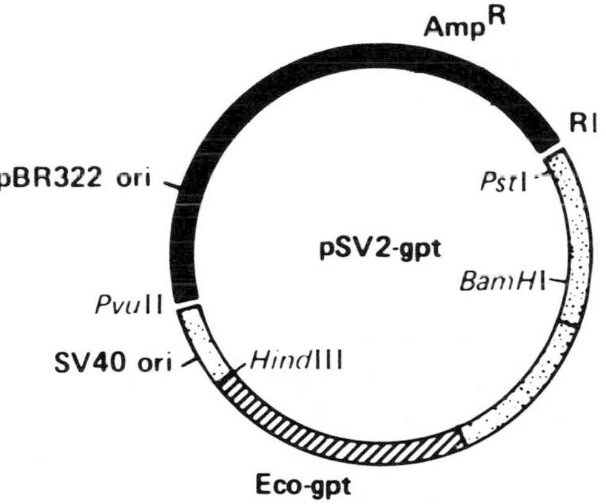

FIGURE 3 The pSV2-*gpt* vector used for transfecting eukaryotic cells. The vector is a plasmid, a closed circle of double-stranded DNA. Contained within the vector are sequences, indicated by the black region, derived from the plasmid pBR322. These consist of the origin of replication and the β-lactamase gene, which provides resistance to the antibiotic ampicillin. The gene providing a marker selectable in eukaryotic cells, in this case, *gpt,* the expression that enables eukaryotic cells to grow in mycophenolic acid, is indicated by the hatched area and was derived from the bacterium *E. coli.* DNA segments derived from the virus SV40 are indicated by stippling. These include the SV40 ori, which contain the early promoter used for expressing the *gpt* gene and splice and poly A addition sites located 3′ of the *gpt* gene. The *Eco*R1, *Pst*I, and *Bam*HI restriction endonuclease cleavage sites are located in regions of DNA not necessary for function in prokaryotic or eukaryotic cells and provide convenient sites for insertion of genes of interest. [From S. L. Morrison and V. T. Oi (1984). *Annu. Rev. Immunol.* **2,** 239–256, with permission.]

marker in eukaryotic cells. When the objective is to produce proteins for subsequent analysis and use, stable transfectants are produced that contain the vectors stably integrated into the chromosome of the cell.

It is essential that the vectors be propagated as plasmids in bacteria to obtain DNA in sufficient quantities for genetic manipulation. To achieve this objective, some plasmids contain the origin of replication and the β-lactamase gene from the plasmid pBR322, which confers resistance to ampicillin on the bacteria propagating these plasmids. Other vectors contain the chloramphenicol acetyl transferase gene, which confers resistance to chloramphenicol. With both vectors it is easy to obtain large quantities of DNA for *in vitro* manipulation.

Because only rare animal cells (10^{-3}–10^{-6}) incorporate and stably express exogenous DNA, it is also necessary to include biochemical markers that can be used to select this minor cell population. The most commonly used markers are drug-resistant markers, connected to a viral (SV40) promoter, splice junction, and poly(A) addition signal so that they can be expressed in eukaryotic cells. Many selectable genes are used, including (1) the phosphotransferase gene from Tn5 transposon (designated *neo*) and (2) the xanthine–guanine phosphoribosyl transferase gene (*xgprt* or *gpt*). The product of the *neo* gene inactivates the antibiotic G418, an inhibitor of protein synthesis in eukaryotic cells. Biochemical selection with *gpt* expression is based on the fact that the enzyme encoded by this gene uses xanthine as a substrate for purine nucleotide biosynthesis, whereas the homologous eukaryotic enzyme uses only hypoxanthine. Thus, when the conversion of inosine monophosphate to xanthine monophosphate is blocked by mycophenolic acid, cells provided with xanthine can survive only if they express the bacterial *gpt* gene. Biochemical selection with G418 or mycophenolic acid depends on two entirely different mechanisms; therefore, they can be used in vectors to select independently for the expression of exogenous recombinant DNA gene segments.

When the objective is to use gene transfection to produce a functional protein it is important that a high level of production be obtained. Gene amplification is one approach used to increase expression levels, and a selectable marker that has been very useful for gene amplification is dihydrofolate reductase (DHFR). DHFR catalyzes the conversion of folate to tetrahydrofolate which is required for purine, amino acid, and nucleoside biosynthesis. The folic acid analog methotrexate (MTX) binds and inhibits DHFR, causing cell death. Cells survive in increasing concentrations of MTX by increasing their synthesis of DHFR by gene amplification.

To create functional antibody molecules, the genes encoding H and L chains must be transfected into the same cell, and both polypeptides must be synthesized and assembled. Both the H- and the L-chain genes can be inserted into a single vector and then transfected; this approach generates large, cumbersome vectors that are difficult to manipulate. A second approach is to transfect sequentially the H- and L-chain genes, using different drug-resistant markers to select for the expression of the different vectors. Alternatively, both genes can be introduced simultaneously on different DNA fragments using either electroporation or protoplast fusion. The latter approach is usually the most efficient in creating complete antibody molecules.

V. OBTAINING VARIABLE REGIONS FOR EXPRESSION

Variable regions needed to produce the desired antibody can be obtained as cDNA or genomic clones. By obtaining variable regions from human antibody-producing cell lines, it is possible to produce totally human antibodies. This approach is appropriate for rescuing low-producing human cell lines and for obtaining human immunoglobulins with the desired isotype. Classically variable regions have been obtained from cell lines identified as producing antibodies of the desired specificity and this continues to be the most common approach. However, it is now possible to identify and select bacteriophage that express variable regions of desired binding specificity. Moreover, *in vitro* mutagenesis can be used to obtain antibodies with higher binding affinities.

For a gene to be expressed in mammalian cells, it must be provided with an appropriate promoter and enhancer. The promoter is the region in which transcription of the gene is initiated; the enhancer is a regulatory element that increases the level of expression. Both the promoter and the enhancer may be obtained from either viruses or eukaryotic genes. Both promoters and enhancers can be tissue specific (i.e., they will function only in certain cell types). Therefore, the choice of promoter and enhancer must be matched with the cell type in which the genes will be expressed.

The organization of the antibody H- and L-chain genes facilitates the isolation of the desired expressed variable regions as genomic clones. Hundreds of different H- and L-chain variable regions are present in the genome. However, as described earlier, for a variable region to be expressed it must be positioned next to a J segment (for the L chain) or a DJ segment (for the H chain) (Fig. 2). Therefore, in an antibody-producing cell, the expressed variable regions can be distinguished from the hundreds of nonexpressed variable regions because of the proximity to a J segment. Molecular probes are available for the J segments, and the expressed variable region can be identified using a J-region probe without any prior information as to the sequence of the expressed variable region. An advantage in using genomic clones is that the variable region is obtained with its own promoter.

Expression of variable regions derived by cloning cDNA (i.e., DNA complementary to messenger RNAs) provides an alternative approach. The cDNAs are cloned into a plasmid and introduced into bacteria; a J-region probe is used to identify bacteria containing antibody genes. A variation on this approach has been devised using the polymerase chain reaction (PCR) technique, which allows the amplification of only the desired variable regions.

The number of available variable regions can limit the number of applications for which the recombinant antibodies can be used. Increasing attention has focused on the use of the polymerase chain reaction (PCR) as a way to more efficiently obtain variable regions. Clearly, the easier it is to acquire new variable regions, the more will be available for functional studies.

Initially, PCR primers for the amplification of specific variable regions were designed using known amino acid sequences. However, the real challenge is to be able to clone variable regions from expressing cells without prior knowledge of their sequence. Because there are relatively few constant regions, designing primers for the 3′ end of the variable region is straightforward. The design of primers for the 5′ end is more challenging but is in fact facilitated by the nature of the antibody molecules. Because there is considerable conservation of the framework residues, it is possible to synthesize a family of relatively few degenerate primers which can be used to clone the majority of the different variable regions. However, because the primers are to the framework region, this approach may introduce amino acid substitutions and these may alter antibody affinity. A second approach is to use a set of redundant primers that prime in the relatively conserved leader sequence (signal sequence). In this approach, no amino acid changes result in the cloned antibody because the leader sequence is removed from the mature antibody molecule. Vectors have been described for the expression of PCR product in both mammalian and bacterial cells.

Two approaches can be used to construct vectors that express immunoglobulin cDNA clones. In one approach, the variable regions obtained from H-chain cDNAs are substituted for the corresponding variable region in a genomic clone. This approach depends on the presence of the appropriate restriction endonuclease cleavage sites to permit the DNA exchange; these sites may either occur naturally or be engineered into the vectors and cDNA clones.

In a second, more general approach, a heterologous transcription unit is created. An entire immunoglobulin gene is created by joining a cDNA variable region with a cDNA constant region. The immunoglobulin genes are then inserted into a transcription unit containing the necessary signals. The choice of promoter and enhancer elements will depend on the cell type in which the gene is to be expressed; frequently, viral sequences are used. Using this approach, variable regions from mouse hybridomas have been expressed with both human H- and L-chain constant regions.

Oligonucleotide synthesis is becoming a more efficient and less costly process. The technology already exists for synthesizing complete immunoglobulin domains. It is conceivable that synthesis will replace cloning as the preferred method for acquiring variable regions of the appropriate sequence and combining specificity.

VI. PRODUCTION OF CHIMERIC ANTIBODIES

In Greek mythology, the chimera is a she monster with a lion's head, a goat's body, and a dragon's tail. Chimera has come to characterize molecules put together from diverse gene segments not normally found associated. Chimeric antibody genes are assembled from diverse gene segments not normally found associated. Chimeric antibodies can be intraspecies, where variable and constant regions are from the same species, or interspecies, where, for example, the variable region is of mouse origin and the constant region is of human origin. The term *chimeric antibody* also refers to hybrid gene segments derived from different antibody genes or in which a segment is assembled from more than one source. Finally, a chimeric antibody can include antigen-binding specificities joined to nonantibody protein structures.

Among the most useful chimeric antibodies are *mouse/human chimerics,* in which the variable region is derived from a murine hybridoma cell line and the constant regions are human. By using the variable regions from the murine antibodies, a wide range of antigen-binding specificities can be used. The resulting chimeric antibodies are superior to completely murine antibodies for use in humans because they have reduced immunogenicity and are better able to interact with the human immune system. Mouse/human chimeric antibodies also have advantages over the available human antibodies, which have a limited range of antigen-binding specificities, and are often of an inappropriate isotype. One procedure for the production of chimeric antibodies in mammalian cells using genomic clones is outlined in Fig. 4.

Genomic clones of immunoglobulin variable and constant region genes are easy to manipulate using genetic engineering techniques because of their division into separate coding regions, the exons. The distinct domains of an immunoglobulin polypeptide are each encoded by a discrete exon. The intervening DNA sequences (introns) separating each domain provide ideal sites for manipulating the antibody gene shown in Fig. 1; because the intervening sequences are removed by splicing when mRNA is made, alterations within them will not affect the structure of the protein. In addition, the RNA splice junction between variable and constant region exons in both H- and L-chain genes is similar, making it is easy to manipulate the Ig structure by deleting, exchanging, or alterating the order of exons. The exon structure of the immunoglobulin gene can also be exploited to construct antibody-cassette expression vectors, which are then assembled in various ways. Moreover, chimeric molecules can be constructed using cDNA fragments.

Results in many systems have indicated that it seems to be a general rule that the specificity of binding of a murine variable region is unchanged when it is expressed associated with a human constant region. It would also seem likely that human variable regions can be expressed associated with different constant regions without changing their combining specificity.

VII. ISOLATING SPECIFIC ANTIBODIES USING BACTERIOPHAGE

Expression of variable regions in bacteriophage has been developed as an alternative for obtaining specific

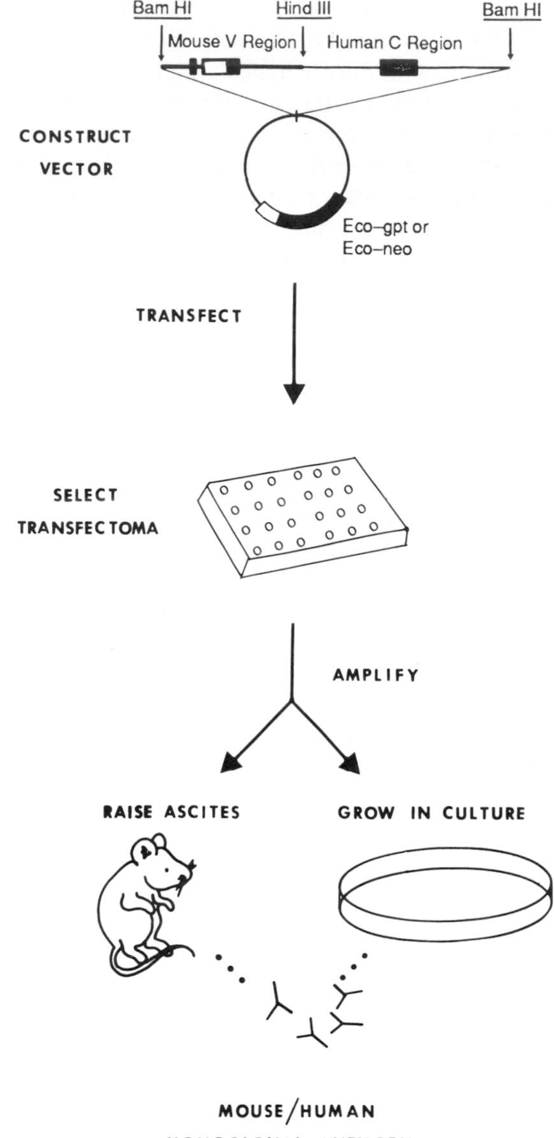

FIGURE 4 Steps in the production of a genetically engineered antibody molecule. The first step is to construct the gene of interest in the appropriate expression vector. The gene segments can be obtained from genomic or cDNA clones. In this example, the variable region from a murine hybridoma is joined to a human constant region; both were obtained by genomic cloning. The vectors are then transfected into an appropriate recipient cell; if a complete antibody molecule is to be produced, both heavy- and light-chain genes must be expressed. The transfected cells must be treated with the appropriate drug selection so that the rare cells in which the transfected genes are expressed can survive and nonselected cells will perish. The transfected cells producing the antibody of interest are then expanded and used to either raise ascites in mice or grown in tissue culture; the genetically engineered antibodies can be isolated from the ascitic fluid or culture supernatant. [From S. L. Morrison (1985). *Science* **229**, 1202–1207, with permission.]

antibodies. Although the initial expression systems were based on bacteriophage λ, those using filamentous phage have been most extensively used. One important advantage is an expanded ability to obtain human antibodies. Depending on the phage protein used for fusion, Fv or Fab fragments can be expressed on the surface of filamentous bacteriophages (f1, M13, and fd) as either single or multiple copies of the antibody of interest. The bacteriophage-expressed V region repertoires can be obtained from naive or immune lymphocytes or assembled *in vitro*. The functional V domains enable the specific phage to bind antigen. Rare phage of the desired specificity can be isolated using affinity chromatography with multiple rounds of binding and expanding the bound phage, yielding specific phage even if the desired specificity was present on less than 1 in 10^6 of the original phage in the library.

Although initial studies isolated specific clones from a library of the rearranged heavy and light chains prepared using immunized mice, more recent experiments have isolated specific antibodies from libraries prepared from peripheral blood lymphocytes of unimmunized donors. Experimental approaches which enable the construction of highly diverse libraries make it feasible to prepare extremely large libraries which should permit the isolation of high-affinity human antibodies specific for virtually any antigen from a single library. In addition, higher affinity antibodies can be selected following random mutation of the phage library.

Although binding specificities can be expressed as either single chain fusions or as Fabs, the use of Fab fragments offers advantages because heavy and light chain libraries can be produced independently and reassorted. The use of two vectors also facilitates the construction of hierarchical libraries in which a fixed heavy or light chain is paired with a library of partners. By introducing an amber mutation between the antibody chain and the coat protein it is possible to either display the antibody on phage using suppressor strains of bacteria or produce soluble fragments using nonsuppressor strains. The cloned variable regions can also be transferred to other vectors for the expression of complete antibody molecules.

VIII. PRODUCTION OF HUMAN ANTIBODIES USING MICE

Mice have been engineered to produce specific human antibodies in response to antigenic challenge. This has been accomplished by inserting elements of the human heavy- and light-chain locus into mice in which production of the endogenous murine heavy and light chain was disrupted. The resulting mice can synthesize human antibodies specific for human antigens and can be used to produce hybridomas, making human antibodies. The mice have been produced so that several different human variable regions for both the light and the heavy chain are available for expression, although the number of different available variable regions is less than in the normal human. Some mice have the ability to isotype switch, i.e., to express the specific variable region with different human constant regions, making it possible to obtain antibodies of the same specificity with differing biologic properties.

IX. USE OF RECOMBINANT ANTIBODIES TO STUDY STRUCTURAL CORRELATES OF ANTIBODY EFFECTOR FUNCTIONS

Functions such as complement fixation, ADCC, and Fc receptor binding are mediated by the H-chain constant region, different isotypes of which display different biologic properties as shown in Table I. Although some structures important for these activities have been defined, the exact molecular correlates of many antibody effector functions remain unknown. With the use of recombinant chimeric antibodies, the study of structure–function relations within the antibody molecule can now be approached systematically. It is now possible to study chimeric antibodies in which the same variable region is joined to different human constant regions. They can be used to determine the amino acid sequences responsible for specific effector functions. The following is an example.

The human IgG subclasses share extensive sequence homology throughout their constant regions with the exception of the so-called hinge region, which shows considerable variation in length and amino acid composition. One question of interest is the extent to which the hinge region modifies effector functions. It has been suggested by several groups that hinge length (i.e., the number of amino acids residues in the hinge region) determines the degree of segmental flexibility and affects complement and Fc receptor binding by IgG molecules. IgG3 has an extended hinge encoded by four exons, the last three of which are identical; it is flexible and fixes complements well. In contrast, human IgG4 has a short hinge encoded by a single exon; it is a rigid molecule and does not fix complement.

To assess the role played by the hinge, a panel of IgG3 molecules of differing hinge length was constructed and the hinge regions exchanged between IgG3 and IgG4. The experiments demonstrated that the presence of at least one of the hinge segments is essential for complement activation, consistent with the idea that an intact H_2L_2 molecule is required for maintaining complement binding activity. However, a single hinge exon is sufficient to endow IgG3 with its full capacity to activate complement, whereas lengthening the hinge to seven exons results in a mole-

TABLE I
Some Properties of Human Immunoglobulins

	IgG				IgA				
	IgG1	IgG2	IgG3	IgG4	IgA1	IgA2	IgM	IgD	IgE
Average concentration in normal serum (mg/ml)	8	4	1	0.4	3.5	0.4	1	0.03	0.0001
Molecular mass (kDa)	150	150	180	150	150–600	150–600	900	180	190
Half-life in serum (days)	23	23	8	23	6	6	5	3	2.5
Active in complement fixation-classical pathway	+++	±	+++	–	–	–	+++	–	–
Binds to high-affinity Fc receptor	+++	0	+++	+	–	–	–	–	–
Sensitizes human mast cells for anaphylaxis	–	–	–	–	–	–	–	–	+
Present in musocal secretions	–	–	–	–	++	++	±	–	–

cule with impaired ability to activate complement. Providing IgG4 with the hinge of IgG3 does not result in an antibody that can activate complement, indicating that additional structural aspects of the IgG4 molecule interfere with its ability to activate complement.

Chimeric antibodies were used to study the role of glycosylation. All IgG molecules contain a consensus glycosylation sequence (Asn-*X*-Thr-, where *X* represents any amino acid) in the C_H2 domain. The presence of this carbohydrate is important for some of the effector functions of Ig, including Fc receptor binding and complement activation. When *in vitro* site-directed mutagenesis was used to remove that glycosylation site from C_H2 of chimeric mouse/human antibodies, the resulting carbohydrate-deficient antibodies, although properly assembled, secreted, and capable of binding antigen, were more sensitive to proteases than their corresponding wild-type IgGs, and some had a shorter serum half-life in mice. Thus, an absence of carbohydrate profoundly changes some properties of the antibody molecule while leaving others intact. The precise influence of carbohydrate structure on antibody function remains to be determined.

A property of antibody molecules critical for their *in vivo* function is the ability to bind to the cellular Fc receptor. Chimeric IgG1 and IgG3 bind the high-affinity Fc receptor (FcγR1) with a high affinity; chimeric IgG4 binds with an approximately 10-fold lower affinity, and binding by chimeric IgG2 is not detectable. Using site-directed mutagenesis, residue 235 of mouse IgG2b has been pinpointed as interacting with this Fc receptor.

X. GENETICALLY ENGINEERING THE ANTIBODY VARIABLE REGION

A potential limitation to the *in vivo* use of chimeric mouse/human antibodies is the immune response that may arise to the murine variable regions. A way to minimize this response would be to make the mouse variable regions more human. When the three-dimensional structure of antibody variable regions is compared, they are found to be similar. They all contain framework regions of similar structure that determine the position of complementary-determining residues (CDRs); these in turn form the antibody-combining site. This conservation of structures makes it feasible to transfer the CDRs from one variable region to another. The CDRs from mouse monoclonal antibodies can be substituted into the framework of a human-antibody, with the resulting chimeric antibodies continuing to bind their specific antigen. The possible applications of this "variable region grafting" remain to be evaluated.

XI. CATALYTIC ANTIBODIES

One application of *in vitro* engineering techniques is the ability to produce Abs with catalytic properties. Although both enzymes and antibodies exhibit binding specificity, enzymes bind and stabilize the transition state of a reaction whereas antibodies normally bind a compound in its ground state. However, by immunization with a hapten, which is a transition-state analog, an enzyme-like binding site can be constructed complementary to the transition state. The resulting antibodies should accelerate a reaction by binding and stabilizing the transition state, the highest energy entity in the reaction pathway. Using monoclonal antibodies it is possible to distingiush antibodies that are catalytic from those which merely bind hapten and to produce the desired antibodies in quantities sufficient for exhaustive study. Although the use of transition-state analogs has been the predominant strategy for *in vitro* production of monoclonal catalytic antibodies, haptens have also been designed to use the binding energy of the antibody to reduce the entropy of activation. Haptens have also been designed which incorporate a metal cofactor. Catalytic residues and cofactor binding sites can be deliberately introduced into antigen-combining sites by site-directed mutagenesis. The ability to screen highly diverse phage libraries will also facilitate the development of better catalytic antibodies. Although considerable ingenuity has been demonstrated in the design of haptens to elicit catalytic antibodies by various means, it is too soon to predict the ultimate usefulness of catalytic antibodies as therapeutic or industrially important reagents. However, they provide useful tools in studying enzyme–substrate interactions. Although no catalytic antibodies have yet been able to accelerate reactions as well as enzymes, the past few year have seen the catalysis of more energetically difficult reactions. New strategies such as bifunctional catalytic antibodies with metal-binding sites and with transition-state complementarity will likely produce better catalytic antibodies.

XII. ANTIBODY FUSION PROTEINS

Fusion of a nonantibody protein with an antibody can be achieved in several different ways. Nonantibody sequences can be substituted for the variable region so that the fused moiety acquires antibody-associated properties such as effector functions or improved pharmacokinetics. These molecules have been called "immunoadhesins" because they contain an adhesive molecule linked to the immunoglobulin Fc effector domains. Alternatively, an enzyme, toxin, growth factor, or biological response modifier can be substituted for or joined to the constant region of the antibody. These molecules retain the binding specificity of the antibody and, depending on the position of the substitution, different antibody-related effector functions and biologic properties will be retained. Such molecules have potential use in immunoassays, in diagnostic imaging, and in immunotherapy.

Immunoadhesins in which CD4, the target molecule recognized by HIV, replaced the variable region of the heavy chain were effective in blocking cell killing and virus production. Immunoadhesins made using cytokine receptors were shown to be effective in blocking the actions of the cytokine. Immunoadhesins in which a cytokine was used to replace the variable region of the heavy chain were able to target antibody-mediated cytotoxicity to cells bearing the cytokine receptor.

The antibody-combining specificity can be used to provide specific delivery of an associated biologic activity yielding an antibody-targeted pharmacological reagent. Joining a tissue-type plasminogen activator (t-PA) to a fibrin-specific antibody resulted in a thrombolytic agent that is more specific and more potent than t-PA alone. Joining tumor necrosis factor (TNF) to an antitransferrin receptor (TfR) antibody resulted in a fusion protein with TNF cytotoxic activity toward cell lines with the TfR. Human lymphotoxin (LT) joined to the variable region of an antiganglioside antibody was cytolytic for cells sensitive to LT but not for resistant cells. Interleukin (IL)-2 is a cytokine involved in the generation of an effective immune response. The activities of IL-2 include the stimulation of T cells to proliferate and become cytotoxic as well as the stimulation of natural killer (NK) cells and the generation of lymphokine-activated killer (LAK) cells, both of which respond with increased cytotoxicity for tumor cells. IL-2 was fused with an antibody-combining specificity in an attempt to specifically deliver IL-2 and to achieve local immune activation without systemic toxicity. Studies in several laboratories indicate that these cytokine–antibody fusion proteins may indeed be effective in eliciting an antitumor immune response.

One of the challenges for effective diagnosis or therapy of human disease is the efficient and specific targeting of the active agent to the desired locale. One region of the body particularly difficult to target is the brain because drug delivery to the brain is limited by the poor transport of water-soluble drugs through the brain capillary endothelial wall which make up the blood–brain barrier (BBB) *in vivo*. In order to obtain required nutrients and factors from the blood, the BBB has specific receptors which transport compounds such as insulin, transferrin, and insulin-like growth factors 1 and 2 from the blood to the brain. These receptors provide potential vehicles for transport into the brain. Studies have indicated that antibodies fused to these growth factors are much more effective in reaching the brain than antibodies lacking associated growth factors.

Several enzymes have been joined to antibody-combining sites, including the nuclease from *Staphylococcus aureus* and the Klenow fragment of *E. coli* DNA polymerase I. The antibody continued to bind antigen and the enzymes were catalytically active, albeit with reduced specific activity.

XIII. CHIMERIC ANTITUMOR ANTIBODIES

Murine mAbs that recognize tumor-associated antigens are potentially useful for recognizing tumors for diagnostic imaging and for therapy via ADCC, CDC, or antibody-conjugated anticancer agents. However, the use of these murine antibodies has been limited by the development of a human antimouse immunoglobulin antibody response and/or failure of the antibody to possess the desired biologic properties (e.g., serum clearance pharmacokinetics and the ability to interact optimally with the human immune system). Mouse/human chimeric antibodies composed of variable regions derived from murine hybridomas and human constant regions should be superior for administration to humans because of their reduced immunogenicity, altered pharmacokinetics, and ability to function more effectively with human effector cells.

Variable regions from a number of murine mAbs reactive with tumor-associated antigens have been cloned, and chimeric antibodies have been prepared using human constant region genes. In all cases, the resulting chimeric proteins retained the binding specificity of the murine mAbs. Chimeric antitumor antibodies have been shown to exhibit the biological properties appropriate for their respective human immunoglobulin subclasses. Several human monoclonal and antitumor mAbs have been developed, but most are of the IgM isotype, limiting their usefulness. These could be changed to one of the human IgG isotypes, using the approach outlined earlier.

In some instances the substitution of human constant regions has resulted in enhanced antitumor activity or has imparted biological activity to a nonfunctional murine antibody possessing desirable binding specificity. For example, the conversion of antitumor antibodies to chimeric antibodies with a human γ_1 constant region results in a chimeric antibody that mediated ADCC in the presence of human effector cells at a concentration 100 times lower than that required for the murine antibody. Similarly, a murine mAb that recognizes the CD20 antigen expressed in both normal and malignant B cells lacks cytolytic activity; conversion of the antibody to chimeric mouse/human IgG1 results in an antibody possessing the antigen-binding specificities of the parental antibody as well as the capability to mediate ADCC with human effector cells and CDC with human complement.

In the management of cancer, the most immediate advantage of chimeric mAbs may be the reduction of the human antimurine antibody response, which is mostly directed toward the murine Fc region of the Ig molecule. The human antimurine antibody response creates two problems: the potential of an allergic response with anaphylaxis, and the rapid clearance of the administered antibody, which would prevent the vast majority of the mAb from reaching the tumor site. Thus, in virtually all the previously reported human therapy trials using multiple administrations of murine mAbs, only the first and/or perhaps the second mAb administrations efficiently reached the tumor site. The use of chimeric antibody diminishes this problem.

One of the major advantages in the use of genetically engineered mAbs is the ability to modify the Ig molecule to alter pharmacokinetics. We may wish to slow down plasma clearance of a mAb so that it will have the opportunity to mediate ADCC or complement-mediated cytolysis or to speed up the plasma clearance of an mAb conjugated to a radioisotope, in which the circulating conjugate may cause damage to normal cells. Genetically engineered Igs provide several ways to alter pharmacokinetic properties. These include (1) large alterations in size by the addition or deletion of domains and more subtle alterations in size using smaller deletions, (2) construction of Fv molecules, (3) alterations in glycosylation, and (4) mutations in sequences controlling clearance rates. [*See* Pharmacokinetics.]

Finally, antitumor mAbs may be modified so that they can act as more efficient vehicles for the delivery of antitumor drugs, toxins, radionuclides, or biologi-

TABLE II

Some Potential Applications of Genetically Engineered Chimeric Antibodies

Genetic modification	Potential uses
Variable regions attached to new constant region derived from either same or different species	Produce isotype variants within a species; can be used to make the available IgM human antibodies more useful by converting them to IgG antibodies Can be used to make mouse/human chimerics with the variable region from mouse and the constant region from humans. The resulting chimeric antibodies should have the following properties; reduced immunogenicity when used *in vivo* in humans, increased ability to interact with the human immune system, and improved pharmacokinetics
Variable regions modification before joining to a constant region	Humanize mouse variable regions by positioning mouse-binding regions (complementarily determining regions) within human framework regions *In vitro* site-specific mutagenesis to produce variable regions with altered binding constants
Modification of constant regions by domain shuffling or deletion or *in vitro* mutagenesis	Improved biologic properties Improved pharmacokinetics and tissue distribution
Antibody sequences joined to nonantibody sequences	Joining to drugs or toxins for their more efficient delivery Join to enzymes for either *in vivo* application or to make better reagents for *in vitro* applications Insert biologic response modifiers to increase their effectiveness or to improve targeting Insert sequences to facilitate labeling with radioisotopes

cal response modifiers. This can be achieved by either directly ligating the antitumor agent into the Ig molecule or by the ligation of efficient linkers for drugs, radioisotopes, or bioresponse modifiers into the Ig molecule.

XIV. ADDITIONAL APPLICATIONS AND FUTURE PROSPECTS

Many potential applications of chimeric or genetically engineered antibodies exist in addition to use as anticancer agents (Table II). Monoclonal and chimeric antibodies have been produced to cell surface antigens, which are important for immune interactions. These antibodies have great potential as immune response modifiers to facilitate organ transplantation and address problems of autoimmunity. It is also possible that antibodies may be developed to approach the problem of immune nonreactivity such as is seen in AIDS. Chimeric antibodies also have potential applications in the treatment of infectious diseases. They should be far superior to antisera derived from heterologous sources (e.g., horses) and should address the heterogeneity and limited availability of antisera derived from human sources. Chimeric antibodies also have great potential for use as vaccines devised to exploit the antiidiotypic network. [*See* Idiotypes and Immune Networks.]

BIBLIOGRAPHY

Better, M., Chang, C., Robinison, R., and Horwitz, A. H. (1988). *Escherichia coli* secretion of an active chimeric antibody. *Science* **240**, 1041.

Boulianne, G. L., Hozumi, N., and Shulman, M. J. (1984). Production of functional chimeric mouse–human antibody. *Nature (London)* **312**, 643.

Brüggemann, M., William, G. T., Bendon, C. J., Clark, M. R., Walker, M. R., and Jefferis, R. (1987). Comparison of the effector functions of human immunoglobulins using a matched set of chimeric antibodies. *J. Exp. Med.* **166**, 1351.

Haber, E., Quertermous, T., Matsueda, G. R., and Runge, M. S. (1989). Innovative approaches to plasminogen activator therapy. *Science* **243**, 51.

Jones, P. T., Dear, P. H., Foote, J., Neuberger, M. S., and Winter, G. (1986). Replacing the complementarity determining regions in a human antibody with those from a mouse. *Nature (London)* **321**, 522.

Köhler, G., and Milstein, C. (1975). Continuous cultures of fused cells secreting antibody of predefined specificity. *Nature (London)* **256**, 495.

Lui, A. Y., Robinson, R. R., Hellstrom, K. E., Murray, E. D., Chang, C. P., and Hellstrom, I. (1987). Chimeric mouse-human IgG1 antibody that can mediate lysis of cancer cells. *Proc. Natl. Acad. Sci. USA* **84**, 3439.

Morrison, S. L., Johnson, M. J., Herzenberg, L. A., and Oi, V. T. (1984). Chimeric human antibody molecules: Mouse antigen binding domains with human constant region domains. *Proc. Natl. Acad. Sci. USA* **81**, 6851.

Mulligan, R. C., and Berg, P. (1981). Selection for animal cells that express the *Escherichia coli* gene coding for xanthine-guanine phosphoribosyltransferase. *Proc. Natl. Acad. Sci. USA* **78**, 2072.

Neuberger, M. S., Williams, G. T., and Fox, R. O. (1984). Recombinant antibodies possessing novel effector functions. *Nature (London)* **312**, 604.

Skerra, A., and Plückthun, A. (1988). Assembly of a functional immunoglobulin Fv fragment in *Escherichia coli*. *Science* **240**, 1038.

Southern, P. J., and Berg, P. (1982). Transformation of mammalian cells to antibiotic resistance with a bacterial gene under control of the SV40 early region promoter. *J. Mol. Appl. Genet.* **1**, 327.

Sun, L. K., Curtis, P., Raksowicz-Szulczynska, E., Ghrayeb, J., Chang, N., Morrison, S. L., and Koprowski, H. (1987). Chimeric antibody with human constant regions and mouse variable regions directed against a carcinoma-associated 17-1A antigen. *Proc. Natl. Acad. Sci. USA* **84**, 214.

Tao, M.-H., and Morrison, S. L. (1989). Studies of aglycosylated chimeric mouse-human IgG: Role of carbohydrate in the structure and effector functions mediated by human IgG constant regions. *J. Immunol.* **143**, 2595.

Tan, L. K., Shopes, R., Oi, V. T., and Morrison, S. L. (1990). The hinge region: Influence on complement activation, C1q binding and segmental flexibility in chimeric human Igs. *Proc. Natl. Acad. Sci. USA* **87**, 162.

Winter, G., Griffith, A. D., Hawkins, R. E., and Hoogenboom, H. R. (1994). Making antibodies by phage display technology. *Annu. Rev. Immunol.* **12**, 433.

Genetic and Environmental Influences on Psychopathology

DAVID L. DiLALLA
Southern Illinois University at Carbondale

STEVEN J. VANDENBERG[1]
University of Colorado

GLOSSARY

Dizygotic twins Fraternal twins; individuals who developed as the result of two separate ova being fertilized at the same time and who share, on average, 50% of their genes

Linkage Co-occurrence of loci in close proximity on the same chromosome, leading to nonindependent assortment

Monozygotic twins Identical twins; individuals who developed from a single zygote and who share 100% of their genes

Path analysis Statistical analysis of effect of each variable on a second variable, independent of other variables; paths between variables are presented in visual diagrams

Penetrance Proportion of individuals of a given genotype who manifest a specific trait

Pleiotropy Effect of a single gene on two or more different traits

Quantitative genetics Study of inheritance of characteristics of individuals, which differ in degree and not kind, that are quantitative rather than qualitative differences

Restriction fragment-length polymorphism Genetic differences between individuals in the length of small segments of DNA visible after DNA is cut by a restriction enzyme

OVER THE PAST 25 YEARS, RESEARCH IN behavioral genetics has supported the idea that genetic factors play a role in the etiology of many forms of psychopathology, with environmental factors also playing a crucial role in development of disorder. Although it is generally agreed that genetic influences on psychiatric disorders are unlikely to be reflected by simple Mendelian patterns, current research on some disorders has begun to investigate chromsomal regions that may be related to genes of large effect.

I. WHAT IS MEANT BY GENETIC EFFECTS ON BEHAVIOR?

The nature versus nurture debate has given rise to many spirited discussions in psychology and related fields about the origins of human behavior. The extremes of each end of the debate might be summarized by this question: "Is human behavior best viewed as 'hard-wired' in the genes, or are newborn human infants 'blank slates' upon which the effects of experience are written?" Clearly, neither of these positions does justice to the complexity of human behavior and recent evidence points to an intricate pattern of interrelations between genetically influenced behav-

[1]Deceased.

ioral tendencies and environmental forces that together shape human development.

The role of genetic factors in the etiology of many types of abnormal behavior has been supported by a variety of techniques that will be discussed in this article. Throughout our discussion, it is important to keep in mind that genes do not "code" directly for behavior; there is not a one-to-one correspondence between genes and the disorders of interest. Rather, genes provide the blueprint for the body to create amino acids, building blocks for proteins that have important roles in other physiological or hormonal processes (e.g., synthesis of neurotransmitters). For all of the psychological disorders to be discussed here, the role of environmental factors is also extemely important, even though the precise pathways of environmental influence are still unclear. Many researchers describe the relationship between genetic and environmental factors in terms of a "risk" or "diathesis–stress" model. This means that genetic factors may confer risk for development of a particular disorder, but that environmental factors (which could be either protective or destructive) strongly influence whether or not the risk for the disorder is expressed. It should be kept in mind that environment in this context is a broad term that includes prenatal insults and intrauterine biochemical and hormonal conditions, as well as postnatal social and nutritional influences.

Investigation of genetic influences on human behavior has, at times, been rather controversial. Much of the controversy is the result of mistaken attempts to explain differences *between* groups of people by referring to results supporting heritability of individual differences *within* a given population being studied. A brief classroom example illustrates this distinction. Assume that we have selected two genetically heterogeneous groups of pea plants and have planted seeds from one group in enriched soil and the seeds from the other group in soil that was deficient with respect to nutrients. If, after 6 weeks, we measured the height of the plants, we would undoubtedly observe an advantage in favor of the plants from the enriched environment. If we had previously undertaken an analysis of the relative contribution of genetic factors to plant height in a sample of similarly genetically heterogeneous peas that had been planted in benign conditions, we would likely observe a strong genetic contribution to individual differences in height. However, it would be *incorrect* for us to conclude based on this finding that the height difference between the enriched and deficient groups is the result of genetic differences. In human behavioral genetics, the methods described in the following provide information *only* about factors that influence differences among individuals *within* the population under study. Such studies should not be interpreted as providing an explanation for differences observed between groups of individuals.

II. METHODS OF INVESTIGATION

Behavioral genetic research on psychopathology involves human as well as animal populations. Research impossible to undertake with humans can be conducted under highly controlled settings using specially bred laboratory animals with the goal of subsequently applying findings to the human case. For example, there is a large body of animal literature on genetic influences on sensitivity and preference for alcohol and nicotine. Such research has provided an analogue for the study of human problems of addiction to alcohol, nicotine, and other drugs.

Human studies in behavioral genetics typically employ one of several basic research designs. In family "pedigree" studies the prevalence of a particular disorder in relatives of the "proband" (the original clinical case) is studied. A family pedigree is constructed and single-gene hypotheses derived from Mendelian models can be tested to determine whether dominant or recessive, autosomal or sex-linked inheritance can be accepted.

For most behavioral abnormalities, it is now thought that such simple single-gene models are not likely to be correct. Polygenic models (two or more genes influencing the disorder or trait) are more likely to apply. An assumption of this approach is that the liability for the disease is normally distributed in the population. When an individual's liability exceeds some threshold, the disease becomes apparent. Whether the threshold is reached is due partly to genetic and partly to environmental factors. Figure 1 illustrates a single-threshold model for onset of disease.

Genetic and environmental effects on noncategorical characteristics of individuals (e.g., continuously distributed variables such as personality traits or cognitive abilities) can also be investigated using the family method. By assessing similarities and differences among family members who differ with respect to their degree of genetic relatedness, estimates of the importance of genetic and environmental influences on behavior can be made. Such studies generally must include large numbers of multigenerational families

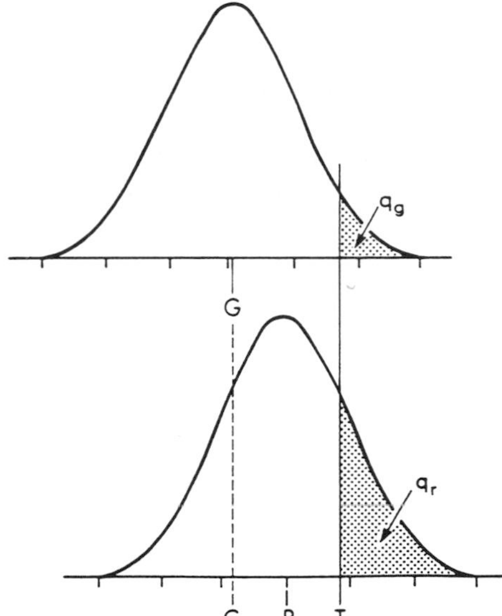

FIGURE 1 Single-threshold model of inheritance of liability toward illness. T is the disease threshold. Top distribution represents general population. G, mean liability; q_g refers to affected individuals. Lower curve represents liability for relatives of affected individuals with mean liability of R and q_r affected relatives. [Modified from D. S. Falconer (1981). "Introduction to Quantitative Genetics," 2nd Ed. Longman, New York.]

in order to tease out the relative contributions of genetic and environmental factors.

Linkage studies also take advantage of the pedigree approach. Linkage studies are based on the identification of restriction fragment-length polymorphisms (RFLPs), which are used to map the genome. RFLPs are small pieces of DNA that are evident after DNA is cut by an enzyme known as a restriction endonuclease. Differences in length of the restriction fragments are inherited according to Mendelian patterns. If a single gene for a disorder is embedded within (or is extremely close to) a particular RFLP, the transmission of the disease gene can be tracked in families. Linkage studies usually capitalize on large multigenerational families in which there are a large number of cases of the disorder of interest. The analysis involves sampling DNA of family members to determine whether there is a high probability that a specific RFLP on a particular chromosome is linked with the presence of the disorder. [*See* DNA Markers as Diagnostic Tools.]

A mainstay in the methodological arsenal of the behavioral genetic researcher is the twin method. The simplest method of twin analysis consists of a direct comparison of concordance rates (percentage of pairs where both twins are affected with the disorder under study) of monozygotic/identical (MZ) and dizygotic/fraternal (DZ) twin pairs, realizing that on average DZ twins share half their genes, whereas MZ twins are genetically identical. The basic premise is that any difference between MZ twins must be due to environmental effects. Similarly, to the extent that genetic factors are important in the development of the disorder under study, MZ twins will more often be "concordant" for the disorder than will DZ twins. As for family studies, twin studies can also shed light on genetic and environmental influences on continuously distributed characteristics of people. Here, to the degree that MZ twins tend to be more similar to each other than DZ twins are to each other—and assuming that MZ and DZ twin pairs generally experience environmental factors equally relevant to development of the target trait or disorder—the greater MZ similarity must be due to genetic factors. A broad estimate of the degree of heritability (usually denoted h^2) can be obtained by doubling the difference between the MZ and DZ correlations for a given measure. Heritability gives an indication of the percentage of variability among individuals that is the result of genetic variability. Analogously, the percentage of variance caused by environmental factors is sometimes referred to as environmentality and can be broadly derived as $1 - h^2$.

A number of other complex statistical techniques based on path analysis have been developed for analyzing twin as well as other types of behavioral genetic data. These techniques allow us to focus simultaneously on a number of variables that may influence a trait or behavior. This is particularly useful in the case of comorbidity, the co-occurrence of two disorders in the same individual. These methods have also been extended to include repeated measures of target variables to produce information on longitudinal problems, including estimates of the ways in which genetic and environmental effects may change during the course of development.

Another often-used technique in behavioral genetic research is the study of adoptees and their biological and adoptive relatives. This method has at its core the natural "experiment" wherein children are removed from the care of their biological parents early in life and are reared by persons to whom they are genetically unrelated. Genetic predisposition and influence of the rearing environment are thus neatly separated. Any similarities between adopted children and their adoptive parents are due to environmental effects or cultural transmission, whereas similarities between adopted children and their biological parents represent the effect of shared genes or preseparation (including prenatal) environmental influences. This assumes "random" placement of children into adoptive homes so that adoptive parents are uncorrelated with biological parents. To the extent that adoptive parents are systematically chosen so that they are similar to their adopted children's biological parents, selective placement has occurred. The effects of selective placement can be assessed by including in the statistical models variables indicative of selective placement such as race, social status, or correlates of these and testing whether their inclusion changes the fit between the model and the observed data.

A complication to all the methods we have discussed is assortative mating or marriage. In the polygenic models that have been derived from basic Mendelian laws, the basic assumption is that every combination of alleles is equally likely to occur. Because it has become obvious that husbands and wives are often more similar than would be expected from this rule for many important characters such as intelligence and some personality traits, techniques have been devised to take this assortment into account. Falconer's introductory book on quantitative genetics provides an entry into this literature.

III. CLASSIFICATION OF ABNORMAL BEHAVIOR

One of the pervasive difficulties encountered in the study of genetics of behavioral disorders is satisfactorily defining the phenotypes of interest. Unlike some genetic disorders such as Down's syndrome, which have relatively clear boundaries and can be precisely identified, many behavioral abnormalities have fuzzy, overlapping boundaries. Even more confusing, the symptoms of various disorders can sometimes mimic each other, so-called phenocopies.

To facilitate diagnosis, treatment, and research, a number of diagnostic conventions for mental disorders have been created. In the United States, the most widely used diagnostic standard is the "Diagnostic and Statistical Manual of Mental Disorders," fourth edition (DSM-IV), of the American Psychiatric Association. The DSM-IV is, in the main, consistent in terminology with the "International Classification of Diseases," tenth edition (ICD-10), published by the World Health Organization. Both diagnostic systems represent a consensus opinion from the psychiatric and psychological community regarding classification of mental disorders. This consensus changes over time as a result of accumulation of new scientific evidence.

Changes in cultural norms, as well as political considerations, may influence what is deemed to constitute mental illness. In most societies, for example, a distinction is made between criminals and mentally ill persons. However, it is difficult, if not impossible, to draw a clear line between these groups. Similarly, the psychiatric and psychological community once considered homosexuality to be a mental disorder. Later, homosexuality was considered to warrant diagnosis only if the individual's sexual orientation caused distress to him or her. In the most recent revision of the DSM, there is no diagnostic category related to homosexuality.

Scientific norms also change over time and may vary in different regions of the world. The accepted definition of schizophrenia used today, for example, is significantly different (and less broad) than the definition used in the United States 30–40 years ago. Analogously, American and European definitions of many forms of psychopathology have often diverged. Hence, a reader of the scientific literature on genetic and environmental influences on psychopathology must pay careful attention to the ways in which the target disorders were defined.

IV. DISCUSSION OF SPECIFIC DISORDERS

A. Schizophrenia

As codified in the DSM-IV, schizophrenia represents a pervasive disturbance in personality, a generally characteristic disordering of thought processes that includes delusions (often of paranoia and of being controlled by an outside force), bizarre sensory perceptions (most often in the form of auditory hallucinations), and inappropriate emotional responses. Onset of symptoms usually occurs during late adolescence to early adulthood, with a smaller proportion of cases having a late adulthood onset. Epidemiological studies have shown the lifetime prevalence of schizophrenia to be approximately 1%. [*See* Schizophrenic Disorders.]

The boundaries of schizophrenia are by no means rigidly defined. In particular, there has been considerable debate about how to conceptualize the affective psychoses (depressive disorders with psychotic components) with respect to schizophrenia. In the United

States, use of the diagnostic category schizo-affective disorder testifies to the difficulty in drawing a line between the two conditions, especially during early episodes of the disorders. Additionally, many individuals present with psychotic symptoms that do not meet diagnostic criteria for schizophrenia, schizo-affective disorder, or bipolar affective disorder, but bear clear similarity to the symptoms of schizophrenia. These concerns have led many researchers to speak of the "schizophrenias" in recognition of the likelihood that there is a spectrum of disorders related to schizophrenia.

Genetic heterogeneity is one potential explanation of a schizophrenia spectrum. Given the assumption that multiple genetic factors are related to risk for schizophrenia, individuals with more of the risk-relevant genes would be more likely to exhibit the disorder than individuals with fewer of the risk-relevant genes. However, individuals below the threshold for developing schizophrenia might still exhibit some mild-to-moderate psychotic symptoms. The concept of genetic reaction range, formulated by I. I. Gottesman, is also relevant to the idea of there being a spectrum of schizophrenia-related disorders. Reaction range refers to the limits of potential phenotypes (observable outcomes) being fixed by the genotype, whereas the actual phenotype is dependent on environmental influences. An important implication is that the same genotypes can lead to different phenotypes (as is the case in MZ twins discordant for schizophrenia) and different (but closely related) genotypes can lead to the same phenotype.

Research supporting a genetic component in the etiology of schizophrenia has come from a variety of sources. Table I presents a summary of family studies on the risk of schizophrenia in the relatives of individuals diagnosed with schizophrenia. These results highlight the familial nature of schizophrenia and show that risk to relatives drops off sharply as we move from first- to second- and third-degree relatives. It is also worth noting that these data show that the vast majority of relatives of individuals with schizophrenia do not develop the disorder.

TABLE I
Average Lifetime Risk[a] for Developing Schizophrenia by Relatedness to Affected Individual[b]

Relationship to individual with schizophrenia	Risk (%)
Genetically unrelated	
Spouse	2
Third-degree relatives	
First cousin	2
Second-degree relatives	
Uncle/aunt	2
Grandchild	5
Half-sibling	6
First-degree relatives	
Child	13
Sibling	9
Sibling with one schizophrenic parent	17
DZ twin	17
Parent	6
MZ twin	48

[a]Summarized across all studies from 1920 to 1987. Note: General population risk = 1%.

[b]Adapted from I. I. Gottesman (1991). "Schizophrenia Genesis: The Origins of Madness." Freeman, New York.

I. Adoption Studies

In L. L. Heston's pioneering study of children adopted away from mothers with schizophrenia (Table II), a higher than expected proportion of those children developed schizophrenia despite having been reared in homes separate from their biological mothers. In another study, the adopted-away offspring of both men and women who had schizophrenia were followed up. "Cross fostering" studies have also been conducted, which investigate children adopted from biological parents with schizophrenia to nonschizophrenic adoptive parents, as well as children unlucky enough to have been adopted from nonschizophrenic parents and placed with adoptive parents one of whom would later develop schizophrenia. All of these studies converge on the conclusion that sharing a rearing environment with a parent who has schizophrenia does not explain the observed familial aggregation of schizophrenia. In all the studies, having a biological

TABLE II
Risk of Schizophrenia in L. L. Heston's Study of Children Adopted Away from Schizophrenic Mothers and from Nonpsychiatric Controls[a]

	Mother schizophrenic	Mother nonpsychiatric
Number of children	47	50
Mean age at follow-up	35.8	36.8
Morbid risk (%)	16.6	0

[a]Modified, with permission, from I. I. Gottesman and J. Shields (1982). "Schizophrenia: The Epigenetic Puzzle." Cambridge Univ. Press, Cambridge, England.

parent with schizophrenia conferred significantly higher risk on adopted-away offspring than would be expected among members of the general population. In most studies, the rate of risk to children adopted away from parents with schizophrenia is similar to that observed in "intact" family studies of children with parents diagnosed with schizophrenia.

2. Twin Studies

The first twin studies of schizophrenia provided strong evidence for a genetic influence, given the substantially higher concordance rate of MZ (average concordance of 65%) as compared with DZ twins (average concordance of 12%). These older twin studies have stood the test of time remarkably well, despite methodological advancements in the twin method as applied to psychopathology. Table III summarizes concordance rates for MZ and DZ twins from several recent studies. The effect of genotype is clear, but the important role of environment cannot be denied when we note that less than 50% of the genetically identical cotwins of individuals with schizophrenia themselves develop the disorder. Finally, a dramatic demonstration of the role of genetic factors in schizophrenia is the finding reported by I. I. Gottesman and A. Bertelsen that children of unaffected members of MZ twin pairs discordant for schizophrenia are as likely to develop schizophrenia as are the children of the MZ twins diagnosed with schizophrenia (Table IV). This highlights the continued influence of the "unexpressed" genotype for schizophrenia.

Critics of the twin method have noted that the higher concordance for schizophrenia in MZ twins may be due to MZ twins being more likely to share a pathological environment than DZ twins. Were this the case, we would expect an overrepresentation of MZ twins in samples of individuals with schizophrenia. However, reviews of major studies of schizophrenia, including studies of twins reared together and apart, have found no such overrepresentation of MZ twins or of twins in general in samples of persons with schizophrenia.

3. Linkage Studies

Although family, twin, and adoption studies strongly implicate the role of genetic factors in the etiology of schizophrenia, the mode of inheritance has not been identified. Some scientists believe that a single gene is responsible for the disorder, although the majority of researchers argue against this and believe that multiple genes are more likely to be involved. In a polygenic system, each gene is assumed to exert a small additive

TABLE III
Probandwise Concordance Rates for Schizophrenia[a] in Recent Twin Studies[b]

Investigator country/year	MZ pairs		DZ pairs	
	Total	Concordance (%)	Total	Concordance (%)
Tienari Finland, 1963/1971	17	35	20	13
Kringlen Norway, 1967	55	45	90	15
Fischer Denmark, 1973	21	56	41	27
Gottesman and Shields United Kingdom, 1966/1987	22	58	33	15
Weighted Average of above four studies		**48**		**17**
Kendler and Robinette[c] United States, 1981	164	31	277	6

[a]Schizophrenia and schizophreniform psychosis.
[b]Adapted from I. I. Gottesman (1991). "Schizophrenia Genesis: The Origins of Madness." Freeman, New York.
[c]Not included in weighted average because sample was initially "prescreened" for physical and emotional health as part of military service requirement of twins. Other studies summarized used nonscreened population-based samples.

TABLE IV
Risk of Schizophrenia and Schizophreniform Psychosis in Offspring of Schizophrenic and Normal Twins[a]

	Total offspring	Affected offspring	Age-corrected morbid risk (%)
MZ twin parents			
Schizophrenic (n = 11)	47	6	16.8 ± 6.6
Normal cotwins (n = 6)	24	4	17.4 ± 7.7
DZ twin parents			
Schizophrenic (n = 10)	27	4	17.4 ± 7.7
Normal cotwins (n = 20)	52	1	2.1 ± 2.1

[a]Modified, with permission, from I. I. Gottesman and A. Bertelsen (1989). Confirming unexpressed genotypes for schizophrenia. *Arch. Gen. Psychiatry* **46**(10), 867.

influence on risk for the disorder. The effects of each gene of the system are also assumed to be mediated by environmental influences. Thus, even with a relatively simple polygenic system, an extremely large number of factors can influence expression of the genotype.

Arguments in favor of polygenic etiology of schizophrenia include the following: (1) prevalence rates for relatives of individuals with schizophrenia do not fit Mendelian expectations (including the relatively small number of cases where offspring are produced by individuals who both have schizophrenia); (2) degree of disability ranges from mild to severe, including "subthreshold" effects, which can be detected in non-affected individuals; (3) there are proportionately more affected relatives of severely ill individuals as compared with mildly ill individuals; (4) as the number of affected family members increases, risk to other members of the family increases; (5) there is a sharp decline in risk as degree of genetic relatedness to the affected individual decreases (e.g., brother versus first cousin); and (6) there is generally a distribution of affected cases on both paternal and maternal sides of the family.

A number of researchers have maintained a search for single genes for schizophrenia, or genes that may have major effects in the development of the disorder. This research is generally conducted using multigenerational families with a relatively large number of individuals affected with schizophrenia. Over the past decade there have been a series of tantalizing reports of potential linkage for schizophrenia. In 1988, one group of researchers led by H. Gurling and R. Sherington reported in *Nature* strong evidence of linkage to chromosome 5. In the same issue of the journal, however, a second group of researchers led by J. R. Kidd and J. L. Kennedy reported equally strong evidence *against* linkage using some of the same chromosome 5 RFLPs. More recently, T. J. Crow and colleagues reported in the *British Journal of Psychiatry* evidence of linkage to sex chromosome autosomal markers, however, this finding has also not been replicated to date. Similarly, suggestions that linkage might be found to chromosomal areas related to dopamine receptor or transport genes have not been supported. The linkage literature to date indicates no definitive support for linkage to any chromosomal area, although the findings do not rule out the possibility that genes of small-to-moderate effect may be found in some of the areas studied.

B. Affective Disorders

The predominant clinical feature of the affective disorders is disturbance of mood in the form of either depression or mania. For bipolar affective disorder (sometimes referred to as manic–depressive illness), an individual must have experienced a full-blown manic episode characterized by symptoms such as uncharacteristically euphoric mood, increased activity, impulsivity, grandiosity, and distortions in perceptual processes (e.g., hallucinations). Experience of depressed mood is not required, although affected individuals often experience cyclic periods of depression and mania. Individuals with unipolar forms of depression experience periods of depressed mood, disturbance in attention and concentration, feelings of helplessness and hopelessness, disturbance in appetite, loss of interest in usual activities, and psychomotor retardation. In some cases, distortions of perceptual processes are also present. Results from the Epidemiological Catchment Area (ECA) program indicated lifetime prevalence ratings ranging from 6.1 to 9.5% for any affective disorder; more recent studies have noted 12-month prevalence to be roughly 11% in the general population. Women generally have higher rates of depression than do men. [*See* Mood Disorders.]

It has been long recognized that depression often "runs" in families and that first-degree relatives of affected individuals tend to have higher than expected rates of affective disorder. This speaks to the familiality of depression but does not illuminate potential genetic influences, as it is plausible that environmental effects within families could lead to depressive symptomatology. Studies of twins and adoptees have convincingly shown that there is a genetic component in the etiology of unipolar and bipolar forms of depression. Concordance rates from recent twin studies of affective disorder are presented in Table V. The higher degree of concordance for MZ as compared to DZ twins implicates the influence of genetic factors. However, twin studies have also indicated that genetic factors contribute more substantially to the development of bipolar affective disorder than to unipolar disorder [see Katz and McGuffin (1993) for details]. In particular, shared environmental influences (e.g., family environment) appear to be stronger for unipolar depression. Though less extensive, adoption studies also provide support for the influence of genetic factors on depression. For example, a 1977 study reported by J. Mendlewicz and J. Rainier on the adoptive and biological parents of adoptees with bipolar affective disorder found the prevalence of depression to be 31% for biological parents (similar to the prevalence of depression among biological parents of non-adopted individuals with bipolar illness) versus 12% for adoptive parents (similar to the prevalence of depression among adoptive parents of nondepressed adoptees). Similarly, R. Cadoret's 1978 study indicated higher rates of affective disorder among adopted-away children of parents with an affective disorder, as compared to adopted-away children of biological parents who had other psychiatric diagnoses. [*See* Depression.]

Recently there has been great interest in tracking down the depression gene or genes. However, a number of difficulties are inherent in this pursuit. The results of twin studies cited earlier testify to the incomplete penetrance of any genes for depression. Were penetrance complete and environmental influence negligible, MZ twins would have full concordance for the disorder. Another difficulty is that the relation between unipolar and bipolar forms of depression is unclear. The clinical presentation of persons diagnosed with depression often is varied, and it seems likely that there are environmentally based "phenocopies"—forms of depression that meet diagnostic criteria but that are caused by environmental stressors. It also seems probable that at least some forms of depression are influenced by polygenes, sets of genes that together predispose an individual toward depression. Finally, recent evidence supports the notion that depression may be heterogeneous at the genetic level. There have been reports of linkage for bipolar disorder (to chromosomes 11 and 6 and to the X chromosome), but these findings have not been widely replicated to date.

In summary, there is compelling evidence in support of the role of genetic factors in the development of depression. It appears that the genetic influence is stronger for bipolar affective disorder than it is for unipolar depression, which shows mild-to-moderate genetic effects. There is evidence for moderate shared-environmental influences on etiology of unipolar depression. Factors such as stressful life events have been implicated in this regard, but there is, as yet, no definitive research on specific environmental factors that contribute to onset of depression among individuals who may have some genetically influenced tendencies toward the disorder.

C. Antisocial Behavior

The conventional wisdom regarding antisocial behaviors, including delinquency and criminality, has been that they are societal problems principally determined

TABLE V
Concordance Rates for Affective Illness in Monozygotic and Dizygotic Twins[a]

	Monozygotic		Dizygotic	
Study	Pairs	Concordance (%)	Pairs	Concordance (%)
Gershon *et al.* (1975)[b]	91	69	226	13
Bertelsen *et al.* (1977)[c]	69	67	54	20
Torgersen *et al.* (1986)[c]	37	51	65	20
McGuffin *et al.* (1991)[c]	62	53	79	28

[a]Adapted from R. Katz and P. McGuffin (1993). The genetics of affective disorders. *In* "Progress in Experimental Personality and Psychopathology Research" (L. Chapman, J. Chapman, and D. Fowles, eds.), Vol. 16, pp. 200–221. Springer, New York.

[b]Summary of six previous studies using pairwise concordance rates.

[c]Probandwise concordance rates.

by environmental influences. Many persons have found the idea of genetic influences on delinquent or criminal behavior to be distasteful, and some investigators have rejected the idea of genetic effects for ideological reasons. To be sure, some of the early studies of genetic influences on criminality suffered from major methodological shortcomings. However, recent additions to the literature provide evidence that cannot be simply ignored. [*See* Crime, Delinquency, and Psychopathy.]

In twin and adoption studies of delinquency and criminality, as in other areas, definition of the phenotype of interest is a major stumbling block to good research. Sampling of affected individuals is also a problem because individuals with Antisocial Personality Disorder (APD) are generally unlikely to self-refer for psychological treatment, so antisocial individuals from clinical samples often present with a variety of other problems. An added difficulty is the high base rate of delinquent behavior in adolescence, which makes it difficult to detect genetic influences on antisocial behavior. Finally, there may be etiological heterogeneity leading to various types of offenders. For example, individuals may be delinquent in adolescence and continue criminal behavior in adulthood; adolescents may exhibit delinquency but not become criminal as adults; and there may be adult criminals who did not engage in delinquent behavior during adolescence.

Genetic studies of delinquency during adolescence do not support the role of genetic etiological factors. For example, average concordance rates for delinquency across twin studies are in the range of 0.87 for MZ twins and 0.72 for DZ twins. The high concordance rates for both types of twins speak to the high base rate for delinquency, and the degree of concordance for DZ twins in particular points to the importance of shared environmental influences. However, some studies have found heritabilities as high as 70% for specific forms of antisocial behavior (e.g., aggressive behavior) as opposed to broadly defined delinquency. Moreover, there is growing evidence that conduct problems in early and middle childhood (often a precursor of later antisocial behavior patterns) are influenced by genetic as well as environmental factors. One explanation for these findings is that for many adolescents delinquency represents a developmental phase strongly influenced by the social environment, whereas some delinquents are influenced, at least in part, by genetic factors that increase the likelihood of conduct problems during childhood; these individuals may continue to engage in delinquent behavior during adolescence and criminal behavior during adulthood.

Studies related to genetic effects on adult criminality are much stronger than those for delinquency, including the results of methodologically sound twin and adoption studies. Twin studies show substantially higher criminality concordance rates for MZ (approximately 0.5) as compared with DZ (approximately 0.2) twins, leading to heritability estimates of roughly 55%. The effect of shared environment also has been shown to be substantial, in the range of 20%. Factors such as peer interaction and socioeconomic status would undoubtedly play a major role with respect to the latter finding.

Adoption studies show that having a biological parent who was criminal places adoptive offspring at higher risk for exhibiting criminal behavior during adulthood. At greatest risk are those individuals for whom both biological and adoptive parents are criminal. The adoption studies also point to the substantial role of environment in shaping criminal behavior. For example, adopted-away children whose biological parents had no criminal history were still found to have higher rates of criminality than the population average (Table VI).

To sum up this rather controversial topic, delinquency per se does not appear to be heritable, although conduct problems during early and middle childhood, as well as aggressive behavior in adolescence, do tend to show some heritability as well as clear environmental influences. Peer and sibling effects may be particularly potent in this regard. For criminality (often used as a marker for antisocial tenden-

TABLE VI
Percentage of Male Adoptees Who Have a Criminal Record[a]

	Biological father	
Adoptive father	Criminal	Noncriminal
Criminal	36.2	11.5
Noncriminal	21.5	10.5

[a]Modified, with permission, from B. Hutchings and S. A. Mednick(1975). Registered criminality in the adoptive and biological parents of male criminal adoptees. *In* "Genet. Res. Psychiatry," Genetic Research in Psychiatry, Procedures of The Annual Meeting of The American Psychopathological Association (R. R. Fieve, D. Rosenthal, and H. Brill, eds.), pp. 105–116. Johns Hopkins Univ. Press, Baltimore.

cies), numerous well-designed studies point toward a definite genetic etiological influence. These studies also highlight clear environmental influences on delinquent and criminal behavior; only a small percentage of adopted-away children of criminal parents exhibit antisocial behavior. It may be that the critical influences of genetic factors may be on personality traits such as extroversion, thrill-seeking versus harm-avoidance, tendencies to attribute "hostile" intent to others, and degree to which individuals are sensitive to the effects of reward and punishment. As noted at the beginning of this article, it should be emphasized that the findings summarized above *only* relate to explanation of within group variability. The findings are *not* relevant to the explanation of any observed differences between race or ethicity groups.

D. Anxiety and Phobic Disorders

Anxiety and phobic disorders encompass a wide range of emotional difficulties related to the experience of severe tension, worry, nervousness, and other negative emotional states. Many individuals who experience anxiety disorders also meet diagnostic criteria for depressive disorders, leading many researchers to speculate about common biological, social, and cognitive etiological pathways. The number of individuals affected by anxiety disorders is quite large; a recent study by R. Kessler and colleagues in the United States indicated a 1-year prevalence rate for adults of 17% for experience of at least one anxiety disorder.

Family studies consistently have shown that family members of individuals diagnosed with an anxiety disorder have a significantly higher likelihood of being diagnosed with an anxiety disorder than do individuals from the population at large. Additionally, family studies provide support for differentiation of some of the specific anxiety disorders. For example, family members of individuals with Panic Disorder are at increased risk for Panic Disorder but do not show higher rates of Generalized Anxiety Disorder.

As we have seen, however, family studies cannot disentangle effects of genetic and environmental factors. Twin studies of anxiety disorder are not as numerous as those undertaken in the area of schizophrenia. A study by S. Torgersen found significantly higher MZ twin concordance for Panic Disorder as compared to DZ twins, but this pattern did not hold for Generalized Anxiety Disorder. At the same time, an extremely large twin study conducted by K. Kendler and colleagues showed moderate effects for genetic factors on Generalized Anxiety Disorder. A twin study by G. Carey and I. I. Gottesman also implicated the effects of genetic factors on phobic disorders and obsessive compulsive disorder. Overall, the findings in this area suggest some familial/genetic influences on development of anxiety disorders, but the effects appear to be modest and there have been conflicting findings regarding influences on specific forms of anxiety disorder. Social factors such as distorted early parent–child attachment patterns and stressful life events have also been regarded as important to the development of anxiety disorders. It may be that there are broad genetic factors that place particular individuals at risk, perhaps via a general tendency to experience negative emotional states, and that risk status interacts with important environmental factors to determine whether or not an anxiety disorder develops.

E. Alcohol/Substance Abuse and Dependence

Abuse of alcohol and other drugs causes a myriad of personal, interpersonal, and societal problems. Although it is clear that social factors are extremely important in the onset and course of substance abuse problems, research in behavioral genetics has also shown that genetic factors play a role in the complex etiology of substance abuse and dependence. The bulk of this research has been conducted on alcohol abuse.

Family studies have long shown intergenerational patterns of alcohol abuse and dependence, but such studies might equally implicate social or environmental factors that "run" in families. An extensive twin study conducted by M. McGue and colleagues has shown that MZ twins tend to be more often concordant for alcoholism than DZ twins, suggesting the etiological genetic factors. However, this pattern is absent among female twins, which suggests possible different etiological pathways for men and women with respect to alcoholism. In several large twin studies of nonalcoholic twins, genetic factors as well as shared environmental factors have been shown to exert an important influence on the frequency and intensity of socially oriented drinking.

Results from adoption studies have shown that adopted children whose biological parents were diagnosed as alcoholic but whose adoptive parents were not still were more likely than individuals in the population at large to develop alcohol problems. At the same time, individuals brought up in a home with an alcoholic adoptive parent, but without alcoholic biological parents, were not at increased risk for developing alcohol problems.

In summary, the family, twin, and adoption literature on alcohol abuse and dependence shows that there are some genetic factors that relate to the development of problem drinking behavior, and that ways in which these influences operate may be very different for men and women. Unlike many other forms of psychopathology, the role of shared environment (e.g., peer interactions shared by twin siblings as well as twin imitation effects) has been shown to be important in development of alcohol abuse and dependence.

There has been great debate regarding the form that a genetic etiological influence is likely to take. Animal research has shown clear genetic influences on sensitivity to alcohol, and such factors might also operate in the human case. Others have noted that personality processes are the likely mediators of genetic effects on alcohol abuse, although research in personality to date has generally not supported the construct of the addictive personality. Understanding genetic influences on alcohol abuse is also complicated by the frequent co-occurrence of other forms of psychopathology with substance abuse. In particular, depressive disorders and antisocial personality characters have been shown to go along with alcohol abuse and dependence. One possibility is that the genetic influence is actually on an "underlying" disorder such as depression, which subsequently leads to alcohol abuse or dependence. Some recent research from large prospective twin studies has suggested that there may be some genetic influences that are jointly shared by depression and alcohol abuse, but that there are also unique genetic influences on each disorder.

F. Personality Disorders

Personality disorders are characterized in the DSM-IV as long-standing and pervasive patterns of personality that are maladaptive or distressing and that cause functional impairment. Many researchers have approached personality disorders from a dimensional perspective, arguing that personality characteristics exist along a continuum and that, at some point along the continuum, the thoughts, emotions, and behaviors of the individual become maladaptive and merit classification as personality disorder. Other researchers see some personality disorders as reflecting "subthreshold" manifestations of other more serious psychological disorders. For example, Schizotypal Personality Disorder (SPD) includes a number of symptoms that are similar to, but not as extreme as, those observed among individuals with schizophrenia. It has been suggested that individuals diagnosed with SPD may have experienced a number of the genetic and environmental risk factors associated with schizophrenia (or factors that protect against the expression of schizophrenia) so that they do not develop schizophrenia but *do* exhibit salient psychological symptoms. Indeed, it has been observed that there is a higher than expected incidence of SPD among biological relatives of individuals diagnosed with schizophrenia.

A large number of family, twin, and adoption studies have indicated moderate genetic influences on the development of personality traits. Generally, somewhere between 40 and 60% of observed personality differences among people have been calculated to be the result of genetic variability. Environmental influences that are shared by people (e.g., siblings who grow up in the same environment) seem to exert little influence on personality similarity. However, unique environmental effects have been shown to exert a strong influence on personality characteristics. To the degree that personality disorders reflect extreme expression of normal personality characteristics, the findings summarized here would predict moderate effects of genetic factors on personality disorders. However, given substantial heterogeneity within personality disorder diagnoses and high overlap between personality disorder diagnoses, it would be incorrect to presume that there is specific genetic influence on personality disorder broadly defined. More likely, there are important genetic and environmental influences on *personality characteristics* that, when maladaptively expressed, are described as reflecting a personality disorder.

V. CONCLUDING REMARKS

Family, twin, and adoption studies have provided converging evidence of genetic contributions to the development of most major forms of psychopathology. These studies also highlight the crucial role of environmental influences on development of mental disorder. For many behavioral disorders, it is environmental influences unique to individuals that appear to be most potent, although shared environmental influences (peer effects, sibling effects) appear to be important for development of depression, conduct problems, antisocial behavior, and alcohol abuse and dependence.

Identification of specific environmental triggering factors or protective factors has proven to be rather difficult, though there are hints that some of the important environmental insults may occur very early in life or even prenatally. In part, the difficulty results from the unanticipated finding that many of the measures typically thought to index "environment" also appear to be influenced in part by genetic factors. This highlights anew the complexity we encounter when trying to unravel the relative effects of nature versus nurture. In fact, nature and nurture may not be as independent as has sometimes been thought, but may be linked together by way of what behavioral geneticists call gene–environment correlations. Simply stated, this means that the environments that we are exposed to (or that we choose) may themselves be influenced by behavioral tendencies that are genetically mediated. It is clear, however, that better understanding of specific environmental factors that either increase or decrease risk for expression of mental disorder will vastly augment our ability to provide preventive interventions for individuals known to be at risk for developing a disorder.

It will also continue to be important to study potential genetic relationships between mental disorders and personality characteristics, as personality may play an important mediating role with respect to development of disorder. Better understanding of the nature of the relationship between personality and psychopathology will also allow us to better address questions such as whether personality distortions cause psychopathology or whether it is the expression of psychopathology that causes personality to appear to be "deviant." Investigating personality and psychopathology simultaneously in a genetically informative study allows analysis of whether specific genetic factors are related to psychopathology alone, or to personality alone, or whether genes that influence some disorder appear to concurrently influence broad personality characteristics. Understanding these issues will take us a giant step toward unlocking the mystery of mental disorder.

BIBLIOGRAPHY

American Psychiatric Association (1993). "Diagnostic and Statistical Manual of Mental Disorders," 4th Ed. American Psychiatric Association, Washington, D.C.

Ehrman, L., and Parsons, P. A. (1976). "The Genetics of Behavior." Sinauer Associates, Sunderland, MA.

Falconer, D. S. (1981). "Introduction to Quantitative Genetics," 2nd Ed. Longman Scientific and Technical, New York.

Fuller, J. L., and Thompson, W. R. (1978). "Foundations of Behavior Genetics." Mosby, St. Louis.

Gottesman, I. I. (1991). "Schizophrenia Genesis: The Origins of Madness." Freeman, New York.

Katz, R., and McGuffin, P. (1993). The genetics of affective disorders. *In* "Progress in Experimental Personality and Psychopathology Research" (L. Chapman, J. Chapman, and D. Fowles, eds.), Vol. 16, pp. 200–221. Springer, New York.

Plomin, R., Defries, J. C., and McClearn, G. E. (1980). "Behavioral Genetics: A Primer." Freeman, San Francisco.

Torrey, E. F., in collaboration with Bowler, A. E., Taylor, E. H., and Gottesman, I. I. (1994). "Schizophrenia and Manic–Depressive Disorder: The Biological Roots of Mental Illness as Revealed by the Landmark Study of Identical Twins." Basic Books, New York.

Vandenberg, S. G., Singer, S. M., and Pauls, D. L. (1986). "The Heredity of Behavior Disorders in Adults and Children." Plenum, New York.

Genetic Counseling

VIRGINIA L. CORSON
Johns Hopkins University School of Medicine

GLOSSARY

Carrier Individual who possesses the gene that determines an inherited disorder and who is usually healthy at the time of study

Chromosome Tightly coiled structures in each cell that contain the genetic blueprint coded by individual genes

Gene Units of information that determine physical characteristics and biochemical properties of the body

Multifactorial Determined by multiple genetic and environmental factors

Recessive Mode of inheritance in which a nonworking gene is inherited from each carrier parent

GENETIC COUNSELING IS A COMMUNICATION process that addresses the human problems associated with the occurrence, or the risk of occurrence, of an inherited disorder or a birth defect in the family. Medical facts, including the diagnosis, prognosis, and available management, are discussed, as well as the role of heredity and the risk of recurrence. Reproductive options are investigated as the family is encouraged to pursue a course of action that seems appropriate to their goals and values. The counselor strives to facilitate the family's adjustment to a disorder or their risk of a genetic problem while offering psychological and emotional support. Genetic counseling services have become an integral part of health care as a result of an increasing awareness of the clinical significance of the principles of human genetics.

I. INDICATIONS FOR COUNSELING

A. Affected Family Member

Approximately 3% of newborns have some birth defect (e.g., a cleft lip or heart defect) that will require medical intervention. One baby in 200 is born with a chromosome abnormality (e.g., Down syndrome). The parents of these children with mental retardation, physical abnormalities, or a suspected genetic disorder will seek genetic counseling to obtain further information about the etiology and implications of their child's problem. Some genetic conditions do not have manifestations until later in life, and an evaluation will be suggested during the teenage or adult years as symptoms arise. In addition, relatives in the extended family will have questions regarding the implications of a disorder and may seek counseling after a diagnosis has been made in one family member. [*See* Birth Defects; Chromosome Anomalies.]

Although traditional genetic counseling is not widely sought for common disorders of adulthood (e.g., diabetes, heart disease, cancer, or emotional illness), the understanding of genetic factors contributing to their occurrence continues to grow, and counseling will expand for these indications. Through early diagnosis of clinical symptoms and the development of predictive genetic markers, relatives of affected individuals can seek appropriate medical care or modify pertinent environmental influences.

B. Prospective Parents

Increasing numbers of couples contact a genetics center with concerns about risks to the fetus or their

TABLE I
Estimated Risk of All Chromosome Abnormalities in Live Births for Specific Maternal Ages

Maternal age	Frequency of chromosome abnormalities
25	1/476
27	1/455
29	1/417
31	1/385
33	1/286
35	1/179
37	1/123
39	1/81
41	1/49
43	1/31

future offspring. Many are referred to discuss the mother's age-related risks for chromosome abnormalities as seen in Table I. Pregnant women in their mid-30s and older are routinely offered prenatal testing to rule out Down syndrome and other chromosome abnormalities. Other couples may have a previously affected child or some other family history concern for which prenatal diagnosis is available. Pregnant women exposed to potentially harmful drugs or infections seek information about possible risks to the fetus. Some couples are referred for counseling and additional testing in follow-up of abnormal prenatal test results obtained by their obstetrician. Finally, some couples have experienced multiple pregnancy losses and are seen to evaluate a possible genetic basis for their reproductive difficulties.

C. Ethnic Background

Individuals of particular ancestries are at increased risk to be carriers of certain recessive genetic disorders. Screening tests and prenatal diagnosis are available to these persons. Eastern European (Ashkenazi) Jews and French-Canadians are at increased risk to be carriers for Tay-Sachs disease, a degenerative, neurological disorder that is usually lethal by 3–5 years of life. Through measurement of the enzyme hexosaminidase A in a blood sample, carriers can be identified and counseled. Blacks are at greater risk to be carriers for sickle-cell anemia and thalassemia, two serious blood disorders. Thalassemia carriers are also more frequent in persons of Greek, Italian, Middle Eastern, Indian, and Southeast Asian backgrounds.

II. HEALTH CARE SETTING AND PROVIDERS

As the demands for counseling services increased in the 1970s, genetics clinics were established in university medical centers, where specialists trained in human genetics were able to see patients and pursue research into understanding the basis of inherited problems. Satellite clinics affiliated with these medical centers were begun in more rural communities or in areas without easy access to a university center. Additional facilities arose in private community hospitals or physicians' offices, largely in response to the growing need for prenatal diagnostic services.

A team of trained individuals is best suited to address the broad range of problems encountered in human genetics and to provide the laboratory support needed for diagnostic testing. Physicians trained in medical genetics are responsible for the medical evaluation and treatment plans of affected patients. Genetic counselors, health professionals with specialized graduate degrees, usually coordinate the clinic activities, participate in the communication process, and assume most of the prenatal genetic counseling responsibilities. Other members of the team may include personnel from the genetics laboratories, nurses, and medical social workers. Specialists from other disciplines such as ophthalmology, cardiology, orthopedics, and radiology are consulted for diagnostic testing and treatment of some patients.

The other professional important to the genetic counseling process is the primary care physician, who refers patients and manages their medical follow-up. These pediatricians, obstetricians, and internists must consider the possibility of a genetic diagnosis in a significant fraction of their patient population.

III. COUNSELING PROCESS

A. Obtain Family History

One of the cheapest and most important tools of the geneticist is the family history, or pedigree. Valuable clues to a diagnosis can be revealed through obtaining information about other family members. Multiple miscarriages or early infant deaths can be indicators of chromosomal or metabolic conditions. Ethnic background can be important, as some problems are more common in certain populations, and consanguinity (mating within the family) points suspicion toward a recessive inheritance pattern. Though sym-

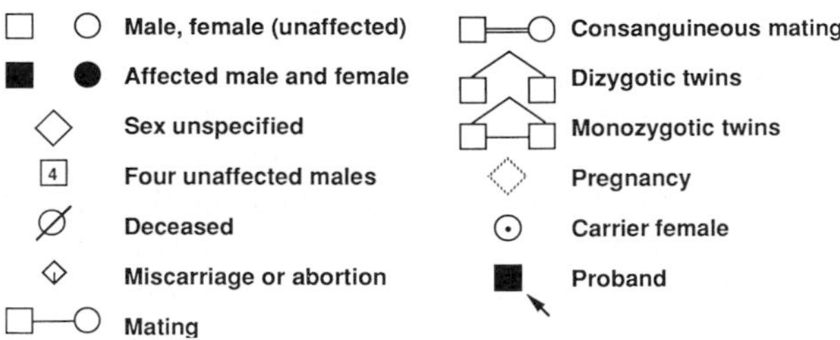

FIGURE 1 Symbols used in constructing a pedigree.

bols for pedigree construction are still being standardized nationally, the conventions seen in Fig. 1 are commonly used and are appropriate in any clinical setting.

B. Review Medical Records and Clinical History

Appropriate medical records regarding the individual under evaluation as well as other pertinent family members should be reviewed by the genetics team. If indicated, a physical examination is performed, and additional factors such as developmental milestones and the mother's pregnancy history are discussed.

C. Evaluate for Diagnosis and Interpret Laboratory Data

Based on the family history and clinical presentation, possible diagnoses will be considered for the individual under evaluation. Laboratory studies or specialist consultations may be indicated to substantiate further or to rule out a particular diagnosis. For prospective parents, the indication for referral will be addressed and options for evaluation through parental testing or prenatal diagnosis presented.

D. Discuss Prognosis, Management, and Inheritance

After a diagnosis is made, the discussion will focus on the significance of this disorder for the family. Expectations about future capabilities and limitations of the affected individual are addressed, as well as plans for medical interventions and therapies. Genetic implications are communicated as they apply to different family members, and carrier detection is offered to appropriate individuals. Issues such as the variation in severity of any diagnosis must be addressed also.

E. Provide Psychosocial Support

The occurrence, or even the risk of occurrence, of a genetic disorder or birth defect can be devastating for the family. Emotional sequelae must be acknowledged and addressed as families deal with their anger, denial, guilt, and depression. Marital problems can be exacerbated, and difficulties with the other children may arise. The counselor's awareness of this process will facilitate dialogue with the parents through reassurance that their emotions are normal reactions and part of a healthy adjustment. Appropriate resources (e.g., support groups and literature references) can be helpful during this period of time. Other economic and community services will be offered as needed.

F. Examine Reproductive Options

Couples at an increased risk for a genetic disorder face a number of reproductive alternatives. Some couples will decide to have children after considering various factors important to their situation. For many genetic conditions, prenatal diagnosis is available and will offer an acceptable alternative. Amniocentesis, the withdrawal of fluid from the sac surrounding the fetus, is most commonly performed at 15–18 weeks gestation with a risk of miscarriage estimated to be 0.3%. Chorionic villus sampling (CVS), an aspiration of early placental tissue, is a newer procedure performed at 10–12 weeks gestation and is thought to have a slightly greater risk of miscarriage than amniocentesis. Other options for prenatal testing include ultrasound for evaluation of the fetal anatomy and maternal serum α-fetoprotein (AFP), human chorionic gonadotropin (HCG), and unconjugated estriol (uE3) screening for open spine defects and Down syndrome. Through these techniques, a number of chromosomal, single-gene, or multifactorial disorders can be diagnosed. If an abnormality is detected through prenatal diagnosis, the options of pregnancy termination or improved obstetric management at delivery can be offered.

Some couples will decline future childbearing after weighing the risks and the burden that the affected child and/or future children could place on their lifestyle. Adoption may be considered by some of these families. For reasons related or unrelated to the genetic diagnosis, a percentage of couples will separate and change marital partners, thus lowering their genetic risks. Under special circumstances, donor sperm or eggs may be used to lower a genetic risk when prenatal diagnosis is not available or is unacceptable.

IV. PRINCIPLES OF EFFECTIVE COUNSELING

A. Counseling Guidelines

When a counseling session begins, it may be beneficial to determine the counselee's preconceptions or knowledge about the disorder in question. Misconceptions about the etiology, inheritance, or prognosis can interfere with the communication process. Both parents should be encouraged to attend so that each has the benefit of hearing the information presented and the opportunity to ask questions. Facts should be explained biologically with the aid of diagrams whenever possible and at a level appropriate for the family's comprehension. Information should be presented when facts are available and when the parents are ready for additional implications of a diagnosis.

The counselor needs to address the feelings of parental guilt that often accompany the birth of a handicapped child. Parents can often be reassured that specific environmental events that occurred during the pregnancy were not responsible for an abnormality. For couples at risk for a specific recessive disorder, the knowledge that *all* individuals are carriers for 5–10 deleterious genes can be helpful in alleviating the feeling that they are different from others.

Much information will be shared during an initial counseling session, and in many cases some form of follow-up contact with the family will be essential to their long-term comprehension. A written summary, a return visit, or a telephone contact with the genetic counselor can be effective mechanisms to reinforce the primary discussion.

B. Risk and Burden

The impact of a diagnosis of a genetic disorder on family units varies according to their perceptions of the burden. Emotional and financial implications of the disease contribute to the stress felt by parents and siblings. For many families, the long-term care requirements of a progressively degenerating disorder carry demanding economic and psychological strains. In contrast, the sudden, unexpected loss of a congenitally malformed child shortly after birth may be accompanied by intense pain, which will diminish over time. The loss is as real, but the emotional and financial demands not as great.

As with burden, perception of the risk magnitude is open to individual interpretation. Some persons will view a 1% risk as "high," whereas others will view that as "low." Many have difficulty in comprehending risk factors in general, and it can be helpful to use an odds likelihood as well as a percentage. Thus, a 3% recurrence risk can be described as one chance in 33, or a 25% risk as one chance in four. The alternative risk that a problem will *not* occur is another important perspective to offer, so that families can put their likelihood for a favorable outcome in perspective. For single-gene disorders, the recurrence risk is the same for each pregnancy.

Reproductive decisions will be made on the basis of this perceived burden and the recurrence risk of the disorder. Genetic counseling can aid in the decision process through identification of these two factors and an exploration of the couple's perceptions of risk and burden.

V. GOALS

Through this communication with genetic counselors and medical geneticists, individuals can achieve an increased understanding of the specific disorder of concern in their family. Genetic implications of the diagnosis are addressed, and risks to various family members outlined. Personal decision making is supported by the genetics team as families face questions of reproductive planning, carrier testing, and treatment for affected individuals. Options for childbearing (e.g., adoption or prenatal diagnosis) are discussed openly for the family's consideration. Ultimately, the goal of genetic counseling is to facilitate increased coping with the realities, or the prospects, of genetic disease and its impact on the family.

BIBLIOGRAPHY

Filkins, K., and Russo, J. F., eds. (1990). "Human Prenatal Diagnosis," 2nd Ed. Marcel Dekker, New York.

Harper, P. S. (1993). "Practical Genetic Counselling," 4th Ed. Butterworth, London.

Marks, J. H., Heimler, A., Reich, E., *et al.* (eds.) (1989). "Genetic Counseling Principles in Action: A Casebook," Birth Defects Original Article Series 25, Vol. 5. Alan R. Liss, New York.

Thompson, J. S., and Thompson, M. W. (1991). "Genetics in Medicine," 5th Ed. Saunders, Philadelphia.

Weaver, D. D. (1992). "Catalog of Prenatally Diagnosed Conditions," 2nd Ed. Johns Hopkins Univ. Press, Baltimore.

Genetic Diseases

MAX LEVITAN
Mt. Sinai School of Medicine

GLOSSARY

Autosomes Kinds of chromosomes normally present doubly in both males and females; in humans there are 22 pairs of autosomes

Chromosomal aberration Change in the number or position of the genes or chromosomes

Dominant Trait or gene that is expressed even if there is only one gene for it at the pertinent locus on a pair of homologous chromosomes

Genetic diseases Pathological conditions caused or influenced by abnormalities in the genetic material

Holandric inheritance Inheritance determined by genes on the Y chromosome

Locus Position of a gene on its chromosome

Maternal inheritance Heredity of traits determined by DNA on cytoplasmic organelles; being passed on almost exclusively via the ovum and not dependent on meiosis, their transmission does not follow Mendelian principles

Mutant Gene or individual that manifests a changed characteristic as a result of mutation

Mutation Change in one or more nucleotides of DNA; in its broadest usage it also includes chromosomal aberrations

Polygenic inheritance Transmission of traits that involve the additive contributions of many loci

Recessive Trait or gene that is not expressed unless both loci of a pair of homologous chromosomes are alike

Simple inheritance Inheritance in which characteristics are attributable to the effects of a single locus

X chromosome Chromosome normally present doubly in the female, but singly in the male, in humans and many other organisms

X linked Genes, or characteristics determined by them, whose loci are on the X chromosome

Y chromosome Chromosome type normally present only in males

GENETIC DISEASES ARE THOSE PATHOLOGICAL conditions that are caused or influenced by an abnormality in the genetic material. If the root cause is the change in one or more nucleotides of a molecule of DNA, that is, in a single gene, it is usually referred to as a mutation; the gene, and often the individual manifesting it, is referred to as a mutant. When the cause stems from a disturbance in the number or position of the genes or chromosomes, it is usually referred to as a chromosomal aberration. The position of a gene on its chromosome is its locus.

A mutated gene that is part of one of the 22 kinds of chromosomes normally present doubly in both males and females, and the associated genetic disease, is referred to as autosomal. However, if it is part of the X chromosome, the one that in humans is normally present doubly in the female but singly in the male, it is referred to as X linked (formerly called "sex linked"). A few genes are known on the Y chromosome, normally present only in males (i.e., holandric inheritance), and on the mitochondrial cytoplasmic organelles; the latter (and associated diseases) are maternally inherited, being transmitted via the ovum.

I. SIMPLY INHERITED DISEASES

When the genetic disease is attributable to change in a single gene locus it is said to be simply inherited. A simply inherited disease is autosomal recessive if the affected person has abnormal autosomal genes for the same locus on the chromosomes received from both parents, autosomal dominant if the affected individual needs to have only one abnormal gene at the pertinent autosomal locus, or X linked if the gene in question is on the X chromosome.

Simply inherited autosomal recessive diseases generally involve genes that normally control the production of enzymes, organic catalysts that enable cells to carry on metabolic processes rapidly at the relatively low temperatures of the human body. When only one such abnormal gene is present, the amount of the enzyme is usually decreased, but enough enzyme is produced via the normal gene present to enable the metabolic process to go on normally.

V. A. McKusick's catalog lists 1730 autosomal recessive phenotypes, of which about 600 can be considered well-established genetic diseases. Most are very rare, occurring in one per 50,000 births or less, but some are more common, particularly in certain ethnic groups. In persons of northern European descent, for example, cystic fibrosis occurs in about one per 2500 births. By contrast, its incidence in Black populations is about one per 17,000 births, and even less in Asians. Similarly, the frequency of Tay-Sachs disease (type I gangliosidosis) is about one per 4000 births in Ashkenazi Jews, about 10,000 times its frequency in Sephardic Jews or Gentiles.

Known enzyme deficiencies lend themselves to prenatal diagnosis by growing embryonic cells obtained from the pregnant woman by various techniques and determining whether they can produce the pertinent enzyme. Molecular genetics has extended these procedures to detect with increasing accuracy the presence or absence of recessive genes (e.g., the ones involved in cystic fibrosis). Prenatal diagnosis is especially useful when a previous child has been born with the recessive defect. Since each subsequent child has one chance in four of being similarly affected, the parents often would hesitate to risk having further children. Prenatal diagnosis demonstrating that the fetus is probably normal can allay these anxieties. The contrary finding, that the fetus is probably affected, poses the psychological and ethical dilemma between selective abortion and the birth of a defective child.

The genes responsible for autosomal dominants generally affect the structural proteins of cells (e.g., hemoglobin, collagen, and myosin). Usually a person producing less than the normal amount of such a protein, because one of his or her genes for it is abnormal, manifests a detectable abnormality.

Although McKusick's catalog lists 4458 autosomal dominant phenotypes, about 650 can be considered diseases, for many of the dominants determine normal variants of serum proteins and cell-surface molecules that act as receptors for circulating substances or provide antigenic specificities to the cells. The latter are the basis of blood groups, such as ABO and Rh. Some of these variants can, however, have pathological consequences. Hemolytic disease of the newborn, for example, may result from certain combinations of Rh, ABO, Duffy, Kidd, and other blood groups in parents and child. Another large group causes physical abnormalities, such as various forms of polydactyly (extra fingers or toes) or brachydactyly (short fingers), that may be noted by a physician (or dentist, if they affect the teeth) but generally have no pathological sequelae.

Sometimes autosomal dominants cause relatively mild diseases in single dose, but rather severe ones when present doubly. Therefore, the more serious diseases are, in effect, recessive. A prominent example is the gene for S hemoglobin. Persons having two such genes (SS) are affected with sickle-cell anemia. Those having one S gene and one normal gene (SA) have milder conditions, but when two SAs mate they may produce SS offspring. Similarly, persons affected with certain thalassemias, many of them so severe as to be lethal early in life, have double doses of genes that result in reduced production of various globins, the protein molecules that form a major component of hemoglobin. Persons with single doses of the same genes, on the other hand, rarely come to clinical attention.

Another serious condition, familial hypercholesterolemia (FH), results from mutations at the locus for production of the cells' receptors for low-density lipoprotein (LDL), the so-called "bad lipoprotein." Cells with one normal and one abnormal gene are less able to capture cholesterol from the blood for its normal uses to build cell membranes, secrete steroid hormones, or produce bile salts. This results in a greater tendency for abnormal accumulations of LDL and its deposition in the linings of critical vessels (e.g., the arteries of the heart), and this increases the likelihood of myocardial infarction (i.e., heart attack). About one in 500 people have single doses of these defective genes, making familial hypercholesterolemia perhaps the most common of all simple Mendelian disorders. The rarer persons with double dose almost invariably develop very high cholesterol levels in the blood and fatal cardiac effects at an early age.

Of the 412 apparent X-linked phenotypes about 150 qualify as diseases, though many (e.g., red–green color blindness, testicular feminization, and several types of deafness) are not seriously debilitating. However, among the diseases are two serious clotting defects (i.e., hemophilias) and several forms of muscular dystrophy and immunodeficiency.

II. POLYGENIC INHERITANCE

The role of genes in common diseases such as diabetes, hypertension, peptic ulcer, bipolar (manic–depressive) mental disorders, Alzheimer's disease, and schizophrenia is currently not well understood, partly because of considerable heterogeneity, that is, conditions with different etiologies may be lumped under the same clinical rubric. Furthermore, many involve complex interactions of genetics and environment. Although occasional reports announce that single important genes governing these traits have been discovered, most geneticists believe that the underlying physiology of these conditions depends on the additive contribution of many genes. These produce a wide spectrum of variation, with those above (or below) a threshold number labeled as affected. Such a multiple factor or polygenic hypothesis probably also accounts for the genetic aspect of birth defects such as spina bifida, cleft palate with or without cleft lip, congenital dislocation of the hip, pyloric stenosis, clubfoot, Hirschsprung disease, and some congenital heart defects and mental deficiencies.

III. CHROMOSOMAL ABERRATIONS

Multiple congenital defects, some mimicking those determined by polygenes, are characteristic of most disorders that result from chromosomal aberrations. These may be divided into two classes: (1) aberrations of number, the affected individual having more or fewer chromosomes per cell than the normal 23 pairs, and (2) aberrations of chromosome structure, situations in which one or more genes are missing, are present in more than their usual numbers per cell, or are present in unusual locations. Aberrations of number are usually due to errors in cell division, whereas errors of structure usually stem from unrepaired chromosomal breakage. However, many aberrations of structure have pathological effects only because they lead to offspring with deficient or excessive numbers or parts of chromosomes. That a deficiency of genes could be harmful is understandable, but the basis for the ill effects of an excess is not well understood. Apparently normal well-being depends on an evolutionarily developed "balance" among the genetic materials that may be disrupted by an excess as well as by a deficiency. [*See* Chromosome Anomalies.]

Aberrations of number are most viable if they involve the X or Y chromosome. Indeed, females with an extra X chromosome (47,XXX) and males with an extra Y (47,XYY) show few, if any, abnormalities. 47,XYY's do tend to be quite tall, and that may be responsible for their greater than random tendency to be involved in criminal activities. (This does not mean that tall men are usually XYY, nor does it mean that XYY are usually criminals.) Males with extra X chromosomes (e.g., 47,XXY; 48,XXXY; 49,XXXXY) exhibit the Klinefelter syndrome; they tend to be infertile and to exhibit various degrees of mental deficiency. Females with more than one extra X chromosome (e.g., 48,XXXX; 49,XXXXX) are likely to be mentally defective and have other symptoms. Females with only a single X chromosome (45,XO) tend to have the Turner syndrome. Its prominent features are short stature, absence of secondary sex characteristics, and ovaries consisting of streaks of connective tissue. There is some evidence that the viable XO's lost the other sex chromosome after fertilization, so that they are or were mosaics of normal and abnormal cells, whereas nonmosaic 45,XO's (the other sex chromosome having been absent from the egg or sperm) usually die *in utero*.

Deficiency of a whole autosome is invariably lethal. Spontaneous abortuses do sometimes contain all the chromosomes in triplicate (triploids) or quadruplicate (tetraploids). Individuals with a single extra autosome (trisomies) appear more frequently. Those that come to term invariably have multiple congenital anomalies and, unless they are mosaics that have many normal cells as well, usually die within a few weeks or months after birth. The exceptions are trisomies for the smallest autosome, labeled number 21. Trisomy-21 is also called Down's syndrome, as the English physician J. L. H. Down first described it clearly; in the literature it is also (unfortunately) called "mongolism" or "mongolian idiocy." It appears in about one per 700 births, more often from older mothers than younger ones. Although also generally beset with congenital anomalies, including varying degrees of mental deficiency, 21-trisomics that survive the perinatal period have good chances of viability and, in the case of females, may even be fertile, with the potential of normal offspring as well as trisomics.

Many partial trisomies and deficiencies have been described. As their effects vary considerably, depending on the nature or amount of the chromosome involved, few clear syndromes of them have been delineated to date. It is known, however, that defects formerly attributed to gene mutations are often due to deletions of whole genes or their critical parts. This is particularly true of many of the thalassemias.

Partial trisomies (sometimes also called duplications) and deficiencies may result from misalignment of normal chromosomes during germ cell production. Deficiencies may also be produced when a chromosome breaks in two places on the same side of its spindle attachment point (i.e., centromere) and reheals without incorporating the substance between the breaks; these are the ones often referred to as "deletions." If the two breaks are on different sides of the centromere and the two terminal pieces are lost, fusion of the two broken ends results in a ring chromosome. Many partial trisomies and deficiencies are, however, the by-products of sperm or ovum production in carriers of inversions or translocations. Such carriers are normal but may have abnormal offspring. An inversion is produced when a chromosome breaks in two places and the substance between the breaks is reincorporated in reverse order. Reciprocal translocations, found in about 2 per 1000 births, occur when two chromosomes suffer one break each, and the chromosomes exchange the broken-off pieces. In the rarer insertional translocation or transposition, the chromosome fragment resulting from two breaks in one chromosome is inserted at a third break. The latter break may be in the same chromosome, in which case it is often called a shift, or it may be in another chromosome. Some Down's syndromes result from carriers of reciprocal translocations involving chromosome 21.

Specific structural aberrations are often associated with diseases. Several have proved useful not only for better understanding the genetic basis of certain cancers but also for locating the genes involved. For example, a translocation, originally thought to be a deletion known as the Philadelphia chromosome because it was first discovered there, is quite consistently associated with chronic myelocytic leukemia, and a detectable deletion of a piece of chromosome 11 has been significant in research on Wilms' kidney tumor. Similarly, the genetics of mental retardation has been greatly advanced by the discovery of the "fragile X chromosome."

IV. GENETICS OF TUMORS AND CANCERS

Susceptibility to tumors or cancers has long been recognized as a genetic disease, but the exact role of heredity has not been clear. A number of syndromes characterized by the development of neoplasia seem to be simply inherited. One of the best understood is xeroderma pigmentosum, a skin disorder resulting from mutants of loci that normally produce enzymes used to repair DNA that has been damaged by ultraviolet light. Persons with single doses of these genes often exhibit a large degree of freckling but do not develop the malignancies. Several other recessive disorders predispose to leukemias or lymphomas, but they may result in tumors at other sites as well. Simply inherited dominant forms most often involve tumors of endocrine glands, skin, or mucosa. Being dominants, the family history can often alert physicians to relatives at risk and, by early detection of cancerous or precancerous lesions, prevent more serious effects. This has been particularly successful in several forms of multiple intestinal polyposis. Similar considerations apply to apparent "cancer families," whose members seem unusually susceptible to a wide variety of neoplasms.

Some of the most frequent cancers, for example, of the breast, uterus, stomach, and colon, do not fit models of simple inheritance because most of the cases are sporadic, rather than familial, that is, they occur in no other relatives. However, recent research has established that the same genes may be involved in the sporadic as well as the familial cases.

Prevailing opinion is that most, possibly all, neoplasia develop because both alleles of certain loci involved in the regulation of cell growth become defective, allowing unregulated cell growth. This is often referred to as the "two-hit" theory. The genes that have been incriminated in this process appear to fall into three classes: (1) genes that normally act as tumor suppressors, sometimes called "antioncogenes"; (2) oncogenes, so named because, like oncogenic viral genes with similar structures, they are capable of causing the transformation of certain tissue cultures from regular monolayer growth to growth in multilayered jumbles resembling the way cancer cells grow in culture; and (3) genes that are responsible for maintaining correct DNA reproduction or repairing if it goes awry.

According to the two-hit theory, in the familial cases a defective gene in one of these classes is inherited, which is why familial tumor susceptibilities often seem to be autosomal dominants. However, its presence does not become apparent unless its allele also mutates in the affected organ sometime during later development (a "somatic mutation"). No defective gene is inherited in the sporadic cases, so that two somatic mutations, often many years apart, must occur at the locus for the cancer to develop. This helps to explain why familial cancers tend to start earlier in life and are more often bilateral than sporadic ones of the same tissue. This contrast was first well established for retinoblastoma, one of the most frequent childhood malignancies. The pertinent locus belongs to the tumor suppressor class, as do those associated with, among others, two kidney neoplasia, Wilms' tumor and von Hippel-Lindau disease, and two forms of neurofibromatosis.

Advances in this field are proceeding rapidly as more and more pertinent genes are located, identified, and cloned, and their normal functions determined. Particularly significant are the recent discoveries of breast cancer genes *BRCA1* and *BRCA2* and genes for adenomatous polyps (FAP and APC) and one type of colon cancer (DCC). [*See* Cancer Genetics.]

BIBLIOGRAPHY

Borgaonkar, D. S. (1994). "Chromosomal Variation in Man: A Catalog of Variants and Anomalies," 7th Ed. Alan R. Liss, New York.

Knudson, A. G. (1993). Antiocogenes and human cancer. *Proc. Natl. Acad. Sci. USA* **90**, 10914–10921.

Levine, A. J. (1995). The genetic origins of neoplasia. *J. Am. Med. Assoc.* **273**(7), 592.

Levitan, M. (1988). "Textbook of Human Genetics," 3rd Ed. Oxford Univ. Press, New York.

McKusick, V. A., with the assistance of Francomano, C. A., Antonarakis, S. E., and Pearson, P. L. (1994). "Mendelian Inheritance in Man: Catalogs of Autosomal Dominant, Autosomal Recessive, and X-Linked Phenotypes," 2 vols., 11th Ed. Johns Hopkins Univ. Press, Baltimore.

Milunsky, A. (ed.) (1992). "Genetic Disorders and the Fetus: Diagnosis, Prevention, and Treatment," 3rd Ed. Johns Hopkins Univ. Press, Baltimore.

Genetic Imprinting

MAX LEVITAN
Mt. Sinai School of Medicine

GLOSSARY

Angelman syndrome Rare congenital disorder characterized by severe motor and mental retardation, enlarged mandibles, and often by seizures, jerky movements, hand flapping, spamodic outbursts of laughter, and absence of speech; also known as "happy puppet syndrome;" associated with the same 15q deletion as the Prader-Willi syndrome

Huntinton chorea Dominantly inherited disease having a degenerative effect on the brain, resulting, usually with late onset, in involuntary spasmodic movements of limb and facial muscles, mental deterioration, and eventual death

Lyon hypothesis Random inactivation of most or all of one of the X chromosomes of the female, thus accounting for dosage compensation of X chromosomes in mammals

N-myc Proto-oncogene of the *myc* family, commonly found in neoplasms, including many neuroblastomas and some lung cancers; when amplified it is associated with shorter survival

Prader-Willi syndrome Rare congenital disorder characterized by short stature with small hands and feet, mild to moderate mental retardation, hypogonadism, poor feeding, and obesity beyond infancy; associated with a deletion of a section of the long arm of chromosome 15

GENETIC IMPRINTING REFERS TO THE DIFFERENT expression of chromosomes, parts of chromosomes, or single genes depending on which of the two sexes they are inherited from. To achieve imprinting, some genetic materials can be modified during gamete production or early embryonic development in one of the two sexes, so the traits determined by the imprinted genes are expressed differently than would be expected under typical Mendelian inheritance.

I. IMPRINTING IN HUMAN PATHOLOGY

Genetic imprinting may be illustrated by one of the first human conditions in which the phenomenon was detected: multiple glomus tumors. Usually a benign neoplastic disorder, this is often inherited as an autosomal dominant. In certain families, all of the affected individuals may be traced to an affected male; none of the offspring of affected females shows the trait. Presumably the pertinent gene is inactivated during oogenesis but not during spermatogenesis. However, the normal sons of affected females may have children with the tumors, indicating that the gene may be reactivated during a subsequent spermatogenesis.

Strong evidence for involvement of genomic imprinting has appeared in a number of other inherited tumors, such as paraganglioma (carotid body tumors), retinoblastoma (and associated osteosarcoma), neuroblastoma, and the Beckwith-Wiedemann syndrome (and embryonal tumors associated with it: Wilms' tumor, adrenocortical carcinoma, hepatoblastoma, or embryonal rhabdomyosarcoma).

Imprinting is also involved in the Prader-Willi syndrome (see Glossary). This syndrome is strongly associated with deletion of a small section of the long arm of chromosome 15, but only if the deletion is on the chromosome 15 derived from the *father* of the affected child. If a similar deletion is on the chromosome 15 derived from the mother, a different condition appears, the Angelman syndrome (see Glossary). It is not known whether the same genes in the chromosome 15 region are involved in the different imprintings of the two syndromes.

A growing body of evidence in both mice and humans points to methylation of cytosine residues in the context of cytosine–guanine (CpG) dinucleotides as the mechanism of imprinting. Such methylation, especially if it occurs in the promoter regions of genes, can nullify the ability of the genes to be transcribed. Certain genes that can be imprinted will be methylated in the production of sperm, others in the production of ova, and they can be reactivated by demethylation when they pass through gametogenesis in the opposite sex. It is still not known why certain alleles are subject to imprinting while others are not, and why they are more likely to be imprinted in one sex than the other.

The methylation hypothesis is strengthened by evidence from mouse genetics that some imprinted genes become unmethylated, that is, lose their imprinting, during the morula (12–16 cells) and early blastula preimplantation stages of embryonic development, but become methylated once again, that is, regain their imprinting, in the subsequent implantation, late blastocyst stage, and the rest of their development and their adult life. This correlates well with the fact that an enzyme that is critical for methylation, DNA methyltransferase, is in short supply during the same stages as imprinting is temporarily lost. [*See* DNA Methylation in Mammalian Genomes.]

II. IMPRINTING AND THE LYON HYPOTHESIS

According to the Lyon hypothesis, named for its discoverer, the British mouse geneticist Mary Lyon, most or all of one X chromosome becomes inactivated in the early female embryo. Many consider this phenomenon to be a form of imprinting, inasmuch as the same basic mechanism, methylation of the inactivated genes, is at work.

III. IMPRINTING AND AMPLIFICATION

Imprinting can involve amplification of genes rather than inactivation, that is, as the gene passes through gametogenesis in one of the sexes, sections of it become duplicated and the gene thereby enlarged. For example, in neuroblastoma, one commonly finds an increased number of DNA segments containing the *N-myc* protooncogene, and such amplification is correlated with poor prognosis of the disease. In an overwhelming proportion of cases it is the paternal *N-myc* gene that is amplified, suggesting that imprinting is responsible. A similar phenomenon occurs in Huntington chorea, an autosomal dominant condition that usually does not become manifested until middle age or beyond. (One of its most famous victims was the folk singer Woody Guthrie.) Earlier manifestations, often even in childhood, is associated with amplification of certain segments of DNA in the gene, but the amplifications, if they occur, are only in Huntington chorea genes inherited from fathers. [*See* Gene Amplification; Human Genetics; Oncogene Amplification in Human Cancer; Oogenesis.]

BIBLIOGRAPHY

McKusick, V. A., Francomano, C. A., and Antonorakis, S. E. (1990). "Mendelian Inheritance in Man: Catalogs of Autosomal Dominant, Autosomal Recessive, and X-Linked Phenotypes," 9th Ed. Johns Hopkins Univ. Press, Baltimore.

Nicholls, R. D. (1993). Genomic imprinting and candidate genes in the Prader-Willi and Angelman syndromes. *Curr. Opin. Genet. Dev.* **3,** 445–456.

van der Mey, A. G. L., Maaswinkel-Mooy, P. D., Cornelisse, C. J., *et al.* (1989). Genomic imprinting in hereditary glomus tumours: Evidence for new genetic theory. *Lancet* **2,** 1291–1294.

Tycho, B. (1994). Genomic imprinting: Mechanism and role in human pathology. *Am. J. Pathol.* **144,** 431–443.

Genetic Maps

MICHAEL DEAN
Frederick Cancer Research and Development Center

I. Mapping in Experimental Organisms
II. Human Genetic Maps

GLOSSARY

Allele One of the forms of a gene

Centimorgan (cM) Measure of genetic distance equal to 1% recombination. In humans, 1 cM is, on the average, 1 million base pairs

Dominant Allele that shows its phenotypic effect even in the presence of another (recessive) allele. Codominant alleles are expressed simultaneously

Eukaryote Organism composed of a cell or cells with membrane-bound nuclei and organelles and multiple chromosomes; includes all plants and animals

Genome Full complement of genetic material of an organism. The DNA sequences contained in all the chromosomes

Genotype Genetic makeup of an individual, including silent alleles. The genotype determines the phenotype, the expressed characteristics of an individual

Heterozygote Individual who has two different alleles for the same gene. Homozygotes have identical alleles

Interference Effect of recombination at one position on the recombination frequency at an adjacent position. The value can be positive or negative

Linkage Proclivity for two genes on the same chromosome to be inherited together. The distance is measured in percent recombination or centimorgans

Meiosis Division of the nucleus that results in the production of haploid germ cells (sperm and eggs)

Microsatellite DNA sequence consisting of a tandem repeat of two to six base pairs; also known as a simple-sequence tandem repeat (STR). The most common repeats are CA/GT repeats. Most microsatellites are polymorphic in length and are therefore useful genetic markers

Molecular cloning Isolation and insertion of a DNA segment into a vector that can be propagated in large quantity. Typically fragments from complex organisms are cloned into bacteria

Operon Group of adjacent genes that perform a coordinated process, usually sequential reactions in a metabolic pathway

Recombination (crossing-over) Shuffling of genetic material that accompanies sexual reproduction and results in new combinations of genes in the progeny

Restriction enzyme Bacterial protein that cleaves DNA at specific positions, allowing physical maps to be constructed. It is the workhorse of recombinant DNA and genetic engineering

Somatic cell hybrid Mixed cell line formed by the fusion of cells from two different species, typically containing the full chromosome complement of one species and a partial set of the other

IN THE 1860s GREGOR MENDEL DESCRIBED THE first principles of genetics, and in the early 1900s analysis of chromosomes from the fruit fly *Drosophila melanogaster* led to the concept of genetic mapping. A genetic map is the blueprint of an organism. It is the linear arrangement of the genes of that species, constructed by measuring the distance between individual genes or genetic markers. The making of a genetic map is one of the first steps toward the complete characterization of the genetic material of an organism, the determination of the total nucleotide sequence of the species. Mapping of genes that cause disease or unusual phenotypes is a crucial step toward the isolation of each gene. Genetic maps have also proved to be valuable in the study of species biology. Today, genetic maps have been constructed for dozens of species from bacteria to humans.

ENCYCLOPEDIA OF HUMAN BIOLOGY, Second Edition, VOLUME 4.

I. MAPPING IN EXPERIMENTAL ORGANISMS

A. Bacteria

Most of the gene mapping to date has been performed on experimental organisms. Many of the organisms studied have fairly simple genetic structures, which allow the ready performance of genetic experiments. The most complete genetic map of a free-living species is that of the bacterium *Escherichia coli.* This organism has been studied in great detail by both biologists and geneticists. Many of the genes of *E. coli* have been identified. Each gene encodes a single protein with a specific function, such as the ability to utilize a certain sugar molecule or synthesize an amino acid. Because this bacterium grows rapidly, and its genetic material is relatively small and can be easily manipulated, maps could be quickly constructed.

The genetic map of *E. coli* is displayed in Fig. 1. A crucial discovery in the development of genetics in *E. coli* was the finding that bacteria can mate or exchange genetic material. Mating occurs when a "male" bacterium binds to another recipient bacterium. Male bacteria contain an extrachromosomal segment of DNA, the F factor, which encodes a structure known as a pilus present at the bacterial surface, through which genetic material can be transferred. In this process, known as conjugation, the F factor itself is transferred. Bacterial strains were identified in which the F factor was incorporated into the bacterial genome. When these bacteria conjugate, parts of the donor's DNA are transferred with the F factor. It was found that for a given strain this process always starts

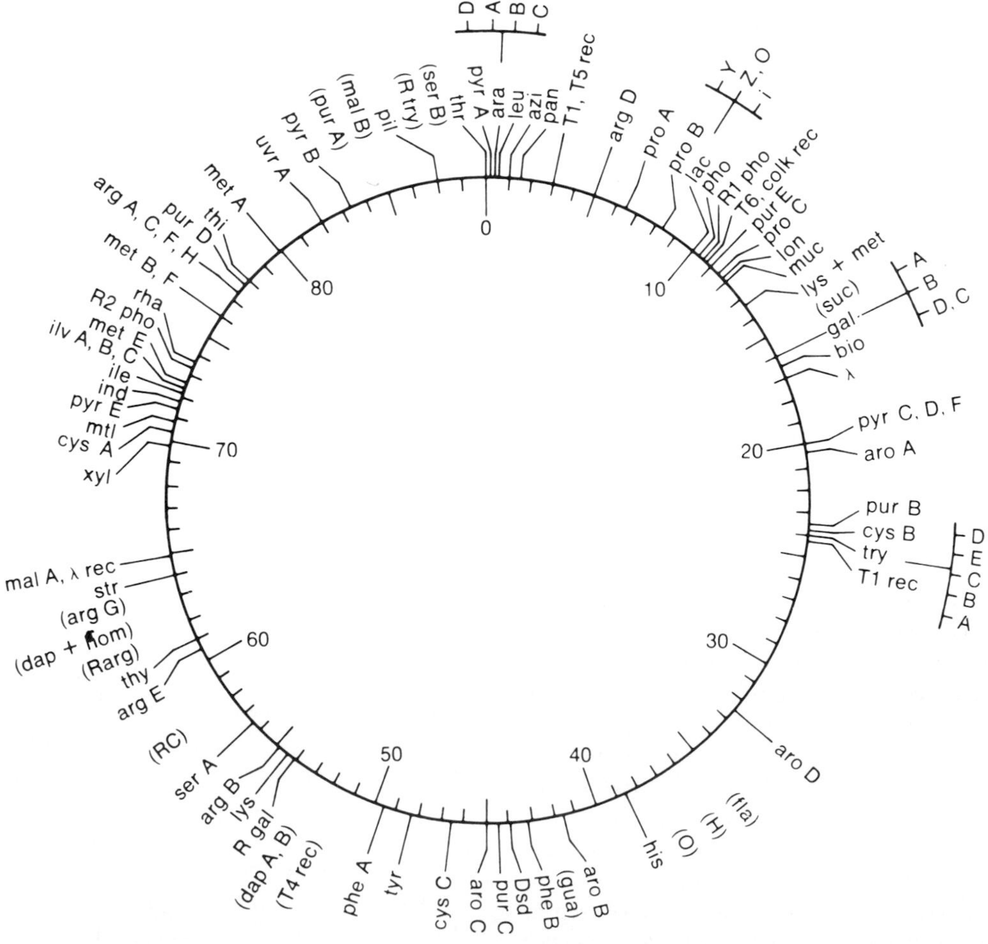

FIGURE 1 A genetic map of *E. coli.* The map is displayed in units of minutes. The position of several operons is shown outside of the central map. [From A. Taylor and M. Thoman (1964). *Genetics* **50,** 667, with permission.]

at the same place in the genome (the point of integration of the F factor) and always proceeds in the same direction at a constant pace. By interrupting the conjugation process at different times (it takes 90 min to transfer the whole genome), the order of genes can be determined. That is, the genes closest to the F factor are always transferred first, whereas distal genes take longer.

Additional methods for mapping bacterial genomes also involve the transfer of segments of naked DNA, a process termed *transformation*, or the use of viruses to carry bits of genetic material, termed *transduction*. In either case the principle is the same. Genetic markers that are close to each other are transferred together more often than markers that are far apart. By determining the frequency in which pairs of markers travel together, a map of the genes can be built. The entire genome of *E. coli* has been molecularly cloned and the fragments mapped by restriction enzymes. The completion of the ultimate map of *E. coli*, the complete nucleotide sequence, is over halfway complete. To date, the genomes of certain viruses, bacteria and unicellular eukaryotes have been completely sequenced.

One important outcome of the mapping of genes in *E. coli* was the discovery that many genes whose proteins are made at the same time and work together in a biochemical pathway are near each other in the genome. Blocks of genes that are coordinately regulated are called *operons*.

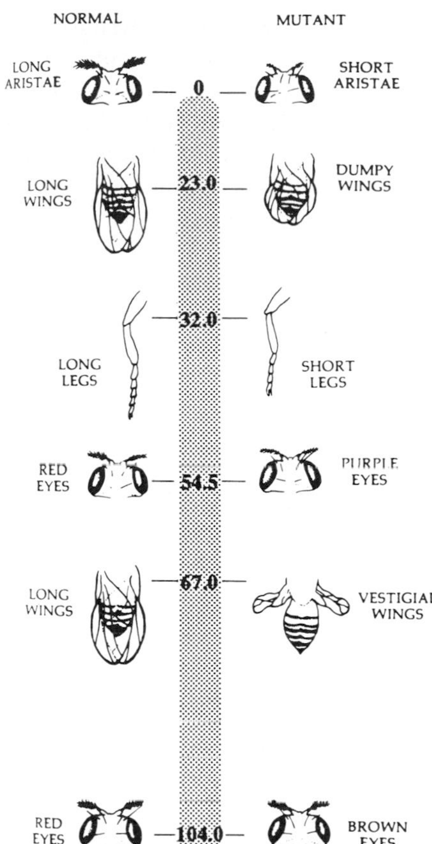

FIGURE 2 Map of several genes on *Drosophila melanogaster* chromosome 2. The position of several genes is shown along with the genetic distance between them, calculated by the recombination frequency. [From H. Curtis and N. S. Barnes (1989). "Biology." Worth Publishers, New York, with permission.]

B. Mapping in Eukaryotic Organisms

Genetic maps of higher organisms are constructed by following the inheritance of genetic markers from parent to offspring. Many of the concepts of genetics were developed by analyzing the inheritance of visible traits (for instance, the color of the eyes, the length of the bristles) in the fruit fly *Drosophila melanogaster*. T. H. Morgan determined in 1911 that certain traits were inherited together more often than expected by chance, suggesting that they were physically linked. A. H. Sturtevant used this approach to build the first map of a eukaryotic chromosome. Figure 2 shows the genetic location of several traits found on a portion of the *Drosophila* second chromosome, along with a depiction of these traits. The genetic positions of several mutations that cause visible traits in the fly are shown along with a depiction of those traits.

We now know that the DNA of all eukaryotes is organized into chromosomes. Most individuals are diploid [i.e., they carry two copies of each chromosome (except the sex chromosomes), one inherited from each parent]. During meiosis, when sperm and egg cells are generated, there is the opportunity for genetic exchange between the chromosome pairs. This process is known as *recombination* or *crossing-over* and leads to the reassortment of alleles (different forms of a gene). Markers that are close together on a chromosome rarely recombine, and those farther apart recombine more frequently. This feature allows the order of genes to be determined by measuring the recombination rate of pairs of markers. Thus a genetic map of a sexual organism is generated by determining the recombination frequency of a set of genes on a chromosome. [*See* Chromosomes; Meiosis.]

Recombination, however, is not a simple process. Although it does give definitive evidence of the order of markers, it does not occur evenly over a chromosome, and it occurs at different rates in the two sexes.

Drosophila offers an extreme example, in which all recombination occurs in females and none in males. A crossover in one part of a chromosome can also affect recombination further down the chromosome, a phenomenon known as *interference*. An example would be three genes, A, B, and C, in which there is 20% recombination between A and B and 30% recombination between B and C. We would expect that a double recombinant with recombination between A and B and also between B and C would occur at a frequency of 6% ($0.20 \times 0.30 = 0.06$). However, the frequency of the double recombinant may be only 2%. In other words, when a recombination happens between A and B, crossing-over between B and C is interfered with or vice versa. This may be a merely physical phenomena, for example, a limitation in the bending of the DNA that carries the genes. However, the degree of interference varies from species to species and region to region, suggesting that the control of recombination is complex.

The assembly of the maps of more than 100 organisms has begun, from viruses to mammals, including humans and plants. Well-studied creatures are yeast, the simplest and best-understood eukaryotic cell; corn, in which many of the principles of segregation were discovered as well as the presence of mobile genetic elements similar to the bacterial F factor; *Caenorhabditis elegans,* a small transparent worm whose every cell can be followed visually from birth to adulthood; and the domestic cow, an important food resource for much of the world. The field of genetics is diverse and touches all the other biological sciences.

C. Mapping the Lab Mouse

Although the studies on bacteria, flies, molds, and plants laid the groundwork for the discipline of genetics, researchers sought an organism that would provide a more useful model for human biology. The clearly superior choice has been the laboratory mouse, *Mus musculus*. Although we might not think we have a great deal in common with a rodent, consider that we are both mammals. Both species give birth to live young, lactate, and grow hair. Mice and humans both have complex brains, intricate behavior patterns, and the ability to learn. We both have much the same organ systems, with quite similar metabolic pathways. Mice, like humans, develop cancer and suffer from viral infections, neurological disorders, and birth defects.

The key to the establishment of mouse genetics was the discovery of pure (inbred) strains with visible genetic traits, whose inheritance could be followed. For centuries the Japanese, mostly as a hobby, bred and collected mice with interesting colors of fur. These lines formed stocks for many of the present strains of experimental animals. By creating purebred strains and crossing them, the genes that determine coat color have been mapped to linkage groups (i.e., chromosomes). During the past several decades, similar mapping has been achieved for a host of traits. Careful observation of traits arising by mutation and the maintenance of such strains, mainly at the Jackson Laboratory in Bar Harbor, Maine, have led to the collection of hundreds of mutant animals. Mouse models for human diseases such as muscular dystrophy, blindness, albinism, colon cancer, obesity, and diabetes have been developed. In addition, there are strains with high susceptibility to cancer and resistance to various viruses and bacteria. For the most part, these phenotypes are controlled by single genes, which have been located on the genetic map of the mouse. One by one these interesting genes are being isolated and characterized, advancing our understanding of mammalian biology.

Mouse genetics has profited enormously from the development of molecular biology. By using cloned DNA fragments to probe specific parts of the genome, any gene or DNA segment can be placed on the mouse genetic map. The principle is illustrated in Fig. 3. Restriction enzymes can be employed to break mouse DNA into fragments of characteristic lengths, which can then be detected with radioactively labeled cloned DNA segments. Sometimes the length of the fragments detected are different between different strains of mice, allowing the inheritance of that particular DNA segment to be followed. This difference is referred to as a restriction fragment-length polymorphism (RFLP). RFLPs have been employed to great advantage in the interspecies backcross. This type of experiment starts with two different subspecies of mice that can breed and form fertile offspring. Animals from the first (F1) generation of mating are then bred (backcrossed) to animals from one of the parent species, and the offspring of this mating are analyzed. In the example shown in Fig. 3, spretus mice are mated to the black/6 (B1/6) inbred strain and the female F1 animals backcrossed to B1/6 mice. Because of the relatively large genetic distance between spretus and B1/6, differences in the size of restriction fragments are frequently detected. The offspring inherit a B1/6 fragment from their father, because he is pure B1/6. They will inherit either a spretus or B1/6 band from the mother, in approximately a 1 : 1 ratio. If the inheri-

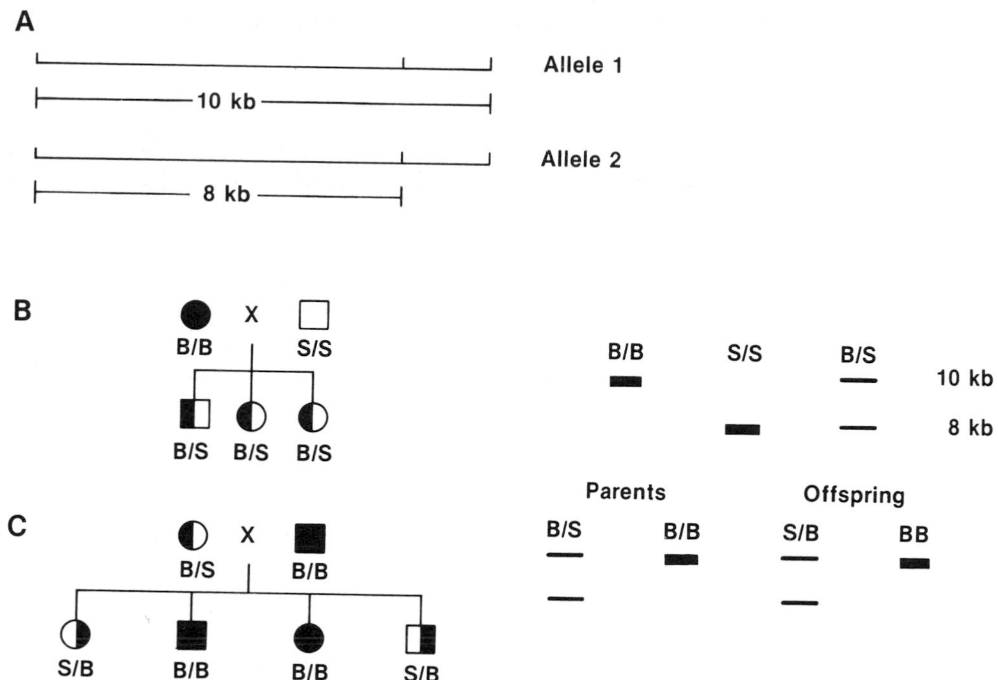

FIGURE 3 Restriction fragment-length polymorphism (RFLP) and the interspecies backcross. (A) An illustration of an RFLP detected using a specific probe and the restriction enzyme X. Two alleles are detected; allele 1 is 10 kb and found in B1/6; allele 2 is 8 kb and is specific to spretus. (B) Generation of the F1 animals for a backcross. Female (round symbols) inbred B1/6 mice are mated to male (square symbols) spretus mice (a closely related species of mice). The resulting offspring inherit one chromosome from each of their parents and are all genetically identical. Also shown is the pattern of an RFLP in these animals in which the 10-kb allele is present in Black/6 and the 8-kb allele in spretus. (C) The backcross. Female F1 animals are mated back to male Black/6 animals. All animals inherit a Black/6 chromosome from the father and either a Black/6, spretus, or a recombined chromosome from the mother.

tance of several markers is determined in the same set of animals, we should find either that they segregate randomly (are inherited together only about half the time) or that they display linkage (are inherited together more often). By determining the proportion of animals that display recombination between two markers, the genetic distance between them can be computed. An example is shown in Table I. Ten backcross animals were typed with probes for two genes, A and B. The data are expressed as S or B to designate whether the animal inherited the spretus or the B1/6 form of the gene from the heterozygous parent. For 8 of the 10 animals, the same form of the gene was inherited. This suggests that the two genes are linked and the recombination distance between them is 20%, two recombinations of 10 meioses. By studying several markers from the same chromosome, a map can be constructed.

Many mammalian genes have been identified based on their ability to code for specific proteins. The DNA that encodes a number of these genes has been isolated. In some cases, alterations in the DNA sequence of the gene (mutations) that cause disease has been identified. Because physical mapping with DNA fragments does not rely on the identification of a visible trait, a large number of markers can be followed in

TABLE I
Backcross Mapping of Two Genes[a]

	Backcross animal									
	1	2	3	4	5	6	7	8	9	10
Gene A	S	S	S	B	S	B	B	B	B	S
Gene B	S	S	S	B	S	S	B	S	B	S

[a]RFLPs for two genes were used to type 10 spretus(S)/B1/6(B) backcross animals. Two recombinations are observed between genes A and B.

the same set of animals. This technique also allows any gene to be mapped, whether or not mutations in the gene cause a discernible phenotype. By combining the mapping of mutants with the physical mapping of DNA segments, the position of the genes that cause mutations can be accurately placed, an important step toward the isolation of the actual gene. The number of genes placed on the mouse genetic map is increasing at an ever-faster pace, as are the number of exciting discoveries from this effort.

II. HUMAN GENETIC MAPS

A. Overview

The genetics of humans is the most complex and difficult to study. It cannot follow the principle of mouse genetics, which depends on the generation of pure strains of individuals, crosses between selected individuals, and the generation and selection of useful mutants. Human genetic analysis must depend on the use of the experiments that nature provides. Human geneticists have made use of genetic diseases and other heritable traits in large families and closed societies with some degree of inbreeding. The reward has been the characterization of hundreds of genetic diseases in humans, the identification and molecular cloning of many of the genes involved in disease, and the ability to make accurate predictions and diagnoses for affected families with several disorders. [*See* Genetic Diseases.]

The earliest human genetic markers used were proteins present in multiple forms in the population. Probably the most familiar are the proteins of the ABO blood group, which are found on the surface of blood cells. There are three alleles in this system (A, B, and O) and four possible phenotypes (A, B, AB, and O) (Table II). O is actually the absence of this protein on the cell (a null allele), and A and B are dominant over O. That is, if we inherit the A allele from our mother and the O allele from our father, the red blood cells will express the A protein on the surface and will be blood type A. A and B are codominant, so if we inherit both alleles, both proteins are expressed and blood type is AB. From the genotypes of the parents, we can predict the possible genotypes of their offspring and follow the inheritance of the alleles in each child.

TABLE II
ABO Blood Groups

Blood group	Genotype
O	O/O
A	A/A, A/O
B	B/B, B/O
AB	A/B

The pool of human markers expanded with the identification of isozymes, enzymes present in multiple allelic forms. However, the number of isozymes is rather small, and the size of the human genome is rather large (23 chromosome pairs and 3 billion nucleotides). Although the principle of using protein markers to detect linkage with human genetic diseases was proposed by J. B. S. Haldane as early as 1937, human genetic mapping did not start in earnest until the 1970s, not through genetics but by using somatic cell hybrids. Somatic cell hybrids are constructed by fusing the cells of two different species, in this case of humans and a rodent, typically mouse or hamster. The fused cell contains the chromosomes of both species but tends to gradually lose those of one of the species. In human–rodent hybrids, the human chromosomes are lost in a fairly random fashion. By isolating a series of cell lines after this loss, a panel of hybrids can be generated, each containing a few or a single human chromosome. Each cell line can then be characterized by karyotyping (visually identifying the chromosomes) to determine which chromosomes it has retained. By detecting the presence or absence of a gene in a panel of hybrids, that gene can be assigned to a chromosome. For instance, the enzyme lactate dehydrogenase A was found to be present in all hybrids that contained human chromosome 11 and in none of the hybrids that did not. Somatic cell hybrids have been used to map genes in many mammalian species, and the techniques is still used as a first step in the mapping of genes. However, somatic cell hybrids do not allow the accurate determination of the position of the gene on the chromosome or the ordering of genes. An alternative approach is to use X rays to fragment the DNA before forming the hybrid. This breaks up the DNA into smaller segments, and these radiation hybrids can be used to make physical maps.

B. Linkage Analysis

The ordering of genes takes advantage of the segregation of markers within families. Many disease genes have been mapped to the X chromosome by observing the segregation of the disease. X-linked recessive disorders are inherited mostly by males. A female, who has two X chromosomes, is a carrier when she has one disease gene allele and one normal allele. Half of her sons will inherit her X chromosome, which carries the mutation, and because they do not receive another X chromosome with the normal allele to compensate, they will suffer the disease. Her daughters will receive one of her X chromosomes, and a normal X chromosome from the father. The result is a disease that affects mostly males and is carried by females. Diseases such as hemophilia A, Duchenne's muscular dystrophy, and color blindness are examples of the genetic price males pay for having only one X chromosome. [*See* Hemophilia, Molecular Genetics; Muscular Dystrophy.]

The first linkage of human genes was described by J. Bell and J. B. S. Haldane in 1937. By analyzing the inheritance of two X-linked diseases (hemophilia and color blindness) within a set of pedigrees, they were able to determine the genetic distance between them. The major difficulties in mapping human genes arise because the individuals for study are selected from an outbred population. This means that no assumptions can be made about the genetic makeup of a particular person. One specific problem is that of determining the phase or arrangement of a group of markers on a chromosome. In the example in Fig. 4A, two parents and an offspring are shown, along with the genotypes for three genes (A, B, and C), each with two alleles (A, a; B, b; C, c). The goal is to determine which alleles the child inherited from each parent, and the arrangement of the alleles in the parental chromosomes (i.e., the phase). Consider first genes A and B. Both the mother and the child are homozygous for A and heterozygous for B. They must each have a chromosome that is A-b and another that is A-B. This is designated by the line drawn between the markers. The father, however, is heterozygous for both A and B, and we cannot tell the phase of his alleles. We also cannot tell which alleles the child inherited and from which parent, except that she must have inherited the father's A allele.

The situation gets worse when we add gene C, in which each person is heterozygous. We get no information about the phase of the C alleles. Even though there is information, it cannot be used to tell which chromosomes were passed on. However, the phase can be sorted out by the addition of just one grandparent (Fig. 4B). Because the grandfather is homozygous for all three genes, he must have two copies of an A-B-C chromosome. His son must have inherited one of these, and therefore his other chromosome must be a-b-c. The daughter clearly received an A-B-C chromosome from her father and an A-b-c chromosome from the mother. Thus by including, whenever possible, information from grandparents, the phase of the markers can often be determined.

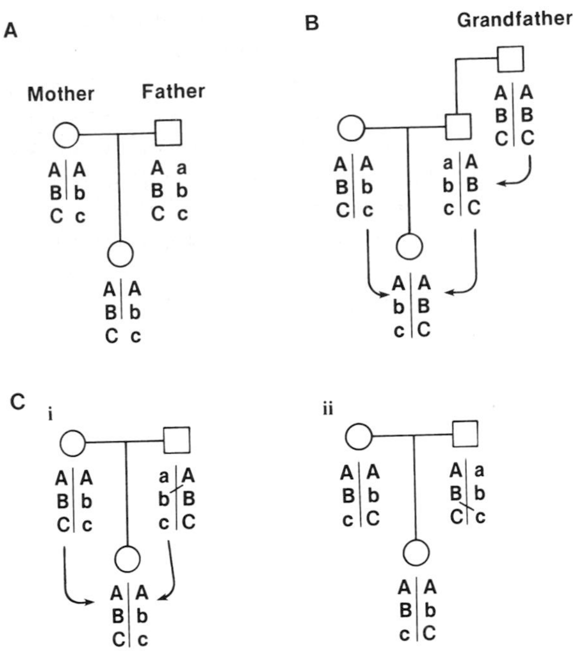

FIGURE 4 Human linkage analysis. (A) The genotypes for three genes (A, B, C) each with two alleles (A, a; B, b; C, c) in a nuclear family. Females are depicted by circles and males by squares; horizontal lines between individuals designate mates, and vertical lines point to offspring. Vertical lines between alleles separate alleles that lie on different homologous chromosomes. (B) Further information provided from a grandparent. Arrows show the path of inheritance of chromosomes. (C) Potential recombination events. Shown are a few of the possible recombinations that could have occurred and yielded the same data. (i) A recombination in the father's chromosomes between genes A and B. (ii) Crossover between genes B and C.

This conclusion was based on the assumption that no recombination events have occurred in any of the chromosomes passed on to the daughter. However, Fig. 4C shows that several single recombinations could have occurred and still produced the same data. In (i), a recombination in the father between A and B produces an A-b-c chromosome, and the child could then have received the mother's A-B-C chromosome, quite a different result than what we predicted in Fig. 4B. Panel (ii) shows another possibility, and many more involve multiple recombination events. The point is that with such data we cannot be completely sure of what happened. How can we make a map when the inheritance of each chromosome cannot be unambiguously followed?

Making genetic maps of human chromosomes depends on the use of certain mathematical and statistical treatments that take into account the uncertainty associated with the data. One of these is the calculation of the log of the odds (LOD) score. Basically these manipulations allow us to calculate the probability that two genes are linked and that a group of linked genes are arranged in a certain order (a genetic map). This can never be absolutely certain, but when the odds become great enough, even the most skeptical become convinced. Usually odds of 1000 to 1 are sufficient to convince the geneticist.

The LOD score is the logarithm of the odds that two genes are linked. Thus a LOD score of 3 means that it is 1000 times more likely that two genes are linked than that they are unlinked. In practice, the LOD score is calculated by following the coinheritance in a set of families of a genetic marker and a genetic disease. Each time the marker and the disease gene are inherited together, the LOD score goes up, and for each apparent recombination, the score decreases. When all the data are added up, if the score is at or below zero, the two can be considered unlinked. The process is repeated with additional markers until a LOD score above 3 is reached, which is taken as proof that the marker and disease gene are linked. If the chromosome location of the marker is known, the disease gene has been mapped to a chromosome. LOD scores less than −2 are considered as evidence that two loci are unlinked.

C. Huntington's Disease

Huntington's disease (HD) is an autosomal dominant disorder of the nervous system, which illustrates both the power of human genetics and the difficult social issues raised by such work. It has been known that the disease is inherited since it was described by George Huntington in 1872. Those who inherit this genetic defect suffer from a progressive deterioration of the nervous system, which typically begins when they are in their 40s.

Huntington's disease patients have a 50% chance of transmitting the disease to each of their children, although most patients do not know that they have the disease until they have passed their childbearing years. Research into the biochemical basis of the disease has been frustrated by the complexity of the brain and the lack of a suitable animal model. In the early 1980s, several groups began to pursue a genetic approach, to search and find the gene using linkage analysis. With the gene in hand, the protein involved could be identified and studied. This process has come to be known as positional cloning, in which a disease gene is isolated based on its location as determined by linkage analysis or chromosomal alteration. In 1983, linkage was detected between a DNA marker on the short arm of chromosome 4 and Huntington's disease. A key to this success was the use of a large, extended Venezuelan family suffering from this disease. This was the first major success of the use of linkage to locate the disease gene without any other clues as to its location.

The discovery of closely linked markers allowed predictions to be made in HD families as to who does or does not carry the defective gene. Figure 5 illustrates a hypothetical family in which a woman with the disease has had four children. She is heterozygous for a closely linked marker (1,2), whereas her husband is homozygous for (2,2). Her oldest daughter is also affected with the disease and has inherited the 1 allele of the marker from her mother. Assuming that no recombinations have occurred, the 1 allele must be on the same chromosome as the HD gene. At this point, the other children can be typed for the marker, and a prediction can be made as to their carrier state. (Remember that this can only be a prediction because of the chance of recombination. The accuracy of the

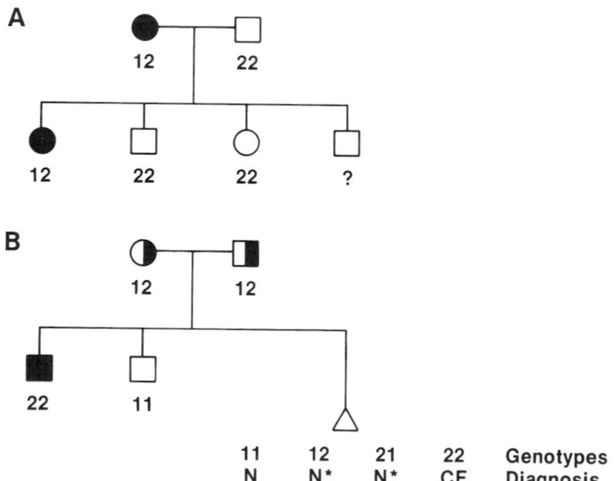

FIGURE 5 Genetic diagnosis of two human diseases. (A) A hypothetical pedigree of a family with Huntington's disease, a dominant disease. The affected individuals are shown by solid symbols, with the genotypes for a closely linked marker shown underneath. 1,2 are the two alleles of a closely linked marker. The disease was passed from the mother to the oldest daughter. The youngest son is the individual requesting diagnosis. (B) Pedigree of a cystic fibrosis family. Because this is a recessive disease, both parents must be carriers (half-solid symbols). The oldest son is affected, and his brother is predicted to be homozygous normal. The fetus is shown by the triangle, along with the four possible genotypes and their predicted outcomes. N* is a healthy carrier.

prediction will be determined by the genetic distance between the marker and the disease gene.)

It is easy to see that this new ability to predict the carrier state of people early in life creates a complex dilemma. A negative result predicts that the individual will be free of disease and can have children without fear of passing the gene to them. A positive result foretells a life that will involve progressive suffering and loss of contact with the outside world. Should we want to know the answer? The testing is always accompanied by extensive counseling to help patients cope with this information. [*See* Genetic Counseling.]

After a 10-year search, the HD gene has been identified. The defect in affected individuals is not a classic mutation, but the expansion of a 3-base-pair microsatellite within the coding region. It is hoped that this research will also provide powerful insight into the workings of the nervous system. [*See* Huntington's Disease.]

D. Cystic Fibrosis

Cystic fibrosis (CF) is the most common fatal genetic disease in the Western world, affecting about one in every 2000 children born to Caucasian parents. It is 10 times rarer in Blacks and almost unheard of in Asians. Like Huntington's disease, CF is a complex disorder, and its molecular basis has eluded investigators for decades. However, the past several years have seen a flurry of discoveries on CF, which have led to a much clearer understanding of the disorder. Clinically the disease involves two major organs (the lung and the pancreas). In most CF patients, the pancreas fails to secrete digestive enzymes, leading to an inability to absorb fats and deficiencies in fat-soluble vitamins. The lungs of CF patients become increasingly congested with a thick mucus, leading to bacterial infections and eventually death. The most consistent clinical symptom for diagnosis is an abnormally high concentration of sodium and chloride ions in the sweat of CF individuals. The levels are usually three times higher than in normal individuals. In the lungs, pancreas, and sweat ducts there are secretory cells, and the process of secretion is largely controlled by the transport of ions into and out of the cell. In CF patients, the regulation of ion transport is defective, leading to obstructions in the pancreas, congestion in the lung, and an ion imbalance in the sweat.

Despite the detailed knowledge gained by studying cells from CF patients, the defective protein was not identified by this approach. Several groups collected samples from families affected by CF and began using linkage analysis to try to locate the mutated gene. Progress was slow until 1985, when four successive reports of linkage were announced. The two closest markers were each found to lie on opposite sides of the gene, at a recombination distance of 1%. This allowed, for the first time, the ability to perform genetic testing and prenatal diagnosis on CF families. An example is shown in Fig. 5B. A couple has a child with CF. Because the disease is recessive, the child must have inherited a defective gene from each parent; they are both carriers. Each parent is heterozygous for a closely linked marker, and the CF child is homozygous for allele 2 of the marker. In this situation, allele 2 is predicted to lie on the CF chromosome in both parents. The patient's brother is clinically normal. However, he could be a CF carrier, and carriers have no clinically detectable traits. The genetic analysis shows that he inherited the 1 allele from both parents and is therefore predicted to be homozygous for the normal chromosome. This analysis can be extended to any subsequent pregnancies in this family. Fetal DNA can be obtained by chorionic villus sampling at 8–10 weeks of gestation or by amniocentesis at 16–18 weeks. The four possible genotypes are shown in Fig. 5B. The fetus has a 1 : 4 chance of being homozygous normal, a 2 : 4 chance to be a carrier, and a 1 : 4 chance of having CF. Again, these are predictions, with the accuracy dependent on the distance between the marker and the mutation. However, with close markers on both sides of the gene, the accuracy is greater than 99%, because they allow the detection of any recombinant in the interval between them.

This discovery was a tremendous breakthrough for some families with a history of CF. Because the disease is recessive, however, most carriers do not know that they carry a defective gene, and thus most CF children are born to completely unsuspecting parents. Remember that the marker does not test for the mutation itself; it can only be used to trace the inheritance of a mutation that is known to be present in the family. To test for carriers in the general population and to learn more about the disease, the gene and the mutation(s) had to be found. This breakthrough was accomplished in September 1989.

By examination of the inheritance of RFLP markers in hundreds of families, it was determined that the CF gene was between the MET gene and the marker D7S8 (Fig. 6A). This distance between MET and D7S8 was determined to be approximately 1000 kilobases. By successively cloning sequences toward the gene from both sides, eventually the gene was reached and identified. The predicted protein, the cystic fibrosis transmembrane ion regulator (CFTR), has a struc-

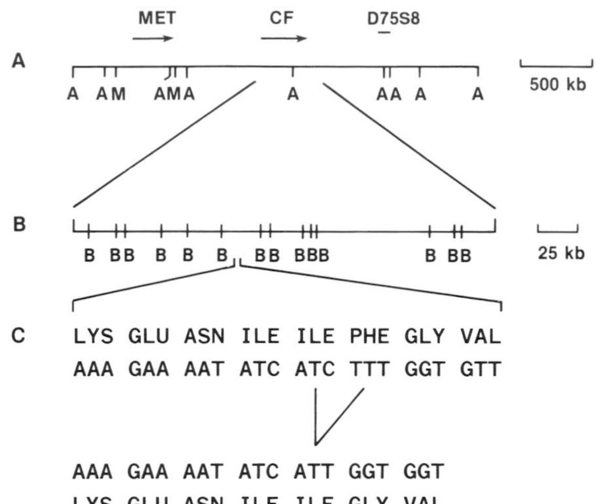

FIGURE 6 Physical maps of the cystic fibrosis locus. (A) A long-range restriction map generated by pulsed-field gel electrophoresis. MET, an oncogene; D7S8, a DNA fragment detecting an RFLP. A, M, cutting sites of restriction enzymes NaeI, MluI. (B) Restriction map of the CF gene. B, Bam HI restriction enzyme site. (C) Sequence of the portion of the CF gene containing the most common mutation. The nucleotide and amino acid sequence of a portion of the CF gene is shown along with the three-nucleotide deletion that causes the most common form of the disease.

ture similar to that of a family of transport proteins localized in the cell membrane, and functions as a chloride channel. Figure 6C shows the nature of the mutation present on the majority of CF chromosomes. The defect is due to the deletion of three nucleotides within the coding region of the protein. The mutation occurs in a region of the molecule that appears to be involved in the interaction of the protein with ATP. This binding is thought to provide the energy needed for the protein to perform its transport functions. Thus the mutation is rather subtle, leading to the production of a slightly smaller protein that is disabled in function. The situation is similar to the case of sickle-cell anemia, which involves a substitution of a single amino acid in a globin gene, which carries oxygen. The substitution leads to a reduction in oxygen binding and to structural abnormalities in the cell.

Figure 6 displays three types of genetic maps. Part C shows the nucleotide sequence of a portion of the gene. The nucleotide sequence is the most detailed type of map, as it displays the complete information encoded by the genome. Part B shows a restriction enzyme map of a large portion of the gene. Restriction maps are generated by digesting DNA with restriction enzymes, resolving the fragments by gel electrophoresis, determining the size of the fragments, and determining the order of the fragments in the original segment of DNA. This type of map is useful for elucidating the size and structure of genes. Part A shows the cleavage position of several restriction enzymes that cleave infrequently in mammalian DNA, every 100,000 base pairs or more. These large fragments can be separated by a technique known as pulsed-field gel electrophoresis, which uses pulsing fields of current to increase the resolution of large DNA molecules. In the CF locus, this method was used to link together distant genetic markers, and it is useful for determining the distance between genes. [*See* Cystic Fibrosis, Molecular Genetics.]

E. Future Prospects

There now exist genetic maps for each human chromosome with a resolution of 1–2 cM. International efforts have also generated collections of large cloned human DNA segments in the form of yeast artificial chromosomes (YACs). Ordered collections of smaller clones have been produced that completely cover some of the smaller chromosomes, and efforts to determine the complete sequence of these chromosomes are under way. Similar projects for yeast and *C. elegans* have produced the complete or nearly complete sequences. Thus in several years the complete sequence of the human genome will be known, as well as the sequence of several important model organisms. These efforts will permit a rapid advance in our understanding of human biology. [*See* Genome, Human.]

In the meantime, rapid progress is being made on the study of many disease genes. The genes for common forms of muscular dystrophy as well as many forms of inherited cancer have been isolated by molecular and genetic techniques. Research is also yielding important clues into the role of inheritance in obesity, epilepsy, schizophrenia, asthma, and other complex diseases. Although the process leading to potential treatments and cures is slow, genetic analysis provides the crucial first step toward identifying the causes of these complex disorders. [*See* DNA Markers as Diagnostic Tools.]

BIBLIOGRAPHY

Dean, M. (1988). Molecular genetics of cystic fibrosis. *Genomics* **3**, 93–99.

Guapay, G., *et al.* (1994). The 1993–94 Généthon human genetic linkage map. *Nature Genet.* **7**, 246–335.

Lalouel, J.-M., and White, R. (1987). Chromosome mapping with DNA markers. *Sci. Am.* **258**, 40–48.

O'Brien, S. J. (ed.) (1993). "Genetic Maps," Vol 6. Cold Spring Harbor Laboratory, Cold Spring Harbor, NY.

Genetic Polymorphism

ANDRE LANGANEY
Université de Genève

DANIEL COHEN
Centre d'Etude du Polymorphisme Humain

ANDRES RUIZ LINARES
Universidad de Antioquia

GLOSSARY

Admixture Effect on genetic pools of the interbreeding of two different populations

Allele One of possible alternative forms of a polymorphic gene

Base pair Unit of information of the DNA molecule; base pairs are the "letters" of the genetic code

Chromosome Linear structure carrying the genes inside the cell nucleus

DNA Deoxyribonucleic acid, the chemical constituent of the genes

Founder effect Sampling effect, in which a limited number of emigrants will take away only a part of the genes of their population of origin and will have different frequencies of the same genes

Gene pool Total genetic makeup of a population

Genes Units of genetic information transmitted from parents to children according to Mendelian laws

Genetic drift Random change of gene frequencies from generation to generation in a single population. Genetic drift is fast in small populations and slow in large populations

Genome All the genetic information of an individual. The human genome is made of 23 pairs of chromosomes carrying 3 billion base pairs of DNA

Genotype Genetic constitution of an individual

Locus Position of a gene in a chromosome

Mutation Random change of DNA structure. Some mutations change gene products, whereas others (silent mutations) do not

Phenotype Physical attributes of an individual

Recombination Random association of parental genes during meiosis (cell divisions producing sex cells)

THE MOST REMARKABLE OBSERVATION OF POPULation biology is that in any sex-reproducing population, all individuals are different from one another except, in some respects, identical twins. This variation within populations and between individuals was studied by Charles Darwin more than one century ago. However, ignoring the biology of sex and genetics, Darwin did not succeed in explaining it. Nevertheless, he understood that natural variation of populations and individuals formed the basis of natural selection, enabling some lineages to transmit their "types" more often than other individuals.

Today, the variation observed between individuals and between populations of the same biological species is termed polymorphism. Polymorphism is the fact that for any characteristic, individuals of the same species will show different types (*poly* = several, *morph* = type). Polymorphism is less general a word than variation, which is used for changes within and between species.

The human species is one of the most polymorphic among living species. Most characteristics vary from one individual to another within populations, and some of these characteristics have different ranges of variation from one population to another. Human polymorphisms can be classified into two types, according to their pattern of variation.

I. QUANTITATIVE POLYMORPHISM

The first type of polymorphism is the variation of the outside appearance of an individual: face and body shape, skin and hair color, height or other body measurements, and so on. This type of polymorphism, easily recognized by the human sight or simple measurements, can be said to be "quantitative." This means that the variation between individuals is continuous and that the measures made in a population will be real numbers of a continuous interval of variation. For instance, somebody is not short, medium, or tall, but more or less tall. People do not have black, white, or yellow skin, but they are more or less brown, the extremes being possible.

There is no doubt that variation between individuals or between populations for quantitative polymorphisms is, at least partly, under genetic control. For example, family heredity of skin color within population, or in cases of admixture, proves that some genes are responsible for being more or less dark- or light-skinned. But despite intensive genetic investigations for more than a century, no one has yet identified the genes that make somebody more or less something. Many sophisticated statistical approaches have not prevented "quantitative genetics" from remaining, to a large extent, a "black box" of modern biology.

So the most interesting result about quantitative polymorphism is its pattern of variation within and between human populations, whatever its genetic background may be. First, for all quantitative characteristics, there is enormous interindividual variation within populations. Measurements of a quantitative polymorphism (e.g., height or skin color) usually vary according to a normal (Laplace–Gauss) probability distribution, which means that measures around the mean of the population are more common than those that lay far aside; but these still exist. For example, height measures vary currently up to 10 inches away from a population mean. This makes that mean differences between populations are commonly lower than extreme differences between individuals within a population. Also, the distribution of measurements for a quantitative polymorphism usually overlap between different populations. This is especially true in the case of skin color, which can be shown to vary continuously worldwide from the lightest white-skinned people to the darkest blacks, through all "yellow-skinned" or brown intermediate grades.

In some cases, the variation seen in quantitative characteristics has been significantly correlated to particular environmental factors, even if the ultimate reason for this geographical pattern remains puzzling. For example, it is well known that the mean skin color of human populations varies according to their latitude of origin. All dark-skinned populations come from the tropics, whereas all populations from temperate or arctic areas are light-skinned; gradients of skin color are observed from the equator to the pole in Africa and Europe, in Asia, and in North and South America. Rigorously speaking, mean skin color is more highly correlated with sunlight than with latitude.

Height and body shape distributions are more complicated, being commonly correlated with the type of environment rather than with latitude. People are short and stocky in the arctic and in high altitude (Andes or Himalayas), but also in equatorial forests. They are middle-sized in temperate forest and savanna, and their mean height increases in the cold temperate zone, whereas they are commonly tall and thin in the hot deserts of Africa.

II. QUALITATIVE POLYMORPHISM

It has been well known, since the beginning of the century, that a successful blood transfusion is conditioned by the ABO blood groups and that these blood groups are determined by Mendelian genes. Every human individual belongs to only one of the four main ABO blood groups (A, B, AB, and O). Thus, the ABO blood groups are a qualitative polymorphism in which an individual belongs or not to a category. Such a polymorphism is also called discontinuous or, in a mathematical sense, discrete. For an individual, the characteristic is "present" or "absent" and is not measurable with continuous real numbers as for quantitative polymorphism. At the population level, measurements can be made of the frequencies of the various possible types (e.g., the four ABO blood groups) or, when they are computable, the frequencies of the alleles that determine the possible genotypes (e.g., alleles A, B, and O for the four ABO blood groups).

In the past 25 years, and particularly during the last 10 years, the rapid growth of understanding in areas such as immunology, biochemistry, and molecular genetics has enabled the discovery of hundreds of new qualitative genetic polymorphisms represented by red cell blood groups, seric blood groups, cell histocompatibility groups, and protein and DNA molecular variants. The number of alleles at each of these polymorphic loci then determines the number of phenotypic classes to which each human individual can belong.

Current estimates of the number of functional genes in humans vary from 40,000 to more than 100,000, that is, about 100 million DNA base pairs among the 3 billion that constitute the human genome. Although it is difficult to confirm, it is usually thought that at least a fifth of the functional human genes are polymorphic, with two or more gene variants detectable at the protein level. This means that the number of theoretically possible genetically different human individuals is much greater than 3 to the exponent 8000 (3 = number of genotypes for loci with two alleles and 8000 = a minimal estimate of the number of polymorphic functional genes). Such a number is much greater than the estimated number of chemical atoms of the entire universe! And it is a considerable underestimation since many genes have more than two alleles and also because one would have to take into account "silent mutations," which change the DNA without changing the gene product or which change noncoding sections of the DNA (a large part of our genome).

This means that human polymorphism is almost infinite in every human population. In any generation, existing individuals are only a tiny sample of the enormous possible diversity. Furthermore, because of recombination during meiosis (the cell division that produces germ cells), two humans never have the same genetic constitution, except for identical twins. Each parent transmits for every genetic system one gene out of the two that he or she received from his or her own parents. So, each germ cell is produced by a "recombination lottery" among billions of possible combinations of alleles at thousands of loci. Thus, during meiosis a parent never produces the same ovocyte or the same sperm, and then each fecundation, combining an original ovocyte with an original sperm, produces a completely new human being. Therefore, each of us has a unique genetic constitution that appeared at conception and will disapper at death. Each human being possesses a kind of genetic identity card in his or her DNA.

The main consequence of this genetic polymorphism is that from generation to generation, human populations are made of different sets of individuals, which means that human populations change from generation to generation. Human evolution is not a scientist's dream or hallucination, it is a fact confirmed by our precise knowledge of genetics. [*See* Human Evolution.]

III. THE ORIGIN OF HUMAN GENETIC POLYMORPHISMS

Human genetic diversity is amplified by genetic recombination but it is rooted in mutations, due mostly to chance errors during gene duplications, that generate new alleles. Because modern humans are likely to belong to a young species (about 1000 centuries old), the majority of present human genetic variants were produced by old mutations that occurred in our ancestors. It has been demonstrated that human beings share some of their genes (those basic in cell biology) with such distant species as fruit flies or even wheat or fungi. Many of our genes are common with those of other vertebrates. Most of them are shared by mammals and even more by other primates, our nearest animal cousins. Sophisticated molecular biology techniques permit us to estimate the age of some of our genes, and the results vary from more than 1 billion years to a few generations for some rare mutations. Recently generated alleles are usually rare in human populations. Nevertheless, we share a large part of our blood group or enzyme alleles with the nonhuman primates, and most of our genes differ only by neutral mutations from those of the other mammals.

IV. VARIATION BETWEEN HUMAN POPULATIONS

Genetic differences between human populations have been found to be much lower than could be expected based on physical appearance.

For most of the studied loci, the same alleles are present in almost all human populations. Differences between populations relate to changes in the frequencies of the same alleles and usually not the presence or absence of an allele, although there are exceptions, usually in systems with numerous alleles. Also, if it is easy to tell that an allele is present in a population, it is almost impossible to be certain that it is absent in another population. Economic cost prevents the study of large samples of individuals, and small samples "miss" rare alleles. Every time that a population

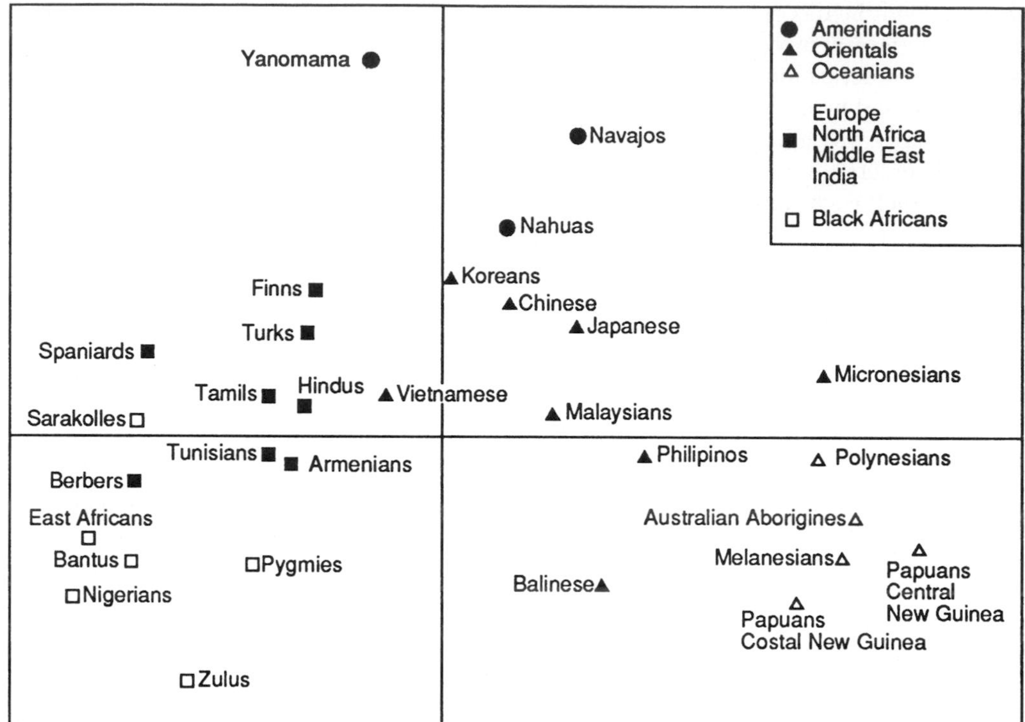

FIGURE 1 Genes and geography: a genetic map of gene frequencies of the human major histocompatibility system (HLA). Without introducing any geographic data, this analysis gives back a map of the world: geography is the major factor of human genetic differentiation. (Data from A. Sanchez-Mazas, Laboratoire de Génétique et Biométrie, University of Geneva.)

has been studied using large samples, alleles that were reputed to be characteristic of other populations were found in the population studied. For example, large samples of European blood donors commonly show, at low frequencies, alleles that had been thought to be present only in black African or Asian populations.

When we study the variation of allele frequencies throughout the world, the first observation is that this variation is continuous from the west of Europe to East Asia and farther into America via the Bering Strait and toward Oceania through Southeast Asia (Figs. 1 and 2). The same can be said from South Africa up to Scandinavia. We cannot find large differences in gene frequencies between neighboring populations, except in small isolated populations (because of founder effects and strong genetic drift) or when a long-range migration has brought together populations formerly isolated for a long time.

Thus, genetic differentiation of human populations is mostly geographic (Fig. 3). The "genetic distance" between populations usually follows the geographic distance between their location and sometimes the structure of communication paths (the shape of continents, the isolation by mountains, oceans, glaciers, and so on). Detailed continental studies have also pointed out that there is a correlation between genetic diversity and language diversity, whereas there are few, if any, correlations between genetic diversity and quantitative polymorphisms. For example, New Guinea Papuans physically look like some African Bantous, however, they are quite different from a genetic point of view. Papuan genes are definitely Asian genes, closer to those of the Chinese than to those of the Africans, whatever their skin color. [*See* Population Genetics.]

V. POLYMORPHISM AND PREHISTORY OF MODERN HUMANS

The oldest fossils of anatomically modern humans have recently been shown to be about 100,000 years old and were found in the Middle East (Qafzeh) and

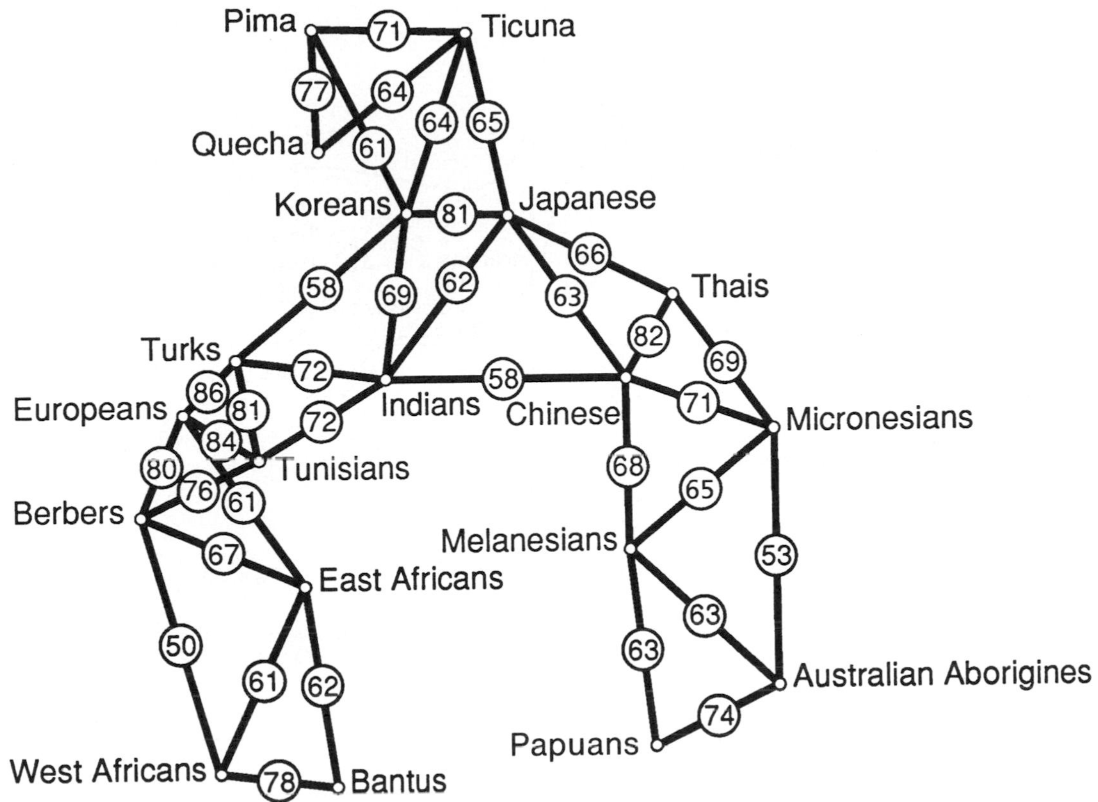

FIGURE 2 The human genetic network: populations are linked according to percentage of gene frequency they share for three very polymorphic genetic systems (Rhesus blood groups, Gm seric groups, and HLA). (Data from A. Sanchez-Mazas and A. Langaney, Laboratoire de Génétique et Biométrie, University of Geneva.)

East Africa (Omo Kibish). Available genetic data on contemporary human populations indicate a limited genetic differentiation between populations. Such results support a recent common origin for all modern human populations, perhaps between 100,000 and 150,000 years before the present, followed by important migrations during prehistoric and historic times. This theory is supported by independent data from immunology and molecular biology and is compatible with fossil and archaeological data. Most human population geneticists think that the ancestors of modern humans originated about a thousand centuries ago in a single population, which could have dwelled somewhere between East Africa and western Asia. Then, this population split and spread throughout the Old World around 70,000 BP, reaching Oceania and America at a later time. Peripheral populations of southern and western Africa, northern and western Europe, eastern Asia, Oceania, and the Americas possess different subsets of the gene pool commonly observed from East Africa to India. Thus, it is likely that when small groups of hunter-gatherers reached these areas, the original gene pool changed, either because they did not carry all alleles with them (the founder effect) and/or because of genetic drift, which is the process of random variation of allele frequencies in populations. This determined the differences in gene frequencies seen between contemporary human populations. On the other hand, a thousand centuries is not enough time to obtain noticeable frequencies of new mutations. Thus, it is believed that founder effects, genetic drift, and migration have been the main factors of modern human genetic evolution.

The high level of discordance between qualitative polymorphism and quantitative polymorphism must be stressed. Although qualitative polymorphism evolves according to human pedigrees and migrations, quantitative polymorphism seems to evolve rapidly, adapting to local environmental conditions. Body shape and related measurements can vary in a few generations, as they have in Europe since the Middle Ages. Skin color seems to be able to change drastically

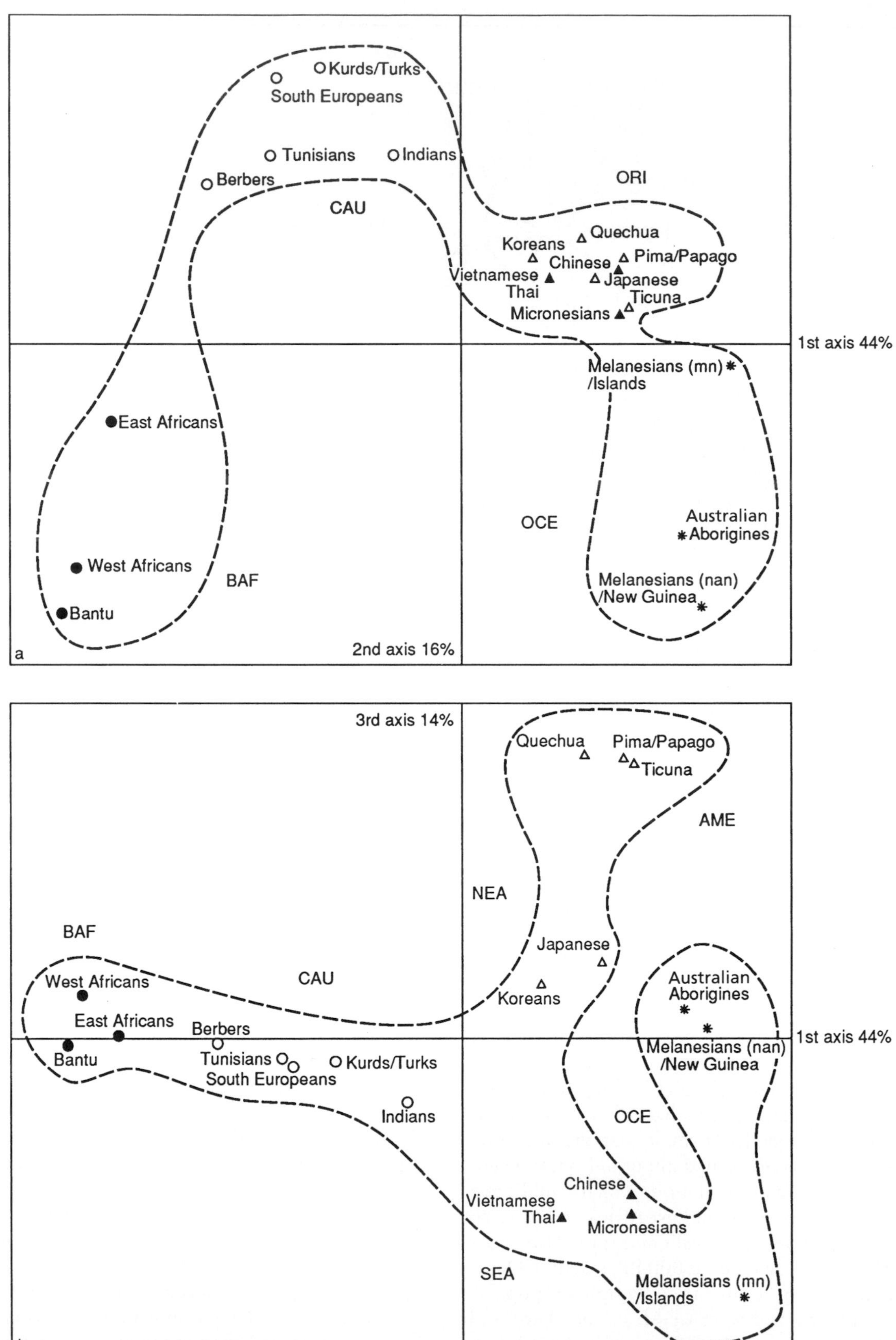

FIGURE 3 Multivariate analysis of the human genetic network. BAF, Black Africans; CAU, caucasoids; ORI, orientals; NEA, Northeast Asia; SEA, Southeast Asia; AME, Amerindians; OCE, oceanians (same polymorphisms as in Fig. 2). [From A. Sanchez-Mazas and A. Langaney (1988). *Hum. Genet.* 78, 161–166, with permission.]

in less than 50,000 years, as is seen between Asians and Melanesians who presumably shared ancestors at about that time. Thus, the external characteristics of the human body (shape, size, color, etc.), which commonly define racial groups, have little to do with the history of human genealogy. Therefore, we must consider that race is mostly a social construct, not a genetic reality.

BIBLIOGRAPHY

Cavalli-Sforza, L. L., Piazza, A., and Menozzi, P. (1994). "The History and Geography of Human Genes." Princeton Univ. Press, Princeton, New Jersey.

Excoffier, L., and Langaney, A. (1989). Origin and differentiation of human mitochondrial DNA. *Am. J. Hum. Genet.* **44,** 73–85.

Excoffier, L., Pellegrini, B., Sanchez-Mazas, A., Simon, C., and Langaney, A. (1987). Genetics and history of sub-Saharan Africa. *Am. J. Phys. Anthropol.* **30,** 151–194.

Goldstein, D. B., Ruiz Linares, A., Cavalli-Sforza, L. L., and Feldman, M. W. (1995). Genetic absolute dating based on microsatellites and the origin of modern humans. *Proc. Natl. Acad. Sci. USA* **92,** 6723–6727.

Langaney, A. (1988). "Les Hommes: Passé, Présent. Conditionnel." Armand Colin, Paris.

Ruiz-Linares, A., Minch, E., Meyer, D., and Cavalli-Sforza, L. L. (1995). Analysis of classical and DNA markers for reconstructing human population history. *In* "The Origins and Past of *Homo sapiens sapiens* as Viewed from DNA" (K. Hanihara and S. Brenner, eds.) World Scientific Publishers, Singapore.

Sanchez-Mazas, A., and Langaney, A. (1988). Common genetic pools between human populations. *Hum. Genet.* **78,** 161–166.

Genetics and Mental Disorders

KATHLEEN RIES MERIKANGAS
JOEL DAVID SWENDSEN
Yale University School of Medicine

NEIL RISCH
Stanford University

I. INTRODUCTION

Although progress in the characterization and assessment of psychiatric syndromes has strengthened the reliability and validity of current nosology, phenotypic imprecision has often been considered the major culprit in the inconsistencies of psychiatric genetic research. Much of the criticism leveled against early psychiatric genetic investigations, and indeed biological psychiatric studies in general, has concentrated on the subjectivity of psychiatric diagnosis. The accepted solution to this problem has gradually developed over the past quarter of a century, reflecting a trend toward explicit "operational" definitions, which provide highly acceptable reliability. However, the introduction of operational definitions for psychiatric research appeared to do little to overcome a more difficult obstacle, that of biological validity. The lack of validity and indistinct boundaries of psychiatric disorder classification are the rate-limiting steps in discovering their etiology, and it is unlikely that the etiologic secrets of the major psychiatric disorders will be unlocked without accurate and valid identification of the syndromes themselves. Although progress in classification is advancing, the findings and conclusions of all psychiatric investigations must therefore be qualified against imperfect biological validity.

With the inherent problems of psychiatric nosology in mind, information about the magnitude of psychiatric disorders in the general population may be used to provide an initial frame of reference for investigating patterns of familial aggregation in clinically ascertained samples. Table I presents the lifetime prevalence rates from several large-scale epidemiologic surveys of psychiatric disorders. The notably higher rates reported by the National Comorbidity Survey are likely to be due to the application of interview methods that minimize biases in retrospective recall, and suggest that almost one-half of the population will meet diagnostic criteria for at least one psychiatric disorder over their lifetime. The high rates of disorder must also be interpreted within the context of psychiatric disorder severity; even the most common forms of illness are often severe enough to prevent basic functioning in social, occupational, and family domains. Taken together, these characteristics offer compelling reasons to investigate the causes of psychiatric disorders with the ultimate goal of developing better models of prevention and treatment.

II. STUDY PARADIGMS IN GENETIC EPIDEMIOLOGY

With its roots in the methods of population and clinical genetics, as well as chronic disease epidemiology, investigations in genetic epidemiology are typically based on one of four research paradigms. The basic strategy of these paradigms is either to hold the environment constant while allowing genetic factors to

ENCYCLOPEDIA OF HUMAN BIOLOGY, Second Edition, VOLUME 4.

TABLE I
Epidemiologic Surveys of Psychiatric Disorder

Disorder	Lifetime prevalence rates (SE) by investigation[a]			
	ECA	Edmonton	CPES	NCS
Anxiety disorder	15.5 (0.7)[b]	11.2 (0.6)[b]	10.5 (0.9)[b]	24.9 (0.8)
Affective disorder	7.9 (0.6)	10.2 (0.6)	14.7 (1.0)	19.3 (0.7)
Psychosis (nonaffective)	1.7 (0.3)	0.6 (0.1)	0.4 (0.2)	0.7 (0.1)
Substance use disorder	16.7 (0.8)	20.6 (0.8)	21.0 (1.3)	26.6 (1.0)
Any disorder	32.6 (1.0)	33.8 (0.9)	36.6 (1.5)	48.0 (1.1)

[a]ECA, epidemiologic catchment area (rates averaged across three sites); CPES, Christchurch Psychiatric Epidemiology Study; NCS, National Comorbidity Survey.
[b]Includes somatoform disorders.

vary or the reverse. Each approach is characterized by inherent strengths and limitations, but progress in all areas continues to clarify the etiology of psychiatric disorder.

A. Family Studies

The observation that some disorders aggregate in families serves as prerequisite evidence suggesting a possible genetic component. The basic family study approach involves identifying individuals with a particular psychiatric disorder (the proband) and then determining the rates of disorder in the proband's relatives. These morbidity statistics can then be compared to the rates of disorder in families of unaffected individuals (controls). The common indicator of familial aggregation is the relative risk ratio, computed as the rate of disorder in families of affected persons divided by the corresponding rate in families of controls.

Although family studies are an important starting point of genetic epidemiology, data from family studies can be difficult to interpret for several reasons. Like all research in psychiatry it is dependent on the diagnostic classification system, and some disorders may show markedly different patterns of familial transmission, depending on minor changes in diagnostic threshold. However, a problem even more specific to family studies is that important environmental factors are also "familial." Key environmental variables such as social support, chronic and acute life stress, economic status, community environment, and many others tend to vary along family lines and are known to have independent effects on mental health. For this reason, family studies may look beyond basic familial aggregation to examine specific patterns of transmission that more clearly identify the genetic influences. These specific patterns of transmission within families may vary according to whether the genes are dominant or recessive, autosomal or X-linked, or multifactorial (including nongenetic factors).

Although the family study design has typically been employed to elucidate the degree and mode of transmission of most disorders, there are numerous other purposes for the application of such data. The major advantage of studying diseases within families is that the assumption of homotypy of the underlying factor eliminates the effects of heterogeneity that are present in comparisons between families. Family studies can therefore be employed to examine the validity of diagnostic categories by assessing the specificity of transmission of symptom patterns and disorders as compared to between family designs. Data from family studies may also provide evidence regarding etiologic or phenotypic heterogeneity. Phenotypic heterogeneity is suggested by variable expressivity of symptoms, whereas etiologic heterogeneity is demonstrated by homotypic expression of different etiologic factors between families. Moreover, the family study method permits assessment of associations between disorders by evaluating specific patterns of cosegregation of two or more disorders within families.

B. Twin Studies

The twin study method compares concordance rates for monozygotic twins (who share the same genotype) with those of dizygotic twins (who share an average of 50% of their genes in common). To support a genetic etiology, the concordance rates for monozygotic twins should be significantly greater than dizygotic twins and, consistent with the concept of familial aggregation, the degree of concordance between co-twins of either type can also be used to provide information about the magnitude of genetic or environmental effects. However, the problem of same-environment confounds has also been raised against twin study paradigms. The possibility of environmental factors that may covary with zygosity is therefore an important consideration.

Although the traditional application of the twin design focuses on the estimation of the heritability of a trait, there are several other research questions for which the twin study may be of value. Differences in concordance rates between monozygotic and dizygotic twins may be investigated at the level of symptoms or symptom clusters in order to study the validity of symptom complexes. Varying forms or degrees of expression of a particular disease in monozygotic twins may be an important source of evidence of the validity of the construct or disease entity. For example, monozygotic twins are not only concordant for depression more often than dizygotic twins, but they may be concordant for specific depression subtypes, underscoring the heterogeneity of these disorders and need for nosology that reflects these entities.

C. Adoption Studies

Family and twin studies are genetically informative because they hold the environment "constant" while examining the rates of disorder across different levels of genetic relationship. An alternative approach is to vary the environment while comparing individuals across degrees of genetic similarity. Adoption studies are part of this latter approach in that psychiatric similarity between an adoptee and his or her biological versus adoptive relatives is examined. An alternative design compares the biological relatives of affected adoptees with those of unaffected, or control adoptees. This approach is the most powerful for identifying genetic factors by minimizing the degree of familial aggregation that can be explained by same-environment confounds. However, adoption studies are also characterized by certain characteristics that may bias results. Biological parents of adopted children have moderately higher rates of psychopathology, alcoholism, or criminality than other parents, and adopted children may themselves be at greater risk for psychiatric disorders. Although such criticisms may be valid reasons to carefully interpret the rates of disorder found in these studies, they do not negate the value of adoption studies to clarify genetic and environmental effects (particularly for disorders showing specificity of transmission).

Estimates of heritability derived from adoption studies may also be used to examine the validity of different phenotypic definitions. For example, the early work on schizophrenia was particularly influential but was criticized by some because of the breadth of the phenotypic definition (which included vague categories such as "latent" and "uncertain" schizophrenia). The reassessment of some of these earlier studies to reflect modern criteria yields far lower rates of schizophrenia or schizotypal personality disorder, but provides better separation between the relatives of schizophrenics versus controls. Thus, a narrower and more reliable definition of the disorder led to an increase in the genetic effect, thereby validating its definition.

D. Genetic Marker Studies

Association and linkage studies comprise the basic paradigms for the identification of biological markers for psychiatric disorders. To be considered a biological marker, a trait must be associated with increased rates of illness, should be observable during phases of illness or recovery, and should be independent of treatment. Genetic markers are a class of biological markers that show Mendelian patterns of transmission, are assignable to a specific location on the chromosome, and are polymorphic with at least two alleles having a gene frequency of at least 1%. Association studies examine whether a disorder is related to a particular marker or allele for the general population (between families) by comparing association rates in unrelated affected individuals with controls. Linkage studies examine the association between genetic markers and disease genes within families and are based on the principle that two genes that lie close in proximity on a chromosome are transmitted to offspring together. The finding of association or linkage does not necessarily mean the linked gene or trait is implicated in causing the disorder, merely that it can be used to identify nearby potential psychiatric disorder genes.

It is now more than two decades since the first linkage study for a psychiatric disorder was performed. Despite the initial excitement generated by the successful application of molecular genetics technology to linkage studies of severe medical diseases, the lack of positive findings in psychiatry has been a sobering experience. Furthermore, although diverse markers have permitted the identification of disease loci for several severe forms of illness, the lack of specificity of the many commonly studied markers prevents their use in identifying the susceptibility to psychiatric disorders. Nevertheless, the identification of new and potentially valuable polymorphic markers across the entire human genome is being accomplished through the efforts of the Human Genome Initiative, and our understanding of the pathophysiology of some of the psychiatric disorders is being enhanced through neuroimaging, pharmacologic challenge, neuroimmunology, and animal studies. It is therefore expected that the application of genetic studies will ultimately be fruitful in elucidating vulnerability factors for the development of major psychiatric disorders. [*See* DNA Markers as Diagnostic Tools.]

III. CHARACTERISTICS OF PSYCHIATRIC DISORDERS THAT IMPEDE GENETIC STUDIES

Family, twin, adoption, and genetic marker studies compose the basic research paradigms of genetic epidemiology. However, the application of these paradigms is complicated by several factors germane to psychiatry itself. One of the most far-reaching impediments to genetic research is the reliability and validity of diagnostic categories. For example, twin studies of male and female alcoholics have revealed a significantly higher heritability for alcohol dependence than for alcohol abuse. Although narrow definitions of alcoholism may provide a more valid phenotype for future genetic analyses, other disorders may require broader definitions, and the appropriateness of using various thresholds is rarely clear in advance of conducting epidemiologic investigations. Research in this domain is therefore necessarily dependent on diagnostic definitions that are in the constant process of refinement.

Another barrier to research in genetic epidemiology is the strong association among certain disorders. *Comorbidity* between psychiatric disorders appears to be the rule rather than the exception: numerous studies of clinical samples have demonstrated the large proportion of patients who simultaneously meet diagnostic criteria for more than a single disorder, and multiple diagnoses within individual subjects appear to be quite frequent in epidemiological surveys of the general population. Comorbidity therefore confounds the study of "pure" disorder etiology, but also poses important questions as to the specificity of risk factors and the appropriateness of diagnostic boundaries.

Cohort effects comprise another limiting factor, as it is often unclear if such effects are artifacts or true differences. For any disease that requires a particular environmental exposure for its development, the disease frequency may differ dramatically according to the variation in the degree of exposure to the particular environmental agent. Many psychiatric disorders have shown striking cohort effects or increasing rates over time, but the source of these changes, or indeed their validity, has not been clarified.

Finally, *nonrandom mating* patterns have been well established among persons with psychiatric disorders. When coupled with the high population prevalence of these conditions, spouse concordance for psychiatric disorders leads to a high frequency of families with bilineal transmission.

IV. EVIDENCE FOR THE ROLE OF GENETIC FACTORS IN THE ETIOLOGY OF PSYCHIATRIC DISORDERS

Although there are formidable challenges to applying genetic epidemiology to the field of psychiatry, progress continues to be made in each of the four paradigms described previously. The following pages will present a review of research in the genetic epidemiology of the major classes of psychiatric diagnoses: schizophrenia, affective disorders, anxiety disorders, and substance dependence. Summaries of linkage studies will be limited to schizophrenia and bipolar illness for which there have been the most substantial number of investigations.

A. Schizophrenia

More is known about the genetic basis of schizophrenia than perhaps any other psychiatric disorder, with genetically informative studies stemming from early this century. Concerning the familial aggregation of schizophrenia, Gottesman pooled data from approximately 40 family studies between 1920 and 1987 and concluded that there is considerable support for the claim that schizophrenia is familial. The risk of schizophrenia and related disorders to first-degree relatives (siblings, parents, children) was on average 9.3 times greater than the risk of schizophrenia to the general

TABLE II
Evidence for Genetic Factors in Schizophrenia

Type of investigation	Standard comparison	Number of studies reviewed	Average relative risk ratio	Ratio range
Family	Relatives of probands vs relatives of controls	9	8.9	2.7–18.5
Twin	MZ probandwise concordance vs DZ probandwise concordance	12	4.4	2.2–12.0
Adoption	Adoptee–biological relatives vs adoptee–adoptive relatives	4	4.3	1.9–10.6

population. However, because early family studies used outdated diagnostic criteria and widely differing methodologies, the conclusions that can be drawn from pooling earlier data are limited. In a more recent review, Kendler and Diehl examined only recent family studies that have included a control group, direct in-person interviews, and blind diagnoses of relatives (see Table II). Based on the average risks to first-degree relatives of proband and control groups across studies, a remarkably similar relative risk of 8.9 is observed. The conclusion that schizophrenia is highly familial is augmented by twin and adoption study support for a genetic etiology. As shown in Table II, a review of 12 twin studies demonstrates an average probandwise monozygotic concordance rate that is over four times the rate of dizygotic twins. Similarly, adoption studies using traditional paradigms and modern diagnostic criteria (if available) demonstrated that the average risk to first-degree relatives was 15.5% compared to 3.6% for controls (giving a grand mean relative risk of 4.3). Very recent adoption studies continue to support the genetic transmission of schizophrenia and related disorders, and have produced findings of similar magnitude.

With such clear evidence of a large genetic role in schizophrenia, much recent research has focused on identifying specific genetic markers for this disorder. Although many association studies of schizophrenia have been performed using diverse polymorphisms, the majority of work in this area has focused on linkage studies. Fueled by initial findings of associations of specific cytogenic abnormalities with schizophrenia, many recent DNA marker studies have particularly focused on regions of chromosomes 5, 6, 8, 11, 22, and sex chromosomes. A summary of the linkage studies of schizophrenia is presented in Table III. The initial excitement generated by the report of linkage between schizophrenia and a marker on the long arm of chromosome 5 was followed by eight studies that failed to replicate the positive finding. Likewise, linkage studies of chromosome 11 and the X were not replicated. More recently, suggestive evidence was re-

TABLE III
Linkage Studies of Schizophrenia and Bipolar Disorder

Disorder	Chromosomal region		Current status
Schizophrenia	5_{q11-13}		Exclusion
	6_{p24-22}		Unconfirmed
	8_{p23-23}		Unconfirmed
	$22_{q12\ 13.1}$		Unconfirmed
	$11_{p15.5}$		Exclusion
	11_{q23}		Exclusion
	XY_{pter}		Unconfirmed
	X_{q27-28}		Exclusion
Bipolar disorder	4_{p16}		Unconfirmed
	5_{q35}		Unconfirmed
	6_{p24}		Unconfirmed
	11	(HRAS)	Unconfirmed
		(INS)	Unconfirmed
		(TH)	Unconfirmed
		(TYR)	Exclusion
		(DRD2)	Exclusion
		(DRD4)	Exclusion
		(D2)	Exclusion
	12_{q23}		Unconfirmed
	13_{q13}		Unconfirmed
	16_{q11}		Unconfirmed
	18_{p}		Unconfirmed
	18_{q}		Unconfirmed
	21_{q22}		Unconfirmed
	X	(CB)	Unconfirmed
		(Factor IX)	Unconfirmed
		(G6PD)	Unconfirmed
		(Xg)	Unconfirmed
		(Xq)	Unconfirmed
		(Dx)	Exclusion
		(F9)	Unconfirmed

ported for a marker on chromosome 6 in the HLA region; numerous studies have tested this region but the results still remain equivocal. Thus, to date, no consistently replicated finding of a gene related to the risk of schizophrenia has emerged.

Although considerable evidence exists that schizophrenia has a strong genetic component, it is also important to underscore that many of the studies in genetic epidemiology indicate that schizophrenia is strongly dependent on nongenetic factors. For example, as the average concordance rate for monozygotic twins is approximately 50%, an individual may have a strong genetic vulnerability to the disease but not manifest the illness due to numerous psychological and environmental influences (including individual differences in the experience of stress, social support, and other diverse factors). Taken together, these investigations support a diathesis-stress model of the disorder whereby a genetic vulnerability serves as a key etiologic factor that is dependent, at least in part, on environmental factors for its ultimate expression. [*See* Schizophrenic Disorders.]

B. Mood Disorders

Mood disorders are composed of a heterogeneous group of syndromes, of which major depression and bipolar disorder (manic-depression) are basic subtypes. Table IV summarizes evidence for genetic factors in bipolar disorder and unipolar depression as reviewed by previous investigators. The overall conclusion from this research is that both major depression and bipolar disorder have important genetic components. Controlled family studies show a 5-fold risk to relatives of depression, and greater than a 10-fold risk to relatives of bipolar patients. The concordance rate for bipolar monozygotic twins is over five times that of dizygotic twins, and depressed twin concordance shows less dramatic but still notable differences. The relative risks based on the few existing adoption studies also confirm that the familial recurrence cannot be attributed solely to environmental factors.

As family, twin, and adoption studies offer a compelling case for the role of genetic factors in mood disorders, especially for bipolar illness, it is not surprising that much effort over the past decade has focused on linkage with DNA markers. Encouraged by reports that bipolar disorder is more common among females than males, the X chromosome served early on as the object of much attention (see Table III). In particular, linkage of bipolar disorder with color blindness (CB) and G6PD deficiency have been found in numerous pedigrees, and significant lod scores have been reported for several other loci on this chromosome. However, as shown by Table III, no single marker on the X chromosome has demonstrated consistent linkage to bipolar disorder. Subsequent positive linkage findings for chromosome 11 rapidly shifted attention to other potential markers with the main areas of interest encompassing loci for the Harvey-RAS oncogene (HRAS), insuline gene marker (INS), and tyrosine hydroxylase (TH) loci. Despite early encouraging findings, however, all subsequent attempts at replication have failed to generate lod scores greater than 3. The more recent linkage efforts continue to examine not only these markers but also loci on numerous other chromosomes. Among the most promising findings of these newer investigations involve loci on chromosome 18, although nonreplication has already been reported.

In summary, despite sizable lod scores linking bipolar disorder to numerous chromosomal loci, reviews of linkage studies continue to conclude a lack of consistent or replicated findings. The widely discrepant

TABLE IV
Evidence for Genetic Factors in Mood Disorders

Type of investigation	Standard comparison	Number of studies reviewed	Average relative risk ratio	Ratio range
Family	Relatives of probands vs relatives of controls	4 (bipolar)	10.8	3.7–17.7
		6 (unipolar)	5.0	1.5–18.9
Twin	MZ probandwise concordance vs DZ probandwise concordance	5 (bipolar)	5.3	3.6–13.4
		2 (unipolar)	1.5	1.2–1.9
Adoption	Adoptee–biological relatives vs adoptee–adoptive relatives	1 (bipolar)	2.6	—
		2 (unipolar)	3.3	1.7–4.8

findings for bipolar disorder, while in some respects disappointing, have nonetheless served an important purpose from the perspective of understanding genetic contributions to this disorder. It is now almost certain that the majority of cases of bipolar disorder cannot be attributed to a single major locus, and the conceptualization of bipolar disorder as a genetically complex and heterogeneous illness is now justified. This conclusion may eventually have important implications concerning phenotypic descriptions and nosology as linkage analyses provide useful information about the biological validity of bipolar disorder subtypes. For example, the nonsignificant linkage findings for X chromosome loci may be due to the inclusion in analyses of both bipolar I disorder and bipolar spectrum, and that linkage to specific loci may not be excluded when examining bipolar I alone. The general failure of this research to generate replicable or convincing findings has nonetheless encouraged some individuals to call for a dramatic increase in the lod score threshold for declaring linkage. However, while the large number of markers being tested across the genome may slightly increase the rate of spurious findings, the current threshold of 3.0 has generally proved robust to false positives. Furthermore, recent analyses demonstrate that if increases in the lod score threshold are warranted, such changes would indeed be minor (i.e., from 3.0 to 3.3 for parametric lod score analysis).

The question therefore remains as to what to do with the growing literature demonstrating nonreplication or otherwise small lod scores. Although researchers may become increasingly skeptical of reports of linkage at current thresholds of significance, at least two reasons for the contribution of such publications have been articulated. First, statistically significant linkage, albeit in a very small subset of families, may indeed represent true genetic contributions to the illness. Second, data from properly defined and executed studies, although unable to produce convincing results when taken singularly, may be accumulated and analyzed to provide more information about predisposing genes. It is important to note, however, that publication of such linkage studies may only be useful if such data are made available to all investigators and if more comprehensive data are presented in the publications (rather than just summary statistics). [*See* Affective Disorders, Genetic Markers; Mood Disorders.]

C. Anxiety Disorders

At present, relatively few studies have examined anxiety disorders from the perspective of genetic epidemiology, and there are virtually no data from certain paradigms (such as adoption studies). However, existing research indicates that most anxiety disorders aggregate in families, and several investigations have offered specific support for genetic etiology (examples of three anxiety disorders are shown in Table V). Perhaps the most consistent support for the role of genetic factors has been found for panic disorder. A review of six controlled family studies using direct interviews provides an average relative risk of 9.4, and new investigations continue to report high levels of aggregation. Although there has been some inconsistency reported by twin studies of panic disorder, studies applying modern diagnostic criteria demonstrate considerably higher rates for monozygotic twins compared to dizygotic twins. Furthermore, current estimates derived from the Virginia twin registry show panic disorder to have the highest heritability of all anxiety disorders at 44%.

Genetic factors are implicated in other anxiety

TABLE V

Evidence for Genetic Factors in Anxiety Disorder

Type of investigation	Standard comparison	Number of studies reviewed	Average relative risk ratio	Ratio range
Family	Relatives of probands vs relatives of controls	6 (panic)	9.4	4.2–17.8
		2 (social phobia)	3.1	3.1–3.2
		2 (OCD)	1.0	1.0–1.1
Twin	MZ probandwise concordance vs DZ probandwise concordance	2 (panic)	2.4	2.2–2.5
		1 (social phobia)	1.6	—
		1 (OCD)	1.9	—

disorders, although comparatively few investigations have been completed. For example, social phobia aggregates in families and twin studies show a higher concordance for monozygotic twins (see Table V). Other phobias (i.e., specific phobia, agoraphobia) have also been shown to be familial, with the three phobia subtypes having similar relative risks and specificity of transmission. More recent data from the Virginia twin study report the estimated total heritability for phobias to be 35%. The aplication of genetic epidemiology to understanding other anxiety disorders has been limited not only due to a dearth of controlled studies, but also because of uncertainty about the appropriateness of phenotypic descriptions. For example, the few family studies of obsessive–compulsive disorder (OCD) that have used standardized assessments with narrow operationalized criteria have demonstrated no increased risk to relatives (see Table V). In contrast, an examination of obsessional symptoms in co-twins of OCD probands revealed an increased risk to monozygotic twins over dizygotic twins. Discrepancies such as these will be clarified as more controlled family, twin, and adoption studies are carried out and as the validity of narrow versus broad definitions are established.

Concerning marker studies of anxiety disorders, the high heritability rates seen for panic disorder have made it the natural focus of research in this area, and many clinical and neurobiological challenge studies have served as a guide by implicating the adrenergic system. However, linkage studies have excluded the possibility that panic disorder was due to mutations in adrenergic receptor loci on chromosomes 4, 5, or 10, and other work has similarly excluded linkage with GABAA receptor genes. Although surveys of the human genome using hundreds of markers have not yielded evidence of linkage, there is reason to be optimistic as the Human Genome Project (and the identification of numerous highly polymorphic markers) will soon lead to major increases in the precision of the human genome map and future research.

D. Substance Use Disorders

The majority of studies concerning the genetic epidemiology of addictive behaviors has focused on alcoholism rather than on drug-related problems, although a shared etiology is suspected by some researchers. Table VI summarizes family, twin, and adoption studies of alcoholism. Not only does alcohol abuse and dependency aggregate in families (comprising a sevenfold risk to first-degree relatives), but twin and adoption studies indicate that this aggregation is partly due to genetic factors. At present, the evidence for a genetic predisposition to alcoholism is stronger for men than for women (but generally significant for both). More recent investigations continue to offer consistent support not only for familial aggregation, but also clarify previously weak areas of evidence by demonstrating a greater concordance for monozygotic over dizygotic female twins. Although the heritability of alcoholism (narrowly defined) has been estimated at 59% by some researchers, the genetic information derived from these twin studies is complex and the heritability of alcoholism (at least in males) appears to be greater when the individual has a comorbid psychiatric diagnosis.

Similar to other domains, the search for specific markers through association and linkage studies in alcoholism has produced equivocal findings. The ma-

TABLE VI
Evidence for Genetic Factors in Alcohol Abuse and/or Dependence

Type of investigation	Standard comparison	Number of studies reviewed	Average relative risk ratio	Ratio range
Family	Relatives of probands vs relatives of controls	5	7.0	2.5–20.1
Twin	MZ probandwise concordance vs	6 (male)	1.6	1.1–2.2
	DZ probandwise concordance	5 (female)	1.1	0.6–1.4
Adoption	Adoptee–biological relatives vs	5 (male)	2.4	1.0–3.6
	adoptee–adoptive relatives	4 (female)	2.4	0.5–6.3

jority of association studies to date have focused on dopamine (DRD2) and serotonin receptor genes as well as the aldehyde dehydrogenase locus. Although several investigations have replicated significant associations between alcoholism and these markers, the majority of investigations are preliminary, nonconfirmatory, or have revealed potential sampling biases that may independently explain observed associations.

Although genetic marker studies for alcoholism (and psychiatric disorders in general) are still in their infancy, results from other paradigms support the conclusion that genetic factors play a moderate role for male drinking problems and at least a modest role for females. In addition, the role of environmental factors in the etiology of alcoholism has been supported by numerous studies from a genetic epidemiologic perspective. For example, adoption study paradigms have shown not only that a disturbed adoptive family environment interacts with a genetic predisposition for alcoholism to affect the risk for the disorder, but that the adoptive family environment may predict alcohol abuse or dependency independent from genetic vulnerability. Finally, although the majority of work in this area has focused on alcoholism, research on other substance use disorders is growing at a fast pace. Current work indicates that these disorders are familial and that they have complex etiologies involving both genetic and environmental components. [*See* Addiction.]

V. SUMMARY AND FUTURE STEPS

Psychiatric disorders not only affect large percentages of the general population, but their severity and impact highlight the crucial need for etiologic investigations from diverse perspectives. Advances in standardized definitions of major psychiatric disorders have dramatically enhanced our ability to reliably characterize behavioral phenotypes for genetic studies. The application of family, twin, adoption, and genetic marker paradigms to this domain has offered new opportunities for understanding the respective roles of genetic and environmental factors for numerous classes of disorders. However, previous research has been impeded by several factors, notably including a lack of biological validity of diagnostic categories as well as the genetic complexity of psychiatric disorders. Investigations of biological markers, particularly linkage studies, have to date failed to produce consistent and replicable findings regarding the possible genetic mechanisms underlying diverse forms of mental disorder. The application of genetic study paradigms to guide definitions of the thresholds and boundaries of psychiatric syndromes (and to refine the precision of phenotypic definitions) will enhance the identification of homogeneous subtypes, thereby increasing the power of future molecular studies in elucidating their genetic basis.

ACKNOWLEDGMENTS

This work was supported in part by an NIMH Research Training Fellowship (5T32MH14235) (Dr. Swendsen), Research Scientist Development Award K02 MH00499 and K02DA00293 (Dr. Merikangas), and R01 HG00348 (Dr. Risch). This article is adapted in part from K. Merikangas and J. Swendsen (1997). The Genetic Epidemiology of Psychiatric Disorders, *Epidemiol. Rev.*

BIBLIOGRAPHY

Cutrona, C. E., Cadoret, R. J., Suhr, J. A., Richards, C. C., Troughton, E., Schutte, K., and Woodworth, G. (1994). Interpersonal variables in the prediction of alcoholism among adoptees: Evidence for gene-environment interactions. *Comprehens. Psychiat.* **35**, 171–179.

Goldman, D. (1995). Candidate genes in alcoholism. *Clin. Neurosci.* **3**, 174–181.

Gottesman, I. I. (1991). "Schizophrenia Genesis." Freeman, New York.

Gurling, H. (1996). The genetics of schizophrenia. *In* "Bailliere's Clinical Psychiatry" (G. Papadimitriou and J. Mendlewicz, eds.), Vol. 2, pp. 15–47. Bailliere Tinda, Philadelphia.

Kendler, K. S., and Diehl, S. R. (1995). Schizophrenia: Genetics. *In* "Comprehensive Textbook of Psychiatry" (H. I. Kaplan and B. J. Sadock, eds.), Vol. VI. Williams & Wilkins, Baltimore.

Kendler, K. S., Neale, M. C., Heath, A. C., Kessler, R. C., and Eaves, L. J. (1994). A twin-family study of alcoholism in women. *Am. J. Psychiat.* **151**, 707–715.

Kendler, K. S., Walters, E. E., Neale, M. C., Kessler, R. C., Heath, A. C., and Eaves, L. J. (1995). The structure of the genetic and environmental risk factors for six major psychiatric disorders in women. *Arch. Gen. Psychiat.* **52**, 374–383.

Kessler, R. C., McGonagle, K. A., Zhao, S., Nelson, C. B., Hughes, M., Eshleman, M. A., Wittchen, H. U., and Kendler, K. S. (1994). Lifetime and 12-month prevalence of DSM-III-R psychiatric disorders in the United States: Results from the National Comorbidity Survey. *Arch. Gen. Psychiat.* **51**, 8–19.

Lander, E., and Kruglyak, L. (1995). Genetic dissection of complex traits: Guidelines for interpreting and reporting linkage results. *Nature Gen.* **11**, 241–247.

McGue, M. (1994). Genes, environment, and the etiology of alcoholism. *In* "The Development of Alcohol Problems: Exploring the Biopsychosocial Matrix." (R. Zucker, G. Boyd, and J. Howard, eds.). U.S. Department of Health and Human Services, Research Monographs, Rockville, MD.

McGuffin, P., Owen, M. J., O'Donovan, M. C., Thapar, A., and

Gottesman, I. I. (1994). "Psychiatric Genetics." American Psychiatric Press, Washington, D.C.

Merikangas, K. R. (1982). Assortative mating for psychiatric disorders and psychological traits. *Arch. Gen. Psychiat.* **39**, 1173–1180.

Merikangas, K. R. (1990). The genetic epidemiology of alcoholism. *Psychol. Med.* **20**, 11–22.

Merikangas, K. R., and Angst, J. (1995). Comorbidity and social phobia: Evidence from clinical, epidemiologic, and genetic studies. *Eur. Arch. Psychiat. Clin. Neurosci.* **244**, 297–303.

Merikangas, K. R., and Kupfer, D. J. (1995). Mood disorders: Genetic aspects. *In* "Comprehensive Textbook of Psychiatry" (H. I. Kaplan and B. J. Sadock, eds.), Vol. 2, pp. 1102–1116. Williams & Wilkins, Baltimore.

Pickens, R. W., Svikis, D. S., McGue, M., and La Buda, M. C. (1995). Common genetic mechanisms in alcohol, drug, and mental disorder comorbidity. *Drug Alcohol Depend.* **39**, 129–138.

Risch, N., and Botstein, D. (1996). A manic depressive history. *Nature Gen.* **12**, 351–353.

Risch, N., and Merikangas, K. R. (1996). The future of genetic studies of complex human diseases. *Science* **273**, 1516–1517.

Weissman, M. M. (1993). Family genetic studies of panic disorder. *J. Psychiatr. Res.* **27**, 69–78.

Woodman, C. L., and Crowe, R. R. (1996). The genetics of anxiety disorder. *In* "Bailliere's Clinical Psychiatry" (G. Papadimitriou and J. Mendlewicz, eds.), Vol. 2, pp. 47–57. Bailliere Tinda, Philadelphia.

Genetic Testing

B. S. WEIR
North Carolina State University

GENETIC TESTING OF BIOLOGICAL SAMPLES PROvides a means of identifying the person from whom the samples came. The genetic profile of a sample is compared to that of a known person, and if the profiles match, that person is not excluded as a source of the sample. Such matching is used in forensic science, in parentage disputes, and in the identification of remains. Until matching profiles can be accepted as establishing a common source, it is helpful to attach numerical weight to the match. Methods for determining these numbers are reviewed.

I. INTRODUCTION

People differ because of differences in the DNA present in virtually all cells in the body. To the extent that people are genetically unique, therefore, many different types of biological samples can be used for identification. Genetic testing is used here to refer to the process of determining whether a biological sample could have come from a specific person. Genetic comparisons are made between *known* and *query* types, where the query type may be from a bloodstain and the known type from a blood sample drawn from the person thought to have left the stain. If the genetic typing of the sample is very detailed, it may be as powerful in identifying the source of the query sample as is the pattern of whorls and ridges in a fingerprint; the term "DNA fingerprint" was initially applied to the results of genetic tests. The current term is "DNA profile" and this refers to the set of alleles seen at the typed loci so that the result of testing is simply a multilocus genotype. If the profile in a query sample is not the same as that from a known person, the sample could not have come from that person. Otherwise the possibility of the person having contributed the sample needs to be considered, and often numerical statements are made to quantify this possibility.

Genetic testing finds uses in many settings, both human and nonhuman. The most widespread use is in parentage testing, where a man may be alleged to be the father of particular child. By comparing the genotypes of the child and its mother, the genetic contribution of the father of the child can often be determined. If the alleged father has the set of paternal alleles, then he is not excluded on genetic grounds from being the father. Analogous methods are used in missing person cases. The genetic profile of a bloodstain thought to be from the missing person can be compared to the profiles of family members of that person. This use of relatives, including parents, siblings, or children, is often referred to as *reverse paternity*. The need for such indirect identification arises in wartime or other disasters when the remains of victims do not allow the use of fingerprints, facial features, or dental records.

In a forensic setting, genetic testing of blood or semen stains at the crime scene may be used to link a suspect to the crime, or bloodstains on a suspect's clothing may be used to link the suspect with a victim. DNA recovered from the back of postage stamps may link a suspect to a letter bomb, and saliva from a cigarette butt may link a suspect with a crime scene. It is also possible to associate people with crimes by genetic testing of plant or animal material. The genetic profile from cat hairs on a suspect's clothing may match that from an animal in the victim's home, or the profile of leaves in a suspect's trunk may match that of a tree where the crime was committed. Genetic testing has also been used to investigate suspected paths of transmission of the HIV virus between people, as well as to identify proprietary lines of domestic plants and animals.

II. TESTING PROCEDURES

Although no two humans, identical twins excepted, are thought to be the same at the DNA level, it is impossible to determine the complete DNA sequence for either known or query samples. Humans differ at about 0.1% of their DNA, so that a region of several thousand base pairs would need to be examined to have a reasonable chance of distinguishing between people. Although DNA sequencing is being used for regions of a few hundred basepairs, better strategies will make use of sequencing chips. The complete human mitochondrial genome, with about 16,500 bp, can be attached to a chip and used to compare with a query sequence. Individual base differences can be detected, and two sequences that have different departures from the chip sequence could not have arisen from the same person. It is not yet known just how variable the human mitochondrion is, even though the D-loop region is being used to identify war remains. It is clear, however, that mitochondrial sequences will not be of use in distinguishing between maternal relatives, who are expected to share the same mitochondria. [*See* Human Genome; Mitochondria.]

Instead of DNA sequencing, or comparisons to long sequences, genetic testing currently is based on highly variable regions of the human genome. The variation exploited is that of differing numbers of *repeat units* of tens of bases (minisatellites) or two to four bases (microsatellites) at certain *genetic markers*. These entities have also been termed variable number of tandem repeats (VNTR, the longer repeat units) and short tandem repeats (STR, the shorter repeat units). A means of estimating the length of a variable region is needed, and the set of pairs of lengths, one pair per region, serves as the genetic testing profile. There can be several thousand different lengths for one VNTR marker, although these are generally collapsed into 20 to 30 classes, and up to 20 or so STR lengths. For any one region, a few percent of any population will match a specific profile, but the proportion drops as the number of regions increases. Even for markers with only two lengths, there are three possible pairs and over 3 billion combinations if 20 markers are used. Identification becomes feasible even though less than 0.1% of the genome is employed.

For questions of identification, such as those arising in forensic science, comparisons are made between two profiles that may have a single source. For parentage issues, however, there is the additional complication of transmission of genetic material from parent to child and the possibility of mutation at the marker locus. The true father may not have a particular paternal allele because mutation had caused the allele he transmitted to the child to be different from both the alleles he had received from his own parents. Accounting for mutation could be as simple as not declaring an exclusion on the basis of a single allele or could be as complicated as using allele-specific mutation rates in paternity calculations. [*See* Mutation Rates.]

Mutation is also an issue in identifying remains from distant relatives. Because of the greater abundance of mitochondrial than nuclear DNA, it is customary to identify bones with mitochondrial sequences both from the bones and from maternal relatives of the deceased person. To allow for greater discrimination, the most variable region of the mitochondrion is used but this is precisely the region where mutation is greatest. Using maternal relatives separated by several generations increases the chances of mutational differences even when a correct identification has been made.

Even though the VNTR systems have the disadvantage of requiring sufficient DNA for Southern transfer detection, and the attendant need for radioactive labels and lengthy procedures, they offer so much discriminatory power for genetic testing that they are likely to remain in use. Attention needs to be paid to the means of distinguishing variants. [*See* DNA Markers as Diagnostic Tools.]

III. BINNING VARIABLE NUMBER OF TANDEM REPEAT LOCI

VNTR loci are characterized by the estimated lengths of a region of DNA, and these lengths are typically between a few hundred base pairs and about 20 kb in length. Although the lengths depend on the number of repeat units the individual has for that marker, the lengths are estimated from migration distances of DNA fragments on an electrophoretic gel. These lengths are therefore continuous, and a complete statistical analysis would necessarily also be continuous. It is convenient, however, to discretize the lengths and invoke discrete analyses. Not only are these analyses of discrete "alleles" simpler to perform and explain, but they also allow publication of tables of allelic frequencies. Two main discretizing, or binning, strategies are in use.

A. Floating Bins

The most natural binning strategy makes use of empirical measures of accuracy of length estimates. Two samples of DNA from the same source are analyzed, and the maximum difference 2α between two lengths of the same-sized fragment is noted. From then on, any two fragments that have a proportional difference in estimated lengths less than 2α will be said to match. Specifically, if the two estimated lengths are e_1 and e_2:

$$|e_1 - e_2| \leq 2\alpha \frac{e_1 + e_2}{2}$$

so that

$$\frac{1-\alpha}{1+\alpha} e_1 \leq e_2 \leq \frac{1+\alpha}{1-\alpha} e_1$$

or, approximately,

$$(1 - 2\alpha)e_1 \leq e_2 \leq (1 + 2\alpha)e_1.$$

To estimate the probability of an evidentiary allele with estimated length e, a sample of people from the population is surveyed, and all those bands with estimated lengths in the range $e \pm 2\alpha e$ will be declared to match the allele. The proportion of such bands in the sample provides an estimate of the probability. It should be noted that if this procedure is applied to all alleles in the sample, the total of the resulting estimates will be more than one.

B. Fixed Bins

A simpler binning strategy divides the range of possible fragment lengths into a small number of nonoverlapping regions. One possibility is to use the "allelic ladder," which is a set of fragments of known lengths whose migration distances on a gel serve as a calibration method to relate migration distance to physical length. Typical ladders have about 30 such lengths. An evidentiary allele is assigned to one of these fixed ranges, or bins, and the proportion of alleles in a sample that also fall into the bin provides an estimate of the allele probability.

A difficulty with the fixed bin approach is that bins do not correspond to matching intervals. The bin may contain sample alleles that would not be said to match the evidentiary allele and may not contain sample alleles that would be said to match. In practice, however, the estimated probabilities for both fixed and floating bin methods are very similar if the fixed bins are about the same width as the matching interval (i.e., about 4α in width).

IV. INTERPRETING PROFILE MATCHES

Eventually, the interpretation of matching genetic profiles will simply be a statement by a forensic scientist that the match indicates a common source for the two profiles, in the same way that a fingerprint expert will declare that matching prints were made by the same person. Challenges to such claims will center on the possibility of error in the declaration of a match rather than the possibility of a coincidental match. That point has not yet arrived, and the history of fingerprints at the beginning of the 20th century suggests that it will take a few years. Until then, it is helpful to attach a numerical statement to the finding of matching profiles. There is likely to be a longer-term need for quantification in questions involving relatives or evidentiary profiles of more than one contributor. Numerical calculations are very conveniently expressed as likelihood ratios.

A. Likelihood Ratios

For both identification and parentage issues, there is a need to interpret genetic evidence E that does not exclude some explanations concerning the source of the evidence. There will often be competing explanations: C and its complement $\overline{C}$. For each explanation, the probability of the evidence can be calculated and these two probabilities are compared. If $\Pr(E|C)$ denotes the probability of the evidence if C is true, then the likelihood ratio L is

$$L = \frac{\Pr(E|C)}{\Pr(E|\overline{C})}$$

and is used in statements such as "The evidence E is L times more likely if explanation C is true than if explanation $\overline{C}$ is true."

The key feature of the likelihood ratio is that it refers to the probability of the evidence, even though it is the probability of an explanation being true that is of primary importance. The genetic testing evidence can address only the former. Any prior belief in the truth of the explanations can be expressed as *prior odds*

$$\text{prior odds} = \frac{\Pr(C)}{\Pr(\overline{C})},$$

and Bayes' theorem leads to posterior odds of

$$\text{posterior odds} = L \times \text{prior odds}.$$

If C is the explanation that the suspect is the person who left the bloodstain at the crime scene, and the prior odds of this were a million to one against, then a likelihood ratio of 100 million would produce posterior odds of 100 to one on C. A common error in the presentation of genetic testing evidence is to confuse the likelihood ratio and the posterior odds. They are not equivalent. Statements such as "The chances are at least 1-in-170-million that anybody else's DNA besides Simpson's could be contained in a blood drop found near the bodies of Nicole Brown Simpson and Ronald Goldman, testified Robin Cotton, director of Cellmark Diagnostics in Maryland" (Associated Press, November 14, 1996) misstate both Dr. Cotton's testimony and the meaning of the evidence.

A common situation is where genetic testing reveals a profile of type A, say, in both the crime sample and a sample from a person S suspected of having bled at the crime scene. Explanation C may be that S is the person who bled at the scene, and explanation $\overline{C}$ that some other person P caused the stain. The likelihood ratio can be manipulated as follows:

$$\begin{aligned} L &= \frac{\Pr(E|C)}{\Pr(E|\overline{C})} \\ &= \frac{\Pr(P \text{ of type } A, S \text{ of type } A|C)}{\Pr(P \text{ of type } A, S \text{ of type } A|\overline{C})} \\ &= \frac{\Pr(P \text{ of type } A|S \text{ of type } A, C)\Pr(S \text{ of type } A|C)}{\Pr(P \text{ of type } A|S \text{ of type } A, \overline{C})\Pr(S \text{ of type } A|\overline{C})}. \end{aligned}$$

If the testing system is reliable, a match is certain when C is true, so $\Pr(P$ of type $A|S$ of type $A, C) = 1$. Furthermore, the probability that S has profile type A does not depend on whether C or $\overline{C}$ is true, so $\Pr(S$ of type $A|C) = \Pr(S$ of type $A|\overline{C})$. The likelihood ratio reduces to

$$L = \frac{1}{\Pr(P \text{ of type } A|S \text{ of type } A, \overline{C})}$$

which shows that what is needed is the probability that P is of type A when it is known that S is of type A.

The simplest assumption is that the profiles of two different people, S and P, are independent. In that case the likelihood ratio is just the reciprocal of the probability that an unknown person has the matching profile. If this probability is 10^{-6}, the likelihood ratio is a million.

B. Estimating Frequencies

The last section pointed to the need for profile probabilities. The very success of genetic testing for discriminating between people makes it difficult to estimate these probabilities. The less likely it is that an innocent suspect has the same profile as the actual perpetrator, the less likely it will be that the profile is seen in any sample from the population. How then can such samples be used to provide probability estimates? Suppose the profile is based on a set of genetic markers $A, B, \ldots$, and contains alleles $A_i, A_j, B_k, B_l, \ldots$. The probabilities of a random person having one-locus genotypes follow from the Hardy–Weinberg law as

$$\Pr(A_i A_i) = p_i^2$$

$$\Pr(A_i A_j) = 2p_i p_j,\ A_i \neq A_j,$$

where p_i is the probability for allele A_i. The product rule can be extended to several loci

$$\Pr(A_i A_j B_k B_l \ldots) = \Pr(A_i A_j)\Pr(B_k B_l) \ldots$$

The product rule, within and between loci, requires an assumption of independence of alleles. Tests for independence are available but are unlikely to have large power unless sample sizes are very large. For neutral genetic markers on separate chromosomes, it is unlikely that any departures from independence would cause the product rule to seriously underestimate the probability of a profile.

It might be objected that profile probabilities of less than 1 in 6 billion, say, have no meaning because there are less than 6 billion people in the world. Such objections ignore the meaning of the number, which is a probability of finding the profile in a random member of the population. The number does not purport to be the proportion of currently living people with that profile. Because the number is derived as the product of allele frequencies, the population for which it could be regarded as a proportion is the hypothetical infinite population formed by complete random mating of a population with the same allele frequencies as those used in the calculations. Confusion with the size of the total population is avoided by concentrating on the need for a conditional probability: what is the probability of somebody having this profile given that it has been seen in the suspect?

C. Single Band Patterns

Apart from the difficulty in assigning probabilities to VNTR fragment lengths, there is a problem caused by the occurrence of single-band patterns for some loci. This may be due to the individual being homozygous for that single band, but is more likely due to the coalescence of bands of similar size on the electrophoretic gel or to one band not being detected. The cause in any particular case will not be known, and a convention has been established of assigning a probability of $2p$ to single-band patterns, where p is the probability of that one allele. This is as opposed to using p^2 from the Hardy–Weinberg result for homozygotes.

D. Confidence Statements

Even if the product rule is accepted as being reasonable, there is uncertainty in a profile probability if it is estimated from a population sample and the uncertainty diminishes as the sample size increases. Ad hoc rules have been suggested for the amount of uncertainty. A report from the U.S. National Research Council (1996) compared profile probability estimates from different samples to suggest that a reasonable confidence limit would extend from one-tenth the estimate to 10 times the estimate. Construction of confidence intervals by bootstrapping shows that the range depends on the value of the probability and involves a factor of 3 rather than 10 when the probability is of the order of 10^{-6}.

E. Population Structure

The likelihood ratio for expressing evidentiary strength of a matching DNA profile was shown earlier to reduce to a statement involving the conditional probability of a person having the profile given that someone else is observed to be of that type. Two people do not have independent profiles if they belong to the same population. In practice, it may not be known whether an unknown perpetrator P belongs to the same population as the suspect S, and it may therefore be prudent to allow for this possibility. The issue is of particular importance if the two people belong to the same subpopulation and if allele frequency estimates are available only for the whole population. In that case, a conditional probability formulation that applies as an average over all subpopulations is appropriate. Such a formulation for a single locus with alleles A_i that have probability p_i is

$$\Pr(A_iA_i|A_iA_i) = \frac{[2\theta + (1-\theta)p_i][3\theta + (1-\theta)p_i]}{(1+\theta)(1+2\theta)}$$

$$\Pr(A_iA_j|A_iA_j) = \frac{2[\theta + (1-\theta)p_i][\theta + (1-\theta)p_j]}{(1+\theta)(1+2\theta)}, \quad A_i \neq A_j.$$

These conditional probabilities are usually greater than the Hardy–Weinberg product rule values. Knowing what value to assign θ is difficult if there are no observations from the various subpopulations, but values such as $\theta = 0.01$ or 0.03 are regarded as being conservative in the sense that they will generally provide overestimates of the probabilities. Higher probabilities lower the likelihood ratio and so reduce the strength of the evidence in support of C versus $\overline{C}$.

F. Relatives

The dependence between profiles of members of the same population is due to common evolutionary history. People with more immediate ancestors are said to be related, and relatives are more likely to share genetic profiles than are unrelated people. Brothers, for example, have a 50% chance of receiving copies of the same allele from each of their two parents, and so they have a 25% chance of receiving copies of the same allele from both parents. This leads to them having the same genotype. There is an additional chance of them being of the same if they receive the copies of different parental alleles which are nevertheless themselves of the same type. This probability depends on allele frequencies in the population. Following standard population-genetic methods, some common cases for the conditional probabilities with

TABLE I
Conditional Probabilities for Relatives

Genotype	Relationship	Conditional probability
A_iA_i	Full brothers	$(1 + p_i)^2/4$
	Father and son	p_i
	Half brothers	$p_i(1 + p_i)/2$
	Uncle and nephew	$p_i(1 + p_i)/2$
	First cousins	$p_i(1 + 3p_i)/4$
	Unrelated	p_i^2
A_iA_j	Full brothers	$(1 + p_i + p_j + 2p_ip_j)/4$
	Father and son	$(p_i + p_j)/2$
	Half brothers	$(p_i + p_j + 4p_ip_j)/4$
	Uncle and nephew	$(p_i + p_j + 4p_ip_j)/4$
	First cousins	$(p_i + p_j + 12p_ip_j)/8$
	Unrelated	$2p_ip_j$

which one relative has the genotype given that the other has it are shown in Table I.

A suspect may acknowledge having a profile matching that in a crime stain, but may raise the possibility that the stain is from his brother. Providing the brother has not been typed, the likelihood ratio for the explanation C: the suspect left the stain, versus $\overline{C}$: the suspect's brother left the stain, is $4/(1 + p_i + p_j + 2p_ip_j)$ when the stain has profile A_iA_j. This quantity cannot be greater than four, whereas the value $1/2p_ip_j$ for unrelated people in different populations can be very much larger if the allele probabilities are small. [*See* Population Genetics.]

G. Mixed Stains

Situations often exist where genetic testing indicates contributions from more than one person because more than two alleles are seen for at least one marker. This can arise when a sample contains material from both victim and perpetrator, for example, or when there are multiple perpetrators. The main difference from the case of single stains is that the evidence may not be certain under both explanations C and $\overline{C}$. An example which shows such a case is when a crime sample reveals alleles A_1, A_2, A_3 at a marker locus. The circumstances of the crime dictate that the stain contained material from two perpetrators, but only one suspect has been identified. That person has type A_1A_2. Explanation C is that the suspect and one other person were the contributors, and $\overline{C}$ is that both contributors are unknown. Under C, the unknown person could be any one of types A_1A_3, A_2A_3, or A_3A_3 and the probability of the evidence is the sum of the probabilities of these types. This probability is less than one. Under $\overline{C}$, the two unknown people could be any of six possible pairs: A_1A_1, A_2A_3 or A_2A_2, A_1A_3 or A_3A_3, A_1A_2 or A_1A_2, A_1A_3 or A_1A_2, A_2A_3 or A_1A_3, A_1A_2 or A_1A_3, A_2A_3 and the six probabilities combine to give $\Pr(E|\overline{C})$.

In general, the evidence consists of a set $\{e\}$ of alleles at a locus and a subset $\{u\}$ of these alleles must have come from unknown people. If there are x unknown contributors, then what is needed is the probability of finding the alleles $\{u\}$, but no alleles not in $\{e\}$,

TABLE II
Likelihood Ratios for Mixtures

E	Suspect	Likelihood ratio: General	Likelihood ratio: p^2 rule	Likelihood ratio: $2p$ rule
$abcd$	ab	$\frac{P_1(cd\|abcd)}{P_2(abcd\|abcd)}$	$\frac{1}{12p_ap_b}$	$\frac{1}{12p_ap_b}$
abc	bc	$\frac{P_1(a\|abc)}{P_2(abc\|abc)}$	$\frac{p_a + 2p_b + 2p_c}{12p_bp_c(p_a + p_b + p_c)}$	$\frac{1 + p_b + p_c}{4p_bp_c(3 + p_a + p_b + p_c)}$
abc	a	$\frac{P_1(bc\|abc)}{P_2(abc\|abc)}$	$\frac{1}{6p_a(p_a + p_b + p_c)}$	$\frac{1}{4p_a(3 + p_a + p_b + p_c)}$
ab	ab	$\frac{P_1(\phi\|ab)}{P_2(ab\|ab)}$	$\frac{(p_a + p_b)^2}{2p_ap_b(2p_a^2 + 3p_ap_b + 2p_b^2)}$	$\frac{p_a + p_b + p_ap_b}{2p_ap_b(2 + 2p_a + 2p_b + p_ap_b)}$
ab	a	$\frac{P_1(b\|ab)}{P_2(ab\|ab)}$	$\frac{2p_a + p_b}{2p_a(2p_a^2 + 3p_ap_b + 2p_b^2)}$	$\frac{1 + p_a}{2p_a(2 + 2p_a + p_b + p_ap_b)}$

among x people. This probability is

$$\Pr(\{u\}|\{e\}) = T_0^{2x} - \sum_i T_{1i}^{2x} + \sum_i \sum_j T_{2jk}^{2x} + \ldots,$$

where T_0 is the sum of the probabilities of all the alleles in $\{e\}$, T_{1i} is T_0 minus the probability of the ith allele in $\{u\}$, T_{2ij} is T_{1i} minus the probability of the jth allele in $\{u\}$ $(j \neq i)$, and so on. Some common values for these probabilities are shown in Table II, where a, b, c, d indicate distinct alleles at a locus.

V. CONCLUSION

The possibility of being able to identify a person from a bone fragment, a hair root, a discarded tissue, or a fingernail has made genetic testing a marvelous tool. Crimes have been solved, families have been reunited, and inheritance disputes have been settled. As more and more genetic markers are used in testing, there will be less and less doubt as to the source of a sample. Until it is generally accepted that matching profiles establish a common source it will be helpful to attach a numerical statement to the evidence of a match; such statements are best derived as likelihood ratios.

BIBLIOGRAPHY

Burke, T., Dolf, G., Jeffreys, A. J., and Wolf, R. (eds.) (1991). "DNA Fingerprinting: Approaches and Applications." Birkhäuser, Basel.

Coleman, H., and Swenson, E. (1994). "DNA in the Courtroom: A Trial Watcher's Guide." GeneLex, Seattle.

National Research Council. (1996). "The Evaluation of Forensic DNA Evidence." National Academy Press, Washington, D.C.

Wambaugh, J. (1989). "The Blooding." William Morrow, New York.

Weir, B. S. (ed.) (1995). "Human Identification: The Use of DNA Markers." Kluwer, Dordrecht.

Weir, B. S. (1996). "Genetic Data Analysis II." Sinauer, Sunderland, MA.

Genomes in the Study of Primate Relations

JONATHAN MARKS
Yale University

GLOSSARY

Evolution Descent and divergence with modification

Homology Relationship between parts of two different organisms, due to their descent from a common ancestor

Homoplasy Similarity between parts of two different species, not attributable to common ancestry, but to parallel evolution (e.g., adaptation to a similar environment or a similar mutation)

Natural selection Differential survival and reproduction of organisms with particular characters. Selection for the mean value or "normal" character is called stabilizing or purifying selection; selection away from the mean or norm is called directional selection

Phylogeny Biological history of organisms, focusing on descent

RESEARCH COMPARING THE GENETIC STRUCTURE of species has been of great use in the study of physical (or biological) anthropology, the study of our place in nature. Genetic studies provide an independent test of phylogenetic hypotheses derived from anatomical studies. These are most often useful when anatomical studies have been ambiguous. Nevertheless, it is not easy to interpret comparisons among genomes, since their evolutionary rates and modes are diverse.

I. INTRODUCTION

For most of the twentieth century, there had been a considerable diversity of opinion as to the relationships of humans to the great apes (chimpanzee, gorilla, and orangutan). Genetic studies based on the intensity of immunological cross-reactions of blood serum, which inferentially measured the similarity of antigenic proteins among species and are thus a reflection of genetic similarity, were performed in the early 1960s by Morris Goodman. These showed that the closest genetic similarity existed among the human, chimpanzee, and gorilla—and that the category "great apes" was therefore phylogenetically artificial, because the chimpanzee, gorilla, and orangutan were not genetically most similar to one another, and therefore probably not one another's closest relatives. Thus, a study of molecular relationships was able to resolve a phylogenetic question derived from anatomical ambiguities.

According to the modern theory of evolution, evolutionary change is the genetic transformation of populations of organisms through time and across space. Anatomies (i.e., phenotypes) are the external products of genomes, to a considerable extent. Whereas anatomies are highly responsive to environmental conditions and can be molded by natural selection in similar ways in different lineages, genomes, for the most part, are not. Thus, the phylogenetic "noise" of parallel evolution should be reduced if one compares genomes rather than phenotypes. Genetic phylogenies are therefore sometimes thought to represent truer histories than anatomically based phylogenies. This is quite naive, however, as the evolutionary processes operating within the genome are now known to be quite complex, are poorly understood, and often yield patterns that are difficult to interpret.

Evolutionary studies of the genome transcend the difficulty of parallel evolution (i.e., homoplasy) caused by similar responses to the environment in different lineages, yet have several difficulties of their own. First, only four nucleotides are possible at any given position (a deletion could be considered a fifth), so the universe of possibilities is limited—a mutation from A to G at any given position is likely to occur many times in different lineages. Second, the genome is filled with redundancy, which can complicate genetic comparisons among species, especially in gross comparisons such as DNA hybridization (see Section VI). Any DNA segment, therefore, has many homologs in another species. In anatomical evolution this is called serial homology; in genomic evolution it is called paralogy. [*See* Genome, Human.]

The concept of homology is critical in evolutionary studies. Two structures are homologous if they are descended from a common ancestor. The adult α1-globin gene in humans (Fig. 1) is homologous to the adult α1-globin gene in gorillas; that is, they are both descended from a common ancestor: the α1-globin gene of a Miocene ape such as *Proconsul africanus*. The human α1 gene, however, is also homologous to the α2 gene alongside it; that is, they are both descended from a common ancestor, an ancestral single α-globin gene present in an Oligocene catarrhine such as *Aegyptopithecus zeuxis*. In the evolution of the monkeys and apes, a duplication of the α-globin gene occurred. As shown in Fig. 1, several duplications had already preceded that one. These duplicates are said to be paralogous DNA segments, or paralogs; the particular α-globin genes across species are said to be orthologous DNA segments, or orthologs.

Homologies within species far outnumber those across species in the comparison of any pair of genomes. There are, for example, at least 14 known

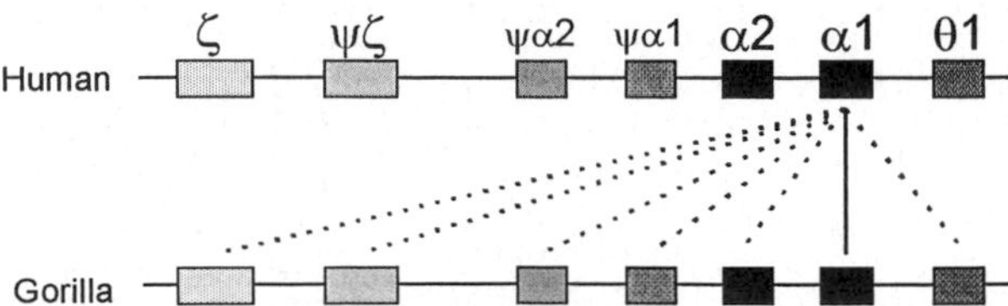

FIGURE 1 Patterns of homology in the genome. In the α-globin gene cluster of the human, each gene is paralogous to every other. The human α1-globin gene is orthologous to one gene in the gorilla (solid vertical line) and paralogous to several others (dotted lines).

homologies of the human blood globin genes. In humans there are also hundreds of thousands of homologies of the *Alu* repeats, short (i.e., about 300 bp) DNA sequences interspersed in the DNA of primates.

Another difficulty in the study of phylogeny from genomes is the problem of typology (i.e., judging the characteristics of a population by a single individual). Anatomists often have knowledge of the variation in the characters they study, in order to gauge whether or not the character in question actually differentiates two populations from one another, or is merely a reflection of the individuality of the specimen being examined. Genomic studies, particularly DNA sequencing, are sufficiently labor-intensive that one is often forced to generalize about the genome of a species, despite the knowledge that the information has been obtained only from a single individual and there are likely to be polymorphisms in the population.

II. GENOMIC EVOLUTIONARY PROCESSES

A number of diverse processes are now known to be operating within the genome, creating diversity within and across species. These modes of mutation can analytically be divided into three kinds: point mutations, genome mutations, and chromosomal mutations. Generally, point mutations tend to accumulate at a fairly regular tempo over time and are the classic mutations familiar to those who have studied genetics in the twentieth century. Genomic mutations are responsible for the pervasive paralogy in the genome, and both these and chromosomal mutations seem to accumulate irregularly in a lineage.

A. Point Mutations

Point mutations are changes that can occur to a single nucleotide or to adjacent nucleotides in the DNA molecule. Most often, these are the substitution of one nucleotide for another, but they might also be the deletion of nucleotide or the insertion of a nucleotide. These could have a major phenotypic effect, or none at all, depending on where the mutation occurs. A point mutation can occur within a gene or between genes (i.e., in intergenic DNA). Within a gene the mutation can occur within coding or noncoding regions. Within coding regions the mutation can result in the coding of a different amino acid or can be a silent mutation, with the same amino acid inserted in the protein.

Natural selection operates on organismal phenotypes and allows favorable phenotypes to proliferate at the expense of alternative phenotypes. Because all organisms are the products of eons of natural selection, most random changes to an organismal phenotype are injurious and are therefore "weeded out" by natural selection. When we compare DNA sequences among species, therefore, we find that coding sequences are always the most similar, as they are most strongly subject to natural selection. Genic noncoding DNA is next most similar across species, and intergenic DNA is the most divergent, presumably virtually invisible to natural selection, as it produces no obvious phenotype.

This highlights an important difference between evolution at the anatomical and genomic levels. In anatomical evolution one expects to find *similarity* among related species (since heredity is fundamentally conservative), and one studies their differences, which are explained by directional natural selection, or adaptations to diverse environments. On the other hand, since so little of the genome interacts with the environment, the accumulation of DNA divergence over time is guided principally by genetic drift, a statistical property of heredity in populations. One therefore expects to find *differences* among DNA sequences of related species, and is instead struck by similarity, which is then attributed to purifying natural selection, indicative of a coding sequence with an important function. Indeed, the homeo box, which may be crucial in the developmental genetics of complex organisms, was recognized for precisely this reason: a high degree of conservatism between flies and humas, which indicated great functional significance for the region.

B. Genomic Mutations

Although point mutations formed the bedrock of evolutionary genetic theory throughout most of the twentieth century, molecular genetics has uncovered a series of other mutational modes that act to differentiate the genomes of individuals, and ultimately the genetic compositions of populations. These are illustrated in Fig. 2 and discussed in the following.

Strand slippage involves a mistake on the part of the DNA polymerase molecule during DNA replication, causing the insertion or deletion of a short series of nucleotides. This occurs most often during a run of identical nucleotides, or short stretches of tandem repeats, and results in the deletion or insertion of a block of DNA—one or a few tandem repeats, or nucleotides.

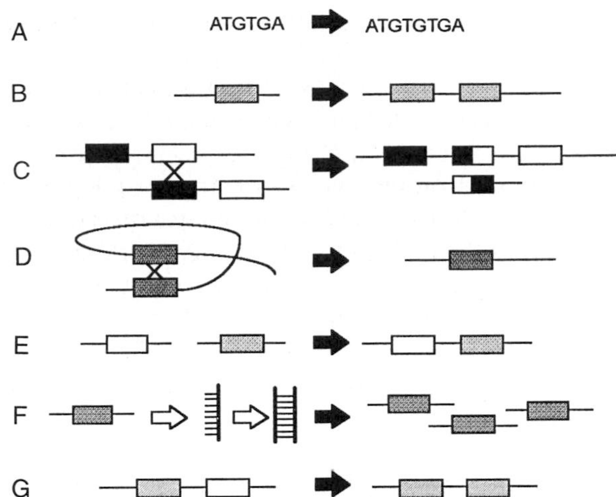

FIGURE 2 Modes of genomic mutation. (A) Strand slippage. (B) Tandem duplication. (C) Unequal crossing-over. (D) Intrachromosomal recombination. (E) Transposition. (F) Retrotransposition. (G) Gene correction.

Tandem duplication involves the insertion of a copy of a DNA segment adjacent to the original. This "rubber stamp" process is the ultimate origin of multigene families.There are three evolutionary fates for a gene duplicate. First, it can continue to make the same gene product as the original, augmenting the function of the first gene. Second, it can accumulate mutations that alter the gene product and its regulation, so long as the original remains intact. In this way the copy can come to take on a specialized function in a particular tissue or at a particular developmental stage. Third, the copy can accumulate mutations that render it inactive; in this case it is called a pseudogene.

Unequal crossing-over is a meiotic consequence of repeated DNA segments clustered together. Though crossing-over is part of the normal process of meiosis, recombining paternal and maternal alleles on a single chromosome, this ordinarily occurs between orthologous DNA segments. If there are tandemly duplicated blocks of DNA, the meiotic machinery can temporarily mispair the chromosomes long enough for a crossing-over to occur between paralogous DNA segments. The evolutionary consequence is a duplication and reciprocal deletion in different gametes.

Intrachromosomal recombination can occur as a consequence of tandemly duplicated segments. A loop formed by a single DNA strand folding back on itself can permit the cross-over process. This would delete the DNA in the loop from the chromosome.

Transposition is the insertion of a piece of DNA into a new location. This is a common property of the relationship of viral DNA to eukaryotic genomes. Here, the genetic element itself is movable, or transposable.

Retrotransposition involves the transcription of an RNA molecule from a DNA template, reverse transcription of the RNA into DNA, and insertion of the DNA copy elsewhere in the genome. The genome can thus be flooded by such RNA-derived copies, and this is the apparent source of the *Alu* repeats. Additionally, some processed pseudogenes appear to be formed in this manner.

Gene correction or conversion is a poorly understood process by which duplicate copies of a DNA segment are kept identical to one another. Usually, these duplicates are adjacent, as in the $\alpha2$- and $\alpha1$-globin genes, but they can also lie on different chromosomes (e.g., the genes for ribosomal RNA).

C. Chromosomal Mutations

The third mode of change to be considered is that which can occur to entire chromosomes, or large segments, and is thereby visible by the light microscope. Balanced changes do not affect the overall quantity of genetic material but merely rearrange it; these occur and are readily visible among the phylogenetic relatives of humans. Most significantly, they do not appear to have an effect of the organismal phenotype. Balanced chromosomal changes seem to reorder large blocks of genetic material, without generally affecting the function of the genes involved. Unbalanced changes, on the other hand, add or delete genetic material and usually result in pathological phenotypes, unless the DNA in question is genetically inert. These chromosomal changes are of four principal kinds (a fifth, polyploidy, the duplication of entire chromosome sets, does not play a role in mammalian evolution, but is significant in the evolution of other groups, and probably occurred in the very remote ancestry of our lineage). The four types are discussed next. [*See* Chromosome Anomalies.]

Inversions result from the breakage of a chromosome in two places, a reversal of polarity of the segment, and subsequent fusion of the breaks. These can often be inferred when comparing the chromosomes of the apes and humans.

Translocations result from the breakage of two different chromosomes and the subsequent improper rejoining of the segments. These are the most frequent chromosomal aberrations detectable in human populations, but are rarely detectable across primate species, since they usually have negative effects on the fertility of the bearer when heterozygous. The homologs of chromosomes 5 and 17 have been translocated in the gorilla.

Fusions and fissions reduce or enlarge the characteristic number of chromosomes per cell. In the human lineage a fusion of two small chromosomes has occurred, creating a large chromosome (i.e., chromosome 2) and reducing the human complement of chromosomes from 24 pairs to 23 pairs.

Heterochromatin amplification or reduction is the expansion or contraction, respectively, of darkly staining material, usually associated with highly redundant satellite DNA. In humans these tend to vary in size, but not location, whereas across species they can be quite labile. In humans this material exists at the centromere of all chromosomes, as well as below the centromere of chromosomes 1, 9, and 16 and at the distal end of the Y chromosome. Chimps and gorillas, by contrast, have this heterochromatic material at the tips of most of their chromosomes. Because this probably represents nonfunctional DNA, it has an evolutionary role in spite of being an unbalanced change.

III. HUMAN EVOLUTION AT THE GENOME LEVEL

Comparisons of the genome have yielded some fascinating, and sometimes counterintuitive, inferences about the recent genetic evolution of humans. In the 1960s and 1970s a number of genetic studies (generally based on protein similarity, as a reflection of gene function and structure) showed that, contrary to the phenotypes, in which humans are clearly different from chimpanzees and gorillas, the genomes of humans were not clearly different. Indeed, the genetic distances that could be measured invariably showed that the distances among the human, chimpanzee, and gorilla were both very small and sufficiently similar to one another as to preclude a clear phylogenetic inference. The split among the three species thus came to be regarded as a genetic "trichotomy," the orangutan lying outside this.

A simple deduction from these observations is that when a small portion of the genome is examined, the sampling error is probably great enough to render any apparent "breaking" of the trichotomy dubious. Indeed, some genetic features can be invoked to link the human and the chimpanzee (e.g., the form of human chromosome 9 and its homologs), the human and the gorilla (e.g., the form of the Y chromosome), or the chimpanzee and the gorilla (e.g., heterochromatin at the tips of the chromosomes). The few existing DNA sequences of these species show some nucleotide substitutions that appear to link each pair of species. The DNA sequence from the β-globin cluster overall appears to favor a human–chimpanzee link, whereas the DNA sequence for involucrin (a protein made by a class of skin cells) appears to favor a chimpanzee–gorilla link, as do the phenotypic data.

More significant than the branching sequence, however, is the recognition that all of the human uniquenesses—habitual bipedalism, brain expansion and language, culture, canine tooth reduction, loss of fur, concealment of ovulation in females, elaboration of the precision hand grip (e.g., holding a pencil), etc.—were accompanied by very little in the way of genetic change. In fact, all of these phenotypic ways in which humans differ from chimpanzees were accompanied by the same amount of genetic change that accompanies the phenotypic ways *gorillas* differ from chimpanzees; and it is difficult to name many of these, aside from body size.

There is no question that genetic change causes evolutionary change, so the human uniquenesses are caused by genetic changes, but these genetic changes are relatively quite few in number. Yet not one was probably a macromutation: a pathological hairless child with a big head and language, born to an australopithecine mother. Rather, the changes in the phenotype per generation were almost certainly very minor. Since the linkage among genotypes, development, and phenotypes—the question of how one gets a phenotype from a genotype—is still one of the major unsolved problems in human biology, this inference is not threatening to the theoretical structure of evolutionary biology. Phenotypic changes, especially large ones, generally occur over the span of many generations and involve changes in the action of several genes.

Because the structure of the functioning gene products appears to differ so little among humans, chimpanzees, and gorillas, it is inferred that the causes of the phenotypic differences lie in the regulation of the genes. Thus, the human uniquenesses are presumed to be attributable to largely the same gene products, but turned on and off at slightly different times, affecting the relative growth rates of body parts, and ultimately resulting in the classic human phenotype.

Another point worthy of consideration is that if the genetic difference between a human and a chimpanzee is small, and much smaller than the difference between the two species phenotypically, then the genetic differences between any two human groups must be much smaller still, and probably also much less striking than any phenotypic differences between these groups.

We can make some generalizations about the modes of evolution in the primate genome. First, the size of the mutation is not correlated with the magnitude of its phenotypic effect. Thus, a single nucleotide difference between two genomes might have no effect at all or might have a significant and perceptible effect. This is easily deducible from the study of genetic pathologies within the human species. When we extend this observation across species, it helps us account for the paradoxical observation that chimpanzees and gorillas are so extraordinarily similar to us genetically, but are very different from us phenotypically.

Second, the genomes of species can change from those of their relatives in quite significant ways in short periods. The functional significance of such changes is unclear, as they involve principally nongenic DNA. Thus, despite the extraordinary similarity of the functional genic DNA of the chimpanzee to the human already noted, chimpanzee's have been measured to have 10% more DNA per cell than humans (Table I). The significance of this DNA, if any, is unknown.

Third, some of these genomic processes can be effective in creating reproductive barriers between populations, and therefore can be more important in speciation than in adaptation. A major role has been hypothesized for tandem duplications of repetitive DNA in speciation, a process that has been called "molecular drive."

TABLE I
Genome Size of Old World Monkeys and Apes, Relative to Humans

Group	Percentage of human genome size
Human (*Homo*)	100
Chimpanzee (*Pan*)	110
Gorilla (*Gorilla*)	102
Orangutan (*Pongo*)	117
Gibbons (Hylobatidae)	79.2–85.2
Cercopithecidae	
Colobinae	103–123
Cercopithecinae	
Cercopithecini	100–149
Papionini	98–107

Fourth, these modes of genetic evolution are qualitatively different and are therefore difficult to compare quantitatively. Are two genomes that differ by four chromosomal inversions more similar than two genomes that differ by two translocations and the emergence of a novel repetitive sequence?

Indeed, attempts to quantify genomic differences can be misleading. For example, if we count the haploid chromosome number as a measure of chromosomal similarity, we would find that the human (n = 23) and the chimpanzee (n = 24) are as different as the black gibbon (*Hylobates concolor; n* = 26) and the siamang (*Hylobates syndactylus; n* = 25). However, this would mask the fact that all of the human chromosomes can be almost perfectly aligned against homologs from the chimpanzee, whereas few of the chromosomes from the two *Hylobates* species can be matched to one another. However, by other standards of genetic distance, the two gibbon species are no more divergent from one another than is the human from the chimpanzee. [*See* Human Genome and Its Evolutionary Origin.]

IV. RATES OF MOLECULAR EVOLUTION

The case of the two gibbon species having highly divergent chromosomes while being nevertheless closely related is an apparent paradox. Should not closely related species be similar in their genetic makeup?

In general, of course, they are. Nevertheless, there are numerous cases of discordant evolutionary rates for some of these genomic evolutionary processes, especially for chromosomal changes. Probably most extreme is a situation from the artiodactyls, in which two species of a deer known as the muntjak, similar enough to form viable hybrids, nevertheless have a diploid number of $2n$ = 46 in one species and $2n$ = 7 in the male and $2n$ = 6 in the female of the other species.

The rate of evolution of chromosomes is thus not well related to the rate of evolution of phenotypes. Two different but closely related species can have karyotypes that are virtually identical (as in the case of many of the baboons and the macaques), radically different (as among the gibbons and the guenons), or only slightly different (as in the great apes and the humans).

The reason for this variation in evolutionary rates is probably that changes in chromosome morphology affect the pairing of chromosomes in meiosis, and hence fertility. These changes are therefore likely to play a part in speciation (i.e., the formation of reproductively isolated communities) in some groups, whereas the phenotypes are directly concerned with adaptation to the environment. In groups that form new species via a chromosomal mechanism, chromosomal diversity between closely related species is relatively high, whereas among groups that form new species by other genetic means, the chromosomes are more similar.

Nucleotide substitutions are more useful for quantifying evolutionary divergence, since one can (in theory) simply count the number of differences among homologous parts of the genome. Here, most detectable nucleotide substitutions are involved in neither adaptation nor speciation; they are neutral, or nearly so. The rate of accumulation of neutral mutations within a given lineage should be roughly constant over time. Thus, given the amount of genetic difference between two species and an estimate of the rate at which neutral substitutions accumulate, one should be able to judge how long the two species have been accumulating mutations separately. That is, the number of nucleotide differences discernible between homologous DNA regions of two species should be indicative of the number of neutral nucleotide substitutions that have occurred over that region, which should be indicative of the time the two species have been diverging. This is the concept of the "molecular clock."

Like most generalizations in science, the molecular clock is valid only to a first approximation. One immediate difficulty is that to implement it in the comparison of two genomes from different species, one must ignore other evolutionary modes (e.g., strand slippage). Often in the comparison of several species' genomes, however, one encounters fairly clear-cut examples (e.g., a string of five adenines in a row in one species, homologous to six adenines in another species and eight adenines in yet another species). A second difficulty is that gene sequences are the least neutral parts of the genome, and thus have selection operating on them to a greater extent than do intergenic DNA sequences. Thus, one should examine intergenic DNA preferentially for these purposes. A third difficulty is that the more divergent two species are, the greater the likelihood that a single nucleotide difference reflects more than one nucleotide substitution. This is the problem of "multiple hits" at the same nucleotide site, which must be statistically accounted for.

For all of these caveats, the molecular clock appears

to work, however crudely. Taking the foregoing factors into consideration, a study of nearly 11 kb from a pseudogene in the β-globin gene cluster shows that the human DNA sequence is about 1.7% different from those of a chimpanzee and a gorilla; the same regions from these three species are about 3.5% different from the homologous region of an orangutan; the region from these four species is about 7.9% different from that of an Old World monkey, the rhesus macaque (a cercopithecid); and the region from these five species is about 11.9% different from that of a New World primate, the spider monkey. Although the rates of molecular change have varied somewhat, there is, nevertheless, a strong association between the amount of detectable genetic divergence and the recency of common ancestry.

V. EVOLUTIONARY ANALYSIS OF GENOMIC DATA

An important component in using the amount of difference to infer phylogenetic relationships is to show that the rate of change is similar in the various lineages. This is accomplished by means of the relative rate test, shown in Fig. 3.

Consider three species: A, B, and C. The amount of difference between A and B is 2, and the amount of difference between B and C and between A and C is 4 (Fig. 3A). We would be tempted to infer that A and B are more closely related to one another than either is to C (Fig. 3B). However, this is valid *only* if the rates of evolution are similar in all of the lineages.

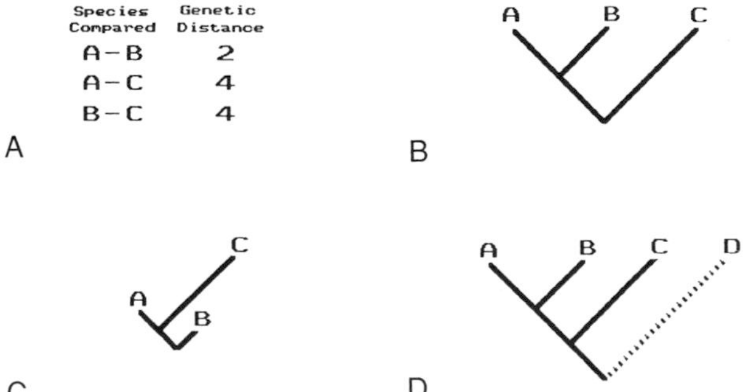

FIGURE 3 The relative rate test. (A) A set of genetic distances measured between pairs of species. (B) Their probable relationships, assuming equal evolutionary rates. (C) Their possible relationships, if the character being measured is evolving more rapidly in species C than in A and B. (D) Comparison of A, B, and C to an outgroup, D, can detect disparities in evolutionary rates across lineages.

If, on the other hand, C were evolving rapidly, then the amount of difference between C and anything else would be very high, regardless of the branching sequence. Thus, the hypothetical data just given are compatible with A and C being closest relatives, if C is accumulating differences much more rapidly than either A or B (Fig. 3C).

The relative rate test invokes a fourth species outside of those being compared (call it D) and asks how different A, B, and C are from D. If C is evolving rapidly, then it should be more different from D than A or B is. If the genetic distances between A and D, B and D, and C and D are all very similar, then we can infer that there is no relative difference in the evolutionary rates among A, B, and C.

An alternative strategy to inferring phylogenetic relationships on the basis of similarity is to count the specific DNA changes themselves. Then, assuming that a nucleotide substitution is a fairly rare event, a tree is constructed that invokes the fewest nucleotide substitutions to account for the data. This is known as a parsimony method (i.e., the tree with the fewest substitutions is the most parsimonious), in contrast to a phenetic method, such as the one just described, in which overall similarity or difference dictates relationship.

Parsimony methods have their limitations as well. There is no guarantee that the most parsimonious tree is the correct one, for there is no guarantee that any particular nucleotide substitution is sufficiently rare as to preclude its having occurred independently in different lineages. Indeed, mutational "hot spots" do occur, and there appears to be a small but significant amount of parallel evolution (i.e., homoplasy) that confuses the fine-scale determination of branching sequences from DNA sequence data.

Since there is no guarantee that the most parsimonious DNA tree is the correct one, for fine-scale branchings one must often judge among the few most nearly parsimonious trees to find the one that is most concordant with other data sets. In other words, the closer the split among three species, the more difficult it is to tell which two are the most closely related, by any method.

What is quite extraordinary through all of these handicaps is that, for the most part, using a fairly small genetic sequence (e.g., hemoglobin) one can reconstruct with a high degree of accuracy the same phylogenetic relationships inferred on the basis of phenotypes. Where the phenotypic and genomic trees conflict in their fine structures, it is often difficult to tell which is wrong. Is the phenotypic tree mistakenly using a similar adaptation in distantly related species to infer phylogeny incorrectly? Or is the DNA segment under study not evolving in the most parsimonious manner, or perhaps not representative of the genome as a whole?

VI. DNA HYBRIDIZATION

Another method for inferring phylogenetic relationships, based on a larger sample of the genome, uses the thermal stability of imperfectly paired DNA strands as a guide to how different the two strands are. If the two DNA strands come from different species, the DNA molecule is said to be a heteroduplex. Since mutations have accumulated along the lineages since the species diverged from one another, a heteroduplex molecule is held together with fewer hydrogen bonds than a homoduplex DNA molecule (i.e., one whose two strands are derived from the same species). Therefore, it should require less energy to break the bonds holding the heteroduplex together, and the strands should dissociate from one another at a lower temperature than the two well-paired strands of homoduplex DNA. In practice, this method has not proven as reliable as had been hoped, in large part because of the complexity of the genome. As the genomes of two species are compared *en masse,* each DNA segment may hybridize either to its specific ortholog or to any of several paralogs in the other species (see Fig. 1). The thermal stability of orthologous DNA measures the divergence of species, whereas the thermal stability of paralogs measures the antiquity of the gene duplications.

It proves difficult to separate these two measurements experimentally. In closely related species, the pairing of orthologs is strongly favored over paralogs, since the species (represented by the orthologs) are very similar, but the duplicated genes (paralogs) are very different. Nevertheless, variation in the extent of paralogous hybridization can generate artifactual differences in the DNA comparisons, which makes the measurement of small differences between the genomes of species imprecise. In distantly related species, orthologous pairing is not as strongly favored, since the species are very different, as are the duplicated genes. Consequently, more of the hybrid DNA sample is composed of paralogous hybrids, which are not what the experiment purports to measure. The technique thus is effective only on a more restricted range of phylogenetic problems than was thought in the 1980s.

DNA hybridization has certainly supported the close relationship among humans, chimpanzees, and gorillas, inferred on the basis of narrower genetic similarities. Indeed, from the available data, it appears that the DNA analyzed from these three species are no more than 1–2% different from one another.

VII. MITOCHONDRIAL DNA AND HUMAN ORIGINS

There is, in addition to the DNA located in the cell's nucleus, a minute amount of DNA in the mitochondria (i.e., cytoplasmic organelles involved in generating ATP for metabolic energy). In humans the mitochondrial DNA (mtDNA) is a compact molecule of about 16,500 bp in length, with none of the redundancy and excess characteristic of the nuclear genome.

mtDNA has several interesting properties. First, it does not undergo recombination; there is no union of maternal and paternal alleles, as in nuclear chromosomes. Second, mtDNA is inherited clonally through the maternal line. At the time of fertilization the zygote inherits mtDNA only from the egg, not from the sperm. Therefore, a child is a mitochondrial clone of its mother and unrelated to its father, in contrast to the situation in the nuclear genome, in which it is equally closely related to both. Third, it is relatively easily isolated and is therefore readily amenable to study. Fourth, it evolves at a rapid pace—in humans about 10-fold faster than nuclear DNA—which makes it amenable to the study of relatively short spans of evolutionary time.

Applied to the trichotomy, mtDNA data have yielded results as ambiguous and mutually contradictory as other molecular data. However, this rapidly evolving DNA can be used to study a series of more recent evolutionary events, the diversification of human groups.

Humans contain far less mtDNA diversity than their closest relatives, which implicates a demographic event having secondarily cut back the detectable levels of mtDNA variation in humans. This could conceivably have been a "founder event,'" wherein all contemporary people are descendants of a small population of ancestors.

Further, comparing the amounts of diversity encountered within continents of the Old World, far more extensive diversity is consistently encountered among Africans than among people from other continents. This in turn may imply that mtDNAs have been evolving longer in Africa than elsewhere. Thus, the root of the human mtDNA family tree should be placed in Africa. The deepest branch of that tree was dated at about 200,000 years ago, on the basis of the amount of difference detectable and the clock rate at which mtDNA diversity appears to be accumulating.

These conclusions were in considerable harmony with an interpretation of the fossil material advocated by several paleontologists. This interpretation sees modern humans as having evolved 100,000–200,000 years ago in Africa, then emigrating and supplanting anatomically archaic populations of Asia and Europe. Other interpretations of the fossil record have humans on each continent evolving largely *in situ,* although in genetic contact with one another. The pattern of mtDNA diversity would here be explained as a consequence of variations in population size and structure.

BIBLIOGRAPHY

Cavalier-Smith, T. (ed.) (1985). "The Evolution of Genome Size." John Wiley & Sons, New York.

Dover, G. A. (1982). Molecular drive: A cohesive model of species evolution. *Nature (London)* **299,** 111.

Godfrey, L., and Marks, J. (1991). The nature and origins of primate species. *Yearb. Phys. Anthropol.* **34,** 39.

Green, H., and Djian, P. (1992). Consecutive actions of different gene-altering mechanisms in the evolution of involucrin. *Mol. Biol. Evol.* **9,** 977–1017.

Li, W.-H., and Graur, D. (1991). "Fundamentals of Molecular Evolution." Sinauer Associates, Sunderland, MA.

MacIntyre, R. J. (ed.) (1985). "Molecular Evolutionary Genetics." Plenum, New York.

Marks, J. (1983). Hominoid cytogenetics and evolution. *Yearb. Phys. Anthropol.* **25,** 125.

Marks, J. (1992). Beads and string: The genome in evolutionary theory. *In* "Molecular Applications in Biological Anthropology" (E. J. Devor, ed.), pp. 234–255. Cambridge Univ. Press, New York.

Pellicciari, C., Formenti, D., Redi, C. A., and Manfredi Romanini, M. G. (1982). DNA content variability in primates. *J. Hum. Evol.* **11,** 131.

Relethford, J. (1995). Genetics and modern human origins. *Evol. Anth.* **4,** 53.

Sarich, V. M., and Wilson, A. C. (1967). Rates of albumin evolution in primates. *Proc. Natl. Acad. Sci. USA* **58,** 142.

Gerontechnology

AD VAN BERLO[1], HERMAN BOUMA[1], JAN EKBERG[2], JAN GRAAFMANS[1], FRED A. HUF[1], WIM G. KOSTER[1], PIRKKO KYLÄNPÄÄ[2], HEIDRUN MOLLENKOPF[3], RITVA ROUTIO[2], JAN RIETSEMA[1], and COR VERMEULEN[1]

[1]*Institute for Gerontechnology, Eindhoven University of Technology*
[2]*National R&D Centre for Welfare and Health (NAWH), Helsinki*
[3]*Wissenschaftszentrum Berlin für Sozialforschung (WZB)*

GLOSSARY

ADL Activities of daily living, as confined to self-care in everyday life for everybody in the house environment; components are dressing, washing, cooking, eating, toilet use, and in-house mobility. The concept is particularly used in case of actual or expected handicaps

Antioxidant Any substance that can delay or inhibit oxidation in the human body (e.g., vitamins E and C, flavonoids)

Biomarker Measure that is indicative for an exposure or effect in an organism

Demography Composition of the population according to many criteria, primarily age and gender (female/male), but also level of education, profession, financial situation, family composition, type of housing, etc.

Environment Internal biological and psychological environment and external social, technical, and other surroundings with which humans interact and within which they function. The concept is a general one and has to be operationally defined for specific situations

Epidemiology Study of incidence and distribution of diseases in human populations

Free radical Often highly reactive chemical molecules that contain one or more unpaired electrons

Gerontechnology Multidisciplinary field of study devoted to technological means to create a preferred living and working environment, good health, and suitable care for adults, especially those over the age of 50

Gerontology Multidisciplinary scientific field of study devoted to biological, psychological, social, and sociological aspects of the process of aging beyond the age of 50

Oxidative stress Situation in which oxidants or radical production exceeds the capacity of the defense system

GERONTECHNOLOGY IS DEFINED AS THE STUDY of technology and aging for the benefit of a preferred living and working environment and adapted health care for aging adults, particularly persons over the age of 50. In many countries, the decisive influence of technology on the daily living and working environment has necessitated using insights from the sciences of aging, that is, gerontology and geriatrics, for steering research and development toward products and services that senior citizens need and can use. Demographic developments such as a high and increasing number of elderly make such effort mandatory. Gerontechnology will be explained and illustrated in the areas of housing, mobility, information and communication, activities of daily living (ADL), and nutrition.

I. INTRODUCTION

Gerontechnology is an application-driven field of inquiry directed at promoting by technical means a good quality of life at home and at work, prevention from disease, and suitable care for citizens over the age of 50. The first international congress on the topic was

ENCYCLOPEDIA OF HUMAN BIOLOGY, Second Edition, VOLUME 4.

held in 1991. Four principal concepts define this rather new field:

1. In many countries, the technological environment is a dominant factor in a majority of the daily activities and experiences of people of all ages. For example, most means and infrastructure for private and public transport rely on technology.
2. Technology can be directed toward serving purposes and intentions of senior citizens as these develop with age and toward the physical, social, and mental capacities that are basic for realizing such intentions.
3. Insights into the aging process as provided by gerontology and by participation of the elderly help define the needs of the elderly and environments in which the products and services have to function.
4. The absolute and relative increase of the number of elderly in many countries constitutes a demographic shift that can be partly accommodated by gerontechnology.

Relevant notions must be made operational so that they can be assessed qualitatively and quantitatively. For example, a preferred living environment is usually one in which a person can function independently, and the degree of independence in carrying out certain tasks can then be assessed. A good health status is clearly desirable and indices for that can be devised. A work ability index has been developed by J. Ilmarinen, and for the home an ADL index is available. Measures are also necessary for specific tasks such as visual acuity as an index for visual performance.

Five types of benefit of gerontechnology can be distinguished:

1. *Prevention* of age-associated loss in physiological, psychological, and social functioning (primary prevention) or in its unwanted consequences (secondary prevention). An example would be a monitoring device for daily physical effort.
2. *Enhancement* of performance in specific tasks and environments. Examples are adaptable housing or user-friendly communication technology.
3. *Compensation* for declining abilities such as higher illumination in visual tasks or task redesign for quicker responses.
4. *Assistance* for informal and professional caregivers such as devices for lifting and for transferring patients.
5. *Improvement* of research in gerontology such as image processing in radiology or better logistics in the health care sector.

The target group is all citizens above the age of 50 years, most of whom are healthy and live an independent life. Active involvement of senior citizens is believed to be indispensable for real progress. Handicapped persons and persons in institutional care are a subgroup to which much technological effort has already been devoted (e.g., wheelchairs, speech interfaces for control).

The following sections are devoted to five active research areas: housing and living, mobility, ADL, communication and information, and nutrition. In the final section, some general trends are discussed.

II. HOUSING/LIVING ENVIRONMENT

Independent living can be supported by "smart" homes. An adaptable smart home is an integrated house in which space is conceived to support interactive relationships. The aim is designing and building for everyone, including elderly, children, and disabled people, for example, barrier-free accessibility and architecture that take account of the diverse capabilities of human beings. An adaptable smart home is built or renovated in such a way that it is easy and inexpensive to adapt to the changing needs of the inhabitants. For the elderly, this entails that devices assisting the activities of daily living can easily be installed later on.

A representative example is the first 12 adaptable smart homes constructed in Finland as part of the European project ASHoRED (Adaptable Smarter Homes for Residents that are Elderly or Disabled). The homes were built to comply with adaptability requirements as well as disability requirements. The devices were chosen based on the expressed needs of the actual inhabitants. Also, provisions were made for convenient transportation between the adaptable smart homes and the service centers as well as the caregivers.

A suburb of adaptable smarter homes with some 1250 homes, offices, shops, and recreation areas has recently been completed in the Finnish suburban area of Joensuu/Marjala. Each home is adaptable to the needs of the inhabitants, as well as to needs of young people, a family with children, and elderly inhabitants or disabled persons. When life situation changes, the home will be adaptable to the new situation, according to the principle that future needs, technologies, and networks are already taken into account in the planning and that solutions can easily be introduced.

The inhabitants can live, move around, and work as independently as they wish. The flexible work and telework concepts implemented in the area will support everyone, especially people who have difficulties reaching their working site. These services are based on the Integrated Services Digital Network (ISDN), which forms the base for the telecommunication infrastructure. Public services like the ISDN-based telecommunications services, such as information services or teleshopping, have also been designed in such a way that they will be accessible to all. The homes are equipped with a heavier power or telecommunications network than the actual regulations require. The cables and cable ducts and how they were installed make future structural rearrangements possible. No specialized home for disabled people is built in this suburb because each home is built in such a way that disabled people can live there. The appropriate physical and biological climate within the home is an important goal of gerontechnology.

III. MOBILITY AND TRANSPORTATION

Mobility is necessary to overcome spatial distances in order to conduct activities inside and outside the home. The extent of mobility depends on:

- individual capacities, needs, and resources (health or frailty status; interests, attitudes, knowledge, and fears; economic resources; family and friendship networks);
- environmental factors such as housing conditions, geographical and climate conditions, the built-up environment, the availability of services, and traffic density; and
- trip-related factors such as purpose of the trip, distance of destination, and availability of barrier-free access to private or public transport facilities.

Indoor mobility constitutes an important precondition for independent living and housekeeping. Mobility becomes more difficult with advancing age because of the increasing risk of a loss of function of the senses or the declining motor performance of the elderly, but also increasingly through unsatisfactory means for private and public transport. Outdoor mobility is the decisive link between the individual and his or her social and built-up environment. It is a fundamental prerequisite for social participation and for independent access to essential commodities and consumer goods.

The majority of trips are short-distance trips made as pedestrians or in some countries by bicycle, and about the same share is carried out by car (as driver or passenger). Whether public transportation is used depends greatly on the area and the quality of the offering. Public transportation use by the elderly is usually higher than that by the population as a whole.

Modern transportation technologies have the potential to enhance mobility significantly. With public traffic enterprises, the needs of the elderly and handicapped are being taken into account more and more in the technical development of buses and trains: wheelchair lifts and various low-floor modifications are increasingly available when purchasing new or replacing old vehicles. Private transportation is made easier through power steering and brakes and through larger and simpler displays placed closer to the center of the driver's field of vision. Comfort of cars can also be improved by larger doors, adaptable seats, and easily accessible luggage trunks. For dealing with various limitations to mobility, from a vague insecurity while walking up to a total inability to move, there are a multitude of technical aids. A differentiated offering has developed in recent years, especially in wheelchairs, where adequate adjustment to individual handicaps is less a technical than a financial problem.

A. Public and Private Transport

In most industrial countries, the traffic systems are dominated by individual transportation, the preferred means being the car. Elderly persons whose physical powers are waning frequently need a car for coping with everyday life. However, there are great differences in the level of ownership of private cars between the elderly and the population as a whole, even though the elderly are frequently dependent on a car. The low level of car ownership results in a more limited mobility radius for elderly people. The share of elderly persons who pursue activities outside the house is significantly lower than that of the general population, even though the amount of activities of older car drivers decreases only slowly with advancing age. The possession of a driver's license and a private motor vehicle is still strongly age dependent at this time, yet car use by older persons is expected to increase as younger generations gradually join the elderly ranks.

This rise in individual transportation will have strong and differing effects. On one hand, the easier accessibility to private cars will contribute to an increase in the mobility of the elderly. On the other hand, the concentration of traffic, especially in cities, can also lead to the elderly using less private transportation. In addition, the shift to individual transportation increases the risk that support for public transportation will be eroded. If important societal needs such as services, shopping possibilities, and cultural and sports offerings are oriented more and more strongly toward the growing share of motorized members of society, then the danger exists that inequality will expand between persons who can own and use a private car and persons who, for economic, physical, or other reasons, cannot use such transportation.

Furthermore, an increasing involvement of the elderly in accidents must be reckoned with in the future. The risk of an accident increases in relation to the amount driven (and thus the exposure to danger) with increasing age. In general, older road users are less involved in accidents than younger drivers, but once involved they are more susceptible to injury. Thus, all measures leading to a reduction in the severity of accidents (air bags, side protection, speed limits, etc.) are of special advantage for this age group.

IV. ACTIVITIES OF DAILY LIVING

Productive physical interaction with the immediate daily environment is a prerequisite for independent living and autonomy. Activities of daily living include all actions a person undertakes regarding personal care, from bathing to meal preparation. More complex and demanding tasks in daily life, such as vacuum cleaning, ironing, and doing light housework with the use of tools or appliances, are called instrumental activities of daily living (iADL). Of the independent elderly (65+ years), about 13% have difficulty with performing at least one out of five ADLs (bathing, transferring from bed or chair, dressing, toileting, and feeding oneself) or walking. This number is over 17% for the iADLs such as shopping, preparing meals, doing light housework, handling money, using a telephone, or getting about the community. Taken together, at least one out of five persons over 65 years has problems in performing tasks and cannot complete them without the help of another person or a special assistive product. The most problematic activities are bathing and light housework.

Products to assist the activities of daily living have to meet criteria derived from good ergonomics practice, that is, safety, health, comfort, and efficiency. A viable market for ADL products will emerge when:

- the distribution is well organized to increase availability;
- the information is widely disseminated so people become more aware;
- the performance of the products is adequate and effective;
- the price is in accordance with the functionality (affordability); and
- they are not stigmatizing and are thus more acceptable.

Some authorities believe that the burden of disability among the elderly will become relatively less as medical care and living conditions improve. Yet the absolute number of disabled elderly will increase because of the "graying" of the population. Therefore the need for better products and environments will increase.

Regarding the optimization of products, considerate effort is still needed. Only slowly are the researchers, designers, and producers gaining more insight into the day-to-day encounters of older consumers and their experiences in these daily situations. Involving elderly persons in codesign, codiscovery, and situated product assessment will likely uncover some of the flaws that are present in current products. User involvement in the research and design processes is become more accepted, and only through this route will the promises of "universal design," "transgenerational design," or "design for all" become real.

The biggest gains in the near future can be expected in the kitchen and bathroom environment, where design criteria related to safety and comfort almost automatically will lead to prevention of accidents and reduction of limitations in ADL performance. The increasing level of awareness and the progress in adaptable and life-span residential design will then lead to a well-balanced mixture of both low-tech and high-tech solutions toward the creation of a comfortable, friendly, and safe environment for daily living.

V. COMMUNICATION AND INFORMATION

Modern technology has opened the possibility to (1) make telecommunication services and terminals accessible to elderly and disabled people and (2) use telecommunication and teleinformatic services to support elderly and disabled people in their activities of daily life. For elderly people, some simple solutions already exist that can be used to alleviate minor mobility, sight, or hearing difficulties. Hands-off facilities, tilting key pads, and pull-down seats are examples of cheap add-on solutions for public telephones. Mobile telephones, cordless telephones, or a device that takes the phone "off the hook" are suitable for people who have difficulty reaching the telephone. Memory phones with stored numbers require only that a single button be pressed. Pressing a single button is also used when using home security telephones and alarm telephones in distress situations. The call to the service center can also be made automatically by the home security system. The videotelephone has been shown to be very useful in this context, affording the service personnel better contact with the distressed caller.

Telephones with adjustable ringing tone, adjustable amplification, or handset receiver coils for the inductive coupling to hearing aids are some examples of solutions for hard-of-hearing people. Large key pads are a solution for people with dexterity problems or poor vision. Information systems like Minitel and Word Wide Web may become very useful for disabled customers. A user interface capable of removing graphics and pictures may be needed for visually disabled persons who prefer information in a synthetic voice form. Global positioning systems (GPS) are already being combined with mobile telephones, and may be used to help, for example, elderly people walking in the woods to find their way home.

Emerging multimedia communication will help the elderly if it is designed to (1) send text, voice, and pictures separately so that the receiver system can present the information in the mode that is most suitable for the user and (2) send the information in redundant forms. For instance, the text should also be understandable without the pictures. Ergonomic interfaces are of primary importance for providing elderly with suitable access to complex devices or systems.

VI. NUTRITION AND BIOMOLECULAR ASPECTS

Good nutrition is of central importance to the health and optimal functioning of people of all ages. Many physiological functions slow down with age and the application of gerontechnology to nutrition is focused on the development of products that will help to reduce or to control diet-related chronic diseases. As stated in the Introduction, prevention of age-associated loss in functioning is one of the benefits of gerontechnology. The roles of gerontechnology with respect to nutrition are not restricted to the improvement of food composition and social environments for the elderly, but include the development of protective diets and biomedical tests to measure the necessity and effects of changing dietary habits. Lack of a univocal theory of aging causes problems in designing the right biomedical tests, but some practical gerontechnological applications can be derived.

Presently there are several theories of aging on the molecular level:

- the immunological theory: The immunological response becomes less and less effective and one becomes a victim of infections or general breakdown. (Food irradiation partly offers a technological solution.)
- the hormone theory: Hormones play a vital role in puberty and menopause and may do so in aging and dying as well. (Restricted energy intake improves the hormonal status. Foods that are low in calorie content, but with essential nutrients, can slow down the aging process.)
- the entropy theory: The tendency of systems to run down toward disorganization.
- the cell death theory: There is a restricted number of divisions for a cell before divisions cease. (Food technology must be directed to the suppression of age-related diseases, rather than the aging process.)
- the genetic mutation theory: Mutations in DNA result in faulty instruction for protein synthesis. This theory shows overlap with the following one.
- the free radical theory: Radicals, often very

reactive species with one or more unpaired electrons, are produced in oxidation reactions, resulting in damaging processes.

Since the latter may very well be an important underlying mechanism in the other theories, an increasing number of scientists support the free radical theory.

Radicals are by-products in the oxidative metabolism of nutrients. Radicals can react with all types of biomolecules, such as lipids (cell injury by membrane damage, atherosclerosis by low-density lipoprotein damage), proteins (decreased enzyme activity), DNA (mutations causing cancer), and carbohydrate (receptor alterations). The human body has several mechanisms for defense against free radicals, such as the enzymes superoxide dismutase (elimination of the superoxide anion radical) and glutathion peroxidase (repair of lipid peroxidation).

In an aerobic environment, aging is a continuous exposure to radicals and insufficient defense of the organism causes dysfunctioning of cells and failure of organ systems. This could result in one or more of the mentioned diseases, depending on the site of radical action. During aging, radical damage increases and accumulates and the likelihood also increases that the defense mechanisms cannot handle the continuous radical generation. Food ingredients can help in this situation, and it is a challenge for technologists to produce foods with adequate radical-scavenging capacity. Both biochemical and epidemiological studies have indicated that food components with antioxidative activity may have an important protective effect against cancer. The evidence of the protective action of fruits and vegetables can no longer be neglected. The same holds for vitamin E and cardiovascular diseases. Nowadays, special attention is paid to neurological disorders, since neural cells can also be damaged by radicals. Studies of fruit flies with improved genes against free radicals may lead to a medical intervention that could prolong human life and improve its quality. However, simply raising the antioxidant content in foods is not the solution for the prevention of age-related dysfunctioning, since radicals are needed and produced in the body phagocytosic defense mechanisms. Future research on the protective role of antioxidants in aging and a broad variety of age-related diseases is promising. Current research is directed to markers for radical stress and associated consequences, which will improve our understanding of aging processes and the gerontechnological contributions to individual health monitoring and intervention programs. [*See* Food Antioxidants; Free Radicals and Disease.]

VIII. EXPECTED DEVELOPMENTS

Gerontechnology is emerging as a multidisciplinary effort to fulfill the needs of senior citizens by improving their technological environment. The scientists of gerontology (social sciences, life sciences, medical sciences) must increasingly collaborate with scientists of technology in well-defined projects. Because of the demographic swift and the increased self-confidence of the elderly, public authorities are espected to actively support such programs and the buildup of international and national communication networks.

Senior citizens must increasingly involve themselves in getting ther needs established, in evaluating technological products and services, and in the early testing of prototypes. This requires the creation of technological centers for the elderly as an infrastructure for industry and for gerontechnological research and development. The majority of healthy elderly will claim more attention in this area, just as handicapped people have been the main beneficiaries until now.

Technological researchers are expected to shift their design and motivation from technology-push to customer-pull principles. Industrial designers will increasingly apply "design for all" principles, since what is required for the elderly is often helpful for middle-aged or younger customers. In the academic world, universities will increasingly organize multidisciplinary education in gerontechnology at undergraduate, graduate, and postgraduate levels.

For all the developments mentioned, efforts are already well under way to make them happen. Although the initiative is now concentrated at a few research centers in a limited number of countries, we expect the efforts to spread in both width and depth, reaching full swing as the twentieth turns into the twenty-first century.

BIBLIOGRAPHY

Bouma, H., and Graafmans, J. A. M. (eds.) (1992). "Gerontechnology." IOS Press, Amsterdam.

Breen, J. (1993). Senior citizens on the road. *In* "International Symposium: Senior Citizens on the Road—Promoting Safety and Mobility," October 1993, Bonn. MS.

Halliwell, B., and Gutteridge, J. H. C. (1989). "Free Radicals in Biology and Medicine." Clarendon Press, Oxford, England.

Langseth, L. (1995). "Oxidants, Antioxidants and Disease Prevention." ILSI Publications, Brussels.

Leon, J., and Lair, T. (1990). Functional status of the noninstitutionalized elderly: Estimates of ADL and iADL difficulties. *In* "National Medical Expenditure Survey Research Findings 4." Agency for Health Care Policy and Research, Rockville, MD.

Manton, K. G., Corder, L. G., and Stallard, E. (1993). Estimates of changes in chronic disability and institutional incidence in the U.S. elderly population of 1982, 1984 and 1989 NLTCS. *Gerontol. Soc. Sci.* 48 (4), S153–166.

Maycock, G., Lockwood, C. R., and Lester, J. F. (1991). The accident liability of car drivers. *In* "TRL Research Report 315." Crowthorn, Berkshire, MA.

Mollenkopf, H., and Hampel, J. (1994). "Technik, Alter, Lebensqualität." Kohlhammer, Stuttgart.

Oever, W. P. J. M. van den, and Graafmans, J. A. M. (eds.) (1993). "Perceived Needs of the Elderly about Mobility." Akontes, Knegsel. The Netherlands.

Pirkl, J. J. (1994). "Transgenerational Design: Products for an Aging Population." Van Nostrand–Reinhold, New York.

Gerontology

ROBERT N. BUTLER
BARBARA KENT
Mount Sinai Medical Center

GLOSSARY

Age-dependent disease Disease that appears to be directly related to chronological aging (e.g., coronary artery disease, osteoporosis, Alzheimer's disease, and Parkinson's disease)

Age-related disease Disease related to aging without a direct time relationship (e.g., cancer: 50% of all cancers arise after age 65, 80% after 50, but there are suggestions of a decreased incidence after 80)

Aging Time-dependent biological changes occurring throughout the life span and leading ultimately to death. "Chronological age" refers to time since conception or birth. Physiological age is a ranking determined by individual performance based on normative group performance over the life span

Average life expectancy Years of life from birth to death, given current mortality rates from disease and accident; length of life of 50% of any birth group

Biomarker Biological property that changes with age and can be used to predict life span. A biomarker of aging is a biological measurement that changes with age

Life span Number of years from birth to death; average length of life for longest-lived individuals; it is species specific

Longevity Length of life

Senescence Postreproductive period

"GERONTOLOGY" IS A TWENTIETH-CENTURY TERM applied to the study of the biology of aging by Elie Metchnikoff in 1907. Throughout human history, how and why we live as long as we do have been questions of great fascination. The concepts of aging and longevity and the cessation of life (i.e., death) have shaped our conception of social organization, economics, religion, and philosophy. The biological tenets to explain aging, however, have remained elusive. Recently, the techniques of recombinant DNA, hybridoma, and transgenic animals promise application of the scientific method to the study of the basic mechanisms of aging. Today gerontology encompasses the study of aging from biological, social, psychological, economic, and, indeed, all perspectives. The biology of human aging is emphasized here.

Human aging is determined by the interplay of intrinsic aspects of aging and longevity with environmental factors. Appreciation of environmental influences on the kinds and rates of biological change in humans comes from systematic longitudinal studies of diverse human populations. Indeed, we know that the antecedents of health and disease include the environment, broadly defined to encompass human behavior as well as the physical and cultural environment, genetic factors, and aging. Of these latter three, research on aging has received the least attention, yet aging profoundly affects the character of health and disease with the passage of time. It is important in gerontology to separate, at least conceptually, aging and disease. Many diseases become more prevalent with age, and in the past some disease-related changes were erroneously attributed to aging. Challenges in studying normal human aging are proper screening and the elimination of disease from normative groups.

In 1825, Benjamin Gompertz, an English actuary, reported that the *force of mortality* (mortality rates) double every 7 years of life, after 30 years of age until very old age. Although the Gompertz equation was

ENCYCLOPEDIA OF HUMAN BIOLOGY, Second Edition, VOLUME 4.

based on human data, it has been found to apply to animals as well.

In the 1980s, studies in humans found that at very advanced ages the likelihood of dying does not increase at the same rate as specified by the Gompertz equation. In fact, it slows so that as one gets closer to maximum life span (115 years), the probability of dying actually decreases. Statistical analyses of the life spans of medflies and *Drosophila,* populations that can be studied by the millions, also demonstrated such plasticity as they neared the end of life.

I. BIOLOGICAL THEORIES OF AGING

A. General

Evidence of biological aging can be found in almost all forms of life, from the simplest to the most complex. Aging changes are well documented on a descriptive level: function x decreases, substance y accumulates, or structure z changes appearance in a characteristic way. Interestingly, time-related changes found in single-celled organisms seem to be preserved in evolution and are exhibited during aging in higher forms of life, including humans. The accumulation of the aging pigment lipofuscin is an example. Not only is the presence of lipofuscin a function of age, but the rate of accumulation in cellular vacuoles is proportional to the aging rate of the species. Thus, neurons from a 3-year-old rat (i.e., close to the end of the normal life span) are found to be 95% filled with the aging pigment, whereas humans must age many more years, to 70 or 80, before such quantities of lipofuscin accumulate in the neurons.

There are many other similarities in aging among species, including changes in cell membrane fluidity, cross-linking of macromolecules, mitochondrial size changes, and loss of fecundity. The underlying mechanisms of aging remain elusive, however. There is no unified single theory of aging, and it remains impossible to separate intrinsic aging processes from external (e.g., environmental and toxic) events. A number of theories of aging are currently under scrutiny. These theories can be roughly divided into two types: determinant and stochastic, or random.

B. Determinant

The determinant theories hold that aging is an orderly programmed process, orchestrated by genetic information, which is played out at a predetermined time to govern the steps in the aging process and, ultimately, the time of death. The evidence that genetics plays a role in aging is overwhelming. An obvious example from nature is the species-specific life span. By definition species are kept reproductively distinct by virtue of the uniqueness of their genetic pool. The variance in life span between individuals within a species is small compared to the variance in life span among species. Within a group of mammals, for example, there is great diversity in life span, the shortest-lived species (the shrew) living about 1.5 years and the longest-lived (the elephant and the human) living to over 80 years.

This observation can be brought into sharper focus using the inbred rodent strains that have strain-specific life expectancies and life spans. The individuals of an inbred strain are genetically identical and might differ from another strain by only a few alleles, or in some cases only one allele. Nevertheless, these minute genetic differences are reflected in significant changes in the mortality statistics and life spans among strains. Along the same line, hybrid vigor is a genetic phenomenon that commonly affects the aging process. That is, the F_1 generation between two inbred strains usually has a longer life span than either of the parent strains. Selective breeding techniques have led to strains of aging-accelerated mice. The problem in studying accelerated aging in these mice is the danger that the life-shortening genetic change is really a pathology-producing change in disguise.

Many other lines of evidence support the involvement of genes in the aging process. The fact that, in most species, the female lives longer than the male suggests either protection by a double dose of the X chromosome or lack of exposure to some deleterious effect of the Y chromosome. In addition, it is commonly found that ancestors of people living to 90–100 years had extended longevity, suggesting a heritable component to human aging. Likewise, in a fashion analogous to the inbred mice, identical twins are more similar in life span than fraternal twins.

These observations notwithstanding, there is no particular group of people bound by culture, environment, or race that is longer lived than another. The reports of the mid-1960s and early 1970s of genetically isolated groups both in the Georgian area of Russia and in the Andes in Ecuador reaching very long life spans have not proved to be correct. Rather, it was found that age is venerated in these societies and that overestimating one's age is a common occurrence.

There are broad questions such as whether genetic

mechanisms really apply in aging. Sir Peter Medawar observed that there is no evolutionary significance to aging. He considered aging to begin after the reproductive period and therefore concludes that there is no selective pressure for aging. It could be argued, however, that in many of the longer-lived species, and certainly in humans, the time involved in the rearing of the offspring creates a selective advantage for longer life. An offspring whose parents live long enough to nurture and protect it has a better opportunity of living to adulthood and continuing the gene pool than one orphaned because of the lack of longevous genes. Also contrary to Medawar's observation is the idea of the inertia of good health programmed by the genes. According to this hypothesis, the more successful the genes are in regulating the developmental phase of life leading up to the point of perpetuation of the species, the longer will be the postreproductive period. In other words, genes conferring robustness and good health during development could have a selective advantage, allowing individuals with these genes to start the postdevelopment period in better health, so that the senescent period takes longer. [*See* Genes.]

Further credence for a determinant theory of aging comes from experiments using tissue culture, carried out by Leonard Hayflick in the 1960s. Events of the cell cycle are under exquisite genetic control. In order for cells to grow in tissue culture, they must continually divide by repetition of the cell cycle. Given the proper balance of nutrients and factors in the medium, fibroblasts can be grown in culture until they reach confluence. At this point they can be divided in half and replated on fresh culture vessels, where they again divide until confluence is reached. This process, called doubling, can be repeated until no more doublings occur and cells stop cycling and die. Cells are then said to have reached their replication, or Hayflick, limit.

The Hayflick limit has interesting gerontological implications because of the correlation between the maximum number of potential doublings of a group of fibroblasts and the life span of the donors. There is a direct proportionality between fibroblast doubling potential and life span for animals whose maximum life span ranges from 4 years (mice fibroblasts, with a low doubling potential) to 80–110 years (humans, with twice the number of doublings) to 175 years (the longest-lived species, the Galápagos turtle, with the greatest number of doublings). Of even more relevance is the observation that fibroblasts from young members of a species have a greater doubling potential than older individuals of the same species, suggesting the winding down of a clock, accumulation of a toxic substance, or depletion of an essential factor. Cells that normally divide in the body demonstrate a Hayflick limit. Cytosurgical techniques in which the nucleus of a young cell is put in an enucleated old cell and vice versa have led to the localization of factors controlling replication to the nucleus. Indeed, when cell cycle events go awry, as in cells transformed by oncogenes, the Hayflick limit is never reached (i.e., the cells become immortal).

The biology of cancer and aging raises significant and interesting questions. There are 30 known oncogenes, some of which are, in a sense, "antiaging" genes, or gerontogenes. The alteration in senescent human fibroblasts seems to be repression of specific protooncogenes along with multiple changes in gene expression. These observations support the view of a determinant role in the process of terminal differentiation and cell senescence. Cancer, then, might be considered "failed aging."

More evidence on the mechanisms involved in the loss of replicative ability in cells as they age comes from the study of telomeres, the repetitive gene sequences at the ends of chromosomes. Telomeres function to aid and preserve chromosomes during cell division; indeed, chromosomes cannot maintain their integrity without their terminal telomeres. The replicative ability of human cells in culture is directly proportional to the length of telomeres on the chromosomes, so that cell cultures with shorter telomeres undergo fewer divisions than those cultures with longer telomeres. Furthermore, cells from people exhibiting premature aging syndromes possess shorter telomeres with the decreased replicative capacity of their cells. Analysis of terminal restriction fragments of the telomere has shown that a number of base pairs are lost each time the cell divides. During cell division, DNA polymerases broker duplication of chromosomal DNA but are not effective at the terminus, where the telomeres reside. It is probable that the loss of telomeric sequence with each division until there is no more accounts for cells in culture reaching the Hayflick limit. Immortal cells, like cancer cells, are those that divide indefinitely in culture and therefore have no Hayflick limit. These cells exhibit long telomeres, which do not decrease in length with cell division. Germ cells such as sperm and ovaries also retain long telomeres. Molecular analysis uncovered a protein that is present in immortal cells but lacking in cells with finite doubling capacity. This protein, telomerase, which is like a polymerase with an imbedded RNA sequence, is thought to be responsible for the duplication of telomeric sequences during cell division. It is present in cells that lack finite doubling capacity and absent in those that do reach a senescent replicative limit.

C. Stochastic

The stochastic theories of aging espouse the idea that aging is caused by a gradual accumulation of randomly occurring deleterious events. Although no one can deny the role of the genetic determinants in biological aging, proponents of the stochastic theories ascribe greater importance to the random effects of environment and random molecular events in the cell in causing aging. Generally, aging is thought of as a "wear-and-tear" process, leading ultimately to death. The stochastic theories can be subdivided into several interrelated aging theories, depending on the importance prescribed to the damaging element or to the part of the organism effected.

Denham Harman formulated the free-radical theory of aging in the mid-1950s. Free radicals are molecules to which an extra electron is attached. They are unstable and can react with almost any biomolecule to perpetuate free-radical reactions and cause damage. According to this theory, free radicals, generated during the course of normal metabolism or acquired from ultraviolet radiation or xenobiotics, cause random damage to cells and biomolecules, and this damage accumulates with time, leading to senescence and eventual death of the organism.

Several features of this theory make it appealing. First, environmental sources of free radicals are plentiful, and studies of lower organisms show a strong inverse relationship between free-radical exposure and life span. Of more interest from a biological point of view is the observation that 5% of the oxygen taken in to support aerobic metabolism is converted to free radicals (i.e., superoxide radicals). A corollary to this hypothesis is that animals that use large quantities of oxygen per gram of tissue have shorter life spans than those that use less. Indeed, an inverse relationship between specific metabolic rate of a range of mammals and maximum life span potential is found. Thus, the hyperactive little shrew whose heart rate can approach 780 beats per minute uses a relatively large amount of oxygen per gram of tissue and has a life span of around 1.5 years. The long-lived elephant has a metabolic rate about one-thirtieth that of the shrew.

Free radicals are too reactive to be measured directly, but their presence can be implied by measuring markers, or "footprints," of free-radical damage. Lipofuscin is thought to be a free-radical footprint generated from free-radical reaction with phospholipids through the intermediate malondialdehyde. Indeed, ethane and pentane gas in expired air indicate lipid peroxidation secondary to free-radical damage. DNA is also a target molecule for free-radical damage. Because DNA repair is a high priority to the cell, the best indication to date of free-radical damage to DNA must be measured indirectly by the accumulation of thymidine dimers in the urine. There is an inverse relationship between these urinary markers of free-radical damage and life span.

Free radicals are generated in cells by a number of intracellular processes in addition to respiration. Prostaglandin synthesis, cytochrome *P*-450, and the inflammatory reaction of macrophages are a few of the sources of intracellular free radicals. Both endogenous and exogenous antioxidants limit damage from free radicals. The genome of animal cells encodes a protein, superoxide dismutase, that scavenges superoxide free radicals and dismutes them to hydrogen peroxide. Hydrogen peroxide is also quite reactive and is probably most responsible for cellular lipid peroxidation. An additional genetically encoded protein, catalase, causes the degradation of hydrogen peroxide to water. When hydrogen peroxide is allowed to come into contact with iron, hydroxyl radicals are formed. They are an extremely reactive species with no known scavenger. Other free-radical scavengers include vitamin C, vitamin E, β-carotene, and urate.

The quantity of superoxide dismutase in cells from many different species varies directly with life span, suggesting that defense against free radicals impedes the aging process. This relationship does not hold up for other intracellular defense enzymes. However, autooxidation occurs much more rapidly in tissues from short-lived animals than from ones with longer life spans, indicating that overall protection from free radicals correlates with longer life. Experiments were performed in which groups of mice were fed substances thought to act as free-radical scavengers. In one of these studies, vitamin E was found to increase the life expectancy of a group of mice, but did not affect the maximum life span. It was felt that vitamin E might forestall age-dependent diseases, but does not have the hoped-for ability to extend life.

Most age-dependent diseases have a free-radical component in their etiology. Emphysema is a good example. Here, the compliance changes of the small airways seen in patients with genetic- or cigarette smoking-induced emphysema are similar to those found in older individuals. Presumably, the free radicals produced as a result of both smoke irritation with subsequent phagocytic activity in the lung and the direct inhibition of antiproteases from free radicals in smoke act in concert to speed up the aging process in the lung; that is, the higher dose of free radicals to which smokers expose themselves increases the rate of progressive accumulation of damage. That cigarette smoking exacerbates so many of the diseases of aging is, in itself, evidence in support of the free-radical theory. Other age-dependent diseases in which free radicals may play a primary role include atherosclerosis, cataractogenesis, arthritis, Parkinson's disease, Alzheimer's disease, and osteoporosis.

The only reproducible method for extending life span in mammals is by caloric restriction. Known as the McCay effect, this has been demonstrated only for rodents. It might apply to longer-lived species, but experiments have not been performed to prove this. When mice or rats have limited access to calories, but are not malnourished or vitamin deficient, they live longer than *ad libitum*-fed controls of the same strain. Caloric restriction can significantly increase life span and therefore is thought to fundamentally affect the aging process.

Interestingly, decreased caloric intake also delays many of the physiological changes with age, as well as the age-dependent diseases thought to be related to free-radical damage. Among other salutary effects, caloric restriction slows lipofuscin accumulation, delays the time of onset of immune and reproductive lossees, increases hepatic catalase activity, increases cell membrane fluidity, and delays the time of onset of cataracts and cancer. A tempting explanation for these observations is that caloric restriction retards aging and age-dependent diseases by decreasing free-radical generation or by increasing their dismutation. The hypothesis that caloric restriction extends life span by decreasing the specific metabolic rate has been tested and does not seem to hold up. In fact, there are no differences in metabolic rate between calorie-restricted and *ad libitum*-fed rats. Current work at the molecular level indicates an increase in endogenous free-radical scavengers in calorie-restricted rodents.

Other stochastic theories of aging are also based on progressive damage at the cellular level, but this might or might not require free radicals in the initiation event. L. Orgel's error catastrophe theory, proposed in 1963, has been for the most part disclaimed, but its heuristic value remains important. This theory holds that as cells age, errors in DNA, from cross-linking or free-radical damage, accumulate and contribute to an ever-growing repertoire of modified proteins until a fatal buildup of defects in cellular machinery leads to an error catastrophe and cell death. In testing this theory, biochemists have discovered that errors in DNA are so faithfully repaired that there is rarely any change in transcription translation through the life of the cell. The amino acid sequences in proteins from young and old cells are identical. With modern techniques to study conformational changes in proteins, however, it is becoming clear that proteins in older cells fold differently from their younger counterparts. This difference in conformation is postulated to be the result of longer transient times in the older cell.

Experiments using radioactively labeled protein markers in red blood cells showed a change in the way proteins are catabolized in older cells. Lysosomes of older cells are not as efficient at picking up and degrading proteins, thereby allowing them to accumulate in the cell. Not only does this allow time for proteins to assume new shapes, it also might overwhelm the cytosolic ubiquitin-dependent pathway for the degradation of smaller proteins. Therefore, while young and old cells synthesize proteins equally well, the ability of the cell to rid itself of proteins (the disposal system) could be more problematic in older cells. The altered shape of proteins, arising from longer cell transit times, could have further implications in the autoimmune diseases accompanying aging.

II. BIOLOGICAL THEORIES OF LONGEVITY

Biological theories of longevity emphasize the effect of coping paradigms on ultimate life span. For example, in clones of one-celled paramecium bombarded by ultraviolet light and allowed to recover under blacklight, treatment with sunlight induced an enzyme to repair the DNA damage. This gave rise to the idea of longevity assurance. A plethora of studies in many species ensued, showing a positive correlation between life span and the ability of an organism to repair damaged DNA. Likewise, cells from older animals have decreased repair abilities compared to cells from younger animals. In tissue culture, cells nearing their Hayflick limit are unable to repair ultraviolet DNA damage as well as cells in earlier doublings. These studies support the importance of the repair of DNA in the maintenance of life. [*See* DNA Repair.]

A single gene capable of increasing life span has been identified in the nematode *Caenorhabditis elegans*. This tractable worm has long been a favorite model for studying the genetics of aging because of its two chromosomes mapped in exquisite detail and its similarity to other aging models; that is, the McCay effect holds true in *C. elegans* and there is timed somatic cell death during the life course of the animal, similar to the thymic involution seen in higher animals. By screening mutants for longer life, a colony was found that outlived controls by 70% in both life expectancy and maximum life span. Called the age-I mutant, the genetic locus has been mapped and offers exciting possibilities for the use of molecular biological techniques in discerning how the gene slows the aging process and prolongs life.

III. HUMAN AGING

Commonsense observations have made clear the inevitability of decline and death in humans. It has only been since the 1960s, however, that the various origins of decline have been systematically differentiated. It is now known that much that had been attributed to human aging per se is due, instead, to a variety of other conditions, such as social adversity, disease, personality, poor conditioning, and toxins (e.g., cigarette smoke and alcohol). When criteria to exclude disease are carefully applied in the selection of a group of young and old people (i.e., a cross-sectional study) to be studied physiologically, declines previously thought to be unavoidably linked to age (e.g., the ability of the heart to increase cardiac output during exercise) are no longer demonstrable. Such studies, along with longitudinal studies now in progress, are leading to a revision of many stereotypes attributed to aging. Instead of a linear decline in function with age, a new rectangularized relationship is emerging in which individuals remain healthy and robust until close to the end of the life span. The length of the human "health span" as this century comes to a close is approaching the human life span.

The twentieth century has seen an extraordinary increase in life expectancy: some 25 years in the industrialized world. This is nearly equal to what had been attained in the preceding 5000 years of human history. This new longevity is obviously a social achievement, not a function of biological evolution. The need to better understand aging because of the rising increase in the elderly population due to this longevity revolution was one of the driving forces for the 1974 Research on Aging Act, which created the National Institute on Aging. Established in 1975, the National Institute on Aging has led to a markedly increased investment in aging research in the United States.

Gerontological studies have pointed the way to interventions to extend the health span. Exercise is touted to mitigate many of the deleterious effects of aging. For example, exercise and calcium are used in the maintenance of bone density to postpone the onset of osteoporosis. Indeed, gerontology, in its applications in medicine and geriatrics, is poised to intervene more dramatically in human aging. The purpose of research into the biology of aging is to control, prevent, or reverse the factors that contribute to aging. Some age-related factors go beyond the standard definitions of disease and include overall functional status and factors that predispose to disease (e.g., osteopenia, age-dependent decreased bone density, which can result in osteoporosis or disease). This illustrates a problem in distinguishing the borderline between age and disease. The constant rise in systolic blood pressure with age at some point becomes a pathological entity called hypertension. Likewise, impaired glucose tolerance that seems to be a normal concomitant of age might cross the line to maturity-onset diabetes. [*See* Bone Density and Fragility, Age-Related Changes; Hypertension.]

Normal aging is difficult to define for the human population. Even when measurable diseases are excluded as carefully as possible, there is a wide diversity in function with increasing age. In fact, the one constant in most clinical studies of functional aging changes is the increase in variance in the measured parameter with age. Whether it be oxygen consumption, renal clearance, glucose tolerance, or systolic blood pressure, to name a few that are measured cross-sectionally as a function of age, the overall pattern of the average can be in a deleterious direction, but invariably elderly individuals are found whose functions are no different from those of 20-year-olds. Did these elderly people not age as fast as their contemporaries, or did they start at a more advantageous level in youth for the measured function? Longitudinal studies are needed to answer this question.

Because aging is a ubiquitous biological phenomenon, many gerontologists have looked for explanations at the level of the cell to build arguments from single-celled organisms that apply to more complicated ones. The fundamental aging process in more complex organisms might not occur at the cellular level, but might be the manifestation of reaching limits in critical systems, which determines aging and life span.

One critical aspect of human aging is the decline in immune function. Starting with thymic involution beginning at puberty, there is a decrease in T cell function with age and a concomitant decrease in the control of antibody production. As the immune system declines with age to perhaps 20% of its peak function in adolescence, there is the prospect for intervention, should means of restoring or maintaining immune function be established. On the other hand, the decreased ability of the elderly to respond to antigen might be protective. Autoimmune diseases such as rheumatoid arthritis might be diminished by senescence of the immune system. Clearly, the pros and cons of interfering with immune senescence must be carefully evaluated. [*See* Autoimmune Disease; Polymorphism of the Aging Immune System.]

The glucocorticoid cascade hypothesis of aging suggests that the emergence of hyperadrenal corticoidism, largely caused by stress as encountered during life, might be responsible for a host of age-associated problems. In initial observations using a rodent model, the recovery of circulating levels of corticosterone after a stress in older rats was found to be dramatically delayed from the recovery in younger animals. Older animals were also found to have higher basal levels of adrenocorticotropic hormone (ACTH), with an accompanying decreased sensitivity of the adrenal gland to ACTH stimulation. Insensitivity to negative feedback control and resultant hypersecretion occur throughout the adrenocortical axis with age. This can be explained by the observation that specific groups of neurons in the hippocampus of aged rat brains are found to be deficient in corticosteroid receptors. It seems that, during life, stresses accumulate until at some point the down-regulation of corticosteroid receptors is great enough to interfere with hippocampal feedback inhibition, resulting in the hypersecretion of corticosteroids. Additional support for the glucocorticoid cascade hypothesis is found in an examination of the age-related diseases that have a significant component of hyperadrenocorticism (e.g., immunosuppression, muscle atrophy, arteriosclerosis, osteoporosis, and steroid diabetes).

The neuroendocrine system has been studied in relation to aging from the perspective of female reproductive capacity, developing a model of a negative feedback defect in some ways similar to the glucocorticoid hypothesis. Using the decrease in ovarian follicular number and its correlation with acyclicity as reproducible biomarkers of aging in rodents, the unexpected result is that transplantation of young ovaries into old mice ovariectomized in youth caused the initiation of estrous cycles, whereas young ovaries transplanted into old female mice who had aged with their ovaries intact did nothing to reinstitute the estrous cycle. It is postulated that ovarian secretion of estradiol is necessary for aging of the hypothalamopituitary axis necessary for the negative feedback control of the estrous cycle. [*See* Neuroendocrinology.]

Supporting this postulate is the observation that when young ovariectomized mice are given a short (i.e., 3-month) elevation of estradiol via their drinking water, cycling is maintained with difficulty after reimplantation of young ovaries, whereas young ovariectomized mice not treated with estradiol resume normal cycling when reimplanted with young ovaries. Further, several markers of reproductive aging, including glial hyperactivity in the arcuate nucleus, are accelerated by chronic exposure of young animals to estradiol and delayed in chronically ovariectomized animals. Glial hyperactivity is usually associated with damaged neurons, so this could indicate a cumulative effect of estradiol to slowly terminate hypothalamic negative feedback control over the estrous cycle. The deterioration of negative feedback homeostatic control systems due to accumulated damage from circulating steroids in other models of aging is similar.

Some of the theories of aging are more general, whereas some are more specific, for human aging. None is sufficient to explain aging adequately. The theories are not mutually exclusive; in fact, they are closely interrelated. Modern gerontology, then, has as its challenge the necessity of dissecting aging from disease and testing the theories of aging and longevity until a coherent and scientifically tenable explanation for biological aging emerges. The payoff for understanding human aging, of course, will be the enormous impact on health, health expectations, and our ability to help each individual reach maximum potential in terms of ability to contribute to society and to have meaning in life regardless of age. To conclude, pioneer American gerontologist Nathan Shock once said, "As we learn more and more about diseases and come to be able to control them, there won't be much left to old age except aging."

BIBLIOGRAPHY

Finch, C. E. (1990). "Longevity, Senescence, and the Genome." Univ. of Chicago Press, Chicago.

Finch, C. E., and Hayflick, L. (eds.) (1985). "Handbook of the Biology of Aging." Van Nostrand–Reinhold, New York.

Kent, B., and Butler, R. N. (eds.) (1988). "Human Aging Research: Concepts and Techniques." Raven, New York.

Schneider, E. L., and Rowe, J. W. (eds.) (1996). "Handbook of the Biology of Aging," 4th Ed. Academic Press, San Diego.

Warner, H. R., Butler, R. N., Sprott, R. L., and Schneider, E. L. (eds.) (1987). "Modern Biological Theories of Aging." Raven, New York.

Gerontology, Physiology

BERTIL STEEN
Göteborg University, Sweden

GLOSSARY

Aging Normal, physiological processes regarding structure and function with advancing age from the stage when maturity is reached

Biological age Age of an individual in relation to genetically determined life span

Biological aging Aging processes related to biology as opposed to psychological and social aging

Chronological age Duration of time passed since the individual was born

Functional age Age of an individual in relation to individuals of the same chronological age

Gerontology Multidisciplinary scientific field devoted to the study of biological, psychological, and social aging

THIS ARTICLE DEALS WITH STRUCTURAL AND functional changes inevitably occurring with advancing age from the stage when maturity is reached until the death of the individual.

Conceptually, biological aging is a normal process and belongs to the field of physiology. Thus, this presentation is part of the gerontological arena, not of geriatric medicine. However, the manifestations of aging are influenced by environmental factors such as diet, physical activity, smoking and abuse of alcohol, and disease, and the extent of such manifestations as well as the speed with which they appear are, thus, to a certain degree influenced by public health preventive activities.

Symptoms and signs seen in advanced age constitute a mixture of the manifestations of normal aging and those of disease, which makes it difficult to differentiate between health and disease in the higher age groups. Knowledge of gerontology is, therefore, a prerequisite to study and practice geriatric medicine.

I. PHYSIOLOGICAL AGING IN GENERAL

Most functions of the human body show different phases throughout life. After conception and during early life, there are phases best characterized as growth, maturity, and improvement. After these comes a phase of aging that goes on until the death of the individual.

Based upon mainly cross-sectional population studies, this phase of aging has been described as continuous and accompanied by inevitably and constantly decreasing functional levels, this decline beginning shortly after maturity is reached.

Some functions clearly reflect this. For example, tissue elasticity in structures such as the ocular lens deteriorates very early in life. Other examples are perceptual speed (manifesting itself in, e.g., increasing reaction time) and rate of oxygen consumption at rest; both of these functions show an almost linear decline throughout life beginning as early as age 20–25.

However, recent longitudinal studies have revealed a much more complicated picture. Some functions seem to be quite uninfluenced by age, such as the oxygen-carrying capacity of the blood. Others do not change markedly until about 70–75 years of age, when the decline begins to be more obvious. This point where the functional curve will start to decline more rapidly is very different for different individuals

ENCYCLOPEDIA OF HUMAN BIOLOGY, Second Edition, VOLUME 4.

and for different functions. Furthermore, the age-related changes may be very small in healthy individuals even at higher age groups, whereas deterioration seems to be much more marked in individuals with certain diseases. This is especially obvious regarding mental aging, in which memory and verbal and logical ability have been shown to be rather unchanged up to age 80 in healthy individuals, whereas patients with cardiovascular disease show marked deterioration. [*See* Aging and Language.]

The variability regarding different functions around the chronological age-related average values is very marked and seems to increase throughout life. In other words, chronological age can be used with more accuracy in childhood and early adult life than in later life, where functional age is much more important. At the chronological age of 70, some functions show distributions of up to 20 years upward and downward in healthy individuals. This means, for example, that some healthy 70-year-olds have functional levels that are as high as those of some healthy 50-year-olds. Part of this variability certainly has genetic causes, but physical and mental training also play roles in this respect.

At least in developed countries, normal aging has fewer and less serious consequences to the average elderly person in spite of the fact that the prevalence rates of most diseases, somatic as well as mental, show a very marked increase from age 60 and upward. However, some kinds of functional decline have practical consequences for most elderly people, such as the aging processes relating to visual acuity.

In general, those functions that presuppose some stress (such as mental functions demanding speed and physical functions demanding maximal performance) suffer more from age than the corresponding functions at rest. Furthermore, functions requiring coordination (such as neuromuscular precision) tend to be more dependent on age.

II. AGING AT THE CELLULAR LEVEL

Tissue and organ aging are preceded by basal aging processes at the molecular and cellular levels. Many studies in this field relate to the division potential of cells, where it has been said that the limited number of generations of normal cells is an expression of cellular aging. The conclusion from such studies is that the genome reorganization occurring during the division cycle creates a shift in cell functions, most markedly seen in postmitotic cells such as neurons and myocardial cells.

There is a decrease of the amount of functionally important DNA fractions in aging cells, while the total amount of DNA seems to be unchanged. Furthermore, damage occurs continuously in the DNA molecules, triggered in part by free radicals and peroxides. The efficacy of DNA repair processes in the cell is related to the life span of the species and has also been shown *in vitro* to be lower with advancing age. A decreasing efficacy of the genetic programs can result in an increased production of defective proteins. [*See* DNA Repair.]

Cellular aging cannot be described as a general decrease of cellular activity, since the production of some substances increases whereas that of others decreases. Some authors claim that cellular aging is related more to lack of capacity to coordinate between synthesis and degradation of cell products than to changed production per se.

Intracellular deposition of waste products is characteristic in postmitotic cells, such as neurons and myocardial cells. One example is lipofuscin in secondary lysosomes, which can occupy a large proportion of the volume of the cells. Others are lipids, hyaline granulae, and vacuoles.

Free radicals, normally produced in aerobic metabolism, are toxic and might contribute to changes occurring with advancing age. However, there are protective mechanisms, such as enzymes (e.g., selenium-containing glutathione peroxidase). It is still unclear whether an increased administration of vitamin E, selenium, or other antioxidative substances influences the aging process in humans.

The mechanisms of cellular aging are still unknown, but the bulk of evidence points toward a combination of different processes.

III. AGING IN ORGAN SYSTEMS

A. Central Nervous System

The distinction between normal aging of the brain and cerebral disease such as the age dementia syndromes—among them Alzheimer's disease—is not clear-cut, and the diagnosis of age-related cerebral disease is therefore difficult. Structural changes occur in both normal aging and disease and are often not *qualitatively* but only *quantitatively* different. [*See* Alzheimer's Disease; Dementia in the Elderly.]

With increasing age, brain weight decreases by about 10% between age 20 and 80 in healthy individuals. However, variation is considerable. The atrophy is predominantly cortical and is accompanied by a compensatory enlargement of the ventricles. Other structural age-related changes are loss of neurons, compensatory gliosis, accumulation of lipofuscin, senile plaques, neurofibrillary tangles, and granulovacuolar degeneration. The brain content of lipids, carbohydrate, protein, and minerals remains the same or declines only slightly, and the water content increases with age. Such changes, which to some degree correlate with the degree of neuronal loss and damage at the synaptic level, may make the brain increasingly susceptible to toxic substances and disease. A marked reduction in radioimmunoassayable peptide levels has been observed with advancing age. Among the substances particularly vulnerable to aging and influenced during disease are the glycolytic enzymes hexokinase and phosphofructokinase, and the different neurotransmitters. It has also been suggested that sympathetic hyperactivity in old age may interfere with cognitive function. [*See* Brain.]

Some deterioration of extrapyramidal function seems to occur with advancing age, which may be the explanation for the motor changes of gait and gestures sometimes resembling those of parkinsonian patients. Deterioration of most sensory systems is well known in the literature; so too are sleep disturbances, which often involve changes of the biological rhythm.

The psychological effects of normal cerebral aging are influenced by environmental processes. They can be influenced by mental training and, furthermore, a large proportion of the psychological differences that have been found between younger and higher age groups can be partly explained by cohort changes.

Perceptual speed decreases with advancing age, and reaction time grows longer. This depends almost entirely on an increase in time required for central processing. Cognitive function very often shows no or very slight decrease with advancing age, and if so then late in life. However, the importance of general health to cognitive function is marked, and a "terminal decline" has been described during the two years before death. Most studies show a decreasing creativity during senescence. Short-term memory does not seem to show any obvious deterioration with advancing age, whereas short-term learning shows some impairment with age due in part to longer central processing time. [*See* Learning and Memory.]

B. Cardiovascular System

Structural age-related changes in the heart include an increase of the amount of collagen and fat, especially in the septal and atrial regions, and sclerosis of valves. Lipofuscin accumulates with aging, leading to "brown atrophy." The sinus node shows a reduced number of cells and a relative increase of connective tissue and fat. Heart volume increases physiologically with age, at least up to age 75. In the vascular system, deteriorating elastin function and an increase of the amount of collagen lead to increasing vascular stiffness.

The influence by age on resting heart rate is very slight. There is, however, an obvious decrease in maximal heart rate response to physical exercise caused by a decreased automaticity of the sinus node. The disposition toward atrial arrhythmias in the elderly is increased. Cardiac output during exercise is lower than earlier in life in both the supine and sitting positions, whereas resting cardiac output is influenced in the supine position only. The most important explanation is a reduction of stroke volume at rest and of heart rate in exercise. An effect of these structural and functional changes is a decrease of oxygen consumption capacity during physical activity. Contributing factors have been suggested, such as the known resistance to β-adrenergic-mediated responses to exercise and the increased vascular input impedance. [*See* Exercise and Cardiovascular Function.]

The structural changes in the vascular system give rise to an increase of total peripheral resistance. Systolic and diastolic blood pressure will increase with advancing age as well as the pulse pressure. These blood pressure changes are seen at rest, but increase with physical exercise. Arterial blood pressure is also influenced by changes of water and electrolyte balance as well as of humoral factors.

Thus, mean arterial blood pressure increases with age in most people despite the fact that the decrease in cardiac output and the increase in luminal volume of the great arteries work in the opposite direction.

C. Renal System

As in most organs, there is also a reduction of cell mass in the kidneys. The loss of renal mass is particularly obvious in the renal cortex, and there is also a loss of glomeruli. These changes are paralleled by a reduction of renal blood flow in both absolute figures and relative to cardiac output.

The glomerular filtration rate decreases on the order of 1% annually from age 50. Since skeletal muscle mass also decreases throughout life, there are no significantly higher levels of creatinine in the serum despite the reduction of glomerular filtration.

Maximal tubular water reabsorption capacity also decreases with age. The ability of the kidneys to handle a water or an acid load is lower in high age than earlier in life. Furthermore, the reactions to such challenges grow more sluggish with advancing age.

D. Respiratory System

Structural changes in the lungs with advancing age include deterioration of elastic properties, for example, in the alveolar walls, and decrease of alveolar volume and surface area. Furthermore, changes in surrounding structures such as calcified costal cartilage and the narrowing of intravertebral space influence the breathing process.

Total lung capacity remains rather unchanged throughout life, whereas the residual volume after maximal expiration increases, leaving a smaller vital capacity for active gas exchange. A major reason for this is the increasing tendency for expiratory small duct and alveolar collapse as a result of changed elastic properties. An additional factor is the increasing stiffness of the thoracic structures.

In spite of these structural and functional changes, alveolar ventilation seems adequate at higher ages, as mirrored by an unchanged carbon dioxide pressure. However, the partial pressure of oxygen decreases significantly with age because of the expiratory collapse of alveoli, and as a consequence of that an uneven distribution of ventilation relative to circulation—again an effect of deteriorating elastic properties. A minor explanation may be decreased alveolar diffusion capacity from air to blood.

The age-related respiratory changes have a minor effect on the average elderly person. However, they may add risks in pulmonary disease, such as pneumonia. It must also be added that smoking may dramatically change the physiological properties of the respiratory system. [*See* Tobacco Smoking, Impact on Health.]

E. Gastrointestinal System

Salivary secretion shows a small decrease with age. Tasting ability declines and a shift between different taste sensations may occur.

Age-related esophageal dysfunction includes inability to relax, especially in the lower esophageal sphincter, and the appearance of nonperistaltic muscle waves. These derangements might give rise to dysphagia and nutritional difficulties.

An age-related mucosal atrophy is common throughout the gastrointestinal tract. The consequences are moderate with the exception that, in about 25% of 60-year-olds, gastric acid production is virtually absent. Also, the production of intrinsic factor and pepsin decreases with age.

Structural and functional changes with age have been described in the liver, where the organ weight decreases somewhat. However, the metabolic capacity is affected very little.

A decrease of the absorption capacity regarding fat, protein, and calcium, and a reduced ability of the calcium absorption system to adapt to low dietary calcium intake with advancing age, has been reported. The age-related changes in the colon are few and very moderate.

In general, the physiological aging of the gastrointestinal system yields very slight effects, especially when the large reserve capacity is taken into consideration.

F. Endocrine System

Although some endocrine derangements in disease may resemble aging processes, no bulk of evidence connects aging per se to endocrine deficiencies. The prevalance of endocrine disorders certainly increases with age, but age does not seem to have a causative bearing on such changes.

Hyperplasia is seen in most endocrine organs with advancing age, with the exception of parathyroid glands.

There is no support for a changed hypophyseal function with age. Thus, the production of gonadotropins has the ability to increase in the feedback system from the gonads.

Pathological derangements of thyroid function are quite common in the elderly. Normal aging processes include fibrotic changes in the thyroid gland with an increase of collagen and lymphocytes. However, these structural changes seem not to give rise to any marked thyroid dysfunction. Also regulation of the thyroid function seems to be well preserved with advancing age. It may be concluded that an observed significantly deranged thyroid function in the elderly most certainly depends on disease rather than aging.

Production of cortisol seems to decrease with advancing age, whereas the cortisol concentration in blood remains unchanged.

Regarding endocrine pancreas function, fasting blood sugar levels increase and glucose tolerance decreases with age.

Gonadal function deteriorates in females and males; males show decreasing levels of testosterone. At menopause the production of estrogens from the ovaries is discontinued. However, steroids—for example, androstenedione—are still produced. An obvious consequence of the hormonal changes in females is atrophy of genitalia, which may produce irritation and infection.

G. Immune System

Aging is accompanied by a decreasing ability to cope with environmental stress. There are many causes relating to this incompetence. An increasingly immunoincompetent system plays a role in this context, since the immune system protects the body against foreign invasion by viruses, bacteria, and fungi, and the growth of other invasive antigens such as cancer cells.

Macrophages seem not to be adversely affected by aging in their handling of antigens. There is strong evidence suggesting that aging may affect thymus-derived T cells, which modulate antibody response. An altered function of T lymphocytes, leading to a deterioration of both cell-mediated and antibody-mediated immunological defense, has been described.

However, such changes seem not to exist in all individuals, and there are longitudinal population studies showing very slight differences by age in these respects.

In general, fewer changes occur in humoral than in cellular immunity. Most evidence suggests that the immunological decline with age is primarily related to changes in the T-cell component of the immune system.

H. Skin

The alteration of the appearance of the skin in elderly people is due to a combination of aging processes per se and environmental damage due, for example, to sun exposure.

The dryness, wrinkling, laxity, and uneven pigmentation seen in most of the elderly stem from changes in the epidermis and dermis. A decreasing ability to proliferate has been described in epidermal cells, and there is a flattening of the dermoepidermal junction. The epidermal turnover rate diminishes on the order of 50% from early to later life. The ability of DNA repair in skin fibroblasts seems to decrease with age, and the skin repair rate (relating to situations such as wound healing and regeneration of blister roofs) declines with age.

The number of active exocrine sweat glands and sweat production decrease. Sebum production decreases as well, paralleling the decrease of gonadal androgen, despite the fact that there is no atrophy of sebum glands.

The number of melanocytes decreases, and the ability to produce vitamin D after ultraviolet light exposure seems to deteriorate with advancing age. Loss of dermal thickness is characteristic of the aging skin.

The aging skin is also vulnerable to pressure sores and trauma. Prurigo is common owing to dryness of the skin, even in the absence of skin disease. Thermoregulation is often impaired because of deranged vascular response to heat or cold, decreased sweat production, and loss of subcutaneous fat.

I. Muscular System

Total muscle mass decreases with advancing age owing to a decreasing number of muscle fibers, whereas there are very small changes in fiber size. Some observations indicate that there is a marked decrease of number of motor units but more fibers in existing motor units, pointing to an ability for dynamic reorganization of fibers.

There seem to be very insignificant changes with age regarding the proportion of "slow" to "fast" muscle fibers, although some studies indicate a reduction in "fast" fiber size. A reduced size of type 2 fibers is associated with denervation and inactive muscles.

As a result of the structural changes, both isometric and dynamic muscle strength decline; coordination and flexibility also decrease with age on the same order as the loss of muscle mass. Some data show that especially isometric muscle performance is impaired to a great extent. This is of practical importance since correcting movements where isometric muscle activity is involved is essential to avoid falls.

Physical training improves both muscular strength and coordination even in the higher age groups, and maximal aerobic power has been shown to increase with training at age 70–75. This is of special geriatric importance since muscle function plays an important protective role regarding the risk for falls and hip fractures.

J. Bone

Peak bone mass is achieved by age 25–30; thereafter an age-related endosteal bone loss starts, though it begins earlier in females than in males. There is great variation in the rate of decrease between individuals. Males seem to lose 0.3–0.5% of bone mass per year, trabecular bone loss being slightly more marked than that of cortical bone. The corresponding figure for females is higher and is close to 1% per year, and young women have about 30% higher mineral content in the skeleton compared to elderly women, whereas the corresponding difference for men is only about 15%.

The amount of bone mineral is related to the strength of the skeleton and, therefore, to the risk of getting fractures. However, other protective factors are involved, such as muscle strength. [*See* Bone Density and Fragility, Age-Related Changes.]

The border between normal aging of the skeleton and osteoporosis is not clear-cut. Osteoporosis and hip fractures currently show an increase in many developed countries. Many factors may contribute to accelerated bone aging and osteoporosis. Such primary risk factors include, apart from age, sex, and race (African-American people being less susceptible), a sedentary life-style, smoking, alcohol abuse, and environmental conditions favoring falls. Adequate intake of dietary calcium is also essential, although the level of recommended intake is still a matter of dispute.

K. Metabolism and Body Composition

Energy metabolism is lower in advanced age than earlier in life. The major explanation for this is probably the decreasing number of cells in the organs, loss of metabolizing tissue, and reduced physical activity, although some intracellular age-induced changes on the enzyme level might also be responsible. Thus, energy expenditure is lower in old age than earlier in life. This seems to be related less to a decrease of basal metabolic rate than to a decrease of energy expenditure relating to physical activity, and diet-induced and temperature-induced thermogenesis. [*See* Energy Metabolism.]

Decreasing number of cells in the organs and increasing disuse of skeletal muscle tissue with age result in a decreasing body cell mass. At age 70, skeletal muscle has been shown to have lost about 40% of its maximal weight in early adult life as compared to about 20% for the liver, and about 10% for the kidneys and lungs.

Most data have shown an increasing amount of body fat with age, especially when expressed as a proportion of body weight. In the eighth decade of life, however, body fat seems to decrease, earlier in females than in males. Subcutaneous fat is deposited more on the trunk than on the extremities in old compared to young individuals, and there is an increase of deep adipose tissue relative to subcutaneous fat with age.

Body water decreases with advancing age in most studies. However, this trend might be less significant in the highest age groups since it might be counteracted by the fact that a change in cardiac and renal function might increase the amount of body water. However, the most important reason for the well-known decrease of body weight during the eighth decade of life has been shown to be a decreasing amount of body water, especially extracellular water.

IV. FUNCTIONAL AGING AND FUNCTIONAL CAPACITY

As mentioned earlier, the functional age of an elderly individual may differ very much from chronological age. Therefore, the importance of assessment of functional capacity of elderly people is obvious. Functional capacity comprises levels and profiles of different functions in an individual in relation to chronological age and the average functional capacity of people of the same chronological age. Such functions should include not only biological and medical parameters but also psychosocial capacity.

Attempts to measure functional age and functional capacity have been done using wide batteries of tests. Certain sensory, psychomotor, and motor functions have been claimed to be particularly useful in this context. Also the assessment of functioning in activities of daily living has shown to be useful.

Since the combination of many different age-related functional deteriorations is obviously of more importance to the average elderly person than effects by age regarding single functions, the attempts to follow functional capacity with age seem to be an urgent field of investigation. Attention has also been paid to the manifestations of disease in the overall judgment of the functional capacity of the aged individual.

Interesting observations in ongoing longitudinal and sequential studies of representative elderly populations seem to indicate a slightly higher degree of performance regarding some physical and mental functions in "young elderly" (age 65–75) now, compared to people of the same age just one or two decades ago. These cohort differences are, however, not a universal phenomenon for all aging functions, and this trend has to be confirmed in further studies.

BIBLIOGRAPHY

Bergener, M. (ed.) (1987). "Psychogeriatrics: An International Handbook." Springer Publishing, New York.

Brocklehurst, J. C. (ed.) (1985). "Textbook of Geriatric Medicine and Gerontology," 3rd ed. Churchill Livingstone, Edinburgh/London/Melbourne/New York.

Chernoff, R., and Lipschitz, D. A. (eds.) (1988). "Health Promotion and Disease Prevention in the Elderly." Raven, New York.

Folkow, B., and Svanborg, A. (1993). Physiology of cardiovascular aging. *Physiol. Rev.* **73**, 725–764.

Horwitz, A., Macfadyen, D. M., Munroe, H. N., Scrimshaw, N. S., Steen, B., and Williams, T. F. (eds.) (1989). "Nutrition in the Elderly." Oxford Univ. Press, Oxford, England.

Libow, L. S., and Sherman, F. T. (eds.) (1981). "The Core of Geriatric Medicine." Mosby, St. Louis.

Masoro, E. J. (1986). Physiology of aging. *In* "Geriatric Dentistry" (P. Holm-Pedersen and H. Löe, eds.). Munksgaard, Vojens, Denmark.

Pathy, M. S. J. (ed.) (1991). "Principles and Practice of Geriatric Medicine," 2nd Ed. Wiley, Bath, England.

Schneider, E. L., and Rowe, J. W. (eds.) (1996). "Handbook of the Biology of Aging," 4th Ed. Academic Press, San Diego.

Terry, R. D. (ed.) (1988). "Aging and the Brain." Raven, New York.

Glutathione

OWEN W. GRIFFITH
Cornell University Medical College

GLOSSARY

Electrophile Molecule having an electron-deficient atom that can react with certain electron-rich atoms (i.e., nucleophiles). Reaction of electrophiles with nucleophilic atoms of cellular proteins and nucleic acids generally destroys or prevents the normal functioning of the protein, nucleic acid, or other molecule attacked

Free radical Molecule having a single unpaired electron. Such molecules are often nonspecifically reactive with a variety of cellular constituents. Unsaturated membrane lipids are particularly susceptible to free-radical damage

K_m Michaelis constant. In characterizing the effect of substrate concentration on the rate of an enzyme-catalyzed reaction, the K_m represents the substrate concentration necessary for the reaction to proceed at one-half of its maximal rate (assuming all other substrates are at saturating concentrations). K_m has units of concentration, and each substrate has a characteristic K_m. Enzymes typically have a high affinity for substrates with low K_m's

Nucleophile Molecule having an atom with nonbonded electrons able to react with electrophiles. Nucleic acids, proteins, and other cell constituents contain numerous nucleophilic atoms often essential to their structure or function

Oxidation–reduction Oxidation is a chemical process in which an atom loses electron density or entire electrons. Reduction is the opposite: a gain in electron density or electrons. Biological oxidation–reduction reactions always occur simultaneously; one reactant is oxidized and the other is reduced. In the aerobic environment of mammalian cells, oxidative reactions are driven directly or indirectly by the energetically favorable reduction of oxygen to water. Reduction of biological molecules, however, often requires the input of additional metabolic energy

Peroxide Molecule containing two oxygens joined by a single bond [e.g., hydrogen peroxide (H–O–O–H) or organic peroxides (R–O–O–H, where R– is an organic molecule)]. Peroxides are moderately strong oxidizing agents

Thiol/disulfide Thiols (i.e., sulfhydryls) are reduced sulfur compounds containing an –SH group. Thiols are readily oxidized to disulfides, which contain the –S–S– structure (i.e., $2\ R{-}SH + H_2O_2 \rightarrow R{-}S{-}S{-}R + 2\ H_2O$). Disulfides can be symmetrical (e.g., R–S–S–R) or mixed (e.g., R–S–S–R′, where R ≠ R′). Although disulfides resemble peroxides in some respects, they are more stable and thus less reactive

GLUTATHIONE (GSH) IS A TRIPEPTIDE (γ-GLUTAmylcysteinylglycine) (Fig. 1) present in all human tissues, generally in high (i.e., millimolar) concentrations. Two aspects of GSH structure are of particular importance: (1) Whereas glutamate in proteins and most peptides is bonded through its α-carboxyl, the glutamic acid residue of GSH is joined to cysteine through its γ-carboxyl; this unusual γ-glutamyl bond prevents the hydrolysis of GSH by ordinary proteases and peptidases. (2) The sulfhydryl (i.e., –SH) group of the cysteinyl residue of GSH accounts for about 90% of the total low-molecular-weight thiols in most tissues; GSH is thus the most prevalent and one of the most reactive nucleophiles in tissues.

GSH is synthesized in the cytosol of all cells from its constituent amino acids by the sequential action of two ATP-dependent enzymes. Within cells GSH plays an important role in the detoxification of a variety of compounds, including free radicals, organic peroxides, hydrogen peroxide, and reactive electrophiles. GSH is not degraded within cells, but is trans-

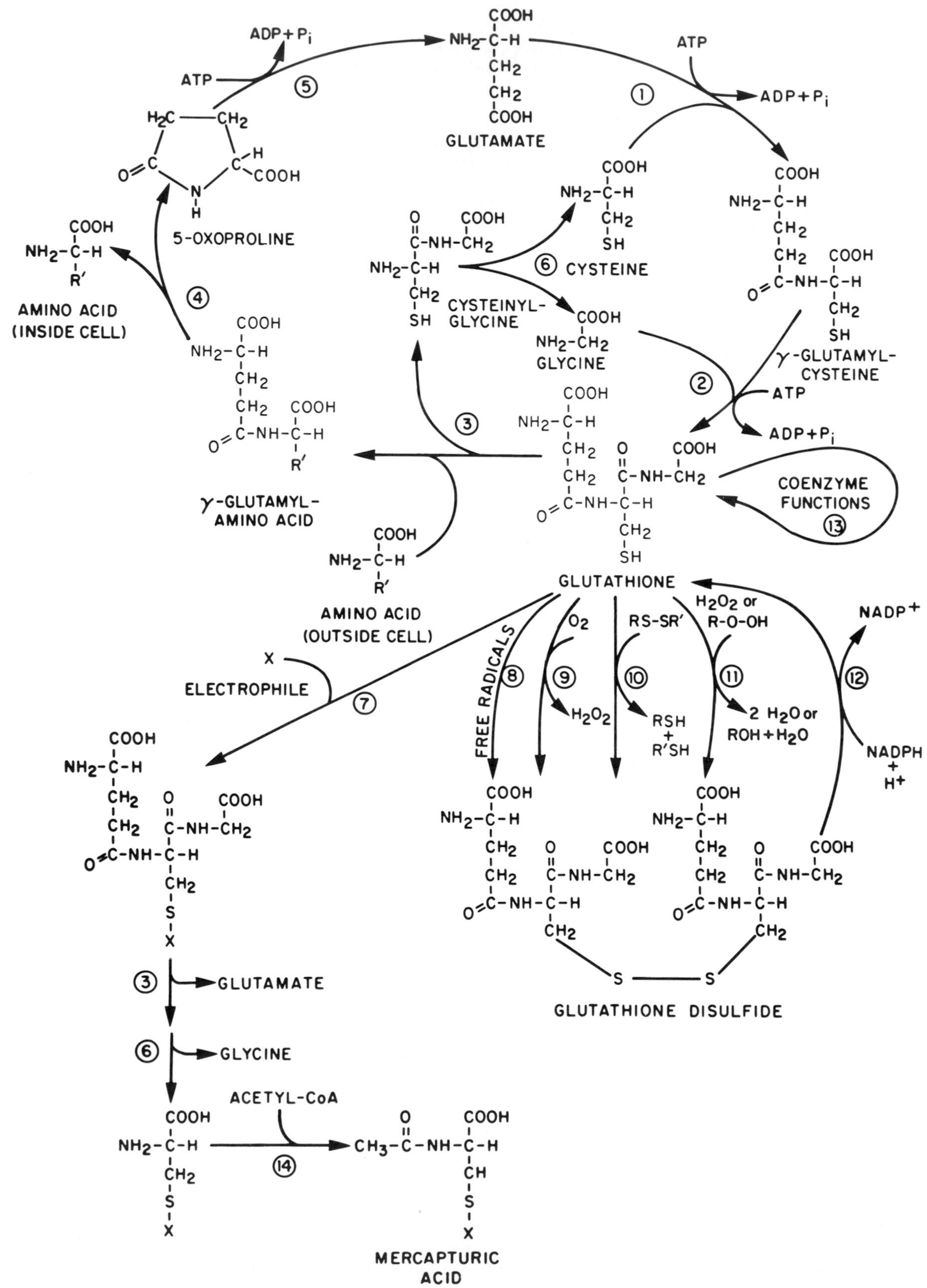
ADP + P_i
ATP
5
GLUTAMATE
ATP
ADP + P_i
1
5-OXOPROLINE
AMINO ACID (INSIDE CELL)
4
CYSTEINYL-GLYCINE
6
CYSTEINE
GLYCINE
γ-GLUTAMYL-CYSTEINE
2
ATP
ADP + P_i
3
γ-GLUTAMYL-AMINO ACID
AMINO ACID (OUTSIDE CELL)
COENZYME FUNCTIONS
13
GLUTATHIONE
X
ELECTROPHILE
7
FREE RADICALS
8
O_2
9
H_2O_2
RS-SR′
10
RSH + R′SH
H_2O_2 or R-O-OH
11
2 H_2O or ROH + H_2O
NADP+
12
NADPH + H+
GLUTATHIONE DISULFIDE
3
GLUTAMATE
6
GLYCINE
ACETYL-CoA
14
MERCAPTURIC ACID

ported intact across the plasma membrane to the extracellular space. Following transport, GSH turnover is initiated on the external surface of certain cells by the γ-glutamyl transpeptidase-mediated cleavage of its γ-glutamylcysteine bond. Cysteinylglycine, a dipeptide product of this reaction, is subsequently hydrolyzed to cysteine and glycine by any of several peptidases.

Tissues vary greatly in the amount of γ-glutamyl transpeptidase present. Liver, a tissue richly supplied with cysteine by both the diet and *de novo* synthesis, makes and secretes substantial amounts of GSH into the plasma, but is deficient in γ-glutamyl transpeptidase; it degrades little of the GSH it releases. Instead, GSH released from the liver is taken up by the blood and transported throughout the organism. Peripheral tissues with significant amounts of γ-glutamyl transpeptidase (e.g., kidney, lung, pancreas, and lymphocytes) degrade not only the GSH released by their own cells, but also that delivered by the blood. Tissues able to degrade GSH take up the amino acids and peptides released (i.e., glutamate, cysteine, glycine, and certain dipeptides). In this fashion the hepatic synthesis and peripheral tissue degradation of GSH provide for the interorgan transport of cysteine, an amino acid often limiting in cellular nutrition.

Inherited disorders affecting most of the enzymes involved in GSH metabolism have been reported. Although all of the disorders are rare, studies of affected patients have significantly elucidated the metabolism of GSH in humans. Pharmacological manipulations of GSH levels and metabolism are also possible. Therapies designed to increase GSH levels offer the possibility of augmenting the protective functions of GSH. Therapies that decrease GSH levels are of interest in the sensitization of tumor cells to radiation and chemotherapy and in the treatment of certain parasitic disorders.

Since mammalian GSH metabolism has been examined most extensively in mice and rats, much of the specific information presented is based on studies in these animals. To the extent that studies of human enzymes or metabolism have been reported, the findings are similar to those in rodents. Where available, data from human studies are presented and are identified as such.

I. CHEMISTRY

A. Structure and Physical Properties

GSH is a tripeptide composed of the protein amino acids L-glutamate (Glu), L-cysteine (Cys), and glycine (Gly) (the structures of these amino acids, GSH, and γ-glutamylcysteine, an intermediate in the synthesis of GSH, are shown in Fig. 1). At physiological pH, both of the carboxyl groups and the α-amino group of GSH are ionized; the tripeptide is thus an anion with a net charge of -1 (the pK_a values of GSH, determined by nuclear magnetic resonance spectroscopy, are 2.05, 3.40, 8.72, and 9.49 for glutamyl α-carboxyl, glycyl carboxyl, amino, and thiol groups, respectively). GSSG, the disulfide formed from GSH, is a dianion at physiological pH. Because of their small size, absence of hydrophobic groups, and presence of multiple ionic groups, both GSH and GSSG dissolve rapidly and extensively in water; solutions greater than 1 *M* are easily prepared at neutral pH. In solution, GSH and GSSG show considerable conformational mobility. Bonds rotate freely, and bond lengths and angles are normal.

High water solubility, permitting significant intracellular or extracellular accumulations without precipitation, is a biologically important property of GSH and GSSG. Although cysteine is also very soluble, it is readily oxidized in aerobic environments to its disulfide, cystine, which is insoluble. When accumulated to concentrations causing precipitation (e.g., >0.6 m*M*), cystine is toxic (cf. the inherited disorder cystinosis, in which intralysosomal cystine accumulation results in cystine crystallization, lysosomal destruction, and, in severe cases, death). It is probable that GSH has been selected by evolution, in part, to provide a safe means for the intracellular storage of cysteine; mechanisms for releasing cysteine from GSH are discussed in Section II.

FIGURE 1 Reactions of GSH synthesis, turnover, and metabolism. The circled numbers correspond to the following enzymes or metabolic processes: 1, γ-glutamylcysteine synthetase; 2, glutathione synthetase; 3, γ-glutamyl transpeptidase; 4, γ-glutamylcyclotransferase; 5, 5-oxoprolinase; 6, cysteinylglycinase; 7, glutathione-*S*-transferase; 8, free-radical inactivation by GSH; 9, enzymatic and nonenzymatic oxidation of GSH by oxygen; 10, enzymatic and nonenzymatic GSH-dependent transhydrogenations; 11, glutathione peroxidase; 12, glutathione reductase; 13, various reactions requiring GSH as a cofactor; and 14, cysteine conjugate *N*-acetyltransferase. P_i, inorganic phosphate.

B. Sulfhydryl Chemistry

The protective functions of GSH are dependent on the chemistry of its sulfhydryl group. As indicated in Fig. 1, GSH can be reversibly oxidized to GSSG. Both free radicals and peroxides are detoxified by reactions that result in GSH oxidation. Because GSSG is rapidly reduced back to GSH by an NADPH-dependent enzyme (see Section III), GSH plays an essentially catalytic role in the detoxification of free radicals and peroxides. In this respect it should be noted that GSH, in contrast to many other thiols (e.g., cysteine), is only slowly oxidized by molecular oxygen in the absence of specific catalysts (Reaction 10, Fig. 1). The chemistry of the GSH sulfhydryl thus facilitates its role as a protectant from cytotoxic oxidants, while minimizing its spontaneous reaction with molecular oxygen, a ubiquitous and essential constituent of mammalian cells.

The sulfur atom of sulfhydryls is electron rich. Because the electrons are by in large easily distorted orbitals (i.e., are polarizable), sulfhydryls are among the strongest and most reactive nucleophiles found in biological systems. The sulfhydryl of GSH is typically reactive and readily attacks a variety of potentially toxic electrophiles. These reactions, catalyzed by glutathione—*S*-transferases (see Section III), result in the chemical joining of GSH to the electrophile (Reaction 7, Fig. 1); the products are chemically stable, thus nontoxic, sulfides. In contrast to the reductive detoxification of free radicals and peroxides, detoxification of electrophiles irreversibly consumes GSH. The sulfide product is metabolized to release glutamate and glycine, but the nutritionally valuable cysteine residue is not recovered.

II. BIOSYNTHESIS AND TURNOVER

A. Enzymology of GSH Synthesis

GSH is synthesized in all mammalian cells by the sequential action of two cytoplasmic enzymes. The first, γ-glutamylcysteine synthetase, catalyzes the ATP/Mg^{2+}-dependent formation of L-γ-glutamyl-L-cysteine from L-glutamate and L-cysteine (Reaction 1, Fig. 1); cleavage of ATP (K_m = 0.2 m*M*) to ADP and inorganic phosphate (P_i) drives the reaction to completion. Under physiological conditions the enzyme is highly specific for L-glutamate (K_m = 1.6 m*M*) and is relatively specific for L-cysteine (K_m = 0.3 m*M*). L-α-Aminobutyrate, a minor metabolite of L-methionine, is present *in vivo* and is a moderately active substrate of the enzyme (K_m = 1.0 m*M*); when L-α-aminobutyrate reacts in place of L-cysteine in the GSH biosynthetic pathway, the ultimate product is L-γ-glutamyl-L-α-aminobutyrylglycine, known as ophthalmic acid, due to its initial discovery in the lens. Although less extensively studied, the structural and catalytic properties of γ-glutamylcysteine synthetase isolated from the human erythrocyte appear to be identical to those described for the rat kidney enzyme.

GSH synthesis is completed by glutathione synthetase, an enzyme catalyzing the ATP/Mg^{2+}-dependent synthesis of GSH from L-γ-glutamyl-L-cysteine ($K_m <$ 50 μM) and glycine (K_m = 33 μM) (Reaction 2, Fig. 1). ATP (K_m = 0.76 m*M*) is again cleaved to ADP and P_i, driving the reaction nearly to completion. Under physiological conditions the enzyme is highly specific for the natural substrates, although, as noted, L-γ-glutamyl-L-α-aminobutyrate is converted to ophthalmic acid. *In vitro,* the enzyme is relatively nonspecific with respect to the L-γ-glutamyl moiety of synthetic γ-glutamylcysteine analogs, but does not accommodate many structural modifications to the L-cysteinyl moiety.

B. GSH Levels and Their Regulation

Intracellular GSH levels range from about 0.5 to 10 μmol/g in a wide range of mouse and rat tissues. In healthy animals each tissue maintains a characteristic GSH level, although some tissues, particularly liver, show a diurnal fluctuation in response to feeding. In many tissues GSH levels change rapidly postmortem, and few human tissues have been examined under conditions likely to preserve *in vivo* concentrations. Nevertheless, data available suggest that human GSH levels are comparable to or a little lower than those in rodents (Table I). In all species, extracellular GSH levels are much lower than intracellular levels. Plasma, cerebrospinal fluid, and urine concentrations are all in the micromolar range (Table I).

Intracellular GSH levels are controlled and determined by the tissue-specific relative rates of synthesis and utilization. Since essentially all γ-glutamylcysteine formed is quickly converted to GSH, the rate of GSH synthesis is normally regulated by the rate of the γ-glutamylcysteine synthetase reaction. This reaction is, in turn, controlled by the amount of enzyme present, by the amount of GSH already in the cell (GSH is a nonallosteric feedback inhibitor; K_i = 2.3 m*M*), and by cysteine availability. Many diets are

TABLE I
Glutathione Levels

Tissues, cells, or physiological fluids	GSH concentration		
	Mouse	Rat	Human
Tissues (μmol/g of tissue)			
Brain	2.08 ± 0.15		
Heart	1.35 ± 0.10		
Lung	1.52 ± 0.13		
Spleen	3.43 ± 0.35		
Liver	7.68 ± 1.22	4.51 ± 0.27	3.55 ± 0.89
Pancreas	1.78 ± 0.31	1.66 ± 0.07	
Kidney	4.13 ± 0.15	2.56 ± 0.09	
Small intestine mucosa	2.94 ± 0.16		
Colon mucosa	2.11 ± 0.19		
Skeletal muscle	0.78 ± 0.05	0.75 ± 0.08	
Cells			
Erythrocytes (μmol/ml of cells)	2.0 ± 0.2		1.9 ± 0.3
T lymphocytes (nmol/10^6 cells)			0.6–1.0
Mixed peripheral blood leukocytes (nmol/10^6 cells)			1.4–2.2
Physiological fluids (μM)			
Plasma	37.0 ± 2.5	22–27	4.98 ± 0.65
Urine	2–5		Negligible
Cerebrospinal fluid	5.05 ± 0.25		Negligible

relatively deficient in the sulfur-containing amino acids cyst(e)ine (i.e., cysteine and cystine) and methionine. Because cysteine biosynthesis depends on sulfur derived from methionine, diets deficient in cyst(e)ine, methionine, or both ultimately limit the supply of cysteine available for GSH synthesis. Correspondingly, several studies in animals and humans indicate that tissue GSH levels can be increased modestly by giving L-cysteine or L-cysteine precursors parenterally or in the diet (see Section IV,B). In human infants, in whom L-cysteine synthesis from L-methionine is limited by the immaturity of some liver enzymes, increased dietary intake of L-cysteine might significantly increase tissue GSH levels.

In the absence of pathological stresses, GSH utilization is controlled primarily by the rate at which GSH is transported out of the cell. Rates of transport vary greatly among tissues (e.g., high in liver, kidney, lung, and leukocytes, but very low in erythrocytes). In all tissues the rate of GSH transport out of the cell is balanced with the synthetic rate in order to maintain intracellular GSH levels at the concentration typical of the tissue. It is noteworthy that the flow of GSH out of a cell represents transport down a sizable concentration gradient; GSH transport to the extracellular space is thus thermodynamically favored, whereas uptake of intact GSH by cells is not. In fact, most studies reported to date indicate that intact GSH is not significantly taken up into any cell type, even if the concentration gradient is made favorable by increasing the extracellular concentration. A consequence of the unidirectionality of GSH transport is that extracellular GSH must be degraded if its constituent amino acids are to be recovered.

Studies with erythrocytes and liver suggest that the transport of GSH and GSSG out of the cell is mediated and controlled by separate mechanisms. Because GSSG is normally present in only trace amounts intracellularly, its rate of transport usually does not significantly affect cellular GSH homeostasis. However, under conditions of oxidative stress, the intracellular concentration of GSSG increases, and, in at least some tissues, its rate of transport out of the cell increases proportionately. Under this circumstance the rate of GSH synthesis and the total rate of glutathione transport out of the cell might not be in balance; intracellular total glutathione levels (i.e., GSH plus 2× GSSG) might decrease substantially. Although loss of cellular glutathione can have serious consequences, it has been suggested that the rapid transport of GSSG out of cells effectively protects the intracellular environment from this product of oxidative stress.

C. Extracellular Turnover

As noted in the definition paragraph, the unusual γ-glutamyl bond in GSH renders the tripeptide resistance to most proteases and peptidases. *In vivo,* GSH degradation is initiated only by γ-glutamyl transpeptidase, an enzyme bound to the external surface of the plasma membrane of many, but not all, cells. Particularly high levels of γ-glutamyl transpeptidase are found in certain epithelial tissues, including the brush border membrane of the kidney proximal tubule, the microvilli of the intestinal lumen, the choroid plexus, and the ciliary body; at each of these locations, it is important that GSH be degraded so that its constituent amino acids can be recovered by the whole organism or specific tissue. γ-Glutamyl transpeptidase catalyzes both transpeptidation and hydrolytic reactions involving the γ-glutamyl bond of GSH [Eqs. (1) and (2), respectively; Eq. (1) is shown as Reaction 3 in Fig. 1].

$$\text{GSH} + \text{amino acid} \rightarrow \gamma\text{-glutamylamino acid} + \text{cysteinylglycine} \quad (1)$$

$$\text{GSH} + H_2O \rightarrow \text{glutamate} + \text{cysteinylglycine} \quad (2)$$

The amino acid indicated in Eq. (1) can be any of a wide variety of L-amino acids; cystine, glutamine, and methionine are particularly active. Some small peptides can also act as acceptors in place of the amino acid. Glycylglycine, for example, reacts at a high rate and is used to assay the enzyme *in vitro*. Even a second molecule of GSH can react in place of the amino acid; in this case the product is γ-glutamyl-GSH. This reaction has been shown to occur *in vivo* in both humans and rodents. Since γ-glutamyl transpeptidase catalyzes hydrolysis and transpeptidation reactions involving the γ-glutamyl bond of GSH, GSSG, γ-glutamylamino acids, γ-glutamyl-GSH, and glutamine (the last is "γ-glutamyl-ammonia" an amino acid present in high concentration in plasma), the extracellular metabolism of GSH and related γ-glutamyl compounds is considerably more complicated than is indicated in Fig. 1 or Eqs. (1) and (2); nevertheless, the products shown in Eqs. (1) and (2) are quantitatively the most important.

Cysteinylglycine and cystinyl-bis-glycine, its corresponding disulfide (see below), are hydrolyzed to their constituent amino acids (i.e., cysteine, cystine, and glycine) by both extracellular and intracellular peptidases; the extracellular enzymes cysteinylglycinase and aminopeptidase M are particularly active toward cystinyl-bis-glycine and cysteinylglycine, respectively. Hydrolysis of cysteinylglycine is shown as Reaction 6 in Fig. 1.

As noted in Section I, GSH is relatively resistant to spontaneous oxidation to GSSG. In contrast, cysteine and cysteinylglycine are easily oxidized. Such oxidation, catalyzed nonenzymatically by trace metals and catalyzed enzymatically by thiol oxidases, occurs in the extracellular environment to form cystine, cystinyl-bis-glycine, and various mixed disulfides. These disulfide products can, in turn, oxidize GSH to GSSG [Eq. (3)]. There is also specific oxygen-dependent enzymatic oxidation of GSH itself within the kidney tubule and of some other sites. At present the physiological role, if any, of extracellular thiol oxidation is unknown. One consequence of such oxidation, however, is that the extracellular thiol/disulfide ratio (typically 2–0.5, depending on the thiol and disulfide considered) is much lower than the intracellular ratio (GSH/GSSG >100 in the cytosol of most cells). Since the breaking and forming of disulfide bonds within and between proteins are affected by the thiol/disulfide ratio, it is possible that the relatively low extracellular thiol/disulfide ratio facilitates the formation of specific protein–protein or protein–ligand disulfide bonds.

$$2\ \text{GSH} + \text{cystinyl-bis-glycine} \rightarrow \text{GSSG} + 2\ \text{cysteinylglycine} \quad (3)$$

The several products of the extracellular γ-glutamyl transpeptidase, dipeptidase, and thiol oxidation reactions are taken up by cells of the surrounding tissue. Separate transport systems mediate the uptake of γ-glutamylamino acids, glutamate, cysteine, cystine, cysteinylglycine, and glycine. Of the several products mentioned, only γ-glutamyl-GSH is thought not to be taken up; it is either excreted in the urine or is further metabolized by γ-glutamyl transpeptidase to yield products that can be taken up. Disulfides taken up by the cell (e.g., cystine and cystinyl-bis-glycine) are reduced to the free thiols by intracellular GSH (Reaction 10, Fig. 1; see Section III). Cysteinylglycine is intracellularly hydrolyzed, presumably by a nonspecific peptidase.

D. Intracellular Metabolism of γ-Glutamylamino Acids: The γ-Glutamyl Cycle

Mammalian tissues vary markedly in their ability to take up γ-glutamylamino acids; tissues with high levels of γ-glutamyl transpeptidase (e.g., kidney and pancreas) tend to be most effective in γ-glutamylamino acid uptake. Within the cytosol γ-glutamylamino acids are metabolized by γ-glutamylcyclotransferase to form 5-oxo-L-proline (i.e., pyroglutamate) and free amino acid (Reaction 4, Fig. 1). The free amino acid formed is incorporated directly into the intracellular amino acid pool and is thus available for a variety of metabolic purposes. 5-Oxo-L-proline, however, is metabolized only by 5-oxoprolinase (Reaction 5, Fig. 1), an ATP/Mg^{2+}-dependent enzyme that hydrolyzes 5-oxo-L-proline to L-glutamate; ADP and P_i are coproducts.

In 1970 A. Meister and M. Orlowski realized that the reactions of GSH synthesis and turnover constitute a metabolic cycle, the γ-glutamyl cycle, which could function as a mechanism for transporting amino acids into the cell. Thus, GSH synthesized within the cell is transported out of the cell and reacts there with an extracellular amino acid in a γ-glutamyl transpeptidase-catalyzed reaction to form the corresponding γ-glutamylamino acid; uptake and intracellular hydrolysis of the γ-glutamylamino acid effectively transport the amino acid into the cell. Hydrolysis of 5-oxoproline and resynthesis of GSH, along with recovery of cysteinylglycine (or cysteine plus glycine), complete the cycle. Subsequent studies using specific inhibitors of various reactions of the cycle have clearly established that amino acids are transported by the γ-glutamyl cycle in γ-glutamyl transpeptidase-rich tissues (e.g., kidney).

Although the cycle might be important in the transport of specific amino acids (e.g., cystine) that are excellent γ-glutamyl transpeptidase substrates, it is noted that the γ-glutamyl cycle is neither the only nor the quantitatively most important mechanism of amino acid transport in any tissue examined to date. The γ-glutamyl cycle, defined so as to account for hydrolysis as well as transpeptidation of GSH by γ-glutamyl transpeptidase, does fully account for GSH turnover in γ-glutamyl transpeptidase-rich tissues. For γ-glutamyl transpeptidase-poor tissues (e.g., liver), GSH turnover occurs by an interorgan γ-glutamyl cycle (e.g., liver and kidney).

III. BIOLOGICAL FUNCTIONS

A. Glutathione Reductase and Maintenance of Intracellular Thiols

Although several physiological and pathological processes oxidize GSH to GSSG, the intracellular GSH/GSSG ratio is, as noted, typically greater than 100. The enzyme responsible for the rapid and nearly complete reduction of GSSG is glutathione reductase, a ubiquitous flavoprotein present in both the cytosol and the mitochondrial matrix (Reaction 12, Fig. 1). As indicated in Eq. (4), GSSG reduction requires the stoichiometric oxidation of NADPH to $NADP^+$; $NADP^+$ is then reduced back to NADPH by any of several cellular dehydrogenases. In most cells glucose-6-phosphate dehydrogenase and 6-phosphogluconate dehydrogenase, enzymes involved in the pentose phosphate pathway of carbohydrate metabolism, are believed to be the most important sources of the NADPH needed to maintain a reduced GSH pool. Inherited partial deficiency of glucose-6-phosphate dehydrogenase is a moderately common human disorder that typically affects the erythrocytes in particular. Under conditions of oxidative stress (e.g., as induced by the antimalarial drug chloroquine), the rate of GSSG formation is high, and the erythrocytes of affected patients cannot maintain a high GSH/GSSG ratio. Failure to maintain a high GSH/GSSG ratio compromises the membrane-protective functions of GSH and leads to erythrocyte destruction; hemolytic anemia is observed clinically.

$$\text{GSSG} + \text{NADPH} + \text{H}^+ \rightarrow 2\ \text{GSH} + \text{NADP}^+ \quad (4)$$

Human tissues contain a number of thiols and disulfides, in addition to GSH and GSSG. Cysteine, cystine, cysteinylglycine, cystinyl-bis-glycine, and γ-glutamylcysteine have been mentioned; there are, in addition, a variety of thiol-containing enzymes and other proteins as well as the coenzymes lipoic acid and coenzyme A. These compounds are maintained predominantly in the reduced form, but are not subject to direct reduction by NADPH. Instead, disulfides such as cystine are reduced by GSH, as shown in Reaction 10, Fig. 1. Equations (5a) and (5b) show in detail the mechanism of cystine reduction by GSH (CyS–SG is the mixed disulfide between cysteine and GSH). Note that the overall process, catalyzed by thiol transhydrogenases, reduces cystine to two molecules of cysteine, while oxidizing two molecules of GSH to GSSG. Because GSSG is readily and completely reduced back to GSH by glutathione reductase, the equilibrium positions of Eqs. (5a) and (5b) are well to the right *in vivo;* the cysteine/cystine ratio, like the GSH/GSSG ratio, is also high intracellularly.

$$\text{Cystine} + \text{GSH} \rightarrow \text{cysteine} + \text{CyS-SG} \quad (5a)$$

$$\text{Cys-SG} + \text{GSH} \rightarrow \text{cysteine} + \text{GSSG} \quad (5b)$$

It is important to emphasize that neither the GSH/GSSG ratio nor the cysteine/cystine ratio is infinite or invariant. Drugs or other conditions producing oxidative stress can significantly decrease the ratios. Even under normal conditions, GSSG and cystine are present intracellularly, and the activity of some enzymes is regulated by the GSH/GSSG ratio. Among those enzymes affected by GSH/GSSG ratios within normal physiological limits are hydroxymethylglutaryl-coenzyme A reductase and phosphofructokinase, enzymes catalyzing rate-limiting reactions in cholesterol biosynthesis and glycolysis, respectively. It is apparent that, for a limited number of enzymes, alterations in the GSH/GSSG ratio represent an important mechanism of metabolic control.

B. GSH and Detoxification of Peroxides

Hydrogen peroxide (i.e., H–O–O–H) is a coproduct of several intracellular enzymes (e.g., superoxide dismutase, D-amino acid oxidase, and monoamine oxidase) and is produced during mitochondrial respiration by the incomplete reduction of oxygen (oxygen is normally reduced by a four-electron process to water, but is reduced by a two-electron process to H_2O_2 up to 5% of the time). Organic peroxides (R–O–O–H) are formed mainly from unsaturated fatty acids, usually by reaction of a fatty acid free radical with molecular oxygen. Both hydrogen peroxide and organic peroxides are subject to spontaneous homolytic cleavage to form highly destructive free radicals (e.g., $H–O–O–H \rightarrow 2\,HO^\cdot$, where $HO^\cdot$ is a hydroxide radical). Furthermore, in the presence of certain metals, particularly iron, hydrogen peroxide and superoxide (i.e., O_2^-, also a normal cellular metabolite) react to form hydroxide radical and singlet oxygen, the latter also a very destructive species. Although peroxide formation is apparently an inescapable consequence of aerobic life, it is obviously important that the cellular levels of peroxides and superoxide be kept at a minimum to prevent their conversion into highly damaging species.

Several defenses have evolved. Superoxide is destroyed by superoxide dismutase with the formation of ordinary oxygen and hydrogen peroxide [Eq. (6)]. Catalase, a peroxisomal enzyme, catalyzes the dismutation of hydrogen peroxide [Eq. (7)], but is inactive toward organic peroxides. Both hydrogen peroxide and organic peroxides are reduced by glutathione peroxidase, a selenium-dependent enzyme present in both the cytosol and the mitochondrial matrix (Reaction 11, Fig. 1). In addition, some isoforms of glutathione *S*-transferase (see Section III,D) also catalyzes the GSH-dependent reduction of organic peroxides; hydrogen peroxide is not reduced. It is apparent that the reduction of peroxides by either glutathione peroxidase or glutathione *S*-transferase causes the concomitant oxidation of GSH to GSSG; as discussed previously, GSSG is reduced back to GSH by glutathione reductase. Because glutathione peroxidase is quantitatively more important than catalase in reducing physiological levels of hydrogen peroxide, cellular defenses against both hydrogen peroxide and organic peroxides depend heavily on the intracellular GSH, glutathione reductase, and a continuing supply of NADPH. At pathologically high levels of hydrogen peroxide, the role of catalase becomes proportionately more important, but does not supplant that of glutathione peroxidase.

$$2\,O_2^- + H^+ \rightarrow O_2 + H_2O_2 \tag{6}$$

$$2\,H_2O_2 \rightarrow O_2 + H_2O \tag{7}$$

C. GSH and Prevention and Repair of Free-Radical Damage

Free radicals are intermediates in several biological reactions essential for life. As already mentioned, other free radicals (e.g., $OH^\cdot$) are highly reactive and tissue-damaging species capable of causing cell death. Ionizing radiation, either from the environment or administered as a treatment of certain cancers, splits a variety of chemical bonds to yield free radicals. Because cells contain high concentrations of water, a common effect of ionizing radiation is the homolytic cleavage of H_2O to $HO^\cdot$ and $H^\cdot$. Cytotoxic-free radicals are also formed metabolically. Carbon tetrachloride is metabolized to form $Cl_3C^\cdot$, for example. In general, free radicals in which the unpaired electron is on carbon or oxygen are highly reactive and cytotoxic species. Such species might add to the carbon–carbon double bonds of membrane lipids, producing lipid radicals and ultimately lipid peroxides. Free radicals such as $HO^\cdot$ and $Cl_3C^\cdot$ might abstract $H^\cdot$ from important biomolecules such as nucleic acids and proteins, leaving those molecules as unstable reactive free radicals.

GSH plays a role in both the prevention and the repair of free-radical damage. Because sulfur-centered free radicals are less reactive than most carbon- and oxygen-centered radicals, direct reaction of GSH with, for example, $HO^\cdot$ yields HOH and $GS^\cdot$, a much less cytotoxic species. Combination of two $GS^\cdot$ species yields GSSG, which is reduced to two GSH, as described in this section. Mammalian cells might also contain GSH-dependent enzymes capable of detoxifying free radicals. The role of GSH and glutathione peroxidase in reducing the lipid peroxides formed by the reaction of lipid free radicals with molecular oxygen was described in Section III,B. [*See* Free Radicals and Disease.]

D. GSH Reaction with Electrophiles: Mercapturic Acid Synthesis

Reactive electrophiles are formed in the course of normal metabolism and during the metabolism of various drugs and other foreign molecules. Many such electrophiles are reactive and add nonenzymatically to any of a wide variety of cellular nucleophiles, including essential nucleic acids and proteins. Since such reactions generally destroy the functionality of the nucleic acid or protein affected, the cell has evolved enzymes that specifically facilitate the reaction of such electrophiles with GSH. Because GSH is an expendable strongly nucleophilic molecule present in high concentrations, reaction with GSH spares more valuable nucleic acids and proteins from electrophilic damage. Reaction of GSH with electrophiles occurs spontaneously and is also catalyzed by a family of related enzymes, the glutathione *S*-transferases (Reaction 7, Fig. 1). The enzyme-catalyzed reaction is quantitatively important for most electrophiles. The product is no longer reactive and is transported out of the cell. Extracellularly, it is metabolized similarly to GSH; the γ-glutamyl moiety is removed by γ-glutamyl transpeptidase, and the bond between the substituted cysteine residue and glycine is cleaved by a dipeptidase. The resulting *S*-substituted cysteine (i.e., CyS-electrophile) is then taken up by various tissues (e.g., kidneys), and the amino group of the cysteine residue is acetylated in an acetyl-coenzyme A-dependent reaction. The resulting product, an *S*-substituted-*N*-acetyl-L-cysteine, is referred to as a mercapturic acid and is excreted in the urine (Fig. 1; pathway indicated by Reactions 7, 3, 6, and 14).

It is noteworthy that GSH, or at least its essential cysteine residue, is irreversibly lost through the formation of a mercapturic acid; in this important respect mercapturic acid formation differs from all of the other protective functions mentioned. Among the common drugs metabolized to mercapturic acids is acetaminophen (e.g., Tylenol), which is activated in the liver to an electrophile which then reacts with GSH. Although the hepatic supply of GSH is more than adequate for normal doses of acetaminophen, massive doses of the drug produce the reactive metabolite in amounts that deplete hepatic GSH. In the absence of GSH, reaction of the activated drug with other cellular nucleophiles occurs and leads to hepatic damage and, in severe cases, death.

E. GSH as a Coenzyme

GSH acts as a coenzyme in several enzymatic reactions (grouped as Reaction 13, Fig. 1). In the glyoxylase reaction, glyoxylase I first catalyzes an internal oxidation–reduction reaction of the hemimercaptal formed nonenzymatically between methylglyoxal and GSH to produce *S*-lactoylglutathione (i.e., CH_3–CO–CHOH–SG → CH_3–CHOH–CO–SG). Glyoxylase II then catalyzes the hydrolysis of this product to D-lactate and GSH. The *cis–trans* isomerization of maleylacetoacetate to fumarylacetoacetate, a reaction involved in the catabolism of the amino acid tyrosine, similarly requires GSH as a coenzyme. Formaldehyde dehydrogenase, an enzyme reversibly interconverting formaldehyde and formic acid, and two enzymes involved in the formation of certain prostaglandins also exhibit a specific requirement for GSH. Houseflies require GSH to detoxify the insecticide 1,1′-(2,2,2-trichloroethylidene)bis[4-chlorobenzene] (DDT); the latter reaction apparently does not occur in humans. [*See* Enzymes, Coenzymes, and the Control of Cellular Chemical Reactions.]

IV. MEDICAL ASPECTS

A. Inherited Disorders of GSH Metabolism

Inherited deficiencies affecting several of the enzymes of the γ-glutamyl cycle have been reported; none is common. In 1970, systemic deficiency of glutathione synthetase was the first such disorder reported. The affected patient was a 19-year-old mentally retarded boy of normal height and weight, exhibiting signs of organic cerebral damage with spastic quadraparesis and cerebellar disturbances. The biochemical defect was discovered during attempts to elucidate a severe acidosis that developed following a surgical procedure. Although the acidosis was effectively controlled with bicarbonate, further studies established that the patient excreted 24–35 g of 5-oxo-L-proline per day in his urine. Once it was determined that glutathione synthetase was severely deficient in the patient, it became apparent that the near absence of GSH released γ-glutamylcysteine synthetase from feedback control and permitted a substantial overproduction of γ-glutamylcysteine. Since this product could not be converted to GSH, it was metabolized intracellularly to 5-oxoproline and cysteine by γ-glutamylcyclotransferase. Production of 5-oxoproline exceeded the capacity of 5-oxoprolinase, and the excess spilled into the plasma, causing acidosis. 5-Oxoproline was eventually excreted in the urine, leading to the clinical characterization of the disorder as 5-oxoprolinuria. To date 16 patients with systemic glutathione synthetase deficiency have been identified; a smaller number are affected by a deficiency involving only the erythrocytes, a relatively benign condition.

Two patients exhibiting nearly complete deficiency of γ-glutamylcysteine synthetase have been reported. These patients reportedly exhibit hemolytic anemia, spinocerebellar degeneration, peripheral neuropathy, myopathy, and amino aciduria. Erythrocyte GSH levels are less than 3% of normal levels, but other tissues exhibit significantly higher GSH levels (e.g., 25% of normal levels in skeletal muscle). The enzyme deficiency is thus not complete.

Two patients with apparently generalized deficiency of γ-glutamyl transpeptidase have also been described. As expected, these patients cannot degrade GSH released by various tissues (mainly the liver); such GSH accumulates in the plasma, is filtered at the kidney glomerulus, and is excreted in the urine. One patient excreted 850 mg of GSH per day. Further studies established that in neither patient is the enzyme deficiency complete; residual γ-glutamyl transpeptidase catalyzes the extracellular synthesis of γ-glutamylcystine (from GSH and cystine) and γ-glutamyl-GSH (from two molecules of GSH). These products are also found in the urine. Although both patients are moderately retarded, the disorder has been sought mainly among patients of mental institutions, and it is not yet established that γ-glutamyl transpeptidase deficiency accounts for the mental deficit.

Three individuals with 5-oxoprolinase deficiency have been reported. Although the enzyme defect is apparently not complete, the affected individuals excrete up to 9 g of 5-oxoproline per day; this value, equivalent to 70 mmol/day, represents a minimum estimate of the daily γ-glutamyl cycle flux in humans. In contrast to patients with glutathione synthetase deficiency, the individuals with 5-oxoprolinase deficiency have normal tissue GSH levels. The condition appears to be benign.

Because glutathione reductase is a flavin-dependent enzyme, acquired deficiency of glutathione reductase is possible when diets deficient in riboflavin are consumed. Moderate glutathione reductase deficiencies due to inadequate flavin availability are, in fact, common, but appear to be without clinical effect. Severe but incomplete deficiency of the glutathione reductase protein has been documented in the erythrocytes of only three patients, all siblings. One patient suffered a hemolytic crisis after eating fava beans (a source of oxidative stress); two suffered cataracts, possibly due to failure to maintain an adequately high GSH/GSSG ratio. Other signs and symptoms were not reported.

Complete deficiency of glutathione peroxidase has not yet been reported, although there are certain Jewish and Mediterranean populations in which tissue levels are about one-half of normal. Because selenium is necessary for the synthesis of active enzyme, acquired deficiency among individuals consuming a selenium-poor diet is common in some parts of the world. No pathological consequences clearly attribute to these incomplete inherited or acquired deficiencies have been reported. [*See* Selenium in Nutrition.]

B. Pharmacological Control of GSH Metabolism: Therapeutic Implications

The multiplicity of enzymatic, transport, and protective functions attributed to GSH has stimulated efforts to develop agents useful in the deliberate pharmacological manipulation of GSH levels and GSH-dependent metabolisms. As noted, various drugs (e.g., acetaminophen) and toxic agents (e.g., many aromatic chemicals) cause tissue damage when their detoxification requires GSH in quantities greater than those present in the tissue. High intracellular levels of GSH also minimize radiation damage and could facilitate the repair of radiation-damaged DNA.

With these facts in mind, several approaches to increasing tissue levels of GSH have been developed. Because GSH synthesis is often limited by the availability of cysteine, GSH levels can be moderately increased by direct administration of L-cysteine or L-cysteine precursors [e.g., *R*-thiazolidine-4-carboxylate (i.e., L-thioproline), *N*-acetyl-L-cysteine, and L-2-oxothiazolidine-4-carboxylate]; the latter two compounds are enzymatically converted by L-cysteine intracellularly by *N*-acetylase and 5-oxoprolinase, respectively. More recently, Meister and colleagues have established that, in contrast to GSH, glutathione ethyl ester (i.e., glycyl carboxylate esterified) and related compounds are transported into many mammalian cells. Esterases within cells hydrolyze the carboxylate ester to produce free GSH within the cytosol.

Since tissue GSH levels can be markedly increased in animals given GSH esters, human trials of these compounds are likely in the near future. In this context it should again be noted that GSH itself is not effectively taken up by intact cells. Parenteral or oral administration of GSH does not, therefore, directly increase cellular GSH levels. GSH is, however, broken down extracellularly (see Section II,C) to provide cysteine, which can be taken up and used for GSH synthesis. In this respect GSH is similar to cysteine and other cysteine precursors.

Whereas the maintenance of normal to supranormal tissue GSH levels is normally of benefit to the organism, the ability to selectively decrease tissue GSH levels can facilitate certain therapeutic interventions. In particular, a variety of *in vitro* studies have established that radiation therapy and certain chemotherapies of tumors are more effective if tumor GSH is first depleted. Several regimens designed to effect such depletion have been developed. One of the least toxic and most effective approaches relies on the administration of buthionine sulfoximine (BSO), a specific and potent inhibitor of γ-glutamylcysteine synthetase. Following administration of BSO, GSH synthesis is blocked, whereas GSH utilization, primarily transport of GSH out of the cell, continues unabated; in tissues with moderately high rates of GSH turnover, GSH depletion occurs within a few hours. Human tumors grown as xenografts in athymic nude mice have been shown to be sensitized to both radiation and various chemotherapies following administration of BSO.

It is also noteworthy that drug resistance in tumors (i.e., the clinical observation that doses of anticancer drugs initially causing tumor remission are not effective in recurrent tumors) has, in some cases, been correlated with increased levels of GSH in the recurrent tumor. Preliminary studies in human ovarian carcinoma and other tumors indicate that the administration of BSO abolishes increased GSH levels and restores drug sensitivity. Finally, studies with mice infected with *Trypanosoma brucei brucei*, an animal model of the human parasitic disease African sleeping sickness (i.e., trypanosomiasis), indicate that treatment with BSO depletes trypanosomal as well as host GSH; treated mice exhibit extended survival and are, in some cases, cured. Further studies have suggested that GSH-depleted trypanosomes were unable to control the oxidative stresses inherent in their normal metabolic reactions; GSH-depleted cells of the host (i.e., the mouse) survived because they had alternative defenses against oxidative stress and depended on metabolic reactions that generated lower levels of cytotoxic oxidants.

ACKNOWLEDGMENT

Studies performed in the author's laboratory were supported in part by National Institutes of Health Grant DK26912. I thank Ernest B. Campbell and Michael A. Hayward for expert technical assistance in these studies.

BIBLIOGRAPHY

Beaudet, A. L., Sly, W. S., and Scriver, C. R. (eds.) (1995). "The Metabolic and Molecular Bases of Inherited Disease," 7th Ed. McGraw-Hill, New York.

Griffith, O. W. (1985). Glutathione and cell survival. *In* "Cellular Regulation and Malignant Growth" (S. Ebashi, ed.), pp. 292–300. Jpn. Sci. Soc. Press, Tokyo.

Griffith, O. W., and Friedman, H. F. (1990). Inhibition of metabolic drug inactivation: Modulation of drug activity and toxicity by perturbation of glutathione metabolism. *In* "Synergism and Antagonism in Chemotherapy" (T.-C. Chou and D. Rideout, eds.). Academic Press, San Diego.

Meister, A. (1989). Metabolism and function of glutathione. *In* "Glutathione: Chemical, Biochemical and Medical Aspects, Part A" (D. Dolphin, O. Avramović, and R. Poulson, eds.), pp. 367–474. Wiley (Interscience), New York.

Meister, A., and Anderson, M. E. (1983). Glutathione. *Annu. Rev. Biochem.* **52**, 711–760.

Packer, L., and Simon, M. I. (eds.) (1995). Biothiols: Glutathione and Thioredoxin: Thiols in Signal Transduction and Gene Regulation. *In* "Methods in Enzymology," Vol. 252. Academic Press, San Diego.

Glycogen

NEIL B. MADSEN
University of Alberta

GLOSSARY

Active site or catalytic site Location on an enzyme protein where the substrate(s) bind and are catalytically transformed into product(s)

Allostery Control mechanism in enzyme activity in which a site other than the active site is occupied by an allosteric effector molecule, which may be an activator or an inhibitor

Amino terminus Beginning of a polypeptide chain that is occupied by an amino acid with a free amino group, its carboxyl group joined to the amino group of the next residue

Coenzyme Small molecule that binds at the active site of an enzyme and participates in the catalytic process

Conformation One of many shapes or structures that a protein molecule may assume

Feedback inhibitor Compound that may be the last in a metabolic pathway and that inhibits an early enzymatic step in the pathway, usually via allostery

Metabolite Any small molecule that undergoes enzymatic transformation in the cell

Polypeptide chain Amino acids combine by the removal of water between the carboxyl group of one and the amino group of the next (forming peptide bonds), yielding long polypeptide chains

Subunit Most enzymes, which are proteins, are dimers or tetramers of polypeptide chains, each folded into a globular shape termed a subunit, which may or may not be identical

T and R states T (taut) state is considered the enzymatically inactive conformation of an allosteric enzyme, whereas the R (relaxed) state is the active conformation

GLYCOGEN IS THE MAJOR STORAGE FORM OF glucose in animal organisms. Glucose molecules are joined together to form long chains with many branch points, resulting in the buildup of huge spherical molecules that are found in large amounts in muscle and liver tissues. In muscle the breakdown of glycogen provides anaerobic (i.e., in the absence of oxygen supply) energy for short, intense bursts of muscular activity (e.g., sprints) and contributes aerobic (i.e., with oxygen available) energy for long sustained activity (e.g., marathon running). Liver glycogen is a reserve source of blood glucose during short periods of fasting (e.g., overnight). During the synthesis of glycogen, several enzymes are employed to transform free glucose to phosphorylated and activated forms, and finally the enzyme glycogen synthase adds the glucose units to the growing chains, whereas another enzyme inserts the branch points. When glycogen breakdown is required, the enzyme phosphorylase cleaves the glucose chains, using phosphate, while a debranching enzyme removes the branch points. Glycogen phosphorylase and synthase are under reciprocal and coordinated controls so that glycogen synthesis and degradation may not occur simultaneously. Thus low blood sugar levels cause hormonal changes that activate, via several intermediate steps, liver phosphorylase and inactivate the glycogen synthase. In muscle the breakdown of glycogen is coordinated with contraction because both processes are stimulated by calcium, which is released in response to nerve impulses. The structure of phosphorylase has been determined in great detail, and it is possible to explain the molecular mechanisms for the various physiological controls that regulate this key enzyme of glycogen metabolism.

I. THE GLYCOGEN MOLECULE

A. Discovery and Chemistry

Claude Bernard, the great French physiologist, discovered glycogen in 1857 during the course of his studies on sugar metabolism and showed that the liver synthesized the substance and used it to maintain the correct level of blood sugar, enunciating also the principle of the maintenance of equilibria for parameters such as blood sugar. Figure 1 is an electron micrograph of a single glycogen particle or molecule isolated by gentle means from rabbit liver. Such rosette-like structures are typical of liver glycogen and have a molecular weight of as much as 500 million, whereas the glycogen molecule from muscle is the size of one of the small spheres making up the liver particle and has a molecular weight of no more than 10 million. All the enzymes needed for the synthesis and degradation of glycogen, as well as those concerned with regulation, are found bound to the glycogen particles. Glycogen is exclusively composed of the sugar glucose and a small amount of a protein known as glycogenin on which the formation of glycogen begins but which we will not consider further. Because each glucose residue in glycogen has a molecular weight of 162, we can see that there can be from several thousand to more than a million glucose residues in one glycogen molecule. As noted in the following section, some tissues store large amounts of sugar in the form of glycogen, but so much sugar as individual glucose molecules would cause the cells to burst from excessive osmotic pressure, whereas a single glycogen molecule exerts no more osmotic pressure than a single glucose molecule.

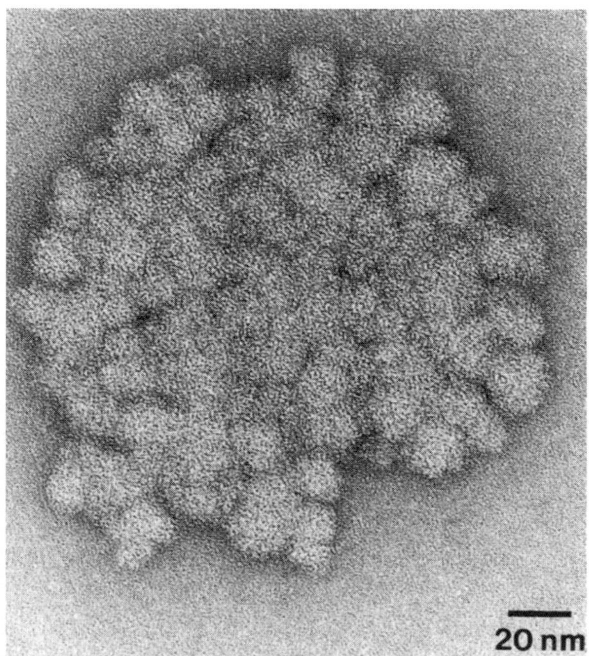

FIGURE 1 Electron micrograph of a single glycogen particle from liver, stained with 2% uranyl acetate. (Courtesy of Dr. D. G. Scraba and Mr. R. D. Bradley, University of Alberta.)

Glycogen is a large branched polymer of glucose. As shown in Fig. 2, long chains of glucose are formed by the elimination of water between carbon 1 of one glucose and carbon 4 of another, forming α-1,4 links. Every four to five glucose molecules in the chain, a branch is formed by an α-1,6 link. Figure 2 shows two outer branches of the molecule with the 4-hydroxy groups available for further elongation. In principle, there would be one reducing 1-hydroxy group per glycogen molecule, but in fact this is coupled to a tyrosine residue of the initiator protein, glycogenin. Continued lengthening of the glucose chains with the introduction of branch points builds up a large tree-like structure, which is illustrated schematically as "mature glycogen molecule" in Fig. 3.

B. Occurrence in Human Tissues and Biological Roles

Tissues containing glycogen include liver, skeletal muscle, heart, kidney, and brain, with most others having at least some. The liver of a well-fed man is approximately 5% glycogen, the skeletal muscle 1.3%, and the brain only 0.1%, so that liver and muscle glycogen are the most significant physiologically. Although the content of glycogen in the liver is higher than in muscle, because of the different tissue masses, the liver can store only 80–100 g of glycogen compared with about 450 g in the total muscle.

In skeletal muscle, the role of glycogen is to provide a source of phosphorylated glucose units (especially 6-phosphate), which can enter the glycolytic pathway and provide energy for muscular contraction. For example, during a sprint, there is sufficient glycogen in the muscles to provide energy for only 20 sec of maximal activity, and most of the energy is provided by the breakdown of glycogen and the anaerobic production of adenosine triphosphate (ATP) required for muscle contraction, because the extreme energy demands to not permit time for external sources (e.g.,

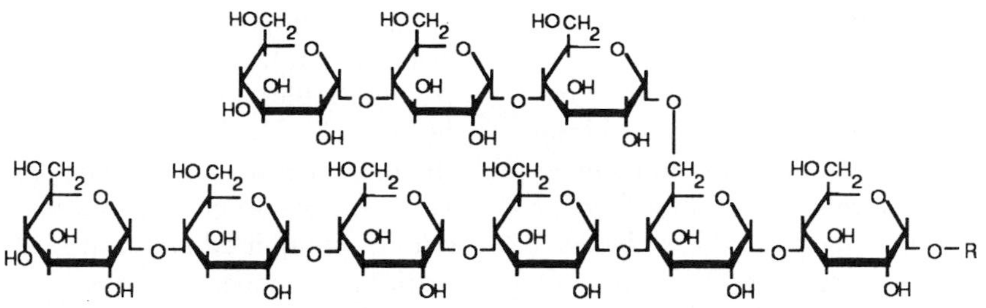

FIGURE 2 Two outer branches of a glycogen molecule, showing how the glucose units are joined by α-1,4 links (horizontal rows), with a branch point consisting of an α-1,6 link connecting the two branches. The two branches terminate in nonreducing ends, whereas R represents the rest of the molecule.

blood glucose and fatty acids), which produce ATP through a more elaborate pathway, to play a role. Marathon runners, however, employ an aerobic metabolism and in the later stages derive 60% of their energy from oxidation of fatty acids as well as blood sugar derived from the breakdown of liver glycogen. Muscle glycogen stores are depleted slowly in long-distance running, but exhaustion is associated with its complete utilization. The regulation of glycogen utilization is obviously under exquisitely sensitive

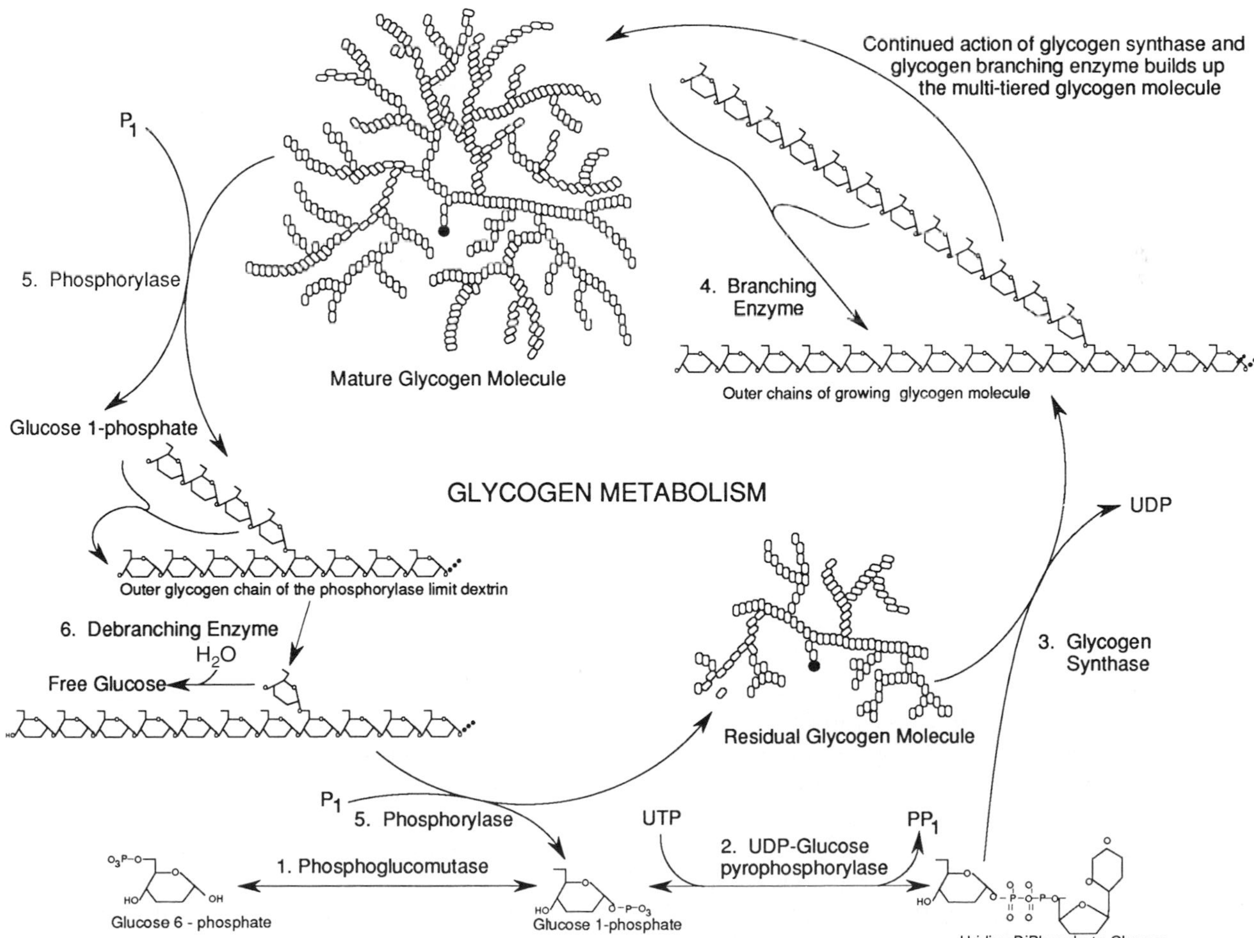

FIGURE 3 Scheme showing the actions of enzymes that synthesize and degrade glycogen.

control to meet a variety of demands, as discussed in Section III. [*See* Adenosine Triphosphate (ATP).]

As first suggested by Claude Bernard, the liver has the chief responsibility for the maintenance of blood sugar levels at a safe minimum and also contributes to the reduction of high levels. When blood sugar levels rise after a meal, the liver rebuilds its glycogen stores to maximal levels. During short fasting periods (e.g., overnight) these stores are depleted to replenish the blood glucose used by other tissues such as the brain. For example, a 24-hr fast lowers the glycogen content from 5% to 0.7%. During longer fasts, some tissues begin to metabolize fatty acids and proteins for energy, but the liver then has to make glucose from protein to supply the brain. The glycogen content of other tissues (e.g., the heart) can provide an emergency supply of energy during brief periods of oxygen deprivation.

II. METABOLISM

A. Enzymes Involved in the Formation of Glycogen

1. Phosphoglucomutase is the enzyme that links the glycolytic pathway to glycogen metabolism. Glucose 6-phosphate, formed in the glycolytic pathway by hexokinase, is diverted from the degradative pathway by its conversion to glucose 1-phosphate by this first enzyme. The reaction is shown in Fig. 3, and it is freely reversible under physiological conditions, because when the enzyme operates long enough on either of the glucose phosphates, it reaches a chemical equilibrium in which there is 95% of the 6-phosphate and 5% as the 1-phosphate. Phosphoglucomutase can operate in both the formation and the degradation of glycogen.

2. UDP-glucose pyrophosphorylase takes the glucose 1-phosphate formed in the previous reaction and reacts it with uridine triphosphate (UTP) to form pyrophosphate (PP_i) and UDP-glucose (uridine diphosphate glucose), as shown in Fig. 3. Although this reaction is freely reversible, it is rendered irreversible under physiological conditions by the hydrolysis of one of the products, PP_i, carried out by the pyrophosphatase ubiquitous in cells. The other product, UDP-glucose, is an activated form of glucose, which is well suited as the donor of glucose for the synthesis of glycogen and other large sugar molecules.

3. Glycogen synthase adds glucose units from UDP-glucose to the 4-hydroxy groups of the glucosyl termini of preexisting glycogen molecules known as primers, causing the formation of long chains of glucose molecules joined in α-1,4-glycosidic linkages, as shown in Fig. 3. The original primers are short glucose chains joined to the tyrosine residue of the initiator protein. Because glycogen synthase catalyzes the first unique, committed reaction leading to glycogen formation, it is under rigorous metabolic and hormonal control, as we will discuss in Section III.

4. Branching enzyme forms branches in the long linear glucose chains resulting from the previous reaction. Long linear chains become insoluble, but with the introduction of branches, a compact, highly soluble spherical molecule comes into being. The number of terminal residues is also increased enormously, so that both the formative and degradative enzymes have more sites on which to work, resulting in faster rates. Branching enzyme exhibits a precise specificity, taking a block of six to seven glucose units from a growing chain, breaking the α-1,4 link, and transferring the block to the 6 position of a glucose residue in an adjacent chain to form a new α-1,6-linked branch, as illustrated in Fig. 3. The new branch is from four to six residues from the next interior branch. It is interesting to note from thermodynamic data that the formation of branches is favored over the reverse reaction, and indeed this enzyme does not play a role in the degradation of glycogen, the removal of the α-1,6 links being effected by a separate enzyme with a different mechanism.

B. Enzymes Involved in the Degradation of Glycogen

1. Phosphorylase was discovered in 1936 when Carl and Gerty Cori demonstrated that the addition of inorganic phosphate to muscle extracts resulted in the conversion of glycogen to a new hexose monophosphate, which they proved to be glucose 1-phosphate. They also showed that the latter compound could be converted into glucose 6-phosphate by muscle and that the phosphorylase reaction was stimulated by adenosine monophosphate (AMP), perhaps the earliest demonstration of allostery. By 1940, they and their colleagues had crystallized phosphorylase (placing it among the first crystalline enzymes) and elucidated the chemistry of the phosphorolysis reaction that it catalyzes. As illustrated in Fig. 4, the enzyme carries out the cleavage of the terminal glycosidic bond of the glucose chains in glycogen, using inorganic phosphate (phosphorolysis) instead of water, which is used in many cleavage reactions (hydroly-

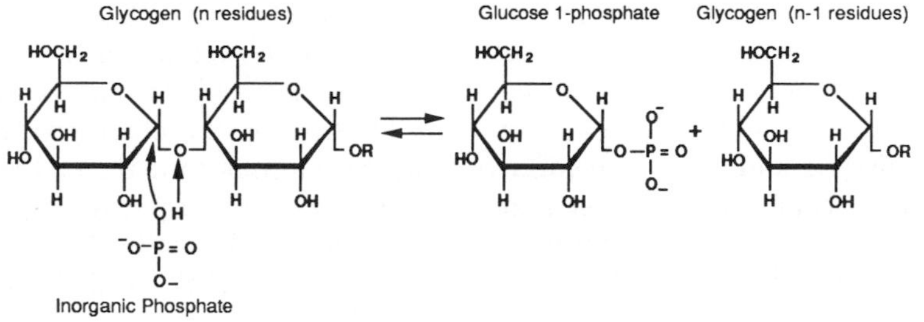

FIGURE 4 The reaction catalyzed by glycogen phosphorylase. The carbon numbering scheme used for glucose is indicated for glucose 1-phosphate.

sis), thus incorporating the phosphate at the C-1 position of the released glucose unit. Although the equilibrium constant of the phosphorolysis reaction is only 0.28 (meaning that only 28% of added phosphate would be converted to glucose 1-phosphate), under physiological conditions the reaction is driven in the direction of glycogen degradation because the ratio of inorganic phosphate to glucose 1-phosphate is 100 or more, so that a high glycogen concentration could not be maintained without the stringent control of phosphorylase, as discussed in Section III. Furthermore, because of the design of its active site, phosphorylase cannot cleave the final four glucose units from an α-1,6 branch, as shown in Fig. 3. Because these branch points are symmetrically arranged in concentric tiers in the spherical glycogen molecule, the action of phosphorylase results in a glycogen with short outer chains, known as a *limit dextrin.*

Phosphorylase contains in its active site a tightly bound coenzyme that is essential for its function. This is pyridoxal phosphate, a derivative of vitamin B_6, and because there is so much phosphorylase in muscle, more than 80% of the vitamin B_6 stores in muscle are bound to this enzyme. Nevertheless, a deficiency of this vitamin shows up as disorders of protein metabolism, in which it plays major roles, rather than of glycogen metabolism. The manner in which pyridoxal phosphate functions in phosphorylase is unique for this coenzyme, involving the interaction of its phosphate with that of the substrate, whereas, in other enzymes, different parts are involved.

2. Debranching enzyme acts on the phosphorylase limit dextrin in the manner depicted in Fig. 3. It transfers a block of three glucose units from one outer branch (the one joined to another branch by an α-1,6 link) to the outer branch containing the "side chain." It is thus an α-1,4 $\rightarrow$ α-1,4-glycosyl transferase. This action leaves a single glucose unit joined by an α-1,6 link to the "main chain." The debranching enzyme then hydrolyzes this link, releasing a free glucose molecule, and thus exhibiting α-1,6-glucosidase activity. It may be noted that both the transferase and glucosidase activities of the debranching enzyme are carried out by a single monomeric polypeptide chain of 160,000 molecular weight, the first example of a "double-headed" enzyme in higher organisms. As shown in Fig. 3, the action of the debranching enzyme on the phosphorylase limit dextrin results in a smaller glycogen molecule with long outer branches on which phosphorylase may act further. Successive alternating actions of phosphorylase and debranching enzyme will obviously reduce the initially large glycogen molecule to the small size found in liver after a fast or in muscle after prolonged and exhausting exercise.

3. Glucose 6-phosphatase is the final enzyme in the pathway for complete degradation of glycogen to free glucose, and removes the phosphate group after phosphoglucomutase has changed the glucose 1-phosphate, arising from glycogen, into glucose 6-phosphate. It is especially important in liver, where its hydrolysis of glucose 6-phosphate permits the export of glucose to the blood, thus fulfilling the liver's role in maintaining blood sugar during exercise and fasting. It is also found in the kidney and intestines but not in muscle and brain, because the latter tissues are the major users of glucose.

C. Glycogen Storage Diseases

A number of genetically determined diseases characterized by accumulations of glycogen in liver, muscle, or other organs have been traced to inherited deficiencies of the various enzymes of glycogen metabolism. The first and most famous of these, characterized by the massive amount of normal glycogen in the liver, was described by E. von Gierke in 1929. It causes low blood sugar and failure to thrive in infants unless carefully controlled by dietary measures.

In 1952, Carl and Gerty Cori showed that glucose 6-phosphatase was missing from the livers of these patients, allowing no release of glucose from the degradation of glycogen. In another disease, the debranching enzyme is missing, and normal glycogen with short outer branches accumulates in liver and muscle. Absence of the branching enzyme results in a normal amount of glycogen in liver and spleen, but it has long outer branches and results in liver failure during infancy. A final example concerns the absence of phosphorylase from muscle. Patients with this problem are unable to perform vigorous exercise because of painful cramps. It is obvious that there would be problems because the vital energy derived from glycogen degradation is not available, but the immediate cause of the cramps appears to be the accumulation of the breakdown product of ATP (the immediate energy source for contraction) [i.e., adenosine diphosphate (ADP)], because the latter cannot be recycled back to ATP.

III. REGULATION OF GLYCOGEN METABOLISM

A. Metabolically Interconvertible Enzymes

1. Covalent attachment of a phosphate by phosphorylase kinase to a specific serine residue on phosphorylase makes this enzyme active under physiological conditions. The phosphorylase that was discovered in 1936 required AMP for activity and is now termed *phosphorylase b*. In 1942 a new form was isolated and crystallized, designated *phosphorylase a*, which does not require AMP for activity, although it binds it even more tightly. Binding of AMP improves activity of the enzyme at low substrate concentration. Eventually it was discovered that both the *a* and *b* forms are dimers of a 97,400 molecular weight subunit containing 842 amino acids in a single polypeptide chain. Phosphorylase *a* contains a phosphorylated serine at position 14 (counted from the amino terminus), in each of its subunits. This is the only chemical difference between the *a* and *b* forms, but there are changes in the three-dimensional structure of the protein, which will be discussed in Section IV,B. The phosphoserine is close to the binding site for the AMP and may be regarded as a covalently bound activator. Both phosphoserine and AMP are on the opposite side of the dimer from the catalytic active site, which can be defined by the presence of the coenzyme pyridoxal 5′-phosphate, bound to lysine 680. Phosphorylase *b* is subject to allosteric inhibition by glucose 6-phosphate, which may be regarded as a feedback inhibition by the chief product of glycogen breakdown in muscle (and the first metabolite in glycolysis). Energy control of phosphorylase *b* is exerted via the allosteric inhibition by ATP (which binds to the same site as does AMP) and the allosteric activation by AMP, as well as the substrate activation by inorganic phosphate. The latter two compounds indicate low energy levels, whereas ATP indicates a high energy level. Phosphorylase *a* has "escaped" allosteric control because it no longer requires AMP nor is it inhibited significantly by ATP or glucose 6-phosphate. Phosphorylase activity can thus be turned on or off by hormonal or nervous controls regardless of the levels of metabolites, permitting a more sophisticated whole-organism regulation to be imposed on the cellular level.

2. Phosphorylase kinase is a complicated enzyme consisting of four copies of each of four different types of subunits, termed α, β, γ, and δ. The entire molecule can thus be denoted $\alpha_4\beta_4\gamma_4\delta_4$ and has a total molecular weight of 1.3×10^6. The function of each subunit has been defined by many years of experimental work, and the catalytic activity has been assigned to the γ subunit. The δ subunit was found to be the calcium binding protein calmodulin, and its function explains the obligate requirement of the kinase for calcium ion, which confers partial activity on the enzyme. Full activity is achieved only when phosphate groups are incorporated into specific serine residues in both the β and α subunits, so that these two subunits are concerned with regulating the activity via intersubunit-transmitted conformational changes. The enzyme that phosphorylates the β and α subunits is the cyclic AMP (cAMP)-dependent protein kinase.

3. cAMP-dependent protein kinase is activated, as implied, by cAMP, which is formed in response to hormonal signals, as described in the following: The protein kinase is composed of two regulatory subunits and two catalytic subunits, forming an inactive tetramer, which can be denoted R_2C_2. When cAMP binds to the regulatory subunits, the catalytic subunits dissociate as free active monomers, which can phosphorylate the phosphorylase kinase. In addition, the cAMP-dependent protein kinase also phosphorylates specific serine residues on glycogen synthase *a*, forming synthase *b*, but in this case, unlike the phosphorylase case, the phosphorylated form of the synthase is physiologically inactive. We will return to this enzyme in Section III,B.

4. cAMP is known as a "second messenger" because it is formed in response to hormones (the first messengers) binding to specific receptors on the outside of the cell membrane. The hormones do not enter the cell but cause conformational changes to be transmitted across the membrane (via proteins that span the membrane) to the enzyme adenylate cyclase on the inside of the cell membrane, causing it to carry out the transformation of ATP into cAMP, as diagrammed in Fig. 5. Among the better-known hormones, adrenalin is secreted in response to emotional states such as fright or anger and it binds to specific receptors on muscle cells where it initiates the chain of events that leads to the breakdown of glycogen, as detailed in Section III,B. Glucagon, however, is secreted in response to low blood sugar, and it binds to specific receptors on liver cells, which again leads to the breakdown of glycogen into the form of glucose that is released into the bloodstream.

B. Glycogen Cascade

1. The glycogen cascade is diagrammed in Fig. 6. The term *cascade* is used because a series of events cascade one into the next to connect the primary hormonal or nervous signals with the ultimate effect on the breakdown or synthesis of glycogen. We have described most of the components of the glycogen cascade, and in this section we will discuss its operation and the principles involved.

2. Hormonal stimulation of glycogen breakdown is initiated by the binding of hormones to their specific receptors. Glucagon binding to its receptors on the liver cell, or epinephrine to the muscle cells, activates adenylate cyclase and causes its conversion to cAMP. A separate enzyme, known as a diesterase, removes the cAMP when the hormonal stimulus is removed. The newly formed second messenger, cAMP, binds to the regulatory subunits of the protein kinase, releasing the active catalytic subunits, which can now cause the phosphorylation of the β and α subunits (in that order) of phosphorylase kinase. It should be noted that the phosphorylated phosphorylase kinase is not active unless sufficient calcium ion is present. In the liver, the calcium level is probably raised by hormonal action, and also the phosphorylated kinase is more sensitive to calcium and can act with the normal low cellular level. In muscle it is unlikely that there is much phosphorylase kinase activity until calcium is released from intracellular stores by the action of the same nerve impulse that caused muscular contraction. Thus we see a vital link between contraction and the breakdown of glycogen, which provides energy for it. Even in the absence of hormonal stimulation, calcium can cause sufficient stimulation of phosphorylase kinase to promote glycogen breakdown, a good provision because epinephrine release is not normally associated with many forms of exercise (e.g., walking up stairs), but a large amount of energy is required. In Fig. 6 we attempt to show that nervous stimulation of muscles causes not only the contraction but also the breakdown of glycogen to provide energy for the contraction, and the unifying link between the two processes is the release of calcium ion.

The next step in the glycogen cascade is the action of phosphorylase kinase to transform phosphorylase *b* to *a* by phosphorylation of serine 14 of the phosphorylase. We have already discussed the fact that this causes the physiological activation of phosphorylase, so that it can now carry out the conversion of glycogen into glucose 1-phosphate, leading to glucose 6-phosphate in muscle and free glucose in liver.

3. Amplification of signal is one of the most interesting features of the glycogen cascade. Four different enzymes are interposed between the hormonal signal and the phosphorolysis of glycogen, and the first three of these act, directly or indirectly, to activate the next enzyme in the sequence. Because an enzyme is a biological catalyst and each molecule acts on many molecules of its substrate per second, each time an enzyme

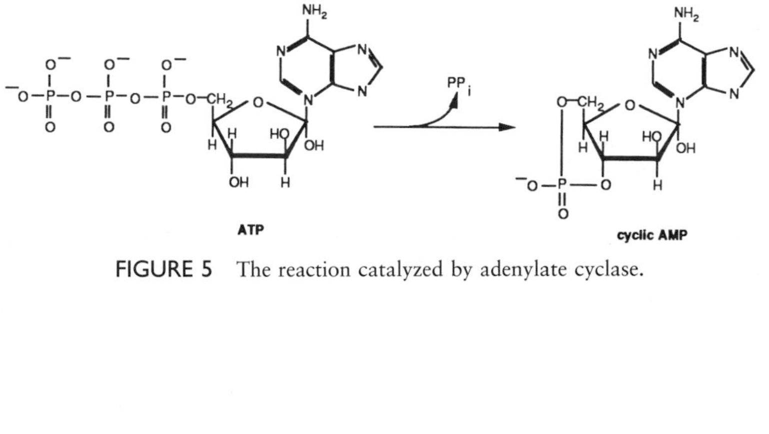

FIGURE 5 The reaction catalyzed by adenylate cyclase.

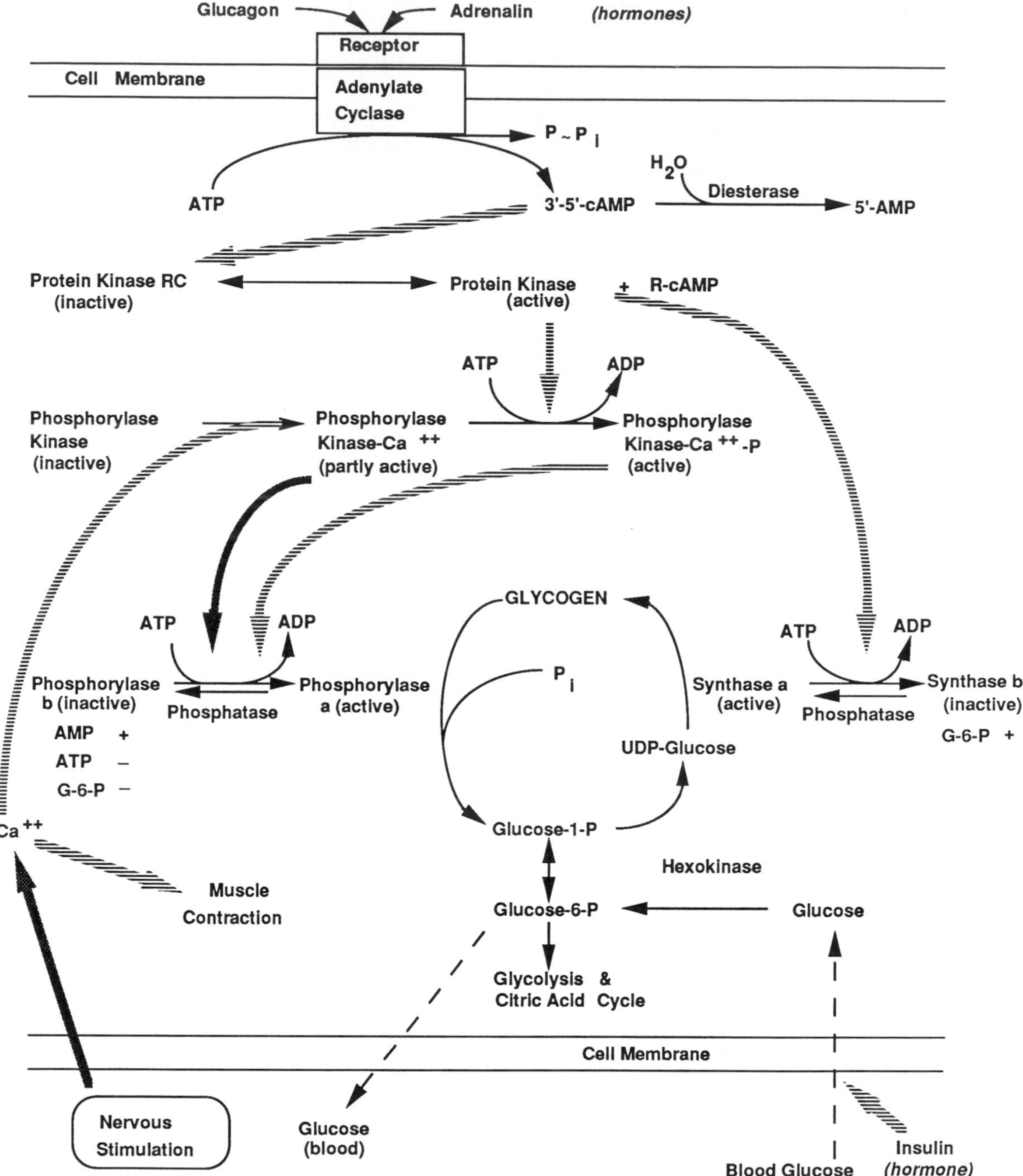

FIGURE 6 Scheme to illustrate the control of glycogen metabolism, including the glycogen cascade (turn on of glycogen breakdown in response to glucagon, or epinephrine, and calcium) and the effect of insulin.

in the cascade is "turned on," it causes an amplification of the signal, much as turning on a radio tube or transistor results in a current far more powerful than the activating current. The threefold repetition of this type of amplification in the glycogen cascade allows a small concentration of hormone in the blood, less than 1×10^{-8} *M*, to activate 2×10^{-7} *M* protein kinase, thence 4×10^{-6} *M* phosphorylase kinase and finally 1×10^{-4} *M* glycogen phosphorylase (concentrations in muscle tissue). In response, the rate of glycogen breakdown caused by strenuous muscular exercise (e.g., sprinting) increases by about 1000-fold, from 0.05 μmole/min/g of tissue in resting muscle to 60 μmole/min. A concomitant feature of the glycogen cascade is that it takes a little time for the maximal rate of glycogen breakdown to be achieved, perhaps 2–3 sec, because the signal has to be passed through several steps.

4. Reciprocal effects on the breakdown and synthesis of glycogen are built into the regulation of glycogen metabolism and are illustrated to some extent in Fig. 6. The first and most important of these is that cAMP-dependent protein kinase phosphorylates not only phosphorylase kinase but also glycogen synthase, resulting in the inactivation of the latter under physiological conditions. This feature makes obvious sense because it prevents the resynthesis of glycogen from glucose phosphates when the latter are needed elsewhere. Phosphorylase kinase can also act on glycogen synthase to ensure further that it is switched off when glycogen breakdown is required.

5. Insulin also has a profound effect on glycogen metabolism, an effect usually opposite to the two hormones epinephrine and glucagon, which turn on glycogen breakdown. In contrast to glucagon, insulin is released from the pancreas in response to high blood sugar, such as occurs after a meal. In some tissues (e.g., muscle), it facilitates the transport of glucose across the cell membrane, as illustrated in Fig. 6. In liver it induces the formation via protein synthesis of a tissue-specific form of hexokinase, leading to the rapid phosphorylation of glucose to glucose 6-phosphate. Finally, it activates the protein phosphatase, which acts on both phosphorylase and glycogen synthase, resulting in the inactivation of the former and the activation of the latter. It may be seen that all of these effects of insulin are consistent with the increase in glycogen synthesis, which has long been known to accompany insulin secretion. It may be added that although insulin leads to inactivation of the protein phosphatases that hydrolyze phosphate off the serine residues of proteins, hormones that raise cAMP levels cause the indirect inhibition of these phosphatases, another example of the antagonism between the two types of hormones. [*See* Insulin and Glucagon.]

IV. STRUCTURE–FUNCTION RELATIONS IN PHOSPHORYLASE

A. Structure of the Molecule in the Crystal

The powerful technique of X-ray crystallography permits the determination of protein structure to atomic resolution, 2 Å or less, so that the position of the atoms can be determined in three-dimensional space. Phosphorylase is the only one of the enzymes concerned with glycogen metabolism to have had its structure determined, and this was a formidable task because there are 842 amino acid residues in each subunit, plus the coenzyme and a phosphate, meaning that there are almost 7000 nonhydrogen atoms (determining the position of hydrogens in proteins is beyond our capability at this time). The phosphorylase structure has considerable general interest because it is under so many different kinds of control and binds so many different types of molecules, and we can gain insights into these features by examining the data derived from X-ray crystallography.

Color Plate 2 shows a computer graphics representation of the normal dimeric form of the phosphorylase *a* molecule obtained by X-ray crystallography, with each sphere representing one nonhydrogen atom but slightly larger than usual to represent the hydrogens we know are present. In this representation of the dimer, one subunit is colored red and its identical neighbor is blue. The two subunits are related by a twofold axis of symmetry such that if we take any feature of one and rotate it by 180° about the center, the identical feature is generated on the other subunit. Note how intimately the two subunits interact at what we call the subunit interface. It is hard to visualize how we could alter the shape of one subunit without forcing the other to also change shape, thus conserving symmetry. In fact, this intuitively obvious physical fact is the basis of allosteric cooperativity. When a small molecule binds to a specific site on one subunit and forces a change in conformation, the other subunit must also change shape, making it easier for a second molecule to bind to its specific site.

It is possible to determine the location of the binding sites for small molecules and proteins, as well as details of their interactions, by X-ray crystallography. For example, the binding of AMP to phosphorylase *a* was determined by soaking a crystal in a solution of AMP, thus allowing the nucleotide to diffuse through the solvent channels between the protein molecules and bind to its specific sites, one per subunit. When the crystal was analyzed by X rays and an electron density map calculated, the extra density caused by the AMP was found to be superimposed on that because of the native protein. By this means, the binding sites of the substrates, activators, and inhibitors of phosphorylase could all be located. On Color Plate 2 we indicate the entrance to the active site, the actual site where the glucose 1-phosphate or phosphate substrates bind, close to the phosphate of the pyridoxal 5′-phosphate, located deep within the subunit. Incidentally, this location of the active site explains why phosphorylase can only remove glucose units from glycogen to within four from a branch point, because it takes four units to reach from the outside to the site of action. Glucose also binds within this active site, occupying much the same position as the glucose part of glucose 1-phosphate, and so it inhibits by competing with the substrate. Also indicated on Color Plate 2 is the binding site for glycogen, determined by the binding of short glucose polymers. This is the site by which phosphorylase binds tightly to some of the terminal branches of the large glycogen molecule, but it acts on other terminal branches that can go in and out of the active site while the enzyme remains bound to the glycogen. We know from kinetic studies, not crystallographic, that glycogen synthase also has binding sites for glycogen separate from its catalytic sites.

We call the side of the phosphorylase dimer shown in Color Plate 2a the catalytic face of phosphorylase because it contains the active sites and also it is this side that binds to the substrate glycogen. If we turn the molecule around by 180°, we look at the opposite side, which we term the *control face* because on this side is found the binding site for AMP as well as the serine 14 that becomes phosphorylated. The control face of the dimer is depicted in Color Plate 2b, in a different molecular graphics coding than that used in Color Plate 2a. On this model we indicate the site where AMP binds when present, the phosphate of serine 14 in phosphorylase *a*, and the amino-terminal peptide that contains the serine 14.

B. Regulation of Phosphorylase Studied at Atomic Resolution

We have mentioned that glucose inhibits by competing with glucose 1-phosphate or inorganic phosphate at the active site. The crystals grown in the presence of glucose contain an inactive form of phosphorylase *a*, not only because of the presence of glucose but also because glucose stabilizes a loop of the polypeptide chain, which must move to allow the phosphate substrates to bind. The conformation of phosphorylase *a* induced by glucose is the inactive T state. Substrates and AMP counteract glucose and stabilize the active R state. Phosphorylase phosphatase binds to the R state but cannot act on the serine 14 phosphate because the latter is tucked into a fold in the protein. In the glucose-inhibited T state, the phosphate of serine 14 is pushed outward enough by the binding of glucose to allow phosphorylase phosphatase to act on it. In the liver, when the glucose level rises after a meal, this change provides a mechanism to inactivate phosphorylase *a* by converting it to phosphorylase *b*, ensuring that the buildup of glycogen in response to increased glucose will not be hindered. Furthermore, the phosphorylase phosphatase, after inactivating the phosphorylase, can then activate glycogen synthase. These metabolite controls on liver glycogen metabolism are in addition to the hormonal controls discussed earlier, and the exact relation between the two is still being worked out. However, via X-ray crystallography, we can see the changes in the molecule induced by the competing inhibitors, activators, and substrates, and thus provide a molecular explanation for the regulation of the phosphatase. It is tempting to speculate that the inhibition by glucose may have been a primitive feedback regulation that has evolved into a regulation of the interconversion of active and inactive phosphorylases.

It has been mentioned before that glucose 6-phosphate, which can be considered the metabolic end product of glycogen breakdown in the muscle, inhibits phosphorylase *b* by competing for the same site as does AMP. Phosphorylase *b* normally exists in the inactive T state, but AMP causes a conformational change leading to the active R state, which binds substrate much better, whereas substrate improves the binding of AMP. By X-ray crystallography we can see the communication between the AMP and catalytic sites, some 30 Å apart, as evidenced by changes in the connecting protein structure. ATP also inhibits phosphorylase *b* by binding to the AMP site and preventing the latter from binding and activating. Thus we have a molecular mechanism for the control of phosphorylase by energy demands, as discussed in Section III,A.

The most sophisticated regulation of phosphorylase is the conversion between the *b* and *a* forms by covalent phosphorylation and dephosphorylation, respectively. Because both forms of phosphorylase have had their structures determined independently to high resolution by scientists in two different laboratories, who then compared their results, we can assess the effects of phosphorylating serine 14 on the structure and hence interpret the effects on function. In phosphorylase *b*, the first 16 residues from the N terminus cannot be "seen" crystallographically, and this, with other evidence from protein chemistry, suggests that this segment of polypeptide chain is disordered and waving about, free from the main body of the protein. In phosphorylase *a*, residues 5–16 are ordered and lie across the interface between the two subunits, as pointed out in Color Plate 2, with the negative phosphate of serine 14 making charge interactions with positive arginine residues from both subunits. This extra contact between the subunits accounts for their lesser tendency to dissociate, and their tighter coupling of symmetry-related changes. In addition, the binding site for AMP is better formed, so that it binds AMP 100 times more strongly, and there are changes at the active site, which lead to tighter binding of substrates. The proportion of phosphorylase *a* in the active R conformation is 100 times more than for phosphorylase *b*, accounting for its "escape" from allosteric controls, discussed earlier, and the energy for this conformational shift is derived from the neutralization of positive charges by the phosphate on serine 14, allowing the amino-terminal peptide to bind the subunits more tightly in a more favorable conformation. The X-ray crystallographic studies on phosphorylase are beginning to provide answers on the mechanisms for regulation of enzymes via both covalent modification and allostery, and we can look forward to more detailed explanations in the future.

BIBLIOGRAPHY

Boyer, P. D., and Krebs, E. G. (eds.) (1986). "Control by Phosphorylation. The Enzymes." 3rd Ed., Vol. XVII, Chaps. 1, 3, 8–12. Academic Press, New York.

Madsen, N. B. (1990). Glycogen phosphorylase and glycogen synthetase. *In* "A Study of Enzymes" (S. A. Kuby, ed.), Vol. II. CRC Press, Boca Raton, FL.

Newsholme, E. A., and Leech, A. R. (1983). "Biochemistry for the Medical Sciences," Chaps. 5, 7, 9, 11, and 16. John Wiley & Sons, New York.

Sprang, S. R., Acharya, K. R., Goldsmith, E. J., *et al.* (1988). Structural changes in glycogen phosphorylase induced by phosphorylation. *Nature* **331**, 215–221.

Stryer, L. (1988). "Biochemistry," 3rd Ed., Chap. 19. Freeman, New York.

Glycolysis

EFRAIM RACKER[1]
Division of Biological Sciences, Cornell University–Ithaca

GLOSSARY

Allostery A control mechanism in enzyme activity (or other protein functions) by occupation of an allosteric effector at a site in the protein other than the catalytic site

Fermentation Any anaerobic process that breaks down organic compounds, resulting in the generation of ATP

High-energy phosphate compounds Conventional term for phosphorylated compounds (e.g., acetylphosphate or phosphoenolpyruvate) that have a large heat of hydrolysis, ΔH (8–12 kcal/mol), and are thermodynamically suited for a transfer of the phosphoryl group to ADP by phosphotransferases

Homeostasis Mechanism in which the steady-state concentration of a biological compound (e.g., blood sugar) is kept constant, usually with the expenditure of energy

Oxidative or Photophosphorylation Process catalyzed by mitochondria, bacteria, or chloroplasts in which the transport of electrons either from food substrates or from H_2O (using light energy) generates a proton-motive force (proton gradient plus membrane potential) that is used to synthesize ATP from ATP and P_i

Pasteur effect An important control mechanism of feedback inhibition in which ATP, the bioenergetic product of glycolysis, inhibits the utilization of sugar

Product inhibition Control mechanism in which the product of an enzyme-catalyzed reaction inhibits the enzyme

Rate-limiting step Reaction in a pathway which acts as a pacemaker (e.g., an enzyme that increases the rate of metabolism when its activity is increased); there can be more than one pacemaker in a pathway

"Tight coupling" Expression used to describe the compulsory linkage of oxidative processes during glycolysis or during electron transport to the synthesis of ATP; represents the basic economic principle in bioenergetics

Nature's economy is clever, simple and beautiful. ATP is made only when it is needed. Man's economy is crazy. Products that are not needed, are advertised on TV and given to me as Christmas presents (e.g., electric backscratchers).

GLYCOLYSIS IS A CELLULAR PATHWAY THAT CATAlyzes the conversion of glucose to lactic acid. The function of this pathway is to convert the energy stored in glucose into adenosine triphosphate (ATP), which is the most universal and important energy currency in living cells. Each molecule of glucose converted to two molecules of lactic acid yields two molecules of ATP. Heating glucose in a test tube in the presence of sodium hydroxide will also yield lactic acid, but the energy is released as heat and cannot be used by cells to synthesize proteins, nucleic acids, and other cellular ingredients. Moreover, glycolysis yields only lactic acid, whereas heat yields a mixture of com-

[1]Deceased.

pounds. Glycolysis is a complex pathway involving about a dozen enzymes and several cofactors. Students of evolution who propose that glycolysis was the primary mode of energy generation when life on earth emerged should be reminded of Renee Dubos, who once pointed out that if this were true, glycolysis must have sprung from the forehead of Zeus, like Minerva, fully armed. More likely, gradients of sodium or proton ions were used as primary sources of energy when we first started. In the course of evolution, however, glycolysis has emerged as the major pathway in the generation of ATP. In the absence of air, glycolysis is not a very efficient process, but in the presence of air, pyruvate, one of its end products, instead of being reduced to lactate, is burned into CO_2 and water via the Krebs cycle by the very efficient pathway of oxidative phosphorylation in mitochondria, which are specialized organelles present in all eukaryotic cells. By generating a proton gradient (i.e., a difference in hydrogen ion concentration across a membrane) during oxidation, glycolysis allows the production of 32 molecules of ATP per molecule of glucose metabolized.

Glucose itself is an intermediate in the flow of energy in the living world. The energy of the sun, transmitted as light, is utilized by plants and other photosynthetic cells in other organelles, the chloroplasts, or in bacteria to cleave water and to generate via an electron transport chain, similar to that found in mitochondria, a proton motive force. This proton-motive force, consisting of a proton gradient and an electrical potential across the membrane, is utilized by an ATP-driven proton pump acting in reverse to synthesize ATP from ADP and inorganic phosphate (P_i). This ATP then catalyzes energy-requiring processes in the cell. Thus, in living cells there are three major pathways for the synthesis of ATP: photophosphorylation, oxidative phosphorylation, and glycolysis or fermentation. Fermentation is a broader term for the breakdown of a variety of sugars and other organic molecules, leading to the production of ATP without the direct participation of oxygen. The fragments generated by fermentation contain less energy than the initial (e.g., carbohydrate) molecule, and this energy difference is utilized to synthesize ATP from ADP and P_i. The products of fermentation vary among living cells, of which yeast is perhaps the most popular, fermenting glucose or sucrose to ethyl alcohol and living happily ever after.

Space limitations do not permit an in-depth discussion of the history of the elucidation of the glycolytic pathway. It is a fascinating account, though, of the glories and failings of the human mind, from which students can derive many lessons. It was believed in the early decades of this century that methylglyoxal was a key intermediate in the formation of lactic acid from glucose. This widely accepted hypothesis, which served to block further progress, was eliminated by showing that the removal of glutathione, a necessary cofactor for the conversion of methylglyoxal to lactic acid by dialysis from a crude yeast extract, failed to affect glycolysis. The elucidation of the individual steps of the glycolytic pathway served as a role model for subsequent analyses of multistep pathways, including that of the oxidation of pyruvate in mitochondria (formulated by Sir Hans Krebs). Mitochondrial oxidative phosphorylation is quantitatively the most important mechanism in animal cells for ATP synthesis. By oxidizing the carbons of pyruvate stepwise one by one to water and CO_2 via the many intermediates of the Krebs cycle, nature invented a machine more efficient than most made by humans.

Before discussing the pathway of glycolysis, with its intricate catalysts, described in minute detail in textbooks of biochemistry, the Pasteur effect should be discussed: This is a very important feature of glycolysis regulation that is studiously deleted from modern textbooks. In 1861, Pasteur published the historical discovery that, per gram of glucose utilized, more yeast is formed in the presence of air than in its absence. These experiments demonstrated that, as in New York City, "life without air" is possible, but expensive. It was the first demonstration that aerobic metabolism (via oxidative phosphorylation) is more efficient than anaerobic glycolysis. Pasteur also observed that, in the absence of air, glucose was more rapidly used per gram of yeast than in air, thereby compensating for the inefficient fermentation process. The inhabition of glucose utilization by oxygen was called the "Pasteur effect" and subsequently was found to be a major regulatory mechanism that serves for the preservation of food throughout the living world. The mechanism of the Pasteur effect will be referred to again at the end of this article. A brief recapitulation of the Pasteur effect is shown below.

1. Glucose is used up more slowly aerobically than anaerobically.
2. However, yeast grows faster aerobically than anaerobically.
3. From this, it must be concluded that the aerobic cell converts glucose into biological energy (ATP) more efficiently than does the anaerobic cell.

4. Cells must therefore possess a mechanism by which glucose utilization is suppressed under aerobic conditions.

I. GLYCOLYTIC PATHWAY: AN INTRODUCTION

Some bacteria, such as the pneumococcus that causes pneumonia in humans, use glycolysis as their only source of energy. Colin MacLeod, an outstanding American biologist who participated in the revolutionary discovery of nucleic acids as the transmitters of genetic information in pneumococci, used to marvel that his beloved pneumococcus utilized glucose by the same enzymatic pathway as the brain of Plato. But instead of oxidizing the end product of glycolysis, the less inspired fermenting bacteria must get rid of the waste products by devising a specific transporter that excretes lactate together with a proton. They must also dissipate the heat generated by this inefficient mechanism. Most bacteria are better off though because they also have oxidative or photosynthetic pathways. Some have additional mechanisms of ATP formation, for example, by the phosphoroclastic cleavage (in which P_i cleaves a carbon–carbon bond) of pyruvate or of fructose-6-phosphate yielding acetylphosphate, which is used as the donor in the formation of ATP from ADP. [*See* Adenosine Triphosphate (ATP).]

For didactic purposes, glycolysis has been divided into five stages, discussed in the following sections.

II. STAGE I: THE OVERTURE

Table I shows that the first stage involves the expenditure of two ATP molecules, which transform glucose into fructose 1,6-bisphosphate. It may seem a funny way to conduct the business of making ATP by first spending it. Quite a bit of ATP energy is wasted in heat during these two steps. The hydrolysis of the sugar phosphate esters ("low-energy phosphates") liberates much less energy than the hydrolysis of ATP (a "high-energy phosphate compound"). In other words, the "energy market value" of simple phosphate esters is much lower than that of the phosphate anhydride in ATP. Why have such wasteful steps survived during the evolution from the pneumococcus to Plato? What are the fringe benefits derived from these steps to justify the waste?

As it turns out, these ATP expenditures are good investments and represent good economy. First, as we shall see, the investment loss is not as bad as it may seem because the phosphorylation steps induce changes in the sugar molecules that permit later, in a single pathway, the recovery of ATP by clever oxidation steps.

A second major benefit to the cells is that by phosphorylation the sugar becomes a carrier of the negatively charged phosphate groups. Bernard Davis, a well-known Harvard biochemist, once wrote an article entitled "On the Importance of Being Ionized." Like Oscar Wilde, Davis is a man of simple tastes; he is always satisfied with the best. And for a cell the best is to make sure that the food that is injested is not allowed to leave. Charged ions do not readily traverse the phospholipid bilayers of the plasma membrane without a passport—a specifically designed transporter, such as the previously mentioned proton–lactate channel or the phosphate transporter. In contrast, uncharged molecules such as glycerol or ethyl alcohol are readily lost from the cells if they are not rapidly metabolized. This as taken advantage of during World War I by diverting yeast to make glycerol from sugar and by using the excreted glycerol for the production of explosives.

The first step in stage I is the conversion of glucose to glucose-6-phosphate by an enzyme called hexokinase. The discovery of this enzyme by Otto Meyerhof is believed to be the first reconstitution experiment in history. Meyerhof added to an extract from muscle that was capable of degrading glycogen, but not glucose, to lactate, a protein fraction from yeast, incapable of fermentation. Together, the two preparations glycolyzed vigorously. Thus, the muscle extract, which does not need hexokinase for generating glucose-6-phosphate from glycogen, served as an assay in the purification hexokinase present in the yeast fraction. Where there is an assay and a will, there is a way to purification. Years later, simpler assays were devised and crystalline hexokinase was isolated.

The second step in stage 1 of glycolysis is the conversion of glucose-6-phosphate to fructose-6-phosphate by an enzyme called glucose-phosphate isomerase. The third step is the phosphorylation of fructose-6-phosphate to fructose 1,6-bisphosphate.

The enzyme that achieves the phosphorylation of fructose-6-phosphate into fructose 1,6-bisphosphate, with the expenditure of yet another ATP, is called 6-phosphofructokinase. This protein contains two special regions: One acts as a catalyst, i.e., it brings

TABLE I

Flow of Glycolysis[a]

Stage 1: From Glucose to Fructose 1,6-Bisphosphate[b]

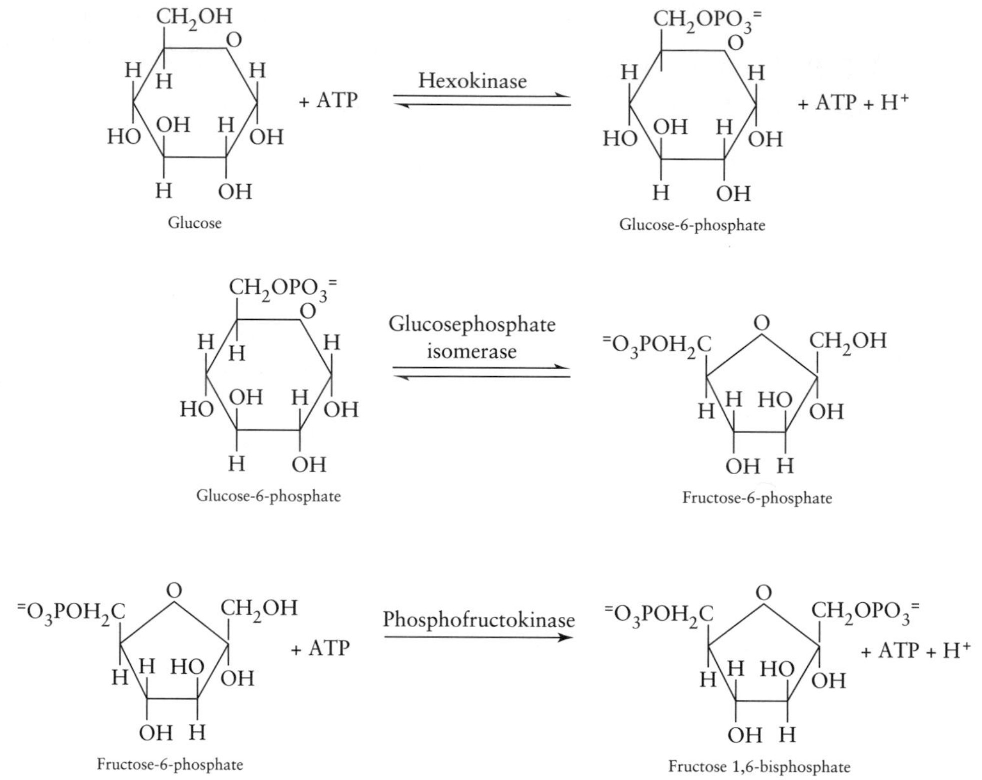

Glucose + ATP $\xrightleftharpoons{\text{Hexokinase}}$ Glucose-6-phosphate + ATP + H^+

Glucose (CH_2OH; H, H, H, HO, OH, H, OH, H, OH)

Glucose-6-phosphate ($CH_2OPO_3^=$; H, H, HO, OH, H, OH, H, OH)

Glucose-6-phosphate $\xrightleftharpoons{\text{Glucosephosphate isomerase}}$ Fructose-6-phosphate

Glucose-6-phosphate ($CH_2OPO_3^=$; H, H, H, HO, OH, H, OH, H, OH)

Fructose-6-phosphate ($^=O_3POH_2C$, O, CH_2OH; H, H, HO, OH, OH, H)

Fructose-6-phosphate + ATP $\xrightarrow{\text{Phosphofructokinase}}$ Fructose 1,6-bisphosphate + ATP + H^+

Fructose-6-phosphate ($^=O_3POH_2C$, O, CH_2OH; H, H, HO, OH, OH, H)

Fructose 1,6-bisphosphate ($^=O_3POH_2C$, O, $CH_2OPO_3^=$; H, H, HO, OH, OH, H)

Stage 2: Fructose 1,6-Bisphosphate to Glyceraldehyde-3-phosphate[c]

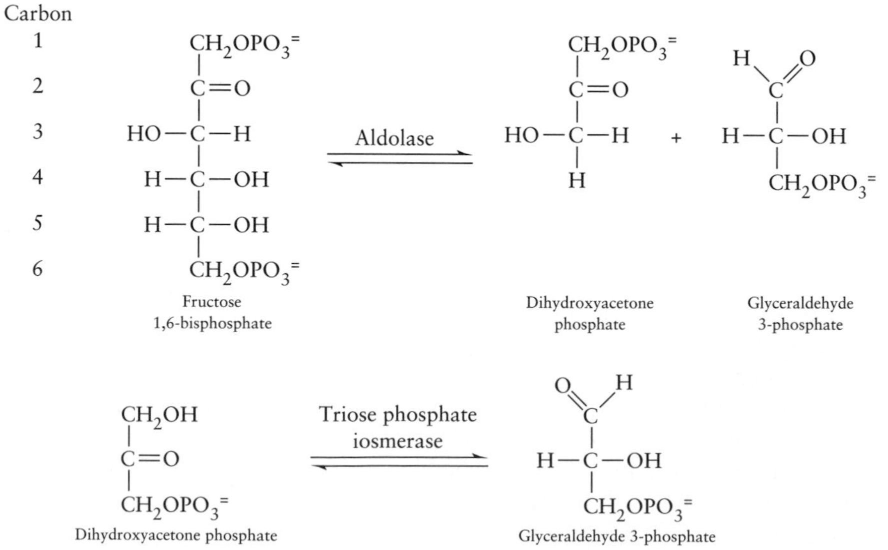

Carbon	Fructose 1,6-bisphosphate
1	$CH_2OPO_3^=$
2	$C{=}O$
3	$HO{-}C{-}H$
4	$H{-}C{-}OH$
5	$H{-}C{-}OH$
6	$CH_2OPO_3^=$

Fructose 1,6-bisphosphate $\xrightleftharpoons{\text{Aldolase}}$ Dihydroxyacetone phosphate + Glyceraldehyde 3-phosphate

Dihydroxyacetone phosphate: $CH_2OPO_3^=$ — $C{=}O$ — $HO{-}C{-}H$ — H

Glyceraldehyde 3-phosphate: $H{-}C{=}O$ — $H{-}C{-}OH$ — $CH_2OPO_3^=$

Dihydroxyacetone phosphate $\xrightleftharpoons{\text{Triose phosphate iosmerase}}$ Glyceraldehyde 3-phosphate

Dihydroxyacetone phosphate: CH_2OH — $C{=}O$ — $CH_2OPO_3^=$

Glyceraldehyde 3-phosphate: $O{=}C{-}H$ — $H{-}C{-}OH$ — $CH_2OPO_3^=$

Stage 3: From Glyceraldehyde 3-phosphate to Phosphoglycerated[d]

$$\underset{\text{Glyceraldehyde 3-phosphate}}{\begin{array}{c}\mathrm{CHO}\\|\\\mathrm{H{-}C{-}OH}\\|\\\mathrm{CH_2OPO_3^{=}}\end{array}} + \mathrm{P_1} + \mathrm{NAD^+} \xrightarrow[\text{dehyrogenase}]{\text{Glyceraldehyde 3-phosphate}} \underset{\text{Diphosphoglycerate}}{\begin{array}{c}\mathrm{O{=}COPO_3^{=}}\\|\\\mathrm{H{-}C{-}OH}\\|\\\mathrm{CH_2OPO_3^{=}}\end{array}} + \mathrm{NADH_2}$$

$$\underset{\text{1,3-Diphosphoglycerate}}{\begin{array}{c}\mathrm{O{=}C{-}OPO_3^{=}}\\|\\\mathrm{H{-}C{-}OH}\\|\\\mathrm{CH_2OPO_3^{=}}\end{array}} + \mathrm{ADP} \underset{\text{kinase}}{\overset{\text{Phosphoglycerate}}{\rightleftharpoons}} \underset{\text{3-Phosphoglycerate}}{\begin{array}{c}\mathrm{O{=}C{-}O^-}\\|\\\mathrm{H{-}C{-}OH}\\|\\\mathrm{CH_2OPO_3^{=}}\end{array}} + \mathrm{ATP}$$

Stage 4: From Phosphoglycerate to Pyruvate[e]

$$\underset{\text{3-Phosphoglycerate}}{\begin{array}{c}\mathrm{O{=}C{-}O^-}\\|\\\mathrm{H{-}C{-}OH}\\|\\\mathrm{H{-}C{-}OPO_3^{=}}\\|\\\mathrm{H}\end{array}} \overset{\text{Phosphoglyceromutase}}{\rightleftharpoons} \underset{\text{2-Phosphoglycerate}}{\begin{array}{c}\mathrm{O{=}C{-}O^-}\\|\\\mathrm{H{-}C{-}OPO_3^{=}}\\|\\\mathrm{H{-}C{-}OH}\\|\\\mathrm{H}\end{array}}$$

$$\underset{\text{2-Phosphoglycerate}}{\begin{array}{c}\mathrm{O{=}C{-}O^-}\\|\\\mathrm{H{-}C{-}OPO_3^{=}}\\|\\\mathrm{H{-}C{-}OH}\\|\\\mathrm{H}\end{array}} \overset{\text{Enolase}}{\rightleftharpoons} \underset{\text{Phosphoenolpyruvate}}{\begin{array}{c}\mathrm{O{=}C{-}O^-}\\|\\\mathrm{C{-}OPO_3^{=}}\\\|\\\mathrm{H{-}C}\\|\\\mathrm{H}\end{array}} + \mathrm{H_2O}$$

$$\underset{\text{Phosphoenolpyruvate}}{\begin{array}{c}\mathrm{O{=}C{-}O^-}\\|\\\mathrm{C{-}OPO_3^{=}}\\\|\\\mathrm{CH_2}\end{array}} + \mathrm{ADP} \underset{\text{kinase}}{\overset{\text{Pyruvate}}{\rightleftharpoons}} \underset{\text{Pyruvate}}{\begin{array}{c}\mathrm{O{=}C{-}O^-}\\|\\\mathrm{C{=}O}\\|\\\mathrm{CH_3}\end{array}} + \mathrm{ATP}$$

Stage 5: Reduction of Pyruvate to Lactate[f]

$$\begin{array}{c}\mathrm{COOH}\\|\\\mathrm{C{=}O}\\|\\\mathrm{CH_3}\end{array} + \mathrm{NADH_2} \underset{\text{dehydrogenase}}{\overset{\text{Lactate}}{\rightleftharpoons}} \begin{array}{c}\mathrm{COOH}\\|\\\mathrm{HCOH}\\|\\\mathrm{CH_3}\end{array} + \mathrm{NAD^+}$$

[a]The sugar structures are presented in the ring form as well as in the open chain form. In nature, we find a mixture of both forms.
[b]Stage 1. The overture: Conversion of glucose to fructose 1,6-bisphosphate.
[c]Stage 2. Cleavage of the sugar molecule: Cleavage of fructose 1,6-diphosphate, which contains six carbon atoms, to dihydroxyacetone phosphate and glyceraldehyde-3-phosphate, each having three carbon atoms. This step is followed by the conversion of dihydroxyacetone phosphate to glyceraldehyde-3-phosphate.
[d]Stage 3. Energy harvest 1: Oxidation of glyceraldehyde-3-phosphate to 3-phosphoglycerate and formation of the first ATP.
[e]Stage 4. Energy harvest II: Conversion of 3-phosphoglycerate to pyruvate and formation of the second ATP.
[f]Stage 5. Epilogue: Reduction of pyruvate to lactate or to other fermentation products in microorganisms such as ethyl alcohol.

the two substrates—fructose-6-phosphate and ATP—close together and allows them to interreact to yield ADP and fructose 1,6-bisphosphate. The second site binds a variety of cellular molecules called effectors, such as ATP, ADP, fructose 2,6-bisphosphate, phosphocreatine, citrate, or inorganic phosphate. As a result of this binding, the shape of the protein molecule changes and the activity of the catalytic site may be either decreased or increased. In other words, the second (noncatalytic) site is a controlling or allosteric site. Phosphofructokinase is one of the most important allosteric enzymes in glycolysis because an inhibition of its action diminishes the utilization of sugar and thereby allows the preservation of energy reserves. By lowering the activity of 6-phosphofructokinase, fructose-6-phosphate and glucose-6-phosphate accumulate. The latter compound, in turn, inhibits the activity of hexokinase, a phenomenon called product inhibition, another simple widely used mechanism in metabolic regulations. Thus, the cells break down only the amount of sugar that is needed for making the amount of ATP energy required for living. The most important compound acting as a negative allosteric effector of phosphofructokinase is ATP. When too much ATP is generated and not enough P_i and ADP are available, phosphofructokinase is inhibited. The enzyme also responds to positive allosteric effectors, compounds that turn the enzyme on again. P_i, ADP, AMP, and cAMP are such positive effectors. When the cell needs a lot of energy and breaks down ATP, it generates ADP and P_i, which in turn activate the machinery of glycolysis that replenishes the ATP level.

III. STAGE 2: CLEAVAGE OF FRUCTOSE 1,6-BISPHOSPHATE TO DIHYDROXYACETONE PHOSPHATE AND GLYCERALDEHYDE-3-PHOSPHATE

The formation of fructose 1,6-bisphosphate in the previous stage represents the clever synthesis of an almost symmetrical molecule that is cleaved by the enzyme aldolase into two interconvertible triosephosphates. This is a clever process because it allows the cell to deal eventually with a single substrate, glyceraldehyde-3-phosphate. If the second expenditure of ATP was not invested, the cleavage of the hexose monophosphate would yield one triose and one triosephosphate and would require two separate sets of enzymes to deal with the two different substrates.

Note that carbons 3 and 4 of the fructose 1,6-bisphosphate molecule become carbon 1 of glyceraldehyde-3-phosphate, carbons 1 and 6 become carbon 3, and carbons 2 and 5 become carbon 2. This feature is important for investigations of the fate of the individual carbons by using glucose labeled at a single carbon site with radioactive carbon 14.

IV. ENERGY HARVEST I—STAGE 3: OXIDATION OF GLYCERALDEHYDE-3-PHOSPHATE TO 3-PHOSPHOGLYCERATE AND FORMATION OF ADENOSINE TRIPHOSPHATE (ATP)

In organic chemistry, the oxidation of an aldehyde to an acid is usually obtained by the intervention of some oxidizing agent, such as bromine. The energy resulting from the oxidation step, however, is lost as heat.

The enzyme glyceraldehyde-3-phosphate dehydrogenase is more ingenious. The simplest formulation of its mode of action is as follows (see Table II): The enzyme (E) that contains a sulfhydryl group (SH) interacts with the aldehyde group of glyceraldehyde-3-phosphate, yielding a product shown in Table II that is oxidized to an enzyme-bound thiolester. In the oxidation step (step 2), NAD^+ serves as the hydrogen acceptor to form $NADH_2$ ($NADH + H^+$). This is the key reaction of glycolysis. Instead of dissipating as heat, the energy of oxidation is preserved by formation of the energy-rich thiolester substrate–enzyme intermediate. A thiolester substrate–enzyme intermediate. A thiolester is a "high-energy" compound because it releases about 10 kcal of heat when hydrolyzed by water. The cell preserves the energy of oxidation of the aldehyde to the acid by allowing P_i to enter instead of water (step 3). This process, called phosphorolysis (instead of hydrolysis), results in the formation of a new high-energy intermediate, 1,3-bisphosphoglycerate. With the help of an enzyme called phosphoglycerate kinase, the phosphate attached via an oxygen at carbon 1 is transferred to ADP to form ATP (phosphotransfer). The steps of oxidation, phosphorolysis, and phosphotransfer catalyzed by glyceraldehyde-3-phosphate dehydrogenase and phosphoglycerate kinase (see Table I, stage 3) are "tightly coupled." Only when the phosphorylated product is removed by phosphoglycerate kinase can another molecule of aldehyde be oxidized. This control mechanism makes certain that virtually no oxidation takes place, unless both P_i and ADP are present

TABLE II
Oxidation of Glyceraldehyde-3-phosphate to 1,3-Diphosphoglycerate[a]

$R = -CHOH-CH_2OPO_3^{=}$

Step 1	$E-SH + C(=O)(H)-R$ Enzyme + glyceraldehyde-3-phosphate	$\longrightarrow$	$E-S-C(OH)(H)-R$ Enzyme–substrate product
Step 2[b]	$E-S-C(OH)(H)-R + NAD^+$ Enzyme–substrate product	$\longrightarrow$	$E-S-C(=O)-R + NADH_2$ Enzyme thiolester
Step 3[c]	$E-S-C(=O)-R + P_i$ Enzyme thiolester	$\longrightarrow$	$E-SH + R-C(=O)-OPO_3^{=}$ Enzyme + 1,3-diphosphoglycerate
Step 4[d]	$R-C(=O)-OPO_3^{=}$ + ADP 1,3-Diphosphoglycerate	$\longrightarrow$	$R-C(=O)-OH$ + ATP 3-Phosphoglycerate

[a]E, enzyme; SH, sulfhydryl group.
[b]Oxidation.
[c]Phosphorolysis.
[d]Phosphotransfer.

to form ATP. We see here how simply and ingeniously nature regulates its energy budget and how the Pasteur effect is achieved. When there is an excess of ATP, but little ADP and P_i, there is no oxidation and the glycolytic machinery comes to a standstill. When ATP is utilized (e.g., during muscular contraction) for biosynthesis or for transport processes, P_i and ADP are liberated and gear the glycolytic machinery that regenerates ATP into motion.

V. ENERGY HARVEST II—STAGE 4: CONVERSION OF 3-PHOSPHOGLYCERATE TO PYRUVATE AND FORMATION OF ATP

The product of the oxidation and ATP-forming steps is phosphoglycerate. In a series of consecutive steps listed in Table I, the phosphate is moved from carbon 3 in 3-phosphoglycerate to carbon 2 in phosphenolpyruvate and H_2O is pushed out of the molecule. These interconversions, catalyzed by the enzymes phosphoglyceromutase and enolase, yield phosphoenolpyruvate, another high-energy compound. This is, in turn, used by the enzyme pyruvate kinase to transfer phosphate to ADP to form ATP, with formation of pyruvate. The principle of the reaction leading to ATP formation is basically the same as in stage 3, in which an oxidation of an aldehyde and a reduction of NAD^+ takes place. In stage 4, the oxidoreduction occurs within the three-carbon molecule. Carbon 2 of 2-phosphoglycerate becomes more oxidized by losing hydrogen, whereas carbon 3 is reduced to a methyl group. The pyruvate kinase reaction is another key point in glycolytic control because it is dependent on the availability of ADP. It is allosterically inhibited by ATP and is stimulated by fructose 1,6-bisphosphate. In some cells, the enzyme is phosphorylated and regulated by protein kinases that inhibit enzyme activity. Names for enzymes such as pyruvate kinase and phosphoglycerate kinase are confusing because, during glycolysis, the reactions they catalyze occur in the forward reaction, i.e., from 1,3-diphosphoglycerate to phosphoglycerate with transfer of phosphate to ADP, whereas most kinases (e.g., hexokinase) transfer the phosphate from ATP to the substrate. But the kinase term is simpler and may serve as a reminder that the reactions are reversible and in fact can operate in reverse when glycogen is formed from pyruvate.

VI. EPILOGUE—STAGE 5: REDUCTION OF PYRUVATE TO LACTATE: FORMATION OF OTHER FERMENTATION PRODUCTS IN MICROORGANISMS

Pyruvate, the product of stage 4, is reduced in the presence of lactate dehydrogenase by $NADH_2$, which was generated during the oxidation of glyceraldehyde-3-phosphate and now returns to NAD^+ and lactate is formed. Thus, the hydrogen cycle through NAD^+ closes and oxidoreductions are balanced. Two molecules of lactic acid contain the same number of oxygens and hydrogens as one molecule of glucose. The oxidoreductions take place internally, without need for molecular oxygen. In yeast, no lactate is formed. Instead, carbon 1 of pyruvate is released as CO_2, while the remaining hydrogen moves to carbon 2 to give rise to acetaldehyde containing two carbons (CH_3CHO). Acetaldehyde is then reduced by $NADH_2$ to yield ethyl alcohol (CH_3CH_2OH), the end product of alcoholic fermentation, and once more oxidation reduction balance is achieved. The oxidized form of the coenzyme (i.e., NAD^+) is regenerated by forming lactate in one case, alcohol in the other.

When things "get sour" in the winery, for example, when the fermentation vat gets contaminated with certain bacteria, acetaldehyde is oxidized to acetic acid (also known as vinegar). Pasteur cured this wine disease by eliminating the infectious bacteria. Some bacteria make formic acid, a compound also produced by ants. Some microorganisms can produce compounds such as glycerol and acetone, used in the manufacture of explosives. As mentioned earlier, during World War I, Karl Neuberg in Germany found a way to force yeasts to accumulate glycerol (CH_2OH $CHOH$ CH_2OH) instead of ethanol by chemically trapping acetaldehyde and forcing dihydroxyacetone to serve as the hydrogen acceptor for $NADH_2$. Chaim Weizmann in England developed another fermentation process in which acetone (CH_3COOH_3) was produced and was also used during World War I for the production of explosives. As in World War II, scientific discoveries can be used to enhance the viciousness of war. Some bacteria accumulate acetoin (CH_3 $CHOHCOCH_3$), which has been used as a starting material for the manufacture of synthetic rubber. Pharmaceutical companies use bacterial fermentation processes to manufacture drugs when it is either too difficult or too expensive to use chemical synthesis. Just as atomic energy, fermentation can be used for good and evil purposes.

Let us now recapitulate what happens to the energy balance during glycolysis (Table III). For each molecule of glucose, two molecules of lactate are formed. Forming fructose 1,6-bisphosphate from glucose requires expenditure of two ATP molecules. Each of the two triosephosphates, which are made by splitting fructose 1,6-bisphosphate, yields one ATP during the oxidation of glyceraldehyde-3-phosphate in the presence of ADP and P_i. The energy balance is now even; the "wasted" ATP is recovered. In stage 4, from two molecules of phosphoglycerate, two molecules of phosphoenolpyruvate are made, generating two molecules of ATP in the presence of ADP. Two ATPs for each glucose fermented have now been accumulated. Because glycolysis is a tightly coupled process, these two extra ATP molecules must be reconverted into ADP and P_i if glycolysis is to continue. ADP and P_i are cofactors of glycolysis present in a cell in relatively small amounts and must continuously be regenerated (Fig. 1). Various work processes (e.g., transport, mus-

TABLE III
Balance Sheet[a]

Expenditure		Earning
	Glucose	
2 ATP →	↓	
	Fructose 1,6-bisphosphate	
	↓	
	2 Triosephosphate	
	↓ + 2 P_i + 2 ADP	→ 2 ATP
	2 Phosphoglycerate	
	↓	
	2 Phosphoenolpyruvate	
	↓ + 2 ADP	→ 2 ATP
	2 Pyruvate	
	↓	
	2 Lactate	

[a]Overall: glucose + 2 P_i, + 2 ADP → 2 lactate + 2 ATP + 2 H_2O.

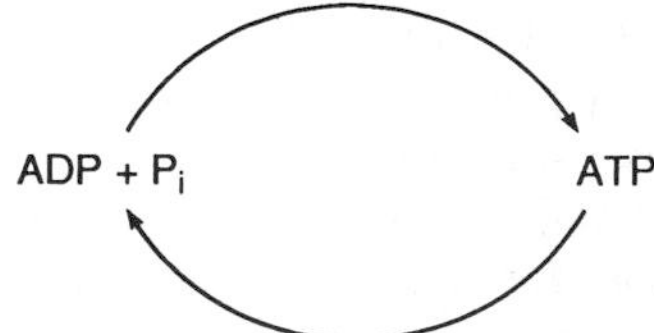

FIGURE 1 The P_i + ADP ↔ ATP cycle in glycolysis and oxidative phosphorylation.

cle contraction, and syntheses) contribute to this "ATPase" activity and thereby activate the glycolytic pathway.

From a thermodynamic point of view, the yield of two ATPs per glucose represents an overall efficiency of about 50%, which is not bad compared with man-made energy transformers. Moreover, many fringe benefits are derived from the glycolytic pathways. It provides, by diversion of intermediates, the glycerol moiety for fats, ribose and deoxyribose for nucleic acids, and some amino acids for protein synthesis. No wonder evolution has conserved this pathway and the human brain shares it with pneumococcus and with many other microorganisms and plants.

Although the basic reactions of the glycolytic pathway are indeed identical in diverse cells, there are important differences in the distribution of the individual enzymes that participate in the pathway. By varying the proportions of the catalysts, nature altered these pathways in such a manner that it appears as if different qualitative mechanisms exist. For example, sodium fluoride, which is added to toothpaste to prevent tooth decay, is a poison when too much is ingested. We know the exact location of action of fluoride. By forming a complex with the enzyme enolase and Mg^{2+}, a cofactor required for enzyme activity, it inhibits the conversion of phosphoglycerate into phosphoenolpyruvate. Some bacteria contain so little enolase that the entire glycolytic pathway is controlled by the activity of this enzyme. We say that in this case enolase is one of the "rate-limiting" steps or that it is a "pacemaker" of the pathway. Other cells, for example, some tumors, contain a huge excess of enolase, over 10 times that required. Enough fluoride can be added to such cells to poison 90% of enolase activity without interfering with the rate of glycolysis. The same amount of fluoride would wipe out glycolysis in a cell that has enolase as a pacemaker.

This basic principle of varying the relative concentration of the glycolytic enzymes is used by nature to vary the pattern and the pacemakers of glycolysis even in different tissues of the same organism. For example, the brain of mammals contains a huge amount of hexokinase, which is needed to assure a steady utilization of glucose, the principal nutrient of brain cells. Brain cells work all the time and therefore need a steady supply of energy. However, muscle cells have different requirements. When muscles work, they need a lot of energy, but they need much less during rest. As a consequence, our muscles store a reservoir of energy in the form of glycogen, a network of interconnected glucose molecules. Muscles therefore contain a large amount of an enzyme which, in the presence of P_i liberated by ATP hydrolysis during muscular contractions, breaks down glycogen to glucose-1-phosphate. The latter is converted into glucose-6-phosphate and thus enters the glycolytic pathway. Hexokinase activity in muscle is quite low, but there is enough to catalyze the slow but continuous conversion of glucose to glycogen during rest periods. [*See* Glycogen; Muscle, Physiology and Biochemistry.]

Sometimes the pattern of glycolysis is altered by inserting an additional enzyme which diverts an intermediate in another direction. For example, the liver contains an enzyme called glucose-6-phosphatase which diverts glucose-6-phosphate by hydrolyzing it to glucose and inorganic phosphate, with liberation of heat. This seems an awful waste of energy, yet the overall metabolic story makes good sense. One of the functions of the liver is the homeostasis of the glucose level of blood. This is just a fancy way of saying that the organ acts as a regulator, keeping the blood sugar level constant. For this purpose, the liver stores glycogen and has a very active glycogen-degrading enzyme, just like muscle; but, in addition, it has glucose-6-phosphatase, which is not present in muscle. When hormones such as adrenaline, which is secreted by the adrenal glands, reach the liver cell, the glycogen-degrading enzymes are activated by a remarkable complex series of events involving a secondary messenger called cAMP. As a result, glucose-1-phosphate is formed, converted to glucose-6-phosphate, and finally glucose is released into the blood.

VII. OXIDATIVE PHOSPHORYLATION

Evolution did not stop at the glycolytic pathway; it created oxidative phosphorylation, an even more complex and much more efficient machinery for energy generation. The respiratory chain of the mitochondria oxidizes the products such as pyruvate, fatty acids, and amino acids that arise from the metabolism of carbohydrates, fats, and proteins in our food. These oxidation processes generate the electrons that are transported via the respiratory chain of mitochondria or of bacteria and are tightly coupled to phosphorylation and to the generation of ATP from ADP and P_i. As in the case of glycolysis, without these two cofactors oxidations stop, a phenomenon called respiratory control. Because both glycolysis and oxidative phosphorylation are tightly coupled and require ADP and P_i, there is a fierce competition for these rate-limiting cofactors. When respiration wins out, glycolysis is inhibited and the Pasteur effect is observed; when glycolysis is overwhelming, as in some cancer cells, the Crabtree effect is observed, an inhibition of respiration. But the Pasteur effect is more complex than inhibition of the oxidation of glyceraldehyde-3-phosphate because of ADP–P_i limitation. This limits the oxidation of glyceraldehyde-3-phosphate and the formation of lactate, but how is the control of glucose utilization achieved? There is a remarkable set of allosteric control mechanisms superimposed at the sites of hexokinase and phosphofructokinase action. The latter is probably among the most regulated enzymes known in biochemistry. The checks and balances are more complex than the controls exerted by Congress and the judiciary branch on the executive branch of our government. The first set of allosteric controls was mentioned earlier: ATP inhibits, whereas P_i, ADP, and AMP stimulate. Other negative controls include phosphocreatine, citrate, and fructose 1,6-bisphosphate. A more recently discovered allosteric effector of phosphofructokinase is fructose 2,6-bisphosphate, which is a potent activator of the enzyme that has reawakened the interest of biochemists in "old hat" glycolysis. The control of the steady-state concentration of fructose 2,6-bisphophate is in itself staggering, involving several enzymes and cAMP. It is likely that fructose 2,6-bisphosphate plays a role in the high rate of aerobic glycolysis in some cancer cells. The inhibition of phosphofructokinases triggered by high ATP and low P_i and ADP results in the accumulation of fructose-6-phosphate and glucose-6-phosphate. The latter is a powerful feedback inhibitor of hexokinase, thereby affecting the utilization of glucose. This inhibition of hexokinase is perhaps also responsible for a diminished influx of glucose via the glucose transporter. This speculation is based on somewhat thin evidence provided by experiments on a phenomenon called catabolite repression, glucose utilization via glycolysis represses other catabolic pathways (e.g., the utilization of galactose). In yeast, this repression is shown to involve one form of hexokinase.

Without belittling the importance of these control mechanisms operating at hexose levels and several other controls catalyzed by protein kinases and phosphatases (see the Bibliography), the bottom line of the Pasteur effect is the competition of glycolysis and oxidative phosphorylation for ADP and P_i. The classical demonstration over five decades ago that dinitrophenol, a proton ionophore and uncoupler of oxidative phosphorylation, releases the inhibition imposed on glycolysis by respiration implicates ADP and P_i as the key regulators in the Pasteur effect. Dinitrophenol not only uncouples respiration from phosphorylation, but also stimulates the mitochondrial ATPase by releasing it from its inhibition by electron transport, to which it is normally tightly coupled.

Glycolysis is altered in diseases. There is a pathology of glycolysis in diseases of liver, muscle, or red blood cells, in which the utilization of glucose or glycogen is impaired because of a hereditary alteration of an enzyme. In most cancer cells, aerobic glycolysis is greatly enhanced, but the mechanism is still uncertain. Now, with the tools of modern molecular biology, it can be shown that transfection of a normal cell with a single oncogene (e.g., *ras*) increases the rate of glycolysis four- to fivefold. Transfections with single oncogenes should allow us to focus on a more specific field of inquiry, namely, the pathway of signal transduction induced by an oncogene. The study of such pathological situations may enhance our knowledge of the normal events in the areas of energy control and expenditures.

A great deal is known about the energy budget of glycolysis and oxidative phosphorylation, but our knowledge of the expenditure budget is pathetic. There are two major ways of generating ATP in animal cells, but thousands of ways of spending it. These expenditures, resulting in the formation of ADP and P_i, represent the sum of the "ATPase" reactions that take place in the cell and that keep glycolysis and respiration going. As noted earlier, these processes are required not only for ATP generation, but for the production of intermediates for many synthetic pathways operating in living cells. From a bioenergetic point of view, our ignorance of expenditures is tragic, but not serious because we are rich. Our potential of ATP generation by far exceeds our needs as long as nutrients are supplied and the bioenergetic machineries are not damaged. We need to know more about our expenditures in order to understand the various ATPase activities that contribute to the rate of glycolysis and are likely to be involved in the high aerobic glycolysis characteristic for most cancer cells.

BIBLIOGRAPHY

Pilkis, S. J., Raafad El-Maghrabi, M., and Claus, T. H. (1988). *Annu. Rev. Biochem.* 57, 755–783.

Racker, E. (1976). "A New Look at Mechanisms in Bioenergetics." Academic Press, New York.

Racker, E. (1985). "Reconstitutions of Transporters, Receptors, and Pathological States." Academic Press, Orlando, FL.

Stryer, L. (1995). "Biochemistry," 4th Ed. Freeman, New York.

Van Schaftingen, E. (1987). *Adv. Enzymol.* **59**, 315–395.

Voet, D., and Voet, J. G. (1995). "Biochemistry." Wiley, New York.

G Proteins

RAFAEL MATTERA
Case Western Reserve University

GLOSSARY

Agonists Synthetic analogs that interact with receptors, eliciting the biological effects normally caused by the physiological ligand (i.e., hormone or neurotransmitter)

Antagonists Synthetic analogs that interact with the receptors with high affinity and saturability, but do not produce the biological effects of the physiological ligands

cDNA Complementary DNA from a messenger RNA template rather than a DNA template

Exon Gene sequence encoding amino acids present in the gene product

Hormones Molecules that are secreted by a particular set of cells and travel through the bloodstream to produce their biological effects in distant cells (effector or target cells)

Intron Noncoding (or intervening) sequence in a given gene

Neurotransmitters Molecules involved in the conduction of electrical impulses from one nerve cell to another or from one nerve cell to muscle cells

Patch-clamp recording Measurement of single ionic channels in small (1–10 μm^2) patches of biological membranes tightly sealed to glass microelectrodes and subjected to adjustable electric fields

Receptors Target-cell molecules that bind hormones or neurotransmitters, with high affinity and saturability, and mediate their actions

RNA splicing Removal of the intron sequences present in the primary RNA transcript, carried out by a complex of RNA-processing enzymes

Signal transduction Conversion of information carried to a cell by chemical or sensory stimuli, and recognized by specific receptors, into a new form, such as the change in the activity of cellular proteins and/or the intracellular concentration of ions and metabolites

Toxic-catalyzed ADP-ribosylation Covalent modification of a substrate protein produced by the enzymatic transference of ADP-ribose from NAD donor molecules to specific amino acids in the acceptor molecule

G PROTEINS ARE HETEROTRIMERIC MOLECULES LOcated in cell membranes that transduce and amplify the signal carried to cells by hormones, agonists, light, taste stimuli, or odorants, thus inducing changes in the activity of cellular enzymes. G proteins possess quaternary structure, defined by the interaction of three different subunits: α, β, and γ, in order of decreasing molecular weights. The heterotrimers bind and hydrolyze guanosine triphosphate (GTP); this catalytic activity (GTPase) controls the deactivation of the protein.

G proteins become activated upon interaction with specific hormone–receptor complexes, in the presence of Mg^{2+} and GTP. Nonhydrolyzable analogs of GTP, such as guanosine-5-O-(3-thiotriphosphate) (GTPγS) or guanosine-5′-yl imidodiphosphate [GMPP(NH)P], or ligands such as AlF^{4-}, can induce stable activation of G proteins in the absence of hormone–receptor complexes. Activation by GTP analogs induces a conformational change that (at least in solution) causes dissociation of the oligomeric G protein into free α subunits and $\beta\gamma$ dimers; both free α and $\beta\gamma$ subunits have been shown to interact with specific effectors. An important property of some G proteins is that they are substrates for ADP-ribosylation catalyzed by bacterial toxins. ADP-ribosylation of specific G protein substrates by cholera toxin (CT) abolishes the GTPase activities associated with these molecules, causing their permanent activation. ADP-ribosylation of substrates mediated by one of the exotoxins of *Bordetella pertussis* (PT) uncouples them from their specific hormone–receptor complexes, with impairment in the signal transduction.

I. GUANINE NUCLEOTIDE-BINDING PROTEINS IN BIOLOGY

The G proteins constitute a subgroup of a family of proteins that can bind and hydrolyze GTP. Nature has assigned an important "switch" role to the GTPases, involving them in critical biological functions such as initiation, elongation, and termination of protein synthesis, microtubule assembly, and signal transduction. These molecules have different conformations and activities depending on whether they have GDP or GTP bound to them. The GDP-bound forms can interact with other molecules that catalyze the exchange of GDP for GTP. The GTP-bound forms are able to interact with, and change the activity of, a third class of molecules (effectors). The GTPase activity converts the protein to the GDP-bound form; this form has decreased ability to interact with the effector and is able to interact with the nucleotide exchanger in order to resume its effector active conformation (Fig. 1). In some cases, the biological effects can occur without GTP hydrolysis, such as translocation in protein synthesis (mediated by the elongation factors G or 2), tubulin polymerization, and signal transduction by heterotrimeric G proteins. In others, such as termination of protein synthesis, GTP hydrolysis is required for the manifestation of the biological effect. This article focuses only on the heterotrimeric GTP-binding proteins involved in signal transduction (referred to as G proteins).

II. SIGNAL TRANSDUCTION IN BIOLOGY: CONCEPT OF AMPLIFICATION

The concept of signal transduction first appeared when studying the regulation of adenylyl cyclase by hormones and the effects of light in photoreceptor cells. The introduction of a third element, coupling the formation of a hormone–receptor complex to the change in the activity of an intracellular effector, has the potential of providing an amplification effect (see legend of Fig. 1). In this way, one molecule of agonist can interact with approximately ten molecules of G_s (the stimulatory guanine nucleotide-binding regulatory component of adenylyl cyclase), which in turn interact with the enzyme adenylyl cyclase.[1] Also, each molecule of light-activated rhodopsin interacts with approximately 500 molecules of transducin (T). Each molecule of transducin activates one molecule of a phosphodiesterase enzyme that hydrolyzes guanosine-3′,5′-monophosphate (cGMP) into guanosine-5′-monophosphate (GMP). The initial amplification provided by the G proteins is subsequently magnified by the activated enzymes: each molecule of cyclase or phosphodiesterase catalyzes the conversion of a large number of nucleotide molecules.

III. CELLULAR FUNCTIONS REGULATED BY G PROTEINS

A. Phototransduction

Transducin interacts with the γ subunit of cGMP-dependent phosphodiesterase. This subunit is inhibitory to the catalytic activity of the $\alpha\beta$ dimer of the phosphodiesterase. The dimer catalyzes the conversion of cGMP into GMP. The decrease in the levels of cGMP results in closing of the Na^+ channels; in this way, a light signal is transduced into an electrical one.

B. Adenylyl Cyclase Activity

Adenylyl cyclases catalyze the conversion of ATP into cAMP. This family of enzymes (eight members have

[1]The G protein subunits are indistinctly referred to by using either the subunit symbol with the appropriate subscript (e.g., α_{i1}) or G and the corresponding subscript, followed by the subunit symbol (e.g., $G_{i1}\alpha$).

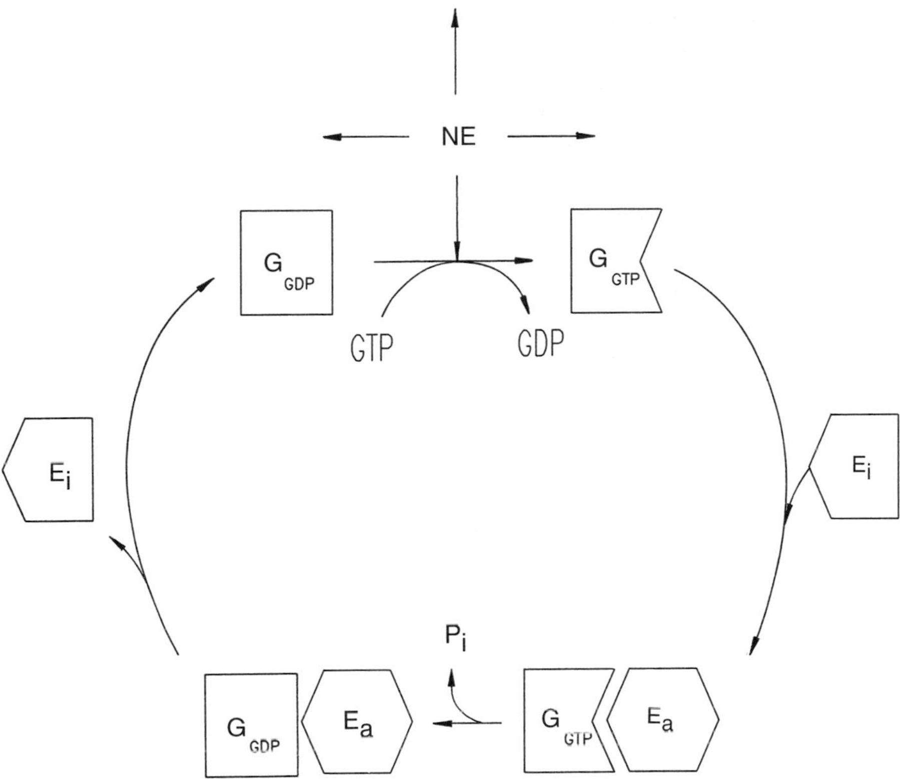

FIGURE 1 The interaction between a GTP-binding protein (G), a GDP/GTP exchanger (symbolized by NE, for nucleotide exchanger), and the inactive and active forms of an effector (E_i and E_a, respectively). One NE molecule (e.g., photolyzed rhodopsin in the activation of the retinal G protein transducin) catalyzes the nucleotide exchange of several hundred GTP-binding proteins (illustrated as multiple arrows emerging from NE). The GTP-binding proteins present two different conformations, depending on their interaction with either GDP or GTP. In this example, the GTP-bound form is able to interact with E_i. As a consequence of this interaction, the effector changes tertiary structure, thus acquiring its active conformation. The activation of the effector by G_{GTP} is sometimes the consequence of a change in the quaternary structure of the effector (the number of subunits present in the oligomeric enzyme). The GTPase activity, which is either constitutively regulated by G or activated by the interaction with a separate molecule (ribosomal proteins or GAPs), shifts the G protein back into the G_{GDP} conformation. This form has low affinity for the effector, resulting in the release of E_i and the availability of G_{GDP} for new interaction with NE. Note the main features of this biological device: (1) the amplification of the NE signal and (2) the self-inactivation of the amplifier, which provides a mechanism for quick blockage of the response, unless the NE is still present to push the system into additional rounds. P_i, inorganic phosphate.

been cloned so far) is subjected to dual regulation by stimulatory and inhibitory receptors. The signal carried by stimulatory receptors is transduced by Gs. The interaction of this protein with ligand-occupied stimulatory receptors induces dissociation of the α subunit from the $\beta\gamma$ dimers. The GTP bound α_s subunit activates the catalytic subunit of adenylyl cyclase. The signal elicited by ligand-occupied inhibitory receptors is transduced by inhibitory G proteins (G_i's). Activation of the G_i's reduces, in turn, adenylyl cyclase activity. The myristoylated α subunits of G_{i1}, G_{i2}, and G_{i3} (but not G_o) inhibit some forms of adenylyl cyclase (see Section VI,A for myristoylation). In addition, $\beta\gamma$ dimers of G proteins can stimulate some forms of adenylyl cyclase (types II and IV) and inhibit others (type I).

The cAMP, one of the most important intracellular messengers, binds and activates the cAMP-dependent protein kinase. This enzyme phosphorylates serine or threonine residues in specific protein substrates that change their activity as a result of the phosphate transfer. The phosphorylation reactions catalyzed by the cAMP-dependent protein kinases constitute the most important biological effect of cAMP. [*See* Protein Phosphorylation.]

C. C-Type Phospholipases

A variety (but not all) of Ca^{2+}-mobilizing receptors interact with either pertussis toxin-sensitive or -insensitive G proteins called G_p's, which in turn activate phosphoinositide-specific C-type phospholipases (PLCs). Recent experiments have identified the pertussis toxin-insensitive G proteins G_q, G_{11}, and G_{16}, as functional G_p's (acting mainly on PLC-β1 and PLC-β3). The pertussis toxin-sensitive activation of PLC (present in only a few cell types) has been attributed to the interaction of $\beta\gamma$ dimers (released from pertussis toxin-sensitive G proteins) with PLC-β2. The superfamily of phosphoinositide-specific C-type phospholipases includes several isoforms that have been purified/cloned and classified (based on their structure and regulation) into three groups termed β (regulated by G proteins; includes PLC-β1, PLC-β2, PLC-β3, and PLC-β4), γ (regulated by receptors with protein tyrosine kinase activity; includes PLC-γ1 and PLC-γ2), and δ (their regulation is poorly understood; includes PLC-δ1 and PLC-δ2). The C-type phospholipases catalyze the rupture of the phosphodiester bond of phosphoinositides, minor lipid components of biological membranes. The reaction produces inositol phosphates (soluble compounds) and a membrane-bound lipid: diacylglycerol. These compounds have profound influences on the state of the cell: the inositol phosphates produce a rise in the concentration of cytosolic Ca^{2+} due to the mobilization of the cation present in intracellular stores and also to the influx of extracellular cation. Ca^{2+} is another extraordinarily important messenger; changes in its levels can drive fundamental processes such as secretion and smooth muscle contraction. The diacylglycerol activates the Ca^{2+}- and phospholipid-dependent protein kinase C (protein kinase C or PKC). This kinase catalyzes the transfer of phosphates to specific serine and threonine residues on protein substrates. Among the PKC substrates in membranes, we can mention the Na^+/H^+ exchanger, Ca^{2+}-ATPase, glucose transporter, and several ion channels.

D. A2-Type Phospholipases

Agonist-mediated release of arachidonic acid appears to involve different enzymes with A2-type phospholipase activity (PLA_2's). These enzymes catalyze the release of fatty acids esterified in the sn-2 position of phospholipids. Arachidonic acid, together with its metabolites (prostaglandins, thromboxanes, and leukotrienes), represents a key intracellular messenger controlling stimulation of secretion from anterior pituitary, pancreatic-β, and mast cells; release of Ca^{2+} from intracellular stores (independent of inositol phosphate formation); activation of PLC, adenylyl cyclase, guanylyl cyclase, and protein kinase C; increase in DNA synthesis; and regulation of K^+ channels. The regulation of some isoforms of PLA_2 is complex and depends on several factors, including their phosphorylation by upstream kinases, the concentration of intracellular Ca^{2+}, and the activity of pertussis toxin-sensitive G proteins. It is still a matter of debate whether the role of G proteins in the regulation of PLA_2 activity is independent from their participation in the control of PLC activity and intracellular Ca^{2+} levels. Recent observations in cells that either do not express $G_{i2}\alpha$ or overexpress dominant negative mutants of this protein support a role for this pertussis toxin substrate in the regulation of PLA_2 activity independent of that of PLC.

E. Ionic Channels

Evidence for direct interaction of ionic channels with receptor-activated G proteins has been obtained recently. This represents an effect independent from the activation of ionic channels resulting from the action of second messengers or protein kinases activated by them (cAMP-dependent protein kinase, PKC, cGMP, etc.). Evidence for direct interaction was obtained from experiments in isolated membrane patches, using incubation conditions designed to minimize secondary effects. Examples of these direct interactions are the coupling of muscarinic agonist-gated K^+ channels in heart by G_i-like proteins (see Section IV), K^+ channels in hippocampal pyramidal neurons (G_o), the regulation of voltage-gated dihydropyridine-sensitive L-type Ca^{2+} channels (G_s), ATP-sensitive K^+ channels, cardiac Na^+ channels (G_s), and amiloride-sensitive sodium channels from renal epithelium. [*See* Ion Pumps.]

F. Regulation of Intracellular Vesicle Traffic

Experiments performed with permeabilized cells and patch-clamp recordings using dialyzed single cells indicate that exocytosis could be directly controlled by G proteins. This regulation seems to be independent of the changes in the levels of intracellular Ca^{2+} or PKC activity that are brought about by G protein regulation of phosphoinositide-specific phospholipase C. It is still unclear whether this is indeed a direct coupling of a G protein to the fusion between secretory granules and plasma membranes, or whether it is due to the G protein-mediated release of another second messenger (e.g., arachidonic acid). More recent experiments have linked G_{i3}, together with low-molecular-weight GTPAses (Rab family), in both constitutive secretion and regulated exocytosis. At the same time, G_s and members of the Rab family are involved in the regulation of endosome fusion.

G. Olfaction

The signal carried by at least some odorants results in activation of adenylyl cyclase in olfactory cilia. This transduction is very likely mediated by $G_{olf}\alpha$, a G protein α subunit exclusively expressed in olfactory tissue, which shares 88% identity with the α subunit of G_s ($G_s\alpha$). The increased levels of cAMP gate a Na^+/K^+ conductance in the ciliary plasma membranes of the olfactory sensory neurons, resulting in membrane depolarization. [*See* Olfactory Information Processing.]

H. Taste Transduction

A G protein α subunit specifically expressed in buds of all taste papillae appears to mediate transduction triggered by taste stimuli in vertebrates. This molecule, termed gustducin, is highly homologous to the α subunits of the transducin heterotrimers expressed in photoreceptor cells.

I. Regulation of Kinase Activity

$\beta\gamma$ dimers directly interact with different kinases, including the β-adrenergic receptor kinases 1 and 2, Raf-1 and phosphoinositide 3-kinase γ, recruiting them to the plasma membrane or cytoskeletal fraction and allowing their interaction with substrates and regulators. The $\beta\gamma$ dimers also play a role in Ras-dependent activation of mitogen-activated protein kinases through a pathway involving increased protein tyrosine phosphorylation and activation of phosphoinositide 3-kinase, shc adapter protein, Grb2 (growth factor receptor-bound protein) and SOS (homolog of *Drosophila Son of sevenless*).

IV. PRIMARY STRUCTURE OF G PROTEIN SUBUNITS

A. α Subunits

The primary structure of approximately 20 α subunits of G proteins has been determined in the past 10 years. Comparison of the analogous subunits among different species shows minimal differences. Four main subgroups (α_s, α_i, α_q, and α_{12}) may be defined within the G protein α subunit family based on relative sequence similarity, sensitivity to ADP-ribosylation catalyzed by cholera and/or pertussis toxin, and the identity of the effector(s) they are coupled to (Table I).

The α_s subgroup include several α_s isoforms and α_{olf} which are cholera toxin substrates and activate adenylyl cyclases. Four isoforms of α_s have been sequenced: α_{s1a}, α_{s1b}, α_{s2a}, and α_{s2b}, encoding proteins of 379, 380, 394, and 395 amino acids, respectively. The four isoforms are thought to be derived from the alternative splicing of a single RNA precursor. α_{olf} is expressed only in olfactory tissue and displays 88% similarity with α_s.

The cDNAs encoding different pertussis toxin substrates—named α_{i1}, α_{i2}, α_{i3}, α_{o1}, and α_{o2}—have also been cloned. Their designation was based on the known effects of pertussis toxin blocking hormonal inhibition of adenylyl cyclase ("α_i" for "inhibitory") and in the abundance of pertussis toxin substrates other than α_i in tissues such as brain ("α_o" for "other"). Although the primary structure of these molecules contains an arginine residue homologous to that susceptible to CT-mediated ADP-ribosylation in α_s isoforms, the α_i's do not undergo this modification (unless forced by interaction with specific agonist–receptor complexes) and are therefore considered only PT substrates. Two α subunits of transducin have been cloned (α_{t1} and α_{t2}); these proteins are 80% homologous, contain 350 and 354 amino acids, and are exclusively expressed in retinal rod and cone photoreceptor cells, respectively. The transducins are substrates for both pertussis and cholera toxins but display higher similarity with α_i proteins. Another α subunit that may be included in this subgroup is α_z: although this protein cannot undergo PT-mediated ADP-ribosylation, it displays approximately 70% similarity with the α_i's and it has been shown to inhibit adenylyl cyclase activity.

Two additional sets of PT-insensitive α subunits have recently emerged: $\alpha_{q/11}$ and $\alpha_{12/13}$. The α_q, α_{11}, and α_{16} subunits have been linked to the PT-insensitive

TABLE I
Diversity in G Protein α Subunits[a]

Subfamily	α subunit	Toxin specificity	Cellular distribution	Effectors
α_s	α_s splice variants	CT	Ubiquitous	Adenylyl cyclase (+), L-type Ca^{2+} channels (+)
	α_{olf}	CT[b]	Olfactory tissue	Adenylyl cyclase (+)
α_i	α_{t1}, α_{t2}[c]	CT, PT	α_{t1}, retinal rods α_{t2}, retinal cones	cGMP-dependent phosphodiesterase (+)
	Gustducin	CT, PT (potential)	Lingual taste buds	cAMP-dependent phosphodiesterase (+) (proposed)
	α_{i1}, α_{i2}, α_{i3}	PT	Ubiquitous	Adenylyl cyclase (−), atrial K^+ channels (+)
	α_o splice variants	PT	Brain, retina, platelets	K^+ channels in hippocampal pyramidal cells (+); opioid, $GABA_B$, M_1 muscarinic, and somatostatin receptor-regulated Ca^{2+} channels in neuronal and endocrine cells (−)[d]; PLC in *Xenopus* oocytes (+)
	α_z	—	Brain, retina, platelets, and red cells	Adenylyl cyclase (−)
α_q	α_q, α_{11}	—	Ubiquitous	Phosphoinositide-specific phosphodiesterases (PLC-β_1 = PLC-β_3 ≫ PLC-β_2)
	α_{16}	—	Hematopoietic cells	Same as for α_q, α_{11}
α_{12}	α_{12}, α_{13}	—	Ubiquitous	Arachidonic acid release (+); α_{13} participates in chemokinesis and developmental angiogenesis

[a]CT, cholera toxin; PT, pertussis toxin; (+) and (−), activation and inhibition, respectively, of the effector.
[b]Inferred by homology with G_s.
[c]Only T_1, the transducin molecule present in retinal rods, has been purified and characterized as both a PT and a CT substrate. T_2 has been identified by molecular cloning and localized in retinal cones by immunocytochemistry.
[d]Specificity for α_o splice variants as well as for different $\beta\gamma$ combinations has been observed.

activation of PLC-β1 and PLC-β_3; α_{12} has been linked to arachidonic acid release and α_{13} to chemokinesis and angiogenesis. Figure 2 compares the sequences of human α_{s2a}, α_{t1}, α_{i1}, α_{i2}, α_{i3}, α_o, α_{olf}, and α_z.

B. β Subunits

SDS-PAGE analysis of transducin purified from rod outer segments shows an apparently unique β subunit with a molecular mass of 36 kDa (termed β1). Heterotrimers other than transducin contain mixtures of β subunits of 35 and 36 kDa. Molecular cloning experiments have found five forms of β subunits (β_1–β_5; Table II) with primary structures displaying 52–90% similarity. The β_5 subunit, a 39-kDa polypeptide almost exclusively expressed in the brain, is the more divergent, displaying only 52% identity with the other subunits. The β_1 and β_3 subunits (both of approximately 36 kDa, the latter being slightly higher), although rather ubiquitous, show a cell-dependent pattern in the retina, where they specifically localize in rod and cone photoreceptor cells, respectively. The β subunits belong to the large family of the GH-WD 40 repeat proteins (containing repeats of approximately 40 amino acids with characteristic glycine–histidine and trytophan–aspartic acid pairs). Other members of this family of eukaryotic proteins are important in replication of DNA, regulation of transcription, polyadenylation and RNA splicing, signal transduction, and neurogenesis. The primary structures of human β_1 and β_2 are depicted in Fig. 3.

C. γ Subunits

Screening of cDNA libraries has resulted in the cloning of cDNAs corresponding to six forms of γ subunits (γ_1–γ_5 and γ_7). The γ subunit of transducin (γ_1) contains 74 amino acids, is exclusively expressed in the retina, and is uniquely hydrophilic. On the other hand, γ_3 is mainly expressed in the brain, the remaining forms (including the abundant γ_2) being more ubiquitous (Table II). The primary structure of the γ_1 and γ_2 subunits is depicted in Fig. 4.

The presence of multiple forms of β and γ subunits raises the possibility of preferential interaction of some β and γ subtypes with the different α subunits. Initial studies have revealed that although β_1 can bind to both γ_1 and γ_2 subunits, β_2 can bind only γ_2 whereas β_3 is unable to interact with either γ_1 or γ_2.

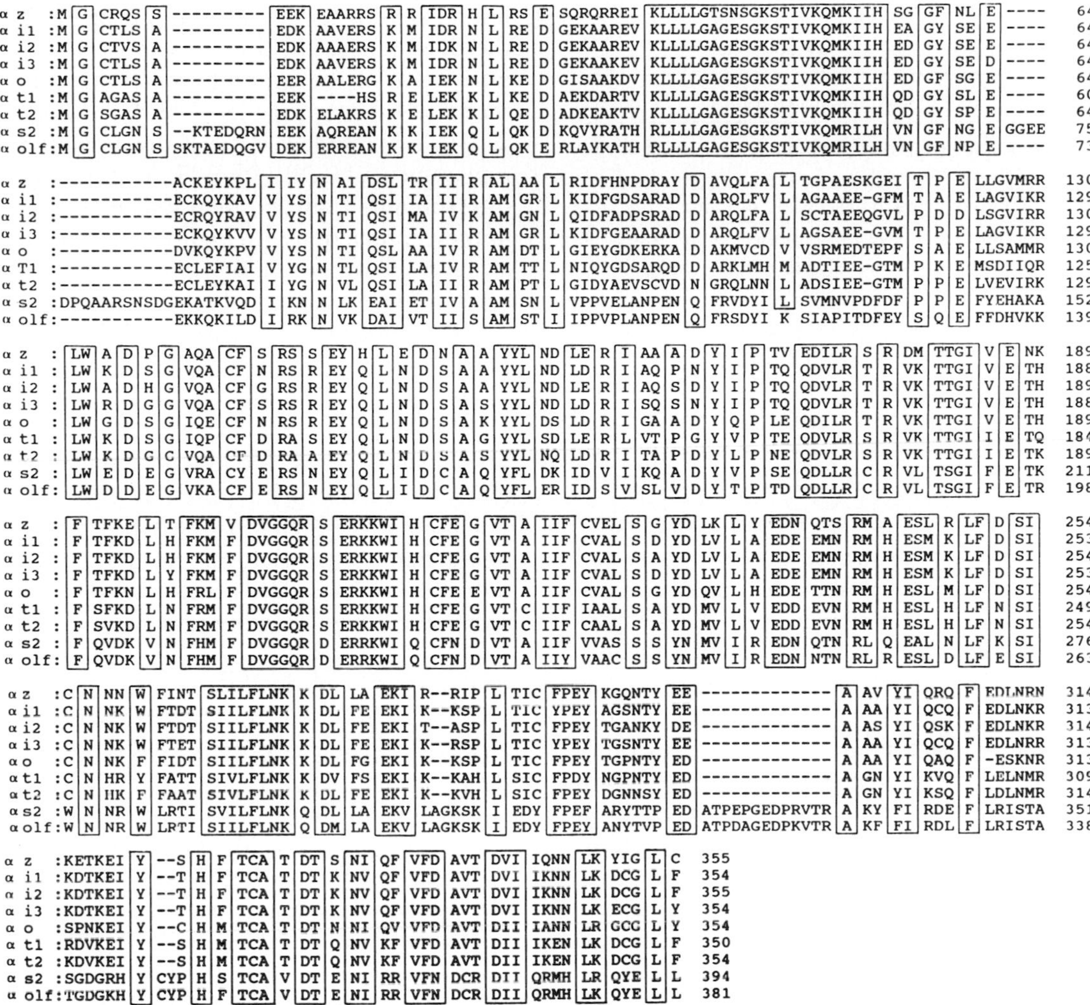

FIGURE 2 The sequences of the human α_z, α_s, α_{i2}, α_{i3}, and α_o, bovine α_{t1}, α_{t2}, and α_{i1}, and rat α_{olf} are given. Variation of the corresponding proteins between species is minimal: the human α_s is 99.7% identical to the corresponding bovine and rat proteins; comparison of the sequences of each of the different types of α_i in human, bovine and rat also shows that they are at least 98% identical. These proteins are the products of nine separate genes. The α sequence shown here (α_{s2a}, 394 amino acids) was deduced from one of the four splicing products of the α_s RNA primary transcript. Single-letter nomenclature for amino acids is as follows: A, alanine; C, cysteine; E, glutamic acid; F, phenylalanine; G, glycine; H, histidine; I, isoleucine; K, lysine; L, leucine; M, methionine; N, asparagine; P, proline; Q, glutamine; R, arginine; S, serine; T, threonine; V, valine; W, tryptophan, Y, tyrosine. Boxed areas are regions of identity or conservative substitutions. Sequences have been spaced to obtain the highest degree of homology. Groups of conservative substitutions are: (a) D, N, E, and Q; (b) R, H, and K; (c) F, Y, and W; (d) S, T, P, A, and G; (e) M, I, L, and V.

V. RELATIONSHIP BETWEEN STRUCTURE AND FUNCTION OF G PROTEINS

A. α Subunits

The significant functional domains in α subunits correspond to the regions involved in the interaction with the $\beta\gamma$ subunits ($\beta\gamma$ site), with guanine nucleotides, with effectors, and with the hormone–receptor complexes (r site). The predicted domains of α_s (human α_{s2a}) involved in interactions with guanine nucleotides (phosphate groups and guanine ring), $\beta\gamma$ dimers, Mg^{2+} ion, and receptors are depicted in Fig. 5. The assignment of domains is based on the effects of mutations on the interaction of G_s with receptors and effectors and by inference from the crystal structure of α_t.

The three-dimensional structure of a G protein α subunit has been recently elucidated. Furthermore,

TABLE II
Diversity in Gβ and Gγ Subunits

Subunit	Cellular distribution
β_1	Ubiquitous (in retina predominantly located in rod photoreceptor cells)
β_2	Ubiquitous, with the exception of retina
β_3	Ubiquitous (however, in retina is expressed in cone but not in rod photoreceptors)
β_4	Ubiquitous
β_5	Brain
γ_1	Retina
γ_3	Mainly in brain
γ_2, γ_{4-7}	Ubiquitous

comparison of the crystal structures of the GDP- and GTP-bound forms of the α subunit of transducin (α_t) has shed light on the conformational changes produced by the GDP/GTP exchange and the identity of the switch regions involved in these changes. The folding of the α subunits reveals three important domains: (1) the GTPase domain, common to other members of the GTPase superfamily (with five helices surrounding a six-stranded β sheet), (2) the helical domain distinctive of α subunits of G protein heterotrimers (entirely α-helical, and (3) an N-terminal helix projecting from the rest of the α-subunit. The first two domains form the cleft where the nucleotide and Mg^{2+} bind. It has been proposed that the helical domain represents the physical barrier that causes the low spontaneous rate of nucleotide exchange in inactive α subunits. The changes triggered by nucleotide exchange are relatively minor and are limited to three switch regions: I (residues 200–210), II (residues 222–242), and III (residues 254–265) (numbering corresponds to the α_{s2} molecule; see Fig. 5). Switches I and II are conserved in other low-molecular-weight GTPases; switch III is exclusive for the α subunits of G proteins. Critical residues involved in the interaction with the γ phosphate of the GTP molecule are R201 and T204 (in switch I) and G226 (in switch II). Changes in switch III result from the propagation (through polar residues, mainly E259) of the perturbations originally induced in switch II. Some residues in switch II (and perhaps I) appear to stabilize the interaction of the β subunits with the NH_2 terminus of the α subunits.

The interaction between Tα and T$\beta\gamma$ subunits occurs at two distinct interfaces: The most extensive contact involves switch regions I and II of the GDP-bound α with the top of the β-propeller domain in the β subunit (see below); the second one, involves interaction of the amino-terminal domain of α with the side of the β-propeller (GH-WD 40 repeats 1 and 2). The first interface is markedly disrupted following replacement of GDP with GTP, leading to dissociation of α subunits from $\beta\gamma$ dimers. The residues involved in these interactions are highly conserved in different α and β subunits, arguing against any obvious differences in the formation of specific $\alpha\beta\gamma$ heterotrimers. Several domains are discussed in the following sections.

1. Region 1–25

This region is involved in the interaction with $\beta\gamma$ dimers (see above). The direct interaction of α with γ subunits has been proposed based on the direct interaction of γ_2, but not β_1, with α_o, and the protective effect of γ_2 on the proteolysis of the 21 amino-terminal residues of α_o by trypsin.

2. Region 47–54 (GAGESGKS)

This region is involved in the interaction with phosphates. The sequence GXXXXGK is considered a consensus element for the interaction with phosphates and is present in all the G protein α subunits that have been sequenced. Most α subunits of G proteins contain the aforementioned GAGESGKS sequence; a few show differences in this region, such as GTSNSGKS (α_z) or GTGESGKS (α_q). These substitutions appear to decrease the rate of release of GDP. Also, substitution of a G residue for V in position 49 (G49V mutation) reduces GTPase activity in α_s. The assignment of residues based on the crystal structure of α_t indicates that E50, G52, K53, and S54 (together with R201 and T204 in switch region I and G226 in switch region II) participate in hydrogen bonding of the α and β phosphates of GTP. The Mg^{2+} ion is coordinated by nonbinding oxygens of the β and γ phosphates and the side chains of S54, T204 (switch II), and D223 (switch III).

3. Region 223–227 (DVGGQ)

This region is also involved, or affects at a distance, the interaction with guanine nucleotides. Reciprocally, this region seems to undergo conformational changes elicited by the interaction of the nucleotide molecule with the 47–53 region. The Q227L mutation (found in pituitary tumors and also introduced by site-directed mutagenesis) dramatically reduces GTPase activity. The G226A mutation in α_s, present in the H21 mutant of the S49 lymphoma cell line, is

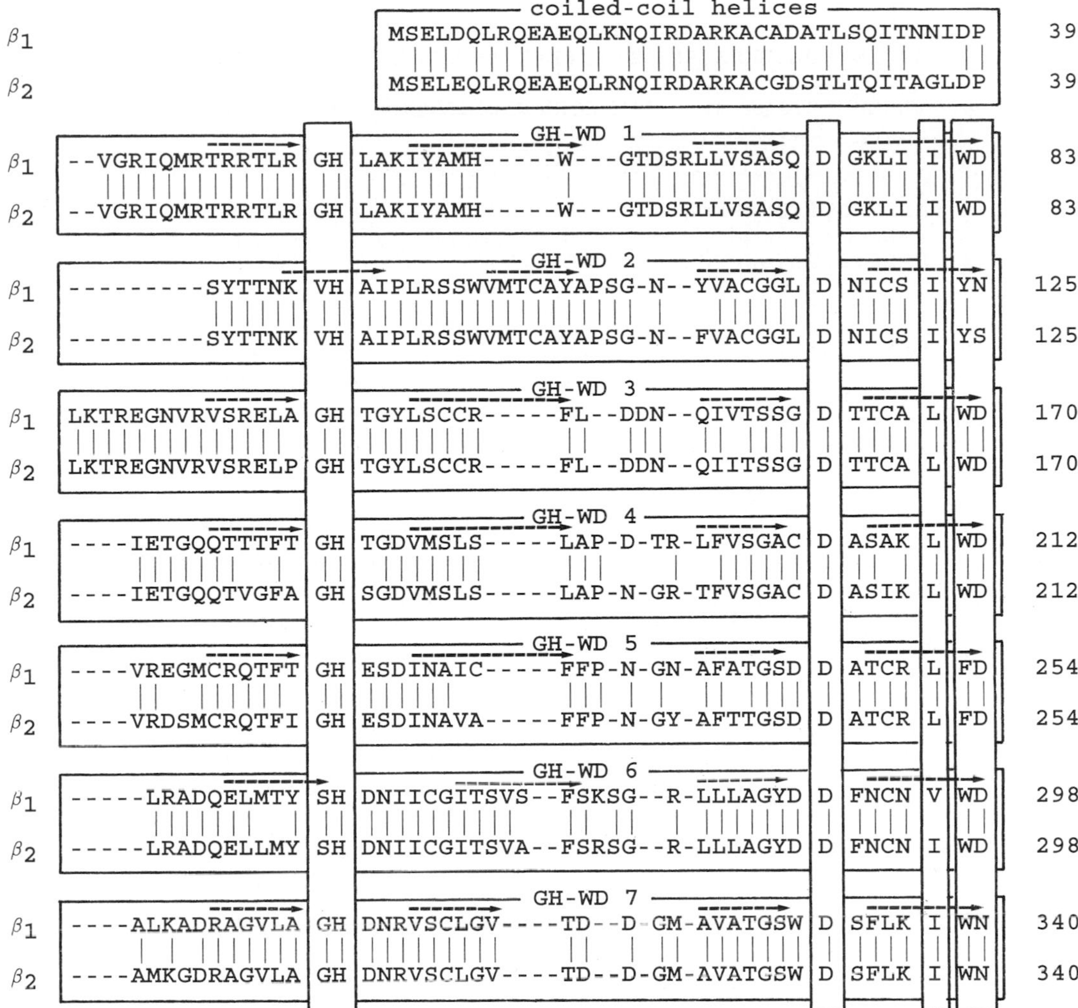

FIGURE 3 Comparison of the primary structures of the human β_1 and β_2 subunits. Vertical lines indicate identical residues. The amino-terminal domain is predicted to form an amphipatic α-helix involved in coiled-coil interactions with Gγ. Gaps in the sequences were introduced to enhance the display of the seven GH-WD 40 repetitive segments. Arrows indicate the position of the antiparallel strands that form the seven β-sheets ("propeller blades"). The first arrow in each GH-WD 40 repeat represents the outermost strand of a β-sheet; the next three arrows in each repeat correspond to the inner strands of the following blade.

able to couple to hormone–receptor complexes but is unable to change the activity of the effector. Since this residue is conserved in G protein α subunits that interact with other effectors (it is unlikely to be located in the effector site), it is possible that this region changes conformation after sensing the interaction of GTP with region 47–53. The D residue in position 223 is involved in the coordination of Mg^{2+} (see earlier). This region is adjacent to the arginine residue in position 201 (switch I), which acts as acceptor during CT-mediated ADP-ribosylation of G_s. This covalent modification, as well as the R201C mutation found in pituitary tumors (see Section X), inhibits GTPase activity.

4. Region 292–296 (NKQDL)

This region is involved in the interaction with the guanine ring. The sequence NKXD has been defined as a consensus sequence involved in guanine nucleotide specificity.

5. Region 374–394

Several lines of evidence suggest that the 21 residues in the carboxy terminus of the protein are involved

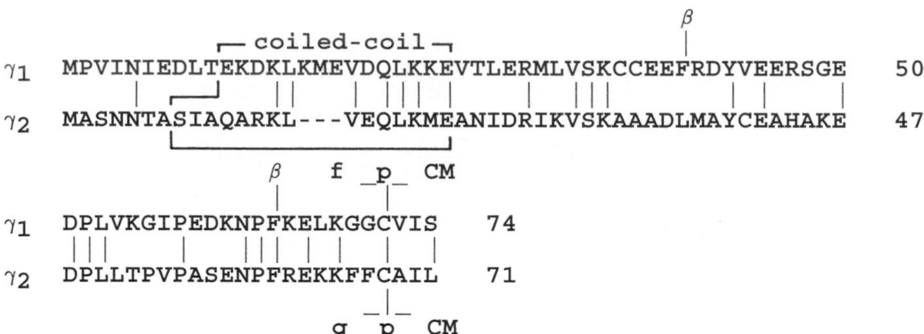

FIGURE 4 Comparison of the primary structures of bovine γ_1 and γ_2. The gap was introduced to obtain the highest degree of homology. Vertical lines show the position of identical residues. f, farnesylation; g, geranylgeranylation; p, proteolytic cleavage; CM, carboxymethylation.

in the interaction with receptors (R site). The R389P α_s substitution, found in the *unc* mutant of the S49 lymphoma cell line (no coupling of receptors to G_s is detected in these cells), has provided experimental support for this hypothesis.

B. β Subunits

The amino-terminal domain of Gβ subunits (Fig. 3) forms an amphipathic α-helix that is involved in coiled-coil interactions with the amino-terminal helix of γ subunits. This amino-terminal region contains the highest density of residues involved in contact with γ subunits (including cystein at position 25 in the β_1–β_4 subunits). In addition, β subunits contain a "propeller-like" structure with seven "blades" (consisting of 4 antiparallel β-strands) that spring outwardly from a central core. This structure classifies the G protein β subunits as members of the β-propeller family of proteins. The GH-WD 40 repeats (Fig. 3) start at the turn between the third and fourth (outermost) β-strands in a given blade, ending at the corresponding position of the following one. The adjacent blades of the propeller are stabilized by interactions between the characteristic W and H, S/T and distal D residues in each GH-WD 40. As opposed to the

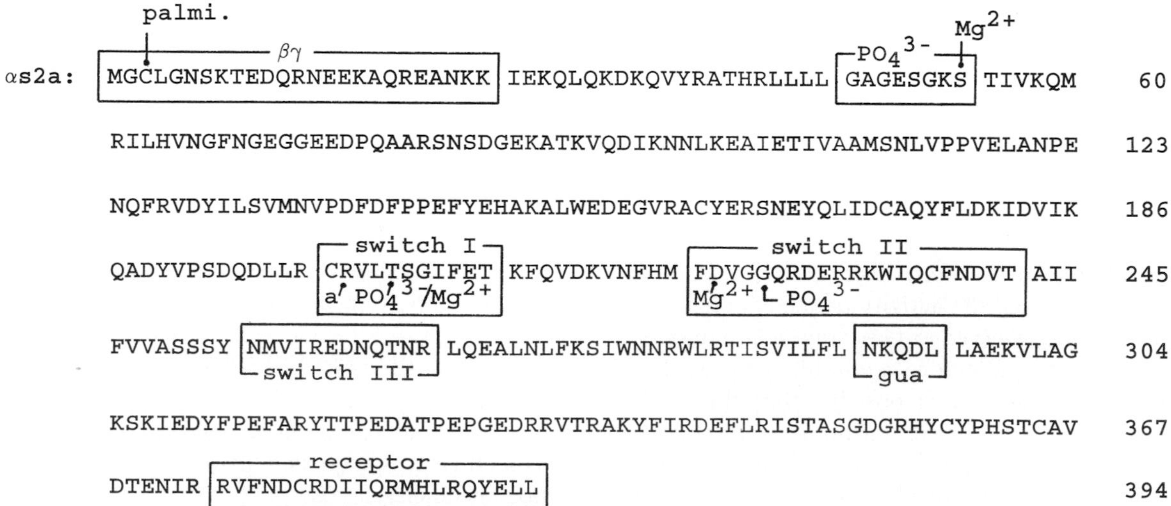

FIGURE 5 Primary structure of human G_{s2a} showing the domains involved in the interaction with $\beta\gamma$ subunits, phosphate groups (PO_4^{3-}), Mg^{2+}, guanine ring (gua), and hormone–receptor complexes, as well as the switch regions that are sensitive to the GDP/GTP exchange (switch regions I, II, and III, respectively). p, site of palmitoylation; a, arginine residue that functions as substrate during CT-mediated ADP-ribosylation. See text for details.

significant conformational changes that occur following GTP binding to α subunits (see above), minimal modifications take place on the $\beta\gamma$ dimers. Thus, the $\beta\gamma$ dimer presents a more rigid conformation and its activation depends only on dissociation from the α subunits. The GH-WD 40 repeats located in the carboxy-terminal region of Gβ subunits interact with the pleckstrin homology (PH) domains (approximately 100 amino acids long) present in several molecules relevant in signaling (such as the β-adrenergic receptor kinases 1 and 2 and PLC isoforms) and cytoskeleton organization (β-spectrin). Regulation by $\beta\gamma$ appears to require a QXXER motif together with additional undefined sequences in the target molecules. A list of effectors regulated by $\beta\gamma$ dimers is shown in Table III.

C. γ Subunits

Posttranslational modification (isoprenylation) in the carboxy terminus of the γ subunits (see Section VI,C) is critical for their interaction with the amino terminus of α subunits, but does not affect the assembly of $\beta\gamma$ dimers. On the other hand, the removal of the three carboxy-terminal residues in the γ subunits impairs the formation of $\beta\gamma$ dimers. The crystal structure of T$\beta\gamma$ dimer shows that Tγ contains two helices and lacks tertiary structure. Tγ binds to Tβ in an extended conformation through both the formation of a coiled-coil between the amino-terminal helices of Tβ and Tγ and multiple hydrophobic interactions (including those of the phenylalanine residues at positions 40 and 64 in γ_1).

D. Structure of Heterotrimers and Interaction with Membranes and Receptors

The amino-terminal region of Tα and the carboxy-terminus of Tγ, containing lipophilic modifications (see Section VI), are separated by only 18 A in a common face of the crystal structure of the heterotrimer. This flat and predominantly neutral surface, which contrasts to the others that are negatively charged, also contains the receptor-coupled carboxy terminus of Tα and is responsible for orienting the heterotrimer to the membranes.

TABLE III
Effectors Coupled to $\beta\gamma$ Complexes[a]

Effector
Type I adenylyl cyclase (−)
Types II and IV adenylyl cyclase (+)
β-adrenergic receptor kinase (recruitment to membrane by direct interaction with $\beta\gamma$)
GIRK1 (atrial muscarinic agonist-gated inwardly rectifying K^+ channel) (+)
N-type and P/Q-type calcium channels (−)
Pheromone-induced mating in *Saccharomyces cerevisiae* (STE20 gene product) (+)
Phosphoinositide-specific phosphodiesterases PLC-β_2 > PLC-β_1 = PLC-β_3 (+)
Phosphoinositide 3-kinase γ
Phosducin, a regulator of cGMP cascade in photoreceptor cells (desensitization of light response)
Raf-1 serine/threonine protein kinase (recruitment to membrane by direct interaction with $\beta\gamma$)
Ras-dependent regulation of mitogen-activated protein kinases (+) (trough increased protein tyrosine phosphorylation, phosphoinositide 3-kinase, shc, Grb2, and SOS)

[a](+) and (−), activation and inhibition of the effector, respectively. Nine unique $\beta\gamma$ dimers (formed by combinations of β_1 and β_2 with γ_1, γ_2, γ_3, γ_5 and γ_7) displayed similar activities when assayed for the inhibition of type I adenylyl cyclase or activation of type II adenylyl cyclase and PLC-β isoforms (with the exception of the less potent $\beta_1\gamma_1$).

VI. COTRANSLATIONAL AND POSTTRANSLATIONAL MODIFICATIONS OF G PROTEINS

A. α Subunits

The α subunits of most, but not all, of the G proteins undergo myristoylation (Table IV). This modification consists in the cotranslational acylation of the amino-terminal glycine with myristic (tetradecanoic) acid. The absence of a serine residue in position 6 of the α_s deduced primary structure, which is present in all the other α subunits, is probably responsible for its lack of reaction with the myristoyl Co-*N*-myristyltransferase complex. This acylation increases the affinity of α subunits for $\beta\gamma$ dimers and seems critical (with the exception of α_s and, perhaps, α_t) for the attachment of the α subunits to the plasma membrane. The absence of acylation in the native α_t probably reflects absence of enzyme–substrate interaction in the photoreceptor cell.

Most α subunits of heterotrimeric G proteins also undergo the posttranslational incorporation of palmitic (hexadecanoic) acid. The palmitate group is incorporated into cysteine residues (located at the amino terminus of the Gα subunits) through labile thioester

TABLE IV
Cotranslational and Posttranslational Modifications of G Protein Subunits

Modification	Substrates
α subunits	
Myristoylation (C14:0)	Amino-terminal glycine residues in α_{i1}, α_{i2}, α_{i3}, α_o, α_t[a], α_z (members of the α_s, α_q and α_{12} families are not substrates)
Palmitoylation (C16:0)	Cysteine residues located in the second position from amino terminus in α_{i1}, α_{i2}, α_{i3}, α_o, α_s; cysteine in eight and/or ninth position in $\alpha_{q/11}$; cysteine residues in α_{12} (probably Cys 11) and α_{13} (Cys 14 and/or Cys 18) (α_t is not a substrate)
Arachydonoylation (20:4)	α_i, α_z, α_q, α_{13} subunits expressed in platelets
Phosphorylation	GDP-bound form of T_α is phosphorylated *in vitro* by PKC, agonist-occupied insulin, and insulin-like growth factor receptors
	α subunits of G_i-like proteins and G_s behave *in vitro* as substrates for PKC; α_{12} (but not α_{13}) is phosphorylated by PKC; both the tyrosine kinase pp60$^{c\text{-}src}$ and the epidermal growth factor receptor phosphorylate α_s
β subunits	
Phosphorylation	β subunits (β_1 and also β subunits in HL60 cells) may function as phosphorylated intermediates transferring γ-phosphate from GTP to GDP bound to α subunits
γ subunits	
Farnesylation ($C_{15}H_{26}O$)	Cysteine residues in the fourth position from carboxy terminus in γ_1
Geranylgeranylation ($C_{20}H_{34}O$)	Cysteine residues in the fourth position from carboxy terminus in nonretinal γ subunits
Proteolytic cleavage	Peptide bond between the third and fourth residues from the carboxy terminus
Carboxymethylation	Carboxy-terminal cysteine resulting from above-mentioned endoproteolytic cleavage

[a]α_t is predominantly modified at the NH_2-terminal glycine by lauric acid (C12:0) and two unsaturated derivatives of myristic acid (C14:1 and C14:2).

bonds. This modification also plays an important role in the attachment of α subunits to the plasma membranes. The biolability of the thioester bond, together with the rapid turnover of this modification, suggests an important point of control of G protein function. In this context, it has been observed that activation of G_s by β-adrenergic receptors results in increased depalmitoylation of α_s and translocation to the cytosol.

Studies in platelets revealed that all α subunits examined (α_i, α_z, α_q, α_{13}) undergo posttranslational arachidonoylation (in addition to palmitoylation); the arachidonate appears to be linked through a thioester bond at the same position(s) where palmitoylation occurs.

CT-mediated ADP-ribosylation of G_s in GH3 pituitary cells has been shown to increase the degradation of this protein, in spite of significantly increasing G_s-mediated stimulation of adenylyl cyclase. This observation suggests that ADP-ribosylation could mark the G protein for increased degradation by cellular proteases. Three endogenous ADP-ribosyl transferases have been purified from eukaryotic cells; these enzymes (A,B,C) can specifically transfer the ADP-ribose moiety to arginine, diphtamide, and cysteine residues, respectively. The ADP-ribosyltransferase C activity, purified from human erythrocytes, uses as specific substrate the α subunit of G_i-like proteins. This modification impairs the epinephrine-induced inhibition of adenylyl cyclase in platelets. It is then possible to speculate that ADP-ribosylation of G proteins by endogenous ADP-ribosyl transferases might represent a physiological posttranslational modification, affecting the stability and the coupling capability of these proteins.

Several lines of evidence indicate that G proteins undergo phosphorylation. The α subunits of G_i-like proteins purified from rabbit liver and recombinant α_s forms have been shown to behave as substrates for phosphorylation mediated by PKC. Also the GDP-bound form of Tα, but not the GTP-associated form, is phosphorylated *in vitro* by PKC and by hormone-occupied insulin and insulin-like growth factor I receptors. Activation of PKC by phorbol esters in multiple cell lines impairs GTP-dependent hormonal inhibition of adenylyl cyclase and activation of phosphoinositide-specific phosphodiesterase. Some of these effects may be explained by the inhibition of activity of G_i-like proteins as a result of PKC-mediated phosphorylation. Other effects of the PKC activity on the adenylyl cyclase system are due to phosphorylation of its catalytic component in some cell types.

B. β Subunits

Histidine residues in β subunits appear to function as phosphorylated intermediates transferring the γ-phosphate from GTP to GDP (presumably bound to α subunits).

C. γ Subunits

The carboxy termini in γ subunits of G proteins undergo posttranslational modification consisting of isoprenylation of a cysteine residue located at the fourth position from the carboxy terminus, proteolytic removal of the last three amino acids, and carboxymethylation of the resulting carboxy-terminal cysteine (Fig. 4). The isoprenylation reaction consists of the covalent attachment of an isoprenoid (either C15 or C20, depending on the particular γ subunit). The incorporation of this lipidic structure to the carboxy terminus of the γ subunits is critical for the anchoring of the $\beta\gamma$ dimers to the plasma membranes. Although the incorporation of the isoprene moiety in γ's does not seem relevant for interaction with β subunits (contrary to proteolytic cleavage, which interferes with this process), the presence of an isoprenylated γ subunit in the $\beta\gamma$ dimers is critical to subsequent formation of heterotrimers with the cognate α subunits. Methylation of γ_1 increases the efficiency of coupling of transducin with receptors.

VII. STEPS IN THE ACTIVATION/ DEACTIVATION CYCLE OF G PROTEINS

In the basal state, the predominant G protein species is the heterotrimer with GDP bound to it. The rate of release of GDP in the presence of unoccupied receptors (basal conditions) is low (although not zero). The intrinsic GTPase activity present in the Gα subunit catalyzes the hydrolysis of GTP into GDP and P_i (inorganic phosphate). Though the P_i is readily released, the GDP remains bound for a much longer time. This sequential release makes the GTPase reaction irreversible. The α_{GDP} displays high affinity for the $\beta\gamma$ dimer, and the heterotrimer is therefore regenerated. Consequently, in the unstimulated steady state only a small fraction of the total number of G protein molecules are in the effector active conformation (GTP-bound). Addition of poorly hydrolyzable analogs of GTP (or AlF^{4-}) results in the accumulation of α subunits in their active conformation, owing to the slow nucleotide exchange occurring in the absence of receptor agonist. Activation of G proteins by agonist–receptor complexes is catalytic as opposed to stoichiometric. For instance, one molecule of occupied receptor interacts with approximately 10 molecules of G_s (inactivation of receptors does not reduce the V_{max} but rather the rate of activation). Upon interaction with agonist–receptor complexes, the rate of release of GDP is increased by over an order of magnitude ("agonist" arrow in Fig. 6). In addition, receptors increase the rate of activation of G proteins by guanine nucleotides. The dissociation of α_{GTP} from the $\beta\gamma$ dimers provides the irreversible step that is necessary in a continuous cycle.

The GTPase activity of purified G proteins is approximately 4–6 hydrolytic cycles per minute ($t_{0.5}$ = 4–7 sec). This is extremely low and inconsistent with the kinetics of some G protein-coupled phenomena (such as visual transduction), suggesting the existence of GTPase-activating proteins for the heterotrimeric G proteins similar to those described for other GTP-binding proteins. In this context, it has recently been observed that effectors such as cGMP-dependent phosphodiesterase and PLC-β act as GTPase-activating proteins for their cognate G proteins (E1 in Fig. 6). A family of regulators of G protein signaling (RGS) that increase GTPase activity of certain Gα subunits has also been recently identified.

Mg^{2+} plays an important role in receptor-driven activation of G proteins. The effect of this divalent cation is threefold: it is required for formation of high-affinity agonist–receptor complexes that are coupled to G proteins, it increases the rate of activation of G proteins by GTP analogs, and it is also required for GTPase activity. Receptor occupancy by agonists decreases the concentration of Mg^{2+} required for G protein activation (from 10 mM to approximately 10 μM). As a result, the concentration of free Mg^{2+} in the cytosol (0.5 mM) becomes sufficient to drive full activation. In this way, the agonist–receptor complexes act as "Mg^{2+} switches" for their coupled G proteins.

VIII. EFFECTS CAUSED BY BACTERIAL TOXIN-MEDIATED ADP-RIBOSYLATION OF G PROTEINS

An important property of the G proteins is that they are substrates for ADP-ribosylation catalyzed by bacterial toxins. The exotoxins produced by some bac-

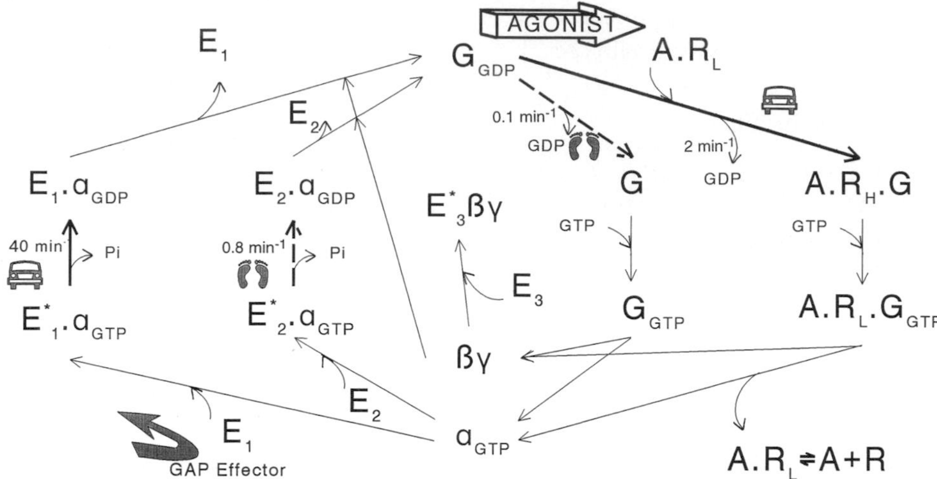

FIGURE 6 Cycle of activation/deactivation of G proteins under resting (downward dashed arrow in the inner hexagon) or agonist-stimulated conditions ("agonist" arrow) and coupling to effectors with or without GTPase activating protein (GAP) activity (E_1 and E_2, respectively). The slow and fast steps are illustrated by dashed and solid arrows (with bare feet and car symbols, respectively). The inner hexagon illustrates slow cycling: unoccupied receptors and coupling with effectors lacking GAP activity; the outer hexagon exemplifies fast cycling: agonist stimulation and interaction with GAP effectors. The rate constants of catalysis for GTP hydrolysis in the presence or absence of GAP effectors are taken from studies by E. M. Ross and colleagues on the interaction of $G_{q/11}\alpha$ with PLC-β1. A.R_L and A.R_H symbolize the low- and high-affinity forms, respectively, of the agonist (A)–receptor (R) complex; E_3, a $\beta\gamma$ effector; P_i, inorganic phosphate.

terial strains (such as *Vibrio cholerae* or *Bordetella pertussis*) display two enzymatic activities: NAD glycohydrolase and ADP-ribosyl transferase. The former implies the excission of NAD^+ into nicotinamide and ADP-ribose; the latter involves the transference of the ADP-ribose moiety to the lateral chains of specific residues of some G protein α subunits (each toxin displays a specific residue/α chain pattern). ADP-ribosylation of substrates by cholera toxin abolishes the GTPase activities associated with these molecules, causing their permanent activation. ADP-ribosylation of substrates mediated by one of the exotoxins of *Bordetella pertussis* uncouples them from their specific hormone–receptor complexes, with impairment in the signal transduction.

IX. STRUCTURE OF G PROTEIN GENES AND CHROMOSOMAL LOCALIZATION

The human α_s gene is approximately 20 kilobases long and is composed of 13 exons and 12 introns. The $G_{i2}\alpha$, $G_{i3}\alpha$, and $T_1\alpha$ genes are composed of 8 exons and 7 introns in their coding region. The positions of the splice junctions on the sequence of cDNAs of $G_{i2}\alpha$ and $G_{i3}\alpha$ are identical; this organization seems to be conserved in the $G_{i1}\alpha$ and $G_o\alpha$ genes. Also 3 out of the 12 splicing sites present in the $G_s\alpha$ gene are also present in the $G_i\alpha$ genes. This suggests that all of these genes evolved from a common ancestral precursor. [*See* Genes.]

The localization of genes encoding G protein subunits in human chromosomes has been reported. $T_2\alpha$, $G_{i3}\alpha$, and $G\beta_1$ genes map to chromosome 1 (1p13 for the first two). $T_1\alpha$ and $G_{i2}\alpha$ genes are located on chromosome 3 (3p21 for both). Genes encoding $G_{i1}\alpha$ and $G\beta_2$ are found on chromosome 7. The $G\beta_3$ gene is present in chromosome 12. $G_o\alpha$, $G_s\alpha$, and $G_z\alpha$ genes are distributed on chromosomes 16, 20, and 22, respectively. The $G_{11}\alpha$ and $G_{16}\alpha$ genes are located on chromosome 19 (19p13 for both). Physical linkage of the genes encoding $G_{11}\alpha$ and $G_{16}\alpha$, $G_{i3}\alpha$ and $T_2\alpha$, and $G_{i2}\alpha$ and $T_1\alpha$ indicate that these three pairs arose by tandem gene duplication prior to the divergence of rodents and primates. [*See* Chromosomes.]

X. REGULATION OF G PROTEIN GENE EXPRESSION

Chronic administration of corticosterone to normal and adrenalectomized rats showed that the steady-state levels of $G_s\alpha$ and $G_i\alpha$ mRNA in cerebral cortex are subjected to positive and negative modulation, respectively, by glucocorticoids. Similar variations in the corresponding levels of proteins are observed. $G_o\alpha$ and G protein β subunits do not appear to be regulated by the glucocorticoid. These effects could be due to (1) direct regulation of subunit gene expression, (2) changes in the synthesis and/or catabolism rate of the messenger RNAs, or (3) an indirect effect elicited by glucocorticoid-sensitive mediators. The identification of consensus hexamer sequences for the interaction with steroid hormone receptors in $G_{i1}\alpha$ supports the first possibility. The prolonged hormone treatments required for the observation of the effects would be compatible with an indirect effect.

Thyroid hormones exert a negative regulation of the levels of β subunits and their corresponding mRNAs in fat cells, as measured in normal, hypothyroid, and hyperthyroid rats. Hypothyroid rat adipocytes also contain increased levels of pertussis toxin substrates and $G_{i1}\alpha/G_{i2}\alpha$ immunoreactive material, without significant changes in the levels of $G_{i2}\alpha$ mRNA. These results seem consistent with the enhanced inhibition of adenylyl cyclase activity produced by specific agonists in fat cells from hypothyroid rats. [*See* Thyroid Gland and its Hormones.]

Expression of β subunits in fat cells also seems to be positively regulated by glucocorticoids, as opposed to the lack of effects observed in cerebral cortex.

The effects of chronic exposure of different cells to agonists suggest a mechanism of coordinate regulation of expression of receptors and their cognate G proteins, together with a reciprocal regulation of the levels of G proteins that trigger opposite effects on the activity of a given effector (such as G_s and G_i-like proteins). These observations have been recently extended by studies demonstrating the coordinate up-regulation of M2 muscarinic receptors, $G_{i2}\alpha$, and acetylcholine-sensitive K^+ channel by cholesterol (or cholesterol precursors) in embryonic chick ventricular myocytes.

XI. CELLULAR AND TISSUE DISTRIBUTION OF G PROTEINS

The α subunits of G_s, the G_i's, G_q, and G_{11}, G_{12}, and G_{13} are ubiquitously expressed and are now considered to be encoded by housekeeping genes (Table II). Examples of α subunits with specific organ localization are $T_1\alpha$, $T_2\alpha$, $G_{olf}\alpha$, and gustducin. Immunochemical studies have shown that the highest concentrations of $G_o\alpha$ are present in the central nervous system, followed by the pituitary gland and sciatic nerve. Other peripheral organs containing G_o (less than 2% of cerebrum concentration) are urinary bladder, stomach, intestines, and heart atrium.

The messenger RNAs encoding for $G_o\alpha$, $G_s\alpha$, and $G\beta_1$ subunits have been mapped in brain by *in situ* hybridization. The localization of these probes varies significantly. $G_s\alpha$ and $G\beta_1$ mRNAs show a similar extensive distribution throughout the brain, particularly in large neuronal cell bodies. On the other hand, $G_o\alpha$ mRNA is localized in fewer areas, including claustrum, habenula, hippocampal pyramidal cells, endopiriform nucleus, granule cells of the dentate gyrus, and the cerebellar Purkinje cells. The distribution of G_o mRNA in brain is consistent with that of the corresponding protein as determined by immunohistochemistry. The localization of $G_s\alpha$ mRNA in brain differs from the distribution of adenylyl cyclase, as determined by binding of labeled forskolin. This diterpene has been shown to interact with the catalytic subunit of the adenylyl cyclase. It is therefore possible that most of the brain G_s is involved in the coupling with effectors other than the adenylyl cyclase. This possibility has received experimental support with the finding that G_s can interact with voltage-gated Ca^{2+} channels.

The intracellular distribution of these messenger RNAs shows that the $G_s\alpha$ message is predominantly located in the cytoplasm as opposed to the mainly nuclear signal detected for $G\beta_1$ mRNA. It is possible that the transport of β subunit mRNA from nucleus to cytoplasm can regulate the synthesis of β subunits and the function of the different heterotrimers. *In situ* hybridization studies suggested that the relative abundance of mRNA encoding G protein subunits in brain is $\alpha_s > \alpha_o > \alpha_i$-like proteins. However, the steady-state levels of protein show a different proportion: α_o is approximately 5 times more abundant than the α_i's and 10 times more abundant than α_s. This would indicate differences in the translational efficiency of the mRNAs or in the turnover of the corresponding proteins.

Measurement of the concentration of G proteins in biological membranes has been carried out by different techniques, such as ADP-ribosylation catalyzed by bacterial toxins, binding of GTPγS followed by protein purification, immunoblotting, and extraction/reconstitution assays. Quantitations based on the transfer of ADP-ribose depend critically on the assay conditions, and generally detect only a fraction of the available α subunits, thus underestimating their true concentration. Some examples of concentration measurements based on immunochemical techniques are:

G_s in S49 lymphoma cells: 20 pmol/mg (0.2% of plasma membrane protein).
G_o in fibroblasts and adipocytes: 10 and 40 pmol/mg, respectively (0.1 and 0.4% of membrane protein).
$\beta_1 + \beta_2$ subunits in undifferentiated and differentiated 3T3-L1 cells (fibroblast-like and adipocyte-like, respectively): 70 and 150 pmol/mg, respectively (0.25 and 0.5% of membrane protein).
β_1 subunit in bovine cerebral cortex: 1.1 nmol/mg of protein (4% of membrane protein).
β_1 subunit in human erythrocytes: 11 pmol/mg of protein (0.04% of membrane protein).

Binding of GTPγS to bovine brain membranes and to fractions obtained in the purification of the G proteins shows that G_o constitutes approximately 1% of the brain membrane protein.

XII. DISEASES INVOLVING DEFECTS IN G PROTEINS

A. $G_s\alpha$

The CT-mediated ADP-ribosylation of G_s in intestinal cells causes a prolonged increase in intracellular cAMP, which in turn results in the large effluxes of salt and water in the gut of patients with the diarrhea of cholera.

A dominant inherited deficiency of α_s, due to the presence of an abnormal allele, is responsible for the resistance to parathyroid hormone (PTH) and other hormones present in type I pseudohypoparathyroidism (PHP-Ia). These patients show a 50% reduction in G_s activity in skin fibroblasts, lymphoblasts, and renal cells, together with olfactory dysfunction. The identification of α_{olf} and the observation that bulbectomy decreases the levels of its mRNA (but not α_s mRNA) suggest that changes in the levels of this protein might be responsible for the impairment in olfaction in PHP-Ia patients. [*See* Parathyroid Gland and Hormones.]

Two localized somatic mutations in the $G_s\alpha$ gene (R201C and Q227R) have been detected in approximately 40% of growth hormone-secreting human pituitary tumors. These mutations block the GTPase activity of the α subunit, which explains the constitutive activation of adenylyl cyclase characteristic of the tumor cell phenotype. The tumor genome contains both mutant and nonmutant alleles, indicating the dominance of the genetic alteration. The conceivable demonstration that expression of the mutant proteins in pituitary somatotrophs induces their malignant transformation defines the $G_s\alpha$ gene as a new oncogene (termed gsp). Similar mutations in $G_s\alpha$ have been identified in thyroid adenomas and in some thyroid carcinomas. [*See* Oncogene Amplification in Human Cancer.]

Mutations in the R201 of $G_s\alpha$ were also detected in several tissues of patients displaying the multiple endocrinopathies that characterize the McCune-Albright syndrome; this pattern appears to originate from an autosomal dominant mutation occurring during early embryogenesis and resulting in a variable mosaic of cells containing normal and mutant $G_s\alpha$.

The known inhibitory effects of cAMP on the proliferation of nonendocrine tissues appear to be the underlying reason limiting the tumorigenic potential of activating $G_s\alpha$ mutants to endocrine cells.

B. $G_{i2}\alpha$

Mutations blocking the GTPase activity of $G_{i2}\alpha$ were identified in some adrenal cortical and ovarian tumors.

C. Other Subunits

The induction of malignant transformation in fibroblast cell lines expressing activating mutations of $G_q\alpha$ and $G_{12}\alpha$ underlies the oncogenic potential of these subunits.

D. Other Pathologies Associated with G Protein Dysfunction

There is conflicting information regarding alterations in G proteins in diabetes mellitus. There is agreement, however, that the disease is associated with an increase in the liver cAMP content. Some groups have reported that the expression of rat liver G_i is abolished in streptozotocin-induced experimental diabetes. On the contrary, others have measured increased levels of liver G_s, with no significant changes in G_i, in similarly treated animals. The spontaneously diabetic BB/Wor rat, a model for insulin-dependent human diabetes, has increased levels of liver G_s and adenylyl cyclase activity. It has also been reported that the diabetic encephalopathy is associated with an altered balance between G_s and G_i/G_o activities, which is observed 14 weeks after the induction of the diabetic state. Clearly more experimentation is required to determine the alterations in G proteins directly produced by the absence of circulating insulin.

G protein dysfunctions have also been observed in genetic hypertension and congestive heart failure.

ACKNOWLEDGMENTS

The author wishes to thank Carlos Obejero-Paz for help in the preparation of the figures. Work in the author's laboratory is supported in part by NIH Grant GM 46552.

BIBLIOGRAPHY

The format and purpose of this encyclopedia excludes the citation of references in the text. For extensive review of the field and as a guide to the original research articles and recent experimental approaches, see the following articles:

Dickey, B. F., and Birnbaumer, L. (eds.) (1993). "GTPases in Biology," Vols. I and II. Springer-Verlag, Berlin.

Iyengar, R. (ed.) (1994). "Methods in Enzymology," Vols. 237 and 238. Academic Press, San Diego.

Neer, E. J. (1995). Heterotrimeric G proteins: Organizers of transmembrane signals. *Cell* **80**, 249–257.

Neer, E. J., and Smith, T. F. (1996). G protein heterotrimers: New structures propel new questions. *Cell* **84**, 175–178.

Growth, Anatomical

STANLEY M. GARN
University of Michigan

GLOSSARY

Aging Decreases primarily in size, functional capacity, or performance in onset and timing differing for different organs and tissues

Cusps Pointed tips of the teeth that are the earliest manifestations of odontogenesis

Development Changes in relative sizes of different body units caused by changes in localized growth rates; also increases in complexity caused by changes at the cellular level

Environmental effects Effects of nutritional level and disease processes on the tempo of growth, maturity timing, and size and proportions; also includes effects of toxins, penetrating radiation, trace elements, and atmospheric and waterborne pollutants

Epiphyses Bony caps on the ends of "long" (tubular) bones that first appear as secondary ossification centers and ultimately unite (or "fuse"), bringing linear growth to an end. The iliac crest also has its own epiphysis

Growth Increase in size primarily caused by an increase in the number of cells, although growth in size may also be caused by an increase in size of individual cells (hypertrophy)

Hypoplastic markings On the enamel surfaces of the teeth, resulting from disturbances in enamel formation during dental development

Incremental or "growth" lines Seen in tooth cross sections, regularly spaced and attributed to cyclical changes in growth during dental development

Lean body mass or lean body weight "Lean" tissue mass as contrasted with the mass of fatty tissue

Life cycle Entire range of defined or definable stages in the life span including the period of the embryo and the fetal period (prenatally) through infancy, childhood, adolescence, adulthood, and senescence

Maturation Attainment of defined steps or events leading toward the specified end points in size, complexity, or function (*vide* skeletal maturation, dental maturation, sexual maturation)

GROWTH, DEVELOPMENT, AND MATURATION are all simple concepts taken from the common language. *Growth* is by definition an increase in size caused by an increase in number of cells. *Development* involves changes in proportions caused by differential growth and an increase in complexity. *Maturation* includes the steps toward attainment of ultimate size, peak performance, or gametogenesis.

Growth in size may be gained through increases in the size of individual cells. Alternatively, a large number of smaller cells may come to occupy less space than a smaller number of larger cells. Development may include calcification of bony elements previously formed in cartilage, or simultaneous changes in dozens of organs. The maturation of the dentition, sexual maturation, and osseous maturation involve different and partially independent processes.

Growth of tissues may involve simultaneous gains and losses as tissues and discrete structures alter their form and "remodel." Development may include programmed cellular death. The maturation of different organs and tissues involves different growth rates and relative timing, so that neural, skeletal, and sexual units peak at different ages. Brain growth ends early but continues to increase in the complexity of neural connections and in storage of information. Skeletal units ("bones") may undergo involution (or loss) at some sites while still gaining at other sites.

I. GROWTH IN THE EMBRYO

Growth begins when one sperm successfully enters an ovum. This large fertilized cell then divides and subdivides into a larger number of smaller cells. There is a reduction in mass but an increase in cell number and complexity. Even at this early stage of embryogenesis, directionality is present. One end of the cell mass will become the head. Sidedness (laterality) is also established for the entire duration of life. [*See* Fertilization.]

By the time the embryo is 25 mm long, a surprising number of body parts have attained a recognizable appearance. The individual segments of the bones of the hand, although still tiny in size, already exhibit adult proportions, one to the other. However, the teeth are but rudimentary dental "germs" until later, and much of their development takes place in postnatal life.

During much of prenatal time the conceptus is only a tiny fraction of the mother's weight and body mass, with minimum demands on maternal tissues. During the entire 269 days of pregnancy, only 25 g of calcium is transferred to the conceptus, less than 4% of the mother's calcium stores. Most of this is added late in fetal development, and most of the gains in fetal weight are added late, at which time the demands on the mother and her nutrient and caloric reserves may then be heavy. Studies on neonates conceived and born during famines show the lasting effects of maternal undernutrition on the conceptus.

However, the fact that different organs and tissues form and complete at different times places great emphasis on the health of the mother and on her environment at these critical periods. Although the placenta provides a barrier for the protection of the fetus, it is only partial. Rubella (German measles) can affect the conceptus and bring about abnormalities of development, as can many viruses and toxins. Penetrating radiation at higher levels can seriously disturb embryogenesis. The workplace and the home now hold dangers for the conceptus not present in previous eras, yet improvements in food supply and distribution also minimize some of the hazards that existed in earlier hunting or agricultural communities. [*See* Embryo, Body Pattern Formation; Fetus.]

II. FIRST YEAR OF GROWTH

The human infant, and the infants of most other species, is best described as "an eating machine with a tremendous capacity for growth." So the human infant in Westernized countries quadruples or quintuples birth size in the course of 1 year, synthesizing an enormous amount of tissue. It does so by consuming food energy in vast amounts that may exceed 10% of its body weight per day.

Growth rates are highest in the first 3 months after birth, continuing the rapid growth of late prenatal time. Growth rates then decrease and decrease again at around the 9-month marker. By 9 months of age, energy needs for growth are sufficiently lower, and the infant (and the parents) may sleep through the night without a nocturnal feeding.

During infancy there is a large increase in muscle mass, and as the muscles grow under neural control, activities increase along a predetermined schedule. Rolling over, then hitching, and then creeping and crawling are the behavioral markers of neuromuscular growth, leading to the ability to stand at the end of the first year of life. Fat stores markedly increase through the ninth month, then plateau. The 1-year-old may be larger and with more lean tissue but only marginally heavier than at the 9-month horizon.

III. GROWTH DURING CHILDHOOD: THE PERIOD OF STRETCHING OUT

The period of growth we call "childhood" exemplifies the effect of changing relative growth rates on size and proportions. Brain growth (and neural growth in general) slows to a small annual pace, so the child no longer appears as large-headed and big-eyed as the infant. Changing relative growth rates of the axial and appendicular skeletal units increase the legs relative to the trunk so that the child is no longer so short-legged. Although the weight of fat and the percentage of fat both increase during childhood, it is the stretching out of the body that comes to our attention.

During infancy, the first of the permanent teeth emerges through the gums. Later, the roots of the deciduous teeth resorb, and they are shed and replaced by the successional teeth, which are larger for the most part and occupy more space in the jaws than their deciduous predecessors. With growth and remodeling of the facial skeleton, the face becomes longer and deeper, relative to the skull, and more adultiform in appearance.

Most of the postnatal ossification centers become radiographically visible during childhood, and their subsequent remodeling serves to provide useful measures of bone age ("skeletal age"). Individual tubular bones gain in widths and circumferences as well as in length, and the mineral mass gains volumetrically. However, blood-forming tissues continue to expand, so the medullary cavities also expand. Consistent with the increase in body mass and the synthesis of new tissues, caloric intakes also increase. So it is that undernutrition, diarrheas, and limiting protein may slow growth during childhood, increasing the size disparity between the poor and the affluent. Conversely, overnutrition is growth-promoting so that obese boys and girls have more lean tissue as well: obese children increase stature beyond the average and are usually advanced in postnatal ossification.

During childhood, weight is gained and stature increases at relatively constant rates, so that massed-data growth charts provide useful indications for clinical assessment. However, boys and girls reflect parental size in their growth progress, so parent-specific growth charts rather than massed-data percentile values are more appropriate for the children of tall or short parents.

Unusual body proportions during childhood typify many of the congenital malformation syndromes and so merit clinical attention. Growth failure or slow growth during childhood may indicate malabsorption states, emotional-disturbance dwarfism, or abnormalities of the growth-regulating hormones. Intervention, as with growth hormone, is now practicable, and even the genetically short child may be accorded a larger adult size if therapy is begun in time.

IV. SEX DIFFERENCES IN SIZE AND DEVELOPMENTAL TIMING

Long before birth the female is already more advanced than the male in many aspects of development and holds this advancement through skeletal maturity. The female is more advanced than the male in ossification timing, the timing of tooth formation and emergence, the timing of epiphyseal union, and the age at which gametogenesis begins. For some ossification centers, the conception-corrected sex difference in ossification may be as much as 25%: for some teeth, the conception-corrected sex difference in eruption timing may exceed 7%. Female advancement in sexual maturation is especially evident at ages 12–14, and it is reflected in dating behavior and (later) in the difference in spousal ages.

Long before birth and throughout life, the male is the larger of the two sexes. The male is heavier, also, because of a larger lean body mass. In adulthood, the lean body mass of the male exceeds that of the female by 3 : 2; the skeletal mass and mineral mass of the adult male exceeds that of the female by 4 : 3. Because males require more food energy to grow a larger body mass and to maintain it, the energy requirement of the male exceeds that of the female at all ages, and this is compounded (after sexual maturation) by higher voluntary and involuntary activity levels of the male and a higher basal oxygen consumption per unit of lean tissue. In fat weight, however, the female equals that of the male or may exceed it, so percent fat is higher in the female, providing an energy reserve against nutritional and reproductive stresses.

Even at those ages when the sex difference in length and weight is small, males are larger than females of the same developmental or "skeletal" age. Comparing the two sexes at the same developmental level (i.e., the isodevelopmental level), boys average some 7% taller than girls, a value comparable to the adult sex difference.

Why the female is advanced over the male developmentally is not clear. The earlier notion that attributed the difference to the second X chromosome present in females and not in males is not in accordance with knowledge of the XXX ("superfemale") or the XXY or XYY "male." Although the developmental advancement of the female over the male for the first two decades of life might (in theory) be associated with a shoter life span, exactly the reverse is the case.

V. GENETIC DETERMINANTS OF GROWTH AND DEVELOPMENT

Bone lengths, muscle volumes, and tooth dimensions all show strong evidence of genetic control, as demonstrated by sibling and twin correlations. So do such discrete developmental events as ossification timing and formation and emergence timing of the 52 deciduous and permanent teeth. Even birth weights follow family line, as documented by mother–daughter comparisons. Mother–daughter correlations for the age of menarche (first menstruation) are systematically positive, approximately 0.25. The order of ossification of the postnatal ossification centers also follows family line, comparing parents and children at the same ages.

These family-line resemblances in dimensions, timing, sequence, and growth rates are equally demonstrable in the poor and the rich, in poorly nourished populations and in Westernized countries. Genetic control is superimposed on environmental determinants, so that taller boys and girls are disproportionately the progeny of taller parents even though the "tall" children in a famine area population may be shorter than the short boys and girls in an affluent society.

Genetic control of individual bone lengths and localized growth rates is also demonstrable in congenital malformation syndromes, including many types of dwarfism. Animal breeders select comparable genetically determined extremes to produce "breeds" or varieties of particular economic value or as household pets. Terriers and bulldogs and Pekinese are the result of intensive inbreeding. These animal examples serve as useful models for some human extremes, especially the achondroplasias. Genetically small animals may evidence unusual variants of growth-regulating hormones, again of value to the understanding of their human analogues.

VI. NUTRITIONAL MODIFICATION OF GROWTH RATES, SIZE ATTAINMENT, AND BODY PROPORTIONS

From the fetal period through the third decade, growth rates and size attainment are affected by available energy, first supplied through the maternal circulation and (in the postnatal period) by food digested and absorbed. A small or abnormal placenta may limit fetal growth, as may maternal undernutrition, or a small maternal body size, or a low weight gain during pregnancy. Conversely, a large maternal size and a large weight gain during pregnancy favor prenatal growth. The "parity effect" (describing the tendency of later-born neonates to be larger in size) reflects the larger body mass and especially the larger fat weight of older mothers.

In the postnatal period, the caloric balance is a major determinant of growth rate, size attainment, maturity timing, and ultimate size. Obese infants, children, and adolescents are faster-growing and earlier to mature. The caloric density of the infant formula as well as the volume consumed directly affects growth rates. Undernourished boys and girls grow less rapidly, mature later, and are of smaller adult size. Food intolerances and malabsorption syndromes limit growth, as is evident in lactose intolerance, wheat-gluten sensitivity, and numerous allergic states. Emotional-deprivation dwarfism may result from reluctance to eat, refusal to eat, or (even in subteen girls) fear of fatness.

Nutritionally deprived children not only grow less rapidly but attain final size with shorter legs relative to trunk. They are more "paedomorphic" (i.e., more child-like in proportions). Although better-nourished children do attain sexual maturity and cease growth earlier in the third decade, the larger growth rate (cm/year) during the growing period more than compensates for the earlier cessation of growth. Even before the fifth year, well-nourished boys and girls are 3–8 cm taller or longer than their poorly nourished developing world peers, and the dimensional advantages of a more positive caloric balance continues during the entire growing period.

VII. OSSIFICATION OF THE ROUND BONES AND EPIPHYSES

The epiphyses at the ends of the tubular bones and the individual round bones are all fully developed during embryogenesis. However, most of them ossify in postnatal life, at which time they become radiographically apparent, and may be followed (in radiographs) through maturity. Postnatal ossification is earlier for every such epiphysis or for round bone in girls than in boys, and the sex difference in ossification timing exceeds 10% overall in girls. This is consistent with the female developmental advancement in general.

Epiphyseal union (union of the epiphyses to the shafts of long bones) is also earlier in the female, and it is the operational reason why girls cease linear growth earlier. However, there is considerable individual variability in ossification timing and timing of epiphyseal union. This is genetic in part, as shown in twin, sibling, and parent comparison. The timing of union is also nutritionally mediated, as shown in the obese and the lean, and in malabsorption syndromes.

There are also population differences in ossification timing that transcend socioeconomic differences.

The order or sequence of ossification of different bony nuclei also differs from family to family. The order of epiphyseal union also differs between families. Individuals may therefore differ much from the "textbook" order or sequence of ossification or union. Bone-to-bone differences in the timing of epiphyseal union and relative bone growth underlie ultimate differences in the relative length of different long (tubular) bones.

VIII. DENTAL DEVELOPMENT

The 20 deciduous or "primary" teeth and the 32 permanent or "secondary" teeth of human beings and other anthropoids develop in quite different fashion from the bony units, reflecting their different historical origins and different embryogenesis. All the teeth form and calcify from top to bottom (crown to root) beginning with the enamel cusps and terminating with closure of the apices of the roots. Each tooth spends the first part of its history within the bony jaws, then emerging as the roots elongate and continuing "eruption" until it meets its opponent from the opposite jaw.

Statistically, the teeth are less variable in their developmental timing than are bony units. Teeth are also far less affected by extremes of nutrition and by hormonal diseases such as hypothyroidism or hypopituitarism. Nevertheless, teeth may bear permanent markings such as incremental lines in the dentin and hypoplastic markings in the enamel as a result of injurious prenatal events, birth trauma, postnatal disorders, or antibiotic therapies during their formations. Like the spicules or bony projections within the marrow cavities of tubular bones, such incremental lines and enamel defects are permanent historical records long preserved in the teeth.

Because the teeth are resistant to nutritional extremes and show lower developmental variability, tooth formation and emergence timing provide a useful approach to age determination in forensic medicine. However, the teeth are quite variable as to number, and the last tooth of each morphological class (i.e., the second incisor, the second premolar, and the third molar) is most likely to be missing (agenesis) because of failure to form. The third molar ("wisdom" tooth) may be congenitally missing in the majority of individuals in some Asiatic populations, but it is rarely missing in native Australians.

As with other calcified structures, there are consistent sex differences in dental developmental timing, females being more advanced than males in all teeth, except for the third molar. There are also consistent population differences in tooth formation and movement, later in those of European ancestry than in Africans, Amerindians, or Eastern Asiatics.

IX. ADOLESCENT GROWTH

Just before the 10th year, especially in Westernized girls, a new set of growth-regulating mechanisms becomes operative, heralded by increased production of follicle-stimulating hormone (FSH), luteinizing hormone (LH), and several other hormones of pituitary origin that control the ovarian cycle and sexual development in both sexes. This is the onset of "adolescence" or of "puberty" (although the two terms have different meanings), and it reflects the responses of many different tissues and organs to hormones of gonadal (testicular and ovarian) and adrenal origin. Adolescence, for want of a better term, actually continues through the teens in human beings and is not complete until well into the third decade! [*See* Puberty.]

Functionally, the body is converted from an eating and growing organism to one capable of gametogenesis, sexual attraction, and (in the female) pregnancy and lactation. So the secondary sexual characteristics develop, in both sexes, including distinctive patterns of subcutaneous fat distribution and distinctively different patterns of body hair and facial hair distribution. There are also distinctively different patterns of sebaceous gland activity and the production of sexually attractant odorous substances (pheromones). Within the female body the uterus enlarges and its lining begins to undergo cyclic cellular losses and regeneration, as part of the menstrual cycle. In the male, the testes enlarge to produce more testosterone from the Leydig cells and to produce sperm from the Sertoli cells. The prostate enlarges to produce prostatic fluid.

There is, of course, a major increase in muscle mass in both sexes, more so in the male than in the female. However, all muscle groups are not equally affected; those of the legs are least differentiated during sexual development, and those of the back and neck far more. The supporting bones increase in length, until their linear growth is terminated by epiphyseal union. These tubular bones also gain in outer diameter (by subperiosteal apposition) and at their inner surfaces (by endosteal or inner surface apposition). So the skeletal mass increases and continues to increase for many years in both sexes, in human beings, in the primates, and in mammals in general.

Most boys and some girls are characterized by an adolescent "spurt" in linear growth, which may (at maximum) approximate 14 cm/year. At this time, and in the rapidly growing adolescent, growth is almost visible, at 0.5 mm/day. This is the time of greatly increased food ingestion to fuel the rapid synthesis of tissue, and this is a time when the mineral intake may lag behind requirements for bone growth. However, many girls and some boys do not demonstrate a textbook adolescent spurt. As a general rule the spurt is greater for those early to mature and lesser for those who are late to mature. The magnitude of the adolescent spurt and its duration is also a familial characteristic; both may be attenuated by inadequate nutrition during adolescence.

During adolescence the skeleton of the arms and legs grows more rapidly than the spine, so that the adolescent becomes more "leggy," like the adult. However, adolescents in undernourished populations retain a more child-like appearance, whereas those in Westernized nations are more old-looking (gerontomorphic) in proportions (i.e., with a long appendicular skeleton relative to the axial skeleton or trunk).

X. SECULAR TREND IN SIZE AND MATURATIONAL TIMING

During the past century and a half, Westernized populations have shown increased growth rates, larger body size, and an earlier age at first menstruation. Such generational or *secular* changes have not yet appeared in some developing world populations, but the secular trends in size and maturational timing in Japan since 1946 have been of outstanding magnitude. Increases in size from one generation to the next are also known for colony-reared primates as compared with wild-captured specimens. The trend toward larger size is not restricted to human beings.

With increased body size at all ages and longer legs and arms, the spatial needs of human beings have also increased, not only for school desks, airplane seats, beds, and mattresses, but also for coffins and cemetery vaults. Bigger men, women, and children need more food to fuel them, more fiber to clothe them, and more energy to cool and warm them, and they excrete more bodily wastes. The trend toward larger body size is therefore of both economic and ecological importance.

Although various explanations have been proposed for the secular trends in size and timing, including out-mating (i.e., interbreeding between genetically unrelated groups) leading to increased gene diversity in the offspring, the secular trend can be reversed under famine stress or wartime deprivations or when developing world economies falter. The secular trend was reversed in Germany after World War I and after World War II, in Russian cities under wartime stress, and in Jerusalem during the seige. This shows that nutrition plays a predominant role.

Hands and feet and legs and arms have shown disproportionate secular increases, so that bodily proportions as well as body size have changed over time. Faces have become longer and deeper although slightly narrower, as shown in parent to adult-child comparisons.

Since 1942, investigators have repeatedly suggested that the secular trend is at an end. However, additional data and larger samples confirm its continuation, particularly in the progeny of less affluent parents and in populations formerly characterized by small body size. Roman and Milanese boys and girls now exceed British stature norms, and formerly short peoples, such as the Japanese, may emerge as the taller peoples of the world, as economic conditions change.

XI. CONTINUING GROWTH IN ADULTHOOD

Although most growth charts end at age 18, linear growth continues for some years, through the mid-20s and often beyond. The lean body mass continues to increase through age 30. The skeletal mass increases through the end of the fourth decade. In most individuals in Westernized countries, fatty tissue (obesity tissue) increases through the seventh decade, through the increase in number and size of fat cells.

The skeletal volume, as apart from the skeletal mass, continues to expand throughout life. Tubular bones become larger, although not longer. Round bones (like the carpals in the hand) increase in size. The head grows in dimensions, so that the hat size increases. The facial skeleton continues to grow. Because the ears and the bony nose also grow throughout adulthood, there is a continuing change in appearance. The continued growth of skeletal volume can be advantageous, compensating in part for the diminution in the skeletal mass. The capacity for continued growth at the outer (subperiosteal) surface of tubular bone may be disadvantageous and possibly a contributing factor in osteoarthritis. The increase in fat mass or fat weight during adulthood may also be disadvantageous with respect to blood pressure and many chronic diseases. Growth does not necessarily end at age 20 or 25. For some organs and tissues it may continue throughout life.

BIBLIOGRAPHY

Bogin, B. (1988). "Patterns of Human Growth." Cambridge Univ. Press, Cambridge, England.

Daughaday, W. H. (ed.) (1979). "Endocrine Control of Growth." Elsevier, New York.

Falkner, F., and Tanner, J. M. (eds.) (1986). "Human Growth: A Comprehensive Treatise," 2nd Ed., 3 vols. Plenum, New York/London.

Goose, D. H., and Appleton, J. (1982). "Human Dentofacial Growth." Pergamon, New York.

Prader, A. (1984). Biomedical and endocrinological aspects of normal growth and development. *In* "Human Growth and Development—Proceedings from the Third International Congress on Auxology" (J. Borm, R. Hauspie, A. Sand, C. Susanne, and M. Hebbelinick, eds.). Plenum, New York.

Tanner, J. M. (1981). "A History of the Study of Human Growth." Cambridge Univ. Press, Cambridge, England.

Growth Factors and Tissue Repair

SHARON LEA AUKERMAN
GLENN F. PIERCE
PRIZM Pharmaceuticals

GLOSSARY

Angiogenesis Process of forming new blood vessels (neovessels)

Autocrine feedback loop Synthesis of a growth factor, which can stimulate the same cell

Cell activation Induction of new gene expression in quiescent cells, in advance of proliferation or differentiation

Chemotaxis Directed migration of specific cell types in response to a biochemical gradient generated by a protein or peptide

Contraction Closure of a wound by forces generated from within to bring the edges together

Extracellular matrix Structural framework between cells, which provides the connective tissue milieu for cellular functions

Fibroblasts Connective tissue cells responsible for synthesis of collagen and other extracellular matrix molecules

Granulation tissue Extracellular matrix and cells that fill open (excisional) wound defects

Growth factors Polypeptides of broad specificity and multiple cellular sources that mediate essential activities such as cellular chemotaxis, activation, proliferation, and differentiation

Macrophages Cells derived from blood-borne monocytes, which secrete multiple growth factors and other cytokines

Mitogen Factor that stimulates cellular proliferation

Reepithelialization Process of covering an open defect through keratinocyte migration and proliferation

THE PROCESS OF TISSUE REPAIR FOLLOWING injuries represents a complex series of physiologic, cellular, biochemical, and molecular events. These events are time-dependent and result in a cascade of numerous biological activities, which ultimately lead to a fully healed wound. Soft tissues include the epidermis and dermis, subcutaneous tissue, fascia, muscle, ligament, and tendon. The repair processes in these tissues are similar and involve the production of a scar to replace the damaged tissue. The sequence of events in soft tissue repair is also similar to that observed for hard tissue repair, such as bone fracture healing and cartilage repair, although the cells involved in hard tissue repair (i.e., osteoblasts and chondroblasts) possess activities and differentiated functions unique to bone and cartilage, respectively.

The field of pharmacologic modulation of soft tissue repair is in its infancy. Although the soluble, insoluble, and cellular mediators have not been precisely elucidated, the application of pharmacologic concentrations of purified polypeptide growth factors, cytokines, and extracellular matrix molecules has nonetheless resulted in the acceleration of normal repair and the reversal of deficient repair in a wide variety of dermal wound healing models in animals. However, early clinical results using these purified factors have been less than encouraging and their role in future therapies for acute and chronic wounds remains to be established.

I. INTRODUCTION: WOUND HEALING AND GROWTH FACTORS

Recent identification of probable mediators such as polypeptide growth factors have permitted an increased understanding of the mechanisms and sequence of normal tissue repair. In addition, understanding the differences between normal and abnormal or deficient wound healing is improving through experiments in which the normal tissue repair process is specifically interrupted. Abnormal repair may occur in individuals who receive drug therapy or radiation or have underlying circulatory or metabolic diseases, which affect their capacity to repair wounds, and may reflect defects at the cellular and mediator levels.

Polypeptide growth factors are a class of molecules recently discovered that are responsible for inducing epithelial, endothelial, or mesenchymally derived cells such as monocytes and fibroblasts to migrate, proliferate, and differentiate. They were initially detected in cellular extracts or conditioned media from normal cells such as platelets, or tumor cell lines, and were purified based on their abilities to stimulate quiescent target cell lines to proliferate. Once the activity was isolated, the proteins were sequenced, cloned, and expressed to permit more detailed studies of their biological activities.

As a class, polypeptide growth factors are released from (1) platelets, the first cells to enter injured areas; (2) macrophages, which subsequently enter wounds; and (3) fibroblasts, which are responsible for restoring structure to the injured area. Thus, growth factors are secreted and utilized by the cells responsible for repairing wounds. Most growth factors belong to families of related molecules, and each growth factor family has a different mechanism of action and different target-cell specificities, thus they are effective agents in different types of soft tissue wounds (Table I). Each family contains related proteins bearing no homology to proteins within other growth factor families. For example, the FGF (fibroblast growth factor) family has ten members, whereas the TGF-β (transforming growth factor-β) family has five members. [*See* Transforming Growth Factor-Beta.] Although members within a family may demonstrate some unique functions, they generally interact with cells via the same receptor (or closely related receptors), triggering identical or similar cellular responses. When growth factors such as PDGF (platelet-derived growth factor), EGF (epidermal growth factor), TGF-α, insulin-like growth factor-II (IGF-II), and basic fibroblast growth factor (bFGF, FGF-2) bind to their receptors, they trigger a receptor tyrosine kinase activity, in which the receptor is autophosphorylated on intracellular tyrosine residues. Other intracellular second messengers are subsequently activated, including G proteins, protein kinase C, and inositol phosphates, which culminate in translocation of the membrane receptor signal to the nucleus and specific early gene (e.g., c-myc, c-fos) transcriptional activity. Thus, intracellular signal transduction events initiate molecular and cellular activities considered essential for tissue repair, including cellular chemotaxis, proliferation, and differentiation.

II. NORMAL SOFT TISSUE REPAIR

A. Cellular and Extracellular Events

Normal repair occurs in an orderly sequence following injury (Table II). Following coagulation and the degranulation of platelets, an acute inflammatory phase results from rapid influx of bloodborne white blood cells, including neutrophils and monocytes (which differentiate into wound macrophages), into the wound. These cells are required to clear the wound of bacteria and debris and represent the initial nonspecific immune response to injury. In addition, the macrophages can be activated to secrete a number of growth factors, which induce fibroblasts to migrate into the wound. These fibroblasts are also activated by macrophage-derived growth factors to synthesize and secrete additional growth factors and extracellular matrix molecules, such as fibronectin and collagen, which are structural molecules that are required during the second phase of repair.

The acute inflammatory phase, also called the lag phase, of tissue repair occurs over the first 2–4 days, depending on the size of the wound (Table II). During this time, neutrophils and monocytes-macrophages predominate in the wound. Wound macrophages differentiate from circulating monocytes, which have been attracted to the wound by various chemotactic factors released during clot formation and platelet degranulation (such as PDGF and TGF-β). By Day 4, fibroblasts from surrounding dermis and underlying connective tissue migrate into the wound and become the predominant cells during the remainder of the repair process. Fibroblast content generally peaks during the first week, then gradually returns to baseline levels over time (i.e., months), as the wound and scar mature. The period of peak fibroblast content within

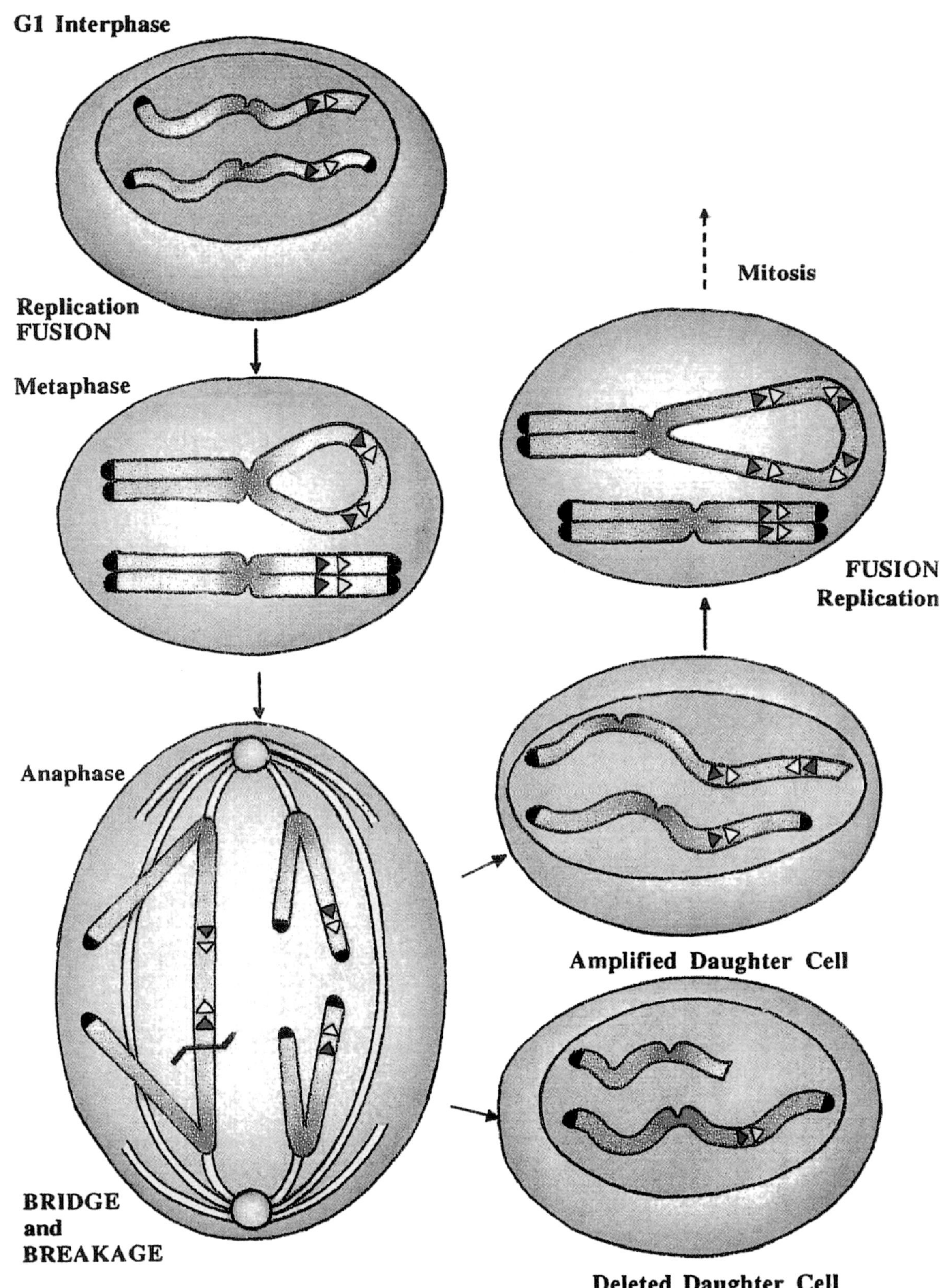

COLOR PLATE 1 Bridge–breakage–fusion cycles. The sister chromatids produced after replication of a broken chromatid fuse at the location of the break, generating a dicentric chromatid. At anaphase, the centromeres of the dicentric chromatid move to opposite poles of the mitotic spindle, creating a bridge which is later broken. Each daughter cell receives a broken chromatid which again replicates and forms another bridge, perpetuating the BBF cycles until the broken chromatid is healed. [*See* Gene Amplification.]

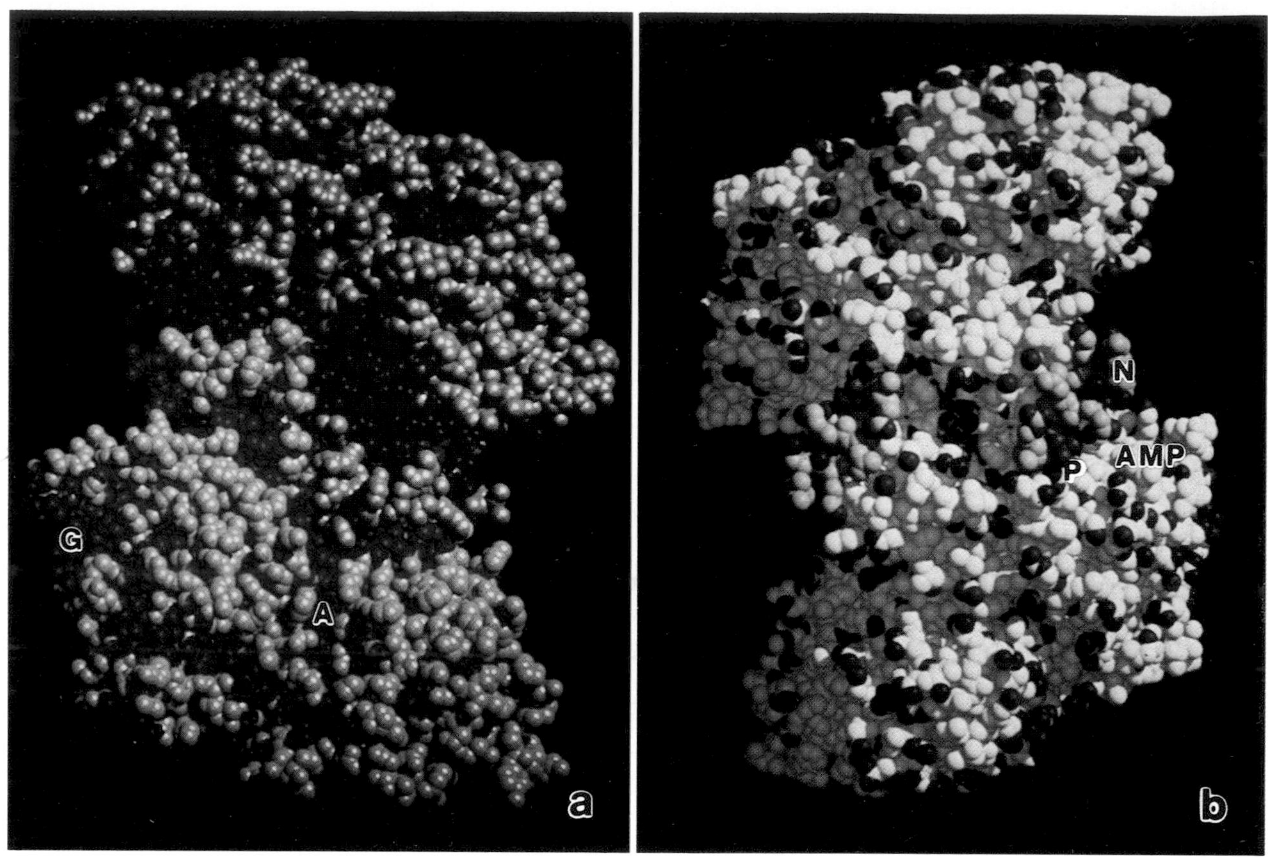

COLOR PLATE 2 Computer-drawn space-filling models of the phosphorylase *a* dimer. (a) The catalytic face with the subunits in different colors and A indicating entrance to active site, G the glycogen-binding site. (b) The control face with N indicating the amino-terminal peptide, P the phosphate of serine 14, and AMP the binding site for AMP. (Drawings are courtesy of Dr. R. Read and Dr. A. Muir, Department of Biochemistry, University of Alberta.) [*See* Glycogen.]

TABLE I
Major Growth Factor Families Important in Soft Tissue Repair

Growth factor family[a]	Important family members	Major sites of synthesis	Target cells	Activities
PDGF	AA, AB, BB	Platelets, macrophages, fibroblasts, endothelial cells, smooth and skeletal muscle cells	Neutrophils, macrophages, fibroblasts, smooth muscle cells	Chemotaxis, gene activation, proliferation
TGF-β	β1, β2, β3	Nearly all cells, largest sources are platelets, bone	Nearly all cells	Regulates proliferation and differentiation of multiple cell types
EGF	TGF-α, EGF	Platelets, macrophages, keratinocytes	Epithelial cells, endothelial cells, fibroblasts	Proliferation, gene activation
FGF	bFGF, aFGF, KGF	Macrophages, fibroblasts, endothelial cells, central nervous system	Endothelial cells, epithelial cells, fibroblasts, chondroblasts	Proliferation, chemotaxis, gene activation, capillary formation
IGF	IGF-I, IGF-II	Liver, mesenchymal cells, central nervous system	Nearly all cells	Proliferation, metabolic control
VEGF	121, 165, 189, 206	Tumors, macrophages	Vascular endothelial cells	Angiogenesis

[a]PDGF, platelet-derived growth factor; TGF, transforming growth factor; EGF, epidermal growth factor; FGF, fibroblast growth factor; KGF, keratinocyte growth factor; IGF, insulin like growth factor; VEGF, vascular endothelial growth factor.

the wound is considered the collagen synthesis phase of repair and lasts from approximately 3 days to 2–4 weeks postwounding, depending on the size of the wound (Table II). During this time, a high rate of new collagen type I synthesis occurs in soft tissue wounds and accounts for the increased physical strength of the wound. The wound strength increases in a linear fashion during the first 10 weeks postwounding and is directly correlated with increased net collagen synthesis. [*See* Collagen, Structure, and Function; Inflammation; Macrophages; Neutrophils.]

The final phase of repair, beginning approximately 2–4 weeks after wounding and extending to one year, is the remodeling phase (Table II). During wound remodeling, collagen cross-linking, collagenolysis, and collagen synthesis occur simultaneously and result in increased structural integrity within the wound. Cross-linking of new collagen is responsible for the significant gains in wound strength achieved during this phase. Collagen breakdown induced by collagenase activity coupled with new collagen synthesis permits increased interwoven collagen bundle formation

TABLE II
Cascade of Biological Activities in Soft Tissue Repair

Repair phase	Time	Cell influx	Biological activities
Coagulation	Immediate	Platelets	Initial growth factor release
Acute inflammatory	Days 0–4	Neutrophils	Phagocytosis, protease secretion
		Monocytes-macrophages	Phagocytosis, growth factor secretion
Collagen synthesis	Days 3–30	Fibroblasts	Growth factor secretion, glycosaminoglycans, fibronectin and collagen synthesis
		Endothelial cells	Growth factor secretion, angiogenesis
		Epithelial cells	Reepithelialization
Remodeling	Day 14 to 1 year	No influx—fibroblasts present	Collagenase-mediated remodeling, collagen synthesis, collagen cross-linking

within the scar. Together, these processes result in a mature collagen-containing scar, which is 80–90% as strong as unwounded dermis after one year. The processes that stop wound healing once repair is complete have not been identified.

Wound contraction is also an important component of soft tissue repair. Wounds heal by a combination of new granulation tissue formation (scar) and contraction of wound edges. Contraction of wounds is thought to be mediated by a specialized cell having characteristics of smooth muscle cells and fibroblasts, the myofibroblast. The myofibroblast appears to differentiate from wound fibroblasts, contains increased intracellular actin bundles, and is prominent during active wound healing. In loose-skinned animals, wound contraction is critical for survival following wounding. In humans, however, contraction may result in disfigurement and excess scarring (e.g., in burn patients), thus the generation of new tissue to fill an open defect frequently would be a preferable outcome.

B. Subcellular Events

The three major phases of wound repair offer a useful framework for following the sequence of healing but do not permit one to evaluate the mechanisms responsible for the repair process. Polypeptide growth factors have been recently shown to mediate many *in vitro* functions considered important for wound healing, such as cellular chemotaxis, proliferation, and activation to a differentiated phenotype. The recent availability of purified recombinant DNA-produced growth factors has permitted their systematic evaluation in models of normal and deficient soft tissue repair. Evaluating the influence of polypeptide growth factors on repair processes *in vivo* has permitted the dissection of the cellular and molecular activities required to initiate and sustain a tissue repair cascade. Growth factors may mediate many of the aforementioned cellular activities considered critical for normal repair of both incisional (i.e., a cut) and excisional (i.e., open granulating) wounds (Table III).

Monocytes-macrophages are required for normal wound healing. Macrophages synthesize and secrete numerous cytokines and growth factors, including interleukin-1 (IL-1), PDGF, TGF-β, bFGF, and TGF-α, and all have proven effects in different models of soft tissue repair (Table III). In addition, growth factors such as PDGF and TGF-β stimulate macrophage chemotaxis and induce the secretion of more growth factors. Therefore, through directed cellular recruitment, gene activation, and positive autocrine feedback loops, growth factors such as PDGF mediate macrophage activities required for wound healing. Thus, PDGF augments and exaggerates the acute inflammatory response induced in normal incisional and exci-

TABLE III
Wound-Healing Cell Types and Associated Growth Factors

Cell type	Growth factors secreted	Probable growth factor-mediated activities
Platelets	PDGF	Neutrophil, monocyte, fibroblast chemotaxis and activation; fibroblast proliferation
	TGF-β	Fibroblast chemotaxis, activation
	EGF	Keratinocyte chemotaxis, activation, proliferation
Neutrophils	None detected	None identified
Monocytes-macrophages	PDGF	Augment, sustain acute inflammatory phase; recruit fibroblasts
	TGF-β	Fibroblast extracellular matrix synthesis
	TGF-α	Same as EGF
	bFGF	Neovessel formation
	IL-1	Induces PDGF synthesis in fibroblasts
	VEGF	Neovessel formation
Endothelial cells	PDGF	Same as above
	bFGF	Autocrine stimulation
Keratinocytes	TGF-α	Autocrine stimulation
	TGF-β	Possible inhibition of keratinocyte proliferation
Fibroblasts	PDGF	Autocrine stimulation, collagenase secretion Extracellular matrix synthesis
	TGF-β	Collagen synthesis
	KGF	Keratinocyte proliferation

sional wound repair. TGF-β is less effective than PDGF in augmenting the acute inflammatory phase because it does not produce the degree of macrophage influx observed with PDGF.

Macrophage products such as TGF-β and PDGF are also potent chemotactic agents for fibroblasts and may mediate directed fibroblast influx into wounds (Table III). Furthermore, TGF-β and PDGF activate fibroblast gene expression and proliferation. Activated fibroblast genes include structural proteins such as fibronectin in addition to TGF-β and PDGF. Thus, positive autocrine feedback loops, similar to those identified in macrophages, permit the amplification of an initial exogenous growth factor signal and may initiate a cascade of activities leading to the increased synthesis of extracellular matrix proteins required for repair. [*See* Transforming Growth Factor, Beta.] Both PDGF and TGF-β have profound, but distinct, effects on extracellular matrix (Table IV). PDGF treatment accelerates and augments glycosaminoglycan synthesis within wounds, which establishes the provisional matrix required for cell migration and collagen deposition. TGF-β is the only growth factor identified that directly stimulates the transcription of procollagen type I. TGF-β also stabilizes procollagen type I messenger RNA, further increasing synthesis of the protein. Thus, although both growth factors induce new collagen synthesis within wounds, TGF-β circumvents the acute inflammatory phase, directly increases new collagen formation, and accelerates its maturation into mature bundles to a greater extent than PDGF. [*See* Extracellular Matrix.]

In contrast, bFGF treatment of wounds does not appreciably affect the inflammatory phase and actually decreases new collagen synthesis. bFGF induces increased collagenase secretion by endothelial cells, a requirement for vascular invasion and neovessel formation within wounds. bFGF is a potent angiogenic agent *in vitro* and directly stimulates the proliferation and migration of endothelial cells, as well as their differentiation into capillary sprouts, tubes, and functional neovessels. In contrast, PDGF and TGF-β indirectly induce supporting neovessel formation within wound granulation tissue but do not trigger the marked angiogenic response observed with bFGF. Both PDGF and bFGF, as well as EGF, accelerate reepithelialization of open wounds, whereas TGF-β actually inhibits keratinocyte proliferation.

Collagenase is an enzyme that degrades collagen and is required for wound remodeling and ultimate structural integrity of the wound. TGF-β decreases collagenase gene transcription *in vitro* and increases transcription of a specific collagenase inhibitor, tissue inhibitor of metalloproteases (TIMP). In contrast, PDGF stimulates collagenase gene expression and, therefore, may enhance wound remodeling (Table IV). PDGF may have a greater role than TGF-β during the remodeling phase of repair, when more collagenase activity is required to further increase the structural integrity of the wound. Therefore, although both growth factors augment soft tissue repair, they have unique and specific mechanisms of action. Thus, TGF-β stabilizes extracellular matrix and PDGF enhances matrix remodeling; importantly, both growth factors ultimately augment wound integrity without inducing excessive scar formation.

Normal soft tissue repair allows for reepithelialization of all types of wounds; however, the truly differentiated functions of the damaged tissue are restored only in partial-thickness injuries (i.e., injuries that penetrate but do not extend through the dermis). Retained epithelial elements such as glands and hair follicles repopulate the damaged dermis and allow full dermal regeneration. In contrast, in full-thickness injuries (i.e., those that extend through the dermis) and injuries involving muscle, tendon, or ligament, the differentiated function of the tissue is not restored, and newly synthesized scar tissue results in a permanent loss of the differentiated cell phenotype and function. Thus, an injury to muscle results in scar formation and not myocyte proliferation and differentiation into a contractile organ. Therefore, the inability to fully regenerate tissues results in scar formation. The determinants that govern full tissue regeneration are unknown but may be at the level of soluble mediators such as growth factors or at the cellular level (i.e., stem cells). This is presently an area of active investigation.

In summary, exogenous growth factor therapy of experimental wounds accelerates and augments the

TABLE IV
Influence of Growth Factors on Wound Extracellular Matrix

PDGF:	↑ Collagenase ↑ Glycosaminoglycans ↑ Procollagen type I = ↑ Matrix remodeling
TGF-β:	↑ Procollagen type I ↓ Collagenase ↑ Collagenase inhibitor (TIMP) ↑ Plasminogen activator inhibitor = Stabilized matrix
bFGF:	↑ Proteases, collagenases = Endothelial cell sprouts and buds, neo vessels

TABLE V
Influence of Growth Factors on the Major Biological Processes Required for Soft Tissue Repair[a]

Granulation growth factor	Reepithelialization	Neovessel formation	Tissue
PDGF	+	+	+
TGF-β	0/−	+	+
EGF, TGF-α	+	0	0
IGF	0	0	0
bFGF	+	+	0/+
KGF	+	0	0
VEGF	0	+	0

[a] +, positive effect; 0/−, no or slight inhibitory effect; 0/+, no or slight positive effect; 0, no effect.

three processes required for repair, namely, reepithelialization, neovessel formation, and granulation tissue (extracellular matrix) synthesis, through direct and inductive activities (Table V).

III. DEFICIENT SOFT TISSUE REPAIR

In many pathologic states, individuals have compromised wound healing despite the redundant nature of the repair cascade. Drugs such as glucocorticoids and chemotherapeutic agents can adversely affect tissue repair processes. Tissue ischemia, local irradiation, or total-body irradiation also diminish tissue repair capacities. Individuals with circulatory diseases such as arterial insufficiency and venous stasis disease are especially susceptible to chronic nonhealing ulcers on their lower extremities. Other individuals with nonhealing ulcers include persons with diabetes and neurosensory and neuromuscular defects. Thus, many individuals have compromised tissue repair due to multiple kinds of underlying diseases, although a central pathogenic component involves the sustained loss of viable tissue and repair functions from inadequate oxygenation and nutrition. As understanding of the functions of the proteins responsible for normal wound healing increases, therapy with purified polypeptide growth factors and matrix molecules may provide utility in states of compromised repair. In addition, FGF and VEGF (vascular endothelial growth factor), angiogenic factors that can induce neovessel formation and thereby increase oxygenation of tissues, may have a critical role in the repair process and in the subsequent maintenance of the healed wound. PDGF, bFGF, aFGF (acidic FGF), KGF (keratinocyte growth factor), IGF, and TGF-β, among others, have been tested in several animal models of compromised repair. These growth factors are capable of reversing decreased repair, although analysis of their activities in these models has revealed important differences in their mechanisms of action.

Glucocorticoids exert a profound inhibition of tissue repair through induction of a near total monocytopenia, resulting in the loss of wound macrophages and a direct inhibition of collagen synthesis by fibroblasts. TGF-β reverses the glucocorticoid inhibition of incisional healing, likely by substituting for the macrophage requirement of normal tissue repair. In contrast, PDGF is not as effective in reversing the deficient repair. Because PDGF is a powerful wound macrophage chemoattractant and activator of TGF-β synthesis, PDGF appears to require the macrophage to mediate repair.

Tissue ischemia is a central pathogenic element in the induction and maintenance of chronic dermal wounds in patients with multiple underlying diseases, ranging from atherosclerosis to diabetes. Ischemia results in lack of oxygenation and nutrients to the tissue, which is then susceptible to minor trauma. Ischemic wounds do not heal because the normal phases of repair do not occur at the levels required to sustain a cascade of biological activities because of the lack of oxygen, nutrients, and cellular influx into the wound. The roles of growth factors in antagonizing repair defects of ischemic wounds are only beginning to be investigated.

PDGF reverses an excisional (open) wound repair defect induced after experimentally rendering a rabbit ear ischemic. PDGF-treated wounds demonstrate enhanced extracellular matrix deposition (granulation tissue formation) and supporting neovessel formation. Ischemic wounds treated with the potent angiogenic agent bFGF also show enhanced neovessel formation within wound granulation tissue, although the healing rate of bFGF-treated wounds is not as accelerated. bFGF is unable to induce new granulation tissue comparable to PDGF; thus, the local increase in neovessels in the absence of normal large vessel flow is not sufficient to reverse deficient repair. Because PDGF does not stimulate endothelial cells directly to undergo differentiation into neovessels, PDGF is likely influencing repair through the enhanced migration and activation of macrophages within the wound, which then release direct angiogenic agents such as bFGF.

In contrast, when KGF, a member of the FGF family, is applied to ischemic wounds it accelerates epithelial healing and stimulates granulation tissue. In normal wounds, KGF stimulates only epithelialization, and does not influence matrix deposition. In ischemic wounds, KGF may be inducing epithelial cells to produce other growth factors, such as TGF-α, which can stimulate granulation tissue formation.

IV. USE OF SPECIFIC GROWTH FACTOR THERAPY FOR TISSUE REPAIR

A. Soft Tissue Repair

The complexity and clinical variability of wound healing have historically limited pharmacologic approaches to accelerate repair. Thus, the treatment of wounds has been dominated by dressings or other nonpharmacologic devices whose clinical efficacy either has not been established or is unsubstantiated by adequate clinical trials. The widespread interest in pharmacological augmentation of tissue repair began in the mid-1980s, but human clinical trials have been limited to this decade (Table VI). There are numerous clinical indications which could potentially be treated with growth factor therapy (Table VII).

TABLE VI
Growth Factors and Cytokines in Preclinical and Clinical Development

Growth factor/cytokine	Preclinical effect	Clinical trial results
PDGF	↑ Neovessels ↑ Reepithelialization ↑ Granulation tissue	↑ Healing in pressure and diabetic ulcers; under FDA review
TGF-β2	↑ Collagen-containing granulation tissue	↑ Healing in venous stasis ulcers; repair of macular holes not successful
TGF-β3	↑ Healing ↓ Scarring	Chronic ulcers, in progress
Basic FGF	↑ Neovessels in a provisional matrix, ↑ Reepithelization	Pressure ulcers—biological effect; under regulatory review in Japan Diabetic and venous stasis ulcers, no effect
IL-1β	↑ Healing in infected open wounds	Pressure ulcers, phase I study in progress
Acidic FGF	↑ Neovessels ↑ Matrix	Diabetic and venous stasis ulcers, in progress
EGF	↑ Reepithelialization	Donor graft sites, chronic ulcers, minimal effects

TABLE VII
Potential Clinical Applications of Growth Factors in Soft Tissue Repair

Category	Clinical application
Burns	Thermal burns Chemical burns
Surgery	Surgical incisions Dental surgery Eye surgery
Ulcers	Decubital (pressure sores) Venous stasis Diabetic Duodenal (intestine) Gastric Oral mucositis
Other indications	Skin grafting Acute and chronic liver disease Inflammatory bowel disease Lung fibrosis

PDGF has been extensively characterized in preclinical wound-healing models. It increases breaking strength of incisional wounds in both normal and impaired healing models. Many of the activities of PDGF are mediated through macrophages, resulting in a cascade of tissue repair processes such that a single topical application can increase incisional healing for at least 7 weeks. In excisional wound models, PDGF markedly increases granulation tissue, stimulates angiogenesis and epithelialization, and reverses the defect present in ischemic wounds.

Based on preclinical results, human studies have been conducted in patients with pressure sores and diabetic foot ulcers. In a small group of young paraplegics, a marked acceleration in healing rates was observed when aqueous PDGF was applied daily to ulcers for 28 days. However, even the control group experienced a marked reduction in ulcer size, presumably owing to the optimal hospital-based care given to all patients and to their young age. In another clinical study, PDGF was delivered in a gel to diabetic foot ulcers daily for 20 weeks. The percentage of wounds achieving complete closure was significantly increased in the treated group compared with the untreated group (48 versus 25%, $p = 0.01$); however, the results were not as impressive as would be anticipated based on the preclinical studies. In summary, strong preclinical data support PDGF as a therapeutic agent for nonhealing wounds. Moreover, although human studies indicate a biological effect of PDGF leading to cell recruitment, activation, and generation of new extracellular matrix, the degree of clinical benefit remains to be determined.

TGF-β1 was the first growth factor shown to accelerate normal wound healing, using a rat incisional model. This work has been extended using a variety of impaired and nonimpaired incisional and excisional wound models. However, the immunosuppressive and fibrotic properties of TGF-β have led investigators to exercise caution when initiating clinical studies. Early phase I studies in venous stasis ulcers are now complete, but no efficacy data from double-blinded, placebo-controlled trials have been reported. A phase I study has also just started for TGF-β therapy in oral mucositis (oral ulcers), a common side effect produced by cancer chemotherapy.

The best-studied member of the FGF family, basic FGF, is the prototype angiogenic factor, which stimulates endothelial and smooth muscle cells and induces proliferation of a variety of epithelial cells and fibroblasts. Although basic FGF is found bound to extracellular matrix, how it is released from cells is under investigation because of its lack of a signal peptide sequence for secretion. It may be released by cells damaged in injured tissue, but a novel energy-dependent export process from living cells has recently been identified. Animal studies have confirmed a marked angiogenic response. In excisional wounds, basic FGF increased granulation tissue. This increase was characterized by tremendous vascularity and provisional glycosaminoglycan-containing extracellular matrix, decreased collagen content, and increased epithelialization. However, basic FGF was inactive in ischemic excisional wounds. This lack of effect was reversed by treating the animals having ischemic wounds in hyperbaric oxygen chambers (raising tissue pO_2 to 300 mm Hg).

Clinical studies with the fibroblast growth factors have largely utilized basic FGF. Unfortunately, in diabetic foot ulcers the results have not been significant. Acidic FGF, a homologous isoform with a similar spectrum of targets, is undergoing clinical tests in diabetic and venous stasis ulcers.

EGF is probably the most extensively studied growth factor in medical science as it was one of the first growth factors ever to be purified. Although EGF has been tested in numerous models of normal and deficient wound healing, preclinical results suggested that it was not as effective as other growth factors (i.e., PDGF) at reversing deficient repair. Clinical studies in patients with chronic skin ulcers demonstrate EGF to be minimally effective in accelerating repair.

Novel approaches to the treatment of chronic wounds have included the application of extracellular matrix to fill in the wound cavity. However, matrix molecules such as hyaluronic acid, bovine collagen, and a modified hyaluronic acid containing Arg-Gly-Asp sequences, which enhance cell migration and attachment in cell culture studies, have not shown clinical promise. Composite grafts containing keratinocytes and a dermal substitute are currently under FDA review and may prove useful in healing full-thickness wounds. On a different note, as an alternative to using purified growth factors, investigators are now exploring gene transfer of growth factor genes to wound sites.

Other types of soft tissue repair processes do not necessarily require the enhanced extracellular matrix deposition and granulation tissue formation induced by PDGF and TGF-β and may benefit from specifically targeted growth factor therapy. These wounds would include partial-thickness dermal injuries such as burns, vascular surgery, corneal and retinal injuries, and acute or chronic wounds in internal (epithelial) organs, where epithelial-specific (i.e., EGF, bFGF, KGF) or endothelial-specific (i.e., bFGF, VEGF) therapy might be more appropriate.

B. Epithelial Repair

Partial-thickness dermal injuries are produced when the wound bed does not extend clear through the epidermal layer. [*See* Skin.] Examples of such wounds are skin graft donor sites, where skin is harvested from an uninjured area to cover full-thickness injuries, or when thermal or chemical burns occur. EGF clearly stimulates epidermal repair in animal excisional wounds and thermal injury models and may also stimulate dermal repair. When applied to donor graft sites in thermal injury patients, EGF caused a 1.5-day acceleration in healing, but this improvement is not clinically significant.

KGF is markedly up-regulated in wounds and, in contrast to all other growth factors studied to date, stimulates all epithelial elements within skin, including hair follicles, sebaceous glands, and sweat glands. KGF stimulates repair in part through stimulating adnexal elements that remain in the wound to repopulate the surface with keratinocytes. KGF has demonstrated activity in a number of preclinical models of tissue repair, including second-degree burns; however, the magnitude of its effect may not be of clinical benefit.

There have been numerous studies of growth factors and their potential role in the healing of corneal wounds (abrasions, incisions) and retinal wounds such as macular holes. Many of the same growth factors studied in soft tissue repair have been implicated in repair processes of the eye as well. The best studied of the growth factors in ocular models are EGF and TGF-β. Clinical trials with TGF-β in patients with macular tears have concluded, and although it was partially effective in repair of macular tears, it did not demonstrate a statistically significant effect in phase III clinical trials.

Numerous recent studies with purified growth factors have begun to investigate the role of these factors in repair of acute or chronic injury in internal (epithelial) organs. KGF is synthesized by fibroblasts and is believed to be a paracrine mediator of epithelial cell proliferation and differentiation in many systems besides the skin, such as the lung, stomach, pancreas, and intestine. For instance, KGF can specifically stimulate type II pneumocytes, the cells in lung alveoli that make surfactant, and may be effective in treating lung disease induced by toxins to type II pneumocytes such as radiation or increased oxygen. KGF is now in phase I clinical testing in cancer patients with oral mucositis. aFGF and bFGF have also been tested in several models of acute injury such as gastric ulcers. bFGF has even begun clinical testing in patients with poorly healing duodenal ulcers, but recent clinical studies proving a role of the bacterium *Helicobacter pylori* in the etiology of the disease suggest that antibiotic therapy may be the most effective treatment for duodenal ulcers.

C. Angiogenesis

The loss of functional blood vessels is a major cause of morbidity and mortality in humans. Growth factors such as aFGF, bFGF, and the more recently described vascular endothelial growth factor, a distant member of the PDGF family, are potent angiogenic agents *in vitro* and *in vivo*. They are capable of stimulating endothelial cell chemotaxis, proliferation, and induction of endothelial sprouts, tube formation, and capillary bud development *in vitro*. Although these activities are considered important for functional neovessel formation, the roles of growth factors in this process remain largely unknown. The mechanisms governing the differential formation of capillaries, veins, or arteries are also unknown.

Vascular endothelial cell growth factor is a potent inducer of angiogenesis and vasculogenesis in the embryo. Neutralizing antibodies to VEGF inhibit tumor growth and metastasis, implicating neovascularization as an essential process in tumor growth. Although VEGF has not been shown to accelerate healing in normal wound models, it appears to function better than basic FGF in ischemic wound models, since VEGF is up-regulated by hypoxia.

Functional neovessels would be required clinically in cases of ischemia and would include patients with diseased coronary arteries and diseased large central and peripheral vessels such as the aorta and femoral arteries, and in patients with microvascular diseases such as arteriosclerosis of peripheral distal vessels. Investigators have demonstrated that aFGF, bFGF, and VEGF can accelerate development of collateral circulation in ischemic limbs and myocardium in preclinical models, and more recently, in early clinical trials. Significantly, PDGF was demonstrated to form *de novo* arterial anastomoses in severed vessels in animal models.

D. Muscle, Tendon, and Ligament Repair

Wounds in muscle, whether the result of trauma to a skeletal muscle or a myocardial infarction that destroys cardiac muscle, heal via the same processes that govern dermal repair. Thus, muscle injuries heal through cell recruitment, extracellular matrix deposition, and eventual scar formation and not through myocyte-mediated repair. [*See* Muscle, Physiology and Biochemistry; Skeletal Muscle.] Growth factors such as bFGF are mitogenic for skeletal muscle satellite cells, the precursors of skeletal muscle myoblasts and mature myocytes. PDGF is synthesized by skeletal and smooth muscle cells and stimulates chemotaxis and proliferation in smooth muscle cells. These observations raise the possibility of the PDGF-mediated positive autocrine feedback loop within muscle cells, similar to that described earlier for macrophages and fibroblasts.

As with vessel repair, adequate preclinical models have not evolved to permit a quantitative analysis of growth factor therapy in muscle repair. A model of smooth muscle repair, an incision in the stomach of rabbits, was recently developed and has shown augmented repair with TGF-α therapy. These results, although preliminary, are of importance because most of the gastrointestinal tract is lined with smooth muscle, and inadequate repair following surgery (i.e., intestinal resections) is a major source of morbidity and mortality in patients.

Tendons and ligaments are modified collagen-containing soft tissue structures that have unique tensile and elastic properties. [*See* Connective Tissue.] When injured, as in the case of injured muscle, scar tissue forms, which frequently may inhibit or diminish the function of the tendon or ligament. An adequate inducer of repair that does not stimulate inflammation and subsequent scar formation has not yet been identified.

V. RELATIONSHIPS BETWEEN TISSUE REPAIR, DEVELOPMENT, AND OTHER PHYSIOLOGIC AND PATHOLOGIC PROCESSES

In many respects, the cells and activities required for tissue repair resemble the processes that occur in such diverse circumstances as embryogenesis, oncogenesis, metastasis, and atherosclerosis and in inflammatory diseases such as rheumatoid arthritis and pulmonary fibrosis. Growth factors released from mesenchymal cells are considered to play critical roles in inducing and sustaining some pathologic activities. Inflammation outside of an appropriate setting such as tissue repair involves recruited and activated macrophages and fibroblasts and likely results in a similar cascade of redundant growth factor and extracellular matrix activities, as observed in wound healing. In pathologic states, components of this cascade may lead to inappropriate cellular proliferation, fibrosis, granulation tissue formation, and scarring, resulting in such diverse diseases as intimal hyperplasia (i.e., an atherosclerotic plaque), restrictive pulmonary fibrosis, synovitis, liver cirrhosis, glomeruloephritis and other chronic inflammatory diseases.

Increased levels of PDGF have been observed in the placenta and uterus during pregnancy, suggesting an autocrine physiologic role for PDGF in trophoblast and smooth muscle cell growth, respectively. Other growth factors (i.e., bFGF and IGF-II) have been found in the placenta. These growth factors may gain access to the fetoplacental circulation and could potentially influence fetal growth and differentiation.

Importantly, growth factors such as bFGF,TGF-β, and PDGF have been implicated in critical mesoderm inductive processes in early morphogenesis. They are maternally encoded and therefore are present within the egg at the time of fertilization. Their critical roles in embryonic stem cell migration and differentiation result in the absence of detectable genetic growth factor deficiencies; such deficiencies likely would be lethal. Later in pregnancy, growth factors such as bFGF and TGF-β have been localized to discrete cell populations at specific times of organ or tissue growth.

Thus, synthesis and secretion of growth factors under appropriate circumstances likely permit critical homeostatic functions such as would healing, placental and uterine growth, and embryonic differentiation to occur, whereas aberrant control of growth factor synthesis and inappropriate secretion may result in fibroproliferative diseases or tumor growth and metastasis.

VI. CONCLUSIONS

Significant and rapid progress has been made in dissecting and elucidating the complex redundant sequence of soft tissue wound healing. However, the exact roles of growth factors in mediating this process require considerably more investigation before therapeutically useful targeted intervention can be achieved. Growth factor-mediated *in vitro* activities are not necessarily replicated within the complex *in vivo* milieu, where other cells and growth factors may either mask or inhibit activities or enhance inductive effects. Recent availability of recombinant proteins has permitted initial clinical evaluations of growth factor therapy to proceed in persons with deficient tissue repair. Only after specific time- and cell-dependent roles of individual growth factors have been elucidated in the repair process can appropriate therapeutic use of growth factors, either alone or in combination, occur. Early clinical trials with single purified growth factors have demonstrated underwhelming results compared to the preclinical model successes. It might be that real clinical benefits may be realized only when combinations of growth factors are used in various patient populations.

Future experiments will further identify the unique roles of specific growth factors in tissue repair, development, normal physiologic processes, atherosclerosis, inflammation, and malignancy. The precise regulation of growth factor gene expression within tissues may represent the critical difference between states of normal and uncontrolled cell growth. Thus, tissue repair models represent a useful paradigm in which to explore the diverse biological activities of growth factors.

BIBLIOGRAPHY

Andree, C., Swain, W. F., Page, C. P., Macklin, M. D., Slama, J., Hatzis, D., and Eriksson, E. (1994). *In vivo* transfer and expression of a human epidermal growth factor gene accelerates wound repair. *Proc. Natl. Acad. Sci. USA* **91**(25), 12188–12192.

Arrick, B. A., and Derynck, R. (1992). Growth regulation by transforming growth factor-β. *Cancer Treat Res.* **63**, 255–264.

Banai, S., Jaklitsch, M. T., Shou, M., Lazarous, D. F., Scheinowitz, M., Biro, S., Epstein, S. E., and Unger, E. F. (1994). Angiogenic-induced enhancement of collateral blood flow to ischemic myocardium by vascular endothelial growth factor in dogs. *Circulation* **89**, 2183–2189.

Beck, T. W., Magnuson, N. S. and Rapp, U. R. (1995). Growth factor regulation of cell cycle progression and cell fate determination. *Curr. Top. Microbiol. Immunol.* **194**, 291–303.

Boyle, W. J. (1992). Growth factors and tyrosine kinase receptors during development and cancer. *Curr. Opin. Oncol.* **4**(1), 156–162.

Brown, D. M., Hong, S. P., Farrell, C. L., Pierce, G. F., and Khouri, R. K. (1995). Platelet-derived growth factor BB induces functional vascular anastomoses *in vivo*. *Proc. Natl. Acad. Sci USA* **92**, 5920–5924.

Danilenko, D. M., Ring, B. D., and Pierce, G. F. (1996). Growth factors and cytokines in hair follicle development and cycling: Recent insights from animal models and the potentials for clinical therapy. *Mol. Med. Today* **2**, 460–467.

Housley, R. M., Morris, C. F., Boyle, W., *et al.* (1994). Keratinocyte growth factor induces proliferation of hepatocytes and epithelial cells throughout the rat gastrointestinal tract. *J. Clin. Invest.* **94**, 1764–1777.

Lindner, V., Majack, R. A., and Reidy, M. A. (1990). Basic fibroblast growth factor stimulates endothelial regrowth and proliferation in denuded arteries. *J. Clin. Invest.* **85**, 2004–2008.

Mason, I. J. (1994). The ins and outs of fibroblast growth factors. *Cell* **78**, 547–552.

Peters, K. G., DeVries, C., and Williams, L. T. (1993). Vascular endothelial growth factor receptor expression during embryogenesis and tissue repair suggests a role in endothelial differentiation and blood vessel growth. *Proc. Natl. Acad. Sci. USA* **90**, 8915–8919.

Pierce, G. F., and Mustoe, T. A. (1995). Pharmacologic enhancement of wound healing. *Annu. Rev. Med.* **46**, 467–481.

Pierce, G. F., Yanagihara, D., Klopchin, K., *et al.* (1994). Stimulation of all epithelial elements during skin regeneration by keratinocyte growth factor. *J. Exp. Med.* **179**, 831–840.

Takeshita, S., Zheng, L. P., Brogi, E., Kearney, M., Pu, L. Q., Bunting, S., Ferrara, N., Symes, J. F., and Isner, J. M. (1994). Therapeutic angiogenesis. A single intraarterial bolus of vascular endothelial growth factor augments revascularization in a rabbit ischemic hind limb model. *J. Clin. Invest.* **93**, 662–670.

Unger, E. F., Banai, S., Shou, M., Lazarous, D. F., Jaklitsch, M. T., Scheinowitz, M., Correa, R., Klingbeil, C., and Epstein, S. E. (1994). Basic fibroblast growth factor enhances myocardial collateral flow in a canine model. *Am. J. Physiol.* **266**, 1588–1595.

Hair

C. R. ROBBINS
Colgate Palmolive Co.

GLOSSARY

Alopecia Loss of hair; it may be localized or diffuse and over any part of the skin that contains terminal hair

Cortex Inner portion of a hair fiber; it contains the major part of the fiber mass and consists of spindle-shaped cells bonded together by intercellular cement

Cuticle (scale) Outer covering of hair fibers, consisting of several layers of flat, overlapping, scale-like structures that envelop each hair fiber

Dermal papilla Germinative tissue at the base of a hair bulb (in the follicle) that produces hair proteins

Hair follicle Microscopic canal leading from the skin surface to the inner dermis. One or more hair fibers grow out of a hair follicle

Hypertrichosis Excessive hairiness that may occur on the limbs, the trunk, the face, or the head. It may be endocrinologic in origin, induced by drugs, or of unknown origin

Keratin Group of highly specialized structural proteins containing the amino acid cystine found on the surface of mammals for protection and/or warmth; for example, hair, nails, claws, horn, scales, and feathers are very rich in keratin

Medulla Loosely packed region of cells containing vacuoles or air pockets, running along the axis of hair fibers, generally located near the center of thick, but not thin, animal hairs

Pilosebaceous unit Unit or system comprising a hair follicle and sebaceous gland

Terminal hair Long, thick hairs that are usually pigmented

Vellus hair Very short (about 1 mm maximum length), fine hairs containing no pigment

HUMAN HAIR IS A KERATIN-CONTAINING TISSUE that grows from large cavities called follicles extending from the surface of the skin through the epidermis to the dermis. Hair is characteristic of all mammals and provides protective, sensory, and sexual attractiveness functions. In humans, it grows over a large percentage of the skin surface. Hair fibers grow in a cyclical manner consisting of three distinct stages called anagen, catagen, and telogen. Structurally, a fully formed hair fiber contains either two or three different morphological units called cuticle, cortex, and medulla. Hair fibers are enveloped by layers of flat, overlapping cuticle scales. Within the cuticle layers is the cortex, the major part of the fiber mass. The cortex consists of fibrous cell remnants containing highly oriented α-helical proteins embedded in a highly cross-linked, less organized matrix. Thicker hairs often contain a third loosely packed porous region, the medulla, located near the center of the cortex.

I. DEVELOPMENT, TYPES, AND FUNCTIONS

A. Development of Hair and Types

Human hair contains keratins, highly specialized proteins produced in the bulb of the follicles. Hair keratins have several common features with other keratins, for example, those of nails, claws, scales, and feathers.

Keratin is not a single substance but a complex mixture of proteins characterized by a high concentration of the amino acid cystine.

Hairs grow from large cavities or sacs called follicles that originate in the subcutaneous tissue (Fig. 1). Rapid cell division occurs at the base of the hair bulb, beginning in the dermal papilla, located in the zone of differentiation and biological synthesis. These cells migrate upward as they move away from the base of the bulb, whereupon they elongate, dehydrate, and eventually die. Finally, these cells move into the zone of keratinization, where the fiber structure is stabilized by the formation of disulfide cross-links from cysteine residues. As the fiber continues to move upward toward the skin surface, it enters the region of the permanent hair fiber. The permanent hair fiber is a completely formed hair and consists of fully structured, dehydrated, cornified cells and intercellular binding matter.

In humans, prenatal hairs originate from the Malpighian layer (stratum germinativum) of the epidermis usually in the third or fourth month of fetal life. Prenatal hairs are called lanugo and are either lightly pigmented or contain no pigment at all. Lanugo hairs are usually shed prior to birth or soon thereafter.

Soon after birth, hairs of the eyelashes, the eyebrows, and the scalp begin to grow at a more rapid rate and the amount of pigment they contain usually increases. Smaller finer hairs called vellus hair grow over the rest of the body. These fine, short (about 1 mm long), unpigmented hairs grow on those parts of the body that appear "hairless," such as the "bald" scalp and the forehead. Terminal hairs are long and thick, and generally finer and shorter on the scalps of children than on those of adults.

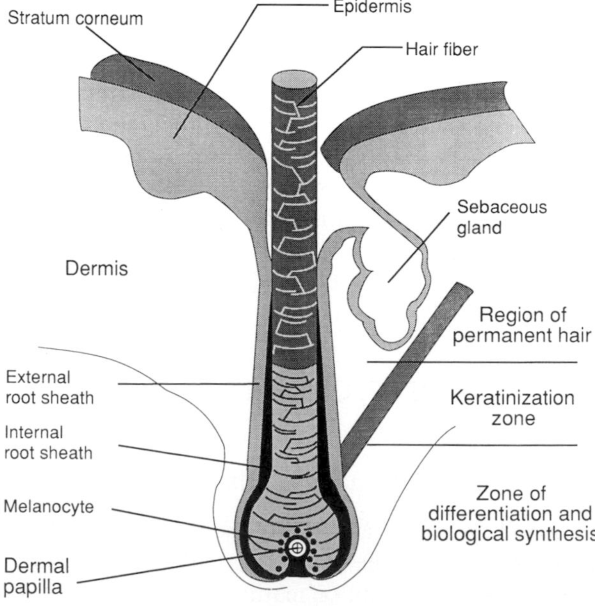

FIGURE 1 Schematic of an active hair follicle illustrating regions of synthesis.

Hair in the axillary, pubic, and beard areas (for males) becomes longer and thicker at the onset of puberty. Those hairs that develop more fully at puberty are called secondary terminal hair. With further aging, hairs become less pigmented (graying) and finer and ultimately hair loss or thinning occurs.

In humans, either terminal or vellus hair generally grows over most of the skin. Those truly hairless regions include the palms of the hands, the soles of the feet, the undersurface of the fingers and toes, the margin of the lips, the areolae of the nipples, the umbilicus, the immediate vicinity of the urogenital and anal openings, the nail regions, and scar tissue.

Terminal hairs, at one stage or another of the life cycle, grow on the scalp, eyelash area, eyebrow area, axillary and pubic areas, the trunk and the limbs, and the beard and mustache areas of males. Vellus hairs grow on all other areas except for the truly hairless areas mentioned.

B. Functions of Hair

Human hair provides limited protective, sensory, and sexual attraction or adornment functions and in an evolutionary sense may be becoming vestigial. Scalp hair provides warmth by insulating the scalp and it can enhance sexual attractiveness. Hair also protects the scalp against light radiation and mechanical impact. Hairs provide protective and/or adornment functions on other parts of the body as well. Eyebrows protect the bony ridges of the eye sockets and inhibit sweat and other liquids from running into the eyes. Eyebrows and eyelashes are both important as adornment. Eyelashes protect the eyes from sunlight and from foreign objects and assist in communication. Nasal hairs help to filter air as it enters the respiratory system.

All hairs are attached to sensory nerve endings, therefore, hairs can function as sensory receptors. For example, one can sense an object by touching hair without even touching the skin.

II. HAIR GROWTH

A. General Features of Hair Growth

The dermal papilla, at the center of the bulb (see Fig. 1), is responsible for controlling the growth cycle of hairs. Basal layers, which produce hair cells, nearly surround the bulb. Melanocytes, which produce hair pigment, are also within the bulb, and blood vessels feed and nourish the growing hair fiber near the base of the bulb.

All human hair fibers have three distinct stages in their life cycle, called anagen, catagen, and telogen (Fig. 2). *Anagen* is the period during which the hair fiber grows. The growing period of hair varies not only among individuals, but also depending on the region of the body. Anagen for scalp hair usually lasts longer than for hair growing on other body regions, usually 2 to 6 years, compared to less than 6 months for hair on the finger.

Catagen is a transition period when metabolic activity slows down in the dermal papilla. During catagen, the bulb shrinks and moves toward the epidermal surface. Catagen usually lasts only 2 to 3 weeks. *Telogen* is the resting stage when growth stops. At telogen, a new bulb begins to grow beneath the shrunken follicle. The old fiber is pushed out and is eventually shed and replaced by a new hair fiber. Therefore, the observation of hairs in the shower after shampooing is generally nothing to be alarmed about; it is usually normal hair fallout. Estimates of the length of telogen are in the range of 6 to 12 weeks. Since hair grows in a mosaic pattern on the scalp, in any given area hairs are in various stages of their life cycle.

Terminal human hair fibers grow at slightly different rates on different body regions. For example, hair grows at approximately 16 cm/year on the vertex or the crown area of the scalp, but at a slightly slower rate (about 14 cm/year) in the temporal area and at a much slower rate in the beard region (about 10 cm/year).

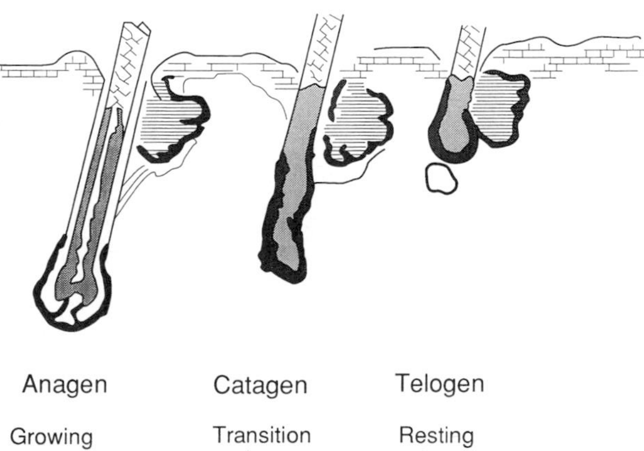

FIGURE 2 Schematic illustrating the life cycle of a human hair fiber. [From C. R. Robbins (1994). Reprinted with permission of Springer-Verlag, New York.]

Since human scalp hair grows at a rate of approximately 16 cm per year and anagen generally lasts 2 to 6 years, scalp hair often grows to a length of 1 m before it is shed. In long hair contests, lengths greater than 1.5 m are frequently measured.

B. Abnormalities in Hair Growth

In rare cases, for unknown reasons, anagen of scalp hair persists for decades. Thus, human scalp hair much longer than 3 m has been documented.

Hypertrichosis or excessive hairiness may be localized or diffuse. The most common type of hypertrichosis is called essential hirsutism or idiopathic hypertrichosis of women. This is the condition in which terminal hairs grow on women in those areas where hairiness is considered a secondary sex characteristic of males, for example, the trunk, the limbs, or the beard or mustache areas. Such women generally do not show endocrinologic abnormality.

Endocrinopathic hirsutism is an uncommon condition resulting from excessive synthesis of hormones with androgenic properties. This condition produces masculinization of females, one symptom of which is excessive growth of terminal hairs in regions that are normally "hairless" in females. Classic examples of this disease are sometimes exhibited in circus side shows, for example, Pastrana, the Mexican bearded lady, who appeared in London in the 1850s.

Hypertrichosis can also be induced by drug therapy. Streptomycin, estradiol, oxandrolone, or minoxidil, when taken internally, can produce an excessive growth of hair on the limbs, the trunk, or the face.

C. Hair Loss

Alopecia (hair loss) may occur over any body region such as the scalp, face, trunk, or limbs. However, alopecia of the scalp, for cosmetic reasons, has received the most attention. Several types of alopecia are commonly seen on the scalp, all involving the follicle. The most common forms are usually linked to the endocrine system or to stress. Some of the different types of stress associated with hair loss are: physical, psychological, disease or illness, or chemically induced stress (including drugs or the immune system).

Hair growth and hair loss are often linked to the endocrine system. The most common form of hair loss is genetically involved, thus the term androgenetic alopecia. Androgenetic alopecia or common baldness is a normal aging phenomenon and occurs in both sexes. It occurs in about 4 out of 10 men and in about 1 out of 10 women. In women, the hair loss is generally more diffuse. Longer hair in women also tends to help cover up the hair loss. In males, hair loss occurs much faster in the vertex and frontal areas of the scalp, thus the term "male pattern baldness."

The rate of conversion from anagen to telogen stages determines whether hair loss or growth occurs. Androgens produced by the adrenals and the sex glands help to control this phenomenon. These hormones are transported via the bloodstream to the pilosebaceous units. Testosterone is converted to dihydrotestosterone (DHT), a more active form of this hormone. The hormones migrate into hair cells in the bulb of the follicles and link to specific receptor proteins, thus forming the active agent in hair protein synthesis. The fact that testosterone induces alopecia in the vertex and frontal areas of the scalp, but induces hair growth in other regions at different stages in life (e.g., the axilla and pubic areas during puberty), suggests that different pilosebaceous units can be programmed to respond to androgens to either induce baldness or grow hair. This activity is a specific local tissue response controlled by the specific receptor protein–hormone combination.

A normal scalp contains 175 to 300 terminal hairs per square centimeter and about 85% of these are in anagen, 1 to 2% in catagen, and the remainder in telogen. In androgenetic alopecia, progressive miniaturization of hair follicles produces smaller vellus-type hair from terminal hairs. As hair loss progresses, the percentage of hairs in anagen decreases while the percentage of hairs in telogen (at rest) increases. A small reduction in the number of follicles also occurs in the bald scalp.

Both during and after pregnancy, shedding rates normally decrease. However, some women during pregnancy report thinning of scalp hair. It is likely that these effects are related to endocrine changes.

Alopecia areata is believed to be related to the immune system, for example, autoimmunity. This disease generally occurs as patchy baldness on an otherwise normal scalp, although sometimes hair from other body regions is affected. When the entire scalp is involved, the condition is called alopecia totalis. If terminal hair loss occurs over the entire body, which is rare, it is called alopecia universalis. Emotional stress has been shown to be an initiating cause of areata. Topical application of steroids is sometimes used to treat this condition.

Trichotillomania is a term used to denote alopecia induced by physical stress, that is, physically pulling or twisting a localized area of hair until noticeable thinning develops. This type of hair loss sometimes occurs in children who unconsciously pull or twist a region of hair. A similar type of hair loss may also occur in adults.

Telogen effluvium indicates sudden but diffuse hair loss due to an acute physical or psychological stress. This condition usually lasts only a few months and is reversible. Drugs used in chemotherapy often induce alopecia, however, this type of hair loss is usually reversible.

III. STRUCTURE OF THE PERMANENT REGION

A. General Structural Features

For practical scientific reasons, terminal human hair fibers are assumed to have a circular cross section. However, some hairs (particularly kinky hair) deviate a great deal from circularity. In fact, the cross sections of hair fibers are better defined as ellipses; for scalp hair the major to minor axis ratios from three racial groups are given in Fig. 3. With the approximation of circularity, the diameter of human scalp hair varies from approximately 30 to 120 μm, being finer in prepuberal children than in adults. The average diameters of hair fibers from the scalp, axilla, and thigh are similar, whereas hair from the pubic region is about 30% thicker. Cross sections of beard hair are

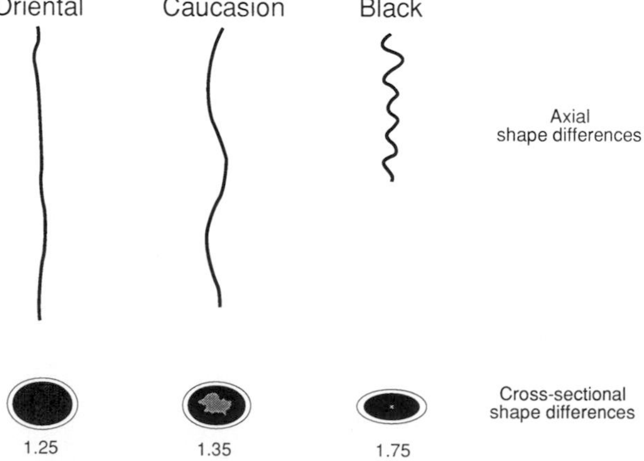

FIGURE 3 Some characteristics of scalp hairs of different racial groups. [From C. R. Robbins (1994). Reprinted with permission of Springer-Verlag, New York.]

also more irregular than for scalp hair. The degree of curl or axial shape of human hair also varies considerably. On average, scalp hair tends to be straightest for Asians and most curly for Blacks, although pubic hair and beard hair tend to be more curly than scalp hair.

B. The Cuticle

Cross sections of the permanent region of human hair fibers generally show three distinct regions (Fig. 4). The outermost region, consisting of several layers of flat, overlapping, scale-like structures, is called the cuticle. The cuticle surrounds the major part of the fiber mass called the cortex. In thicker hairs, generally near the center of the cross section, a third region, the medulla, is sometimes present.

The cuticle is chemically resistant and serves as a protective covering for hair. Cuticle scales are translucent and for scalp hair do not contain pigment. They are attached at the root end and resemble shingles on a roof (Fig. 5). Each scale is about 0.5 to 1 μm thick and about 45 μm long. The cuticle of human scalp hair is about 5 to 10 scale layers thick near the scalp. The number of scale layers often varies among different animal species and can sometimes assist in species identification in forensic studies; for example, wool fiber generally contains 1 to 2 scale layers and human hair 5 to 10, but furs may contain up to 20 scale layers.

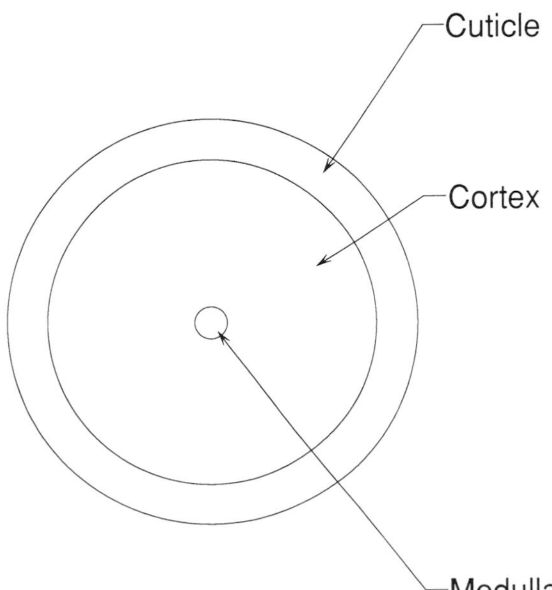

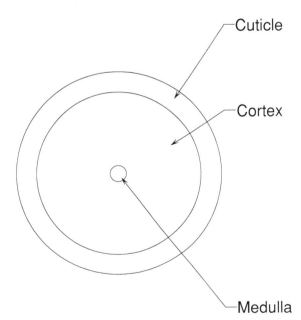

FIGURE 4 Schematic of a cross section of the permanent region of a hair fiber illustrating the three morphological regions.

A cuticle scale, the residue of a highly deformed, flattened cell, is still surrounded by a structure originating from the cell membrane. Internally, each scale has three layers (Fig. 6): the A-layer, the exocuticle, and the endocuticle. The A-layer is chemically resistant and contains a very high concentration of the amino acid cystine (~30%), which cross-links neighboring polypeptide chains through disulfide bonds to build a rigid structure. The exocuticle is also rich in cystine (~15%), whereas the endocuticle has very little cystine (~3%).

As the permanent hair fiber emerges through the skin it contains smooth, unbroken scale edges (Fig. 5, upper left). However, mechanical effects such as combing and brushing and chemical action from cosmetics (e.g., permanent waves, bleaches, and hair straighteners) and from sunlight degrade the cuticle (Fig. 5, upper right and lower left). In some extreme cases, severe mechanical damage can produce split ends (Fig. 5, lower right).

C. Intercellular Matter

The intercellular region, sometimes called the cell membrane complex, consists of cell membranes together with adhesive matter that binds both cuticle and cortical cells. The cell membrane complex is primarily proteinaceous matter and structural lipid that is highly resistant to alkalinity and to reducing agents.

D. The Cortex

The cortex constitutes the major part of the fiber mass. It is approximately 90% of the dry weight of hair and consists of spindle-shaped cells oriented parallel to the fiber axis (see Fig. 6). Each cortical cell is approximately 6 μm thick and 100 μm long. Cortical cells contain helical proteins embedded in a matrix rich in cystine cross-links. Each cortical cell is composed of smaller filamentous structures called macrofibrils (see Fig. 6), which in turn consist of highly organized filaments called microfibrils surrounded by a less organized matrix that is highly cross-linked via cystine.

The microfibrils contain even smaller filaments called protofibrils that are composed of α-helical proteins. The α-helical proteins of the cortex of hair produce its well-known wide-angle X-ray diffraction pattern that provided the basis for the pioneering protein structure work of L. Pauling and R. B. Corey.

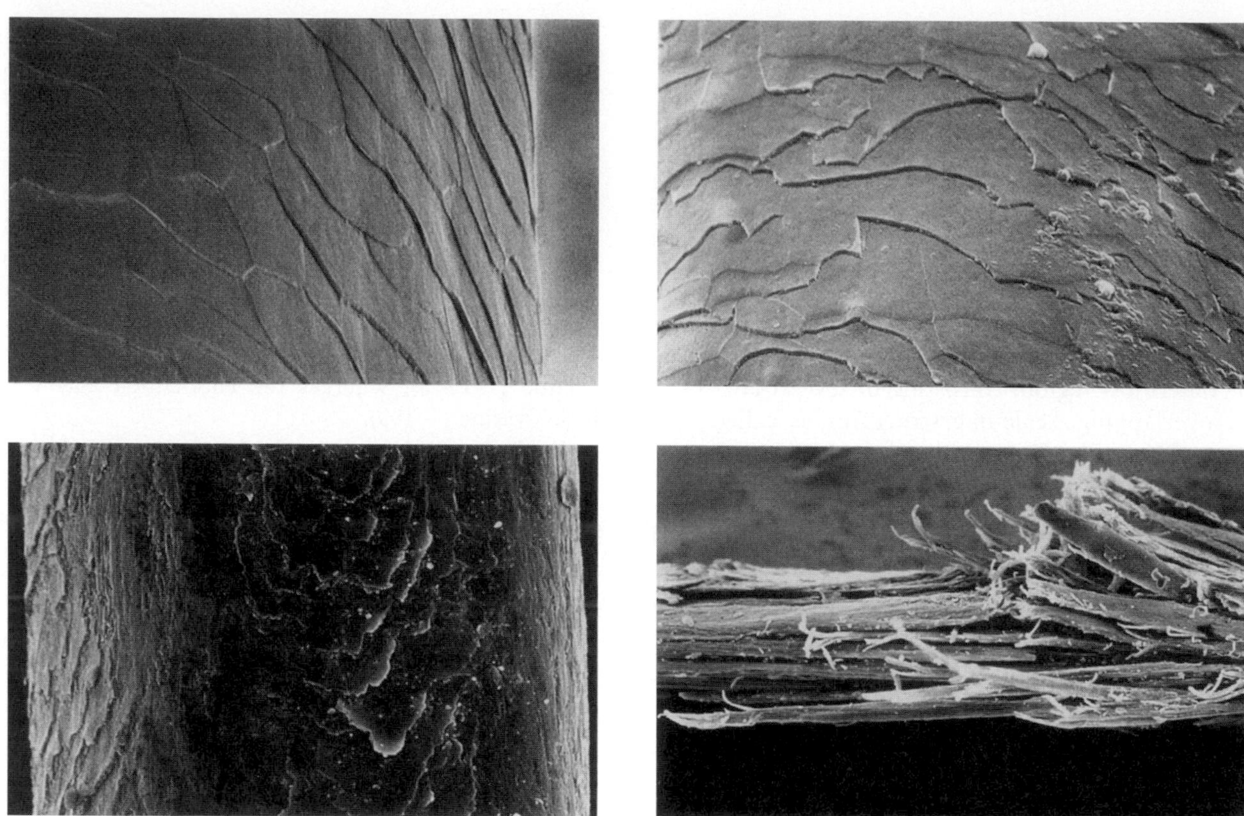

FIGURE 5 Scanning electron micrographs illustrating changes in surface architecture of hair caused by weathering. Upper left: near scalp, note smooth cuticle scale edges and scale surfaces; upper right: 6 inches from scalp, note broken scale edges; lower left: 12 inches from scalp, note broken scales and worn cuticle; lower right: 18 inches from scalp, note split hair fragments.

E. The Medulla

The medulla region is present only in thicker hairs, such as thick human hair, horse tail or mane, and porcupine quill, and is generally absent in most fine human hair. When present, it generally constitutes only a small percentage of the fiber mass. The medulla consists of a porous, loosely packed region of cells with air pockets. It may be continuous along the fiber axis, or it may be present or absent at different spots along the fiber axis. Some thicker animal hairs contain multiple medullas. Two medullas or a double medulla may be present in thicker human hair. The medulla enhances the insulation properties of hair fibers in fur-bearing animals, but otherwise provides no known function in human hair.

IV. CHEMISTRY OF HAIR

A. General Chemical Components

Morphologically, human hair contains two and sometimes three different types of cells and intercellular binding material. Chemically, human hair contains four different types of components: minerals, pigments, lipid, and proteins.

B. Minerals

A large number of elements have been reported in human hair. Some of the elements found in hair, other than C, H, O, N, and S, are: Ca, Mg, Sr, B, Al, Na, K, Zn, Cu, Fe, Ag, Au, Hg, Pb, Sb, Ti, W, Mo, I, P, and Se. The primary sources of these elements in hair are from metabolic irregularities, sweat deposits, and the environment. The most important environmental sources of trace elements in hair are cosmetics, the water supply (bathing, washing, and pool water), and air pollution.

Certain heavy metals in the body tend to accumulate in hair at concentrations well above those present in blood or urine. For example, levels of some toxic metals such as cadmium, arsenic, mercury, and lead in hair have been shown to correlate with the amounts of these metals in internal organs. Human hair has been used as a forensic tool in heavy metal poisoning

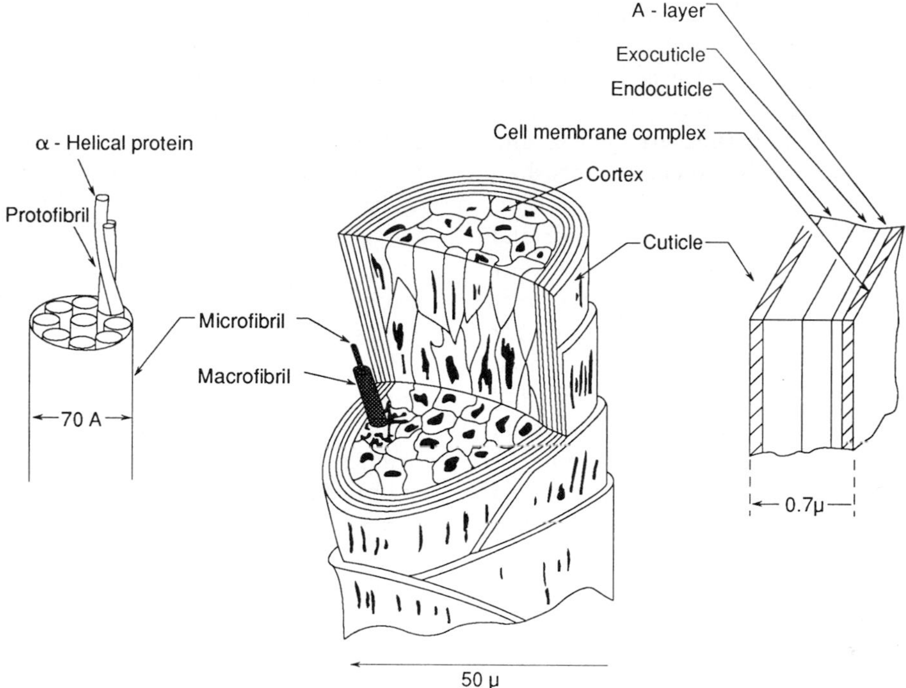

FIGURE 6 Stereogram of the human hair fiber structure illustrating substructures of the cuticle and cortex. [From C. R. Robbins (1994). Reprinted with permission of Springer-Verlag, New York.]

and can be used as a diagnostic tool as well. High levels of cadmium have been reported in dyslexic children and there is an unusually high ratio of potassium to sodium in hair of persons suffering from celiac disease.

Of metals derived from cosmetics or consumer products, lead has been demonstrated in hair treated with dyes based on lead acetate. Zinc and selenium can be detected in hair treated with antidandruff products containing these metals. Many shampoos or toilet soaps contain potassium, magnesium, or sodium ions that can deposit or bind to carboxylate groups in hair. Metals such as lead, cadmium, and copper in hair can derive from air pollution.

C. Hair Pigments

The cortex of human hair also contains remnants of cell nuclei and pigment granules. The pigments of human hair are the brown-black melanins (eumelanins) and the red melanins (pheomelanins). In scalp hair the pigment granules exist only in the cortex and the medulla, but not in the transparent cuticle. Hair from other regions of the body, such as the beard area, sometimes contains pigment in the cuticle as well as the cortex and the medulla.

The formation of hair pigments takes place in the melanocytes (melanin-producing cells) starting with the amino acid tyrosine. The reaction is enzymatically controlled and dopaquinone is believed to be a common intermediate for both the brown-black and the red pigments. Oxidative polymerization of dopaquinone in the absence of cysteine produces brown-black pigments, but in the presence of cysteine produces red pigment. Melanosomes or pigment granules are injected into newly differentiating cells near the base of the bulb. Thus the pigments are within the cells as they undergo keratinization.

The amount of pigment in the permanent hair fiber varies among racial groups. Generally, there is more pigment in the hair of Blacks and Asians than in the hair of Caucasians. Melanocytes as well as keratinocytes are dormant during telogen (resting phase), but they resume activity during anagen. Graying results when melanocytes become less active during anagen.

D. Hair Lipids

Lipids exist in human hair as free lipid or as structural lipid. The structural lipids of human hair are generally associated with the cell membrane material and have not been fully characterized. The free lipid of human hair is primarily of sebaceous origin and its concentration is governed by several factors, including androgens (which control sebaceous gland activity), shampooing frequency, and rubbing against objects such as combs, brushes, and pillows.

Free lipid exists both on the surface of hair and inside the fibers. Shampoos generally remove only surface lipids. The free lipid of human hair is very complex chemically and it is primarily responsible for the condition known as oily hair. It consists of about 50% free fatty acids (saturated and unsaturated) and neutral lipids, including triglycerides, free cholesterol, wax esters, paraffins, and squalene. The free lipid of oily hair differs from that of dry hair in both amount and composition. Thus oily hair is more fluid and contains a higher percentage of wax esters, a higher ratio of unsaturated to saturated fatty acids, and a larger percentage of cholesterol esters than lipid from dry hair. [*See* Lipids.]

Free lipid from children's hair generally contains higher concentrations of paraffinic hydrocarbons and lower concentrations of squalene and cholesterol than that from adult's hair.

E. Hair Proteins

At low humidities when the water content is low, hair contains more than 90% proteins, mostly structural proteins with high concentrations of the amino acid cystine, up to 18% in human hair. Hair also contains hydrocarbon, hydroxyl, primary amide, basic amino acids, and carboxylic amino acid side chains, all at high frequencies or concentrations.

It must be remembered that hair fibers are a complex tissue composed of several different structural regions that are chemically and physically different. For instance, they vary greatly in the concentration of cystine. Therefore, they respond differently to cosmetic or chemical treatments.

Some of the proteins of hair are hygroscopic, therefore, hair responds to changes in relative humidity by either picking up water or losing it. The water content of hair varies from a theoretical zero at 0% relative humidity (RH) to about 32% of its own weight at 100% RH. The binding of water to hair and its content are critical to many hair properties; for example, hair fibers swell with increasing water content, fiber friction increases, and static charge decreases as the water content of human hair increases. The response of human hair to RH is so sensitive that hair has been used as a hygrometer.

All cuticle cells are rich in those amino acids that resist helix formation such as cystine and proline. The cortex contains microfibril and matrix regions. The microfibrils contain α-helical proteins and tend to be richer in the amino acids alanine, leucine, and glutamic acid, which favor helix formation. In contrast, the matrix is richer in cystine and proline and other amino acids that tend to resist helical formation.

The medulla of porcupine quills has been analyzed, showing that medullary proteins are rich in basic and acidic amino acids and poor in cystine content.

When considering the reactivity of human hair to environmental factors and to cosmetic treatments, one may view its reactions in terms of the most reactive functional groups of its proteins. However, one must keep in mind that hair is not a homogeneous material. The best illustration of this fact is the difference in reactivity of the surface of hair to simple acids and bases. Hair proteins of the surface contain a greater frequency of acidic side chains relative to basic side chains as evidenced by its isoelectric point of pH 3.7. This fact demonstrates that the surface of hair has a greater capacity for combination with alkalies than with acids. However, when titrating whole fiber with acids and bases, we find that the entire hair has a greater capacity for acid than for alkalies. [*See* Proteins.]

F. Cosmetically Treated Hair

Several types of cosmetics react with functional groups of hair proteins, altering them in the process. For example, hair is generally bleached cosmetically using strong oxidizing agents consisting of alkaline hydrogen peroxide either alone or combined with persulfate salts. The function of a bleaching system is to lighten hair by degrading hair pigments, but the strong oxidizing agents used also oxidize cystine to cysteic acid, and other amino acid groupings of the hair proteins that are sensitive to oxidation are chemically modified. Changes in surface friction, wet tensile properties, and swelling behavior occur as a result of these changes of the fibers.

Permanent waving and chemical straightening also change the chemical behavior of hair. During permanent waving, hair is treated with a reducing agent that opens disulfide bonds, thus allowing molecular shifting to occur under stress:

$$\text{Hair}-\text{S}-\text{S}-\text{Hair} + 2\ \underset{\text{reducing agent}}{\text{R}-\text{SH}} \rightarrow \text{R}-\text{S}-\text{S}-\text{R} + 2\ \underset{\text{reduced hair}}{\text{Hair}-\text{S}-\text{H}}$$

The stress is induced by bending or curling the hair (permanent waving) or by combing it straight (chemical straightening) while it is in the chemically reduced state. The hair is then treated with a mild oxidizing agent to re-form disulfide bonds after the hair has taken on a new shape:

$$2\ \text{Hair}-\text{SH} \rightarrow \text{Hair}-\text{S}-\text{S}-\text{Hair}$$

Another form of hair straightening involves treatment of hair with strong solutions of alkalies that attack cystine and break open the disulfide cross-linkages to allow the hair to be straightened.

Hair dyes may be classified into various groups: permanent or oxidation dyes, semipermanent dyes, temporary dyes or color rinses, and metallic dyes. Each of these types of dyes has its specific chemistry that may be used to assist in identification in forensic analysis.

Shampoos, conditioners, and hair fixatives (hair sprays, gels, glazes, and mousses) react primarily at or near the fiber surface. These cosmetics can induce temporary and in some cases permanent changes to the surface properties of hair.

Figure 5 illustrates physical changes that may be observed in some human scalp hairs with time. Since scalp hair grows at a rate of about 15 cm/year, the tip ends of 30-cm-long hair have been exposed to environmental effects such as sunlight, wind, rain, and shampoos for a period of about 2 years. We generally refer to these effects collectively as weathering of hair. Chemical changes have been demonstrated in weathered hair from the action of sunlight. Physical changes have also been demonstrated from combing and brushing of hair. The ultraviolet (uv) region of the spectrum oxidizes cystine to higher oxidation states. Changes to other amino acids have also been demonstrated in hair fibers exposed to uv radiation. These chemical weathering changes generally occur more readily near the fiber surface.

V. SOME ABNORMALITIES AND DISORDERS

The louse is a blood-sucking, wingless insect that parasitizes human beings. One species generally attacks the head and another the body. Pediculus capitis, the parasite that is found on scalp hair, is the most common. The adult louse is a small gray-white insect, only about 3 mm long. It grasps hairs with its legs and lays eggs that are cemented to hairs. The eggs hatch into nymphs, which grow into adults. The empty nit or egg sacs are left on the hair fiber surfaces, producing gross fiber distortions.

Monilethrix or beaded hair is a congenital, heredity disease resulting in abnormal human scalp hair. This disease is easy to diagnose because individual hairs look beaded, that is, composed of thin and thick segments along the length of the fiber axis. Hair length in monilethrix rarely exceeds a few centimeters because the fibers are fragile and dry.

Pili torti or twisted hair is also a rare congenital deformity. Pili torti is characterized by flattened fibers with multiple twists. Sometimes this condition produces hairs of normal length, but frequently the hairs are twisted and broken.

Pili annulati or ringed hair is a rare hereditary condition characterized by light (silvery) and dark bands or "rings" along the fiber. This visual effect is produced by regions with and without a medulla. Annulati generally does not present a medical problem because the hairs of this condition often appear otherwise normal.

Trichorrhexis nodosa occurs more often in facial hair than in scalp hair. This condition consists of bulbous nodes producing the appearance of thin and thick segments along the fiber axis resembling monilethrix. However, these nodes are actually partial fractures that crack under stress, forming "broom-like" breaks. These fractures are actually fragmented cortex in which individual separated cortical cells appear like splintered wood.

BIBLIOGRAPHY

Robbins, C. R. (1994). "Chemical and Physical Behavior of Human Hair," 3d Ed. Springer-Verlag, New York.

Rogers, G. E., Reis, P. J., Ward, K. A., and Marshall, R. C. (1989). "The Biology of Wool and Hair." Chapman & Hall, London/New York.

Zviak, C. (1986). "The Science of Hair Care." Dekker, New York.

Headache

OTTO APPENZELLER
New Mexico Health Enhancement and Marathon Clinics Research Foundation

GLOSSARY

Aura Warning of impending attack (e.g., flashing lights, black spots appearing before the eyes, bright zigzag lines)

Bite plates Plastic covering inserted between upper and lower teeth

Extracranial vasodilatation Increase in diameter of blood vessels outside the boney cavity of the head

Focal neurologic deficits Failure of part of the nervous system to function

Hypoxia Lack of oxygen

Metabolism Chemical processes of living cells and organisms necessary for maintenance of function

Migraineur Person suffering from migraine

Neuron Nerve cell

Neurotransmitters Chemicals used for communication between nerve cells

Oxidative metabolism Where oxygen is used for metabolic processes

Pulsatile pain Wavelike increases in pain coinciding with heartbeats

Refractory period Time after an attack during which a new attack does not occur

Trigeminal perivascular nerves Nerves surrounding blood vessels that take origin in the trigeminal nerve cells, which also subserve sensory function

Trigeminal vascular system Nerves surrounding some blood vessels supplying blood to the brain

HEADACHE IS AMONG THE MOST COMMON AFflictions of human beings, but it is seldom the most dangerous. It ranges from slight nondisabling pain in the head or neck to severe pulsating localized or generalized head pain. Most headaches, migraine and tension headache being the most common, are neither symptoms of serious disease nor disease themselves. Rather, recurring headaches, like migraine and tension headaches, are a manifestation of stress, a peculiarly human response to a perceived or actual threat to the headache sufferer's well-being. Only occasionally are headaches a sign of serious disease. These rare headache types include those associated with tumors of the brain, inflammation, or ruptured blood vessels within the cranial cavity.

Most people experience headache at some time in their lives, in part because humans are uniquely constructed to suffer headaches. Unlike other animals, humans have a denser innervation in the head, face, and neck. This suggests a possible evolutionary purpose for the malady, as it may protect the head and its content from damage. The ancestors of humans depended for their lives on the functions (sight, hearing, and oral communication) of the richly innervated head and its organs, also suggesting headache's evolutionary purpose, for when the head or its organs were threatened, life was threatened as well. The highest density of pain fibers (i.e., nerves that conduct impulses to the brain that are perceived as painful) is found in tissues of the head, such as the skin and the lining of the sinuses making up much of the foundation of the face. (Brain tissue itself is insensitive to pain.) Pain is perceived more acutely and more often in or around the head than in any part of the body. Historical evidence (e.g., ancient skulls containing holes drilled before death) suggests that early humans not only experienced headaches but sought remedies for them. It is thought that the holes served to relieve painful pressure within the head.

Modern humans are probably at greater risk for headache than their counterparts in earlier times because they are, arguably, exposed to more stress, a precipitating factor in most headaches. An estimated 40 million Americans suffer from headache; the numbers are similar in other Western countries but are lower in developing countries and in Japan. Headaches, including migraine, are more common in females of all ages but tend to decline in both sexes after the age of 65 yr. Estimates of the incidence of headaches (i.e., the percentage of individuals suffering from head pain) vary with age and range from 70% of men and 84% of women aged 18–24 yr to 26% of men and 50% of women aged 75–79 yr. Headache sufferers in each sex and age group consistently experience more depression and anxiety, particularly about their own health, which they monitor obsessively and frequently imagine is at risk.

I. ANATOMY OF HEADACHE

Most of the information processing necessary for the perception of external stimuli of any kind and of pain occurs within the head, that is, at the site of the most highly developed sense organ—the brain. Because of this, the head has a very rich supply of nerves and is exquisitely sensitive to painful stimuli.

The brain is the most important and well-protected organ. It lies encased in several layers, the outermost being the skin. The skin itself consists of several distinct layers, all of which contain a large number of nerve endings. These endings receive and transmit the conscious sensations, including temperature, touch, pressure, and pain. When this rich nerve network is excessively stimulated, pain is perceived and protective measures are taken. [*See* Brain.]

The largest sensory nerve in this network is the trigeminal. Its sensitive endings are spread not only in the skin of the face and head but also within the muscles subserving chewing, in the lining of the mouth, and within the lining of the nasal cavity. Trigeminal nerve fibers also surround blood vessels within the cranial cavity that supply blood to the brain. These pain-sensitive fibers, which constitute a perivascular nerve fiber network known as the trigeminal vascular system, control the diameter of cranial blood vessels. They are important in headache because, by causing blood vessels to enlarge, they can cause pulsatile head pain.

Within the envelope of skeletal muscles is the boney brain cavity, which is rigid but pierced by numerous holes through which nerves and blood vessels enter and exit. The bones themselves are not sensitive to pain, but their covering of tough membrane and their numerous openings have pain-sensitive tissues. The bones of the cranial cavity and the face contain numerous air cavities, or sinuses, which reduce the weight of the skull without changing its strength. The sinuses are lined with membranes richly supplied with nerve endings and blood vessels that protect these cavities from air pollutants. When these membranes become irritated, they can give rise to pain.

Within the cavity of the skull, the brain rests on several membranes that absorb shocks and provide support. The membranes are composed of a tough outer membrane and finer inner membranes, the latter richly supplied with blood vessels and their accompanying nerves. Between the tough outer membrane and the one directly apposed to brain tissue, the so-called pia mater (the "tender mother"), a fluid layer provides further cushioning from undue displacement or blows to the head. The brain itself does not respond to pain because it lacks appropriate nerve endings or receptors sensitive to painful stimuli. Nevertheless, it can induce headache when it is displaced or swollen, the displacement or swelling thus exerting pressure on blood vessels, veins, and the brain's covering, all of which contain pain-sensitive nerve endings.

II. CLASSIFICATION

Painful symptoms that arise in or about the head are usually called headaches. The main groups recognized from an earlier system of classification (devised in 1962) are still valid in the revised classification (1988). They are vascular headaches, muscle contraction or tension headaches, traction and inflammatory headaches (due to displacement of pain-sensitive structures of the head), and psychogenic headaches. These have no definitive origin in any of the pain-sensitive structures but may be related to biochemical changes in the brains of victims. Other headache types may be lumped into a miscellaneous group (neuralgia, postconcussion headache, and other types). Many headache types respond to a slowdown, relaxation, or removal from the triggering situation, often exemplified by the relief suffers get from isolation and lying down. Although each group is characterized by its symptoms, most, if not all, have symptoms in common, such as nausea, vomiting, and light and sound sensitivity. Fortunately, most headaches are mild and respond to simple measures, but some require interruption of activities and may last from 2 to 24 or more hours. Even though the pain is often intense and frightening, particularly if it is accompanied by signs of nervous system dysfunction, such as paralysis, numbness, or visual phenomena, as may occur in migraine attacks, these headaches are usually benign and transient. Only perhaps 5% are due to serious disease.

A. Vascular Headaches

A large number of pains in the head, ranging from those associated with hangover and hunger to migraine and cluster headache, are in the vascular headache category. All of these pains can be traced to a transient abnormality of blood vessels that may not lie at the root of the disorder but that accompanies the headache. These headaches are by definition throbbing; their pain accompanies the pulsations of the heart. Although a given attack may be localized, at least initially, to one or the other side of the head, successive headaches may occur on different sides. These headaches often are triggered by specific stimuli, and in particular cold foods (ice cream headache) and drinks.

1. Migraine

Migraine and cluster headaches, one subclass of vascular headaches, differ from other such headaches in that they are chronic and are usually life-long afflictions. Unlike occasional toxic vascular headaches (such as hangovers), migraine has a congenital aspect, but the hereditary basis of these headaches has not yet been elucidated. The changes in diameter of cerebral blood vessels resulting in migrainous attacks can occur in anybody, provided external and internal stressors have come together in the right amount and in the right sequence. Nevertheless, the physical and behavioral constitution of people suffering from frequent migraine is different from that of people who may have only one attack during a lifetime.

Each attack consists of at least two phases: the first is associated with a decrease in blood flow to specific areas of the brain, and the second with a marked increase in flow, in both the brain and scalp, and severe pulsatile pain. In some migraines, the phase of falling blood flow brings about dysfunction of the brain (neurologic deficits) resulting from the decrease in oxygenation of circumscribed areas of the brain. These attacks are known as migraine with aura (formerly classic migraine). Others, known as migraine without aura (formerly common migraine), lack neurologic deficits and consist only of pulsatile pain. The nature of a patient's migraine, the site of the pain, and the kind of focal neurologic deficits often change over a lifetime. Migrainous attacks are often accompanied by nausea, ushered in by sleepiness, excessive vigilance, or seemingly boundless energy. After the attack a refractory period lasts 1 or more days.

In childhood migraine sufferers, boys are more often affected by migraine than girls, but the proportion becomes reversed after the onset of puberty. In adults, women are by far the commonest victims of the disorder. Children may also have symptoms, such as colic, motion sickness, or periodic vomiting, without actual headache. Now classed as childhood periodic syndromes, they may be precursors to later migraine attacks.

a. Hypoxic, Neurogenic, and Vascular Theories of Migraine Causation

Numerous theories about the causation of migraine headaches have been proposed and discarded. None has taken into account and explained all the many triggers for headache, but with modern techniques for noninvasive assessment of oxygen supply and demand in the human brain, a comprehensive theory is likely to emerge. Existing theories (hypoxic, neurogenic, and vascular), as yet unproved, merit further investigation.

b. Hypoxic Theory of Causation

One hypothesis proposes that lack of sufficient oxygen supply to the brain is the cause of migrainous attacks. Oxygen is essential to brain function, which is dependent on neurotransmitter activity; when oxygen supply to the brain is diminished, brain function is impaired. This hypothesis does not state precisely why oxygen is lacking, but two possibilities exist: some migraine sufferers may have a physical impairment that prevents adequate oxygen supply to their brains; others may have impairments that interfere with the energy supply to nerve cells and consequently with oxidative metabolism. To support this hypothesis, many similarities have been found between manifestations of migraine and those caused by abnormalities in oxidative metabolism. Thus, this theory proposes that triggering mechanisms for migraine may be a reduction either in the supply of oxygen or in the supply of the primary fuel for brain metabolism, glucose. (Glucose is the only nutrient available for brain cell survival; it requires oxygen for its utilization by brain cells.) Situations that increase the demand for oxygen can also cause attacks, and an inappropriate matching between brain metabolism and blood flow, which brings oxygen and glucose to the brain, may be a trigger as well. Last, metabolic processes that result in biochemical abnormalities may also trigger attacks. Because the hypoxic theory examines several possible headache triggers, it deserves further clinical investigation.

Investigations outside the clinic (at high altitude) support the hypoxic theory. At high altitude, unless they acclimate first, people who do not ordinarily suffer from headaches invariably develop headaches, partly due to hypoxia. Their headaches have many of the same features of migraine, including pounding pain, accompanying nausea, and aggravation of the pain by exertion. Mountain-climbing migraine sufferers learn through experience the altitudes that trigger their migraines because they are doubly affected at high altitude, which contributes a predictable factor (hypoxia) in triggering their attacks.

The experience of patients whose migraine is triggered by eating certain foods supports the hypoxic theory as well. These foods change the availability of brain neurotransmitters by altering the electrical discharge of nerve cells. The neurotransmitters can change the requirements of oxygen of some cells by increasing the firing rate of their electrical impulses. This in turn can promote electrical conductance along certain pathways in the brain involved in the perception of pain.

The hypoxic theory is also supported by the experience of patients suffering from diseases (e.g., strokes) that affect the patency and distensibility of blood vessels in the brain and, thus, oxygen supply to local areas of the brain. In these cases, the metabolic requirements of the brain arguably may not have been met with adequate oxygen, because the damaged blood vessels block the flow of blood (and oxygen) to the brain. Because many fits stop normal respiration and thus oxygen uptake by the blood, people who have disease of the brain (e.g., epilepsy) that manifests itself by fits may have associated headaches. [*See* Epilepsy; Stroke.]

The hypoxic theory may also explain some aspects of tension headache, brought about through the intimate relationship of blood vessels and nerves within the muscles surrounding the head. The contraction of muscles, which are initiated by nerve electrical impulses, together with chemical release of substances at the nerve–muscle junction, induces individual muscle fiber contraction and shortening. Muscle contraction initiated by stress and prolonged (without intermittent relaxation of the muscles) by the additional stress of irritated, contracted muscles stimulates the perivascular nerve fibers and reduces the flow of blood. In turn, the lack of oxygen produced by the decreased flow causes the release of a variety of chemicals that stimulate pain-sensitive endings in the muscles. Direct pressure on nerves by contracting muscles or by unusual positions of the head and neck and neck bones also contributes to this pain.

c. Neuronal Theory of Causation

The well-known relationship of migraines to stress, fatigue, and anxiety, all manifestations of brain nerve cell activity, and the frequency of attacks on weekends and holidays, when many drink alcoholic beverages (which impair brain cell function), suggest that abnormalities of nerve cells may be of primary importance in causing attacks. Similarly, the sensitivity of many sufferers to bright lights, which may trigger attacks within seconds, suggests a reflex, entirely neuronal causation.

Supporting evidence for a neuronal mechanism is the occurrence of deficits of function in discrete brain areas in some sufferers, especially patients who have had frequent attacks spread over many years. Abnormalities on brain imaging scans, only comparatively recently available, have also been found.

Furthermore, a weak association between migraine and epilepsy is well documented. Because both of these disorders are common, however, most instances of the two occurring together are coincidental.

The postulate that migrainous attacks are the result of primarily neural dysfunction implies that, because attacks are paroxysmal, this abnormality in neural function occurs only in paroxysms, and that all vascular phenomena that accompany the attacks are secondary to the abnormality in nerve cell function. The most widely held and longest surviving explanation for the neuronal mechanism is that migraine is an episode of the "spreading depression" first described by A.A.P. Leao, a Brazilian physiologist. This spreading depression is a transient disruption of neuronal electrical activity, which can be elicited in experimental animals and spreads like a wave through some brain areas. The depression seen in experiments travels at a speed of 3 mm/min. The characteristic initial visual phenomena of migraine (bright zigzag lines with surrounding dark areas) have also been calculated to spread at the same rate. When cerebral blood flow was measured during attacks of migraine in some patients, an initial fall in blood flow was seen to spread, like a wave, at a rate of 3 mm/min. But when similar attacks were induced in patients whose migraine was not accompanied by focal neurologic deficits (e.g., visual phenomena), no such abnormalities in cortical blood flow were noted. In spontaneously occurring attacks similarly studied, the expected reduction of cerebral blood flow in those with focal neurologic deficit was found, but this decrease in flow extended well into the headache phase known to be accompanied by extracranial vasodilatation, and an increase in intracranial flow was recordable during the later stage of the attacks.

d. Vascular Theory of Causation

The many vascular phenomena that accompany migraine, including increased pulsation of blood vessels supplying blood to the head and in the scalp and face, the transient improvement of pain when these blood vessels are compressed, and the effect of substances either causing blood vessel dilatation and increasing pain, or causing vasoconstriction and decreasing pain, have suggested that the primary cause of migraine is not in abnormal neuronal function but primarily in changes in blood vessel diameter. Although the neural hypothesis does not deny the importance of vascular phenomena in the genesis of pain, this explanation itself does not account for peripheral effects outside the brain that accompany migraine attacks, but these, perhaps, could be activated through centers in the posterior part of the brain, the so-called brain stem, which does control to some extent the caliber of blood vessels throughout the body.

A migrainous attack may be considered an adaptive or reactive phenomenon. This phenomenon in turn depends on changes in the threshold of responsiveness of the brain to a number of stressful and precipitating factors. Factors such as hormonal balance, changes in the level of neurotransmitters of the brain, lack of oxygen, jet lag, changes in sleep patterns, and changes in blood vessel diameter are important triggers. If these triggers are acting in concert in the right sequence and in sufficient numbers, an attack of migraine occurs. Thereafter, exhaustion of the brain compels rest and recuperation until the setting becomes right for a subsequent attack. The threshold at which attacks occur varies from person to person and is probably genetically determined to some extent, but it also varies during various times of the life cycle and fluctuates with physical or mental stress, broadly defined. This concept allows for the occasional attack in otherwise ordinary people never subject to headache and accounts for those who have frequent and incapacitating recurrences of migraine and a family history of migraine.

2. Cluster Headache

Cluster headache, like migraine, is considered a "vascular" headache because pain during attacks is brought about by changes in cerebral blood flow and blood vessel diameter. The intensity of pain in cluster headache is unique. Unlike migraineurs, who lie still during their attacks, cluster patients pace back and forth in agony. Also different from migraine is the near certainty cluster patients have about the timing of their attacks during a cluster period. These attacks occur in series that may last for weeks or months (the cluster periods) and are separated by intervals of complete freedom that may last months or even years. Patients are equally uncertain about when a cluster period will end. The cyclic nature of this disease remains unexplained. The prevalence of attacks at night suggests a relationship to sleep phases and perhaps to biologic clocks, but cluster periods themselves do not seem to follow any patterns, although in some patients they are more likely in autumn. The attacks are much shorter in duration than migrainous attacks, lasting at most 3 hr, but usually only 20–40 min. During a cluster period, attacks can occur just once every other day or as often as eight times daily. Characteristically, these attacks are associated with autonomic nervous system dysfunction ipsilateral to the side of the head pain, which is strictly unilateral. Conjunctival injection, a red eye, tearing, congestion of the nose, nasal discharge, and forehead and face sweating indicate the autonomic nervous system involvement, together with a smaller pupil and droopy eyelid and sometimes swelling of the eyelid ipsilateral to the pain.

Cluster males (the condition is far commoner in men than in women) are thought to have a characteristic appearance. They often have a ruddy complexion, thick facial skin, visibly dilated small blood vessels across the bridge of the nose and cheeks, and a prematurely deep-furrowed forehead, together with a broad chin and skull. They are often tall and have a rugged appearance and have a higher prevalence of hazel-colored eyes than other men. Few of them are blue-eyed. There is no pattern of inheritance for cluster headache, although these physical characteristics suggest that heredity may play some role. Personality factors are found to be similar to those of migraineurs: difficulty in handling stress is also characteristic of cluster patients. In addition, male sufferers smoke more and drink more coffee and alcohol (in between cluster periods only) than do ordinary men. During a cluster period, however, they all abstain from alcohol, because it invariably precipitates an attack.

Like migraine, cluster headache is treated with drugs that affect blood vessel diameter. One special type of cluster headache, mainly affecting women, strikes 10–18 times a day and may remain uninterrupted by pain-free periods for up to 20 years, so that victims are unable to participate in ordinary life activities. This type of headache, called chronic paroxysmal hemicrania, responds to the anti-inflammatory drug indomethacin and may completely disappear with the use of this drug. This remains the only type of headache that can be "cured" permanently by the use of medication, even though it is not known how this type of pain is produced, and it remains the rarest type of the so-called benign headaches.

3. Toxic Vascular Headache

Anyone who has experienced fever has usually also experienced a toxic vascular headache, known as fever headache. Infection by bacteria or viruses causes release of fever-producing substances, the so-called endogenous pyrogens. These substances, after they enter the blood and pass to certain areas of the brain, cause release of prostaglandins (vasodilators), which also affect the brain's thermostat and lead to an increase in body temperature by shivering and blood vessel constriction or other reactions, such as metabolic heat production by the liver. When the higher body temperature has finally been reached, an increase in oxygen demand and fuel supply, caused by the increased temperature, in turn demands an increased blood supply, which is accompanied by dilatation of blood vessels and headache. Such headache is dull, deep, often aching, generalized, and throbbing in its early stage, although it may occasionally be restricted to a specific area at the back of the head. As it does with other vascular headache, any jolt, sudden movement, or strain aggravates fever headache.

B. Muscle Contraction or Tension Headaches

Tension headaches associated with continuous contraction of muscles around the head and neck are by far the commonest kind of headache. Typically, the pain is dull, lasts several hours, and recurs over periods of days, weeks, or years. It is often felt on both sides of the head, in the temple, and at the back of the head; the pain is sometimes described as a hatband around the crown. These headaches are often aggravated by physical or emotional stress or both, but in all cases severe disabling pain is real and not imagined. Excessive muscle contraction around the neck and scalp also occurs in other kinds of headaches.

C. Traction and Inflammatory Headaches

An important category of clinical pain is that caused by traction or displacement of pain-sensitive structures of the head or inflammation. The headache is due to disease or injury to the head, and if ignored or untreated may lead to permanent damage and even be life-threatening. Although such headaches may initially respond to painkillers, delay is often dangerous, and sufferers should seek medical attention. Such headaches, grouped because of their ability to cause permanent damage, have diverse causes. The traction headaches are the result of mechanical forces that distort by pressure or traction pain-sensitive structures within the cranium. The forces are mostly due to brain tumors, blood clots, or collections of pus. If such lesions are sufficiently large, and particularly if they change size relatively rapidly, as blood clots and deposits of pus can, they cause pain. On the other hand, some slow-growing tumors can become large before pain becomes severe.

The second subgroup of these disease-related headaches is caused by inflammation, the site of the inflamed tissue usually giving the name to the underlying disease: meningitis, phlebitis, and arteritis. In meningitis, the meninges, or coverings of the brain, become inflamed. In phlebitis, from the Greek word for vein, the large veins draining blood from the brain are often involved; in arteritis, an inflammation of arteries, the vessels at the temple are often at fault. Arteritis is a disorder that most often strikes the elderly and may be associated with loss of sight or multiple strokes. It often causes an intense and deep-seated ache, with tender and rigid blood vessels on the surface of the skull.

1. Temporal-Mandibular-Joint Dysfunction

A widely touted cause of headache is temporal-mandibular-joint dysfunction (TMJ). The importance of abnormal jaw muscle contraction or arthritis in the jaw joints, causing imbalance of the bite and thus headache, is debated. TMJ typically is initially treated with bite plates prior to grinding teeth for better alignment. Some specialists, who do not generally believe in this type of causation, do not treat patients with headache by bite plates. Others use these devices to treat a variety of headaches and report frequent success. Bite plates are also used by athletes, who report improvement in performance when the plate is used and hard clenching of teeth during muscular exertion is prevented mechanically by the device.

2. Headache Caused by Degenerative Disease of the Cervical Spine

Degenerative disease of the cervical spine causing narrowing of joint spaces can also cause headache, because the disease causes imbalance in muscle contraction, which in turn may put pressure on the nerve roots exiting from the spinal cord. The pain these patients experience is indistinguishable from the pain of tension headaches, but, unlike tension headache pain, may respond to massage, traction on the neck, or surgery.

3. Inflammatory Headache Caused by Sinus Infection

Inflammation of the sinuses rarely causes chronic headaches but may cause pain in the face if acute infection is present. Most patients with long-standing sinusitis suffer from ordinary tension-type headache. That is not to say that such inflammation may not on occasion give rise to serious neurologic problems manifested by headache. A chronically blocked and inflamed sinus cavity, for example, can sometimes lead to inflammation of veins that drain blood from around the sinuses into the cranial cavity, and infection thus spreads, causing meningitis, which may be followed by brain abscess. [*See* Inflammation.]

D. Psychologic/Psychogenic Headaches

1. Emotion, Stress, and Mental Illness-Associated Headache

Chronic head or face pain is often associated with the muscle contractions of repeated and involuntary facial grimacing, nearly always a symptom of psychologic stress or mental illness. Patients suffering from such illnesses frown, squint, and grind their teeth (particularly at night) for prolonged periods of time. The contracted muscles become tired and release substances that cause chronic pain, which may be aggravated by malalignment of the jaw and improved by a bite plate.

2. Delusional or Conversion Headaches

Even though psychologic factors can aggravate the preceding headache types, the pain is real: it has a physical basis. On the other hand, a complaint of headache can also express emotional disturbances, usually associated with sexual conflicts, repression, anxiety, or depression, and is often related to changes in an individual's life that are not wanted or cannot be controlled. Though patients suffering from these headaches are convinced of the organic nature of their illness, they often are rather indifferent to the pain they complain of so bitterly. Such people typically cannot establish normal social and situational contacts; they often experiment with drugs, including painkillers and tranquilizers, and usually present a complicated management problem because of their frequent associated drug dependency. Such patients often require prolonged behavioral manipulations.

E. Miscellaneous Headache

1. Neuralgia

Several varieties of neuralgia, or pain arising from nerves around the head, are dreaded because of the intensity of the pain. Though these disorders usually are not life-threatening, they make existence miserable. The most common is trigeminal neuralgia or tic douloureux, which is related to the trigeminal nerve system, the most widespread sensory system of the head. Its victims are usually 40 years old or older. The pain is an intense aching, burning, or stabbing, lasting only a few seconds but coming in bursts, typically on one side of the head. Characteristically, the pain of trigeminal neuralgia is triggered by stimuli that affect the trigeminal sensory system, including cold air, touch, chewing, and sometimes clearing the nose. Individual sufferers become vividly aware of specific trigger zones and stimuli. Glossopharyngeal neuralgia is much rarer than tic douloureux. Clinically, the pain is similar, but the sensitive trigger zones are located in the back of the throat or the tongue, from which the pain radiates toward the ear in terrifying, painful stabs. Swallowing is the most common trigger for this neuralgia.

2. Postconcussion/Posttraumatic Headache

Occasionally after head injuries resulting in lacerations of the scalp, scar tissue forms around the small sensory nerve fibers that are abundant in the head region. Scarred nerves can produce occasional stabs of pain, initiated by touch or pressure, that later evolve into continuous contraction of scalp muscles. The result is indistinguishable from tension-type headaches. A clue to this headache mechanism is often the presence of an extremely sensitive spot related to a scar. The injection of a local anesthetic and sometimes steroids may interrupt the vicious cycle of abnormal and spontaneously arising nerve impulses in the constricted or scarred nerve, thereby relieving the scalp muscle contraction and pain.

a. Postconcussion Syndrome

A constellation of symptoms following head injury, consisting of headache, dizziness, irritability, difficulty in concentration, and alcohol intolerance, is labeled the postconcussion syndrome, headache being the most troublesome and prominent of the symptoms. This constellation of complaints is common after head injury, but its cause and treatment continue to arouse controversy.

b. Posttraumatic Headache

A brain concussion changes the levels of neurotransmitters (substances that are released or taken up by neurons for communications with other neurons) or neuromodulators (which have a similar function but take effect more slowly). The electrical brain activity is therefore affected by the impact, which changes levels of neurotransmitters, thus interfering with the regulatory capacity of the central autonomic nervous system. Subsequently the peripheral autonomic nervous system and particularly the trigeminal vascular systems are affected. The resultant abnormalities in cerebral blood flow following concussion are well demonstrated and may lead to very serious consequences. Occasionally this alteration in vasomotion leads to primary traumatic brain damage and more diffuse secondary disorders of the brain. These secondary changes in the brain after trauma do not seem to be pathogenetically releated to posttraumatic headache, because the brain itself is insensitive, but they point to a relationship between brain concussion and vascular disorders occurring immediately after the injury. The resulting decrease in blood flow to the brain and change in diameter of blood vessels, known to be sensitive to pain, can cause a headache that is clinically very similar to migraine. It is pulsatile and is often aggravated by exertion and by bending forward. Unlike typical migraine, however, the pain is not concentrated in one side of the head; it is either diffuse or frontal. Headache persisting for years after head injury, however, cannot easily be attributed to the original impact, and its causes are mysterious. Most authorities suggest that 30–50% of patients develop headache after injury, and because head injuries are common, the posttraumatic headache, persistent and incapacitating, forms the bulk of posttraumatic handicaps (serious neurologic deficits are less common in these patients). No clear correlation between the severity of the injury or damage to the brain and the incidence of posttraumatic headache has emerged. It has even been suggested that posttraumatic headache is less troublesome after major cerebral injury, and more disabling after minor concussion. Psychologic factors, such as anxiety or the desire for unemployment compensation, that are related to the trauma undoubtedly play a role in many posttraumatic headaches. Some posttraumatic headaches mimic ordinary tension headaches; others have many, if not all, of the features of migraine, including the accompanying nausea and light and sound sensitivity. A third, distinctive, type often follows injury to the neck. The cause is thought to be the autonomic innervation in the sheaths of major arteries to the head. Some rare patients experience pain and tenderness of the neck for weeks after the injury, later giving way to severe pain with sweating on the injured side in temple and frontal areas, accompanied by dilatation of the pupil, nausea, and light sensitivity. The pain may last for hours or days and responds to specific drugs.

3. Headache Associated with Drug Use or Drug Withdrawal

Drug-related headaches usually indicate that a chemical gained access to the body either through the skin or through swallowing or inhaling, causing a chemical imbalance. Such "toxins" may also give rise to classic attacks of migraine or cluster headache, but in many sufferers these headaches occur in the absence of focal neurologic deficits and in close relationship to the absorption of the offending substances. The pain is, as in other vascular headaches, pulsatile, and it may occur after ingestion of excessive amounts of alcohol or in particularly susceptible individuals after eating hot dogs, Chinese food, or frozen dinners, or after inhaling air pollutants. It also may occur in association with fever or infectious diseases such as influenza. Blood vessels are affected by triggers that cause relaxation of smooth muscle cells, thus causing dilatation of the vessels. The vessels dilate and, in turn, the pain-sensitive nerves surrounding them signal, in the form of pain, the abnormal diameter to the brain. Such headaches often respond to treatment with aspirin or similar drugs and often to physical manipulation, such as the application of ice bags to the painful area, a well-recognized treatment for hangover headaches. The application of ice bags tends to constrict the painful and swollen blood vessels, at least on the surface of the scalp, thus providing a nonchemical relief for the painful pulsatile vasodilation. In tension headaches, however, chemical treatment is sometimes warranted. Alcohol, the culprit in hangover headaches, can be used to advantage in treating tension headaches when it is consumed in moderation because alcohol temporarily blocks stress and allows the headache sufferer to relax the constricted muscles causing the pain.

Hangover headaches depend not only on the kind of alcohol consumed but also on how quickly and in what quantities it is drunk. Psychologic effects are also important because, characteristically, the pain does not start until blood alcohol levels decrease following breakdown in the liver and subsequent excretion. Moreover, because of intangible psychologic effects, the actual blood alcohol levels reached during any drinking bout have little relation to the subsequent hangover headache.

Most habitual coffee drinkers (those who routinely consume three or more cups of coffee a day) will suffer withdrawal headaches when coffee is unavailable. Caffeine constricts blood vessels, and in moderation may be useful for some migraine attacks, for example, or for hangover headaches. But, in those users who consume large amounts, the sudden withdrawal of the constricting effects of caffeine, as often occurs over weekends, may lead to sudden and unaccustomed dilatation of blood vessels and headache. Such withdrawal headaches are rarely recognized for what they are and may sometimes account for what is known as the letdown headache of migraineurs, which also tends to occur at weekends at a time when many sufferers from caffeine withdrawal headaches are incapable of maintaining their blood vessels in a constricted state through ready access to the coffee urn.

4. Headache Associated with Substances Ingested or Inhaled

A Chinese restaurant syndrome has been traced to the liberal use of monosodium glutamate (MSG), which causes blood vessel dilatation, headache, and many other symptoms, such as chest pain. MSG is also liberally found in frozen TV dinners, canned meat, and powdered soups. In susceptible persons, MSG, even in small quantities, may cause headache. If taken in large quantities, its vasodilatory capacity causes pain, even in subjects not usually prone to headache.

Inhalation of substances that trigger headache is not uncommon. These include nitrates and nitrites. Both compounds have a nitrogen atom and an oxygen atom and are used for the preservation of foods and for preservation of color in meats. Nitrates are part of the nitroglycerin molecule, an explosive sometimes used as a treatment of cardiovascular diseases. If used in medicine, it may be absorbed through the skin and cause vasodilation and complicating headache, a frequent affliction of workers who handle nitrate in munition factories and on other jobs requiring the use of explosives. Such headaches are often accompanied by flushing of the face, an increase in heart rate, and lightheadedness.

Many air pollutants, and occasionally perfumes, may trigger headaches. Dyes and paints cause headache in some people; carbon monoxide from exhaust fumes or tobacco smoke in poorly ventilated rooms cause headache in others.

Solvents, paints, and paint removers are volatile vasoactive substances and may quickly cause severe headache because of their dilatory effect on blood vessel diameter.

5. Headache Associated with Infection

Infections of the eyes, ears, nose, mouth, teeth, and sinuses cause pain in the affected structures. A natural reaction to pain in these structures is stiffening of neck muscles and contraction of facial muscles, which in turn cause tension headache, making the infection-related pain doubly painful. Thus, pain in association with such inflammatory or other diseases of these parts of the head and neck have many of the associated features of tension headache, plus local pain in the primary structures affected.

6. Headache Associated with Metabolic Disorder

Patients having abnormal metabolism such as hypoglycemia (low blood sugar) or kidney disease may have a worsening of headache if they suffered before the onset of the metabolic disorder. Others have a new onset of headache in close relationship to the beginning of their metabolic disorder; the headache usually disappears within about 7 days after the abnormal metabolic state is corrected.

Headaches in this category include those associated with the hypoxia of altitude exposure, an altitude-induced metabolic disorder, or that occurring in association with pulmonary disease, which interferes with transfer of oxygen from the air to the blood. Some patients become hypoxic because of prolonged periods of apnea (lack of breathing) during sleep and also develop daytime headache. Occasionally, an increase in the arterial blood content of carbon dioxide without accompanying hypoxia can cause headache.

Those patients who become hypoglycemic, those on dialysis for kidney disease, and occasionally those without hypoglycemia who are fasting or who are undergoing plasma exchange may develop headache. These headaches usually disappear with correction of the metabolic abnormality induced by dialysis or plasma exchange.

7. Miscellaneous Headache without Structural Lesions

Tight head coverings can cause external compression headaches, also known as "goggle headaches."

Sexual activity can give rise to so-called coital headache, which may be dull or explosive and which occurs at the time of orgasm or immediately thereafter, when the headache sufferer resumes an upright posture.

8. Headache Associated with Exercise

Acute headache associated with exercise, straining, coughing, sneezing, laughing, or stooping can be a symptom of life-threatening intracranial disorders, but more often it results from poorly understood benign causes related to exercise. Exertional headache must be distinguished from the more common aggravation of an established headache by exercise. Exercise aggravates vascular headaches of the migrainous type, but in some migraines and many cluster headaches it can be useful in aborting the attack. Benign exertional headache is more common in men than in women and is more frequent in older individuals. Benign exertional headache not associated with intracranial lesions is usually bilateral, frontal, occipital, or generalized. It is often abrupt in onset and severe; it may be sharp, stabbing, or pulsatile and last for several minutes after the exercise. The cause of benign exertional headache in nonathletes is obscure.

Disease-associated exertional headaches are rare, but they can accompany a number of serious disorders that increase the pressure within the skull. In headache associated with lesions within the head, headache may be induced by any kind of exercise, but its severity is related to the degree of effort. In some patients the headache is pulsatile and associated with visual disturbances; it may be mistaken for attacks of migraine induced by exertion.

Migrainous attacks can occur after athletic effort of any kind but tend to be more frequent at high altitude. Many such effort-induced migrainous attacks include only some of the symptoms. They may have just visual phenomena, usually occurring immediately after the exertion and remaining without subsequent pain. Nausea and severe pulsatile, unilateral pain indistinguishable from the classic variety of migraine may also occur. However, effort migraine is most common in poorly trained individuals, and dehydration, excessive heat, low blood sugar, and unaccustomed altitude place such individuals at particular risk.

III. CONCLUSION

Common headache is annoying but not life-threatening. It may cause disability if severe and frequent. Because most headache is aggravated by stress, the necessity to attend to the pain usually removes the sufferer temporarily from stressful situations. In this sense, headache may protect from prolonged and threatening situations that seriously jeopardize function. Most headache can be managed but not permanently cured. Only a few headache types have specific curative remedies.

BIBLIOGRAPHY

Blau, J. N. (ed.) (1987). "Migraine: Clinical, Therapeutic, Conceptual and Research Aspects." Chapman & Hall, London.

Dalessio, D. J., and Silberstein, S. D. (eds.) (1993). "Wolff's Headache and Other Head Pain," 6th Ed. Oxford Univ. Press, New York.

Rose, J. C. (ed.) (1989). "New Advances in Headache Research," Smith-Gordon, London.

Health Care Technology

HAROLD C. SOX, JR.
Dartmouth Medical School

GLOSSARY

Effectiveness Results of using a technology under the usual conditions of medical practice

Efficacy Results of using a technology under ideal circumstances

THE OFFICE OF TECHNOLOGY ASSESSMENT OF THE United States Congress defines health care technology as the "techniques, drugs, equipment, and procedures used by health care professionals in delivering medical care to individuals, and the systems within which such care is delivered." In other words, health technology encompasses the materials and methods used in the diagnosis, treatment, and prevention of illnesses and diseases.

I. PROBLEMS IN USING HEALTH TECHNOLOGY

The goal of using health technology is to apply it to the patients who will benefit. There are several reasons why physicians have difficulty achieving this goal consistently.

The outcome of using a health technology is inherently unpredictable for several reasons. First, the individual's biological characteristics that determine response are often unknown to the physician. Second, the true state of the patient is often unknown at the time of a decision to use a technology. Consequently, some patients are treated unnecessarily, yet are subjected to the risks and costs of the technology. Third, the technology itself is imperfect. Diagnostic tests may fail to identify patients with disease and may incorrectly identify as diseased patients who are not diseased. Treatments do not cure everyone and may have undesirable side effects.

Health technology is often very costly. Whether this factor is important in the decision to use a technology bdepends on who must pay. The individual with comprehensive health insurance is likely to be indifferent to cost, whereas the insurance company will be concerned, particularly in the case of extraordinarily costly technologies such as organ transplantation. Different areas of the United States vary widely in the per capita rate of coronary bypass surgery or surgery for benign prostate disease, for example. A crucial question is whether greater expenditures buy measurable differences in the quality of life.

The health effects of many health technologies are poorly understood. Measures of outcome have often been limited to intermediate end points of the disease (e.g., blood pressure in studies of the treatment of high blood pressure) rather than primary end points such as quality of life or length of life. Studies have not included measures such as quality of life, functional status, and satisfaction with care. Many studies of new technology do not have a concurrent control group that receives the current standard intervention. This problem is particularly true of diagnostic tests and procedures. The advantage of using a very costly technology as compared with using a less costly technology is often unknown because many studies do not examine the relationship between cost and effectiveness. Because of these deficiencies, physicians are often unable to answer their patients' questions about the likely effect of a proposed treatment or test.

Physicians cannot always use what they know about a technology in their decision making because other factors dominate their thinking. Patients sometimes urge their physician to use a technology that is unlikely to benefit them, which creates a conflict between the physician's role as the patient's advocate and the physician's responsibility to a society that is concerned about the costs of health care. In the United States, the rise of managed care has produced a new form of ethical conflict, one that occurs when a managed care plan creates incentives for physicians to do as little as possible. Furthermore, physicians sometimes are influenced by concern about malpractice litigation when deciding whether to use a technology that has a small chance of benefiting the patient.

II. METHODS FOR EVALUATING HEALTH TECHNOLOGY

There are several sources of evidence on which to base a judgment about the effectiveness of technology. These sources differ in the extent to which they conform to two requirements of an ideal study: validity and general application. A valid study is one whose conclusions are true. Validity is most likely to occur in a study of a new technology when it is possible to ascribe any effects on clinical outcomes to one and only one variable, the technology itself. To achieve this goal in a clinical study, one must control a number of extraneous factors that might affect the outcome of care. Studies of care under carefully controlled, ideal circumstances are called *efficacy studies.* The problem with such studies is that the conclusion, however valid, may apply only to the patient population that was used for the study. By restricting the study population to a narrowly defined group, one sacrifices general applicability in order to achieve validity. Ultimately, physicians need to know if a technology works in their patients. Studies of patient care under usual workaday circumstances are called *effectiveness studies.* In such studies, the study population is representative of usual practice, and the physicians are community physicians. Valid effectiveness studies are infrequent because community practice is a difficult venue in which to perform a study that controls for extraneous variables. Thus, most studies of health technology sacrifice validity to achieve general application or they compromise general applicability to achieve validity.

The following sections discuss several methods for evaluating health technology.

A. Randomized Clinical Trial

The term "randomized clinical trial" refers to the practice of random assignment of patients to receive the intervention under study or a comparison intervention. In a study of a new drug for high blood pressure, the patient receives either the new drug or a placebo—a biologically inert substance that is packaged to look just like the new drug. Assignment to drug or placebo is random. If the study groups are large enough, random assignment assures that they differ only in exposure to the new drug and are identical with respect to all other variables, *known and unknown,* that might affect the outcome of care.

Blinding, a method for preventing the study participants from knowing the treatment assignment, is a desirable feature of a randomized trial. In a study that is single-blind, the patient does not know the treatment, but the doctor does know. In this case, preconceived notions of what the study should show might lead to unconscious bias in recording the findings that comprise end points of the study. In a double-blind study, neither the doctor nor the patient knows what the patient is receiving. A double-blinded study minimizes the chance that preconceived ideas will influence the way in which the patient or the physician reports the data.

There are two types of errors in interpreting the results of a randomized trial. The first is called an alpha, or Type I, error. A Type I error means that a clinical investigator assumes that a difference exists when it does not exist. For example, the investigator assumes that a 40% rate in the intervention group as compared with a 36% rate in the control group means that the intervention is effective. In other words, a Type I error is a false-positive trial result. The commonly reported "p value" for an observed difference between control and intervention groups is the probability that the observed difference is due simply to chance variation in two samples from the same universe. A low p value means that the observed difference is very unlikely to be due to chance and probably represents a true effect of the intervention. The second type of error is called a beta, or Type II, error. A Type II error is a false-negative trial result: there was a true difference, but the study failed to detect it. A Type II error occurs when random variation in sampling, because of small study populations, causes two samples from different universes to appear to have been drawn from the same universe. Studies try to enroll enough patients to reduce the probability of a Type II error to 20% or less.

There are several advantages of the prospective clinical trial with randomized controls. The most important is that, given a sufficient number of participants, the intervention is the only way in which the experimental and control groups can differ. This characteristic means that the clinical investigator can confidently attribute differences in outcome to the intervention. The clinical trial is a well-accepted method for assessing a health technology; well-designed trials on topical clinical problems are, in the aggregate, quite influential. Clinical trials generate new data; a major clinical trial of coronary artery surgery led to nearly 100 journal articles in less than a decade.

Randomized clinical trials have a number of shortcomings. Clinical trials in chronic disease can be very costly, particularly if standardized patient care must be provided during many years of follow-up observations. If the number of outcome events is too small, a clinical trial may have a false-negative result, a Type II error. Thus, clinical trials are unsuited for the study of rare diseases, diseases with a low rate of outcome events, or interventions that are expected to have a small effect. Sometimes the intervention becomes obsolete while the investigators wait to accumulate enough trial end points to draw valid conclusions. To maximize the chance of seeing an effect, a clinical trial may exclude patients who are likely to experience a trial end point for reasons unrelated to the intervention. For example, a study of coronary artery surgery might exclude patients with known cancer because they are likely to die of cancer before having a full opportunity to experience a trial end point. Such studies sacrifice general applicability to maximize the chance of observing an effect. Finally, a randomized trial result favoring one treatment over another may not apply to all patients. Unless the study populations is very large, it may not be possible to identify the subgroups that get the most benefit.

B. Meta-analysis

Meta-analysis, or "after-analysis," is a method for evaluating the weight of evidence that has accumulated from many studies. The basic premise is that combining several similar creates a pooled population that is large enough to minimize the chance of misinterpreting the results and making a Type I or Type II error.

Meta-analysis is performed in several steps:

• Define a literature search strategy: Typically, one uses an electronically stored literature data base such as MEDLINE, which the United States National Library of Medicine maintains. MEDLINE stores keywords from the title, the abstract, and sometimes the full text of articles. The keywords define the strategy for searching the data base for relevant articles. By publishing the search strategy, the author makes it possible for other investigators to duplicate the search.

• Define criteria for excluding studies: The exclusion criteria define the population to which the conclusions of the meta-analysis will apply. Criteria may be broad (no articles in a foreign language; no studies of animals; no review articles) or relatively narrow. The authors of a meta-analysis must state their exclusion criteria.

• Identify poorly designed studies: The conclusions of poorly designed studies may be less valid than those of well-designed studies. The authors may exclude such studies or analyze them separately to see if study quality influences the results. Often, the best studies of a technology show a smaller effect on health outcomes than poorly designed studies.

• Test for homogeneity of studies: In the ideal meta-analysis, all studies examine a sample from the same universe of patients and use the same intervention in exactly the same way. In practice, studies usually differ. Since it is unsound to combine studies that are very dissimilar, the analyst needs a method for examining a series of studies and deciding whether the results of the individual studies are sufficiently alike to justify combining them. The investigator may exclude studies by inspection or identify them by testing for the homogeneity of the pooled population. Typically, one uses a statistical test to decide if the results of individual studies differ more than would be expected if each study population was a sample from the same universe. Significant heterogeneity indicates that differences among reported results are not due to random variation alone.

• Combining results: There are many methods for combining results. Results of individual studies are usually expressed in terms of the relative risk of an outcome in the intervention as compared with the control group. The Mantel–Haenszel estimation of the combined relative risk is a commonly used method for interpreting the results of several studies.

• Perform a statistical test of the significance of differences in the treatment effect size in the intervention group and the control group.

• Perform sensitivity analysis: In a sensitivity analysis, the investigator tests the consistency of the main results in different subpopulations of studies. For example, the analysis of the entire group of studies may show a large difference between control and intervention group. If, after deleting poorly designed studies, the difference between control and intervention group is small and insignificant, conclusions based on the differences seen in the total population of studies may be invalid.

C. Administrative Data Set Analysis: A Special Form of Cohort Study

Among the defects in prospective, controlled studies, two stand out: small sample size and highly selected study populations that sometimes bear little resemblance to the everyday practice of medicine. Recently, epidemiologists have turned to a new source of data: data sets formed as a by-product of running a hospital or medical practice, for example, a hospital's claims for reimbursement for services rendered to a patient. When a patient who is enrolled in the U.S. government's Medicare system is discharged from the hospital, the hospital submits a claim for payment. This claim contains a code number for the patient's principal diagnoses and codes for major procedures that were performed during the hospitalization. These data are stored in electronic form in a Medicare computer file along with the dates of service and information about the patient, such as age and gender. The investigator forms a cohort of patients whose common feature is hospital admission for a specified diagnosis but who differ according to which intervention they received while in the hospital. Another Medicare file contains information on enrollees' vital status. The investigator tries to compare the interventions by comparing survival at a specified interval after discharge from the hospital. For example, one can compare survival at one year after a hospital admission in which narrowing of the blood supply to the brain was a coded discharge diagnosis and a patient either had an operation to relieve the narrowing or did not have the operation.

There are a number of advantages to analyzing insurance claims to measure the effectiveness of health technology. The most important is the representativeness of the study population, which is the entire membership of the insurance program, and of the providers, which include all participating hospitals and physicians. The breadth and size of this population are in sharp contrast to those for randomized trials, which are often small and narrowly selected. A second advantage is complete follow-up information. Randomized trials sometimes lose contact with patients and can say nothing about their clinical outcome. In some cases, the reason for being unable to contact the patient is the death of the patient, which is an important study end point. One can reliably discover if a patient is alive or dead from administrative data sets because the next of kin must report the death of a patient who is a beneficiary of the Social Security System. A third advantage is the very large number of patient records in a claims data base. In a very large data base, there may be many records from patients with a rare disease or a disease with infrequent end points. A final advantage of claims data bases is low cost: the data are collected for administrative purposes, and research entails little additional cost.

Claims data bases have one very important disadvantage, which is common to other forms of cohort analysis. Consider the problem of interpreting longer survival in a group of patients who had a surgical procedure for narrowing of the arteries of the brain. Did the patients survive longer because of the procedure or because of other factors, such as overall good health, that led their surgeon to encourage them to have surgery? If many of the untreated patients had another disease, such as cancer, that limited the benefit that could be derived from surgery, their surgeon will not encourage them to have the operation. If many of them went on to die of cancer, survival would be shorter in the untreated patients even though the treatment might actually be harmful. These extraneous factors can be taken into account by comparing only patients who did not have cancer, and these comparisons can often be done without losing statistical power because of the large number of patients in a claims data base. However, this method allows one to correct only for factors that are known and that are recorded in the data base. The advantage of randomization is that all confounding factors, known and unknown, are taken into account. A second disadvantage of claims data bases is the relative paucity of clinical data that would allow one to examine clinically important relationships by analyzing illness outcomes in subgroups of patients. This problem should resolve as technical advances lead to greater ease in recording complex clinical data in electronic form. A final disadvantage of current data bases is the lack of information on the patients' ability to function in daily life.

The limited amount of clinical information in administrative data sets is one of two important shortcomings of this method. In fact, the Health Care Financing Administration is collecting an extensive clinical data set on a randomly chosen sample of Medicare patients. However, even this admirable effort cannot solve the serious problem of making valid inferences about cause and effect from studying a cohort of patients. To decide that a treatment is the cause of improved survival, one must exclude the possibility that another factor (a so-called "confounding variable") is responsible for the effect. Typically the confounding variable and the putative causal variable (the treatment) vary together (they are covariates), so that it is difficult to know which is responsible for the effect. To overcome this problem, investigators use a set of techniques known collectively as multivariate analysis. The common feature of these techniques is a set of predictor variables (typically characteristics of the patients and their medical care) and a dependent variable (typically a measure of the outcome of illness). Multivariate analysis will show the effect of each predictor variable on the dependent variable *independently of the influence of other predictor variables.* Multivariate analysis is a very powerful technique, but it has an important limitation. It can only analyze the effect of a variable that is in the data set. The contrast with a randomized clinical trial helps to understand this point: if the study groups are large enough, random assignment assures that they differ only in exposure to the intervention and are identical with respect to all other variables, *known and unknown,* that might affect the outcome of care. A multivariate analysis of a cohort study cannot claim to include as predictor variables all potentially confounding variables, both known and unknown.

D. Consensus Conferences

A consensus conference is a panel of individuals who have been convened to render a joint opinion. For example, the panelists may be asked to read an article that describes the current state of the art. Often the article is a comprehensive, critical review of past studies and a distillation of the evidence into draft recommendations for good medical practice. In many conferences, the panelists use a closed ballot to express an independent opinion, the results of the balloting are discussed, and a second vote is taken. This method, which is a variation of the Delphi technique, obtains an expression of group opinion while minimizing the risk that a few highly vocal individuals will unduly influence the proceedings. This method also identifies areas of controversy and areas of agreement.

There are several advantages to consensus conferences. They are relatively inexpensive, as compared with clinical studies that test hypotheses by obtaining new data, and they often lead to a description of the current practices of expert clinicians. Consensus conferences can be quite influential because the opinion of experts carries considerable weight with many clinicians.

Yet consensus conferences have important shortcomings. The validity of the conclusions is limited by the quality of the data. Some conferences give the participants only a day or two to complete their deliberations and write detailed recommendations. A thorough discussion of the evidence is difficult under these conditions. Participants who are under intense time pressure to complete a task are more likely to endorse current practice than to consider new ideas. Finally, there is no opportunity for nonparticipants to comment on the recommendations. Individuals whose opinion is influenced by factors other than scientific evidence have the opportunity to sway group opinion, and there is little time to negotiate a sound, evidence-based compromise solution. When all of these conditions are present and, in addition, the conferees are renowned and the conclusions are well publicized, the conference may retard the search for truth rather than advance it.

E. Decision Analysis and Cost-Effectiveness Analysis

Decision analysis can provide an answer in circumstances where a randomized trial cannot be performed or will take a long time to produce a result. It is a method for simulating the events that follow a medical intervention by taking into account the uncertainty about when events will occur or whether they will occur at all. The method can also take into account patients' feelings about the outcomes that they might experience. Decision analysis uses data from published studies and is not a substitute for a randomized trial, which makes observations on patients' response to illness and treatment.

A model is a representation of reality. A model of a decision problem typically represents events in a tree structure. Events that are under the physician's control are represented by decision nodes. Unpredictable events (such as death following a surgical operation) are depicted by chance nodes, with a branch for each event that might occur. Each branch at a chance node is labeled with the probability that it will occur. The probabilities are obtained from published studies, from analysis of claims data bases, or from an expert clinician. Figure 1 is a decision tree to help decide between surgery or medical management of chronic pancreatitis, a disease in which inflammation of the pancreas causes abdominal pain. By calculating a weighted average value at each chance node starting with the tips of the tree on the right and working back to the left, one calculates an expected, or average, outcome for medical treatment and for pancreatic surgery. The average length of life spent in good health ("quality-adjusted life expectancy") is the measure of outcome in this model. The basic principle of expected value decision making is to choose the decision alternative that has the largest expected, or average, value.

Cost-effectiveness analysis is a general term for expressing the efficiency with which a health care intervention transforms the resources that it requires into clinical outcomes. One way to calculate cost-effectiveness is to use an expected value decision model. The model calculates expected cost and a measure of clinical outcome, such as life expectancy. When comparing two decision alternatives, the difference in cost between the two alternatives (marginal cost) divided by the difference in outcome (marginal outcome) is the cost per unit of clinical outcome (cost-effectiveness). Thus, the model might compare an improved treatment with the usual treatment by calculating the quotient of the additional cost of the new treatment and the additional length of life provided by it.

The advantages of decision analysis, as compared with randomized clinical trials, include low cost and speed. Flexibility is another advantage. The investigator can bring a decision analysis up to date by substituting new values for probabilities or changing the structure of the model as new data or treatment alternatives become available. A decision analysis can generate an answer that serves as a clinical policy that can be applied to many patients. A decision model can also calculate a patient-specific result if the parameters of the model are specific to a particular patient.

The role of decision analysis in the hierarchy of methods for assessing technology is to provide new insights about old data by casting them in the perspective of making a decision. However, the parameters of the model are obtained from prior observations of patients, and the validity of the conclusions is limited by the quality and applicability of these data. Often the model represents events about which there is little or no information. Uncertainty about such events need not be a weakness of the analysis. In a technique called sensitivity analysis, the model recalculates the expected value of each decision alternative as the investigator substitutes first one value and then another value for the uncertain variable, eventually covering the entire range of possible values. If the preferred decision alternative is the same in all cases, the variable is not an important determinant of the decision. Sensitivity analysis identifies the variables that are really important to decision making and focuses the research agenda on obtaining better information about these variables.

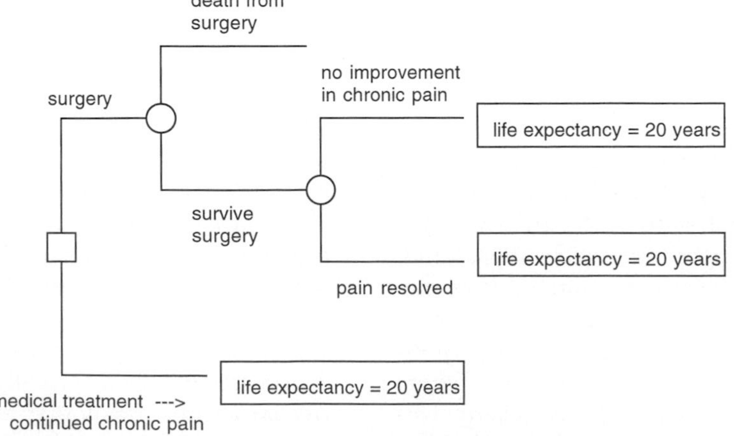

FIGURE 1 A decision tree for the management of chronic pancreatitis. The choices are a surgical procedure or medical treatment. The outcomes are perioperative death, resolution of pain, and continued pain.

III. SYSTEMS FOR EVALUATING HEALTH TECHNOLOGY IN THE UNITED STATES

The systems for evaluating health technology are in a state of change, and professional organizations participate because they have a responsibility to educate physicians. They evaluate technology in order to provide their membership with current information about how to use technology effectively. The American College of Physicians is an example of a professional society with a very active program of technology assessment. The federal government supports the U.S. Preventive Services Task Force, an expert panel that makes recommendations about preventive services. Those who pay for health care are motivated to discover which medical practices are appropriate and should therefore be reimbursed. The Blue Cross/Blue Shield Association is an example of an organization that evaluates technology. With the rise of managed care, practicing physicians have become active in developing practice guidelines, often using the guidelines of professional organizations as templates.

IV. THE SPREAD OF NEW TECHNOLOGY

Several factors appear to promote the adoption of new technology. To begin with, the theory underlying the technology should be consistent with prevailing medical theory. A technology that is easily learned, such as the use of a new drug, may be more quickly adopted than a complex procedure that requires manual dexterity and long practice. A technology that deals successfully with a common and urgent clinical problem is especially likely to be adopted. Furthermore, a strong and influential advocate in the professional community speeds adoption of new technology. The practice setting can also have an influence, especially if it is one in which physicians control the decision to acquire a new technology. In light of these complex influences, the effect of a careful evaluation of the technology may have less effect than might be expected, especially if the study results receive little publicity.

V. FUTURE DIRECTIONS

The patient's role in deciding: Patients are taking a greater role in deciding what care they will receive, and the study of technology now includes formal evaluation of patients' preferences for the outcome states they may experience. Videotaped programs inform the patient of these outcomes and the probability that they will occur.

New measures of effectiveness: The clinical outcomes that serve as the end points for a study are the measures of effectiveness. Past studies have used survival or a measure of disease activity as the sole study end point. As competing technologies become more powerful, they often differ little in their effects on disease and survival, and the choice between them will depend on outcomes such as the ability to function in society physically, emotionally, and socially. Current studies of technology are starting to use these secondary measures.

New methods to disseminate information about technology to physicians: The most important unsolved problem in technology evaluation is how to influence the physician to adopt a new practice style. The marketplace is providing the motivation to adopt practices that improve the quality of care and reduce costs. Health care organizations invest time and effort into analyzing the details of patient care and striving to improve it. "Improving quality while taking costs out" has become a catch phrase in U.S. health care.

The role of the computer: Soon, all of a patient's clinical data will be stored in a computer rather than in the traditional paper medical record. This development will mean a very rich data source for clinical studies of technology. The computer will probably soon play a role in guiding the selection of technology for a patient, since the computer can present the results of technology assessments to the physician at the time that a decision is required; this form of intervention is known as "point of service decision support."

BIBLIOGRAPHY

Feeny, D., Guyatt, G., and Tugwell, P. (eds.) (1986). "Health Care Technology: Effectiveness, Efficiency, and Public Policy." The Institute for Research on Public Policy, Montreal.

Fineberg, H. V., and Hiatt, H. H. (1980). Evaluation of medical practices: The case for technology assessment. *N. Engl. J. Med.* **303**, 1086–1091.

Kassirer, J. P., Moskowitz, A. J., Lau, J., and Pauker, S. G. (1987). Decision analysis: A progress report. *Ann. Int. Med.* **106**, 275–291.

L'Abbe, K. A., Detsky, A. S., and O'Rourke, K. (1987). Meta-analysis in clinical research. *Ann. Int. Med.* **107**, 224–233.

National Academy Press (1985). "Assessing Medical Technologies." National Academy Press, Washington, D.C.

National Academy Press (1989). "Medical Technology Assessment Directory: A Pilot Reference to Organizations, Assessments, and Information Resources." National Academy Press, Washington, D.C.

Sox, H. C., Jr. (ed.) (1990). "Common Diagnostic Tests: Use and Interpretation," 2nd Ed. American College of Physicians, Philadelphia.

U.S. Preventive Services Task Force (1995). "Guide to Clinical Preventive Services," 2nd Ed. Williams & Wilkins, Baltimore.

Health Psychology

HOWARD S. FRIEDMAN
University of California, Riverside

GEORGE C. STONE
University of California, San Francisco

GLOSSARY

Behavioral contracting Procedure in which provider and client mutually negotiate a set of goals for behavior change on the part of client and explicit rewarding or punishing consequences for fulfilling or failing to fulfill the contract

Cognitive processes Manipulation of information by the brain (i.e., interpreting, storing, retrieving, comparing, linking, etc.)

Conditioning Fundamental process, presumably in the central nervous system, whereby changed patterns of response to specific situations are brought about by temporally contingent relations between environmental events and behavior; it occurs in many phyla of multicellular animals

Holism Idea, in this context, that animal organisms demonstrate emergent properties not predictable from knowledge of their component subsystems

Long-term versus short-term memory Symbol memories of humans and higher mammals. Long-term memory retains information for years, short-term for a few seconds, without rehearsal; long-term retains vast amounts, short-term about seven items

Pain behavior Behavior typically performed when painful stimuli are applied: verbal and nonverbal complaints, impaired functioning, and relief-seeking actions, such as self-medication

Psychoneuroimmunology Interdisciplinary field of research that demonstrates and seeks to explain influences on the immune system by cognitive and emotional processes

Psychosexual development Development of sex-related beliefs, attitudes, desires, and behaviors. In psychologies deriving from the work of Freud, an elaborate sequence centering on parent–child relations has been proposed

Retrospective versus prospective design In psychological epidemiology, design that seeks to establish causal relations from recall or records of a group characterized by a consequent event, versus following a group possessing the antecedent attribute to see if the consequent occurs

AS DEFINED BY THE DIVISION OF HEALTH PSYCHOLogy of the American Psychological Association,

> Health Psychology is the aggregate of the specific educational, scientific, and professional contributions of the discipline of psychology to the promotion and maintenance of health, the prevention and treatment of illness, the identification of etiologic and diagnostic correlates of health, illness and related dysfunction, and the analysis and improvement of the health care system and health policy formation.

I. ANTECEDENTS

A. Clinical Psychology

Studying the relation of mental and physical problems, the field of clinical psychology has developed to apply psychological knowledge and techniques to health problems. Until recently almost all the work of clinical psychologists was concentrated on "mental health problems," in which the behavior or emotional experience of the client/patient defined the issue to be diagnosed and treated. Although the first psychological clinic was established in the United States in 1896, the major growth of clinical psychology began after the Second World War, when a massive program was initiated and supported by the National Institute of Mental Health to prepare clinical psychologists to treat returning veterans and members of the armed forces, concentrating their effort on problems of psychotherapy, schizophrenia, mental retardation, and alcoholism. In the 1970s, emphasis began shifting to all the other issues of health and illness and efforts began to redress the balance.

B. Medical Psychology

Unlike clinical psychology, "medical psychology" never developed into a regulated field of practice nor did it give rise to major professional organizations. Perhaps as a consequence, several different kinds of activity have been going by this name. In Great Britain and associated countries, medical psychology was almost synonymous with "psychiatry"—the treatment of mental illness. Throughout most of Europe, and to some extent in the United States, the term refers to the psychological aspects of medical care, emphasizing the experience of the patient and the "humanizing" of the doctor–patient relation. Medical psychologists of this kind mostly teach physicians-in-training to be attentive to these aspects of their work. Yet a third usage, found in the United States and elsewhere, refers to the practice of clinical psychology in medical settings or with patients suffering from problems of bodily disease.

C. Behavioral Medicine

The most widely accepted definition was an outcome of the Yale Conference on Behavioral Medicine, held in 1977, and a second conference sponsored by the Institute of Medicine in 1978:

> Behavioral medicine is the interdisciplinary field concerned with the development and integration of behavioral and biomedical science knowledge and techniques relevant to health and illness and the application of this knowledge and these techniques to prevention, diagnosis, treatment, and rehabilitation.

This definition is broader than that of health psychology, because it includes contributions from other behavioral sciences such as sociology and anthropology and also from biomedical sciences. Implicit in the definition, however, is a concentration on the biomedical aspects of health problems, an emphasis that has generally been characteristic of both research and practice in the field. Psychologists who refer to their practice as "behavioral medicine" usually employ behavioral techniques to address a medical problem per se, with a view to altering some aspect of bodily function either directly or indirectly. Those who call themselves "medical psychologists" may engage in such work also, but they are more likely to include concern for the individual patient's efforts to cope with the illness and treatment as a human problem—a problem of living—and perhaps to engage in psychological diagnostic procedures to help fit medical procedures to the particular person to whom they are to be applied. Conceptually, both psychologists who engage in behavioral medicine and those who go by the name medical psychologist are health psychologists.

D. Psychosomatic Medicine

The term "psychosomatic medicine" refers to the role of internal conflict in causing mental disease. Active development of the psychosomatic concept began in the 1920s, when physicians, mostly of psychoanalytic persuasion, began to collect case histories and speculate about the relation of the emotions and symbolic representations of inner conflicts to bodily disease. To varying degrees, the ideas of three major theorists were interwoven to create sometimes fanciful explanations for several "psychosomatic diseases" (e.g., asthma, stomach ulcers, and rheumatoid arthritis). Sigmund Freud's concepts of psychosexual development were linked with Ivan Pavlov's discoveries about the processes of conditioning, and physiological mechanisms were proposed using Walter Cannon's "wisdom of the body" approach to physiology.

The concept of psychosomatic diseases reached its peak in the 1940s and early 1950s. Then increasing methodological rigor revealed that the relations were more general and less symbolically colorful than the early workers had portrayed them. Emphasis shifted from specific psychosomatic diseases to a recognition of the great importance of psychological phenomena in most disease processes. This idea was expressed as early as 1950 by George Engel, a quarter of a century before he coined the heuristic term *biopsychosocial model* to refer to it. This viewpoint now pervades behavioral medicine and guides the work of most health psychologists who work at the mind–body interface. The term *psychosomatic medicine* persists to refer to a special emphasis on holism and on symbolic influences within the body of behavioral medicine.

E. Beginnings of Health Psychology

In the 1970s, psychologists, noting the lack of work on most aspects of health and illness, began to engage in new kinds of research and to develop new forms of intervention. From the learning laboratories came techniques, known as biofeedback, that allowed people to gain some voluntary control over health-relevant behavior, such as the regulation of heart rate or skin temperature, that had traditionally been considered involuntary. Social psychologists began to study a phenomenon known as "noncompliance," that is, the failure of patients to follow physicians' recommendations for which they had often paid dearly in time, trouble, and money. As more and more psychologists began to work on the many problems they encountered in the new environments of the health system, they organized, first as a task force, then as a Section on Health Research in one of the divisions of the American Psychological Association, and finally, in 1978, as an independent Division of Health Psychology. Similar actions were taking shape in other parts of the world. A Task Force for Health Psychology was established in the Interamerican Society of Psychology in 1982 and a European Society of Health Psychology in 1987. A truly international enterprise has grown up, emphasizing health as much as illness and prevention as much as cure, based on a universal recognition of health as a primary human value.

II. SCOPE OF THE FIELD

Health psychology represents the overlay of psychological theory, data, and methods on the problems of health. These problems include not just the diagnosis and treatment of disease and injury, but identifying and responding to environmental hazards to health; fostering health-promoting behavior on the part of the public; supporting persons in their appraisals of their symptoms and their decisions as to what should be done about them; minimizing the discomfort and ardors of medical treatments that patients must undergo and helping them to cope with the irreducible stresses and to adhere to their treatment regimens; providing programs and facilities for those with residual disabilities after treatment; educating health professionals; and planning, financing, and staffing appropriate health care facilities and organizations.

Figure 1 diagrams some of the relations in the health system in which psychological processes influence health outcomes.

To recapitulate, health psychology relates to all phases of the health process from hazard abatement and avoidance to rehabilitation after illness or trauma. In principle, it relates to all aspects of the health system (e.g., financing, training of personnel, design of organizations, and health care delivery). It is also concerned with all parties to health transactions, including persons at risk, their families, health care providers, organizational managers and support staff, and policy makers.

III. MAJOR RESEARCH ISSUES

Causal relations between health and behavior can be organized in either direction: behavior can have health effects, and health can affect behavior. Pathways by which behavior can change the health status of an individual can be broadly grouped as follows:

1. Behavior can lead to physical trauma, broken bones, contusions, concussions, etc. In some cases, the relation is truly accidental, but there are also persons who deliberately expose themselves to situations of high risk for various psychological reasons. Some earn their livelihoods in hazardous occupations; some crave excitement and a sense of danger. On the positive side, some persons take precautions against injury by wearing seat belts or protective helmets, making plans and preparations for natural disasters, and supporting regulations or legislation to remove or mitigate hazards in the environment.

2. Behavior can lead to the penetration of the body by toxic or pathogenic materials. The exposure may be deliberate, as in substance abuse and high-risk sexual behaviors, or unintended, as when someone buys a home on a site once used for the disposal of toxic wastes or in an area infested by insect vectors for serious diseases. Behavior taken to avoid exposure to or damage by such materials is properly considered preventive health behavior. Oral hygiene reduces the damage to teeth and gums of the bacteria that are naturally present in the mouth.

3. Behavior can affect the bodily condition of the individual through the choice of the amount and kinds of foods ingested, the amount of physical exercise engaged in, and the care taken in conjunction with

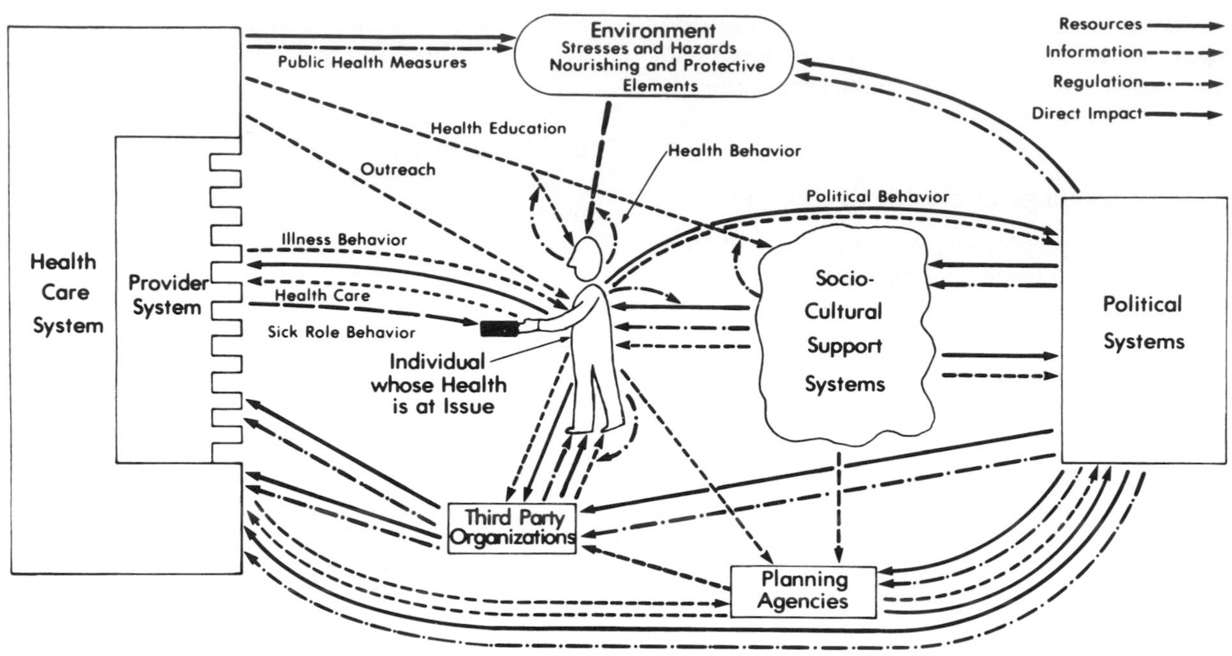

FIGURE 1 The health system.

such exercise to avoid injuries from excessive strain or improper postures and movements. Much of the current emphasis on "healthful life-styles" is directed toward behavior in this category.

4. Behavior can greatly influence the level of activity in the body's stress systems: the autonomic nervous system that regulates glands and organs, and the immune system that responds to pathogenic intrusions and novel substances that may appear in the body. Stress can be increased by life choices and by the ways in which one perceives or construes life events. Stress can also be reduced by certain practices, including various forms of meditation or contemplation. It may also be possible to reduce stress through the adoption of different frameworks for interpreting life circumstances.

In relation to each of these pathways, health psychologists attempt to understand the ways in which the risks are perceived and to devise and evaluate methods for communicating accurate understanding of the risks. They study decision-making processes and the determinants of risk-taking behavior. In relation to the stress pathways, they study the psychophysiological processes involved and the appraisal and coping mechanisms associated with greater or lesser physiological response to comparable stresses.

An individual's health status can also have profound effects on his or her behavior. For example:

1. Duration and mutability: Most conditions of disease, illness, or injury cause some degree of physical limitation for shorter or longer periods. In some cases it is of only a few days duration and in others it is present from birth and lasts for a lifetime. Closely related to this characteristic of the limitation is the degree to which it can be expected to change. Some problems are transitory and can be expected to pass without special (psychological) intervention. Some are totally unchangeable, and some can be changed to varying degrees by varying expenditures of effort, time, and other resources.

Health psychologists study the impact of duration of the limitation and of its time of onset on the total reaction of the person and on the design of interventions to limit the degree of incapacitation.

2. Increasing versus decreasing limitations: Limitations on behavioral capacities from health causes vary greatly in their time courses. Some appear suddenly in full force, as in the case of strokes. Others, such as Alzheimer's disease, come on gradually over a period of years. Reactions of patients and of those around them are different as a result of variations in this attribute. There may also be substantial difficul-

ties arising from the removal of limitations. Persons recovering from heart attack often retain psychologically based limitations that are not warranted by the degree of their recovery of cardiac function. Persons relieved of long-term obesity or facial disfigurement through surgical procedures may undergo catastrophic disturbances of their life adaptations and require significant psychological intervention to reestablish equilibrium.

Much more research has been focused on those issues in which psychological factors influence health than on those in which the health status of the individual has psychological sequelae. There are, however, a great many instances of each.

A. Prevention of Injury and Illness

As costs of health care continue to grow, recognition is increasingly being given to the idea that prevention costs less than curing—in monetary terms as well as in terms of human suffering. Furthermore, a good case can be made that much of the increase in life expectancy of the past century is due not to curing disease but rather to preventing disease, through improved sanitation, improved nutrition, less crowded housing, and childhood inoculations. Similarly, many of the greatest contributors to current morbidity and premature mortality involve preventable behaviors—tobacco use, alcohol and drug abuse, poor diet, and failure to take prophylactic measures. Nevertheless, substantial difficulties need to be overcome in developing effective programs of prevention.

Before one can develop methods for preventing health problems, one must understand how they come about. The pathogens, toxins, and other hazards of the environment must be causally linked to the injuries and diseases that we seek to prevent. Geneticists, epidemiologists, physiologists, structural engineers, and many others, including psychologists, contribute to our understanding of etiology (i.e., the causes of illness) and to the discovery of interventions that could mitigate the harmful effects of the hazards identified. Without exception, however, when a cause has been discovered and a potentially effective intervention has been designed, someone's behavior will have to be changed before prevention can be accomplished. In many cases, it is the behavior of the person whose health is at risk that must change. People must wear seat belts, stop smoking, change their sexual practices, and change the ways they clean their infants or build their houses. Changing behaviors of health care providers can also be difficult, even in an era characterized by technological progress. In other cases, it is the behavior of persons in corporate entities that must change, reducing the release of toxic substances into the environment, the sales of dangerous products, or the imposition of health-damaging working conditions. Sometimes changes involve the passage of legislation or the adoption of administrative regulations, in which case the behavior of legislators and other policy makers must change, and perhaps so must the behavior of voters. In every case, however, the ultimate change process occurs in the behavior of individual humans, and the psychological processes of behavior change are involved.

A special characteristic of preventive behaviors makes them difficult to establish and maintain. In most cases, it is necessary to replace behavior that provides more immediate gratification with behavior that may have little or no reward other than the knowledge (or belief or hope) that it may reduce the likelihood of some unfavorable health condition or event at some time in the future. Psychological research on prevention is mainly concerned with discovering ways to overcome barriers to behavior change so that well-established principles and procedures can be applied.

B. Stress and Illness: Psychophysiology

For more than 50 years there has been increasing understanding of the ways in which our individual appraisals of what happens to us give rise to responses in our nerves and glands that influence our bodies in many ways. Among other things, these responses alter our digestive processes; our metabolism of sugars, fats, and other substances; the functioning of our hearts and blood vessels; and the growth and distribution of the blood cells that help to protect us from invading bacteria and toxic materials such as plant pollens.

Stress is the state of an organism when reacting to challenging new circumstances, be they an invading virus or a social stressor like a divorce. In addition to biological changes, these stress-inducing challenges are usually associated with motivational and emotional changes like anger, anxiety, or depression. Increasing evidence is documenting the biological pathways through which these psycho-emotional states may increase the likelihood of disease onset or disease progression. Particular attention is focusing on atherosclerosis and coronary heart disease, inflammatory diseases like arthritis, and cancer progression. Activation of the sympathetic nervous system (sympathetic adrenal medullary axis) affects blood pressure, heart function, and catecholamine levels (and hence metabolism). Activation of the hypothalamic–pituitary–adrenocortical axis affects blood levels of cortisol and related stress hormones, with likely effects on the immune system. So, for example, increasingly sophisticated attempts are being made to assess the contribution of chronic hostility to hypertension, and the possible effects of severe emotional repression on the progression of cancer.

Between the conditions and events that occur in our environments, as they might be recorded by cameras and scientific instruments, and the physiological mechanisms of our body, there are psychological processes known as "stress appraisal." We all know that an event that may be devastating to one person may be handled with equanimity by another. Psychologists seek to characterize the properties of situations and of individuals that give rise to these differences. The psychological essence of the stressful situation is that it threatens the integrity of the individual, however that may be conceived. People attempt to cope with stressful situations in two major ways. These have been termed *emotion-focused* and *problem-focused* coping, depending on whether the individual tries to reduce the distressing emotions aroused by the threat or attempts to mitigate the source of the stress. Depending on the nature of the stressful situation (i.e., whether it can be altered by human effort), one or the other of these methods may be more effective in minimizing the impact on the person. Interactions between these psychological mechanisms and the physiological mechanisms of the stress response are extraordinarily complicated. [*See* Psychoneuroimmunology; Stress-Induced Alteration of GI Function.]

C. Personality and Illness

The idea that certain kinds of people were especially prone to develop particular diseases was, as previously noted, a major preoccupation of the field of psychosomatic medicine in the middle of this century. As more sophisticated kinds of statistics were applied and larger groups of patients studied with better sampling methods, the inductions that had been put forward proved difficult to substantiate. There are, however, continuing indications that differences in personality are related to health.

The most consistent evidence linking personality to physical health comes from studies of hostility and its associated emotional response style. Interest in the study of hostility stemmed from research on the Type A Behavior Pattern (TABP), which had suggested that emotion-based aspects of the TABP, such as hostility and excessive competitiveness (assessed using coded verbal and nonverbal behaviors from a structured interview), were more strongly predictive of physical health than paper-and-pencil measures of the construct. In support of this, measures of hard work did not predict ill health if there was no emotional disturbance. Hostile individuals may or may not exhibit overt anger and aggression. In particular, individuals who are alienated and cynically hostile tend to be suspicious and resentful and especially prone to disease.

Although one might think that optimistic and cheerful individuals would be physically healthier than pessimistic individuals because of their more positive outlook on life, research to date has not been able to establish a simple and firm link between optimism and physical health. Optimists tend to report fewer symptoms than pessimists, they enjoy quicker and better recovery after surgery, and they may use more adaptive coping strategies. However, this does not necessarily predict disease incidence or mortality risk because being too optimistic and carefree may put one at risk for a series of unhealthy behaviors.

Neuroticism (being anxious and depressed) is associated with self-reported symptoms, but not always with more objectively measured health outcomes such as mortality risk, although there appear to be pathological components of neuroticism that are associated with more objectively defined health outcomes. For example, diseased or disease-prone people are much more likely to be dysphoric, alienated, and anxious than they are to be enthusiastic, balanced, friendly, and well integrated into their communities. It is not clear exactly why this is the case. In an attempt to focus more attention on the positive aspects of mental health that are associated with good physical health, the latter characteristics (like enthusiasm and emotional equilibrium) have been termed a "self-healing personality." Overall, there is little doubt that personality traits and emotional response patterns should be considered risk factors for disease—just as cholesterol, smoking, and blood pressure have traditionally been considered risk factors. Further, we know that children with certain personal characteristics grow up to be adults who engage in health-compromising lifestyles whereas others live health-promoting lives. For example, a number of longitudinal studies of children and adolescents have predicted smoking and drinking behavior from aggressive, weak, or unstable personalities.

D. Decisions in Health Care

Every phase of the health care process is fraught with decisions. Persons at risk and patients must decide about their health behaviors, what risks they will take, whether their symptoms require expert care and, if so, where to seek it, and whether to follow the recommendations made by the expert consulted. On the other side, a major component of the expertise of health professionals resides in their capacity to decide about the patient's diagnosis and what treatment to recommend.

For many years, psychologists have been studying the ways in which people make decisions, and health psychologists have been active in applying and extending their findings to the decisions of the health system. Much of this work has been based on a model that treats the person as rational—seeking to optimize outcomes under conditions of limited information. Within this model are two principal issues: (1) How one can best estimate the probabilities of various states of the world (e.g., that a person does or does not have a particular disease), and the probabilities of various outcomes given a particular state of the world and a particular course of action (e.g., the probability that a cardiac catheterization will yield important diagnostic information, given the patient's presenting symptoms, versus the probability that the patient will die during the procedure); and (2) how the values of various outcomes can be compared. How does one weigh the dollar cost of a treatment plan against the possibility that it will remove a particular symptom or reduce the chance of a serious health problem at some time in the future?

The input of psychologists is needed whenever human behavior is a part of the picture. For example, it has been determined that the choice between a surgical and a pharmacological approach to a certain kind of hypertension depends on one's estimate of the probability that patients will faithfully follow their prescribed drug regimens. The problem of assigning numerical values to human values for purposes of formal decision making is one that psychologists and philosophers have struggled with without much success. In the end, it simply comes down to a careful measurement of human preferences. However, such measurement is no easy matter. Policy makers decide how the preferences of various persons and groups should be weighted, but psychologists are needed to deal with difficulties in getting accurate, stable, and consistent expressions of an individual's preferences.

For most of the day-to-day decisions that are made, however, patients and health professionals do not call on statisticians, philosophers, or policy makers for help. Sometimes they use approximations to the rational decision model, subject to limitations of the human mind in knowledge, long-term and short-term memory, and their understanding of the model itself. Often, irrational factors enter in. Important among these are the suppression of information that gives rise to discomfort when contemplated, and the rejection of mental calculations that are lengthy and effortful. Health psychologists study the operation of such tendencies in patients and professionals and seek ways to overcome them.

IV. INTERPERSONAL ISSUES IN HEALTH CARE

Much of health care depends on successful interpersonal interactions. For success to be achieved, a patient must enter the health care system, truthfully and accurately describe his or her symptoms and behaviors, understand health or treatment recommendations, follow (cooperate with) the recommendations, and integrate them into ongoing life. Physicians sometimes call these matters the "art" of medicine, but health psychologists consider these matters to be social science, and susceptible to rigorous study.

A. Provider–Patient Interaction

One useful way of construing the medical interaction is as a joint problem-solving situation. The patient brings the problem and presents it to the provider. The provider uses his or her expert knowledge to elicit further information, achieve a diagnosis, decide on possible forms of treatment, and present one or more options to the patient. The patient, in most situations, accepts or rejects the recommendation and follows or fails to follow the prescribed plan. The results of this sequence of actions are then evaluated and, if necessary, additional information is sought or new options considered. There are many possible variations on this basic plan. A widely accepted classification, first proposed in 1956 by T. S. Szasz and M. H. Hollender, divides them into three models: (1) active provider–passive patient; (2) guidance–cooperation; and (3) mutual participation. These models differ in the degree to which there is continuing two-way communication throughout the transaction and the degree to which responsibility for decisions is shared. Although psychologists and other social scientists have tended to favor the mutual participation model on philosophical grounds, research has made it clear that different models are appropriate for different situations. One important determinant is the expectation by the patient and the preparation for assuming responsibility. Problems arise when provider and patient are applying different models to the transaction without knowledge that they are doing so.

B. Social Support

A second aspect of interpersonal relations in health care that has been much studied by health psychologists is that of "social support." People turn to each other for support. Some people are surrounded by networks of relatives and friends; others live alone and may have few or no significant persons in their lives. In general, people with strong networks of social support, particularly men with loving spouses, are healthier, live longer, adhere to treatment regimens better, and cope better with chronic or life-threatening illness. Early studies tended to focus on the quantity of support (i.e., the size of social networks). Increasingly, it is recognized that the quality of the support is at least as important; some families, typically counted as support, might be better considered as sources of stress. Quality of support is much more difficult to appraise than quantity. It may also be the case that social support sometimes results from good health and a healthy personality rather than being a cause of good health.

C. Informed Consent

An ethical principle increasingly important in the health care system is that of "informed consent." In its strongest form, this doctrine holds that no procedure should be performed on persons until they have full information about the possible consequences and an opportunity to reflect on the implication of this information for their own lives and those of others. Only then should consent be given, and if it is not, the procedure should not be undertaken.

Several psychological problems arise in conjunction with this concept. How much information does one need to make the best possible decision? How can a person without medical training understand highly technical information? How can one anticipate what it would be like to be in a condition one has never experienced or known anyone to have experienced? What should be done when a patient asks the health care provider to make the decision? Should we insist on informing people who prefer not to have the information? What happens when a person with inadequate understanding of probability is informed about rare but serious side effects of a treatment—do they make better or worse decisions? For that matter, how do we compare the quality of decisions made? Psychologists have a good deal of information about "postdecisional cognitive processes" that point to major difficulties in assessing satisfaction or regret concerning the consequences of actions taken. Health psychologists cannot provide definitive answers to many of these questions, which mostly involve significant value issues, but they can help to determine the probabilities of alternative outcomes of the consent process based on analyses of the situations and the participants.

D. Cross-Cultural Issues in Health Psychology

Cultural beliefs and traditions are associated with a range of health-related behaviors, from diet to meditative practices. There is a wealth of evidence that certain religious groups, such as Mormons and Seventh Day Adventists, are generally healthier than the population of the United States as a whole. This difference can be readily explained by their lower use of alcohol and tobacco and, in the case of Seventh Day Adventists, their lower consumption of saturated animal fats. Japanese who live in Japan have lower incidence of cancer of the bowel and a higher incidence of cancer of the stomach than persons of Japanese ancestry in the United States who have adopted mainstream American culture.

Cultures also differ in their "display rules" as to the degree to which emotions may be expressed in social intercourse. Some psychological theories have proposed that suppression of emotion can have deleterious health consequences. Thus, a culture's display rules for emotional effects might have health effects. Beliefs of different religious groups about the origins and meaning of suffering may influence the seeking of health care and the experience of pain. The rapidly growing subdiscipline of medical anthropology addresses these and related issues of the interaction among cultural beliefs, conceptions of disease, and appropriate health care delivery.

V. MAJOR FORMS OF INTERVENTION

In terms of activities performed by health psychologists, the field is at least as much a form of practice as it is a domain of research. The practice of psychology spans a wide range of approaches from precisely controlled techniques addressed to some specific component of an individual person's behavior to participation in the design of large-scale systems. In the broadest terms, psychological practice has one of two purposes: It may attempt to describe or predict—to *assess*—how individuals act or feel in specified situations, or it may attempt to *change* the ways in which they behave or feel. Although these two purposes are conceptually related, the procedures by which they are accomplished are altogether different.

A. Assessment in Health Psychology

Assessment, in the context of psychological practice, attempts to answer three kinds of questions: (1) What needs to be done to improve the condition of the client(s)? (2) Which of the possible approaches to the problem thus defined is most likely to work with this particular person or group? (3) Are the selected interventions accomplishing the kinds of change that was intended? The fundamental problem of assessment in health psychology is that of describing psychologically relevant attributes of persons and environments that can predict what interventions will lead to improvements in the health of the target persons or populations. Because health psychology encompasses all aspects of psychology, the range of its assessment techniques is correspondingly broad. A few examples must suffice as illustrations.

1. Assessment of Health Status

Basic to many problems is the need to specify the "health status" of individuals. This has proven to be difficult because health is not a uniquely valued state, but a collection of statuses on multiple dimensions. A person who is able to carry out the activities of daily living is healthier, all other things being equal, than the one who is not. Similarly, one who is free of pain or other troublesome symptoms is healthier. A person who can run far and fast, jump high, and lift heavy weights is generally considered healthier than one who cannot, even though neither may expect or care to engage in such activities. A body that can resist disease or injury is healthier. A community whose members engage in socially valued work is usually said to be healthier than one characterized by crime, violence, and vandalism.

Although there is general consensus as to the healthy direction on most of these dimensions, there are substantial differences in the weight assigned to each in an overall judgment of health. Health psychologists in collaboration with other social scientists have employed their techniques for measuring preferences to bring maximum clarity and precision to what remains a difficult and cloudy area.

2. Measurement of Pain

A persistent and difficult problem in health care has been that of evaluating patients' pain reports, particularly in those conditions in which the pain is chronic and cannot be correlated with any tissue damage. On the basis of their tradition of working with subjective experience, psychologists have been able to make some contribution to the recognition that such pain is "real" and deserving of professional attention. Behavioral methodologies developed to objectify the subjective have also provided some assistance in the understanding and treatment of pain conditions. In many pain clinics this has taken the form of shifting emphasis from the treatment of pain to the treatment of "pain behavior." [*See* Pain.]

3. Psychological Risk Factors

Research demonstrating psychological differences between those who succumb to many different kinds of health problems, from relapse in smoking programs to myocardial infarctions, has given rise to the concept of "psychological risk factors." There is a logical progression from retrospective discoveries to prospective verification to the development of instruments to assess risk and ultimately to the attempt to modify the behavior that indexes risk and the psychological characteristics that give rise to such behavior. Two current areas of active interest are the investigations of the coronary-prone personality and the "addictive personality." Having identified a coronary-prone person or an addictive personality, it may be possible to reduce his or her risk through psychological interventions.

4. Assigning Patients to Treatments

An application beginning to be exploited is the use of psychological testing to assign patients to medical treatments. For example, treatment outcomes for pain patients assigned to anesthesiological versus behavioral treatments have been predicted on the basis of scores from the Minnesota Multiphasic Personality Inventory, a widely used psychological test. Treatment programs for chronic, severe asthma patients have been designed on the basis of prognoses for rehospitalization, medication compliance, and other aspects of reaction to medical care based on a battery of psychological tests. Conclusive validation of such assignments is difficult, so that progress in this area can be expected to be slow.

5. Assessing the Quality of Life

Advances in medical technology have made it possible to sustain life in many situations and conditions in which death would have been almost certain in previous times. With this capacity has come, inevitably, the question of whether there are circumstances in which it is undesirable to do so. Are there times when the quality of life preserved is "worse than death"? Much more difficult to face is the question of whether there are qualities of life that are not worth the money it costs to sustain them. When the question is posed in this way, most people are reluctant to agree that there are. However, because the resources applied to health care inevitably compete with other applications of those resources, both within the health system and beyond it, choices must be made. Arguably, maximizing quality of life throughout a population is the best basis for allocating money for health care. Whether it is or not is a question that must be debated by philosophers, theologians, and people-at-large. But if it is, another arguable proposition, that the essence of quality of life is the integration of feeling states over time, becomes relevant. Psychological assessment includes the appraisal of feeling states (e.g., joy, pain, sadness, pleasure). Categories are still unstable and measurement is crude, but if human preference is the measure, psychologists are the ones to develop the instruments.

B. Behavior Change Applied in Health Psychology

Although some people feel a certain ethical unease at the idea of interventions designed to change another's behavior, the socialization and education of children, the design of wage and salary systems, and the writing of persuasive books are all examples of behavior change methods applied with little or no concern for the consent of the target of change. However, just as the surgeon's ethical standards are different from those of the swordsman, so the psychologists' should be different from those of the huckster or the law enforcement officer. Concepts of informed consent in health care should be as fully applied to behavior change as they are to any other form of intervention.

There are three major avenues to changing behavior, defined by the level of the person-in-environment system that is addressed. At the most basic level, the laws of conditioning, worked out by a long series of researchers from Pavlov to B. F. Skinner and their students, are adapted for use with human beings. The psychologist works directly or indirectly, through the agency of the individual whose behavior is to be changed, to control the association between behavior and events in the immediate environment. "Biofeedback" and "behavior modification" are names applied to techniques based mainly at this level. At the second level, verbal and other symbolic input to the individual is used in an effort to alter the internal, symbolic representations of the person to be changed. Examples of this approach are counseling, education, and persuasion. The third approach is through modifying social and physical contexts within which behavior occurs in the expectation that these changes will interact with the motivational and learning systems of individuals to change their behavior. Actual methods used in attempting to change behavior are rarely pure application of techniques from a single level. Examples of several applications in health psychology follow.

1. Biofeedback

For many years a distinction was made in psychology between voluntary and involuntary behavior. In general, actions of striated muscles were said to be voluntary and those of the smooth muscles and glands were involuntary (i.e., not subject to conscious control). Beginning in the 1960s, evidence began to accumulate that many bits of behavior previously thought to be involuntary could be changed if information about the ongoing state of the organ was provided to the person in visual or auditory form. By means of this biofeedback a person could gain some control over such things as heart rate, brain rhythm, and stomach acidity. In many cases, the pathways by which such control was achieved were obscure and sometimes varied from person to person. Attempts have been made to use biofeedback to alter many different bodily conditions. At this time there is substantial evidence for its use in controlling skin temperature of the hands, which provides relief in Reynaud's disease, a disturbance of peripheral circulation, and in migraine headache. The specificity of the effect has been questioned, however, because in many situations relaxation training without biofeedback produces equally good results. The most clear-cut evidence of the specific value of biofeedback is in conditions in which control of skeletal muscles has been inadequately learned or impaired through illness or injury. Well-documented effects have been found in torticollis (twisted neck), muscle spasms, scoliosis (crooked back), sphincter control to eliminate urinary or fecal incontinence, and rehabilitation of muscle control after stroke or injury. [*See* Biofeedback.]

2. Behavior Modification

The nervous system is organized in such a way that behavior leading to pleasurable or satisfying states of the organism becomes more likely and behavior that leads to pain or distress becomes less likely. The techniques known collectively as behavior modification make use of these fundamental properties of individuals to produce change. They may proceed by increasing clients' recognition of the situations in which an action is rewarded (produces pleasure or satisfaction), punished (produces pain or distress), or has no appreciable consequence of either kind. One of the techniques used is called "modeling" (i.e., essentially demonstrating the desired behavior and its rewarding consequences). Often the clarification is largely verbal, in which case this approach is not much different from the counseling techniques described in the next section. The other major technique of behavior modification is to alter the rewarding or punishing consequences of health-related behaviors. In some cases a monetary reward can be linked to performing a desired action, such as appearing for an appointment on time, or to a result presumed to follow from a recommended action, such as losing weight or controlling blood sugar level (in the case of diabetics). Some programs have used punishments, usually in the form of forfeiting all or part of a deposit made at the beginning of a program such as weight control or smoking cessation, for failure to achieve specified goals.

Often a process of "behavioral contracting" is employed in which the health care provider and the client together work out a set of contingencies. Patients are often taught to design and apply reinforcing contingencies for themselves in "self-control" programs. In such programs, and in most others, even when behavior modification is not the designated mode of approach, the rewards and punishment of social approval and disapproval play an important part in changing behavior.

3. Counseling

Counselors are persons trained to facilitate behavior change primarily by talking with their clients. Most health psychologists who engage in counseling also subscribe to the view that behavior is controlled to a large extent by rewards and punishments (including highly personal and subjective satisfactions such as satisfaction with "knowledge of a job well done"), but they place much greater emphasis on the idea that it is not the actual relations of actions to consequences that determine behavior but the ways in which these relations are represented in clients' memories. Internal representations are limited in their accuracy by the limited experience of the individual and by the failure of the categories within which experience is recorded to correspond fully to the attributes of actions and situations that actually determine outcomes. Distortions arise, often serious ones, as a result of experiences in which strong emotions have occurred. Such traumatic experiences (or thoughts) can affect the handling of information received later in such a way that errors of representation are perpetuated or aggravated.

Counselors work by encouraging their clients to express and explore the emotions associated with the health issues that confront them and then to evaluate their own representations in light of the consensual representations of others, revealed to them through the words of their counselors and such reading or other inputs as the counselor may recommend. Counseling is more useful the more emotionally laden the health problem being faced and the more perplexing the values that must be balanced against each other in choosing a course of action. Helping patients cope with severely disabling or terminal illness is one of the areas in which counseling is most often employed. Counseling can often be helpful when drastic changes in behavior are required to accomplish some health objective, although in such cases behavior modification techniques will usually be needed as well. (Many health psychologists are adept at both.)

The expertise of health counselors resides in both their skills in the counseling process and their knowledge of the probabilities and the value issues associated with the particular health problems their clients are confronting. Counseling skills, even without specific knowledge about the health condition, can help clients make use of the information and advice they are receiving from other sources. Knowledge of the health issues without the counseling skills is not sufficient for counseling. Considerable disagreement exists regarding how much and what kind of education and training are required to produce a skilled counselor.

4. Education

Education is a term with broad and varied meanings. In one sense it refers to any activities systematically conducted with a view to increasing the knowledge, skills, or competence of those educated. Most of the behavior change activities described here could be incorporated within such a broad definition. A common narrow definition refers to a change process involving verbal transmission of information that appeals primarily to human rationality. Often the educator disclaims any intent to produce *specific* changes in beliefs or behavior, such as might be the result of "propaganda," "indoctrination," or "coercion." "Persuasion" incorporates both the rational and irrational. In its most rational expression, persuasion seeks to organize the material for maximum effectiveness in accomplishing some specific change. At the other extreme it uses, covertly, knowledge of motivating factors to achieve change that could not be gained simply by the presentation of information. A striking example is the use of psychological "reactance," the tendency of humans to resist apparent efforts to constrain their freedom of thought, to back them into desired positions by suggesting a graded series of contrary views for them to resist. The statements "You really have to be more concerned about your business than about your family" and "Your family is probably more concerned about the income you produce than the effect of your work on your health" are reactance-inducing items that have been used in a series to induce a person to seek an appointment with his or her physician. [*See* Persuasion.]

Most individual educational efforts in the health system are directed to patients and to health care personnel. However, they may be addressed to family members or other affiliates of patients and occasionally to individuals involved in the planning or administration of health-related activities. Education of persons in groups is more often directed to health care workers and professionals. For most of this century, psychologists have participated in teaching health care providers about the psychology of the patient and about provider–patient interactions.

The involvement of health psychologists in educative and persuasive interventions is more often than not as critics and evaluators of the efforts of others. Carefully controlled studies show that simply increasing the information available to individuals, even when they have incorporated it to the degree that they can reproduce it on demand, is not an effective way of changing behaviors deeply entrenched by habit, tradition, or preference. Educators, journalists, and media specialists are usually more artful at composing effective presentations of information, but psychologists can be useful in tailoring presentations for special audiences and in analyzing the details of what is working and what is not. Sometimes they can point to irrational barriers between knowledge and action and propose methods of penetrating them. "Health education," however, is often designed and conducted by health professionals who not only lack the expertise to do it effectively, but do not realize that such expertise exists. Health psychologists, working with them, can sometimes be the agents of educating the would-be educators.

5. Systems and Community Approaches

Some interventions by health psychologists are directed to groups of people who live or work together. In such cases interactive properties of "systems" of people are taken into account. The whole field of "systems analysis" has developed to describe, objectify, and quantify the idea that when one acts on one of a set of interconnected elements (i.e., on a system), there are compensating adjustments throughout the system. Systems vary enormously in size and complexity. Three levels of human systems (i.e., those in which the elements are humans) that have been addressed by health psychologists are families, health care organizations, and communities.

1. The work of health psychologists with families has developed largely in collaboration with the rebirth of the family doctor in the form of the specialist in family medicine. Another factor has been the importance recently placed on life-styles, obviously and demonstrably much influenced by family patterns. Yet another is based on the growing emphasis on social support as a major influence in the response to stress, the formation of health goals, and the adherence to intended health regimens. Families are often the primary source of social support. Simply recognizing the importance of families is only the beginning, however, of the approach to families being developed by psychologists and other social scientists. Drawing on general systems theory, group dynamics, dynamic psy-

chology, and anthropology, a "family system theory" is growing up that takes families as a unit of analysis. The impact of stress on family units, their decision-making processes, and their mobilization to deal with crises of extended and life-threatening illness are studied to develop interventions directed to the family as whole.

2. Psychologists working in health care organizations have focused most of their attention on two issues: how to increase the satisfaction of clients/patients with the care they receive, and how to match the utilization behavior of patients to their need for health care to make the best use of health care resources. Much of the effort in the area of patient satisfaction has been devoted to developing methods for assessing it. Many studies, particularly the earlier ones, have been limited in their discovery of correlates of patients' evaluation of their health care by the fact that patients tend to give high marks to their physicians and their services, even when other evidence suggests that substantial amounts of dissatisfaction exist. Furthermore, it has not proven easy to obtain differential ratings of the quality of different aspects of the care received. However, when health psychologists applied sophisticated techniques of questionnaire development to the problem, it was possible to break through these barriers and to obtain information that could guide administrators and teachers in their efforts to improve the quality of care.

In early studies of factors influencing patient satisfaction it was often found that interpersonal aspects of the provider–patient encounter dominated, with technical quality taking a distinctly secondary place. Many contemporary discussions of the topic continue to rest on this generalization. However, careful analyses using better methods for gathering patients' appraisals reveal that patients are sensitive to providers' competence in performing the technical tasks of health care. Their own effective participation in the health care process is more affected by the latter factors than by purely interpersonal processes.

The organizational issue of utilization is complex. In essence, it is the problem of matching health care services to human needs in such a way as to make the best possible use of available resources. There are three aspects of the problem: underutilization, overutilization, and getting people to the right place, including the problem of referral. In general, poorer segments of the population have greater health needs and receive fewer health care services than the more affluent. This maldistribution may be reduced, but it is not eliminated even when the services are offered free of charge. However, a small proportion of persons uses a relatively large proportion of health services. For example, in a nationwide survey conducted in the United States in 1980, it was found that about 5% of the population made 26% of all outpatient visits and accounted for more than 22% of outpatient expenditures. The problem of misdirection of the initial request for service affects both under- and overutilization. In underutilizing groups, entry into the health care system will often result in a referral to another site, and such referrals result in a completed appointment at the proper facility less than half of the time. By contrast, persons who are thought to seek more from the health care system than it can properly deliver—the overutilizers—actively seek referrals and make many self-referrals to a great variety of service providers in usually unsuccessful efforts to gain relief from symptoms.

Health psychologists have not been much involved in this set of problems as yet. Sociologists and demographers have tried to explain underutilization and psychiatrists have studied overutilizers, whereas economists have studied the impact of the pricing of services on these behaviors. Problems of making successful referrals are usually approached as instances of interaction and communication.

3. At the community level, health psychologists have focused on the use of community organizations and other resources to decrease health-damaging behaviors and increase behaviors favorable to health. Examples of such programs are those conducted by Stanford University behavioral scientists in several communities in northern California and by Finnish researchers in the county of North Karelia, Finland. With follow-up periods of 10 or more years, changes in health behaviors and in physiological and epidemiological indicators of cardiovascular risk and morbidity have been demonstrated. Large-scale projects of this kind have been conducted in Australia, South Africa, and Switzerland, as well as several areas in the United States. Cardiovascular risk and substance abuse are the topics of most studies thus far conducted. These are multidisciplinary projects. Psychologists join with physicians and other health professionals, health educators, and others in their planning, design, implementation, and evaluation phases. Psychologists' expertise in individual and community dynamics can spell the difference between success and failure in attaining project goals.

6. Public Health Interventions

In practice, public health workers today often make use of educational approaches on a large scale, as described in the previous section. They support media campaigns to get people to wear seat belts or laws mandating warning messages on packages of tobacco products and in advertising for such products. In terms of a defining philosophy, however, the public health approach primarily relies on actions that do not require the active engagement of the public whose health is to be protected. A classic example comes from the work of a London physician, John Snow, during a cholera epidemic in the mid-nineteenth century. Careful mapping of cases led him to conclude that a particular public water pump was the source of contamination, even though the mechanisms of cholera infection were unknown at the time. Snow did not try to convince people not to use water from the pump; he used his authority as a government physician to have the pump handle removed. Other examples of the "passive" approach that requires little or no participation by members of the public are draining swamps to control malaria, engineering highways and protective structures to reduce deaths and injuries from highway accidents, and promoting mandatory inoculation to provide life-long protection from infectious diseases.

The two major instruments for changing behavior via the public health approach are environmental change and regulation. The health psychologist can contribute to both of these by anticipating reactions to the measures taken. There is almost no intervention that does not require some degree of acquiescence or cooperation from individuals, and psychological knowledge applied in the selection, design, and implementation of environmental or regulative changes can increase their effectiveness.

VI. PSYCHOLOGY OF THE ADDICTIONS: AN EXAMPLE

The integration of the various concerns and activities of health psychologists as they relate to a major class of health problems can be exemplified by a consideration of the situations in which some behavior has become intractably excessive. Until recently, problems of alcohol abuse, drug addiction, smoking, obesity, pathological gambling, and pathological hypersexuality have been considered separately, using different theoretical approaches and independently developed bodies of intervention technique. Now there is a recognition that the underlying psychological mechanisms have much in common so that a coalescence of research and practice is under way.

A consensus is emerging that these conditions are complex biopsychosocial phenomena in which predisposing constitutional factors, environmental influences, psychological mechanisms of learning and defense, and moral and ethical beliefs all play important parts. In some patterns of excessive behavior, the physiological processes of tolerance (i.e., requiring increasing amounts of a substance to achieve the same bodily effects) and of withdrawal symptoms that are uncomfortable, painful, and even life-threatening assume much larger roles than in others. These are the processes that have been strongly associated with the term *addiction.* Even in those conditions in which these physiological processes figure most strongly, in the case of the opiates, barbiturates, and alcohol, psychobiological processes of learning and the generalization of overlearned habits to ever-greater numbers of eliciting or discriminative stimuli give rise to the intractability of the conditions and form the basis for relapse. In the remainder of this section, the complexly determined pattern of intractably excessive performance of some behavior is referred to as an addiction.

The addictions have several essential features in common. First, some behavior that elicits a pleasurable response soon after the behavior has occurred becomes so frequent that it causes serious harmful consequences to the body of the addicted individual or in the social network of that individual. The forces that maintain the behavior are so strong that even though the affected person fully understands the harm that is being done, it is extremely difficult to change the behavior pattern. Commonly, sporadic efforts to change are increasingly overwhelmed by patterns of defensive denial, rationalization, and other behaviors that protect the addicted person from full recognition of the need for change. When addicted persons enroll in programs to change behavior, usually about a quarter of them succeed in eliminating or substantially reducing the unwanted behavior, but unless there is some continuing factor at work, most of these individuals will revert to the addicted pattern within a few months to a year.

A. Societal Aspects of Addiction

Specification of what constitutes an excess of an appetitive behavior is a social definition, although biological and psychological factors are deeply involved in its establishment. What is considered pathological excess in one culture or in a particular time period may be well within the range of the acceptable in others. When certain behaviors (e.g., the use of particular substances such as opium or cannabis) are totally proscribed by a society, there is no legally acceptable level of the behavior, although there may still be a socially tolerated level. The distribution of the amount of the behavior performed by the members of a population is skewed because most persons are moderate, with ever fewer engaging as the level of the behavior increases. Unless there is rigid enforcement of a legal or religious proscription of the behavior, the mode or peak is usually somewhat above total abstinence.

The costs to society of the addictions are enormous. Two kinds of costs are recognized: internal costs that are borne by the addict (including, in some analyses, costs to family members) and external costs that are borne by employers, taxpayers, victims of accidents, purchasers of insurance policies, and others. Estimates made in the late 1980s of combined internal and external costs identify direct costs for facilities and treatment of alcohol and other drug abuse, exclusive of tobacco, at about $2 billion per year. In the same period, it was estimated that total costs, arising mainly from lost productivity and unemployment, amounted to nearly $200 billion per year. These costs are offset to a greater or lesser degree by "savings" to society because substance abusers have shorter lives and therefore collect less in the way of pensions, make less use of nursing homes, and so on. Some investigators have attempted to place a dollar value on expected years of life lost to the substance abuser. A common method of doing so is in terms of lost income as the substance abuser is withdrawn from the work force. A more psychological approach has drawn the estimate from data regarding the amount that people are willing to pay for a small change in life expectancy.

Analyses of this kind are mainly carried out by economists. Psychologists can contribute, however, through their knowledge of methods for assessing human preferences, thus refining subjective costs of pain, suffering, discomfort, and lost life. Applications of these analyses are found in legislation and regulations that set excise taxes, proscribe sale of certain commodities, or establish schedules of premiums or copayments in health insurance plans. Involvement of psychologists in planning such social interventions can lead to greater accuracy in anticipating public response to them.

In terms of health effects, smoking has been named as the single greatest preventable source of illness and mortality in the developed world. A major epidemiological study in a midwestern state concluded that alcohol abuse contributed to 12% of all years of life lost, mostly because of consequences of accidental injuries. Obesity that exceeds 130% of the average or recommended weight is associated with premature mortality that is 1.3 or more times that of persons with normal weight. The harm that results from excesses of gambling and sexual behavior are difficult to document in terms of physical illness and mortality figures. Often their deleterious effects have greater impact on families and other associates than on the addicted person. For these reasons, some critics claim that these last two conditions should not be considered in the category of illnesses, but rather as socially, perhaps arbitrarily, defined deviance. However, there are excessive rates of suicide and mental illness in these conditions. [*See* Nonnarcotic Drug Use and Abuse; Obesity; Tobacco Smoking, Impact on Health.]

Health psychologists contribute to the scope and accuracy of the epidemiological studies that document these effects by studies of methods for phrasing questions and structuring questionnaires that appraise both the problematic behavior and the health status of respondents. They are also involved in making use of the results of such studies in developing educational and treatment programs for preventing and treating addictive behavior.

B. Individual Aspects of Addiction

The addictions are peculiarly psychological disorders, in that the problem arises specifically from a particular pattern of behavior. In almost every kind of health problem, behavioral factors are involved in the etiology, treatment, and recovery from the disease, but in the addictions behavior is the central aspect. This is not to say that physiological and social factors are unimportant. In every facet of addiction, biopsychosocial interaction is pervasive. These interactions are traced through the stages of addiction in the sections that follow.

1. Predisposing Factors

The occurrence of excessive appetitive behaviors is not uniform over populations but is associated with ethnicity, education, income, family membership, and other sociodemographic variables. Although the epidemiological evidence on which this statements rests is of variable quality for the several forms of the addictions, it has provided the basis for efforts to attribute predispositions based on genetic, familial (including psychodynamic), and socioenvironmental factors. By far the most extensively studied condition is that of excessive use of alcohol. It has been possible to accumulate conclusive evidence that risk factors for alcoholism can be associated with each of these domains of variables. Details of mechanisms are, for the most part, not well established. Genetic factors have also been sought with some claimed success in the case of excessive eating, and recent studies point to a possibly genetic abnormality of norepinephrine metabolism in pathological gamblers. Repeated proposals that endocrine abnormalities may be involved in hypersexuality have not been substantiated as yet. Postnatal predisposing factors have their impact through learned aspects of the behavior and are treated in subsequent sections.

2. Taking Up the Addictive Behavior

Addictive behavior is learned behavior. Therefore, opportunities to observe others engaging in such behavior usually precede the actual performance of the behavior. The observation may occur in real life, in the behavior of parents or peers, or it may be encountered first in television, films, or written materials. Studies of the initiation of these behaviors, which in their excess become problematic, have mostly been conducted with adolescents. In this group, the most strongly associated factor is almost always the degree to which same-aged associates, especially close friends, engage in the behavior. Parental behavior, attitudes, and permissiveness all make important contributions. A variable sometimes called "tolerance for deviance," which is closely related to the belief as to whether the behavior is wrong, is also important. Personality factors, such as rebelliousness, stress on independence, low sense of psychological well-being and self-esteem, and lower academic motivation and achievement, are all associated with relatively early initiation of substance use, sexual behavior, and gambling. Investigation of these topics is one of the most active areas of research in health psychology.

3. Active Phase of the Addiction

Except for legally proscribed behaviors, the addictions consist of socially defined excesses of behavior exhibited by most members of a population to some degree. Even when a behavior is totally forbidden by law, sizable proportions of a population may engage in it on one or a few occasions. The processes by which persons change from occasional, or moderate, controlled degrees of the behavior to a level deemed excessive are multiple and complex. At the simplest level, basic principles of learning dictate that the more a behavior occurs, the more situations become occasions for the performance of the behavior. (In technical terms, by response generalization, more and more situations come to be discriminative stimuli in whose presence the behavior is highly probable.)

For most persons, both the rewarding effect and the generalization process are opposed and checked by social and physiological consequences of occasional excess or by the anticipation of such consequences, learned vicariously. A significant factor in establishing the equilibrium point at this first stage, as in the initiation process, is the behavior and attitudes of other persons in the settings that the individual frequents. Poorly understood differences among individuals in the positive, rewarding effects of the behavior, the propensity to inhibit behavior in anticipation of punishment or disapproval, and the impact of belief systems on behavior influence the progression into a second stage in which guilt, shame, and deceit become prominent. These new factors introduce individually characteristic degrees of anxiety arising from self-awareness of the problematic nature of the behavior. The rewarding effects of the addictive behavior may now be enhanced by their capacity to reduce anxiety and to distract attention from the awareness of transgression of social or internalized ethical norms. Thus a complicated positive feedback situation is created in which the knowledge of excess becomes a factor in continuing and increasing it.

When the addiction has reached this stage of self-perpetuation, it begins to transform the life patterns of the individual. Family and friends are likely to become alienated, and the addict becomes isolated or affiliated with other addicts in a way of life that is centered on and dominated by the excessive behavior pattern. The life course of the alcoholic or the person addicted to other drugs often ends prematurely as a result of drug-related deterioration of the body or from illness or injury brought on by intoxication and loss of domicile. Compulsive gamblers and those who are pathologically hypersexual may become drug addicts or alcoholics as the social structure of their lives is destroyed. Some of the eating disorders, notably anorexia nervosa and bulemia, are serious health problems that can result in death.

4. Resolution without Treatment

Many persons who are at one time in their lives actively addicted to some form of excessive behavior eventually return to a more balanced way of living. Authoritative estimates of the percentages who do so are hard to come by, and they vary substantially among the several forms of addiction. The quality of the data varies as a function of the covertness of the behavior and the social investment in its study. Probably the best data are available for the tobacco addiction. These data were almost all collected during the period after the intensification of public health efforts to get people to stop smoking in the 1960s. This fact reminds us that quantitative generalizations from one form of addiction to another, from one cultural context to another, and from one social era to another are not possible.

In the case of smoking, somewhat more than half the persons who report they were at one time regular smokers will also report that they have now been abstinent for 1 year or more. Other data indicate that relapse is relatively rare after such a period of abstinence. Elimination of the addiction is not often achieved on the first effort, however. Perhaps 4–6% of those who attempt to quit smoking succeed on any given attempt. Of those who do succeed in quitting, it appears that more than 9 of 10 quit on their own without having been involved in any formal treatment programs. Recent, methodologically sound research on persons with serious alcohol addictions indicates that about half of them are abstinent or are free of drinking problems 4 years after initial contact even if they have received no treatment in the interim. (A third or more are free of alcohol problems if they have received treatment.) Even in the case of addiction to heroin, which is certainly one of the most intractable of all the addictions, it has been reported that during a 10-year period as many as 40% of those previously addicted will become abstinent. Accurate figures for the number of heroin addicts who become abstinent without participating in an effective treatment program at some time are not available.

Although the mechanisms for freeing oneself from an addiction vary considerably in detail, there are some patterns in common. The shifting equilibrium that prevails between the positive and negative consequences of the addictive behavior during active addiction is disrupted through some event or series of events that bring the addict to a "turning point." Death of a close associate, loss of job, disruption of family, or sometimes just the realization that one is living in total subjugation to a dependency leads to a crystallized intention to change. In the traditions of alcoholism, it is said that one must reach "rock bottom" before the turn to recovery can be made. However, data indicate that the turning point can occur at any degree of addiction and that recovery is more likely and quicker the less intense the addiction.

Many devices are tried in the course of the difficult process of change. Some persons are successful with a process of "tapering off" (i.e., gradually reducing the amount of the addictive behavior until its strength is lessened). They may restrict the times and settings when they engage in it and thereby gradually reverse the process of generalization referred to earlier. Others, influenced more by the perceived hazards of relapse, prefer abrupt and total change to abstinence—"cold turkey." Addicts committed to change often adopt patterns of avoiding the situations in which the urge to engage in the addictive behavior is particularly strong. Smokers may give up a morning cup of coffee or an afternoon cocktail, because these are strongly associated with the pleasure of cigarettes. Particularly important is finding new interests, activities, and desires to fill the void created by the removal of all the behaviors associated with the addiction. Episodes of relapse are almost universal, and the manner in which they are handled by the addict is critical for the ultimate success of the recovery process.

Although present understanding of the origins of, development of, and independent recovery from the addictions is substantial, continuing psychological research is adding further knowledge of the mechanisms that underlie the phenomena and is providing details of the variations to be found among the several kinds of addictions and as a result of variations in the social settings and individual characteristics of the addicts. This knowledge is used to design treatment programs for interrupting the patterns and restoring addicts to more normal functioning.

5. Treatment of the Addictions

There is an enormous literature on the treatment of addiction. In the case of obesity and hypersexuality, surgical and hormonal approaches can be effective in some cases. However, in the vast majority of cases, some method for producing behavior change must be used. Incarceration in a prison or hospital is usually effective for its duration, but relapse is virtually certain unless other change techniques are employed during the period of confinement. Psychological programs for change are offered by proponents of all the major theoretical approaches to the understanding of human behavior. Behaviorists emphasize the elimination of the habitual aspects of addictive behavior and the substitution of new habits. Cognitivists concentrate on clarification of the consequences of the addiction, the clarification of values, the formation of new intentions, and reappraisal of the addict's potential for change ("self-efficacy" beliefs). Psychodynamicists focus on the underlying conflicts that have contributed to the susceptibility to addiction preparatory to subsequent re-educative processes. Socially oriented therapists may try to help the addict establish supportive social networks made up of nonaddicted persons, and they may involve family members in the treatment process. In fact, most therapies involve elements from all of these approaches.

Each of these levels of address can contribute to the change process in individuals who have reached the turning point. All except the most stringently habit-focused can aid the ambivalent addict to come to the turning point. Claims of success in treatment vary widely. Short-term change may be achieved in 50–80% of those who complete a treatment, but the long-term success rate is much lower because of relapse. Careful studies of long-term (1 year or more) outcomes usually find 20–30% abstinence in the case of substance abuse. Many more addicts demonstrate significant reduction in the amount of the addictive behavior in which they engage, often with significant lessening of adverse consequences. These accomplishments are better than those achieved by addicts on any given independent effort to change. Extensive studies of heroin and alcohol treatment programs indicate that they can be cost-effective, saving more in terms of identifiable costs to society than is spent on them. Success rates are higher the more broadly the problem is addressed, and no one of the approaches mentioned appears to be substantially better than the others. The most basic conclusion regarding recovery from addiction is that it must be regarded as a complex process extended in time rather than a single event.

6. Postrecovery Condition

Relapse after short-term success in eliminating an addiction is likely. Research emphasis has therefore shifted to discovering methods for preventing relapse. Investigation of the circumstances under which it occurs provides a basis for warnings to ex-addicts. A common finding is that high stress is frequently involved, so training in stress management is often offered. Formal programs of aftercare can substantially reduce the relapse rate. Even an agreed on schedule of follow-up contacts can be a significant factor in maintaining recovery.

A major controversy in the psychology of the addictions has been the question of whether it is possible for a former addict to continue in the controlled performance of the formerly excessive behavior. Of course, in the case of excessive eating, it is essential to do so. This fact has been pointed to as an explanation of the belief that excessive eating is among the most difficult of the addictions to eliminate. Data on long-term outcomes of weight reduction programs are scarce, however. When the problem behavior is illicit or illegal, controlled continuation is an unstable behavior pattern, because engaging in the behavior tends to be incompatible with normal life patterns. Alcohol and tobacco are two substances that are not essential to life, but whose use is legal and socially acceptable. Some persons are able to develop long-term patterns of well-controlled use of these substances after periods of strong addiction. The preponderance of evidence indicates that most of those who attempt to sustain such a pattern will revert to the pattern of addiction at some stressful time in their lives. This question has not, however, been adequately studied.

VII. CONCLUSION

In every aspect of the health system, problems arise that involve human reactions to complex situations that affect health outcomes. Human emotions, beliefs, decisions, intentions, communications, and interactions influence the processes of abating and avoiding hazards in the environment, interpreting symptoms, seeking and participating in health care, and recovering maximum ability to function effectively after injury or illness. These psychological factors also determine the actions of those who provide care, develop and manage health organizations and public health programs, and formulate health policies, regulations, and legislation. They influence the behavior of those who conduct, interpret, and support research on health problems.

Contemporary psychology, in conjunction with adjacent fields of information science, neuroscience, and social science, has much to offer in the understanding of the role of such factors and in the application of this understanding to the design of effective interventions. To realize the potential contributions, psychologists must immerse themselves in the problems and settings of the health system. Those who do so are properly called health psychologists. Their number is rapidly increasing throughout the developed and developing world.

BIBLIOGRAPHY

Adler, N., and Matthews, K. (1994). Health psychology: Why do some people get sick and some stay well? *Annu. Rev. Psychol.* **45**, 229–259.

Adler, N. E., Boyce, T., Chesney, M. A., Cohen, S., *et al.* (1994).

Socioeconomic status and health: The challenge of the gradient. *Am. Psychologist* **49**(1), 15–24.

Baum, A., and Fleming, I. (1993). Implications of psychological research on stress and technological accidents. *Am. Psychologist* **48**(6), 665–672.

Belar, C. D. (1988). Education in behavioral medicine: Perspectives from psychology. *Ann. Behav. Med.* **10**(1), 11–14.

Cohen, S., and Williamson, G. M. (1991). Stress and infectious disease in humans. *Psychol. Bull.* **109**(1), 5–24.

Dimsdale, J. E., and Baum, A. (eds.) (1995). "Quality of Life in Behavioral Medicine Research." Lawrence Erlbaum Associates, Hillsdale, NJ.

Friedman, H. S. (ed.) (1990). "Personality and Disease." John Wiley & Sons, New York.

Friedman, H. S., Hawley, P. H., and Tucker, J. S. (1994). Personality, health, and longevity. *Curr. Dir. Psychol. Sci.* 3(2), 37–41.

Friedman, H. S., Tucker, J. S., Schwartz, J. E., Tomlinson-Keasey, C., *et al.* (1995). Psychosocial and behavioral predictors of longevity: The aging and death of the "Termites." *Am. Psychologist* **50**(2), 69–78.

Sarason, B. R., Sarason, I. G., and Pierce, G. R. (eds.) (1990). "Social Support: An Interactional View." John Wiley & Sons, New York.

Shumaker, S. A., Schron, E. B., Ockene, J. K., Parker, C. T., Probstfield, J. L., and Wolle, J. M. (eds.) (1990). "The Handbook of Health Behavior Change." Springer, New York.

Stone, G. C., Weiss, S. M., Matarazzo, J. D., Miller, N. E., *et al.* (eds.) (1987). "Health Psychology: A Discipline and a Profession." Univ. of Chicago Press, Chicago.

Stroebe, M. S., Stroebe, W., and Hansson, R. O. (eds.) (1993). "Handbook of Bereavement: Theory, Research, and Intervention." Cambridge Univ. Press, New York.

Traue, H. C., and Pennebaker, J. W. (eds.) (1993). "Emotion Inhibition and Health." Hogrefe & Huber Publishers, Gottingen, Germany.

Heart, Anatomy

EDWARD CHEN
Massachusetts General Hospital and Harvard Medical School

GLOSSARY

Aorta Exits from the left ventricle through the aortic valve and transports oxygenated blood to the rest of the body

Atrioventricular valve Divides the atrium and the ventricle

Atrium Chamber of the heart that receives blood from the vena cavae or pulmonary veins

Mediastinum Middle portion of the thoracic cavity that contains the heart, great vessels, trachea, and esophagus

Pericardium Sac that encloses the heart and the roots of the great vessels

Pulmonary artery Exits from the right ventricle through the pulmonary valve and transports deoxygenated blood to the lungs

Semilunar valve Divides the ventricle and the outflow tract of the great vessel (pulmonary artery or aorta)

Sino-atrial node Site in the atrial muscle where contraction of the heart is initiated

Ventricle Chamber of the heart that pumps blood to the aorta or pulmonary artery

THE HEART IS A MUSCULAR ORGAN THAT WEIGHS approximately 300 g; this varies with sex, height, and skeletal structure. It acts as a double pump in which the right side receives deoxygenated blood from the body and pumps it to the lungs; the left side receives oxygenated blood from the lungs and pumps it to the rest of the body. In order to sustain life without irreversible damage, the heart must pump continuously for the duration of the individual's life. With each beat, the heart pumps approximately 70 ml of blood which, over a lifetime, adds up to several million gallons of blood.

I. PERICARDIUM

The pericardium is a sac that encloses the heart and the roots of the great vessels and is filled with about 30 ml of clear fluid. The fibrous layer of pericardium is attached to the diaphragm and sternum. The fibrous pericardium limits the movement of the heart and inserts into the outer coats of the great vessels of the heart (aorta, pulmonary artery, superior and inferior vena cavae, and the pulmonary veins). The serous portion of the pericardium is further divided into parietal and visceral layers. The visceral layer is also called the epicardium which directly covers the heart. The clear pericardial fluid is located in the pericardial cavity between the parietal and the visceral layers of the serous pericardium (Fig. 1). This fluid functions as a lubricant which allows movement of the heart within the mediastinum.

II. CARDIAC CHAMBERS

A. Right Atrium

The right atrium receives deoxygenated blood from the body via the superior and inferior vena cavae. It also receives drainage from the coronary sinus which carries the deoxygenated blood of the heart (Fig. 2). Located on the medial wall of the right atrium is the

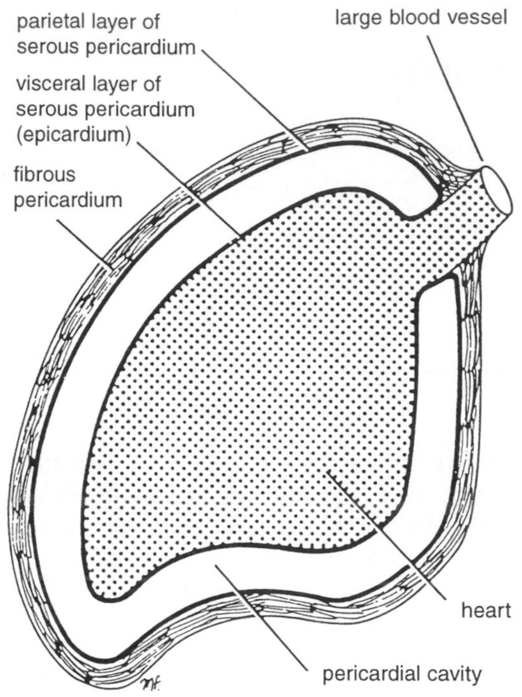

FIGURE 1 Layers of the pericardium (from Snell, 1995).

fossa ovalis which marks the foramen ovale. This previously patent opening allowed blood from the right atrium to travel into the left atrium, thus bypassing the lungs. At birth this opening closes. Valves are located at the orifice of the inferior vena cava and the coronary sinus. The anterior wall of the right atrium consists of an appendage, or auricle, as well as a number of parallel muscular ridges called pectinate muscles (Fig. 3). The smooth muscular portion is divided from the pectinate muscles by the crista terminalis. The right atrium communicates with the right ventricle through the atrioventricular orifice.

B. Right Ventricle

The right ventricle receives deoxygenated blood from the right atrium through the tricuspid valve and pumps it to the lungs via the pulmonary arteries. Although this chamber is thicker than the right atrium, it is actually thinner than the left ventricle. In cross section, it is crescent-shaped, and a significant portion of the ventricle is lined with a series of irregular muscular ridges called trabeculae carneae. A moderator band is a muscular band that extends from the ventricular septum to the anterior papillary muscle. The septum divides the left ventricle from the right ventricle,

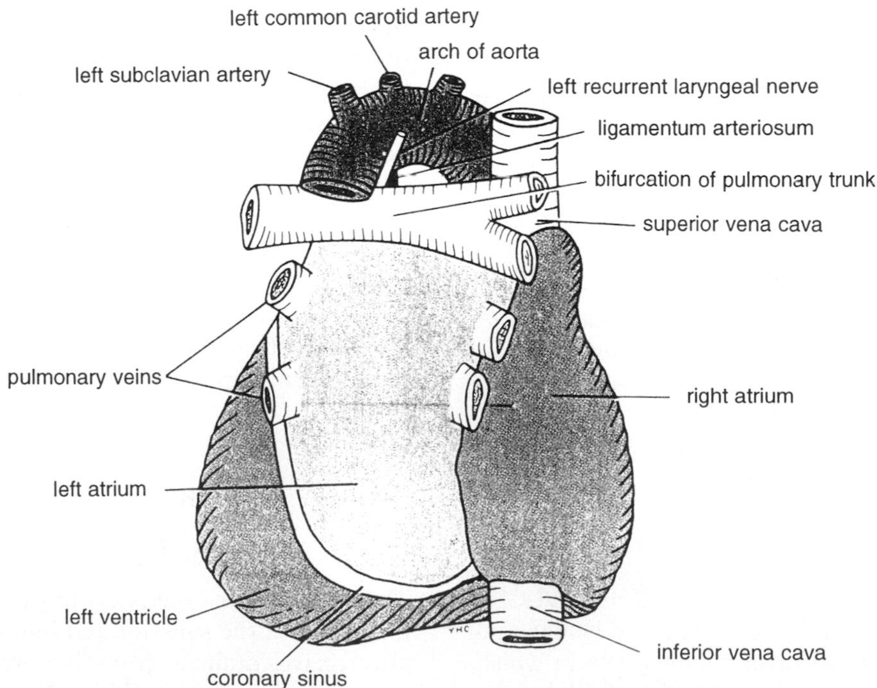

FIGURE 2 Posterior surface of the heart (from Snell, 1995).

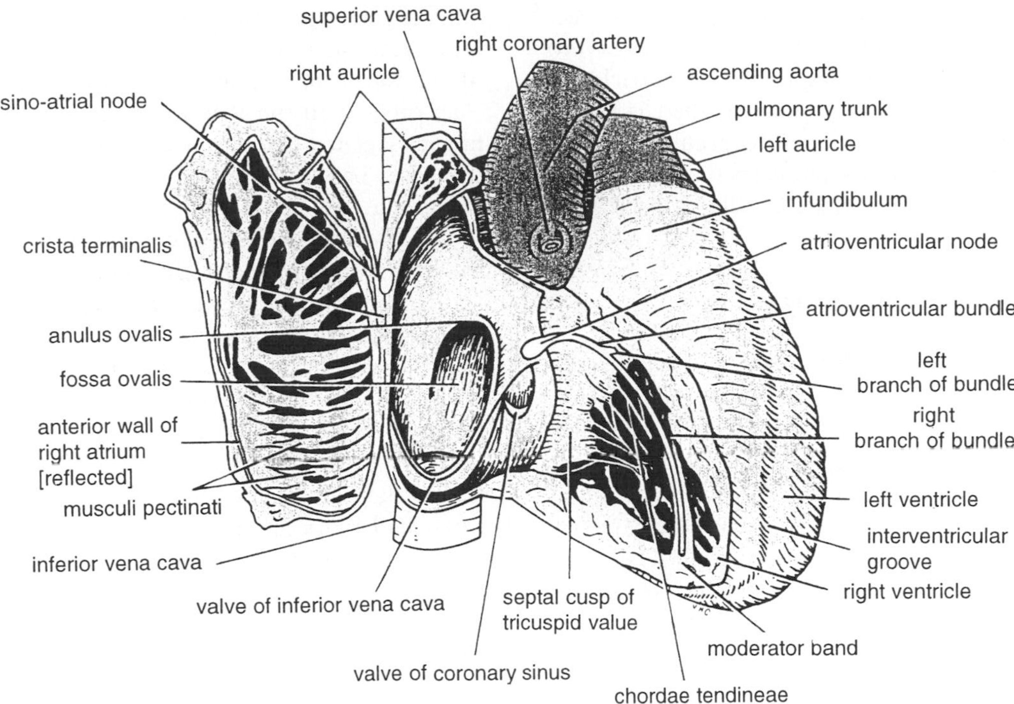

FIGURE 3 Interior of the right atrium and right ventricle (from Snell, 1995).

and the papillary muscles are responsible for proper function of the atrioventricular valves. The crista supraventricularis is a muscle band that divides the inflow and outflow tracts of the right ventricle (Fig. 3).

C. Left Atrium

The left atrium receives oxygenated blood from the lungs by way of four pulmonary veins (Fig. 2). With respect to the right atrium, the left atrium is actually posterior and to the left. Like the right atrium, the left atrium also has an auricle, or appendage, and the pectinate muscles are the only trabeculated areas of the atrium. Unlike that of the right atrium, there is no crista terminalis in the left atrium. The rest of the left atrium inside is smooth. Along the septum is the valve of the fossa ovalis, the previous site of the foramen ovale. The left atrium communicates with the left ventricle through the atrioventricular orifice.

D. Left Ventricle

The left ventricle receives oxygenated blood from the left atrium through the mitral valve and pumps it to the rest of the body via the aorta (Fig. 4). The muscular wall of the left ventricle is much thicker than the wall of the right ventricle, and the intraventricular pressure of the left ventricle is many more magnitudes greater

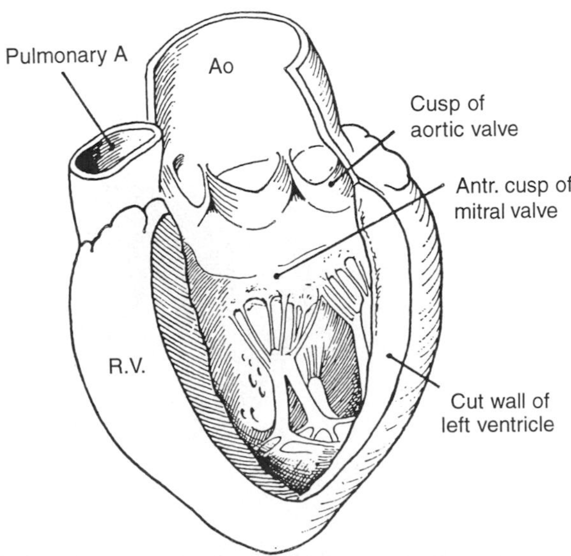

FIGURE 4 The interior of the left ventricle (from Ellis, 1997).

than the pressure in the right ventricle. In cross section, it is concentric, and it subsequently bulges into the right ventricle, thus giving the right ventricle the crescent shape. Like the right ventricle, there is a large area of irregular muscular ridges (trabeculae carneae) and there are two sets of papillary muscles which attach to the mitral valve. [*See* Cardiac Muscle.]

III. CARDIAC VALVES

The valves of the heart fall into two types: the atrioventricular (tricuspid and mitral) valves and the semilunar (pulmonary and aortic) valves (Fig. 5). The valve is composed of leaflets, or cusps, that are translucent and delicate. Tendinous cords called chordae tendineae connect the edges of the cusps with the papillary muscle attached to the ventricular wall. The valve allows only one-way flow of blood from the atrium to the ventricle. When the ventricle contracts, the pressure of the blood forces the cusps to come together and close the valve. The chordae maintains the position of the leaflets and prevents them from being pushed into the atrium during the contraction of the ventricle.

The four valves (tricuspid, pulmonary, mitral, and aortic) are located near each other. In cross section, the aortic valve is in the center between the mitral and the tricuspid valves (Fig. 5). The pulmonary valve is anterior and to the left of the aortic valve. The rings of the mitral and tricuspid valves are connected to each other with the septum of the heart to form the fibrous skeleton of the heart. The fibrous rings support the bases of the valve leaflets and prevent excessive stretching.

The tricuspid valve is the atrioventricular valve of the right atrium and right ventricle. It has three leaflets: anterior, posterior, and septal (Fig. 5). They are attached to the chordae which is then connected with the papillary muscles of the right ventricle. Of note is the close location of this valve with the conduction system of the heart. The pulmonary valve connects the right ventricular outflow tract with the orifice of the pulmonary artery. It is also composed of three leaflets: right, left, and anterior (Fig. 5).

The mitral valve is the atrioventricular valve of the left atrium and left ventricle. Unlike the others, this valve is composed of two leaflets: anterior and posterior (Fig. 5). Its name is related to the two-piece headdress (miter) worn by bishops. The anterior cusp is larger and is closely related to the aortic orifice. The chordae tendineae attach these cusps to the two large papillary muscles (anterolateral and posteromedial) that extend from the inner wall of the left ventricle. Because of the larger surface area of the two leaflets

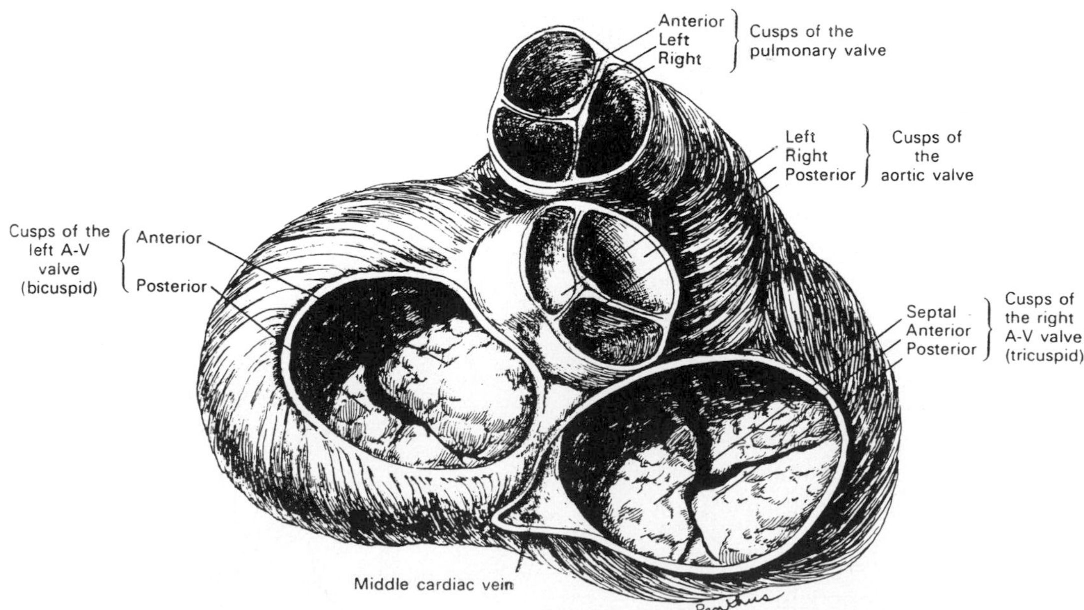

FIGURE 5 The atrioventricular (bicuspid/mitral and tricuspid), pulmonary, and aortic valves of the heart viewed from above after removal of the atria, pulmonary trunk, and aorta (from Gray, 1985).

relative to the area of the mitral orifice, there is some degree of coaptation when the mitral valve is closed.

The aortic valve is located between the left ventricular outflow tract and the aorta at the aortic orifice. It has three leaflets: right, left, and posterior (Fig. 5). The sinuses of Valsalva are the outpouchings in the aortic vessel behind each aortic cusp. The sinus of the right cusp is the origin of the right coronary artery, and the sinus of the left cusp is the origin of the left coronary artery. The posterior cusp is often referred to as the noncoronary cusp. This valve is in fibrous continuity with the anterior leaflet of the mitral valve and the septum.

IV. CONDUCTION SYSTEM

In the absence of disease states and structural abnormalities, the heart contracts in a rhythmic fashion at about 70 beats per minute. This is a spontaneous process that results in the contraction of the two atria first, followed by the contraction of the two ventricles. There is a delay between contraction of the atria and ventricle so that the blood from the atria can empty into the ventricles before the ventricle contracts.

The path of the conduction system consists of the sino-atrial (SA) node, atrioventricular (AV) node, atrioventricular bundle, and the Purkinje fibers (Fig. 3). The SA node is at the upper end of the crista terminalis of the right atrium. The SA node is often referred to as the "pacemaker" of the heart because the electrical impulse is initiated here. It then travels to the AV node, which is located in the septum dividing the right and left atria near the orifice of the coronary sinus. This impulse then continues into the atrioventricular bundle (bundle of His) where it divides into two bundle branches, one for each ventricle. It then continues as a series of fibers of the Purkinje plexus. The result is a coordinated contraction of the atria and ventricles.

V. CORONARY ARTERIES

The coronary arterial system is divided into left and right which arise from the portion of the aorta just above the aortic valve (Fig. 6). The right coronary artery (RCA) originates from the sinus of the right aortic leaflet and runs between the main pulmonary artery and the right atrial appendage. It then travels

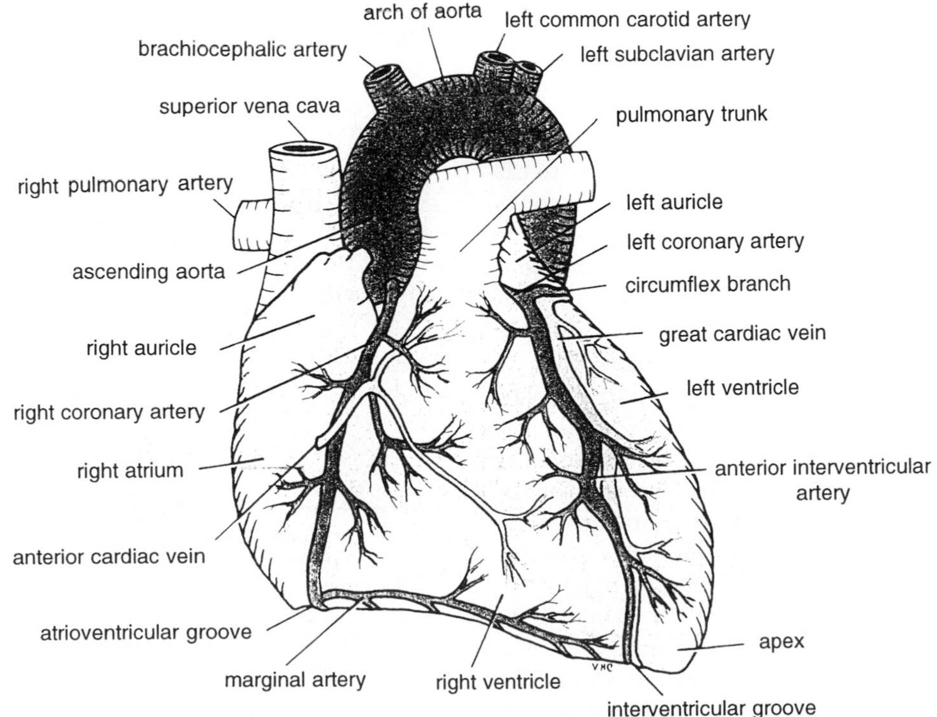

FIGURE 6 Anterior surface of heart and great blood vessels (from Snell, 1995).

along a groove located between the right atrium and the right ventricle along the exterior portion of the heart. The RCA gives rise to several branches that supply blood to the right atrium and ventricle. A marginal branch supplies the right ventricle. The RCA then loops around the heart to anastomose with the left coronary artery. The left coronary artery originates from the sinus of the left aortic leaflet and runs between the main pulmonary artery and the left atrial appendage. It quickly divides into an interventricular branch and circumflex branch. The interventricular branch is often referred to as the left anterior descending artery (LAD). Its branches supply the right and left ventricles. The circumflex branch travels along a groove located between the left atrium and the left ventricle and supplies blood to the left ventricle.

Variations in the arterial supply of the heart are common. There can be differences in origin, length, and number of branches for each artery. The term "dominance" refers to the arterial supply of the posterior interventricular, or posterior descending, artery (PDA). The PDA runs along the groove that divides the left and right ventricle underneath the heart. The heart is termed "right dominant" if the right coronary artery supplies the PDA, and the heart is labeled "left dominant" if the left coronary artery branches into the PDA.

VI. GREAT VESSELS

The great vessels comprise the conduits that either return deoxygenated blood from the systemic circulation or pump oxygenated blood from the heart to the rest of the body. The superior vena cava brings venous blood from the head, neck, and upper limbs to the right atrium, and the inferior vena cava returns venous blood from the portions of the body below the diaphragm (Fig. 6). This venous blood travels from the right atrium to the right ventricle and then enters the main pulmonary artery, or pulmonary trunk. The main pulmonary artery then divides into the right and left pulmonary arteries which lie in close proximity with the aorta as well as the main branches of the trachea, namely the left and right mainstem bronchi. After gas exchange takes place in the lungs, oxygenated blood returns to the heart via two pulmonary veins from each lung and enters the left atrium. After it passes through the left atrium, mitral valve, and left ventricle, this blood then enters the aorta. The aorta is the large artery of the thorax that is divided into the ascending aorta, aortic arch, and descending aorta. Along the way the aorta divides off into smaller arterial branches that supply the major organs and peripheral tissues. [*See* Cardiovascular System, Anatomy.]

VII. CLINICAL IMPLICATIONS

Various factors can affect the structure and function of the heart. The pericardial sac can become inflamed or the amount of pericardial fluid surrounding the heart can increase. Both can result in impaired filling of the heart to the point where insufficient blood is ejected to the rest of the body. The cardiac chambers can either be defective secondary to errors in fetal development (congenital) or be altered by the aging process in which the chamber size may enlarge. In both situations, the path of blood flow results in inefficient pumping of blood from the heart. Each cardiac valve configuration can change following an inflammatory process, resulting in either a narrowing (stenosis) or incompetence (regurgitation) of the valve. Again, the heart will pump blood inefficiently. The conduction system can be affected by a myriad of factors, including drugs and inflammatory processes such that conduction can either increase or slow down dramatically, thus affecting how much blood is pumped over a given period of time. With genetic, environmental, and aging factors each playing a role, the patency of the major coronary arteries can narrow to the point where blood supply to major parts of the heart is jeopardized, resulting in a compromise in oxygen supply, or ischemia. This imbalance of oxygen supply and demand can progress to an infarction where there is actual damage to the heart muscle. Finally, the walls of the aorta can weaken to the point where there is an undesirable stretching of the vessel under a high pressure system that can result in rupture of the vessel. [*See* Cardiovascular System, Physiology and Biochemistry.]

BIBLIOGRAPHY

Cotran, R. S., Robbins, S. L., and Kumar, V. (1994). "Pathologic Basis of Disease," 5th Ed. Saunders, Philadelphia.

Ellis, H. (1997). "Anatomy for Anaesthetists," 7th Ed. Blackwell Scientific Publications, London.

Gray, H. (1985). "Anatomy of the Human Body." Lea & Febiger, Philadelphia.

Kirklin, J. W., and Barratt-Boyes, B. G. (1993). "Cardiac Surgery," 2nd Ed. Churchill-Livingstone, New York.

Snell, R. S. (1995). "Clinical Anatomy for Medical Students," 5th Ed. Little, Brown, and Co., Boston.

Heart–Lung Machine

HANS H. J. ZWART
Wright State University

GLOSSARY

Bubble trap Specialized blood filter that can also trap a limited amount of air bubbles from the blood

Bypass flow Amount of blood pumped per minute

Cannula Short piece of tubing inserted into major blood vessel

Cardioplegia Production of cardiac arrest by infusing the coronary arteries with potassium solution, often combined with cooling

Cardiopulmonary bypass Procedure to substitute the function of heart and lungs

Cell saver Device to collect and purify blood spilled in the operative field

Crystalloid solution Solution of simple electrolytes such as sodium chloride

Doppler sound probe Device that emits sound waves and senses echoes of those waves. Air bubbles distort normal echoes from blood and can thus be detected

Emboli Foreign particles in the blood, traveling through the circulation and lodging somewhere in the tissues

Heat exchanger Device to control the temperature of the blood

Left heart bypass Bypass from left heart to aorta

Oxygenator Device to add oxygen and to remove carbon dioxide from the blood

Right heart bypass Bypass from right heart to pulmonary artery

THE HEART–LUNG MACHINE IS AN APPARATUS that is used to temporarily replace the function of both the heart and the lungs. It was designed primarily for use in heart surgery since most heart operations can only be carried out on the arrested heart. The heart–lung machine is made up of an artificial heart and a lung-part, connected with each other and with the patient by a system of cannulas and tubes. The system also incorporates filters and equipment to control the temperature of the blood. The blood is drained from the right side of the heart to the lung-part of the machine, where oxygen is added and carbon dioxide is removed. The pump part thus transports the blood back to the aorta or one of its branches.

I. COMPONENTS OF THE HEART–LUNG MACHINE

A. Overview

The patient's blood flows to the heart–lung machine via a $\frac{1}{2}$-inch-diameter plastic tube called the venous line. The line is connected on the patient's side to one or two cannulas inserted into the right atrium of the heart and, on the machine side, to a venous reservoir that is usually open to air. The air blood level is 50–100 cm below the patient's heart and the sole propellant for blood flow through the venous line is gravity. One-quarter-inch plastic suction tubes are also connected to the reservoir to return to the heart–lung machine blood suctioned out of the surgical field. One of these suction tubes may be temporarily used to decompress the heart. Medications and fluids supplied to the patient via the heart–lung machine are introduced into the blood circuit via the venous reservoir as well. All fluids and substances drain from the reservoir to an oxygenator either by gravity or by means of a pump, depending on the type of oxygenator. The oxygenator usually incorporates a heat exchanger to help keep the blood temperature at the desired level. Finally, the blood is pumped back to the patient via a $\frac{3}{8}$-inch-inside-diameter tube called the arterial line. On the patient's side the line is connected to the blood return cannula, which is usually inserted into the ascending aorta with the tip pointing downstream. Still another set of tubes serves to perfuse and cool the heart itself. This so-called cardioplegia system consists of $\frac{3}{16}$-inch plastic tubing connected on the patient's side to a needle inserted into the ascending aorta just downstream from the aortic valve. On the machine side, the cooling tube is connected to a separate apparatus specifically designed for cardiac cooling.

B. Artificial Heart Components: Blood Pumps

The pump in the heart–lung machine circuit must be capable of propelling blood at a rate of up to 8 liters/min against a pressure as high as 150 mm Hg. During a typical run, the flow rate is 4–5 liters/min at a blood pressure of 60–80 mm Hg. Depending on the type of oxygenator used, the arterial pump is located either upstream or downstream from the oxygenator. In circuits with a high-resistance artificial oxygenator, two pumps may have to be used, one upstream and one downstream from the artificial lung.

The pump most commonly used is of the *roller type,* in which one, mostly two, or sometimes three rollers squeeze a flexible tube, containing the blood, within a rigid housing. The rollers rotate, isolating a certain volume of blood within the tube, and transporting it from the inlet to the outlet. It is of simple construction, easy to operate, inexpensive, and reliable. In case of power failure, it is easy to rotate the rollers by hand using a special handle fitted on the axis of the pump. The tube containing the blood is disposable, so the cleaning process is simple. Extensive laboratory testing has provided evidence that the pump causes little damage to the blood, provided it is not used for periods of longer than 4–6 hr at flow rates of about 5 liters/min. The disadvantage of the pump is the occurrence of spellation, that is, dislodgment of small plastic particles from the wall of the tubing, which occurs mainly after prolonged use. These particles can travel into the arterial circulation and lodge in any part of the body. Other members of the group of rotating pumps are the Archimedean screw pump and the finger pump. The Archimedean pump consists of a rotor, in the form of a helical screw, which revolves within a housing with a double internal helical thread. During pumping, the fluid pockets between rotor and housing steadily progress along the thread toward the outlet. Finger pumps transport the blood by sequential compression of a straight rubber tube by multiple fingers. These two types of pumps were only of historical interest until recently, when the Archimedean screw was skillfully engineered into a promising mechanical circulatory device to assist in closed-chest left heart bypass (see Section III,C).

During many years the roller pump has served reliably in millions of operations lasting 2–5 hr. More recently though, the need for prolonged application of bypass led to further refinements of the heart–lung machine, and a less traumatic pump type was introduced. This *centrifugal* or *impeller pump* imparts velocity to blood with a propeller contained in a disposable plastic translucent chamber, without touching it, and rotates at a high rpm. The chamber is mounted on a pump house and the propeller rotation is effected by an electromagnetic field. This type of pump increases output as more volume is presented at the inlet; it is very sensitive to pressure at the outlet, decreasing output as the pressure rises. The pump cannot generate much suction at the inlet. These characteristics make this pump type well suited for prolonged use, such as days or weeks. There is less damage to the blood than in roller pumps, and no spellation. At present, centrifugal pumps are expensive and therefore are not used widely for routine heart surgeries. As production costs decrease, the pumps will probably be used for short-term applications as well.

Finally, reciprocating pumps comprise the last family. These pumps imitate the natural heart: the blood enters the chamber and is ejected by a decrease of the chamber volume. They typically need an inlet and an outlet valve. These pumps are used primarily in artificial heart research. To expel the blood, the volume of the pump chamber is decreased either by a rigid plate connected to a plunger or by circumferential compression within a rigid outer chamber by gas or by a liquid. The most commonly used artificial heart pump so far was of the reciprocating type.

C. Artificial Lung

The purpose of the artificial lung (oxygenator) is to transfer oxygen into the patient's blood and to eliminate carbon dioxide from it. This is a formidable task and it was the major limiting factor of the original heart–lung machines in the 1950s, because the gas transfer was to be carried out reliably and gently. Oxygen and carbon dioxide are carried by the hemoglobin molecule, which is present within the red blood cell. The ventilating gases have to transfer through the blood plasma to the red cell, must cross the cell membrane, and diffuse through the cell fluid to the hemoglobin molecules. Diffusion of oxygen through the layers is much slower than that of carbon dioxide. To speed up the gas exchanges the blood exposed to the ventilating gas should be in a thin layer, the blood transit time through the oxygenator should be long, and mixing of the blood should be efficient. The artificial lung should be able to oxygenate venous blood fully at a flow rate of up to 6 liters/min. To reach this goal, the oxygenator has to transfer 250–500 ml of oxygen per minute into that volume of blood and remove an equal amount of carbon dioxide.

The earliest clinical oxygenator model was of the vertical screen type. Venous blood from the patient was delivered to the top of a vertical screen. The blood flowed down the screen in a thin film and was exposed to a high flow of oxygen. The oxygenation capacity could be varied by increasing the number or the surface area of the screens. A large volume (3.5 liters) was necessary to fill the device (priming volume). This was all very problematic because blood was used to prime the machine.

A less bulky oxygenator of the same type was developed in which blood was distributed as a film on a number of rotating disks mounted in a cylinder. Blood filled only the bottom of the cylinder; the rest was ventilated with oxygen. While rotating, the disks picked up a thin layer of blood and exposed it to the oxygen. The capacity of these oxygenators could be adjusted by adding more disks and by varying their speed of rotation.

In still another type of oxygenator, oxygen was simply blown into the blood via a showerhead-like inlet. As a result, the blood starts to foam and it is in the foam bubbles that gas transfer takes place. The initial problem of these so-called bubble oxygenators was the inability to defoam the blood before it was returned to the patient. It should be noted that even small gas particles in blood can cause major damage to all organs of our bodies by stopping flow through small blood vessels. Effective debubbling was eventually achieved by exposing the foam to surfaces sprayed with silicone. Bubble oxygenators were accepted because they were easy to regulate and had a small priming volume. Once they were produced as disposable units, they became very popular. However, even the late models cause too much blood damage, especially when used over prolonged periods of time.

The oxygenator type that reproduces most closely the normal lung is the membrane oxygenator. In this artificial lung, the blood and gas phases are separated by a membrane. Its development depended on membrane technology. Early oxygenators were plagued by membrane rupture and inefficient gas transfer, but as the membrane quality improved, oxygenators of this type have become the artificial lung of choice even in short-term perfusions.

Membrane oxygenators have many other advantages. The priming volume is low and since there is little direct contact between gas and blood, blood damage is minimal. The entire heart–lung machine blood circuit is closed to air, promoting sterility and safety, and further decreasing blood damage. A newer variety of the membrane oxygenator is the capillary oxygenator. The blood is contained within or around multiple small rubber tubes rather than between sheets of membrane, reducing the priming volume even further. The modern membrane oxygenators are all disposable.

D. Other Parts

Since in the heart–lung machine circuit blood is exposed to room temperature, its temperature must be controlled. Once the patient is on the machine, the normal blood temperature of 37°C is allowed to drop to around 30°C, decreasing the oxygen need of the body. This gives an extra margin of safety to the use of the heart–lung machine. Certain complicated operations necessitate cessation of all blood perfusion through the body, which can be done safely only if the body temperature is decreased to around 10°C. Decreasing the temperature is easier and quicker than increasing it. Modern heat exchangers can increase blood temperature by about 0.5–1.0°C/min at a blood flow rate of 5 liters/min. Thus, the heart surgeon has to notify the perfusionist to start rewarming the patient some 20–30 min ahead of terminating the surgery. The heat exchangers are usually built into the artificial oxygenator and are thus disposable as well.

Filters are incorporated to ensure that no foreign particulate matter is being injected into the patient. They are essential in suction lines through which excess blood is removed from the operative field and returned to the venous reservoir. This blood may contain small tissue fragments, dissolved fat, blood clots, and even small pieces of sutures. Filters are also frequently employed in the blood return tube. The pore size of the filters varies from 25 μ in the venous filters to 40 μ in the arterial lines. A bypass circuit is always provided around the arterial line filter in case the latter clogs up. The modern arterial filters also serve as bubble traps, but only some 100–200 ml of gas can thus be prevented from entering the patient. Larger amounts would escape the bubble trap and enter the patient.

One or two tubes are provided in the heart–lung machine circuit to enable the surgeon to collect excess blood from the chest wound and return it to the blood circuit. This so-called *cardiotomy suction* is of great importance. Sudden accidents occurring during surgery could lead to dangerous blood loss if it were not promptly returned to the patient. The suctioning, however, causes most of the blood damage: it is the primary port of entry through which foreign particles and infectious sources enter the blood circuit of the heart–lung machine. The surgeon must realize these dangers and use the cardiotomy suction sparingly and wisely. The suction lines are mounted in one or two roller pumps, which return the blood to the venous reservoir. While the aorta is clamped and the heart is stopped, one of the suction lines is often used to decompress the heart by connecting it with the perfusion cannula in the aorta. Any blood accumulating within the heart is thus returned to the heart–lung machine circuit.

The roller pump produces a near-continuous outflow into the arterial blood circulation of the patient, whereas the natural heart produces a pulsatile flow with a difference of 20–50 mm Hg between highest and lowest pressure (pulse pressure). Whether the pulsatile flow or the continuous flow is better for the patient has been a matter of dispute ever since heart surgery has been practiced. For the 2- to 4-hr duration of routine heart surgery, advantages of pulsatile over nonpulsatile flow cannot be demonstrated. It is difficult to produce pulsatile flow with the roller pump, but several pulsating devices have been designed and are used in surgery of any duration; centrifugal pumps are especially suitable. Most surgeons prefer pulsatile flow if it is readily available.

Knowledge of the rate of bypass flow is of vital importance to the heart–lung machine operator. When roller pumps are used, the flow is simply a function of the rpm of the pump. If a centrifugal pump is used, the flow is usually measured by a Doppler sound probe in the arterial line or with an electromagnetic probe.

Several chemical constituents of the blood must also be measured frequently. They include hemoglobin, potassium, glucose, activated clotting time (ACT), and arterial blood gases. In most hospitals, blood samples from the patient are taken to the laboratory every 20 min to obtain these parameters. The laboratory returns the values by telephone. Depending on the quality of the lab, it takes 5–20 min to obtain these vital measurements. Equipment is now available that can provide on-line measurements of many of these values. A small tube delivers blood from the heart–lung machine to the lab machine. There it is exposed to all manner of analysis without being contaminated, so that it can be returned to the heart–lung machine. These laboratory machines are very expensive and therefore are not widely in use.

E. Ancillary Equipment

Modern heart surgery requires additional equipment used in concert with the heart–lung machine. A fairly recent addition is an apparatus to cool the heart. Once the patient is on cardiopulmonary bypass, it is safe to stop the heart. Cardiac arrest is needed to facilitate the miniature vascular operations on the coronary arteries and/or to open the heart to repair or replace diseased valves.

In the past, the aorta was simply clamped, causing anoxic heart arrest. A much better method is to cool the heart. Cooling alone, however, often leads to arrhythmias including ventricular fibrillation, during which the heart contracts with short, uncoordinated movements. In this state, the oxygen need of the heart increases sharply. To avoid this complication, potassium is added to the cooling solution, resulting in immediate arrest. Over the years, special equipment has been designed to facilitate this so-called hypothermic-cardioplegic arrest. A cooling unit contains a roller pump and a reservoir for cold water. Blood is drained from the arterial line of the heart–lung machine and passed through a coil suspended in the cold water. Once the blood temperature has decreased to some 5–10°C, a crystalloid solution containing a high potassium concentration is added, resulting in a final potassium level of some 20 mEq/liter. The mixture is pumped into a needle inserted into the ascending aorta. Some 300–600 ml of cold mixture are delivered to the heart every 20 min.

A newer apparatus is the so-called cell saver. It provides a special suction line separate from the ones connected to the heart–lung machine, which is used before and after heart–lung machine bypass. The blood is first mixed with heparin, which makes it unclottable, then is passed through a filter and emptied into a reservoir. From there, it is moved to a centrifuge, where the cellular elements of the blood are isolated. These cells are returned to the patient. This machine has led to a further decrease in the use of donated blood in heart surgery.

In patients with poor renal function, an artificial kidney may be combined with the heart–lung machine. The artificial kidney is generally used just before heart surgery. However, an artificial kidney-like apparatus called an *ultra filtration device* is connected in series within the arterial line. This apparatus can be used advantageously during cardiopulmonary bypass. The blood flows between membranes that retain the cells and the larger molecules of the proteins, allowing the removal of large amounts of excess fluid, including small molecules and waste products.

II. THE HEART–LUNG MACHINE IN USE

A. Hemodynamics during Bypass

Before entering the operating room, the heart–lung machine is assembled and filled (primed with fluid). Nowadays blood is not used to prime the circuit but rather crystalloid solutions that closely resemble normal blood plasma in electrolyte contents. Albumin (30–50 g) is added to increase osmotic pressure and to coat the inside of all plastic surfaces, in the hope of decreasing blood damage. The total priming volume is 2000–2500 ml. The fluid is then circulated through the machine by short-circuiting the arterial with the venous line. At this point, a 0.2-μ filter is temporarily inserted into the circuit to sift out all plastic particles that could have been left behind in the devices by the manufacturer. After testing the function of all moving parts the heart–lung machine is wheeled into the operating room and positioned next to the operating table.

At this point the surgeon is usually well into the operation. The tubes that will connect the patient with the machine are handed over to the surgical assistant. The patient is given some 300 units of heparin per kilogram of body weight intravenously. The surgeon places so-called pursestring sutures into the front wall of the aorta and in the right atrial wall. Pursestring sutures, as the name implies, are circle-shaped stitches that can be pulled up with a resulting narrowing of the tissue captured within the circle. (The aortic and venous cannulas are inserted into the bloodstream via the center of the pursestrings.) About 5 min after the heparin has been given to the patient, a blood sample is analyzed for clotting, usually by measuring the ACT and comparing it to the control value. The heparinized values should be some three times the control value of 80–150 sec. During complete bypass the ACT level should be at least 500 sec. At that point the surgeon inserts a short tube (cannula) via the pursestring into the aorta with the tip pointing downstream. The cannula is connected to the arterial line of the heart–lung machine. Taking extreme care not to insert any air bubbles in the system, the surgeon inserts another cannula via a pursestring into the right atrium. This cannula is then connected to the venous line of the heart–lung machine. Two suction lines are provided by the machine. Up to this point, the surgeon has been using the suction tube provided by the cell saver machine. Once the patient is heparinized, this suction is not used any more until the heparin is neutralized. Still another tube system is provided by the apparatus to cool the heart.

With all the tubes in place, the heart–lung perfusion or cardiopulmonary bypass can be started. The venous line is opened first and as venous blood flows by gravity from the right atrium to the heart–lung machine, the perfusionist starts the arterial pump and returns the blood to the patient via the arterial line. The arterial blood pressure of the patient nearly always falls, even though the bypass flow is the same as or higher than the amount of blood that the patient's heart pumped prior to the bypass. The lowering of blood pressure is caused by a reflex causing a marked drop of the vascular tone, and thus decreasing vascular resistance. The arterial blood pressure waveform changes from pulsatile to straight (continuous flow pattern). Often, the vascular resistance recovers spontaneously but, if it does not, the perfusionist will inject drugs to contract the arteries.

During bypass, the arterial blood pressure depends on the bypass flow times the resistance of the arteries. The flow has to be at least 2 liters per meter square of body surface area per minute (i.e., 4–6 liters/min) for an average adult. If the flow is less, the metabolism of the body suffers and the patient will sustain damage. Thus, the primary requirement in bypass is adequate flow. Too low a flow is considered an emergency that must be corrected immediately. With an adequate blood flow, the blood pressure is controlled by changing the arterial resistance. Powerful drugs are available to either increase or decrease it. Most operators keep arterial blood pressure between 60 and 90 mm Hg.

During bypass, the circulation is very sensitive to the blood volume. Loss in blood volume is reflected in the immediate decrease in blood flow. Volume can be lost directly into the operative field and it is the surgeon's responsibility to frequently check for accumulation of blood and to return it to the heart–lung machine via the suction. Urine production, which can be as high as 1000 cc/30 min, is another possible source of volume loss. It is replaced by adding crystalloid solution to the heart–lung machine.

As soon as the patient is on the heart–lung machine, the blood temperature is allowed to drift down to about 30°C in most routine heart operations. At the end of the operation, the surgeon initiates the discontinuation of the bypass. The bypass flow is slowly decreased, taking care not to accumulate more volume in the heart–lung machine. Thus, the natural heart gets back more and more of its original blood supply and, as a result, pumps out more blood. Once it is certain that the heart increases its output, the bypass flow is lowered further, until finally it is stopped altogether. At this point, the heart should have taken over the pump function and put out at least as much as the last amount of blood pumped by the heart–lung machine. If the heart fails to take over the circulation, a rare event, the patient has to be put back on the heart–lung machine. Once the heart–lung machine has been switched off, the surgeon will try to return as much of the priming volume to the patient as is tolerated. The aortic cannula is removed and connected to the venous cannula. The patient is given protamine sulfate to counteract heparin, and the proper level of blood coagulability is established by remeasuring the ACT. Once the priming volume of the heart–lung machine is returned to the patient completely, the venous line is removed as well. At this point, the patient is on his or her own.

B. Metabolism during Bypass: Cooling the Body

Metabolism is evaluated by measuring the rate of oxygen usage by the tissues. At rest, it is approximately 120–150 ml/m^2/min. The heart pumps 3.0–3.5 liters/m^2/min. The metabolism is said to be adequate if sufficient amounts of oxygen are delivered to tissues and carbon dioxide and other metabolites are removed. A number of blood parameters determine this ideal state: the hemoglobin content, the oxygen content, the viscosity, the flow, and pressure and resistance relationships, in addition to the conscious level of the patient and the body temperature.

During bypass the hemoglobin is decreased from a normal of 12–14 to 6–8 mg/100 ml due to dilution of the blood by the nonblood priming volume, which does not contain hemoglobin. The hemoglobin can be increased by supplying the patient with blood transfusions. The oxygen content of the blood is easily optimized with modern oxygenators, resulting in complete saturation of the hemoglobin. The hemodynamic parameters of flow, pressure, and resistance can be controlled readily in routine operations (see Section II,A). The blood viscosity decreases with hemodilution and increases with cooling, but in practice only at the very low temperatures of 10–15°C does it significantly affect bypass flow and tissue perfusion. At the lower temperatures, low flow is compensated by decreased demand for oxygen. Anesthesia alone decreases oxygen demand by 10–25%, depending on the depth. When patients start to wake up during surgery, the oxygen consumption rises immediately. The depth of anesthesia is actually evaluated by measuring the oxygen content of venous blood leaving the patient for the heart–lung machine. If all other factors are satisfactory, a decrease of the venous oxygen level signals the need for more anesthetic.

Probably the most effective way to decrease oxygen demand is to lower the body temperature: the demand is 50% of normal at 30°C; 25% at 22°C; and 12% at 16°C. As the temperature lowers, heart action decreases, and eventually the heart stops. Since blood circulation is controlled while on cardiopulmonary bypass, these effects are not worrisome, and are even encouraged. Thus, from the early days of heart surgery, body cooling usually to about 30°C has been used to create a margin of safety against low bypass flow conditions.

C. Effects of Heart–Lung Machine on Organs: Complications

During cardiopulmonary bypass, all organs are at the mercy of the artificial circulation. If inadequate, one or all vital organs may be damaged, sometimes irreversibly. The injection of foreign particles or air bubbles may also cause severe damage. The heart is probably the strongest organ to resist bypass-induced damage, and the most amenable to treatment. The brain is the most vulnerable. Damage to the brain, kidneys, liver, and lungs is difficult or impossible to influence therapeutically.

As cardiopulmonary bypass is started, the heart receives less blood and consequently pumps less, but, if its function is normal, the intracardiac pressures do not change. The metabolic need of the heart does not change much either even though the heart pumps less blood. The oxygen consumption of the heart depends more on the pressure against which it pumps than on the volume load. If the heart is failing, for instance in valvular heart disease, the start of the heart–lung machine may actually cause heart damage by increasing the arterial pressure. The increasing afterload creates a greater demand for oxygen, which the heart cannot always satisfy. This state of overloading lasts only a short time in the everyday practice of heart surgery, but it may become a problem when prolonged closed chest cardiopulmonary bypass techniques are used.

Once the patient is on full bypass, nearly all the venous blood returned to the heart is diverted and the cardiac output drops low enough that the heart only occasionally empties and causes a pressure wave in the aorta. At this point the heart is cooled. A tube from the cooling apparatus is inserted via a pursestring suture into the aorta close to the aortic valve. The aorta is clamped just downstream from the cooling cannula so that the arterial blood cannot reach the coronary arteries. The cooling machine pumps diluted, cold blood with a high potassium content at a rate of 200–300 ml/min into the coronary arteries (see Section I,E). A total amount of 300–400 ml is delivered every 20 min. The result is cardiac arrest and cooling of the heart down to 10–15°C. Under these circumstances, the heart can safely be without blood supply for 20–30 min. Once the operation on the heart proper is completed, the aortic clamp is removed and warm aortic blood is allowed to reperfuse the coronary arteries. This causes the heart to resume its function. Malfunction may cause mild functional changes and, rarely, massive cardiac failure and death.

The lungs frequently suffer some decrease in function during bypass. Good measurements on pulmonary function are hard to obtain in the immediate postoperative period because the patient does not put forth much of a breathing effort. Blood gas levels usually indicate a defect in oxygen diffusion in the lung, resulting in low partial oxygen tension in the arterial blood. Generally these abnormalities subside in 2–5 days, but in some cases the pulmonary dysfunction progresses to intractable failure. This condition, called *acute respiratory distress syndrome* (ARDS), is life-threatening and requires the use of the respirator for prolonged periods of time. The mortality rate can be as high as 50%. It is of great importance that the patients be frequently encouraged to breath deeply in the early postoperative period. Another complication is *atelectasis,* that is, collapse of air space in the lungs, which arises when chest pain makes the patient suppress deep breathing efforts. It leads to decrease of arterial oxygen content, cardiac arrhythmias, and pulmonary infection. [*See* Pulmonary Pathophysiology.]

The brain is very sensitive to inadequate perfusion. If the blood supply to the brain stops for only a few minutes, irreversible damage occurs and the patient never wakes up, although all of the other vital organs function properly. This complication is rare, especially because the decrease of body temperature down to 30°C gives some margin of safety. *Emboli,* consisting of foreign particles when lodged in the brain, may cause strokes, the severity of which depends on the size of the embolus and its location in the brain. Emboli may originate from the heart–lung machine, from the aortic wall at the site of the arterial cannula, and from other areas of the vascular system, notably ulcerations in the carotid arteries. Heart–lung machine emboli used to be a major problem, but not now. Particles such as air bubbles, plastic fragments, and antifoam coagulates have been located in peripheral tissues. By careful selection of the aortic cannulation site, the second source of embolization can be kept to a minimum. The effect of carotid artery disease is also minimized by screening patients before heart surgery and cleaning out these vessels if necessary. Since such preventive measures have been taken, the incidence of postoperative stroke has dropped to some 0.5%. Psychiatric abnormalities occur at fairly high frequency but subside a few months after the surgery. Depression is quite common; fully developed psychosis is rare. Efforts to study the brain function during surgery by recording the EEG (electroencephalogram) have not given useful results.

The function of kidneys depends on a blood pressure of at least 60 mm Hg. Below that pressure, renal perfusion is still adequate for tissue survival but not for urine production. Kidney dysfunction and even failure may be produced in the immediate postoperative period by medications or by insufficient circulation. Renal failure calls for the use of a temporary artificial kidney. Complete recovery may occur.

Cardiopulmonary bypass causes damage to the blood owing to the blood's contact with foreign surfaces, turbulence, squeezing in the tubes, foaming in the bubble oxygenator, or mixture with tissue debris in the wound.

Red blood cells are destroyed during cardiopulmonary bypass as shown by an increase of hemoglobin in the plasma. Most of the damage is caused by suctioning of the blood from the operating area. If it is extensive, as during multiple valve replacements, the urine must be alkalinized by keeping the partial pressure of the carbon dioxide in the blood low and urine production high. A proportion of red cells are not destroyed but are weakened to the point that they are captured and eliminated by the body itself in the first few days postoperatively, resulting in a drop of hemoglobin without discernable blood loss.

After bypass, white cells are less effective against foreign intruders. In the postoperative period, the white blood cell count frequently increases and remains high for 5–10 days, raising the question of the presence of infections. The platelet count drops approximately 40% during bypass and increases to slightly below normal levels in a few hours afterward. Blood proteins are denaturated and destroyed but not to a dangerous extent. The clotting factors are altered significantly, primarily as a result of the administration of heparin before and protamine sulfate after bypass.

III. APPLICATIONS OF THE HEART–LUNG MACHINE

A. Heart Surgery

During heart surgery the machine is needed to exclude the heart from the blood circulation, stop, and open it. The most frequent operation is the coronary artery bypass operation, in which cardiopulmonary bypass may be needed for 1–3 hr.

In valve surgery, cardiopulmonary bypass must last for a longer time, up to 6 hr. The wound area suction is used extensively in this operation, and thus more blood damage is incurred. In heart transplantation, the heart–lung machine is used for about 3 hr.

During heart surgery for congenital defects, a heart–lung machine miniaturized to adapt to babies is used. Whole-body cooling down to 10°C is generally applied.

B. Other Applications: Lung Surgery and Aortic Surgery

The heart–lung machine has been used in the past for complicated pulmonary operations, but this use is now discontinued. Cardiopulmonary bypass is used in operations on the first part of the aorta from the aortic valve down to the origin of the artery to the left arm. The rest of the intrathoracic aorta can be operated on with or without bypass. If bypass is used, it does not have to include the heart and the lungs. A simple tube from the aortic arch to the descending aorta suffices, although some surgeons use complete heart-lung bypass. The simple bypass has the advantage that generalized heparinization is not needed, an important consideration since descending thoracic aortic surgery is frequently done on patients shortly after major accidents.

C. The Heart–Lung Machine as an Assist Device for Heart and Lung Failure

Early in the development of the heart–lung machine, it was recognized that cardiopulmonary bypass could be used for assistance to the failing heart. However, prolonged bypass periods are needed, causing considerable damage to blood and its cellular elements, even with the advanced equipment presently available. Uncontrollable bleeding is also a problem, which becomes particularly severe if the patients are heparinized.

Cardiopulmonary bypass can be used to help patients who become very unstable just before surgery, especially during complicated procedures in the heart catheterization laboratory. Cannulas are inserted into a femoral artery and a femoral vein and are connected to the heart–lung machine. Bypass flows of 2–4 liters/min can be obtained and with that circulation is usually sufficiently restored. If the increased arterial pressure puts a heavier workload on the heart, threatening heart failure, the left side of the heart must be decompressed. This result can be obtained by applying short periods of external cardiac massage or, more effectively, by draining the left ventricle. The Archimedean screw pump proved very effective in pumping blood from a cannula inserted in the ventricle via the aorta.

Cardiopulmonary bypass using peripheral cannulation can be used for the treatment of cardiogenic shock after myocardial infarction. Since the refinement of equipment and the recognition of the need to decompress the left ventricle, the application of the bypass in this area has been revived during the last few years. Given the high incidence of myocardial infarction, this field of application may grow significantly.

Another area of application of the heart–lung machine is treatment of severe pulmonary failure. The bypass can be lifesaving in some 50–90% of newborn babies; in adults, it is applied in patients who have acute respiratory distress syndrome. In these cases, only the lungs are bypassed, not the heart. The technique is called extracorporal membrane oxygenation

(ECMO). Blood can be drained from and returned to either a femoral vein or an artery. In adults, the blood circuit is usually from vein to vein. The bypass is operated at a low flow of some 500–1000 ml/min, sufficient to normalize the blood gases even though the patient breathes only two to four times per minute. Oxygen transfer is affected primarily in the lungs, whereas the elimination of carbon dioxide occurs in the artificial lung. Such a bypass can be carried on for many days until recovery of pulmonary function occurs. The major complication is profuse bleeding. Equipment improvement may control this complication in the future. Using materials coated with heparin makes heparinization of blood unnecessary, reducing bleeding danger.

Variants of the cardiopulmonary bypass are the *left heart bypass* and *right heart bypass,* in which the patient's own lungs are used for oxygenation. Left heart bypass is of interest because it effectively decompresses the left ventricle; it has therefore been studied extensively for assisting the failing heart. In heart surgery it is used as a last resort in patients who cannot be weaned off the heart–lung machine. The mortality of such patients, however, is understandably high. Implantable left heart bypass pumps have been used in these cases for prolonged assistance. The ultimate permanent application of both the left and right heart implantable pump systems is the artificial heart.

BIBLIOGRAPHY

Akutsu, T. (ed.) (1986). "Artificial Heart III." Springer-Verlag, Tokyo/Berlin/Heidelberg/New York.

Friedman, E. A. (ed.) (1988). "ASAIO Transactions," Vol. 34, No. 3. Lippincott, Philadelphia.

Galletti, P. M., and Brecher, G. A. (1966). "Heart–Lung Bypass: Principles and Techniques of Extracorporeal Circulation." Grune & Stratton, New York.

Ionescu, M. F. (ed.) (1982). "Techniques in Extracorporeal Circulation," 2nd Ed. Butterworths, London.

Reed, C. C. (1985). "Cardiopulmonary Bypass." Texas Medical Press, Stafford.

Heat Shock

MILTON J. SCHLESINGER
Washington University School of Medicine

GLOSSARY

BiP Immunoglobulin heavy-chain binding protein

Chaperonins Family of proteins whose proposed role is to ensure that the folding of certain polypeptides and their assembly into oligomeric structures occur correctly

Glucose-regulated protein (GRP) Number following GRP refers to the protein's subunit molecular weight; thus, GRP94 is the glucose-regulated protein with a subunit molecular weight of 94,000

GRO EL Protein in *Escherichia coli* required for cell division and for bacteriophage replication

Heat shock promoter element Sequence in the DNA, usually in the 5′ nontranscribed region of a heat shock gene, that controls expression of that gene

Heat shock protein (HSP) Number following HSP refers to the protein's subunit molecular weight (cf. GRP)

Heat shock transcription factor (HSF) Protein that binds to the heat shock promoter element and affects expression of a heat shock gene after a heat shock

Rubisco Ribulose bisphosphate carboxylase-oxygenase

HEAT SHOCK REFERS TO THE ACTIVATION OF A SET of evolutionarily conserved genes and the subsequent synthesis of their encoded proteins as a result of an increase in temperature above that normally experienced by a cell. The temperature range required to initiate a heat shock generally does not exceed a 10% change from the physiological temperature; thus, heat shock can occur in organisms that normally live in temperatures ranging from 4°C to >60°C. Many heat shock genes are also activated by other agents that stress the cell such as organic solvents, heavy metals, nutrient deprivation, oxidants, and viral infection. Some heat shock genes are activated during normal growth and development of the organism (Fig. 1).

I. BACKGROUND

The internal temperature of the human body can fluctuate only within a very narrow range without leading to death of the individual. Even within this range, severe damage can occur; however, evolution has built within the cells of the human as well as those of virtually all species in the living world—including the bacteria—a set of biochemical activities that protects the cell during a thermal stress and allows for its recovery later when temperatures return to normal. Not all of the mechanisms employed by a cell for thermal protection are known, but one that has been studied in considerable detail is the activation of a small set of genes almost immediately after the cell senses the temperature shock.

The first observation of a heat-inducible gene activation was the appearance of new puffs in the chromosome of *Drosophila* embryos that had been moved from their normal temperature of 25°C to 30°C. Puffs are sites of intense RNA transcription, and the RNAs from the heat shock puffs are selectively translated to produce high levels of proteins called heat shock proteins. Similar proteins were subsequently found in heat-shocked cells obtained from many different organisms, including human, chicken, plant, yeast, and bacteria (Fig. 2), although the temperatures re-

ENCYCLOPEDIA OF HUMAN BIOLOGY, Second Edition, VOLUME 4.

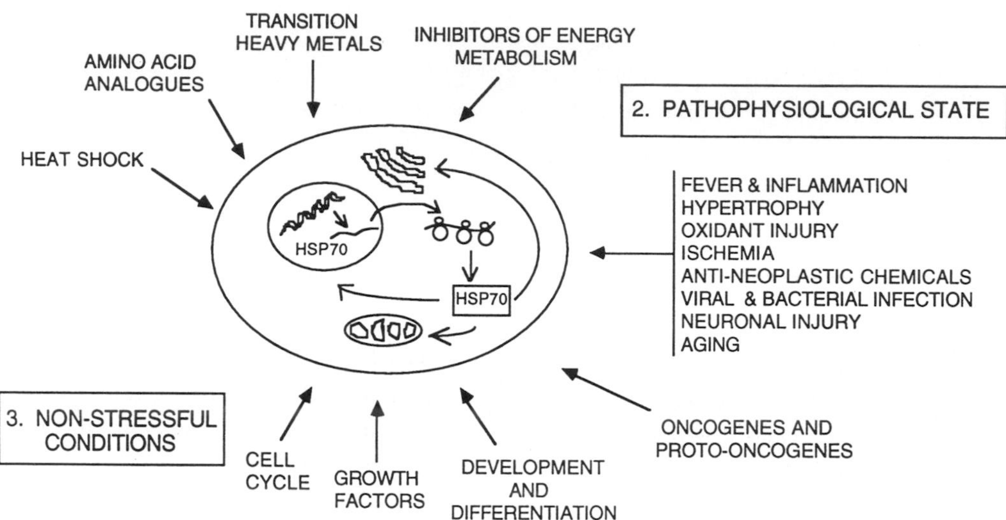

FIGURE 1 Conditions that result in the induction of heat shock gene expression in eukaryotes. [Reproduced, with permission, from R. I. Morimoto, A. Tissières, and C. Georgopoulos (1994). "The Biology of Heat Shock Proteins and Molecular Chaperones." Cold Spring Harbor Laboratory Press, Cold Spring Harbor, NY.]

quired to activate their genes were vastly different. The major proteins fell into four groups based on their size: small proteins of 15,000–30,000 molecular weight, intermediate proteins of 50,000–60,000 and 70,000–80,000, and large proteins of 90,000–100,000. Many features of the structure and function of these proteins are described in this article. However, it is the formation of these proteins and the factors controlling the expression of their genes that provide the operational definition for the heat shock response.

II. HEAT SHOCK GENES AND THEIR EXPRESSION

A. Features of a Heat Shock Gene

Heat shock genes can be distinguished from other protein-encoding genes by a short sequence of DNA, noted as the heat shock promoter element, which is located in the 5′ nontranscribed portion of the gene positioned about 80–200 nucleotide base pairs from the start site of messenger RNA (mRNA) transcription. In eukaryotic cells, the promoter consists of contiguous arrays of an inverted repeat of five nucleotide base pairs containing *n*GAA*n*, where *n* can be any of the four bases (Fig. 3).

Another feature of many heat shock genes is the lack of introns (noncoding sequences) within the protein-coding region of the gene—a property that enables heat shock mRNAs to bypass the normal pre-mRNA splicing activities that are inhibited in a heat-shocked cell. Also, some sequences in the 5′ and 3′ transcribed but noncoding domains of heat shock genes stabilize the mRNA during heat shock and confer a selectivity for its translation by the cell's protein synthetic machinery. [*See* Genes.]

B. Expression of Heat Shock Genes

Activation of all eukaryotic heat shock genes occurs when the heat shock promoter element is occupied by a DNA-binding protein called the heat shock transcription factor (HSF). Transcription factors are proteins of about 70,000 molecular weight that are found as "inactive" monomers or dimers in the cytoplasm of the unstressed cells. Several isoforms, which are encoded by distinct genes, appear to be activated by distinct forms of stress or by developmental signals. Activation involves oligomerization, phosphorylation, and translocation to the cell nucleus, where the

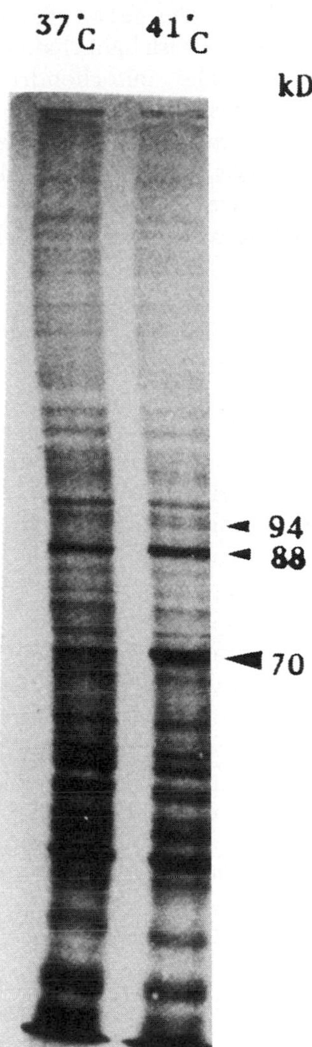

FIGURE 2 Synthesis of heat shock proteins in human endometrium incubated at 41°C. Tissue samples were given radioactive amino acids and incubated for 2 hr at either normal or heat-shocked temperatures. Proteins were extracted and separated by electrophoresis in polyacrylamide gels. The pattern shown is an autoradiogram; heat shock proteins are indicated. [Reproduced, with permission, from A. Ron and A. Birkenfeld (1987). Stress proteins in the human endometrium and decidua. *Hum. Reprod.* **2**, 277–280.]

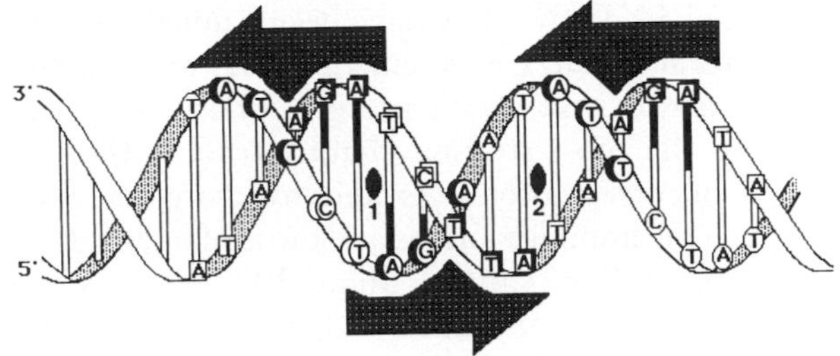

FIGURE 3 A segment of double-stranded DNA containing a heat shock promoter element (*n*GAA*n*). Shaded large arrows are heat shock transcription factors bound to DNA. [Based on a model, with permission, presented by O. Persic, H. Xiao, and J. T. Lis (1989). Stable binding of drosophila heat shock factor to head-to-head and tail-to-tail repeats of a conserved 5 bp recognition unit. *Cell* 59, 797–806.]

HSFs bind to selective regions, that is, the heat shock promoters, of the DNA that also contain other DNA-binding proteins. The structure of the DNA-binding domain in the HSF from *Kluyveromyces lactis* has been determined to 1.8 Å and consists of a variation of the helix–turn–helix motif found in other DNA-binding proteins. The resulting complex leads to transcription of the heat shock gene. In the *Drosophila* chromosome, an RNA polymerase is already bound to and transcribes a short region of the heat shock gene in the nonstressed cell, but it is blocked in its ability to proceed beyond a specific region of the gene. Binding of the heat shock factor relieves this inhibition and allows for the complete transcription of the gene. [*See* DNA and Gene Transcription.]

The transcription factors are different in bacteria. In *Escherichia coli,* the heat shock transcription factor is an isoform of the sigma factor, the regulatory subunit of the DNA-dependent RNA polymerase. The heat shock sigma factor is made during normal cell growth but is degraded very rapidly and its steady-state level is low. Heat shock blocks degradation of this protein, resulting in higher levels of the factor, which then replaces the normal sigma factor in the RNA polymerase complex. This new complex binds selectively to heat shock promoter elements, thereby transcribing only heat shock genes.

The biochemical mechanisms that sense heat shock and produce the specific alterations in heat shock transcription factors are unknown. In bacteria, some evidence indicates that a thermal unfolding of proteins is the trigger. In the eukaryotic cell, the thermal-sensitive element could be the heat shock transcription factor itself or a protein bound to it. The latter might be a heat shock protein.

III. HEAT SHOCK PROTEINS AND THEIR FUNCTIONS

A. General

The precise number of heat shock proteins in an organism is not known, but in the human it is unlikely to exceed 50. Only a few have been studied in detail, but several features have emerged that are common to the major heat shock proteins of many species. Perhaps the most striking characteristic is the high degree of conservation in protein structure that exists between the human proteins and those of highly divergent species. For example, about 50% of the amino acid sequences in the human 70,000 molecular weight heat shock protein is found in a heat shock protein of similar size in *Escherichia coli,* and antibodies raised against a chicken heat shock protein of this size recognize the human and the yeast protein.

A second important property of the major heat shock proteins is that they can be grouped into distinct protein families. Individual members of a family are closely related in structure but are distinct with regard to their location in a cell as well as by the conditions controlling their synthesis and turnover. Features of several major heat shock protein families are noted in the following sections.

B. The HSP70 Protein Family

The protein whose synthesis appears most rapidly and most intensely in the human cell after heat shock has a molecular weight of 70,000. There is, however, a basal level of this protein in nonstressed cells, and the protein also appears at specific stages during the normal cell division cycle as well as after infection of human cells with viruses. Thus, expression of the human *hsp70* gene is regulated by multiple promoter elements, including one responding to the heat shock transcription factor. HSP70 is usually present as a dimer in the cytoplasm, but most of it migrates to the nucleus during a heat stress where it binds tightly to proteins in the nucleolus—the site of ribosome assembly.

The human HSP70 family contains four other proteins that are slightly larger in molecular weight. One of these, HSP72, is found only after a heat shock. Another of similar size is present at high levels in normal cells and is identical to a protein that disassembles clathrin-coated vesicles. The third and fourth members of the family are localized to the endoplasmic reticulum and the mitochondria of the cell. The former is a glycoprotein called glucose-regulated protein 78, GRP78, and is identical to BiP, a protein that was first identified by its binding to the immunoglobulin heavy chain in pre-B cells that did not make the immunoglobulin light chain. In the absence of light chains, immunoglobulin heavy chains denature; however, BiP prevents this inactivation. Later, when light chains are produced, the heavy chains are released from BiP and complex with light chains to form active immunoglobulins. The mitochondrial protein is termed GRP75 and is closely related to a similar protein found in yeast mitochondria, where its function is essential for survival of the organism. Synthesis of these GRPs is induced by agents that affect protein glycosylation or increase intracellular calcium concentrations.

The functions of HSP70 have been surmised from a variety of activities detected for several of the HSP70 family members from different organisms. Some of these were just described. In yeast, an HSP70-like protein is required to unfold certain cytoplasmic proteins destined for import into mitochondria or for transport into the endoplasmic reticulum and ultimately secretion from the cell. In *Escherichia coli,* the HSP70 homologue (DNA K) participates in the initiation of DNA replication for bacteriophages and for the host cell at high temperatures.

In all of these examples, the HSP70 protein first forms a tight complex with a partially folded polypeptide or a complex of polypeptides and then—in a reaction driven by adenosine triphosphate (ATP)—modifies the bound protein so that it either unfolds or the complex dissociates (Table I). In fact, HSP70 has a weak ATPase activity and the amino-terminal domain of the protein contains an ATP-binding site. The atomic structure of this ATP-binding site has been solved by X-ray diffraction analysis and resembles that of similar ATP-binding sites in other proteins. During heat shock, HSP70 is postulated to bind to protein complexes that are particularly sensitive to temperature denaturation, to nascent polypeptide chains, or to polypeptide subunits that have not yet folded to a thermostable conformation. In these complexes, the proteins are protected from denaturation, but when the cell returns to its normal physiological state, HSP70 is released, thereby allowing the protein to resume its normal structure and function.

C. The HSP60 Protein Family

The HSP60 proteins are the most highly conserved in structure among all of the heat shock proteins. Tetradecamers of the 60,000 molecular weight subunit with a sevenfold axis of symmetry are localized to the inner compartment of the cell's mitochondria, where they catalyze the refolding of polypeptides newly imported across the mitochondrial membrane. These polypeptides are encoded by nuclear genes and are synthesized on cytoplasmic polysomes. They con-

TABLE I
Heat Shock Protein Complexes

Heat shock protein	Target protein	Result after ATP action
HSP70 family member	Clathrin-coated vesicles	Dissociation
	Premitochondrial and presecreted proteins	Unfolding
	DNA replication–initiation complex	Dissociation
	Newly imported, unfolded proteins in the lumen of the endoplasmic reticulum	Dissociation
	Unfolded forms of tumor-suppressor proteins	?
	Nucleolar proteins (pre-mRNA splicosomes; pre-rRNA ribosomes)	Dissociation
	Steroid hormone receptors	Dissociation
HSP60 family member	Large rubisco subunit	Folding
	Bacteriophage structural proteins	Assembly
HSP90 family member[a]	Steroid hormone receptors	Activation of the receptor
	Phosphokinases (oncogene products)	Activation of the kinase
	Eukaryotic initiation factor 2 kinase	Activation of the kinase
	Actin, tubulin	?

[a]Requires binding of ligands other than ATP.

tain amino acid sequences that target them to receptors on mitochondrial membranes, but to be transported through the mitochondrial membrane, they must first be unfolded—a reaction catalyzed by an HSP70 family protein. Upon emerging into the lumen of the mitochondria, a refolding occurs, catalyzed by the HSP60-type protein. The term chaperonin has been invoked to identify this kind of interaction. In plants, the chaperonin protein is the rubisco-binding protein. It is abundant in the chloroplast, where it participates in formation of the multimeric complex of enzymes that fix carbon dioxide into organic material. The bacterial protein, called GRO EL, is needed for assembly of bacteriophages and for normal cell growth. Antibodies directed against the protein isolated from tetrahymena recognize homologous proteins in *Escherichia coli* and in humans.

D. The HSP90 Protein Family

Members of the HSP90 group are highly phosphorylated, abundant cytoplasmic proteins. About 50% of their sequences are conserved among yeast, trypanosomes, fruit flies, chickens, and mammals. A bacterial form is 40% related in sequence to that of other species. One member of the family, GRP94, is localized to the eukaryotic cell's endoplasmic reticulum.

A number of other cellular proteins form a stable complex with HSP90. These include viral and cellular protein kinases, steroid hormone receptors, tubulin, and actin. When complexed with HSP90, the kinases and receptors are inactive. Binding of hormone releases the HSP90 from the receptor complex and unmasks the DNA-binding domain of the receptor.

The HSP70, HSP60, and HSP90 protein families share a common function in that all form dissociable complexes with a variety of cellular polypeptides (summarized in Table I).

E. The HSP15–30 Protein Family

Proteins in the 15,000–30,000 molecular weight range are synthesized in almost all cells after heat shock and are the most abundant heat shock proteins in plants. Many of these proteins form very large polymers, which appear as heat shock granules in the cytoplasm. Some of these granules, particularly in stressed plant cells, are highly regular in structure and contain specific small RNAs. Upon recovery from stress, the granules dissociate. A portion of the amino acid sequence for many of the small heat shock proteins is closely related to a sequence in bovine lens α-crystallin, and this may be the region responsible for polymerization.

Many of the small heat shock proteins become phosphorylated when cells are stimulated by a variety of agents, many of which are associated with cell growth and differentiation. Changes in the extent of phosphorylation are distinct from activation of transcription of the small HSP genes noted during heat shock. In normal cells, the small heat shock protein functions as an inhibitor of actin polymerization and is found associated with cellular structures active in microfilament remodeling.

Some of the genes encoding these proteins in fruit flies are activated during late larval and early pupal stages of development, possibly by the hormone ecdysone. In yeast, the protein is synthesized when cells stop growing and when they initiate sporulation.

F. The Ubiquitin System

All cells contain enzymes or enzymatic pathways for degrading intracellular polypeptides, proteins whose activities must be carefully regulated during normal cell function as well as proteins that become misfolded and denatured as the result of stress such as a heat shock. In the eukaryotic cell, a universal system for degrading intracellular proteins to small peptides and amino acids employs a heat-stable cofactor called ubiquitin, a polypeptide of 76 amino acids whose sequence is almost invariant from plants to humans. A set of enzymes activates ubiquitin in the presence of ATP and covalently attaches it to lysine groups on the protein destined for degradation. Multiple ubiquitin molecules are subsequently added and the polyubiquitinated protein is recognized by a large multicomponent complex containing proteases. Proteins targeted for ubiquitination have either charged or very bulky hydrophobic amino acids at their amino terminus or have an abnormal, misfolded domain within the body of the protein. The latter would be most likely present during a heat shock.

Genes encoding several of the enzymes involved in the ubiquitin pathway are activated by heat shock. These include a gene encoding ubiquitin itself and two of the enzymes that transfer the activated ubiquitin to the target protein. Other cellular components utilize ubiquitin, and some of these are also affected by a heat shock (see Section III,G).

Ubiquitin is not present in prokaryotes; however, the gene (*lon*) that encodes the ATP-dependent protease La in *Escherichia coli* is activated by heat shock. La probably functions like the ubiquitin system to remove aberrant proteins formed during stress in bacteria.

Thus, proteins damaged (i.e., unfolded) by heat shock have two alternative fates: they may be rescued and repaired by a heat shock protein, or they may be proteolytically degraded.

G. Other Proteins

In addition to the proteins already discussed, the synthesis of the glycolytic enzymes enolase, glyceraldehyde-3-phosphate dehydrogenase, and phospho-glycerate kinase is stimulated by a heat shock. Thermal stress also increases the formation of some extracellular and cell-surface proteins such as thrombospondin, C-reactive protein, and a collagen-binding protein.

A summary of how the major HSPs interact in a stressed cell is presented in Fig. 4.

IV. EVENTS IN A HEAT-SHOCKED CELL

The activation of heat shock genes is only one of several changes that occur in the cell in response to a stress. Many of the cell's metabolic activities are affected, and virtually every organelle in the cell shows some modification. There are also changes in the cell's morphology and cytoskeleton. As expected, the extent of these alterations is closely related to the intensity and duration of the stress. This section considers some of the more profound changes.

Among the cytoskeletal elements—microfilaments, microtubules, and intermediate filaments—it is the latter that is most sensitive to stress. Upon a mild heat shock, the intermediate filament network collapses to a perinuclear cytoskeletal ring but reappears in its extended form shortly after removal of the stress. Formation of heat shock proteins appears to be necessary for this recovery. With more severe stress, actin bundles are found in the nucleus and cytoplasmic microtubules are destroyed.

The morphology of the mitochondria is also sensitive to a heat shock stress and the normally shaped elongated tubules of this organelle collapse into structureless "globs" that cluster around the nuclear membrane (see Fig. 5). The collapse and subsequent recovery of mitochondria after a stress occur under those same conditions noted for the changes in the intermediate filament network and suggest that the latter acts as a tether for the mitochondria.

Metabolic activities in the nucleus are among the most sensitive to stress. DNA synthesis slows dramatically, ribosomal RNA precursors accumulate, and splicing of mRNA is inhibited. Histones are modified by phosphorylation and deubiquitination—changes that are normally found during condensation of the chromatin.

The protein synthetic machinery is not as sensitive but polysomes dissociate immediately after a brief severe heat shock. Most proteins are stable to temperatures in the physiological range of a heat shock; however, an increase in the level of ubiquitin-conju-

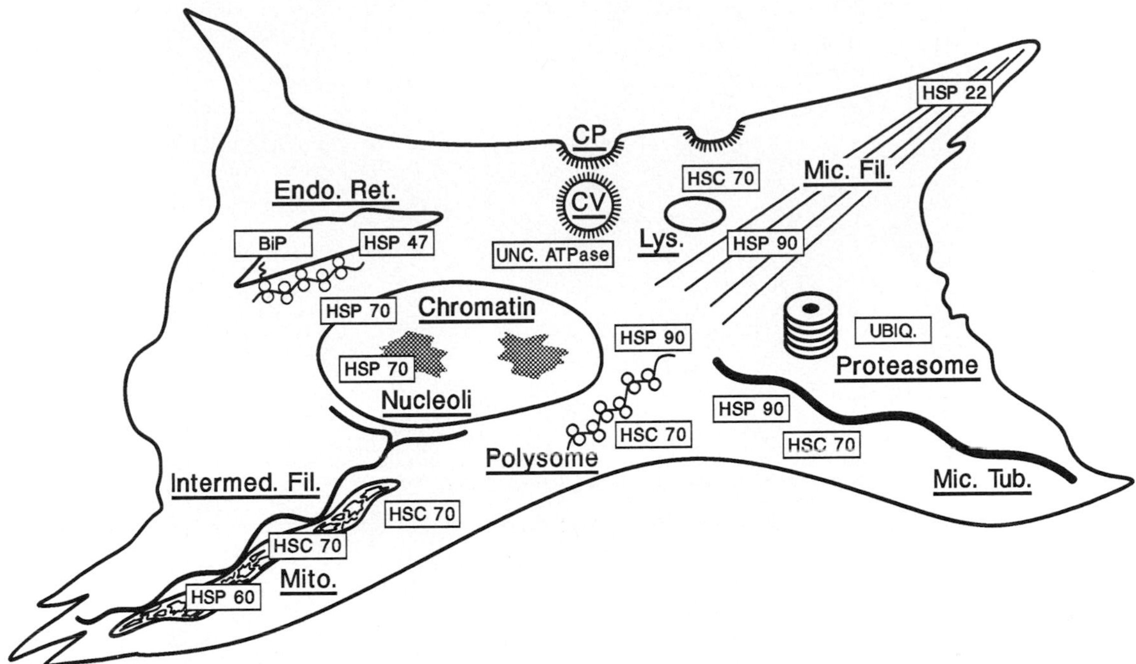

FIGURE 4 A heat shock protein view of the cell. The major heat shock proteins are noted in boxes associated with various cellular organelles (nucleus, cytoplasm, mitochondria, endoplasmic reticulum, lysosome) and cytoskeletal elements (microfilaments, microtubules, and intermediate filaments). CP, coated pits; CV, coated vesicles; UNC., uncoating; UBIQ., ubiquitin; Lys., lysosome; Mic. Fil., microfilaments; Mic. Tub., microtubules; Intermed. Fil., intermediate filaments; Mito., mitochondria; Endo. Ret., endoplasmic reticulum.

gated proteins indicates increasing numbers of denatured proteins in the heat-shocked cell.

At the plasma membrane, there is a decrease in sodium–potassium ATPase activity. With severe stress, internal cell membranes are affected: the Golgi stack and other cytoplasmic veiscles are disrupted, and the endoplasmic reticulum is swollen and fragmented. In addition, the cell shifts to the glycolytic pathway for ATP generation and accumulates lactic acid. Levels of calcium also increase in the heat-shocked cell.

Proteins are modified in a variety of ways after a heat shock (e.g., the ribosomal subunit S6 is dephosphorylated).

V. STRESSORS OTHER THAN HEAT SHOCK

In addition to heat shock, a variety of other kinds of environmental abuses reprogram the cell's protein synthetic machinery to produce heat shock proteins or related family proteins. Table II lists many of these together with the types of stress proteins elicited. Whether or not these different stressors act through a common mechanism to signal the activation of heat shock genes is unknown; however, heat shock genes have multiple promoter elements, and some of these are used for factors specific to a particular stressor. Even where a heat shock element is the effective promoter, activation of the heat shock transcription factor could be different for distinct stressors.

VI. THERMOTOLERANCE

Damage caused by a severe heat shock can be reduced if the cell or organism is first subjected to a milder heat stress. This phenomenon is called thermotolerance, but it is not limited to either thermal stress or thermal protection. For example, a mild stress by ethanol will protect a cell from heat shock damage, and a mild heat shock protects cells from ethanol.

The molecular mechanisms leading to stress tolerance have not been completely defined; however, a strong correlation exists between synthesis and persis-

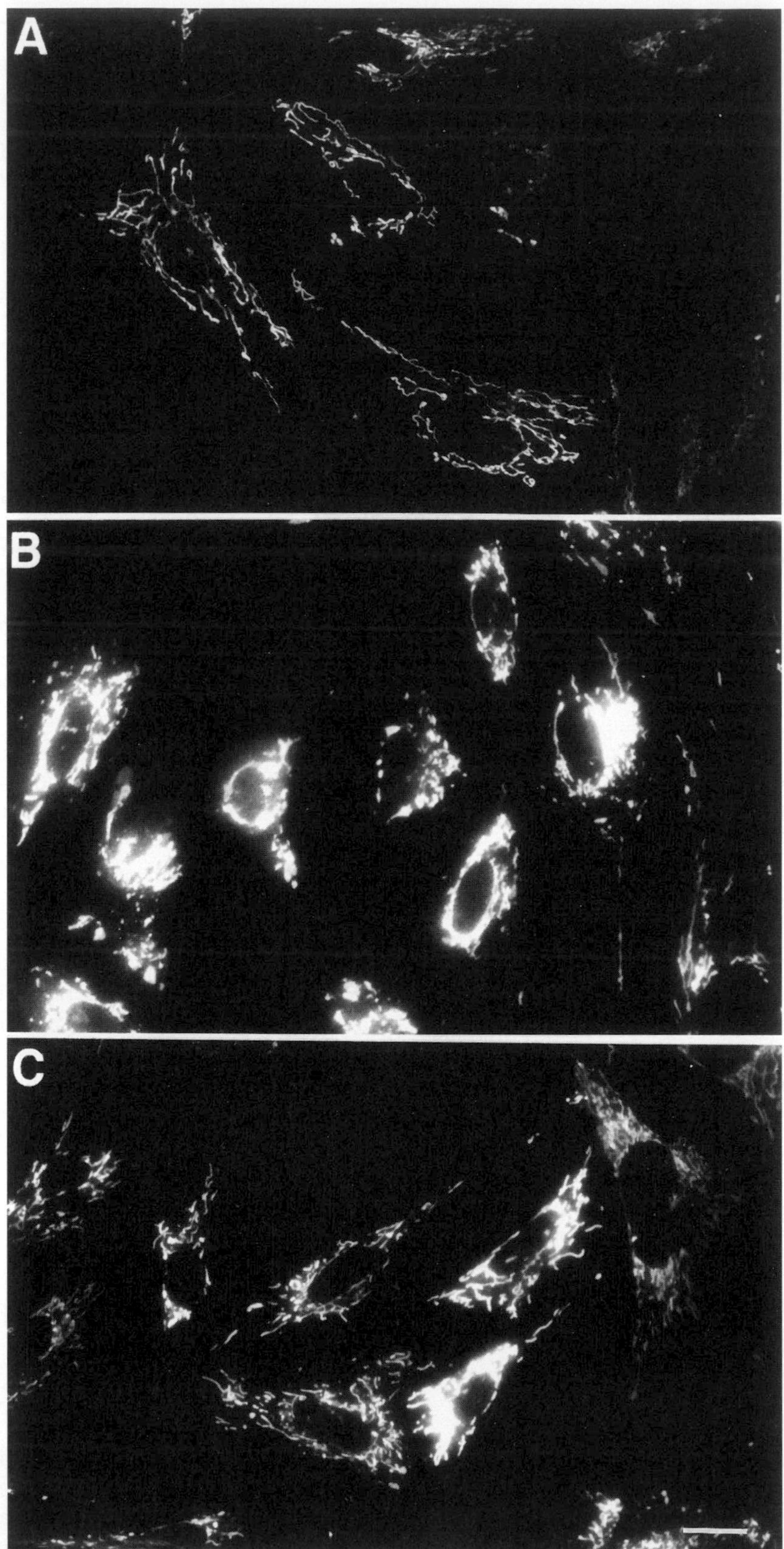

FIGURE 5 Mitochondria in normal, heat-shocked, and recovered chicken embryo fibroblasts (CEF). Live CEF were stained with rhodamine 123. (**A**) Normal cells. (**B**) Cells heat-shocked for 1 hr at 45°C. (**C**) Cells heat-shocked for 1 hr at 45°C and incubated at 37°C overnight. Bar = 25 μm.

TABLE II
Nonheat Shock Inducers of Heat Shock Protein Synthesis

Type	Class	Examples
Stressors that induce mainly the heat shock proteins	Oxidizing agents and drugs affecting energy metabolism	Arsenite, peroxide
	Transition series metals	Mercury, copper
	Chelating drugs	Salicylate
	Amino acid analogues	Canavanine, azetidine
	Inhibitors of gene expression	Puromycin
	Steroid hormones	Dexamethasone
	Virus infection	Herpes simplex
	Others	Ethanol, wounding, recovery from anoxia
Stressors that induce mainly the glucose-regulated proteins	Glucose deprivation	Deoxyglucose, tunicamycin
	Membrane permeants	Calcium ionophores

tence of heat shock proteins and thermotolerance. Agents that induce tolerance also induce the heat shock proteins, and some tissue culture cell lines, selected for resistance to heat shock damage, constitutively synthesize high levels of certain heat shock proteins such as HSP70. Furthermore, tolerance cannot be induced during stages of tissue development where heat shock protein induction is blocked, nor can tolerance occur when cells are stressed by growth in the presence of analogues of amino acids. In the latter case, the analogues interfere with the formation of active forms of the heat shock proteins. There are, however, examples where thermotolerance is achieved in the absence of heat shock protein synthesis and other cases where thermotolerance does not occur despite high levels of heat shock proteins. Thus, although heat shock proteins may be necessary for thermotolerance, they may not be sufficient. [*See* Thermotolerance in Mammalian Development.]

The phenomenon of stress tolerance is currently being tested as a potential mechanism for preserving cell viability prior to several types of medical operations, such as organ transplants and cardiovascular surgery.

VII. HEAT SHOCK PROTEINS AND THE IMMUNE RESPONSE

The immune system of higher vertebrates consists of a complex, interactive set of cells whose activities lead to secretion of circulating proteins (the immunoglobulins) and to activation of specific types of cells (the T lymphocytes), which together protect the organism from a wide variety of potentially toxic materials and from microorganisms whose growth can cause severe damage and death. Among the latter are various parasitic and nonparasitic protozoa, fungi, bacteria, and viruses whose life cycles involve growth and development at widely different temperatures. When many of these invade the human body (e.g., by transmission from insects), the microorganism experiences an immediate heat shock and responds by producing heat shock proteins. The immune response to infection by many microorganisms elicits immunoglobulins and T lymphocytes that recognize heat shock proteins as the major foreign antigens. As a result, the organisms are destroyed and the infection is limited. Because the structure of heat shock proteins is so widely conserved, a broad cross-protection to many different types of invasive microorganisms could occur early in the life of the individual. These same proteins exist in normal human cells; however, they are present at low levels and confined inside a cell and, thus, are hidden from the immune recognition system. In abnormal, stressed, or damaged cells, there are higher levels of these proteins, and some may be present at the cell surface. This situation would lead to immune-mediated destruction of these cells—a phenomenon referred to as immune surveillance. In fact, a major mouse tumor antigen is identical to HSP86, a member of the HSP90 family of heat shock proteins. Immune recognition of human heat shock proteins that are induced by a localized inflammation or a systemic febrile response could also lead to autoimmune diseases. In fact, individuals suffering from the latter possess antibodies and T lymphocytes recognizing heat shock proteins. [*See* Immune Surveillance, Immune System.]

BIBLIOGRAPHY

Hendrick, J. P., and Hartl, F.-U. (1993). Molecular chaperone functions of heat-shock proteins. *Annu. Rev. Biochem.* **62,** 349–384.

Hershko, A., and Ciechanover, A. (1992). The ubiquitin system for protein degradation. *Annu. Rev. Biochem.* **61,** 761–807.

Kaufmann, S. H. E. (1991). Heat shock proteins and the immune response. *Immunol. Today* **11,** 129–136.

Morimoto, R. I. (1993). Cells in stress: Transcriptional activation of heat shock genes. *Science* **259,** 1409–1410.

Morimoto, R. I., Tissières, A., and Georgopoulos, C. (1992). "Stress Proteins in Biology and Medicine." Cold Spring Harbor Laboratory Press, Cold Spring Harbor, NY.

Morimoto, R. I., Tissières, A., and Georgopoulos, C. (1994). "The Biology of Heat Shock Proteins and Molecular Chaperones." Cold Spring Harbor Laboratory Press, Cold Spring Harbor, NY.

Nover, L. (1991). "Heat Shock Response." CRC Press, Boca Raton, FL.

Parsell, D. A., and Lindquist, S. (1993). The function of heat-shock proteins in stress tolerance: Degradation and reactivation of damaged proteins. *Annu. Rev. Genet.* **27,** 437–496.

Schlesinger, M. J. (1994). How the cell copes with stress and the function of heat shock proteins. *Pediatr. Res.* **36,** 1–6.

Schlesinger, M. J., Ashburner, M., and Tissières, A. (1982). "Heat Shock: From Bacteria to Man." Cold Spring Harbor Laboratory Press, Cold Spring Harbor, NY.

Welch, W. J. (1993). How cells respond to stress. *Sci. Am.* **268,** 56–81.

HeLa Cells

TCHAW-REN CHEN
The American Type Culture Collection

GLOSSARY

Cell line Cells that grow serially in a sterile glass or plastic container (*in vitro*). It can be a strain from normal cells with finite life span, a cell line derived subsequently from a cell strain that undergoes transformation, or the newly derived cell population from transformed and/or malignant cells with infinite life span *in vitro*

Cryopreservation Maintaining the viability of cells by storing at low temperature. Usually liquid nitrogen is used for the storage purpose

DNA fingerprint profile Pattern profile of polymorphic DNA restriction fragments revealed by using specific probes targeting at hypervariable sequences in a cell genome. Each profile is unique to each individual

Multidrug resistance Simultaneous resistance to multiple drugs that are not necessarily related structurally and functionally to the selecting drug

Phenotype Observable properties of an organism resulting from interactions between genotype and environment

Selection medium Specially designed medium to selectively grow a particular cell type, such as somatic hybrid cells and mutant cells, while eliminating both parental cells simultaneously

Somatic cell hybrids Newly formed cells from fusions of somatic cells. The donor cells for fusion may be from either the same or different species

Universal donor Clonal mutant cell population that possesses two or more selectable markers for selection. These cells are used to select new hybrid cells by fusing with the other donor cells in which no selective mutant markers are required for selection in the specific selection medium

THE HeLa CELL WAS ISOLATED FROM THE CERVIcal carcinoma of a black female, Henrietta Lacks (HeLa), in 1951. Initially, the biopsy was diagnosed as epidermoid carcinoma. However, slides from the same biopsy were recently reexamined and found to be adenocarcinoma. There are three HeLa cell lines currently deposited at the American Type Culture Collection (ATCC): the original HeLa (ATCC CCL 2) and clonal isolates HeLa 229 (CCL 2.1) and HeLa S3 (CCL 2.2).

I. INTRODUCTION

HeLa cells were the first human cell line established *in vitro,* and are probably the most extensively studied human cell line of the last three decades. These studies relate to cell and molecular biology, virology, oncology, somatic cell genetics, and interactions with microbial organisms. Since they are well adapted to *in vitro* growth, intrusion by HeLa cells into an exotic culture often results in total replacement of the initial cell population by HeLa cells. This contamination problem became so serious that, for a period, all established cell lines were suspected to be or to contain HeLa cells. Strict quality control and characterization were required to ensure the identity of cell lines. These requirements, in turn, stimulated research for a better means of cell line quality control and characterization, identified many other contaminated cell lines, and ultimately reduced cases of contamination with HeLa and other cell lines. Although cell line contamination has not been totally alleviated, incidences using misidentified cell lines have declined dramatically.

II. HISTORY OF HeLa CELL LINES

On February 1, 1951, a 31-year-old black woman suffering from intermenstrual spotting was admitted as an outpatient to the gynecology department of Johns Hopkins University Hospital. The biopsy obtained was initially identified as carcinoma. On February 8, 1951, George Gey began growing the cells *in vitro* and the established cell line was named HeLa after the patient. Recent reexamination of the original slide from the same biopsy, however, identified the specimen to be an unusual type of a very aggressive adenocarcinoma of the cervix. Although the patient's malignancy was still in an early clinical stage and was promptly treated with radium therapy, she died in August, 1951.

For nearly 1 year, HeLa cells were maintained in a plasma clot culture in roller tubes. During 1952, these cells were grown serially as monolayer cultures in a medium composed of chicken embryonic extract (2% or 5%) and balanced salt solution (Hanks' 48% or 45%) supplemented with either human placental serum, human adult serum, or human ascitic fluid (50%). Currently, HeLa cells are serially propagated in a standard culture medium supplemented with fetal or newborn calf serum. They are also successfully grown in a chemically defined medium without serum content (serum-free medium). Cells are cryopreserved in liquid nitrogen and can be reconstituted into live cultures for repeated use. The newly thawed cultures maintain the same cell properties as held previous to cryopreservation.

III. CHARACTERISTICS OF HeLa CELL LINES

Of the three HeLa cell lines in the ATCC repository, the HeLa designated as CCL 2 exhibits characteristics closest to the original HeLa reported in earlier studies. The characteristic differences between HeLa and two isolates are shown in Table I.

Genetic characteristics of HeLa cells have been studied extensively. These cells express M, N, S, s, Tj^a, and HLA (A28, A3, BW 35) antigens and have the enzyme phenotypes listed in Table II. The fast-moving type A variant of the enzyme, glucose-6-phosphate dehydrogenase (G6PD, E.C. 1.1.1.49), occurs almost exclusively in the negroid population. This G6PD-A variant is rarely found in Caucasians, with the exception of inhabitants of Mediterranean areas.

TABLE I
Some Characterizations of HeLa Cell Lines[a]

Characteristic	HeLa (CCL 2)	HeLa 229	HeLa S3
Number of serial subcultures from tissue of origin	Unknown: 90–102 from culture received by W. F. Scherer, 1952	Unknown: 78–88 from culture sent to W. F. Scherer in 1952	Unknown
Morphology	Epithelial-like	Epithelial-like	Epithelial-like
Virus susceptibility	Susceptible to poliovirus type 1 and adenovirus type 3	Susceptible to adenovirus type 3 and vesicular stomatitis (Indiana strain) virus; this strain is 2–3 logs less sensitive to poliovirus type 1–3 than HeLa CCL 2	Susceptible to poliovirus type 1 and adenovirus and vesicular stomatitis virus
Chromosome number	Mode 82 Range 78–86		Mode 68 Range 51–74
HeLa marker chromosomes[b]	1 copy each of t(1p3q) and t(3p5q), 3–5 copies of i(5p), 1–2 copies of der(19)		1 copy each for all four markers described for HeLa in this table
Tumorigenicity	In nude mouse		In cheek pouch of hamster
Submitted to ATCC by:	W. F. Scherer	J. T. Syverton	T. T. Puck

[a]From American Type Culture Collection (ATCC).
[b]After Chen (1988). See Table III for detailed description of these marker chromosomes.

TABLE II
Enzyme Phenotypes in HeLa Cells

Enzyme	Phenotype	Enzyme code (E.C.)
Adenylate kinase	AK(1-1)	2.7.4.3
Adenosine deaminase	ADA(1-1)	3.5.4.4
Acid phosphatase	AcP(AB)	3.1.3.2
Phosphoglucomutase	PGM(1-1)	2.7.5.1
6-Phosphogluconate dehydrogenase	6PGD(A)	1.1.1.44
Glucose-6-phosphate dehydrogenase	G6PD(A)	1.1.1.49
Esterase D	ESD(1)	3.1.1.1
Peptidase D	PEP-D(1)	3.4.13.9

By contrast, Caucasians possess the slow-moving type B variant (G6PD-B). Therefore, G6PD-A is commonly used as one of the best genetic markers to identify HeLa and other cell lines of negroid origin. Likewise, G6PD-B is used as an equally distinct marker for Caucasoid-derived cell lines.

HeLa cells have the modal number of 82 chromosomes per cell. By special staining methods, the banding pattern of a chromosome can be delineated. These bands are the manifestation of longitudinal differentiation of chromosome arms and display as crossbars vertical to the longitudinal stretch of arm(s). The band pattern is unique and consistent to each chromosome, and is useful for identifying individual chromosomes and chromosome arms. In a HeLa cell, there are about 20 structurally altered (marker) chromosomes. Among these are t(3p5p), der(19)t(13;19) (q14;p13), and i(5p), which are invariably found in HeLa cells and other cell lines confirmed to be HeLa contaminants (Table III). Thus, these marker chromosomes are generally used for identifying cells in the HeLa cell line family. Since HeLa was initiated from a female patient, the absence of Y chromosome can also be used to identify the HeLa cell line family.

IV. CELL LINE CONTAMINATION

S. M. Gartler first suspected that many human cell lines established later were HeLa cells rather than new cell lines as claimed by the originators. This is because many cell lines derived from Caucasians exhibited the G6PD-A instead of the G6PD-B phenotype. Later the 1–1 pattern of phosphoglucomutase (PGM, E.C. 2.7.5.1) was identified for HeLa cells, thus giving an additional genetic marker for cell identification. Other enzyme phenotypes found in HeLa cells are listed in Table II.

Gartler's suspicion was confirmed in later studies in which other HeLa-specific characteristics were examined. Hitherto, more than 100 cell lines were identified or suspected to be HeLa contaminants (Table IV). However, as became clear later, enzyme patterns alone were not sufficient to claim unambiguously that a culture was HeLa cells.

Comparison of DNA fingerprint profiles between individuals is now available and is a powerful method for unbiased identification of cell lines. Using Jeffrey's multilocus probes, the probability that the same fingerprint profile will occur in two individuals is estimated at 2.9×10^{-17}, which is unlikely to occur in the human population on earth. The DNA fingerprint profile for HeLa cells is well defined by many laboratories, including the ATCC. Studies conducted on cell lines believed to be HeLa contaminants all revealed the same DNA fingerprint profile, as expected. However, at present, DNA fingerprint profiling still cannot routinely distinguish the coexisting clonal populations found in the same individual and, likewise, sublines in the HeLa cell line family. [*See* DNA Markers as Diagnostic Tools.]

In this respect, cytogenetics can be more informative if coexisting clones exhibit chromosome changes. The three ATCC HeLa cell lines and other HeLa contaminant derivatives, such as the widely used KB, D79, and HEp-2, can be identified clearly by cytogenetics. Therefore, one must be aware that biochemical or molecular approaches cannot totally replace cyto-

TABLE III
HeLa Chromosome Markers[a]

Designation	Description
t(1p3q)[b]	Translocation of short arm (p) of chromosome No. 1 (N1) and long arm (q) of N3 at centromere
t(3p5q)	Translocation of N3 p arm and N5 q arm at centromere
i(5p)	Translocation of two N5 p arms to form isochromosome
der(19)t(13;19) (q14;p13)	Translocation of distal N13 q arm at q14 position to distal N19 p arm at p13 position

[a]From Chen (1988).
[b]This marker chromosome is absent in some HeLa cells.

TABLE IV

Cell Lines with Characteristics Unique to HeLa Cells[a]

Designation	Source
HeLa (adenocarcinoma cervix)	ATCC
HeLa (=CCL 2)	ATCC via A. Deitch A. Mukerjee V. Klement from Flow Labs., Inc. G. Gey Unlisted Four individuals, unlisted Grand Island Biological Co. N. Differante
HeLa 229 (=CCL 2.1)	ATCC
HeLa S3	G. Nette from E. Robbins L. Levintow
HeLa S3g	M. Griffin via G. Melnykovych
HeLa S3k	K. Kajievara via G. Melnykovych
KB (carcinoma, oral)	Unlisted
KB (=CCL 17)	ATCC S. Mak V. Klement from MBA H. Sussman E. Priori Commercial, unlisted
H.Ep.-2 (carcinoma, larynx)	Unlisted
H.Ep.-2 (=CCL 23)	ATCC Individual, unlisted P. Dent V. Klement from MBA M. Webber K. McCormick
H.Ep.-2 (clone)	K. V. Ilyin
AV3 (amnion)	Unlisted
AV3 (=CCL 21)	ATCC
AV3 (103)	I. Keydar from ATCC
AV3 (F-49-1)	P. Peebles from ATCC
L132 (=CCL 5) (lung)	ATCC
L132 (G-38-7)	P. Peebles from ATCC
Chang liver (liver)	Unlisted
Chang liver (=CCL 13)	ATCC R. Chang Individual, unlisted
HBT3 (carcinoma, breast)	P. Arnstein from R. Bassin
HBT-E (16c, clone of HBT-3)	R. Bassin
HBT-39b (carcinoma, breast) (clone 6)	P. Arnstein from E. Plata
HEK (kidney)	Commercial, unlisted J. Rhim from C. Pfizer, Inc. C. Pfizer, Inc.
HEK/HRV (HEK, virus transformed)	S. Aaronson
MA160 (prostate)	The originators
MA160	P. Price, MBA M. Vincent, MBA
Prostate (=MA160)	Unlisted
Detroit 6 (=CCL 3)	Child Research Center of Michigan or ATCC
Minnesota EE (esophageal epithelium)	Individual, unlisted
Minnesota EE (=CCL 4)	ATCC
Intestine 407 (jejunum, ileum) (=CCL 6)	ATCC
Intestine 407	Commercial, unlisted
Intestine 407 (=HEI = CCL 6)	G. Spahm from ATCC
SA4 (TxS-HuSa$_1$) (liposarcoma)	C. Pfizer, Inc.
SA4	D. Morton
RT4 (carcinoma, bladder)	J. Leighton via N. Abaza
Detroit 30A (carcinoma, ascitic fluid)	W. D. Peterson, Jr.
Detroit 98 (=CCL 18) (sternal marrow)	ATCC
Detroit 98s (=CCL 18.1)	ATCC
Detroit 98/AG (CCL 18.2)	ATCC
Detroit 98AT-2 (=CCL 18.3)	ATCC
Detroit 98/AHR (=CCL 18.4)	ATCC
FL (=CCL 62) (amnion)	ATCC
CaOV (carcinoma, ovary)	N. P. Mazurenko
J96 (leukemic blood)	T. A. Bektemiov
JIII (monocytic leukemia) (=CCL 24)	Commercial, unlisted Unlisted ATCC
T-9 (transformed normal diploid)	O. G. Andzaparidze
DAPT (astrocytoma, piloid)	A. O. Bykovsky
AO (amnion)	A. O. Bykovsky
KP-P$_1$ (carcinoma, prostate)	P. Lee via M. Glovsky
ElCo (carcinoma, breast)	R. Patillo
HCE (carcinoma, cervix)	D. Brown
CMP (adenocarcinoma, rectum)	Unlisted
CMPII C2	D. Rounds via J. Kim
JHT (placenta)	J. Cho via J. W.-Peng
OE (endometrium)	The originators P. Di Saia via L. Milewich
SH-2 (carcinoma, breast)	The originators G. Seman via R. Miller
SH-3 (carcinoma, breast)	The originators G. Seman via R. Miller
ESP$_1$ (Burkitt lymphoma, American)	P. Price from E. Priori E. Priori
EB33 (carcinoma, prostate)	F. Schroeder
D18T (synovial cell)	D. A. Peterson
M10T (synovial cell)	D. A. Peterson
Detroit 6 (sternal marrow)	Unlisted Commercial, unlisted
Detroit 6 (=CCL 3)	ATCC
Detroit 6 (clone 12) (=CCL 3.1)	ATCC
NCTC2544 (=CCL 19) (skin) (epithelium)	ATCC
NCTC3075 (=CCL 19.1)	ATCC
WISH (amnion)	Individual, unlisted
WISH (=CCL 25)	ATCC
Girardi heart (heart) (=CCL 27)	ATCC
TuWi (=CCL 31)	ATCC
Wong-Kilbourne (conjunctiva) (=CCL 20.2)	ATCC

[a]From Nelson-Rees *et al.* (1981).

genetics for the identification of interspecific, intraspecific, and intraindividual clonal cell populations.

W. A. Nelson-Rees proposed four tangible characteristics to identify HeLa cells: (1) G6PD-A, (2) PGM 1-3 type, (3) absence of Y chromosome, and (4) the three HeLa-specific markers described earlier. Over 100 cell lines established by different laboratories were shown to have these four HeLa characteristics and thus confirmed to be in the HeLa cell line family (see Table IV). Some researchers, however, are less than candid in accepting this fact. They have pointed out that some obvious chromosome differences exist among HeLa-confirmed cell lines. Karyotypes of aneuploid cell lines change slightly among only a certain group of chromosomes in long-term subcultures. By contrast, chromosome changes observed in some of those cell lines of HeLa contaminants exceed that expectation.

However, we have not as yet found another cell line that simultaneously possesses the three HeLa chromosome markers in cells. Thus, the probability of obtaining these chromosome markers simultaneously from two individuals may be extremely low. Other chromosome distinctions, such as low number of copies (0–1) for normal chromosome 3 (N3) and high number of copies (3–4) for N17, add to the repertoire of the HeLa-unique chromosome pattern. This evidence, together with enzyme data and other cellular properties (e.g., epithelioid cell morphology), demonstrates that characteristics of HeLa cells are indeed well defined. Hence, cell lines that simultaneously bear all those genetic markers should be classified as sublines to the HeLa cell line family.

Some sublines in the HeLa family may have specific properties. This is feasible since most of these sublines were derived from different cultures with the HeLa cells being the only common denominator. In this respect, new cell populations may acquire portions of the host genetic materials, rendering a unique property specific only to the cell populations from each specific combination of cell types. These cells are considered as sublines in the HeLa family, but not as an independent cell line derived from a separate individual.

V. GROWTH OF HeLa CELLS

Many different media have been used for culturing HeLa cells. Among those frequently used are Eagle's minimal essential medium (EMEM) or Dulbecco's modified EMEM (DME) supplemented with 10% fetal or newborn calf serum. Cultures are generally incubated at 37°C under 5% CO_2–95% air conditions. If necessary, antibiotics are added to the medium to control microbial contamination. The commonly used antibiotics include penicillin (100 units/ml) and streptomycin (100 μg/ml) in combination, gentamicin (50 μg/ml), or chlortetracycline (50 μg/ml).

HeLa cultures grown strictly in a chemically defined medium without serum supplementation are also available (CCL 2.3). For serum-free media, use Ham's F12, RPMI, or similar media supplemented with insulin, transferrin, hydrocortisone (aldosterone), fibroblast growth factor, epidermal growth factor, and glutamine. These elements are essential for normal cell growth; without them, cell growth may be seriously affected. The growth rate and clonal growth of HeLa cells in a serum-free medium are comparable to those in serum-supplemented standard medium.

VI. STORAGE OF HeLa CELLS

HeLa cells, like other cells grown *in vitro,* can be frozen to store indefinitely. They can also be thawed to grow *in vitro* without affecting their normal growth. For cryopreservation, 1 million or more cells are placed in 1 ml cryopreservation medium consisting of 10% dimethyl sulfoxide (DMSO) in a complete medium. DMSO may be replaced with the same amount of glycerol.

Either a 2-ml-capacity clinical-grade glass ampoule or polypropylene tube with secured screw-top closure may be used as containers for cryopreservation. After cells are added, the container is either sealed (ampoule) or tightly secured (polypropylene tube) and placed either directly in a −70°C freezer overnight or in a programmed freezer for stepwise temperature decreases. It is then stored in deep-freeze conditions. For long-term storage, containers are often immersed in the liquid phase. Otherwise, the containers may be placed in the vapor phase of the liquid nitrogen tank.

VII. HeLa CELL VARIANTS

Many variant cell lines have been isolated from HeLa cells (Table V). For example, HeLa S3 was selected to grow in suspension in a medium lacking divalent cations such as Ca^{2+} or Ca^{2+}/Mg^{2+}. A mutant clone exhibiting both ouabain-resistant and hypoxanthine phosphoribosyl transferase (HPRT)-deficient mutants has been isolated. This clone is known as the universal

TABLE V
Variant Cell Lines Derived from HeLa and KB Cells

Character	Cell lines	Derivation
Nutritional variants		
Glutamine requiring	HeLa I-11a	HeLa I-11
Glutamine independent	HeLa I-11	HeLa S3-1
Carbohydrate	Ribose variants	
	Xylose variants	
	Lactate variants	
Grow in protein- and lipid-free synthetic media	HeLa-P3	HeLa
Quantitatively different in their requirement of serum	HeLa S1	HeLa
Drug resistant		
Actinomycin D	HeLa-R	HeLa
1-β-D-Arabinofuranosylcytosine	KB/araC	KB
8-Azaguanine	S3AG1	HeLa S3
Bromodeoxyuridine (Brd Urd)	HeLa BU-10	HeLa S3
	HeLa BU-15	HeLa S3
	HeLa BU-25	HeLa S3
	HeLa BU-50	HeLa S3
	HeLa BU-100	HeLa S3
Chloramphenicol	296-1	HeLa S3
		HeLa S3
Colchicine, multidrug resistant	KB-8-5	KB
	KB-C1- -C4	KB
Adriamycin, multidrug resistant	KB-A1	KB
Vinblastine, multidrug resistant	KB-V1	KB
Erythromycin	ERY2301	HeLa
Ethylmethane sulfonate	HeLa A6	HeLa S3
Ouabain	D98-OR	D98/AH-2
6-Thioguanine	H23	HeLa
	D98/AH-2	HeLa
Toxin resistant		
Diphtheria toxin	KB-R2	KB
	KB-R2A	KB
Epidermal growth factor–*Pseudomonas* toxin	ET	KB
Altered virus susceptibility		
Poliovirus sensitive	HeLa I-3	HeLa S3
Poliovirus resistant	HeLa S3-1C	HeLa S3
Poliovirus resistant	"R"	HeLa S3
Others		
Alkaline phosphatase lacking	A clonal line of giant HeLa cells	HeLa
Ultraviolet sensitive	S-1M	HeLa S3
	S-2M	HeLa S3

donor and is useful in isolating intraspecific somatic cell hybrids.

For isolation, interline (both interspecific and intraspecific) cell hybrids acquiring wild type from complementation are selectively grown in specially designed selection media. The most widely used selection media are those supplemented with hypoxanthine-aminopterin-thymidine (HAT medium) or alanosine-adenine (AA medium). These selection systems must rely on another selection condition, for example, differential growth rates and/or growth substrata of donor cells, or clonal isolation, to separate hybrid clones from other surrounding cells. Otherwise, donor cells are required to have another mutant property for selection, and selection of hybrid clones depends totally on the successful production of the wild type that corrects adverse mutants by complementation. In instances of the universal donor such as the ouabain^{R}/

HPRT⁻ HeLa clone, the donor mutant HeLa is killed by growth in ouabain-supplemented HAT, and the other donor is killed owing to the toxicity to ouabain. Thus, only hybrid cells with ouabain-resistant and HPRT enzyme properties survive.

The D98AH2 clone, which is HPRT-negative and tumorigenic in nude mice, is often used for studying tumorigenicity by hybridizing with nonmalignant normal cells. This study supports the idea that malignancy is recessive to the normal wild type. Hybrid cells in this respect exhibit nonmalignant phenotype and obviously retarded growth. Cytogenetic studies on these cell hybrids have further suggested that the presence of N11 is essential for the nonmalignant phenotype. The N11 may have the tumor-suppressing gene that suppresses malignant expression of D98AH2 cells.

Multidrug-resistant sublines KB-8-5, KB-C1, C1.5, C2, C2.5, C3, C3.5, and C4 were isolated from the KB, which is another suspected subline in the HeLa family. They have been used to elucidate the molecular basis for multidrug resistance, especially for the function of P-glycoprotein as an energy-dependent efflux pump.

VIII. CONCLUSION

The HeLa cell line was derived from the cervical adenocarcinoma of a black woman and is the first cell line established to grow *in vitro* from human material. HeLa and other cell lines derived from contamination with HeLa are grouped into the HeLa cell line family, which is undoubtedly the most widely used and studied human cell line. The cell line possesses readily recognized signatures, such as the presence of three HeLa-specific chromosome markers, lack of Y chromosome, G6PD type A, and PGM 1-3-type enzyme phenotypes. Other markers, including specific DNA fingerprint profiles, are available to clearly identify this group of cells. Many mutant and variant clones have been isolated and used for studying a variety of subjects. HeLa cells are infamous for their broad involvement in the cell line cross-contamination. However, this notoriety in turn has led to concerted scientific studies, including cell line characterization and quality control, to ensure the authenticity of a cell line. The contribution of HeLa cells to the scientific community is, indeed, unsurpassed by any other cell lines as a research material.

BIBLIOGRAPHY

Akiyama, S. (1987). HeLa cell lines. *In* "Molecular Genetics of Mammalian Cells" (M. M. Gottesman, ed.), "Methods in Enzymology," Vol. 151, p. 38. Academic Press, San Diego.

Akiyama, S., Fojo, A., Hanover, J. A., Pastan, I., and Gottesman, M. M. (1985). Isolation and genetic characterization of human KB cell lines resistant to multiple drugs. *Somat. Cell Genet.* **11**, 117.

Chen, C.-J., Chin, J., Ueda, K., Clark, D., Pastan, I., Gottesman, M. M., and Roninson, I. (1986). Internal duplication and homology with bacterial transport proteins in the mdr 1 (P-glycoprotein) gene from multidrug-resistant human cell. *Cell* **47**, 381.

Chen, T. R. (1988). Re-evaluation of HeLa, HeLa S3, and Hep-2 karyotypes. *Cytogenet. Cell Genet.* **48**, 19.

Gartler, S. M. (1967). Genetic markers as tracers in cell culture. *J. Natl. Cancer Inst.* **26**, 167.

Gey, G. O., Coffman, W. D., and Kubicek, M. T. (1952). Tissue culture studies of the proliferative capacity of cervical carcinoma and normal epithelium. *Cancer Res.* **12**, 264.

Kaelbling, M., and Klinger, H. P. (1986). Suppression of tumorigenicity in somatic cell hybrids. *Cytogenet. Cell Genet.* **41**, 65.

Nelson-Rees, W. A., Daniels, D. W., and Flandermeyer, R. R. (1981). Cross-contamination of cells in culture. *Science* **212**, 446.

Srivatsan, E. S., Benedict, W. F., and Stanbridge, E. J. (1986). Implication of chromosome 11 in the suppression of neoplastic expression human cell hybrids. *Cancer Res.* **46**, 6174.

Taylor-Robinson, D., Sarathehandra, P., and Furr, P. M. (1993). *Mycoplasma fermentans*—HeLa cell interactions. *Clin. Infect. Dis.* **17** (Suppl. 1), S302.

Helminth Infections

CELIA HOLLAND
Trinity College, Dublin

GLOSSARY

Anemia Reduction in the hemoglobin concentration, the hematocrit, or the number of red blood cells to a level below that which is normal for a given individual. Anemia has been defined as the hemoglobin level below −2 S.D. of the mean value in the population. WHO recommends characterizing the hematologic status of populations by determining the frequency distribution of hemoglobin or hematocrit values

Anorexia Loss of appetite

Anthropometric measurements Assessments of protein–energy malnutrition that include weight, height, weight-for-height, mid/upper arm circumference, and skinfold thickness. Assessments are made relative to a standard or reference value from a well-nourished population of the same age. The reference values most commonly used are the Harvard standards or the more recent NCHS–WHO reference values. The NCHS references have been recommended by WHO for use in developing-country surveys

Intensity of infection Number of individuals (determined directly or indirectly) of a particular helminth species in each infected host in a sample

Malabsorption Failure to assimilate essential ingredients from the diet leading to malnutrition. In this context, failure of the intestine to absorb the products of digestion as a consequence of changes in the intestine

PEM Protein–energy malnutrition. Clinical signs and anthropometric measurements are used to diagnose serious protein–energy malnutrition

Steattorhea Steattorhea is characterized by loose, smelly, fatty stools and is a symptom of fat maldigestion, which is accompanied by malabsorption of more than just fat. It is generally defined as fecal fat excretion greater than 6 g per day

D-Xylose absorption This is used specifically as a measure of carbohydrate malabsorption but is often used generally as an indicator of malabsorption. A measurement of urinary xylose excretion is usually made 5 hr after a 5-g oral dose of D-xylose. This is used because it is rapidly absorbed by the small intestine and excreted unmetabolized in the urine

THE IMPACT OF FOUR MAJOR SOIL-TRANSMITTED helminth nematode infections on human nutrition is considerable. The soil-transmitted nematodes are defined as those parasites in which part of life cycle development occurs in the soil (i.e., the development of eggs or larvae prior to their ingestion by or penetration of the definitive host). As far as humans are concerned, the most prevalent species are the human roundworm. *Ascaris lumbricoides,* the human hookworm, *Ancylostoma duodenale* and *Necator americanus,* the human whipworm, *Trichuris trichiura,* and *Strongyloides stercoralis.* All four species are nematode parasites that inhabit the human intestine as adult worms and produce eggs or larvae that pass out in the feces. These species are discussed both because of their high prevalence worldwide and the fact that their distributions frequently overlap with those of malnutrition. In addition, they have tended to be neglected and somewhat underestimated in public health terms.

I. BACKGROUND

In many parts of the world, particularly the subtropics and tropics, human beings suffer the double insult of parasitic infection and malnutrition. Recent global estimates indicate that the soil-transmitted helminths are some of the most common infections in the world (Table I). Nutrient deficiency diseases are equally prevalent (Table II), and intestinal nematode infections have been implicated in the aggravation of protein–energy malnutrition, iron deficiency anemia, and to a lesser extent vitamin A deficiency. Actively growing children who are often heavily infected with several parasite species are particularly at risk of developing malnutrition. [*See* Malnutrition.]

Both gastrointestinal nematodes and malnutrition flourish in an environment dominated by poverty, insufficient sanitation, crowded living conditions, and inadequate health care facilities. Morbidity associated with such parasitic infections is likely to be underestimated, and little systematic study of the symptomatology associated with them has been undertaken. Certainly the fact that they inhabit the intestinal tract makes it highly likely that they will influence nutritional status, but there is a need to establish the importance of their role and, therefore, a realistic picture of their public health significance.

TABLE I

Prevalence of Major Infections in Africa, Asia, and Latin America, 1977–1978[a,b]

Infection	Rate of incidence in millions/year
Diarrheas	3000–5000
Tuberculosis	1000
Ascariasis[c]	800–1000
Hookworm infections	700– 900
Malnutrition	500– 800
Malaria	800
Trichuriasis	500
Amebiasis	400
Filariasis	250
Giardiasis	200
Schistosomiasis	200

[a]The infections are arranged in order of their frequency. Adapted from Z. S. Pawlowski (1984). Implications of parasite–nutrition interactions from a world perspective. *Fed. Proc.* **43**, 256–260. Reproduced with permission.

[b]Data on strongyloidiasis is not available from this source. *Strongyloides stercoralis* has been estimated to infect 56 million people worldwide but this is likely to be an underestimate.

[c]Major soil-transmitted helminths are shown in boldface type.

TABLE II

Gross Estimates of the Total Number of Persons Affected by Three Dietary Deficiency Diseases Worldwide[a]

Deficiency	Prevalence[b]	Age (years)
Protein and energy[c]	500 million	0–6
Iron	350 million	18–45 (women)
Vitamin A[d]	6 million	All

[a]Adapted from M. C. Latham (1984). Strategies for the control of malnutrition and the influence of the nutritional sciences. *Food and Nutrition* **10**, 5–31. Reproduced with permission. Also adapted from Stephenson (1987).

[b]Estimates are gross and do not express the significant variations within and between countries.

[c]Those suffering from stunted growth defined by weight below the 2.5 m percentile of the WHO growth standards.

[d]This is an underestimate because it includes cases from Southeast Asia only.

The intestinal nematode infections are transmitted as a result of inadequate disposal of human feces, thereby disseminating the infective eggs or larvae into the environment. Infection by ingestion of infective eggs (*Ascaris lumbricoides, Trichuris trichiura*) or penetration by infective larvae (hookworm species, *Strongyloides stercoralis*) can occur. Direct ingestion of the larvae, in the case of *Ancylostoma duodenale*, can result in complete development. The development of *Necator americanus* appears to require percutaneous entry with a phase of larval development in the lungs. The eggs or larvae require a certain period of development in the environment under appropriate conditions in order to become infective. All four species can produce large numbers of infective stages, particularly *A. lumbricoides*, which has been estimated to produce an average of 200,000 eggs per day per female worm. On ingestion or penetration, the larval stages undergo a form of migration and eventually end up as adult worms in the small intestine. Their location varies, with *A. lumbricoides* inhabiting the jejunum, *T. trichiura* the cecum, hookworm species the upper small intestine, and *S. stercoralis* the duodenum and the upper jejunum. In heavy infections the parasites may extend their location.

The intensity of a helminth infection is a particularly important parameter for the accurate assessment of the dynamics of infection and the morbidity and mortality associated with that infection. Intensity of infection has been found to have a strong positive correlation with symptomatology and nutritional morbidity. This applies to the eggs and worms of *Ascaris, Trichuris,* and hookworm and the larvae of *Strongyloides*. It is therefore important to know the intensity of infection for each individual in a nutrition–parasite study.

In terms of population dynamics, the distribution of worm intensity is characteristically aggregated or overdispersed within a host community or population. This means that most infected hosts harbor few or no parasites while a small proportion harbor heavy parasite burdens. This distribution has been described for *Ascaris, Trichuris,* and hookworm from a variety of countries worldwide. Further evidence has shown that individuals exhibit a predisposition to heavy or light infection with these three parasite species, although the precise explanations or mechanisms for this predisposition are still unknown. One factor may be host nutritional status, indicating not only that heavy infection can contribute to nutritional morbidity but that poor nutritional status may enhance the risk of harboring a heavy infection.

Strongyloides must be considered differently from the other three species. The process of autoinfection means that the dynamics of infection are not governed purely by the processes of immigration and emigration. Despite this, heavy infections of *Strongyloides* do occur, described as the hyperinfection syndrome, and those individuals who suffer it are predisposed by a number of factors (see Section IV on strongyloidiasis).

The epidemiology of these infections is an important consideration when assessing the relationship between intestinal parasitic infection and malnutrition.

Some attention has been paid to the design of nutrition–parasite studies, and certain recommendations were made by L. S. Stephenson in 1987. Three types of studies have been commonly employed to look at the interaction between host nutrition and parasitism. These are studies in laboratory animals, clinical studies in a hospital setting, and community-based field studies. Community-based field studies, which are often the most useful, are limited by ethical considerations. The choice of population, the study design, and the form of intervention must be evaluated with care. The aggregated distribution of the helminths in a community needs to be taken into account, and selective sampling may be necessary to identify a number of heavily infected individuals. Polyparasitism can also make it difficult to assess the impact of a single parasite species. Host dietary intake is another important consideration, as this will influence the impact of an infection and the ability of a host to respond to losses incurred.

II. HOOKWORM

Adult hookworms live in the upper small intestine, attached to the mucosa by their buccal capsules. Worms feed by drawing a plug of mucosal tissue into their buccal capsules. Blood loss occurs because of the sucking of the worms and from lesions in the mucosa.

The clinical severity of the disease is closely related to the intensity of infection and the condition of the host. Host factors include iron intake and bioavailability, iron resources and needs, and general state of health. The most important nutritional impact of infection is the blood and iron loss as a consequence of the feeding of the adult worms in the intestine. In acute infections, nausea, vomiting, and diarrhea and abdominal pain can occur. In chronic infections, most of the symptoms are those of iron deficiency anemia. In a classic study M. Layrisse and M. Roche demonstrated a highly significant relationship between circulating hemoglobin levels and hookworm egg counts in a rural Venezuelan setting. Hemoglobin levels were significantly lower in subjects passing more than 2000 eggs per gram of feces for women and children and more than 5000 eggs for adult men. Mean hemoglobin level decreased linearly as egg counts increased.

A number of estimates of intestinal blood loss and iron loss of hookworm-infected people are shown in Table III. A person who passes 2000 eggs per gram of feces loses an estimated 1.3 mg of iron for *Necator* and 2.7 mg of iron for *Ancylostoma*. In tropical countries only about 10% of the iron ingested is absorbed. The type as well as the amount of dietary iron consumed has a major influence on the bioavailability and hence the amount of iron a person actually absorbs daily. In most areas of the world where hookworm is prevalent, iron intake is of predominantly vegetable origin. The nonheme iron present in foods of vegetable origin is more poorly absorbed than the heme iron in foods of animal origin. Another source of iron is contaminant and insoluble organic iron, which is relatively un-

TABLE III
Intestinal and Fecal Blood and Iron Losses in Hookworm Infection in Humans[a]

Parameter[b]	*Necator americanus*	*Ancylostoma duodenale*
Intestinal blood loss in ml per worm per day mean	0.03	0.15
Range	0.01–0.04	0.05–0.30
Intestinal blood loss in ml per 2000 epg mean ± SD	4.3 ± 2.02	8.9 ± 2.32
Intestinal iron loss in mg per 2000 epg	2.0	4.2
Fecal iron loss in mg per 2000 epg[c]	1.3	2.7

[a]From Stephenson (1987), pp. 128–160, with permission from Taylor & Francis.

[b]epg = eggs per gram of feces. An infection of 2000 epg, or about 80 worms, is considered a moderate subclinical infection by public health standards.

[c]These values assume a hemoglobin level of 14 g/dl. Fecal iron loss is less than intestinal iron loss because some of the iron lost is reabsorbed and reutilized and is not excreted in the feces. Average fecal iron loss per person per day has been estimated to be on the order of 5–9 mg.

available for absorption but probably accounts for some of the reports of high iron content of meals from some countries where nutritional iron deficiency is endemic.

Studies of the absorption of fat, carbohydrate, vitamin A, vitamin B_{12}, and folic acid have been undertaken in hookworm-infected people. The majority of these investigations have provided little evidence for a malabsorption syndrome associated with hookworm infection and indicated that the structural and functional abnormalities found were more likely to be due to PEM or tropical sprue.

One study that did provide evidence for a link between hookworm infection and malabsorption involved 14 anemic infected patients from Costa Rica. Malabsorption of carbohydrate, fat, and vitamin A was detected in the majority of the cases, and absorption improved after treatment. Biopsies showed abnormal villous architecture, which receded after treatment. Six lightly infected patients and eight people with other parasites showed no evidence of malabsorption. In contrast to these findings, 10 studies from the Indian, African, Asian, and Latin American continents found little or no evidence of malabsorption of fat or carbohydrate in patients with hookworm infection of varying intensity.

Further evidence for the lack of a malabsorption syndrome associated with hookworm infection was provided by a study of 15 Colombian subjects. Patients with severe protein–energy malnutrition but no hookworm infection showed abnormal small bowel absorptive function with flattened villi and a decreased crypt–villous ratio. This returned to normal after protein repletion. Patients with protein–energy malnutrition and anemia secondary to hookworm infection showed similar functional and histological findings that normalized after protein repletion, despite the persistence of their infection and associated anemia.

Patients without protein–energy malnutrition and with severe hookworm infection and anemia failed to show as pronounced biochemical or histological evidence of malabsorption. Findings like these, similar to those described for *Strongyloides*, further emphasize the need for meticulous evaluation of the nutritional status of infected patients.

Hookworm infection has been shown to be associated with reduced food intake and anorexia, although there have been no specific studies undertaken to examine this. Another important, yet ill-studied, aspect of hookworm is its possible association with growth stunting. A significant increase in growth was reported in children treated for hookworm and *S. hematobium*. The degree of growth improvement correlated well with a decrease in hookworm eggs after treatment. Possible causes could include reduced food intake or protein loss as a result of blood loss caused by worms feeding or because of a protein-losing enteropathy in an inflamed intestine. Some studies have shown albumin or protein loss to increase in line with increasing worm burden. It is likely that protein loss does occur but will only cause protein deficiency in severe chronic infections and/or extremely low intakes.

In terms of food intake and its effect on growth, an ingenious "natural experiment" took place involving 26 well-nourished Indian men who played a game on a previously used but recently ploughed defecation field. They manifested intense itchiness of the skin after the game and went on to develop heavy hookworm infection with abdominal pain and vomiting. They lost an average of 7 kg in weight. None of the men had steatorrhea or villous atrophy, and only six of them had abnormal D-xylose values. The authors attributed the men's weight loss to deficient caloric intake due to fear of food precipitating abdominal pain.

Another important aspect of hookworm-induced

iron deficiency anemia is the functional outcome that can affect worker productivity, reproduction, and cognitive performance. A number of studies have demonstrated reduced work capacity among anemic workers and increased work output after iron supplementation. A study among latex tappers in Indonesia demonstrated a correlation between work output and hemoglobin concentration. The work output was significantly less (19%) in anemic tappers than in nonanemic tappers. After treatment, the output of the anemic group increased to that of their nonanemic colleagues.

Severe anemia is associated with increased risk of premature delivery, increased maternal and fetal morbidity, and mortality. Even relatively mild anemia has been found to be associated with premature delivery and a correlation between maternal hemoglobin levels and fetal birth weight has been demonstrated.

There are many indications (but little firm evidence) of a relationship between impaired cognitive and behavioral function and helminth disease, including hookworm infection. Iron deficiency anemia has been found to have deleterious effects on cognitive performance and this outcome of hookworm infection is obviously in need of further study.

III. ASCARIASIS

Ascariasis is a prevalent infection worldwide, particularly in children, who manifest the heaviest infections. The parasite produces many resistant eggs that can survive for long periods of time in the environment.

Most nutritionally significant symptomatology is associated with the presence of adult worms in the small intestine and includes abdominal pain, nausea, diarrhea, anorexia, and nutrient malabsorption. Complications resulting from the aggregation of adult worms in the intestine and the migration of worms to other parts of the body have been described. These can be fatal and are expensive and difficult to treat, particularly in inadequate health care systems. Pathology associated with the larval migration has not received much systematic attention, although cough, asthma, substernal pain, fever, a skin rash, and eosinophilia have been noted.

Extensive studies of *Ascaris suum* in the pig have been undertaken, which can act as a useful laboratory model for *Ascaris lumbricoides* infection in children. Evidence has accumulated that *Ascaris*-infected pigs fed a low-protein diet exhibit reductions in food intake, in weight gain, and in weight gain per amount of food consumed. This reduction can be correlated with worm burden. Decreased nitrogen retention and decreased absorption of lactose and fat have also been described in infected animals. Reduced mucosal lactase activity and histological abnormalities of the jejunal mucosa have been demonstrated. These findings have stimulated an interest in the relationship between nutritional morbidity and ascariasis in growing children who may be malnourished.

A number of clinical studies have shown that *Ascaris* infection is associated with decreased absorption of nitrogen, steattorhea, and abnormal D-xylose absorption. Jejunal biopsies of infected children also showed abnormal villous architecture, including broadening and shortening of the villi, elongation of the crypts, and cellular infiltration of the lamina propria. After treatment there was evidence of reduced fecal nitrogen, and steattorhea and D-xylose malabsorption regressed. Mucosal damage also decreased after treatment.

Lactose maldigestion has also been shown to be associated with *Ascaris* infection. *Ascaris*-infected Panamanian children exhibited a marked decrease in lactose digestion, measured by breath H_2 concentration compared to age- and sex-matched controls. This decrease in lactose digestion was linearly related to worm burden and returned to normal 3 weeks after treatment. Symptoms associated with milk intolerance have been described from infected children and include abdominal pain, diarrhea, and flatulence after drinking milk. Uninfected controls did not manifest these conditions after lactose intake. An assessment of milk intake among *Ascaris*-infected children revealed a significant reduction in lactose consumption compared to controls using the 24-hr recall method. The food weighing method failed to detect this difference. [*See* Lactose Malabsorption and Intolerance.]

Accelerated mouth-to-cecum transit time has also been recorded from infected Panamanian children. This was significantly negatively correlated with worm burden and may be a possible mechanism for decreased nutrient absorption. Evidence has also been provided that ascariasis interferes with vitamin A absorption, and in one study there was a clear improvement in the absorption of vitamin A after deworming.

A number of community-based longitudinal studies have concluded that ascariasis can contribute to growth retardation in preschool children and that periodic deworming can improve growth.

One study based in the Machakos district in Kenya examined children aged 12 to 72 months three times at 3.5-month intervals. Infected and uninfected children did not differ in terms of socioeconomic status or anthropometric measurements at the start of the study. However, there was a decrease in skinfold thickness of *Ascaris*-infected children compared with the controls. Three and a half months after treatment, previously infected children showed significantly higher weight gain and percent expected weight gain and an increase in skinfold thickness compared with the controls. This represented a 33% increase in expected growth rate over the 3.5-month period. Multiple regression analysis revealed that *Ascaris* infection was the most important of 37 possible health, nutritional, and socioeconomic variables in explaining the decrease in skinfold thickness before and the increase in skinfold thickness after treatment.

In Tanzania, improved growth rates were also reported among preschool children who received levamisole at 3-month, intervals over a 1-year period compared to a control group who received a placebo.

Recently, Malaysian children treated for ascariasis every 3 months for a year gained significantly more weight and showed greater skinfold thickness compared with untreated infected children. The difference in weight gain between the treated and untreated children was about 16% of the 1-year increment.

Furthermore, a recent Burmese study demonstrated a significant increment in height for age among children 12, 18, and 24 months after treatment with levamisole compared to a group of children who did not receive treatment.

Several other studies that have examined the relationship between ascariasis and child growth have failed to demonstrate similar findings. In a recent book on the impact of helminth infection on human nutrition, Stephenson reviewed these studies in detail and offered some explanations for the variations in the findings concerning child growth. Aspects of experimental design, including the choice of population, sample size, successful drug treatment, and data analysis, are considered. Successful drug treatment is particularly important, and in a number of studies *Ascaris* had not been completely eradicated from the infected children at the time that growth assessment took place.

Another important factor is the underlying socioeconomic status of the individuals or community under study. In a cross-sectional survey of the relationship between ascariasis and percent weight for age in three Balinese hamlets, the interaction between infection and nutritional status was influenced by other socioeconomic factors. This underlines the importance of evaluating the dietary and socioeconomic variables in a community that may in turn influence the relationship between infection and malnutrition.

IV. STRONGYLOIDIASIS

Strongyloides stercoralis is unusual in that it possesses both free-living and parasitic larval and adult stages. Under certain conditions, not yet fully understood, free-living larval forms can give rise to infective larvae that can penetrate and infect a human host. Unlike the other species of parasites described here, it is the larval form that is used for diagnosis. Female worms, which live in the intestinal mucosa, lay eggs that hatch into larvae in the gut lumen. They are then passed out in the feces. These larvae can be difficult to detect in a routine smear and require the use of a special technique, which is not widely used, and therefore the prevalence of this parasite is likely an underestimation.

Two forms of autoinfection can occur and play an important role in the public health significance of the disease. Some larvae may metamorphose and molt into infective larvae en route down the bowel. These larvae can then directly penetrate the wall of the gut and enter the bloodstream; this is termed *internal autoinfection. External autoinfection* can occur when fecal matter containing infective larvae makes contact with the perianal region and larvae invade the local tissue. These processes are responsible for the perpetration of the infection long after an individual has left an endemic area and can contribute to the severe form of the disease.

Strongyloides infections can be divided into asymptomatic, chronic, and severe. In both chronic and severe forms of the disease, abdominal pain, nausea, vomiting, diarrhea, and weight loss of varying severity occur. These symptoms can lead to a decrease in food intake and an increase in nutrient excretion.

If allowed to persist, strongyloidiasis can cause malnutrition and even death as a consequence of hypovolemic shock due to extensive diarrhea and vomiting. A malabsorption syndrome has been found to be associated with *Strongyloides* infection, including lesions in the duodenum and jejunum. Fibrosis as a consequence of bacterial contamination of ulcers in the small intestine may also contribute to this malabsorption.

Infection with *Strongyloides* can persist for many years as a consequence of autoinfection and occasionally progress to the so-called hyperinfection syndrome or disseminated strongyloidiasis. Acute strongyloidiasis involves a massive buildup of both adult and larval worms in the intestinal mucosa. Parasites can also invade other organs. This condition has a mortality rate of 50–70%. Cause of death includes malnutrition, but bacterial sepsis in individuals immunocompromised by therapy or disease often contributes to the fatal outcome. A number of conditions are thought to predispose individuals to the development of these heavy infections, including malignant neoplastic disease, corticosteroid or immunosuppressive therapy, and malnutrition. Recently disseminated strongyloidiasis has been reported from a number of AIDS-infected patients. It is particularly difficult to establish if malnutrition was present before infection and therefore contributed to it or arose as a consequence of the disease. Depressed cell-mediated immunity, as a consequence of PEM, has been described as a causative agent of the hyperinfection syndrome. But a recent case description of an immunologically competent patient developing disseminated strongyloidiasis indicates that even in uncompromised hosts *Strongyloides* should not be allowed to persist.

Evaluation of the relationship between strongyloidiasis and malnutrition is complicated by the fact that most of the information available comes from case studies of one or more patients, suffering from the hyperinfection syndrome, in a hospital setting. This type of data has obvious limitations, particularly in terms of the evaluation of morbidity at the community level.

A number of studies have focused on the issue of a malabsorption syndrome associated with strongyloidiasis. The findings tend to be somewhat contradictory, and in general the sample sizes were low, no uninfected individuals were included as controls, follow-up after treatment was not always described, and the biochemical, radiological, and histological investigations undertaken varied from study to study. Five studies provided evidence for a malabsorption syndrome associated with *Strongyloides* infection and reported steattorhea, abnormal D-xylose and folic acid absorption, and abnormal jejunal and duodenal biopsies. Some patients manifested diarrhea, vomiting, and weight loss. Several studies that undertook follow-up after treatment reported a regression of steattorhea and that intestinal biopsies returned to normal.

In contrast, a Costa Rican study that included infected and uninfected controls found no evidence of a malabsorption syndrome with normal fat and carbohydrate absorption, despite pathological changes in the jejunum and duodenum. In response to these somewhat contradictory findings, a carefully controlled study of individuals of known nutritional status was undertaken, the premise being that the malabsorption and intestinal changes recorded in previous studies might be secondary to malnutrition and not directly due to the parasitic infection. In malnourished patients with *Strongyloides,* malabsorption persisted in spite of total parasite eradication and disappeared in patients fed a low-protein diet in spite of a continued parasite presence. These findings indicate how important it is to evaluate the nutritional status of an infected host to untangle the relative contributions of infection and malnutrition.

V. TRICHURIASIS

Adult worms of the human whipworm, *Trichuris trichuira,* inhabit the mucosa of the cecum, into which they burrow by means of their anterior stylet. In heavy infections the worms can be found in the wall of the appendix, the colon, and the most posterior section of the ileum.

Children harbor the heaviest infections; intensity declines with adulthood. This decline has been attributed to behavioral changes in adulthood that result in a reduction in exposure or acquired resistance, which may lead to diminished establishment. Severity of infection is dependent on parasite intensity but also on host-related factors such as age, general health status, iron resources, and experience of past infections.

Infections can be classified as light, moderate, and heavy. Light infections tend to be regarded as asymptomatic. Moderate infections produce abdominal pain, diarrhea, vomiting, weight loss, and anemia in malnourished children. Heavy infections, which tend to be confined to endemic areas, are accompanied by bloody diarrhea, abdominal pain, pronounced weight loss, severe anemia, rectal prolapse, and finger clubbing. These symptoms are likely to contribute to a reduction in food intake and an increase in nutrient excretion. The diarrhea described is not of small intestinal origin and is likely to result from inflammation of the colon.

Despite high prevalence worldwide, there has been relatively little measurement of the morbidity associated with trichuriasis at the community level. Heavy trichuriasis has been associated with anemia, chronic diarrhea, and dysentery and protein–energy malnutrition.

It has not been conclusively proved that blood is an important food of *Trichuris,* but several workers have found blood and blood products inside worms at autopsy. Certainly the anterior stylet is responsible for the laceration of tissue and blood vessels, causing bleeding from the mucosa. A number of studies have reported an association between trichuriasis and anemia and an inverse relationship between hemoglobin level and intensity of infection. Only one study has concluded that trichuriasis is a cause of anemia. An estimated blood loss of 0.005 ml per day per worm and a blood loss per child of 0.8–8.6 ml per day has been reported. This is higher than that attributed to losses from uninfected persons (0.2–1.5 ml per day) but considerably less than that described for hookworm. Inconsistencies in the findings of a number of the studies may be explained by differences in the intensity of infection and the selection of study subjects. Individual variation in iron nutritional status and iron requirements may also be important. More community-based studies are needed to examine the relationship among trichuriasis, blood loss, and anemia in children with heavy and light infections, some of whom are anemic.

Evidence that trichuriasis can contribute to PEM has been provided from Malaysia, South Africa, and St. Lucia. Heavily infected Malaysian children and controls of similar socioeconomic status showed striking differences in nutritional status and clinical signs. All the infected children had chronic dysentery, and many manifested rectal prolapse, edema, anemia, and finger clubbing in contrast to the uninfected children (Table IV). Their growth was also significantly poorer and they had lower hematocrit and serum albumin levels. After treatment, nutritional index, number of red blood cells, and serum albumin significantly increased, and the frequency of rectal prolapse, diarrhea, edema, and clubbing was reduced in the infected children. A South African study revealed a significant increase in weight gain after treatment. The authors concluded that this weight gain was likely to be a consequence of the cessation of diarrhea, an improved appetite after treatment for anemia, and the improved hospital diet.

Finally, heavily infected children from an endemic community in St. Lucia showed greater growth deficits than the rest of the population. These deficits were attributed to growth stunting because low height-for-age was more significantly associated with heavy tirchuriasis than low weight-for-age or arm circumference. The authors concluded that trichuriasis is associated with growth stunting as a consequence of dysentery. These provocative observations indicate the need for further investigations using a longitudinal design.

TABLE IV

Features of *T. trichiura* Infection in Heavily Infected Malaysian Children ($n = 67$)[a,b]

Feature	Incidence (%)
Blood in stools	73
Edema	18
Clubbing	12
Rectal prolapse	51
Refeeding syndrome	26
Pica	67
Previous attendance as in-patients for diarrhea	44

[a]Adapted from R. H. Gilman, Y. H. Chong, C. Davis, G. Greenburg, M. K. Virik, and H. B. Dixon (1983). The adverse consequences of heavy *Trichuris* infection. *Trans. Roy. Soc. Trop. Med. Hyg.* 77, 432–438. From Stephenson (1987), pp. 161–201, with permission from Taylor & Francis.

[b]Visualization of *T. trichiura* worms in the rectum by anoscopy was used to diagnose heavy infection. The control group of children ($n = 73$) had 0% for all categories.

Some workers are of the opinion that the contribution of the whipworm to childhood morbidity, and in particular malnutrition, has been significantly underestimated.

VI. CONCLUSIONS

In general, further investigations are needed of the relationship between helminth infections and malnutrition, with a particular emphasis on carefully controlled, longitudinal, community-based studies.

Evidence is plentiful that hookworm is an important cause of anemia in many countries. It now seems unlikely that hookworm infection causes a malabsorption syndrome. The contribution of hookworm infection to reduced food intake, anorexia, and possible growth stunting merits further attention.

A number of researchers have now demonstrated convincingly that successful treatment for ascariasis can improve growth and macronutrient absorption in malnourished individuals in an environment where ascariasis and malnutrition are endemic. The first evidence of an improvement in height-for-age among children treated for ascariasis over a 2-year period has been provided from Burma.

Careful attention needs to be paid to the socioeconomic background of the communities under study.

Reductions in appetite and food intake may be one of the explanations for the impact of ascariasis on child growth. As in the case of hookworm, there is a lack of systematic studies on the influence of ascariasis on food intake and anorexia. Reduced food intake has been demonstrated in *Ascaris*-infected pigs. Precise measurement of food intake in individuals living within their own communities is notoriously difficult to undertake, but an attempt to do so is urgently needed.

Community-based observations on the epidemiology of strongyloidiasis are generally lacking, and this makes an evaluation of the influence of this parasite on nutritional status difficult. Some evidence of a malabsorption syndrome associated with *Strongyloides* has been found, but more studies are needed employing larger sample sizes, individuals of known nutritional status, and controls.

Evidence presented on trichuriasis indicates that the impact of this parasite on the nutritional status of children has not received adequate attention. The syndrome of chronic dysentery, rectal prolapse, anemia, poor growth, and clubbing associated with heavy *Trichuris* infection is of obvious public health significance and indicates that the identification of heavily infected individuals is important. The observation that heavy *Trichuris* infection is associated with growth stunting is also significant and needs further investigation using a longitudinal design.

In terms of control, some effective anthelmintic drugs are now available for *Ascaris, Trichuris,* and hookworm. Despite this, the prevalence of these infections remains staggeringly high in the countries of the subtropics and tropics. Mass chemotherapeutic programs have proved to be expensive and difficult to implement in such countries. In addition, these diseases have not always received the attention they deserve in public health terms. [*See* Chemotherapy, Antiparasitic Agents.]

The concept of targeted or selected chemotherapy has been discussed, given the aggregated distribution of the parasites concerned. Selection for treatment of the most heavily infected individuals or those perceived to be at particular risk, such as children, could reduce transmission and the number of infective stages in the environment. Limited resources would also be used more cost-effectively. The identification and characterization of these heavily infected individuals is therefore an important priority in epidemiological research.

In addition, the provision of improved sanitary facilities and public health education can only assist in the reduction of the soil-transmitted helminths. A purely biomedical approach to parasite control will not be adequate. There is a need to take account of behavioral and cultural attitudes in the design of control programs. *Ascaris* control programs have been found to act as a useful tool in the introduction of other primary health care programs and priorities.

In the case of *Strongyloides,* mass treatment is not appropriate because of the lower worldwide prevalence. Despite this, because of autoinfection and the hyperinfection syndrome, vigilance is needed. Diagnostic procedures need to be improved, and epidemiological information is generally lacking.

BIBLIOGRAPHY

Anderson, R. M. (1986). The population dynamics and epidemiology of intestinal nematode infections. *Trans. Roy. Soc. Trop. Med. Hyg.* **80**, 686–696.

Cooper, E. S., and Bundy, D. A. P. (1988). *Trichuris* is not trivial. *Parasitol. Today* **4**, 301–306.

Crompton, D. W. T. (1985). Chronic ascariasis and malnutrition. *Parasitol. Today* **2**, 47–52.

Crompton, D. W. T. (ed.) (1993). "Human Nutrition and Parasitic Infection." *Parasitology* **107**, S1–S203. (This is a special multiauthor supplement on the topic.)

Grove, D. I. (1989). "Strongyloidiasis: A Major Roundworm Infection of Man." Taylor & Francis, London/Philadelphia.

Holland, C. V. (1989). An assessment of the impact of four intestinal nematode infections on human nutrition. *Clin. Nutr.* **8**(6), 239–250.

Nesheim, M. C. (1989). Ascariasis and human nutrition. *In* "Ascariasis and Its Prevention and Control" (D. W. T. Crompton, M. C. Nesheim, and Z. S. Pawlowski, eds.). Taylor & Francis, London/Philadelphia.

Nokes, C., Grantham-McGregor, S. M., Sawyer, A. W., Cooper, E. S., Robinson, B. A., and Bundy, D. A. P. (1992). Moderate to heavy infections of *Trichuris trichuira* affect cognitive function in Jamaican school children. *Parasitology* **104**, 539–547.

Pawlowski, Z. S. (1984). Trichuriasis. *In* "Tropical and Geographical Medicine" (K. S. Warren and A. A. F. Mahmoud, eds.). McGraw-Hill, New York.

Schad, G. A., and Warren, K. S. (1990). "Hookworm Disease: Current Status and New Directions." Taylor & Francis, London/Philadelphia.

Stephenson, L. S. (1987). "Impact of Helminth Infections on Human Nutrition." Taylor & Francis, London/Philadelphia.

Hematology and Immunology in Space

CLARENCE F. SAMS
NASA/Johnson Space Center

MEHDI TAVASSOLI
University of Mississippi School of Medicine

GLOSSARY

Anemia Reduction from the normal red blood cell mass

Erythropoiesis Production of red cells in the bone marrow from proliferation, differentiation, and maturation of pluripotential stem cells. Daily production is 1/120th of red cell mass, which reflects the survival of red cells in the circulation (120 days)

Erythropoietin Hormone produced by the kidney in response to low pO_2 that acts on erythroid-committed progenitor cells in the bone marrow to stimulate erythropoiesis

Oxygen-carrying capacity Capacity of blood to transport oxygen from lungs to tissues. This capacity can be altered by a large number of physiological and pathological factors, including the concentration of hemoglobin, molecular alterations in hemoglobin, rheology of blood, affinity of hemoglobin for O_2, and red cell metabolism

Packed cell volume (PCV or hematocrit) Ratio of red cell volume to whole blood measured in a small sample. In practice it serves as an estimation of red cell mass

Red cell mass Mass of circulating red cells within the vascular space, measured directly by dilution of a ^{51}Cr-labeled red cell sample or derived indirectly from the measurement of ^{125}I-labeled plasma space and hematocrit. Expressed as mg/kg body weight

Red cell survival Survival of a red cell from the time of its delivery into the circulation (in the form of a reticulocyte) to its removal from the circulation in about 120 days. Because the removal of aged red cells follows an exponential curve, the survival is usually expressed as "half-life," which is 60 days. However, because the measurement is usually determined by labeling cells with ^{51}Cr and the radioactivity is subject to physical decay as well as physical elution, normal ^{51}Cr half-life is considered to be 28–30 days

Reticulocyte Youngest form of red cell that can enter blood; it lacks a nucleus, but still contains certain cytoplasmic elements necessary for hemoglobin synthesis that, on staining, give the cell a reticulated appearance (thus the name). These elements are lost within 24 hr, after which the cell takes the appearance of an ordinary red cell. The number of reticulocytes as a proportion of total circulating red cells serves as an index of bone marrow activity in erythropoiesis

SMEAT Ground-based control study designed to mimic a full 56-day Skylab flight except for the conditions of microgravity and possible cosmic rays. SMEAT served as a control for the subsequent physiological alterations noted during the spaceflights.

THE MOST CONSISTENT IMMUNOHEMATOLOGIC finding during orbital flights has been a substantial loss of red cell mass, often known as "anemia of space flight." However, anemia, as usually defined, is based on the concentration of red cells or hemoglobin in a given volume of whole blood. Total blood volume consists of the red cell mass plus the plasma volume. In clinical and experimental conditions, anemia is recognized by the measurement of packed cell volume (PCV or hematocrit) or hemoglobin concentration. In these situations a loss of red cell mass occurs usually without or with a negligible loss of plasma volume. Thus, the oxygen-carrying capacity of blood (PCV or hemoglobin concentration) falls, leading to anemia. During orbital flights, the loss of red cell mass is often associated with a decrease in plasma volume. This is the result of fluid shifting from the lower to upper body in the absence of gravitational forces. Volume receptors then act to reduce plasma volume. Parallel decreases in plasma volume render simple concentration measurements unrepresentative of the true red cell mass: hematocrit and hemoglobin concentration remain "normal," while the total red cell mass is reduced. Nonetheless, because of its currency, the term "anemia" of space flight is used here interchangeably with reduction in red cell mass. [*See* Hemoglobin.]

I. METHODOLOGY

A. Method of Detection

Because PCV and hemoglobin concentration are not representative measures for the anemia of spaceflight, the reduction of red cell mass is generally detected by isotope dilution. A sample of blood is obtained, red cells are labeled with radioactive ^{51}Cr, and the labeled cells are then reinfused in the subject's circulation. After some 20 min, when dilution equilibrium has been attained, another blood sample is removed, and its radioactivity is compared with that of the original reference sample. Simple calculations then permit the determination of red cell mass. Plasma volume can be similarly determined using ^{125}iodine-labeled albumin. Determinations are made at several points before flights (to ensure the stability of the readings) and repeatedly after flights (to determine the postflight pattern). Comparisons are then made between immediate preflight and postflight readings. By drawing a few other blood samples over an appropriate period and without the use of additional ^{51}Cr, the red cell mass technique can also be used to gain information on red cell life span. Alternatively, red cell life span can be determined by the incorporation of ^{14}C-labeled glycine into hemoglobin. These techniques can be performed during flight. However, because most flights are of short duration and because of concern that excess blood-drawing may in itself lead to the loss of red cell mass, in-flight determinations are not the rule. The amount of blood-drawing has been kept to a necessary minimum, and it has not been a factor in the loss of red cell mass.

B. Controls

Before the initiation of orbital flights, it was thought that physiological alterations as a result of exposure to microgravity could be predicted from bed rest studies. In its reduced requirement for muscular exercise, bed rest can indeed mimic microgravity. There was no indication, however, that a loss of red cell mass occurred during bed rest, although actual measurement of red cell mass was not done. To provide a baseline, a definitive experiment was initiated during the Skylab program (1973–1974). Known as the Skylab Medical Experiment Altitude Test (SMEAT), this was a ground-based experiment that simulated all aspects of a full 56-day Skylab mission. The physical facility, atmospheric pressure and composition, crew activity, diet, and the timetable of events were all representative of actual flight. Only the gravitational effects, the mental stress of being in space, and possibly the effects of cosmic radiation differed between SMEAT and the actual Skylab flight. Immunohematologic findings of SMEAT were:

- no significant changes in the red cell mass (2.7% ± 0.4, which is not significant);
- no significant shortening of red cell life span;
- slight but insignificant (1.6%) increase in plasma volume (decreases have been reported in most orbital flights);
- no significant changes in red cell glycolytic enzymes;
- normal blood cell counts;
- no major alterations of chromosomes and genetic materials;
- no significant immunologic alterations.

These findings provide a firm basis for the interpretation of actual flight data, so that the immunohematologic alterations observed during actual flights could likely be attributed to the effect of microgravity.

II. MAGNITUDE OF LOSS

In all orbital flights where red cell mass has been determined, losses have been noted. The consistency is seen in both the U.S. and Russian programs, which are similar in their essential features. Table I summarizes the results obtained. The most profound deficit was present in Gemini flights, where the atmosphere of the spacecraft cabins consisted of 100% O_2 at a partial pressure of 258 mm Hg (far greater than the

TABLE I
Red Cell Mass Losses during American and Russian Flights in Which Measurements were Made Under Comparable Methodologies and Conditions[a]

Flight	Duration of flight	No. of subjects	Percent decrease
American flights			
Gemini 4	4	2	13
Gemini 5	8	2	21
Gemini 7	14	2	14
Apollo 7	11	3	3
Apollo 8	7	3	1
Apollo 9	10	3	7
Apollo 14	10	3	4.7
Apollo 15	12	3	10.1
Apollo 16	12	3	14.2
Apollo 17	13	3	11.2
Skylab 2	28	3	14
Skylab 3	59	3	12
Skylab 4	84	3	7
Spacelab 1	10	4	9.3
SLS-1	9	3	11
Russian flights			
Soyuz 13	8	2	3
Salyut 3	16	2	14
Salyut 5	18	2	14
Salyut 4	30	2	16
Salyut 5	49	2	31
Salyut 4	63	2	20
Salyut 6	96	2	26
Salyut 6	140	2	14
Salyut 6	175	2	18

[a]All data were obtained directly by ^{51}Cr labeling except those from *Gemini 4*, where the data were calculated from plasma volume and hematocrit.

normal 160 mm Hg). Such hyperoxia can lead to a loss of red cell mass via two different mechanisms: peroxidation of red cell membranes, leading to shortening of red cell life span (hemolysis), and suppression of the production of erythropoietin and a resulting decrease in red cell production. Both mechanisms were documented. The ^{51}Cr survival of red cells in the Gemini crews was significantly reduced (half-life 20 days, normal 28–30 days). This reduction was associated with reduced plasma α-tocopherol levels and red cell membrane lipids, indicating oxidative injury and subsequent hemolysis. Moreover, no compensatory reticulocytosis was observed, suggesting suppression of red cell formation. During subsequent flights, when the atmosphere of cabins was changed to a mixture of O_2 and N_2 to provide a pO_2 of 150–185 mm Hg, the shortening of red cell survival was alleviated. Table II compares red cell survival in Gemini and subsequent flights. In *Spacelab 1*, postflight survival of red cells was entirely normal.

TABLE II
Red Cell Survival[a]

	Preflight	Flight	Postflight
Gemini 4, 7	—	20	—
Apollo 7–8	25	28	25
Apollo 14–17	24	23	27
Skylab 2–4	26	(123)[b]	24

[a]Means of ^{51}Cr half-life in days; normal is 28–30 days. Most studies were done in two members of the crew.
[b]^{14}C-glycine red cell life span in days.

III. PATTERN OF LOSS

Almost all the red cell mass deficit occurs during the first few weeks of flight as a linear function of flight duration. Red cell mass then appears to stabilize as physiological adjustment to the new conditions occurs. The cause of the initial loss and the subsequent adjustment may be the fluid shift that occurs in response to microgravity conditions. This fluid shift results in the removal of plasma from the gravity-dependent vascular spaces and the eventual establishment of a new equilibrium with a normal hematocrit and reduced plasma volume. Thus, the loss of red cell mass is not progressive and does not interfere with flights of very long duration. This is in contrast to the body's loss of calcium during spaceflight, which is progressive. Calcium loss and radiation exposure are thought to be the factors that limit the duration of human exposure to microgravity. [*See* Space Travel, Biochemistry and Physiology.]

IV. MECHANISMS OF LOSS

Spaceflight-induced reductions in the circulating red cell mass are usually attributed to either a more rapid elimination of red cells from the circulation or a decrease in their input into the circulation. The rapid appearance of the deficit, within 2 weeks, suggests that red cells are being lost from the circulation. The life span of red cells in the circulation is relatively long (normally 120 days). Although normalization of cabin atmospheric O_2 eliminated hemolysis due to hyperoxia, there is evidence that some random hemolysis may occur. This can be caused by exposure to cosmic rays, particularly HZE particles (particles with high charge Z and high energy E). Sequestration and subsequent destruction of red cells by the reticuloendothelial system may also be a major factor and can lead to reduction in red cells within the body. Consistent with this interpretation is the increase in serum ferritin (Table III). The ferritin increase parallels the reduction in red cell mass, indicating that iron is being released from sequestered red cells and stored in the form of ferritin for subsequent synthesis of hemoglobin. Other parameters of iron kinetics are also consistent with the changes in ferritin levels.

Evidence for erythropoietic suppression is derived from reticulocyte studies and marrow differential cell counts. Reticulocytopenia, or at least reticulocyte counts not high enough to compensate for the degree of red cell deficit, has been a consistent finding during space flights. Table IV shows serial reticulocyte counts in the crew members of the three Skylab and *Spacelab 1* missions. In every case, total circulating reticulocytes fall on the day of spacecraft recovery as compared with preflight. Morphological study of the marrow shows reduced erythropoietic activity as well as a 20% reduction in bone marrow cellularity. Russians have also reported decreased concentration of red cell progenitors and contraction of the pool of younger cells in the bone marrow, indicating a suppression of red cell production. There is no evidence that nutritional deficiencies contribute to the loss of red cell mass.

Decreases in erythropoietin levels have been observed during flight. Data from *Spacelab-1* and *SLS-1* indicate significant reductions in erythropoietin levels within 24 hr after reaching orbit. These reduced levels inhibit the maturation and release of red cells from the marrow. The production of red cells appears to be arrested at multiple points during the maturation of red cell precursors, and the induction of programmed cell death (apoptosis) by the low erythropoietin levels may be occurring. The arrest of red cell production late in the maturation cycle provides a potential mechanism consistent with the rapid reestablishment of reticulocyte production upon landing.

Hence, the microgravity-induced fluid shift causes a reduction in plasma volume and a perceived excess of red cells. This reduces the production of erythropoietin hormone, leading to the suppression of bone marrow erythropoiesis and consequent reticulocytopenia. In addition, an immediate suppression of new red cell

TABLE III
Serum Ferritin during *Spacelab 1* Flight (Mission Duration: 10 Days)[a]

Flight day:	1	7	10	11	18	22
Flight	103	134	133	145	96	84
Control (simulation)	99	82	86	84	61	67

[a]Data given as percent of preflight mean value.

TABLE IV
Mean Reticulocyte Counts ($\times 10^9$/Liter)

			Day after landing			
Mission	Duration	Preflight	0	1	7	14
Skylab 2	28	34	15	18	27	29
Skylab 3	59	36	23	—	66	82
Skylab 4	84	45	40	53	64	83
Spacelab 1	10	64	24	54	48	64

release from bone marrow occurs, which, coupled to the normal, age-related destruction of red cells, causes a steady decline in the red cell mass. Random hemolysis may also contribute to the loss of red cell mass. A new equilibrium is eventually established that is maintained throughout longer flights. This scheme is diagrammatically shown in Fig. 1, but it must be considered tentative. As more data become available, this scheme could be modified.

V. RECOVERY

Red cell mass invariably recovers after landing, indicating that the physiological control mechanisms for red cell production are not damaged. As expected, the recovery of red cell mass is heralded by reticulocytosis (see Table IV), indicating the resetting of the regulatory mechanisms for erythropoiesis. Curiously, after short-duration flights, reticulocytosis is somewhat delayed, and the appropriate response is not seen for several weeks. This might be attributable to the dormancy of bone marrow stroma, which is needed for regulation of erythropoiesis. After flights of longer duration, reticulocytosis usually occurs within 1 week after landing, and the recovery of red cell mass is typically swift and uncomplicated.

VI. FUNCTIONAL CAPACITY OF SPACE-BORN RED CELLS

Salyut 6 was an orbiting space station that supported several successive missions of increasing duration. These missions provided an unusual opportunity to study the functional capacity of space-born red cells. Some of the crewmen in these missions remained aboard for periods of 140 and 175 days. Because the life span of red cells is approximately 120 days, circulating red cells in these crew members were entirely replaced by red cells produced in space. The oxygen-carrying capacity of these space-born cells was completely normal, suggesting that the cellular proliferation in the weightless state leads to the production of normally functioning cells. Similarly, no major alterations have been reported in other proliferating cell systems, such as the skin or the gastrointestinal mucosa. However, changes in the immune cells or subtle changes in other cell systems cannot be ruled out.

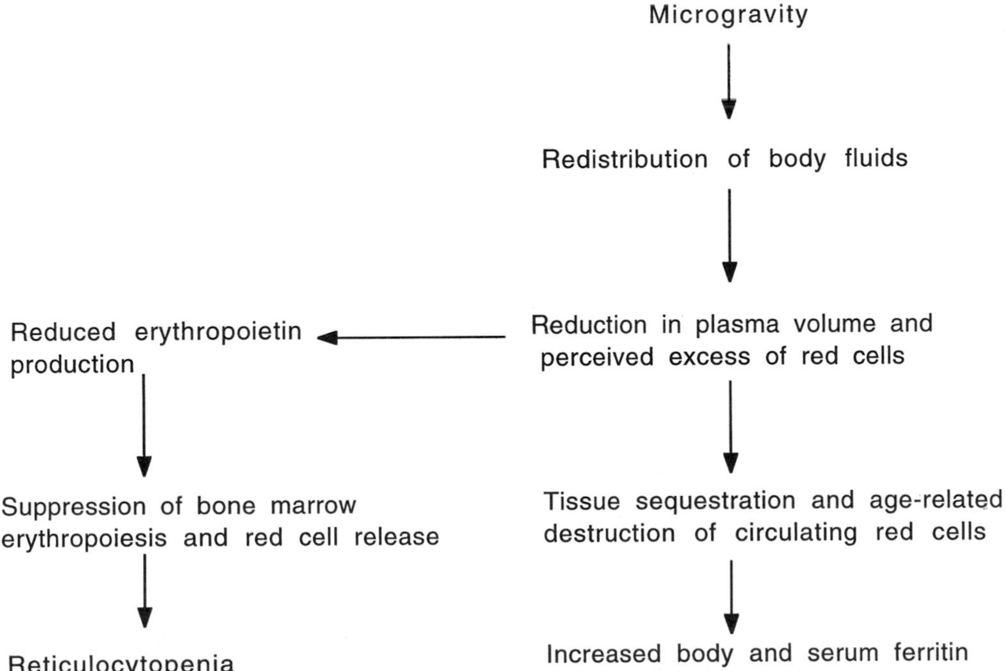

FIGURE 1 Possible mechanism of how microgravity can lead to the suppression of erythropoiesis in bone marrow and consequently to red cell mass deficit.

VII. OTHER HEMATOLOGIC AND IMMUNOLOGIC FINDINGS

The effects of spaceflight on the immune system and the influence of these changes on crew health remain controversial. Consistent alterations in circulating immune cells have been reported upon landing, though the relevance of these changes to those occurring during flight is currently unclear. Postflight changes in circulating white blood cells include granulocytosis, lymphopenia, and variable alterations in the lymphocyte subpopulations and the monocytes. Decreased

TABLE V

Changes in Hematologic, Immunologic, and Serum Parameters Immediately after Landing as Percent of Changes from Preflight Measurements in the Four Members of the Crew of *Spacelab I*

Parameter	Percent change on landing (±SEM)	
Blood volume		
Red cell mass, ml/kg	−9.30	1.60
Plasma volume, ml/kg	−5.98	4.30
Blood volume, ml/kg	−10.50	0.87
Erythrocyte hematology		
Erythrocyte count, $\times 10^{12}$/liter	−5.83	1.03
Hemoglobin, g/dl	−3.4	0.3
Hematocrit, liter/liter	−3.2	0.8
Mean corpuscular volume, fl	2.0	1.0
Mean corpuscular hemoglobin, pg/cell	8.3	0.9
Mean corpuscular hemoglobin concentration, g/dl	5.1	1.4
Erythrocyte production		
Reticulocyte number, $\times 10^{9}$/liter	−61	8
Reticulocyte production index	−60	8
Reticulocyte RNA, % of cytoplasm	24.7	3.1
Erythropoietin, units/ml	−72	10
Iron		
Transferrin, mg/dl	−10.7	7.5
Serum iron, μg/dl	−32	6.6
Unbound iron-binding capacity, μg/dl	−13.7	5.9
Total iron-binding capacity, μg/dl	−17.5	6.1
Saturation of transferrin, %	−17	4
Platelets and leukocytes		
Platelet count, $\times 10^{9}$/liter	12.0	10.7
Leukocyte count, $\times 10^{9}$/liter	17	16
Neutrophils, $\times 10^{9}$/liter	34	14
Lymphocytes, $\times 10^{9}$/liter	−0.4	13
Monocytes, $\times 10^{9}$/liter	40	50
Eosinophils, $\times 10^{9}$/liter	80	100
Serum chemistry		
Osmolality, mOsm/liter	0.6	0.3
Sodium, mEq/liter	−0.4	0.2
Potassium, mEq/liter	−1.5	6.5
2,3-DPG, mol/g Hb	−9.3	6.9
ATP, mol/g Hb	−6.9	14.2
Serum proteins		
Total serum protein, g/dl	−3.0	1.7
Albumin, g/dl	−0.7	3.9
Alpha-1 globulin, g/dl	10	10
Alpha-2 globulin, g/dl	10	10
Beta globulin, g/dl	−3	8
Gamma globulin, g/dl	−30	5
Haptoglobulin, mg/dl	12	8
Ferritin, ng/ml	62	36

numbers of natural killer cells and reduced cytotoxic function are also observed during the days immediately after landing. Other changes in lymphocyte function include reduced proliferation in response to mitogens and altered production or secretion of cytokines. Delayed-type hypersensitivity (DTH) testing of crew members during flight has confirmed that a reduction in cell-mediated immunity occurs while in orbit. The relationship between the *in vivo* DTH tests and the postflight *in vitro* results will require further study to elucidate. However, the available data suggest that (1) a dysregulation of the immune system occurs during flight, (2) additional changes in circulating immune cells occur during reentry and reambulation, and (3) immunologic parameters return to preflight values within the first days to weeks after flight.

These changes likely result from the psychological and physical stresses encountered during launch, in flight, and during reentry. Such stress can activate the sympathoadrenal system, inducing changes in neuroendocrine hormones and a subsequent immunomodulation. Consistent with this hypothesis is the observed elevation in norepinephrine and epinephrine levels after flight. Increases in cortisol levels are not the cause of the immune changes, since serum cortisol levels are slightly reduced in Space Shuttle crew members following spaceflight.

Other changes in serum proteins and other blood parameters do not appear to have a major influence on crew health. A slight increase in α-2-macroglobulin and significant increases in IgA and IgM immunoglobulins have been noted after Shuttle flights. Slightly elevated C3 and C4 complement levels were also present. These changes are suggestive of potential alterations in humoral immune function, although the functional consequences of such alterations are unclear. Other changes in serum proteins are either generally related to the loss of red cell mass or insignificant at the level of our present understanding. Table V is a representative data set of hematologic, immunologic, and serum parameters that were determined for the four members of the crew of *Spacelab-1*. [*See* Lymphocytes.]

BIBLIOGRAPHY

Huntoon, C. L., Whitson, P. A., and Sams, C. F. (1993). Hematologic and immunologic functions. *In* "Space Physiology and Medicine" (A. E. Nicogossian, C. L. Huntoon, and S. L. Pool, eds.), 3rd Ed., pp. 351–362. Lea & Febiger, Malvern, Pennsylvania.

Leach, C. S. (1992). Biochemical and hematologic changes after short-term space flight. *Microgravity Quart.* **2**, 69–75.

Leach, C. S., Chen, J. P., Crosby, W., Johnson, P. C., Lange, R. D., Larkin, E., and Tavassoli, M. (1988). Hematology and biochemical findings of Spacelab 1 flight. *In* "Regulation of Erythropoiesis" (E. D. Zanjani, M. Tavassoli, and J. L. Ascensao, eds.), pp. 415–453. PMA Publishing Corp., New York.

Meehan, R. T., Whitson, P. A., and Sams, C. F. (1993). The role of psychoneuroendocrine factors on space flight-induced immunological alterations. *J. Leukoc. Biol.* **54**, 236–244.

Talbot, J. M., and Fisher, K. D. (1986). Influence of space flight on red blood cells. *Fed. Proc.* **45**, 2285–2290.

Tavassoli, M. (1982). Anemia of space flight. *Blood* **60**, 1059–1067.

Tavassoli, M. (1986). Medical problems of space flight. *Am. J. Med.* **81**, 851–854.

Udden, M. M., Driscoll, T. B., Pickett, M. H., Leach-Huntoon, C. S., and Alfrey, C. P. (1995). Decreased production of red blood cells in human subjects exposed to microgravity. *J. Lab. Clin. Med.* **125**, 442–449.

Hemispheric Interactions

MALCOLM JEEVES
University of St. Andrews

GLOSSARY

Cerebral specialization Neuropsychological studies of brain-damaged people, patients with the forebrain commissures surgically divided, and normals have shown that each cerebral hemisphere is selectively superior for processing and analyzing certain kinds of sensory input and for performing motor output; for example, in most right-handed people, while the left hemisphere is better at handling speech, writing, language, and calculation, the right hemisphere is better at spatial construction, singing, and playing musical instruments

Forebrain commissures Axons of some cortical neurons cross from one cerebral hemisphere to the other; commissural fibers are distinct bundles of these axons clearly identifiable using autoradiographic techniques; they cross the midline in one of the forebrain commissures, namely, the anterior commissure, corpus callosum, and dorsal and ventral hippocampal commissures

Splenium Commissural fibers from a given cortical region tend to occupy a distinct location in one of the forebrain commissures; the splenium is in the posterior part of the largest of the forebrain commissures, the corpus callosum, and cross-connects areas of the cortex involved in visual processing

GIVEN THAT HUMANS HAVE TWO CEREBRAL HEMISPHERES, it has long been debated whether or not this implies the possibility of two minds within one brain. That two hemispheres acting together are better than one is in accord not only with our intuition but with empirical data. How the two hemispheres work together to ensure the unified activity of the mind has for the past four decades been the subject of intense experimental investigation. Questions arise readily. What sorts of interactions take place to ensure the moment-by-moment, efficient, unified functioning of a system made up of two partially independent modules? What mechanisms make possible the efficient sharing of raw sensory data, yet also, as required, ensure sensory isolation and response inhibition? We consider in turn hemispheric integration, hemispheric inhibition, hemispheric facilitation, individual differences, and the developmental aspects of each of these questions. Normally, the brain acts as a unified whole despite the fact that the two cerebral hemispheres in most people are functionally different and anatomically distinct. Moreover, each hemisphere appers to represent the world differently. Although not a clear-cut dichotomy, most left hemispheres emphasize a digital linguistic representation, whereas most right hemispheres specialize in analogue perceptual representations. These complementary functional systems were revealed most clearly following the now classic split-brain studies of Roger Sperry and his colleagues.

I. INTERHEMISPHERIC CONNECTIONS

The two cerebral hemispheres are intimately cross-connected through three transverse commissures, referred to collectively as the forebrain commissures (Table I). The axial or midline parts of the body tend to have the most numerous commissural connections, whereas those that deal with sensory and motor func-

ENCYCLOPEDIA OF HUMAN BIOLOGY, Second Edition, VOLUME 4.

TABLE I
Hemispheric Interactions Mediated through the Forebrain Commissures

Integration	ensures sharing of lateralized sensory input and coordination of bilaterally controlled motor output
Inhibition	(1) ensures the development of topographic sensation and precise motor control by suppressing the contribution from uncrossed ipsilateral pathways (2) ensures hemispheric independence, thus allowing implementation of modularity of representation and processing in the normal brain
Facilitation	(1) in the intact cortex, the corpus callosum exerts a facilitatory, or modulating, influence on the neural activity of both hemispheres (2) this modulatory action may actively participate in the functional reorganization that takes place after brain injury
Attention and meta-control	ensures normal maintenance and switching of attention and distributes arousal to task-relevant regions of the brain

tions of extreme distal parts of the extremities have relatively few. The higher-order association areas, in the frontal lobe and at the parieto-temporo-occipital junction, have variable densities of contralateral connections. After conception, if development proceeds normally, the first crossing is of the fibers that will form the anterior commissure at about the 50th day. The second crossing, the hippocampal commissure, appears soon thereafter at the end of the second gestational month. Finally, the corpus callosum begins its development at about the 12th gestational week. The corpus callosum is formed in all its parts 18–20 weeks after gestation but does not reach full functional maturity until myelination is completed at around 10 years of age. [*See* Brain.]

At birth, there is exuberant growth of the fibers in the commissural systems. In cats and monkeys, the extent of this is indicated by the elimination of 70% of these fibers in the mature callosal system. Estimates of the total number of fibers in the corpus callosum have been revised steadily upward; most recent estimates are from 500 to 800 million. The area of the corpus callosum predicted the number of fibers in humans when small- to intermediate-size fibers were included in the counts (Aboitiz *et al.*, 1992). The relation of area to number of fibers in the human corpus callosum may reflect functional differences between individuals (see the following).

In addition to the forebrain commissures, there are diencephalic and mesencephalic interhemispheric connections. These include hypothalamic, supraoptic, and habenular commissures, the massa intermedia (mediating thalamic connections), the posterior commissure, and the collicular commissure.

Different topographical regions of the forebrain commissures contain fibers relating to different functional specializations (Fig. 1). While somatic sensory functions are localized to the central portion of the body of the callosum, the splenium deals mainly with visual function. This topographical organization is confirmed and illustrated by clinical and behavioral evidence from studies of patients with discrete lesions of the forebrain commissures. The anterior commissure cross-connects the olfactory areas on the two sides of the brain as well as parts of the temporal lobes.

The cross-connection of the two hemispheres through the commissures has, perhaps naturally, focused on the likely information transmission and information transfer functions of the system. However, the fact (see Section II) that patients in whom the commissures have been cut are indistinguishable from normals in activities of daily living warns against overemphasizing this role for the commissures to the exclusion of other possibilities. And why such a massive structure as the corpus callosum should be present if other significant roles are not carried by it is hard to conceive.

Recently several hypotheses about the functions of the corpus callosum have been proposed that, on the face of it, seem mutually exclusive. Some emphasize an excitatory, others an inhibitory, role in modulating hemispheric interactions. On closer scrutiny it is plausible that the observed heterogeneity of interhemispheric connections could allow for such a variety of functions within the one system.

II. HEMISPHERIC INTEGRATION

When the forebrain commissures are cut in humans, a dramatic disconnection syndrome occurs. Detailed laboratory studies of such split-brain patients indicate the coexistence of two seemingly independent cognitive systems within the same brain; thus, one hemisphere is unable to communicate with the other hemisphere. The split-brain syndrome, however, has relatively few effects on everyday life. Thus, normally, the forebrain commissures and particularly the corpus

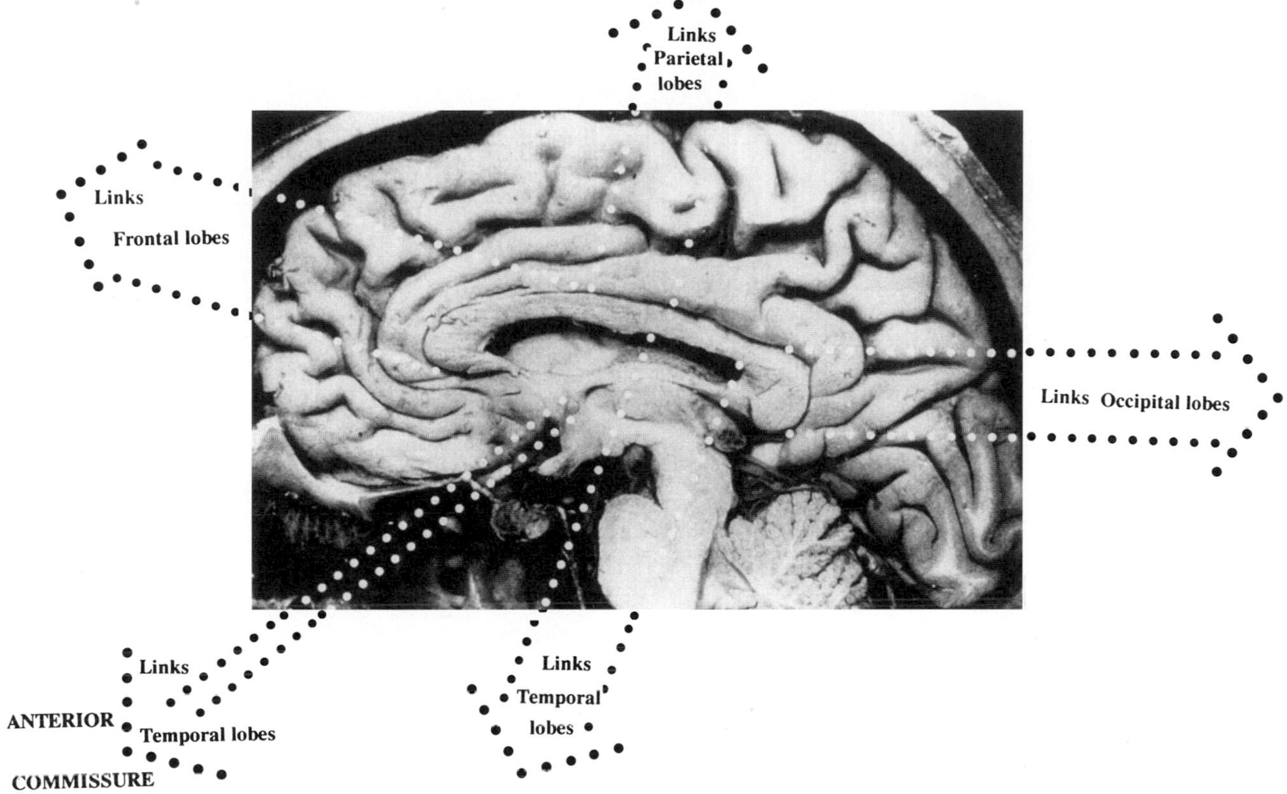

FIGURE 1 The human corpus callosum and the anterior commissure seen *in situ* contain the majority of the fibers cross-linking the two cerebral hemispheres. Different topographical regions of the forebrain commissures contain fibers relating to different functional specializations. Somatic sensory and motor functions are localized to the central portion of the body of the corpus callosum; the splenium deals mainly with visual-related functions.

callosum ensure that the processing operations of the two hemispheres can be integrated. For example, in the tactile modality, if a familiar object is placed in the right hand of a split-brain patient, he has no difficulty in naming it; however, if the same task is given but now with the object in the left hand, the patient is usually unable to name it, because the information from the left hand goes to the right hemisphere but owing to the commissure section is then denied access to the brain's left hemisphere, where in most right-handed people the language system is found. Similarly, in the visual modality, if the picture of an object is flashed briefly to the right side of where the patient is fixating, the information goes to the left side of the brain. A split-brain patient has no difficulty in naming an object thus presented. If, however, a picture is flashed to the left side of the visual field so that the information goes to the right side of the brain, the split-brain patient is unable to name correctly what he has seen. The same is true if letters of the alphabet or words are presented to the right or left visual fields. Thus, it was concluded that the two hemispheres communicated and passed information for specialized processing back and forth through the corpus callosum (Fig. 2).

In recent years, the necessity of the forebrain commissures for hemisphere integration, as just outlined, has been modified somewhat. Studies have shown that some rudimentary forms of visual information pass from one hemisphere to the other even though the corpus callosum, the anterior commissure, and the hippocampal commissure have all been cut. Researchers conclude that subcortical structures may have the functional capacity to subserve the integration of information at a visuospatial and also higher-order cognitive level; however, they noted that such integration as achieved was far from perfect.

The debate about how much information can be transferred between the hemispheres through subcortical pathways continues. It seems clear, however, that

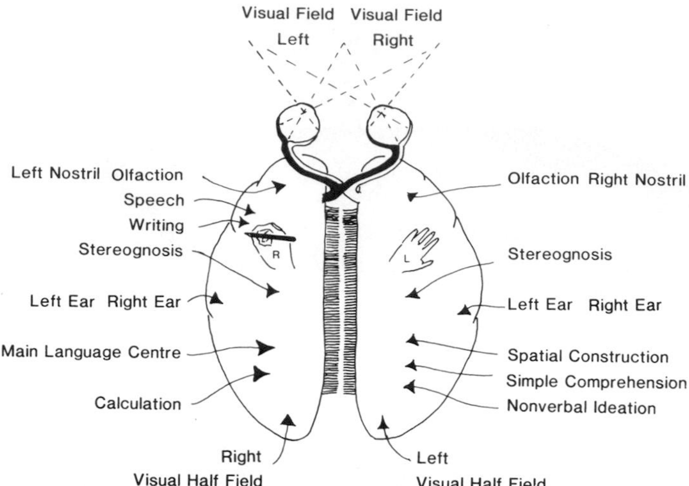

FIGURE 2 A schematic representation of some of the specialized functions of the cerebral hemispheres. The corpus callosum is shown as sectioned down the midline, thus abolishing most routes for hemispheric interactions as described in the text.

despite the limitations on information transfer in split-brain patients, they can identify numbers and letters sent to the mute right hemisphere and also can integrate low-level visuospatial information across the midline. This suggests that certain sorts of information can be transferred and cross-integrated without the corpus callosum.

The possibility for two hemispheres to share information even though the neocortical commissures have been cut gains further support from earlier studies of people who were born without the corpus callosum and, in some instances, also without the anterior commissure. Extensive studies on this small group of patients showed that in everyday living they manifest no obvious difficulties; like the surgical split-brain patients, their deficits became evident only with careful laboratory testing. So long as the anterior commissure is present, they seem able to share information between the two cerebral hemispheres although not as efficiently as a person with a normal intact brain. The compensatory mechanisms making possible their seemingly normal performance probably vary depending on the sense modality being tested. Thus, visual transfer may be effected through the anterior commissure, whereas tactile transfer may be made possible through the enhanced elaboration and development of uncrossed sensory pathways.

Recent work on animals indicated that in mice in which the corpus callosum did not develop the putative callosal axons that failed to cross to the other hemisphere did not reroute through the anterior commissure. If the same thing were to apply in humans, it would curb present enthusiasm for attributing too great a compensatory role for the anterior commissure in the absence of the corpus callosum. In general, compensatory mechanisms are better for those modalities that are more bilaterally organized, for example, temperature and audition, than for those that project mainly to contralateral cortical structures, for example, touch and vision (Lassonde and Jeeves, 1994).

The importance of the integrative action of the neocortical commissures in ensuring efficient hemispheric interaction is further underlined by the many reports demonstrating the so-called bilateral processing advantage. This refers to the finding that when two stimuli, normally visual, must be compared and a judgment made as to whether they are the same or different, if one stimulus is sent separately to each hemisphere, the task is accomplished faster and with fewer errors than if both stimuli go to one hemisphere. To a point, this finding is counterintuitive, because in the bilateral presentation condition one stimulus must be transferred through the forebrain commissures to the other hemisphere for comparison to be made. From a physiological point of view, such transfer involves information transmission with the consequent possibility of degradation of the signals, which in turn might be expected to result in poorer performance than when both stimuli arrive within the same hemisphere and no such interhemispheric transfer is called for. Because the net effect in the bihemisphere condition is improved performance, it further underlines the efficiency of the integrative processes through the forebrain commissures.

The advantage conferred by having the stimuli to be compared appearing in both visual fields rather than only one does not, however, always emerge. The general finding is that as the difficulty of the task increases, so the benefits of bihemispheric input emerge and increase. Yet that conclusion must be qualified since some studies (e.g., A. Belger) suggest that the between-hemispheres advantages are obtained only on difficult tasks for which a dispersal of processing load is possible, that is, both hemispheres have some capacity for processing that type of information.

III. RECIPROCAL HEMISPHERIC INHIBITION

When clear differences in hemispheric functioning were first reported, physiologists pointed out that such cerebral specializations most likely could come about only as a result of inhibitory processes mediated through the forebrain commissures. This view was subsequently taken up by psychologists who argued that, in the course of normal development, the inhibitory processes mediated by the forebrain commissures ensured the specializations of function of the two cerebral hemispheres observed at maturity (see Fig. 2) (e.g., special linguistic capacities develop in the left hemisphere so it inhibits and prevents similar developments in the right hemisphere). The anatomical and physiological substrate for this process is assumed to be the forebrain commissures and primarily the corpus callosum; however, this view has been challenged. The results of detailed studies of people born without the corpus callosum failed to find the predicted bilateral representation of function, which should, according to the inhibitory hypothesis, occur in the absence of the corpus callosum from birth. Thus, though it seems unlikely that the normal role of the corpus callosum is to bring about hemispheric specialization, fine-tuning of hemispheric specialization still may be afforded by the forebrain commissures. If this is true, then while the brains of acallosals are lateralized, the extent of lateralization may be less than that found in the normal population (Lassonde and Jeeves, 1994). [*See* Cerebral Specialization.]

Because hemispheric interaction in the form of inhibition may not be necessary to ensure normal hemispheric specialization, it does not mean that other important inhibitory roles are not mediated through the corpus callosum. Researchers have suggested that inhibition, at the cognitive level, is called for to enable component mental operations elaborated concurrently in both hemispheres to be protected from interfering cross-talk between the hemispheres until the operations are fully developed and ready to adopt specific relationships with one another within the total action program. In a somewhat similar vein, others have suggested that hemispheric independence is a common and ubiquitous state in the normal brain and that a process of callosal inhibition could ensure such independence.

With the increasing availability of magnetic resonance imaging techniques to visualize brain structures, attempts have begun to relate measures of the size and shape of the corpus callosum in normal healthy adults to indices of callosal functioning. One study involving 60 healthy subjects examined possible interhemispheric inhibitory mechanisms. Their behavioral task was dichotic listening. Nonsense syllables were presented simultaneously to both ears and those to the right ear going to the left hemisphere were, in common with earlier findings, reported more accurately. The differences between right ear and left ear performances were greater in right-handers than left-handers. Since performance on this task is thought to depend entirely on the left hemisphere (where the right ear input goes to), the negative correlations they found between right ear performance and total area of the corpus callosum, as well as with the anterior one-third of the corpus callosum, were interpreted by the investigators as due to the right hemisphere inhibiting the left hemisphere through the corpus callosum. Other investigators believe that with a larger corpus callosum, excitatory signals (see Section IV) that are sent from the right hemisphere only appear to be inhibitory because they disrupt ongoing left hemisphere processing. Certainly the heterogeneity of callosal connections allows for both inhibitory and excitatory influences.

It has been argued that in the normal person, the uncrossed ipsilateral pathways for sensory input and motor control are not allowed to compete with the crossed pathways, and this occurs through the suppressive action or the inhibition through the corpus callosum. Because this is absent in an acallosal patient, the competition between crossed and uncrossed pathways remains, resulting in reduced sensory and motor performance. These ideas were based on the reduced ability of adult acallosals to make fine sensory discriminations and to exercise precise motor control.

These findings have been confirmed and extended. The inhibitory role can account for the development of dominance of the contralateral over the ipsilateral pathways in the sensory and motor systems, which appear to be essential for finely tuned sensation and motor control. [*See* Motor Control.]

IV. HEMISPHERIC FACILITATION

Evidence indicates an important facilitative role of the corpus callosum. Studies of patients born without the corpus callosum, as well as of children and adults who had undergone partial or total callosotomy for control of intractable epilepsy, demonstrated deficits that were not limited to *inter*hemispheric processing but were evident also in *intra*hemispheric processing; therefore, the corpus callosum seems to exert a facilitatory action modulating neural activity in *both* hemispheres. In the absence of the corpus callosum, this is assumed to be reduced or absent and, thus, the deficits in intrahemispheric as well as interhemispheric processing occur. One implication of this hypothesis on the modulating role of the corpus callosum is that in the absence of the callosum or following damage to it, neither hemisphere will likely achieve its full potential. Furthermore, clinical reports indicate that the corpus callosum, through its modulating action, may actively participate in the functional reorganization that takes place after brain injury. Restitution of language functions occur only when the forebrain commissures remain intact. Thus, this type of hemispheric interaction, labeled modulatory action, may be one important way in which an intact cerebral hemisphere can help to compensate for loss or impairment of functions when the other hemisphere is damaged.

Evidence from two other lines of investigation of brain processes is consistent with a facilitative role in hemispheric interactions through the corpus callosum. These are, first, studies of regional cerebral metabolism that indicate high and positive correlations between homologous brain sites in humans and, second, the coherence of EEGs in the two hemispheres. Coherence is a measure of the covariance of activity for given EEG frequencies. Such coherence measured during sleep is significantly reduced in infants born without the corpus callosum as compared with normal infants. This is interpreted as indicating an integrative-facilitative role for the corpus callosum in normal children, thus producing a coherent pattern of activity in homologous sites in the two hemispheres.

V. ATTENTION AND META-CONTROL

When split-brain patients were briefly shown figures that were made up of the left half of one stimulus joined at the vertical midline with the right half of another stimulus, two things were evident. First, they did not notice that they were looking at composite figures; second, they seemed to select, attend to, and visually complete only one half of the figures shown. This suggested that with the callosum cut and normal hemispheric interactions disrupted, they could not allocate attention normally. Taken together with the observation that split-brain patients also show deficits in vigilance, the results suggest that the corpus callosum plays a key role in maintaining normal sustained attention, a view supported further by recent developmental data (see Section VI). It has been suggested that the neural basis for these attentional effects is that the corpus callosum plays a role in maintaining bilateral arousal, possibly by a positive feedback loop between the hemispheres.

One of the leading researchers in the early work on the split-brain patients, after studying how such patients went about a task that could be done either verbally or visually, noted that the right hemisphere (left visual field presentations) dominated performance despite its lower capacity on the task. She wrote that "depletion of higher control that occurs with division of the corpus callosum not only reduces perceptual awareness for one half of space but often does so in a maladaptive fashion, so that the more capable hemisphere is not employed in cognitive processing" (Levy, 1985, p. 24).

These and other similar studies suggest that overall or meta-control of attention is a process that is disrupted when the forebrain commissures are cut and that as a consequence such patients often use the less-skilled hemisphere to perform a task. What emerges from studies of split-brain patients suggests that meta-control processes mediating hemispheric interactions through the corpus callosum may be important in distributing arousal to task-relevant regions of the brain in normal healthy subjects also.

VI. DEVELOPMENTAL ASPECTS OF HEMISPHERIC INTERACTIONS

As noted earlier, the corpus callosum is one of the last paths of the nervous system to mature. It is generally accepted that the callosal fibers are not fully myelinated (and therefore mature) until around 10 years of age. Therefore, as a system, the forebrain commissures are among the last to achieve full functional capacity. It has been further suggested that hemispheric function during development, and presumably up until about 10 years of age, reflects the increasing interhemispheric influences discussed earlier, because these are mediated largely through the maturing corpus callosum. Attempts to demonstrate this developmental progression physiologically and behaviorally have met with increasing success. An early study using cortical evoked potential techniques reported that relay times across the corpus callosum decreased linearly with logarithm of age from 3.5 years to puberty. More securely based timing techniques using behavioral measures of manual response times to simple visual stimuli presented either side of a central fixation point have indicated that whereas in young adults the mean interhemispheric transmission time is 2 to 3 msec, in 6-year-old children it is 6 to 8 msec. At the other end of the life span, where it is known that degenerative processes in the nervous system have occurred, studies have shown that in 70-year-olds the interhemispheric transmission times are increased to 6 to 8 msec.

Another way of studying such developmental changes is exemplified by a study in which children were given the task of judging whether the textures of two pieces of cloth, lightly touched by a finger, are the same or different. The judgments were made by either two different fingers on one hand or two different fingers on different hands. They found that whereas all the children showed an expected improvement in overall performance with increasing age, the younger preschool children had greater difficulty when the two judgments were made by different hands than when they were made by the same hand. With the two-handed condition, information must cross the forebrain commissures, whereas with the one-hand condition, no such crossing is involved. Other researchers have studied the ability of children of different ages to indicate correctly when the tip of a finger has been lightly touched. They show the experimenter which finger was touched by touching it themselves with the thumb of that hand (this is the uncrossed condition). Alternatively, they may be asked to indicate in the same manner, on the other hand, the finger that was touched on the first hand (this is the crossed condition). In the latter case, a callosal crossing is involved, in the former it is not. Younger children found this crossed condition much harder relative to the same-hand, uncrossed condition when compared with older children. The importance of the development of fully efficient interhemispheric integration is evident when one remembers that complex tasks demand that the human brain enlists and coordinates the special abilities of both hemispheres. The question naturally arises of what happens if, for whatever reason, the corpus callosum is not fully functional in a developing child. A study of the efficiency of children of different ages in transmitting information from one hemisphere to the other showed that dyslexic children resemble normal children younger than them both in overall level of correct response and in the ratio of crossed (same hand) to uncrossed (two hands) errors in a task such as that just described. Typically, the reading-disabled children approximate normal children 4 to 6 years younger than themselves. When the abilities of children to select, sustain, divide, and focus attention, as described, were carried out, the results suggested that as hemispheric interaction improves with age, so also does sustained attention. However, the callosal deficit may simply be a particular instance of a more general deficit, which is prominent in the preceding studies, because the corpus callosum is simply the largest of an extensive set of cortico-cortical connections. Pictures of brains using magnetic resonance imagery link underdevelopment of the corpus callosum with ill-defined learning difficulties in children.

Thus, hemispheric interactions through the forebrain commissures seemingly play a variety of important roles to ensure the development to full capacity of the potential for unified functioning of the human brain.

VII. CONCLUSIONS

Hemispheric interaction is not a simple matter. Several distinguishable interhemispheric functions are mediated through the forebrain commissures. Of these, hemispheric integration is crucial for the unified functioning of the differently specialized cerebral hemispheres. An inhibitory role of the commissures is equally important, not only during development, but also at maturity to make possible the independent functioning as and when required of a particular cerebral hemisphere without interference from the other hemisphere. A facilitatory function has been suggested, which ensures the fine-tuning of what is happening in both hemispheres and guarantees that both hemispheres achieve their full potential. Regarding development, there are two distinguishable functions, both of which depend on a properly functioning neo-commissural system. First, there is the sharing function, alternatively called information transmission or integration, which increasingly guarantees the sharing of information as and when appropriate between what is happening in each cerebral hemisphere. The shielding function is equally important. It is the way in which one hemisphere may, through the callosum, inhibit the activity of the other hemisphere so that it does not interfere with the ongoing activity of the first hemisphere. Third, there is the role of the system in exercising meta-control so that arousal is appropriately distributed to task-relevant brain regions.

BIBLIOGRAPHY

Aboitiz, F., Scheibel, A. B., Fisher, R. S., and Zaidel, E. (1992). Fiber composition of the human corpus callosum. *Brain Res.* **598**, 143–153.

Clarke, J. M., Lufkin, R. B., and Zaidel, E. (1993). Corpus callosum morphometry and dichotic listening performance: Individual differences in functional interhemispheric inhibition? *Neuropsychologia* **31**, 547–557.

Hoptman, M. J., and Davidson, R. J. (1994). How and why do the two cerebral hemispheres interact? *Psychol Bull.* **116**(2), 195–219.

Jeeves, M. A. (1986). Callosal agenesis: Neuronal and developmental adaptations. *In* "Two Hemispheres—One Brain" (F. Lepore, M. Ptito, and H. H. Jasper, eds.). Liss, New York.

Lassonde, M. (1986). The facilitatory influence of the corpus callosum on intrahemispheric processing. *In* "Two Hemispheres—One Brain" (F. Lepore, M. Ptito, and H. H. Jasper, eds.). Liss, New York.

Lassonde, M., and Jeeves, M. A. (1994). *In* "Callosal Agenesis: A Natural Split Brain?" Advances in Behavioural Biology, Vol. 42. Plenum, New York/London.

Levy, J. (1985). Interhemispheric collaboration: Single-mindedness in the asymmetric brain. *In* "Hemispheric Function and Collaboration in the Child" (C. T. Best, ed.), pp. 11–31. Academic Press, San Diego/New York.

Ramaekers, G., and Njiokiktjien, C. (1991). *In* "Pediatric Behavioural Neurology," Vol. 3. Suyi, Amsterdam.

Zaidel, E., Clarke, J. M., and Suyenobu, B. (1990). Hemispheric independence: A paradigm case for cognitive neuroscience. *In* "Neurobiological Foundations of Higher Cognitive Function" (A. Scheibel and A. Wechsler, eds.). Guilford, New York.

Hemoglobin

M. F. PERUTZ
Medical Research Council Laboratory of Molecular Biology, Cambridge

GLOSSARY

Bohr effect Influence of pH on the oxygen affinity of hemoglobin, discovered by Danish physiologist Christian Bohr early in this century

2,3-Diphosphoglycerate Ester of glyceric acid and phosphate, present in red blood cells, which lowers the oxygen affinity of hemoglobin; abbreviated DPG

Hydrogen bond Bond between charged or dipolar groups, mediated by a hydrogen atom carrying either a whole or a fractional positive charge

Hydrophobic Refers to groups that carry no net charge or permanent dipole (e.g., aliphatic or aromatic hydrocarbons) and are water repellent

Ligand Molecule that binds to the heme iron (e.g., oxygen, carbon monoxide, or nitric oxide)

Polar Refers to groups carrying either a net positive or negative charge or a permanent electric dipole; such groups attract water molecules, which are themselves dipolar, and they are therefore soluble in water, or hydrophilic

Salt bridges Hydrogen bonds between acidic and basic groups (i.e., between one group carrying a net positive charge and another carrying a net negative charge)

HEMOGLOBIN IS THE PROTEIN OF THE RED BLOOD cells that allows vertebrates to transport oxygen from the lungs to the tissues and helps the return transport of carbon dioxide from the tissues back to the lungs. Like all proteins, hemoglobin is made of small organic molecules called amino acids, strung together in a linear sequence called a polypeptide chain. There are 20 different amino acids, and their sequence in the chain is genetically determined. The hemoglobin molecule is made up of four polypeptide chains: two α chains of 141 amino acid residues each and two β chains of 146 residues each. The α and β chains have different sequences of amino acids, but fold up to form similar three-dimensional structures. Each chain harbors one heme, which gives blood its red color. The heme consists of a ring of carbon, nitrogen, and hydrogen atoms called porphyrin, with an atom of iron at its center (Fig. 1). A single polypeptide chain combined with a single heme is called a subunit of hemoglobin, or a monomer of the molecule. In the complete molecule four subunits are closely joined, as in a three-dimensional jigsaw puzzle, to form a tetramer.

In red muscle there is another protein, called myoglobin, similar in constitution and structure to a β subunit of hemoglobin, but made up of only one polypeptide chain and one heme. Myoglobin combines with the oxygen released by red blood cells, stores it, and transports it to subcelluar organelles called mitochondria, where the oxygen generates chemical energy by the combustion of glucose to carbon dioxide and water.

Myoglobin is the simpler of the two molecules. This protein, with its 2500 atoms of carbon, nitrogen, oxygen, hydrogen, and sulfur, exists for the sole purpose of allowing its single atom of iron to form a loose chemical bond with a molecule of oxygen (O_2).

An oxygen-free solution of myoglobin or hemoglo-

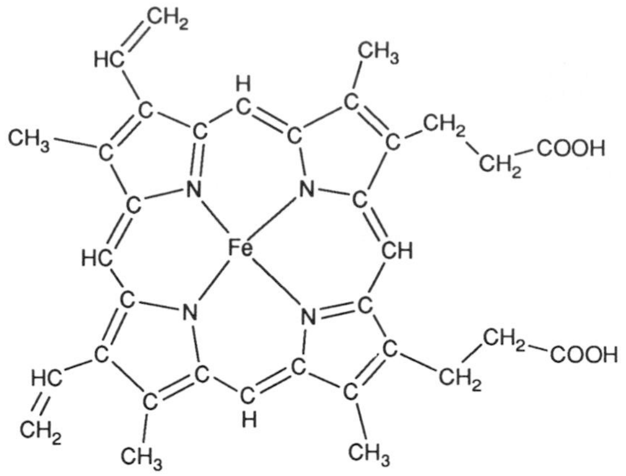

FIGURE 1 Iron–protoporphyrin IX. The $-CH_3$ (methyl) and $-CH-CH_2$ (vinyl) groups attached to the porphyrin ring are hydrophobic (i.e., water repellent) and are buried in the interior of the globin. The $-CH_2-CH_2-COOH$ (propionate) groups are hydrophilic (i.e., water attractant) and dip into the surrounding water.

bin is purple, like venous blood; when oxygen is bubbled through such a solution, it turns scarlet, like arterial blood. If these proteins are to act as oxygen carriers, then hemoglobin must be capable of taking up oxygen in the lungs, where it is plentiful, and giving it up to myoglobin in the capillaries of muscle, where it is less plentiful. Myoglobin, in turn, must pass the oxygen on to the mitochondria, where it is still more scarce.

I. OXYGEN EQUILIBRIUM

The equilibrium between myoglobin and oxygen can be represented by the equation

$$[\mathrm{Mb}] + [\mathrm{O_2}] \overset{K}{\rightleftharpoons} [\mathrm{MbO_2}],$$

where the brackets indicate concentration and K is the equilibrium constant. If we define Y, the fractional saturation with oxygen, by the equation

$$Y = \frac{[\mathrm{MbO_2}]}{[\mathrm{Mb}] + [\mathrm{MbO_2}]},$$

then the equilibrium is best represented by a graph in which the logarithm of $Y/(1 - Y)$ is plotted against the logarithm of the partial pressure of oxygen. In myoglobin it forms a straight line at 45° to the axes. The intercept of the line with the horizontal axis drawn at $Y/(1 - Y) = 1$ gives the equilibrium constant K. This is the partial pressure of oxygen, at which exactly one-half of the myoglobin molecules have taken up oxygen. The greater the affinity of the protein for oxygen, the lower the pressure needed to achieve half-saturation and the smaller the equilibrium constant. The 45° slope remains unchanged, but lower oxygen affinity shifts the line to the right and higher affinity shifts it to the left.

If the same experiment is done with blood or with a solution of hemoglobin, a different result is obtained. The curve rises gently at first, then steepens, and finally flattens out as it approaches the myoglobin curve. This strange sigmoidal shape signifies that oxygen-free molecules (i.e., deoxyhemoglobin) are reluctant to take up the first oxygen molecule, but their appetite for oxygen grows with the "eating." Conversely, the loss of oxygen by some of the hemes lowers the oxygen affinity of the remainder. The distribution of oxygen among the hemoglobin molecules in a solution therefore follows the Biblical parable of the rich and the poor: "For unto every one that hath shall be given, and he shall have abundance; but from him that hath not shall be taken away even that which he hath." This phenomenon suggested some kind of communication among the hemes in each molecule, and physiologists therefore called it heme–heme interaction. The term "cooperativity" is now preferred.

The equilibrium curve of hemoglobin begins with a straight line at 45° to the axes, because at first oxygen molecules are so scarce that only one heme in each hemoglobin molecule has a chance of catching one of them, and all of the hemes therefore react independently, as in myoglobin. As more oxygen flows in, the four hemes in each molecule begin to cooperate and the curve steepens. The tangent to its maximum slope is known as Hill's coefficient (n), after the English physiologist A. V. Hill, who first attempted a mathematical analysis of the oxygen equilibrium.

The normal value of Hill's coefficient in the blood of healthy human subjects under standard physiological conditions is 3.0, which means that in this part of the curve the fractional saturation of hemoglobin with oxygen increases with the third power of the partial pressure of oxygen. The curve ends with another line at 45° to the axes, because oxygen has now become so abundant that only a single heme in each molecule is likely to be free. In this situation there can be no cooperativity, and all hemes in the solution once more combine with oxygen independently, as in myoglobin.

II. COOPERATIVE EFFECTS: PHYSIOLOGICAL PURPOSE

Hill's coefficient and the oxygen affinity of hemoglobin depend on the concentrations of several chemical factors in the red blood cell: protons (hydrogen ions; i.e., hydrogen atoms without electrons, whose concentration, measured as pH, reflects the acidity of the solution), carbon dioxide, chloride ions, and an ester of glyceric acid and phosphate, called 2,3-diphosphoglycerate (DPG). These are known as heterotropic ligands, as opposed to oxygen or carbon monoxide, which are called homotropic ligands. Increasing the concentration of any of these heterotropic ligands shifts the oxygen equilibrium curve to the right, toward lower oxygen affinity, and makes it more sigmoidal (Fig. 2). The cooperative binding of oxygen and the influence of the heterotropic ligands on the oxygen equilibrium curve are known collectively as the cooperative effects of hemoglobin. Strangely, none of the heterotropic ligands influences the oxygen equilibrium curve of myoglobin, even though the chemistry and the structure of myoglobin are related closely to those of the individual chains of hemoglobin. The explanation emerged only recently (see the following).

What is the purpose of these cooperative effects? Why is it not good enough for the red blood cell to contain a simple oxygen carrier such as myoglobin? The answer is that such a carrier would not allow enough of the oxygen in the red blood cell to be unloaded to the tissues, nor would it allow enough carbon dioxide to be carried to the lungs by the blood plasma.

The cooperativity of oxygen binding and release, the effects of the heterotropic ligands, and the effect of temperature conspire to maximize the difference in fractional saturation with oxygen between arterial and venous blood. The partial pressure of oxygen in arterial blood is normally 90–100 mm Hg; in mixed venous blood, 35–40 mm Hg. Under standard conditions (i.e., pH 7.4, pCO_2 of 40 mm Hg, a DPG concentration of 5 m*M*/liter of packed red blood cells, carbon monoxyhemoglobin of 1%, and a temperature of 37°C) in whole blood from apparently healthy subjects, these partial pressures correspond to oxygen saturations of 98–100% in arterial and 66–73% in venous blood. Under these circumstances only one-quarter to one-third of the total oxygen carried is released in the tissues; the fractional saturation with oxygen in muscular veins during heavy exercise is believed to be lower, so that more oxygen would be delivered to the muscles.

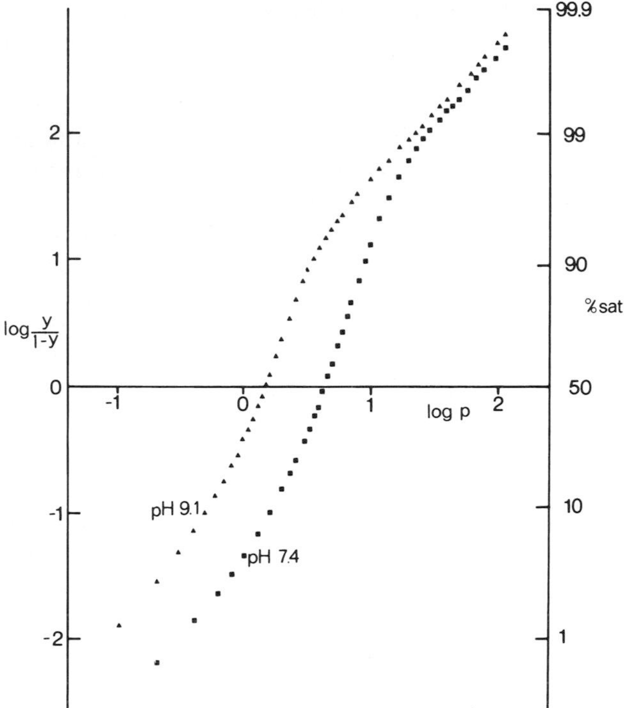

FIGURE 2 Double-logarithmic Hill plots of the oxygen equilibrium curves of normal human hemoglobin at two different pHs. The right-hand curve was measured at physiological pH, the left-hand one at alkaline pH. Its lower part is shifted to the left, toward higher oxygen affinity. %sat, percentage of saturation with oxygen; p, partial pressure of oxygen. (From J. V. Kilmartin, unpublished results.)

The more pronounced the sigmoidal shape of the equilibrium curve, the greater the fraction of oxygen that can be released. Several factors cooperate to that purpose. In the tissues, oxidation of nutrients by the tissues liberates lactic acid and carbonic acid; these acids, in turn, liberate protons, which shift the curve to the right, toward lower oxygen affinity, and make it more sigmoidal, which allows more oxygen to be released. DPG has the same effect. The number of DPG molecules in the red cell is about the same as the number of hemoglobin molecules, 280 million, and probably remains fairly constant during circulation; a shortage of oxygen, however, causes more DPG to be made, which further lowers the oxygen affinity and therefore helps to release more oxygen. The human fetus has a hemoglobin with the same α chains as the hemoglobin of an adult human, but different β chains that have a lower affinity for DPG. This gives fetal hemoglobin a higher oxygen affinity and facilitates the transfer of oxygen from the maternal to the fetal circulation.

If protons lower the affinity of hemoglobin for oxygen, then the laws of action and reaction demand that oxygen lowers the affinity of hemoglobin for protons. Hence, the liberation of oxygen causes hemoglobin to combine with protons and vice versa; at physiological pH about two protons are taken up for every four molecules of oxygen released, and two protons are liberated again when four molecules of oxygen are taken up (Fig. 3). This reciprocal action is known as the Bohr effect and is the key to the mechanism of carbon dioxide transport. The carbon dioxide released by respiring tissues is too insoluble to be transported as such, but it can be rendered more soluble by combining with water to form a bicarbonate ion and a proton. The chemical reaction is written

$$CO_2 + H_2O \rightarrow HCO_3^- + H^+.$$

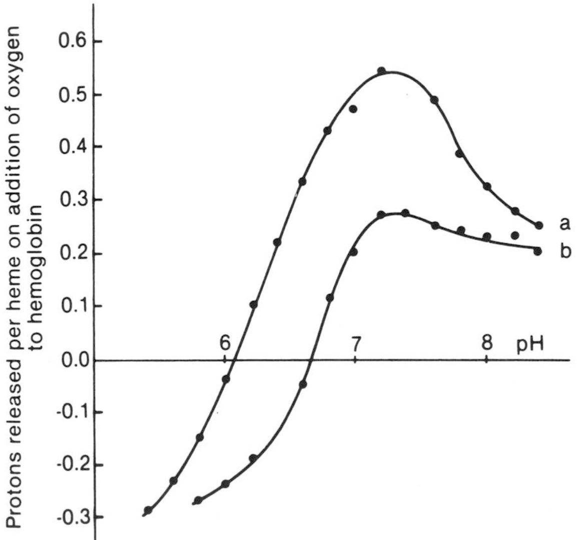

FIGURE 3 Discharge of protons upon uptake of oxygen by hemoglobin, known as the Bohr effect, after Danish physiologist Christian Bohr. (a) Native human hemoglobin. (b) Human hemoglobin from which the C-terminal histidines of the β chains, which are two of the principal residues responsible for the Bohr effect, have been cleaved. [From J. V. Kilmartin and J. F. Wootton (1970). Inhibition of Bohr effect after removal of C-terminal histidines from haemoglobin β-chains, *Nature* **228**, 776.]

In the absence of hemoglobin this reaction would soon be brought to a halt by the excess of protons produced, like a fire going out when the chimney is blocked. Deoxyhemoglobin acts as a buffer, mopping up the protons released in the tissues and thus tipping the balance toward the formation of soluble bicarbonate. In the lungs the process is reversed. There, as oxygen binds to hemoglobin, protons are cast off, driving carbon dioxide out of the solution so that it can be exhaled.

III. ALLOSTERIC PROTEINS

Many proteins besides hemoglobin exhibit cooperative effects. Most of these are enzymes that catalyze (i.e., speed up) chemical reactions in the living cell. Such enzymes consist of several subunits, each containing a site that catalyzes the same reaction. The combination of substrate (i.e., the molecule that undergoes the reaction) with one of the subunits raises the substrate affinities of all of the subunits, just as the combination with oxygen of one of the four subunits of hemoglobin raises the oxygen affinity of all of them. Chemical compounds that bear no resemblance to the substrates themselves often regulate the substrate affinity of such enzymes, just as hydrogen ions or DPG regulate the oxygen affinity of hemoglobin.

In most of these proteins, the cooperative effects arise through a transition between two or more alternative structures with different substrate affinities. These structures are distinguished by the arrangement of the subunits and the number and kinds of bonds between them. If there are only two alternative structures, the one with fewer and weaker bonds between the subunits would be free to develop its full catalytic activity or oxygen affinity. It is therefore called R, for "relaxed." The activity would be damped in the structure with more and stronger bonds between the subunits; this form is called T, for "tense." In the absence of oxygen or substrate, nearly all of the protein molecules have the T structure; if the protein molecules are saturated with oxygen or substrate, nearly all of them have the R structure. The greater the proportion of molecules in the R structure, the greater the affinity of the solution for oxygen or substrate.

Cooperativity arises by a change in the relative population of protein molecules in the T and R structures as the reaction progresses. For example, in hemoglobin the steep slope of the oxygen equilibrium curve near half-saturation of the solution with oxygen arises when the change in the relative population of the two forms as a function of oxygen saturation is greatest. Proteins that exhibit cooperativity and change their structures in response to chemical stimuli are called allosteric. Compounds other than oxygen or substrates that change the equilibrium between T and R states, such as DPG in red blood cells, are called allosteric effectors. In hemoglobin the heterotropic ligands all act as effectors that shift the allosteric equilibrium toward the T structure. This is the reason why they shift the oxygen equlibrium curve of hemoglobin to the right, toward lower oxygen affinity. They do not affect the oxygen equilibrium curve of myoglobin because it is a monomer, and allosteric effects arise only if several monomers combine, as in hemoglobin.

IV. THREE-DIMENSIONAL STRUCTURE

It has become customary to speak of the primary, secondary, tertiary, and quaternary structures of proteins. Primary refers to the amino acid sequence; secondary to the local conformation of the polypeptide chain, such as the α helix or pleated sheet; tertiary to the fold of a single polypeptide chain, as in myoglobin; and quaternary to the assembly of several chains or subunits (e.g., the hemoglobin tetramer).

The fold of the polypeptide chain is the same in myoglobin and in the α and β chains of hemoglobin. It is made up of seven or eight α-helical segments and an equal number of nonhelical ones placed at the corners between them and at the ends of the chain (Fig. 4). The helices are named A–H, starting from the amino terminus, and the nonhelical segments that lie between helices are named AB, BC, CD, and so on. The nonhelical segments at the ends of the chain are called NA at the amino terminus and HC at the carboxyl terminus. Residues within each segment are numbered from the amino terminus: A1, A2, CD1, CD2, and so on. Evolution has conserved this fold of the chain, despite great divergence of the sequences: the only residues common to all hemoglobins are the proximal histidine F8, which attaches the heme to the globin, and phenylalanine CD1, which wedges the heme into its pocket (Fig. 5). Most globins also have a histidine on the distal (i.e., oxygen) side of the heme. In myoglobin and in the α subunits this histidine forms a strong hydrogen bond with the bound oxygen, but in the β subunits the bond is either weaker or absent.

Myoglobin has ionized side chains distributed all over its surface, but the surfaces of the α and β globin chains have nonpolar patches that allow them to com-

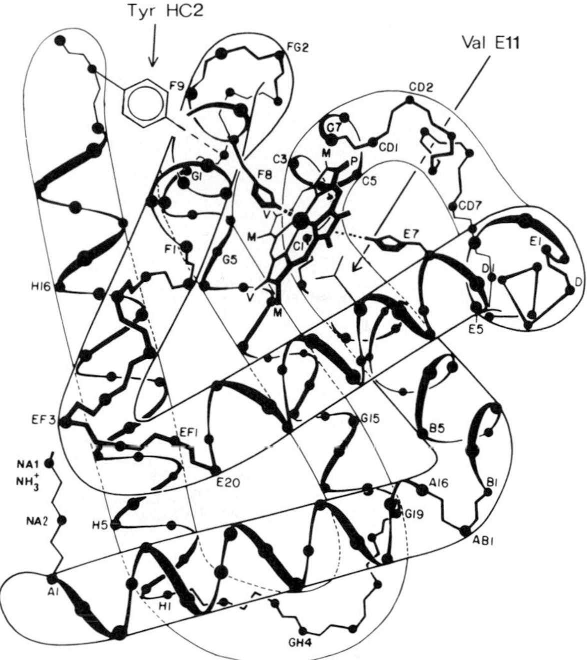

FIGURE 4 Secondary and tertiary structure of the hemoglobins showing α carbons and coordination of the hemes. Shown are the proximal histidine F8, which links the heme iron to the globin; the distal histidine E7 and valine E11, which make contact with the bound oxygen; and tyrosine HC2, which links the carboxyl terminus of the chain to helix F by a hydrogen bond. The exact numbers of residues in the different segments are the same in all mammals, but vary in other vertebrates and especially invertebrates. The lettering of the helical and nonhelical segments is explained in the text. M, V, and P, methyl, vinyl, and proportionate side chains, respectively, of the heme.

bine to form the $\alpha_2\beta_2$ tetramer. To make a model of the tetramer, each of the subunits must first be joined to its partner around a twofold symmetry axis, which brings one subunit into congruence with the other by a rotation of 180°. One pair of chains is then inverted and placed on top of the other to make a tetramer in which the four subunits are arranged at the corners of a tetrahedron (Fig. 6). The twofold symmetry axis that relates the pairs of α and β chains runs through a water-filled cavity at the center of the molecule. This cavity widens upon transition from the R structure to the T structure to form a receptor for the allosteric effector DPG between the two β chains and also for chloride ions.

The heme is made up of one atom of ferrous iron at the center of protoporphyrin IX, which is synthesized in the bone marrow from glycine and acetate (see Fig. 1). It is wedged into a pocket of the globin with its hydrocarbon side chains interior and its polar

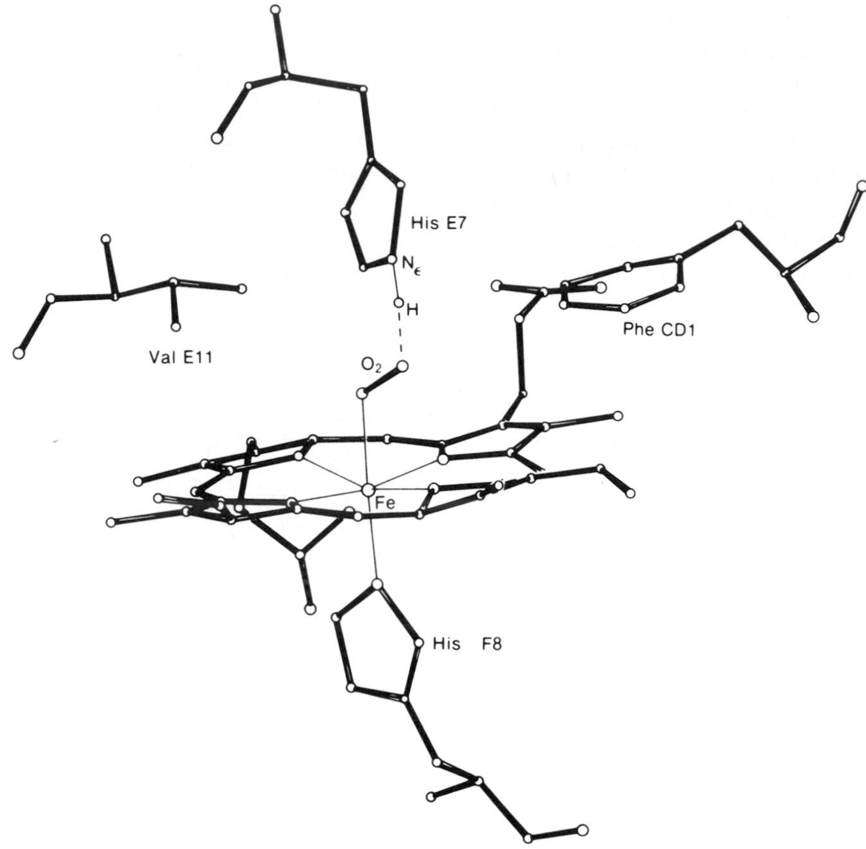

FIGURE 5 Arrangement of proximal and distal histidines in oxymyoglobin, showing the hydrogen bond between N_ϵ of the distal histidine and the bound oxygen. His, histidine; Val, valine; Phe, phenylalanine.

propionate side chains exterior; it is in contact with about 20 side chains of the globin, all hydrophobic apart from the two histidines.

V. TRANSITION FROM THE DEOXY STRUCTURE TO THE OXY STRUCTURE

Upon transition from the deoxy (T) structure to the oxy (R) structure, one $\alpha\beta$ dimer rotates relative to the other by 12–15° (Fig. 7). The rotation is accompanied by a shift of ~1 Å along the rotation axis and makes the two $\alpha\beta$ dimers move relative to each other along the $\alpha_1\beta_2$ and $\alpha_2\beta_1$ contacts. The two contacts form a two-way switch that ensures that the $\alpha\beta$ dimers click back and forth between no more than two stable positions, so that any stable intermediates in the reaction of hemoglobin with ligands must have either the quaternary R or the T structure. The T structure is constrained by additional bonds between the subunits, which oppose the changes in tertiary structure needed to flatten the hemes upon combination with oxygen. These bonds take mainly the form of salt bridges (i.e., hydrogen bonds between oppositely charged ions), such as

$$-NH_3^+ \begin{matrix} \cdots O- \\ \\ \cdots O \end{matrix} C- \quad \text{or} \quad -NH^+ \begin{matrix} \cdots O- \\ \\ O \end{matrix} C-$$

One pair of salt bridges is formed by the carboxyl-terminal arginines of the α chains, and another by the carboxyl-terminal histidines of the β chains. In addition, four pairs are formed by DPG with cationic groups of the β chains (Fig. 8).

Chloride ions were believed to form additional salt bridges in the T structure, but in fact they act differently. The channel that runs through the center of

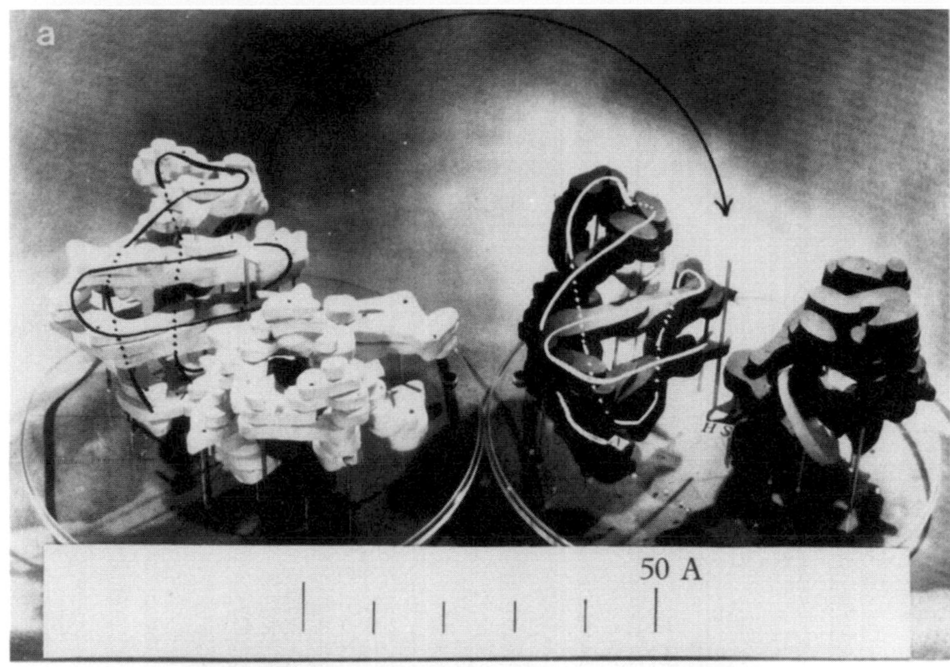

FIGURE 6 Assembly of hemoglobin tetramer. (a) A pair of α subunits (white) and a pair of β subunits (black) are placed on either side of the twofold symmetry axis (central rod with white sign). The pair of α subunits is then inverted and placed on top of the β subunits to form the complete tetramer (b). Note the hemes in separate pockets. Rotation by 180° about the twofold symmetry axis brings the molecule to congruence with itself. The letters refer to the helical and nonhelical segments shown in Fig. 4 and explained in the text.

the T structure is lined predominantly with positively charged, basic amino acid side chains. Their mutual electrostatic repulsion destabilizes the T structure and biases the allosteric equilibrium toward the R structure. Chloride ions diffuse into the channel and stabilize the T structure by neutralizing the positive charges without being bound to any one of them. This stabilization of the T structure lowers the oxygen affinity.

Hydrogen ions, DPG, chloride ions, and carbon dioxide all stabilize the T structure. They lower its oxygen affinity and retard the T $\rightarrow$ R transition. For example, at low concentrations of the allosteric

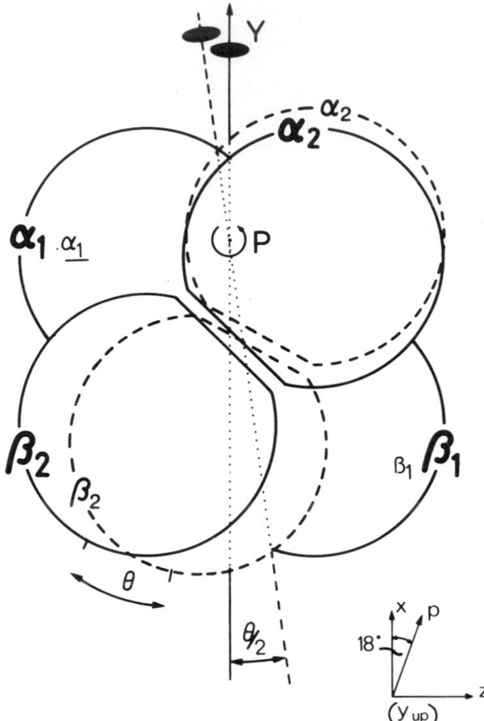

FIGURE 7 The change in quaternary structure that accompanies the ligation of hemoglobin. Bold symbols and solid lines in the diagram refer to deoxyhemoglobin; light symbols and dashed lines refer to oxyhemoglobin. The binding or release of oxygen causes little movement of the α_1 and β_1, or α_2 and β_2, subunits relative to each other. The oxygenated and deoxygenated $\alpha_1\beta_1$ dimers have been superimposed. The position of the oxygenated $\alpha_2\beta_2$ corresponds to that obtained by moving the deoxygenated $\alpha_2\beta_2$ dimer as follows: rotating it about an axis *P* (which is perpendicular to the twofold symmetry axis, marked *Y*, of both the oxygenated and deoxygenated molecules, and to the picture plane) by an angle θ = 12 to 15° and shifting it along the axis *P* by 1 Å into the page. [From J. Baldwin and C. Chothia (1979). Haemoglobin: The structural changes related to ligand binding and its allosteric mechanism. *J. Mol. Biol.* **129**, 175.]

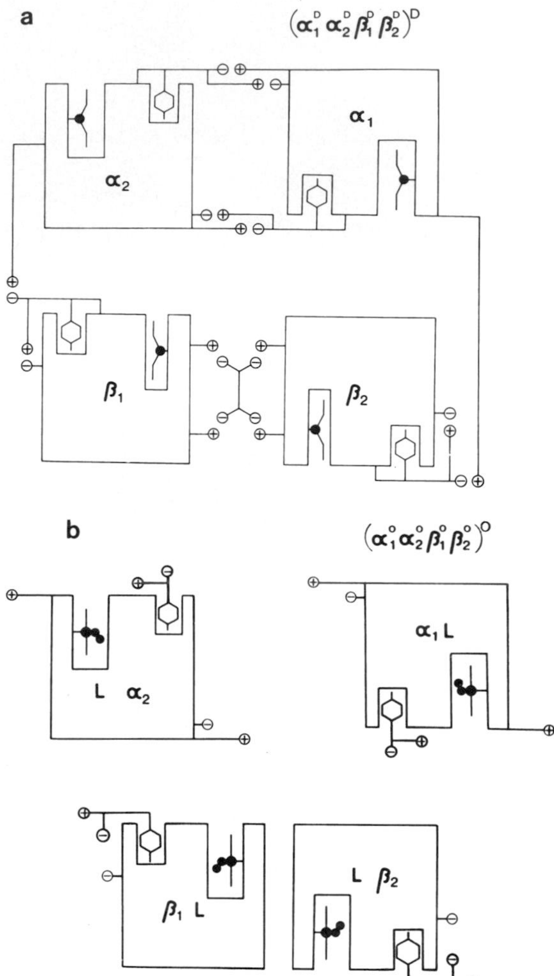

FIGURE 8 (a) Salt bridges by carboxyl-terminal residues, marked by juxtaposed positive and negative charges, in the deoxyhemoglobin (T) structure. The molecule with the four negative charges between the β chains of the T structure is 2,3-diphosphoglycerate. It must be released before transition to the R structure (b), where the gap between the β chains becomes too narrow to accommodate it, and the salt bridges are broken. The solid circles in (a) represent the iron atoms; in (b) the iron atoms and oxygen molecules attached to them. The lines extending from the solid circles represent the porphyrins. The hexagons represent the tyrosines HC2 that precede the carboxyl-terminal amino acid residues.

effectors, most of the hemoglobin molecules in a solution may click from the R structure to the T structure when the second oxygen molecule is bound. At high effector concentration they may click over only when the third oxygen molecule is bound. All effectors except chloride ions act by fortifying the salt bridges of the T structure. In this way they promote the release of oxygen to the tissues. The allosteric effectors leave the oxygen affinity of the R structure unchanged.

The transition from the T structure to the R structure is triggered by stereochemical changes at the hemes. In deoxyhemoglobin the porphyrins are domed, and the iron atoms are displaced by 0.4 Å from the plane of the porphyrin nitrogens, regardless of the quaternary structure of the globin. Upon binding oxygen the iron atoms move toward the porphyrin. These remain domed in the T structure, but flatten upon transition to the R structure. As a result, upon transition from deoxyhemoglobin (oxygen free) in the quarternary T structure to oxyhemoglobin (hemoglobin saturated with oxygen) in the quaternary R structure, the iron atoms and the proximal histidines move toward the mean planes of the porphyrins by 0.5–0.6 Å (Fig. 9). In deoxyhemoglobin in the quaternary T structure, combination of the β hemes with oxygen is blocked by the distal valine; in the R structure this block is removed by tilt of the β hemes relative to helix E. The doming and undoming of the porphyrins, the movements of the irons and the proximal histidines, and, in the β subunits, the movement of the distal residues relative to the heme are seen as the only perturbations that could set in motion the changes in quaternary structure.

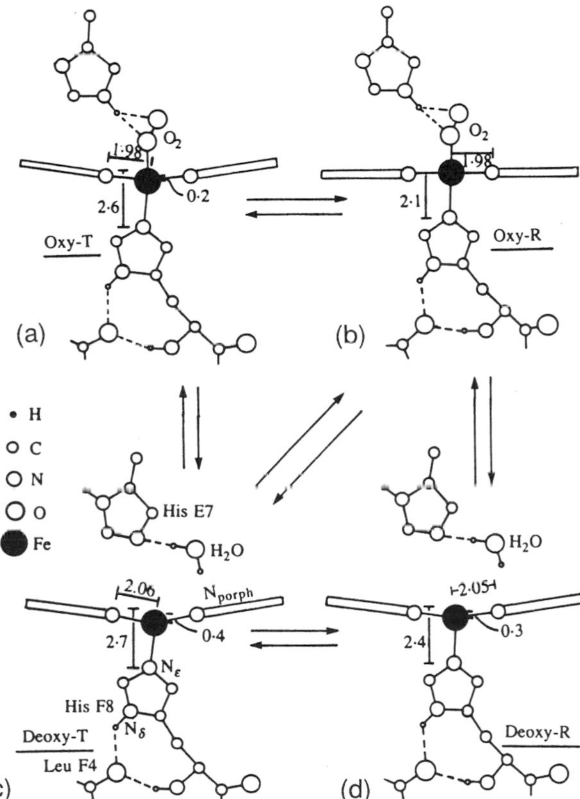

FIGURE 9 Changes in the heme stereochemistry on binding of oxygen by the α subunits in the R and T structures. (a) Upon uptake of oxygen by the T structure, the heme remains domed and the iron remains displaced from the porphyrin plane, due to the constraints by the tightly packed side chains of the globin around the heme; on the other hand, upon dissociation of oxygen from the R structure (d), the porphyrin becomes domed, because R structure, being relaxed, does not force the heme to remain flat. (c) In the oxygen-free T structure the porphyrin is domed and the iron atom is displaced from the porphyrin plane. (b) In the oxygenated R structure the porphyrin is flat and the iron lies in the porphyrin plane. The numbers give distances in Ångstrom units. [Reprinted with permission from M. F. Perutz, G. Fermi, B. Luisi, B. Shaanan, and R. C. Liddington (1987). Stereochemistry of cooperative effects in hemoglobin. *Acc. Chem. Res.* **20**, 309. Copyright 1987 American Chemical Society.]

VI. EMBRYONIC AND FETAL HEMOGLOBINS

The transport of oxygen differs in the embryonic, fetal, and adult stages of development. The early embryo obtains oxygen from the maternal interstitial fluid and uses a hemoglobin known as $\zeta_2\varepsilon_2$. The developing fetus obtains its oxygen via the placenta, using a hemoglobin known as F_{II} or $\alpha_2\gamma_2$. This has the same α chain as the adult form, but its β chain, known as γ, differs from the adult form in 39 positions. In fact, two different kinds of γ chain are made; they differ only in position H14(136), where one chain has a glycine and the other has alanine. There is also a minor component, known as hemoglobin F_I, in which the amino-terminal valines of the γ chains are acetylated. The oxygen affinities of hemoglobins F_{II} and F_I are higher than those of the two adult forms [hemoglobin A ($\alpha_2\beta_2$) and A2 ($\alpha_2\delta_2$)], which facilitates the transfer of oxygen across the placenta from the adult to the fetal circulation. Their higher oxygen affinity is due to their lower affinity for the allosteric effector, DPG, that is, at equal concentrations of hemoglobin and DPG, the latter lowers the oxygen affinity of fetal hemoglobin less than that of adult hemoglobin.

VII. HEMOGLOBIN DISEASES

Hemoglobin genes are subject to mutations that alter the structure of the globin. Several hundred variants of human hemoglobins have been isolated and chemically characterized. In most of them the abnormality consists of the replacement of one pair of identical amino acid residues by another. All of these replacements are consistent with single-base substitutions in the DNA coding for the globin chains. Some variant hemoglobins have residues deleted or inserted; in others the chains are cut short or elongated, and yet others contain hybrids of β and δ or β and γ chains. Though many of the abnormal hemoglobins are without effect on their carriers' health, others give rise to symptoms. In over 30 abnormal hemoglobins the stereochemical causes of these symptoms are now known.

Among the hemoglobin diseases caused by single-amino-acid substitutions, sickle-cell anemia is the most serious, affecting the largest number of people. Many other substitutions give rise to hemolytic anemia when they affect the stability of hemoglobin, causing unfolding of the globin chains and the formation of clumps of unfolded hemoglobin molecules in red blood cells. These anemias are generally less serious, first, because they do not lead to the blocking of blood vessels that causes the crises characteristic of sickle-cell anemia and, second, because each of these substitutions is sufficiently rare to exclude its occurrence in the father and the mother of the same child. Many substitutions raise the oxygen affinity of hemoglobin, which diminishes the amount of oxygen delivered to the tissues. A lack of oxygen is registered by a sensor in the kidneys that responds by releasing a hormone called erythropoietin. This hormone stimulates the synthesis of red blood cells in the bone marrow, with the consequence that such patients tend to suffer from polycythemia (i.e., an excess of red blood cells), which could be severe enough to cause discomfort and occasional symptoms. On the other hand, their high erythrocyte count and high oxygen affinity make them better adapted than other people to life at high altitudes. There are far fewer substitutions that lower the oxygen affinity. Those that do occur often manifest themselves by cyanosis (i.e., bluish color of the skin due to insufficient oxygenation of the blood) and sometimes lead to diminished red blood cell synthesis. [*See* Sickle Cell Hemoglobin.]

Another group of hemoglobin diseases, known as thalassemias, arises from failures to synthesize either the α or the β globin chains. Individuals with only one sickle cell or thalassemia gene and one normal hemoglobin gene are generally healthy, because their red blood cells contain enough normal hemoglobin to supply them with oxygen, but people with two defective hemoglobin genes tend to be severely crippled. In 1949 the British geneticist J. B. S. Haldane first spotted that these diseases are most frequent in areas where malaria is prevalent. This has been confirmed by studies in many parts of the world. For reasons that we do not yet understand, the altered hemoglobin inhibits the multiplication of the malaria parasite in the red blood cells of infants carrying one sickle cell or one thalassemia gene. These infants are therefore more resistant to malaria than are normal infants and stand a better chance of surviving to adult age. [*See* Malaria.]

It seems that the mutations causing either sickle-cell anemia or thalassemia arise spontaneously in human populations. In the absence of malaria, selective pressure penalizes the carriers of the altered hemoglobins and they produce fewer children, but if malaria is present, fewer children with normal hemoglobin than children who carry one defective hemoglobin gene survive to reproductive age. The high incidence of thalassemia in malarial islands of Melanesia and its rarity in malaria-free islands are particularly impressive, since people have inhabited these islands for no more than 3000 years; Darwinian selection must therefore have operated in historical times. It is the best example of evolution by natural selection in humans.

VIII. GENETICALLY ENGINEERED HEMOGLOBIN AS A BLOOD SUBSTITUTE

Scientists have long tried to find ways of using solutions of hemoglobin as substitutes for blood in transfusion, but they failed for two reasons. Red blood cells contain very concentrated hemoglobin solutions in which all the hemoglobin molecules are tetramers, but hemoglobin freely dissolved in the blood would be more dilute. In dilute solution the hemoglobin molecules split into halves containing one α and one β subunit. The half-molecules are quickly excreted by the kidneys, turning the urine red, and would therefore be lost to the blood. In addition, they appear to poison the kidneys. Furthermore, DPG would be too dilute to have any effect, and without it the oxygen affinity becomes too high for effective delivery of oxygen to the tissues.

K. Nagai has invented a method of linking the two halves of the molecule so that they cannot fall apart. He noticed that the beginning of one α chain lies close to the end of its partner α chain. He therefore suggested connecting them by a bridge made of one residue of glycine, so that the two α chains now form one continuous chain. He also suggested lowering of the oxygen affinity by substituting asparagine 109β by lysine, a substitution known to lower the oxygen affinity in an abnormal human hemoglobin. This hemoglobin has a normal sigmoid oxygen equilibrium curve and remains in the circulation for several days before being broken down. There are high hopes that it will prove suitable for transfusion.

BIBLIOGRAPHY

Antonini, E., and Brunori, M. (1971). "Hemoglobin and Myoglobin and Their Reactions with Ligands." North-Holland, Amsterdam.

Bunn, H. F., and Forget, B. G. (1986). "Hemoglobin: Molecular, Genetic and Clinical Aspects." Saunders, Philadelphia.

Dickerson, R. E., and Geiss, I. (1983). "Hemoglobin." Cummings, Menlo Park, CA.

Fermi, G., and Perutz, M. F. (1981). "Haemoglobin and Myoglobin: Atlas of Biological Structures" (D. C. Phillips and F. M. Richards, eds.). Oxford Univ. Press (Clarendon), Oxford, England.

Imai, K. (1982). "Allosteric Effects in Haemoglobin." Cambridge Univ. Press, Cambridge, England.

Ogden, J. E. (1992). Recombinant hemoglobins in the development of red-blood-cell substitutes. *Trends Biotechnol.* **10,** 91–96.

Perutz, M. F. (1979). Regulation of oxygen affinity of hemoglobin. *Annu. Rev. biochem.* **48,** 327.

Perutz, M. F. (1990). Mechanisms regulating the reactions of human hemoglobin with oxygen and carbon monoxide. *Annu. Rev. Physiol.* **52,** 1–25.

Perutz, M. F., Fermi, G., Luisi, B., Shaanan, B., and Liddington, R. C. (1987). Stereochemistry of cooperative effects in hemoglobin. *Acc. Chem. Res.* **20,** 309.

Hemoglobin, Molecular Genetics and Pathology

DAVID WEATHERALL
Institute of Molecular Medicine, University of Oxford, John Radcliffe Hospital

GLOSSARY

Enhancer DNA sequence that enhances the transcriptional activity of genes and may regulate their expression in particular tissues

Exon Segment of the coding sequence of a gene

Intron Intervening sequence within a gene that contains noncoding sequences, which do not appear in the processed messenger RNA

Promoter Region on the DNA molecule to which RNA polymerase binds and initiates transcription

Pseudogene Sequence with homology to a particular gene but that contains one or more mutations that prevent its normal function

Trans-activating factor DNA-binding protein that interacts with specific regulatory regions within or close to structural genes

HEMOGLOBIN IS THE OXYGEN-CARRYING PROTEIN of the red cell. Like all mammalian hemoglobins, the human hemoglobins vary in structure during different periods of development, an adaptive process designed to meet differing oxygen-transport requirements. All human hemoglobins have a tetrameric structure, consisting of two pairs of different globin chains, each associated with one heme molecule. Hemoglobin is an allosteric protein, that is, its configuration undergoes alterations that are essential for its normal function. Its oxygen-binding properties are reflected in a sigmoid oxygen dissociation curve, which means that it can bind oxygen tightly in the lungs and release it rapidly when it encounters a low partial pressure of oxygen in the tissues. Furthermore, the oxygen affinity of hemoglobin can be modified according to physiological needs, the curve shifting to the left or right in response to pH, temperature, and carbon dioxide levels. Some of these adaptive changes are the result of binding small molecules, notably 2,3-diphosphoglycerate. All of these allosteric functions require the interaction of two unlike pairs of globin chains. Thus, the red blood cell precursors must synthesize these globin subunits in a synchronous manner, and a regulatory system has evolved that ensures that different structural hemoglobins are produced at appropriate times during fetal and adult life; fetal hemoglobin has a higher oxygen affinity than adult hemoglobin, an adaptive response to the oxygen requirements of the fetus. Apparently, by a series of gene duplications followed by mutations that have modified the function of particular forms of hemoglobin, we have arrived at our present state of evolution, in which we have different hemoglobins adapted specifically to varying physiological needs at particular phases of development.

I. STRUCTURE AND HETEROGENEITY OF HUMAN HEMOGLOBIN

Human adult hemoglobin is a heterogeneous mixture of proteins consisting of a major component, hemoglobin A, and a minor component, hemoglobin A_2, constituting about 2.5% of the total. In intrauterine life, the main hemoglobin is hemoglobin F. The structure of these hemoglobins is similar. Each consists of two different pairs of identical globin chains. Except for some of the embryonic hemoglobins (see the following), all normal human hemoglobins have one pair of α chains: in hemoglobin A these are combined with β chains ($\alpha_2\beta_2$), in hemoglobin A_2 with δ chains ($\alpha_2\delta_2$), and in hemoglobin F with γ chains ($\alpha_2\gamma_2$).

Human hemoglobin shows further heterogeneity, particularly in fetal life. Hemoglobin F is a mixture of two molecular species, which differ by only one amino acid residue, either glycine or alanine at position 136 in their γ chains; they are designated $\alpha_2\gamma_2^{136\,\mathrm{Gly}}$ and $\alpha_2\gamma_2^{136\,\mathrm{Ala}}$. The γ chains containing glycine at position 136 are called $^{G}\gamma$ chains; those that contain alanine are called $^{A}\gamma$ chains. At birth, the ratio of molecules containing $^{G}\gamma$ chains to those containing $^{A}\gamma$ chains is about 3 : 1; this ratio varies widely in the trace amounts of hemoglobin F present in normal adults. The fetal–adult hemoglobin switch, which relates γ to β chain production, starts before birth and, except in some pathological states, is complete by the end of the first year of life.

Before the eighth week of intrauterine life, there are three embryonic hemoglobins: Gower 1 ($\zeta_2\varepsilon_2$), Gower 2 ($\alpha_2\varepsilon_2$), and Portland ($\zeta_2\gamma_2$). The ζ and ε chains are the embryonic counterparts of the adult α chains, and β, γ and ζ chains, respectively. ζ chain synthesis persists beyond the embryonic stage of development in some of the genetic disorders of hemoglobin production; so far, persistent ε chain production has not been observed. During fetal development, there is an orderly switch from ζ to α chain and ε to γ chain production, followed by β and δ chain production after birth. In some inherited hemoglobin disorders, γ chain production is persistent into childhood and adult life. [*See* Hemoglobin.]

II. GLOBIN GENE CLUSTERS

Each different globin chain is the product of a specific gene. The α and ζ genes form a cluster on chromosome 16, whereas the γ, β, and δ gene cluster is on chromosome 11. The different human hemoglobins together with the arrangement of these gene clusters is shown in Fig. 1.

Although there is some individual variability, the α gene cluster usually contains one functional ζ gene and two α genes, designated $\alpha 2$ and $\alpha 1$. It also contains four pseudogenes—$\varphi\zeta$, $\varphi\alpha 1$, $\varphi\alpha 2$, and θ—that is, gene loci with homology to the α or ζ genes but with mutations that render them functionless. They are thought to be evolutionary remnants of once-active genes. The θ gene is remarkably conserved among different species. Although it appears to be expressed in early fetal life, its function is unknown; it seems unlikely that it can produce a viable globin chain.

Each α gene is located in a region of homology approximately 4 kb (kb = 1000 nucleotide bases) long, interrupted by two small nonhomologous regions. It is thought that the homologous regions have resulted from gene duplication and that the nonhomologous segments may have arisen subsequently by insertion of DNA into the noncoding regions around one of the two genes. As is the case for most mamma-

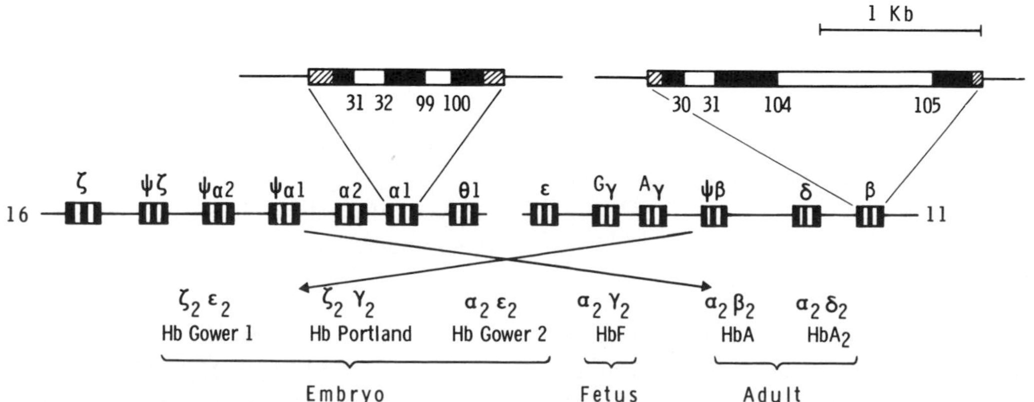

FIGURE 1 The human hemoglobins and the globin gene clusters on chromosomes 11 and 16. The extended figures of the $\alpha 1$ and β genes show the exons in dark shading, the introns in open boxes, and the flanking regions in lined shading. 1 Kb = 1000 nucleotide bases. Hb, hemoglobin.

lian genes, the α genes are divided into coding regions (exons) separated by intervening sequences (IVS), or introns. All the globin genes have two introns and three exons. The exons of the two α-globin genes have identical sequences. The first intron in each gene is identical, but the second intron of $\alpha 1$ is 9 bases longer and differs by 3 bases from that in the $\alpha 2$ gene. Despite their high degree of homology, the sequences of the two α-globin genes diverge in their 3′ untranslated regions 13 bases beyond the TAA stop codon (see the following). These differences provide an opportunity to assess the relative output of the two α genes. Apparently, the production of $\alpha 2$ messenger RNA exceeds that of $\alpha 1$ by a factor of 1.5–3.

The $\zeta 1$ and $\zeta 2$ genes are also highly homologous. The introns are much larger than those of the α-globin genes, and, in contrast to the latter, IVS 1 is larger than IVS 2. In each ζ gene, IVS 1 contains several copies of a simple repeated 14-bp sequence, which is similar to sequences located between the two ζ genes and near the human insulin gene. There are three base changes in the coding sequence of the first exon of $\zeta 1$, one of which gives rise to a premature stop codon, thus turning it into an inactive pseudogene.

The region separating and surrounding the α and α-like structural genes has been analyzed in detail. This gene cluster is highly polymorphic. There are five so-called hypervariable regions in the cluster, one downstream from the $\alpha 1$ gene, one between the $\zeta 2$ and $\zeta 1$ genes, one in the first intron of both ζ genes, and one 5′ to the cluster. These regions have been sequenced and found to consist of varying numbers of tandem repeats of nucleotide sequences. Taken together with numerous single-base restriction fragment-length polymorphisms (RFLPs; i.e., nucleotide variants that alter the patterns of DNA fragments after treatment with restriction enzymes), the genetic variability of the α gene cluster reaches a heterozygosity level of approximately 0.95. Thus, identifying each parental α-globin gene cluster in the majority of persons is possible. This heterogeneity has important implications for tracing the evolutionary history of the α genes.

The arrangement of the β-globin gene cluster on the short arm of chromosome 11 is ε, ${}^{G}\gamma$, ${}^{A}\gamma$, $\psi\beta$, δ, β. Each of the individual genes and their flanking regions has been sequenced. Like the $\alpha 1$ and $\alpha 2$ gene pairs, the ${}^{G}\gamma$ and ${}^{A}\gamma$ genes share a similar sequence. In fact, the ${}^{G}\gamma$ and ${}^{A}\gamma$ genes on one chromosome are identical in the region 5′ to the center of the second intron, yet they show some divergence 3′ to that position. At the boundary between the conserved and divergent regions, there is a block of simple nucleotide sequence, which may be a "hot spot" for the initiation of recombination events that have led to unidirectional gene conversion (i.e., matching of the two genes) during evolution.

Like the α-globin genes, the β gene cluster contains a series of single-point RFLPs, although in this case no hypervariable regions have been identified. The arrangements of RFLPs, or haplotypes, in the β-globin gene cluster fall into two domains. On the 5′ side of the β gene, spanning about 32 kb from the ε gene to the 3′ end of the $\psi\beta$ gene, three common patterns of RFLPs exist. In the region encompassing about 18 kb to the 3′ side of the β-globin gene, three common patterns also exist in different populations. Between these regions, there is a sequence of about 11 kb in which randomization of the 5′ and 3′ domains occurs, and, hence, where a relatively higher frequency of recombination may occur. Recent studies indicate that β-globin gene haplotypes are similar in most populations, although they differ markedly in individuals of African origin; these results suggest that these haplotype arrangements were laid down very early during evolution and are consistent with data obtained from mitochondrial DNA polymorphisms, which point to the early emergence of a relatively small population from Africa, with subsequent divergence into other racial groups.

III. STRUCTURE OF THE NONCODING REGIONS

The regions flanking the coding regions of the globin genes contain a number of conserved sequences that are essential for their expression. The first is the ATA box, which serves to accurately locate the site of transcription initiation at the CAP site, usually about 30 bases downstream, and which also appears to influence the rate of transcription. In addition, there are two so-called upstream promoter elements; 70 or 80 bp upstream is a second conserved sequence, the CCAAT box, and further 5′, approximately 80–100 bp from the CAP site, is a GC-rich region with a sequence that can be either inverted or duplicated. These promoter sequences are also required for optimal transcription; mutations in this region of the β-globin gene cause its defective expression. The globin genes have also conserved sequences in their 5′ flanking regions, notably AATAAA, which is the polyadenylation signal site. The promoter regions and adjacent sequences for each of the globin genes have been sequenced. They show many different binding sites for both hemopoietic-line-specific DNA-binding proteins such as GATA-1 and NF-E2, together with sites for a number of binding proteins that are expressed ubiquitously in many different cells.

IV. EXPRESSION AND REGULATION OF THE GLOBIN GENES

The primary transcript of the globin genes is a large messenger RNA precursor containing both intron and exon sequences. During its stay in the nucleus, it undergoes a good deal of processing, which entails modifying the 5′ end and polyadenylation of the 3′ end, both of which probably serve to stabilize the transcript. The introns are removed from the messenger RNA precursor in a complex two-stage process, which depends on certain critical sequences at the intron–exon junctions.

Very little is known about the regulation of globin gene transcription. The methylation state of the genes clearly plays an important role in their ability to be expressed; in human and other animal tissues, the globin genes are extensively methylated in nonerythroid organs and are relatively undermethylated in hemopoietic tissues. A change occurs in the methylation pattern and chromatin configuration around the globin genes at different stages of human development. Increasing evidence indicates that these genes come under the influence of so-called trans-acting factors, DNA-binding proteins that may be developmental stage-specific. The promoter sites are involved in efficient transcription of the globin genes. In addition, however, other sequences apparently are involved, particularly in tissue-specific expression. For example, evidence indicates the existence of so-called enhancer sequences, which may act by coming into spatial apposition with the promoter sequences to increase the efficiency of transcription of particular genes. Several enhancer sequences for the human globin genes have been defined, although their precise role in the regulation and specificity of expression remains to be determined.

It is now clear that the globin gene complexes are under the control of remote cis-acting sequences. About 15 kb upstream from the ε-globin gene there is an important regulatory region called the β locus control region (β-LCR). Naturally occurring deletions of this region completely inactivate the whole β-globin gene complex. This region is characterized by a set of developmentally stable, DNase I-hypersensitive sites, HS1, 2, 3, and 4, distributed over approximately 15 kb. This region contains many binding sites for both hemopoietic-specific and nonspecific DNA-binding proteins. The expression of the human α-globin gene is similarly dependent on an upstream regulatory element, referred to as HS-40 because it is characterized by an erythroid-specific DNase I-hypersensitive site 40 kb upstream from the embryonic ζ gene. If this site is lost by deletion, the entire α-globin gene cluster is inactivated. Although the precise mechanism of action of these key regulatory regions has not yet been determined, it seems likely that they become opposed to the promoters of the different globin genes at different stages of development.

Globin chain synthesis is directed in the cytoplasm of red cell precursors by the processed messenger RNA. Translation is initiated by the binding of a specific initiator transfer RNA, which carries the amino acid methionine to the initiation codon AUG. The initiation complex includes the two subunits that make up the ribosomes together with a number of proteins called initiation factors and with guanosine triphosphate and adenosine triphosphate. The first amino acid incorporated into globin chains is methionine, which is later cleaved. The globin messenger RNA is then translated in a stepwise fashion, during which time amino acids are added to the growing chain until the ribosomes reach the termination codon UAA. At this time, the ribosomal subunits drop off the messenger RNA and are reutilized for further protein synthesis. The completed peptide chain, which starts to fold into its complex secondary and tertiary configuration and which probably binds heme while still on the ribosomes, then associates with its partner chains to form the definitive hemoglobin molecule. α and β chain synthesis is almost synchronous, although a slight excess of α chains is degraded by proteolytic enzymes in the red cell precursor.

V. DEVELOPMENTAL CHANGES IN GLOBIN GENE EXPRESSION

Knowledge about the developmental regulation of the globin genes is equally incomplete. β-globin synthesis commences early during fetal life, at approximately 8–10 weeks of gestation. Subsequently, it continues at a low level (ca. 10% of the total non-α-globin chain production) up to about 36 weeks of gestation, after which it is considerably augmented. At the same time, γ-globin chain synthesis starts to decline, so that at birth approximately equal amounts of γ- and β-globin chains are produced. Over the first year of life, γ chain synthesis gradually declines, and by the end of the first year, this amounts to <1% of the total non-α-globin chain output. In adults, the small amount of hemoglobin F is confined to an erythrocyte population called F cells.

How this series of developmental switches is regulated is unknown. They are not organ-specific but are synchronized throughout the developing hemopoietic tissues. Although environmental factors may be involved, some form of "time clock" is built into the hemopoietic stem cell. At the chromosomal level, regulation apparently occurs in a complex manner involving the interaction of developmental stage-specific trans-activating factors with sequences in the $\gamma\delta\beta$-globin gene cluster.

VI. MOLECULAR PATHOLOGY

The gene disorders of hemoglobin are the most common single-gene disorders in the world population and affect millions of individuals. They result from either mutations that alter the structure of a globin chain or those that cause a drastic reduction in the rate of production of one or more globin chains.

Most of the structural hemoglobin variants result from a single-base change in one or another of the globin genes, causing the production of a hemoglobin variant with a single amino acid substitution. Structural variants with shortened or elongated subunits are more rare. Though many hemoglobin variants are harmless, some give rise to diseases of varying severity. The best-known example of a disease due to a structural hemoglobin variant is sickle-cell anemia. Sickle-cell hemoglobin results from the substitution of glutamic acid by valine in the sixth position of the β-globin chain. This causes abnormal aggregation of hemoglobin molecules and, hence, a sickling deformity of the red blood cell. This in turn leads to shortening of the red cell survival and aggregation of sickled erythrocytes in small blood vessels with subsequent death of tissue due to a reduced oxygen supply. Some hemoglobin variants cause molecular instability; the hemoglobin molecule precipitates in the red blood cell, causing its premature damage in the circulation. Others interfere with oxygen transport.

The genetic disorders of hemoglobin that are characterized by a reduced rate of production of the α- or β-globin chains are called thalassemias. There are many forms of thalassemia, the most common being α- and β-thalassemias, which are due to a reduced rate of production of the α- or β-globin chains. This leads to imbalanced globin chain synthesis with the precipitation of the chain that is produced in excess, damage to the developing red blood cell, and, hence, a varying degree of anemia. The α-thalassemias are classified into the α^0-thalassemias, in which both pairs of α-globin genes are lost ($\alpha\alpha/--$), and the α^+-thalassemias, in which one of the pair of α genes is lost ($\alpha\alpha/-\alpha$). In the homozygous state for α^0-thalassemia, no α genes remain; the condition causes intrauterine death. The compound heterozygous state for α^0- and α^+-thalassemia results in a clinical disorder called hemoglobin H disease, which is associated with moderate anemia and globin chain imbalance with the production of β_4 molecules, or hemoglobin H. Less commonly, the α-globin genes in α-thalassemia are not deleted but are inactivated by point mutations similar to those found in β-thalassemia.

Over 150 different mutations have been described in the globin genes of patients with β-thalassemia. In this case, deletions are rare and the majority result from point mutations that produce premature termination codons within exons, mutations that cause a shift in the reading frame of the genetic code and, hence, premature termination of chain synthesis; a wide variety of different mutations that involve the critical splice junctions and, hence, cause abnormal splicing of messenger RNA; and point mutations that lie near the promoter boxes and cause defective transcription of the globin genes. The promoter mutations and some of the splicing mutations may produce quite mild defects in β-globin production and a less serious form of the disease.

It is now possible to identify these different mutations in fetal DNA and to offer rapid prenatal diagnosis of the thalassemias. This is of particular importance because these are the most common single-gene disorders in humans and have a very high frequency in the Mediterranean region, the Middle East, the Indian subcontinent, and many parts of Southeast Asia.

BIBLIOGRAPHY

Bunn, H. F., and Forget, B. G. (1986). "Hemoglobin: Molecular, Genetic and Clinical Aspects." Saunders, Philadelphia/London.

Stamatoyannopoulos, G., Nienhuis, A. W., Leder, P., and Majerus, P. W. (1994). "The Molecular Basis of Blood Diseases," 2nd Ed. Saunders, Philadelphia/London.

Weatherall, D. J., Clegg, J. B., Higgs, D. R., and Wood, W. G. (1995). The hemoglobinopathies. *In* "The Metabolic Basis of Inherited Disease" (C. R. Scriver, A. L. Beaudet, W. S. Sly, and D. Valle, eds.), 7th Ed. McGraw–Hill, New York.

Hemophilia, Molecular Genetics

EDWARD G. D. TUDDENHAM
Clinical Research Centre, United Kingdom

JULIE HAMBLETON
University of California, San Francisco

GLOSSARY

Coagulation Process whereby blood is transformed into a solid clot

Exon Protein-coding region of gene

Factor VIII Protein cofactor of coagulation, deficient in hemophilia A

Factor IX Inactive zymogen precursor of factor IXa; deficient in hemophilia B

Factor IXa Activated form of factor IX, which activates factor X to factor Xa

Factor Xa Enzyme that activates prothrombin

Factor XI Inactive zymogen precursor of factor XIa; deficient in hemophilia C

Factor XIa Activated form of factor XI, which activates factor IX

Fibrinogen Protein that gives rise to blood clot when acted upon by thrombin

Hemophilia Literally, "love of blood"; medically, any bleeding disorder due to coagulation factor deficiency, usually inherited

Intron Intervening sequence of gene between exons (i.e., spliced out during messenger RNA processing)

Platelet Blood cell involved in coagulation

Prothrombin Inactive zymogen precursor of thrombin

Splice junction Sequences of DNA at junction of intron and exon that are essential for correct splicing

Thrombin Terminal enzyme of the coagulation process

von Willebrand factor Protein that carries factor VIII in the blood and promotes platelet adhesion; deficient in von Willebrand's disease

HEMOPHILIA COMPRISES A DIVERSE GROUP OF DISorders in which deficiency of a clotting factor leads to an inheritable tendency to bleed. The bleeding tendency may be mild, occurring only after injury, or severe, occurring spontaneously depending on the factor and its level of deficiency. Diagnosis of hemophilia depends on specific testing of the coagulation system for deficiency of particular coagulation factors. Treatment is by means of replacement of the relevant factor. Inheritance varies for different factors according to their chromosomal location and physiological behavior. Mutations in the genes for the 10 clotting factors involved in hemophilia are now being discovered and are highly diverse, including all of the known types of mutation affecting gene function. Treatment of hemophilia with plasma-derived clotting factor concentrates produced a high rate of infection with viruses prior to institution of detergent-solvent and purification procedures. The currently available factor products now undergo viral inactivation and purification steps or are derived from bioengineered recombinant DNA.

I. HISTORY OF HEMOPHILIA

Failure of the clotting mechanism produces very dramatic clinical problems, which have impressed medical observers from early times. A village in which

ENCYCLOPEDIA OF HUMAN BIOLOGY, Second Edition, VOLUME 4.

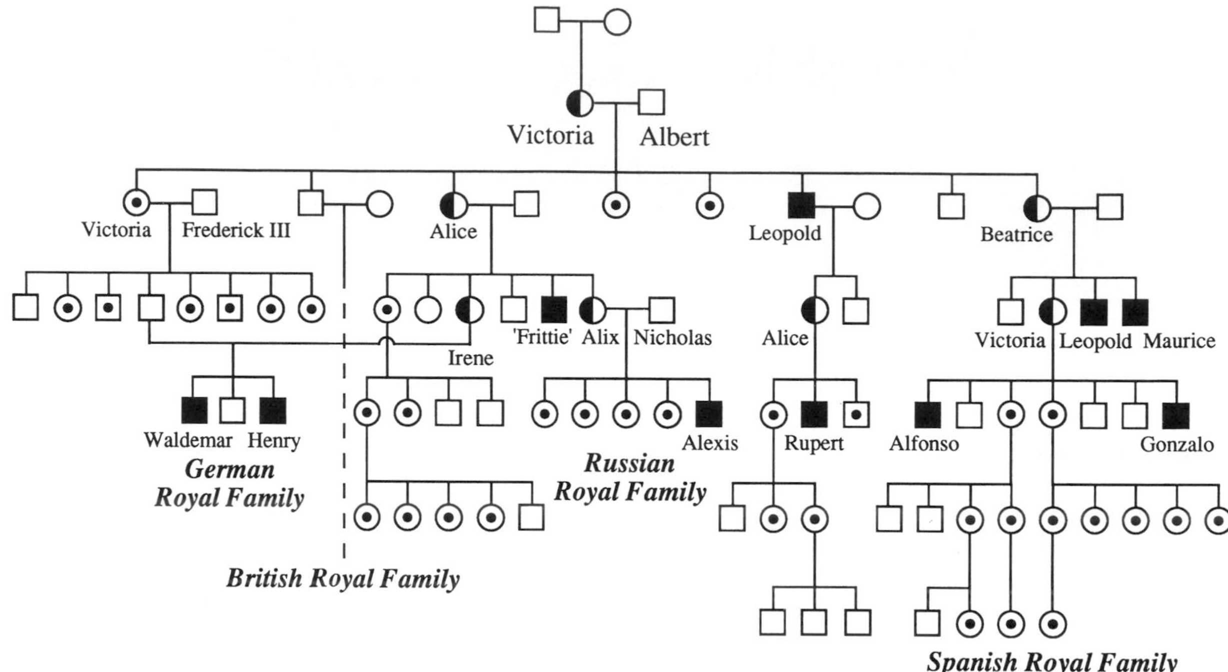

FIGURE 1 Queen Victoria's antecedents and descendants show that she was a carrier of sex-linked hemophilia, most probably arising as a new mutation in the gamete she received from her elderly father. Two of her daughters were definite carriers and passed the hemophilic gene to three other European royal families. Her son Leopold was a hemophiliac. Whether the royal hemophilia was due to factor VIII or factor IX deficiency is not known and will probably never be determined because the trait died out in this family before the availability of specific tests. The present British royal family descends through a normal male and, therefore, cannot carry the gene. □, normal; ■, hemophiliac; ⊡, possible hemophiliac; ○, normal; ◐, carrier; ⊙, possible carrier.

many men and boys suffered bleeding after injury was described by Khalif Ibn Abbas, the great Moorish surgical writer in the tenth century. In 1803, John Otto described in the Medical Repository of New York a family with hemophilia affecting males and transmitted by apparently unaffected carrier females. Many other cases and family reports appeared during the nineteenth century and the clinical features and patterns of inheritance of sex-linked hemophilia became well known. Queen Victoria was a carrier of hemophilia, which she transmitted through her daughters to three of the royal households of Europe (Russian, German, and Spanish; Fig. 1). Delayed coagulation of hemophilic blood was demonstrated toward the end of the nineteenth century, but the prevailing theory of blood coagulation at that time was inadequate to account for the deficiency in hemophilia. In 1936, researchers partially isolated a principal from normal blood able to correct the clotting time of sex-linked hemophilic blood. Around 1950, two kinds of sex-linked hemophilia emerged (then called hemophilia A and B), and they are now known to be due to deficiency of coagulation factor VIII and factor IX, respectively. Treatment of these two prevalent bleeding disorders with concentrates prepared from normal blood began in the 1960s, transforming the lives of hemophiliacs, who up to then had suffered early onset of crippling due to bleeding in the joints and a very short life expectancy due to catastrophic bleeding in the brain or after injury. Factor VIII and factor IX were purified to homogeneity in the 1970s and cloned and sequenced in the early 1980s, leading to discovery of the underlying mutations in hemophilia A and B.

II. NORMAL BLOOD COAGULATION

Figure 2 diagrams the processes whereby exposure of blood to surfaces outside the circulation, as occurs whenever a blood vessel is breached, leads through a linked series of enzymatic interactions to the production of fibrin clot. The blood clotting factors circulate as inactive precursors, either of enzymes (called zymo-

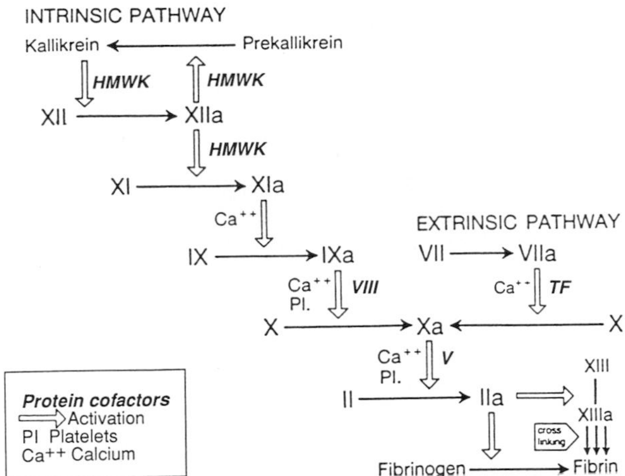

FIGURE 2 Blood coagulation cascade. The intrinsic pathway is initiated when blood contacts extravascular surfaces such as collagen. The extrinsic pathway is initiated by contact with tissues bearing the cell-surface receptor tissue factor (*TF*). "a" indicates the activated form of a clotting factor.

gens) or of cofactors (called procofactors). Both the so-called intrinsic pathway, starting with factor XI, and the extrinsic pathway, starting with plasma factor VII, are required for normal blood clotting. The way in which factor XI or factor VII is activated is unclear and remains controversial. In test-tube experiments, factor XI is activated by factor XII in concert with two other proteins, but all three of these contact factor proteins [factor XII, high-molecular-weight kininogen (HMWK) prekallikrein] are apparently dispensable for normal coagulation, because individuals who lack them have no problem with bleeding. Factor XIa activates factor IX. Tissue factor is present at the surface of many cells in the body and is a highly efficient cofactor for the activation of factor X by factor VIIa. No individual lacking tissue factor has been described, and such a deficiency would probably not be compatible with life. Factor VIII is a procofactor, which is itself activated by thrombin and then becomes efficient in promoting activation of factor X by activated factor IX, a step that occurs on the surface of platelets. Thus, the two processes of initiation/activation converge on factor X, generating factor Xa. Factor Xa activates prothrombin, and for this it requires the assistance of a cofactor V and platelet surface. Factor V circulates as a procofactor requiring to be activated by thrombin. (From where the initial thrombin for activation of factors V and VIII is derived is uncertain.) Finally, thrombin removes short peptides from fibrinogen (fibrino-peptides A and B), yielding fibrin monomers. Fibrin monomers spontaneously polymerize by end-to-end and staggered side-to-side association, producing the visible and structurally solid fibrin clot. The clot is stabilized by another enzyme that cross-links fibrin monomers covalently. This factor is designated factor XIII and, like factors V and VIII, requires to be activated by thrombin.

III. DEFECTS OF COAGULATION—THE HEMOPHILIAS

A. Factor VIII Deficiency—Hemophilia A (Classic Hemophilia)

Hemophilia A is the most common severe bleeding disorder in humans, affecting approximately 1 in 5,000 male births, or 1 in 10,000 of the general population. The gene for factor VIII is situated at the tip of the long arm of the X chromosome, not far from the gene for factor IX and close to the genes for color vision and for glucose-6-phosphate dehydrogenase (Fig. 3). The severity of bleeding in hemophilia A depends on a residual amount of clotting factor in the patient's blood. If none can be measured, the patient has a severe tendency to bleed with apparently spontaneous episodes of bleeding into his joints and muscles as often as three or four times a week. Paradoxically, the healing of small cuts and scratches is normal in hemophilia A because blood platelet function is normal in these patients and able to plug small holes in the smaller blood vessels. If untreated, bleeding into the joints causes severe pain and immobility but spontaneously resolves in a few days to weeks. The consequence of such bleeding is rapid and progressive articular damage leading to deformity and fixation of the joints most affected, the large load-bearing joints—knees, ankles, elbows. Any joint in the body can be affected and bleeding into the muscles can give rise to obstruction of blood flow, leading to death of tissue and contraction (Fig. 4). Mild hemophilia occurs in patients with residual factor VIII >5% of the normal level. Episodes of bleeding into joints and tissues affects these patients only after significant injury (Table I). Hemophilia A follows the typical inheritance of an X-linked recessive disorder. Females who carry a defective factor VIII gene are themselves protected from bleeding by its normal counterpart on their other X chromosome (Fig. 5). Occasionally, owing to the process of random inactivation of one X chromosome per cell, which occurs early in female embryogenesis, most of the normal

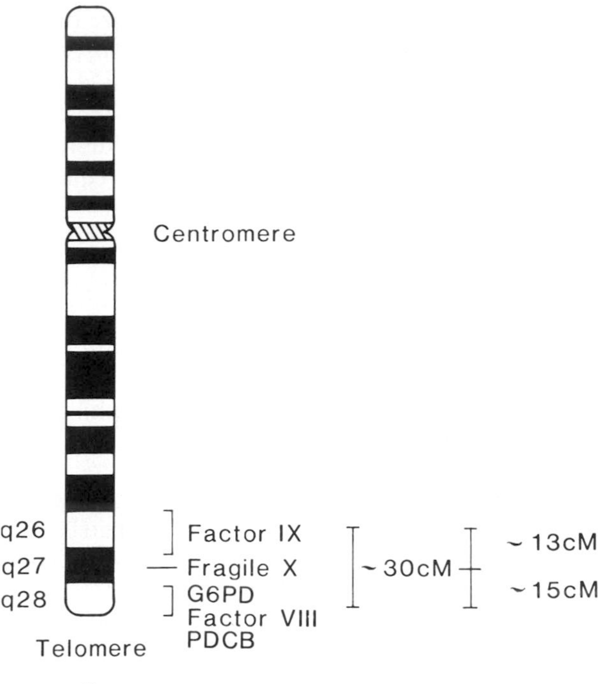

FIGURE 3 The X chromosome showing selected gene locations near the tip of the long arm. Fragile X is associated with mental retardation, glucose-6-phosphate dehydrogenase (G6PD) with hemolytic anemia, and protan deutan color blindness (PDCB) with color blindness. Factor VIII and factor IX genes are mutated in hemophilia A and B, respectively. The scale of distances is given in centimorgans (cM). One cM represents approximately 1 million base pairs of DNA.

X chromosomes are inactivated and, thus, a female carrier may have mild or moderate hemophilia. On average, half of the sons of a carrier female are affected by hemophilia A and half of her daughters are carriers. A male with classic hemophilia or hemophilia B has only normal sons and only carrier daughters.

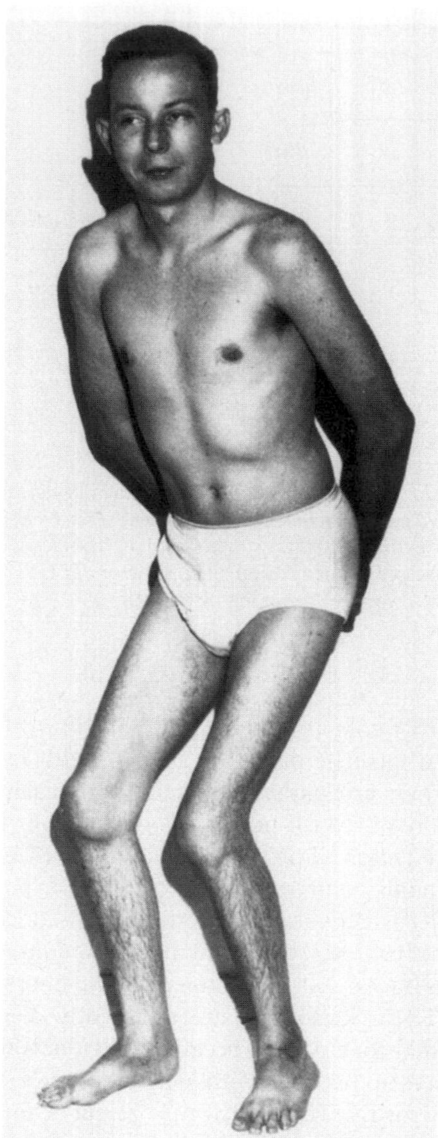

FIGURE 4 Hemophiliac who grew up before treatment with factor VIII was available. He suffered repeated joint and muscle bleeding, which led to deformity and fusion of the joints and wasting of the muscles.

B. Factor IX Deficiency—Hemophilia B

Hemophilia B cannot be distinguished from hemophilia A by examining the patient or taking his history, and the inheritance of the two disorders is the same. For example, which disorder was transmitted by Queen Victoria (see Fig. 1) is unknown. Specific blood tests are necessary to reveal a deficiency of clotting factor IX. The frequency of this disorder is somewhat less than that of hemophilia A, affecting about 1 in 30,000 males.

C. von Willebrand's Disease

von Willebrand's disease resembles hemophilia A in that there may be low levels of factor VIII, but an additional defect is found in failure of platelets to adhere to surfaces such as collagen that are exposed after blood vessel injury. This leads to prolonged bleeding from cuts because platelet plugs do not form in small capillaries after injury. The explanation for this combined defect of coagulation and platelet func-

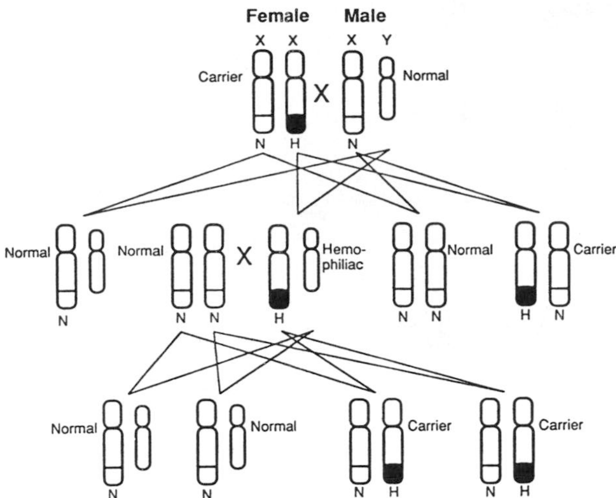

FIGURE 5 Inheritance of X-linked hemophilia. Females who carry a defective factor VIII or IX gene are protected from its effects by the normal gene on the paired chromosome. Males have only one X chromosome and, therefore, suffer from the defect.

TABLE I
Clinical Severity of Hemophilia[a]

Level of factor VIII or IX (% of normal)	Bleeding tendency
<2	Severe: frequent apparently spontaneous bleeds into joints and other internal organs.
2–10	Moderately severe: bleeding after minor trauma, with occasional spontaneous bleeds.
10–30	Mild: bleeding only after major trauma or surgery.

[a]Applies to hemophilia A and B, and not to von Willebrand's disease nor to factor XI deficiency (see text).

tion puzzled researchers for many decades. It is now clear that the deficient blood protein in this disorder—von Willebrand factor—has two functions: (1) to support and protect factor VIII in the circulation, preventing its rapid destruction, and (2) to promote adhesion of platelets under conditions of high shear rate. Patients with von Willebrand's disease lack von Willebrand factor owing to a defect of the von Willebrand factor gene located on chromosome 12. Most often, the disorder is transmitted as an autosomal dominant with a moderate reduction of both factor VIII and von Willebrand factor levels. These patients suffer bleeding from small vessels—the vessels where shear rate is high and von Willebrand factor platelet-adhesion promotion is most important. This gives rise to bruising, nose bleeds, bleeding from other mucous membranes of the intestinal tract, and heavy menstrual bleeding in women. Bleeding in the joints and muscles is rare because the factor VIII levels are usually above 10%. Some surveys indicate that the incidence of mild von Willebrand's disease may be as high as 1% of the general population. Because the bleeding is mild, diagnosis is often delayed into the second or third decade of life. Rare individuals have inherited a recessive gene for von Willebrand factor deficiency from both parents and completely lack the factor from their circulation and have very low levels of factor VIII. These patients suffer both from the type of bleeding seen in hemophilia A and from the superficial bleeding characteristic of platelet-adhesion defect.

D. Factor XI Deficiency—Hemophilia C

Hemophilia C has a most curious racial distribution, being common among Ashkenazi Jews and rare among all other racial groups, including the Sephardic Jews. Up to 8% of Ashkenazis (Jews of the Eastern Diaspora) have a defect in synthesis of factor

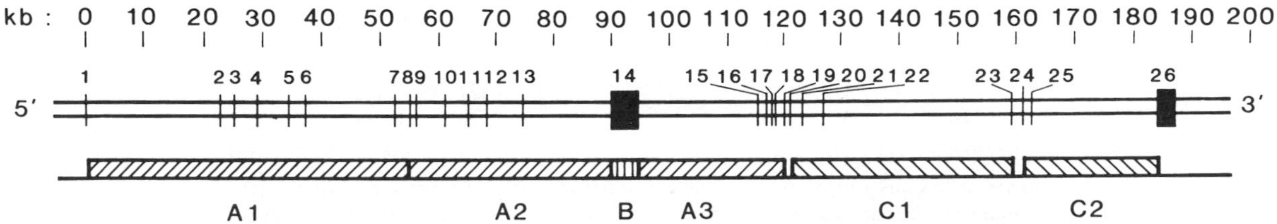

FIGURE 6 The factor VIII gene. Top line: scale in kilobase pairs of DNA. Middle line: diagram of gene; exons shaded dark, introns unshaded. Bottom line: factor VIII protein. A1–A3 domains are homologous to the copper-carrying protein ceruloplasmin. The B domain is removed when factor VIII is activated by thrombin. C1 and C2 domains are homologous to the slime mold protein discoidin.

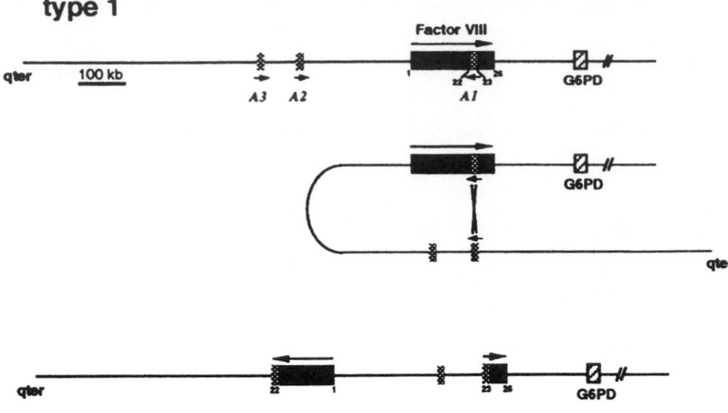

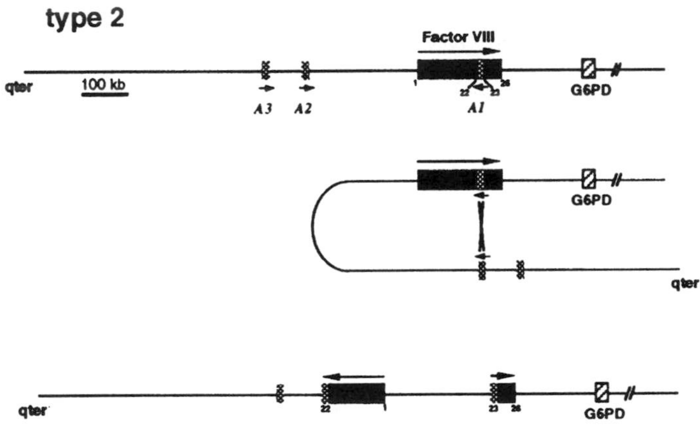

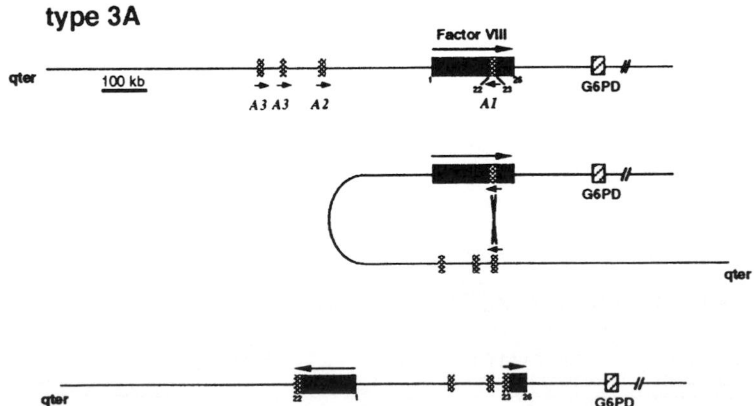

FIGURE 7 Schematic representation of the mechanisms of the different factor VIII gene inversions caused by intrachromosomal crossing-over between homologous sequences A. The mechanisms of types 1 and 2 are as proposed. The mechanisms of inversions 3A and 3B are two of several alternatives to explain the Southern blot patterns. The diagram at the bottom of the figure schematically shows the results of Southern blot analysis in the different types of factor VIII inversions. [Reproduced, with permission, from S. E. Atonarakis *et al.* (1995). *Blood* **86**(6), 2206.]

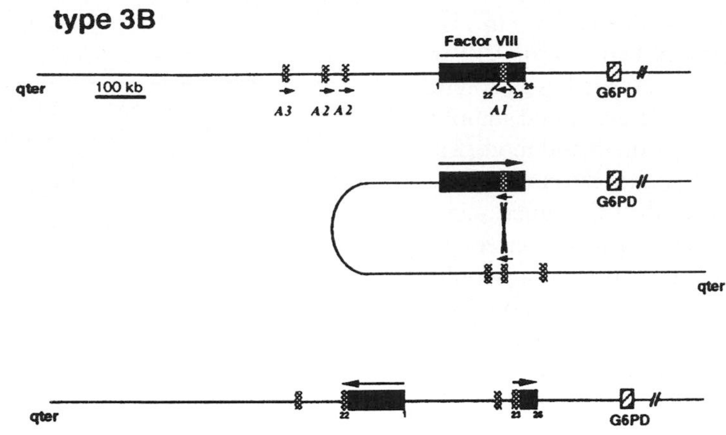

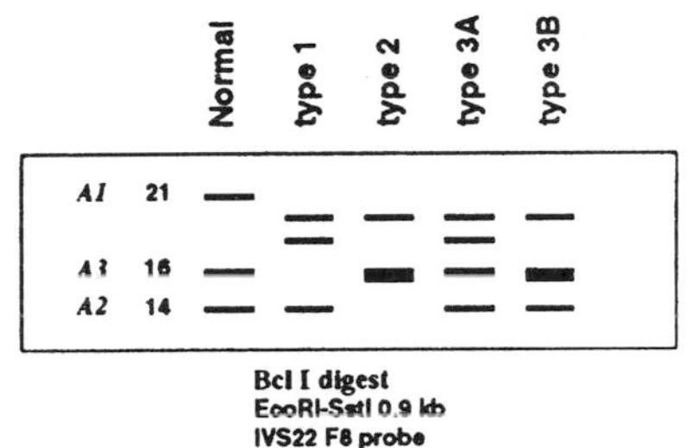

FIGURE 7 *(Continued)*

XI with moderately reduced levels of this contact factor. They generally have little or no clinical problem from bleeding except after severe trauma, and many such individuals have no problems with bleeding at all. Because the gene frequency is so high, homozygous cases are quite common, and these individuals have little or no factor XI in their circulation but still only suffer with a mild bleeding tendency, quite unlike the situation for factor VIII and factor IX deficiencies.

E. Deficiency of Other Clotting Factors

Most of the other clotting factors represented in Fig. 2 (factors I, II, V, X, VII, and XIII) are rarely found to be deficient in individuals with a hemophilia-like bleeding tendency. Because the inheritance in each case is autosomal recessive, usually only one generation is affected. Curiously, deficiency of factor XII or of prekallikrein, or HMWK, is not associated with hemophilic bleeding.

IV. MUTATIONS THAT CAUSE HEMOPHILIA

A. Factor VIII Gene Defects in Hemophilia A

The factor VIII gene spans a large region of the X chromosome, being about 190,000 base pairs in length. The protein-coding regions of the gene are divided into 26 exons (shown as shaded bars in Fig. 6). The processed messenger RNA for factor VIII is about 9000 base pairs in length and specifies a protein of 2351 amino acids. The gene is expressed mainly in liver cells, which produce the protein and modify it in various ways before releasing it into the circulation. Various kinds of mutation can affect the functioning of genes, and with such a very large gene an extremely large number of small or large mutations that would affect its function can be predicted. Furthermore, because hemophilia A is often lethal to affected males and greatly reduces a sufferer's chances of having children, mutations that cause it will be rapidly lost from the population and the incidence of the disorder must be maintained by a constant input of new mutations. This gives rise to the prediction made in 1935: a high proportion of cases of hemophilia is due to recent mutation and, hence, these mutations are likely to be diverse. The most common mutation found in upward of 50% of unrelated patients with hemophilia A is an inversion of a portion of the factor VIII gene (Fig. 7). The inversion is mediated by an unequal crossing-over of a particular DNA sequence that is copied three times within the factor VIII gene. This intrachromosomal exchange between homologous sequences in the gene results in a portion of the gene whose sequence is inverted, and thus no intact factor VIII protein is produced.

Point mutations and deletions in the factor VIII gene have also been described. Because of the extremely large size of the gene, there are technical problems of localization of point mutations, but those that have been found are indeed highly diverse. The simplest category of mutation to identify is a large deletion of the gene, and this has been found in about 5% of cases. Table II lists the size and extent of various deletions of the factor VIII gene that have been found in hemophilia A patients. All except one of the patients with these deletions have severe hemophilia A. The exception is a case in which exon 22 has been deleted. It so happens that the splice junctions between exons 21 and 22 and exons 22 and 23 are in-frame, so that removal of the exon does not interrupt the reading frame of the messenger RNA. It simply removes 52 amino acids from near the carboxy terminus of the protein. Apparently, the shortened protein retains about 5% activity in the circulation. Some, but not all, of these individuals have made antibodies to factor VIII, which was given to them as treatment for bleeding. Another type of mutation is the substitution of a single base creating either a stop signal or the substitution of an amino acid that affects the function of the protein, Note that substitution of an amino acid that had no effect on factor VIII function would not be detected clinically but simply be part of the natural variability as found, for example, in blood groups. Point mutations that create a premature stop codon have been found to occur preferentially at certain sites where the nucleotides cytosine and guanine occur together in the order CG followed by an A coding for arginine (Table III). Table IV lists such mutations identified in patients with severe hemophilia A. The proposed mechanism for these mutation "hot spots" is based on the fact that mammalian DNA is often methylated at cytosines followed by a guanine. Spontaneous deamination of methyl cytosine would convert the base to a thymidine, which, being a normal constituent of DNA, is not recognized by repair mechanisms. Where CG is followed by an A coding for arginine, the conversion will create a TGA codon for termination of translation. For example, the finding of four independent mutations at codon 2209 in the factor VIII coding sequence converting an arginine to stop is far too frequent to have happened as a result of chance alone given the numbers of patients screened. A similar mechanism operating on the opposite strand of DNA will convert the G to an A (its pair having been converted to a T on the lower or antisense strand). Such mutations have indeed been found associated with either severe or moderate hemophilia depending presumably on the effect of the substitution of a glutamine (CAA) for an arginine

TABLE II
Deletions in the Factor VIII Gene Causing Hemophilia A

Exons deleted	Factor VIII level (% of normal)	Antibodies to factor VIII
1–26	<1	No
1–22	<1	Yes
1–5	<1	No
3	<1	No
6	<1	No
14[a]	<1	No
22	5	No
26[b]	<1	No
15–18	<1	Yes
23–25	<1	No
23–26	<1	Yes
24 and 25	<1	No

[a]Two independent cases.
[b]Four independent cases.

(CGA) or a cysteine for an arginine at various sites. These examples are also listed in Table V. The mutations in which arginine occurs at a site where thrombin activates factor VIII by cleaving the primary chain of amino acids are of great theoretical interest for the function of factor VIII in coagulation. In some hemophilia A patients, either one of two critical thrombin-sensitive arginine residues has been replaced by a cysteine owing to the mutation of a C to T in the first base of codon 372 or 1689. Factor VIII protein with 1689 cysteine in place of arginine has a light chain, which is resistant to thrombin, and no functional activity in coagulation. A similar mutation at codon 372 has been found in a patient with moderately severe hemophilia A. This man's factor VIII with 372 cysteine in place of arginine has a thrombin-resistant heavy chain and low residual activity.

All of these point mutations have been discovered by screening, specifically at the CG codons, either with DNA-cutting enzymes, which cut at such sequences, or by using synthetic short stretches of DNA to probe these regions of the factor VIII gene. As other regions of the factor VIII gene are searched for defects in hemophilia, other mutations are expected to be found affecting the function of the factor VIII protein or the processing of the factor VIII message. The factor VIII gene represents a large target area of the human genome, and the accumulation of knowledge on the

TABLE III

Notation for Amino Acids

Amino acid	Three letter	One letter
Alanine	Ala	A
Cysteine	Cys	C
Aspartic acid	Asp	D
Glutamic acid	Glu	E
Phenylalanine	Phe	F
Glycine	Gly	G
Histidine	His	H
Isoleucine	Ile	I
Lysine	Lys	K
Leucine	Leu	L
Methionine	Met	M
Asparagine	Asn	N
Proline	Pro	P
Glutamine	Gln	Q
Arginine	Arg	R
Serine	Ser	S
Threonine	Thr	T
Valine	Val	V
Tryptophan	Trp	W
Tyrosine	Tyr	Y

TABLE IV

Point Mutations in the Factor VIII Gene Causing Hemophilia A

Codon[a]	Mutation	Substitution	Factor VIII level in patient's blood (%)	Number of unrelated cases
272	GAA → GGA	Glu → Gly	2	1
336	CGA → TGA	Arg → Stop	<1[b]	4
372	CGC → TGC	Arg → Cys	4[b]	2
1689	CGC → TGC	Arg → Cys	<1 or 4[b,c]	3
1941	CGA → TGA	Arg → Stop	<1	2
2116	CGA → TGA	Art → Stop	<1	1
2116	CGA → CCA	Arg → Pro	<1	1
2147	CGA → TGA	Arg → Stop	<1	2
2209	CGA → TGA	Arg → Stop	<1	4
2209	CGA → CAA	Arg → Gln	<1	3
2307	CGA → TGA	Arg → Stop	<1	1
2307	CGA → CAA	Arg → Gln	9	1

[a]Codons numbered 1–2332 in the sequence of mature factor VIII.

[b]Nonfunctional factor VIII protein was present in these cases with failure to activate due to mutation at thrombin cleavage site.

[c]Unexplained variation in residual activity of different cases.

TABLE V
Mutations in the Factor IX Gene Causing Hemophilia B

(Location[a]) codon	Mutation	Substitution/effect	Factor IX level in patient's blood (%)	Region of gene or protein	Number of unrelated cases
(−20)	T → A	Alters gene regulation	Increases at puberty	Promoter	1
(−6)	G → A	Alters gene regulation	Increases at puberty	Promoter	1
(13)	A → G	Alters gene regulation	Increases at puberty	Promoter	2
(6704)	GTA → GGA	Destroys exon C splice donor	<1	Promoter	1
−4	CGG → CAG	Arg → Gln	<1	Propeptide release	7
−4	CGG → TGA	Arg → Trp	<1	Propeptide release	1
−1	AGG → AGC	Arg → Ser	<1	Polypeptide release	1
7	GAA → GAC	Glu → Asp	<1	Gla domain	1
11	CAA → TAA	Gln → Stop	<1	Gla domain	1
27	GAA → AAA	Glu → Lys	<1	Gla domain	1
29	CGA → TGA	Arg → Stop	<1	Gla domain	2
29	CGA → CAA	Arg → Glu	5	Gla domain	1
33	GAA → GAC	Glu → Asp	10	Gla domain	1
47	GAT → GGT	Asp → Gly	10	Aromatic stack	1
50	CAG → CCG	Glu → Pro	<1	Aromatic stack	1
55	CCA → GCA	Pro → Ala	10	EGF[b]	2
60	GGC → AGC	Gly → Ser	10	EGF	3
114	GGA → GCA	Gly → Ala	?	EGF	1
120	AAC → TAC	Asn → Tyr	<1	EGF	1
145	CGT → TGT	Arg → Cys	10	Activation site	3
145	CGT → CAT	Arg → His	10	Activation site	8
173	CAA → TAA	Gln → Stop	<1	Activation peptide	1
180	CGG → CAG	Arg → Gln	<1	Activation site	1
180	CGG → CAG	Arg → Trp	?	Activation site	2
182	GTT → TTT	Val → Phe	<1	Protease	1
(20,566)	GGT → GTT	Destroys exon f splice donor	<1	Protease	1
194	TGG → TGA	Trp → Stop	<1	Protease	1
222	TGT → TGG	Cys → Trp	<1	Protease	1
248	CGA → TGA	Arg → Stop	<1	Protease	3
248	CGA → CAA	Arg → Gln	5	Protease	1
252	CGA → CTA	Arg → Leu	?	Protease	1
252	CGA → TGA	Arg → Stop	<1	Protease	2
257	CAC → TAC	His → Tyr	?	Protease	1
260	AAT → ACT	Asn → Thr	10	Protease	1
291	GCT → ACT	Ala → Thr	10	Protease	1
291	GCT → CCT	Ala → Pro	?	Protease	1
296	ACG → ACT	Thr → Met	<1	Protease	1
307	GTA → GCA	Val → Ala	10	Protease	1
311	GGA → CGA	Gly → Arg	5	Protease	1
333	CGA → CAA	Arg → Gln	1–5	Protease	4
336	TGT → CGT	Cys → Arg	5	Protease	1
338	CGA → TGA	Arg → Stop	<1	Protease	4
363	GGA → GTA	Gly → Val	?	Protease	1
390	GGA → GTA	Ala → Val	1–5	Protease	2
396	GGA → AGA	Gly → Arg	<1	Protease	1
397	ATA → ACA	Ile → Thr	5–10	Protease	7
407	TGG → AGG	Trp → Arg	<1	Protease	1
411	AAA → TAA	Lys → Stop	<1	Protease	1

[a](*N*), nucleotide position from capsite; −m, codon number of prepropeptide; −m, codon number of mature protein.
[b]EGF, epidermal growth factor-like domain.

random pattern of mutations across this region causing hemophilia should help us to understand more of the mechanisms by which such errors in DNA arise. The discovery of the mutation hot spots at CG dinucleotides is an example of a result from this natural laboratory.

B. Factor IX Gene and Hemophilia B

The factor IX gene is located on the X chromosome on the centromere side of the factor VIII gene at a probable distance of 30 million base pairs. As shown in Fig. 8, eight exons are within a gene spanning 33,500 base pairs. The protein for which this gene codes—factor IX—is shown in Figure 9, a diagram using the single-letter amino acid code (see Table III). The primary amino acid sequence consists of 461 amino acids as the nascent chain emerges from the ribosome into the endoplasmic reticulum. This undergoes several modifications before it is released into the circulation from the liver cells in which it is synthesized. Certain glutamic acid residues are modified by the addition of an extra COOH group to their gamma carbon, and these are indicated in the figure as a gamma symbol (γ). This step requires the presence of vitamin K and explains why antagonists of vitamin K, such as warfarin, produce a bleeding tendency. The signal for carboxylation to occur is contained within the prepro leader sequence, which itself is cleaved from the main protein by a processing protease during secretion from the cell. Furthermore, an aspartic acid residue at position 64 is hydroxylated in the β position (indicated by a β). Two asparagine residues have sugar molecules added.

By comparing the sequence of factor IX with that of other proteins, some fascinating similarities have emerged. The overall structure of factor IX very closely resembles that of three other proteins of blood coagulation: factor VII, factor X, and the anticoagulant protein C. The region labeled catalytic domain resembles other protein-cutting enzymes on the general pattern of trypsin. A similar domain is found in all the proteolytic enzymes of coagulation. Quite surprisingly, two domains resemble epidermal growth factor, and a domain labeled aromatic stack is thought to be involved in protein–protein interaction in coagulation. Activation of factor IX by factor XIa, the preceding enzyme in the intrinsic clotting pathway, occurs through cleavage between residues 145 and 146, and 180 and 181, releasing the activation peptide, on which the two asparagine-linked sugar chains are located (marked with lozenges on Fig. 9).

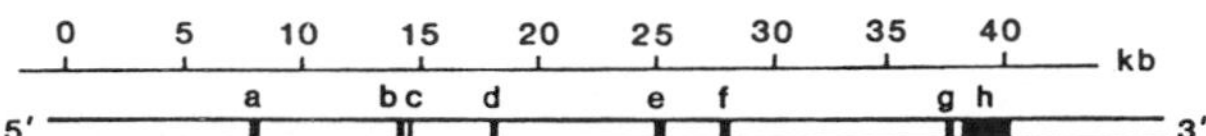

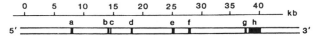

FIGURE 8 The factor IX gene. Top line: scale in kilobases. Middle line: diagram of gene; exons shaded, introns unshaded.

As with hemophilia A, hemophilia B mutations are rapidly lost from the population because of their highly adverse effect on reproductive fitness; therefore, new mutations account for a large proportion of cases and are diverse at the molecular level. About one-third of patients with hemophilia B have a non- or poorly functional factor IX molecule in their circulation. Some of these were the first mutations to be solved at the molecular level in hemophilia. One such factor is factor IX Chapel Hill (named after the location of the laboratory that carried out the research) and is caused by the substitution of histidine for arginine at residue 145. This interferes with activation by factor XIa, and factor IX Chapel Hill has low activity. The mutation responsible for the amino acid substitution affects the middle residue of codon 145, converting CGT to CAT. It is an example of the type of mutation discussed earlier, occurring at CG dinucleotides thought to be caused by deamination of methyl cytosine. Also shown in Figure 9 are nine other substitution (termed missense) mutations producing moderate or severe hemophilia B. The substitution of arginine by serine at position −1 in the prepro sequence led to failure of release of the N-terminal peptide and the abnormal factor IX molecule, which is nonfunctional. Table V lists more exhaustively other point mutations found in the factor IX gene in hemophilia B. Among these are examples of mutations affecting gene regulation, often referred to as hemophilia B Leyden. This class of mutations affects the way in which factor IX gene is controlled and leads to a condition characterized by low levels of factor IX in childhood that rise gradually toward normal levels around puberty. It seems that the mutations affect the response of the gene to testosterone, such that the gene is poorly expressed until the testosterone levels rise at puberty when the gene is turned on. Other point mutations affect the critical junctions between exon and intron, such that there is failure of correct processing of messenger RNA as seen in hemophilia B Oxford 1 and 2.

Deletion of part or all of the factor IX gene has been found in other patients, many of whom have made antibodies to factor IX after treatment for bleed-

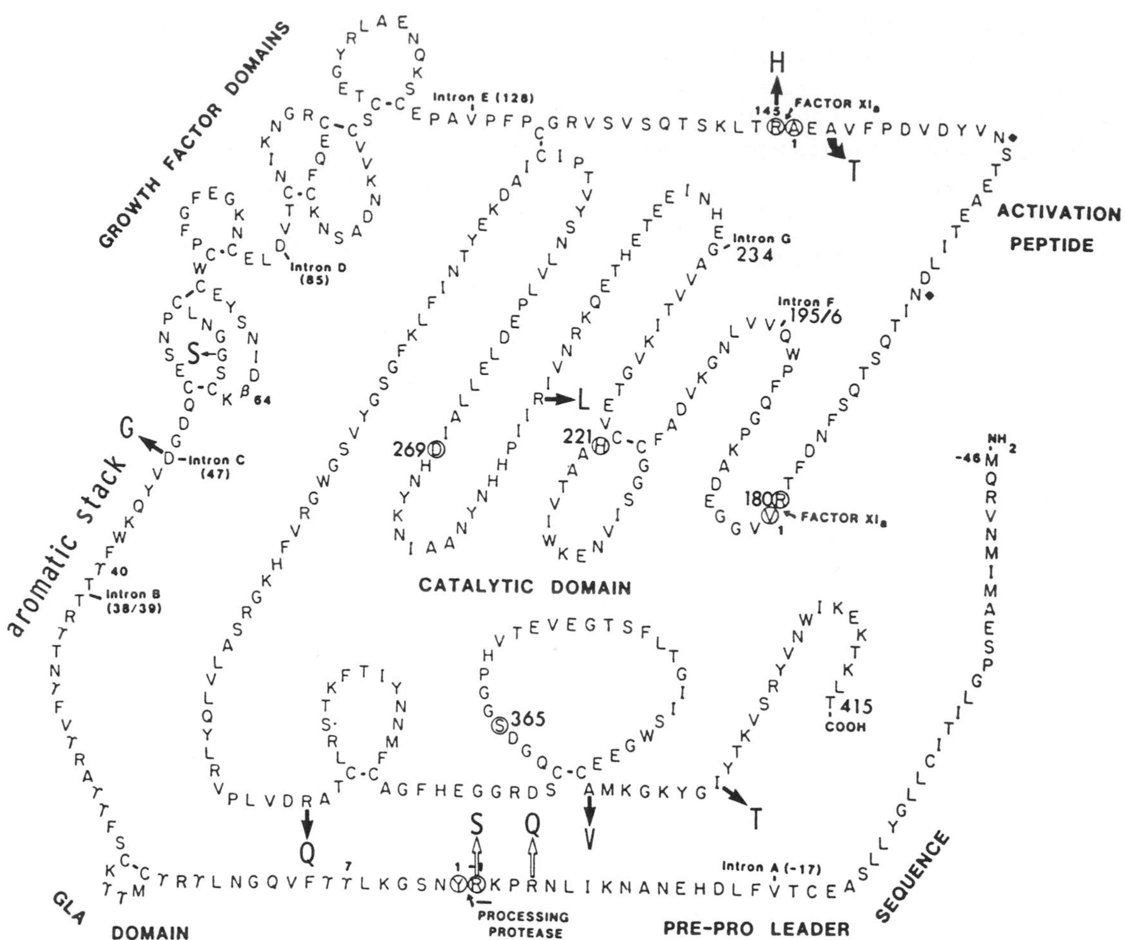

FIGURE 9 Factor IX protein (amino acid sequence single-letter codes as in Table III). Point mutations with substitution of a single amino acid producing hemophilia B; →Q, →S, etc. A ↘ T protein polymorphism. Arrows denote change in sequence 2° to mutation or polymorphism. [Modified, with permission, from Yoshitake, S. *et al.* (1985). Nucleotide sequence of the gene for human factor IX (antihaemophilic factor B). *Biochemistry* **24,** 3736. Courtesy of the American Chemical Society.]

ing. Some of these are listed in Table VI. Disorders of the factor IX gene in hemophilia B are proving to be a natural laboratory for study of the mechanisms of mutation and the correlation between structure and function of the factor IX protein.

C. von Willebrand Factor and von Willebrand's Disease

von Willebrand factor circulates in plasma as a series of multimers (Fig. 10) of a basic protomer with molecular weight approximately 240,000. During biosynthesis in endothelial cells lining the blood vessel, the proteins are assembled first into dimers through carboxy-terminal disulfide interchange and then into

TABLE VI
Deletions in the Factor IX Gene Causing Hemophilia B

Exons deleted	Antibodies to factor IX	Number of unrelated cases
a–h	Yes	12
a–h	No	2
a–c	Yes	1
e, g, h	Yes	1
f–h	Yes	1
e and f	No	1
d	No	1
d	Yes	1

multimers of higher order through N-terminal disulfide interchange in a step that depends on the presence of a very long propeptide. This peptide is excised from von Willebrand factor multimers during assembly and secreted separately but has no known function other than to promote multimer assembly. The importance of multimer assembly is attested by the fact that some individuals with von Willebrand's disease have von Willebrand factor in their circulation that is not properly multimerized and of low molecular weight and has very poor function in platelet adhesion although able to support factor VIII. The complete amino acid sequence of the von Willebrand factor monomer has been established and consists of a single chain of 2050 residues. The sequence of the propeptide was obtained from complementary DNA sequence and consists of 763 amino acids that are cleaved from the promoter during synthesis and assembly. The gene for von Willebrand factor encoded on chromosome 12 has only recently been cloned and mapped and consists of 52 exons. von Willebrand's disease is highly diverse in terms of clinical expression and the presence of different types of structural defects in von Willebrand factor.

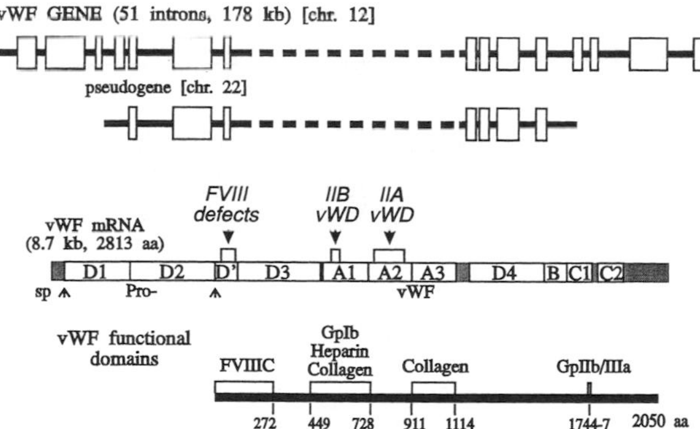

FIGURE 10 Schematic of the human von Willebrand factor (vWF) gene, mRNA, and protein. The vWF gene and pseudogene are depicted at the top, with boxes representing exons and the solid black line, introns. The vWF mRNA encoding the full prepro-vWF subunit is depicted in the middle as the stippled bar and lettered boxes. The locations of signal peptide (sp) and propeptide (pro-) cleavage sites are indicated by arrowheads, and the lettered boxes denote regions of internally repeated sequence. The approximate localizations for known vWF functional domains within the mature vWF sequence are indicated at the bottom. Numbers underneath the domains refer to amino acid residues within the mature vWF subunit. The clusters of mutations responsible for type IIA vWD, type IIB vWD, and the FVIII binding defects are indicated. aa, amino acids; chr, chromosome. [Reproduced with permission, from D. Ginsburg and E. J. W. Bowie (1992). *Blood* 79(10), 2507.]

D. Mutations Causing von Willebrand's Disease

An increasing number of mutations in the von Willebrand factor gene have been identified and are associated with the diseases. A small subgroup of patients with homozygous recessive von Willebrand's disease develop antibodies to the protein after infusion, and these individuals have deletions of all or part of their von Willebrand factor gene. The mutation in a patient with von Willebrand's disease in which there was failure of proper multimer formation has been localized to substitution of valine by asparagine at residue 844. It was shown by mutagenizing a normal von Willebrand factor clone at this precise position that the resulting von Willebrand factor made *in vitro* in a tissue culture system failed to form multimers, thus proving that this mutation was indeed the cause of the clinical condition, and coincidentally that residue 844 must be involved in multimer assembly. Other identified mutations are in a region of the von Willebrand factor gene wherein the protein product binds to specific receptors on platelets. These mutations render the von Willebrand factor ineffective in promoting platelet aggregation. Another form of von Willebrand's disease involves a mutation in a region where binding of von Willebrand factor to factor VIII occurs. The loss of this interaction due to the mutation results in rapid clearance of factor VIII from the circulation.

E. Factor XI Gene and Hemophilia C

Factor XI protein circulates as a dimer of two identical subunits, each consisting of a single chain of 607 amino acids. The gene for factor XI is 23,000 base pairs in length and contains 15 exons. The chromosomal location of the factor XI gene is 4_q35. Jewish patients with factor XI deficiency invariably have reduced amounts of circulating factor XI protein in parallel with the reduction of coagulant activity. Three mutations responsible for factor XI deficiency in Jews have now been identified, as shown in Table VII. This information raises a very interesting question in evolutionary genetics. Prior to the discovery of this molecular heterogeneity, it had always been proposed that factor XI deficiency in this racial group was due to the founder effect and subsequent genetic drift in a closed breeding population. Because the polymorphism clearly is due to at least three separate mutational events, some heterozygote advantage to partial factor XI deficiency is much more likely; however,

TABLE VII
Mutations in the Factor XI Gene Causing Hemophilia C

Type	Site	Mutation	Effect on expression	Frequency[a]
I	Junction of exon 14 and intron N	GT → AT	Destroys splice donor of exon 14	$\frac{1}{12}$
II	Exon 5, codon 117	GAA → TAA	Glu → Stop	$\frac{5}{12}$
III	Exon 9, codon 283	TTC → CTC	Phe → Leu	$\frac{6}{12}$

[a]Results of preliminary survey among Ashkenazi Jews.

the nature of this advantage can only be conjectural at present. One possibility is that some environmental factors, perhaps dietary and/or cultural, would promote a high incidence of thrombosis in this group, protection from which is provided by a partial coagulation factor deficiency.

F. Mutations Associated with Other Coagulation Factor Defects

Abnormal fibrinogen, which clots poorly (dysfibrinogenemia), has been shown in some cases to be due to a mutation at or close to the thrombin cleavage site of the fibrinogen α chain. Some cases of factor X deficiency have been associated with deletions of the factor X gene. So far, the molecular basis of factor VII deficiency and of factor V deficiency has not been reported, but it is to be expected that a range of mutations will be found in different families with these various coagulation factor deficiencies.

V. TREATMENT OF HEMOPHILIA

The mainstay of treatment for hemophilia A and B for the past 25 years has been concentrates derived from pooled normal human plasma. Twenty years ago, these concentrates were relatively impure and, until 10 years ago, were not treated in any way to inactivate viruses. As a consequence, the population of patients with hemophilia A and B became successively infected with hepatitis B, hepatitis non A non B (one form of which has recently been identified as hepatitis C virus), and human immunodeficiency virus. The drastic consequences of these infections are now being seen in end-stage liver failure and acquired immunodeficiency syndrome. In the past 7 to 10 years, severely heat-treated or detergent-treated concentrates, which are probably free of infective lipid-enveloped viruses, have become available, and very recently highly purified clotting factor proteins have become available based on monoclonal antibody purification schedules. Simultaneously, genetically engineered factor VIII is now available, and genetically engineered factor IX is undergoing clinical trials. Over the next decade, treatment of hemophilia A probably will depend more and more on genetically engineered replacement therapy. Eventually, it is hoped that gene

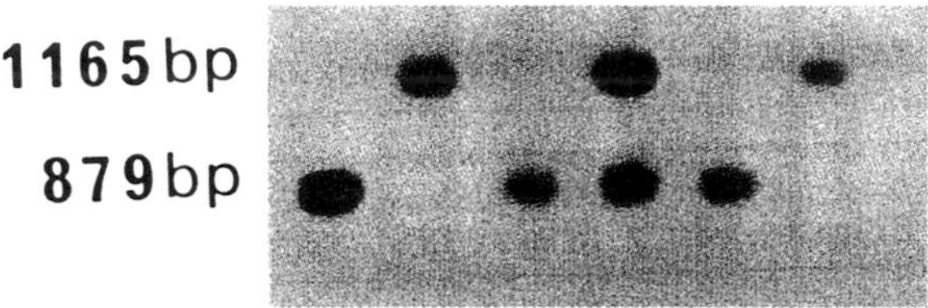

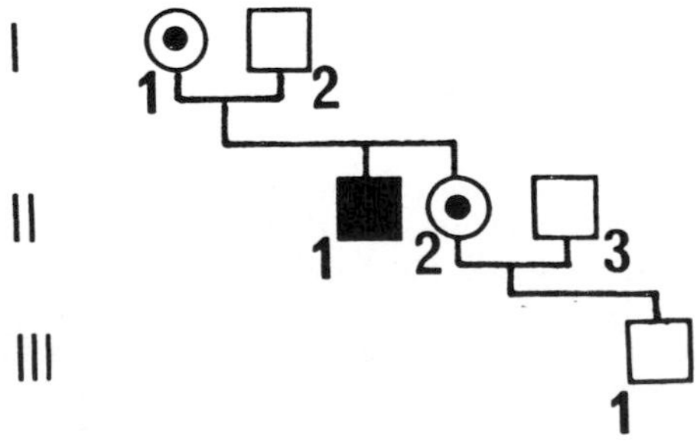

FIGURE 11 Use of a gene probe to diagnose hemophilia from chorion villus biopsy sample at 8 weeks of gestation. In this family, hemophilia A is associated with the lower band on the Southern blot. Because the male fetus (III 1) has the upper band, he cannot have hemophilia A, as proven 7 months later. [Reprinted, with permission, from J. Gitschier *et al.* (1985). Antenatal diagnosis and carrier detection of haemophilia A using a factor VIII gene probe. *Lancet* **1**, 1093.]

therapy by introduction of suitably modified coagulation factor genes into somatic tissues will provide an effective cure or amelioration for the bleeding tendency in hemophilia.

VI. GENETIC COUNSELING

Every male patient with hemophilia A or B has a number of female relatives who are definite or possible carriers of his gene. These females often request information on their carrier status and advice on risk of producing hemophilic offspring. Based on the new knowledge of the genetic loci involved, it is now possible to provide definitive carrier status information in a high proportion of cases. For hemophilia B, now that the precise mutation can be localized in all cases, carrier prediction can be provided with certainty for all female relatives of a patient with this disease. For hemophilia A, genetic linkage analysis can provide such diagnosis in about 50% of cases, but in other cases, because of an unknown level of a new mutation linkage, analysis is only partially informative. An evaluation looking for inversion of factor VIII should be done, as this is the most common cause of hemophilia A. Antenatal diagnosis using gene probes can be performed on biopsy samples from chorion villus material as early as 8 weeks of gestation (Fig. 11). Thus, carriers of hemophilia can make the choice to bear only normal sons. [*See* Genetic Counseling.]

BIBLIOGRAPHY

Antonarakis, S. E., Rossiter, J. P., Young, M., Horst, J., deMoerloose, P., Sommer, S. S., Ketterling, R. P., *et al.* (1995). Factor VIII gene inversions in severe hemophilia A: Results of an international consortium study. *Blood* **86**(6), 2206.

Colman, R. W., Hirsh, J., Marder, V. J., and Salzman, E. W. (eds.) (1994). "Hemostasis and Thrombosis. Basic Principles and Clinical Practice," 3rd Ed. Lippincott, Philadelphia.

Ginsburg, D., and Bowie, E. J. W. (1992). Molecular genetics of von Willebrand disease. *Blood* **79**(10), 2507.

Kane, W. H., and Davie, E. W. (1988). Blood coagulation factors V and VIII: Structural similarities and their relationship to hemorrhagic and thrombotic disorders. *Blood* **71**, 539–555.

Lakich, D., Kazazian, H. H., Antonarakis, S. E., and Gitschier, J. (1993). Inversions disrupting the factor VIII gene as a common cause of severe hemophilia A. *Nature Genet.* **5**, 236.

Tuddenham, E. G. D. (ed.) (1989). "The Molecular Biology of Coagulation. Baillière's Clinical Haematology," Vol. 2, No. 4. Baillière Tindall, London.

Tuddenham, E. G. D. (1989). von Willebrand factor and its disorders: An overview of recent molecular studies. *Blood Rev.* 3(4), 1–12.

Tuddenham, E. G. D. (1989). Inherited bleeding disorders. *In* "Postgraduate Haematology" (A. V. Hoffbrand and S. M. Lewis, eds.), 3rd Ed., Chap. 23. Heinmann Medical Books, Oxford, England.

Vehar, G. A., Lawn, R. M., Tuddenham, E. G. D., and Wood, W. I. (1989). Factor VIII and factor V: Biochemistry and pathophysiology. *In* "The Metabolic Basis of Inherited Disease" (C. R. Scriver, A. L. Beaudet, W. S. Sly, and D. Valle, eds.), Chap. 86. McGraw–Hill, New York.

White, G. C., and Shoemaker, C. B. (1989). Factor VIII gene and hemophilia A. *Blood* **73**, 1–12.

Zimmerman, T. S., and Ruggeri, Z. M. (eds.) (1989). "Coagulation and Bleeding Disorders." Marcel Dekker, New York.

Hemopoietic System

KENNETH DORSHKIND
University of California, Riverside

GLOSSARY

Chemotaxis Orientation of a cell along a chemical concentration gradient or movement in the direction of the gradient

Cytokine Cell product, usually a glycoprotein, that has effects on the growth, differentiation, activation, and/or survival of another cell population

Hemopoiesis Process of blood cell formation that takes place in the bone marrow or medullary cavity

Pluripotent hemopoietic stem cell Immature cell that can self-renew and generate additional stem cells and has the potential to differentiate and generate progeny of all hemopoietic lineages

Progenitor cell Cell with limited proliferative capacity and a restricted differentiative potential

Retrovirus RNA virus; after infecting a cell, its RNA is copied into DNA by the enzyme reverse transcriptase, and the DNA becomes part of the genome.

THE MAJORITY OF BLOOD CELLS HAVE A RELAtively short life span, and new cells must be continually produced to replace them. This process depends on a highly regulated series of events in which immature precursors progress through several stages of growth and differentiation, resulting in the formation of mature blood cells of a particular type. Blood cell development is known as hemopoiesis, and this occurs in cavities within the bones. The soft tissue that occupies the cavities is the bone marrow, and it is composed of the various hemopoietic populations and the supporting cells that form the environment in which blood cell production occurs. These supporting tissues are collectively referred to as the stroma, and they play an important role in the regulation of blood cell production.

I. BLOOD CELL TYPES AND THEIR FUNCTIONS

A. Classification of Blood Cells

Hemopoietic cells are classified into two broad categories: myeloid cells and lymphoid cells. Myeloid populations include red blood cells, or erythrocytes, as well as members of the granulocytic, monocytic, and megakaryocytic lineages. The two principal lymphoid cells are B and T lymphocytes. Additional terms, such as leukocytes or white blood cells, are also used and refer only to granulocytes, monocytes, and lymphocytes. In any case, mature myeloid and lymphoid cells are easily detected upon examination of the marrow, and some of their distinguishing characteristics are described in the following sections. In addition to the fully differentiated cells, immature progenitors of each of the different blood cell types are also present.

B. Lymphoid Cells

Lymphocytes are found in the blood and various lymphoid tissues such as the spleen and lymph nodes. These mononuclear, nonphagocytic white blood cells are key components of the body's immunological defense system and are classified as T cells or B cells. [*See* Lymphocytes.]

Two main subpopulations of T cells have now been identified based on functional criteria and their expression of cell-surface antigenic determinants; these include helper T cells and cytotoxic T cells. Helper cells are important in the regulation of the immune response and can affect the function of other T cells or B lymphocytes through direct contact with those cells or via the release of soluble factors. Cytotoxic T cells function to eliminate tumor cells or virus-infected cells. T cells are the exception to the rule that all blood cells are produced in the bone marrow. T-cell development occurs in the thymus, and precursors present in the marrow are believed to migrate to that tissue and provide a continual source of immature cells from which thymocytes are generated. The maturational state of these marrow emigrants is unknown, but they could be a prothymocyte committed to T-cell development or the pluripotent hemopoietic stem cell itself.

The other main type of lymphocyte is the B cell. B lymphocytes are produced in the bone marrow and are the class of lymphocyte responsible for the production of immunoglobulin (Ig) or antibody molecules. Each newly produced B cell expresses an immunoglobulin on its cell surface that can specifically recognize a particular foreign material known as an antigen. Following binding of the antigenic substance to the Ig molecule, the B cell is stimulated to synthesize identical Igs, which are secreted. All of the immunoglobulin molecules secreted by any one B cell or its progeny have the same antigenic specificity. Whether or not, analogous to T cells, subpopulations of B cells exist is an active area of interest. Some investigators believe that those human and murine B cells that express the CD5 cell-surface antigen constitute a separate B-lymphocyte lineage. [*See* B-Cell Activation, Immunology.]

Natural killer (NK) cells are an additional population best considered with the lymphocytes. These cells are present in the spleen, lymph nodes, peripheral blood, and peritoneal cavity and are capable of killing a target, usually a tumor or virus-infected cell, without prior sensitization. NK cells are also known as large granular lymphocytes owing to the presence of granules in their cytoplasm. To what lineage NK cells belong is unclear, and this has been an area of considerable investigation. NK cells originate in the bone marrow from hemopoietic precursors. [*See* Natural Killer and Other Effector Cells.]

C. Myeloid Cells

The majority of blood cells can be classified as myeloid cells and include monocytes–macrophages, several different types of granulocytes, megakaryocytes, and erythrocytes. Macrophages are mononuclear cells derived from monocytes and are specialized for phagocytosis. These cells can ingest particulate antigens or microorganisms and digest these materials with hydrolytic enzymes they produce. Macrophages vary in size and shape depending on their state of activation. In general, however, they have a cell surface from which projects numerous membrane processes, a spherical or oval nucleus, and a cytoplasm with granules that contain numerous hydrolytic enzymes. Macrophages play a key role in the immune response by presenting antigen to lymphocytes and through the secretion of various factors that can modulate the function of lymphocytes and other cells in the body. [*See* Macrophages.]

The granulocytes are a second category of myeloid cells. These cells receive their name because of the presence in their cytoplasm of granules that contain various enzymes or active proteins. Based on their appearance after hematologic staining, granulocytes are classified as neutrophils, eosinophils, and basophils. Neutrophils can be recognized by their multilobed nucleus and numerous cytoplasmic granules. The latter are classified as primary granules, which develop first during neutrophil differentiation and contain hydrolases, lysozymes, myeloperoxidase, and cationic proteins, and secondary granules, which constitute up to three-quarters of the granules and contain lysozymes and lactoferrin. These agents function to destroy or break down substances phagocytized by the cells. Neutrophils are sensitive to chemotactic agents and can migrate to the site of an inflammatory response at which they phagocytize and kill bacteria and other infectious agents. Eosinophils are bilobed cells that are characterized by the presence of large reddish-orange-staining granules that contain peroxidase and other enzymes. Because their numbers increase during parasitic infections, they may be involved in defense against those organisms. Eosinophils may play a role in various allergic reactions. Basophils have a constricted nucleus and a cytoplasm that contains coarse bluish-black-staining granules. The granules in these cells contain heparin, histamine, and serotonin. These substances are released following stimulation of the cells and contribute to allergic responses such as smooth muscle contraction and increased blood vessel permeability. Basophils have also been implicated in defense against parasites. [*See* Neutrophils.]

Megakaryocytes or thrombocytes are large cells that can have multiple nuclei that form as a result of polyploidy. The function of megakaryocytes is to produce platelets, and this occurs as a result of budding of cytoplasmic fragments from the cells. Platelets function in coagulation, hemostasis, and thrombus formation.

Erythrocytes, or red blood cells, are nonnucleated biconcave disks that are able to transport oxygen to the tissues. This is because the cells contain hemoglobin, a protein that combines readily with oxygen.

II. HEMOPOIETIC HIERARCHY

A. Introduction

If a sample of bone marrow is smeared on a slide and stained with an appropriate hematologic reagent, a marked heterogeneity of cell types can be observed. The mature cells described earlier can be easily recognized as well as other populations not readily identifiable. These are hemopoietic cells of various lineages and at different stages of development. The maturational state of these cells can be determined using a variety of criteria that include their histologic appearance, expression of defined cell-surface or cytoplasmic antigenic determinants, responsiveness to growth and differentiation factors, and proliferative and differentiative potential. All of these parameters have provided useful information that has permitted a scheme of hemopoiesis to be formulated.

B. The Pluripotent Hemopoietic Stem Cell

At the head of the hemopoietic hierarchy is the pluripotent hemopoietic stem cell (PHSC). This is defined as a cell that has the ability to divide and generate additional stem cells (a property known as self-renewal) as well as to differentiate and give rise to the blood cell types already described. The PHSC is a rare cell and may account for $<0.1\%$ of the nucleated cells in the bone marrow. Their incidence is even lower in other tissues such as the spleen. Despite this, these cells clearly exist.

One study that provided evidence for the existence of a multipotential stem cell analyzed blood cells in women who were heterozygous for the enzyme glucose-6-phosphate dehydrogenase (G-6-PD). All blood cells that derive from any one pluripotent hemopoietic stem cell would express one of the two forms of G-6-PD type. When cells from G-6-PD heterozygous females who had developed chronic myelogenous leukemia (a disease that originates in a single cell) were analyzed, cells from most of the hemopoietic lineages were shown to express the same G-6-PD type. This provided strong evidence that these populations were the progeny of a common stem cell that exhibited a marked differentiative capability. Another classic study that demonstrated the existence of the PHSC involved the induction of chromosomal markers in bone marrow cells by subjecting them to a low, sublethal dose of ionizing irradiation. This induces random chromosomal translocations in a high proportion of the surviving cells, and because each is unique, it can be used as a cytogenetic marker to detect cells derived from the marked cells. This cell population (called clonal) was used to repopulate X-irradiated mice, in which the bone marrow is destroyed. At various times following reconstitution, cytogenetic analysis was performed on the myeloid, B, and T cells that derived from the cells used to repopulate the animal. The study showed that in some of the mice, myeloid, B, and T cells all bore the same unique chromosomal marker, providing very strong evidence that they were all progeny of a single pluripotent stem cell in which the marker had been created.

Another series of experiments that demonstrated the existence of the PHSC repopulated mice with bone marrow cells that were infected with a retroviral vector carrying the bacterial neomycin resistance gene (which allows for the selection of the cells carrying the gene). The retrovirus genome integrates randomly into the genome of the marrow cells, and the viral integration site creates in each cell a unique change (called restriction fragment-length polymorphism), which can be recognized by preparing DNA from the cells and cutting it with an appropriate restriction enzyme. These fragments can be visualized by a suitable laboratory technique. At appropriate periods after reconstitution of irradiated mice, myeloid, B, and T cells were isolated from the animals, and the cells' DNA was analyzed. These studies demonstrated that neomycin-resistant B cells, T cells, and myeloid cells could be detected in which the retroviral vector had integrated at the same site. The most straightforward explanation for these results was that these cells were the progeny of a pluripotent hemopoietic stem cell that had been infected by the retrovirus.

A clonal analysis of hemopoietic cells in these mice was also performed at various times after reconstitution. This investigation showed that some hemopoietic stem cells stably contributed to all hemopoietic lineages in a reconstituted mouse over time, whereas others underwent temporal changes. For example, in some animals, myeloid and lymphoid progeny of a particular stem cell were observed at the initial sampling time, but only one or the other cell type was present at a later point. In additional animals, new stem cell clones were observed to have emerged, based on the observation that lymphoid and myeloid cells that bore the same unique viral integration site were detected at the time of a second sampling but not an earlier point following reconstitution. These studies are of importance, because they document that the contribution of a pluripotent stem cell to the various lineages can vary with time.

The foregoing experiments have provided evidence for the existence of the pluripotent hemopoietic stem cell, but they do not allow it to be directly studied in isolation. A major goal of experimental hematologists is to isolate and grow pure stem cell populations. This would present great advantages for studying how the growth and differentiation of that population are regulated and for identifying and characterizing genes that are temporally expressed during blood cell development. A variety of approaches have been used to purify stem cells from the marrow based on their physical characteristics and expression of cell-surface determinants. Recent work has indicated that murine stem cells express a unique combination of cell-surface determinants that allow their isolation. Markers expressed on human pluripotent hemopoietic stem cells, such as CD34, can also be used for their isolation.

C. Multipotential Precursors

The differentiative potential of multipotential precursors is more limited than that of a pluripotent stem cell. Two types of mulitpotential precursors have been described. A multipotential myeloid precursor is defined as a cell that can generate all hemopoietic cells except lymphocytes. The evidence that these restricted populations exist is based on several criteria. First, in the preceding experiments using retrovirally marked cells, patterns of repopulation were observed in which only myeloid but not lymphoid cells bore the same viral integration site. A second piece of evidence is based on classic experiments that demonstrated that bone marrow cells will form macroscopic colonies on the spleen 8–13 days following their injection into lethally irradiated mice. The cell that initiates these colonies is known as the spleen colony-forming unit (CFU-S; S stands for spleen) and the colonies are called spleen colonies.

Based on the observation that lymphoid cells are rarely, if ever, detected as progeny of the CFU-S and the colonies contain primarily myeloid cells, the assay is generally accepted as detecting a multipotential myeloid-restricted precursor. Spleen colonies form at various times after injection of bone marrow cells into the irradiated mice. Those that appear on Day 13 are generally thought to be derived from a multipotential myeloid-restricted precursor. Spleen colonies also appear at earlier times, but they may be derived from more differentiated progenitor cells described in Section II,D.

A common lymphoid precursor is a second type of restricted stem cell. By definition, such a cell would have the potential to generate both B and T lymphocytes but not myeloid cells. Several laboratories have provided indirect evidence for the existence of a common lymphoid precursor, but to date, a pure population of such cells has not been isolated. Whether lymphocytes are the progeny of a common lymphoid precursor committed to the B- and T-cell differentiative pathway or are direct descendants of the PHSC remains an unanswered question.

D. Progenitor Cells

Progenitor cells have a limited proliferative capability and a restricted differentiative potential. For each of the blood cell populations described earlier, a progenitor cell committed to that particular lineage exists. There are two principal means by which progenitor cells can be detected. The classic method takes advantage of the ability of progenitors to form lineage-restricted hemopoietic colonies in semisolid medium. The growth of these colonies depends on the presence of a growth factor that potentiates their formation. Progenitor cells can also be identified by their expression of lineage-specific antigens.

1. Myeloid Progenitors

Investigators in Israel and Australia were the first to demonstrate that bone marrow or spleen cells could form colonies in a semisolid medium such as agar following culture under appropriate conditions. The colonies that grew contained granulocytes and/or macrophages, and subsequent analyses indicated they had arisen from a single progenitor cell. Distinct progenitor cells for most of the myeloid lineages have now been identified, and their formation of colonies in semisolid medium depends on the presence of a particular lineage-specific growth and/or differentiation factor. For example, as shown in Fig. 1, it is now known that a bipotential granulocyte–macrophage

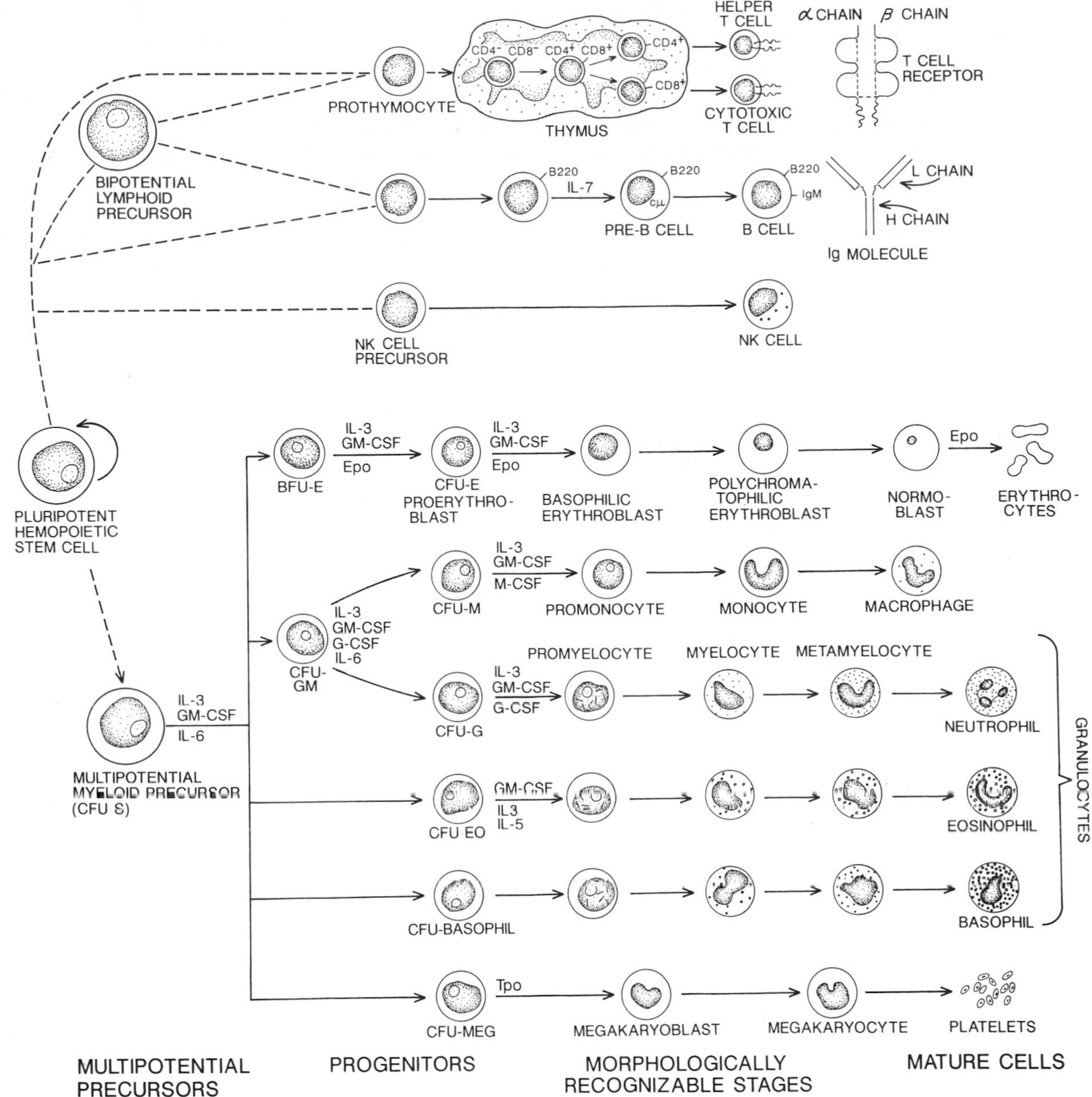

FIGURE 1 A scheme of the cells of the hemopoietic system showing the various lineages and maturational states that have been defined. Dashed lines indicate pathways in which lineage relationships remain unresolved. The arrow on the pluripotent hemopoietic stem cell indicates that the cell is capable of self-renewal. Some of the cytokines that regulate the growth and/or differentiation of blood cells are indicated. Their placement in the diagram is intended to indicate the point at which they act, but this is only an approximation. The reader must consult primary literature for the precise, up-to-date effects of these and other cytokines.

progenitor cell exists that can differentiate into cells of one or both of these lineages.

Progenitor cells are not necessarily of the same precise maturational state, and multiple cells may be included in a progenitor cell compartment. This point is best illustrated by analysis of cells of the erythroid lineage. Two stages of erythropoiesis have been defined based on the ability of erythroid progenitors to form colonies that appear at different times following initiation of cultures. An erythroid progenitor known as the erythroid burst-forming unit (BFU-E) is an immature erythroid progenitor and the erythroid colony-forming unit (CFU-E) is a more diffrentiated cell. Both the BFU-E and the CFU-E meet the definition of progenitor cells, because they are lineage-restricted in their differentiative potential. Figure 1 also shows the other myeloid progenitors that have been identified, and heterogeneity within these compartments also likely exists.

2. Lymphoid Progenitors

The bone marrow contains precursors that can generate B and T cells, but the development of these two types of lymphocytes occurs in different sites. B lymphopoiesis proceeds in the marrow of adult mammals, whereas primary T-lymphocyte development occurs in the thymus.

Many researchers now believe that T-cell production depends on the migration of a bone marrow-derived T-cell precursor to the thymus. Whether this is a lymphoid stem cell, a progenitor committed to the T-cell developmental pathway (the prothymocyte), and/or the pluripotent stem cell itself is not entirely clear. In any case, once the T-cell precursor enters that tissue, it interacts with the thymic stroma and a series of events occur that result in the expression of the T-cell-receptor and T-cell surface antigens. Major efforts are under way to identify the anatomical sites within the thymus in which various differentiative events occur, to define the lineage relationships, which result in the production of the various T-cell subpopulations, and to further clarify the molecular events that result in the expression of the genes that encode the T-cell receptor. It is generally accepted that a cellular intermediate that expresses both the CD4 and CD8 antigens develops during the intrathymic stage of differentiation, and this gives rise to T-cell subpopulations that include the CD4+ T-helper cells and the CD8+ cytotoxic T cells. These newly produced T cells express the heterodimeric T-cell receptor, which most commonly consists of an α and a β chain (Fig. 1), although it may also be formed by a γ and a δ chain. These chains are expressed during T-cell differentiation as a result of a complex process in which genes that encode segments of these molecules rearrange and become expressed in developing thymocytes. The T-cell receptor on each newly produced T lymphocyte has a unique site at its amino terminus, known as the variable region, that recognizes antigen in association with cell-surface determinants encoded by the major histocompatibility locus.

B-cell development occurs in the bone marrow, and maturation of cells in this lineage results in the expression of the Ig molecule as well as other nonimmunoglobulin cell-surface antigens. Distinct stages of human and murine B lymphopoiesis have been defined through the use of antibodies that recognize these molecules (Fig. 1). For example, the CD45R (B220) molecule is a 220,000 molecular weight glycoprotein present on murine B-cell progenitors. Similar molecules may be present on human B-cell progenitors.

The Ig or antibody molecule is formed by the assembly of two identical heavy-chain and two identical light-chain glycoproteins. The heavy-chain genes are expressed first, and following translation of heavy-chain messenger RNA, heavy chain of the immunoglobulin M class is expressed in the cytoplasm of a cell population termed a pre-B cell. Light-chain genes are subsequently expressed, and as light-chain protein appears, the complete Ig molecule is assembled in the cytoplasm of the pre-B cell and immediately appears on the surface of what is now termed a newly produced B lymphocyte. As with the T-cell receptor, expression of the Ig molecule depends on the rearrangement of gene segments that encode the heavy- and light-chain molecules. The amino terminus of the heavy and light chains forms a unique site specialized for recognition of an antigen.

Colony assays for B-cell progenitors, analogous to those for erythroid and myeloid populations, are a recent development that provides an additional means to identify B-cell precursors.

E. Morphological Identification of End Cell Populations

Before the development of colony assays or the availability of antibodies that recognize antigens expressed by hemopoietic precursors, morphology was the principal means by which immature cells in the various lineages were identified. However, by the time a cell of a particular lineage is histologically recognizable, it is more mature than progenitors that are detected in colony assays. For example, it is possible to distinguish immature erythroid cells such as the erythroblast. This cell, which is thought to correspond to a mature CFU-E, has a basophilic cytoplasm and a nucleus containing multiple nucleoli. More differentiated progeny of the erythroblast can also be defined based on cell size, the nuclear chromatin pattern, and cytoplasmic basophilia (Fig. 1). Similarly, various stages of myelopoiesis can be recognized under the light microscope. One of the earliest recognizable precursors of granulocytes is the promyelocyte. This cell, which is certainly more differentiated than the progenitors that initiate colonies in semisolid medium, has a large nucleus with a mucleolus. Peroxidase-positive granules are present in the cytoplasm, and a well-developed Golgi apparatus is present. From the promyelocyte stage on, cells in the granulocytic series are distinguished based on the formation of specific granules in their cytoplasm, and once these appear, the cells are classified as myelocytes. Cells are classified as metamyelocytes following the development of the distinguishing nuclear characteristics noted in Section I,C.

Some appreciation of the appearance of the various hemopoietic cells remains important. This is particularly true for the clinician, because diagnosis of certain types of hematologic dyscrasias such as leukemia is still based on the morphological appearance of the cells present in the marrow or peripheral blood.

F. Transcriptional Control of Hemopoiesis

The process by which hemopoietic precursors commit to the development of a particular blood cell lineage is the focus of current investigation in many laboratories. This event appears to be dependent upon the highly regulated and coordinated expression of multiple genes at particular times during development; and one way for the precursor cell to accomplish this is by the expression of one or a few cell-specific transcription factors. Transcription factors bind to DNA and regulate gene expression.

Studies in this area have been facilitated by the development of transgenic techniques that allow the generation of mice that lack expression of a specific gene. Several laboratories used this technology to "knockout" the genes encoding particular transcription factors in order to assess their role during hemopoietic development. The analysis of these mutant mice revealed that failure to express some transcription factors results in a complete absence of blood cell formation, while the lack of expression of others is more selective. For example, mice in which the gene that encodes the TAL-1 transcription factor has been "knocked out" do not produce any blood cells and die *in utero*. This result suggests that the TAL-1 transcription factor acts at the pluripotent stem cell stage. On the other hand, other transcription factors are required for development of specific hemopoietic lineages. Thus, mice lacking the E2A transcription factor fail to develop B lymphocytes and are immunodeficient, but all other blood cell types can be detected.

III. ORGANIZATION OF HEMOPOIETIC TISSUES

Blood cell development takes place in association with a fixed framework of supporting elements known as stromal cells. An understanding of the location of these stromal populations and the architecture of the marrow cavity is best appreciated in the context of blood cell circulation to the bone.

The principal blood supply to a typical long bone is the nutrient artery. This enters at midshaft through the nutrient foramen and travels in the central portion of the medullary cavity in parallel with the longitudinal axis of the bone. This centrally running artery sends off branches, which radiate toward the inner or endosteal surface of the medullary cavity, and at that point these branches anastomose with tributaries from other arteries that supply the bone. These subendosteal vessels become continuous with a series of venous sinusoids that deliver their contents toward a central sinus in the center of the marrow cavity. Blood from the central sinus exits the bone via the nutrient foramen and enters the systemic circulation.

Stromal cells are present in the intersinusoidal spaces and form the framework on which hemopoiesis proceeds. These cells, which have been referred to as reticular cells by morphologists, have numerous cytoplasmic projections that form extensive connections with one another and the processes of stromal cells in the adjacent intersinusoidal spaces. Stromal cell processes also have extensive contact with the endothelial cells, which form the lining of the sinuses. Developing blood cells in the intersinusoidal spaces are in close contact with the stromal cells, and this association led early morphologists to propose that the stroma can regulate blood cell production. As blood cells mature, they must ultimately exit the medullary cavity and enter the circulation. They do so by passing from the intersinusoidal spaces and through the endothelial cells to enter the venous sinusoids.

Under normal, steady-state conditions, the bone marrow is the only site in which the simultaneous production of myeloid cells and B lymphocytes occurs. This suggests that local tissue influences critical for that process are operative in the medullary cavity. It is generally accepted that the various stromal cell types and their products are one source of these regulatory signals, and the term hemopoietic microenvironment has been used to describe the localized sites in the medullary cavity in which blood cell production proceeds. Some investigators consider that components of the hemopoietic microenvironment are able to induce a pluripotent hemopoietic stem cell to differentiate along a particular lineage. However, this is not universally accepted, and it remains unclear if the microenvironment acts directly to induce the commitment and differentiation of multipotential stem cells or, instead, provides the permissive signals that allow the further development of a precursor that has become randomly committed to a particular lineage through its own intrinsic genetic programming. These controversies aside, it is evident that a number of growth and differentiation signals exist that affect hemopoiesis.

IV. REGULATION OF BLOOD CELL PRODUCTION

A. Introduction

The identification of the environmental signals that regulate blood cell production and the characterization of their effects on hemopoietic targets remain challenging areas of investigation. Such knowledge is of relevance to the basic understanding of a fundamental developmental process and has clinical relevance as well. When regulatory controls do not function properly, a number of hemopoietic abnormalities can occur. Much of what is known about the regulation of hemopoiesis is based on studies of murine tissues and cells, and it is reasonable to expect that similar regulatory controls are operative in human marrow.

The environmental cues a differentiating hemopoietic cell receives can be delivered via interactions with other cells, through the association of the cell with components of the extracellular matrix, or by the binding of a soluble mediator to a receptor on the cell membrane.

B. Cell–Cell Interactions

Morphological examination of marrow sections reveals that hemopoietic cells are tightly clustered in the intersinusoidal spaces in association with each other and the supporting stromal elements. This obviously creates the potential for direct cell–cell interactions between these populations and has suggested that hemopoietic and stromal cells might associate with one another via recognition molecules that permit direct membrane binding. Such molecules are now being identified by several laboratories that have prepared monoclonal antibodies to stromal cells. These antibodies apparently block the interactions between stromal cells and blood cells. These reagents will be of particular value in functional studies and in purifying and characterizing cell interaction molecule(s)

The endothelial cells that line the marrow sinusoids also play a role in regulating the passage of cells from the marrow into the systemic circulation. A number of elegant ultramicroscopic analyses have demonstrated that blood cells actually pass through the endothelial cell wall on their way into the sinusoids. What signals may be transmitted from the endothelial cell to the migrating blood cells remain to be defined, but the potential for this exists, because endothelial cells are known to secrete various factors with effects on hemopoiesis.

An important means of regulating passage of cells from the marrow may be mediated by the reticular and endothelial cells. The processes of reticular cells are associated with the abluminal surface of the endothelial cells and the degree of association can be variable. It has been hypothesized that an extensive reticular-endothelial cell association may serve to retain cells in the marrow, while a retraction of reticular cell process from the endothelium may facilitate blood cell emigration.

C. Cell–Matrix Interactions

The extracellular matrix (ECM) is a meshwork of polysaccharides and proteins that form an organized framework in the tissues. Fibroblasts in the tissues secrete much of ECM. The two main ECM components include polysaccharide glycosaminoglycans that are linked to a protein backbone and fibrous proteins. The latter include structural components such as collagen and elastin and adhesive proteins such as fibronectin and laminin. The matrix was initally believed to serve as a packing material in the tissues, but studies now indicate that it plays an important role in the development, migration, proliferation, shape, and metabolic function of cells. [*See* Extracellular Matrix.]

Hemopoietic cells have been reported to have receptors for various ECM components to which they may bind. What particular events may be signaled by such binding remains to be determined. Also, evidence indicates that hemopoietic growth factors can bind to ECM components and that the ECM thereby com-

TABLE I
Selected Cytokines and Their Effects on Hemopoietic Cells[a]

Cytokine	Source	Effects
IL-1	Macrophages, monocytes, fibroblasts, endothelial cells	Synergy with CSFs, stimulation of CSF and IL-6 secretion by stromal cells
IL-2	T cells	Stimulates thymocyte and T-cell growth
IL-3	T cells	Stimulation of proliferation–differentiation of myeloid-restricted stem cells and erythroid, monocytic, granulocytic, and megakaryocytic progenitors
IL-4	T cells, stromal cells	Pre-B-cell maturation, potentiates responsiveness of myeloid progenitors to CSFs
IL-5	T cells	Stimulation of eosinophil progenitors
IL-6	T cells, macrophages, stromal cells	Acts as CSF to stimulate proliferation of myeloid progenitors
IL-7	Stromal cells	Stimulation of B-cell progenitor–proliferation; stimulates thymocyte proliferation; augments proliferation of mature T cells to other cytokines
M-CSF	Macrophages, endothelial cells, stromal cells	Stimulates proliferation–differentiation of macrophage progenitors
G-CSF	T cells, stromal cells	Stimulates proliferation–differentiation of granulocyte progenitors
GM-CSF	T cells, stromal cells	Stimulates proliferation–differentiation of granulocyte–macrophage progenitors
Tpo	?	Megakaryocyte production

[a]The cellular sources of a particular factor and the effects are not necessarily restricted to the parameters noted. This is only a selected tabulation of sources and actions of these cytokines. In addition, these are not the only cytokines with effects on hemopoietic progenitors.

partmentalizes these cytokines and helps to present them to developing blood cells.

D. Soluble Mediators

Experiments in which a conditioned medium was used to stimulate the formation of hemopoietic colonies in agar demonstrated that some soluble factors could regulate the growth and/or differentiation of hemopoietic cells. It is now known that the molecules that potentiate the formation of the granulocyte and macrophage colonies are members of a family of cytokines known as the colony-stimulating factors. Additional factors that can affect the growth, differentiation, survival, and activation of hemopoietic cells have now been identified and include erythropoietin, which acts on erythroid cells, and, more recently, a variety of interleukins with effects on lymphoid and myeloid cells (Table I). The considerable progress that has been made in the identification and characterization of the actions of these multiple cytokines is due in part to the application of molecular approaches to cytokine analysis. Techniques for the cloning of the genes that encode these mediators have become relatively routine, and because of this, high concentrations of essentially pure recombinant factors are now available for *in vitro* and *in vivo* experiments. [*See* Cytokines in the Immune Response.]

Among the best characterized factors are the family of glycoproteins known as the colony-stimulating factors (CSFs). Three of the four CSFs were named according to their initially described effects on the growth and differentiation of a particular progenitor cell population and include macrophage (M-CSF), granulocyte (G-CSF), and granulocyte–macrophage (GM-CSF). Though the primary targets of these mediators are indicated by their name, their actions are not restricted to those lineages. For example, GM-CSF targets include eosinophils, megakaryocytes, and multipotential myeloid precursors. A fourth CSF is known as multi-CSF or interleukin-3 (IL-3). IL-3 targets include granulocytes, macrophages, eosinophils, megakaryocytes, and multipotential precursors. The various hemopoietic cells noted earlier each express several hundred CSF receptors, and occupancy of as few as 10% of them can simulate a biological response by the cells. The CSFs have a high specific activity and are active at concentrations of 10^{-10}–10^{-12} molar. Although a major action of CSFs is the stimulation of cell proliferation, they also have effects on the survival, differentiation, maturation, and activation of their targets.

Red blood cell production depends on the hormone erythropoietin. This is produced in the kidney, and its level of production is sensitive to the oxygen levels in the blood. A lack of oxygen or shortage of red blood cells results in an increase in erythropoietin synthesis and a stimulation of red blood cell production in the bone marrow. Recently, the gene that encodes thrombopoietin (Tpo), which regulates megakaryocyte growth and differentiation, has been cloned.

Another class of molecules with effects on the growth and differentiation of hemopoietic cells is the interluekins. This name was chosen to indicate that the source of these cytokines is leukocytes. In addition to IL-3, at least 13 other interleukin molecules have now been described. Interleukin-1 (IL-1) is primarily a macrophage–monocyte product, although other tissue cells such as fibroblasts and endothelial cells can produce it. IL-1 exhibits potent synergistic effects with other CSFs. Other interleukins with hemopoietic effects include interleukins-4, -5, -6, and -7. For example, IL-6 was originally defined by its ability to promote the growth of Ig-secreting B cells. This molecule is now recognized to act as a CSF to stimulate granulocyte progenitors and multipotent myeloid precursors. IL-7 is the first cytokine available in recombinant form that has been shown to affect B-cell precursors in the bone marrow. The gene for IL-7 was cloned from a bone marrow stromal cell line. IL-7 can stimulate the proliferation of B-cell progenitors. Its effects are not restricted to B-lineage cells, as both thymocytes and peripheral T cells are sensitive to its actions. IL-7 apparently does not have any differentiative function, and this would suggest that additional molecules with such actions on B-lineage cells exist.

Cytokine biology is a dynamic and complex area, and it is difficult to discuss all of the subtleties of this field. However, several important facts regarding the actions of these regulatory molecules must be appreciated and can be summarized as follows.

1. The effects of any one cytokine are often pleiotropic (i.e., multiple). For example, the interleukins are critical to the mediation of the immune response and many were initially described by their effects on mature lymphocytes. The fact that most of these factors have potent effects on the growth and differentiation of hemopoietic cells of various lineages was a secondary discovery.
2. The same factor may exhibit differential effects on cells within the same lineage. IL-4 has been proposed to stimulate or potentiate the differentiation of mature B cells while acting as an inhibitor of the proliferation–differentiation of early B-cell precursors.
3. The potential exists for additive and synergistic interactions between various cytokines. IL-1 and the CSFs are known to synergize and IL-4 has been shown to potentiate the response of myeloid progenitor cells to the actions of CSFs. Precisely how the interactions between two cytokines result in an augmented cellular response remains to be defined. If a cell expresses receptors for both cytokines, then the result of dual binding to these could easily explain an additive response. The explanation for a synergistic response is not as straightforward. In the case of IL-1–CSF synergy, it is possible that one molecule, such as IL-1, could stimulate an increase in the number of CSF receptors expressed by the cell and, thus, potentiate the response to that cytokine.
4. The effects of any particular hemopoietic factor are not necessarily mediated through a direct action on the hemopoietic target. Many of these molecules indirectly stimulate bystander cells to secrete a second molecule, which in turn acts on the hemopoietic progenitor. Numerous cells such as fibroblasts, bone marrow stromal cells, and endothelial cells are known to express IL-1 receptors. One effect of IL-1 binding to them is that they are stimulated to secrete a number of other mediators that include the CSFs. These in turn may then act on the hemopoietic target alone or synergistically with the IL-1 molecules that originally induced their production.
5. Certain hemopoietic cells can secrete cytokines to which they respond. For example, macrophages are sensitive to the effects of M-CSF and may also produce that mediator. This raises the possibility that autocrine (self-regulating) and paracrine (regulating neighboring cells) regulatory mechanisms are operative.
6. It is becoming clear that control of cell production in the hemopoietic system is also governed by various negative regulatory molecules that can have inhibitory effects on the growth and differentiation of cells. One such molecule, transforming growth factor-β, has been shown to affect the production of lymphoid and myeloid lineage cells.

It is important to note that virtually nothing is known about factors that regulate the self-renewal and differentiation of pluripotent hemopoietic stem cells.

V. INTERFACE BETWEEN BASIC AND CLINICAL STUDIES

Because hemopoietic cells are easily obtained and manipulated, the hemopoietic system is an excellent model for investigating the parameters that regulate cell growth and differentiation. The development of culture systems that facilitate these studies has been a major technological advance, and systems that allow long-term, stromal cell-dependent murine myelopoiesis or lymphopoiesis have been described and are in widespread use. These cultures are being used to study the growth of human hemopoietic cells as well.

The accessibility of bone marrow, the relative ease in manipulating it, and the development of techniques for its transplantation have contributed to great interest in applying the basic knowledge obtained about hemopoietic cells to clinincal medicine. For example, the experiments in which retroviral vectors were used to infect stem cells have potential use far beyond simply establishing lineage relationships. There is now considerable interest in using this approach to correct genetic defects that affect blood cell development or function. Thus, hemopoietic stem cells could be harvested from an individual and infected with a retrovirus that contains a normal copy of the defective gene. These cells would then be used to repopulate that individual's hemopoietic system with "normal" cells. This therapy depends on perfecting techniques for retroviral infection of stem cells and for ensuring that the exogenous gene is expressed in a regulated manner.

Another example of how the basic studies of the hemopoietic system have rapidly overlapped with clinical applications is evident from the rapid progress in cloning the genes for the various cytokines. Virtually unlimited quantities of factors such as erythropoietin. GM-CSF, and G-CSF are now available and are being administered to patients with promising results. Although considerably more basic research in the mechanisms of action of these factors remains to be conducted, information regarding their actions is being obtained from these clinical trials. Thus, not only can basic information about the hemopoietic system be applied to human medicine, but the results obtained from these applications can contribute to the fundamental understanding of blood cell production.

BIBLIOGRAPHY

Abramson, S., Miller, R. G., and Phillips, R. A. (1977). Identification of pluripotent and restricted stem cells of the myeloid and lymphoid systems. *J. Exp. Med.* **145**, 1567.

Dorshkind, K. (1990). Regulation of hemopoiesis by bone marrow stromal cells and their products. *Annu. Rev. Immunol.* **8**, 111.

Keller, G., Paige, C., Gilboa, E., and Wagner, E. F. (1985). Expression of a foreign gene in myeloid and lymphoid cells derived from multipotent haematopoietic precursors, *Nature* **318**, 149.

Kincade, P. W., Lee, G., Pietrangeli, C. E., Hayashi, S. I., and Gimble, J. M. (1989). Cells and molecules that regulate B lymphopoiesis in bone marrow. *Annu. Rev. Immunol.* **7**, 111.

Klein, J. (1982). "Immunology: The Science of Self–Non-self Discrimination." John Wiley & Sons, New York.

Jordan, C. T., McKearn, J. P., and Lemischka, I. R. (1990). Cellular and developmental properties of fetal hematopoietic stem cells. *Cell* **61**, 953.

Metcalf, D. (1988). "The Molecular Control of Blood Cells." Harvard Univ. Pess, Cambridge, MA.

Metcalf, D., and Moore, M. A. S. (1971). "Haemopoietic Cells." North-Holland, Amsterdam.

Nicola, N. A. (1989). Hemopoietic cell growth factors and their receptors. *Annu. Rev. Biochem.* **58**, 45.

Shiudasani, R. A., and Orkin, S. H. (1996). The transcriptional control of hematopoiesis. *Blood* **87**, 4025.

Spangrude, G. J. (1989). Enrichment of haemopoietic stem cells: Diverging roads. *Immun. Today* **10**, 344.

Herpesviruses

BERNARD ROIZMAN
University of Chicago

HERPESVIRUSES SHARE TWO FUNDAMENTAL properties that differentiate them from other viruses. All herpesvirus particles are architecturally identical even though they may differ considerably with respect to their biological properties and the size, structure, and composition of their genomes. In addition, all herpesviruses investigated to date are able to remain latent, i.e., in a nonmultiplying form, in their natural hosts.

I. INTRODUCTION

Most of the herpesviruses infecting humans and the diseases they cause have been known for many years. Herpes simplex viruses (HSV) were first isolated during the second decade of this century, but only in 1962 did it become known that there are two distinct serotypes: type 1 (HSV-1) and type 2 (HSV-2). The latest human herpesvirus was discovered in 1994. Interest in the biology of these agents has been spurred by a general resurgence of interest in sexually transmitted agents.

A. Definition and Taxonomy of Herpesviruses

Herpesviruses are recognized on the basis of the structure of their virus particle, the *virion* (Fig. 1). From the center out, the typical herpesvirus consists of a *core* consisting of the viral chromosome coiled in the form of a ring (toroid); a shell of proteins called the *capsid* and consisting of 162 protein complexes or *capsomeres;* a variable, asymmetric layer of proteins designated as the *tegument;* and a membrane called the *envelope* to differentiate it from cellular membranes from which it is derived (Fig. 1). The infectious particle consists of approximately 30 to 40 species of viral proteins.

The herpesviruses constitute the family Herpesviridae, which is composed of over 100 different species. Members of this family have been isolated from, turtles, fish, birds, marsupials, and just about every mammal in which they were sought. Although in the past they have been named after the disease they produce [e.g., varicella-zoster virus (VZV)] or their discoverers [e.g., Epstein–Barr virus (EBV)], the rules adopted in 1981 require that, with few exceptions, herpesviruses must be named after the family of the host in ascending numerical order [e.g., canid herpesvirus No. 1, the only canine herpesvirus known; and human herpesvirus No. 8 (HHV8), the latest herpesvirus isolated from humans].

Herpesviruses differ considerably with respect to several characteristics: (1) the animal host range, (2) the cell host range, (3) the rate of multiplication, (4) the tissue in which the virus remains latent in its native or experimental host, and (5) the size, average base composition, and sequence arrangement of their genomes. Herpesviruses have been provisionally classified into three subfamilies on the basis of their biological properties largely because these are easier to deduce. Thus, *alphaherpesvirinae,* exemplified by the

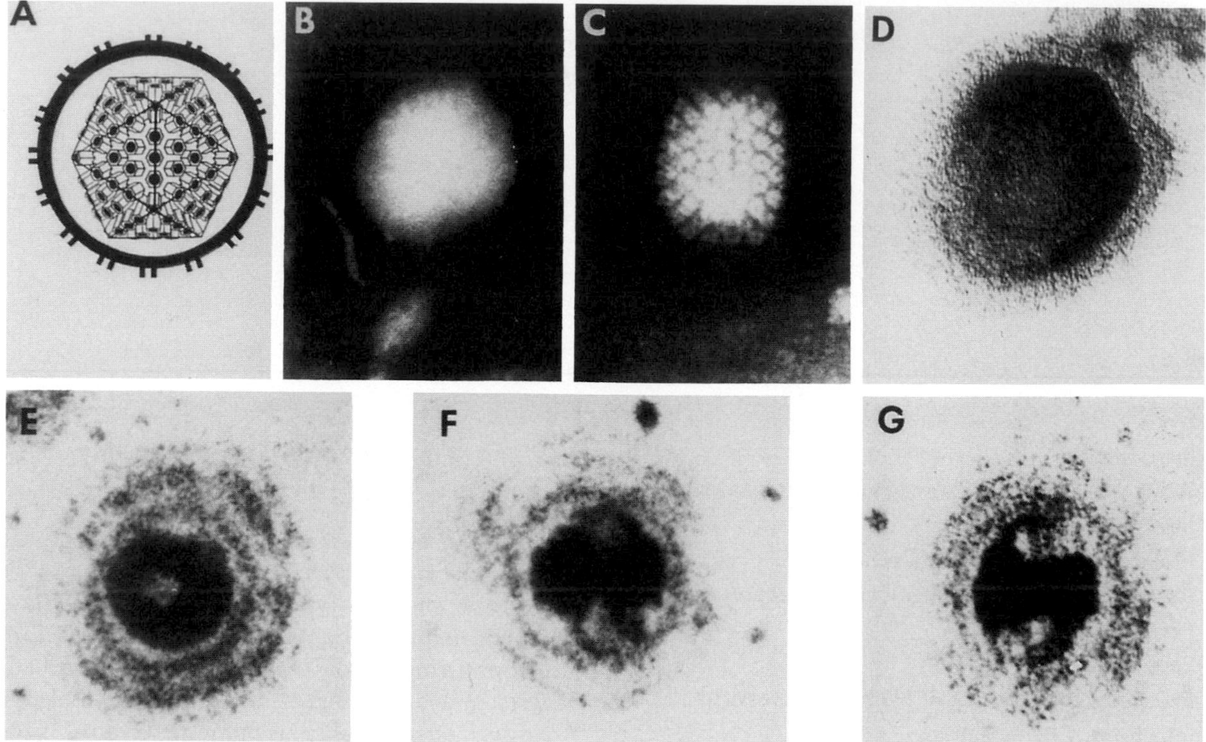

FIGURE 1 The structure of herpesviruses. (A) Schematic representation of a herpes virion. The outer ring represents the viral envelope with glycoproteins (spikes) projecting from its surface. The shell, the capsid, has a diameter of 105 nm and the shape of an icosadeltahedron consisting of 150 hexameric and 12 pentameric protein subunits (the capsomeres). A hole runs through each capsomere along its long axis. (B and C) Electron micrographs of an intact virion and capsid, respectively, stained with an electron-dense stain. The envelope prevents the stain from penetrating the virion and therefore the virion is electron translucent. The electron-dense stain between the capsomeres and in the holes of the capsomeres outlines the structure of the capsid. (D) Stained capsid showing the outlines of the DNA coiled in its interior. The DNA is coiled in the form of a toroid seen as a ring when viewed from the top (E) or bar when viewed from the side (F and G). The stain used on the preparations shown in E, F, and G reacted primarily with the DNA and with a lower intensity with the proteins in the capsid and envelope.

herpes simplex viruses 1 and 2 and varicella-zoster virus, usually multiply in a variety of cells from different species and tissues, spread rapidly from cell to cell in culture, and are associated with neuronal tissues during latency. The major characteristics of *betaherpesvirinae,* exemplified by human cytomegalovirus (HCV), are a restricted host range and a sluggish spread from cell to cell. The *gammaherpesvirinae* (e.g., EBV) grow in B or T lymphocytes and remain latent in these cells. The unambiguous differentiation of herpesviruses into subfamilies based on gene arrangement and conservation, and specific characteristics of nucleotide sequences of their genes is likely to supplant current criteria for classification.

B. Enumeration of Herpesviruses Infecting Humans

Humans become infected with 10 known herpesviruses. Of this number, 9 are human pathogens transmitted from person to person mostly by physical contact (e.g., herpes simplex viruses), but in some instances in the form of aerosols (e.g., varicella-zoster or chicken pox virus). These viruses and their taxonomic designations are HSV-1 and HSV-2 (human herpesviruses No. 1 and No. 2), varicella-zoster virus (human herpesvirus No. 3), human cytomegalovirus (human herpesvirus No. 4), Epstein–Barr virus (human herpesvirus No. 5), human herpesviruses No. 6A and No. 6B (HHV6A, HHV6B), human herpesvirus No. 7 (HHV7), and human herpesvirus No. 8 (HHV8), the latest discovered virus.

Another herpesvirus that occasionally finds its way into the human host is the simian B virus. Although its biology in its native host, the Old World rhesus and maccacus monkeys, is similar to that of HSV-1 in humans, transmission of the virus to humans by bites can cause severe disease and death. Herpesviruses infecting domestic animals do not cross into humans. Conversely, human herpesviruses are not as a rule transmitted to domestic animals, although rare transmission of HSV-2 to horses can cause severe disease.

II. BIOLOGY OF LYTIC AND LATENT HUMAN HERPESVIRUS INFECTIONS

A. Definition of Lytic and Latent Infections

Herpesviruses interact with their hosts in two different ways. In *lytic* infections, the herpesviruses multiply in *permissive* cells and these cells are then destroyed. Diseases caused by herpesviruses are the direct consequences of cell destruction that obligatorily accompanies virus multiplication, although the immune response to viral multiplication may also contribute to tissue destruction. In *latent* infections the viral genome is retained in an episomal state in the nucleus of the cell harboring the virus. The virus may express a small fraction of its genome designed primarily to maintain the cell and the viral genome contained in that cell, but virus multiplication and the ensuing destruction of the infected cell does not take place. The virus may become activated, multiply, and destroy the cell. [*See* Virology, Medical.]

B. Infection and Interaction of Herpesviruses with Their Human Hosts

1. Herpes Simplex Viruses 1 and 2

HSV-1 and HSV-2 are transmitted by the contact of healthy mucosal cells with infected saliva, genital secretions, or tissues from an individual with lesions containing virus. HSV-1 is more commonly transmitted by oral contact whereas HSV-2 is more commonly acquired by sexual contact. Both viruses are readily transmitted to the newborn during birth by contact with infected tissues. Individuals exposed for the first time to either HSV-1 or HSV-2 may experience a mild to severe infection depending on their immune state: the infection is much more severe in newborn and in immunologically compromised individuals than in healthy adults. The virus multiplies in cells at the portal of entry and spreads through nerve endings to the sensory nerves innervating that site. Once in the nerve, the virus is transported through the axon to the nucleus where it remains latent, sometimes for the life of the host. While in the latent state, the virus is shielded from the immune system and, as long as it expresses a small number or no viral gene products, it is not affected by drugs that can suppress viral multiplication during lytic infection. In a fraction of individuals harboring the virus in a latent state, the virus is periodically activated, multiplies, and is then transported through the axon to a site at or near the portal of entry, but most frequently to the mucocutaneous junction of the mouth (fever blister at the junction of the lips and skin), the cornea, or genital organs. The reactivated virus may spread to cells innervated by the infected nerve and cause a blister that ultimately is cleared by the immune system. The molecular mechanisms for the activation of viral multiplication are not known. However, termination of the latent state and virus multiplication follow trauma (e.g., ultraviolet light damage, burns) of tissues innervated by sensory nerves harboring latent virus, emotional stress, menstruation, intake of hormones, etc. Most initial infections are self-limited, but rare cases, especially in neonates, result in encephalitis which can be fatal. Encephalitis in adults is rare and is usually the result of reactivated latent virus.

2. Varicella-Zoster Virus

VZV causes chicken pox, a childhood disease rapidly transmitted from virus shedders to nonimmune contacts. In the nonimmune individual, the infecting virus multiplies at the portal of entry (lung and pharynx) and is disseminated by white blood cells to the skin where it multiplies and causes erythematous skin vesicles characteristic of chicken pox. In a latent form, the virus remains in the host harbored in sensory ganglia. Activation of the virus, which usually occurs in adults and usually only once, results in shingles, a painful disease resulting from the radial spread of infection to cells and tissues innervated by a ganglion harboring the virus. In immunocompromised individuals, primary infection or reactivation of the virus may result in extensive, life-threatening, and difficult to treat lesions. As in the case of HSV-1 and HSV-2 infections, the latent virus is shielded from the immune response of the host and plays a key role in the maintenance of the virus in the human population. The child recovering from chicken pox becomes the reservoir of the virus and maintains it in the largely immune population. This virus becomes available for transmission to the children of the next human generation when it is reactivated in the adult shingles patient.

3. Human Cytomegalovirus

Human cytomegalovirus rarely causes significant disease in healthy, immunologically competent adults, although the virus can remain in latent form. A severe and often fatal disease can result from the reactivation of latent virus or first infection in severely immunocompromised adults. Infection of a pregnant woman can result in transplacental transmission of the virus, and moderate to extensive tissue damage to the fetus is evident postpartum in the form of organ malformations, mental retardation, etc. Newborns infected during birth as well as infants infected during gestation may secrete the virus in urine and saliva for long intervals of time and show moderate to severe sequelae of disease, including impairment of hearing. Not much is known about the mechanisms of latency, the reactivation of virus multiplication, or of the cells harboring the virus.

4. Epstein–Barr Virus

The Epstein–Barr virus is associated with a number of clinical syndromes. In Europe and North America it is the causative agent of infectious mononucleosis, a self-limiting polyclonal lymphoproliferative disease of young adults resulting from transmission of the virus contained in infected saliva by kissing. The virus multiplies in nasopharyngeal tissues; it is readily isolated from throat washings and from B lymphocytes circulating in the blood. The immune system of most healthy individuals can cope effectively with infectious mononucleosis. The virus persists, however, in a latent state in B lymphocytes. In many communities, infection occurs at an early age and, except in immunocompromised individuals, the infection can be inapparent. Some children are genetically unable to respond to the EBV infections and develop a fatal polyclonal lymphoma after infection. The presence of latent EBV is a common feature of a monoclonal B-cell lymphoma described first by Burkitt and occurring in some areas of Africa and New Guinea known for a high rate of infection with the malarial parasite. The B-cell lymphoma is most likely a sequel of chromosomal translocation and the role of EBV is not clear. EBV is also a marker of cancer cells constituting a nasopharyngeal carcinoma prevalent in some genetically predisposed Chinese populations in southeast Asia. Although the presence of a secretory antibody to EBV structural proteins is an important marker of nasopharyngeal carcinoma, the role of the virus in the etiology of the cancer is not known. [*See* Epstein–Barr Virus.]

5. Human Herpesvirus No. 6

This virus was initially isolated from lymphocytes of healthy adults as well as from individuals infected with the human immunodeficiency virus 1. Serologic reagents and nucleotide sequence analyses indicate the existence of two closely related but nonidentical viruses designated HHV6A and HHV6B. The virus grows best in T lymphocytes. Gene arrangements indicate that these viruses are closely related to CMV and belong to the betaherpesvirinae subfamily. Studies on the prevalence and distribution of antibodies to the virus in the human population indicate that infection with this virus occurs early in life and is widespread. HHV6B appears to be the causative agent of roseolla, a childhood exanthem that occurs in afebrile children after a short febrile episode. The virus may cause central nervous system infections and severe manifestations in immunocompromised individuals. The prevalence of HHV6A is less common. Like other herpesviruses, HHV6 very likely remains latent in the human host. The tissue in which it remains latent, the mode of transmission, and the long-term effects of infection with this virus are not yet known.

6. Human Herpesvirus No. 7

Human herpesvirus No. 7 was initially isolated from cultures of T lymphocytes of an apparently healthy donor. The virus is commonly isolated from human saliva. It infects children of a somewhat older age than HHV6 and has not been linked to specific human diseases. Limited sequence analyses suggest that it also belongs to the betaherpesvirinae subfamily.

7. Human Herpesvirus No. 8

The latest discovered herpesvirus has been deduced to exist on the basis of cloning of DNA fragments containing genes related to but not identical to the Epstein–Barr virus from Kaposi's sarcoma and from body cavity lymphomas associated with AIDS. The role of this virus in either Kaposi's sarcoma or the body cavity lymphomas is not known.

III. MOLECULAR BIOLOGY OF HERPESVIRUS

A. Structure and Sequence Arrangements of Viral Genomes

The herpesvirus genomes vary in length, from approximately 120,000 to approximately 240,000 bp and in base composition, from 32 to 76 mol of guanine + cytosine percent. The variability in both size and base composition is extraordinary; it reflects the evolutionary divergence of viruses well integrated and evolving with their hosts through a large portion of the evolution of life on earth. With the exceptions of HSV-1 and HSV-2 and of HHV6A and HHV6B, the human herpesviruses identified to date are no more related to each other than they are to viruses isolated from other species. Thus HSV and EBV are each much more closely related to other herpesviruses isolated from lower animals than they are to each other. These observations suggest that each virus evolved a niche in its human host independently of other herpesviruses. Notwithstanding the difference in the sequence arrangements and base composition of their genomes, a significant fraction of genes are conserved. Structurally, the virus particles that harbor and protect the genomes of different viruses cannot be differentiated from each other visually, although they differ significantly with respect to immunologic specificity, the exact size, and number of viral proteins.

The herpesvirus genomes studied most extensively are those of the HSV, EBV, CMV, VZV, and HHV6A and B. The significant features of the chromosomes are as follows.

1. In the virus particle, the genomes are linear and double stranded. A single nucleotide protrudes at each end (3′ extension).

2. In all viral genomes, reiterated DNA sequences are interspersed among unique, nonreiterated sequences. Large repeats divide the herpesvirus genomes into major domains. In the case of HSV genomes, it is convenient to describe the genome as consisting of two covalently linked components designated as large (L) and small (S). Each component consists of unique sequences flanked by inverted repeats. The inverted repeats of the L component are approximately 9 kbp long and can be further subdivided into the terminal *a* sequence and the internal *b* sequence; the sequence *ab* and its inverted repeat, *b′a′*, flank a long stretch of a quasi unique sequence designated as U_l. The inverted repeats of the S component, each 6.4 kbp long and designated as *a′c′* and *ca*, flank the quasi unique sequence of this component designated as U_s. The size of terminal *a* sequence varies from approximately 200 to 500 bp depending on the strain and is the only shared sequence by the two components. The *a* sequences at the termini of the L component can be present in several copies that are directly repeated and adjacent to each other.

The two components L and S can invert relative to each other. The frequency of inversion is so high that viral DNA extracted from viruses produced in a few thousand cells and representing the progeny of a single infectious virus particle consists of four equimolar populations of viral genomes differing solely in the relative orientation of the L and S components. HSV-1 and HSV-2 genomes are identical in size and are colinear with respect to the arrangement of their genes. The CMV genome (225 kbp) is significantly larger than the genomes of the other herpesviruses, but the arrangement of its reiterated sequences is similar to those of HSV-1 and HSV-2 genomes. The genomes of VZV differ from those described earlier in that only one set of the unique sequence (U_s) is flanked by inverted repeats and only that unique stretch of DNA inverts relative to the other longer (U_l) stretch of unique sequences. DNA extracted from these viruses or from infected cells consists of only two isomeric populations. The EBV genome is representative of still another group of herpesviruses. Some representatives of this group contain a sequence at the termini that is repeated in tandem (directly) many times. Other internal stretches of viral DNA may also be repeated in tandem numerous times. In the absence of large stretches of sequences that are repeated in an inverted orientation, the isomerization of viral genomes does not occur. In the genomes of HHV6A, HHV6B, and HHV7, a significant portion of the genome at one terminus is repeated in the same orientation at the other terminus.

3. A significant feature of the herpesvirus genomes is their polymorphism. The variability in the number of repeats of a particular reiterated sequence probably reflects both intragenomic and intergenomic recombinational events. They are of lesser interest than polymorphism in the restriction endonuclease cleavage sites. The observation that restriction endonuclease cleavage sites are stable and do not change on passage in cell culture and from one individual to his or her contact has established restriction endonuclease fingerprinting as an important tool for tracing the transmission of virus from person to person. The significant conclusion drawn from the application of this epidemiologic tool is that the increase in the rate of genital herpesvirus infections seen since the early 1970s reflects a "frenzy of transmission" (i.e., an increase in the number of sexual contacts between infected and uninfected individuals rather than the consequence of an epidemic caused by a new virus sweeping the continent). [*See* Genetic Polymorphism.]

B. The Herpes Virion: Architecture and Composition

The herpesvirus envelope is derived from cellular membranes, but contains viral rather than cellular proteins. The number of viral proteins varies. HSV-1 and HSV-2 contain at least 10 glycosylated proteins on the surface (glycoproteins B, C, D, E, G, H, I, K, L, and M) and an unknown number of proteins in or on the underside of the envelope. The glycoproteins D, B, H, and L are the only ones essential for infection and multiplication in cells in culture. Of these proteins, B and H appear to be conserved among several herpesviruses. At different stages of maturation the capsid contains from 5 to 7 proteins. Another 10–15 proteins are located in the tegument, between the envelope and the capsid.

IV. REPRODUCTIVE CYCLE IN PRODUCTIVE INFECTION OF PERMISSIVE CELLS

A. Reproductive Cycle

The sequence of events in the replication of HSV-1 and HSV-2 is depicted in Fig. 2 and with minor variations is applicable to all herpesviruses. The key feature of the herpesvirus reproductive cycle is a highly ordered, tightly regulated sequence of viral gene expression that lasts 18 (HSV) to 72 (CMV) hr. The process may be summarized as follows.

1. Infection is initiated by the attachment of viral glycoproteins (most likely B or C) to domains of as yet unidentified cell surface proteins. Viral proteins, most likely glycoprotein D with B and H, fuse the envelope with the plasma membrane. The capsids are then transported along the cytoskeleton to nuclear pores. Factors at the environment of the nuclear pore act on viral proteins and trigger the release of viral DNA into the nucleus. The DNA circularizes immediately upon entry into the nucleus. [*See* Nuclear Pore, Structure and Function.]

2. Fusion of the envelope with the plasma membrane results in the release of several viral proteins along with the capsid. These include a protein that turns off the synthesis of host proteins and DNA (virion host shutoff, VHS) and a protein (α *trans* induction factor, αTIF) that is transported into the nucleus and *trans* activates the transcription of the 6α genes, the first set to be expressed after infection. *Trans* activation results from the concurrent binding of αTIF and host transcriptional factors (Oct-1 and C_1) to specific sites on the viral DNA.

3. The 6α proteins perform regulatory functions. One protein, infected cell protein No. 4 (ICP4), is required for the *trans* activation of all genes expressed

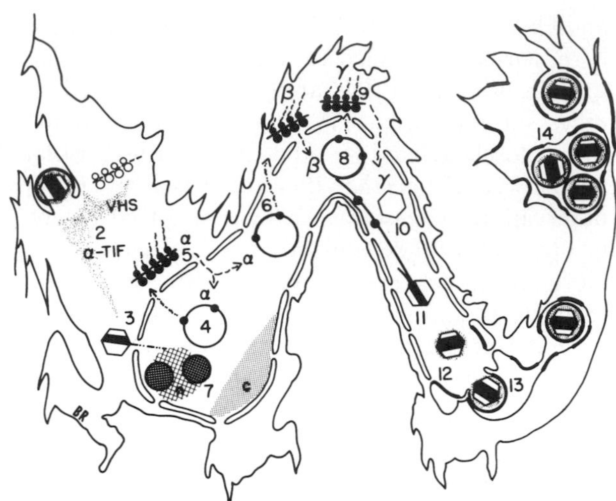

FIGURE 2 Schematic representation of the replication of herpes simplex viruses in susceptible cells. (1) The virus initiates infection by the fusion of the viral envelope with the plasma membrane following attachment to the cell surface. (2) Fusion of the membranes releases two proteins from the virion. VHS shuts off protein synthesis (broken RNA in open polyribosomes). α-TIF (the α gene *trans*-inducing factor) is transported to the nucleus. (3) The capsid is transported to the nuclear pore where viral DNA is released into the nucleus and immediately circularizes. (4) The transcription of α genes by cellular enzymes is induced by α-TIF. (5) The 5α mRNAs are transported into the cytoplasm and translated (filled polyribosome), and the proteins are transported into the nucleus. (6) A new round of transcription results in the synthesis of β proteins. (7) At this stage in the infection the chromatin (c) is degraded and displaced toward the nuclear membrane, whereas the nucleoli (round hatched structures) become disaggregated. (8) Viral DNA is replicated by a rolling circle mechanism that yields head-to-tail concatemers of unit length viral DNA. (9) A new round of transcription/translation yields γ proteins consisting primarily of structural proteins of the virus. (10) The capsid proteins form empty capsids. (11) Unit length viral DNA is cleaved from concatemers and packaged into the preformed capsids. (12) Capsids containing viral DNA acquire a new protein. (13) Viral glycoproteins and tegument proteins accumulate and form patches in cellular membranes. The capsids containing DNA and the additional protein attach to the underside of the membrane patches containing viral proteins and are enveloped. (14) The enveloped proteins accumulate in the endoplasmic reticulum and are transported into the extracellular space.

later in infection. It binds directly to viral DNA at specific sites in both 5′ nontranscribed and transcribed noncoding domains. This protein also represses the transcription of its own gene and that of at least one other (ORF-P) gene. ICP0, the product of the α0 gene, acts as a promiscuous transactivator. The products of two other α genes, α22 and α27, regulate the expression of genes transcribed late in infection. The first set of genes induced by the α proteins, designated as β, specify proteins that replicate viral DNA by a rolling circle mechanism. β proteins include a DNA polymerase, primase, helicase, thymidine kinase, DNase, dUTPase, ribonucleotide reductase, and both single-strand and double-strand specific DNA-binding proteins.

4. Most virion structural proteins are specified by the late or γ genes, a large, heterogeneous group whose expression requires functional α proteins and viral DNA synthesis. These proteins are transported into appropriate compartments. Thus, the capsid proteins assemble into empty capsids in the nucleus whereas some tegument and envelope proteins form patches in membranes of the infected cells. Viral DNA is cleaved from concatemers at the *a* sequence and is packaged into the preformed capsids. After envelopment at the inner lamellae of the nuclear membrane, viral particles are transported in the endoplasmic reticulum through the Golgi to the extracellular space. In cultured cells, a large number of particles may become de-enveloped and enveloped *de novo* in the membranes of the endoplasmic reticulum. It has been suggested that glycoprotein D made in the infected cells blocks reinfection of the cells with newly released virus.

5. In the case of the HSV-1 genome, at least y5 genes can be deleted without affecting the ability of the virus to grow in cells in culture. These genes include $\alpha 0$, $\alpha 22$, and $\alpha 47$, six of the virion glycoproteins (C, E, G, I, KM), the genes specifying thymidine kinase, dUTPase, ribonucleotide reductase, and numerous other proteins whose function is not known. It is highly unlikely that all of these viral genes are truly redundant. At least some of the deleted genes are dispensable only because cells in culture express cellular functions that complement the deletion mutants and allow them to multiply. Other viral genes may be required for alternative modes of entry or replication in cells that lack certain host factors.

B. Fate of Infected Cells

Cells productively infected with any of the known herpesviruses invariably die. In addition to the early shutoff of the host macromolecular metabolism characteristic of some herpesviruses, the infected cells usually exhibit irreversible disaggregation of the nucleoli, margination and pulverization of chromatin, and major alterations in the structure of cellular membranes. Much of the pathology associated with herpesvirus infections is caused by destruction of the productively infected cells.

V. LATENT INFECTION

The hallmark of herpesviruses is their ability to establish latent infections in a specific population of host cells (i.e., in neurons of sensory or autonomic ganglia in the case of HSV-1 and HSV-2, in dorsal root ganglia in the case of VZV, and at least in B lymphocytes in the case of EBV). In latently infected cells, the viral DNA is maintained in an episomal (closed circular) form, only a fraction of viral genes is expressed, and the functions of these genes are to restrict viral gene expression and to maintain both the copy number (approximately 10–200 copies per cell) and the state (episomal, not integrated into host chromosomes) of the viral DNA. Latent infections of cells capable of dividing or with a limited life span (e.g., the latent infection of B lymphocytes with EBV) are very different from those of neuronal cells with HSV-1 or HSV-2. The EBV genome expresses several genes in latently infected B lymphocytes. The products of some of these genes act on a specific site on the EBV genome, the *ori P*, to induce the replication of the DNA to the desired copy level per cell and to maintain the genome in a closed circular form whereas other genes appear to modify the physiology and longevity of the cell. The sensory neurons do not divide, and although the genome copy number appears to be elevated, there is no evidence for the synthesis of viral proteins that perform similar maintenance functions as those seen in EBV-infected B lymphocytes.

The factors that activate the latent virus to multiply appear to vary. Only approximately half of the individuals harboring latent HSV reactivate the virus. Reactivation follows physical or emotional trauma, hormonal imbalance, or injury of tissues innervated by the neurons harboring the latent virus. Experimentally, activation has been associated with increased prostaglandin synthesis. Activation may occur with such frequency as to appear to be a chronic smoldering infection or as infrequently as once in a lifetime. Activation of the latent varicella-zoster virus (shingles) rarely occurs more than once in a lifetime.

VI. PROSPECTS FOR CONTROL OF HUMAN HERPESVIRUS INFECTIONS: CHEMOTHERAPY

Effective chemotherapy of herpesvirus infections is based on differences in substrate specificities and the structure of viral and cellular enzymes. The most significant drug currently licensed for use against HSV infections is *acyclovir,* which is activated by phosphorylation largely, but not exclusively, by viral thymidine kinase. The triphosphate form is utilized preferentially by the viral DNA polymerase and it acts as a DNA chain terminator. Because its phosphorylation in the absence of the viral thymidine kinase is drastically reduced, its toxicity to uninfected cells is low. Its virtue is also its weakness. Acyclovir and drugs that act in a similar fashion are effective only if the enzymes that activate it are present and if they are administered while viral DNA synthesis takes place. They are not effective against latent viruses. Although daily administration blocks the appearance of symptoms of reactivated virus, the long-term effects of the intake of drugs, which could act as human DNA chain terminators, remain to be established. Drugs similar to acyclovir are being tested for other human herpesvirus infections. There are no drugs or strategies as yet to rid humans of latent herpesviruses. Acyclovir is not effective against all human herpesviruses. New drugs specific for other human herpesviruses have been developed or are in the process of development. Those currently available (e.g., gancyclovir against CMV) are more toxic than acyclovir. [*See* Chemotherapy, Antiviral Agents.]

Prevention of Infection

It is far more desirable to prevent disease by prior immunization than to treat it with drugs. In practice, human herpesviruses present unique challenges that have been met only partially. Central to the practical aspects of disease prevention is that each virus has a specific *portal of entry* (e.g., mouth and gastrointestinal tracts for poliovirus, respiratory tract for influenza virus) and a specific *target organ* in which cell destruction caused by virus multiplication is the cause of the illness. In principle, when the target organ and portal of entry differ, the immunity induced by vaccination will be rapidly stimulated by the virus multiplying at the portal of entry, blocking the dissemination of the virus to the target organ. Vaccination does not, as a rule, block the infecting virus from multiplying at the portal of entry. Furthermore, in the case of herpesviruses, immunity to the virus does not affect latent virus or the activation of its multiplication in the neuron, although the severity of the lesions and the time it takes for the lesions to heal are determined by the immune system. For HSV, the target organ is the portal of entry, inasmuch as the virus multiplying at the portal of entry (cell of the mucous membranes of the mouth or genitals) colonizes the sensory ganglia and is responsible for the recurrent lesions. To block the establishment of latency, the immunity must be so high as to preclude or at least reduce viral multiplication at the portal of entry. In essence, the immunity induced by vaccination must be as high and as sustained as that induced by natural infection and be maintained by periodic reactivation of the latent virus. None of the many anti-HSV vaccines tested to date have met this objective.

The situation appears to be different in the case of VZV. Here the portal of entry and target organ are sufficiently different to render vaccination useful in the case of highly susceptible populations. A vaccine developed by Professor Michiaki Takahashi of Osaka University in Japan has been extensively tested and licensed for use in the United States. There are no vaccines of infection with other herpesviruses.

BIBLIOGRAPHY

More, P. S., and Chang, Y. (1995). Detection of herpesvirus like DNA sequences in Kaposi's sarcoma in patients with and without HIV infections. *N. Engl. J. Med.* **332,** 1181–1186.

Roizman, B., Lopez, C., and Whitley, R. J. (eds.), (1993). "The Human Herpesviruses." Raven Press, New York.

Hippocampal Formation

DAVID G. AMARAL
University of California, Davis

IVAN SOLTESZ
University of California, Irvine

GLOSSARY

Anterograde amnesia Loss of the ability to acquire or recall new information

Axon Output process of neurons that forms connections between brain regions

Cytoarchitectonics Characteristic organization and packing density of neurons; a major criterion used in defining boundaries of brain regions

Dendrites Tree-like processes of neurons that form their receptive surface

Long-term depression Activity-dependent, long-lasting decrease in synaptic efficacy

Long-term potentiation Activity-dependent, long-lasting increase in synaptic efficacy

Synapse Specialized region at the end of an axon where one neuron communicates with another neuron through the release of chemical substances

Temporal lobe One of the four lobes of the human brain, the others being the frontal, parietal, and occipital. The temporal lobe is located on the side of the brain in the region below and behind the temple

Ventricles System of cavities in the brain that are filled with cerebrospinal fluid

THE HIPPOCAMPAL FORMATION IS A BRAIN REGION located in the inner, or medial, portion of the temporal lobe. It comprises four subregions: the dentate gyrus, the hippocampus, the subicular complex, and the entorhinal cortex. These four subregions are linked by prominent connections that tend to unite them as a functional entity. It was originally thought to play a role in olfaction or emotion, but the major behavioral function now associated with the hippocampal formation is memory. Damage to the human hippocampal formation results in permanent anterograde amnesia, that is, the individual cannot store in long-term memory any of his or her new experiences. Most older memories remain intact, though hippocampal damage may also result in some retrograde amnesia. The hippocampal formation is particularly sensitive to a variety of traumas and disease states and is often damaged in anoxia/ischemia, epilepsy, and Alzheimer's disease.

I. LOCATION AND STRUCTURE OF THE HIPPOCAMPAL FORMATION

The most distinctive subregion of the hippocampal formation, and the one from which it takes its name, is the hippocampus. The term *hippocampus* (or sea horse) was first applied in the sixteenth century by the anatomist G. Arantius, who considered the three-dimensional form of the grossly dissected human hippocampus to be reminiscent of this sea creature (Fig. 1). Others likened the hippocampus to a ram's horn, and De Garengeot named the hippocampus "Ammon's horn" after the mythological Egyptian god. The terms hippocampus and Ammon's horn (or Cornu Ammonis) are now used synonymously.

ENCYCLOPEDIA OF HUMAN BIOLOGY, Second Edition, VOLUME 4.

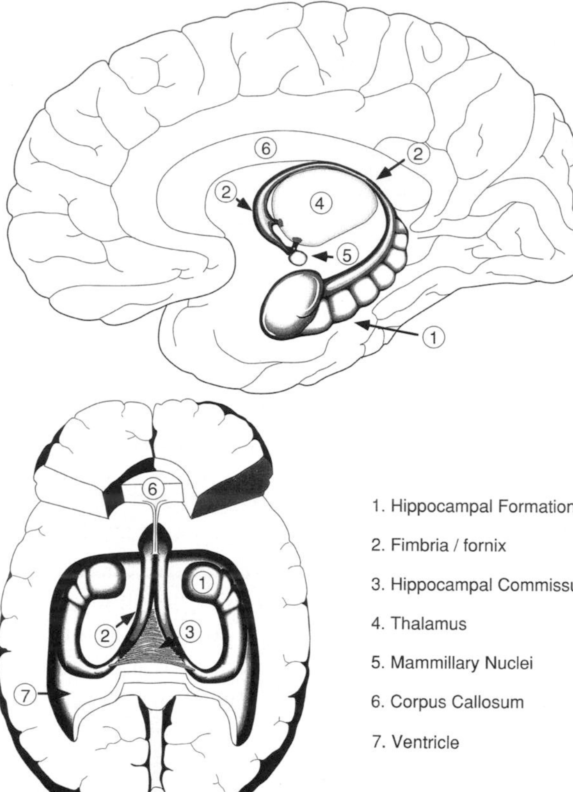

FIGURE 1 The position of the hippocampal formation and related structures in the human brain. The top image represents the medial surface of the human brain. The front of the brain is to the left. The position of the hippocampal formation in the medial portion of the temporal lobe is illustrated. The subcortically directed output bundle of axons, here labeled the fimbria/fornix, can be seen to arc over the thalamus and to ultimately descend into the diencephalon. Connections are made with both the mammillary nuclei and the thalamus. The bottom illustration is a cutaway of the brain as viewed from above and behind. The front of the brain is at the top of the image. The C-shaped structure of the hippocampal formation can be seen to occupy the floor of the lateral ventricles.

The hippocampal formation is located in the medial portion of the temporal lobe, and the hippocampus and dentate gyrus form a prominent bulge in the floor of the lateral ventricle. The hippocampal formation is widest at its anterior extent, where it bends toward the medial surface of the brain. The subtle bumps (or gyri) formed in this region give it a foot-like appearance, and the name *pes* (foot) *hippocampi* has classically been applied to this area. The main portion, or "body," of the hippocampus becomes progressively thinner as it bends posteriorly and upward toward the corpus callosum (see Fig. 1).

The medial surface of the hippocampus contains a flattened bundle of axons called the fimbria (Figs. 1 and 2). Axons originating from neurons in the hippocampus and subicular complex travel in a thin layer called the *alveus* that covers the hippocampus and coalesces in the medially situated fimbria. At rostral levels, the fimbria is thin and flat but it becomes progressively thicker caudally as fibers are continually added to it. As the fimbria leaves the posterior extent of the hippocampus, it fuses with the ventral surface of the corpus callosum and travels anteriorly in the lateral ventricle. The major portion of the rostrally directed fiber bundle is called the body of the fornix. At the end of its anterior trajectory, the body of the fornix descends and is called the column of the fornix. The fornix then divides around the anterior commissure to form the precommissural fornix, which enters the forebrain and terminates mainly in the septal nuclei, and the postcommissural fornix, which terminates in the diencephalon. Near the point where the fimbria fuses with the posterior portion of the corpus callosum, some of its fibers extend across the midline of the brain to form the hippocampal commissure. A variety of gross anatomical terms have been applied to the commissural fibers, but the term *psalterium* (alluding to a harp-like stringed instrument) is most common.

The four subregions of the human hippocampal formation can be differentiated cytoarchitectonically in neuroanatomical preparations in which thin sections of the brain have been stained to show the distribution of neuronal cell bodies (see Fig. 2). The dentate gyrus is the simplest of the subregions and has a trilaminate appearance; two of the layers contain large numbers of cell bodies and the third contains mainly dendrites and axons. The principal cell layer of the dentate gyrus (the granule cell layer) is populated primarily by one class of neuron, the granule cell. The human dentate gyrus contains approximately 9 million granule cells on each side of the brain. The dendrites of the granule cells extend mainly into the overlying molecular layer, where they receive their main input from the entorhinal cortex.

The human hippocampus can be divided into three distinct fields labeled CA3, CA2, and CA1 (see Fig. 2). The distinction of the three fields relates primarily to differences in their connections and more subtle differences in neuronal size and shape. All of the fields of the hippocampus have essentially one cellular layer, the pyramidal cell layer, which is populated by neurons with triangularly shaped cell bodies (the pyramidal cells). Pyramidal cells have dendrites emanating from their top (or apical) and bottom (or basal) sur-

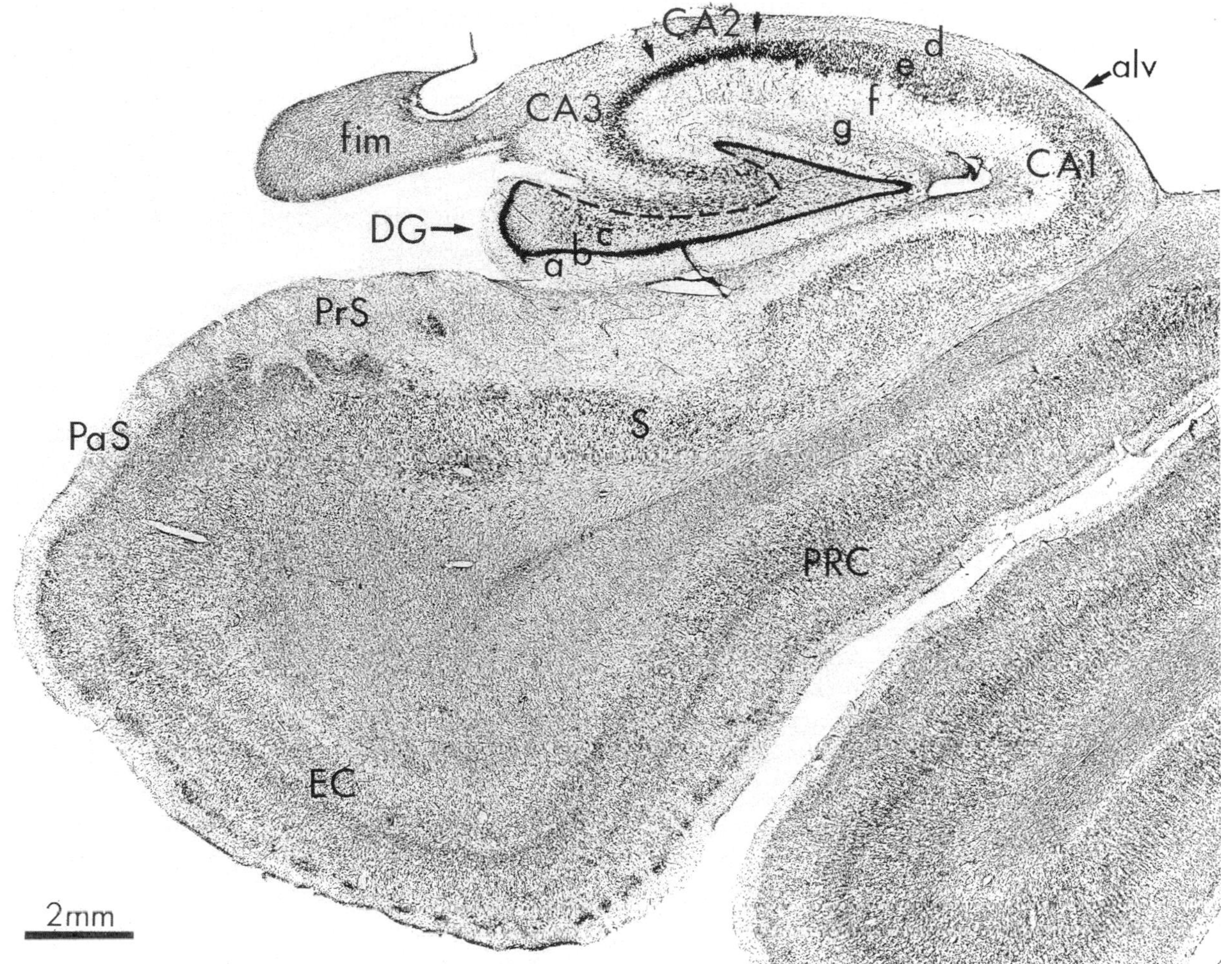

FIGURE 2 A thin section cut through the hippocampal formation of the human brain that is stained for the visualization of neurons. Each dark spot indicates the location of one neuronal cell body. The four major fields of the hippocampal formation can be differentiated in this type of preparation. The dentate gyrus (DG) has three layers: the molecular layer (a), granule cell layer (b), and polymorphic cell layer (c). The hippocampus can be divided into three fields, CA3, CA2, and CA1. The covering of the hippocampus is composed of axons originating from the pyramidal cells and is called the alveus (alv). Some of these axons coalesce to form the fimbria (fim). The next layer, the stratum oriens (d), contains the basal dendrites of the pyramidal cells. The cell bodies of the pyramidal neurons are contained in the pyramidal cell layer (e). The apical dendrites of the pyramidal cells extend into the overlying stratum radiatum (f) and stratum lacunosum-moleculare (g). The subicular complex comprises three distinct regions: the subiculum (S), the presubiculum (PrS), and the parasubiculum (PaS). The last subregion of the hippocampal formation is the entorhinal cortex (EC), which is a multilaminate cortical area that resembles the neocortex. The adjacent perirhinal cortex (PRC) is one source of sensory information to the hippocampal formation.

faces. On each side of the brain, there are about 2 million neurons in the CA3 region, 220,000 in CA2, and almost 5 million neurons in CA1. The relatively acellular layers above and below the pyramidal cell layer all have distinctive names. The outside surface of the hippocampus, located deep to the pyramidal cell layer, is formed by axons of the pyramidal cells and is called the alveus (see Fig. 2). Between it and the pyramidal cell layer is the stratum oriens, which contains the basal dendrites of the pyramidal cells. The region above the pyramidal cell layer contains the apical dendrites of the pyramidal cells and is divided into several strata (stratum lucidum, stratum radiatum, and stratum lacunosum-moleculare). Different connections are formed in each of these strata.

The subicular complex can be subdivided into three distinct fields, the subiculum, the presubiculum, and the parasubiculum (see Fig. 2). As in the hippocampus, one principal cell layer in the subiculum is populated by about 2.5 million neurons on each side. Details concerning the laminar organization of the other components of the subicular complex, the presubiculum and parasubiculum, are complex and not yet well understood. The subicular complex is an important region of the hippocampal formation, however, because it is from this region that the major outputs to subcortical regions such as the thalamus and hypothalamus originate. [*See* Hypothalamus; Thalamus.]

The term *entorhinal cortex* was coined by the early neuroanatomist Korbinian Brodmann to name the region that lies adjacent to the shallow rhinal sulcus in nonhuman brains. More than any other subregion of the hippocampal formation, the entorhinal cortex has undergone substantial regional and laminar differentiation in the human brain. Unlike the other hippocampal subregions, the entorhinal cortex is a multilaminate cortical region that resembles the cytoarchitectonic appearance found in the neocortex. It is distinguished from the neocortex by the presence of clusters of darkly stained neurons in layer II, located just below the surface of the brain. It is also distinct in lacking a layer of small granule cells that typically populate the fourth layer in the neocortex.

II. CONNECTIONS OF THE HIPPOCAMPAL FORMATION

The connections between the four subregions of the hippocampal formation, that is, the intrinsic connections, and between the hippocampal formation and other portions of the brain, that is, the extrinsic connections, have been the topic of extensive neuroanatomical study for more than a century. The fundamental intrinsic circuitry of the hippocampal formation and its major inputs and outputs are illustrated in Fig. 3.

The four subregions of the hippocampal formation are connected by unique and largely unidirectional connections. The entorhinal cortex can, for convenience, be considered the first step in the intrinsic hippocampal circuit. Cells located primarily in layers II and III of the entorhinal cortex project to the molecular layer of the dentate gyrus. The connection between the entorhinal cortex and the dentate gyrus is called the *perforant pathway*. Some of the entorhinal projections also terminate in the subiculum and in the CA1 and CA3 fields of the hippocampus.

The dentate granule cells give rise to axons, the mossy fibers, that form connections with pyramidal

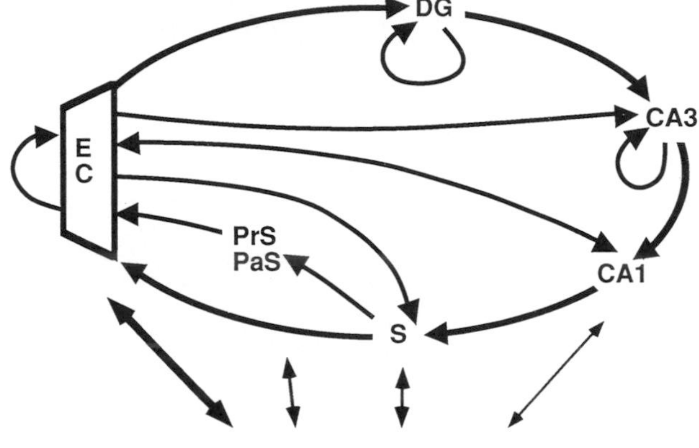

FIGURE 3 Summary of the major intrinsic and extrinsic neural connections of the hippocampal formation. The oval in the center of the illustration represents the hippocampal formation. Arrows indicate the direction of the major intrinsic hippocampal connections. The major subcortical inputs and outputs of the hippocampal formation are listed at the top of the illustration. Several cortical areas that are interconnected with the hippocampal formation are listed on the bottom. The cortical interconnections with the hippocampal formation are primarily made with the entorhinal cortex.

cells of the CA3 region of the hippocampus. The other main constituent of the granule cell layer is the dentate basket cell, which gives rise to a dense plexus of fibers and terminals that surround the granule cell bodies. These basket cells are known to use the inhibitory neurotransmitter γ-aminobutyric acid (GABA). There are other classes of neurons in the dentate gyrus, particularly in its diffuse polymorphic cell layer, that give rise to a variety of feedback and feedforward circuits within the dentate gyrus.

The pyramidal cells of CA3 have axons that project to other levels of CA3 (associational connections) and to subcortical regions, especially the septal nuclei. CA3 cells also originate the major input to CA1; this projection is called the Schaffer collaterals. The CA3 field contains a number of nonpyramidal cells, many of which form local circuits within CA3. The pyramidal cells in CA1 have a pattern of connections that is quite distinct from the one in CA3. What is perhaps most striking is that CA1 pyramidal cells do not project significantly to other levels of CA1, that is, there are virtually no associational connections within CA1. Rather, CA1 pyramidal cells project predominantly to the subiculum. The subiculum, in turn, projects to the presubiculum and parasubiculum, and all three components of the subicular complex project to the entorhinal cortex.

One of the more striking features of the intrinsic circuitry of the hippocampal formation is that it is largely unidirectional. The CA3 field does not project back to the granule cells of the dentate gyrus, for example. Nor do CA1 pyramidal cells project back to CA3. Thus, aside from the initial entorhinal input that reaches all of the hippocampal fields in parallel, information flow from the dentate gyrus through the other fields follows an obligatory serial and largely unidirectional pathway. This is in marked distinction to the situation in most other cortical regions, where connections are usually reciprocated.

A variety of chemical substances (neurotransmitters) mediate the transfer of information within the hippocampal circuitry. As in all other cortical regions, some of these substances excite the neurons onto which they are released and some are inhibitory. The hippocampal formation is particularly rich in the glutamate family of transmitters and receptors. One class of glutamate receptor, the *N*-methyl-D-aspartate (NMDA) receptor, has been implicated both in the modulation of neural activity that accompanies learning and memory and with the pathological activation of neurons that accompanies ischemia-induced neuronal cell death.

The fimbria and fornix form the classic efferent, or output, system of the hippocampal formation, and the human fornix contains about 1.2 million axons. The precommissural fornix, which innervates primarily the septal nucleus and other basal forebrain structures, arises mainly from neurons of the hippocampus and, to a lesser extent, from the subiculum and entorhinal cortex. Axons originating in the subicular complex are mainly connected to the diencephalon, that is, the thalamus and hypothalamus, particularly the mammillary nuclei located in the posterior hypothalamus (see Fig. 1). The fornix and fimbria also contain axons that originate in other structures and that terminate in the hippocampus. An important connection carried in the fornix originates in neurons of the forebrain that contain the neurotransmitter acetylcholine; these axons terminate in many portions of the hippocampal formation.

Connections from one brain region to the same region on the other side of the brain are called *commissural connections*. In nonprimate brains, the hippocampal formations of both sides are heavily interconnected. However, in the primate brain, including humans, the side-to-side interconnections of the hippocampal formation appear to be rather meager. This anatomical observation is consistent with the behavioral finding that damage to the hippocampal formation on one side of the human brain preferentially impairs memory for either verbal (on the left) or spatial (on the right) types of information. This indicates that memory function may be lateralized in the human hippocampal formation.

One fact that has been uncovered about the human hippocampal formation comes from the analysis of patients who have had bilateral damage to the hippocampal formation. These people are unable to learn or retrieve new information about their day-to-day existence. Their ability to recall old memories, however, remains largely preserved. This implies that memory is not stored in the hippocampal formation but perhaps in one of the brain areas that the hippocampal formation communicates with. Thus, it is important to know from what brain regions the hippocampal formation receives its information about ongoing events (sensory information) and to what brain regions the hippocampal formation delivers its processed information. These latter regions would be strong candidates for memory storage sites.

For many years it was thought that the hippocampal formation was not intimately interconnected with the higher processing centers of the neocortex. This view was based largely on neuroanatomical studies conducted in rodents. However, with neuroanatomical studies of the organization of the nonhuman primate hippocampal formation, it has become clear that this area is in receipt of substantial sensory information from the highest levels of the neocortex. Several cortical regions, located mainly in the frontal and temporal lobes, are connected to the entorhinal cortex. It is interesting that the neocortical regions that project to the entorhinal cortex are themselves in receipt of inputs from cortical regions representing several sensory systems. The hippocampal formation can be thought of as being positioned at the top of a pyramid of information processing. Like the *New York Times*, it hears all of the news fit to print to memory. Information is then relayed to the other hippocampal fields via the intrinsic connections. The processing that takes place during the trajectory through the hippocampal formation presumably begins the process of making a memory. It is also quite clear that the processed information is returned to the entorhinal cortex, which then sends return projections back to the neocortex. Thus, the hippocampal formation is privy to high-level sensory information processing that takes place in the cortex and has the connections necessary to send its own processed information back to potential storage sites in the neocortex. [*See* Neocortex.]

III. BEHAVIORAL FUNCTIONS OF THE HIPPOCAMPAL FORMATION

The hippocampal formation has historically been implicated in a variety of functions. In the latter part of the nineteenth century, for example, the neurologist Wilhelm Sommer considered the hippocampal formation to be a component of the motor system because he found that damage to it correlated with the seizure disorder associated with temporal lobe epilepsy. For much of the first half of this century, the hippocampal formation was thought to be primarily related to olfactory function and was considered to be a prominent component of what was called the rhinencephalon, or olfactory brain. But this notion was put to rest by the Norwegian neuroanatomist Alf Brodal, who indicated that the hippocampal formation was as prominent in anosmic species as in species that rely heavily on the sense of smell. In the 1930s, the neurologist James Papez considered the hippocampal formation to be a central component in a system for emotional expression. In his view, the hippocampus was a conduit by which perceptions of ongoing experiences could be collected and channeled to the mammillary nuclei of the hypothalamus, which would produce the physiological components of emotions. There has been little substantiation of Papez's theory, however, and the role of orchestrator of emotional expression is now more closely linked with another prominent temporal lobe structure, the amygdala.

Perhaps the most widely accepted and long-lived proposal of hippocampal function relates to its role in memory. It has been known for nearly a century that damage to certain brain regions can result in an enduring amnesic syndrome that is characterized by a complete, or near-complete, anterograde amnesia. Affected patients are incapable of recreating a record of day-to-day events, although most past memories remain largely intact. It is now clear that damage isolated to the human hippocampal formation is sufficient to produce this form of memory impairment. The most famous example is a patient with the initials H. M. As a young man, H. M. suffered from epilepsy that was so severe as to be life-threatening. In 1956, H. M. underwent a neurosurgical procedure in which the hippocampal formation on both sides of his brain was removed. Though this surgery had a moderating effect on his seizures, from the time of his surgery until the present day, H. M. has not been able to store any new information into long-term memory. In all other respects, however, H. M.'s cognitive functions appear normal. If you were to meet him, for example, you could carry on an intelligent conversation for hours and he would be able to accurately relate information about his life prior to his surgery (he was 27 at the time of the surgery). You would have little indication from his behavior that the surgery had taken place. But, if you left the room for ten minutes and then returned to continue your conversation, he would have no recollection whatsoever of having met you nor of anything that you had talked about during the previous conversation.

Although fairly convincing evidence now exists that the hippocampal formation plays a prominent role in the formation of enduring memories, the mechanism by which it exerts its influence is far from clear. Since damage to the hippocampal formation does not cause a loss of distant or well-established memories, it appears that the hippocampal formation cannot be the final repository for stored information. Rather, it appears that the hippocampal formation must interact with storage sites in other, presumably cortical, regions to consolidate ephemeral sensory experiences into long-term memory.

Electrophysiological studies conducted primarily in rodents have demonstrated that neurons in the hippocampus are preferentially activated by certain aspects of the environment. If one records the neural activity of a single hippocampal cell while a rat is running around in a maze, for example, the cell might be activated only when the rat travels through a certain location of the maze. Data of this type have prompted the suggestion that the hippocampus can form a "cognitive map" of the outside world. In a more general sense, it might be thought that the neurons of the hippocampal formation, acting as an assembly of differentially activated units, can form a representation of ongoing experience. Perhaps the interaction of this hippocampal representation of experience with the more detailed information of the experience located in the neocortex is the route through which long-term memories are formed. One implication of the electrophysiological data is that neurons in the hippocampal formation are not uniquely sensitive to certain types of information. Rather, neurons in the hippocampal formation may act more like random-access memory (RAM) in a computer and are therefore potentially activated by all types of information. Since it would be difficult for evolution to anticipate all the various forms of information that might need to be stored as memory, a generalized memory buffer system would be highly adaptive.

IV. PHYSIOLOGY OF THE HIPPOCAMPAL FORMATION: THE SUBSTRATE OF MEMORY?

What is the biological substrate of memories? The study of the cellular events underlying memory formation, and how these may be enhanced to facilitate memory, is one of the most exciting areas of modern neuroscientific research. Events in the external world are represented in the brain as electrical activity patterns in neurons. One current hypothesis is that memory might be related to an increased efficiency of electrical communication between neurons that represent experiences. Neurons are connected to each other via specialized junctions called synapses. It is these sites of communication that are believed to be modified during learning.

In the late 1940s, D. O. Hebb and J. Konorski proposed that a synapse connecting two neurons is strengthened if the neurons are active at the same time. This hypothesis is referred to as the coincidence-detection rule. In 1973, T. V. P. Bliss and T. Lømo showed that brief trains of high-frequency electrical stimuli applied to the axons of entorhinal cortical neurons that project to the dentate gyrus cause a rapid and long-lasting increase in the efficacy of synaptic transmission between the perforant path fibers and the granule cells of the dentate gyrus. This increased efficacy is referred to as long-term potentiation (LTP). LTP has now been demonstrated in several other parts of the hippocampal formation, as well as in the neocortex. Several properties of LTP make it an attractive model of the cellular mechanisms underlying information storage. First, like memory, LTP can be generated very rapidly. In fact, activity-dependent synaptic potentiation can occur within milliseconds of the stimulation. Second, both LTP and memory are strengthened with repetition and, third, LTP lasts as long as several months. Fourth, LTP is selective since potentiation occurs only at stimulated synapses. Fifth, to induce LTP you must stimulate the brain at levels above a threshold, that is, not all stimulation induces LTP and not all information that we perceive gets encoded into long-term memory. This latter property is referred to as cooperativity. Most importantly, activation of a weak input at the same time as a convergent strong input leads to the potentiation of the weak input. Therefore, LTP is associative. If one thinks about a specific autobiographical memory, it includes information about time, place, sensory perceptions and even emotions. The associativity of LTP might provide the basis for linking these various types of information into a consolidated memory about an episode in our lives.

Though LTP may appear to be a somewhat artificial model of memory formation, it has several commonalities with what is known concerning normal functioning in the hippocampus. For example, the high-frequency synchronized firing of presynaptic neurons that occurs during the induction of LTP is likely to take place normally in the hippocampus during learning. In normal synaptic communication, low-frequency activity in the presynaptic fibers leads to the release of the neurotransmitter glutamate from terminals, which then crosses the synaptic cleft and binds to NMDA and non-NMDA types of glutamate receptors. It is mostly the non-NMDA receptor activation that is responsible for generating synaptic responses under these conditions. During low-frequency synaptic transmission, glutamate has little effect on the NMDA receptor because the NMDA receptor is blocked by magnesium ions. Perhaps the most interesting feature of the NMDA receptor is that the magnesium blockade can be switched on and off depending on the resting voltage of the neurons. When the resting electrical potential of the neuron with the NMDA receptors is made more positive, that is, when the cell is depolarized, the blocking magnesium ions are expelled and the NMDA receptor can then take part in synaptic transmission. During high-frequency, LTP-inducing activity in the presynaptic fibers, this is exactly what happens. The postsynaptic membrane becomes depolarized, and the Mg^{2+} block of the NMDA receptor disappears. Therefore, the NMDA receptor is a "molecular coincidence-detector," becoming active following the synchronous activity in the presynaptic fiber and the depolarization of the postsynaptic neuron's membrane. Unlike most of the non-NMDA receptors, the NMDA receptor is permeable to Ca^{2+}, so high-frequency activity of the presynaptic axon leads to an increase in the intracellular concentration of Ca^{2+}. The increased intracellular Ca^{2+}, in turn, triggers a series of complex biochemical events, ultimately resulting in persistent increase of synaptic strength.

Several classes of protein molecules have been implicated in this process of LTP, including proteases such as calpain, protein kinases such as protein kinase C, and the Ca^{2+}/calmodulin-dependent protein kinase (CaMKII). There is evidence that the biochemical cascade triggered by the high-frequency electrical stimulation that leads to LTP may result in modification of existing proteins, new protein synthesis, and/or changes in gene transcription. These recent discoveries have been greatly aided by the construction of mutant "knockout" mice that lack a gene coding for one of these molecules. During LTP, alterations may also occur in the presynaptic fiber terminals as a result of molecules released from the potentiated postsynaptic cell. The identity of these retrograde messengers is hotly debated and may involve the fatty acid arachidonic acid or the gases nitric oxide or carbon monoxide.

Because what goes up often must come down, there are mechanisms that appear to reverse the increase in synaptic efficacy that takes place during LTP. A decrease in synaptic strength may increase the flexibility of the memory-storing neuronal circuits and allow for the possibility of forgetting. Indeed, several forms of long-term depression (LTD) have been described. One form is induced following prolonged low-frequency stimulation of afferent fibers. This form of LTD is, in many ways, very similar to LTP, because it is specific and requires both the activation of NMDA receptors and the postsynaptic increase in CA. Whereas LTP and LTD enjoy substantial support as the physiological mechanisms for memory and forgetting, there are a variety of other hypotheses concerning the physiology of memory. At least short-term memory, for example, may involve the reverberatory activation of circuits in the hippocampal formation. There is a prominent electroencephalographic wave pattern called "theta activity" that appears to coordinate information as it enters each of the links of the hippocampal circuit.

V. SENSITIVITY OF THE HIPPOCAMPAL FORMATION TO ILLNESS

Various clinical conditions result in morphological alterations of the human hippocampal formation. Interestingly, some of the damage seems to occur as a result of the influx of calcium that may be associated with LTP. The hippocampus appears to be somewhat metastable. It can quickly encode information into a stable memory, but when pushed too far, the cells may be excited to death.

Though the causative factors of hippocampal pathology are not thoroughly understood for most disease states, it is clear that each of the different hippocampal cytoarchitectonic fields is more or less vulnerable to damage. In ischemia and temporal lobe epilepsy, for example, the CA1 field of the hippocampus (the so-called Sommer's sector) suffers the greatest neuronal cell loss. Head injury frequently results in memory deficits and epilepsy, and postmortem examinations of head-injured patients reveal selective damage to the hippocampus, often restricted to a specific part of the dentate gyrus (end-folium sclerosis). In other neuropathological conditions, such as Alzheimer's disease, the entorhinal cortex may suffer greater pathology.

Among the many conditions that produce pathological changes in the hippocampal formation, Alzheimer's disease is probably the most devastating. It is an age-related neurodegenerative illness that results in profound memory impairment and dementia. A number of pathological profiles, including senile (or neuritic) plaques and neurofibrillary tangles, are consistently seen in some of the hippocampal fields, especially the entorhinal cortex. Ultimately, Alzheimer's disease leads to a massive death of neurons in the hippocampal formation and other brain regions. There is good reason to suspect that the initial memory impairment associated with the disease is due in large part to the devastation of the hippocampal formation. [*See* Alzheimer's Disease.]

Temporal lobe or complex partial epilepsy is another neurological disorder in which the hippocampal formation is severely affected. This most common form of epilepsy was first associated with damage to the hippocampal formation in the late 1800s by Sommer, who conducted the first postmortem microscopic examination of a brain from a long-term epileptic patient. Sommer noted a dramatic loss of neurons in the hippocampal formation that was relatively selective and involved a region that in modern terminology would encompass CA1 and part of the subiculum. In approximately two-thirds of the cases of temporal lobe epilepsy, the hippocampal formation is the only structure that shows pathological modifications. During the first part of this century, it was generally believed that hippocampal pathology was a consequence of the epileptic seizures rather than their cause. Currently, however, increasing emphasis is being placed on the idea that disruption of normal hippocampal function may be an initiating factor in temporal lobe seizures. [*See* Epilepsies.]

In a number of other pathological conditions, the hippocampal formation is preferentially damaged. Among these is the loss of neurons, primarily in CA1, consequent to the ischemia associated with cardiorespiratory arrest. As noted earlier, these patients often demonstrate an anterograde memory impairment resulting from the hippocampal damage.

BIBLIOGRAPHY

Amaral, D. G., and Insausti, R. (1990). The hippocampal formation. *In* "The Human Nervous System" (G. Paxinos, ed.). Academic Press, San Diego.

Chan-Palay, V., and Kohler, C. (eds.) (1989). "Hippocampus New Vistas." Alan R. Liss, New York.

Collingridge, G. L., and Bliss, T. V. P. (1995). Memories of NMDA receptors and LTP. *Trends Neurosci.* **18,** 54–56.

Mody, I., and Soltesz, I. (1993). Activity-dependent changes in structure and function of hippocampal neurons. *Hippocampus,* **3,** 99–112.

Morris, R. G. M., Davis, S., and Butcher, S. P. (1990). Hippocampal synaptic plasticity and NMDA receptors: A role in information storage? *Philos. Trans. Roy. Soc. London* (*Ser. B*) **329,** 187–204.

Remple-Clower, N. L., Zola, S. M., Squire, L. R., and Amaral, D. G. (1996). Three cases of enduring memory impairment after bilateral damage limited to the hippocampal formation. *J. Neurosci.* **16,** 5233–5255.

Squire, L. R., Shimamura, A. P., and Amaral, D. G. (1989). Memory and the hippocampus. *In* "Neural Models of Plasticity" (J. Byrne and W. Berry, eds.), pp. 208–239. Academic Press, San Diego.

Histones

GARY S. STEIN
JANET L. STEIN
ANDRÉ J. VAN WIJNEN
University of Massachusetts Medical Center

GLOSSARY

Cell cycle Interval between the completion of mitosis in the parent cell and the completion of the next mitosis in one or both progeny cells. The periods of the cell cycle are sequentially defined as mitosis (prophase, metaphase, anaphase, and telophase), G_1 (the period between the completion of mitosis and the onset of DNA replication), S phase (the period of the cycle during which DNA replication occurs), and G_2 (the period between the completion of DNA replication and the onset of mitosis)

Histone proteins Five principal species of basic chromosomal proteins designated H2a, H2b, H3, H4, and H1 that range in size from 11,000 to 25,000 daltons. Histone proteins complex with DNA to form the primary unit of chromatin structure, the nucleosome

Nucleosome Primary unit of chromatin structure in eukaryotic cells consisting of approximately 200 nucleotide base pairs of DNA and 2 each of the core histone proteins (H2a, H2b, H3, and H4)

Posttranscriptional control Components of gene expression involving regulation mediated at the level of messenger RNA processing within the nucleus and/or cytoplasm, the translatability and/or stability of mRNA, and the assembly or posttranslational modifications of polypeptides

Promoter regulatory elements DNA sequences, generally but not necessarily 5′ (upstream) from the mRNA transcription initiation site, which modulate the specificity and/or level of transcription

Transcription factor Protein with sequence-specific recognition for a gene promoter element that positively or negatively influences the level of transcription

Transcriptional control Component of gene expression involving the synthesis of RNA, utilizing DNA as a template

HISTONE PROTEINS ARE THE MOLECULES RESPONsible for packaging DNA into chromatin, the protein–DNA complex that constitutes the eukaryotic genome. Histone–DNA complexes form the primary unit of chromatin structure, the nucleosome. Equally important, modifications in histone–DNA interactions occur in association with modifications in the expression of specific genes. Human histone genes have been cloned and characterized with respect to the regulation of expression that occurs in proliferating cells at the time when DNA is replicated, providing histone proteins to package newly replicated DNA into chromatin. Regulatory sequences of the histone genes that determine the specificity and levels of transcription, as well as factors that bind to regulatory elements to mediate histone gene expression, have been identified.

I. HISTONE PROTEINS

A. General Properties

The five principal species of histone proteins are designated H2a, H2b, H3, H4, and H1. These low-molecular-weight chromosomal polypeptides (Fig. 1) range in size from 11,000 to 25,000 daltons (Da). They share the common feature of a net positive charge due to a high representation of the basic amino acids arginine, lysine, and histidine, which facilitates the interactions of histones with negatively charged DNA

ENCYCLOPEDIA OF HUMAN BIOLOGY, Second Edition, VOLUME 4.

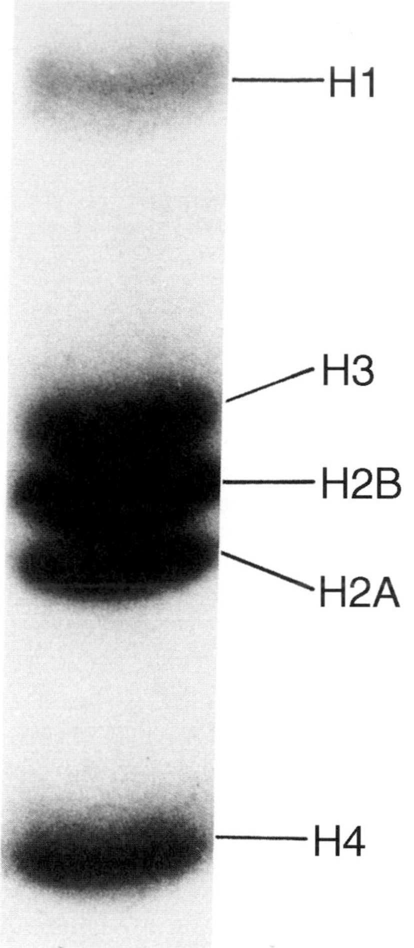

FIGURE 1 The five principal classes of histone polypeptides (H1, H3, H2B, H2A, and H4) fractionated electrophoretically in acetic acid–urea polyacrylamide gels.

molecules. The amino acid sequences of the histone proteins have been highly conserved during evolution, as illustrated by only limited amino acid substitutions in the histones of organisms separated phylogenetically as far as plants and mammals. This retention of the histone amino acid sequences reflects the conserved role of these proteins in chromatin structure and the apparently stringent requirement to support the similarly conserved primary unit of chromatin structure, the nucleosome (see Section I,B).

Despite the conserved nature of the histone proteins, these chromosomal polypeptides are encoded in a multigene family with approximately 20 nonidentical copies of each core (H2a, H2b, H3, and H4) and H1 gene. As a result, the five principal classes of histone polypeptides can be separated into several groups: (1) those that are represented in most cells and tissues and synthesized only in proliferating cells at the time of DNA synthesis (>90%); (2) those that are found in many cells and tissues but are expressed independently of proliferation, either constitutively during the cell cycle or following the completion of proliferation at the onset of tissue-specific gene expression associated with differentiation; and (3) those that are expressed solely in specialized cell types, such as sperm and avian erythrocytes, in which there are highly specific requirements for modifications in the packaging of DNA into chromatin.

Additional heterogeneity of the histone proteins is reflected by a series of posttranslational modifications that include acetylation, methylation, phosphorylation, and ADP-ribosylation. Such modifications alter the distribution of charge in specific domains of the histone proteins and may influence histone–DNA, as well as histone–histone, interactions. In the case of acetylation and phosphorylation, nuclear deacetylases and phosphatases permit the removal of acetate and phosphate moieties from histone polypeptides. These posttranslational modifications are involved in the incorporation of newly synthesized histones into chromatin (e.g., histone–DNA binding or histone–histone interactions) and may provide a basis for changes in the interactions of histones with DNA for remodeling chromatin architecture. An example of a dramatic reorganization in chromatin structure and organization is the condensation of chromatin into discrete and identifiable chromosomes at the onset of mitosis, which is accompanied by changes in histone phosphorylation. A more subtle, yet functionally important, modification in chromatin structure that is reflected by changes in histone–DNA interactions occurs when the expression of specific genes is activated or repressed. [*See* Chromatin Folding.]

B. Chromatin Structure

The magnitude of the problem associated with chromatin structure is illustrated by a requirement for the ordered packaging of $2\frac{1}{2}$ yards of DNA within the confines of the cell nucleus in a manner that supports the selective expression of specific genes required for the biogenesis and maintenance of cellular phenotypes. An equivalent amount of histone and DNA are present in cell nuclei, forming complexes known as nucleosomes, which are the primary unit of chromatin structure (Fig. 2). Each nucleosome consists of a core particle of approximately 140 base pairs of DNA wound around a complex consisting of two H2a, H2b, H3, and H4 molecules and a linker DNA region of approximately 40–60 base pairs. Under the elec-

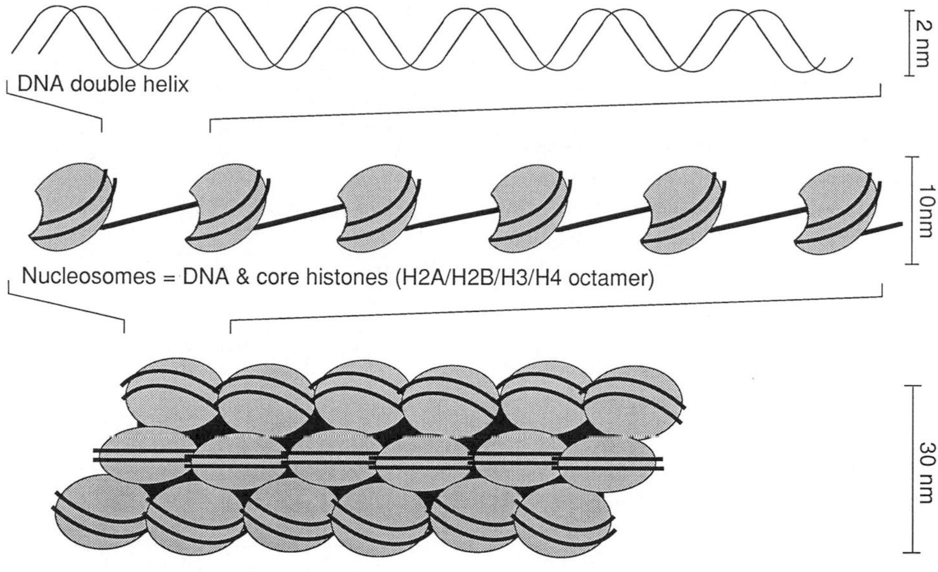

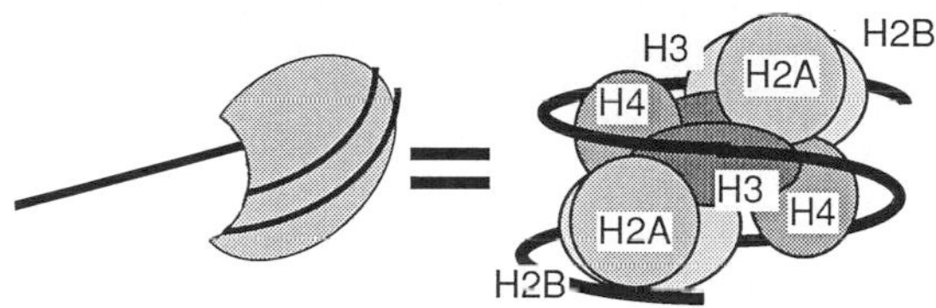

FIGURE 2 *Top:* Three principal levels of chromatin organization. *First:* The 2-nm, deproteinized, double-stranded DNA double helix. *Second:* The organization of DNA into nucleosomes. The beads-on-a-string structure comprises a 10-nm fiber: each bead consists of two each of core histone proteins (H2A, H2B, H3, and H4). The string component of the structure is the DNA. *Third:* The higher-order organization of chromatin structure mediated by association of nucleosomes through linker histone H1 into a 30-nm chromatin fiber. *Bottom:* Organization of the core histone proteins within a nucleosome-octamer.

tron microscope, the nucleosomes appear as a series of beads (protein–DNA complexes) on a string (linker DNA joining the nucleosomes). The H1 histones bind to the linker region and are involved with nucleosome–nucleosome interactions. Though the interactions of histones with DNA and the H3–H4 and H2a–H2b histone interactions within the nucleosomes have been firmly established, this organization only accounts for a packing ratio of seven. Clearly, higher-order structural organization of chromatin is required for accommodating the genome within the cell nucleus, and here our understanding of chromatin structure is minimal. The 10-nm beads-on-a-string structures are packaged as a 30-nm chromatin fiber, and further packaging results in chromatin fibers of 100 nm (see Fig. 2). Additional insight into the interactions of histone with DNA is required, but it is also necessary to further define the role of "nonhistone," sequence-specific DNA binding proteins in directly mediating DNA conformation and in modulating histone–DNA interactions within the nucleus. [*See* DNA in the Nucleosome.]

II. HISTONE GENE EXPRESSION

Historically, the observations that the cellular content of histone proteins doubles during the S phase of the cell cycle (the period of the cell cycle when DNA replication occurs) and that histone protein synthesis occurs concomitantly with DNA synthesis provided the first example of gene expression functionally related to cell growth. As such, investigations in several laboratories over the past two decades have focused on the regulation of histone gene expression as it relates to the complex and interdependent series of events required for cell proliferation. The selective synthesis of histone proteins during the S phase of the cell cycle is mediated by the regulation of histone gene expression at both the transcriptional and posttranscriptional levels. At the transcriptional level, the extent to which mRNA is transcribed from histone genes is regulated. Posttranscriptional regulation involves the control of histone mRNA processing and/or stability.

Multiple levels of control have been established by addressing experimentally the relationship of DNA replication to histone protein synthesis, cellular histone mRNA levels, and histone gene transcription. The rationale for this approach is that a stoichiometric relationship between DNA replication, histone protein synthesis, and histone mRNA levels is indicative of transcriptionally mediated expression, whereas the transcription or presence of histone mRNAs in cells when histone proteins are not synthesized is a direct indication of posttranscriptional control. Differences, and particularly fluctuations, in the relationship between the principal parameters of histone gene expression—mRNA levels, mRNA synthesis, and protein synthesis—suggest that both transcriptional and posttranscriptional regulation are operative.

A functional, as well as temporal, relationship between DNA replication and the expression of human core and H1 histone genes was initially indicated by the constant histone : DNA ratio (1 : 1) observed in a broad spectrum of cells, tissues, and organs and the doubling of cellular levels of histone protein during the S phase of the cell cycle. Direct measurements then confirmed that histone protein synthesis is largely confined to S phase and that inhibition of DNA replication results in a rapid cessation of histone protein synthesis. Cellular levels of histone mRNAs have been measured throughout the cell cycle in continuously dividing cell populations and after stimulation of quiescent cells to proliferate. In all cases the cellular levels of human histone mRNA accumulation reflect cellular levels of both histone protein synthesis and DNA replication. Similarly, inhibition of DNA replication brings about a dose-dependent loss (selective destabilization) of histone mRNAs that parallels decreases in DNA and histone synthesis. Measurements of histone gene transcription indicate enhanced synthesis of histone mRNAs early during the S phase of the cell cycle. A summary of the principal biochemical events associated with histone gene expression during the cell cycle is presented in Fig. 3. [*See* DNA Synthesis.]

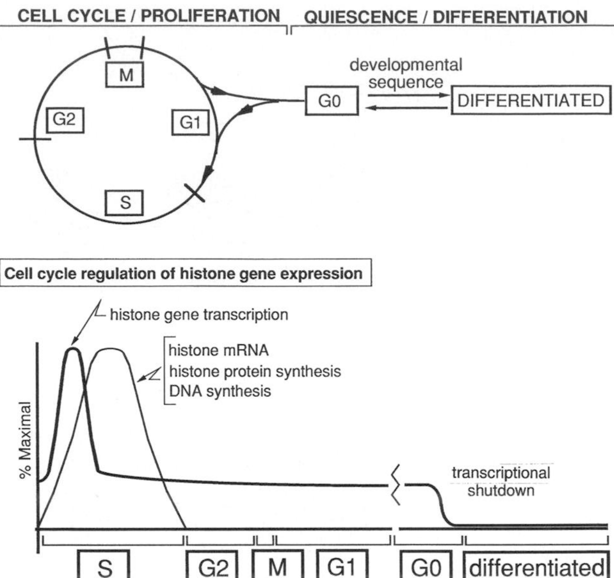

FIGURE 3 Regulation of histone H4 gene expression. *Top:* Schematic representation of the cell cycle (G_1, S, G_2, Mitosis), indicating the pathway associated with the postproliferative onset of differentiation initiated following completion of mitosis. *Bottom:* Representation of data defining the principal biochemical parameters of histone gene expression; histone protein synthesis and the presence of histone mRNA are restricted to S phase cells (DNA synthesis). Constitutive transcription of histone genes occurs throughout the cell cycle with an enhanced transcriptional level during early S phase. These results establish the combined contribution of transcription and mRNA stability to the S phase-specific regulation of histone biosynthesis in proliferating cells, with histone mRNA levels as the rate-limiting step. The transcriptional down-regulation of histone gene expression at the onset of differentiation is shown.

A viable model for regulation of human histone gene expression must therefore account for: (1) a precisely timed increase in transcription of histone genes when DNA replication is initiated; (2) elevated cellular levels of histone mRNA and histone protein synthesis, which are tightly coupled with DNA replication; (3) a rapid and selective destabilization of histone mRNAs concomitant with the termination of DNA replication at the natural end of S phase or following inhibition of DNA synthesis; and (4) complete down-regulation of histone gene transcription and histone mRNA destabilization when cessation of proliferation accompanies the onset of differentiation.

The increased transcription of histone genes early during S phase and the coordinate accumulation of mRNAs for core and H1 histone proteins that closely parallels the initiation of DNA and histone protein synthesis suggest that the onset of histone gene expression is at least in part transcriptionally mediated. In fact, it is reasonable to postulate that throughout S phase the synthesis of histone proteins is modulated by the availability of histone mRNAs. The stabilization of histone mRNAs throughout S phase and the destabilization of histone mRNAs when DNA replication is completed or inhibited are highly selective, and largely posttranscriptionally controlled. The selectivity of histone mRNA destabilization is suggested since both the natural end of S phase and inhibition of DNA replication are associated with a rapid loss of histone mRNAs, whereas only minimal fluctuations, qualitative or quantitative, are observed in other cellular mRNA populations under these conditions. The extremely tight coupling of histone gene expression with the extent of ongoing DNA replication is supported by the coordinate and stoichiometric decreases of histone mRNAs and histone protein synthesis after inhibition of DNA replication. [*See* DNA and Gene Transcription.]

The complete suppression of histone gene expression when cells become postproliferative during differentiation is accompanied by and functionally related to both repression of histone gene transcription and degradation of histone mRNA. This down-regulation of histone genes supports a principal component of physiological requirements for differentiation, a loss of competency for proliferation and cell cycle progression. In summary, the preferential expression of histone genes during the S phase of the cell cycle apparently involves control at both transcriptional and posttranscriptional levels. The initiation of DNA replication in human cells is associated with a 3- to 5-fold increase in histone gene transcription, which after several hours returns to a basal level that is maintained outside of S phase. This increase represents enhancement rather than an activation of histone gene transcription, and hence the mechanisms that regulate histone gene transcription differ from those observed when activation of nonexpressed genes is initiated. In contrast, the selective destabilization of histone mRNAs at the end of S phase or after inhibition of DNA synthesis is posttranscriptionally mediated. Differentiation is functionally linked to transcriptional down-regulation of histone genes.

III. ORGANIZATION AND REGULATION OF HUMAN HISTONE GENES

A. General Organization

The human histone genes are organized into clusters of core alone (H2a, H2b, H3, and H4) or core together with H1 histone coding sequences. Several such segments of human genomic DNA containing histone coding sequences are illustrated schematically in Fig. 4. Chromosome mapping studies based on *in situ* hybridization of metaphase chromosomes with radiolabeled human histone gene probes, Southern blot analysis of DNAs from panels of mouse–human somatic cell hybrids, and hybridization with DNAs from flow-sorted human chromosomes have indicated that the human histone gene clusters are represented on at least chromosomes 1 and 6. Within these clusters, there is generally a pairing of H2a with H2b genes and H3 with H4 genes. This organization of human histone genes is similar to that observed for mouse, rat, and chicken, but somewhat more complex than the simple, tandemly repeated clusters found in lower eukaryotes such as sea urchin and *Drosophila.*

B. Organization of a Cell Cycle-Regulated Human Histone Gene

Despite the clustering of cell cycle-regulated human histone genes, each histone coding sequence is an independent transcription unit. All amino acids of the histone protein are encoded in contiguous nucleotides since these genes lack introns. Also noteworthy are the absence of a polyadenylation site in the 3′ region and the nontranslated leader and trailer segments of the mRNA that are less than 50 nucleotides. This simple organization of the mRNA coding region of the histone gene appears to be optimal for rapid processing of the mRNA transcript and export to the cytoplasm for immediate use as a template for histone protein synthesis.

Regulatory sequences controlling expression of histone genes have been functionally defined by introduc-

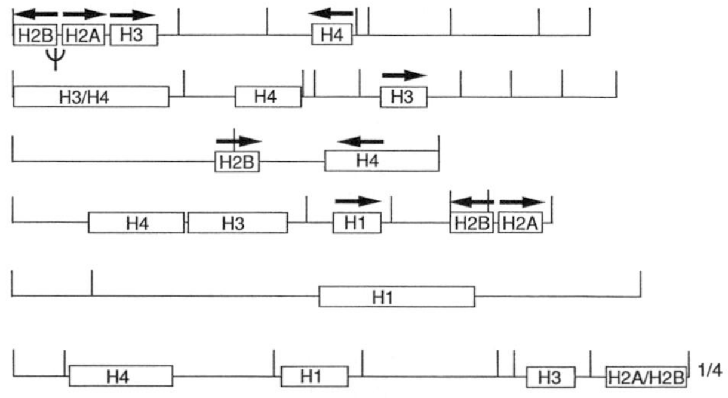

FIGURE 4 The organization of genomic DNA segments containing some of the human histone coding sequences; arrows designate directions of transcription. H2B and H2A pseudogenes are designated by the symbol ψ.

ing a systematic series of mutations that were assayed for consequential effects on histone gene transcription in intact cells. The 5′ flanking regions of histone genes contain consensus regulatory sequences of many genes transcribed by RNA polymerase II (Fig. 5). Proximal regulatory elements include a TATA box approximately 30 base pairs upstream from the transcription start site. As has been observed for all genes that have been analyzed for the localization of promoter regulatory elements, an extensive series of sequences that influence both specificity and level of transcription have been identified in the initial 1000 nucleotides of the 5′ promoter region.

Figure 5 is a schematic representation of the regulatory organization of the initial 1.0 kB of an H4 histone gene promoter. Though this region contains the minimal sequences required for regulated expression, the functional limits of the H4 gene appear to extend considerably upstream. Indeed, cis-acting elements up to −6.5 kB may influence developmental expression of the H4 histone gene *in vivo* in transgenic animals. Two domains of *in vivo* protein–DNA interactions for the H4 histone gene have been established in the intact cell at single-nucleotide resolution. These have been designated H4-Site I (nt −156 to −113) and H4-Site II (nt −97 to −47). Two distal regulatory

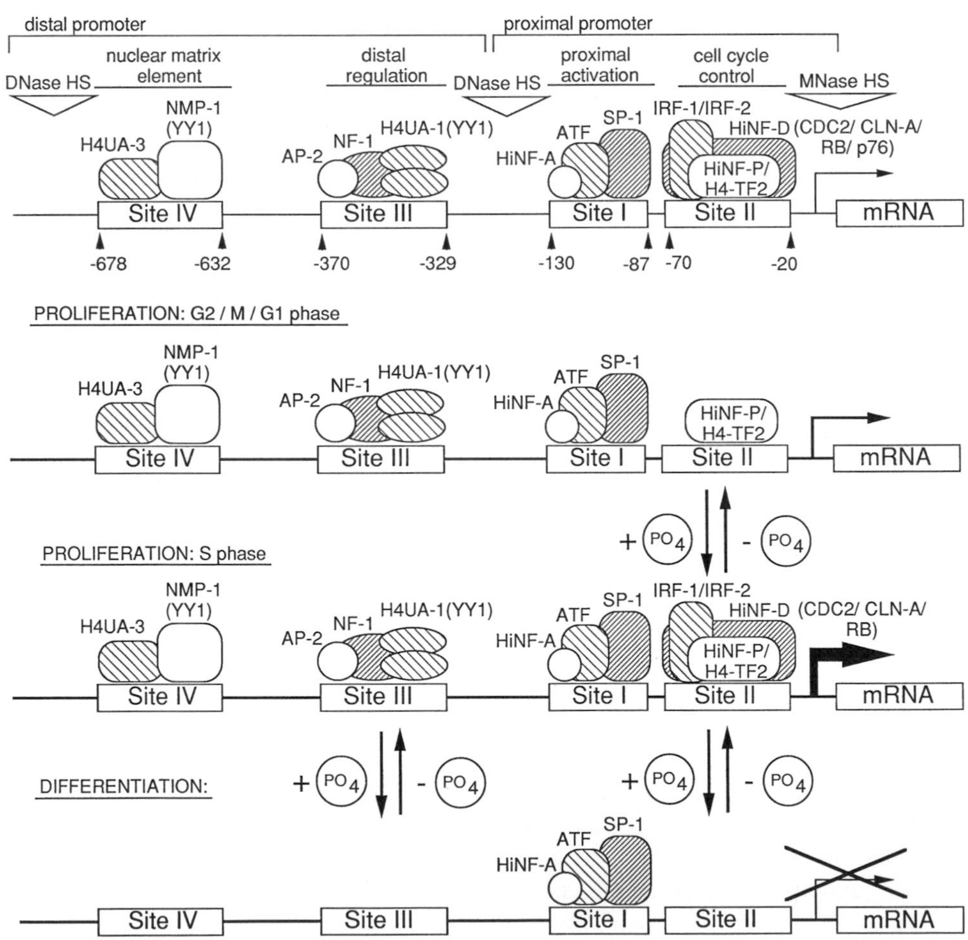

FIGURE 5 Schematic representation of promoter regulatory elements and transcription factors that support histone gene expression. *Top:* The representation and organization of gene regulatory sequences is designated by Sites I–IV. The ovals and boxes represent transcription factors. Proximal and distal cell cycle regulatory elements are designated along with nuclease-sensitive regions (DNase HS, MNase HS). Also shown are sites of histone gene interactions with the nuclear matrix. *Three lower segments:* Phosphorylation-dependent modifications in interactions of transcription factors with histone gene promoter elements both during the cell cycle and following differentiation. These modifications in protein–DNA interactions control the extent to which the histone gene is transcribed, which is indicated by the thickness of the horizontal arrows over the mRNA regions of the gene.

elements designated Site III (nt −370 to −329) and Site IV (nt −678 to −632) have been experimentally established.

The biosynthesis of histone proteins that occurs independent of DNA synthesis and proliferation is supported by histone genes with a sequence organization that controls proliferation-independent expression. Variations from organization of the cell cycle-regulated histone genes are present in both the mRNA coding and flanking sequences. The unique sequence organization that renders these histone genes transcribable in the absence of DNA synthesis and facilitates messenger RNA stability independent of proliferation includes variant promoter elements, introns in the mRNA coding sequences, and 3′ polyadenylation sequences.

C. Regulatory Factors That Control Expression of Human Histone Genes

Three approaches have been pursued to identify proteins that bind in the 5′ flanking regions of cell cycle-regulated human histone genes at regulatory elements that influence transcription. The first experimental approach has been to establish the sites of protein–DNA interactions in the histone gene promoter in intact cells. The second approach has been to assay the binding of factors in fractionated nuclear extracts to isolated sequences upstream from cell cycle-regulated histone genes. The third has been isolation and functional characterization of histone gene promoter binding factors.

In vivo protein–DNA interactions were established by a technique known as "genomic sequencing" in which the guanine (G) sequencing reaction of Maxam and Gilbert is carried out in intact cells. In this way, sites of protein–DNA interactions can be detected at single-nucleotide resolution. Figure 5 shows schematically diagrammed a series of protein–DNA interactions that have been established both in intact cells and by *in vitro* protein binding studies using isolated promoter elements for sequence-specific protein binding. The identity of transcription factors that have been experimentally established as contributing to histone gene expression are indicated. Modifications in transcription factor interactions at histone gene promoter sequences that define the extent to which the gene is expressed or repressed are also shown. The complexity of histone gene promoter organization and the diverse series of regulatory factors that modulate histone gene transcription are becoming increasingly apparent. This complexity in histone gene promoter organization serves as a paradigm for transcriptional competency at the G1/S phase transition point of the cell cycle. The proximal promoter domain H4-Site I is a bipartite cis-activating element that interacts distally with ATF1, a member of the leucine zipper family of transcription factors, and proximally with a zinc finger DNA binding protein Sp1, with both factors capable of synergistically mediating stimulation of transcription. The H4-Site II is a multipartite protein–DNA interaction site for sequence-specific factors HiNF-D, HiNF-M, and HiNF-P (H4-TF2). The proximal region of H4-Site II spans a TATA motif and is sufficient to mediate accurate transcription initiation, presumably by interaction with initiation factor TFII-D. However, the distal region of H4-Site II influences transcriptional competency, as well as the timing and extent of H4 mRNA synthesis *in vivo*. The Site II region binds factors that regulate basal histone gene transcription and are responsible for the cell cycle-dependent enhancement of histone gene transcription at the onset of S phase. There are two principal components of cell cycle transcriptional control at Site II. The first is HiNFD, which includes cyclin A, cdc2, and an RB-related protein. The second is HiNF-M, which is IRF2 (interferon responsive factor 2).

The distal activating elements H4-Sites III and IV encompass regions that stimulate transcription *in vivo* and interact with the heteromeric nuclear factors H4UA-1 and H4UA-3, respectively. Additionally, H4-Site IV overlaps with a putative nuclear matrix attachment site spanning nt −730 to −589. This element interacts with a sequence-specific nuclear matrix protein (NMP-1) that has recently been identified as the YY1 transcription factor. Site IV protein–DNA interactions may influence expression of the H4 histone gene promoter by transient anchorage to the nuclear matrix. The integration of mechanisms controlling the coordinately regulated transcription of multiple histone genes may involve several shared promoter-binding activities, including both ubiquitous and histone gene-specific transcription factors. HiNF-D-related protein–DNA interactions are also represented in H3 and H1 histone gene promoters, suggesting the possibility of coordinate transcription factor interactions regulating several histone gene classes.

Insight into transcriptional control of histone gene expression has been provided by identification of modifications in interactions of promoter binding factors within the initial 1.0 kB of a human H4 histone gene promoter at Sites I, II, III, and IV, and relating these to the extent of gene transcription. Protein–DNA interactions at these regulatory elements during the cell cycle and with the down-regulation of proliferation during differentiation are schematically shown in Fig. 5.

Taken together with the contribution of multiple regulatory elements to the specificity and level of histone gene transcription, this series of promoter–factor interactions reflects the modular organization of the cell cycle-regulated histone genes. A single, rate-limiting step to the transcription of the cell cycle-regulated histone genes has not been established. Rather a series of sequences comprising both positive and negative regulatory elements contribute individually as well as synergistically to determine the extent to which the gene is expressed.

BIBLIOGRAPHY

Heintz, N., Sive, H. L., and Roeder, R. G. (1983). Regulation of human histone gene expression: Kinetics of accumulation and changes in the rate of synthesis and in the half-lives of individual histone mRNAs during the HeLa cell cycle. *Mol. Cell. Biol.* **3,** 539–550.

Holthuis, J., Owen, T. A., van Wijnen, A. J., Wright, K. L., Ramsey-Ewing, A., Kennedy, M. B., Carter, R., Cosenza, S. C., Soprano, K. J., Lian, J. B., Stein, J. L., and Stein, G. S. (1990). Tumor cells exhibit deregulation of the cell cycle histone gene promoter factor HiNF-D. *Science* **247,** 1454–1457.

Marzluff, W. F., and Pandey, N. B. (1988). Multiple regulatory steps control histone mRNA concentrations. *Trends Biochem. Sci.* **13,** 49–52.

Pauli, U., Chrysogelos, S., Stein, G., Stein, J., and Nick, H. (1987). Protein–DNA interactions *in vivo* upstream of a cell cycle regulated human H4 histone gene. *Science* **236,** 1308–1311.

Ramsey-Ewing, A., van Wijnen, A., Stein, G. S., and Stein, J. L. (1994). Delineation of a human histone H4 cell cycle element *in vivo:* The master switch for H4 gene transcription. *Proc. Natl. Acad. Sci. USA* **91,** 4475–4479.

Sierra, F., Lichtler, A., Marashi, F., Rickles, R., van Dyke, T., Clark, S., Wells, J., Stein, G., and Stein, J. (1982). Organization of human histone genes. *Proc. Natl. Acad. Sci. USA* **79,** 1795–1799.

Stein, G., Lian, J., Stein, J., Briggs, R., Shalhoub, V., Wright, K., Pauli, U., and van Wijnen, A. J. (1989). Altered binding of human histone gene transcription factors during the shutdown of proliferation and onset of differentiation in HL-60 cells. *Proc. Natl. Acad. Sci. USA* **86,** 1865–1869.

Stein, G. S., Stein, J. L., and Marzluff, W. F. (eds.) (1984). "Histone Genes." John Wiley & Sons, New York.

van Holde, K. E. (1988). "Chromatin." Springer-Verlag, New York.

van Wijnen, A. J., Aziz, F., Grana, X., De Luca, A., Desai, R. K., Jaarsveld, K., Last, T. J., Soprano, K., Giordano, A., Lian, J. B., Stein, J. L., and Stein, G. S. (1994). Transcription of histone H4, H3 and H1 cell cycle genes: Promoter factor HiNF-D contains CDC2, cyclin A and an RB-related protein. *Proc. Natl. Acad. Sci. USA* **91,** 12882–12886.

Vaughan, P. S., Aziz, F., van Wijnen, A. J., Wu, S., Soprano, K., Stein, G. S., and Stein, J. L. (1995). Activation of a cell cycle regulated histone gene by the oncogenic transcription factor IRF2. *Nature* **377,** 362–365.

Hormonal Influences on Behavior

EDWARD P. MONAGHAN
S. MARC BREEDLOVE
University of California, Berkeley

GLOSSARY

Androgen Class of steroid hormones, the primary source of which is the testes. The major circulating androgen is testosterone

Endogenous Originating from within the organism

Estrogen Class of steroid hormones, the primary source of which is the ovaries. The major circulating estrogen is estradiol

Exogenous Originating from outside the organism

Gonads The sex organs; testes in the male, ovaries in the female

Hormone Chemical secreted by one cell and capable of affecting other cells throughout the organism

Perinatal Around the time of birth

Perineum Genital region and structures occupying the pelvic floor

Sexual dimorphism Structural or behavioral difference between males and females of the same species

Steroids Class of hormones with cholesterol as the common precursor. The primary sources of steroids are the gonads, placenta, and adrenal cortex

HORMONES, WITHOUT DOUBT, CAN INFLUENCE the behavior of animals, including humans. However, the causes of behavior are seldom simple, and it is often difficult to conclude that hormone X causes behavior Y. Rather, through controlled study and manipulation of hormone levels in various situations, we can demonstrate that a given hormone is involved in certain behaviors and whether it increases or decreases the likelihood that the behavior will appear.

I. INTRODUCTION

Before dealing with particular hormones and behaviors, we need to discuss briefly the mechanisms and timing of hormone actions and the scientific approaches to investigating hormonal influences on behavior. Classically speaking, a hormone is a chemical substance that is secreted into the bloodstream by an endocrine gland and that travels to and influences particular target tissues. To produce an effect in the target tissue, the hormone must bind to a specific receptor that is itself a protein. Much of this article deals with steroid hormones, such as androgens and estrogens, that readily diffuse into cells. Steroid receptors are contained within the target cells and, when bound to the appropriate hormone, interact with the cell's DNA to modulate gene activity. The other class of hormones, protein or peptide hormones, usually binds to receptors on the external surface of the cell membrane and triggers a series of internal events that greatly vary from hormone to hormone and even from tissue to tissue. Two points should be emphasized: (1) if a tissue or an individual lacks receptors, the hormone cannot have its effect, and (2) the hormone does not completely change the target, but instead acts by adjusting ongoing processes. [*See* Endocrine System; Steroid Hormone Synthesis.]

When can hormones act to influence behavior? The organizational/activational model for hormone ac-

ENCYCLOPEDIA OF HUMAN BIOLOGY, Second Edition, VOLUME 4.

tion proposes that during early development hormones permanently organize neural substrates such that the individual will be capable of displaying particular behavioral patterns in response to later hormone exposure. As discussed later, this model is best illustrated by sexually dimorphic behaviors (i.e., those behavior patterns more common in one sex than the other). For example, the comparatively high testosterone (T) level in male rats around the time of birth organizes their central nervous system (CNS) (i.e., the brain and spinal cord) such that the rise in T levels during puberty activates male-specific behavior patterns. Treatment of adult female rats with T produces little masculine behavior. Thus early exposure to some hormones can regulate neural development, which will then determine hormone responsiveness later in life.

The study of hormonal involvement in a particular behavior is a three-step process. We start by observing and recording behavior and plasma hormone levels in an intact subject. Next we remove the endocrine gland, which produces an implicated hormone, and note the effect on the behavior. Finally, if removal of the hormonal source affects behavior, we supply, via injection, exogenous hormone and see if the behavior is restored to normal levels. Obviously, such controlled surgical manipulations are not carried out on human subjects. Most basic biomedical research is conducted on animals, and it is important to note that although there are species differences, the information gained in animal studies provides the basic framework to explain human phenomena. This article focuses on human behavior, but references will be made to animal research, which has provided particular insight.

II. SEXUAL BEHAVIOR

A. Introduction

It has often been stated that the further advanced along the "evolutionary ladder" an animal is, the less influence gonadal hormones have on its behavior. Thus hormones seem to play a greater role in the sexual behavior of rats than of primates. This may be an oversimplification. The effects of hormonal alterations may be easier to see in rodents than in primates, partly because of the greater control we have over extraneous variables when dealing with rodents. It is difficult to conduct well-controlled experiments involving primates (especially humans), but recent findings have suggested that gonadal hormones do play an important role in the behavior of "higher animals."

Hormones are essential for the expression of sexual behavior in vertebrates. They are involved in everything from the fetal development of sexual structures to the production of copulatory behavior. It may not be immediately obvious why we mentioned the development of sexual structures when we are concerned with behavior, but looking at behavior without looking at morphology can lead to confusion. Whenever a behavior involves two or more individuals, whether they are rats or humans, the physical appearance of each individual can profoundly affect the resultant behavior. Furthermore, research on animals has shown that the same signals that direct masculinization of the body also direct masculinization of the nervous system and therefore behavior. With this in mind, we will begin by briefly describing sexual differentiation, the process by which a fetus develops as either male or female.

Early during fetal development, the individual's genetic sex, XX or XY, determines the differentiation of the gonads: a genetic female, XX, will develop ovaries, and a genetic male, XY, will develop testes. From that point on, further development of phenotypic sex (the male or female body type) depends not on the sex chromosomes directly, but on hormones secreted by the gonads. In the normal case, the testes of an XY individual release hormones that masculinize bodily structures: T acts to masculinize the external genitalia, and another testicular factor works in concert with T to differentiate the internal structures. It is the absence of these testicular secretions in the XX individual that results in the feminine phenotype. In addition to determining these somatic characters, hormones also act on the CNS of developing embryo to affect the likelihood that the individual will engage in certain behaviors later in life; this is referred to as the organizational role of hormones in the behavior of animals. The early exposure to androgen seems to organize irreversibly the CNS in a masculine fashion. The challenging question, which is largely unanswered, is whether this is also true for humans.

Because it is our assumption that any differences in behavior must be reflected by differences in the nervous system, it is worth while to look at some of the sexual dimorphisms demonstrated in the CNS, especially those found in both human and nonhuman animals. First, we will consider a sexual dimorphic area in the hypothalamus: that region of the brain involved in homeostasis and appetitive behaviors. The preoptic area (POA) of the hypothalamus appears to be important in sexual behavior because lesions there selectively abolish male sexual behavior in a variety of species. Located within this region is the sexually dimorphic nucleus of the POA (SDN-POA). The SDN-POA of both rats and humans is larger in males than in females. Hormonal manipulation in rats has shown that it is the presence of high levels of perinatal T that cause males to have a larger SDN-POA. Although we know that the POA is involved in sexual behavior, the behavioral relevance of the SDN portion of the POA has not been established. [*See* Hypothalamus.]

A second sexual dimorphism of the CNS is located in the spinal cord of both rats and humans. Male rats have a group of large motoneurons called the spinal nucleus of the bulbocavernosus (SNB) that innervate and control muscles of the perineum, specifically those attached to the penis. Both the motoneurons and their muscles are absent in adult female rats. At birth, both sexes have SNB cells and perineal muscles, but perinatal T levels prevent the death of these cells in males. Adult T levels regulate the activity of this neuromuscular system, and as might be expected, it is important in male copulatory behavior. These same perineal muscles in humans are innervated by a homologous spinal area called Onuf's nucleus, which has more motoneurons in males than in females. Continued research involving sexual dimorphisms will lead to a greater understanding of how early androgens influence neural development and therefore how differences in behavior arise in both rats and humans. [*See* Spinal Cord.]

The final aspect of sexual differentiation in humans is the formation of gender identity, which is believed to be completed early in childhood and refers to how an individual perceives oneself (i.e., as either male or female). There are conflicting opinions as to what determines gender identity, but a reasonable hypothesis is that prenatal hormone levels and postnatal environmental factors (experience and learning) interact in the process of psychosexual development. As we are about to see, these three aspects of sexual identity (genotypic sex, phenotypic sex, and gender identity) are not always in agreement.

B. Syndromes Affecting Sexual Development

1. Androgen Insensitivity

Several syndromes alter the course of sexual differentiation. The most striking disagreement between genetic and phenotypic sex is the androgen insensitivity syndrome (AIS). AIS individuals are genetically male (XY) but, because of a genetic mutation, lack androgen receptors. During development the Y chromosome causes their gonads to differentiate as testes that release T, but because there are no receptors present, body tissues fail to respond and therefore a feminine phenotype results. In cases of complete insensitivity, the external appearance of such individuals is quite feminine, and they are often undetected and are raised and behave in a normal female pattern, except that at puberty menstruation fails to appear.

Animal studies have indicated that it is a metabolite of T, estradiol (see Fig. 1), that is responsible for much of the sexual differentiation of brain tissue and that this neural masculinization ultimately affects behavior. Because AIS individuals have receptors for estradiol, we might predict that masculinization of the neural structures would occur despite the feminine appearance. In that case, we would expect problems in the development of gender identity and sexual orientation, but no such problems are reported. The lack of difficulty in psychosexual development of AIS individuals suggests that either these early hormones do not influence the formation of gender identity and sexual orientation or that consistent environmental factors can override hormonal cues. The difficulty in deciding between these two possibilities becomes especially evident when we consider the conflicting conclusions reached in the study of other syndromes.

2. 5-α-Reductase Deficiency

As the name suggests, this syndrome is due to a deficiency in the enzyme 5-α-reductase, which, as illustrated in Fig. 1, plays an essential role in allowing T to masculinize the external genitalia during develop-

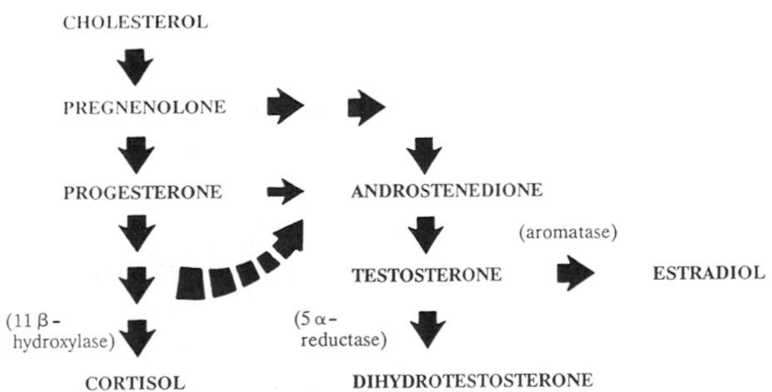

FIGURE 1 Simplified schematic diagram of synthesis of some steroid hormones. Cholesterol is the precursor for all steroid hormones. In gonads, enzymes facilitate synthesis in the direction of androstenedione, a precursor to testosterone. Testosterone can then be converted to dihydrotestosterone (DHT) by 5α-reductase, which is present in many peripheral tissues, or to estradiol by aromatase, which is present in the ovaries and some areas of the central nervous system. One of the major steroids produced by the adrenals is cortisol, and the enzyme 11-β-hydroxylase is important in its production. In the absence of 11-β-hydroxylase (congenital adrenal hyperplasia syndrome), steroid production in the adrenals is shifted in the direction of androgens.

ment. Individuals born with this syndrome are genetically male (XY), but because of the enzyme deficiency do not convert T to the more active androgen, DHT; therefore they are born with external genitalia that are feminine or at least ambiguous: the scrotum resembles labia and the size of the phallus is intermediate between that of a clitoris and a penis. This disorder is common in a small village in the Dominican Republic where it is referred to by the villagers as "guevedoces"—penis at twelve. Reportedly, these individuals are raised as girls until puberty, when testicular T secretions trigger development of the male phenotype (i.e., deepening of the voice, increase in muscle mass, enlargement of the phallus, and descension of testes). After puberty, their gender identity and sexual orientation become masculine. Some researchers have concluded from these cases that the environment (i.e., being treated as a girl throughout childhood) has little effect on adult gender identity and that the presence of T early in development, at puberty, or both completely determines psychosexual development as a male. An alternative explanation of the guevedoces phenomenon is that although the affected individuals look somewhat feminine, the villagers are clearly aware of some differences and treat them subtly differently than normal boys or girls. This enables a smooth assumption of a male gender identity around puberty. It is also possible that hormones have little influence on human gender development, and in a secure setting where such gender shifting is accepted, humans are so adaptable that they can accept a pubertal sex change, especially when the change brings the many social advantages accorded males in that culture.

3. Congenital Adrenal Hyperplasia

The adrenal glands are a source of steroid hormones, the corticosteroids. As shown in Fig. 1, the synthesis of adrenal and gonadal steroids is closely related, but normally only small amounts of androgen are released from the adrenals. However, in individuals with congenital adrenal hyperplasia (CAH), a deficiency in 11-β-hydroxylase results in an abnormally high production of androgen. If this occurs in an XX individual, depending on the extent of the deficiency and the time it arises during development, the genitalia can be masculinized to varying degrees. The appearance of ambiguous genitalia at birth leads physicians to a diagnosis of CAH. The genitalia can be surgically corrected, and maintenance treatment with exogenous steroids can halt excess androgen production.

Because of differences in the time of initiation of hormone treatment and surgical correction and the possibility that the parents communicated ambiguities about the sex of their child, it is quite difficult to draw definitive conclusions about whether this early exposure to T itself affected the later behavior of these females. Despite problems finding appropriate comparison groups, some studies have reported a consistently higher level of early tomboyish behavior and, as adults, a higher incidence of bisexual fantasy and experience in CAH women than in AIS or normal women. However, the majority of CAH females have exclusively heterosexual relations. Thus the studies of CAH women leave open the possibility of hormonal influence on gender development but do not provide evidence for an overpowering role for hormones.

C. Female Sexual Behavior

Thus far we have dealt exclusively with the effect of T on development and behavior. Although some T is secreted by the ovaries of adult females, estrogen (E) and progesterone (P) are the major steroid products of the ovaries. Hormone levels vary throughout a female's menstrual cycle (referred to as "estrous cycle" in nonhumans). E levels are high in the early to midcycle (i.e., before and around the time of ovulation). P levels are high late in the cycle, after ovulation. T levels, although always much lower than E, also peak around midcycle.

With these normal, predictable changes in hormone levels, one might expect experiments on hormonal effects on female sexual behavior to be among the first studies conducted in hormones and behavior. This is true when we consider the "lower" mammals in which the females will mate only when they are in hormonally activated estrus (heat), but human and many other primate females were considered to be sexually receptive throughout their cycle. Thus it was long concluded that their sexual behavior was independent of proximate hormonal stimulation, but scientists have recently taken a closer look at this proposal.

Female sexual behavior can be divided into three categories. Whether or not a female will copulate if approached by a male is referred to as receptivity. Proceptivity describes the actions a female will take to attract a male or to initiate a sexual encounter. Attractivity refers to how well an individual female serves as a sexual stimulus for a male. For example, male monkeys show preferences for sexual partners who are at a particular stage of their estrus cycle; therefore we could say that these females are more attractive. In terms of stimulus value, a given rat, monkey, or woman can have a different level of attractivity depending on her hormonal state.

Although rhesus females will allow males to mount (i.e., are receptive) at any point during their estrus cycle, it was found that around midcycle they were much more likely to work to gain access to a male, indicating that proceptive behavior peaks around the time of ovulation. Rhesus males are more attracted to females (i.e., are more likely to mount and continue until ejaculation) when the females are in midcycle and least likely to show extended mounting late in the female's cycle. We can thus see that although rhesus females can show receptive behavior independent of specific hormone level, other important components of sexual behavior *are* influenced by hormones. Animal studies such as these have led to a reevaluation of hormonal effect on the sexual behavior of women.

Obviously, collecting data with human subjects is more difficult. Scientists usually rely on self-reports, which can be inaccurate and prejudiced. To determine how hormones might influence a woman's proceptive behavior, some researchers have asked that in addition to keeping track of sexual activity the couples record who initiates the encounter; some such studies have reported an increase in female-initiated encounters around midcycle. Another approach is to look at the frequency of masturbation, and again there are reports of a midcycle peak. What hormone is responsible for the increased interest in sexual activity around midcycle? Both E and T levels peak at this time, but several lines of evidence implicate T as the effective hormone.

Most human studies that demonstrate a relation between hormones and female sexual behavior have correlated peak or average hormone level with overall sexual gratification, interest, and activity. No correlation was found between E levels and a variety of sexual behavior/interest measures. However, T levels positively correlated with sexual gratification, arousal, desire, and masturbation frequency. Further implication of T in female sexual behavior comes from hormone replacement therapy in women who have undergone hysterectomy. T, but not E, treatment increased sexual desire and fantasies, although E treatment did augment vaginal lubrication. However, neither hormone had an effect on frequency of copulation. Thus in women, hormones may not affect receptivity, but T seems to be involved in the proceptive or cognitive aspects of sexual behavior.

A discussion of female sexual behavior must address the premenstrual syndrome (PMS). PMS has become a popular topic not only in the general press and scientific literature, but also in the courtroom, where attorneys have used PMS as a defense in criminal cases. Despite all this attention there is little scientific understanding of the syndrome.

PMS encompasses a wide variety of disturbances, including somatic complaints of bloating and breast tenderness; mood changes related to tension, anxiety, and irritability; and behavior changes such as crying, social withdrawal, and spousal fighting. Few clinicians or researchers doubt that many women suffer to varying degrees from some form of PMS, but there is much disagreement as to its treatment and cause. Early reports concluded that PMS was a result of abnormally high levels of prolactin, a peptide hormone released by the pituitary, and claimed that bromocriptine, a prolactin antagonist, alleviated symptoms in a group of patients. Later studies demonstrated that a placebo was equally effective in treating all the symptoms of PMS except breast pain.

Recent well-controlled studies measured a variety of hormonal levels throughout the menstrual cycle and compared women who suffer from PMS and those who show no symptoms; no differences in hormone levels were found. Thus there appears to be no simple hormonal basis for PMS. A possible explanation for the reported effectiveness of various treatments is that acknowledgement of the symptoms and attempts to provide relief serve to alleviate some of the suffering in women who previously had been told nothing was wrong. At present the best explanation for the cause of PMS involves differences in sensitivity or dynamic response to normal hormonal fluctuations between PMS sufferers and other women.

D. Male Sexual Behavior

When looking at the activational effects of T on male sexual behavior, one general rule seems to emerge regardless of the species being considered—there is plenty of T. In normal males, whether rodents or men, individual differences in sexual behavior cannot be predicted from differences in T levels. Although levels vary, even those male rats with relatively low T concentrations have two to three times the amount of hormone necessary to maintain normal copulatory behavior. Behavioral differences between males with intact, functional testes are most likely a result of differences in their neural substrate, i.e., the quantity of dynamics of the hormone receptors, or neural connections, or nonhormonal influences.

Although low versus high levels of T do not explain differences in sexual behavior in male rats or humans, there is considerable evidence that having *no* T influences sexual behavior in both species. Hypogonadal men have *extremely* low T levels and show varying degrees of hyposexuality. Treatment with exogenous T restores normal levels of sexual activity and interest. To determine how T acts to affect male sexual behavior, researchers compared hypogonadal and normal men in their erectile responses. No differences were found in their ability to maintain erections in response to an erotic film. However, unlike normal males, hypogonadal men were unable to maintain erections when sexual fantasy was the only stimulus. These individuals also showed significantly fewer and less intense spontaneous erections. Taken together, these data suggest that the mechanical or physiological component of sexual arousal is not drastically influenced by T but that the hormone works at a motivational or cognitive level. In other words, although T is not necessary to produce erections, it may act to lower the threshold for sexual arousal.

Early in our discussion of hormones and behavior, we mentioned the organizational/activational hypothesis of hormone action. From the review of both male and female sexual behavior, it is evident that hormones do have an activational effect (i.e., the presence of hormones in the adult influences the expression of sexual behavior). Although humans are not totally dependent on the presence of gonadal hormones for the expression of sexual behavior, evidence shows that the presence or absence of gonadal hormones can affect many aspects of sexual interactions in both men and women. However, in contrast to most other mammals, it has been difficult to demonstrate an organizational effect of early hormones on later sexual behavior in humans. The various human syndromes resulting in prenatal hormonal abnormalities do not always produce the alterations in sexual behavior predicted by animal models, and the problems that are observed could be explained by nonhormonal mechanisms. Presently, it is unclear whether humans differ from other mammals in the early action of hormones on the CNS or whether social factors are much more effective in overriding early hormonal organization in humans. [*See* Sexual Behavior.]

III. AGGRESSIVE BEHAVIOR

In almost all mammalian species, hamsters and hyenas being among the few exceptions, spontaneous intraspecies aggression is primarily a male phenomenon. This is one reason why it is widely accepted that in animals, aggression is very much dependent on T levels. Unfortunately, it is somewhat easier to find exceptions to this rule than support for it. In naive animals (those who were never before exposed to an aggressive situation), the presence or absence of T will affect whether they will fight or merely submit. But once an animal has experienced aggressive encounters, his T level has little effect on subsequent responses. On the contrary, success or failure in an aggressive encounter tends to affect T levels more than do T levels affect the probability or outcome of an aggressive encounter. Considering the difficulty in seeing a simple relation between hormones and aggression in animals, it is not surprising that the study of human aggression has proven exceedingly difficult. Many of the antagonistic encounters between people occur on the verbal level: Is this aggression? Animal researchers use tissue damage as the criterion for an aggressive act: Is corporate "back-stabbing" the modern socialized form of intrahuman aggression? The difficulty in answering such questions, and the tendency for many researchers in the area to ignore them, has resulted in a less than satisfying literature. [*See* Aggression.]

Clinical research involving hormone manipulations has been conducted on men convicted of violent crimes. Most of the participants in these studies were sexual offenders, making interpretation difficult because of the combination of sexual and aggressive components of the act. Nonetheless, some interesting and consistent conclusions have been reached. In several studies, participants were treated with drugs that act as antiandrogens either by reducing T levels or by interfering with hormone receptor dynamics. Recidivism in subjects treated with antiandrogens was significantly lower than in those individuals who did not receive drug treatment. While on the drug, the men reported a reduction in the frequency and intensity of fantasies and urges related to sexually deviant/aggressive activities. Researchers have concluded that the antiandrogens act to reduce an individual's motivation to engage in deviant behaviors, thus allowing for the adaptation of socially acceptable alternatives. However, men so motivated to reform that they will accept antiandrogen treatment may be less prone to recidivism even without the drugs.

In normal adult males, no clear correlation has been found between plasma T levels and nonsexually related aggression, although some evidence exists for such a relation in adolescents. Studies of high school-aged males have reported a greater incidence of hostile acts among individuals with high T levels. Such studies show a correlation (i.e., there is some relation between T and aggression in adolescents), but they do not demonstrate that T causes an increase in aggressive behavior. T may increase the motivation to engage in aggressive behaviors or permit the expression of such behaviors in individuals predisposed to aggression due to social factors. Another possible explanation involves the peripheral effects of the hormone, (i.e., the appearance of the individual). Peripheral effects of T include increased muscle mass and body hair; this more "mature" or "intimidating" appearance may influence the way peers or authorities respond to that individual.

The importance of considering the appearance of an individual when assessing behavioral interactions is often neglected by researchers in behavioral endocrinology, who are often eager to jump to conclusions about neural substrates and motivation. For example, reports indicate a higher than expected proportion of XYY individuals involved in violent crimes. These individuals tend to develop faster and have higher T levels than their adolescent peers. Genetic and hormonal explanations have been offered to explain this high level of aggression. An alternate explanation might take into account the appearance of XYY adolescents. Actions that might be interpreted as adolescent pranks when committed by a normal 15-year-old may be viewed as violent when committed by an XYY individual if he looks more like an adult. Thus in humans, if we exclude sexually related actions, it is difficult to see a direct effect of hormones on aggressive behavior.

A great deal of attention has recently been focused on steroids in sports. A controversy has arisen about the role androgens may play in physical performance. As for endogenous levels, androgens reportedly rise before competition and remain high in the winners relative to the losers. However, the hormone levels both before and after a match appear to correlate more closely with the individual's mood than with his or her performance (i.e., the contestants with the highest spirits before a match had higher T levels, and among the winners, those who were most satisfied with their performance had the higher postmatch T levels). As for exogenous steroid use, much clinical research suggests that it does little to improve fitness or performance, despite widespread assumptions to the contrary. Why then is its illicit use so widespread in sports? In addition to a possible placebo effect, some researchers suggest that the dosage used by many bodybuilders is 10–100 times that used in clinical studies and that, at these levels, a significant effect on training may be seen. At such high doses, steroid use is exceedingly dangerous: in addition to the cardiovascular and liver damage that may result, many individuals show affective dysfunctions (e.g., depression, mania, hallucinations, and paranoid delusions). In a recent report, more than 20% of the subjects showed significant behavioral problems while on steroids. Even here we must consider the possibility that the behavioral problems may be caused by the internal conflict such users must feel about knowingly risking their health to achieve recognition or acceptance. Considering that the relation between endogenous steroids and behavior is complicated and often not well understood, the use of exogenous steroids may or may not be useful in producing the desired effects but definitely can have deleterious effects.

IV. COGNITIVE/MOTOR TASKS

There are a variety of tasks in which performance differs between men and women; tests involving dichotic listening, verbal, spatial, or complex manual skills are the most often reported. In addition to developmental and environmental factors that may contribute to these differences, recent reports suggest that hormonal levels at the time of testing can also affect such performance. [*See* Sex Differences, Psychological.]

We will consider two tasks, one involving spatial skills in which men tend to score higher, the other a test of complex manual ability in which women excel. Women's performance on both tasks varies throughout the menstrual cycle and is correlated with the level of E and P. Scores on the manual dexterity test were higher at the midluteal phase when E and P are high than during menstruation when E and P levels are low. In contrst, women were more accurate in spatial judgments during menstruation than during the midluteal phase. Another study examined androgen levels and performance on spatial tasks and found that females with high androgen levels scored higher than those with low levels. The opposite trend was observed in men, those with high androgen levels scoring lower than those with low levels. Given that E and P are present in much higher concentrations in females and much more androgen is present in males, it is interesting to note that among females the performance on the manual dexterity task (female dominated) was highest when E and P levels were high and that females with high androgen levels performed best on spatial tests (male dominated). However, data on men point out once again that simple relations between hormones and behavior are rare. Taken together, both studies suggest that observed differences in nonsexual and nonaggressive behaviors may also be subtly influenced by hormones.

V. INGESTIVE BEHAVIOR

Both the hormones and some of the behaviors we have discussed thus far are important in the survival of the species but are not necessary for the survival of an individual. This enabled us to study how drastic hormonal manipulations affected various aspects of behavior. Unlike gonadal hormones, an individual is dependent on adrenal, thyroid, pancreatic, and pituitary hormones for survival, and it is obvious that behaviors such as drinking and eating cannot be drastically altered without deleterious effects to the individual. As we will see, when considering such behaviors, it is easy to demonstrate simple relations with some hormones, but because the behaviors are so essential, auxiliary mechanisms of control have also evolved.

Two hormones are known to play important roles in water retention and excretion: aldosterone and vasopressin (also known as antidiuretic hormone). Aldosterone is a steroid hormone secreted by the adrenal cortex; it acts on the kidneys to reduce salt excretion. Vasopressin is a peptide hormone secreted by the posterior pituitary; it stimulates the reabsorption of water by the kidneys. The release of vasopressin is under neural control, but aldosterone secretion is dependent on the plasma levels of another hormone, angiotensin. Both vasopressin and aldosterone act to conserve water, and loss of function of either of these hormones leads to increased thirst and therefore increased water consumption in both humans and other mammals.

An important factor in the control of feeding behavior is the level of glucose in the blood. Insulin, a peptide hormone secreted by the pancreas, is a major determinant of blood glucose levels. The presence of insulin in the bloodstream allows muscles and organs to take up and use circulating glucose; in the absence of insulin, only the brain and liver can use this nutrient. An injection of insulin will deplete blood glucose levels and therefore stimulate hunger, which will result in feeding behavior. High blood glucose levels, such as after a meal, stimulate the release of insulin, which will allow its uptake and storage by peripheral cells. However, a small release of insulin can be stimulated by the taste or smell of food, or by some conditioned signal that a meal is eminent. This is why light hors d'oeuvres may increase our appetite—their taste and smell can trigger a pulse of insulin that depletes blood glucose and therefore increases hunger. In individuals afflicted with diabetes mellitus, the pancreas does not produce insulin; if left untreated, these people will increase their food intake and maintain high blood glucose levels, but because their cells cannot use the energy, they grow thin and eventually quite ill. [*See* Insulin and Glucagon; Peptide Hormones of the Gut.]

Several neural and hormonal factors appear to play roles in cessation of feeding behavior. The presence of food in the stomach both stimulates stretch receptors, which directly signal the brain to stop eating, and causes the gut to release peptide hormones, glucagon and cholecystokinin. The precise role these and other hormones play in producing satiety is not yet clear.

In addition to providing necessary energy, feeding behavior serves to maintain and, in some cases, increase body energy stores of fat. The question of how the body monitors and maintains a particular mass of adipose tissue has intrigued researchers as well as the weight-conscious public. A recent theory suggests that the brain monitors the average blood insulin levels over relatively long periods. The amount of food intake, adipose tissue, and activity interact to affect plasma insulin. Average insulin levels may indicate the level of fat storage and may in turn influence long-term feeding behavior.

VI. CONCLUSION

Hormonal effects on behavior are obviously extensive. This article has touched on some of the areas that have been vigorously researched and in which a fair level of understanding has been attained. There are several other areas in which hormones are known to play a role. For instance, the gonadal, adrenal, thyroid, and pituitary hormones have all been implicated in depression, but as is often the case, it is not known whether it is dysfunction in the endocrine glands that contributes to the development of depression or whether the depressive disorder causes abnormal hormonal activity. Thus our discussion leads to a general conclusion that the role of hormones in human behavior is important and extensive but rarely simple.

ACKNOWLEDGMENTS

We thank Nancy Forger and John Dark for their comments on the manuscript.

BIBLIOGRAPHY

Becker, J. B., Breedlove, S. M., and Crews, D. (1992). "Behavioral Endocrinology." Bradford Books.

Davidson, J. M., and Myers, L. S. (1988). Endocrine factors in sexual psychophysiology. *In* "Patterns of Sexual Arousal" (R. C. Rosen and J. G. Beck, eds.). The Guilford Press, New York.

DeVries, G. J., DeBruin, J. P. C., Uylins, H. B. M., and Corner, M. A. (eds.) (1984). "Progress in Brain Research," Vol. 61. Elsevier, New York.

Fishman, R. B., and Breedlove, S. M. (1988). Sexual dimorphism in the developing nervous system. *In* "Handbook of Human Growth and Developmental Biology" (E. Meisami and P. S. Timeras, eds.), Vol. 1. CRC Press, Boca Raton, FL.

Hines, M. (1982). Prenatal gonadal hormones and sex differences in human behavior. *Psychol. Bull.* **92**(1), 56–80.

Reinish, J. M., Rosenblum, L. A., and Sanders, S. A. (eds.) (1988). "Masculinity/Femininity: Basic Perspectives." Oxford University Press, New York.

Rosenzweig, M. R., and Leiman, A. L. (1982). "Physiological Psychology." D. C. Heath and Company, Lexington, MA.

Rosenzweig, M. R., Leiman, A. L., and Breedlove, S. M. (1996). "Biological Psychology." Sinauer, Sunderland, MA.

Sherwin, B. B. (1988). A comparative analysis of the role of androgen in human male and female sexual behavior: Behavioral specificity, critical thresholds, and sensitivity. *Psychobiology* **16**(4), 416–425.

Strunkard, A. J., and Stellar, E. (eds.) (1984). "Eating and Its Disorders." Raven Press, New York.

Svare, B. B. (ed.) (1983). "Hormones and Aggressive Behavior." Plenum Press, New York.

Human Evolution

HENRY M. McHENRY
University of California, Davis

GLOSSARY

Australopithecus Extinct forms of humans who lived in Africa from about 5 to 1.3 million years ago. They were characterized by human-like bipedal posture, brain size about one-third that of modern people, and relatively large cheek teeth. One species of this genus probably evolved into the genus *Homo*

Hominid Organisms belonging to the zoological family Hominidae, which includes both extant and modern forms of humans such as *Australopithecus, Homo erectus,* and the Neanderthals

Hominoid Organisms belonging to the zoological superfamily Hominoidea, including modern and fossil species of humans and apes

Molecular clock Method of computing the time of divergence between evolutionary lineages based on the genetic distance among living species

Pongid Organisms belonging to the zoological family Pongidae, including modern great apes (chimpanzee, gorilla, and orangutan) and closely related fossil species

THE HUMAN EVOLUTIONARY LINEAGE ORIGInated in Africa in the Late Miocene, 8 to 5 million years ago, from an ancestor shared with the African great apes (chimpanzee and gorilla). Probably by 5 but certainly by 3.7 million years ago, our ancestors walked bipedally, had front teeth intermediate in shape between those of modern apes and people, and had a relative brain size about one-third that of *Homo sapiens.* By 2 million years the teeth are basically human-like (except for the large size of the cheek teeth) and the relative brain is about 50% larger than that of its predecessor. The species *Homo erectus* first appears at 1.8 million years ago with a brain nearly twice that of the first hominids. Perhaps at this time, but certainly by about 1 million years ago, our family disperses out of Africa to colonize Asia and Europe. Not until about 300,000 years ago are brains expanded into the modern human range. Anatomically modern *Homo sapiens* appears to have originated in Africa at least 100,000 years ago. By about 35,000 years ago, this form had occupied all of the Old World, including Australia. Although some people may have reached the Americas earlier, the first substantial population there dates to about 12,000 years ago.

This brief outline of current knowledge derives from the remarkably successful cooperation of the molecular biologist, comparative morphologist, geologist, paleontologist, archaeologist, biological anthropologist, geneticist, ecologist, and individuals from many other fields. Such a sketch is, of course, tentative pending further information and interpretation, but a great deal of hard evidence lies behind it.

I. PLACE IN NATURE

A. Taxonomy of *Homo Sapiens*

Homo sapiens is 1 out of about 180 living species within the order Primates, which is 1 out of 18 mammalian orders. The order Primates contains prosimians (such as lemurs), monkeys, apes and humans. Our species is placed in the suborder Haplorhini along with tarsiers, monkeys, and apes. Within Haplorhini we are grouped with Old World monkeys and apes in infraorder Catarrhini, which in turn is divided into two superfamilies, Cercopithecoidea (Old World monkeys) and Hominoidea (apes and people). Tradi-

ENCYCLOPEDIA OF HUMAN BIOLOGY, Second Edition, VOLUME 4.

tionally the Hominoidea are divided into three families, Hylobatidae (gibbons and siamangs), Pongidae (orangutans, gorillas, and chimpanzees), and Hominidae (humans), but research since the 1960s has clearly demonstrated that this classification does not reflect the true phyletic relationships. Several lines of evidence, especially molecular systematics, show that African great apes (gorilla and chimpanzee) are more closely related to humans than to Asian apes (gibbons, siamangs, and orangutans). Some authors believe that the classification should reflect that fact by placing the African apes in the family Hominidae. By this scheme, the Hominidae is divided into subfamilies Ponginae (gorilla and chimpanzee) and Homininae (humans). There are several other recently proposed revisions in the classification, but the most widely used is still the traditional division of Pongidae (great apes) and Hominidae (humans), with the understanding that this does not reflect the phyletic relationship.

B. Genetic Relationships

As early as 1904, G. H. F. Nuttall published "Blood Immunity and Blood Relationships," which explored the genetic relationships among animals using immunology techniques. More than half a century lapsed before M. Goodman showed the power of molecular biology in clarifying our place in nature. Using the method of immunodiffusion, in 1962 and 1963 he published what has become established fact: *Homo sapiens* is much more closely related to African great apes than either is to the Asian great ape. Subsequent tests using a variety of methods, such as amino acid sequencing, microcomplement fixation, DNA hybridizations, electrophoresis, and nucleotide sequencing, confirm the surprisingly close relationship among humans, chimpanzees, and gorillas. The genetic similarity among these three is comparable to that between dog and fox, cat and lion, and sheep and goat, and only slightly less than that between horse and donkey.

By the late 1960s, V. Sarich and A. Wilson accumulated enough genetic comparisons among primates to show that immunologically detected differences in albumin most likely changed at a constant rate of time. By calibrating this change with a widely accepted date for the evolutionary divergence of two lineages, they derived a molecular clock for primate evolution that placed the origin of the hominid lineage at 4.2 million years ago. Subsequent studies by a variety of methods by numerous individuals have shown that the molecular clock is generally correct, but irregularities in rate require significant error ranges. Recent estimates of the origin of the hominid lineage range from about 8 to 4 million years ago.

C. Traits Shared with Apes

Despite the obvious fact that *H. sapiens* is profoundly different from all other animals in many resepcts, there are a number of traits we share uniquely with apes that show our phylogenetic affinity with them. These shared and unique characteristics are most conspicuous in the forelimb and trunk anatomy. Apes and humans share exceptionally short lumbar regions of the spine, the lack of a tail, highly mobile shoulder and elbow joints, broad chests that are flattened from front to back, reduced ulnar olecranon processes, a vertebral column that protrudes into the chest cavity, long clavicles and acromial processes, a broad sternum, a diaphragm that is perpendicular to the spine, an obliquely positioned heart that adheres closely to the diaphragm, and abdominal organs that are closely attached to the posterior wall of the body cavity. These and other traits are probably related to orthograde posture. The molar teeth show detailed similarity in cusp number, fissure pattern, and overall form. There are a few traits that are uniquely shared by humans and the African great apes, such as the presence of a frontal sinus that develops from the ethmoid bone and the fusion of the os centrale in the wrist. In addition, African great apes and humans share certain molecular similarities, such as unique substitutions of amino acids in the myoglobin chain at positions 23 and perhaps 110, two shared substitutions in the fibrinopeptide A and B chains, three transversions, eight transitions, and three deletions in the nuclear DNA sequence, and 16 substitutions in the mitochondrial DNA. [*See* Comparative Anatomy.]

D. Unique Traits of Hominidae

As Lamarck, Huxley, Haeckel, Darwin, and other nineteenth-century evolutionists pointed out, bipedalism was probably the primary change in the origin of our evolutionary lineage. Twentieth-century discoveries of fossils confirm this. Habitually walking on the hindlegs with the forelimbs free for carrying and perhaps wielding weapons is the first fundamental change away from our common ancestors with the apes. This change required a major reorganization of the hindlimb and spinal column away from the pattern common to most mammals. The most conspicuous anatomical differences are the shortened pelvic blade and the reorientation of the foot from a grasping organ to one in which the sole function is propulsion.

Modern human brain size is about three times greater than that of apes of the same body size. From the fossil record it is clear that brain size evolution occurred after the adaptation of bipedalism. Most of the change occurred during the last 2 million years of evolution. Why the brain expanded is the subject of much speculation, but certainly the origin of language is related to this process.

II. EVOLUTIONARY HISTORY

A. The Stock from Which Our Family Arose

The suborder Anthropoidea first appears in the fossil record during Middle Eocene times (about 45 million years ago) in Asia and North Africa. The first substantial fossil evidence of the origin of the evolutionary lineage leading to modern catarrhines (Old World monkeys, apes, and humans) occurs in the geological strata dating to 38 to 31 million years ago in the Fayum deposits of Egypt. These deposits present an extraordinary window on the early stages of the evolution of the primates. The climate was warm, the habitat forested, and a major river moved slowly through the site. Primates constituted a major component of the mammalian fauna. A tarsier-like form is represented (by a jaw fragment) as well as a loris-like creature (known only from a tooth). The most common primates are grouped under the family name of Parapithecidae (*Parapithecus* and *Apidium*), which are not catarrhines (they retain the primitive characteristics of having three premolar teeth). The oldest Catarrhine is *Catopithecus,* which shared many unique cranial traits with late members of its suborder, but had prosimian-like teeth with high cusps, shearing crests, and shallow jaws. The primitive dental features may imply that the origin of Anthropoidea did not involve a dietary shift. In later strata, catarrhine fossils occur with teeth that are more typical of Anthropoidea with low and rounded cusps on the cheek teeth. These include the genus *Propliopithecus,* who was small (2–3 kg), arboreal, quadrupedal, and sexually dimorphic, and *Aegyptopithecus,* who was larger (about 6 kg), diurnal, and arboreal. Its postcrania show an adaptation for slow and deliberate branch climbing and quadrupedal walking. It had strong sexual dimorphism in body size, canines, and skull morphology. The teeth indicated fruit eating. The best-preserved cranium shows a long, almost lemur-like snout, but other facial skeletons are much less prognathic. The brain is relatively small compared to those of modern catarrhines, but catarrhine-like in having relatively smaller olfactory bulbs, larger visual cortex, and a more complex sulcal pattern.

Current paleontological evidence confines catarrhines to Africa until about 17 million years ago. Until this date, Africa and Arabia were separated from Eurasia by the Tethys Sea. Unfortunately there is a gap in the primate fossil record in Africa between 31 and 25 million years. Between 22 and 17 million years there is a rich sample of catarrhines in East Africa during a period when tropical forests extended much farther east and north than they do now, although there are some indications that woodland or brushland grassland habitats existed as well.

The Early Miocene primates of East Africa show the earliest evidence for the divergence between the two catarrhine superfamilies, Cercopithecoidea (Old World monkeys) and Hominoidea (apes and humans). Cercopithecoid fossils are relatively less abundant than those of hominoids. The Early Miocene hominoids are quite different from any living species and are best placed in one or more separate families. There are at least seven genera. The adaptive diversity among these hominoids is considerable. Body sizes range from 4 to 40 kg. Their diets were primarily frugivorous, but some species were folivorous. Most were arboreal quadrupeds, but some may have been capable of forelimb suspensory behavior as well. The postcranium does not, however, reveal many (if any) of the distinctive traits that are shared by all the living hominoids. In fact, the Early Miocene hominoids shared very few derived traits with living hominoids.

Between 17 and 12 million years ago there was a major faunal change in the Old World resulting from the exchange of species between Africa and Eurasia. This event was apparently triggered by the breaking down of the Tethys Sea barrier, which had kept Afro-Arabian faunas isolated. Although forest was much more abundant than today, there is evidence for some open habitat and possibly grassland, particularly in the northern latitudes. There is also evidence of climatic cooling at this time.

The first hominoids outside of Africa appear at this time (17–12 million years ago) in both Europe and Asia. The adaptive and phyletic diversity is much higher than among modern hominoids. Between 15 and 12 million years ago there are at least four genera and nine species of large-bodied hominoids in Eurasia and possibly the African diversity was equally great, although fewer African fossil primates are known during this period. Most of these hominoids have thick molar enamel similar to that seen in modern orangutans and fossil hominids. They were apparently frugivorous, had marked sexual dimorphism, and looked somewhat like modern apes. Although the postcranial fossils are rarer and unassociated, they appear to share the suite of characteristics unique to modern hominoids. The divergence of the evolutionary lineage leading to the orangutan probably occurred by 12 million years ago, as shown by the remarkable discovery of the facial skeleton of *Sivapithecus* in Pakistan. This specimen shares numerous derived traits with the Asian great ape. There is a wealth of new material from Europe and especially China that shows the rich diversity of these ape-like forms in the Miocene compared to the impoverished diversity of modern ape species.

At about 10 or 11 million years ago there is evidence for increasing open habitat in the Old World and another major faunal turnover. By about 7 million years, forest- and woodland-adapted fauna are replaced by more open-country species in South Asia. In Africa the record is less well documented, but certainly abundant forests of the Early Miocene give way to more and more patchy woodland and grassland habitats in the Middle to Late Miocene. The collection of Middle to Late Miocene hominoids has rapidly expanded in recent years, but the picture is still far from clear. There are still no fossils that are clearly linked uniquely to any of the living African great apes, although a palate from 8-million-year-old beds in Kenya shows some gorilla-like traits. There are tantalizing bits of hominoid fossils in the Late Miocene, but nothing that can be linked specifically to living species until possibly 5 and certainly 4 million years ago, when fossils occur that distinctly belong to the human evolutionary lineage.

B. First Bipeds

The molecular clock predicts the divergence of the human and African ape evolutionary lineages to be some time between 5 and 8 million years. During this period and later, the African habitats were changing drastically with increasing seasonality of rainfall and spreading grassland. Areas of tropical rain forest were reduced, creating isolated pockets of forest in a sea of grass. Unfortunately there are few primates known from this period: a hominoid tooth at 9 million years ago, a palate with some gorilla-like features combined with unique traits at 8 million years, an ape-like molar crown at 7 million years, and a mandibular fragment with one molar tooth at 5–6 million years ago. The latter specimen from Lothagam in Kenya has some derived traits that appear to be shared with later hominids, such as a relatively decreasing molar length, entoconid size, and mandibular depth. It is associated with open-country fauna.

By 4.4 million years ago the earliest hominid species, *Ardipithecus ramidus,* can be identifed from deposits in Ethiopia. Overall it is quite ape-like, but it does share a few unique features with later hominids such as broader and less projecting canines, a shorter cranial base, and some details of the elbow. It had thinner molar enamel than later hominids, implying perhaps a diet more similar to that of African apes. It lived in a closed woodland habitat. Unfortunately there are as yet no hindlimb fossils available to ascertain its form of locomotion.

At least by 4 million years ago there is fossil evidence of bipedalism. A well-preserved hominid leg bone from Kanapoi, Kenya, has the special architecture associated with the stresses of bipedalism. It is attributed to the taxon *Australopithecus anamensis,* whose dentition is almost as primitive as that of *Australopithecus ramidus.* By at least 3.6 million years ago the much better known species *Australopithecus afarensis* appears in the record. Bits of this species may be as old as 4 million years, but the richest samples derive from Laetoli, Tanzania (3.6 and 3.8 million years) and Hadar, Ethiopia (3.2 to 2.8 million years).

Laetoli produced the remains of 23 hominids, including jaws, associated dental rows, an infant skeleton, and three sets of footprint trails made by bipeds. The footprints are clearly hominid with distinctly convergent big toes and human-like proportions. The dental remains are very similar to those found in the Afar region of Ethiopia dating between 3.0 and 3.4 million years ago. This Hadar collection is wonderfully complete, with one associated skeleton (A.L. 288-1, "Lucy"), the fragmentary remains of at least 13 individuals who apparently died together at one spot, and numerous other skeletal parts. The combined samples of Laetoli and Hadar yield an excellent picture of what this early human species looked like.

Australopithecus afarensis was fully bipedal as indicated by the Laetoli footprints and the Hadar postcranial skeletons. The pelvic blades are low, the sacrum wide, and the pelvic basin quiet human-like in shape, although it is not identical to that of modern humans. The knees are characteristically human and not ape-like. The toes are relatively shorter than those of any ape, but not reduced as much as they are in modern humans. The forelimbs are relatively quite small and the wrists and hands show no adaptation for ape-like knuckle-walking. The skeleton retains many ape-like traits, but in most fundamental respects it is adapted for bipedality. Its long, curved toes and fingers and many other ape-like features may imply that this first human species was a more adept tree climber than later species of hominid. Although its pelvis and hindlimb are fundamentally human-like, there are basic differences that imply a somewhat different form of bipedality from that seen in modern *H. sapiens*. The pelvic blades, for example, face more posteriorly, the thighs are relatively short, the knees appear to lack a human-like meniscus attachment, the ankle in at least one specimen slopes in an ape-like direction, the foot architecture has many ape-like traits, and the toes are relatively long and curved.

In other characteristics, *A. afarensis* possesses a mixture of hominid and pongid qualities. There is strong sexual dimorphism in body size, with females weighing perhaps as little as 29 kg and males closer to 45 kg or even more. The brain size is about that of a modern chimpanzee (384 cc), which is almost one-third the size of a modern human brain of the same body size. The skull is quite pongid-like with a prognathic muzzle, an unflexed cranial base, and a strong development of the posterior fibers of the temporalis muscle. The canine is considerably reduced from the size seen in modern apes, but it is larger than that of modern humans. The lower first premolar is variable, but in many specimens it is quite ape-like in orientation and size of inside cusp.

A similar form of hominid, *Australopithecus africanus,* occurs between about 3.0 and 2.3 million years in South African cave deposits. In many postcranial parts it is remarkably similar to the Hadar hominids, although there are some differences, particularly in the hand. *Australopithecus africanus* is distinct from the Hadar and Laetoli hominids in having many cranial and dental traits that are more similar to later *Homo* such as a reduced muzzle, a more flexed cranial base, and a bicuspid first premolar. Females appear to be as small as those from Hadar, but males are not quite as large. The average brain size is larger (420 cc), and there are no specimens as small as the smallest one from Hadar. The cheek teeth are relatively larger.

C. Extinct Cousins: The Robusts

From about 2.7 to 1.3 million years ago there lived a variety of hominids referred to as "robust" australopithecines because of their hypertrophied masticatory system. At least three and probably four species are known. The earliest, referred to by some as *Australopithecus aethiopicus,* is known from East African sites dating between 2.7 and 2.3 million years ago. Its cheek teeth are enormous, as are all the supporting structures related to heavy chewing, i.e. massive jaws, strongly buttressed skull, enormous area of attachment for the muscle of mastication. In many ways the skull shares primitive characteristics with *A. afarensis* and pongids, such as a small brain (399 cc), an unflexed cranial base, a prognathic muzzle, and strong development of the posterior fibers of the temporalis muscle. In the later robust species these traits are lost, which makes them appear more *Homo*-like. These species include *A. boisei,* found abundantly in East Africa between 2.3 and 1.3 million, and *A. robustus* of South Africa. Although brain size is about 100 cc larger, the postcrania is fragmented and difficult to associate with these species. But what specimens there are indicate a remarkably human-like form. It is generally assumed that the massive development of the chewing apparatus was an adaptation to a herbivorous diet, including small hard objects such as nuts, seeds and tubers. There is some evidence that at least *A. robustus* ate meat as well.

D. The Appearance of *Homo*

The first abundant evidence of what most investigators would refer to as *Homo* occurs in strata dated to about 2 millions years ago, although fragmentary material is known as early as 2.4 million years. Between 1.9 and 1.6 million years ago there are many specimens that probably belonged to the genus *Homo,* but the variability is higher than would be expected from a single species. Whether or not this variability indicates more than one species has not been resolved. The species name associated with this material is *habilis.* The *Homo*-like traits include an expanded brain (about 50% larger than that of *A. africanus* in relative size) and reduced cheek teeth. The smallest postcranial specimen was probably less than 1 m tall and may have weighed less than 32 kg; the largest probably weighed over 52 kg. Relative to joint size, the hindlimbs are much less robust than those of *Australopithecus.* Stone tools first appear in the archaeological record at 2.5 million years and it is often inferred that *Homo* was responsible for them. Well-preserved living floors occur at least by 1.8 million years and show that these hominids were using tools for several activities, including butchering, plant processing, and wood carving.

The species *H. erectus* first appears at 1.9 million years ago in East and South Africa. Brain size in the larger specimens is twice that of *Australopithecus*. Relative cheek-tooth size was reduced considerably from the large size seen in earlier hominid species. Body size of the larger specimens probably exceeds 60 kg. It may be that body size sexual dimorphism was not as marked in early *H. erectus* as it apparently was in *H. habilis,* but there are as yet no associated female skeletons complete enough to be certain. At about the same time as the origin of *H. erectus,* the Acheulean material culture appears, characterized by bifacially flaked large stone tools ("handaxes"), which are found throughout most of the Old World for the next 1.5 million years. There is a great deal of evidence supporting the hypothesis that the origin of *H. erectus* marked the beginning of major change in the adaptive strategy of Hominidae, although it remained a numerically minor part of the vertebrate fauna collected in Pleistocene beds despite the strong bias of collectors.

E. Colonizing the Old World

By at least 1 million years ago and perhaps as early as 1.8 million years ago, some populations of *H. erectus* had left their African homeland and colonized Eurasia. The best known of the early hominids outside of Africa are those from Java, dating to perhaps 1.8 million years ago, although dating is much less precise than it is for earlier parts of the record and there are only two sites with such ancient dates. The imprecision is a result of the complexity of the geology in the Asian sites.

The *H. erectus* of Java was found in 1893. This famous discovery of Eugene DuBois consisted of a skull cap with an estimated cranial capacity of 940 cc associated with a thigh that was well within the range of variation of *H. sapiens*. This association led DuBois to name the creature *Pithecanthropus erectus* (erect ape-man). Unfortunately, few other specimens came to light until the 1930s. Between 1930 and the outbreak of the Asian part of World War II came a pulse of discovery in Java and China that established *H. erectus* as a well-documented species preceding our own. A wealth of new Javanese and Chinese fossils was discovered and later described that gave a picture of a hominid of intermediate brain size (about 800 to 1000 cc, compared to the average for *H. sapiens* of 1300 cc) with large brow ridges, a low cranial vault, teeth of intermediate size, and robust but modern-looking postcrania. Several fire layers are present in the best-known *H. erectus* site in China (Zhoukoudien), leading most investigators to infer the controlled use of fire. The first appearance of humans in the rest of Eurasia is less well documented. There are archaeological sites that may date back to as early as 1 million years, but the dates are problematical. There are only a sprinkling of human remains in Western Europe before about 300,000 years and they do not resemble *H. sapiens* very closely. Most if not all of the European fossils before 35,000 years are best regarded as Archaic *H. sapiens*.

F. Archaic *Homo sapiens*

The term Archaic *H. sapiens* refers to a heterogeneous collection of Old World hominids between about 400,000 and 35,000 years ago. The term is not precise and there is a need for a formal taxonomic reappraisal. The contrast between what is now called Archaic *H. sapiens* and anatomically modern *H. sapiens* is greater than that between Archaic *H. sapiens* and *H. erectus* except in one important characteristic, namely, brain size. Brain size is usually within the modern human range of variation (1000 to 1700 cc), although there is one exceptionally small specimen (Sale of North Africa with 860 cc). They resemble *H. erectus* with their large faces, robust skeletons, brow ridges, and long low skulls. Earlier specimens are associated with Acheulean culture, but by about 200,000 years, stone tools became more sophisticated, particularly in their manufacture of prepared-core flake tools. The best-known variety of Archaic *H. sapiens* is the Neanderthal.

Neanderthals were a relatively homogeneous group of Archaic *H. sapiens* who occupied Europe and West Asia between about 200,000 and 35,000 years ago. Their facial morphology was unmistakable with exaggerated midfacial prognathism, leading to what must have been enormously protrusive noses. They were the first prehuman fossils known to science, and consequently they have played a major role in the interpretation of human evolution. Now that much more is known about prehistory, Neanderthals are seen in perspective as a relatively isolated extreme variant of Archaic *H. sapiens* that held out against anatomically modern *H. sapiens* until 40,000 years ago, when they disappear.

G. Aanatomically Modern *Homo sapiens*

Anatomically modern *H. sapiens* is distinctly different from the archaic form of the species in having small faces, high foreheads, a true chin, and longer and less robust limbs, especially the distal segments (forearm and shin). They are first known in Africa by about 100,000 years or slightly earlier. There is some evidence that they were in the Middle East by 90,000 years ago, but remains of Archaic *H. sapiens* are much more common in the area until about 40,000 years ago, when they disappear from the record. At this time the culture changes dramatically with the introduction of more finely worked stone blades, a wider variety of tools of all kinds, and art including cave paintings and stone carvings. There is accumulating evidence that the morphological and behavioral changes between Archaic and modern *H. sapiens* are profound enough to warrant changing the taxonomy so as to restrict the species *Homo sapiens* to moderns only. Comparisons of the nuclear and mitochondrial DNA of living humans show that the highest variability among modern humans is in Africa. This fact, combined with fossil and archaeological evidence, supports the African origin of modern *H. sapiens*.

H. People of the Earth

There is no unanimity of opinion concerning the origin and spread of anatomically modern *H. sapiens*. Some scholars emphasize the apparent regional continuity between local forms of fossil hominids and living populations in the same area. For example, several traits characteristic of *H. erectus* in China resemble those of some populations of modern Chinese, such as the form of cheekbones. On the other hand, all modern humans are more similar to one another than to Archaic *H. sapiens* or *H. erectus*, which can be interpreted as indicating a close genetic relationship and a restricted geographical area of origin. Particularly telling is the fact that maternally inherited mitochondrial DNA of all living people is best interpreted as having a single origin in Africa. Had local premoderns outside of Africa contributed to the modern gene pool, presumably that contribution would be detected in the maternal line. Much more work needs to be done in this area before any certain conclusions can be drawn.

Anatomically modern humans had reached Australia by at least 30,000 years ago. Archaeological evidence shows that the colonization of the rest of the Pacific began several thousand years ago from the east, reaching the Marquesas Islands by about A.D. 600 and New Zealand by about A.D. 1200. People arrived in the Americas from northern Asian by at least 12,000 years ago, for which there is abundant evidence. Sites before that date are rare and their authenticity is often challenged. Small populations may have occupied the Americas much earlier, but they remained at low population densities until 12,000 years ago.

BIBLIOGRAPHY

Aiello, L., and Dean, C. (1990). "An Introduction to Human Evolutionary Anatomy." Academic Press, San Diego.

Andrews, P. (1992). Evolution and environment in the Hominoidea. *Nature* **360**, 641–646.

Fleagle, J. G. (1988). "Primate Adaptation and Evolution." Academic Press, San Diego.

Jones, S., Martin, R. D., and Pilbeam, D. (eds.) (1992). "The Cambridge Encyclopedia of Human Evolution." Cambridge Univ. Press, Cambridge, England.

Klein, R. G. (1989). "The Human Career." Univ. of Chicago Press, Chicago.

Martin, R. D. (1990). "Primate Origins and Evolution." Chapman & Hall, London.

McHenry, H. M. (1994). Tempo and mode in human evolution. *Proc. Nat. Acad. Sci. USA* **91**, 6780–6786.

Tattersall, I., Delson, E., and Van Couvering, J. (eds.) (1988). "Encyclopedia of Human Evolution and Prehistory." Garland, New York.

White, T. D., Suwa, G., and Asfaw, B. (1994). *Australopithecus ramidus*, a new species of early hominid from Aramis, Ethiopia. *Nature* **371**, 306–312.

Wolpoff, M. H. (1995). "Human Evolution 1996 Edition." McGraw-Hill, New York.

Wood, B. A. (1992). Origin and evolution of the genus *Homo*. *Nature* **355**, 783–790.

Human Genetics

H. ELDON SUTTON
University of Texas at Austin

GLOSSARY

Allele Alternate form of a gene

Autosome Any chromosome other than a sex chromosome. Humans have 22 pairs of autosomes

Dominant Allele that is expressed when only one copy is present

Genotype Specific set of genes present in an organism, whether or not they are expressed

Heritability That part of the variation of a trait in a population that is due to a variation in genotypes

Linkage Occurrence of genes on the same chromosome. Genes that are close together have reduced recombination in meiosis

Locus Position of a gene or other genetic element on a chromosome

Meiosis Type of cell division in germ cells that reduces the number of chromosomes by one-half, producing gametes that have only one of each kind of chromosome

Mitosis Type of cell division that occurs in somatic cells and in earlier divisions of the germ cell line. Each daughter cell receives an exact copy of the parental set of chromosomes

Phenotype Observable organism, resulting from interaction of the genotype and the environment

Recessive Allele that alters the phenotype only when both chromosomes have the allele

Segregation Separation into different daughter cells of genes that are together in a parent cell

Sex chromosome Chromosome whose variation in number is associated with sex. In humans the X and Y chromosomes are sex chromosomes

GREGOR MENDEL'S STUDIES IN *PISUM SATIVUM* (I.E., flowering peas) led to the conclusion that inherited variations result from combinations of discrete stable determinants (i.e., genes). For most traits these occur in pairs, except in gametes, which receive only one of each pair during meiosis. Fertilization of an egg creates new pairs. The similarity of transmission of genes and of chromosomes in cell division led to the recognition that genes are located on chromosomes.

Genes can occur in many alternate forms (i.e., alleles). Dominant alleles are expressed in the phenotype when a single copy is present. Recessive alleles are expressed only when both members of the pair are similar. These rules describe many of the human pedigrees of autosomal dominant and recessive transmission of traits. Genes on the X and Y sex chromosomes have somewhat different patterns of transmission, because females have two X chromosomes and males have one X and one Y chromosome. Modern studies of the molecular nature of genes and gene action have led to understanding of these events and the occasional apparent deviations that occur. [*See* Genes.]

The inheritance of many traits is more complex. These traits can reflect variation at multiple gene loci as well as in the environment. Nevertheless, the influence of genetic variation can be detected by comparing similarities among close relatives, who share many genes, with similarities among unrelated persons.

Mitochondria are cytoplasmic organelles that have DNA, which codes for some of the proteins and for RNA. Variation in mitochondrial genes is transmitted only by females, who supply the cytoplasm for ova.

Such transmission is known as maternal, or cytoplasmic, inheritance.

I. MENDELIAN HEREDITY

Interest in the transmission of inherited traits precedes modern science by hundreds of years. Indeed, the X-linked pattern of transmission for human traits such as hemophilia and color blindness was known, although the rules were not based on understanding the mechanics or stochastic nature of heredity. It remained for Mendel in 1865 to determine experimentally the rules that apply to all eukaryotic organisms for the genetic transmission of traits.

A. Mendelian Principles

1. Genetic Segregation

Mendel's experiments with *P. sativum* were successful in large part because he selected clearly alternative traits to observe. We now know that such qualitative variations are much more likely to result from variation of a single gene than are quantitative traits. His initial crosses of two true-breeding stocks assured that only one kind of genetic determinant—the gene in modern terminology—was present at most loci of the parental stocks.

The basic observation made by Mendel was that in a cross of two parental stocks differing for a trait (e.g., round versus wrinkled peas), only one trait was observed in the F_1 (first filial) generation. This can be expressed symbolically:

Parents	Round × Wrinkled
F_1	Round

Avoiding another error of many of his predecessors, Mendel carried out another cycle of breeding and found that the determinants of wrinkled peas were neither modified nor lost in the F_1 generation:

F_1	Round × Round
F_2	3 Round : 1 Wrinkled

Further breeding showed that the round peas in the F_2 generation were of two types. One-third produced only round progeny when bred among themselves, whereas two-thirds continued to produce both round and wrinkled peas. Wrinkled peas from the F_2 generation produced only wrinkled progeny when bred among themselves.

From these observations Mendel concluded that heredity depended on the transmission of determinants that occurred in various combinations in individual plants, that the expression of the associated characteristics depended on the combinations of determinants, and that determinants were recovered from combinations without having been altered.

The number of such determinants could be deduced by the ratio of peas in the F_2 generation. Because the same results were obtained irrespective of which parent possessed which characteristic, Mendel concluded that the contribution of each parent was identical. The minimum number of determinants in a gamete (i.e., pollen or ovule) would be one; therefore, the minimum number in a fertilized ovule must be two. The crosses could then be symbolized:

Parents	RR × WW
Gametes	R × W
F_1	RW
Gametes	R, W × R, W
F_2	1RR + 2RW + 1WW

The ratio of round to wrinkled peas in F_2 would be a simple statistical combination, producing three round to one wrinkled if the combination RW produced round peas. This was the ratio observed experimentally. Proposing a greater number of determinants would generate different ratios. Hence, Mendel concluded that the number of determinants in each gamete is one. These observations comprise the law of genetic segregation.

In modern terminology the determinants are genes, and the alternate forms are alleles. The trait that is expressed in F_1 is dominant, and the unexpressed trait is recessive. If the same allele is received from both parents, the organism is homozygous for that locus. If two different alleles are present, the organism is heterozygous. "Genotype" refers to the particular combination of alleles of an individual. The expression of these genes produces the phenotype, the result of the genotype acting with other genes in a particular environment. A cross such as that described in which a single trait is considered is called a monohybrid cross.

The expected genetic ratios in the F_2 generation provide a powerful test of the validity of a particular genetic hypothesis. The genotypic ratios are those that refer to the actual genotypes. In the foregoing exam-

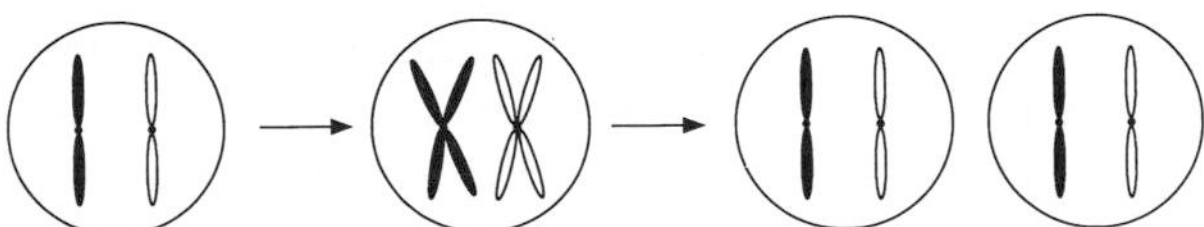

FIGURE 1 Transmission of chromosomes in mitosis. Each daughter cell receives an exact copy of the parental set of chromosomes.

ple, 1 : 2 : 1 is the ratio of RR : RW : WW. The phenotypic ratios refer to the observable differences—3 round : 1 wrinkled in the example. For allelic combinations at some loci, the phenotypes of heterozygotes are distinct from either homozygote, in which case genotypic and phenotypic ratios are the same.

2. Chromosomal Basis of Heredity

In 1902 W. S. Sutton and T. Boveri independently recognized that the distribution of chromosomes in cell division and in gamete formation are the same as for Mendelian genes. They proposed that genetic information is stored on chromosomes and is replicated and distributed in this form, a theory that has been proved amply.

Humans have 46 chromosomes, consisting of 23 pairs. For 22 of the pairs, both members are similar in males and females and are called autosomes. The remaining pair are sex chromosomes, consisting of two X chromosomes in females and one X and one Y chromosome in males. The chromosomes that compose each autosomal pair are homologous (i.e., they have the same array of genes). The X and Y chromosomes have a small region of homology at the tip of the short arms, but they are not otherwise homologous. [*See* Chromosomes.]

During division of somatic cells, each daughter cell receives an exact duplicate of the chromosomes of the parental cell in a type of division known as mitosis (Fig. 1). In meiosis, the cell divisions that produce germ cells, each daughter cell receives only one of each pair of parental chromosomes (Fig. 2). Homologous chromosomes pair physically in meiosis. During this period, crossing-over can occur. This involves a physical exchange of chromosome segments and can occur once or several times in each pair of chromosomes, depending on the size. Crossing-over allows alleles that were originally on different chromosomes to be on the same chromosome. The other crossover product would have the complementary combination of alleles. [*See* Meiosis; Mitosis.]

3. Independent Assortment

Mendel's experiments were based on seven different traits (e.g., round versus wrinkled peas and yellow versus green pods). Identical results were obtained for each of the seven traits. Furthermore, in dihybrid crosses, in which simultaneous segregation of two traits was considered, the results from each trait were independent of each other. That is, in the F_2 generation the likelihood of a round or a wrinkled pea was independent of the likelihood of a yellow or a green pod. Thus, the original parental combinations of alleles were not preserved in the gametes produced by the F_1 plants. This is known as the law of independent assortment.

4. Linkage

We now know that a large number of genes (50,000–100,000 in mammals) are located on a small number of chromosomes (i.e., 23 pairs in humans). Since the chromosomes remain intact in various parts of the cell cycle, not all pairs of genes should assort independently. Those on different chromosomes do. Those that are widely separated on the same chromosome also assort independently as a result of crossing-over

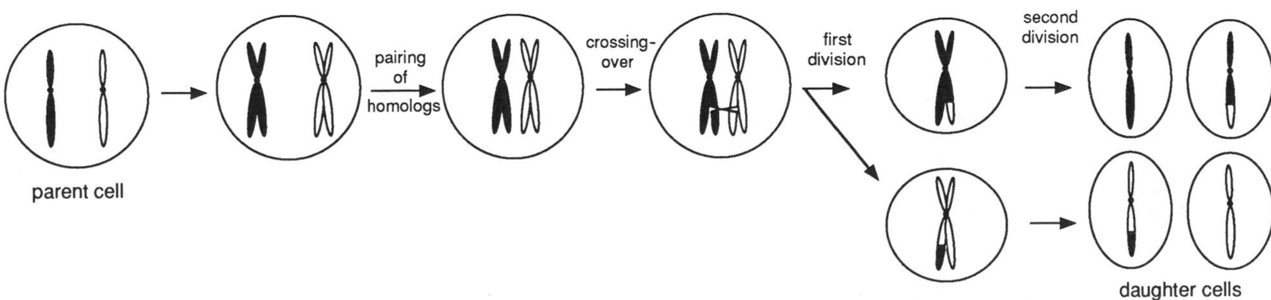

FIGURE 2 Distribution of chromosomes in meiosis. Early in meiosis each chromosome pairs with its homolog. Crossing-over occurs during this period. The chromosomes then separate, and the members of a homologous pair go into different daughter cells. The chromosomes then complete division, and a second cell division occurs. The end result is four potential gametes, each of which has only one copy of each of the original set of chromosomes.

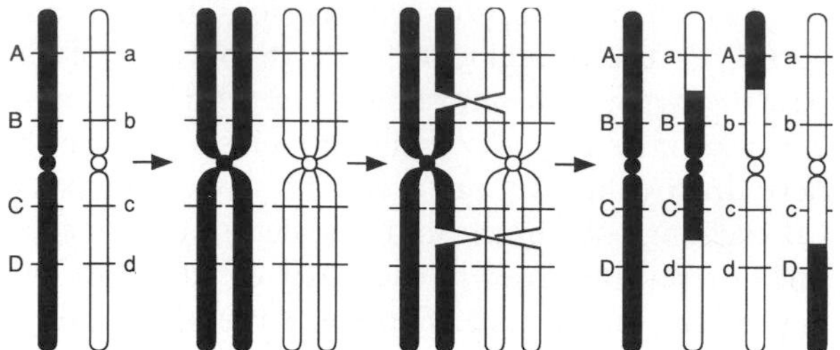

FIGURE 3 Crossing-over during meiosis. Crossing-over permits the recombination of alleles on a particular chromosome. Thus, a chromosome in a gamete is a composite of the original homologous pair of chromosomes from which it is derived.

that might occur between them (Fig. 3). However, if two such loci are close together, the likelihood of crossing-over between them is low, and the parental combinations of alleles are largely preserved in meiosis. Such loci are described as linked. Two or more linked loci constitute a linkage group.

A gamete with new combinations of alleles is described as recombinant, whether the new combination arises from the independent assortment of chromosomes or from crossing-over between loci on the same chromosome. Since the frequency of crossing-over between two loci is approximately proportional to the distance between them, one can express their proximity to each other by the percentage of recombination. For example, if an F_1 individual were formed from the parental gametes *AB* and *ab*, where *A* and *a* are alleles at one locus and *B* and *b* are alleles at another, then the nonrecombinant (parental) gametes produced by F_1 would be *AB* and *ab*; the recombinant gametes would be *Ab* and *aB*. If the recombinants composed 10% of the total, the loci would be 10 centimorgans (map units) apart.

By comparing pairs of loci, one can construct a linear map showing the order of loci and the distances between adjacent loci for each chromosome. Such a map is called a genetic map, based entirely on genetic recombination. It is similar to a physical map in that loci are in the same order, but the rate of crossing-over need not be uniform along a chromosome. As a result the genetic map can show distortions of scale compared to the physical map. [*See* Genetic Maps.]

B. Human Pedigrees

The rediscovery of Mendelian principles in 1900 promptly led to searches for Mendelian traits in other species. The generality of Mendelian heredity was soon established for both plants and animals, including *Homo sapiens*. Indeed, over 4000 Mendelian traits have now been identified in humans, and the rate of discovery continues to increase. A compilation describing all known Mendelian traits is published at regular intervals under the supervision of V. A. McKusick and is maintained as a regularly updated data base that is accessible on Internet.

In analyzing transmission patterns in mammals, it is necessary to distinguish between genetic loci located on autosomes and those on the sex chromosomes. It is often useful to summarize both genotypic and phenotypic information in the form of a pedigree. Some of the conventions used in drawing pedigrees are shown in Fig. 4. The types of pedigrees commonly encountered depend on the relative frequencies of the normal "wild-type" and variant alleles. For the most part, interest is in rare traits that are associated with inherited disease, and discussion here is focused on these situations. [*See* Genetic Diseases.]

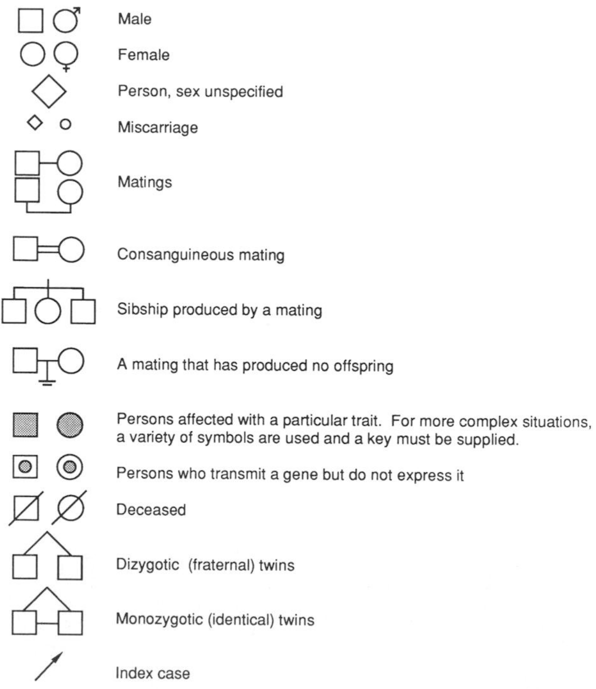

FIGURE 4 Symbols commonly used in depicting pedigrees of inherited traits. Roman numerals are often used to designate generations, I being the oldest. Arabic numerals are used to designate individuals, beginning with 1. Thus, the fourth person in the second generation is II-4. [Reproduced by permission from H. E. Sutton (1988). "Introduction to Human Genetics," 4th Ed. Saunders College, Philadelphia.]

1. Autosomal Dominant Inheritance

The first Mendelian trait described in humans was reported in 1905 by W. C. Farabee. Brachydactyly is characterized by short broad fingers. It is a rare, relatively benign trait that is easily recognized. It also has high penetrance (i.e., all persons with the appropriate genotype express the trait). A pedigree of the original family described by Farabee is shown in Fig. 5.

Analysis of pedigrees is particularly useful for rare dominant traits. Because persons homozygous for the rare allele would be exceedingly rare, there are essentially only two genotypes observed, *Aa* and *aa*, where *A* represents the dominant allele and *a* the recessive. Each genotype thus corresponds to a different phenotype.

Several characteristics of autosomal dominant inheritance are useful in analyzing human pedigrees. (1) Every affected person should have an affected parent. The only exception would be instances in which a new mutant gene appears, but mutation is rare and should not be encountered often. (2) The dominant allele should be transmitted equally to sons and daughters, regardless of the sex of the affected parent. Hence, the number of affected males and females should be equal.

Expression of a trait can, of course, vary between males and females or can be influenced by other factors. Penetrance can also be less than 100%. It is therefore necessary to rule out other explanations when deviations from the ideal transmission pattern occur. The demonstration that a trait is dominantly transmitted by the foregoing criteria requires that it not confer sterility or impose a health burden that is incompatible with reproduction. Were it to do so, the only affected persons would be those who received a newly mutant allele, which they could not transmit.

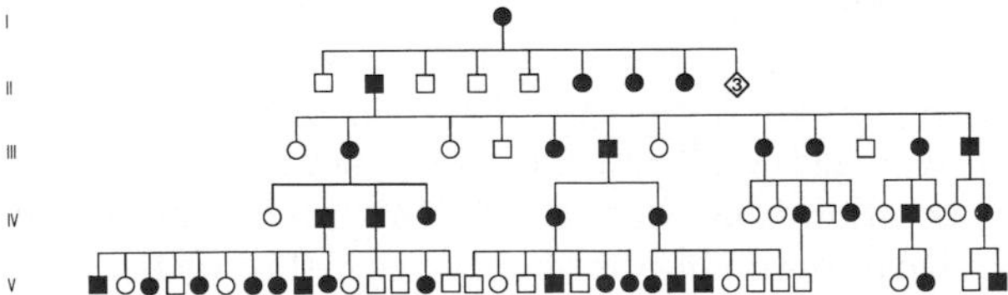

FIGURE 5 Pedigree of brachydactyly, an autosomal dominant trait. This was the first example of Mendelian heredity reported in humans. [From W. C. Farabee (1905). "Inheritance of Digital Malformations in Man," Papers of The Peabody Museum of American Archaeology and Ethnology, Harvard University, Vol. 3, pp. 69–77, Cambridge, MA.]

There are many examples of autosomal dominant inheritance in humans, of which Huntington disease, achondroplasia, and some forms of osteogenesis imperfecta (i.e., brittle bones) are especially well known. [*See* Huntington Disease.]

2. Autosomal Recessive Inheritance

The expression of recessive traits depends on the presence of the recessive allele on both chromosomes. This requires that both parents transmit the allele. Males and females should be equally often affected (i.e., homozygous), although a detrimental trait can be expressed more severely in one sex or the other. Since both parents of a child with a recessive trait must be heterozygous, the typical pedigree for a recessive trait contains a single sibship with affected persons.

A common finding for rare recessive traits is an increase in consanguinity in the parents of affected persons. In such matings both parents are much more likely to be heterozygous for the same recessive allele received from one of the common ancestors than would two unrelated persons. Each child from such a mating would therefore have a 25% chance of being homozygous for the allele. The increase in consanguinity of parents of children with rare traits is often the most compelling argument that the traits are recessively inherited.

Because of the typical pattern of isolated affected sibships, pedigree analysis is generally of little value in proving that a particular trait is inherited as a rare autosomal recessive. Such familial clustering can result from environmental as well as genetic causes. Many deleterious conditions are inherited as autosomal recessive traits. These include phenylketonuria, Tay–Sachs disease, and cystic fibrosis. [*See* Cystic Fibrosis, Molecular Genetics; Phenylketonuria, Molecular Genetics.]

3. Pseudoautosomal Inheritance

There is a small region of homology between the human X and Y chromosomes, in both cases near the ends of the short arms (Fig. 6). Crossing-over regularly occurs in meiosis in this region, leading to the exchange of terminal segments between the X and Y chromosomes. Genes in this region show the same pattern of transmission as autosomal genes, hence the term pseudoautosomal. The older literature uses the term "partial sex linkage" for these genes. So far the only functional genes placed in this region in humans are *MIC2*, responsible for a cell-surface antigen, and *ASMT*, a gene involved in melatonin synthesis.

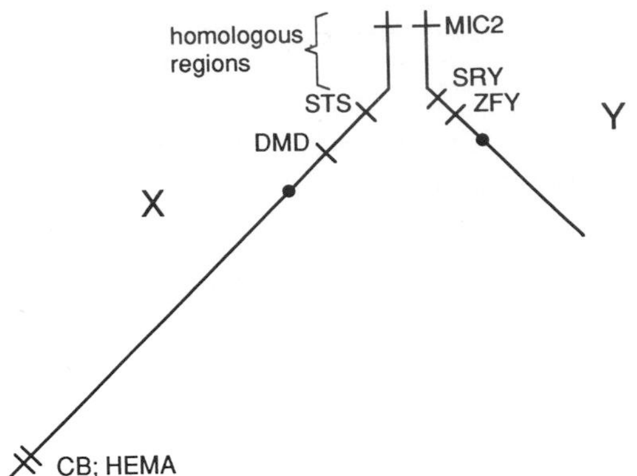

FIGURE 6 Regions of homology and nonhomology of the human X and Y chromosomes. Gene symbols are as follows: *MIC2*, a cell-surface antigen; *SRY*, sex-determining gene Y; *ZFY*, a gene possibly involved in male fertility; *STS*, steroid sulfatase; *DMD*, Duchenne muscular dystrophy; *CB*, deutan color blindness; *HEMA*, hemophilia A.

4. X-Linked Dominant Inheritance

The terms "X linked" and "sex linked" are synonymous. This pattern of transmission differs from the autosomal pattern in that a male transmits his one X chromosome to all of his daughters but none of his sons, whereas a female transmits one of her two X chromosomes to each child, irrespective of the sex of the child. Therefore, a male with a rare X-linked dominant trait transmits it to all of his daughters but none of his sons. A heterozygous female transmits it to one-half of her daughters and one-half of her sons. Since females have twice as many X chromosomes as males, the frequency of an X-linked dominant trait is twice as great among females than among males. Good examples of X-linked dominant transmission, other than of normal traits, are rare.

5. X-Linked Recessive Inheritance

The distinctive pattern of transmission in families was responsible for X-linked recessive traits being the first recognized. The inheritance of hemophilia is noted in the *Talmud*, and the rules of transmission of hemophilia were established in 1820 by C. F. Nasse and became known as Nasse's law.

In order for a female to express an X-linked recessive trait, she must be homozygous for the variant allele. Since males have only one X chromosome, they express the trait when that X chromosome has the variant allele (i.e., they are hemizygous for the trait). An affected male cannot transmit the trait to his sons, since they receive his Y chromosome. He transmits the recessive allele to all of his daughters, who will be heterozygous carriers unless their mother also transmits a recessive allele. Therefore, affected females can occur only if their fathers are also affected. A consequence of this pattern of transmission is the fact that all affected males in a pedigree are related to each other through heterozygous females. Thus, for rare traits an affected male can have affected maternal uncles or an affected maternal grandfather, but not both. Hemophilia, red–green color blindness, and Duchenne muscular dystrophy are common examples of X-linked recessive inheritance. A pedigree of hemophilia among the royal families of Europe is shown in Fig. 7. [*See* Hemophilia, Molecular Genetics; Muscular Dystrophy.]

6. Y-Linked Inheritance

Since chromosomes have genes, a logical supposition is that the Y chromosome has genes. Except for the small portion of the Y chromosome that is homologous to the X chromosome, there appear to be few, if any, genes except that (or those) necessary for sex determination. This is reasonable, considering that no very important genes could be put on a chromosome that is absent in females.

There have been reports of genes that appear to be transmitted according to rules expected for Y linkage. These rules are that a trait should occur only in males and, when present in a male, must be transmitted to all of his sons but none of his daughters. The only such trait that has withstood rigorous examination is maleness itself. The gene for male determination, *SRY*, is located on the short arm of the Y chromosome near the region of X chromosome homology (Fig. 6).

C. Genetic Segregation in Human Families

The power of Mendelian analysis in experimental crosses lies in the strong predictions that it makes of genetic ratios. These predictions depend on some knowledge of parental genotypes, as in a monohybrid cross in which the parents are known to be homozygous. Indeed, if a trait is determined by a single pair of segregating alleles, then the genetic ratio in the F_2 generation must be 1:2:1. This prediction can be tested by observing the phenotypes in the F_2 generation.

Genetic ratios can also be tested in human families. However, great care must be exercised not to violate the assumptions that underlie their use.

1. Problem of Small Family Size

Genetic ratios refer to stochastic (i.e., random), rather than deterministic, events. Therefore, the reliability

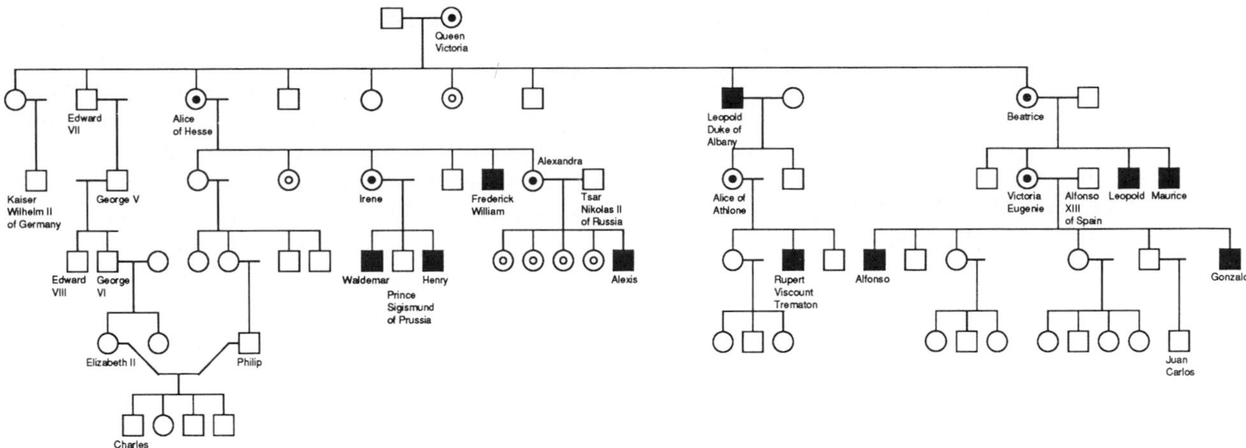

FIGURE 7 Pedigree of hemophilia A in the royal families of Europe. □, normal male; ○, normal female, presumed homozygous; ■, hemophilic male; ⊙, normal female, proved heterozygous; ◎, normal female, at risk, status not established. [Reproduced by permission from H. E. Sutton and R. P. Wagner (1985). "Genetics: A Human Concern." Macmillan, New York.]

of observations depends on the number of offspring produced from a particular cross. Too few observations are associated with large error and might fail to invalidate a false hypothesis. In experimental genetics the offspring from crosses of known genotype can generally be pooled to produce sufficient data to achieve any desired statistical validity. In individual human families the number of offspring is much too small for reliable comparisons with Mendelian predictions. Families can be pooled, but at the risk of mixing different genotypes that are phenotypically similar.

Another problem is the bias introduced when families are ascertained (located) through affected offspring. The family that by chance produces two affected offspring is more likely to be selected in a sample as a similar family that produced only one affected offspring. Had the same family, by chance, produced only nonaffected offspring, it would not have been selected at all.

Nevertheless, it is useful to be able to test segregation ratios in humans. The segregation ratio is important evidence in support of autosomal recessive inheritance, and there are occasions when deviations from the theoretical segregation ratios can disclose significant biological events. Because the nature of the ascertainment biases is often known, corrections can be applied that allow the calculation of correct ratios.

2. Simple Sib Method of Correction

A correction for ascertainment, the simple sib method, is useful in situations in which a rare trait is suspected of being inherited as an autosomal recessive trait. Typically, affected persons occur in isolated sibships, which can contain one or more affected persons as well as nonaffected persons. According to the hypothesis to be tested, all such subships are produced from matings of the type $Aa \times Aa$, where A is the normal dominant allele and a the recessive allele. Offspring should be in the proportion 3 $A/-$: 1 aa. (The dash indicates that the second allele can be either A or a.)

Location of the affected sibships is through an affected member. The procedure for correction assumes that the likelihood of including a sibship in the sample is directly proportional to the number of affected persons. Thus, sibships that happen to include no affected members would be missed. Furthermore, a sibship with two affected members is exactly twice as likely to be located as a sibship with one affected member. These assumptions apply when only a small sample of the possible sibships in the base population is chosen. Another assumption is that sibships with more affected members are not more likely to occur in the population base from which the sample is drawn than are sibships with fewer affected members.

If these conditions are met, the ascertainment bias can be removed by subtracting the index case from the total. Thus, for an autosomal recessive trait a ratio of 75% nonaffected : 25% affected should be found among the sibs of the index cases. The actual results can be tested against these theoretical expectations, using appropriate statistical methods.

Other methods of correction make other assumptions about the methods of ascertainment. With the use of computers, it is possible to develop complex models of segregation to test against observations, which incorporate variables such as penetrance, modifying genes, aberrant segregation, and unequal survival.

D. Factors That Modify Mendelian Expectations

Although the validity of Mendelian transmission is well documented, a number of discoveries alter and extend our understanding of how genes act as well as more complex genetic situations.

1. Dominance and Recessiveness

The term "*dominant*" was used earlier to describe the allele that is expressed in the F_1 generation in a monohybrid cross. Mendel himself found that in experiments with other plants the F_1 phenotype is intermediate between those of the two homozygotes.

As additional systems have been studied, the concepts of dominance and recessiveness have changed. A dominant allele is one that is expressed in the phenotype when only one copy is present. There is no implication that homozygous and heterozygous genotypes have the same phenotype. Often, they do not. Nor is there the suggestion that phenotypic expression of the second allele is suppressed in heterozygotes. Often, the two alleles are expressed independently of each other; the term "codominant" applies to such pairs of alleles. A recessive allele is one that affects the phenotype only when the allele is present on both chromosomes.

At the molecular level, dominance and recessiveness are readily understood. One functional allele can be adequate for normal development and metabolism. A defective allele would then be recessive, since heterozygous combination of that allele with a normal allele would produce a normal phenotype. On the other hand, if one functional allele cannot supply normal cell requirements for the gene product, a nonfunctional allele would be dominant. For some loci a variant allele might produce a product that interferes with normal function, an effect that is likely to be dominant.

The terms "dominant" and "recessive" are useful in describing transmission patterns in families or experimental crosses. They should always be used with the understanding that an isolated allele is neither dominant nor recessive. Rather, the terms have meaning only in the context of the interaction of alleles in a biological system.

2. Allelic Diversity

Genetics is often taught as if each locus had only two alleles. Such pedagogical practice obscures the fact that there might be many alleles. Consideration of the many nucleotides that compose a gene makes it obvious that the number of nucleotide substitutions, deletions, and additions within the coding regions, as well as changes in regulatory regions, should make the potential number of allelic variations enormous.

This is indeed the case. To be sure, many of the variations would be eliminated rapidly by natural selection after a transient existence. However, new alleles are continuously being produced by mutation, many of them detrimental, some normal variants, and perhaps rarely one that is functionally more adaptive than the parental allele from which it arose.

Molecular analysis of many mutant alleles associated with disease has verified great allelic heterogeneity at many loci. The mutant alleles responsible for Duchenne muscular dystrophy result from many independent mutations and are of many different types. Diseases inherited as recessive traits often involve two different nonfunctional alleles in an affected person, described as a compound heterozygote. In some instances a particular mutant allele might be prevalent, even though there are also many other variant alleles with the same effect. In the case of cystic fibrosis (CF), there is one common *CF* allele that has a one-codon deletion, but other alleles have the same phenotypic effect.

The products of different alleles can, of course, be associated with different phenotypes. One example is found in deficiencies of hypoxanthine phosphoribosyltransferase, an X-linked recessive trait. A modest deficiency produces a rare form of gout. Allelic variants that produce no active enzyme cause Lesch–Nyhan syndrome, a condition that involves mental deficiency, compulsive biting of fingers and lips, and other defects. Mutations at a single locus therefore might be responsible for more than one disorder.

The great heterogeneity of mutations suggests that some of the variation in phenotype of inherited traits in outbred populations is due to different combinations of alleles. In small inbred populations, allelic variation is likely to be less, and persons who express an inherited trait are more likely to be genetically homogeneous.

3. Multiple Loci

Although mutations at many loci produce phenotypes that are unique to each locus, this is not universally true. A complication that must always be considered in studying inherited variations in humans is whether all similar phenotypes are caused by mutations at a single locus. There are many examples to the contrary. One of the first to be established was the rare, dominantly inherited condition elliptocytosis (EL). In some families, the *EL* locus was proved to be very close to the Rh blood group locus, subsequently placed on chromosome 1. In other families, the locus was shown not to be near the Rh locus, and this second locus (*EL2*) has been placed on another chromosome. The two loci produce products that affect the same cell function, and either of them, if defective, can interfere with this function.

Individually these loci conform strictly to Mendelian transmission patterns. Should the two loci be segregating within the same kindred, failure to recognize this fact would lead to apparently aberrant segregation.

4. X Chromosome Inactivation

A puzzle that concerned geneticists for many years was the mechanism of dosage compensation of X-linked loci. Most of the loci known to be on the X chromosome have no direct relationship to sex. Females with two X chromosomes have twice the number of copies of most of these genes than do males. However, no differences could be found in function of the genes in males and females.

The solution was found in X chromosome inactivation, a hypothesis first clearly formulated by M. Lyon and demonstrated to apply to mammals, including humans. In each XX zygote (fertilized egg), both X chromosomes are active. However, early in embryogenesis one of the X chromosomes in each cell becomes largely inactive, leaving only one active X chromosome, as in males. Through subsequent cell divisions, the same X chromosome remains inactive, replicating somewhat later than the active X chromosome.

At the time of the original inactivation, the choice of which chromosome is to be inactivated is an independent random event in each cell. Some cells have one chromosome active; some will have the other. Females are therefore a mosaic of cells and tissues that differ with respect to which X chromosome is active. Because inactivation occurs in the early embryo, any particular cell at that time might serve as the progenitor for a large number of cells in the developed individual. On average approximately one-half of the cells have a specific active X chromosome. But chance might favor one or the other X chromosome in a particular tissue. For this reason many females who are heterozygous for an X-linked recessive trait show some expression of the trait. At the cellular level each cell is found definitely to express one or the other allele, but not both.

5. Uniparental Disomy

On very rare occasions, a gamete (egg or sperm) will have two copies of a particular chromosome and, during fertilization, will join with a gamete that has no copies of that chromosome. Such gametes arise by abnormal segregation of the chromosomes in a process called nondisjunction. The fertilized egg that results from this union has the correct number and kind of chromosomes, but both copies of this particular chromosome will have arisen from one parent. This situation is called *uniparental disomy* and can further be designated heterodisomy if the two chromosomes are nonidentical homologs, or homodisomy if they are identical.

In the case of some chromosomes, the two homologs in a zygote must be derived from different parents for normal development to occur. In the case of other chromosomes, the two homologs can come from the same parent. There is a problem, however, in that the offspring may receive two copies of a recessive allele from a parent who is heterozygous. Expressed symbolically, an $AA \times Aa$ mating will produce a child who is *aa*. Often in such cases, the child will be homozygous for alleles at many loci, several of which may be associated with abnormal development. Such a mating would be incapable of producing an *aa* child according to Mendelian rules. Fortunately, the risk for recurrence in the family is near zero.

6. Imprinting

One of the cornerstones of Mendel's Law of Genetic Segregation was the equivalence of males and females in the transmission of genes. This has been extensively documented for the vast majority of nuclear genes, with mitochondrial genes being a conspicuous but understandable exception. There are a few nuclear genes that also bend this rule. It was first shown in mice that, at a few loci, alleles are expressed or not depending on whether they are received from the mother or the father. An allele that is not expressed in one generation may be expressed in the next if transmitted from a parent of the opposite sex. The allele therefore is modified temporarily during the transmission process. The modification does not involve alteration of the DNA sequence, however. This phenomenon is called *genetic imprinting*.

The clearest example of imprinting in humans involves two inherited disorders that are due to mutations on chromosome 15. Prader–Willi syndrome includes mild or moderate mental retardation, with hypotonia in infancy, hypogonadism, and other features. Some 70% of the patients have a small deletion at a particular position on the long arm of chromosome 15, always on the chromosome that was inherited from the father. The disorder can also occur on the very rare occasions when both chromosomes come from the mother (uniparental disomy). These observations indicate that this region of chromosome 15, or perhaps a single gene located in this region, is functional only if the chromosome comes from the father. The second disorder is Angelman syndrome, which involves more severe mental retardation. The disorder is also associated with small deletions of the same region of chromosome 15, but the affected chromosome is always transmitted by the mother. It is apparent that normal development depends on receiving one chromosome 15 from the mother and one from the father.

Not all chromosomes are subject to imprinting. Perhaps only a few loci are. In general, the phenotype associated with deletions is independent of sex of the parent who transmitted it.

7. Unstable Genes

An important characteristic of genes is the fidelity with which they are replicated and transmitted. Mutations do occur, but they are rare, on the order of one per million gametes for most loci. [*See* Mutation Rates.]

Some 10 genes are known that, in some instances, are highly unstable. The common characteristic of these genes is the presence of a sequence of three nucleotides repeated in tandem a number of times. In the first example discovered, the fragile X syndrome (FRAXA), the *FMR1* locus on the X chromosome contains repeated copies of the trinucleotide 5′-$(CGG)_n$-3′. In most normal persons, n varies from 6 to 52. In this range, the gene is very stable, and the value of n does not change from one generation to the next. A few normal persons have an X chromosome with values of n from 50 to about 200. These persons, if they are female, are highly likely to transmit that X chromosome with an increased number of copies of the trinucleotide. If the offspring receives an X chromosome with greater than 200 copies, the *FMR1* gene will not function properly. Expansion may generate *FMR1* genes with greater than 1000 repeats. Persons with 50 to 230 copies are said to have a premutation; greater than 200 copies constitutes a full mutation. Females who are heterozygous for a normal *FMR1* allele and a full mutation are mildly retarded mentally. Males who are hemizygous for a full mutation are more severely retarded and have a typical facies and very large testes (macro-orchidism). The fragile X syndrome is one of the most frequent causes of inherited mental retardation, estimated to occur in 1 in 1500 males and 1 in 2500 females.

The instability of the premutation allele leads to pedigrees that do not show simple Mendelian transmission of the FRAXA phenotype. A male with a premutation will be normal and will transmit that premutation to all his daughters, who will be normal. However, the daughters may transmit either premutations or full mutations to their children. With succeeding generations of transmission by females, the number of trinucleotides increases, increasing the risk that offspring will have full mutations and will be affected. The increase in risk in succeeding generations is called *anticipation.*

Another example of a trinucleotide expansion disorder is Huntington disease. In this case the trinucleotide repeat is 5′-$(CAG)_n$-3′. The numbers of repeats required for a full mutation are less than in the case of FRAXA. Also, the expansion occurs when transmitted from a male rather than from a female. This explains the earlier onset and greater severity that is observed when the Huntington allele is received from the father rather than the mother.

The mechanism for trinucleotide expansion is not understood, nor is the reason for the expansion being dependent on the sex of the transmitting parent.

II. COMPLEX INHERITANCE

Analysis of single-gene variations has been highly successful in humans. There are many other traits of great interest that are more complex in their transmission. Indeed, prior to the rediscovery of Mendel's work, most efforts were directed to the study of these complex traits. Preeminent among the scientists working on complex traits in humans was Francis Galton, who invented several statistical techniques to solve problems in human genetics, techniques that are still in use in many fields.

A. Basis of Complex Traits

With the demonstration that Mendelian heredity applies generally to plants and animals, an obvious question was how one could explain such traits as height, behavior, or crop yields in Mendelian terms. These traits do not occur in simple alternative forms. Rather, they are distributed as continuous variables. Other traits (e.g., cleft lip in humans or the number of toes in guinea pigs) might occur as discrete inherited variables, but still are not due to variation at a single locus.

An answer lay in the hypothesis that these are multigenic traits. These can also be called multifactorial traits, a term that admits the role of environmental, as well as genetic, variation. Combinations of alleles at several loci constitute the genotype that produces a certain phenotype. Different combinations of alleles might alter the phenotype by small increments. Furthermore, a particular phenotype can result from several different combinations of alleles, making it difficult to infer the genotype from the phenotype.

TABLE I
Correlations between Relatives for Total Dermal Ridge Counts[a]

Relationship	No. of pairs	Observed correlation coefficient	Theoretical correlation coefficient
Parent–child	810	0.48	0.50
Mother–child	405	0.48 ± 0.04	0.50
Father–child	405	0.49 ± 0.04	0.50
Father–mother	200	0.05 ± 0.07	0.00
Midparent–child[b]	405	0.66 ± 0.03	0.71
Sib–sib	642	0.50 ± 0.04	0.50
Monozygotic twins	80	0.95 ± 0.01	1.00
Dizygotic twins	92	0.49 ± 0.08	0.50

[a]From S. B. Holt (1961). *Brit. Med. Bull.* **17**, 247.
[b]Midparent is the average of the values of the two parents.

B. Analysis of Complex Traits

Even though complex traits cannot be tested with segregation analysis and cannot be partitioned directly into variation at specific loci, the existence of a genetic component of variation can be tested. The underlying premise is that the alleles that contribute to a multigenic trait are found in greater number in close relatives than in more distantly related or unrelated persons.

1. Intrafamilial Correlations

The method of study introduced by Galton was the comparison of family members for similarity. Galton's studies were performed from approximately 1865 to 1900, between Mendel's original report and its rediscovery in 1900. Thus, Galton knew nothing of Mendelian theory. His analyses were based only on the idea that a child receives hereditary determinants equally from each parent. From this it follows that sibs have one-half of their heredity in common, grandparents and grandchildren have one-fourth in common, half-sibs have one-fourth, first cousins have one-eighth, etc.

To compare the observed similarity among related persons with theory, Galton devised the correlation coefficient, which standardizes the variation on a dimensionless scale and expresses the overall similarity between pairs of persons (e.g., parent and child) for the trait. The correlation coefficient can vary from +1, if the two measures are always the same, to −1, if they consistently vary in opposite directions. If the variation is independent, the correlation coefficient is 0. With the eventual understanding of Mendelian heredity, it was possible to calculate the expected correlations for various degrees of relationship, as shown in Table I, on the assumption that all variation is genetic.

An example of a trait whose variation is virtually entirely due to heredity is given in Table I. Fingerprints are determined in the embryo and do not change after that. The pattern intensity is measured by the number of ridges crossed if a line is drawn from the center of the pattern on each finger to the triradius in the case of a loop, or to both triradii in the case of a whorl. Arches have no triradii and are scored as 0. The sum of the values for the 10 fingers is the overall measure for the person. This measure turns out to be complex in its genetic determination, with no indication of environmental influences. This does not mean, of course, that environment is not important in the development of fingerprint ridges. Rather, it means that variation in environment is not important in variation of ridge counts, whereas genetic variation is.

2. Twin Studies

Galton realized that twins might be especially useful in the analysis of inherited traits. This is because some twins are identical genetically, whereas fraternal twins are two sibs that happen to be conceived at the same time because of double ovulation. These are referred to as monozygotic and dizygotic twins, respectively. By comparing the variation within pairs of monozygotic twins and dizygotic twins of the same sex, one can avoid the effects of different ages and environments that characterize comparisons of single-born sibs.

Twin studies are widely used to assess the magnitude of genetic variation of complex traits. They have other special uses as well. For example, since monozygotic twins are genetically identical, any variation within a pair must be nongenetic. The discordance that occurs for many traits within monozygotic pairs provides an opportunity to identify environmental factors that differ for the members of a pair that can contribute to the appearance of the trait.

3. Adoption Studies

For some traits, particularly behavioral traits, the underlying assumptions of the twin method can be questioned. Monozygotic twins might, in fact, have more similar environments than do dizygotic pairs, because monozygotic twins may be treated more alike. An alternative approach is to compare related persons, one of whom is adopted into a different setting at birth. Typically, adopted children are compared with their biological parents versus their foster parents. To the extent that the adopted child resembles the biological parent, the underlying cause is assumed to be shared genes. A powerful comparison would be monozygotic twins separated at birth, but these are too few in number for most studies.

C. Nature and Nurture—The Meaning of Heritability

Geneticists often wish to express the portion of the variability in a complex trait that can be attributed to genetic variability. This term, "heritability," can be formulated in various ways. Typically, the denominator is the total variation of a trait in the population. The numerator is the genetic variation, which, together with the variation attributed to environment, should equal the total variation.

Every phenotype is the result of a genetic blueprint executed in a specific environment. In this context it is meaningless to talk about whether heredity or environment is more important, but it is meaningful to attribute the variation in a trait to heredity or environment. If a population is genetically uniform for the alleles that determine a trait, then the only source of variation is environment. The trait is not less genetic, but the variation is. Conversely, increasing the genetic heterogeneity might increase the total variation but decrease the fraction due to environment.

There have been many acrimonious debates on the relative contributions of heredity and environment to a particular trait. Most often, these debates are based on misunderstanding of heritability: It is a measure of how much genetic and environmental variation occur in the population, not whether the trait is genetically determined.

III. MATERNAL INHERITANCE

Mitochondria (Mt) are the only cell organelles other than nuclei in animals that have DNA. In plants, chloroplasts also have DNA, and the principles of maternal inheritance also apply. The DNA in Mt is in the form of a ring, similar to the DNA in bacteria. There are many identical copies of the DNA rings in the inner matrix of each Mt. The number of nucleotides in a ring of human Mt DNA is very small compared to a haploid complement of nuclear DNA: 16,569 base pairs versus 3 billion base pairs.

The number of genes coded in Mt DNA is also small, but these include a complete set of transfer RNA and ribosomal RNA genes. In addition, there are genes that code for key electron transport proteins that function in the use of oxygen for the production of ATP. Not all proteins found in Mt are coded in Mt DNA. Indeed, most are coded in the nucleus and find their way into Mt after synthesis. The nuclear-coded proteins can show genetic variation that is transmitted in typical Mendelian patterns.

Proteins coded by Mt genes have very different transmission patterns. There are thousands of Mt in the cytoplasm of an ovum. By contrast, a sperm has a small number in the midpiece. When an ovum is fertilized, few if any of the paternal Mt enter the zygote. Therefore, the Mt in the developing embryo are entirely maternal in origin. These biological facts determine the transmission of traits coded in Mt DNA, known as maternal or cytoplasmic inheritance. Whatever variations are present in the mother are transmitted to every child; fathers make no contribution of Mt DNA. These principles are amply borne out by studies of variants of Mt DNA. A pedigree of a Mt disorder is shown in Fig. 8. Because of the small number of Mt genes, there are only a few disorders that show maternal inheritance.

One characteristic of maternally inherited disorders is the variation in expression from one person to another. This is due to heteroplasmy, that is, the presence of two populations of Mt in each cell, one normal and one mutant. The degree of functional impairment depends on the relative proportions of the two popu-

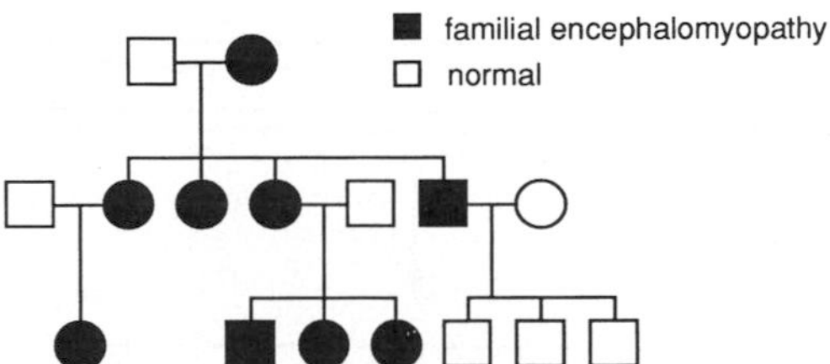

FIGURE 8 Pedigree of familial mitochondrial encephalomyopathy showing maternal transmission of the disorder. All children of affected females are affected regardless of sex. No child of an affected male is affected. [Based on a report by D. C. Wallace, X. Zheng, M. T. Lott, J. M. Shoffner, J. A. Hodge, R. I. Kelley, C. M. Epstein, and L. C. Hopkins (1988). *Cell* **55**, 601–610.]

lations. Since the distribution of Mt at cell division appears to be a chance event (as opposed to the elaborate segregation of chromosomes), some daughter cells receive a greater proportion of normal Mt and some a smaller proportion. If the proportion in the ovum differs substantially from that of the mother's tissues, the expression in the child can vary accordingly.

IV. GENETIC MARKERS

Genes are useful not only for understanding particular traits but also as indicators or markers for other genetic events. Reference has already been made to the cotransmission of alleles at closely linked loci. The transmission of a particular allele at one locus allows the prediction of which alleles are transmitted at nearby loci. This can have important application in genetic counseling. The kinds and frequencies of alleles in different human populations can be used to infer evolutionary relationships about the populations. Genes are also used forensically, as in matching evidence from a crime scene with alleged perpetrators or in establishing paternity.

Historically, such applications were limited to comparing phenotypes: blood groups and certain proteins in blood. With the development of DNA technology, we are no longer limited to the study of functional genes as markers. Indeed, there is a vast amount of DNA variation that can be used, most of which is not known to be associated with functional genes. Yet, DNA variations are transmitted as faithfully as are genes, following Mendelian rules. They are inherited as codominant traits, which is very advantageous. And they all can be detected by means of a few well-established techniques.

A. Restriction Fragment-Length Polymorphisms

DNA can be cleaved by a number of microbial enzymes called restriction endonucleases, and many of these enzymes cleave at characteristic nucleotide sequences. For example, the restriction enzyme EcoRI cleaves the sequence 5′-GAATTC-3′. (Only one of the two strands is shown.) In a large piece of DNA, this sequence might occur many times. Digestion with EcoRI would cut the DNA at each of these sites, producing a number of "restriction fragments." If the number of fragments is not large, they can be separated from each other in an electric field (electrophoresis) on the basis of differences in size and can be observed using a fluorescent DNA-binding dye. If one is interested in identifying a particular fragment, and especially if the number of fragments is large, as it would be if the entire genomic DNA of a mammal is digested, a DNA probe can be used that binds only to the fragment of interest. The probe is a piece of DNA that is complementary to one of the DNA strands and can be detected by its radioactivity or fluorescence. The technique commonly used is known as a Southern blot. Figure 9 illustrates the kind of results that are observed.

There are numerous small variations in DNA sequence among different persons and between the two sets of chromosomes that each person has. These have no functional significance usually, but occasionally a variation occurs in a restriction site. When this happens, the enzyme cannot cleave the DNA at that site. The fragment produced at that position would consist of the two flanking fragments joined together. A probe that detects one of these flanking fragments would therefore show the presence in the population of two different fragment sizes (Fig. 9). The importance of these variations is that they are inherited like any codominant trait. An individual can be *LL* (homozygous for the large fragment), *LS* (heterozygous for the large and small fragments), or *SS* (homozygous for the small fragment).

There are many thousands of these variable sites in the human genome. No two persons in the world have the same restriction fragment-length polymorphism (RFLP) patterns except identical twins, provided, of course, that a sufficiently large number of sites is studied.

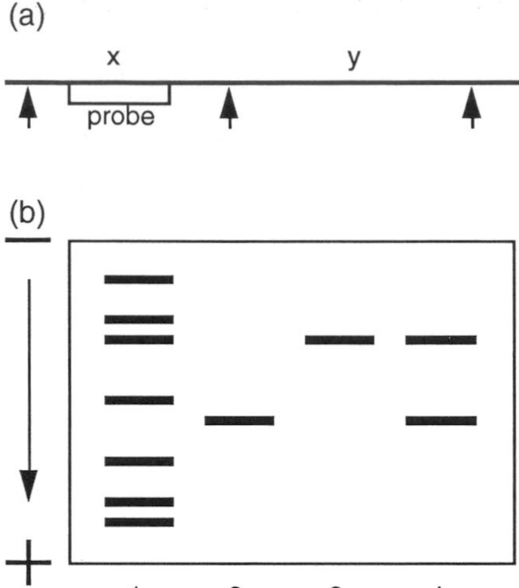

FIGURE 9 Characterization of RFLPs. (a) A segment of DNA with three restriction sites (indicated by arrows). The DNA probe binds to fragment x. (b) Electrophoretic separation of restriction fragments. Lane 1: A mixture of seven fragments separated by size, the smallest migrating fastest toward the positive pole. Lane 2: The pattern that would be produced by digestion of DNA from someone homozygous for the chromosome in (a). The fragment size would correspond to x. Lane 3: The pattern that would be produced if the middle restriction site in (a) were absent. The fragment size would equal $x + y$. Lane 4: The pattern that would be produced by someone heterozygous for the two fragment sizes.

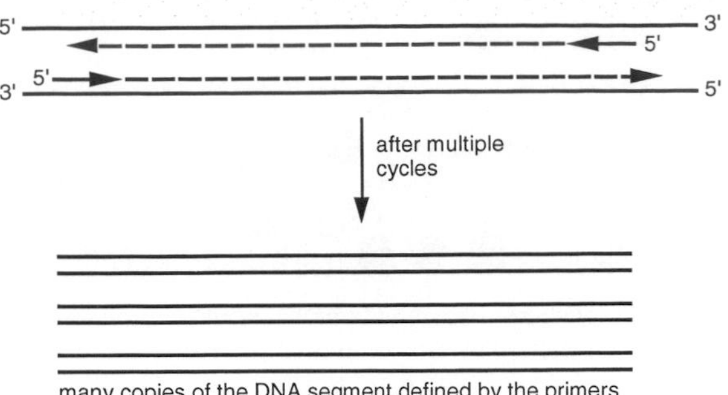

FIGURE 10 The polymerase chain reaction (PCR). The solid lines in the upper diagram are the two DNA strands that contain a segment to be copied. The solid arrows are primers that are complementary to the ends of the segment to be copied. Newly synthesized strands of DNA (dashed arrows) are extended from the 3′ ends of the primers. The extension of the strands is catalyzed by a heat-stable form of DNA polymerase. The system can be carried through a large number of cycles solely by changing the temperature. High temperatures (90°C) cause the strands to separate. Lower temperatures (<50°) allow complementary sequences to pair. Each cycle doubles the number of copies of the segment that terminates in primer sequences.

B. Microsatellites

Another type of frequent DNA variation involves tandem repeats of di-, tri-, and tetranucleotides. A common example would be 5′-$(CA)_n$-3′, when n can be any of a number of values, usually in the range of 5 to 15. Each different number of repeats corresponds to a different "allele" and is transmitted faithfully according to Mendelian rules. Because of the large number of common alleles at these positions, the proportion of persons who are heterozygous for two different alleles is very high, making these microsatellites very useful genetic markers. The name "microsatellite" comes from use of the term satellite DNA to describe the tandem repeats of very much longer segments of DNA.

Detection of microsatellite variants begins with a technique called *polymerase chain reaction* (PCR). DNA polymerase is an enzyme normally used by cells to replicate DNA. In order to initiate replication of a piece of single-stranded DNA, there must first be a primer, that is, a short piece of complementary DNA or RNA that can be extended by the polymerase from the 3′ end. Any stretch of DNA can be amplified in the presence of two primers that define the segment to be copied (Fig. 10). A cycle of copying consists of separating the two DNA strands by heat in the presence of a great excess of primers, allowing the primers to form double-stranded structures with the source DNA, and extending the primers with polymerase. The cycles can be controlled by temperature. A typical cycle lasts only about 2 to 3 minutes, and each cycle doubles the number of DNA copies of the segment defined by the choice of primers. After some 30 cycles, there would be 2^{30} or about one billion copies for each original molecule of DNA. This enormous amplification makes it possible to start with as little as one sperm or one cell from an eight-cell embryo and generate enough copies of DNA from some gene of interest to do genetic diagnosis or, in the case of microsatellites, to count the number of repeats by measuring the size of the fragment by means of electrophoresis.

The PCR/microsatellite technique has been especially useful in criminal cases, where the amount of material left at a crime scene may be very limited: small droplets of blood or semen, a few hairs with roots, skin flakings, etc. As with RFLP markers, no two persons are alike if a sufficiently large series of markers is tested, except for identical twins.

BIBLIOGRAPHY

Gelehrter, T. D., and Collins, F. W. (1990). "Principles of Medical Genetics." Williams & Wilkins, Baltimore.

Jorde, L. B., Carey, J. C., and White, R. L. (1994). "Medical Genetics." Mosby–Year Book Inc., St. Louis.

Mange, A. P., and Mange, E. J. (1990). "Genetics: Human Aspects," 2nd Ed. Sinauer Associates, Sunderland, MA.

McKusick, V. A. (1992). "Mendelian Inheritance in Man," 10th Ed. Johns Hopkins Univ. Press, Baltimore. The "Online Mendelian Inheritance in Man" (OMIM) is maintained as a continually updated data base at The National Center for Biotechnology Information, National Library of Medicine. It can be reached through the World Wide Web at http://www3.ncbi.nlm.nih.gov/Omim/.

Sutton, H. E. (1988). "Introduction to Human Genetics," 4th Ed. Saunders College, Philadelphia.

Human Genome

PHILIP GREEN
Washington University School of Medicine

Revised by
GIORGIO BERNARDI
Institut Jacques Monod, Paris

GLOSSARY

Chromosome Thread-like structure in the cell nucleus consisting of a single DNA molecule (up to several hundred million nucleotides in length), together with associated proteins necessary for maintaining its structural integrity (human cells contain 23 pairs of chromosomes)

Chromosome map Schematic indication of the relative locations of particular DNA sites in a chromosome. In linkage maps, distances between sites are expressed in terms of the rate at which crossing-over between the sites occurs during meiosis; in physical maps, distance is expressed as the number of nucleotides between the sites

Deoxyribonucleic acid (DNA) Molecule that carries genetic information. It consists of two complementary helical strands, each of which is a polymerized sequence of nucleotide subunits

Gene Segment of the DNA molecule encoding the information necessary for the cell to construct a protein or RNA molecule. Genes are the "functional units" of the genome

Nucleotide Molecule consisting of a phosphate group, a sugar group (which together form part of the DNA strand backbone), and one of the bases adenine, cytosine, guanine, or thymine

THE GENOME CONTAINS THE HEREDITARY INFORmation necessary to specify the human organism. This information is encoded in the sequence of some 3 billion nucleotide subunits composing the DNA in the nuclei of human haploid cells. Over 99% of this sequence is identical in essentially all humans and distinguishes us as a species from other organisms; differences at various points in the other 1% ensure that no two individuals, except identical twins, have exactly the same genetic complement and account for the extensive inherited variation seen in the human population.

The genome includes on the order of 100,000 genes, segments of the DNA containing the necessary instructions to manufacture protein molecules in the appropriate cells. Since essentially all biological structures and processes involve proteins, and the structures of all proteins are specified by genes, knowledge of the genome provides a focus for organizing our understanding of human biology. Identifying all genes has accordingly become a major priority of current biological research.

I. GENETIC INFORMATION

A. DNA and Chromosomes

Every human develops from a single cell. One of the central preoccupations of twentieth-century biology has been to understand how the genetic material in this cell encodes the information necessary to specify a complete individual.

The structure of the molecule that carries this information, DNA, was found by James Watson and Francis Crick in 1953 (Fig. 1). It is a double helix in which each of the two helical strands consists of a sequence

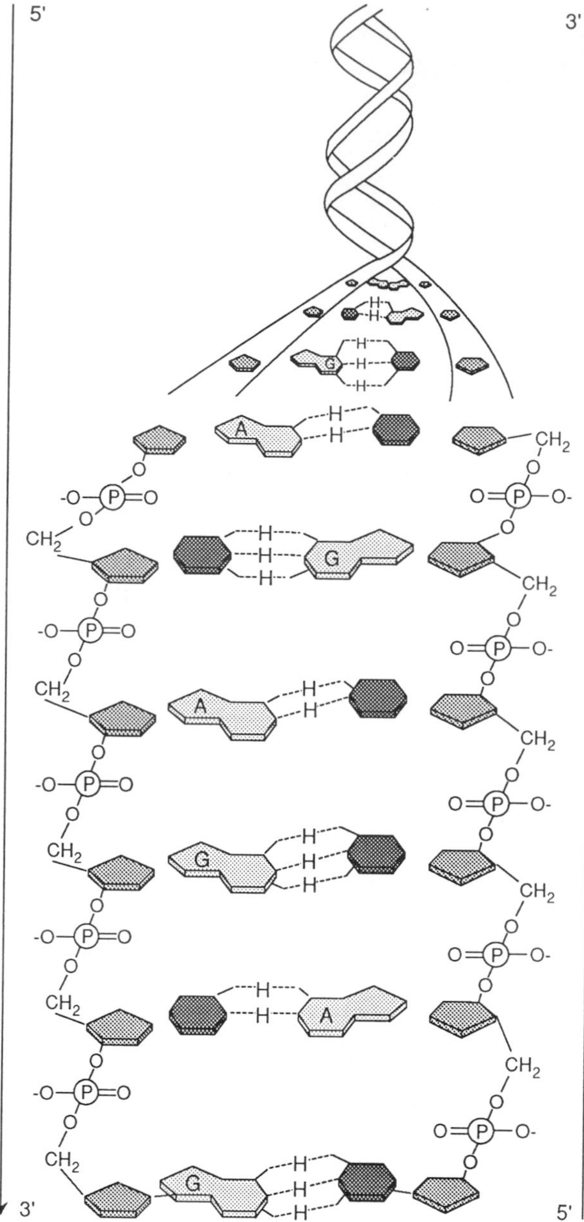

FIGURE 1 Structure of DNA. Note that the two sugar–phosphate chains run in opposite directions. This orientation permits the complementary bases to pair. [From P. W. Davis and E. P. Solomon (1986). "The World of Biology," 3rd Ed., p. 208. Saunders College, Philadelphia.]

of chemical subunits, called nucleotides. These are of four different types denoted A, C, G, and T after the names of their component bases adenine, cytosine, guanine, and thymine. Each base in one strand is paired with a unique complementary base in the other strand via hydrogen bonds: T is always paired with A, and C with G. The bases are covalently attached to a strand backbone formed from alternating sugar and phosphate groups (every nucleotide contributing one of each). Each strand has a direction: any nucleotide has a 5′, or upstream, neighboring nucleotide, as well as a 3′, or downstream, neighbor (the numerical terminology derives from the sites in the sugar group to which the adjacent phosphates are attached). The two strands run in opposite directions. Any particular DNA molecule thus could be described by giving the sequence of letters representing the bases (in 5′-to-3′ order) in one of the strands. It is no exaggeration to say that biology has been revolutionized by the insight that the genetic information is contained in the specific nucleotide sequence of the DNA molecule. [*See* DNA and Gene Transcription.]

The structure of DNA also immediately suggests the means for its replication: Since the base sequence of either strand determines the other strand, two faithful replicas of the molecule can be created by separating the two strands and synthesizing new strands complementary to each. This synthesis is carried out in the cell by the enzyme DNA polymerase, which appends the complementary nucleotides one by one in the 5′-to-3′ direction. DNA replication occurs prior to cell division, one replica then being segregated to each daughter cell. As a result, nearly all of the 10 trillion cells in the mature human organism have DNA complements identical to that in the single-celled zygote from which it started its complex development.

The genome is packaged in the form of chromosomes, each of which consists of a single long DNA molecule together with proteins that maintain its structural integrity. There are 24 different human chromosomes (Fig. 2): the sex chromosomes X and Y and 22 autosomes, composing a total of approximately 3 billion base pairs (bp) of DNA. Most human cells are diploid, containing two copies of each autosome and either two X's (in females) or an X and a Y (in males). Genetic information is transmitted to the offspring via haploid cells (formed from germ-line diploid cells during meiosis) that contain one chromosome from each of the 23 pairs. These haploid cells, the sperm and the egg, each contribute one-half of the genetic material of the zygote formed by their union. [*See* Chromosomes.]

B. Genes

Most information in the genome is in the form of genes, segments of the DNA molecule that encode the instructions necessary for the cell to construct two

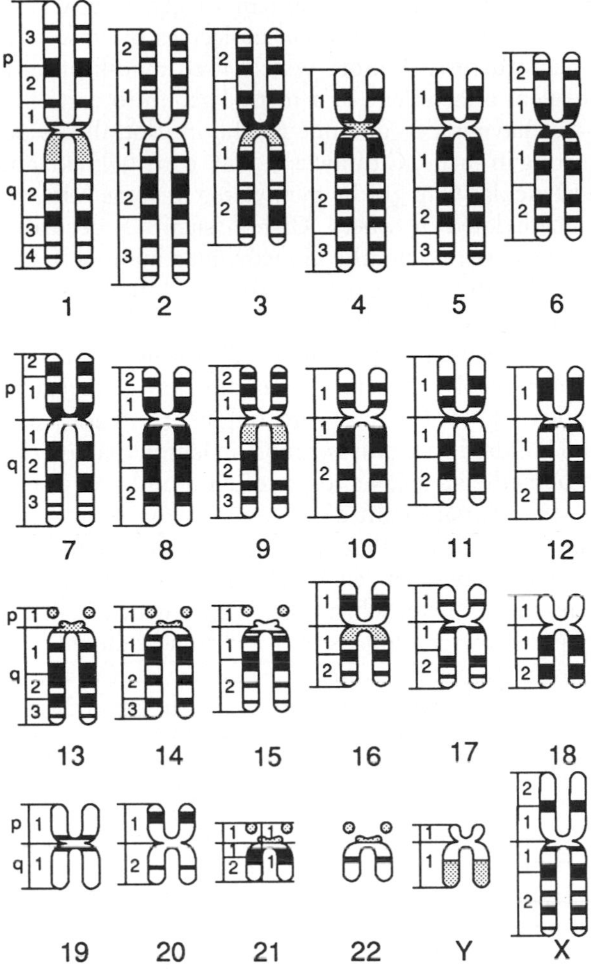

FIGURE 2 The human chromosomes, with designations for the Giemsa-stained metaphase bands (see Section IV,B,2). [Modified from J. J. Yunis (1976). High resolution of human chromosomes. *Science* **191**.]

kinds of macromolecules: proteins and RNA. Proteins, polymers whose chemical subunits are 20 different kinds of amino acids, play crucial functional roles in nearly all cellular processes and are essential components of most cellular structures. The vast majority of the estimated 100,000 human genes encode proteins. RNA is a polymer with four types of nucleotide units similar to those of DNA (but with slightly different backbone sugar groups and the base uracil instead of thymine). It is usually single stranded, but can base-pair with a DNA or RNA strand of complementary sequence in the same way that two DNA strands do. RNA molecules also play a number of important functional and structural roles in the cell, particularly in various stages of the process by which protein-encoding genes are expressed (i.e., the nucleotide sequence information is converted into the protein molecule). [*See* Genes.]

The structural features of protein-encoding genes are best understood in terms of the stages of the expression process. In the first stage, transcription, the enzyme RNA polymerase synthesizes (in the 5′-to-3′ direction) an RNA molecule, the transcript, complementary to a segment of one of the two DNA strands. The genetic information to control this process is contained in the promoter, a part of the gene that typically includes several short sequences recognized by nuclear proteins that interact with the polymerase. One promoter component is usually a short sequence rich in A's and T's, the TATA box, necessary to accurately position the polymerase to start transcription. Other promoter elements, which are usually located within a region of 100 or so nucleotides 5′ to the TATA box, determine the rate at which transcripts are initiated (and thus, indirectly, the number of protein molecules produced in the cell). Some of these elements mediate changes in transcription rate in response to certain signals, such as those conveyed by hormones or growth factors; others determine the types of cells in which transcription occurs.

The transcript starts at a point 25–30 bases downstream from the TATA box and usually extends for several thousand nucleotides farther downstream. A curious feature of most human genes is that the coding region, the part of the gene that encodes the protein structure, is interrupted by one or more noncoding segments called introns. [The coding segments (called exons) often correspond to structural domains in the protein molecule, which suggests that genes might have been assembled over the course of evolution from exon "modules." Known genes vary widely in their number of exons, from one to more than 50.] These intronic sequences are spliced out of the transcript to leave an RNA molecule with a single contiguous coding region, typically consisting of 1000 or so nucleotides. Introns often vastly exceed the coding region in length, totaling more than 1 million nucleotides in some genes. They contain short nucleotide sequences to permit their recognition by the splicing machinery, but are otherwise not believed, in general, to contain significant information of functional importance in gene expression.

The processed RNA molecule (now referred to as mRNA) is then transported out of the cell nucleus to a ribosome, where it is translated into the protein molecule: Starting with a trinucleotide ATG (which encodes the amino acid methionine) near the beginning of the mRNA molecule, each successive nucleotide triplet, or codon, in the mRNA molecule is "read," and the amino acid that corresponds to it (via the genetic code) is attached to the nascent peptide chain. Translation stops, and the completed protein is released from the ribosome, when a termination, or "stop," codon (i.e., TAA, TAG, or TGA) is reached.

In addition to the genes that encode proteins, there are others that are transcribed to yield functionally important RNA molecules not translated into proteins. These "RNA genes" are transcribed by different RNA polymerases from the one that transcribes protein genes, and many of them have promoter sequences contained wholly or partially within the transcribed sequence, rather than 5′ to it.

C. Chromosome Maintenance

The genome also contains sequences necessary for maintaining chromosome integrity and ensuring proper chromosomal replication and segregation to the daughter cells at cell division. The telomeres are specialized structures at the chromosome ends to ensure the proper replication of the end and protect it from cellular enzymes that would otherwise degrade it. Human telomeres include several hundred to several thousand copies of the six-nucleotide sequence TTAGGG, repeated in tandem, and might include other functionally important sequences. The centromere is the chromosomal structure to which the spindle is attached during cell division, to ensure segregation of the chromosomes to the daughter cells. The DNA sequences essential for centromeric function have not been identified, but may include the so-called α-satellite sequences, which consist of several hundred thousand tandem repeats of simple-sequence DNA found in the centromeric region. [*See* Telomeres, Human.]

Chromosomes also have (as yet unidentified) specific sequences to signal the sites at which DNA replication is to start and other sequences at which various protein components of the chromosomes bind the DNA.

D. Augmenting the Genome's Information: Methylation

There are several ways that the information in the genome's nucleotide sequence can be modified or augmented to affect gene expression. Rearrangements of the DNA in some genes (e.g., antibody genes) occur in cells of the immune system to generate a large number of different protein molecules of closely related structure. A more widely used modification is covalent attachment of a methyl group to certain C bases having a 3′ guanine neighbor. Not all C's occurring in such CG dinucleotides are methylated; and a particular C might be methylated in some cells, yet unmethylated in others. Certain such C's occurring near genes are generally unmethylated in those tissues in which the genes are expressed, and methylated in tissues in which the gene is not expressed, suggesting that the methylation is involved in gene expression. Also possibly involved in expression are regions known as "HTF (Hpa II tiny fragment) islands," several hundred base pairs or more in length that are rich in unmethylated C's and are often found near genes.

Methylation might also play a role in X inactivation. Early in female embryogenesis one of the two X chromosomes in each cell assumes a more condensed (i.e., heterochromatic) state, which is stably inherited by the replicas of this chromosome through subsequent cell divisions. The condensed X is known to be more heavily methylated than the other X, and most of its genes are not expressed. The purpose of X inactivation is probably to ensure that X chromosome genes, most of which are irrelevant to sex determination, have the same number of active copies (i.e., one) in female cells as in male cells.

Methylation also appears to be involved in imprinting, a phenomenon in which the expression of certain genes depends on the parent from whom the gene is inherited. In at least some cases, imprinting is associated with methylation patterns that are stably maintained in the development of somatic (non-germline) tissues, but undergo sex-specific alterations in the germ line. Imprinting might underlie the observation that the severity of certain genetic diseases depends on the parental origin of the disease gene.

E. Changes in the Genetic Information

The central role of the genome makes it essential that it be passed down virtually unchanged, both from cell to cell within the organism and from generation to generation. As a result, mechanisms have evolved to ensure that DNA replication is extraordinarily accurate and that damage to the DNA is detected and repaired efficiently. Nevertheless, the action of chemical agents, errors in the DNA replication process, or radiation do occasionally alter (i.e., mutate) the DNA sequence in a particular cell, such that the changed sequence is inherited by the descendants of that cell.

Mutations occurring in germ-line cells might be passed on to the offspring; in contrast, a somatic mutation affects only tissues derived from the mutated cell within the individual in whom the mutation occurs.

Mutations are of several types, including substitution of one nucleotide for another, deletion or insertion of one or a few nucleotides, and, more rarely, deletions, duplications, or rearrangements of more extensive regions (up to several million nucleotides or more in length). A relatively frequent type of nucleotide substitution occurs when a methylated C loses an amine group, which changes it to a T. Thus, CG dinucleotides tend to mutate to TG or (when the mutation occurs on the complementary strand) to CA. This type of mutation has occurred frequently enough during vertebrate evolution that CG dinucleotides are found much less commonly in the genome than would be expected from the frequencies of C's and G's.

Germ-line mutations are important as the raw material for evolutionary change. Any mutation initially occurs in a single individual; it becomes fixed in the species only if, by some subsequent generation, every member of the species is descended from that individual. This can occur if the mutation confers a selective advantage on the individual carrying it; however, statistical considerations show that it also, by chance, occurs occasionally with neutral mutations that have no significant effect on the organism. It is, in fact, possible that a large number of mutations that have been incorporated into the genome during human evolution are of the latter type.

A mutation occurring in a gene might or might not have a significant effect on the structure or function of the gene product, or its expression; when it does, it is likely to be deleterious (because most genes have already undergone millions of years of evolutionary fine-tuning, and most improvements are likely to have been implemented already). In humans these deleterious mutations (if they are not lethal) can cause genetic diseases, over 3000 of which are known. Although most of these are relatively rare, significant genetic components are increasingly being identified for common diseases such as diabetes, some psychiatric disorders, heart disease, and many cancers. Genetic diseases are an important resource for the study of human biology, because they provide the only examples of the effects on the whole organism of altering the structure or expression of a gene. [*See* Genetic Diseases.]

Mutations provide genetic diversity. Their efficient utilization for species evolution depends on the ability to mix the genetic material from different individuals, so that ultimately several advantageous mutations might be combined within a single individual. In part this is ensured by sexual reproduction and independent segregation of chromosomes at meiosis, but an important additional mechanism, crossing-over, is necessary to allow mutations occuring at different positions (i.e., loci) on different copies of the same chromosome to be combined on the same chromosome. Crossing-over occurs during meiosis, when the two homologous chromosomes in each pair are aligned next to each other. Roughly speaking, both chromosomes break at the same position, and each piece is rejoined to the other chromosome so as to create two new copies of the chromosome, which are mosaics of the previous copies. The breaks can occur anywhere along the chromosome, but more frequently in some regions (particularly near the ends of the chromosomes) than others. In male meiosis the X is paired with Y, but crossovers between these chromosomes occur only in the pseudoautosomal region, consisting of about 2 million bp near one telomere of both the X and the Y.

F. "Junk" DNA

Genes and chromosome maintenance information are thought to account for no more than 5–10% of the DNA content of the genome. The function(s), if any, of the remaining 90% (in which we include the introns as well as the intergenic regions) is unknown. Although there are undoubtedly additional types of functionally important sequences remaining to be discovered, the information content in the intergenic regions is not likely to be as dense as in the genes: Comparisons of human intergenic sequences with the corresponding sequences in evolutionarily closely related mammals indicate that mutations are accumulating there at a substantially higher rate than in coding regions, almost certainly an indication that they contain less information for selective forces to maintain. Much of this sequence presumably reflects the continual flux (over millions of generations) of random rearrangements, duplications, and other accumulated mutations, which could lead occasionally to new useful genes, but for a large part are neutral. Presence of a sequence in the genome does not entail functional significance.

There are some recognizable features in this noncoding DNA. Perhaps 50% of it consists of several types of repeated sequences, which are found in numerous copies in the genome. The most prevalent of these is the Alu element, a sequence of some 300 bp found in over 500,000 copies dispersed throughout the genome and accounting for between 5 and 10% of all human DNA. The Alu element is closely related in sequence to the gene for a small RNA molecule (7SL RNA), a component of the cellular machinery to target proteins to particular compartments, and is probably derived from it evolutionarily, but is not believed to serve any similar function within the cell. The Alu element is transcribed and, like small RNA genes, has a promoter wholly included within the transcribed sequence. This could account for its evolutionary spread through the genome: one notion is that RNA molecules are occasionally "reverse-transcribed" into DNA by an enzyme, reverse transcriptase, which synthesizes a DNA strand complementary to the RNA sequence. Following synthesis of the complementary DNA strand, the DNA molecule could then be inserted at the site of a chromosome break. If this process occurs in germ-line cells, it might be incorporated into the genome of subsequent generations. (Reverse transcription and insertion into the genome have also occurred occasionally with the mRNA from protein-encoding genes, but since the promoters of these genes are not contained within the transcript, these "processed pseudogenes" lack promoters and generally are not expressed.)

Alu (and other repeated elements) could thus simply be examples of "selfish," or parasitic, DNA, sequences that have accidentally evolved the capacity to take advantage of the cell's machinery to spread themselves through the genome. On the other hand, these repeated elements could play an important role in facilitating evolutionary change, either by altering the expression of nearby genes or by mediating occasional genome rearrangements (by causing misalignment of the two homologous chromosomes during meiosis).

II. GENOME ORGANIZATION

Many genes are found in gene clusters, or families, whose members are located near each other in the same chromosomal region and have closely related nucleotide sequences. Often, the genes in a family encode proteins with similar, but not identical, functions or differ in their patterns of expression (e.g., the tissues or developmental stages at which they are transcribed). Each cluster has likely arisen during evolution from a single progenitor gene by several rounds of duplication (perhaps caused by misalignment of the two chromosomal regions during meiosis, accompanied by crossing-over within the misaligned region); the duplicate copies then form the raw material for further mutational changes, which can lead to altered expression or function. Many clusters include nonexpressed pseudogene members, the result of duplication events that failed to evolve advantageously and instead acquired inactivating mutations.

In some cases, proximity of the genes in a family could play a functionally important role in their expression. An example is the genes encoding the RNA molecules used as structural components of ribosomes: These ribosomal RNA genes occur in large numbers of adjacent, virtually identical tandemly repeated units near the ends of the acrocentric chromosomes (i.e., those in which the centromere is near one end of the chromosome) 13, 14, 15, 21, and 22 and are coordinately expressed in association with the nucleolus during the construction of ribosomes.

A fact of great interest is that the genome is a mosaic of subregions that differ widely in C/G content, the percentage of their nucleotides that are C's or G's. Each subregion, or isochore, is at least several hundred thousand base pairs in length and is relatively homogeneous in C/G content. The C/G content can be one of a small number of different values ranging from about 40% to about 55%. Genes tend to be located in the most C/G-rich isochores (although some are found in the other isochores). [*See* Human Genome and Its Evolutionary Origin.]

III. GENOME VARIATION: POLYMORPHISM

Genes or other genomic DNA segments that occur in two or more reasonably common variant forms (i.e., alleles) in the human population are said to be polymorphic. The variants represent old mutations that have been inherited by an appreciable fraction of the current population. Most of these mutations are probably neutral, although some gene polymorphisms reflect competing selective advantages of the different alleles. Studies of several gene regions from different individuals suggest that roughly 1 in 100 to 1 in 1000 nucleotide sites is polymorphic, with any two copies of the same region differing at about 1 in every 1000 nucleotides (although it is unclear whether these figures are characteristic of the genome as a whole); by comparison, the genomic sequences of a human and a chimpanzee (the most closely related primate species) in such regions differ at 1–2% of the sites.

Polymorphisms in gene coding regions account for much of the inherited observed variation in the human population, underlying such diverse traits as blood groups, metabolic differences, and (perhaps) aspects of personality. These polymorphisms are nearly always nucleotide substitutions, since most deletions or insertions disrupt the reading frame, leading to an entirely different (and usually nonfunctional) protein product.

An important use of polymorphisms is in distinguishing different copies of the same chromosome. For example, when an individual is heterozygous (i.e., has two different alleles) at a polymorphic site, it becomes possible to track the inheritance of that chromosomal region in the children of that individual. This provides a powerful tool for localizing disease genes through linkage analysis (see Section IV,B,1). Polymorphisms are also useful for studying the origins of human population groups, which often differ in the frequencies of particular alleles.

A disproportionate number of nucleotide substitution polymorphisms occur at CG dinucleotides, reflecting the propensity of methylated C to mutate to T. Another important class of polymorphic loci is the minisatellite, or variable number of tandem repeat, loci. These sites (whose functional significance is unknown) consist of up to several hundred copies of a short (i.e., typically less than 20 base) sequence, repeated in tandem, the alleles differing in the numbers of copies of the repeated element. One class of minisatellite loci consists of tandem repeats of the dinucleotide GT. These are often found near genes and could play some role in their expression.

Some minisatellite loci are highly polymorphic, having multiple alleles that are each relatively rare. These sites provide a powerful tool for distinguishing the DNA of different individuals, and accordingly have important medical and forensic applications. Although minisatellite loci are found throughout the genome, the highly polymorphic ones often tend to be located near the ends of chromosomes, presumably reflecting a higher mutation rate in the telomeric region. [*See* Polymorphism, Genes.]

IV. METHODS FOR STUDYING THE GENOME

Most current knowledge about the genome has been obtained using a variety of powerful tools developed over the last 20 years to detect and manipulate DNA, RNA, and protein molecules. These techniques have made it possible to explore the organization of the genome, to find a number of genes within it, to determine the nucleotide sequences of these genes, and in many cases to analyze their various functional components by testing the effects of changes in promoter elements and coding sequences.

A. Tools for Manipulating DNA *in Vitro*

Methods have been developed to detect, amplify, synthesize, separate, alter, and determine the nucleotide sequence of DNA molecules in the laboratory. Many of these rely on enzymes purified from bacteria, the byproducts of extensive biochemical studies of the mechanisms by which bacterial cells process DNA. Among the more important techniques are:

1. Chemical synthesis of oligonucleotides, short single-stranded DNA molecules (usually less than 100 nucleotides in length) of any desired sequence.

2. The polymerase chain reaction (PCR), a method for amplifying a particular sequence from genomic DNA. Two oligonucleotides (primers) are synthesized corresponding to sequences flanking the region to be amplified, one of them upstream of this region on one strand, the other upstream of it on the other strand. These are then added to a mixture containing the genomic DNA, a thermostable bacterial DNA polymerase, and nucleotides. The mixture is heated to separate the DNA molecules into single strands (by breaking the hydrogen bonds between complementary bases). Upon cooling, the oligonucleotides anneal to each strand; the polymerase then synthesizes complementary strands starting at these oligonucleotides and continuing downstream through the desired region. Each repeated cycle of heating, cooling, and DNA synthesis doubles the number of copies of the sequence between the oligonucleotides, so that in a short time it is possible to amplify this sequence 1 million-fold or more. Since the amount of starting material can be extremely small—the DNA from a single human cell has been used—this provides an extremely powerful method both for detecting whether a particular sequence is present and for making sufficient amounts of it for further analysis or manipulation. Its main limitations are that the nucleotide sequence of the region to be amplified must be at least partially known, and at present it is difficult to amplify regions more than a few thousand base pairs in length.

3. Cleavage of DNA molecules with restriction enzymes, which recognize specific nucleotide sequences

(usually 4–6 bases in length) within a DNA molecule and cut it there. For example, the enzyme *Eco*RI recognizes the nucleotide sequence GAATTC and cleaves between the G and the first A. Since this sequence also occurs on the opposite strand, *Eco*RI "digestion" of DNA produces a set of double-stranded fragments with the single-stranded four-nucleotide overhang AATT at both ends. The overhang on one fragment is complementary to the overhang on any other fragment and can anneal to it; the enzyme DNA ligase can then be used to covalently link two such restriction fragments together or circularize a single linear fragment. Restriction digestion in conjunction with ligation provides a powerful general method for making recombinant DNA molecules composed of pieces from several different sources. Several hundred restriction enzymes of varying sequence specificities have been purified from various bacterial species.

4. Size fractionation of mixtures of DNA molecules (e.g., restriction digestion fragments) by gel electrophoresis. The DNA molecules (which are negatively charged at normal pH) are impelled through a gel by an electrical field; separation according to size results from the fact that large molecules are retarded more by the gel matrix than are small ones. Agarose gels can resolve molecules differing by a few percent in length and ranging in size from several hundred to 50,000 bp. Acrylamide gels can resolve single- or double-stranded DNA molecules of up to several hundred bases in length that differ in length by as little as a single nucleotide.

Large DNA fragments are poorly resolved by ordinary electrophoresis, because they orient themselves for efficient passage through the gel such that retardation by the matrix is no longer proportional to length. Pulsed-field gel electrophoresis circumvents this phenomenon by periodically changing the field direction to reorient the molecules and thereby restore proportional retardation of large fragments. This permits resolution of molecules up to several million base pairs in length. [*See* Pulsed-Field Gel Electrophoresis.]

5. Cloning, a powerful method both for fractionating a complex DNA source into individual pieces and for making as many copies as desired of any particular piece. Cloning is usually a prerequisite to nucleotide sequence determination and to functional studies of a particular DNA segment. The DNA molecules to be cloned are attached (typically by restriction digestion and ligation) to a vector DNA molecule, and the ligated molecules are introduced into microorganism host cells. The vector includes sequences that permit replication of the ligated molecule within the host, and usually also a marker gene to permit selective growth or identification of only those cells that have successfully incorporated the DNA. *Escherichia coli* is commonly used as a bacterial host and, as vectors, either plasmids, which replicate as circular DNA molecules and include antibiotic resistance genes as markers, or bacteriophage, viruses that infect bacteria, are used. DNA fragments of up to about 45,000 bp can be cloned in this way. Larger DNA fragments (of several hundred thousand base pairs in length) can be cloned as artificial chromosomes in yeast. Traditional cloning of small DNA fragments is being supplanted for many purposes by PCR, which can be regarded as *in vitro* cloning.

Clone libraries are collections of clones of fragments from a particular organism's DNA. Libraries fractionate a complex DNA source into manageable smaller pieces, which can be screened (e.g., by hybridization, described next) with a probe for a particular DNA sequence to obtain a clone containing that sequence. Bacteriophage, plasmid, and yeast artificial chromosome (YAC) libraries of genomic DNA are all used in this manner. YAC libraries in particular are proving to be an important tool for analysis of the human genome owing to the large size of the cloned fragments. [*See* Chromosome-Specific Human Gene Libraries.]

The importance of gene coding regions and their low density within the genome make it useful to efficiently identify these sequences. Libraries of cloned coding regions can be constructed by using the enzyme reverse transcriptase to synthesize single-stranded DNA complementary in sequence to the mRNA molecules purified from a particular tissue type; these cDNA molecules are then converted to double-stranded DNA molecules using DNA polymerase, ligated to plasmid or bacteriophage vector sequences, and introduced into the bacterial host. Expression cDNA libraries use specialized vector sequences that allow the bacterial host to transcribe and translate the cloned coding region. These libraries can then be screened using methods (e.g., labeled antibodies) that directly detect the protein, rather than the DNA.

6. Hybridization, a powerful method for detecting a sequence of interest within a complex mixture. It is based on the tendency for two single-stranded DNA molecules of complementary sequence to seek each other out and anneal, or hybridize. A cloned, or synthetic, DNA segment (the probe) is radioactively or fluorescently labeled, and its strands are separated and allowed to anneal with target DNA whose strands

have also been separated. Usually, the target DNA has been fractionated (either electrophoretically or in the form of a clone library) and transferred to a solid support, typically a nylon membrane. (Transfers from electrophoretic gels are known as Southern blots.) The location on the membrane at which the probe anneals indicates the fragment size or clone containing the probe sequence.

7. Determining the nucleotide sequence of a small cloned or PCR-amplified DNA segment. This depends on the ability to generate, for each of the four types of nucleotides, a family of single-stranded DNA fragments that are subsequences of the cloned segment, and have the same 5′ end but terminate at variable 3′ positions corresponding to the various occurrences of that particular nucleotide in the segment. These fragments can be created either (1) by partial digestion with chemicals that cleave DNA strands at occurrences of the nucleotide (Maxam–Gilbert sequencing) or (2) by DNA polymerase-mediated synthesis of fragments complementary to one of the original segment strands, using a reaction mix that contains, in addition to the usual nucleotides, a modified form of one nucleotide that, when incorporated by the polymerase, terminates the nascent strand at that nucleotide (Sanger sequencing). The relative lengths of these fragments (to single-base resolution) are then determined by acrylamide gel electrophoresis; correlating the terminating nucleotide of each fragment with its length yields the nucleotide sequence. In a single gel run it is possible to determine the sequence of a DNA molecule of several hundred nucleotides.

B. Mapping

Mapping (i.e., determining the locations of genes or other DNA segments within the genome) provides useful information for cloning genes, for inferring evolutionary relationships between genes and organisms, and for organizing our knowledge of the genome. There are several mapping techniques in current use. [*See* Genetic Maps.]

1. Linkage maps (historically the first type of genomic map to be constructed, using experimental organisms) delineate the order of polymorphic sites along a chromosome. The map distance between two sites is defined as the average number of crossovers occurring between them per meiosis, the unit of measurement usually being the centimorgan (cM, corresponding to an average of one crossover per 100 meioses; on average 1 cM in the human genome corresponds to a physical distance of about 1 million bp, but there is substantial local variation in this value). Because rates of crossing-over differ in male and female meioses, map distances must be specified separately for each sex.

Linkage maps are constructed from studies of the inheritance of alleles at polymorphic loci in a panel of families. When an individual is heterozygous at each of two different loci, it is often possible to determine whether or not particular alleles at the two loci are coinherited by his or her children. For loci near each other on a chromosome, an allele at one locus is coinherited with that allele at the other locus that happens to lie on the same copy of the chromosome in the parent, except when a crossover occurs between them. Loci on different chromosomes show no allele coinheritance. Thus, inheritance studies allow one to identify when polymorphic loci are on the same chromosome and to determine the relative distances and order for loci on the same chromosome. In comparison to studies of experimental organisms, the analysis of human linkage data is complicated by the facts that the allele phase (i.e., the particular copy of the parental chromosome on which a given allele lies) is often not known, any particular individual is likely to be heterozygous for only a subset of the loci being studied, and it is impractical to study large numbers of families. This has necessitated the development of specialized mathematical and statistical techniques (i.e., multilocus linkage analysis) to extract the maximum amount of information from the data.

Convenient for mapping purposes are restriction fragment-length polymorphisms (RFLPs), DNA sequence variations, which affect the length of particular genomic DNA restriction fragments, either by nucleotide substitution within a restriction enzyme recognition site or by insertion or deletion between sites (e.g., minisatellites). RFLPs are scored by hybridizing a probe that detects the variable fragment to Southern blots of genomic DNA digested with the appropriate restriction enzyme. RFLP linkage maps with an average spacing of 5–10 cM between sites have been constructed for each human chromosome. These maps have revealed several interesting facts: crossing-over occurs about twice as frequently in female meiosis as in male meiosis, with the exception of certain regions (found mostly near the ends of chromosomes) where males have a higher rate. Polymorphisms are unevenly distributed in the genome, the most highly polymorphic loci more often found near the telomeres than in other parts of the chromosomes, and on the autosomes than on the X.

2. Cytogenetic maps are based on the discovery that preparations of metaphase human chromosomes stained with certain dyes (e.g., Giemsa) reveal a characteristic pattern of alternating light and dark bands for each chromosome, a total of about 300 bands in the genome as a whole (Fig. 2). The physical basis for this differential staining is unclear, although Giemsa light (R) bands are known to be relatively more C/G-rich than the Giemsa dark (G) bands. Medical genetics has benefited enormously from the use of cytological staining to identify and classify mutations due to gross chromosome rearrangements, such as translocations (in which a part of one chromosome is joined to another), deletions, and aberrations in chromosome numbers (such as trisomy 21, the basis of Down's syndrome).

This ability to identify individual chromosomes and portions of chromosomes has led to several useful mapping techniques. Somatic cell hybrids, fusions of a human cell with a rodent cell, preferentially lose human chromosomes during replication, so that after several cell divisions only one or a few human chromosomes are retained with the rodent chromosomes. The chromosome containing a particular human DNA segment can often be identified by hybridizing a probe for that segment to DNA prepared from a panel of such hybrids retaining different human chromosome complements. In an important extension of this technique, radiation hybrid mapping, the human chromosomes are fragmented by radiation prior to the cell fusion; hybrid panels constructed in this manner allow the mapping of DNA segments relative to each other within a chromosome.

In situ hybridization of a radioactively labeled human DNA probe to metaphase chromosomes can be used to localize the probe sequence to a specific chromosomal band (on average, a region of about 10 million bases). Recent fluorescent labeling techniques permit the simultaneous visualization of several nearby probes, which allows higher-resolution ordering of probes within a band.

3. Restriction maps indicate the order and distances [usually measured in kilobases (kb), or thousands of nucleotides] between restriction enzyme sites in a particular genomic region. Restriction enzymes with recognition sequences of 4–6 bp generally produce genomic fragments averaging a few hundred to a few thousand base pairs in length and are used to construct high-resolution restriction maps of regions up to 30 kb or so in length. These maps are usually acquired primarily from analysis of restriction digests of small overlapping clones spanning the region in question.

In contrast, long-range restriction maps spanning several million base pairs rely on the use of a small number of restriction enzymes (having recognition sequences of eight or more nucleotides in length and/or containing the rare dinucleotide CG) that produce much larger fragments, with a size of several hundred kilobases or larger. These maps are usually constructed by hybridizing probes from the region to Southern blots of pulsed-field gels to identify the sizes of fragments containing the probe sequence.

4. Overlapping clone maps, sets of DNA clones that cover particular genomic regions, are particularly useful because they provide a replica of the genome in pieces of manageable size. They are usually constructed in either of two ways, both involving the use of clone libraries: fingerprinting, which involves generating characteristic sequence or restriction site data (e.g., the fragment sizes for a particular restriction enzyme) for each clone in the library and identifying overlaps via shared portions of the fingerprint patterns; or library screening, using DNA probes from the region, with overlaps now implied by the presence of shared probe sequences. The latter method includes as a special case chromosome walking. In that method, as clones are added to the map, new probes are generated from the ends of the cloned segment and used to rescreen the library for the next neighboring clones.

Overlapping YAC clone maps are presently being constructed using both methods for a number of genomic regions.

C. How Genes Are Found

Once a protein has been isolated, cloning the corresponding gene is relatively straightforward. If part of its amino acid sequence can be determined, oligonucleotides corresponding to part of the gene sequence can be synthesized (using the genetic code) and used to probe a genomic or cDNA library. Alternatively, if the protein can be detected *in vitro* (e.g., via an antibody that binds the protein), it can often be found in a cDNA expression library without having to determine the amino acid sequence first.

When the protein is not directly detectable—for example, if it is known only through the existence of a genetic disease—finding its gene is much more difficult. The approximate location for disease genes can often be identified using the techniques of linkage mapping to find a polymorphic DNA locus that tends to be coinherited with the disease gene in families with the disease. Having identified the chromosomal location of the polymorphic locus, it is then necessary to search through cloned DNA from this region to find the gene. Most successful searches to date have depended on the existence of more specific clues to location, such as chromosomal rearrangements (i.e., translocations or deletions) disrupting the gene in some individuals with the disease. In the absence of these, it becomes necessary to identify candidate genes within the region of interest (by screening cDNA libraries for transcribed sequences or by identifying sequences that are evolutionarily conserved among several organisms), sequence them, and look for differences between the sequences of affected and normal individuals. This is rendered more difficult by the necessity to distinguish polymorphisms from the disease mutation, but has been successful in a few instances. The availability of overlapping clone maps covering most of the genome should greatly aid such searches.

V. HUMAN GENOME PROJECT

In recent years it has become apparent that a systematic approach to finding all human genes would have many benefits. Present cloning approaches, despite impressive successes, have yielded only a few percent of the estimated 100,000 genes. Availability of the entire genome sequence would greatly accelerate progress: once the sequence of a small part of any particular gene were determined, the rest would be immediately accessible. Furthermore, it is likely that there are many genes or other functionally important regions of the genome whose presence is totally unsuspected, and that therefore might not be accessible to current cloning efforts.

For these reasons, determination of the complete structure of the genome has become the subject of a major initiative that is both international and interdisciplinary in scope, involving input from mathematicians, computer scientists, engineers, chemists, and physicists as well as biologists. As currently envisaged, this genome project will have several goals:

1. A high-resolution (i.e., average spacing of 2–5 cM) linkage map of the genome, to facilitate the mapping of disease genes and provide landmark sites of known chromosomal order for use in constructing physical maps. This will be facilitated by the use of a common panel of families for inheritance studies, which permits researchers from different laboratories to combine the data for different polymorphic sites into a single map. The DNA from such a panel of 60 large three-generation families is currently being distributed by the Centre d'Étude du Polymorphisme Humain.

2. Physical maps of several types, including radiation hybrid maps, long-range restriction maps, *in situ* hybridizaton maps, and overlapping clone maps. These disparate maps will be merged into a single common map giving the positions of sequence tagged sites (STSs), short regions consisting of a few hundred bases of known sequence, for which PCR assays have been developed. A set of STSs of known order, spaced at approximately 100-kb intervals along each chromosome, will make any genomic region immediately accessible (by providing PCR probes for screening genomic libraries). The STS concept provides a common language for physical mapping, much as the use of a shared set of reference families provides a common language for linkage mapping.

3. Determination of the 3 billion-nucleotide genomic sequence. This will most likely depend heavily on sets of overlapping clones constructed in Goal 2 and will require substantial prior technological research to increase the speed and decrease the cost of current sequencing methods. Since polymorphic variation implies that there is no unique human sequence, it would also be desirable to have at least a partial catalog of the polymorphic sites.

4. It is important to recognize that the sequence is not an end in itself, but is only useful insofar as we can identify the genes and find out what the corresponding proteins do in the cell. Gene identification directly from sequence data is difficult because of the interrupted structures of most genes and the fact that a relatively small proportion of the genome is in coding regions. Sequencing the genomes of several "model" organisms (including the yeast *Saccharomyces cerevisiae,* the nematode *Caenorhabditis elegans,* the mouse *Mus musculus,* and the fly *Drosophila melanogaster*) will provide a useful resource for identifying the functionally important regions (many of which will be evolutionarily conserved and thus have similar sequences in several species) and will allow experimental manipulations to test hypotheses about gene function. These and other approaches to determining gene function can be expected to occupy human biologists for many years after the entire human genome sequence is known.

BIBLIOGRAPHY

Bernardi, G. (1995). The human genome: Organization and evolutionary history. *Annu. Rev. Genet.* **29**, 445–476.

Committee on Mapping and Sequencing the Human Genome (B. Alberts, chairman) (1988). "Mapping and Sequencing the Human Genome." National Academy Press, Washington, D.C.

Donis-Keller, H., Green, P., Helms, C., Cartinhour, S., Weiffenbach, B., Stephens, K., Keith, T., Bowden, D., Smith, D., Lander, E., Botstein, D., Akots, G., Rediker, K., Gravius, T., Brown, V., Rising, M., Parker, C., Powers, J., Watt, D., Kauffman, E., Bricker, A., Phipps, P., Muller-Kahle, H., Fulton, T., Ng, S., Schumm, J., Braman, J., Knowlton, R., Barker, D., Crooks, S., Lincoln, S., Daly, M., and Abrahamson, J. (1987). A genetic linkage map of the human genome. *Cell* **51**, 319–337.

Lewin, B. (1994). "Genes V." Oxford Univ. Press/Cell Press, Oxford, England.

Sambrook, J., Fritsch, E. F., and Maniatis, T. (1989). "Molecular Cloning: A Laboratory Manual," 2nd Ed. Cold Spring Harbor Laboratory, Cold Spring Harbor, NY.

Vogel, F., and Motulsky, A. G. (1986). "Human Genetics: Problems and Approaches," 2nd Ed. Springer-Verlag, Berlin/New York.

Human Genome and Its Evolutionary Origin

GIORGIO BERNARDI
Institut Jacques Monod, Paris

GLOSSARY

Recombination A process that gives rise to cells or individuals (recombinants) associating in new ways two or more hereditary determinants (genes) by which their parents differed

Sister chromatids Two chromatids derived from one and the same chromosome during its replication in interphase

Translocation Chromosomal structural change characterized by the change in position of chromosome segments (and the genes they contain)

THE BASIC QUESTION THIS ARTICLE ADDRESSES IS the organization of nucleotide sequences in the human genome and the evolutionary origin of such organization. It shows that, far from being a "gene bag," the genome is highly ordered from the chromosomes down to the nucleotide level. Nucleotide sequences, whether in genes or in the very abundant nongenic segments, obey precise compositional rules. Compositional rules are evident also in chromosome structure in that DNA composition is at least partly responsible for chromosome bands. These structural properties are associated with functional properties and shed new light on genome evolution.

I. INTRODUCTION

A. The Genome

Every living organism contains, in its genome, all the genetic information that is required to produce its proteins (and some ribonucleic acids, RNAs, which are not translated into proteins; see the following) and that is transmitted to its progeny. The genome consists of deoxyribonucleic acid (DNA), which is made up of two complementary strands wound around each other to form a double helix. The building blocks of each DNA strand are deoxyribonucleotides. These are formed by a phosphate ester of deoxyribose (a sugar), linked to one of four bases: two purines [adenine (A) and guanine (G)] and two pyrimidines [thymine (T) and cytosine (C)]. In the DNA double helix, purines pair with pyrimidines (A with T, G with C)and the phosphates bridge the paired building blocks of the two strands to form the double helix. [*See* DNA and Gene Transcription; Human Genome.]

During cell replication, the two strands of the double helix are unwound, and a complementary copy of each is made, producing two identical copies (except for rare mistakes, or mutations) of the parental double helix. The two strands are also unwound at the time when one strand, the sense-strand carrying the genetic information, is copied into a complementary RNA, which differs from the DNA master copy in having ribose instead of deoxyribose and uracil instead of thymine. RNA transcripts of genes are used as templates for the synthesis of proteins.The translation of each RNA transcript into the corresponding protein involves a very complex machinery that makes use of ribosomes (particles made up of two subunits, each containing a ribosomal RNA) and transfer RNAs specific for different amino acids. It will suffice here to mention that subsequent sets of three adjacent nucleotides (or triplets, also referred to as codons) of the transcript specify amino acids that follow each other in the protein chain. Because there are 64 triplets (three of which are termination codons, marking the end of translation) and only 20 amino acids, all amino acids except for methionine and tryptophan are encoded by more than one codon. In other words, several synonymous codons may be used to specify the same amino acid; therefore, the genetic code is said to be degenerate, which means that alternative coding possibilities (synonymous codons) exist for the same amino acid. Differences among synonymous codons are mainly at the third codon positions.

The genome of living organisms differs greatly in size, from 4.2 Mb [megabases, or millions of base pairs (bp)] for a typical bacterium such as *Escherichia coli* to 3000 Mb or 3 Gb (gigabases, or billions of bp) for a eukaryote such as humans. Whereas prokaryotes (bacteria) are characterized by small genome sizes, clustering around the value given for *E. coli*, eukaryotes exhibit larger genome sizes that cover a wide range—from 13 Mb for the yeast *Saccharomyces cerevisiae* to 3 Gb for mammals.

The much larger genome size of higher eukaryotes is not only due to the presence of a larger number of genes (see the following). In fact, the increase in size is mainly due to noncoding sequences that can be both intergenic and intragenic. The latter sequences, called introns, separate different coding stretches, or exons, of most eukaryotic genes. The intron part of the primary transcript is eliminated by splicing, leaving the mature transcript or messenger RNA that encodes a given protein.

Eukaryotes differ from prokaryotes in other respects as well. They have a nucleus that is separated from the cytoplasm by a nuclear membrane. In addition to the nuclear genome, eukaryotic cells also have organelle genomes, which are located in mitochondria and, in the case of plants, also in chloroplasts. Organelle genomes contain a very limited yet essential amount of genetic information (genes) encoding organelle-specific proteins and RNAs. Organelle genomes apparently originated from symbiotic bacteria, which entered proeukaryotic cells. Like the bacterial genomes, organelle genomes are physically organized in a rather simple way. In contrast, in the nuclear genome of eukaryotes, DNA is wrapped around histone octamers to form nucleosomes, which are packaged into chromatin fibers. These fibers are folded into chromatin loops, consisting of 30–100 Kb (kilobases, or thousands of bp) of DNA, which are, in turn, packaged into chromosomes.

B. The Human Genome

The nuclear genome of humans consists of about 3 billion base pairs, whereas the size of the mitochondrial genome is only 16,000 bp. Estimates of the number of (nuclear) human genes range from 30,000 to 100,000. If coding sequences average 1000 bp, they would represent 1–3% of the human genome, 97–99% of which is therefore made up of noncoding sequences. It should be noted that the larger number of genes in humans compared with bacteria is mainly due to the fact that many human genes exist as multigene families, the result of gene duplications during evolution.

Our present knowledge of human coding sequences, in terms of primary structures (or nucleotide sequences), is limited to about 5000 of them. In other words, we only know about 5–15% of our coding sequences, or 0.15% of our genome. Our knowledge of noncoding sequences in terms of primary structure is even more limited. Nevertheless, we know that a sizable part of intergenic noncoding sequences is formed by repeated sequences that belong to several families. Two important families are called LINES and SINES (the long and short interspersed sequences), which are present in about 100,000 and 900,000 copies, respectively. These families of repeated sequences have been studied largely by reassociation kinetics. This experimental approach is based on the fact that, if small DNA fragments (having a size of 300–400 bp) are separated into their complementary strands, the reassociation of the latter proceeds at a rate that depends on the frequency in the genome of the sequences present in the fragments. Single strands from sequences that are present very many times in the genomes will find their complementary strands faster than single strands from genes that are present a few times or even only once in the genome. This technique allows estimates of the relative amounts of repetitive sequences and single-copy sequences, the latter being present only once or a small number of times in the genome.

Another level of knowledge of the human genome concerns chromosomes. Each human germ cell (sperm or oocyte) contains 23 chromosomes. In haploid cells, 22 chromosomes (1 to 22 in order of decreasing size) are autosomes, which are identical in both sexes. The 23rd chromosome, the sex chromosome, is an X chromosome in females and a Y chromosome in males. Somatic cells are diploid; they have two haploid chromosome sets. Female diploid cells have two X chromosomes (one of which is inactive), whereas males have one X and one Y chromosome. During mitosis, chromosomes condense and, at metaphase, they are characterized by specific staining properties. G-bands (Giemsa-positive, or Giemsa dark bands; equivalent to Q-bands, or quinacrine bands) and R-bands (reverse bands; equivalent to Giemsa-negative or Giemsa light bands) are produced by treating metaphase chromosomes with fluorescent dyes, proteolytic digestion, or differential denaturing conditions. [*See* Chromosomes.]

Under standard conditions, Giemsa staining produces a total of about 400 bands that comprise, on the average, 7.5 Mb of DNA. If staining is applied to prophase chromosomes, which are more elongated, up to 2000 bands can be visualized. At this high resolution, one chromosomal band contains, on the average, 1.5 Mb of DNA. Prophase chromosomes can also be studied at meiosis, the process leading to haploid germ cells. Staining meiotic chromosomes (usually at pachytene) produces results that are similar to those just mentioned. Indeed, a pattern of chromatin condensations, the chromomeres, is visualized. Chromomeres and the interchromomeres separating them correspond to high-resolution Giemsa-positive and Giemsa-negative bands, respectively. A number of approaches, ranging from purely genetical to molecular ones, have allowed the assignment of genes not only to individual chromosomes but also to chromosome bands.

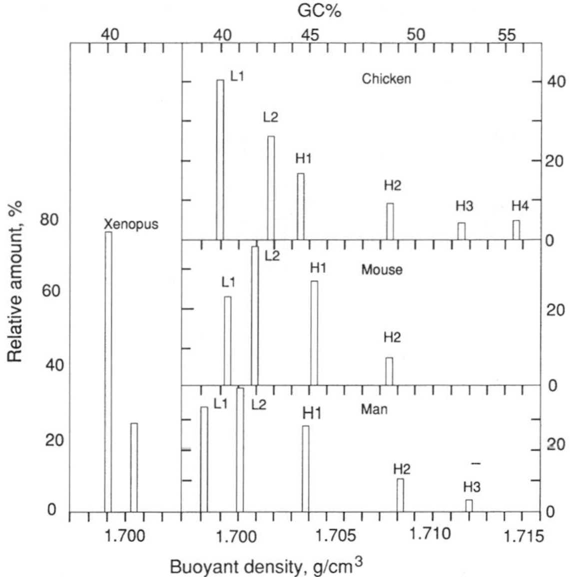

FIGURE 1 Histograms showing the relative amounts and modal buoyant densities in CsCl and GC levels of the major DNA components (L1, L2, H1, H2, H3, H4) from *Xenopus,* chicken, mouse, and human. Satellite (clustered repeated sequences) and minor components (like ribosomal DNA) are not shown. [Reproduced, with permission, from G. Bernardi (1989). The isochore organization of the human genome. *Annu. Rev. Genet.* **23**, 637–661.]

II. ISOCHORES AND GENOME ORGANIZATION

Our present knowledge of the human genome is clustered around two levels of organization: genes, which are in a DNA size range of a few Kb, and chromosome bands, which are in a size range of a few Mb. We know very little about the intermediate size range.

Recent discoveries, however, have not only linked the gene level with the chromosome level of genome organization but also shed light on the functional and evolutionary implications of genome organization. Indeed, the human genome is made up of isochores, large DNA domains (>300 Kb) that are compositionally homogeneous and belong to a small number of families ranging from 35 to 55% GC (Fig. 1; GC is the molar percentage of the two complementary bases G and C in DNA). This point was demonstrated in two ways. (1) The relative amounts of isochore families, as judged by fractionating human DNA fragments according to their base composition, is independent of molecular size of the fragments between 3 and >300 Kb. The fact that breaking down DNA fragments to small sizes does not change the ratios of the compositional families obviously indicates compositional homogeneity in the larger fragments. (2) Hybridization of single-copy sequences to compositional fractions of DNA fragments about 100 Kb in size occurs within a very narrow GC range, about 1%. This means that all DNA fragments that carry a given gene are extremely close in composition despite the fact that they were derived from intact chromosomal DNA by a random breakdown due to the unavoidable physical and enzymatic degradation that occurs during DNA preparation.

It is important to note that isochore families differ in number and relative amounts in different vertebrates (see Fig. 1), because compositional patterns or isochore patterns represent different genome phenotypes. Likewise, compositional patterns can also be

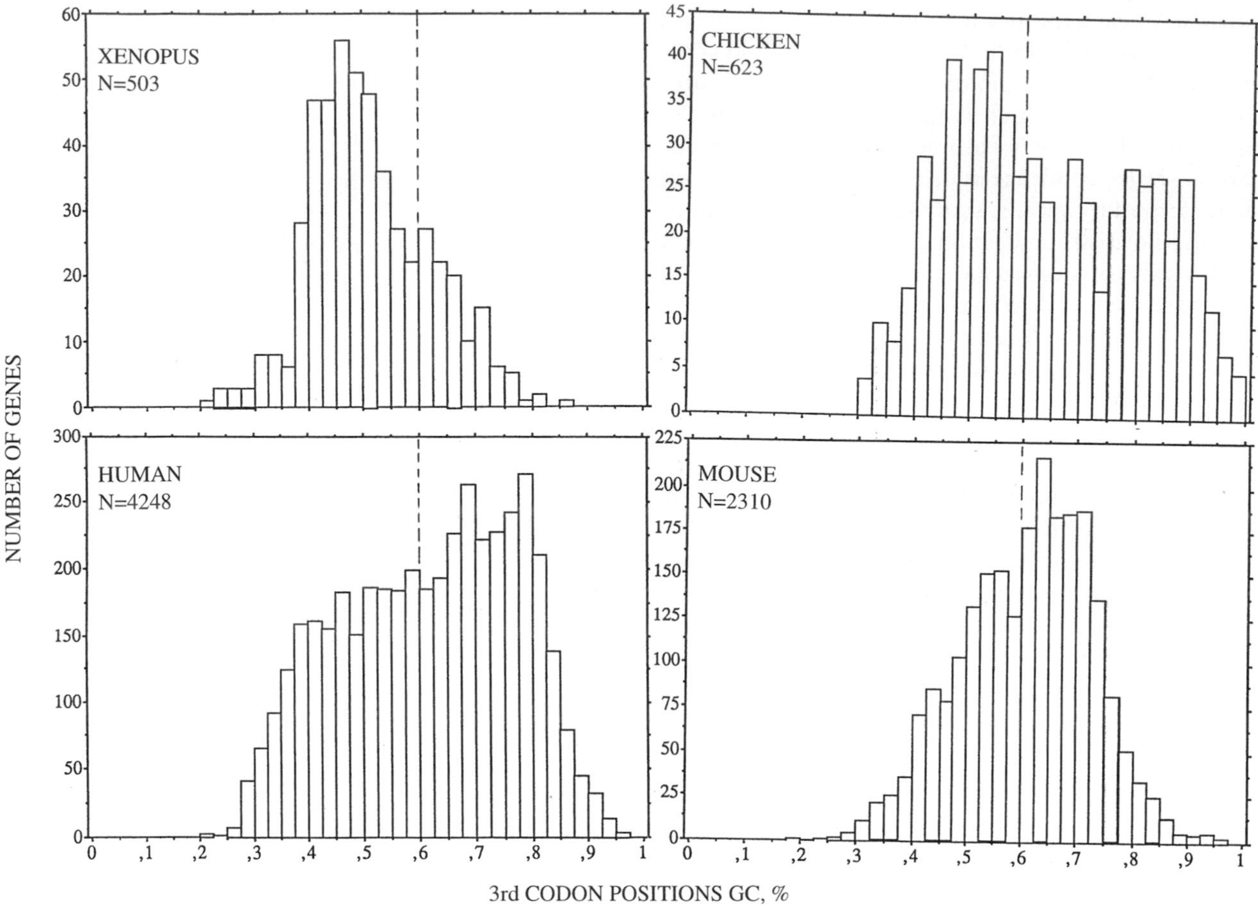

FIGURE 2 Compositional distribution of third codon position from vertebrate genes. (This distribution is the most informative because of its wider spread compared with coding sequences and first or second codon positions.) The number of genes under consideration is indicated. A 2.5% GC window was used. The vertical dashed line at 60% GC is shown to provide a reference. [Reproduced, with permission, from G. Bernardi (1993). The human genome organization and its evolutionary history: A review. *Gene* **135**, 57–66.]

seen by looking at the compositional distribution of coding sequences (or of their third codon positions; Fig. 2).

The discovery of an isochore organization in the human genome (and in the eukaryotic genomes in general) was accompanied by the discovery of compositional correlations (1) between genes and the isochores containing them and (2) between chromosomal bands and isochores. These correlations will be discussed in the two following sections.

A. Isochores and Genes

The localization of a number of genes in compositional fractions of human DNA demonstrates a very important point, namely, that a linear relationship exists between GC levels of coding sequences (and of their first, second, and third codon positions) and GC levels of the DNA fragments (about 100 Kb in size) containing them (Fig. 3). Because the DNA fragments are mainly composed of noncoding intergenic sequences, the compositional correlation just described is in fact a correlation between coding sequences and the noncoding sequences that embed them. This suggests that both coding sequences representing <5% of the genome and the flanking noncoding sequences representing >95% of the genome are subject to similar compositional constraints.

The existence of this compositional relationship is also important from another viewpoint. Indeed, it allows us to understand the genomic distribution of genes over isochores having a different GC content. This can be deduced from histograms showing the compositional distribution of coding sequences or of their different codon positions. In fact, because the coding sequences are compositionally correlated with the vast DNA regions surrounding them, the compositions of coding sequences themselves indicate the locations of these genes in different isochore families. For instance, a coding sequence that is GC-rich will be present in a GC-rich isochore, whereas a gene that is GC-poor will be located in a GC-poor isochore. If one examines the composition of all human coding sequences studied so far (or, even better, of their third codon positions, which are free to change with less or no alteration of the corresponding amino acids), one discovers (Fig. 4) a strong predominance of GC-rich coding sequences, which are mainly located in the GC-richest isochores. The results of Fig. 4 point to an extremely nonuniform distribution of genes in the human genome because the GC-richest isochores are the least represented in the human genome.

It is interesting that this gene distribution is mimicked (1) by the distribution of CpG doublets, the main potential sites of methylation (CpG doublets in GC-poor genes are underrepresented relative to statistical expectations, whereas in GC-rich genes, CpG doublets approach the statistical frequency) and (2) by the distribution of CpG islands, which are sequences >0.5 Kb in size characterized by high GC levels, by clustered unmethylated CpG's, by G/C boxes (e.g., GGGGCGGGGC or closely related sequences), and by clustered sites for rare-cutting restriction enzymes (which recognize GC-rich sequences comprising one or two unmethylated CpG doublets). CpG islands are associated with the 5′ flanking sequences, exons, and introns of all housekeeping genes (namely, of the genes whose activity is required by every cell) and many tissue-specific genes (namely, of genes that are expressed only in some specific tissues, such as liver cells).

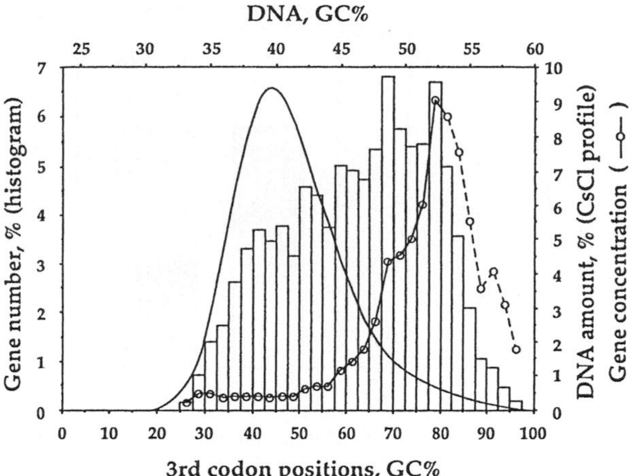

FIGURE 4 Profile of gene concentration in the human genome as obtained by dividing the relative amounts of genes in each 2.5% GC interval of the histogram by the corresponding relative amounts of DNA deduced from the CsCl profile. The apparent decrease in gene concentration for very high GC values (dashed line) is due to the presence of rDNA in that region. The last concentration values are uncertain because they correspond to very low amounts of DNA. [Reproduced, with permission, from G. Bernardi (1993). The human genome organization and its evolutionary history: A review. *Gene* **135**, 57–66.]

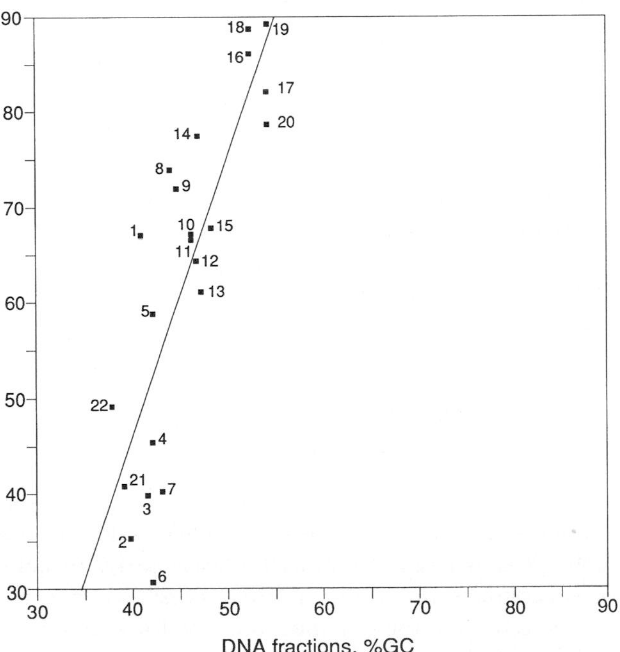

FIGURE 3 Plot of GC levels of third codon positions against the GC levels of compositional DNA fractions in which they were localized. The numbers indicate different coding sequences. [Reproduced, with permission, from B. Aïssani, G. D'Onofrio, D. Mouchiroud, K. Gardiner, C. Gautier, and G. Bernardi (1991). The compositional properties of human genes. *J. Mol. Evol.* **32**, 497–503.]

B. Isochores and Chromosome Bands

Hybridization of compositional fractions of human DNA corresponding to different isochore families has allowed the definition of the relationship of chromosomal bands to isochore families (Fig. 5). T(elomeric) bands are essentially formed by GC-rich isochores (mainly of the H2 and H3 families), whereas R′-bands, namely, the R(everse) bands exclusive of T-bands, comprise both GC-rich isochores (of the H1

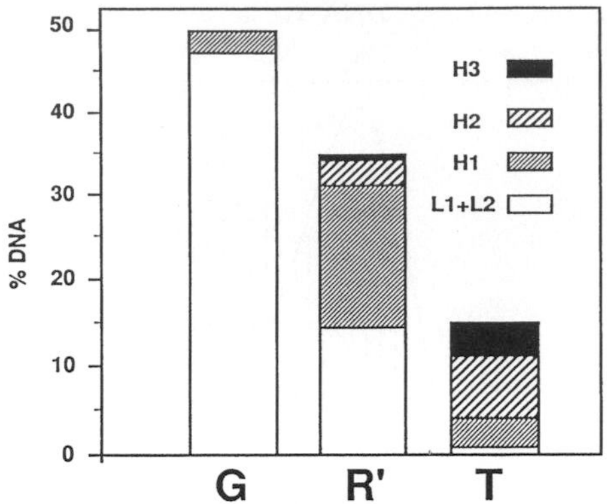

FIGURE 5 A scheme of the relative amounts of isochore families L1 + L2, H1, H2, and H3 in G-bands, R′-bands, and T-bands; R′-bands are R-bands exclusive of T-bands. [Reproduced, with permission, from S. Saccone, A. De Sario, J. Wiegant, A. K. Rap, G. Della Valle, and G. Bernardi (1993). Correlations between isochores and chromosomal bands in the human genome. *Proc. Natl. Acad. Sci USA* 90, 11929–11933.]

family) and GC-poor isochores. Finally, G(iemsa) bands are practically only formed by GC-poor isochores.

III. ISOCHORES AND GENOME FUNCTIONS

Although the functional correlates of the isochore organization of the human genome are still largely open problems, a number of points are already well established as outlined in the following.

A. Isochores and Integration of Mobile and Viral DNA Sequences

Stable integration of mobile and viral DNA sequences is mostly found in isochores of matching composition. Mobile sequences that have been amplified by retrotranscription and translocated to numerous loci of the human genome during mammalian evolution, such as LINES and SINES (see Section I), are predominantly located in isochores of matching GC levels. This indicates that reinsertion is targeted to matching genome environments and/or that integration is more stable within such environments. Incidentally, such reinsertion may be a cause of mutation, if it occurs in genes that are thereby disrupted.

Needless to say, these observations are of interest in connection with the integration of foreign DNA into the genome of transgenic mammals. In this case, the important, yet unresolved, question concerns the effect of genomic compositional context on the expression of integrated sequences. Recent results on transcribed and nontranscribed sequences of human T-lymphocyte virus type I (HTLV-I) suggest that such an effect exists.

B. Isochores, Translocation Breakpoints, and Fragile Sites

Translocation breakpoints are not randomly located on chromosomes. R-bands and G/R borders are the predominant sites of exchange processes, including spontaneous translocations, spontaneous or induced sister-chromatid exchanges, and the chromosomal abnormalities seen after X-ray and chemical damage. Likewise, fragile sites tend to be more frequent in R-bands or near the border of R- and G-bands. Moreover, cancer-associated chromosomal aberrations are also nonrandom, because a limited number of genomic sites are frequently associated with cellular oncogenes or fragile sites. These observations indicate that R-bands and G/R borders are particularly prone to recombination. They also raise the question of the role played in these phenomena by the compositional discontinuities at G/R borders and within R-bands, as well as by the genomic distribution of SINES, CpG islands, and other recombinogenic sequences, such as minisatellites, which are located predominantly in R-bands.

Chromosomal rearrangements have two important consequences: the activation of oncogenes by strong promoters that could be placed upstream of them by the rearrangement, and the formation, in evolutionary time, of reproductive barriers and speciation.

C. Other Functional Aspects of Isochores

Isochores are related to replication and condensation timing in the cell cycle. G-bands replicate late in the cell cycle, whereas R-bands replicate early. In contrast, condensation timing during mitosis occurs early in the cell cycle for G-bands and late for R-bands.

The distribution of genes over the isochores also has some correlation with gene function. Housekeeping genes (including oncogenes) are preferentially distributed in GC-rich isochores, whereas tissue-specific genes are more abundant in GC-poor isochores.

The range of GC values (25–100%) in codon third

positions of human genes is almost as wide as that exhibited by the genes of all prokaryotes. This very extended range implies very large differences in codon usage (namely, in the differential use of synonymous codons for the same amino acid) for GC-poor and GC-rich genes present in the same genome. In particular, at the high-GC end of the range, an increasing number (up to 50%) of codons are simply absent. In turn, GC values in codon third positions are paralleled (although, expectedly, to a more limited extent) by the GC values in first and second positions. The range of the latter values leads to very significant differences in the frequency of certain amino acids encoded by GC-rich and GC-poor genes. For instance, the ratio of alanine + arginine to serine + lysine (namely, of the two amino acids that contribute most to the thermodynamic stability of proteins over the two that do so the least) increases by a factor of four between proteins encoded by the GC-poorest and the GC-richest coding sequences in the human genome.

The distributions of CpG doublets and of CpG islands suggest that the distribution of methylation in the genome of humans and other vertebrates is highly nonuniform, a point of interest in view of the role of DNA methylation in gene function and of the distribution of housekeeping and tissue-specific genes.

The results on CpG islands have an additional functional relevance. In GC-poor isochores, genes are usually endowed with a TATA or a CCAAT box and an upstream control region, whereas in the GC-rich isochores there is often no TATA box but promoters containing properly positioned "G/C boxes" that bind transcription factor Sp1, a protein that activates RNA polymerase II transcription. These GC-rich promotors are apparently associated with all genes located in GC-rich isochores.

IV. ISOCHORES AND GENOME EVOLUTION

As shown in Figs. 1 and 2, the compositional patterns of the human genome are characterized by a wide GC range of both isochore families and coding sequences and by a predominance of GC-rich genes. These patterns are found in all mammals explored so far, with only a slightly narrower distribution in three families of myomorpha (a suborder of rodents that includes mouse, rat, and hamster). Very similar, yet not identical, patterns are found in birds, a vertebrate class that arose from reptiles independently of mammals. In the case of birds, the compositional distributions of both DNA fragments and coding sequences attain, however, higher GC values than in mammals (see Figs. 1 and 2). In sharp contrast, the genomes of the vast majority of cold-blooded vertebrates exhibit compositional patterns that are narrower and do not reach the GC levels of the GC-rich DNA fragments of warm-blooded vertebrates or of the coding sequences contained in them (see Figs. 1 and 2).

The compositional patterns of vertebrate genomes define two modes of genome evolution. In the conservative mode, which prevails in mammals (and in birds), the composition of DNA fragments and coding sequences is maintained during evolution despite a very high degree of nucleotide substitutions (which may be >20% in third codon positions). This compositional conservation appears to require negative selection, operating at the isochore level, to eliminate any strong deviation from a presumably functionally optimal composition. It is clear that if, during mammalian evolution, a number of genes retain a GC level of >90% in their third codon position, the mutations that would have decreased this level must somehow have been eliminated.

In contrast, in the transitional or shifting mode, parallel compositional changes are seen in both isochores and coding sequences. The two major compositional transitions found in vertebrate genomes are the GC increases that occurred between the genomes of reptiles on the one hand and those of birds and mammals on the other. (Apparently, these increases were accompanied by the replacement of TATA and CCAAT boxes by GC-rich promotors.) These compositional changes are due to a directional fixation of point mutations largely caused by natural selection at isochore levels. Of course, selection implies functional differences and, therefore, supports the idea that isochores are functionally relevant structures. Moreover, the compositional relationships between coding and noncoding (particularly intergenic) sequences indicate that the same compositional constraints apply to both kinds of sequences. The selection pressures underlying such constraints cannot be understood if noncoding sequences are "junk DNA," with no biological functions.

V. CONCLUSIONS

Isochores represent a new structural level in the organization of the genome of warm-blooded vertebrates that bridges the enormous size gap between the gene level, including both their exon–intron systems and the corresponding regulatory sequences, and the chromosome level, with its banding patterns. These three levels are correlated with each other, because genes match compositionally the isochores in which they are harbored, while GC-poor and GC-rich isochores are DNA segments located in G- and T-bands, respectively, R′-bands comprising both GC-poor and GC-rich isochores.

Whereas isochores from the genomes of warm-blooded vertebrates belong to a number of families characterized by large differences in base composition, this is not true for cold-blooded vertebrates, in which isochores are characterized by much smaller differences in composition, which correspond to much weaker banding patterns in metaphase chromosomes.

However, isochores are not only structural units but also appear to play functional roles. Some of these, like the integration of mobile and viral sequences or recombination and chromosome rearrangements, are well established. The observations on the gene distributions of the genomes, the relationships of such distributions with gene functions (housekeeping, tissue-specific), with codon usage, and with different kinds of regulatory sequences are also indicative of functional roles for isochores. On the other hand, DNA replication timing and chromosome condensation timing at mitosis seem to be rather correlated with the chromomere–interchromomere organization of chromosomes, independently of the composition of the corresponding DNA stretches, because they are also found in cold-blooded vertebrates.

Isochores are evolutionary units of vertebrate genomes. Their composition may be conserved in spite of enormous numbers of point mutations or may undergo dramatic changes after more modest numbers of point mutations. In the case of the two independent evolutionary transitions from cold-blooded vertebrates to mammals and birds, the compositional transitions occurring in the genome seem to be largely associated with the optimization of genome functions following environmental body temperature changes. Interestingly, these transitions appear to be accompanied by very conspicuous changes in promoter sequences.

In conclusion, two large compositional compartments can be distinguished in the human genome and, more generally, in the genomes of warm-blooded vertebrates (Table I). The first compartment, the paleogenome, is characterized by its similarity to what it was, and still is, in cold-blooded vertebrates: the late-replicating, compositionally homogeneous, GC-poor isochores of early-condensing chromomeres contain relatively rare, GC-poor (largely tissue-specific) genes having TATA box promotors (CpG islands are scarce). The second compartment, the neogenome, is characterized by the fact that it changed its compositional features compared with what it was in cold-blooded vertebrates. In the neogenome, the ancestral, early-replicating, GC-poor isochores of late-condensing interchromomeres were changed into compositionally heterogeneous, GC-rich isochores that contain abundant genes (perhaps including most housekeeping genes) have G/C box promotors (genes and CpG islands are particularly abundant in the GC-richest isochores).

TABLE I
The Human Genome[a]

Paleogenome (G-bands, GC-poor isochores)	Neogenome (T-bands, GC-rich isochores)
Chromomeres	Interchromomeres
Late replication	Early replication
Early condensation	Late condensation
Abundance of LINES	Abundance of SINES
Compositional homogeneity	Compositional heterogeneity
Scarcity of genes	Abundance of genes (esp. in H3)
GC-poor gene (esp. tissue-specific)	GC-rich genes (esp. housekeeping)
Scarcity of CpG islands	Abundance of CpG islands
TATA box promotors	G/C box promotors
Less frequent recombination	More frequent recombination

[a]Modified from G. Bernardi (1989). The isochore organization of the human genome. *Annu. Rev. Genet.* **23**, 637–661.

BIBLIOGRAPHY

Bernardi, G. (1989). The isochore organization of the human genome. *Annu. Rev.Genet.* **23**, 637–661.

Bernardi, G. (1993). The human genome organization and its evolutionary history: A review. *Gene* **135**, 57–66.

Bernardi, G. (1995). The human genome: Organization and evolutionary history. *Annu. Rev. Genet.* **29**, 445–476.

Huntington's Disease

EDITH G. McGEER
University of British Columbia

GLOSSARY

γ-Aminobutyric acid (GABA) Major inhibitory neurotransmitter in the brain

Caudate One of the two nuclei making up the striatum (see Putamen)

Cholinergic Pertaining to neurons that use acetylcholine as a neurotransmitter

Chorea Ceaseless occurrence of a wide variety of rapid, highly complex, jerky movements that appear to be well coordinated but are involuntary

Choreiform Having the characteristics of chorea

Codon Series of three nucleotide bases in a gene that codes for a particular amino acid in the protein formed on translation of the gene; for example, the triplet CAG codes for glutamine

Dementia Loss of mental faculties, particularly memory

Excitotoxin Excitatory amino acid that destroys neurons in brain regions into which it is injected. The exact mechanism is unknown, but these materials appear to overstimulate or excite neurons to such a degree that they die. Most known excitotoxins are not found normally in brain, although they are chemically related to glutamate, which does occur in the brain

Extrapyramidal system Group of subcortical nuclei, also called the basal ganglia, having to do with movement control

Glutamate Major excitatory neurotransmitter in brain; excessive glutamate or some analogs can be excitotoxins

Neurotransmitter One of a group of specific chemicals used by the nervous system to carry messages from one neuron to another

Peptide Compound made up of a chain (polymer) of amino acids

Putamen One of the two nuclei making up the striatum (see Caudate)

Striatum Forebrain structure that forms a major part of the extrapyramidal system; used here as a synonym for caudate/putamen

HUNTINGTON'S DISEASE, FORMERLY CALLED HUNTington's chorea, is a rare, genetic neurodegenerative disorder characterized by involuntary movements, cognitive impairment, and personality change. The choreiform movements cease during sleep. Huntington's disease generally manifests itself in affected individuals in the fourth decade of life and progresses within 10 or 15 years to death. Only palliative treatment is available. The brain is often shrunken, with extensive cortical atrophy, but the most characteristic and consistent pathology is loss of neurons in the caudate and putamen. The mutant gene has been localized to the short arm of chromosome 4 and the specific mutation has been identified as an expansion of a series of a repeated codon. The function of the gene product, however, is still unknown. In many families, DNA probes can now be used to estimate the risk of a fetus carrying the gene. Since injections of excitotoxins into the neostriatum in animals can reproduce much of the striatal pathology of the disease, it has been hypothesized that the genetic defect may be in some aspect of the endogenous excitatory amino acid systems and that neuronal death might be slowed by treatment with an inhibitor of excitation. Both hypotheses await future testing.

I. GENERAL DESCRIPTION

Huntington's disease (HD) occurs equally among men and women and the abnormal movements usually begin between the ages of 37 and 47. They may be preceded by as much as a decade by personality changes. Curiously, the mental symptoms can mimic schizophrenia in the early stages, and some victims have been kept in mental hospitals for several years before onset of the choreiform movements permits the correct diagnosis. In the later phases, however, the severe dementia is unmistakable. A few cases have been reported where choreiform movements are not seen; this is why the name Huntington's disease rather than Huntington's chorea is now preferred. The disease is relentlessly progressive and death usually ensues within 10 to 15 years.

In approximately 10% of HD cases, the onset may occur before the age of 15 and the onset in a few cases may be delayed until the sixth or seventh decade. The progress of the disease is generally more rapid in affected children than in young adults and is unusually slow in the older onset cases (see Section III).

HD is a rare disorder that affects all races, but its incidence can vary widely depending on the inheritance. The incidence in Europe, America, and Australia is around 4–7 per 100,000, and only about one-tenth of that in Japan and China. Pockets with a very high incidence exist, as in northern Sweden, around Lake Maracaibo in Venezuela, and in Tasmania. In many of these instances, all of the cases can be traced back to a common ancestor.

The extent of suffering is increased because the onset of the disease commonly occurs after the birth of children who perpetuate it; both the unaffected spouse and the children suffer from the knowledge that, statistically, 50% of the offspring will develop this devastating, presently untreatable disorder. Moreover, it is difficult for such children to decide whether they should in turn procreate, knowing that, if they are not themselves victims, they will probably not transmit the disorder (see Section III). The existence of DNA probes that can predict, in many families, the likelihood of any individual carrying the gene could theoretically be helpful in genetic counseling, but it is still being debated as to how justified such testing is at this time when no help can be offered to those who test positively.

II. PATHOLOGY

A. Histological Pathology

There is a general atrophy throughout the brain, but the greatest and most typical atrophy occurs in the caudate/putamen. The weight of the total brain is usually reduced by about 20%, whereas that of the caudate/putamen is reduced by more than 50%. Neuronal degeneration occurs in many brain regions, but the neuronal loss in the caudate/putamen is the characteristic pathological feature of the disease (Fig. 1). Many types of neurons with their cell bodies in the caudate/putamen are affected, although the large neurons tend to be better preserved than the small.

B. Chemical Pathology

Attention has been focused on the chemistry of the caudate/putamen since this is the site of the heaviest and most consistent pathology. All, or almost all, neuronal types with cell bodies in the striatum seem to be affected, whereas the same kinds of neurons are all right elsewhere in brain. Many of the identified biochemical losses in other regions, such as the substantia nigra, can be traced to loss of projections from the caudate/putamen (Fig. 2). Hence HD has been called a region-specific disease as contrasted with a neuron-specific disease such as Parkinsonism, where the critical pathology seems to be loss of dopamine neurons throughout the brain. It is worth noting that these nigro-striatal dopamine neurons are preserved in HD.

Positron emission tomography (PET) studies of glucose metabolism in HD show clear, progressive decreases in the caudate/putamen. Whether such decreases in local glucose metabolic rate precede or follow the onset of clinical symptoms remains a matter of some controversy.

III. THEORIES AS TO THE CAUSE

HD is a genetic disorder and the mutation has been shown to involve unstable repeats of a CAG codon within the protein-coding portion of a gene on chromosome 4. This type of mutation was unsuspected until about 1991, but has now been found in the genes responsible for seven neurological disorders, including HD. A great surprise is the instability of the gene. The repeat is variable in both normal individuals and patients, although HD cases have a larger repeat size (30–121) than normals (9–34). But a clinically normal male in the high normal range can have a child with a higher number of repeats who will show the disease. Moreover, there is a correlation between the number of repeats and the age of onset of the illness and, since the number of repeats tends to increase in successive generations, the disease will appear earlier and be more severe.

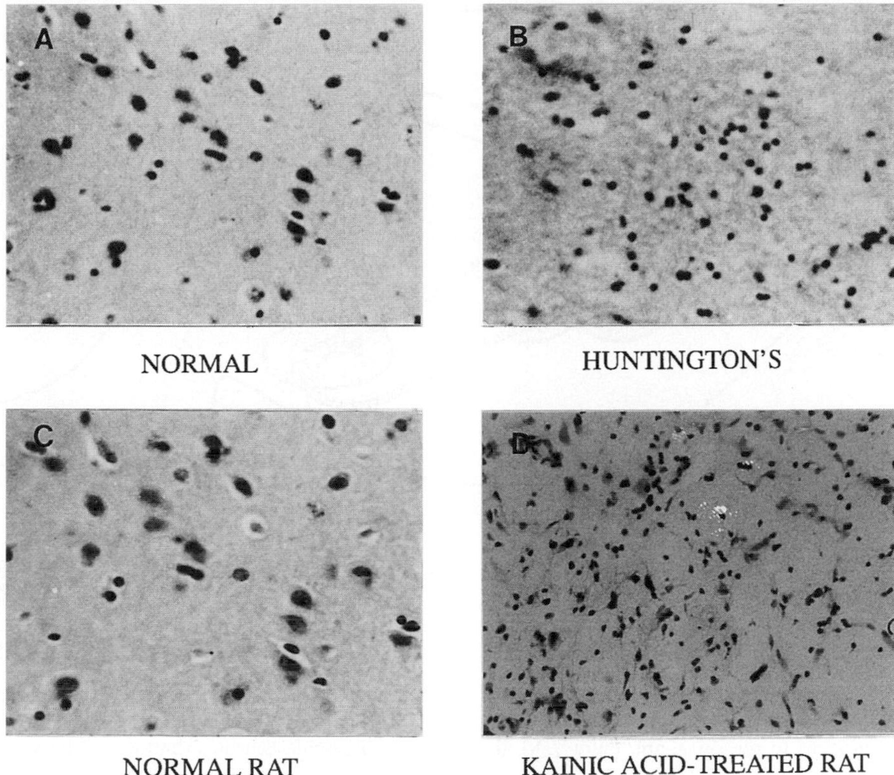

FIGURE 1 Histological sections of the caudate of a neurologically normal human (A), a case of Huntington's disease (B), a control rat (C), and a rat after an intrastriatal injection of kainic acid (D). Note the loss of the larger cells (neurons) in B and D, with preservation or increase in the numbers of very small cells (glia).

Another surprising finding is that the protein coded by the gene appears to be rather widely and diffusely localized in the body and brain, making it difficult to understand the fairly selective nature of HD pathology. The function of this gene product has not yet been identified. It is hoped that such identification will indicate the root cause of the disorder and suggest methods of effective treatment.

In the meantime, considerable interest has been focused on the excitatory amino acid systems in brain because the kainic acid and quinolinic acid animal "models" of HD suggest that excitatory amino acids may be involved in the mechanism of cell loss (Section IV). According to this hypothesis, some genetic defect in an excitatory amino acid system of brain results in progressive neuronal destruction though an excitotoxic mechanism. The caudate and putamen might be particularly vulnerable because of the massive excitatory amino acid projection that they receive from the cortex (Fig. 2). Although unproven, this hypothesis has led to suggestions of possible therapeutic measures to slow the progression of the disorder (see Section V). [*See* Genetic Diseases.]

IV. ANIMAL MODELS

The injection of excitotoxins such as kainic acid or quinolinic acid into the striatum of rats produces local neuronal loss that is histologically similar to that seen in HD (Fig. 1), and extensive biochemical studies have indicated that the lesion-induced changes in many neurotransmitter systems in the basal ganglia are similar to those found in HD (Table I). Bilaterally lesioned rats also show some behavioral abnormalities and pharmacological responses similar to those seen in the human disorder. Thus, for example, the rats show enhanced activity during a rat's normal waking period (Fig. 3A), abnormal locomotion, learning problems, and weight losses reminiscent of those seen in HD. They also show markedly enhanced responses to amphetamine (Fig. 3B) and

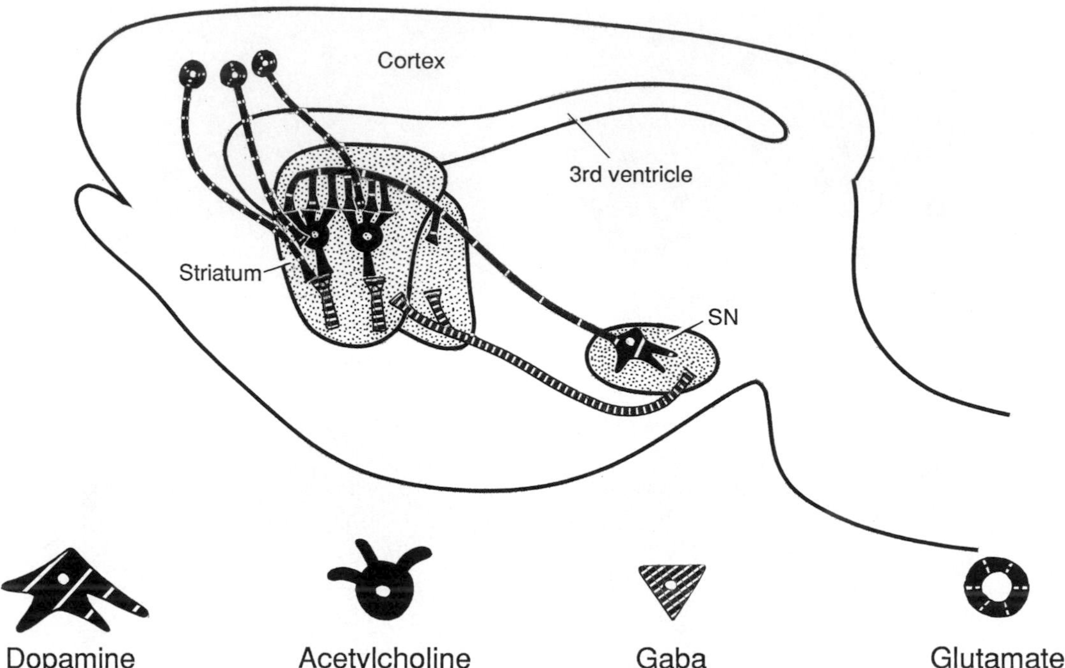

FIGURE 2 Diagrammatic representation of a few of the projections to and from the striatum, as well as a few of the striatal interneurons in a rat brain. Many other neurons using other transmitters exist in this system. For example, descending neurons to the substantia nigra (SN) also use various peptide neurotransmitters such as substance P and dynorphin.

TABLE I
Examples of Some Biochemical Measures on Presynaptic Neurotransmitter Indices in the Striatum and Substantia Nigra in Huntington's Disease and Excitotoxic "Models"

Neurotransmitter	Huntington's disease	Excitotoxic "models"
In the striatum		
Neurotransmitters in striatal neurons		
GABA indices	Decreased	Decreased
Acetylcholine indices	Decreased	Decreased
Substance P	Decreased	Decreased
Enkephalin	Decreased	Decreased
Dynorphin	Decreased	Decreased
Somatostatin	Normal or increased	Normal or decreased[a]
Neurotransmitters in projections to the striatum		
Dopamine indices	Normal or increased	Normal or increased
Noradrenaline indices	Normal or increased	Normal or increased
Serotonin indices	Normal or increased	Normal or increased
In the substantia nigra		
Neurotransmitters in projections from the striatum		
GABA indices	Decreased	Decreased
Substance P	Decreased	Decreased
Dynorphin	Decreased	Decreased

[a]This is an example of a neurotransmitter system in which a quinolinic acid lesion mimics HD better than does a kainic acid lesion.

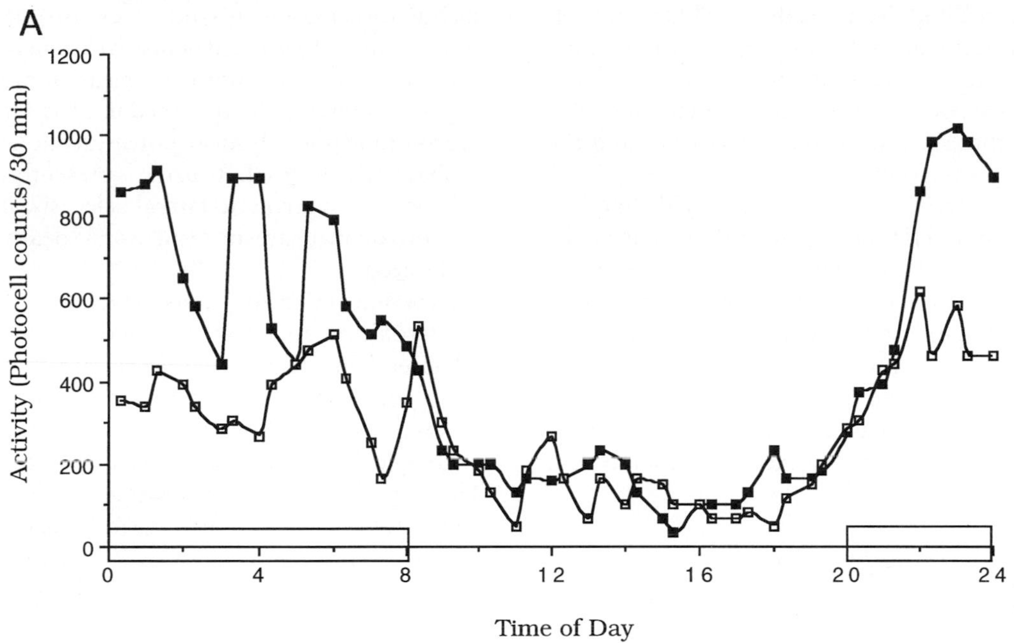

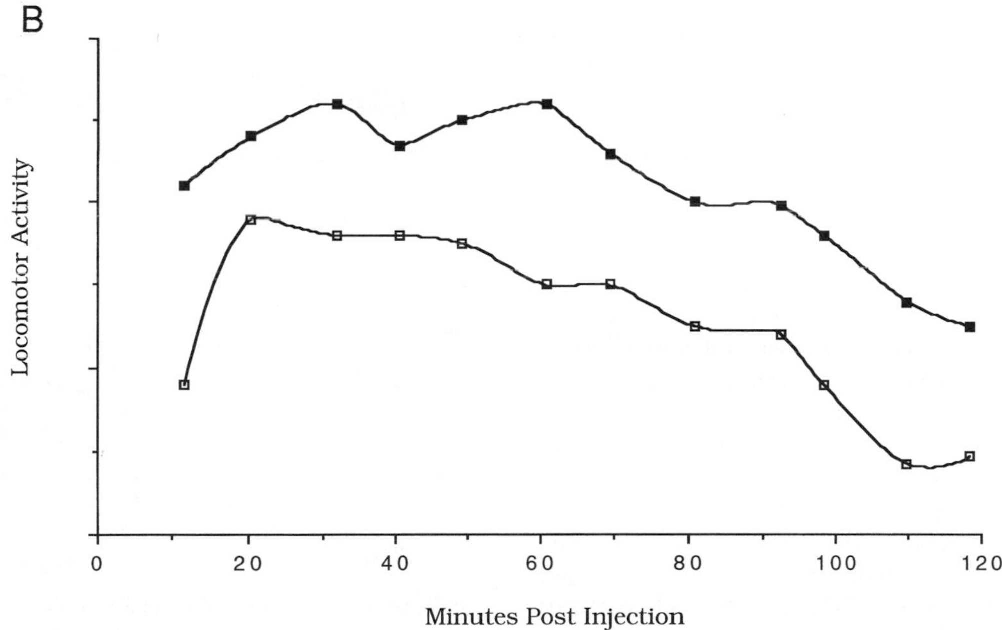

FIGURE 3 Mean activity measurements in groups of kainic acid-lesiond rats (■) as compared with groups of controls (□). (A) Activity over the course of a day, showing nocturnal hyperactivity, the period of light being from 8 to 20 hours. (B) Locomotion after injections of amphetamine.

scopolamine but an attenuated cataleptic response to haloperidol and sedative effects with apomorphine. The hyperactivity in the rats is decreased on treatment with haloperidol or physostigmine. These findings have been interpreted as akin to the pharmacological responses seen in HD.

Rats sacrificed several months after such lesions show considerable striatal atrophy and some neuronal loss in the cortex and thalamus that make the pathology even more similar to that in HD than the changes seen shortly after lesioning. It is clear, however, that the time course of neurotoxicity in such "models" differs from that seen in the human disease, and the genetic factor is missing.

The biochemistry of the quinolinic acid "model" seems to resemble HD more closely than that of the kainic acid "model." Moreover, quinolinic acid is a normal mammalian metabolite of tryptophan and is found in brain; kainic acid is derived from a seaweed and does not occur in animals. Hence, much attention has been focused on the quinolinic acid system, but the general hypothesis, based on these models, merely suggests some abnormality in excitatory amino acid systems in HD brain.

V. TREATMENT

No form of treatment is known to arrest the mental deterioration. Dopamine blockers (such as haloperidol or chlorpromazine) or dopamine depleters (such as reserpine or tetrabenazine) may help to control the involuntary choreiform movements in some cases; such drugs are better known for their tranquilizing action in schizophrenia.

When the early work on biochemical pathology was done (see Section II), some attempts were made to replace the GABA or cholinergic deficiencies. These were without beneficial effect. In view of the complex nature of the biochemical losses in HD (Table I), it seems highly unlikely that an effective replacement therapy can be devised. Side effects would also be significant because of the existence elsewhere in brain of healthy neurons of the affected types. Attention is therefore being focused on therapies designed to prevent or inhibit the cellular destruction. The existence of the excitotoxic "models" of HD (see Section IV) has led to the suggestion that agents that inhibit either the release of neurotransmitter glutamate or the action of excitatory amino acids at postsynaptic receptors might slow progression of the disease. Other agents, such as taurine and naloxone, which antagonize excitotoxicity in animal models, have also been suggested. In theory, none of these agents would have any effect on established symptomatology but they might be particularly useful in delaying the onset of clinical symptoms in persons carrying the gene. The only drug of this type tested so far is the relatively weak and nonspecific $GABA_B$ agonist, Baclofen; it did not appear to slow the progression in established cases. Better agents might be glutamate receptor antagonists or inhibitors of glutamate release, such as are being widely studied for possible use as anticonvulsants or to prevent neuronal loss in ischemia and hypoxia. One such glutamate release inhibitor, lamotrigine, is presently being tested in early HD cases and a second and possibly more powerful one, riluzole, will probably also be tried. Riluzole is presently under clinical test in amyotrophic lateral sclerosis, in which an excitotoxic mechanism of neuronal death is also hypothesized.

Experimental transplants have been done in the excitotoxic rat and monkey "models" of HD. Neonatal or fetal rat or monkey striatal cells are injected into the lesioned area some days after striatal injections of kainic or quinolinic acid. The cells can be shown histologically to live and to develop a blood supply. The transplanted tissue also seems to show near-normal glucose metabolism (Fig. 4). Biochemical measurements indicate some recovery of GABA and cholinergic indices and no loss in weight of the caudate/putamen. Some normalization of the behavior and response to drugs has also been reported after such

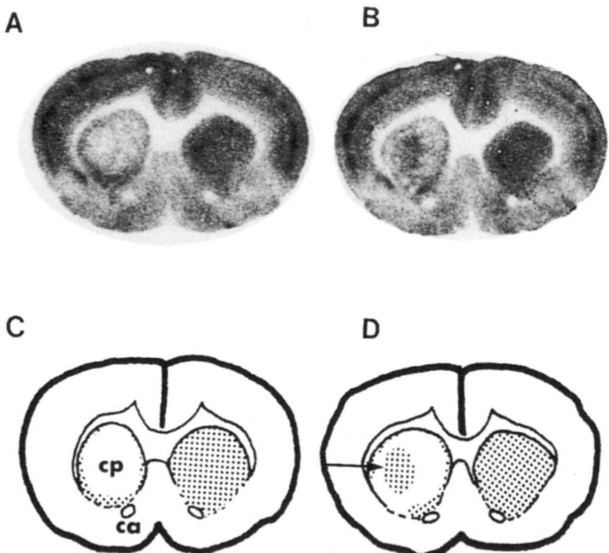

FIGURE 4 Autoradiographic studies of glucose metabolism in brain sections from rats that had been given injections of kainic acid into the left striatum about 11 weeks before sacrificing. Extent of the normal striatal area is dotted in the diagrams. Rat (B, D) received a transplant of striatal cells from a neonatal rat 3 weeks after the kainic acid injection. The extent of the transplant is also indicated by dotting. The darker the shading, the greater the rate of glucose metabolism. Note the lower glucose metabolism in the striatum on the injected side of the rats as compared to the uninjected side. HD cases show similarly low glucose metabolism in the striatum when studied by PET. The transplanted cells (arrow) appear to have nearly normal glucose metabolism. cp, caudate/putamen (striatum); ca, internal capsule.

transplants. A few cases of such transplants in human cases of HD have been reported, but whether they are helpful remains in question.

BIBLIOGRAPHY

Fisher, L. J., and Gage, F. H. (1994). Intracerebral transplantation—Basic and clinical applications to the neostriatum. *FASEB J.* **8**, 489–496.

Koutouzis, T. K., Emerich, D. F., Borlongan, C. V., Freeman, T. B., Cahill, D. W., and Sanberg, P. R. (1994). Cell transplantation for central nervous system disorders. *Crit. Rev. Neurobiol.* **8**, 126–162.

Martin, J. B. (1993). Molecular genetics of neurological diseases. *Science* **262**, 674–677.

Sanberg, P. R., and Coyle, J. T. (1984). Scientific approaches to Huntington's disease. *CRC Crit. Rev. Clin. Neurobiol.* **1**, 1–44.

Sutherland, G. R., and Richards, R. I. (1994). Dynamic mutations. *Am. Sci.* **82**, 157–163.

Hydrocephalus

J. E. BRUNI
M. R. DEL BIGIO
University of Manitoba

GLOSSARY

Cerebrospinal fluid (CSF) Watery, clear, colorless liquid produced largely by the choroid plexus and contained within the cerebral ventricles, central canal of the spinal cord, and the subarachnoid spaces surrounding the central nervous system

CSF shunting Surgical procedure used for the management of hydrocephalus in which an obstruction in the CSF flow pathway is bypassed using an implanted silicone rubber tube that diverts CSF flow to an alternate body location

Hydrocephalus Condition in which there is overproduction, obstructed circulation, and/or impaired absorption of CSF resulting in accumulation of fluid within the cerebral ventricles and/or subarachnoid spaces, progressive ventricular dilatation, raised intracranial pressure, and consequent neurological and functional disturbances

Ventricles Cerebrospinal fluid-filled cavities located within each cerebral hemisphere, the diencephalon, and the brain stem

HYDROCEPHALUS (GR. WATER + HEAD) IS A CENtral nervous system condition characterized by a pathological excess of cerebrospinal fluid (CSF) intracranially due to its overproduction, impaired absorption, or circulation. The condition has been recognized since the time of Hippocrates (460–377 BC), although it took until the nineteenth century for the pathophysiology of the CSF circulation to be finally understood. Since that time, understanding and management of hydrocephalus have advanced such that outcome has also greatly improved.

Hydrocephalus is mainly a childhood disorder occurring congenitally or as a sequelae of intracranial tumors, hemorrhage, or infections. In the adult, hydrocephalus is less common and usually arises from subarachnoid hemorrhage, head injury, tumors, infections, or unknown causes.

In hydrocephalus there is a progressive enlargement of the ventricles, increased intracranial pressure (ICP), white matter edema, diminished cerebral blood flow, and compression of nervous tissue. Because the brain is encased in a bony vault, the pathological increase in CSF volume and ICP that occurs with hydrocephalus is compensated for by a corresponding reduction in the volume of some other intracranial compartment (vascular or neural). The extent of the CSF pressure increase depends on the severity of the obstruction and the ability of the neural and vascular compartments to adapt. If left untreated, deterioration of motor, endocrine, and neuropsychological functions occurs and death may ensue. This article presents a brief overview of the etiology, pathophysiology, and clinical manifestations and management of hydrocephalus.

I. PRODUCTION, CIRCULATION, AND ABSORPTION OF CSF

Cerebrospinal fluid is a clear, colorless fluid that resembles an ultrafiltrate of blood plasma. It is formed

ENCYCLOPEDIA OF HUMAN BIOLOGY, Second Edition, VOLUME 4.

by the choroid plexus of the four cerebral ventricles by a process of active choroidal epithelial cell secretion combined with passive capillary diffusion. [*See* Cerebral Ventricular System and Cerebrospinal Fluid.] A portion (10–30%), however, is also formed at sites other than the choroid plexus. In humans, the total volume of CSF is estimated to be about 140 ml. The rate of formation is approximately 0.35 ml/min or 500 ml/day and the normal pressure measured in the recumbent position by lumbar puncture varies from 25 to 70 mm (2–5 mm Hg) in infants and from 65 to 195 mm water (5–15 mm Hg) in adults.

The circulation of CSF through the ventricular system and subarachnoid space is attributable to pressure waves in the fluid generated by pulsatile arterial–brain expansion and the pressure gradients produced by production and absorption of CSF. From the lateral ventricles, where most of the CSF is produced, it passes into the third ventricle through the interventricular foramina (of Monro). From the third ventricle, the CSF reaches the fourth ventricle via the narrow cerebral aqueduct (of Sylvius). CSF then leaves the ventricular system at the level of the medulla oblongata through three apertures: the midline foramen (of Magendie) and the paired lateral foramina (of Luschka). These apertures communicate with the subarachnoid space surrounding the brain and spinal cord.

The ultimate fate of the CSF in both its intracranial and spinal course is to drain by way of finger-like extensions of the subarachnoid space called arachnoid villi or granulations that project into venous channels (sinuses and lacunae). [*See* Meninges.]

Small amounts of CSF are also absorbed via pial vessels and across the walls of cerebral capillaries within the brain parenchyma. Some absorption, particularly in smaller mammals, occurs via lymphatic channels adjacent to the subarachnoid space that surrounds cranial and spinal nerves. The importance of these alternative routes of absorption in humans, however, is still being debated.

II. CLASSIFICATION OF HYDROCEPHALUS

Advances in the understanding of hydrocephalus over the years have resulted in many different classifications of the disorder. Recent schemes view hydrocephalus as any abnormal accumulation or retention of intracranial brain water. Using this definition, a distinction is then made between the intraparenchymal (cerebral edema) and extraparenchymal accumulation of increased CSF. Extraparenchymal hydrocephalus can then be further subclassified into subarachnoid and intraventricular forms depending on the location of the accumulated CSF volume. Accumulation of fluid within and enlargement of the ventricular system are usually referred to as internal hydrocephalus, whereas enlargement of the subarachnoid space is called external hydrocephalus. Accumulation of fluid within the subarachnoid space and cerebral ventricles as a consequence of cerebral atrophy is given the special designation of hydrocephalus *ex vacuo* but is an entity considered distinct from hydrocephalus.

The more conventional scheme, predicated on the definition of hydrocephalus as an imbalance between production and absorption of CSF, uses levels of obstruction along the CSF pathway as the basis for classification. According to this scheme, hydrocephalus may be obstructive or nonobstructive in form. The only nonobstructive form of hydrocephalus is caused by overproduction of CSF by the choroid plexus as occurs infrequently with tumors (papilloma) of the choroid plexus. Noncommunicating and communicating forms of obstructive hydrocephalus are usually distinguished.

A. Noncommunicating Hydrocephalus

Noncommunicating hydrocephalus is distinguished by an intraventricular obstruction to CSF flow that occurs at or before the outlet foramina of the fourth ventricle. Such an obstruction would cause CSF to accumulate and the ventricular system proximal to the obstruction to enlarge (Fig. 1). Particularly vulnerable sites for obstruction are the narrow cerebral aqueduct (of Sylvius) and the exit foramina of the fourth ventricle. This form of hydrocephalus is referred to as noncommunicating because CSF cannot pass freely between the ventricles and the subarachnoid spaces.

B. Communicating Hydrocephalus

In contrast, communicating hydrocephalus is caused by an obstruction to CSF flow that occurs somewhere in the subarachnoid space, external to the ventricular system proper. In this situation, the ventricles remain in communication with one another and with the subarachnoid space (SAS), hence the designation of communicating hydrocephalus. The SAS over the cortical surface may be enlarged because of impaired flow of CSF as a result of occlusion or fibrosis within the subarachnoid space near absorption sites. The im-

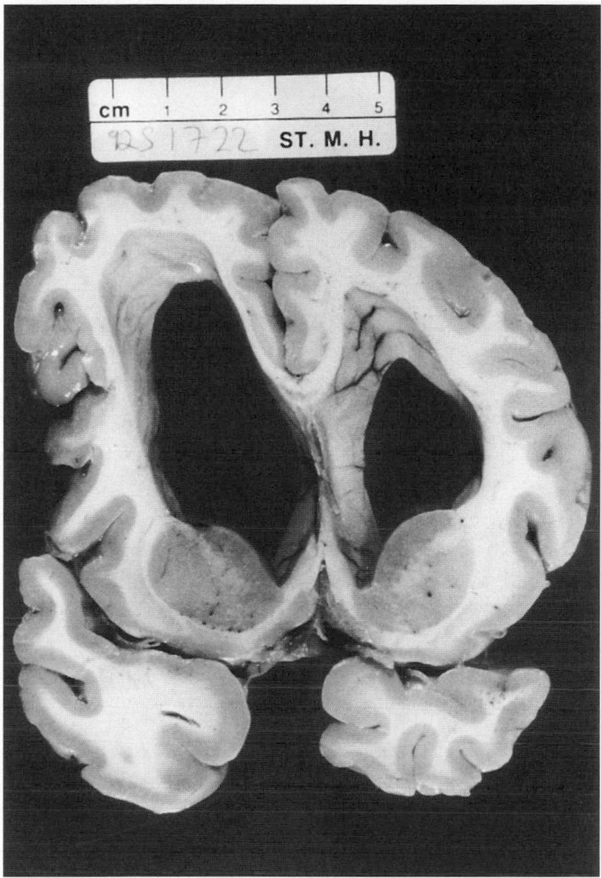

FIGURE 1 Coronal section through the brain of a 21-year-old hydrocephalic male showing the greatly enlarged frontal horns of the lateral cerebral ventricles and consequent compression of the adjacent nervous tissue.

paired flow may occur at the level of the basal cisterns/incisura or it may be due to inadequate absorption at the level of the arachnoid villi or to impaired venous drainage. It may occur as a result of hemorrhage, tumor, or inflammatory exudate in the SAS.

III. ETIOLOGY OF HYDROCEPHALUS

Hydrocephalus is primarily an affliction of children occurring as either a congenital (due to central nervous system maldevelopment or intrauterine infection) or acquired disorder (as a sequela of intracranial tumors, trauma, hemorrhage, or infection). The cause of hydrocephalus is often a significant factor in predicting outcome; both morbidity and mortality. Only the more common causes of the disorder are discussed here. The characteristic feature of progressive hydrocephalus in infancy is a markedly enlarged head, wide cranial sutures, and a bulging anterior fontanelle indicative of increased intracranial pressure. Hydrocephalus is diagnosed using neuroradiologic tests. Ultrasound, a noninvasive imaging technique is used most commonly to diagnose infantile hydrocephalus because it requires an incompletely ossified skull. Computed tomography is also commonly used because it is quick, inexpensive, and noninvasive (Fig. 2). Magnetic resonance imaging, also noninvasive, does not subject the patient to radiation, is used to provide better detail.

A. At Birth and Acquired

Congenital or inherited hydrocephalus occurs in about one per thousand live births but has a much higher incidence in the offspring of consanguineous parentage. When present at birth, it often occurs (in about one-third of cases) as a consequence of stenosis of the cerebral aqueduct (of Sylvius). One form of hereditary hydrocephalus is an inherited autosomal dominant disorder linked to a defective gene on the X chromosome. Males are primarily affected and exhibit varying degrees of other abnormalities, including mental retardation along with the hydrocephalus.

Hydrocephalus also often accompanies a wide range of other developmental malformations of the central nervous system, such as myelomeningocele, Arnold-Chiari hindbrain deformity, and Dandy-Walker posterior fossa malformations.

The use or abuse of certain toxic or pharmacological substances or nutritional deficiencies (vitamins) during pregnancy may be causative of some of the malformations associated with hydrocephalus. As a general rule, the earlier during the period of gestational development that a toxic agent (teratogen) is present, the more severe is the malformation caused.

Hydrocephalus may also result from uterine infections such as toxoplasmosis, which may cause narrowing at the aqueduct. Viral infections during pregnancy have been implicated both directly and indirectly (via the induction of malformations) in the genesis of fetal hydrocephalus. Common agents such as cytomegalovirus, varicella, rubeola, herpes, and mumps have been implicated. They are capable of producing inflammation of the meninges and/or of the ependyma, which in turn may narrow the CSF channels. The ependyma lining of the cerebral ventricles is reported to be a specific target for many viruses.

Hemorrhage into the ventricle from the germinal matrix of the developing nervous system is a lesion

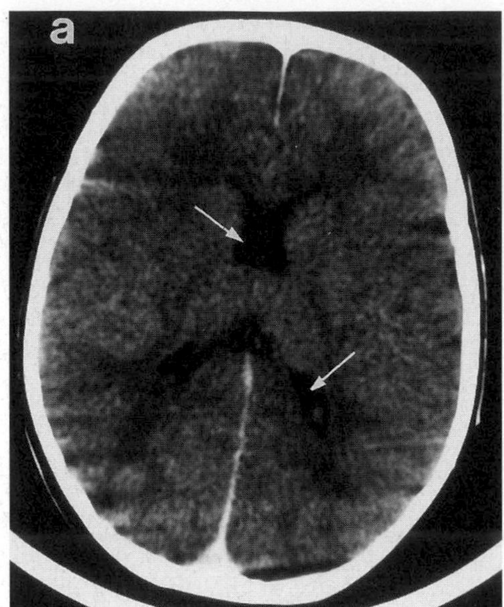

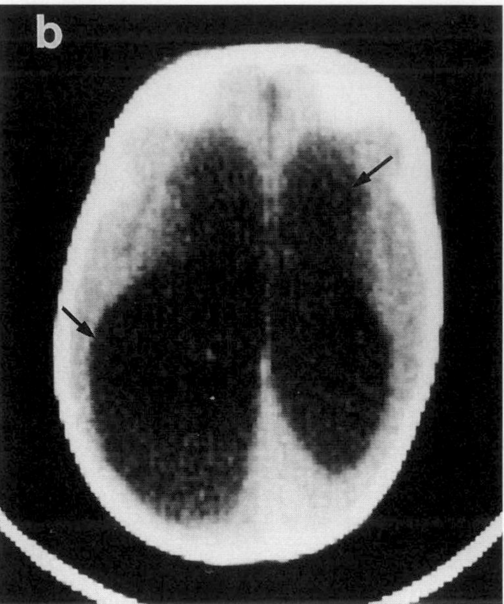

FIGURE 2 Computed tomography scans of the heads of two children, two years of age, taken in the horizontal plane. A normal brain with slit-like lateral ventricles (arrows) is shown in Fig 2a. A hydrocephalic brain with grossly enlarged ventricles (arrows) that are at least 10× greater than the normal volume is shown in Fig. 2b.

peculiar to infants born prematurely, that is, at 24–28 weeks gestation. This has been reported to occur with an incidence that exceeds 40%. The vulnerability of blood vessels in the periventricular matrix zone (subependyma) of the cerebral ventricles during neonatal development predisposes to hemorrhage in preterm infants. This in turn produces a communicating type of hydrocephalus because fibrosis or arachnoiditis can impair CSF flow at the arachnoid villi or outlet foramina of the fourth ventricle. The aqueduct may also become blocked due to an ependymal reaction or by debris, and in this case a noncommunicating form of hydrocephalus could develop.

Bacterial meningitis acquired during the neonatal period may scar the subarachnoid space and also obstruct outlet foramina and impede CSF flow. During childhood, central nervous system tumors (astrocytomas, medulloblastomas, ependymomas) may develop and, depending on their location, physically obstruct CSF flow through the ventricular system. In the case of tumors, their removal may be sufficient to alleviate the hydrocephalus. Overproduction of CSF is also a cause of hydrocephalus, albeit rare, that may occur with tumors (papillomas) of the choroid plexus. Lastly, head injuries at all ages but especially during infancy may also cause hydrocephalus by obstructing the CSF flow or absorption secondary to intraventricular and/or subarachnoid bleeding.

B. Adult

In the adult, hydrocephalus is less common and may result from subarachnoid hemorrhage, head injury, obstruction due to tumor masses or cysts, infections, or unknown causes. Intracranial bleeding into the subarachnoid space can occur as a consequence of trauma, the rupture of aneurysms, or vascular malformations. The ensuing accumulation of blood within the system or basal cisterns obstructs CFS outflow or absorption at the arachnoid villi and can result in hydrocephalus.

A peculiar form of hydrocephalus known as "compensated hydrocephalus" or "normal pressure hydrocephalus" (NPH) also sometimes occurs in adults. In this syndrome, there may be brain atrophy but total intracranial volume remains the same and mean intracranial pressure is within the normal range. Although not completely understood, it has been attributed to obstruction of subarachnoid CSF flow and/or resorption or more recently, white matter damage has been suggested ischemic. Clinically the syndrome is characterized by memory loss, urinary incontinence, and gait apraxia. The treatment involves CSF shunting. Because of different possible causes of the dementia, however, not all patients show improvement following this procedure.

IV. CLINICAL MANIFESTATIONS

As a consequence of enlarging ventricles individuals may suffer a range of neurological dysfunction. To some extent the disturbances are determined by the speed with which the ventricles enlarge. If they enlarge rapidly, for example, following hemorrhage into the cerebral aqueduct and/or fourth ventricle or following shunt malfunction, the main symptoms can be headache, nausea, vomiting, papilloedema, disturbances of vision, and altered or lost consciousness due to rising intracranial pressure and decreasing blood flow to the brain. In these individuals, treatment must be given urgently or else death would likely result.

Less rapid expansion of the lateral and third ventricles results in more subtle intellectual, motor, and emotional disturbances as a result of dysfunction and/or damage to structures surrounding the ventricles. The most common finding in both adults and children is the disturbance of motor function manifested as an unsteady widely spaced gait and impaired hand and arm movements. This is thought to be due to damaged projections from the motor and premotor area as they run alongside the ventricles. The need to urinate frequently also occurs; due possibly to compression of the medial surface of the frontal cortex. Dementia or impaired intellectual function can be seen in adults with hydrocephalus.

Children with hydrocephalus tend to have a lower intellectual capacity. Whereas adults tend to be apathetic, hydrocephalic children often exhibit an outgoing talkative personality sometimes referred to as "cocktail party syndrome." Their speech may be fluent and rapid but with few meaningful ideas (Hagberg and Sjorgen, 1966).

Less common clinical features of hydrocephalus include a movement disorder of unknown cause similar to that found in Parkinson's disease. Patients exhibit a tremor, extreme slowness of movement, and rigidity of the limbs due to increased muscle tone. In children, as the third ventricle enlarges there may be compression of the upper brain stem regions with consequent impairment of upward movement of the eyes. This results in a characteristic "setting sun" sign in which the eyes are cast in a downward gaze. Expansion of the third ventricle can also compress hypothalamic structures responsible for endocrine regulation. Hydrocephalic children may have short stature and delayed puberty because of impaired growth and gonadal hormone production.

V. PATHOPHYSIOLOGY

A distinction must be made between neuropathological abnormalities that cause hydrocephalus and those changes that are caused secondarily by the hydrocephalus. In the first category are the malformations or destructive processes that actually cause hydrocephalus (see Section III).

Brain damage that is the result of enlarged ventricles is dependent on the rate and extent of ventricular enlargement. Damage to the cerebrum may be seen anywhere from the ventricle surface to the cortex. The earliest changes observed are thinning and disruption of the ependyma, the single-cell layer that lines the surface of the ventricles. [*See* Ependyma.] As the ventricles enlarge, the neuropil is progressively compressed, although paradoxically the white matter immediately adjacent to the ventricle becomes edematous. Axons coursing in the immediate vicinity of the ventricles can be compromised; their myelin investments may be disrupted. The brain reacts to this injury by increased production of astrocytes (gliosis), a process that in effect scars the brain. In extreme cases of ventricular enlargement and brain compression, the cortical mantle itself may exhibit atrophic changes such as neuronal loss.

Brain damage in hydrocephalus can be attributed to the following factors: (1) the physical distortion of the brain substance that may stretch axons and impair the flow of nervous impulses, (2) the chronic ischemia and reduction in blood flow associated with the increased fluid volume and intracranial pressure, and (3) the adverse effects of accumulated breakdown products of metabolism that normally would be eliminated from the brain by outflow of CSF.

VI. OPERATIVE TREATMENT

A. Shunt Procedures

The most widely used treatment for hydrocephalus is surgical diversion or "shunting" of CSF from the ventricular system to an alternate body location (Fig. 3). This is accomplished by placing the proximal end of a silicone rubber tube into the lateral ventricle of

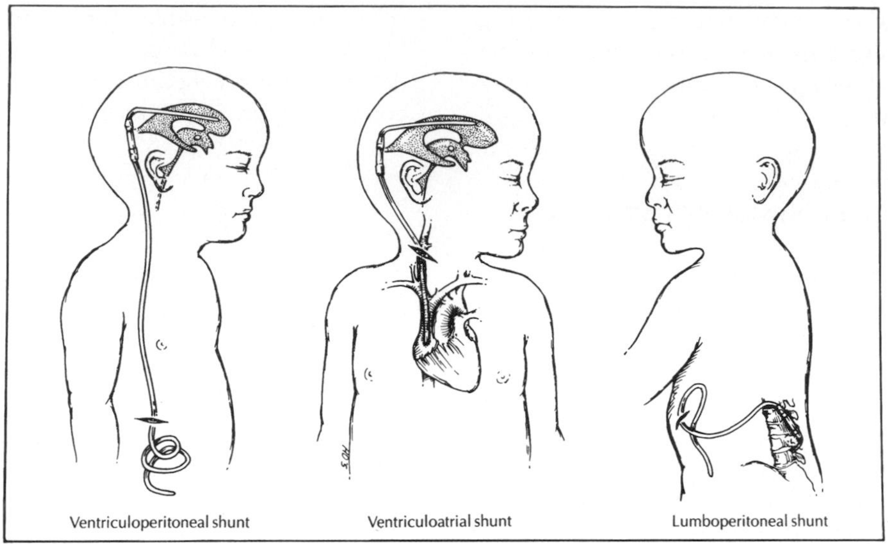

FIGURE 3 Types of surgically implanted CSF shunts. [Reproduced with permission from H. E. James (1992). Hydrocephalus in infancy and childhood. *Am. Family Phys.* 45(2), 733–742.]

the brain and tunneling the tubing underneath the skin to the peritoneal cavity (ventriculoperitoneal shunt) or less commonly the right atrium of the heart (ventriculoatrial shunt) or the pleural cavity (ventriculopleural shunt). Each of these locations provides an adequate space and blood supply where reabsorption of the fluid can take place. Along the course of the shunt tube, a one-way valve is placed that is used to control pressure and flow of CSF (Pudenz, 1981). Over a period of time, such a drainage device allows the ventricles to reduce in size, although they do not necessarily return to normal size. Lumboperitoneal shunts are also sometimes used in cases of communicating hydrocephalus. They are inserted into the lumbar subarachnoid space (cistern) and allowed to drain into the peritoneal cavity.

CSF shunts may lead to complications that can result in some morbidity and mortality. The most common complication encountered is occlusion of the ventricular shunt catheter by ingrowth of brain tissue. Shunts may also become infected, usually due to bacteria that normally live on the skin surface. Less often, shunts that have been in place for many years may develop mechanical problems due to malfunction, breakage, or disconnection. In most cases the patient must undergo a surgical procedure (shunt revision) to replace a portion of the shunt. In the case of infections they must be treated with antibiotics. Shunts may also work too well and overdrain CSF through a siphoning effect. This may result in collapsed (slit) ventricles, and the patients may develop headaches, nausea, lethargy, or, in elderly patients, subdural hematomas. A number of devices have been designed and employed for siphon control with variable success to correct this problem of overdrainage.

B. Treatment without Shunt Procedures

Hydrocephalus may also be treated by procedures known as ventriculostomy or ventriculocisternostomy. Ventriculostomy, one of the earliest procedures employed for temporary treatment of hydrocephalus, bypasses an obstruction by establishing drainage from the ventricle to the exterior. In contrast, ventriculocisternostomy creates a communication between the obstructed ventricular system (usually the third ventricle) and the subarachnoid space. The removal of the choroid plexus, the source of most CSF production, by a surgical procedure known as choroid plexotomy or by endoscopic coagulation has also been attempted.

VII. NON-OPERATIVE TREATMENT

Pharmacological intervention, using drugs known to decrease the rate of formation of CSF, are also employed. Acetazolamide or furosemide block carbonic anhydrase activity and have been given, usually to children, with the intention of decreasing CSF production. They tend not to be effective clinically however because the effects are only temporary and because of undesirable side effects.

Osmotic diuretic agents (mannitol, diamox, urea, isosorbide) administered orally or intravenously have also been used for the nonsurgical management of hydrocephalus. These pharmacological agents temporarily alter intracranial fluid dynamics and may only delay the need for inserting a shunt catheter.

VIII. EXPERIMENTAL MODELS OF HYDROCEPHALUS

A. Congenital Hydrocephalus

In experimental animals, hereditary defects of development that result in disturbed flow of CSF and dilatation of the cerebral ventricles are used as models to study the pathogenesis of hydrocephalus in humans. Regardless of whether hydrocephalus occurs spontaneously or is induced by environmental agents, the cytopathology of the disorder is similar. In most cases, obstructed CSF flow is due to defective development of the ventricular system (usually the cerebral aqueduct) or of the subarachnoid space and frequently occurs in association with anomalies in other tissue/organ systems. Several distinct inherited forms of hydrocephalus have been described in a variety of laboratory (mice, hamsters, rats) and domestic (cattle) animals. Recessively inherited hydrocephalus of variable severity has been reported in several inbred strains of rodents. They all develop a domed-shaped skull during the neonatal period and usually die before reaching maturity.

Congenital hydrocephalus can also be experimentally induced in the offspring of laboratory animals following maternal treatment with chemicals (6-aminonicotinamide, cyclophosphamide, ethylene thiourea, ethylnitrosourea), exposure to assorted pathogens (mumps virus, myxoviruses), by X irradiation of the fetuses, and by feeding vitamin (A, B_{12}) -deficient diets.

B. Induced Obstructive Hydrocephalus

Hydrocephalus has repeatedly been induced in otherwise normal animals by some form of physical obstruction of CSF flow pathways. The most commonly chosen method has been the intracisternal injection of a variety of agents, most commonly kaolin or silicone oils. The former is an irritant that produces leptomeningitis and thus induces hydrocephalus by inflammatory obstruction of the basal subarachnoid spaces. Inflammatory complications have been greatly reduced by the use of inert silicone oils that produce a direct mechanical obstruction to CSF flow and/or absorption.

For the most part, studies of hydrocephalus using such experimental models have shown a consistent array of pathological changes. Their sequence, severity, and extensiveness have been found to correlate with the degree of ventricular dilatation rather than the mode of induction.

BIBLIOGRAPHY

Black, P. McL., and Ojemann, R. G. (1990). Hydrocephalus in adults. *In* "Neurological Surgery" (J. R. Youmans, ed), 3rd Ed., Vol. 2, Chap. 41, pp. 1277–1298. Saunders, Philadelphia.

Bruni, J. E., Del Bigio, M. R., Cardoso, E. R., and Persaud, T. V. N. (1988). Hereditary hydrocephalus in laboratory animals and humans. *Exp. Pathol.* 35, 239–246.

Chapman, P. H. (1990). Hydrocephalus in childhood. *In* "Neurological Surgery" (J. R. Youmans, ed.), 3rd Ed., Vol. 2, Chap. 40, pp. 1236–1276. Saunders, Philadelphia.

Del Bigio, M. R. (1993). Neuropathological changes caused by hydrocephalus. *Acta Neuropathol.* **85**, 573–585.

Gleason, P. L., Black, P. McL., and Matsumae, M. (1993). The neurobiology of normal pressure hydrocephalus. *Neurosurg. Clin. N. Am.* 4(4), 667–675.

Hagberg, B., and Sjorgen, I. (1966). The chronic brain syndrome of infantile hydrocephalus. A follow-up study of 63 spontaneously arrested cases. *Am. J. Dis. Child* **112**, 189–196.

Mori, K., Shimada, J., Kurisaka, M., Sato, K., and Watanabe, K. (1995). Classification of hydrocephalus and outcome of treatment. *Brain Dev.* **17**, 338–348.

Pudenz, R. H. (1981). The surgical treatment of hydrocephalus—An historical review. *Surg. Neurol.* **15**, 15–26.

Hypercalcemia of Malignancy

GREGORY R. MUNDY
A. JOHN YATES
CHARLES A. REASNER
University of Texas Health Science Center

GLOSSARY

Humoral hypercalcemia of malignancy Hypercalcemic syndrome occurring in patients with cancer that is due to circulating factors produced by the tumor cells

Hypercalcemia Increase in the calcium concentration in the blood

Myeloma Neoplastic disorder of plasma cells universally associated with bone destruction, which may be either localized or generalized and often complicated by hypercalcemia

Primary hyperparathyroidism Overproduction of parathyroid hormone by a primary disorder of the parathyroid glands, usually due to a single benign parathyroid gland adenoma

ENDOCRINE SYNDROMES ASSOCIATED WITH CANcer are caused by humoral factors produced by the tumors and are unrelated to direct tumor cell invasion of tissues or to tumor metastases (spread of the tumor to distant sites). These syndromes have also been called paraneoplastic syndromes or ectopic hormone syndromes. They frequently mimic overproduction of hormones by benign tumors of endocrine glands. Among these syndromes, humoral hypercalcemia of malignancy is probably the most common. The hypercalcemic syndrome is similar to that seen in patients with primary hyperparathyroidism. In primary hyperparathyroidism, hypercalcemia is due to excess production of parathyroid hormone (PTH) by one or more of the four parathyroid glands. The endocrine syndromes associated with malignancy are responsible for considerable morbidity and even mortality. They are important for physicians to recognize because they can be frequently treated or prevented by appropriate therapy. Other examples of these endocrine syndromes include Cushing syndrome, inappropriate antidiuretic hormone secretion, hypoglycemia, cachexia, fever, and other nonmetastatic hematopoietic, neurologic, and dermatologic syndromes.

Tumor cells have an amazing capacity to synthesize and secrete peptides that cause systemic effects. Tumors have been described that have produced up to nine ectopic proteins. It is not surprising, therefore, that so many syndromes caused by the systemic effects of tumor peptides have been described.

Hypercalcemia is one of the most serious of these syndromes. It always indicates increased calcium resorption from the skeleton, although there are also usually effects of tumor products on the other major organs of calcium homeostasis, the kidney and the gut. Tumor involvement of the skeleton also causes bone pain and fracture after trivial injury, as well as hypercalcemia. This is a particularly unfortunate complication of cancer, because once tumors have metastasized to involve the skeleton, they are usually incurable, and the only therapy possible is palliative.

I. FREQUENCY OF HYPERCALCEMIA IN PATIENTS WITH MALIGNANCY

With the introduction of routine serum calcium measurements by the autoanalyzer in the mid-1970s, hypercalcemia has become recognized as a common electrolyte abnormality in the general population, as well as in hospitalized patients. Estimates of the incidence and prevalence of primary hyperparathyroidism have dramatically risen since that time. In addition, hypercalcemia has also been noted more frequently in patients with malignant disease. Epidemiologic surveys performed in England and the United States suggest that there are approximately 270 new cases of primary hyperparathyroidism per million population per year and approximately 150 new cases of hypercalcemia occurring in patients with malignant disease per million population per year. Primary hyperparathyroidism and malignant disease are the cause of hypercalcemia in the vast majority of hypercalcemic patients. Other causes of hypercalcemia listed in Table I are much less common, comprising about 10% of all patients.

A relatively small number of specific malignancies are responsible for hypercalcemia in most cancer patients (Table II). Malignant hypercalcemia is most often caused by squamous cell carcinoma of the lung, whereas anaplastic carcinoma or oat cell carcinoma of the lung is rarely associated with hypercalcemia. Carcinoma of the breast of all types is often responsible for hypercalcemia. Twenty to 40% of patients with hematologic malignancies, such as myeloma, develop hypercalcemia. Squamous cell tumors of the head and neck and the upper end of the esophagus are also often associated with hypercalcemia. Many common cancers are almost never associated with hypercalcemia (e.g., carcinoma of the colon and carcinoma of the female genital tract). In contrast, some rare malignancies are frequently associated with hypercalcemia [e.g., cholangiocarcinoma and vasoactive intestinal polypeptide (VIPomas)].

TABLE I
Causes of Hypercalcemia

Causes of Hypercalcemia
Primary hyperparathyroidism
Sporadic (80% adenoma)
Familial (hyperplasia)
Malignant disease
Hyperthyroidism
Immobilization
Vitamin D intoxication
Vitamin A intoxication
Familial hypocalciuric hypercalcemia (FHH)
Diuretic phase of acute renal failure
Chronic renal failure
Thiazide diuretics
Sarcoidosis and other granulomatous diseases
Milk-alkali syndrome
Addison's disease

TABLE II
Relative Frequency of Causes of Hypercalcemia

Cause	No. of patients	Percentage of cases
Primary hyperparathyroidism	111	54
Malignant disease	72	35
Lung	25	35
Breast	18	25
Hematologic (myeloma 5, lymphoma 4)	10	14
Head and neck	4	5
Renal	2	3
Prostate	2	3
Unknown primary	5	7
Others (gastrointestinal 4)	6	8

II. PATHOPHYSIOLOGY

In most patients with cancer, the hypercalcemia is due to a combination of increased calcium released from bone (increased bone resorption) and decreased excretion of calcium by the kidneys (increased renal tubular calcium reabsorption). Frequently, impairment of glomerular filtration occurs, as a result of either underlying disease or secondary to hypercalcemia. The reduced glomerular filtration rate further contributes to the decreased renal excretion of calcium. The combined effects of increased bone resorption and decreased renal excretion of calcium offset the decreased calcium absorption from the gut seen in most patients. The pathophysiology of hypercalcemia of malignancy is best considered in terms of three distinct syndromes: (1) humoral hypercalcemia of malignancy, (2) hypercalcemia associated with osteolytic disease, and (3) hematologic malignancies.

A. Humoral Hypercalcemia of Malignancy

The humoral hypercalemia of malignancy (HHM) syndrome is caused by tumor products that circulate and cause hypercalcemia because of their systemic effects. It is unrelated to local osteolytic bone destruction. Solid tumors are those most often associated with HHM (e.g., carcinomas of the lung, head and neck, kidney, pancreas, and ovary). Occasionally, patients with malignant lymphomas develop HHM. Most patients with HHM have many of the biochemical features of primary hyperparathyroidism, including hypercalcemia and increased urinary excretion of phosphate and cyclic AMP. However, other characteristics of HHM, including decreased calcium absorption from the gut, decreased bone formation, and metabolic alkalosis, are not seen in primary hyperparathyroidism. The variations that occur in different patients with HHM are most likely caused by differences in the tumor-derived factors and the variable response of the host immune cells. Specific factors that have been implicated in the pathophysiology of HHM are discussed in the following sections.

1. Parathyroid Hormone-Related Protein

Albright first suggested that the hypercalcemia of malignancy is caused by a parathyroid hormone-like substance produced by the tumor; the phrase "pseudohyperparathyroidism" was used to refer to this syndrome. Evidence that supported the presence of a PTH-like factor includes the presence of a factor in tumor tissue that stimulates adenylate cyclase activity in renal tubular membranes and osteoblast cells that can be inhibited by synthetic antagonists to parathyroid hormone. The protein responsible for this activity has been identified, and the gene has been molecularly cloned. Eight of the first 13 amino acids of this PTH-related protein (PTH-rP) are identical to those of PTH. The gene encoding PTH-rP is clearly different from the PTH gene and resides on a different chromosome, but probably had a similar evolutionary origin. There is more than one PTH-rP owing to differential use of the information of the gene during formation of messenger RNA (alternate processing) and possibly to proteolytic degradation of the protein after secretion. The PTH-rP is able to bind and activate the PTH receptor.

The PTH-rP seems to mimic all the biologic effects of PTH and, in addition, has its own unique effects. Like PTH, PTH-rP stimulates bone resorption both *in vitro* and *in vivo,* increases renal tubular calcium reabsorption, increases cyclic AMP generation, increases renal phosphate excretion, increases 1,25-dihydroxyvitamin D production, and causes hypercalcemia when given by injection. Despite the similarities between PTH-rP and PTH, the syndromes of HHM and primary hyperparathyroidism are clearly different. These differences are likely due to other factors that are produced by the tumor or host immune cells, which have unique effects on target tissues to modulate the effects of PTH-rP. [*See* Parathyroid Gland and Hormone.]

2. Transforming Growth Factor α

Solid tumors frequently produce polypeptide growth factors that may be responsible for maintaining the transformed phenotype in tumor cells. These autocrine factors have a similar conformation to that of epidermal growth factor and are able to bind to and activate the epidermal growth factor receptor. The best described of these factors is transforming growth factor α (TGFα). *In vitro,* TGFα is a potent stimulator of osteoclasts, whose function is to resorb bone. Its effects on bone are even more potent than those of PTH or PTH-rP. It also causes hypercalcemia when injected into normal mice. There is now considerable evidence that some tumors produce both TGFα and PTH-rP. This is clearly true in the rat Leydig tumor (a testicular tumor), a well-studied animal model of HHM. Because the effects of TGFα and PTH-rP on bone are synergistic *in vitro,* it is not surprising that the rat Leydig tumor is associated with profound increases in osteoclastic bone resorption. However, the relation between these factors may be complex. In some circumstances, TGFα opposes the effects of PTH-rP. For example, TGFα decreases the adenylate cyclase response to PTH-rP (and PTH) in both renal tubular cells and cells with the osteoblast phenotype (which cause bone formation). Moreover, *in vitro,* TGFα stimulates replication of osteoblastic cells but inhibits their differential function.

3. Tumor Necrosis Factor

Tumor necrosis factor (TNF) is a potent bone-resorbing cytokine (locally acting factor) that is secreted by normal activated macrophages. It stimulates the formation of osteoclasts by enhancing both proliferation of progenitor cells and differentiation of committed progenitor cells and also activates mature osteoclasts. Like TGFα and PTH-rP, TNF causes hypercalcemia when infused or injected *in vivo* or whenever cells carrying the gene for TNF are introduced into animals. Although there is no current evidence that solid tumors themselves produce TNF directly, a number of tumors have now been identified that have the capacity to stimulate TNF production by normal host cells. In some circumstances, TNF production is due to tumor production of granulocyte-macrophage colony-stimulating factor (GM-CSF), which in turn acts on host immune cells to provoke TNF release. In nude mice carrying some human tumors, the serum calcium can be lowered by injecting antibodies to TNF, suggesting that tumor cells produce bone-resorbing factors such as PTH-rP and TGFα and that host cells produce bone-resorbing cytokines such as TNF. Hypercalcemia is presumably due to the effects of these factors acting in concert. The hypercalcemia in nude mice caused by the introduction of cells carrying the TNF gene is associated with markedly increased osteoclastic bone resorption, indicating that increased circulating TNF, produced either by tumor cells or by host cells, can cause the syndrome of hypercalcemia of malignancy.

4. Interleukin-1

Interleukin-1 (IL-1) is a multifunctional protein that is present in supernatant media harvested from activated leukocyte cultures. It causes hypercalcemia when injected into normal mice. Both IL-1α and IL-1β appear to do so equally by stimulating osteoclastic bone resorption.

IL-1 causes marked changes in bone resorption. There are increases in osteoclast numbers and active bone resorption surfaces, and marked expansion of the marrow cavity. IL-1 causes increased osteoclast formation in a marrow culture system, which detects the effects of factors on the generation of osteoclasts from human mononuclear precursors. In addition, IL-1 stimulates differentiation of committed progenitors into mature osteoclasts and increases the capacity of the mature osteoclasts to resorb bone. Whether this effect is mediated directly or indirectly through other cells is still not clear.

Interleukin-1 is produced by several solid tumors and at least one lymphoma associated with HHM. Surprisingly, these tumors produce IL-1α, which is the less predominant form produced by normal activated leukocytes.

It is important to note that both IL-1 and TNF act synergistically on bone with PTH, PTH-rP, and TGFα. Like TGFα, IL-1 has been shown to modify the effects of PTH or PTH-rP on target bone cells. Preincubation of osteoblastic cells with IL-1 decreases the subsequent adenylate cyclase response to PTH or PTH-rP.

B. Osteolytic Metastases

Hypercalcemia often occurs in patients with widespread bone metastases associated with extensive osteolytic bone destruction. In these cases it is likely that the tumor cells either directly or indirectly stimulate local osteoclast activity, which is responsible for the bone destruction and hypercalcemia. The most common tumor in this category is breast cancer, in which hypercalcemia is almost always limited to patients with extensive metastatic bone disease. Renal tubular calcium reabsorption is increased in some of these patients via mechanisms that are not clear. Although occasional patients with breast cancer and hypercalcemia have been shown to produce PTH-rP, in the great majority, nephrogenous cyclic AMP is not increased and increased renal tubular calcium reabsorption is likely increased by other mechanisms. Hypercalcemia is estimated to occur in approximately one-third of patients with breast cancer. Breast cancer cells have the capacity to cause the release of ^{45}Ca from devitalized bone *in vitro*, suggesting that the tumor cells may have a direct role in localized bone resorption and the development of hypercalcemia. However, examination by scanning electron microscopy of bone surfaces suggests that osteoclast stimulation by tumor factors is the major mechanism.

Breast cancer cells produce a number of factors capable of stimulating osteoclast activity in culture, particularly TGFα. It is usually present in greater amounts in patients with hormone-independent, estrogen receptor negative cell lines. In the estrogen receptor positive cell lines, TGFα production can be increased by incubating the cells with estrogen. This may be important in the development of acute hypercalcemia seen in patients with breast cancer and widespread metastatic disease who have been treated with antiestrogens. Prostaglandins are also produced by many of these breast cancer cells, and their production is similarly enhanced by estrogens. Breast cancer cells also produce PTH-rP, particularly when in the bone microenvironment. This may be an important mechanism for local bone destruction. Because the mechanism of hypercalcemia in this important tumor is relatively unexplored, other factors produced by tumor cells could also be involved in the bone destruction. [*See* Breast Cancer Biology; Transforming Growth Factor, Alpha.]

C. Myeloma and Other Hematologic Malignancies

In myeloma, hypercalcemia occurs in approximately 50% of patients and is almost always associated with extensive osteolytic bone disease. Local factors produced by myeloma cells activate adjacent osteoclasts, causing the osteolytic bone lesions and, in some patients, hypercalcemia. Hypercalcemia is far more common in patients with impaired glomerular filtration, which itself is a common feature in myeloma and results from a variety of causes. Therefore, it is not surprising that hypercalcemia occurs so frequently.

The major mediator produced by myeloma cells that is responsible for increasing osteoclast activity appears to be lymphotoxin, a polypeptide cytokine released by normal activated T lymphocytes. Human or murine lymphotoxin stimulates osteoclastic bone resorption in organ culture and causes increased bone resorption and hypercalcemia when injected into normal intact mice. Not all the bone-resorbing activity can be blocked by antibodies to lymphotoxin, suggesting that other factors may also be involved. One possibility is the cytokine IL-6, which has the capacity to increase osteoclast formation from committed progenitors *in vitro*. Possibly, production of IL-6 in combination with lymphotoxin could be responsible for the effects observed on bone in patients with myeloma.

Patients with lymphomas occasionally develop hypercalcemia. This can be seen with T-cell or B-cell lymphomas, or histiocyte lymphomas. The mechanisms here are clearly heterogeneous. In some of these patients, increased circulating concentrations of 1,25-dihydroxyvitamin D have been found, probably due to the presence of 1α-hydroxylase activity in the tumor cells, which results in the conversion of 25-hydroxyvitamin D to the active form of vitamin D. However, this is a rare event, and serum concentrations of 1,25-dihydroxyvitamin D are not increased in the majority of patients. Other lymphomas have been shown to produce IL-1α and PTH-rP. [*See* Lymphoma.]

III. TREATMENT

A. General

Medical therapy for hypercalcemia of malignancy is usually successful; many agents will lower the serum calcium satisfactorily. However, no therapy is ideal, and the treatment choice should be individualized for each patient. The selection of an individual form of therapy depends on a number of factors, including the rate of rise of the serum calcium, specific contraindications to a particular form of therapy (such as the presence of renal failure), the extent of the increase in serum calcium, the presence of symptoms, and the pathophysiology of the hypercalcemia. The general principles of therapy for hypercalcemia are (1) effective treatment of the tumor where possible, (2) treatment of any exacerbating cause that has precipitated hypercalcemia, and (3) treatment of dehydration, which is common in patients with the hypercalcemia of malignancy and may lead to a spiral of disequilibrium hypercalcemia and progressive worsening of the serum calcium as fluid losses provoke hypercalcemia, which causes nephrogenic diabetes insipidus and consequently worsens the dehydration.

In patients with malignant disease, hypercalcemia is usually progressive, and the serum calcium may rise rapidly over a few weeks. This occurs because of progressive increases in bone resorption and overall bone destruction. Because the symptoms of hypercalcemia are unpleasant and can be prevented, we feel all patients with the hypercalcemia of malignancy should be treated actively.

B. Nonurgent Therapy for Hypercalcemia

Patients with a serum calcium less than 13 mg/dl at the time of diagnosis usually do not have symptoms related to hypercalcemia or have relatively mild symptoms. A number of agents are satisfactory for this situation. These agents may also be used in patients previously treated for more severe hypercalcemia who have mild residual hypercalcemia.

1. Bisphosphonates

Bisphosphonates, analogues of pyrophosphate, bind to mineralized surfaces and inhibit mineral deposition and osteoclastic bone resorption. They have a direct cellular effect to inhibit the bone resorption process, although their precise mechanisms of action are still unclear. In the United States, etidronate and pamidronate are the only members of this group currently approved for human use. However, it is likely in the next few years that additional members of the bisphosphonate family will be approved. The newer generation bisphosphonates are extremely effective in lowering the serum calcium, usually within 72–96 hr. These drugs are effective in more than 95% of patients with the hypercalcemia of malignancy and are the medical agents of first choice.

2. Plicamycin

Plicamycin was developed as a cytotoxic drug during the 1960s and is an effective antitumor agent in certain rare embryonal tumors. It was found by serendipity that plicamycin also lowered the serum calcium, and it has been widely used since then for the treatment of hypercalcemia of malignancy. It inhibits DNA-dependent RNA synthesis, and it is probably cytotoxic for osteoclasts. In organ culture, its effects are long-lasting and presumably irreversible. It is extremely effective in the treatment of hypercalcemia of malignancy, lowering the serum calcium in greater than 80% of patients. However, it must be given by intravenous infusion over several hours. The infusion should not be repeated until the serum calcium rises again. It is potentially dangerous in patients who have impaired renal function because it has direct nephrotoxic effects and is cleared by the kidney. Other side effects include bone marrow suppression, a nonthrombocytopenic bleeding diathesis, and hepatotoxicity. For these reasons, plicamycin should be reserved for patients with normal renal function when other drugs have been unsuccessful.

3. Calcitonin and Glucocorticoids

Glucocorticoids have been used for many years in the treatment of hypercalcemia of malignancy. When used alone they reduce serum calcium in approximately 30% of all patients but rarely to the normal range. They are most likely to be effective in patients with hematologic malignancies such as myeloma or the lymphomas. They are occasionally effective in patients with breast cancer or lung cancer, but it is difficult to predict which patients will respond. *In vitro,* they are effective inhibitors of bone resorption stimulated by agents such as the cytokines, but are less effective against PTH- or PTH-rP-stimulated resorption. Calcitonin is not particularly convenient as it has to be injected. The many chronic side effects of glucocorticoids are not seen in most patients, who in general have no more than 3–6 months of remaining life owing to the progression of the tumor.

Calcitonin alone in many patients produces a transient fall of serum calcium, which is not maintained. This "escape phenomenon" is seen both *in vitro* and *in vivo.* However, the escape phenomenon can be blocked by glucocorticoids. The combination of the two drugs is effective in patients who require a rapid fall in serum calcium, particularly in patients with hematologic malignancies. It is an attractive form of treatment in patients with fixed impairment of renal function, in whom most other forms of treatment are contraindicated. In many patients, it may be necessary to withdraw calcitonin for 48 hr every week to avoid escape. There are no significant side effects when calcitonin is used in these doses.

4. Phosphate

Oral phosphate has been used for many years in the treatment of hypercalcemia. It is only likely to be effective in those patients in whom the serum phosphorus is less than 4 mg/dl. The mechanism of action is complex. Phosphate inhibits osteoclastic bone resorption, decreases calcium absorption from the gut, and promotes soft tissue calcium deposition. Oral phosphate should not be used in patients who have impaired renal function or who have serum phosphorus levels greater than 4 mg/dl. The major side effect of long-term phosphate therapy is troublesome diarrhea, which often limits its usefulness. Intravenous phosphate should only be used when other forms of therapy have failed. It frequently provokes extraskeletal calcium deposition, particularly in the presence of impaired renal function.

5. Gallium Nitrate

Gallium nitrate, which was originally used as a cytotoxic agent, inhibits bone resorption *in vitro* and produces prolonged lowering of the serum calcium *in vivo.* It seems to be effective in most forms of hypercalcemia of malignancy. It has recently been approved for treatment of hypercalcemia of malignancy and is moderately effective with relatively few side effects.

6. Indomethacin and Aspirin

Indomethacin and aspirin are very effective inhibitors of prostaglandin synthesis and were widely used when prostaglandins were thought to be the major cause of hypercalcemia of malignancy. However, patients rarely respond to them.

C. Emergency Treatment of Hypercalcemia

Emergency treatment for hypercalcemia is necessary in patients with a serum calcium greater than 13 mg/dl. This level of serum calcium is almost always associated with severe symptoms, including thirst, nausea, and altered mental status. Such hypercalcemia is potentially life-threatening in patients with malignant disease in whom the serum calcium may rapidly rise further, particularly if patients become dehydrated. Other situations that can precipitate severe hypercalcemia include thiazide therapy, treatment with estrogens or antiestrogens in patients with breast cancer, or vomiting with associated dehydration and prerenal azotemia.

1. Fluids

Patients with severe hypercalcemia are often dehydrated and may require repletion with 6–10 liters of normal saline over 24 hr. Because severe hypercalcemia is associated with fluid depletion, and calcium and sodium handling by the renal tubules is linked, normal saline should be given to replace fluid losses. Normal saline repletion will lead to a sodium and calcium diuresis, which will help lower the serum calcium. Occasionally, patients will become hypernatremic (excess sodium), and under these circumstances half-normal saline should be substituted for normal saline.

2. Loop Diuretics

Furosemide has been frequently used in the United States for the emergency treatment of hypercalcemia because it promotes a calcium diuresis. However, in the doses used by most clinicians it is not clear that it produces a significant effect beyond that of saline therapy alone. Moreover, loop diuretics at the required doses may be dangerous and may lead to other electrolyte problems, dehydration, and worsening of hypercalcemia. [*See* Diuretics.]

3. Bisphosphonates

The most effective current agents in the emergency treatment of hypercalcemia are the bisphosphonates. Etidronate is effective in approximately 70–80% of patients and will usually lower the serum calcium to the normal range after 72 hr, without serious toxicity. Normalization of calcium occurs in about 50% of related cases. The major side effect is impairment of bone mineralization, which reduces the uptake of calcium by the skeleton and is the reason why etidronate is less effective than newer, more potent bisphosphonates that do not interfere with mineralization. Osteomalacia is not a concern in the short term but may be more important in patients receiving large doses over prolonged periods of time. Pamidronate is effective in approximately 95% of patients and will usually lower the serum calcium to the normal range after 72 hr without serious toxicity. Pamidronate is more effective than etidronate. Within the next few years, there will probably be additional bisphosphonates available that have similar efficacy to pamidronate, but will have the advantage that they can also be taken orally.

4. Calcitonin and Glucocorticoids

The combination of calcitonin and glucocorticoids is often effective in the emergency treatment of hypercalcemia, especially in patients with hematologic malignancies. It is probably the most rapid and safest form of therapy in patients who have hypercalcemia associated with cardiac failure or renal failure. Unfortunately, it is not universally effective. The use of calcitonin together with a bisphosphonate may cause a more rapid lowering of the serum calcium than the use of the bisphosphonate alone.

5. Plicamycin

Plicamycin (mithramycin) is often efficacious in the treatment of severe hypercalcemia but has to be given by infusion and usually takes 24–48 hr to produce a beneficial effect. It should be avoided in patients who have impaired renal function because of its nephrotoxicity.

6. Dialysis

Dialysis has been used occasionally in the emergency treatment of hypercalcemia in patients with renal failure. It is useful only as a transient remedy when other therapies are contraindicated or ineffective.

ACKNOWLEDGMENTS

The authors are grateful to Nancy Garrett and Thelma Barrios for expert secretarial assistance in the preparation of this manuscript. Part of the work described here was supported by Grants CA 40036, RR-1346, AR-28149, and AR-39357.

BIBLIOGRAPHY

Case Records of the Massachusetts General Hospital (Case 15-1971) (1971). *N. Engl. J. Med.* **284,** 839–847.

Garrett, I. R., Durie, B. G. M., Nedwin, G. E., Gillespie, A., Bringman, T., Sabatini, M., Bertolini, D. R., and Mundy, G. R. (1987). Production of the bone resorbing cytokine lymphotoxin by cultured human myeloma cells. *N. Engl. J. Med.* **317,** 526–532.

Grill, V., Ho, P., Body, J. J., Johanson, N., Lee, S. C., Kukreja, S. C., Moseley, J. M., and Martin, T. J. (1991). Parathyroid hormone-related protein—Elevated levels in both humoral hypercalcemia of malignancy and hypercalcemia complicating metastatic breast cancer. *J. Clin. Endocrinol. Metab.* **73,** 1309–1315.

Moseley, J. M., Kubota, M., Diefenbach-Jagger, H., Wettenhall, R. E. H., Kemp, B. E., Suva, L. J., Rodda, C. P., Ebeling, P. R., Hudson, P. J., Zajac, J. D., and Martin, T. J. (1987). Parathyrhoid hormone-related protein purified from a human lung cancer cell line. *Proc. Natl. Acad. Sci. USA* **84,** 5048-5052.

Mundy, G. R. (1988). The hypercalcemia of malignancy revisited. *J. Clin. Invest.* **82,** 1–6.

Mundy, G. R., and Martin, T. J. (1982). Hypercalcemia of malignancy—Pathogenesis treatment. *Metabolism* **31,** 1247–1277.

Mundy, G. R., Ibbotson, K. J., D'Souza, S. M., Simpson, E. L., Jacobs, J. W., and Martin, T. J. (1984). The hypercalcemia of malignancy: Clinical and pathogenic mechanisms. *N. Engl. J. Med.* **310,** 1718–1727.

Mundy, G. R., Ibbotson, K. J., and D'Souza, S. M. (1985). Tumor products and the hypercalcemia of malignancy. *J. Clin. Invest.* **76,** 391–395.

Myers, W. P. L. (1960). Hypercalcemia in neoplastic disease. *Arch. Surg.* **80,** 308–318.

Nakai, M., Mundy, G. R., Williams, P. J., Boyce, B., and Yoneda, T. (1992). A synthetic antagonist to laminin inhibits the formation of osteolytic metastases by human melanoma cells in nude mice. *Cancer Res.* **52,** 5395–5399.

Powell, D., Singer, F. R., Murray, T. M., Minkin, C., and Potts, J. T. (1973). Non-parathyroid humoral hypercalcemia in patients with neoplastic disease. *N. Engl. J. Med.* **289,** 176–181.

Sporn, M. B., and Todaro, G. J. (1980). Autocrine secretion and malignant transformation of cells. *N. Engl. J. Med.* **303,** 878–880.

Suva, L. J., Winslow, G. A., Wettenhall, R. E. H., Hammonds, R. G., Moseley, J. M., Diefenbach-Jagger, H., Rodda, C. P., Kemp, B. E., Rodriguez, H., Chen, E. Y., Hudson, P. H., Martin, T. J., and Wood, W. I. (1987). A parathyroid hormone-related protein implicated in malignant hypercalcemia: Cloning and expression. *Science* **237,** 893–896.

Yates, A. J. P., Gutierrez, G. E., Smolen, P., Travis, P. S., Katz, M. S., Aufdemorte, T. B., Boyce, B. F., Hymer, T. K., Poser, J. W., and Mundy, G. R. (1988). Effects of a synthetic peptide of a parathyroid hormone-related protein on calcium homeostasis, renal tubular calcium reabsorption and bone metabolism. *J. Clin. Invest.* **81,** 932–938.

Yoneda, T., Alsina, M. M., Chavez, J. B., Bonewald, L., Nishimura, R., and Mundy, G. R. (1991). Evidence that tumor necrosis factor plays a pathogenetic role in the paraneoplastic syndromes of cachexia, hypercalcemia, and leukocytosis in a human tumor in nude mice. *J. Clin. Invest.* **87,** 977–985.

Hypertension

MORTON P. PRINTZ
University of California, San Diego

GLOSSARY

Autonomic nervous system Nervous system that carries information from the brain to all the internal organs; this system operates with minimal, if any, conscious effort and is an autonomic responding system; two parts of the autonomic system are recognized: a sympathetic branch and a parasympathetic branch; almost all organs receive nerve fibers of both parts, and optimum organ system function requires both to be operational

Cardiac output Amount of blood pumped by the left ventricle of the heart in 1 min; this is a measure of the pumping action of the heart

Cardiovascular system Refers to the organ system, which includes the heart and all the blood vessels, arteries, veins, and capillaries

Diastolic pressure Minimal arterial blood pressure, which occurs during the rest phase (or diastole) of the heart

Endothelial cells Cells that line the inner face of blood vessels and are in immediate contact with blood

Heart left ventricle The heart consists of four chambers: right and left atria and right and left ventricles; the atria serve to pump the blood into the ventricles, which are the main pumping chambers; blood returns from the body and enters the right atrium and is pumped to the right ventricle and then through the lungs; it returns to the left atrium, which pumps it into the left ventricle; the left ventricle is the pumping chamber that forces the blood out into the arteries toward all the tissues

Neurotransmitters Chemicals that are made and stored in nerve cells and that, when released by a process termed exocytosis, carry information from one nerve cell to another cell; some neurotransmitters excite the target cell, whereas others inhibit; however, all are carriers of information

Systolic pressure Maximal arterial blood pressure, which occurs when the left ventricle of the heart pumps blood into the arteries

Vascular resistance Resistance to the flow of blood out from the arteries into the capillary blood vessels in the organs

HYPERTENSION, OR HIGH BLOOD PRESSURE, REFERS to an elevation of systemic arterial blood pressure that is greater than expected based on population measurements of systemic arterial pressure for individuals of comparable age and sex. Blood pressure and volume are under the control of many organ systems, among which are the heart, nervous system, blood vessels, and the kidney. Hypertension is considered to be a disorder of regulation rather than a simple disease process. "Essential" or primary hypertension is the most common form of the disorder and its cause is still unknown. Once a diagnosis of hypertension is made, patients are classified according to the severity of the hypertension and damage to critical organs such as the heart or kidneys. Therapy of hypertension is individualized and consists of both drug and nondrug approaches. There are several types of antihypertensive drugs, each type designed to influence one or more of the mechanisms of the body that control blood pressure or volume. The high occurrence of this disorder throughout the world, combined with the evidence that untreated hypertension leads to increased mortality from heart disease, stroke, kidney failure, and atherosclerosis, makes hypertension a major medical problem.

I. BLOOD PRESSURE: DEFINITION AND CHARACTERIZATION

A. Definition

Blood pressure, at its simplest, is the pressure of the liquid within blood vessels; however, hypertension refers to elevated pressure in the arteries, the blood vessels that serve as the pathway for blood from the heart to all the tissues. The blood pressure in the arteries is determined by (1) the force provided the blood by the heart during its contraction and ejection of blood into the arterial compartment (blood vessels), (2) the rate of flow of blood out of the arterial compartment and into the tissues via the capillaries, (3) the volume of blood within the vascular compartment, and (4) the tension generated by the walls of the blood vessel resisting the blood pushed into the arteries by the heart. To understand how hypertension could develop by a variety of possible mechanisms, a brief discussion of the physiology of the cardiovascular system is warranted.

B. Vascular Compartments

Humans, like all animals, have a closed and continuous compartment through which blood flows carrying oxygen and nutrients to tissues, and carbon dioxide and waste products from tissues to sites of metabolism and elimination. This compartment, referred to as the vascular compartment, actually consists of three separate but connected compartments: arterial (arteries), venous (veins), and capillary. [*See* Cardiovascular System Anatomy.]

The arterial compartment commences at the outflow from the left ventricle of the heart and extends, in an ever-branching manner, to all the organ systems and tissues. The large vessels of the arterial system are termed conduit vessels, because they serve mainly as feeder vessels to the larger number of smaller vessels ramifying to the tissues. The arterial system is a high-pressure circuit, and the large conduit vessels receive the full pressure and flow of the blood pumped from the left ventricle of the heart. As the vessels ramify to smaller and smaller vessels, the pressure within these vessels declines and is lowest (in the arterial circuit) at the junction with the capillary network of the tissues.

Arterial blood vessels are complex and consist of multiple cell types and three discrete layers. In immediate contact with blood is a single-cell-thick lining consisting of endothelial cells with underlying connective tissue protein and termed the intima of the vessel. The middle of the arterial blood vessel (termed the media) consists of multiple layers of contractile smooth muscle cells. The outer layer (termed the adventitia) consists of connective tissue cells (fibroblasts), elastic and connective tissue proteins, sympathetic nerve endings, and other cell types, which vary according to the size of the vessel and organ system. The media smooth muscle layer probably has the most important role in determining the level of arterial blood pressure in this vascular compartment. This will be discussed later.

The capillary compartment consists of the smallest blood vessels in all three compartments of the vascular system. These vessels consist of a single cell layer of endothelial cells with associated connective tissue proteins. The structure of the capillary system serves to permit rapid movement of materials, in both directions, between the blood contained within the vessel and the adjoining tissue. This vascular compartment lacks both the outer adventitial and medial smooth muscle layers found in the arterial system and, therefore, is primarily a passive circuit for exchange.

The venous compartment commences at the outflow from the capillary compartment and extends to the right atrium of the heart. This vascular system serves two functions: to collect the blood and return it to the heart and to act as a reservoir of blood in the body. The venous compartment is essentially a mirror image of the arterial compartment with small veins, highly ramified, joining into successively larger vessels. Veins are complex vessels just like arteries and consist of an endothelial cell lining, medial smooth muscle, nerves, and multiple types of cells; however, the venous system is a low-pressure circuit. The pressure in the large veins is approximately 3–5 mm of mercury. Accordingly, the smooth muscle layer is much thinner than that found in the arterial compartment for an equivalent-sized vessel. The veins are referred to as capacitance vessels because they have the capacity to expand and hold large amounts of blood. When needed, this blood can be recruited into the arterial and capillary compartments through the activity of the sympathetic nervous system. [*See* Vascular Capacitance.]

The vessels connecting the heart and lung constitute a special compartment of the vasculature. While arteries from the left ventricle of the heart to all the organ systems (except the lungs) carry highly oxygenated blood and veins returning to the right atrium of the heart carry low-oxygenated blood from the tissues, the converse exists in the blood vessels of the lungs (called pulmonary vessels). The pulmonary artery,

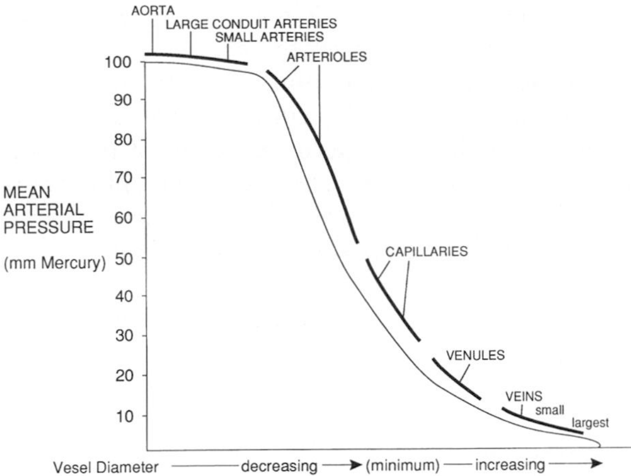

FIGURE 1 Schematic of the distribution of mean arterial pressure through the vascular compartment of humans. As discussed in the text, the maximum decrease in arterial pressure occurs in the resistance vessels, the arterioles, with a further decrease in pressure through the capillaries. Arterial pressure continues to decrease through the venous side of the circulation, reaching a minimum at the entry to the right heart receiving chamber, the right atrium. In contrast to a continuous drop in arterial pressure, blood vessel diameter first decreases to a minimum in the capillary network and then increases again through the venous capacitance vessels.

from the right ventricle of the heart to the lungs, carries venous blood of low-oxygen content, whereas the pulmonary vein, from the lungs to the left atrium of the heart, carries highly oxygenated blood. [*See* Cardiovascular System, Physiology and Biochemistry.]

Pulmonary hypertension is excess blood pressure in the pulmonary artery from the heart to the lung. It is not equivalent to systemic arterial hypertension, the subject of this article, and has, in general, other causes and treatment. For this reason, we will not deal with this special form of hypertension.

C. Levels of Blood Pressure in Vascular Compartments

Blood pressure is highest in the large conduit arteries leaving the left ventricle of the heart (aorta and carotid arteries) and decreases steadily through the arterial, capillary, and venous compartments; however, the largest drop in pressure occurs on the arterial side (Fig. 1). Maintenance of this decreasing gradient of pressure is essential to ensure adequate blood delivery to all tissues and optimum fluid and material exchange in the capillary network and to deliver the blood back to the right side of the heart to be pumped through the lungs and recycled through the body. Disease processes that interfere with any of these three functions will invariably result in a form of hypertension.

D. Determinants of Arterial Blood Pressure

Arterial blood pressure is determined by the pumping action of the heart, by the resistance to flow of the blood out from the arterial vascular bed and into the capillaries, and by the compliance (distensibility) of the walls of the arteries. The arterial pressure varies during the contraction phase (termed systole) and rest phase (termed diastole) of the left ventricle of the heart. When blood pressure is measured, both upper and lower values are quoted (e.g., 120 over 80). The upper value is the maximum pressure measured (systolic pressure), whereas the lower value is the lowest (diastolic pressure). The magnitude of the difference between systolic pressure and diastolic pressure is pulse pressure.

During contraction of the left ventricle of the heart (i.e., systole), blood is ejected into the large arterial conduit vessels and arterial pressure builds rapidly to the maximum systolic pressure (Fig. 2). The magni-

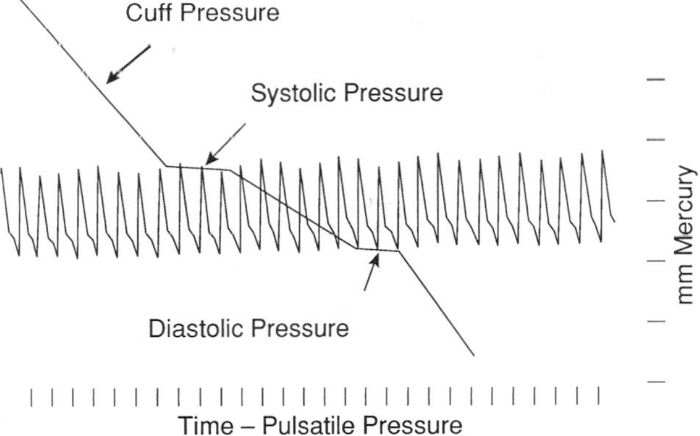

FIGURE 2 The pulsatile pressure profile is illustrated schematically in a large conduit artery. As discussed in the text, the maximum arterial pressure is termed systolic pressure and the minimum is termed diastolic pressure. Also illustrated is the relationship between the sphygmomanometer (cuff) pressure and a patient's actual arterial pressure. When the cuff pressure is greater than the systolic pressure, blood flow is impeded. To obtain a measurement of systolic and diastolic pressures, the cuff pressure is slowly decreased. When the cuff pressure equals the maximum arterial pressure, flow begins through the artery and the sound of flowing blood, heard with a stethoscope, marks systolic pressure. As blood flows through a partially restricted blood vessel, a characteristic sound continues to be heard. Diastolic pressure is that cuff pressure associated with a loss of the characteristic sound of flow through the partially restricted vessel.

tude of systolic pressure is determined by the rate and force of ejection of blood into the arterial vessels by the left ventricle of the heart, by the degree of compliance of the large conduit arteries, and by the rate of outflow from the arterial compartment into the capillary compartment. During diastole, as the left ventricle relaxes, permitting a refilling of the left ventricle by blood from the left atrium, arterial pressure declines to the minimum diastolic pressure (Fig. 2). The rate of decline is governed both by the rate of outflow of blood from the arterial compartment and by the elasticity of the conduit arteries. The latter results in what is termed a windkessel effect, namely, a rebound by the distended vessel wall, which results in pressure being exerted on the blood in the arteries. The windkessel process is very much like the stretched rubberband of a slingshot rebounding and exerting force on the object being propelled. The pressure does not drop to zero during diastole because the heart contracts, repeating the cycle. Under normal conditions, if the heart rate is slowed, the minimum (diastolic) pressure reached would be expected to be lowered.

A critical determinant of both systolic and, especially, diastolic pressure is the rate of outflow from the arterial compartment. This rate with which blood leaves the arterial compartment and enters the capillary compartment is determined by the degree of constriction of resistance elements in small arteries. These resistance elements, which are equivalent to valves regulating the flow of water from a water hose, are small arterial blood vessels (approximately 30–50 μm in diameter). They are termed resistance elements because by varying the degree of constriction of the vessel they set the resistance to blood flow out of the arterial compartment and, thereby, the level of arterial pressure. Constricting the resistance elements leads to increased arterial pressure, whereas relaxing (dilating) the elements leads to decreased arterial pressure. Note that both systolic and diastolic pressures are influenced by these resistance elements.

The smooth muscle of the media of the resistance elements controls the degree of constriction or relaxation of the vessel. Smooth muscle cells within the media layer are spindle-shaped and oriented side-by-side and at an angle to the direction of the vessel. The small resistance vessels and, specifically, their smooth muscle cells are richly supplied by nerves under the control of the sympathetic branch of the autonomic nervous system. The nerves controlling the resistance vessels release a chemical (neurotransmitter) called norepinephrine, which upon binding to a specific protein on the smooth muscle cell (termed the alpha$_1$-adrenergic receptor) causes the smooth muscle cells to constrict. Through coordinated contraction of the many smooth muscle cells within the wall of the vessel, the inner diameter (termed lumen) of the vessel is reduced and constricted. In this manner, blood flow through the lumen is impeded, resistance to flow is increased, and arterial pressure rises. [*See* Smooth Muscle.]

Although systolic and diastolic pressures are important in determining whether or not a patient has true hypertension, to understand the possible causes and treatment of high blood pressure, we must also consider the interrelationships among the heart, blood vessels, and autonomic nervous system. Rather than distinguish the extremes of pressure (systolic or diastolic pressure), we speak about the arterial compartment as a whole and use an "averaged" arterial pressure, namely, mean arterial pressure (MAP). Mean arterial pressure, when averaged over time, is defined by the following relationship involving cardiac output (CO) and total peripheral vascular resistance (TPVR): $\text{MAP} = \text{CO} \times \text{TPVR}$, or mean arterial pressure is the product of cardiac output and total peripheral vascular resistance. TPVR is the total resistance offered by the arterial vascular compartment to the flow of blood out from the arterial compartment. Cardiac output is the amount of blood (in liters) pumped by the left ventricle of the heart over a full minute. This volume of blood pumped in 1 min is determined by the force of contraction of the left ventricle, by the rate of contraction of the heart (i.e., the heart rate), and by the amount of blood contained within the left ventricle chamber during each contraction. The latter is controlled partly by the amount of blood that returns to the heart from the venous compartment, termed venous return, and by the resistance encountered when the heart pumps the blood into the arterial circuit. Because the veins function largely as reservoirs for the blood, changes in both blood volume and the degree of constriction of venous smooth muscle influence the low blood pressure in the veins and the amount of blood returned to the heart. Cardiac output is defined by the following relation involving stroke volume (SV), that is, the volume of blood ejected by the left ventricle with each beat and heart rate (HR): $\text{CO} = \text{SV} \times \text{HR}$. Therefore, mean arterial pressure is determined by the stroke volume, the heart rate, and the total peripheral vascular resistance of the arteries, or $\text{MAP} = \text{SV} \times \text{HR} \times \text{PVR}$.

Because arterial blood pressure is governed by the action of the heart, by the state of resistance to flow by the vessels, by the state of constriction of veins, and by the total volume of the system, high blood pressure could be caused by several different factors; however, numerous control mechanisms operate to compensate for changes in one or more of these factors so as to keep arterial blood pressure constant. For example, if peripheral vascular resistance increases, cardiac output is decreased through homeostatic mechanisms by decreasing heart rate. Likewise, if cardiac output drops, as, for example, might occur if there is sudden loss of blood due to hemorrhage, there is a homeostatic increase in sympathetic nervous system discharge to blood vessels resulting in constriction of smooth muscle in resistance elements of the arterial bed and a concomitant increase in total peripheral vascular resistance and arterial pressure. This homeostatic regulation of arterial pressure is dynamic and operates continuously. The regulatory system responds very rapidly to changes in pressure. For example, changes may be sensed by the system within two heart beats and reflex responses initiated.

Hypertension is considered a disorder of regulation, in part because in hypertensives these homeostatic mechanisms appear to continue to operate but regulate at a higher level of arterial pressure. A "resetting" of the level of homeostasis permits sustained elevated arterial pressure and, ultimately, the development of hypertension. With chronic elemvation of arterial pressure, there is a thickening of the smooth muscle layer of the arteries (possibly so as to resist the increased pressure within the blood vessel). This thickening causes the lumen of the vessel to become smaller, which itself raises the resistance to flow (increases vascular resistance). Over time, the changes in the blood vessels also lead to a decrease in the distensibility or compliance of the vessel. As it becomes less distensible, there is less "give" with each pressure pulse from the heart's contraction, and this also will result in a marked elevation of systolic and diastolic pressures.

II. HYPERTENSION AS A DISORDER IN HUMANS

A. Normative Levels of Arterial Pressure in Humans

Blood pressure is termed a "quantitative trait" because there is not a single value for "normal" arterial blood pressure. Rather, blood pressure depends on age, sex, body weight and behavior or activity of the individual. Blood pressure increases with age and generally is lower in women than in men who are the same age. This tendency toward lower blood pressures in women may be changing in Western societies owing to the increasing involvement of women in professional and business life. The reasons for such a change are unclear but may include increased stress or altered life-style, increased consumption of alcohol or cigarette smoking, or dietary changes. More epidemiological studies are needed to determine if a trend toward more hypertension in women is really developing. Blood pressure is also influenced by body mass or weight and by extent of physical conditioning (exercise, etc.). Therefore, a blood pressure that is considered "normal" in one person may be abnormal in another.

Through measurements of large numbers of adult subjects, all in apparent good health, the well-known values of 120 (systolic) and 80 (diastolic) have entered our vocabulary to imply "normal blood pressure." In general, these values are good approximations for the population of such subjects as a whole; however, biological variability results in a range of "normal" blood pressures with some being on the extreme low or extreme high end. Such individuals are not necessarily hypertensive; therefore, a complete history and physical examination with appropriate tests are essential to define whether or not a patient has the disorder of hypertension.

No single blood pressure value defines when a patient becomes classified as having hypertension. Because of the variability in blood pressure values among the population as a whole, as well as the many factors that go into setting the level of arterial pressure, classification as to when an individual is hypertensive and requires therapy has been controversial. This question is especially important for patients who fall in the borderline hypertensive category (i.e., have diastolic pressures between 90 and 95 mm Hg) because many of the antihypertensive drugs used for treatment may, with long-term usage, have undesirable side effects that must be monitored and prevented. Studies of the adult population have resulted in guidelines. For adults within the ages of 21–60 years and with a weight within the normal range for their age and body build, a diastolic pressure >90 mm Hg, taken in an appropriate manner and with a complete evaluation (see the following), is considered evidence for a possible classification of hypertension. In patients not under any antihypertensive medications, diastolic pressures between 90 and 104 mm Hg define the category of mild hypertension, with 90–94 mm Hg constituting borderline hypertension. Likewise, patients not receiving therapy with diastolic pressures between 105 and 120 mm Hg are placed into the moderate to severe hypertension range. Because drug therapy is judged optimally effective when the diastolic pressures are brought into the normal range (i.e., <90 mm Hg), the foregoing categorization applies only when the patients are free of any antihypertensive medications.

Why is diastolic pressure used to identify and categorize hypertensives? Though elevated systolic pressures (160 mm Hg or more) generally alert health care professionals to the possibility of hypertension and, in hypertensive patients, may provide the physician information about the long-term effects of the disorder in that patient, systolic blood pressure is very sensitive to stress or anxiety. The systolic pressure, as discussed earlier, is determined largely by the action of the heart pumping blood into the arterial vascular bed. On the other hand, diastolic pressure is determined largely by the peripheral vascular resistance and blood vessel compliance. Transient and reversible episodes of elevated arterial pressure would not necessarily constitute a diagnosis of hypertension; however, sustained elevation of arterial pressure would be an indication of either changes in the blood vessel structure (as discussed earlier) or alterations in homeostatic regulation. These changes would be most evident when the heart is in diastole (rest phase). For this reason, diastolic pressure is critical to the diagnosis and decision of when to begin therapy and the type of therapy. When the diastolic pressure rises to >95–100 mm Hg, there is no question but that hypertension, or some other cardiovascular disorder, must be considered and appropriate treatment instituted.

B. How is Hypertension Recognized?

Although many people believe that headaches or rapid heart beat may indicate high blood pressure, these symptoms are not most commonly associated with hypertension. Hypertension has been termed the "silent killer," because a patient may be hypertensive for many years without any evident symptoms. Prior to recent efforts among health professionals to identify hypertensive patients, ophthalmologists or even opticians would discover hypertensive patients when examining the retina (the back inner surface of the eye). The retina is highly vascularized and with sustained, chronic, untreated hypertension the vascular bed becomes damaged, which may be readily observed on examination. Atherosclerosis and diabetes may also lead to changes in retina blood vessels, so retinal examination by itself will not provide a diagnosis of hypertension.

Diagnosis of hypertension is also commonly discovered during routine physical examinations. However, the procedures followed by the physician or health care professional must be rigorous to make a diagnosis. The institution of drug therapy without a thorough examination and analysis (see the following) is not advisable. Why is diagnosis so difficult? The reason is that the measurement of arterial pressure using devices external to the body is not trival! Almost all methods of measurement employ a form of the sphygmomanometer technique, an indirect method to measure pressure using the cuff. This involves measurement of the external pressure required to occlude, totally or in part, blood flow in a readily accessible artery, such as the brachial artery of the upper arm. The procedure involves wrapping an appropriately sized rubber cuff (which is an inflatable air bag like a balloon) around the upper arm and pumping air into the cuff to inflate it, thereby impeding the flow of blood through the major artery (basilar) of the arm. A stethoscope is placed in a location on the side of the cuff away from the heart so as to listen to characteristic sounds (termed Korotkoff sounds) formed by the resumption of blood flow as the cuff air pressure is slowly lowered. In the absence of flow, no sounds are heard. Two sounds are generally listened for. When the first sound is heard, which is the start of blood flow in the vessel as the cuff pressure is lowered, the cuff air pressure at that point defines systolic (maximal) pressure. This air pressure, and therefore blood pressure, is expressed in terms of millimeters of mercury (mm Hg) because the first, and still most accurate, sphygmomanometers use a column of liquid mercury to determine air pressure. When flow sounds are lost (indicating no further cuff restriction on flow in the artery), the cuff pressure is taken as the diastolic (minimal) pressure. Automated devices, which are gaining wider usage, also use an occluding technique as described for the sphygmomanometer.

Though the cuff blood pressure measuring devices found in pharmacies and food stores, which use the finger rather than the arm, may be quite reliable, they cannot be used to exclude or to diagnose hypertension; rather, to make the diagnosis, a systematic and rigorous protocol must be followed and blood pressures should be taken on both arms while the subject is sitting quietly. Most importantly, pressure cuffs must be used that are appropriate for the size and body mass of the individual. The use of too small a cuff may give an anomalously high pressure (more cuff pressure is needed to occlude the arterial blood flow), whereas too large a cuff may give a falsely low value. If a high reading is obtained, before the diagnosis of hypertension can be made, three separate blood pressure measurements should be taken and averaged, and this should be done on at least two separate visits at least 1 week apart. Often, blood pressure is measured both when the patient is upright (sitting or even standing) and when in a lying-down position. There is a well-described phenomenon known as "white-coat hypertension," which is an elevated blood pressure whenever the pressure is taken in a doctor's office by the physician or even the nurse. This is attributed to a short term elevation of blood pressure due to the stress of being examined, and may disappear when the pressure is taken in a less forbidding environment, for example, at home. For this reason, it is essential that measurements be taken over several visits and in a calming environment so as to ensure an accurate assessment of the patient's blood pressure. If hypertension is suspected, a complete examination must be conducted to exclude secondary forms of hypertension (see the next section), including a complete family history to determine if there is a genetic predisposition to be hypertensive (see Section II,D).

C. Types of Hypertension in Humans

The most common form of human hypertension, estimated to be responsible for >80% of all hypertension, is of unknown cause and, therefore, is called primary, or essential hypertension. Some experts feel that human essential hypertension actually reflects a variety of different forms of hypertension, each attributable to a different abnormality in blood pressure and/or volume regulation. In contrast, other experts believe that essential hypertension is one syndrome that arises from a mix of different abnormalities in regulation. Essential or primary hypertension may also include patients who exhibit a salt-dependent form of hypertension as well as obesity-dependent hypertensives, if there is no evident explanation for their hypertension.

A much smaller percentage (estimated to be approximately 10–15% of all hypertensives) have hypertension that is due to a known or identifiable cause. This form of hypertension is termed secondary hypertension. A common form of secondary hypertension is due to diminished or restricted arterial blood flow into the kidney. This results in a form of hypertension that is caused predominantly by activation of the renin–angiotensin system (see Section III.B) and is termed renovascular hypertension. It is estimated that approximately 5–8% of all human hypertension is renovascular in origin. Hypertension may also be due to other causes, some reversible and some irreversible. For example, secondary hypertension may be due to selected pharmacological (i.e., drug-induced) exposures. Although the patients with these unusual forms of secondary hypertension are much fewer than those with essential or renovascular hypertension, because they may reflect a particular group of the population, by age or sex, they may be the cause of a predominant form of hypertension in that group. For example, hypertension was found in young women who were taking selected estrogen-containing oral contraceptives. Because all the subjects were women of comparably young age, this form of hypertension comprised a significant fraction of the types of hypertension in this age group. When the association with the pill was established, modifications in the formula used in those oral contraceptives that had the highest frequency of causing the disorder resulted in a major drop in the incidence of hypertension in this group of patients. (It should be noted that some investigators believe that women who exhibited this form of hypertension are genetically prone to the disorder and may, as they get older, show signs of essential hypertension; however, this view is controversial and not yet proven or disproven. Clearly, many years are needed to determine if a correlation exists.)

Hypertension may also be caused by a relatively rare tumor of the adrenal medulla, termed a pheochromocytoma. Pheochromocytoma patients may exhibit episodes of very high arterial pressure. These periods of hypertension are attributable to the release of large quantities of neurohumoral substances (epinephrine and norepinephrine, among others) from the malignant tumor tissue at unpredictable times.

A form of hypertension called "isolated systolic hypertension" is found in the elderly and reflects markedly elevated systolic pressure (>160 mm Hg) without a diastolic pressure >90 mm Hg. The cause of isolated systolic hypertension is not fully known; however, one explanation is that it may result from reduced compliance of the arterial blood vessel (with age). This would result in less of a windkessel effect and higher systolic pressure with the pulse of blood pumped by the heart's left ventricle. The treatment of isolated systolic hypertension is somewhat different than with the more common form of essential hypertension.

Malignant hypertension is a term used for patients who show high, and frequently irreversible, hypertension. This term in no way implies a cancerous origin to the hypertension; rather, patients with malignant hypertension are generally those who have had hypertension for long periods of time and who fail to respond to drug therapy. Often, malignant hypertension is associated with severe kidney failure and/or marked changes in the vascular compartments. Such hypertensive patients exhibit very high, sustained arterial pressure, both systolic and diastolic, and constitute a medical emergency necessitating bold drug treatment to lower the blood pressure and prevent further organ damage.

D. Genetic Predisposition to Hypertension

Strong evidence indicates that human essential hypertension has a hereditary or genetic component, that is, the potential for developing hypertension can be passed from one generation to another within a family. This means that there are hypertensive-prone individuals; however, evidence is not yet sufficient to indicate whether all hypertensive-prone patients will become hypertensive. In fact, many experts believe that hypertension can still be prevented (or at least significantly delayed in onset) by appropriate dietary and exercise regimens. As part of the original diagnosis, physicians and health care professionals must determine if there is any family history of hypertension or other cardiovascular disorder. (The latter information is important because hypertension may have gone undetected in earlier generations but the end result, heart disease or stroke, may have brought the patient to the attention of the physician.)

In contrast to essential hypertension, very little evidence indicates (or disproves) that secondary hypertension has a hereditary or genetic component. Even drug-induced hypertension may reflect some hereditary aspects because, in the example of oral contraceptive hypertension, only a fraction of all the women on the medications in question became hypertensive. As a rule for good health, whenever a patient knows of the presence of hypertension in their family, they should be on guard as to a possible hereditary relationship.

III. PATHOPHYSIOLOGY OF HUMAN ESSENTIAL HYPERTENSION

The possible cause or causes of essential hypertension are still unknown, and discovering the cause is complicated by the many factors that determine the level of blood pressure. At the very least, disturbances in autonomic nervous activity, in blood volume, in the action of the heart as a pump, or in the distensibility of arterial blood vessels all may result in abnormal levels of arterial pressure. A complete discussion of these factors is beyond the scope of this article; however, a brief introduction is warranted because therapy is often directed at one or more of these possible causative mechanisms.

A. Elevated Activity of the Sympathetic Nervous System

The sympathetic branch of the autonomic nervous system is important in setting the level of activity of the body so as to meet all the needs of the organism, other than those of a purely vegetative nature. For example, the sympathetic nervous system may be activated by such innocent activities as standing up or walking across the room. Blockade of the sympathetic nervous system by drugs may cause postural hypotension, which is a failure to maintain adequate blood pressure when going from a recumbent (lying or sitting) position to a standing position. Individuals with postural hypotension may faint on standing or getting out of bed too quickly. The sympathetic system is also stimulated by swallowing or other consummatory activity. The sympathetic nervous system normally is also involved in regulating metabolism (glucose, fat, etc.), in setting the level of heart rate, peripheral vascular resistance, and blood pressure in proportion to anticipated needs, and in the regulation of body temperature during high-activity periods. Stimulation of the sympathetic nervous system results in an increased heart rate, increased peripheral vascular resistance, and, generally, increased mean arterial pressure. These are but a few of the obvious functions of this nervous system; however, the sympathetic system also regulates hormone systems important in salt (sodium) and water balance. At the other extreme of human behavior, stress, fear, anger, and defense reactions all markedly elevate the activity of the sympathetic nervous system. These behavioral responses may result in either selected or global activation of the sympathetic nervous system.

In the early stages of human essential hypertension, evidence indicates an enhanced sympathetic nervous system activity and, often, an increased heart rate. Under normal circumstances, in response to elevated arterial pressure (with or without increased heart rate), reflex mechanisms are activated that reduce the level of activity of the heart and should thereby lower blood pressure. These reflex mechanisms involve changes in the activity of the sympathetic nervous system and the parasympathetic nervous system. The latter slows heart rate. However, at some point in the development of hypertension, these reflex mechanisms appear to fail and the original level of blood pressure (and, possibly, sympathetic nervous sysem activity) is not regained. Rather, a new blood pressure baseline becomes the point of regulation. If this new baseline is at a higher blood pressure, a resetting of the "set point" for blood pressure has occurred. Blood pressure is then regulated at this new level.

The sympathetic nervous system may also be activated *because* of improper function of an organ system such as the kidney. For example, if there is insufficient filtration and urine formation, increasing the activity of the sympathetic nervous system could overcome the insufficiency. However, if this activation persists, then a new set point in function results. Often this is associated with an increased arterial pressure. This leads to a circle of continuing resetting of arterial pressure and/or sympathetic activity to maintain function. The net effect of chronically elevated sympathetic nervous system activity, which remains unchecked by reflex activity, is elevated heart activity, increased peripheral vascular resistance, increased sodium retention by the kidney, and elevation of the activity of the renin–angiotensin system and other neurohormones.

B. Retention of Sodium and Fluid

There is no question that with established hypertension, abnormalities in kidney function could lead to retention of sodium and, with it, water. Maintenance of salt and water homeostasis is the result of optimum arterial pressure and kidney function, and is also under the control of various peptide hormones and steroids. One peptide hormone system, the renin–angiotensin system, is intimately involved in maintaining sodium and water homeostasis. Renin is an enzyme released from the kidney into the blood in response both to sympathetic nervous system activity and to the level of sodium intake (lowered sodium leads to increased renin release). Angiotensin comes from a large protein (prohormone) that is made in the liver and is continuously released into the blood. Renin in the blood then breaks a chemical bond in the prohormone, releasing a small, inactive peptide called angiotensin I. This peptide is converted to a smaller, but active, peptide called angiotensin II by an enzyme called angiotensin-converting enzyme (ACE). Angiotensin II can constrict blood vessel smooth muscle and, thereby, influences peripheral vascular resistance, and also stimulates the release by the adrenal gland of a steroid hormone called aldosterone. Aldosterone acts directly on the kidney to increase sodium (and water) retention by the body. Angiotensin II also has important actions within the kidney and the intestines to regulate sodium and water conservation. This hormone system has other functions, which include regulating and enhancing the functional activity of the sympathetic nervous system, stimulating heart activity, and possibly affecting growth of some cells.

Because of the importance of the renin–angiotensin system in affecting so many systems that control blood pressure, many studies look at the possible involvement of this hormone system in the cause of some forms of hypertension. In fact, renovascular hypertension is believed to be caused by excess activity of the renin–angiotensin system. This system has also become a major target in the design of antihypertensive drug therapy. Even though renin–angiotensin system activity is elevated in some forms of essential hypertension, little evidence indicates that it is the cause of the hypertension. Nevertheless, ACE inhibitors, which interfere with the formation of angiotensin II, are very effective in lowering blood pressure in many forms of hypertension. In addition, one mechanism whereby beta-blockers (such as propranolol) lower blood pressure in hypertensive patients is believed to be through interference in sympathetic nervous system stimulation of renin release by the kidney and formation of angiotensin II.

Retention of sodium and water can occur in almost all forms of hypertension, even those that appear to be independent of the renin–angiotensin system. For example, one form of hypertension is characterized by a low plasma renin activity but high sodium retention and elevated (expanded) volume. This form of hypertension may reflect abnormalities in kidney function or in hormones that regulate kidney function and excretion. Restriction of sodium (or salt) intake is good advice to all hypertensive patients. Some investigators argue that there is excess salt intake in developed Western societies and that almost everyone should reduce their salt intake. However, the role of sodium intake as a factor in the development of hypertension remains controversial. There is no question that dietary salt intake is much higher than needed in Western societies, and many studies have demonstrated a positive relationship between the amount of salt consumed in the diet and the incidence of hypertension; however, not everyone is hypertensive. Therefore, many investigators believe that some patients are "salt-sensitive." Such individuals may have a hereditary predisposition to retain excess sodium (and water) and, as a consequence or independently, a predisposition to develop hypertension. Almost all investigators agree that a moderate, rather than high salt intake is good advice for adults. Salt sensitivity varies among the races—Caucasian, African-American, and Asian—and this has been considered to be evidence for a genetic influence. Partly because of this variability in salt sensitivity, the most appropriate antihypertensive treatment differs between races.

C. Relationship of Chronic Stress and the Development of Hypertension

One attractive, but as yet unproven, theory is that repeated exposure to stress in everyday life activities may result in abnormal cardiovascular responses in hypertensive-prone individuals. Another theory is that repeated stress episodes lead to irreversible changes in the arterial blood vessels, resulting in a sustained elevation of peripheral vascular resistance and, ultimately, hypertension. As yet no evidence exists to prove or disprove this thesis. Intuitively, repeated exposure to stressful situations would be expected to result ultimately in some adverse effect on the cardiovascular system. However, there is no evidence supporting a hypertensive "type-A" personality, as has been described for patients at risk for a myocardial infarction. Nevertheless, disproving a stress–hypertension link has been as difficult as trying to establish linkage. One reason for this difficulty may lie in the complex interplay of different mechanisms regulating blood pressure and volume.

IV. HYPERTENSION AS A DISORDER

A. When Does Elevated Arterial Pressure Require Treatment?

Hypertension is classified according to the level of diastolic pressure and three categories are relevant to our discussion: borderline hypertension with a diastolic pressure between 90 and 94 mm Hg; mild hypertension with a diastolic pressure between 95 and 104 mm Hg; and moderate to severe hypertension with a diastolic pressure >104 mm Hg. These three categories are loosely defined because some patients are at greater risk of developing severe cardiovascular disorders with any degree of hypertension than are other patients. In addition, essential hypertension is a chronic disorder that extends over many years. It is generally accepted that without any treatment, a patient will progress from mild to severe hypertension in a short time. Even with treatment, many patients will progress, over years, to a higher diastolic pressure. *Therefore, the goal of therapy is to reduce the rate of that progression or prevent it entirely.*

When to initiate therapy and the types of therapy remain controversial. Most physicians and investigators believe that some form of therapy must be initiated when a *definitive* diagnosis of hypertension has been made. Therapy generally takes the form of either antihypertensive drugs (i.e., pharmacological treatment) or nonpharmacological approaches. If a secondary form of hypertension is considered, drug therapy is usually initiated immediately along with efforts to deal with the source of the hypertension. However, with essential hypertension, whether to treat or not depends on the classification of severity, with moderate or severe forms of hypertension necessitating antihypertensive drug treatment along with nondrug therapy. In the case of borderline hypertension and even with mild forms of hypertension, there is an ongoing debate as to the appropriate type of therapy, if any. The primary goal of all therapeutic approaches must be to lower the diastolic (and systolic) pressures into the normal range or, if that cannot be reached, to lower pressure as much as possible. A diastolic pressure <90 mm Hg is usually the target of therapy. To achieve that goal, the therapeutic approach should be individualized to the patient. For example, many physicians will move quickly to institute some form of antihypertensive drug therapy when treating a patient with a family history of hypertension. Because so many drugs are now available to treat hypertension, the type of therapy employed generally reflects the experiences of the physician in treating similar cases of hypertension.

B. Types of Therapy Used to Lower Blood Pressure

There are two general approaches to lowering blood pressure back to the normal range. For essential hypertensive patients who are classified as borderline hypertensives, many physicians will initially recommend nondrug therapy. This frequently includes

TABLE I
Categories of Antihypertensive Drugs

Type of agent	Reason for their use
Diuretic	Diuretics increase urine production by increasing the loss of sodium and water from the body. They have many secondary actions, some undesirable such as increasing loss of potassium. Potassium supplementation, by drugs or diet, is often necessary. Diuretics may be used alone or in combination with other drugs.
Potassium-sparing diuretics	A special group of diuretic drugs that reduce the loss of potassium from the body.
Beta-blocker	Beta-blockers are drugs that partially block the ability of the sympathetic nervous sytem to stimulate the heart and to release renin from the kidney. Some beta-blockers may also work in the brain to reduce sympathetic nervous system activity and secretion of hormones, which may raise blood pressure. There are many types of beta-blockers.
ACE inhibitors	A form of antihypertensive therapy that has become very popular. These agents interfere with the formation of angiotensin II. They also affect other systems that may control blood pressure and kidney function. However, their blockade of angiotensin II formation is believed to be a primary mechanism by which they lower blood pressure.
Alpha-blocker	Alpha-blockers are drugs designed to block the ability of the sympathetic nervous system to contract the smooth muscle of blood vessels and, thereby, decrease blood vessel constriction and peripheral vascular resistance. Since the sympathetic nervous system maintains blood pressure and blood flow to the brain when a person stands by constricting blood vessel smooth muscle, alpha-blockers often cause postural hypotension.
Centrally acting drugs	These drugs are designed to work primarily in the brain to reduce the activity of the sympathetic nervous system, but some may have additional mechanisms by which they lower blood pressure.
Calcium-channel antagonists	These drugs are designed to interfere with the entry of calcium into blood vessel smooth muscle cells. Because calcium entry is essential for contraction of the smooth muscle cell, calcium antagonists reduce peripheral vascular resistance. These agents consist of several different types, some of which are most active on the heart and reduce contractile force of the left ventricle of the heart. For this reason, they are often used in treating heart failure. Calcium channel antagonists also reduce the release of renin by the kidney and, thereby, decrease production of angiotensin II.

weight reduction, restriction of salt intake, stress reduction and changes in life-style, exercise regimens, and discontinuance of alcohol intake or cessation of smoking. All of these approaches constitute nondrug therapy of hypertension (and other cardiovascular disorders as well). Depending on the individual, these therapeutic approaches may be all that is needed. For example, a relatively small reduction in weight has been shown to markedly lower blood pressure in essential hypertensives. In some patients, losing weight resulted in their being able to discontinue much or all of the antihypertensive drugs they were taking. Weight reduction achieves two immediate goals: reduction in the workload placed on the heart and reduction in total blood volume. Both effects are beneficial. Another nondrug approach is stress relaxation and/or modification in life-style. Although it is not fully proven, many investigators, physicians, and patients feel that stress exacerbates high blood pressure and, therefore, stress reduction has potential benefits. Stress reduction would lower sympathetic drive, thereby lowering activation of neurohormones as well as directly decreasing stimulation of the heart. Nondrug therapies such as stress reduction may take several months before evidence of their effectiveness is clear. For this reason, many physicians prefer to start the patient on limited antihypertensive drug therapy to assist in lowering the blood pressure more quickly. Whether or not this is necessary depends on each individual patient.

It is generally considered that with time (years) patients with essential hypertension will show progression of the disorder and require the institution of some form of drug therapy. Because in most patients hypertension progresses eventually to the point where stronger drugs are necessary, a "stepped care approach" with drug therapy is followed. With stepped care, a first drug is used (along with nondrug therapy) to achieve the target goal of a diastolic pressure <90 mm Hg. If (or when) that drug fails to achieve the target goal, either a second drug is added or a substitution of another "type" of antihypertensive drug is tried. Ultimately, after many years in most patients, two or even three different drugs are used to lower the blood pressure. The choice of the first drug differs between therapy in the United States and Europe and undergoes continuing debate. In the United States, a diuretic (see the next section) is frequently the first drug used in hypertensives, whereas in Europe, a beta-blocker is often the drug of choice. ACE inhibitors have also become a very widely used early drug because they are so well accepted by many patients. Whatever drug is used, nondrug therapeutic approaches must be continued because, when effective, nondrug therapy reduces the need or amount of drug therapy and thereby reduces the frequency of occurrence of adverse (or undesirable) side effects of drug treatment. All antihypertensive drugs can cause side effects. The treatment of essential hypertension is a long-term process and patients should be familiar with the drugs they are taking and what side effects may result.

C. Categorization of Antihypertensive Drugs According to Their Target

As discussed earlier, blood pressure may be influenced by a wide variety of systems in the body. Antihypertensive drugs have been developed to affect all of these systems. Table I lists the main groups of drugs in common use.

The ultimate goal of all forms of therapy is to reduce systemic arterial pressure, to reduce the rate of progression to more severe forms of hypertension, and to reduce the rate of damage to those critical organs (heart, brain, kidney, and eyes) as a result of continued high blood pressure. But the drugs themselves are not harmless. Chronic use of thiazide diuretics has been linked to abnormalities in lipid metabolism and potentially to coronary artery atherosclerosis. Similarly, many of the antihypertensive drugs, by lowering blood pressure, reduce the ability of the kidney to excrete sodium and water. These drugs require the simultaneous use of a diuretic to enhance urine production and sodium elimination from the body. Many of the stronger antihypertensive drugs cause postural hypotension (discussed earlier, i.e., a drop in blood pressure upon standing up which may result in fainting). Some beta-blockers can precipitate asthmatic attacks in sensitive patients.

Hypertension is a disorder of regulation and any therapeutic approach must focus on restoring, if possible, the regulation of blood pressure. The treatment of hypertension involves an ongoing series of decisions and evaluations by the physician based on each individual patient. Continued interaction and communication among patient, physician, and health professional are essential to the prevention of the damaging effects of hypertension. In the case of hypertension, as with many diseases or disorders, prevention through changes in dietary intake, exercise, stress reduction, and life-style remain the preferred first step in treatment.

BIBLIOGRAPHY

Abbott, D., Campbell, N., Carruthers-Czyzewski, P., Chockalingam, A., David, M., Dunkley, G., Ellis, E., Fodor, J. G. McKay, D., and Ramsden, V. R. (1994). Guidelines for measurement of blood pressure, follow-up, and counselling. Canadian Coalition for High Blood Pressure Prevention and Control. *Can. J. Publ. Health* **85** (Suppl. 2), S29–S43.

Anastos, K., Charney, P., Charon, R. A., Cohen, E., Jones, C. Y., Marte, C., Swiderski, D. M., Wheat, M. E., and Williams, S. (1991). Hypertension in women: What is really known? The Women's Caucus, Working Group on Women's Health of the Society of General Internal Medicine. *Ann. Internal Med.* **115**, 287–293.

Farquhar, J. W. (1995). The place of hypertension control in total cardiovascular health; Perspectives outlined by the Victoria Declaration. *Clin. Exp. Hypertension,* **17**, 1107–1111.

Kaplan, N. M. (1994). "Clinical Hypertension," 6th Ed. Williams & Wilkins, Baltimore.

McVeigh, G. E., Flack, J., and Grimm, R. (1995). Goals of antihypertensive therapy. *Drugs* **49**, 161–175.

National Heart Lung and Blood Institute (1993). "The Fifth Report of the Joint National Committee on Detection, Evaluation and Treatment of High Blood Pressure (JNC V)," NIH Publication No. 93-1088. National Institutes of Health, Washington, D.C.

Page, I. H. (1987). "Hypertension Mechanisms." Grune & Stratton/Harcourt Brace Jovanovich, Orlando, Fla.

Tobian, L., Brunner, H. R., Cohn, J. N., Gavras, H., Laragh, J. H., Materson, B. J., and Weber, M. A. (1994). Modern strategies to prevent coronary sequelae and stroke in hypertensive patients differ from the JNC V Consensus Guidelines. *Am. J. Hypertension* **10**, 859–872.

Hyperthermia and Cancer

GEORGE M. HAHN
Stanford University School of Medicine

GLOSSARY

Carcinogen Drug or physical agent that induces cancer

Chemotherapy Treatment of disease by chemical agents

Mutagen Drug or physical agent that causes genetic changes in the offspring of an exposed cell or organism

Randomized prospective trial Clinical test comparing two or more treatments; neither the patient nor the physician is usually aware of the treatment assignment, which is made randomly

Transformation Cellular change that converts a normal cell to one that has the potential of growing as a cancer

WHEN THE TEMPERATURE OF A HUMAN RISES above its usual value (37°C), then that individual has had a hyperthermic episode. This episode may be due to one of several causes: fever, heat stroke, or intentionally induced hyperthermia for the treatment of cancer or other diseases. Fever and heatstroke necessarily involve increases in temperature of the whole body; intentionally induced hyperthermia may be whole body or localized. In the latter case, core body temperature may not change by more than a fraction of a degree. Both intentional and accidental hyperthermic episodes induce certain changes in cellular metabolism. Somewhat paradoxically, these may have beneficial or deleterious effects, particularly with respect to the treatment or the induction of cancer.

I. INTRODUCTION

The interior of the human body is designed to function within a limited temperature range. Under nonpathogenic conditions, the core temperature of the average person varies by only fractions of a degree around a set temperature. Although this varies from individual to individual, the set temperature of the vast majority of people is between about 36.5° and 37°C. Throughout the body, temperature is maintained at surprisingly constant levels. Only in the liver and perhaps in other regions of high metabolic activity does the temperature increase slightly. Near the skin, particularly under conditions where convective temperature loss is considerable, a temperature gradient extends to a depth of perhaps 1 cm. Parts of the lung may be at lower temperature, largely because of local evaporation. In most of the rest of the body, however, temperature remains constant, within a fraction of a degree of the set point. Obviously, the body needs a variety of homeostatic systems to maintain such temperature uniformity. These systems operate on a local as well as a global level. It has been demonstrated in many experiments that the hypothalamus is a major controlling element that governs global control. Control on a local level is far less well understood, although it is effected primarily by changes in blood flow in the tissue of concern. [*See* Body Temperature and Its Regulation.]

Several situations, however, cause changes in the body temperature. We will consider here only increases in either local temperatures or increases at the level of the entire human. The global change that most of us are familiar with is that caused by fever. This form of hyperthermia apparently involves hypothalamus directed change in the body's set point. Homeostatic mechanisms are employed to maintain a core temperature around the new set point; this can be as high as 41°C. Possible beneficial functions of fevers have been discussed ever since physicians have attempted interventional medicine. For example, the Greek physician Parmenidis said that given the means to induce fevers, he would be able to cure all illnesses. In more modern times, physicians frequently thought fevers harmful and suppressed them with agents such as aspirin. In very recent years, the pendulum has begun to swing back, and many physicians now believe that fevers have, indeed, beneficial functions and should be suppressed only if they constitute a specific danger to the patient. Nevertheless, very few physicians would believe today that serious illnesses, particularly cancer, can be cured simply by means of fevers. This is because we can now satisfy Parmenidis' wish: We can induce fevers by injecting a variety of substances having pyrogenic properties. Perhaps of most interest among these substances are several bacterial toxins. These have indeed been tried as cancer remedies but only with limited success. Elevated temperatures induced by pyrogens behave very much like fevers; the body adjusts its set point and uses its homeostatic mechanisms to maintain the elevated temperature.

There are, however, other ways of raising the body's temperature. For example, in hot climates, particularly when the body is exposed excessively to sun, core temperature may rise. Mechanisms will then be set in motion to reduce body temperature. In the human, the primary mode to do this is sweating; because of the large amount of heat expended as heat of evaporation, skin temperature is cooled temporarily. Cardiac output is increased, so that the excess heat is rapidly carried to the skin and dissipated during evaporation. If all these maneuvers are insufficient, core temperature continues to rise and severe injury may result from "heatstroke." Exposure of the body to microwaves, either accidental or intentional, will increase core temperature. Again, homeostatic mechanisms will be brought into action to attempt to lower the body's temperature to its set point. Temperature can obviously also be increased locally. The most obvious means of doing so is conduction. If we touch a hot object, local skin temperature will be raised. Obviously, if the object is very hot, pain will quickly cause us to make every attempt to break contact. Here also, the body will attempt to maintain temperature by locally increasing blood flow to carry off the excess heat that has been introduced via conduction. The same phenomenon occurs when heating locally is carried out by absorption of energy emitted by infrared, focused, or locally applied microwaves or ultrasound radiations.

Parmenidis, when he mentioned his desire to cure all disease with heat, presumably included cancer. In his time, malignancies had been described in considerable detail. Is there really any reason to expect elevated temperatures to act as an anticancer treatment? Or is it perhaps the opposite: Does heat induce cancer? This article discusses these two possibilities, and considerable experimental evidence is presented to suggest that both of these, at least in a limited sense, may be correct. As far as treatment is concerned, Parmenidis was obviously overoptimistic. We now know that heat by itself has only a limited influence on the progression of malignant disease. But many studies now indicate very strongly that in conjunction with more conventional treatments such as radiation therapy and chemotherapy, heat has much to offer. Thus, hyperthermia is an anticancer agent. In the fully developed organism, heat by itself may not be carcinogenic; it does, however, increase the carcinogenic efficacy of many agents, including ionizing radiations and many drugs, although this depends to some extent on the order of application. For example, if cells are heated and then radiated, the rate of transformation from a nonmalignant to a potentially malignant state is increased; if the order is inverted, it may be decreased. In the developing organisms, particularly in the fetus of rodents, heat clearly has deleterious, teratogenic effects. It is not unreasonable to suggest that in some cases developmental defects on the cellular level lead to cancer.

II. HYPERTHERMIA AS A TREATMENT MODALITY

Late in the 19th century, several attempts were made to treat malignancies by elevating body temperatures. The rationale for these attempts was based on well-documented occurrences of spontaneous remissions in cancer patients who had episodes of high fevers while suffering from malignant disease. This subject has been reviewed in detail. The reviewers found that one-third of 450 spontaneous remissions of histologically proven malignancies were known to be associated with the development of acute fevers that resulted from concurrent diseases such as malaria or typhus. Similarly, one-third of spontaneous remissions of lymphomas in children were also found to be associated with high fevers. The lymphoma remissions were of short duration, but several "cures" of carcinomas and sarcomas have been described. These observations certainly provided an impetus for physicians to attempt to treat incurable malignancies by elevating body temperatures. This was initially achieved primarily via the injection of bacterial agents or chemical pyrogens. The most prominent of these studies was performed by Coley late in the 19th century. He developed a bacterial toxin ("Coley's toxin") that, when injected into patients, induced fevers that ranged from 38° to 41°C. Some favorable results were obtained in these studies, particularly against osteosarcomas and soft tissue sarcomas. Current knowledge suggests that perhaps the effect of pyrogens and bacterial toxins relates not only to hyperthermia but perhaps also to stimulation of nonspecific immune responses. Following these early studies, progress in hyperthermic treatments of tumors was slow. Possibly, this was related to the development of X-ray therapy as a modality for treating cancers. In any case, it was not until the early 1930s that radiation was combined with hyperthermia. Because of the great difficulties involved in heating tumors, this work was not followed up by many investigators until the early 1960s when a series of experiments showed that hyperthermia by itself can have definitive antitumor effects, and, in combination with radiation therapy, it has curative potential. Both experiments involved localized heating. Regional and even whole-body heating have also been attempted. Because different physical means are employed with the two types of uses of hyperthermia in the clinic, they are described individually below.

A. Localized Hyperthermia

With localized hyperthermia, heating is achieved primarily via use of electromagnetic techniques or ultrasound, although for interstitial applications, heating via conduction of either water or electrically heated implants or catheters is also feasible. Each of these techniques has its own specific advantages and disadvantages. Unfortunately, the disadvantages are often very important. For example, electromagnetic energy is rapidly absorbed in tissue, particularly at the high frequencies where focusing and beam-shaping would permit selective heating of specific tissue volumes. Therefore, deep-seated tumors (e.g., cervical cancers, tumors near the intestines) do not receive enough microwave energy and do not heat readily. Ultrasound penetrates more readily, but cannot traverse tissue–air interfaces or penetrate bone. As a result, only a relatively small number of tumors can readily be heated by ultrasound. Interstitial techniques, while perhaps yielding the best temperature distributions, require surgery and, therefore, are primarily used in conjunction with brachytherapy (i.e., X-ray therapy delivered by implanted sources of radioactive materials).

1. Heat Alone

In one study, small human tumors (<4 cm in diameter, <3 cm deep) were heated by ultrasound. Tumor temperature varied typically ±1° across the tumor volumes. Near bone, even larger deviations were seen. Perhaps these temperature variations do not sound appreciable, but cell killing by heat varies in an exponential fashion, and a 1° change in temperature corresponds to a 50% change in effective dose. Temperatures in the region of 43°C were maintained for about 30 min, and treatments were repeated three times a week for a total of six treatments. Complete disappearance of the tumors was seen in about 10–15% of the patients, but those remissions were of short durations and lasted only about 6 weeks. Very similar results were obtained by several other groups that employed a variety of means to heat the tumors. The conclusion from these studies was that heat alone was not able to eradicate enough tumor cells to cause long-term disappearance of the treated lesions. There were probably two reasons for this. First, uniform tumor heating proved to be difficult if not impossible, perhaps because of variations in localized blood flow. Blood acts as an efficient coolant of tissue. The other possibility is that a fraction of the cells in the treated tumors were heat-resistant and, therefore, exposed to an inadequate thermal "dose." Several of the studies did not comment on one interesting finding, namely that heat seemed to do little, if any, damage to normal, nontumor tissue. The only problem seen was occasional skin blistering or minor necrosis of fatty tissue. Similar comments were made in many of the studies cited in the next paragraphs.

2. Hyperthermia Plus X-Irradiation

There have now been about 25–30 studies that in one way or another compared the antitumor efficacy of radiation only with that of radiation plus hyperthermia. Most of these studies used historical controls, i.e., the responses seen in patients treated with a combination of hyperthermia and radiation were compared with responses that had been previously seen in a more or less matched group of patients. Therefore, not all of these studies are truly randomized trials, and they have frequently been criticized for this reason; nevertheless, looking at the results is worthwhile. These have been surprisingly uniform. Complete response rates (i.e., clinical disappearance of the treated lesion) increased by about 50–75% when heat was added to X-irradiation. The typical target intratumor temperature that investigators attempted to reach was 43–44°C. Some tumors proved extremely difficult to heat, and very likely only a small fraction of these tumors was heated to the target temperature. But even among the tumors that readily heated (and where such heating was documented by adequate thermometry), portions of the tumor remained well below the target temperature. It is reasonable to suggest that treatment failures were concentrated among poorly heated lesions. An interesting aspect of the studies was that the number of heat treatments seemingly were found irrelevant: tumors treated 3–4 times with hyperthermia did as well as those treated 10–15 times. Tumors treated included breast cancer, head and neck lesions, melanomas, and other surface-accessible lesions. Only one study examined the durations of local control and concluded these exceeded tumor-free intervals in historical controls.

Because prospective randomized studies are difficult and expensive to carry out and, furthermore, require a large patient number to yield statistically significant results, several investigators studied the effect of heat on matched pairs of lesions. In these studies, patients with multiple tumors had one of their tumors treated with radiation plus heat, another lesion matched in size with radiation alone, and in some cases a third with heat alone. Such studies, because of the internal controls, require relatively few patients. These studies also yielded uniform results. The tumors treated with a combination of the two modalities responded best; radiation-only treated tumors responded less well; and tumors heated but not irradiated rarely responded with complete remissions.

Only three randomized prospective studies have involved radiation and heat. Two of these showed results similar to those found in the earlier studies that had employed retrospective controls: approximately a doubling of the response rate when hyperthermia was added to radiation therapy. The third study, perhaps the largest of all clinical trials, was undertaken by a consortium of several radiation therapy centers. The results of this multi-institution trial purported to show that the addition of hyperthermia to radiation did not influence the response of treated tumors. Unfortunately, the study suffered from major defects. First, the majority of the tumors treated could not have been heated by available equipment; the study placed no restriction on tumor size. Second, because of the lack of adequate quality control, there was insufficient thermometry so that no estimate could be made on temperature distributions within most of the treated lesions. But even that study, when the tumors were stratified according to size, showed that in the smaller lesions (i.e., in those where adequate heating was possible), the addition of hyperthermia to radiation therapy indeed resulted in statistically significant benefits.

3. Hyperthermia Plus Chemotherapy

Cells in culture are much more sensitive to many drugs if exposure is carried out at elevated temperatures. Similarly, animal tumors are more sensitive to these drugs if chemotherapy is combined with local heating. For these reasons, physicians have also combined systemic drug treatment with localized hyperthermia. Most of the published reports originate in Japan. Most of the descriptions are anecdotal; few, if any, of the trials are randomized. Thus, although almost all data suggest that hyperthermia increases response rates over those seen with chemotherapy alone, no conclusive proof for the added efficacy exists.

B. Regional and Whole-Body Hyperthermia

Most cancers are systemic. This may be because of the nature of the specific disease, such as the leukemias, or it may result from failure to treat localized disease sufficiently early or adequately so that metastases occur, and sites other than the primary one become involved. The use of whole-body hyperthermia is an attempt to deal with this problem. Regional heating, involving heating of entire limbs or other large volumes, is an intermediate application of hyperthermic therapy. Some major differences exist between local and systemic hyperthermia treatments. First, for whole-body exposures, there is an absolute temperature limit of 42°C or less. Exceeding that temperature results in irreversible damage to liver and probably to brain. Second, the physical implementation of systemic heating (and of regional heating) is far easier than that of localized heating.

There are several ways of inducing elevated temperatures throughout the body. The use of toxins and other pyrogens has already been discussed. Although these do induce hyperthermia, the degree of hyperthermia induced tends to vary from patient to patient, and how long the elevated temperature will be maintained is difficult to predict in individual patients. Therefore, quality control becomes exceedingly difficult, and these agents have been abandoned in the clinic as inducers of hyperthermia.

Currently used techniques utilize conduction, absorption of electromagnetic energy, or external heating of blood and reinfusion via an arteriovenous shunt. The last technique is particularly useful for regional heating. Heating that relies on conduction alone is perhaps the slowest method of raising the patient's temperature. With hot water "blankets," for instance (i.e., thin layers of partially heat-conductive materials carrying conduits for rapid transport of heated water), the time required to do so is on the order of 3 hr. Such a long warm-up time may be undesirable for both practical and theoretical reasons. The arteriovenous shunt technique is perhaps the most rapid, but it requires a surgical procedure for the placement of the shunt and hospitalization during treatment. Perhaps the most convenient technique, and one that involves the least patient discomfort, combines heating by conduction with absorption of electromagnetic energy. With all the approaches, however, the most difficult and important aspect is the control of temperature to ensure that it does not rise above the target, usually 41.8°C. Exceeding that temperature for any appreciable length of time can lead to irreversible liver and brain damage.

1. Hyperthermia Alone

The results of exposing patients to whole-body hyperthermia without concomitant chemotherapy have not been particularly encouraging. For example, in one of the earliest studies reported, of 49 patients treated, 31 had either objective or subjective responses. Durations of responses were short; however, the disease progressed in most patients within a few weeks of completion of treatment. At the National Cancer Institute, out of 14 treated patients, 4 had objective responses. These responses lasted for 3–12 months. These results are typical of clinical data obtained in a small percentage of responses (10–30%), and these only of short durations (typically 6 months or less).

2. Heat Plus Chemotherapy

Better results were obtained when hyperthermia was combined with chemotherapy. Some studies have reported on treatments that combine drugs such as adriamycin, mitomycin C, and a nitrosourea with whole-body hyperthermia. A Texas group has used hyperthermia for many years in the perfusion of limbs affected with melanoma, apparently with considerable success. An East German group described the treatment of children with whole-body hyperthermia and chemotherapy and also claimed good results.

III. IS HYPERTHERMIA AN INDUCER OF CANCER?

No direct evidence, either from human or from animal data, suggests that heat by itself can cause cancer. Certainly, epidemiological studies show that in some warm countries, the incidence of skin cancers, particularly squamous cell carcinomas and melanomas, is very high; however, the causative agent here is certainly excessive exposure to the sun. Part of the sun-emitted spectrum of frequencies is in the ultraviolet range. Many studies have demonstrated that a portion of the ultraviolet radiation is preferentially absorbed by DNA, and the damage there, or its faulty repair, leads to changes in the cell (probably involving oncogenes) that finally result in neoplastic disease. The possibility that the increased temperature in the warm countries makes ultraviolet radiation a more efficient carcinogen has not as yet been examined. This concept—heat as a cocarcinogen—is discussed later. [*See* Skin, Effects of Ultraviolet Radiation.]

What about indirect evidence? Here, two sets of data need to be considered. First, heat may be a mutagen. We know that most carcinogenic agents are also mutagenic, and vice versa. If heat can be demonstrated to be a mutagen for mammalian systems, then the thought would have to be entertained: It may also be a carcinogen. Second, if carcinogenesis is considered as an error in development, then data on heat-induced teratogenicity need to be examined.

A. Hyperthermia as a Mutagen or a Transformant

When *Drosophila* cells were exposed to 38°C for 1 hr (a hyperthermia exposure because the normal growth temperature of the fruit fly is 25°C), this treatment doubled the production of lethal mutations. In the only study on human cells, investigators measured 6-thioguanine resistance (a frequently used assay for mutagenesis) induced by heat in human lymphoblasts. A 10-min exposure to 45°C approximately doubled the mutant fraction, causing these workers to suggest that heat behaved like a strong chemical mutagen.

Transformation of mammalian cells *in vitro* has become a powerful technique used to examine the carcinogenicity of various treatments and drugs. The technique consists of exposing density-inhibited cells to the putative carcinogen and then, at various times later, examining the culture for foci of morphologically recognizable transformations. Cells from such transformed clones are then shown to be capable of producing tumors in appropriate hosts, whereas the untransformed cells are incapable of doing so. The process of acquiring this malignant potential is called transformation. Curiously, several studies have shown that heat is not able to transform mammalian cells. Thus, an apparent discrepancy appears between data on mutations and transformations. The explanation may well be that mutations usually involve only a single DNA modification, while it is known that transformation requires at least two such modifications. Thus, heat may be able to do the former, while unable or unlikely to do the latter.

B. Hyperthermia as a Teratogen

The teratogenic literature is much more definite. Heating cells clearly can cause modifications of development that can lead to abnormalities in the offspring. Heating of developing *Drosophilia,* during the pupal stages, results in very specific, abnormal phenotypes of the emerging fruit flies. The type of abnormality relates to the time of heating and was surprisingly specific. For example, if a heat shock is given 38 hr after pupariation, this resulted in growth of abnormal hair in about 90% of the distal wing area, whereas on the proximal part of the wing, <10% of the hair was abnormal. Shifting the heat shock in time by only a few hours reversed this ratio. This curious phenomenon very likely is related to the induced synthesis of heat shock (or stress) proteins.

In other species, a great variety of heat-induced birth defects have been observed. Animals involved range from chickens to monkeys. These experiments have been reviewed recently. In chicken, defects involve the ventril body, head and feet, limbs, and eyes as well as other parts of the birds' bodies. In mice, the major abnormality that has been examined in detail is a neural tube defect, namely failure of the closure of the tube. This failure is frequently associated with the brain outside the cranium at birth.

These studies and many others also point toward the extreme heat sensitivity of the embryos' central nervous system. In guinea pigs, for example, heating at 43°C for 1 hr led to offspring that were slow, clumsy, and frequently did not bond with the mother. Several of the newborn guinea pigs died within the first few days after birth. Autopsy showed that brain sizes were well below those of control animals, which had not been exposed to the elevated temperature. These findings are of considerable interest to medicine because heat, in many ways, acts like alcohol in inducing stress responses. Many of the characteristics found in the newborn guinea pigs delivered from mothers that had been exposed to hyperthermia resemble those found in children suffering from what is called the fetal alcohol syndrome (associated with maternal intake of alcohol). Many biochemical modifications that are induced by heat are also induced by alcohol; it is not unreasonable to hypothesize that it is these biochemical changes that are responsible for malformations in the newborn.

Finally, in humans, several epidemiological studies have related maternal febrile episodes to births of malformed children. Although these studies can hardly be considered as proof that the thermal episode was related to the malformation of the offspring, nevertheless the epidemiological studies on humans, when combined with the direct experimental evidence for many animal systems, are a strong indication that, indeed, hyperthermia is a teratogen in humans. [*See* Thermotolerance in Mammalian Development.]

C. Hyperthermia as a Coinducer

In addition to data cited, hyperthermia is definitely a cocarcinogen and a coteratogen. *In vitro* studies have shown that if carcinogens are applied to cells at elevated temperatures, the rate of transformation is usually increased. This is particularly true if the two treatments are combined. Similarly, if a teratogen is given in conjunction with an increase in body temperature, the probability of malformation in the newborn increases. Thus, heat acts as an agent able to increase the efficacy of many carcinogens and teratogens. In many ways, this is hardly surprising. Temperature increases the rate of chemical reactions and increases the probability of errors in error-prone DNA repair systems.

From these data, the following emerges. Heat by itself is not known to cause cancer; however, it is a mutagen and, as such, may contribute to its induction. In addition, many carcinogens, when applied at elevated temperatures, are more efficient than they are at 37°C. For these agents, heat acts as a cocarcinogen. The teratogenic effects of heat also suggest that elevated temperatures, if these occur at specific times during pregnancy, may contribute to induction of malignancies, although no data or experiments support this hypothesis.

IV. CONCLUSION

Hyperthermia appears to have a role in the treatment of cancer, particularly as part of a multimodality approach. It appears to be most effective against localized tumors. The major obstacle to its wider acceptance is the difficulty of heating lesions, particularly if these are not near the body's surface.

Somewhat paradoxically, hyperthermia may also influence induction of cancer. Again, its major importance may be in enhancing effects of other treatments, i.e., as a cocarcinogen or as a coteratogen.

BIBLIOGRAPHY

Bull, J. M., Lees, D., Schuette, W., Whang-Pang, J., Smith, R., Bynum, G., Atkinson, E. R., Gottdiener, J. S., Gralnick, H. R., Shawker, T. H., and DeVita, V. T., Jr. (1979). Whole body hyperthermia: A phase I trial of a potential adjuvant to chemotherapy. *Ann. Intern. Med.* **90**, 317–323.

Coley, J. B. (1893). The treatment of malignant tumors by repeated inoculations of erysipelas, with a report of ten original cases. *Am. J. Med. Sci.* **105**, 488–511.

Crile, G., Jr. (1963). The effects of heat and radiation on cancers implanted in the feet of mice. *Cancer Res.* **23**, 372–380.

Edwards, M. J. (1986). Hyperthermia as a teratogen: A review of experimental studies and their clinical significance. *Teratogen. Carcinogen. Mutag.* **6**, 563–582.

Engelhardt, R. (1987). Hyperthermia and drugs. *In* "Hyperthermia and the Therapy of Malignant Tumors" (C. Streffer, ed.), pp. 136–203. Recent Results in Cancer Research, Vol. 104. Springer-Verlag, Berlin/Heidelberg/New York.

Hahn, G. M. (1982). "Hyperthermia and Cancer." Plenum Press, New York.

Matsuda, T. (ed.) (1993). "Cancer Treatment by Hyperthermia, Radiation and Drugs." Taylor & Francis, London.

Molls, M., and Scherer, E. (1987). The combination of hyperthermia and radiation: Clinical investigations. *In* "Hyperthermia and the Therapy of Malignant Tumors" (C. Streffer, ed.), pp. 110–135. Recent Results in Cancer Research, Vol. 104. Springer-Verlag, Berlin/Heidelberg/New York.

Pettigrew, R. T. (1975). Cancer therapy by whole body heating. *In* "Proceedings of the International Symposium on Cancer Therapy by Hyperthermia and Radiation" (M. Wizenberg and J. E. Robinson, eds.), pp. 282–288. American College of Radiology Press, Baltimore, MD.

Warkany, J. (1986). Teratogen update: Hyperthermia. *Teratology* **33**, 365–371.

Warren, S. L. (1935). Preliminary study of the effect of artificial fever upon hopeless tumor cases. *Am. J. Roentgenol.* **33**, 75–87.

Hypothalamus

QUENTIN J. PITTMAN
University of Calgary

GLOSSARY

Autonomic nervous system Part of the nervous system that is concerned with involuntary body functions through an action on glands, smooth muscle, and cardiac tissue

Baroreceptors Specialized stretch receptors located in the walls of some major blood vessels that are stimulated by elevations in blood pressure and transmit this information to the brain

Homeostasis Relatively stable state of equilibrium (of the internal body environment)

Neurohypophysial neuron Neuron with a cell body in the hypothalamus and an axon that extends into the posterior pituitary

Nucleus Aggregation of neuronal cell bodies

Tuberoinfundibular neuron Neuron in the medial basal hypothalamus that projects to the median eminence, where it secretes either (a) hypothalamic release or (b) hypothalamic release-inhibiting hormones, which are transported by the blood through capillaries to the anterior pituitary

THE HYPOTHALAMUS IS COMPOSED OF A NUMBER of groups of cells, lying close to the base of the brain and bordering on the third ventricle. An important area of the brain involved in the control of homeostatic functions as diverse as eating, drinking, sleep, thermoregulation, cardiovascular regulation, and hormone secretion, the hypothalamus receives a variety of inputs, both of a neural and of a humoral (blood-borne) nature. Outputs from the hypothalamus result in hormonal, behavioral, and autonomic responses designed to maintain homeostasis. A wide variety of neurotransmitters are utilized within hypothalamic neural networks to accomplish these integrative and control functions.

I. BACKGROUND

Despite its small size (less than 4 g), the hypothalamus exerts enormous control over the body. An appreciation of this influence developed over the first few decades of this century as descriptions emerged of syndromes associated with hypothalamic lesions and dysfunction in humans. Much of our information about the hypothalamus, however, comes from animal experiments. It is such studies that have brought the hypothalamus out of the shadow of the overlying cortical areas to establish its important role in the control of homeostasis. As a result of these studies, the discipline of neuroendocrinology is virtually synonymous with that of the study of the hypothalamus. The identification and structural characterization of many of the chemicals important in brain function have arisen from analyses of extracts of hypothalami, where they are synthesized and stored in great abundance. Finally, the concept of neurosecretion, that is, that a neuron could synthesize and release a hormone (peptide), which is now a commonly accepted fact in the neuropharamacological literature, had its infancy in studies of hypothalamic–pituitary relationships in lower vertebrates.

In order to understand hypothalamic function, it is necessary to have an appreciation of its anatomy.

II. ANATOMY

A. Intrinsic Anatomy

The hypothalamus is an ill-defined area of neural tissue lying on each side of the midline and bordering

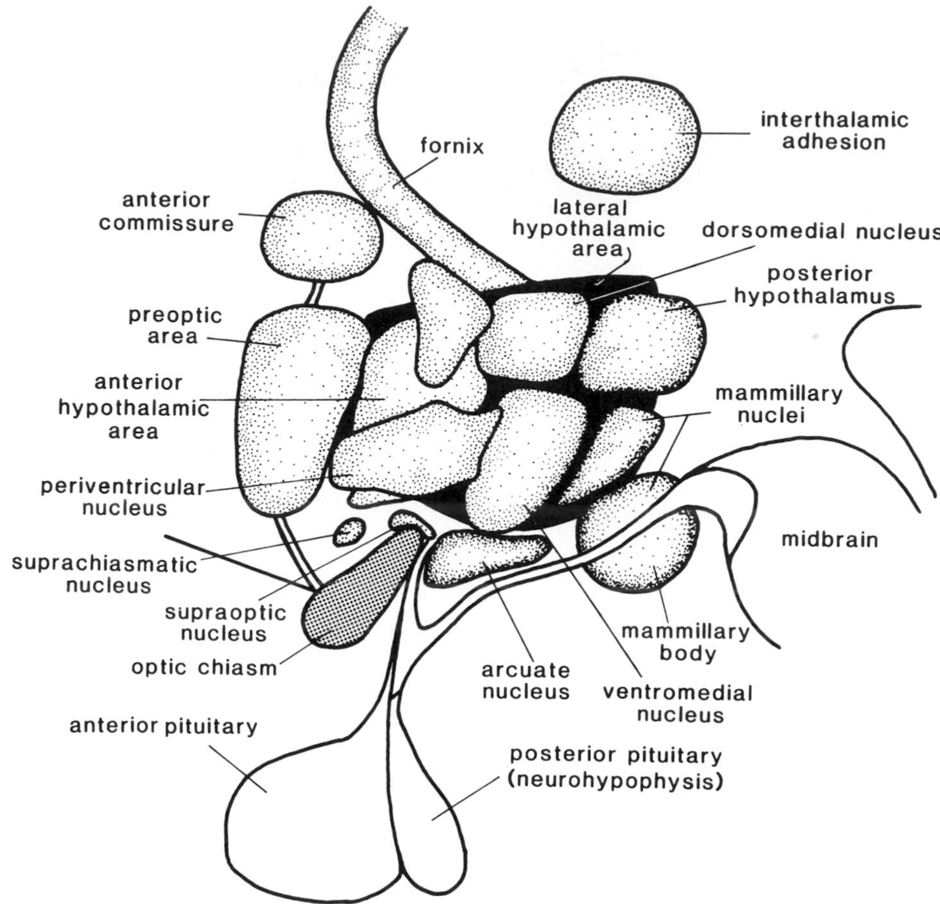

FIGURE 1 Intrinsic hypothalamic nuclei and surrounding structures. This is a sagittal view from the medial aspect of the hypothalamus with anterior on the left.

the walls of the third ventricle. As its name indicates (hypothalamus), it is situated ventral to the thalamus at the base of the diencephalon. The pituitary stalk that joins the hypothalamus to the pituitary exits from the ventral surface; its anterior border is considered the plane of the anterior commissure and its posterior border the mammillary bodies that lie at the junction between the diencephalon and the midbran. This small area of tissue, which in coronal section is barely the size of an adult's thumbnail, is made up of a number of nuclei. These nuclei are packed tightly together and have been named, as indicated in Fig. 1, on the basis of their location within the hypothalamus, their relationship to neighboring structures, and the organization of the afferent and efferent connections. In many (but not all) cases, a particular nucleus is associated with specific functions. At the anterior end of the hypothalamus is the preoptic area, which lies anterior to the optic chiasm. Because a well-defined group of neurons is not apparent from anatomical studies, the term *preoptic area* has been given to this group of cells. Indeed, the preoptic area is often considered to be *anatomically* distinct from the hypothalamus but is so closely related *functionally* to the hypothalamus that it is generally included in any description of hypothalamic anatomy and physiology.

At its posterior edge, the preoptic area gives way to the anterior hypothalamic area, another ill-defined group of cells that often is functionally closely related to the preoptic area. On the medial aspect, near the walls of the third ventricle, the periventricular nucleus makes its appearance and continues posteriorly along the ventricle as a broad sheet of cells. At the anterior–posterior plane of the optic chiasm, small groups of cells (nuclei) with well-defined borders appear; these cells include the suprachiasmatic nucleus and, slightly more posteriorly and laterally, the supraoptic nucleus. The latter nucleus is particularly evident on histological sections because of its large magnocellular cell bodies; this nucleus is important in the control of posterior pituitary function. Dorsal and medial to the supraoptic nucleus and bordering the dorsal aspects of the third ventricle lies the functionally related paraventricular nucleus. Immediately posterior to the optic chiasm is a small nucleus lying immediately above the exit of the pituitary stalk; this nucleus, called the arcuate nucleus, is important in endocrine control, as is the larger, ventromedial nucleus, which lies dorsal and slightly lateral to the arcuate nucleus. As can be seen in Fig. 1, the dorsal and posterior portions of the hypothalamus are occupied by the dorsal medial nucleus and posterior hypothalamus, respectively. A prominent landmark is provided by the ventral projection of the mammillary nuclei, which form a midline structure called the mammillary body. This protuberance (which is evident on the ventral surface) marks the posterior border of the hypothalamus. Running throughout the length of the hypothalamus and abutting laterally upon the subthalamic region and internal capsule is an anatomically indistinct area of tissue called the lateral hypothalamus.

B. Inputs to the Hypothalamus

In light of the predominant role played by the hypothalamus in a wide variety of internal body functions, it is hardly surprising to learn that the hypothalamus is extensively connected to numerous other parts of the nervous system. Some of these pathways are evident as large bundles of nerve fibers easily observed upon gross dissection; others are more diffuse, consisting of small nerve fibers that eluded anatomical description until the advent of more sensitive anatomical techniques. In addition to having neural inputs, the hypothalamus is also in the position to monitor both its local and distant environments through intrinsic sensory neurons that monitor both local conditions in the extracellular fluid and those of their blood supply. Through its afferent and intrinsic pathways, the hypothalamus receives a wide variety of information important to its control of the internal body environment (Fig. 2).

1. Visceral Inputs

The hypothalamus receives extensive information concerning the internal state of the body. Such visceral inputs are relayed to the hypothalamus from a series of ascending fiber tracts that arise in lower brain stem areas. For example, the nucleus tractus solitarius, a brain stem nucleus that receives much of the sensory information arising from the viscera, has a direct projection to various hypothalamic nuclei. A number of brain stem and pontine nuclei that synthesize and utilize noradrenaline as a neurotransmitter (e.g., locus ceruleus and ventral lateral medullary nuclei) give rise to an ascending system of catecholamine fibers that innervate much of the hypothalamus. Many of these fibers ascend in a lateral area of the brain stem in the ventral noradrenergic tract of the medial forebrain bundle. Others ascend through a midline tract called the dorsal longitudinal fasciculus.

2. Somatic-Sensory and Primary Sensory Inputs

Sensory information from the skin is relayed to the hypothalamus through ascending, polysynaptic pathways that arise in the dorsal horn of the spinal cord and cranial sensory nuclei. This information, which projects to the thalamus and higher cortical areas, probably reaches hypothalamic nuclei either from collaterals of ascending fibers or directly from the thalamus. Receptors in the skin transduce a wide variety of sensory signals; of particular importance to the hypothalamus are sensory fibers whose electrical activities change in response to temperature. These fibers carry thermal information to the hypothalamus. Other sensory inputs from the skin carry somatic-sensory information (e.g., touch, pain). [*See* Thalamus.]

Although olfactory input is probably of relatively less importance in humans than it is in lower mammals, many of the connections of olfactory areas to hypothalamic nuclei have been retained in the human. Of particular importance is the input from olfactory areas of cortex to the amygdala, a nucleus lying in the temporal lobe. The amygdala extensively innervates a number of hypothalamic nuclei through two fiber pathways—the stria terminalis and the ventral amygdalo–fugal pathway. Visual sensory information reaches the hypothalamus directly through the retino–hypothalamic tract, which projects to the suprachiasmatic nuclei.

3. Cortical–Limbic Inputs

The limbic system consists of a number of structures that border on the ventricular system of the brain and are thought to be important in processing emotional behavior. The hypothalamus is one of the major destinations of limbic system information. Much of this information is relayed to the hypothalamus from a

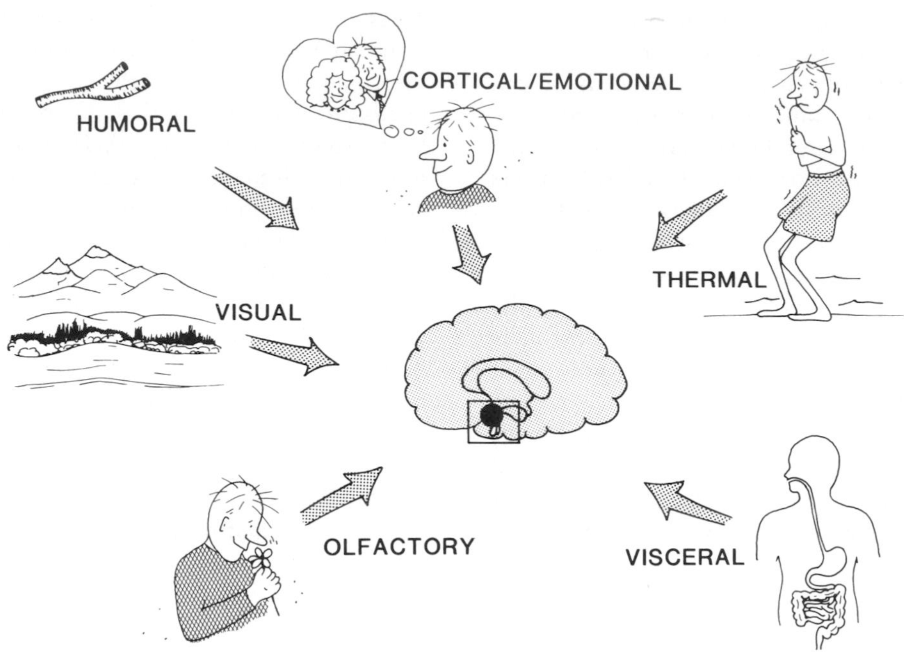

FIGURE 2 Inputs to the hypothalamus.

temporal lobe structure known as the hippocampus. The hippocampus gives rise to a large bundle of fibers called the fornix, which sweeps into the hypothalamus from its anterior dorsal aspect to innervate the posterior hypothalamic nuclei, in particular the mammillary complex. This white matter tract is easily identified on gross dissection of the brain, and its large myelinated fibers dominate histological sections of the hypothalamus. The septal nuclei, which lie anterior and dorsal to the hypothalamus, are also considered part of the limbic system, and they give rise to descending projections, which run through the medial forebrain bundle to innervate the lateral aspects of the hypothalamus and other descending structures. Cortical input to the hypothalamus arises from prefrontal cortex and reaches the hypothalamus either directly through cortical hypothalamic fibers or indirectly via relay through the thalamus. [*See* Hippocampal Formation; Emotional Motor System.]

4. Humoral (Blood-Borne) Inputs

In its position at the base of the brain, the hypothalamus is among the first of neural structures to be perfused by blood ascending in the carotid arteries. Indeed, the Circle of Willis, which distributes blood from the carotids to the various parts of the diencephalon, cortex, and subcortical areas, forms a ring at the ventral aspect of the hypothalamus. The hypothalamus contains sensory neurons of a variety of types that respond to local blood-borne elements. This function is facilitated by a unique type of vascular epithelium in the medial basal aspects of the hypothalamus. Whereas most parts of the brain maintain a well-demarcated functional barrier (the blood–brain barrier) between the blood and local extracellular fluid, the hypothalamus has a "leaky" blood–brain barrier owing to the presence of fenestrated epithelia lining the capillaries. Thus, many substances that circulate (e.g., hormones, nutrients) diffuse freely into the hypothalamus, where specialized neurons can transduce substance levels into electrical activity. [*See* Blood–Brain Barrier.]

It is also possible that the hypothalamus participates in a short, circulatory feedback loop from the pituitary. It has recently been established that blood may flow retrogradely from the pituitary to the brain; this provides a means by which pituitary secretions can quickly reach the hypothalamus, where their levels can be monitored via neurons having specialized receptors. [*See* Pituitary.]

Other information of importance to the pituitary appears to be that of the metabolic state of the body. In addition to having neurons specialized for monitoring metabolically important substances (e.g., glucose), there are neurons throughout the hypothalamus that respond to local temperature. As this temperature is largely a function of the blood perfusing this structure, these neurons are uniquely positioned to monitor core temperature as reflected by the temperature of blood arising from the thorax.

Because of its location bordering the walls of the third ventricle, the neurons of the hypothalamus are also in contact with the cerebrospinal fluid that fills the ventricular system and that can diffuse freely through the ependymal lining of the ventricles into hypothalamic tissue. Although evidence is still fragmentary, it is thought that specialized neurons may exist near the ventricular space that can respond to fluctuations in levels of neurotransmitters and ions that occur in the cerebrospinal fluid.

C. Outputs of Hypothalamus

The anatomical and humoral outputs of the hypothalamus provide this structure with the means to affect many behavioral and physiological functions. An overview of these various controls is given in the following section.

1. Endocrine Outputs

The identification and description of the control of pituitary function have provided one of the more fascinating stories of anatomical and physiological investigations of the past 60 years and have given rise to the entire discipline of neuroendocrinology. The intimate anatomical relationship between the hypothalamus and the pituitary is probably best appreciated when one realizes that the posterior pituitary (neurohypophysis) is actually a ventral growth of the hypothalamus. Magnocellular neurons of the paraventricular and supraoptic nuclei project via the hypothalamo–neurohypophysial tract through the infundibular stalk to end as free nerve endings in the neurohypophysis. The peptides arginine vasopressin and oxytocin (as well as a number of other peptides) are synthesized in these nuclei, transported down the axons of the neurohypophysis, and released into the extracellular space, from where they enter the bloodstream.

In contrast to the organization of the posterior pituitary, the glandular cells of the anterior pituitary receive no direct neural projections. Rather, a series of neurosecretory neurons (tuberoinfundibular neurons) scattered throughout the various hypothalamic nuclei (but concentrated in the arcuate and ventromedial nuclei) send their axons to the base of the brain (median eminence) and release their contents (releasing or release-inhibiting hormones) into a specialized vascular bed. These portal plexus capillaries transport the hypothalamic releasing or inhibiting hormones from the hypothalamus to the anterior pituitary, where they diffuse into this tissue to affect the release of the various anterior pituitary hormones. [*See* Neuroendocrinology.]

In some lower mammals, the pituitary contains an intermediate lobe, which secretes the hormone α-melanocyte-stimulating hormone and is controlled by arcuate nucleus neurons that project directly onto intermediate lobe cells. In humans, the intermediate lobe is present in the fetus but is largely absent, except for a few scattered cells, in the adult.

2. Cortical–Behavioral Outputs

The hypothalamus exerts its control over behavioral functions through mono- and polysynaptic pathways to limbic system structures. Among such projections are those to the septal nuclei, the amygdala, habenula, and medial dorsal nucleus of the thalamus. A particularly important tract providing a relay for hypothalamic information to the limbic system is the mammillo–thalamic tract, which, as its name suggests, arises in the mammillary body and projects to the anterior nucleus of the thalamus. The thalamus, in turn, relays this information to specific limbic structures, in particular, the cingulate gyrus.

3. Autonomic Outputs

Because of limitations in anatomical techniques, and possibly owing to its well-described involvement in endocrine control of the pituitary, descending connections from the hypothalamus were originally thought to be few in nature. Anatomical studies on brains from patients succumbing to problems relating to hypothalamic lesions identified short descending pathways, including the mammillo–tegmental tract, the dorsal–longitudinal fasciculus, and caudal extensions of the medial forebrain bundle. With the advent of more sophisticated anatomical tracing methodology, however, studies in lower mammals have brought to light the fact that the hypothalamus possesses an extensive system of descending fiber tracts that innervate virtually all of the autonomic nuclei of the brain stem and even descend as uninterrupted axons to the most caudal aspects of the spinal cord, where they innervate the sympathetic and parasympathetic preganglionic neurons. A similar system of descending fibers is found in the human. Thus, in terms of size and distribution of their axonal projections, certain neurons of the hypothalamus rival that of the large upper motor neurons (Betz cells) of the cerebral cortex, which exert control over the somatic motor system.

4. Intrinsic Connections

In addition to widespread connections to and from the hypothalamus and other parts of the nervous system, the intrinsic hypothalamic nuclei are also extensively interconnected. Throughout the hypothalamus are many short axon interneurons, which receive information from adjacent nuclei, either directly or from collaterals of descending or ascending fibers.

D. Neurotransmitters

In the mammalian nervous system, neurons communicate with each other largely through the secretion of low-molecular-weight chemicals called *neurotransmitters*, which diffuse across specialized junctions between neurons called *synapses* to affect the electrical activity of neighboring cells. Virtually all the "classic" neurotransmitters (including acetylcholine, noradrenaline, serotonin, dopamine, adrenaline, γ-aminobutyric acid, and glutamate) are well represented in the nerve fibers innervating hypothalamic nuclei. As a number of drugs of abuse and medications are known to act through their interference with such classic neurotransmitter systems, it is apparent that hypothalamic function can be altered by exposure to such agents. Of particular interest over recent years has been the realization that the hypothalamus is virtually a cornucopia of peptide hormones that act as neurotransmitters. Experiments describing the neurosecretory nature of hypothalamic neurons and subsequent electrophysiological studies of hypothalamic "neuroendocrine" neurons provided some of the first evidence that peptides could function as neurotransmitters. There have now been well in excess of 50 different peptides described as having putative neurotransmitter functions within the brain, and virtually all of these are represented within the hypothalamus. However, the task of identifying the roles of these peptides in specific aspects of hypothalamic anatomy and physiology has barely begun. [*See* Peptides.]

III. FUNCTIONS OF THE HYPOTHALAMUS

A. Homeostatis and Control Theory

The great French physician and scientist Claude Bernard first recognized the fact that the internal environment of the body, or "milieu interieur," is maintained in a constant state. Walter Cannon subsequently coined the term *homeostasis* to describe the mechanism by which animals maintain this relatively constant internal environment. The remarkable fact that the internal body environment can be regulated at a constant level in the face of widely varying external conditions of temperature, food availability, metabolic demands, and so on, is due, in part, to the hypothalamus. To achieve such regulation, the hypothalamus must be able to sense the various internal environments of the body, integrate this information, and compare it to a theoretical "set point," and then respond to any perturbations with an appropriate effector signal. The means by which the hypothalamus obtains its afferent information and the pathways by which its output is directed to appropriate destinations have been outlined in the preceding sections. The following discussion examines the various functions controlled by the hypothalamus.

B. Specific Activities

The location of the hypothalamus deep within the brain has not facilitated investigation into the functions of this structure. From the dorsal aspect, one must pass through the entire cerebrum to reach the hypothalamus; from the ventral aspect, the base of the skull limits accessibility. In humans, tumors and cardiovascular accidents have provided the bulk of the experimental material revealing functional correlates to damage or stimulation of particular areas. Even under such circumstances, it has often been difficult to differentiate between a syndrome caused by *damage* to a specific area and that resulting from irritation or stimulation of an area adjacent to, for example, a small hemorrhage. Particularly with respect to the endocrine system, however, a number of diseases that exhibit classic sequelae can now be associated with specific hypothalamic deficits. The information that has been obtained about the human hypothalamus, however, is strongly based on experimental studies in lower mammals. Fortunately, there appears to be remarkable preservation of form and function throughout the various mammalian genera, and experiments in mammals as diverse as the rat and a subhuman primate often give surprisingly congruent results. In these animals, it has been possible to carry out lesion, stimulation, and recording experiments that have revealed important aspects of hypothalamic function. However, even under controlled laboratory conditions, it has been difficult to achieve clear and

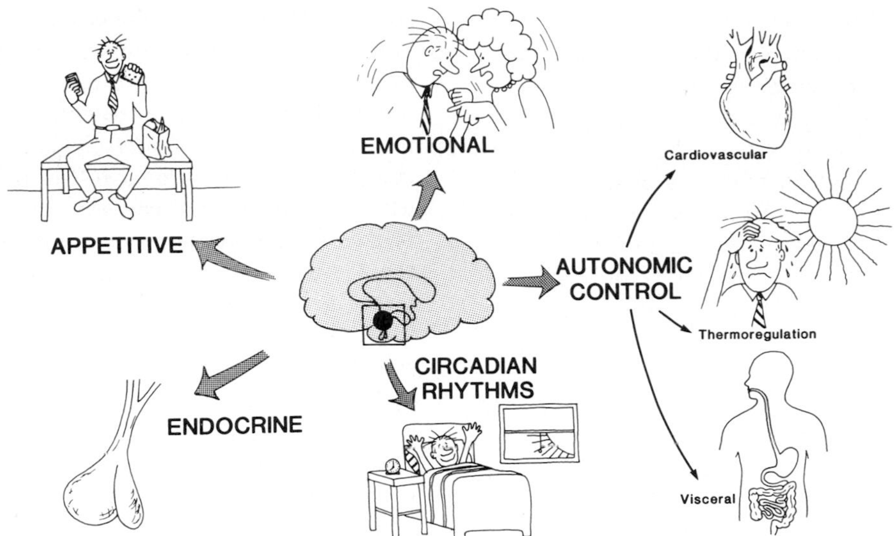

FIGURE 3 Activities under hypothalamic control.

easily interpretable results; the various hypothalamic nuclei are closely situated, and it has been difficult to affect one area of the hypothalamus discretely without impinging on adjacent areas. Nonetheless, with these caveats in mind, a description of hypothalamic functions emerges as outlined in Fig. 3.

1. Appetitive Behaviors

The hypothalamus plays an important role in the control of feeding behavior. It has been known for many years that stimulation of medial parts of the hypothalamus, in particular the ventromedial nucleus and adjacent tissue, will inhibit eating, whereas stimulation of the lateral hypothalamus will cause an increase in food intake in lower mammals. In keeping with these observations, lesions in medial hypothalamic areas occurring both naturally in humans and after experimental placement in lower mammals cause increased feeding; destruction of more lateral hypothalamic areas causes anorexia. Rather than a direct stimulation of the appropriate behavior (eating, chewing, food seeking), the evidence indicates that an actual set point of body weight regulation appears to be changed. These changes in food intake are associated with corresponding changes in metabolism, which suggest that they are part of an integrated response to the regulation of caloric balance. While the actual mechanisms responsible for the regulation of food intake are numerous, the hypothalamus plays an important role. Among other regulatory events is one in which neurons in the hypothalamus respond to changes in blood glucose level with changes in electrical activity. For example, cells in the ventromedial hypothalamus respond with increased activity to high blood glucose levels, whereas neurons throughout the lateral hypothalamus more often decrease their activity in response to elevations in glucose. Although it is undoubtedly simplistic to make the relationship, it may be noted that such changes are appropriate in terms of our understanding of the relative roles of medial and lateral hypothalamus in regulating food intake. Thus, as is expected for a controlled system, both the sensory mechanisms are present and the necessary connections exist from the hypothalamus to the limbic system to elicit the appropriate behaviors. [*See* Feeding Behavior.]

The debate as to the exact cue for signaling thirst as well as the location of cells responsible remains vigorous to this day. It would appear that changes in blood osmolaity and possibly sodium content activate neurons in areas near the hypothalamus (organum vasculosum of the lamina terminalis and subfornical organ), which in turn send strong projections to the hypothalamus. In addition, osmoreceptors are located in the anterior hypothalamus and preoptic area, which directly respond to changes in plasma osmotic pressure (concentration of solutes in the blood). [*See* Thirst.]

The hypothalamus appears to integrate signals from the various brain structures and activates drinking behavior via pathways to the limbic system in a manner similar to that described for eating behavior.

2. Autonomic Control

The pivotal role of the hypothalamus in the control of the autonomic nervous system has long been recognized, and it is often described as the "head ganglion" of the autonomic nervous system. The autonomic nervous system is divided on the basis of its anatomy, pharmacology, and, to a certain extent, its function into two components: the sympathetic (involved in arousal, or "fight or flight," behavior) and parasympathetic (involved in vegetative function) systems. At one time, it was thought that there was a topographical relationship within the hypothalamus to these two systems, with the anterior hypothalamus considered the parasympathetic center and the posterior hypothalamus the sympathetic center. Although it is true that in animals "sympathetic" effects can be activated by electrical stimulation of the posterior hypothalamic areas, and "parasympathetic" responses by stimulation of anterior hypothalamic areas, it is equally true that more discrete stimuli throughout a variety of areas of the hypothalamus can cause generalized autonomic effects. The strong interconnections between various hypothalamic nuclei and brain stem and spinal cord autonomic nuclei provide an anatomical basis for the hypothalamus to exert control of cardiovascular function, thermoregulation, and visceral function. [*See* Autonomic Nervous System.]

a. Cardiovascular System

With respect to the cardiovascular system, it is known that information is relayed from baroreceptors synapsing in the medulla within nuclei that project directly or via polysynaptic pathways to the hypothalamus. Electrical or chemical stimulation of neurons in hypothalamic areas that receive baroreceptor information causes changes in both arterial pressure and in heart rate; such effects are most likely mediated through descending projections to areas of the ventral lateral medulla and to the spinal cord, which control peripheral arterial muscles and cardiac muscle. In addition, a number of circulating hormones influence blood pressure, and their release can be influenced by such descending pathways or via hypothalamic connections to the pituitary (see Section III,B,5).

b. Thermoregulation

Classic experiments carried out in the 1930s established that animals with hypothalamic lesions were incapable of normal thermoregulation and gave rise to extensive studies that have unequivocally placed the hypothalamus as a pivotal structure involved in the neural control of body temperature. Thermal sensory information from the skin and other parts of the body is relayed via multisynaptic pathways to the hypothalamus, where it is integrated with information obtained from intrinsic receptors. Through mechanisms not yet well understood, this information is compared to a hypothetical "set point," and, where deviations from this temperature take place, appropriate effector mechanisms are brought into play to return temperature to normal. These effector mechanisms include the regulation of peripheral vasomotor tone, alterations in sweat secretion, changes in respiratory rate, changes in metabolism, and alterations of behavior and posture appropriate for changes in heat loss or heat gain. In addition to its control over normal thermoregulation, the hypothalamus, and particularly the preoptic/anterior hypothalamic areas, appears to be important in the development of fever. In lower mammals, application of minute quantities of pyrogens (agents that cause fever) directly to these areas causes prompt, experimental fevers; destruction of neurons in these same areas inhibits fever development in these animals. [*See* Body Temperature and Its Regulation.]

c. Visceral Function

In accord with the hypothalamic control of appetitive behaviors described earlier, the hypothalamus also directly influences gut function to provide complementary autonomic effects. For example, it has been shown in rats that electrical or chemical stimulation of a number of hypothalamic areas (e.g., paraventricular nucleus, ventromedial hypothalamus, dorsal hypothalamus) can influence gut muscle function as well as acid secretion from the stomach. It is thought that the pathways from the hypothalamus to brain stem autonomic areas that control acid secretion may be of particular importance in the development of ulcers in response to stress, with the hypothalamus providing the intermediary role in relaying cortical and limbic information to the autonomic nervous system. [*See* Visceral Afferent Systems: Signaling and Integration.]

3. Emotional Expression

In lower mammals, electrical stimulation of several areas in the medial and posterior hypothalamic nuclei causes impressive displays of aggressive and vicious behavior; because it is thought that such behaviors are merely a behavioral output rather than the "thought" associated with anger, this phenomenon is known as sham rage. Destruction of other areas of the hypothalamus (e.g., the ventromedial nucleus) will also lead to such behavioral alterations. In contrast to these aversive behaviors, it is known that other areas of the hypothalamus, in particular the lateral areas, appear to be associated with behaviors best described as "reinforcing" or "pleasurable." That is, animals will actually self-stimulate electrodes placed in the lateral hypothalamus, even to the detriment of other behaviors; presumably, stimulation of neurons in these areas activates pathways leading to cortical areas and are interpreted as pleasurable. [*See* Affective Responses.]

Subjective aspects of fear, rage, and pleasure are also associated with hypothalamic damage or irritation in humans. This evidence is consistent with the idea that the hypothalamus is one relay in the pathway involving limbic and cortical areas that are associated with emotional behavior.

4. Circadian Rhythms

Many animals exhibit behaviors that show a remarkable periodicity that approximates a 24-hr day. These circadian rhythms are thought to be generated by an oscillatory mechanism in the suprachiasmatic nucleus and entrain to the light–dark cycle via impulses arriving from retinal–hypothalamic tracts. In fact, destruction of the suprachiasmatic nucleus in such animals will cause these rhythms to become totally disrupted. Human beings also have a suprachiasmatic nucleus and have been shown to display circadian rhythms in body temperature and endocrine activity. Of particular interest is the hypothalamic control of rhythms of sleep and wakefulness, which appear to be mediated by hypothalamic connections to the reticular activating system of the midbrain. [*See* Circadian Rhythms and Periodic Processes.]

5. Endocrine Control

Through its anatomical connections to the posterior pituitary and to the median eminence, the hypothalamus represents the final common pathway for brain control of pituitary secretion. The availability of specific radioimmunoassays for various peptides and hormones has made it possible to measure these substances in the circulation and to identify the stimuli responsible for their release. With the advances in biochemical characterization of these peptide hormones, and the availability of pure synthetic peptides, it has also been possible to administer these substances and thereby identify target end organ responses.

The major hormones released from the posterior pituitary are the peptides arginine vasopressin and oxytocin. Arginine vasopressin is also known as antidiuretic hormone because of its action within the kidney to enhance water reuptake and concentrate the urine. In addition, this hormone has other actions within the body to regulate blood pressure through its action on arterial smooth muscle and an action on metabolism through its enhancement of glycogen conversion to glucose in the liver. Oxytocin receptors are found on smooth muscle, in this case on uterine smooth muscle and on mammary smooth muscle. Thus, its peripheral action is to cause milk letdown via contraction of the mammary smooth muscle and to cause uterine contractions (during the birth of a baby). Oxytocin receptors are also found on fat cells and on selected other peripheral tissues; oxytocin actions on these structures may explain its presence in males.

Regulation of the secretion of these hormones is brought about through specialized receptors in the hypothalamus and afferent stimuli arising from peripheral receptors. Neurons in the anterior hypothalamus, and possibly the vasopressin neurons themselves, are capable of sensing the osmolality of the blood and, in response to increases in osmolality, cause increased synthesis and release of vasopressin. Reductions in blood pressure and blood volume, signaled via specialized cardiovascular receptors in the periphery, are also potent stimuli for vasopressin release, as are some types of stressful stimuli (e.g., pain and nausea).

Oxytocin is released in response to tactile stimulation of the nipples (i.e., when a baby suckles) and after distention of the birth canal during delivery. Cortical inputs are thought to provide a behavioral influence on oxytocin release; this is sometimes manifested by the spontaneous milk letdown experienced by a mother in response to the cries of her baby. Emotional stress, on the other hand, appears to inhibit oxytocin secretion, thereby making breast-feeding difficult for a mother under such stress.

Hormones responsible for the release or inhibition of release of anterior pituitary hormones are synthesized in tuberoinfundibular neurons scattered throughout the hypothalamus and terminating on the medium eminence. The secretory products of these neurons control glandular cells of the anterior pituitary, which synthesize and release luteinizing hormone, follicle-stimulating hormone, growth hormone, thyroid-stimulating hormone, adrenocorticotropic hormone, prolactin, and β-endorphin. The control of the tuberoinfundibular neurons is beyond the scope of this article. Nonetheless, in terms of general principles, there appear to be negative or positive feedback mechanisms that affect the release of their products. Thus, there are receptors in the hypothalamus that respond to circulating hormones produced both in distant parts of the body (e.g., steroids, such as estrogen produced in the ovary) as well as from the anterior pituitary. In addition, afferent stimuli from both viscera, as well as the cortex and limbic system, can affect release. Secretion of the anterior pituitary hormones is often synchronized to the light–dark cycle, and this circadian control is thought to be exerted via retinal signals entering the suprachiasmatic nucleus. [*See* Peptide Hormones and their Convertases.]

C. Integrative Aspects

The control of hypothalamic function has been discussed for each regulated function, yet it must be emphasized that response to perturbations in our internal body environment is accomplished through recruitment of many of the functions of the hypothalamus. For example, the response to a severe hemorrhage will include autonomic responses designed to elevate blood pressure (e.g., vasoconstriction of peripheral arteries and increased heart rate), hormonal responses mediated via descending autonomic fibers (e.g., adrenaline secretion from the adrenal gland), and pituitary secretion of vasopressin and a number of anterior pituitary hormones. Similarly, the response to cold will include behavioral (e.g., warmth-seeking behavior, putting on warmer clothes), autonomic (e.g., vasoconstriction and shivering), and endocrine (e.g., increased secretion of thyrotropin-releasing hormone) responses.

The importance of these regulatory functions controlled by the hypothalamus cannot be understated. A lesion or deficit in the afferent circuitry, the central organization or the output of the hypothalamus for the control of these homeostatic functions, has serious consequences for the individual. As an "ancient" part of the brain present even in the lowest vertebrates, the hypothalamus has maintained its essential role in supporting bodily functions.

BIBLIOGRAPHY

Appenzeller, O. (1990). "The Autonomic Nervous System," 4th Ed. Elsevier, Amsterdam.

Ganten, D., and Pfaff, D. (eds.) (1986). "Morphology of the Hypothalamus and Its Connections." Springer-Verlag, New York.

Gordon, C. J., and Heath, J. E. (1986). Integration and central processing in temperature regulation. *Annu. Rev. Physiol.* **48,** 595–612.

Loewy, A. D., and Spyer, K. M. (eds.) (1990). "Central Regulation of Autonomic Functions." Oxford Univ. Press, New York.

Nerozzi, D., Goodwin, F. K., and Costa, E. (eds.) (1987). Hypothalamic dysfunction in neuropsychiatric disorders. *In* "Advances in Biochemical Psychopharmacology," Vol. 43. Raven, New York.

Reichlin, S., Baldessarini, R. J., and Martin, J. B. (eds.) (1978). "The Hypothalamus." Raven, New York.

Smith, O. A., and DeVito, J. L. (1984). Central neural integration for the control of autonomic responses associated with emotion. *Ann. Rev. Neurosci.* **7,** 43–65.

Stoddard, S. L. (1991). Hypothalamic control and peripheral concomitants of the autonomic defense response. *In* "Stress, Neurobiology and Neuroendocrinology" (M. R. Brown, G. F. Koob, and C. Rivier, eds.), pp. 231–253. Marcel Dekker, New York.

Swanson, L. W., and Sawchenko, P. E. (1983). Hypothalamic integration: organization of the paraventricular and supraoptic nuclei. *Annu. Rev. Neurosci.* **6,** 269–324.

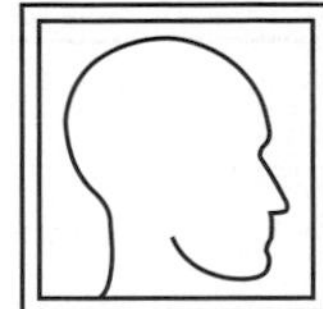

Immune Surveillance

HANS WIGZELL
Karolinska Institute, Stockholm

GLOSSARY

Homograft reaction Rejection reaction against transplanted tissue from a foreign individual where T lymphocytes are the dominating aggressive cells

Immunogenicity Capacity of a substance or cell to induce a specific immune reaction against it

Natural killer cells Aggressive cells with an inborn tendency to kill malignant cells more readily than normal cells

T lymphocytes White blood cells requiring the thymus to mature properly; they constitute a major specific protective force against many infectious diseases

IMMUNE SURVEILLANCE (IS) IS HERE DEFINED AS A mechanism through which arising "potentially dangerous mutant cells" are constantly eliminated and where the mechanism(s) is "of immunological character." The need for IS was considered to constitute the major reason behind the development of the thymus-dependent part of the immune system, with the capacity to eliminate *in situ* malignant cells before they had developed into minitumors. IS would thus have the form of a natural, spontaneous immune reactivity already preexisting before the appearance of the tumor cells. This should then be put in contrast to other, more conventional forms of specific immune reactions, which may become induced against an already established, growing tumor. Of course, in the latter situation, IS clearly has already shown itself as a failure with regard to function.

I. CONCEPT OF IS

The word immune means exempt. It is derived from past observations of a severe infectious disease that swept through a district and caused many people to become sick, with several even dying; however, when the same disease returned at a later time, people who had survived the first epidemic were now exempt, or immune. When the term immune surveillance is used, however, it is not used in the context of protection against infectious diseases but rather as a constant guard against invaders from within (i.e., against cancer cells). Several scientists have suggested that a primary function of the immune system is to protect the multicellular society of the individual against its "asocial" members—the malignant tumor cells.

The first theory along these lines was probably formulated by the famous German immunologist Paul Ehrlich in 1909. He believed that malignant cells did arise throughout the life of the individual but that the vast majority of such cells would be eliminated by the immune system constantly surveilling the tissues against newly arising tumor cells. Accordingly, the appearance of tumors would more or less be a mistake, by which the cancer cells somehow managed to "sneak" through this constant watch-guard mechanism of IS. [*See* Immune System.]

More recent formulations of IS stem from researchers who, in the later part of the 1950s, put IS into a more modern concept of cellular immunology. As the strength of the homograft reaction at that time was just being cognized, this particular form of immune reactivity was proposed to have as its prime biological basis the elimination of malignant cells. When T lymphocytes later became known as the major force for the specific, cell-mediated immune reactions causing rejection of foreign grafts, T cells were considered the most likely candidates for the effector cells of IS. Still later, when natural killer (NK) cells were discovered, with their significant tendency to preferentially react against malignant as compared with normal targets, they also were indicated as potential effectors of IS. [*See* Lymphocytes; Natural Killer and Other Effector Cells.]

II. ATTEMPTS TO VALIDATE IS

The preceding concept of IS against cancer is intellectually highly attractive; however, a number of situations disprove the general validity of the hypothesis. A major assumption when considering the existence of IS is that the presence of a deficiency in the proposed immune effector mechanism for IS would automatically lead to an increased risk of tumor development. Superficially, significant support would seem to exist for IS using the foregoing reasoning. If one assumes that T lymphocytes are the performers of IS, literature abounds with articles showing that immunodeficient individuals do indeed express a significantly increased risk for tumor development. It was initially shown for individuals with genetically defined immunodeficiency disorders being followed by similar observations in transplantation patients with drug-induced immunosuppression. When the acquired immunodeficiency syndrome epidemic began, the HIV-infected individuals, upon entering into immunodeficiency, displayed a most dramatic increase in malignant disorders, reaching cancer frequencies on the order of 25% or more in some cohorts.

These data would initially seem to support the existence of IS. However, a major problem with these findings in relation to IS is the type of cancers arising. If IS is of general significance, one would expect that all forms of tumors would be increased in immunosuppressed conditions; however, this is not the case (Table I). The most common forms of cancer in humans, such as mammary, pulmonary, gastrointestinal, and prostatic tumors, do not appear at increased frequencies in immunodeficient individuals. In contrast, only a limited number of malignancy types dominate among the tumors appearing in immunosuppressed humans. Among these tumors are lymphomas, skin tumors, and a tumor type normally very rare in Caucasian groups, namely, Kaposi's sarcoma. Undoubtedly, the immunosuppression is intimately linked to the increase of these tumor types. For instance, the depth of immunosuppression induced by drug treatment in organ-grafted individuals is directly positively correlated with the increase in percentage of individuals developing malignancies. However, these normally rare tumors occurring at high frequency in immunosuppressed individuals may not be due to a faulty IS. The lymphomas arise, for instance, within an immune system afflicted by damage. Many other systems show that chronic inflammation and repair per se in a tissue may serve as promoters for tumor development. Likewise, the lymphomas that appear are of a type called Burkitt's lymphomas, a tumor intimately associated with the Epstein–Barr virus, known to start replicating in immunodeficient individuals. For Kaposi's sarcoma, a similar situation prevails. Evidence suggests that the immunodeficiency situations linked with this type of tumor cause the production of certain growth factors, one or more of which act as efficient promoters of the cells composing Kaposi's sarcoma. In addition, there is increasing evidence that a previously unknown virus is present in Kaposi's sarcoma development. Again, this would suggest that the damage to the immune system is releasing tumor-promoting factors (viruses, growth factors) that are responsible for the increase in malignancies, with no evidence for IS playing a discernible role. [*See* Lymphoma.]

The third type of tumor with increased frequency in immunodeficient individuals, skin epithelial tumors of squamous or basal cell type, is the only one for which a plausible case for IS containing the carcinoma cells exists. Ultraviolet (UV) radiation at high doses is known to be immunosuppressive and can, at certain

TABLE I
Increased Appearance of Certain Tumor Types[a] in Immunosuppressed Individuals

	Relative tumor incidence (control = 1)[b]		
	Kaposi's sarcoma	B lymphomas	Skin tumors
Organ-grafted	450 (23)	39 (36)	29
Acquired immunodeficiency syndrome	>10,000	>40	increased

[a]Tumors such as pulmonary, mammary, prostatic, and gastrointestinal show no increase in immunosuppressed individuals.
[b]Figures within parentheses indicate (in months) tumor appearance after induction of immunosuppression.

wavelengths, also be shown to induce tumors with high immunogenicity in inbred mice. UV radiation in humans likely will act in a similar manner. The fact that UV-induced tumors in mice are unusually immunogenic can be used to make a case that the ubiquitous UV radiation from the sun creates a vital need for efficient protection against mutagenized premalignant cells in the skin, perhaps via IS; however, this argument needs more proof before it can be accepted as fact.

In animals, the results with regard to consequences of T-cell deficiency are similar to those in humans. In most circumstances, mice lacking thymus and T lymphocytes have normal frequencies of tumors, unless the mouse colony is infected with tumor-inducing viruses. In such a case, the T-cell-deficient mice develop tumors in a very high percentage; however, this does not prove the existence of IS but, rather, a lack of conventional immune protection against a tumorigenic virus, allowing it to cause a more severe infection and, thus, a higher probability to cause cancer.

In summary, immunodeficiencies involving T lymphocytes thus fail to reveal the existence of IS against tumor cells. But the more recently discovered cell type NK cells—with their tendency toward "spontaneous" killing of preferentially malignant cells—could serve as an alternative candidate taking care of IS. Such cells can clearly be shown in experimental animal systems to act *in vivo* against transplanted tumor cells. However, in human beings with relatively select NK cell deficiencies such as the rare Chediak–Higashi's disease, no convincing evidence indicates a general increase in malignancies. Some increase in lymphoid malignancies has been recorded, but as this disease involves damage within the bone marrow cells, this again may have secondary reasons. Likewise, in the animal systems no significant support for IS via NK cells has been forthcoming. Therefore, one can conclude that immunodeficiency disorders involving either T or NK cells have so far failed to provide evidence for the existence of any significant IS against tumor cells.

III. DOES IS AGAINST TUMORS EXIST?

From the foregoing information, one can make certain comparatively firm conclusions. All severe immunosuppression induced by various ways (genetic, drug-induced, acquired via viral infection) result in an increase in cancer disease in the afflicted individuals. However, the identical restricted types of tumors appear (lymphomas, skin tumors, Kaposi's sarcoma) at higher frequencies, whereas all common types of cancer appear at normal rates. The tumors that do arise at enhanced rates in these immunosuppressed individuals may all do so for reasons unrelated to a failure of IS.

At face value, these data would appear to dismiss the existence of an IS mechanism constantly protecting the individual against the appearance of tumors. It may be wise to postpone this final disposal, at least for some time. It is normally considered that the immune system is composed of cells from the bone marrow with a division of labor between subsets of cells. This definition may be too narrow if one returns to the original meaning of the word immune (i.e., exempt). Thus, cell functions may exist in the body that are yet to be defined with "immune" consequences and where such cells are not necessarily derived from the bone marrow. For instance, in the body, tumor cells may reside for prolonged periods of time—years—without starting to grow. This is most easily shown in situations where metastasis appears a decade or more after the primary tumor was removed. One tumor type that comparatively frequently acts in such a manner is the melanoma. Whether normal cells in the vicinity of the "dormant" tumor cells keep the cancer cells from growing or whether there are inborn restraints in the tumor cells that may be released with time is unknown. From the clinic, a significant number of reports indicate the same message. For example, a patient who underwent a leg operation for a primary melanoma some 15 years later develops mammary carcinoma. After surgery of the carcinoma, local X irradiation is given. A few months after irradiation, hundreds of melanoma metastases can be seen growing locally in the skin of the irradiated area. The X irradiation has undoubtedly activated dormant melanoma cells and/or alternatively eliminated some restraining element(s) provided for by the local tissue cells. If the latter is true, is it justified to call this IS despite the fact that it would not cause the death of the tumor cells but merely prevent growth?

Studies on the possible existence of IS against tumors have been a fruitful way to explore how the immune system functions in general, despite the failure to provide its existence in a more general manner. New findings or concepts regarding how the immune system may function will surely be tested in future research on IS.

BIBLIOGRAPHY

Burnet, F. M. (1957). Cancer—A biological approach. *Br. Med. J.* **1,** 779.

Burnet, F. M. (1970). The concept of immunological surveillance. *Progr. Exp. Tumor Res.* **13,** 1.

Ehrlich, P. (1957). "The Collected Papers of Paul Ehrlich" (F. Himmelweit, ed.). Pergamon, Oxford, England.

Mukheiji, B. (1997). Molecular Basis for Tumor Immunity. In "Encyclopedia of Cancer." (Bertino, Joseph, ed.). Academic Press, San Diego.

Old, L. J. (1989). Chapter IV Defense: Structural basis for tumor cell recognition by the immune system. *In* "Progress in Immunology VII" (F. Melcher, ed.). Springer-Verlag, Berlin.

Penn, I. (1986). Cancer is a complication of severe immunosuppression. *Surg. Gynecol. Obstet.* **162,** 603.

Thomas, L. (1959). Reactions to homologous tissue antigens in relation to hypersensitivity. *In* "Cellular and Humoral Aspects of the Hypersensitive States" (F. S. Lawrence, ed.), p. 529. Harper, New York.

Immune System

ANDREW M. LEW
The Walter and Eliza Hall Institute of Medical Research, Australia

GLOSSARY

Antibody A specialized protein (immunoglobulin) secreted by B cells that binds antigen (in the form of toxin or as part of a microbe) leading to neutralization or elimination of the offending pathogen; one antibody usually binds to only one antigen

Antigen Any molecule (usually foreign) that is recognized by antibody or the T-cell receptor

B cell Lymphocyte that makes antibody and initially developed in the bone marrow (human, mouse) or bursa of Fabricius (chicken)

B-cell receptor Surface membrane-bound form of antibody

Immunoglobulin Protein that antibodies are made of (structural term; Porter and Edelman won the 1972 Nobel Prize for determining its structure)

Leukocyte White blood cell; comprises lymphocytes, monocytes, macrophages, neutrophils, eosinophils, basophils, and dendritic cells

Lymphocyte A particular subset of leukocytes endowed with immune recognition

T cell Lymphocyte that has a T-cell receptor and initially developed in the thymus

THE IMMUNE SYSTEM CONSISTS OF A MOBILE NETwork of highly specialized white blood cells (lymphocytes). Each lymphocyte has an individual specificity for a given part of a given molecule of a given microorganism. Because there are so many lymphocytes, the range of possible assailing microorganisms is catered for by the total system. The major trigger for lymphocytes to proliferate is the presence of the particular microorganism that the lymphocytes react against. Therefore, those lymphocytes with that specificity expand as a cohort (clonal expansion) so when the same microorganism attacks a second time, the immune response is faster and more vigorous. Lymphocytes are not only endowed with a powerful weaponry, but they communicate with each other and enlist the aid of other innate defense mechanisms. [*See* Lymphocytes.]

I. THE IMMUNE SYSTEM AS AN ARMY

The immune system defends the body against foreign pathogens, especially infectious microorganisms, and as such is an army whose headquarters and battleground are the body. It is therefore essential that it is highly specific for its target. Many of its features are summarized in Table I and are discussed later in this section. The immune system evolved in vertebrates, arguably because vertebrates are in general long lived and therefore need highly developed immune surveillance. Children with severe immunodeficiencies die of infectious diseases when only a year old, even in our antibiotic era, unless some replacement of their immune system is successful.

A. An Army Needs Lots of Soldiers

Five to 10% of the body's total number of cells are devoted to defense. The immune system has about 2 trillion lymphocytes.

B. An Army Needs Specialized Forces to Perform the Many Varied Tasks

For example, lymphocytes can be grouped into T and B cells. B cells make antibody which can neutralize

ENCYCLOPEDIA OF HUMAN BIOLOGY, Second Edition, VOLUME 4.

TABLE I
Features of the Immune System

Specificity
Memory
Amplification
Proliferation
Recruitment and mobilization
Specialization
Networks of hubs and guard posts
Networks of communication by contact and by chemical messages
Special training grounds

poisons (e.g., tetanus toxin) and can clear away viruses from the circulation. Certain specialized T cells (called helper T cells) help B cells make more antibody and also instruct B cells to make the antibodies more effective. Still other T cells (called cytotoxic T cells) kill cells that harbor virus. Some of these aspects are discussed in Section III.

C. An Army Identifies the Enemy and Strikes Only Enemy Targets

This specialization of workloads goes even further. Each T cell has one specificity and each B cell has one specificity. The offspring of each of these cells forms a cohort with the same/similar specificity, which is termed a clone. Each clone is stimulated to proliferate when bound to the specific antigen. In this way, only in the presence of a particular microorganism are the clones reactive to (and only those clones reactive to the microorganism) expanded. This is the clonal selection theory which earned Burnet the 1960 Nobel Prize. The immune system has the theoretical capacity of generating a trillion specificities but in reality probably generates a million specificities. This enormous repertoire allows it to identify and discriminate any molecular shape (the major reason why antibodies are so useful in diagnostics and as research tools). Cells with the right specificity for a given target are selected to be activated and expanded. Self-reactive cells are eliminated or inactivated, although in rare circumstances this goes awry and autoimmune disease develops.

D. An Army Learns from Previous Exposure to a Certain Enemy and Is More Prepared the Next Time

This phenomenon is called immunological memory. This and the specificity outlined earlier are the very basis of how vaccines work. The secondary immune response is faster, greater, and more avid (antibodies bind more strongly to its target) than the primary response. Someone immunized with a particular vaccine (a nonpathogenic version of a noxious agent) will recognize the noxious agent, react quickly before the agent has a chance to do much harm, and readily overwhelm it.

E. An Army Needs Short- and Long-Range Weapons

The former is provided by close cell contact (e.g., perforin and granzyme B from cytotoxic T cells perforate and kill virally infected cells) and by cytokines (special hormones that mainly act at close range); the latter by antibodies which pervade the body, especially through the circulation. There are also important locally acting antibodies, especially at mucosal surfaces.

F. An Army Enlarges and Mobilizes Its Forces Quickly

Lymphocytes can reproduce offspring cells, can recruit members of its own army from nearby hubs and from the circulation, can recruit other cells, and can amplify its own effectiveness using defense proteins from the circulation (e.g., complement) or defense proteins from other cells.

G. An Army Communicates Efficiently at the Battle Front and to Central Networks

The immune system has an elaborate communication system. At the whole body level, there is a network of lookouts and hubs. Lymph nodes drain the lymph (act as first port of call and are therefore ideal guard posts), and within each lymph node is a complex hub with highly differentiated areas for B-cell proliferation and for T- and B-cell collaboration. At the cellular level, this involves both direct contact of cells (using a host of surface molecules both to establish contact and to signal certain messages) and the use of locally acting hormones (viz. chemical messengers). Because the cells themselves are mobile, they can go to various parts of the body to raise the alarm.

H. An Army Trains Cadets

The bone marrow is the initial training ground for B cells, and the thymus is the initial training ground for T cells. Further training "on the job" occurs in the peripheral lymphoid organs (lymph nodes, spleen, etc.).

II. INNATE VS SPECIFIC IMMUNE SYSTEMS

The immune system that is generally referred to (as described in Section I) is the specific (also called acquired) immune system. There is also a more primitive defense system that does not recognize individual microorganisms, or a particular molecule shape. It lacks the exquisite specificity of the specific immune system and hence is referred to as innate or natural or nonspecific immunity. Although it tends not to be as focused or as vigorous, it is very important, especially for organ systems that are exposed (i.e., have direct connections to the outside), e.g., skin, respiratory system, eye, gut, and bladder.

There are many mechanisms of innate immunity and only a few examples are mentioned here.

A. Anatomical Barriers

The maze of the nasal turbinates is a course filter for what we breathe in through the nose. The cornified epithelium of the skin is a layer of armour that stops many physical, chemical, and infectious agents from penetrating.

B. Flushing Away

If voiding from the bladder is inhibited, there is an increased chance of urinary tract infections. In the windpipe, a layer of mucus paddled by the ciliary beat constantly traps and clears away small particles.

C. Antimicrobial Chemicals

Acid in the stomach can reduce the level of ingested salmonella 1000-fold.

D. Antimicrobial Proteins

Lysozyme (in saliva and other body secretions) efficiently digest the bacterial cell wall. Notably, the exposed avian egg is rich in lysozyme. Interferons α and β are secreted by cells as a response to viral infection and make surrounding cells more resistant to viral infection; interferon γ from T cells in the specific immune system also performs this task, as well as other tasks. A special set of proteins found in blood plasma is complement. It is a cascade of enzymes, hormones, and proteins that attack the cell membranes; this protection afforded by complement is paramount in *Neisseria* infections. This cascade can be activated by microorganisms directly as well as being enlisted by antibody after it has bound antigen.

E. Phagocytes

Macrophages/monocytes and neutrophils are the professional phagocytes, efficiently engulfing foreign organisms and damaged cells. They are particularly efficient at this in the presence of antibody ± complement. Because they contain certain bags of enzymes, these phagocytes can efficiently digest foreign particles such as bacteria. The formation of pus is largely due to an accumulation of neutrophils, as a response to some noxious agent. [*See* Macrophages; Neutrophils; Phagocytes.]

As described in Sections D and E, the specific immune system often recruits the innate immune system to amplify its effectiveness.

III. COMPONENTS OF THE IMMUNE SYSTEM

Another way to describe the specific immune system is that it is a mobile network of a highly specialized set of white blood cells called lymphocytes. Whereas there is only one type of red blood cells (erythrocytes) and they essentially do one job, i.e., carry oxygen around the body, there are several types of white blood cells (leukocytes). For example, neutrophils and macrophages mentioned in Section II,E are leukocytes. The lymphocyte is the major leukocyte in lymph and in the lymphoid organs (see Section IV). It comprises 20–30% of leukocytes in human blood.

The two main types of lymphocytes are B and T cells and these are the cells that confer specificity. This specificity is determined by T- and B-cell receptors (Table II).

The B-cell receptor (antibody bound on the cell surface of B cells) binds antigen and so triggers the cell to make the secreted form of antibody. Both the membrane and secreted forms of antibody recognize conformational (three-dimensional) structures, i.e., see intact antigens. The T-cell receptor, however, is educated in the thymus to react only to the body's own major histocompatibility complex (MHC) molecules and these molecules present only small fragments of

TABLE II
Specificity Determined by B- and T-Cell Receptors

B-cell receptor	B cells recognize antigens as intact molecules. This recognition is mediated by a form of antibody that is found on the B-cell surface. Upon its engagement with antigen along with other signals, B cells start to proliferate and make the secretory form of antibody.
T-cell receptor	T cells recognize antigens as digested products (short peptides) bound within a groove of a major histocompatibility complex molecule of another cell. This recognition is mediated by a special protein structure called the T-cell receptor. Upon its engagement with antigen along with other signals, T cells start to proliferate and produce its wide weaponry (proteins that kill, inflammatory hormones, etc.).

antigens. This phenomenon is called MHC restriction for which Zinkernagel and Doherty won the 1996 Nobel Prize. The T-cell receptor thus sees a structure that is partly peptide and partly MHC. Cytotoxic T cells recognize 8–10 amino acid peptides (derived by digestion of antigens) within the groove of MHC molecules called Class I. Helper T cells recognize 8–24 amino acid antigenic peptides within the groove of MHC molecules called Class II. Both MHC Class I and Class II molecules have a similar shape in that the part T cells see is like two hands joined together, as shown in Fig. 1. The peptide is bound in a groove created by the palms of the hands (in molecular terms, the β-sheet) and the two coils of clenched fingers (two α-helices). One pivotal consequence of this is that whereas B cells and antibody can only see intact proteins on the cell surface or outside the cell, proteins hidden inside cells can be identified by T cells by way of the proteins' breakdown products (peptides) being presented on the cell surface with MHC molecules. This is why T cells and not antibody are effective against intracellular bacteria (like *Mycobacteria tuberculosis*) and viruses once inside a cell (Table III). [*See* T-Cell Receptors.]

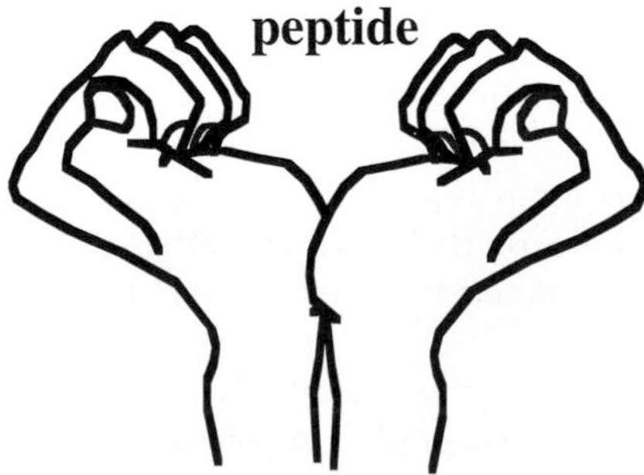

FIGURE 1 The major histocompatibility complex (MHC) molecule binds a proteolytic fragment (peptide) from proteins taken up or produced within cells. The peptide is bound between the two α-helixes of the MHC molecule (depicted as clenched fingers) and a β-sheet platform (depicted as the palms of the hand). The T-cell receptor recognizes both the peptide and part of the α-helixes.

The terms humoral (viz. liquid) immunity and cell-mediated immunity were coined to describe immunity that could be transferred from one animal to another by plasma and by cells, respectively. Humoral immunity correlated with antibody and cell-mediated immunity correlated with T cells.

Functionally, B cells make antibody. Cytotoxic T cells kill virally infected cells and some tumor cells. Helper T cells aid and modify the production of antibodies by B cells and secrete many different types of cytokines (short-acting hormones that usually act locally) which perform numerous functions, including killing (e.g., lymphotoxin, tumor necrosis factor), promoting the growth of other lymphocytes (e.g., IL-4), and activating macrophages (e.g., interferon-γ). The different effector mechanisms of these various cells are reflected in the different microorganisms that they are best at combating (summarized in Table III).

Dendritic cells are found virtually everywhere T and B cells are found. They are also leukocytes and can be of similar lineage to lymphocytes or to macro-

TABLE III
Defenses against Infectious Diseases

Soldiers	Pathogen
B cells (antibody)	Bacteria that live outside cells, e.g., *Streptococcus pneumoniae* or *Haemophilus influenzae* Toxins, e.g., tetanus toxin or diphtheria Viruses before they have invaded inside a cell (i.e., during reinfection or in the bloodstream), e.g., polio or hepatitis B
Helper T cells	Bacteria that live inside cells, e.g., *Mycobacteria tuberculosis* Certain protozoa, e.g., *Toxoplasma gondii* Certain viruses, especially those that lyse cells, e.g., smallpox
Cytotoxic T cells	Certain viruses, especially those that do not lyse cells, e.g., EBV (glandular fever) or HIV

phages. They are ideal cells for presenting antigens to T and B cells, not only because of their wide distribution and locality (e.g., juxtaposed to T and B cells in lymph nodes), but also because of their high levels of MHC molecules and other important surface molecules that help stimulate T and B cells.

Cells in the immune system communicate not only by cytokines (as mentioned earlier), but also by these specialized surface molecules. Some examples of the latter include B7 on dendritic cells which stimulate CD28 on T cells (to secrete IL-2) and CD40L and OX40 on T cells which stimulate CD40 and OX40L on B cells, respectively. Many of these interactions are used for more than one purpose. [*See* B-Cell Activation.]

IV. HIGHWAYS AND HUBS OF THE IMMUNE SYSTEM: A BRIEF ANATOMY

A. Highways

The highways of the immune system are the two major circulation systems of the body: blood (arteries and veins) and lymph (lymphatic vessels). The lymphatic vessels drain from nearly all parts of the body (notable exceptions are the brain and spinal cord) to local lymph nodes. This is one of the reasons why breast cancer can spread so easily to the axillary lymph nodes (in the armpit).

Most lymphatic vessels anastomose (join up) freely. The lymph percolates through one or more lymph nodes to form larger vessels until they collect into two major channels; the thoracic duct and the right lymphatic duct. These enter the blood via the brachiocephalic vein. Lymphocytes, dendritic cells, and colloidal/particulate material thus all drain to lymph nodes from sites of injury and lymphocytes reappear into the site of injury via the circulation. This lymphatic/blood relay is extremely efficient and there is a very high turnover in these highways.

B. Lymphoid Organs Are Hubs of the Immune System

Essentially, all B cells are derived from the red bone marrow (often seen as part of a lamb chop or T-bone steak, or inside a rib; cf. yellow bone marrow is mainly fat). The bone marrow is also the major site of where the secondary antibody response is generated. Essentially, all T cells are so called because they are lymphocytes that receive special training in the thymus. The thymus is located above the heart and is a large organ in the young but is no longer essential in the adult once enough T cells have seeded to the spleen and lymph nodes.

As mentioned in Section I, lymph nodes are both guard posts and hubs of interaction. Sometimes when you get an infection in your arm, your armpit feels tender and your physician may palpate there. Likewise, the area near the angle of your jaw bone may be palpated if an infection of the mouth/throat is suspected. These are sites of lymph nodes which can swell considerably due to enormous lymphocyte proliferation and recruitment from elsewhere. If you buy calf or lamb liver from your butcher, you may see a gray, 1- to 2-cm-long lymph node, shaped like a broad bean and located near one of the big veins. Its looks are deceiving. It is a complex organ containing discrete areas of germinal centers (where B cells proliferate to generate memory cells and the immunoglobulin genes undergo minor mutations to produce more avid antibodies), foci of high antibody secretion, scaffolds of dendritic cells presenting antigen to T and B cells, and zones of T and B cell cross-talk. The spleen is also a lymphoid organ with specialized architecture. Peyer's patches along the intestine are hubs for lymphocytes surveying the gut. There is almost nowhere in the body where lymphocytes are not nearby, some in discrete organs such as tonsils whereas others seem scattered, e.g., just under the skin surface. Even these latter lymphocytes can quickly gather into nearby hubs, e.g., draining lymph node, and after rites of induction return to the site of injury via the circulation. It would not be surprising therefore that there are many homing receptors on lymphocytes (e.g., instructing T cells to go to lymph nodes or other hubs, instructing lymphocytes to leave blood vessels, etc.) and respective receptors (addressins) in the organs they home into.

BIBLIOGRAPHY

Alberts, B., Brat, D., Lewis, J., Raff, M., Roberts, K., and Watson, J. D. (1994). "Molecular Biology of the Cell," 3rd Ed. Garland, New York.

Paul, W. E. (ed.) (1993). "Fundamental Immunology," 3rd Ed. Raven Press, New York.

Roitt, I. M. (1994). "Essential Immunology," 8th Ed. Blackwell, Oxford.

Zinkernagel, R. M. (1996). Immunology taught by viruses. *Science* **271**, 173–178.

Immunoassays, Nonradionucleotide

KENNETH D. BURMAN
Washington Hospital Center and George Washington University Medical Center

GLOSSARY

Chemiluminescent assay Technique of measurement that uses compounds with emitted light as a detection system; in most assays, the greater the unknown concentration, the more light is emitted

Enzyme-linked immunosorbent assay (ELISA) Technique of measurement that generally uses antibodies to detect unknown substances; an enzyme may be attached to an antibody, which when activated will change color; the color density will correlate with the concentration of the unknown substance

Fluorescent assay Technique of measurement that uses emitted fluorescence intensity as a detection system

Fluorescent polarization assay Technique of measurement that uses the deflection or degree of polarization of transmitted light in its detection system

Immunoassay Method of measurement that uses antibodies to detect concentrations of unknown hormones, drugs, or substances

Nonradionuclide assay Any technique or method of measurement of hormones, drugs, or other substances that does not use isotope (e.g., enzyme-linked immunosorbent and chemiluminescent assays)

Radionuclide assay Any technique or method of measurement of hormones, drugs, or other substances that does use isotope (e.g., radioimmuno- and radioimmunometric assays)

Thyrotrophin Hormone secreted by the pituitary gland that stimulates function of the thyroid gland and production of the active thyroid hormones triiodothyronine (T_3) and thyroxine (T_4); specific thyrotrophin receptor molecules are on thyroid gland cells

NONRADIONUCLIDE IMMUNOASSAYS ALLOW THE measurement of unknown concentrations of drugs, hormones, or other substances; these assays do not employ isotopes. The major types of these nonnuclide techniques are enzyme-linked immunosorbent assay (ELISA), chemiluminescence, fluorescence, and fluorescence polarization. Nonnuclide assays can be highly sensitive and, in most cases, easy to perform. Compared with assays employing radionuclides [e.g., radioimmunoassays (RIAs), radioimmunometric], the reagents used for nonnuclide assays are, in general, inexpensive, easy to obtain, have a long shelf-life, and do not require special radioactive disposal or administration. [*See* Radioimmunoassays.]

I. GENERAL COMMENTS

A nonradionuclide immunoassay refers to any assay that uses an antibody for detection or separation but does not employ any type of radionuclide. Several points should be emphasized at the onset. First, no inherent fundamental differences exist in principle between immunoassays that employ isotope and those that do not. The basic chemical interactions are usually quite similar; however, the detection systems in these two assay systems are quite different. Second, RIAs and nonradionuclide assays each have their own set of advantages and disadvantages, which may lead one laboratory to choose one system over another. For example, sensitivity or specificity usually depends on other factors in addition to the detection system, and, as a result, for a particular hormone or drug, an RIA may be more sensitive, whereas in another

ENCYCLOPEDIA OF HUMAN BIOLOGY, Second Edition, VOLUME 4.

circumstance a nonnuclide assay may have advantages. Therefore, the detection system per se (e.g., isotope or nonisotope) is usually only one determinant of an assay's characteristics, and most often practical laboratory or cost considerations will dictate which assay is most useful in a given circumstance.

As a third general comment, one aspect of immunoassays that frequently enhances sensitivity and specificity relates to the use of two or more antibodies in a system, resulting in a "sandwich"-type assay. Usually, these types of assays are more sensitive and specific than classic single-antibody assays, irrespective of whether or not isotope is employed in the detection system. Usually each antibody of the "sandwich" recognizes a different antigenic epitope, increasing specificity. In the following analysis, ELISA assays are used as a prototype for nonnuclide assays, and general and specific principles of these techniques will be considered; then, other important types of nonradionuclide assays will be described.

II. GENERAL PRINCIPLES

A. Customary Assay

In a customary assay, the antigen is allowed to react with an antibody that would be typically labeled with an enzyme (e.g., alkaline phosphatase). Under suitable conditions, the antigen and antibody would be allowed to react, and then a substrate (in the case mentioned here, *p*-nitropenyl phosphate) would be added. This reaction results in a change in the solution color from clear to yellow. Furthermore, the intensity of the yellow color would be directly correlated with the number of antibody molecules that are bound to the antigen (Fig. 1). Control samples are always included, and time at which the substrate is allowed to react is constant in an individual assay. In the preliminary analysis of an assay, the appropriate control antigen and antibody reaction are analyzed such that it is known when the enzyme reaction has plateaued. An assay in which the reaction was still progressing rapidly would be unusual; comparison of one tube to another would be difficult in such a situation. We shall discuss further details of such a reaction later. In such a customary assay, the antigen concentration could be determined by analysis of the color change or by optical density of the solution. A standard curve, usually with several sets of known control samples, must also be run (Fig. 2). In this manner, an unknown sample concentration could be determined by comparison with known concentrations in the standard curve. In nonnuclide immunoassays, as in RIAs, a great deal of preliminary analysis must be performed. The concentration of antibodies, proper enzyme label, and substrate concentration, as well as time of incubation, type of optimal density reader, and type of plate, for example, must all be taken into consideration before a nonnuclide assay can be used reproducibly. This type of assay can be modified by using two antibodies for detection, usually enhancing sensitivity (Fig. 3).

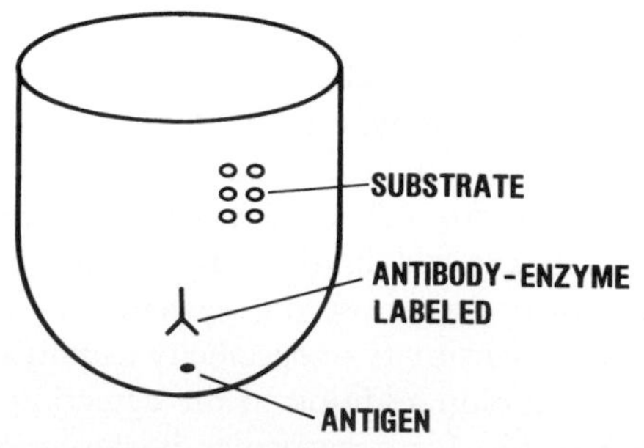

FIGURE 1 Schematic depiction of a customary ELISA assay.

B. Single Sandwich ELISA

In a single-sandwich-type assay, an antibody is allowed to adhere to the bottom of an ELISA well, and the antigen, perhaps in serum or other biological fluid, is then layered onto the antibody. After thorough washing, the second antibody, which is enzyme-labeled, is allowed to react with the solution (Fig. 4). Finally, a substrate is added and color change of the solution is quantitated, and the concentration of the antigen in the biological sample is determined by comparison to a standard curve. Of course, each assay contains several known standards as well. Obviously, the concentration of antigen in the biological sample is directly proportional to the amount of enzyme-labeled antibody bound to the antigen and, as a result, optical density readings are directly related to the color change (Table I). The antibody plated on the bottom of the microtiter plate usually recognizes a different portion, or epitope, of the antigen than does the enzyme-labeled antibody. The use of those two antibodies, which recognize different portions of the antigen, represents the most advantageous part of this assay (Fig. 5). The background is low and the degree of specificity and sensitivity is high, largely because of the two different antisera used. Again, careful attention to each component of the assay is required.

FIGURE 2 Typical ELISA plate using an alkaline phosphatase-labeled antibody. The intensity of the yellow color after the addition of substrate is correlated with the unknown samples' concentration. Note that the standard curve is in columns 3–6 and that the unknown samples are in columns 7–11. The outside lanes are not used for sample determination.

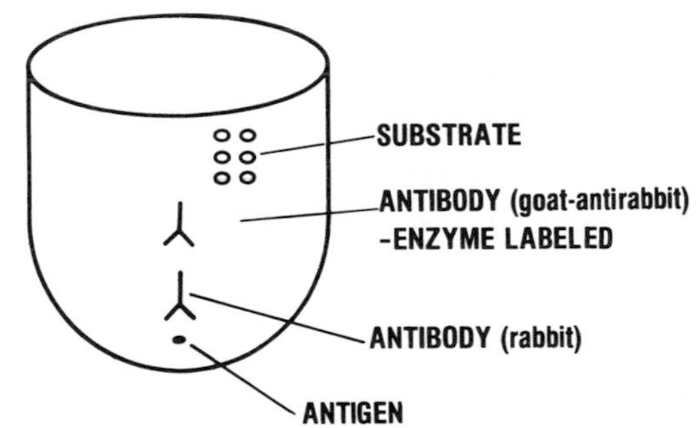

FIGURE 3 Schematic depiction of a modified ELISA assay.

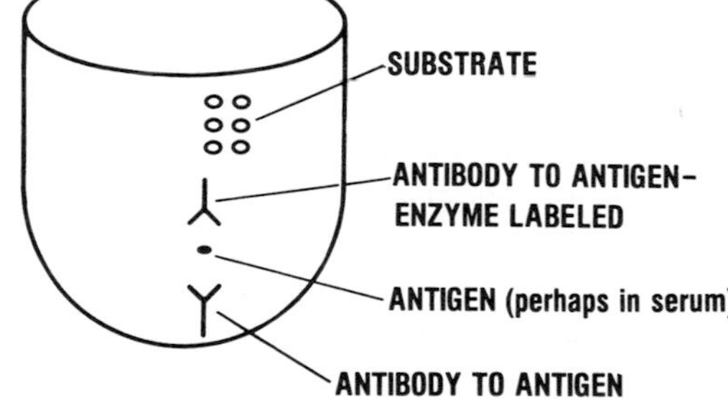

FIGURE 4 Schematic depiction of a single-sandwich ELISA.

C. Double-Sandwich ELISA

The double-sandwich-type assay is similar to those just noted but employs a total of three antibodies. The first antibody is directed against one site on the antigen; this antibody is used on the bottom of the microtiter well. The unknown concentration antigen, located in a biological fluid, is then layered onto the antibody. After thorough washing, a second antibody,

TABLE I
Examples of Different Enzymatic Systems in ELISA Assays

Enzyme	Substrate	Color
Alkaline phosphatase	*p*-Nitrophenyl phosphate	Yellow
Horseradish peroxidase	5-Amino salicylic acid	Red
Biotin	Avidin	Blue

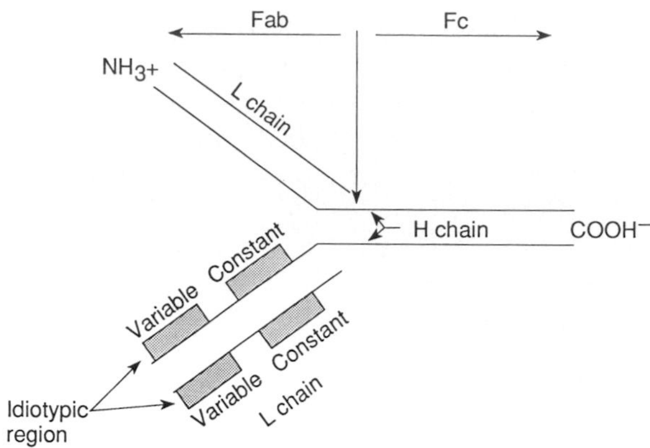

FIGURE 5 Schematic depiction of the structure of an immunoglobulin. There are two heavy and two light chains. The Fc portion of the molecule allows binding relatively nonspecifically, whereas the Fab portion contains a highly specific binding region, the most relevant of which is the variable, or hypervariable, region of the idiotypic region. This region is important in binding to antigens.

which recognizes a different epitope or binding site of the antigen, is then added; this second antibody is not enzyme-labeled. A third antibody (in this case, an enzyme-labeled antibody that recognizes the second antibody) is added, and then a substrate is added. In general, the first antibody is raised in one species (e.g., goat), whereas the second antibody is generated in a different species (e.g., a rabbit). Finally, the third antibody does not recognize the unknown antigen but, rather, is directed against the nonbinding protein (Fc portion) of the second antibody; that is, the third antibody is an antirabbit antibody directed against the Fc portion of rabbit IgG. These types of assays are very specific and have high sensitivity. Because the first (goat) antibody may also react with the third (anti-Fc) antibody, increased background interference may result. Thus, frequently, the first (goat) antibody will be processed to leave the antigen-binding region intact, but the Fc fraction will be removed. Many variations of this type of assay can be used (e.g., several different rabbit antibodies, each directed against different portions of the antigen molecule, can be used and may help to enhance sensitivity and specificity). Semantically, it must also be noted that the descriptive term "double sandwich" may be euphemistically pleasing, but it is a nonspecific term that can be applied to various types of assay.

D. Basic Principles

Several basic principles inherent in the development of an ELISA assay merit reiteration.

1. When an antigen or antibody is chemically linked to an enzyme, the resulting conjugated substance does not lose any of its original antigenicity and antigen-binding capacity. As a result, the characteristics of antigen–antibody binding are not affected.
2. Many substrate molecules can react with only one enzyme molecule. This assay characteristic leads to an enhancement of the detection system and, thus, leads to a highly sensitive assay.
3. Molecules unbound to microtiter wells or unattached to antigen or antibody can be effectively removed from the assay well by vigorous washing. Obviously, this results in minimal background interference and the relationship between optical density reading and assay conjugates becomes direct and generally linear. Usually, albumin or other large proteins are included into the wash buffers to decrease nonspecific binding.
4. The process of adsorption of an antigen or antibody onto the microtiter plates occurs without loss of biological or immunological activity.

III. SPECIFIC CONSTITUENTS

A specific ELISA assay is capable of detecting thyrotrophin (TSH) receptor antibodies. The TSH receptor is a cellular membrane protein of thyroid cells, which binds TSH, a pituitary hormone, and initiates the complex and largely unknown process of thyroid gland stimulation. In Graves' disease, antibodies are formed against the TSH receptor; these antibodies act like TSH and are quite effective in causing prolonged thyroidal stimulation through the TSH receptor. Normally, TSH production and action is under negative feedback inhibition by the circulating thyroid hormones; however, stimulation by antibodies in Graves' disease is not under normal feedback homeostasis and results, in time, in a clinical disorder characterized by excessive secretion and action of the thyroid hormones triiodothyronine and thyroxiine. In the ELISA assay, human tissue containing TSH receptors (e.g., thyroid or guinea pig fat cell membranes) is used to isolate a solubilized membrane-enriched fraction. About 2 μg protein in 100 lambda solution is layered onto flat-bottomed polystyrene microtiter plates at pH 7.2 and allowed to remain at 4°C for about 18 hr. This procedure allows the membranes to adhere to the plates. The precise mechanism by which this occurs is unknown, but it probably relates to charge attraction, because generally proteins are positively charged and plastics are negatively charged.

After application of the antigen, the wells are washed three or four times with phosphate-buffered saline (PBS) (pH 7.2), and then 100 lambda of a 1% human serum albumin solution (PBS, pH 7.2) is added and allowed to remain in the wells for 1 hr at 37°C. This step is extremely important because it reduces nonspecific binding in the assay.

Following washing with a PBS solution (pH 7.2) containing 0.05% Tween 20 (a detergent) and 0.1% albumin, the wells are ready for the biological substance to be measured, for instance, serum. One hundred lambda of serum is added and incubated for 1 hr at 37°C. Various dilutions of the serum are used, depending on the assay used and the concentration of the antigens in the biological substance to be tested, as well as on the level of nonspecific binding or interference found in a particular fluid. Again, the wells are washed diligently and then the detection system is added. To detect TSH receptor antibodies, an anti-human IgG conjugated to the enzyme alkaline phosphatase (100 lambda in a 1 : 1000 titer) is added and allowed to sit in the wells for 1 hr at 37°C. After more washes, 150 lambda of a solution containing *p*-nitrophenyl phosphate (1 mg/ml) in diethanolamine buffer (pH 9.8) (with 0.01 *M* $MgCl_2$) is added and incubated for a sufficient time to obtain good reading and low background as judged from positive and negative controls included in each plate. The incubation time of the activated enzyme in this last step is the most crucial and is based on preliminary studies aimed at optimizing the concentrations and times of each reagent.

The control samples are very important. They must include controls of (1) antibody conjugate alone, with buffer added instead of an unknown solution in an empty well (without antigen); (2) biological fluid with antibody conjugate (in this instance, the wells also do not contain antigen); and (3) antigen with antibody conjugate. These controls should give very low readings. Background values also must be low.

IV. VARIOUS TYPES OF NONNUCLIDE ASSAYS

In the last several years, there has been a virtual explosion in technology resulting in a large variety of other nonnuclide assays. Several of these will be discussed to give an overview of their basic characteristics.

In many areas, chemiluminescent assays are becoming very popular. Their main advantage is that they can exhibit tremendous sensitivity. In chemiluminescent assays, a specific chemiluminescent tracer or compound is used (either linked to antigen or antibody) in the detection system. Chemiluminescence or bioluminescence refers to the property of light emission of certain compounds, chemical or biological, respectively, especially after they are in an excited state. This luminescence may be used in assays in a manner analogous to radioisotopic compounds. These luminescent labels are stable in their basal state for long periods of time, thereby representing an advantage over isotopes and even over various enzyme labels. The greatest advantage of these compounds is the sensitivity they impart to an assay, mainly because of their decreased background "noise." As an example, acridinium esters are used in a commercially available kit as compounds that emit light when treated with hydrogen peroxide and alkali. In an assay, similar to the ELISA assays discussed earlier, the second antibody is conjugated with the acridinium ester. The amount of light emitted is in proportion to the amount of antibody bound. A chemiluminescent assay generally may detect 0.005 μU/ml of TSH, whereas a good ELISA and RIA will detect 0.10 μU/ml and 0.3 μU/ml, respectively. Over the last several years, chemiluminescent assays have been automated and are now routinely applied to the measurement of a wide variety of hormones [e.g., TSH, free T4, parathyroid hormone (PTH), human chorionic gonadotropin (HCG), and estradiol]. Indeed, these chemiluminescent assays have been the standard method of measurement. Increased sensitivity, specificity, ability to automate these procedures, and rapid assay times are each important reasons for their increased application in clinical settings.

Another new type of nonnuclide assay is designated as time-resolved fluorescence. In a commercially available assay, the detection system uses a europium-chelate, which has a prolonged fluorescence delay time, conjugated to the second antibody. After thorough washing and addition of an enhancement solution, the europium is converted into cations that dissociate, form chelates with portions of the solutions, and give rise to fluorescence, which is measured in a time-resolved fluorometer. A major advantage of this type of system is that it has low background interference, because of the prolonged fluorescent decay time, and thus increased sensitivity.

Fluorescent polarization assays also represent an innovative, nonnuclide, commercially available method of assessing the concentrations of various substances. These assays are based on the principle that molecules bound to a fluorophor will emit polarized light in a different manner, depending on whether or not the fluorophor can rotate freely; that is, if the fluorophor is bound to the antibody, the complex is free to rotate; the polarized light directed through this solution will be deflected and becomes less polarized as it exits. In contrast, if the antibody is bound to the antigen, the resultant emitted light will be more polarized. This assay depends on competition between the fluorophor-labeled antigen and unlabeled antigen in biological fluid. It is widely used in a variety of hormone or drug assays.

V. CONCLUSION

Several different types of nonnuclide radioassays have been examined. The basic principles in an assay that utilizes isotopes (e.g., RIA, immunoradiometric assay) are the same as those involved in a nonnuclide assay. There may be inherent advantages in using one assay over another, but the decision to use a particular assay must often involve practical laboratory considerations. Issues such as primary objective, finances, experience in a given area, and the laboratory environment are all important (Table II). Further, we must not forget that a comparison of isotopic versus nonisotopic assays must consider such nonreadily apparent issues as time involved in radiation safety administration, radiation disposal, equipment costs, and stability of reagents.

In this context, however, nonnuclide assays certainly must be considered a tremendous advance over nuclide assays, and we await with expectation further innovative assay developments in the future.

TABLE II
Comparison of Different Types of Assays[a]

	RIA	ELISA	Luminescent
Isotope	++++	0	0
Convenience	+	+++	++
Cost	++	+	+++
Sensitivity	++	+++	++++

[a]These assessments are subjective and apply to each type of assay only in general terms. Individual exceptions exist.

ACKNOWLEDGMENTS

The author thanks representatives from Pharmacia Diagnostics, Fairfield, New Jersey; and Abbott Laboratories, North Chicago, Illinois, for helpful discussions and information regarding nonnuclide assays. Portions of this report have been presented in several immunoassay technique courses that have been sponsored by The Endocrine Society, Bethesda, Maryland. The author also thanks Ms. Estelle Coleman for editorial assistance.

BIBLIOGRAPHY

Baker, J. R., Lukes, Y. G., Smallridge, R. C., Berger, M., and Burman, K. D. (1983). Partial characterization and clinical correlation of circulating immunoglobulins directed against thyrotrophin binding sites in guinea pig fat cell membranes. *J. Clin. Invest.* **72,** 1487.

D'Avis, J. R. E., Black, E. G., and Sheppard, M. C. (1987). Evaluation of a sensitive chemiluminescent assay for TSH in the follow-up of treated thyrotoxicosis. *Clin. Endocrinol.* **27,** 563.

Engvall, E., and Perlmann, P. (1972). Enzyme-linked immunosorbent assay (ELISA). *J. Imunol.* **109,** 129.

Franklyn, J. A., Black, E. G., Betteridge, J., and Sheppard, M. C. (1994). Comparison of second and third generation methods for measurement of serum thyrotropin in patients with overt hyperthyroidism, patients receiving thyroxine therapy, and those with nonthyroidal illness. *J. Clin. Endocrinol. Metab.* **78,** 1368.

Soini, E., and Kojola, H. (1983). Time-resolved fluorometer for Lanthanide chelates—A new generation of non-isotopic immunoassays. *Clin. Chem.* **29,** 65.

Spencer, C. A., Lo Presti, J. S., Patel, A., *et al.* (1990). Applications of a new chemiluminometric thyrotropin assay to subnormal measurement. *J. Clin. Endocrinol. Metab.* **70,** 453.

Sturgess, M. L., Weeks, I., Evans, P. J., Mpoko, C. N., Laing, I., and Woodhead, J. S. (1987). An immunochemiluminometric assay for serum free thyroxine. *Clin. Endocrinol.* **27,** 383.

Symposium on Ultrasensitive TSH Assays (October–December 1988). *Mayo Clin. Proc.* **63,** Parts I–III.

Tseng, Y.-C., Burman, K. D., Baker, J. R., Jr., and Wartofsky, L. (1985). A rapid, sensitive enzyme-linked immunoassay for human thyrotropin. *Clin. Chem.* **31,** 1131.

Voller, A., Bidwell, D., and Bartlett, A. (1980). Enzyme-linked immunoassay. *In* "Manual of Clinical Immunology" (N. Rose and H. Friedman, eds.), 2nd Ed., Chap. 45, p. 359. American Society for Microbiology, Washington, D.C.

Weeks, I., Sturgess, M., Siddle, K., Jones, M. K., and Woodhead, J. S. (1984). A high sensitivity immunochemiluminometric assay for human thyrotrophin. *Clin. Endocrinol.* **20,** 489.

Yolken, R. H. (1978). Enzyme-linked immunosorbent assay. *Hosp. Pract.* **December,** 121.

Immunobiology of Transplantation

WILLIAM E. GILLANDERS
TED H. HANSEN
Washington University in St. Louis

GLOSSARY

Allele Any alternative form of a gene present at a particular genetic locus

Allogeneic Genetically distinct individuals (or tissues) of the same species that express different histocompatibility antigens

Allograft Graft obtained from an allogeneic donor

B lymphocyte Lymphocyte primarily responsible for humoral immunity; precursor of antibody-producing plasma cells. Characterized by the presence of a clonally distributed surface immunoglobulin receptor that constitutes the B-cell antigen receptor

CD4 (cluster of differentiation antigen 4) Cell-surface molecule identifying one of two major subsets of T cells. CD4 T cells recognize antigen in association with MHC class II molecules

CD8 (cluster of differentiation antigen 8) Cell-surface molecule identifying one of two major subsets of T cells. CD8 T cells recognize antigen in association with MHC class I molecules

HLA (human leukocyte antigen) Term used to define antigens of the human major histocompatibility complex

Histocompatibility antigen Antigen responsible for allograft or xenograft rejection. Histocompatibility antigens are encoded by histocompatibility genes

Major histocompatibility complex (MHC) Complex of genetic loci originally identified based on its central role in tissue compatibility and allograft rejection. Genes in the major histocompatibility complex encode class I and class II MHC molecules

Polymorphism Quality or character of occurring in several different forms in the same population. Variants of a polymorphic gene are termed alleles

Rejection Immune process leading to the destruction of an allograft or xenograft

Syngeneic Individuals (or tissues) that have identical genotypes and share all histocompatibility antigens

T lymphocyte Lymphocyte primarily responsible for cell-mediated immunity. Characterized by the presence of a clonally distributed cell-surface T-cell receptor complex responsible for antigen recognition

Tolerance Transplantation tolerance is the long-term acceptance of grafted tissue in the absence of continuous immunosuppression. Transplantation tolerance involves the development of specific nonreactivity to the grafted tissues

Transplantation Process of grafting tissue from one individual to another individual. Transplantation from one individual to the same individual is termed autologous transplantation

Xenogeneic Individuals (or tissues) that are of different species and therefore, by definition, express different histocompatibility antigens

Xenograft Graft obtained from a xenogeneic donor

THE SUCCESS OF MODERN CLINICAL TRANSPLANTATION represents one of the most important advances in medicine during the past 30 years. Currently, solid organ transplantation is often considered a viable clinical alternative, and in many cases the only definitive therapy for isolated end-stage organ failure. Advances in transplantation biology have been fundamental to this success. However, allograft rejection, and the complications associated with nonspecific immunosuppression therapy required to prevent rejection, limit the success of transplantation. Therefore, one of the goals of transplantation research remains the understanding of the cellular, biochemical, and molecular events leading to allograft rejection. A thorough understanding of these processes will aid in the development of new approaches to prevent rejection.

Transplantation has been intricately linked with the field of basic immunology throughout its short history. Early transplantation research resulted in significant contributions to our understanding of basic immunology. The identification of the major histocompatibility complex (MHC), and its subsequent structural characterization, has been central to the basic understanding of T-cell antigen recognition. Now, however, ongoing research in basic immunology may help to resolve two of the most important problems in clinical transplantation. The study of T-cell tolerance induction may help overcome the complications associated with nonspecific immunosuppression, currently used to treat allograft rejection. In addition, the use of nonhuman primates, or miniature swine in xenogeneic transplantation, may alleviate the acute shortage of donor organs. This article will summarize the basic principles of transplantation biology and focus on their importance in clinical transplantation.

I. FUNDAMENTAL ISSUES IN TRANSPLANTATION

Early attempts at tissue transplantation between individuals were uniformly unsuccessful. Although it was technically feasible to transplant skin, and even solid organs between individuals, these grafts underwent necrosis and failed. Examination of failed skin grafts revealed an inflammatory infiltrate termed rejection. Experiments in animals to determine the cause of skin graft rejection revealed that *rejection is a form of specific immunity*. These classic experiments were performed by the British scientist Peter Medawar. The experiments are presented in Table I and can be summarized as follows:

1. A skin transplant between two different strains of inbred mice will be rejected by a naive recipient within 7 to 10 days. This process is known as *first set rejection* and is the result of an inflammatory immune response. As shown in Table I, a naive animal from strain A will reject a skin transplant from a donor strain B animal in a normal fashion as a first set rejection.
2. If a second graft from the same donor strain (strain B) is placed on the original recipient (strain A), it will undergo more rapid rejection, a process known as *second set rejection*. Thus, the rejection of tissue grafts results in memory, a fundamental characteristic of normal immune responses.
3. If, however, the second graft is from a different donor strain (strain C), it will be rejected in a normal fashion, as a first set rejection. Thus, the rejection of tissue grafts is specific, another fundamental characteristic of the immune system.
4. Adoptive transfer of lymphocytes from an animal that has rejected a skin graft to a naive animal transfers the capacity to produce rapid, or second-set, rejection. Thus, a recipient strain A animal rejects a skin graft from a donor strain B animal. Lymphocytes from the recipient animal are then adoptively transferred to a naive strain A animal. This naive animal can subsequently reject a skin graft from a donor strain B animal in an accelerated fashion, or by second set rejection.

These experiments clearly demonstrate that allograft rejection is a form of specific immunity. Subse-

TABLE I
Allograft Rejection Is a Specific Immune Response

Experiment	Recipient	Donor	Prior treatment of recipient	Rejection
1	Strain A	Strain B	None	Slow (first set)
2	Strain A	Strain B	Sensitized by previous rejection of strain B graft	Rapid (second set)
3	Strain A	Strain C	Sensitized by previous rejection of strain B graft	Slow (first set)
4	Strain A	Strain B	Adoptive transfer of lymphocytes from strain A animal that has previously rejected strain B graft	Rapid (second set)

quent research in transplantation biology has focused on the following issues, which will be discussed in greater detail: (1) What distinguishes a foreign tissue graft from self, allowing it to be rejected? In particular, what antigen(s) stimulate graft rejection? (2) What is the basis for the recognition of these antigens? The process of allograft rejection is one of the most aggressive immune responses characterized. (3) How does the immune system, once it distinguishes a tissue graft as being foreign, mediate its destruction? (4) Can the process of graft rejection be specifically prevented, without compromising the ability of the immune system to respond to other unrelated foreign antigens?

II. HISTOCOMPATIBILITY ANTIGENS AND ALLOGRAFT REJECTION

Histocompatibility antigens are the antigens responsible for allograft rejection. They are encoded by histocompatibility genes, and the search for these genes led to the discovery of the MHC, a complex of genetic loci of critical importance in allograft rejection. The structural characterization of the molecules encoded by the MHC has proven to be of fundamental importance in the basic understanding of antigen presentation, as well as alloreactivity. [*See* Major Histocompatibility Complex (MHC).]

A. The Genetics of Histocompatibility

Early researchers studying transplantation used genetic techniques to identify the antigens responsible for allograft rejection and to delineate the fundamental principles of transplantation biology. Central to this analysis was the use of inbred strains of mice, which are the result of sequential brother–sister matings for at least 20 generations. Therefore, every individual animal is genetically identical to all other animals within the same inbred strain. Inbred mice are homozygous at every genetic locus and produce identical homozygous progeny. Early researchers noted that transplants between animals of the same inbred strain are not rejected, whereas transplants between animals of different inbred strains are always rejected. These studies reveal that there is a genetic basis governing histocompatibility. The genes responsible for graft rejection are known as histocompatibility genes. Histocompatibility genes can be defined as genes that encode molecules that are polymorphic within a species and are sufficiently immunogenic to evoke a rejection reaction. These genes are almost always codominantly expressed: the product of alleles on both chromosomes is expressed. The laws of transplantation as described by these early researchers are summarized in Table II. Although it was not fully understood at the time, these laws are the result of classic Mendelian inheritance of multiple, independently segregating histocompatibility genes. Thus, rejection will occur if a histocompatibility gene is present in the donor, but not in the recipient. F_1 hybrid animals, the result of mating two different inbred strains, receive a single dose of all genes present in both parental strains. Thus, F_1 animals accept grafts from either parental strain, but grafts from F_1 animals are rejected by both parental strains.

Selective breeding allowed researchers to identify the histocompatibility genes responsible for graft rejection. Congenic mouse strains are inbred strains that have been derived by selective matings so that they differ from one another at a single independently segregating genetic locus. Congenic strains that differ at a single histocompatibility locus are termed congenic resistant strains. As these congenic resistant mouse strains were being established, it became apparent that there was one histocompatibility locus of overwhelming importance in determining the fate of an allograft, and this locus was termed the major histocompatibility locus. Further analysis of the major histocompatibility locus revealed that it was not a single locus, but rather multiple closely linked loci forming a complex of loci termed the *major histocompatibility complex*. Although initially identified because of their important role in allograft rejection, the molecules encoded by the MHC are of central importance in the normal function of the immune system. These

TABLE II
The Laws of Transplantation

1. Transplants between individuals of the same inbred strain are not rejected.
2. Transplants between individuals of different inbred strains are always rejected.
3. Transplants from a member of an inbred parental strain to an F_1 hybrid offspring are not rejected. For example, an $(A \times B)F_1$ hybrid animal will not reject an allograft from an A or B strain animal.
4. A transplant from an F_1 hybrid animal to one of the inbred parents is rejected. For example, an A or B strain animal will reject an allograft from an $(A \times B)F_1$ hybrid animal.

molecules have been extensively studied and their important features are considered separately in later sections.

Based on the principles discussed here, it is possible to experimentally determine the number of genetic loci that code for histocompatibility antigens. The results of such experiments reveal that there are as many as 30–40 histocompatibility loci in certain strain combinations. The other loci capable of causing graft rejection encode antigens referred to as *minor histocompatibility antigens* to distinguish them from the MHC antigens. Minor histocompatibility antigens were initially characterized by relatively slow rejection responses. Rejection responses to a single minor histocompatibility antigen are relatively weak and are easily overcome by nonspecific immunosuppression. Frequently, however, if two strains differ at multiple different minor histocompatibility loci, rejection can be as rapid as if they differed at a single MHC locus. It was initially believed that minor histocompatibility antigens were cell-surface molecules. However, there is increasing evidence that minor histocompatibility antigens are peptides derived from polymorphic proteins that are processed and presented by MHC molecules. Thus, minor histocompatibility antigens represent any molecule that is polymorphic within a species, as long as the allelic variation can be expressed as a peptide fragment capable of being presented by a MHC molecule. This model helps to explain why it is very difficult to detect antibody responses to minor histocompatibility antigens.

B. The Molecular Structure of MHC Antigens

The MHC was originally identified based on its dominant role in determining the immune response to allografts. However, it is now clear that the MHC is of central importance in the normal function of the immune system. MHC molecules play a central role in *antigen presentation* and in *the discrimination between self and nonself.* A central tenet of basic immunology is that antigen-specific T cells do not recognize foreign antigens alone, but recognize peptide fragments in association with molecules encoded by the MHC. The MHC encodes for two different types of cell-surface glycoproteins, known as MHC class I and class II molecules. MHC molecules function as peptide receptors, binding representative intracellular peptides and transporting them to the cell surface. At the cell surface, MHC–peptide complexes are recognized by antigen-specific T cells. Although each individual MHC molecule binds only a single peptide, MHC molecules have the capacity to bind many different peptides. In fact, investigators have recently confirmed that MHC molecules expressed on the cell surface are associated with thousands of different peptides. The majority of these peptides are derived from endogenous proteins and are perceived as "self" by the host immune system.

There are two central pathways of antigen presentation. Endogenous proteins are degraded in the cytosol, transported to the endoplasmic reticulum, and associate with MHC class I molecules. These complexes are then transported to the cell surface and presented to $CD8^+$ T cells. In a second pathway of antigen presentation, extracellular proteins are processed in an endosomal pathway and associate with MHC class II molecules. At the cell surface, these complexes are presented to $CD4^+$ T cells.

MHC molecules are also fundamental in the process of thymic education and in the discrimination between self and non-self. Developing T cells circulate through the thymus and interact with MHC molecules expressed on the surface of the thymic epithelium. T cells expressing T-cell receptors (TCRs) with high affinity for self-MHC molecules are deleted in the process of negative selection. However, T cells expressing TCRs with intermediate affinity for self-MHC molecules are positively selected, resulting in a TCR repertoire that can effectively recognize foreign antigens in association with self-MHC molecules. The result of this process is the formation of a T-cell repertoire that can effectively discriminate between self and non-self.

MHC molecules are extremely polymorphic in every species that has been analyzed. It is believed that this polymorphism has evolved to ensure that members of the species will be capable of binding many distinct foreign peptide antigens. The importance of MHC molecules in both allograft rejection and normal immune function has led to a concerted effort by researchers to elucidate the structure of these molecules. An understanding of the structure of MHC molecules helps to explain the special nature of allorecognition.

The structures of MHC class I and class II molecules are shown schematically in Fig. 1, and salient features are summarized in Table III. Fully assembled MHC class I molecules consist of a heterotrimeric complex of two separate proteins and a peptide ligand. The complex consists of a polymorphic MHC-encoded α chain, or heavy chain of 45 kDa, noncovalently associated with β_2-microglobulin, a nonpoly-

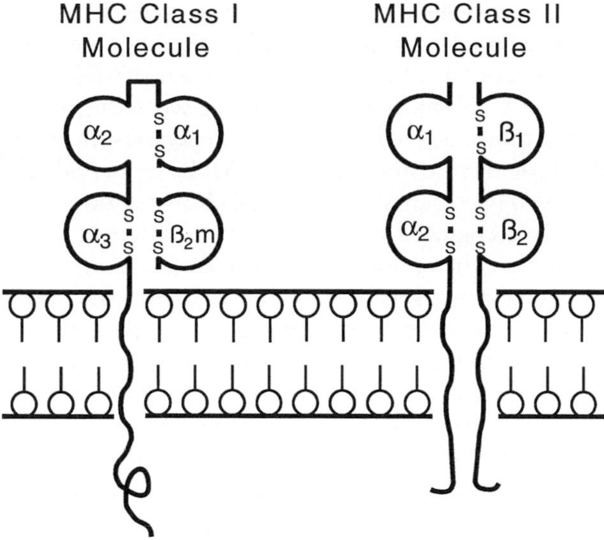

FIGURE 1 Schematic representation of class I and class II MHC molecules. MHC class I molecules consist of a heterotrimeric complex of two separate proteins and a peptide ligand. The complex consists of a polymorphic MHC-encoded α chain, or heavy chain of 45 kDa, noncovalently associated with β_2-microglobulin, a nonpolymorphic light chain of 12 kDa, and an associated peptide ligand of 8–10 amino acids in length. The class I heavy chain consists of five domains: the cytoplasmic domain, the transmembrane domain, and the $\alpha1$, $\alpha2$, and $\alpha3$ domains. The $\alpha1$ and $\alpha2$ regions interact to form the peptide-binding cleft. Class II molecules are composed of two noncovalently associated glycoproteins and an associated peptide of 8–20 amino acids. The α and β chains are both approximately 30 kDa, and each chain consists of four domains: the cytoplasmic domain, the transmembrane domain, and two extracellular domains $\alpha1$ and $\alpha2$, or $\beta1$ and $\beta2$. The peptide-binding cleft of class II molecules is formed by the interaction of both chains involving the $\alpha1$ and $\beta1$ domains.

TABLE III
Summary of the Major Histocompatibility Complex

1. MHC class I molecules
   - Single polymorphic heavy chain (45 kDa)
   - Noncovalent association with β_2-microglobulin (12 kDa)
   - Five domains:
     - $\alpha1$, $\alpha2$, $\alpha3$ extracellular domains
     - Transmembrane domain
     - Cytoplasmic domain
   - Ubiquitous tissue distribution: expressed on all nucleated cells
   - Encoded by HLA-A, -B, and -C loci in humans
   - Present peptide antigens to CD8 T cells
2. MHC class II molecules
   - Two polymorphic chains:
     - α chain (29 kDa)
     - β chain (34 kDa)
   - Each chain contains four domains:
     - $\alpha1$, $\alpha2$, or $\beta1$, $\beta2$ extracellular domains
     - Transmembrane domain
     - Cytoplasmic domain
   - Limited tissue distribution: expressed on macrophages, dendritic cells, B cells, and vascular endothelium
   - Encoded by HLA-D locus in humans (DP, DQ, and DR)
   - Present peptide antigens to CD4 T cells

morphic light chain of 12 kDa, and an associated peptide ligand of 8–10 amino acids. The class I heavy chain consists of five domains: the cytoplasmic domain, the transmembrane domain, and the $\alpha1$, $\alpha2$, and $\alpha3$ domains. The $\alpha1$ and $\alpha2$ regions interact to form the peptide-binding cleft. Class I molecules are expressed ubiquitously on the surface of all nucleated cells.

Class II molecules are composed of two noncovalently associated glycoproteins and an associated peptide of 8–20 amino acids. These two polymorphic glycoproteins are both encoded within the MHC and are similar in overall structure to each other. The α and β chains are both approximately 30 kDa, and each chain consists of four domains: the cytoplasmic domain, the transmembrane domain, and two extracellular domains $\alpha1$ and $\alpha2$, or $\beta1$ and $\beta2$. The peptide-binding cleft of class II molecules is formed by the interaction of both chains involving the $\alpha1$ and $\beta1$ domains. Class II molecules have a limited tissue distribution, with expression limited to so-called professional antigen-presenting cells (APC): macrophages, dendritic cells, B cells, and vascular endothelium.

C. Direct Allorecognition

The process of acute cellular allograft rejection is a complex event, dependent on the coordinated interaction of the many different cell types and soluble mediators of the immune system. Fundamental to this process, however, is the recognition of allogeneic MHC molecules by recipient T cells. The T-cell receptor, present on the cell surface of all T cells, is responsible for specific antigen recognition. The TCR is a disulfide-linked $\alpha\beta$ heterodimeric glycoprotein encoded by a family of variable and constant region genes. The structure of the TCR has recently been resolved and as predicted, shows homology with immunoglobulin molecules (Ig). In a normal MHC-restricted immune response, T cells recognize foreign antigens in the form of peptide fragments bound to MHC molecules. The recent cocrystallographic analyses of TCR with MHC class I/peptide complexes provide a model for these molecular interactions. In this model, the α-helical portions of the $\alpha 1$ and $\alpha 2$ domains of the class I molecule serve to define both the peptide-binding cleft and contact points for interaction with TCR. The CDR1 and CDR2 regions of the TCR provide the primary contacts for the α-helical portions of the MHC molecule, whereas the CDR3 region, encompassing the V(D)J junction, provides the primary contacts with peptide.

Despite these advances in characterizing the molecular basis of a normal MHC-restricted immune response, the recognition of foreign MHC antigens in allograft rejection continues to be investigated. It is clear, however, that alloreactive T cells are capable of recognizing intact donor MHC molecules on the surface of donor antigen-presenting cells in a direct pathway of allorecognition. In other words, T cells are capable of recognizing allogeneic MHC molecules directly without the usual requirement that these antigens be processed and presented by self-MHC molecules. Recognition of allogeneic MHC molecules in this fashion is known as *direct allorecognition.* Direct allorecognition explains an apparent paradox that has long puzzled immunologists. The ability to reject allografts would appear to confer no evolutionary survival advantage as tissue transplantation is not normally encountered in nature. However, the T-cell response to allografts expressing allogeneic MHC molecules is one of the most intense immune responses characterized. The extraordinary strength of T-cell responses to allogeneic MHC molecules is even apparent *in vitro* and can be readily measured in mixed lymphocyte cultures, whereas T-cell responses to foreign antigens presented in association with self-MHC molecules generally require *in vivo* priming before they can be measured *in vitro.* In addition, the number of precursor cells that recognize allogeneic MHC molecules is very high. *In vitro,* as many as 1 in 50 T cells are capable of recognizing a single allogeneic MHC molecule, compared to 1 in 10,000 T cells that typically recognize foreign antigens presented in association with self-MHC molecules. This high precursor frequency explains the intensity of T-cell responses to allogeneic MHC molecules *in vitro* and *in vivo.*

Direct allorecognition is the result of T-cell cross-reactivity. For instance, a T cell that recognizes a viral peptide in association with a self-MHC molecule may also recognize an allogeneic MHC molecule. In short, Allo-MHC = Self-MHC + peptide X. Recent experimental evidence strongly suggests that direct allorecognition is dependent on the endogenous peptides associated with allogeneic MHC molecules. In short hand, Allo-MHC + peptide A = Self-MHC + peptide X. This model can help to explain the high precursor frequency of T cells recognizing allogeneic MHC molecules. MHC molecules are associated with, or present, thousands of different endogenous peptides. Each of these complexes (Allo-MHC + peptide A_1, Allo-MHC + peptide A_2, Allo-MHC + peptide A_3...) can be recognized by different cross-reactive T cells. Note that the endogenous peptides ($A_1 + A_2 + A_3$) recognized in association with allogeneic MHC molecules need not be polymorphic; the complex of Allo-MHC + peptide is recognized based on polymorphic determinants in the Allo-MHC molecule, as well as on conformational changes in the peptide structure related to Allo-MHC binding. This hypothesis is known as the *determinant frequency hypothesis.* Another hypothesis to explain the high precursor frequency of alloreactive T cells is the *determinant density hypothesis.* There is evidence that certain endogenous peptides may be presented on the cell surface by a large number of MHC molecules. This high determinant density can be a powerful stimulus to activate alloreactive T cells with relatively low affinity for the determinant, increasing the potential number of T cells that can respond. The extraordinary strength of T-cell responses to allogeneic molecules can thus be explained by the unique way in which these molecules are recognized by cross-reactive T cells.

Direct allorecognition is not the only way in which allogeneic MHC molecules are recognized. They are also recognized by recipient T cells in a process known as *indirect allorecognition.* In this process, allogeneic MHC molecules are processed by host APC and presented as peptide fragments in the context of self-MHC molecules. Indirect allorecognition is therefore a special case of normal antigen presentation. Direct and indirect allorecognition are depicted schematically in Fig. 2. Recent studies clearly demonstrate that allograft rejection can prime indirect alloresponses *in vivo.* However, the relative importance and precise role of indirect allorecognition in allograft rejection remain to be determined. It is believed that indirect allorecognition may be important in the process of chronic rejection.

B lymphocytes can also recognize allogeneic MHC molecules, and antibodies can mediate allograft rejection. In particular, antibodies are involved in the clinical syndromes of hyperacute and accelerated rejection and may be important in the process of chronic rejection. Therefore, antibody-mediated rejection will be considered in Section II,F. The role of T cells in allo-

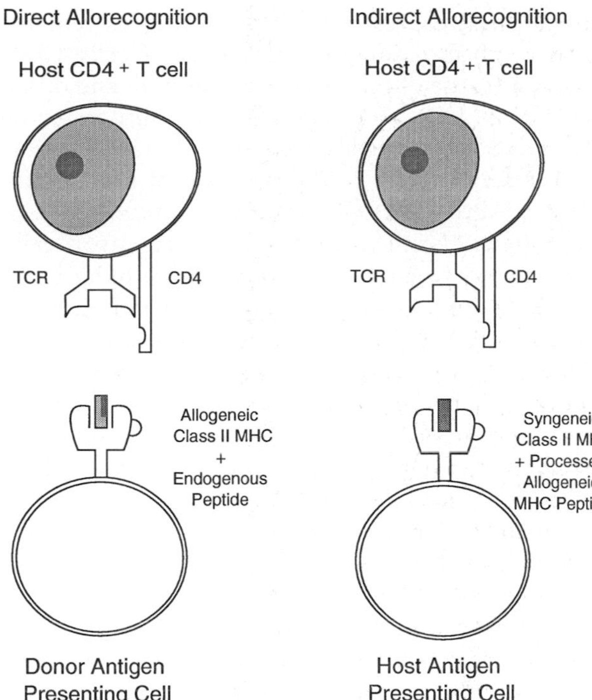

FIGURE 2 Direct and indirect allorecognition. Direct allorecognition is represented schematically on the left. Recipient T cells recognize donor MHC molecules on the surface of donor APCs without the usual requirement that these molecules be processed and presented by self-MHC molecules. Endogenous peptides associated with allogeneic MHC molecules are believed to be involved in direct allorecognition. In indirect allorecognition, allogeneic MHC molecules are processed and presented by syngeneic MHC molecules on the surface of host antigen-presenting cells.

graft rejection will first be examined, as it is generally considered that T cells are responsible for most cases of allograft rejection, particularly acute cellular rejection.

D. Allorecognition *in Vitro:* The Mixed Lymphocyte Culture and the Role of CD4$^+$ and CD8$^+$ T-Cell Subsets

The central importance of T cells in allograft rejection has been confirmed in experimental animal models. Nude mice, which are athymic and lack mature T cells, are unable to reject allogeneic skin grafts. However, adoptive transfer of T cells restores the ability of these animals to reject allografts. In addition, the success of OKT3 in clinical transplantation underscores the importance of T cells in allograft rejection. OKT3 is a murine monoclonal antibody reactive with the CD3 receptor on T cells. Treatment with OKT3, therefore, specifically targets mature T cells. OKT3 is one of the most effective agents used in the treatment of acute cellular rejection. The allogeneic *mixed lymphocyte culture* (MLC) is an *in vitro* model system that has allowed investigators to study the T-cell response to alloantigens and, in particular, to define the role of different T-cell subsets in alloreactivity.

The allogeneic MLC is performed by incubating mononuclear leukocytes from two different individuals or inbred strains together *in vitro*. If there is an MHC antigen disparity between the two individuals or inbred strains, T cells will proliferate in response to the allogeneic MHC molecules. This proliferation can be measured by the incorporation of [^{3}H]thymidine into replicating cells. Mononuclear leukocytes are typically isolated from peripheral blood in humans or from the spleen in rodents. To simplify the interpretation of the allogeneic MLC, one population of cells is irradiated to prevent cell proliferation. The irradiated population of cells serves only as "stimulators," and any proliferation measured can be attributed to the "responder" cell population. By studying the response to single MHC class I or class II antigen disparities, and with the use of monoclonal antibodies specific for different T-cell subsets, it has been possible to define the role of CD4$^+$ and CD8$^+$ T-cell subsets in alloreactivity.

CD4$^+$ T cells express the CD4 coreceptor molecule. The CD4 coreceptor recognizes invariant determinants on the surface of MHC class II molecules, and CD4$^+$ T cells normally interact with MHC class II molecules. The primary function of CD4$^+$ T cells is to secrete cytokines that promote the growth and differentiation of CD8$^+$ T cells and B lymphocytes. For this reason, CD4$^+$ T cells are known as helper T cells. In the MLC, CD4$^+$ T cells recognize allogeneic MHC class II molecules and proliferate strongly. In addition, CD4$^+$ T cells secrete helper cytokines to promote the differentiation of CD8$^+$ T cells. Therefore, alloreactive CD4$^+$ T cells function as do normal CD4$^+$ T-helper cells. CD4$^+$ T cells are responsible for the majority of proliferation in the MLC. If no MHC class II antigen disparity exists between the responder and stimulator populations, very little proliferation is observed. As noted earlier, the surface expression of MHC class II molecules is limited to certain specific cell types. Stimulation of alloreactive CD4$^+$ T cells is therefore limited to these cell types, including macrophages, dendritic cells, B cells, and vascular endothelium.

CD8⁺ T cells are known as cytolytic T cells because their primary function is to lyse cells that express foreign antigens on their cell surface. The CD8 coreceptor, present on the surface of $CD8^+$ T cells, recognizes invariant determinants on the surface of MHC class I molecules. Therefore, $CD8^+$ T cells normally recognize foreign antigens in association with MHC class I molecules. In the MLC, $CD8^+$ T cells are stimulated by allogeneic MHC class I molecules. Although $CD8^+$ T cells do not proliferate significantly, they do differentiate into cytolytic T cells capable of lysing targets expressing specific allogeneic MHC class I molecules. Individual cytolytic T cells are specific for a single MHC class I molecule, although a bulk population of cytolytic cells generated in a MLC will contain cells reactive with all the allogeneic MHC class I molecules expressed by the stimulator cells. Although $CD8^+$ T cells do secrete cytokines after activation, the growth and differentiation of these cells is most effectively promoted by $CD4^+$ T cells present in the same culture. The $CD8^+$ T-cell response is therefore strongest when both MHC class I and class II antigen disparities exist.

Experimental studies in rodent models have confirmed the *in vitro* model outlined here. In these studies, the rejection of MHC class I- or MHC class II-disparate allografts was investigated in recombinant inbred mouse strains. Monoclonal antibodies to the CD4 and CD8 coreceptors were used to define the role of these T-cell subsets in response to MHC class I or class II alloantigens. These studies reveal that $CD4^+$ T cells are central to the rejection of MHC class II-disparate allografts, whereas $CD8^+$ T cells are responsible for the rejection of MHC class I-disparate allografts. [*See* CD8 and CD4: Structure and Function.]

E. Allorecognition *in Vivo:* The Importance of Antigen-Presenting Cells

The process of allograft rejection *in vivo* can be considered in terms of two interdependent processes: (1) sensitization of alloreactive T cells and (2) activation of effector mechanisms leading to the actual rejection of the allograft. Central to the process of sensitization is the interaction between alloreactive T cells and donor-derived antigen-presenting cells. APCs, including dendritic cells, macrophages, and activated B cells, are specialized cells capable of effectively stimulating $CD4^+$ and $CD8^+$ T cells. Although histocompatibility antigens are expressed on the surface of all cells in a foreign allograft, APCs are critical elements of the sensitization process for two reasons. First, *APCs constitutively express MHC class II antigens,* required for the activation of alloreactive $CD4^+$ T cells. $CD4^+$ T cells, because of their role in the growth and differentiation of $CD8^+$ T cells, and their ability to recruit nonspecific effector cells to the site of graft rejection, are central elements in the rejection of most grafts. Second, APCs effectively *provide costimulatory signals* required for the activation of alloreactive T cells. Costimulatory signals are the result of cytokine secretion by APCs and the expression on APCs of accessory molecules capable of interacting with T cells. Accessory molecule interactions considered important include B7 interaction with CD28 and ICAM-1 interaction with LFA-1. Recent evidence suggests that stimulation of T cells in the absence of these second signals may actually lead to functional inactivation, or *anergy* induction.

The importance of donor APCs in allograft rejection is known as the *passenger leukocyte hypothesis.* APCs present in the allograft at the time of transplantation have been known as passenger leukocytes. Almost immediately following transplantation, donor APCs migrate from the graft to recipient lymphoid organs, where sensitization can occur. At the same time, donor APCs are replaced in the allograft by recipient-derived APCs. The importance of APCs in the sensitization process is most clearly demonstrated in animal models, in which the survival of endocrine tissue allografts is markedly or indefinitely prolonged after depletion of donor APCs. Subsequent challenge with a second allograft containing APCs leads to the rapid rejection of both grafts. In addition, the importance of APCs in the sensitization process may explain the variability seen in the rejection of skin and cardiac allografts in rodent models. Skin allografts are typically rapidly rejected, and high doses of immunosuppression are required to arrest the rejection process. Cardiac allografts, however, undergo more delayed rejection, and this process is more responsive to immunosuppressive medications. Langerhans cells, a subpopulation of dendritic cells, are responsible for the intense rejection of skin allografts. Langerhans cells are common in the interstitium of the skin and are only slowly replaced by recipient APCs. However, few APCs exist in cardiac allografts, and these are rapidly replaced by recipient APCs.

F. The Role of Antibodies in Allograft Rejection

Although B-cell recognition of allogeneic MHC molecules is not completely understood, antibodies can mediate allograft rejection and are responsible for the clinical syndromes of hyperacute and accelerated rejection. In addition, antibodies are now believed to play an important role in the process of chronic rejection. However, because B- and T-cell responses often coexist, it is difficult to precisely define the role of antibodies in other forms of allograft rejection. Hyperacute and accelerated rejection are the result of preformed antibodies. Preformed antibodies to blood group or ABO antigens occur naturally and are generally of the IgM isotype. Preformed antibodies reactive with allogeneic MHC molecules do not occur naturally, but are the result of sensitization after pregnancy, blood transfusion, or prior transplantation. Antibodies reactive with MHC molecules are therefore generally of the IgG isotype. ABO antigen matching and the antibody cross-match are used in clinical transplantation to prevent hyperacute rejection and will be discussed in the next section.

III. MODERN CLINICAL TRANSPLANTATION

Clinical transplantation has been developed rapidly since kidney transplants were first performed in the 1950s. One of the first successful series of transplants was performed in identical twins, underscoring the central role of the immune system in transplantation. The subsequent development of effective immunosuppressive medications has dramatically improved the success of renal transplantation and has permitted the development of cardiac, pulmonary, and hepatic transplantation. Advances in surgical technique, donor organ preservation, and histocompatibility antigen testing have contributed to the success of modern clinical transplantation. Although allograft rejection and the complications associated with nonspecific immunosuppression required to prevent allograft rejection continue to limit the success of clinical transplantation, the growing shortage of donor organs is rapidly becoming the single most important problem in transplantation.

A. Antigen Matching in Clinical Transplantation

The success of modern clinical transplantation is dependent on the use of nonspecific immunosuppression. However, in most instances, an effort is made to match the donor and recipient as closely as possible. This involves (1) blood group or ABO matching, (2) screening of the recipient for preformed antibodies reactive with human leukocyte antigens (HLA) on donor cells (the cross-match), and (3) HLA typing of the donor and of potential recipients, with an attempt to match as closely as possible donor and recipient HLA types.

As noted earlier, preformed antibodies to ABO antigens and to HLA are responsible for the syndromes of hyperacute and accelerated graft rejection. These syndromes cannot be treated by immunosuppression, and therefore preoperative ABO matching and an antibody cross-match of the donor and recipient are routinely performed. The *ABO antigen system* was first defined based on the expression of ABO antigens on the surface of red blood cells. IgM antibodies reactive with ABO antigens occur naturally, limiting blood transfusions across ABO barriers. Transplantation across ABO barriers results in hyperacute rejection, making ABO matching critically important. Transplantation across ABO barriers is not attempted. In addition to ABO matching, the recipient is screened for the presence of preformed antibodies reactive with HLA antigens on the allograft. Anti-HLA antibodies do not occur naturally but are the result of exposure to foreign HLA after pregnancy, blood transfusion, or previous transplantation. Prior to transplantation, serum from the recipient is incubated with donor lymphocytes and complement in a procedure known as the *cross-match*. If preformed antibodies reactive with HLA on the donor cells are present, the cells will be lysed after complement fixation. ABO matching, and screening for preformed antibodies have largely prevented hyperacute rejection in clinical transplantation.

For practical considerations, *HLA typing* and prospective antigen matching are performed primarily only in renal and bone marrow transplantation. The limitations of organ preservation techniques do not generally allow prospective HLA typing in hepatic, cardiac, or pulmonary transplantation. Also, elective living-related donor transplantation is currently limited primarily to renal and bone marrow transplantation. The evidence that HLA typing improves renal allograft survival is most clear when considering the survival of two-haplotype-matched living related grafts. Siblings with a two-haplotype match are HLA-identical, but differ at minor histocompatibility loci. One-haplotype-matched siblings share half of their HLA antigens, and unrelated individuals generally share few HLA antigens because of the extreme polymorphism of the MHC. Recent data from the Transplantation Unit at Massachusetts General Hospital reveal that HLA-identical living related grafts have a one-year survival rate of 100%, one-haplotype-matched living related grafts have a one-year survival rate of 92%, and grafts from unrelated cadaveric donors have a one-year survival rate of only 83%. The introduction of cyclosporine (CsA) in the mid-1980s dramatically improved the success of cadaveric renal transplantation. For this reason, improvements in cadaveric renal allograft survival rates associated with HLA matching are difficult to demonstrate unless there is a complete HLA match. Long-term follow-up of renal allograft recipients reveals that antigen matching may be even more important in the prevention of chronic rejection.

HLA typing is dependent on serologic, cellular, and molecular techniques. Serologic techniques utilize defined antisera to identify HLA antigens on the surface of lymphocytes. In a commonly performed test, the lymphocytotoxicity assay, antisera with known anti-HLA specificity are added to lymphocytes from individuals of unknown HLA type. Complement is then added, and lysis is measured. Cell death indicates antibody binding, identifying the HLA antigens present on the surface of the lymphocytes. The lymphocytotoxicity assay is used to type the HLA-A, -B, -C, and -DR antigens. However, serologic typing of other class II antigens is imprecise, and other techniques are required to more accurately identify these antigens. Originally, cellular techniques were used to type class II antigens, but increasingly molecular techniques based on the polymerase chain reaction are used. The mixed lymphocyte culture forms the basis for cellular typing of HLA class II antigens. In a commonly performed assay, lymphocytes from individuals of unknown HLA type are added to irradiated homozygous typing cells, and proliferation is measured. Homozygous typing cells are cells known to be homozygous at the HLA-D locus. Thus, the HLA type of unknown lymphocytes is identified by determining which homozygous typing cells the lymphocytes do not respond to. Recent advances in molecular biology allow HLA antigens to be identified based on their genetic sequence. The genetic region encoding polymorphic residues in class II MHC molecules can be amplified using primers derived from conserved flanking regions with the polymerase chain reaction. The amplified region can then be sequenced, and the HLA-DR, -DP, and -DQ alleles can be typed at the molecular level. As noted earlier, these new molecular techniques are especially useful for typing class II antigens as the cellular techniques are time-consuming and often indeterminate.

B. Effector Mechanisms and the Pathology of Allograft Rejection

Despite an increased awareness of the mechanisms leading to allograft recognition, rejection of allografts remains a major problem in clinical transplantation. Rejection episodes can be divided into several major categories based on the effector mechanisms leading to rejection (antibody-mediated rejection versus cellular rejection), as well as on the clinical circumstances associated with rejection (hyperacute, acute, or chronic rejection). The most extensive clinical experience involves renal transplantation, and the following discussion will outline the pathology of rejection episodes in this organ.

Hyperacute rejection is a clinical syndrome characterized by allograft rejection within minutes to hours after revascularization. Hyperacute rejection is the result of preformed antibodies to donor AB blood group antigens, or preformed anti-HLA antibodies. These antibodies react to antigens on the vascular endothelium of the transplanted graft. Antibody deposition leads rapidly to endothelial cell activation and platelet aggregation. Complement activation leads to endothelial cell injury, further promoting platelet activation and coagulation. Within a short period of time, vascular occlusion results in irreversible tissue ischemia. Characteristic histologic features include microvascular occlusion and endothelial cell destruction.

Acute rejection episodes occur commonly after the first week following transplantation. Acute rejection is characterized by an intense cellular infiltrate. Infiltrating cells include lymphocytes, macrophages, and polymorphonuclear leukocytes. Acute rejection commonly involves infiltration and destruction of the graft parenchyma. In renal transplantation, lymphocytic infiltration of tubular epithelial cells is common. Infiltration of vascular endothelium is also commonly seen. Acute rejection episodes commonly respond to increased immunosuppression, including high-dose steroid therapy and antilymphocyte antibody therapy. Successful treatment of acute rejection is associated with a decrease in the cellular infiltrate. Although acute rejection is less common after the first 6 months following transplantation, it can occur up to years later and is often seen in association with chronic rejection.

Chronic rejection is a slowly progressive condition that is generally unresponsive to increasing doses of immunosuppression. The pathogenesis of this condition is poorly understood, but it is thought to represent the accumulation of immunologic damage over the period of months and years. Chronic rejection involves damage to the vascular endothelium, with concomitant proliferation of intimal smooth muscle cells. This process leads to progressive vascular occlusion and ischemic damage to the graft. Histopathologic changes consist of interstitial fibrosis and the loss of normal cellular architecture. Experience with cadaveric renal transplant recipients reveals the importance of chronic rejection. One-year graft survival in these patients is 82%, but survival is reduced to only 35% at 10 years.

C. Regulation of Graft Rejection: Nonspecific Immunosuppression

The number of histocompatibility loci, as well as the extreme polymorphism within the major histocompatibility locus, makes allograft rejection almost invariable in clinical transplantation. Allograft rejection can be prevented in one of two major ways. First, nonspecific immunosuppression can be used to globally suppress recipient immune responses. Second, the immune response to donor histocompatibility antigens can be specifically inhibited. Modern clinical transplantation is dependent on the use of nonspecific immunosuppression to prevent allograft rejection. This approach, however, is limited by the complications associated with nonspecific immunosuppression, including the predisposition to infection and the development of neoplasms. Because immunosuppressive agents commonly have severe side effects when used at high doses, these agents are often used in combination. In any event, the development of protocols to specifically inhibit the immune response to donor histocompatibility antigens remains one of the primary goals of transplantation research.

Corticosteroids were one of the first agents used to treat allograft rejection. Currently, high-dose intravenous corticosteroid therapy is often used to treat acute cellular rejection. How corticosteroids act to suppress the immune system is incompletely understood. Corticosteroids primarily affect T cells and macrophages, but have no effect on antibody production by B cells. Corticosteroids reduce inflammation, and one of the proposed mechanisms of action is that they interfere with cytokine gene transcription. However, they do have many side effects, including hypertension, hyperglycemia, poor wound healing, avascular necrosis, and psychosis, as well as the predisposition to infection and malignancy shared by other nonspecific immunosuppression medications. Therefore, corticosteroids are often used at low doses in combination with other medications to prevent these side effects. [*See* Anti-Inflammatory Steroid Action.]

Azathioprine is a derivative of 6-mercaptopurine and inhibits purine synthesis. Thus, azathioprine blocks the synthesis of DNA and RNA and inhibits the proliferation of activated T cells. Azathioprine therapy is often associated with myelosuppression, and for this reason it is often used at low dose in combination with other agents.

Cyclosporine (CsA) is one of the most effective nonspecific immunosuppressive medications currently in use. The introduction of CsA during the mid-1980s resulted in dramatic improvements in cadaveric renal allograft survival rates. In addition, CsA has allowed the successful transplantation of both cardiac and hepatic allografts, which were almost universally unsuccessful prior to its introduction. CsA is a cyclic peptide isolated from the soil fungus *Tolypocladium inflatum Gams*. CsA binds specifically to the intracellular protein cyclophilin and inhibits the enzymatic activity of calcineurin, a critical intermediate in the T-cell activation signal transduction pathway. Thus, CsA blocks the transcription of the early T-cell activation genes IL-2, IL-3, IL-4, and IFN-γ. This effectively blocks the IL-2-dependent activation and differentiation of T cells. The specificity of CsA, inhibiting primarily the IL-2-dependent activation of T cells, is associated with a lower incidence of opportunistic infections than other immunosuppressive medications. However, the use of CsA is associated with nephrotoxicity. The nephrotoxicity is a particular limitation in renal transplantation, as CsA toxicity can be difficult to differentiate from rejection episodes. This side effect is dose-related, and for this reason CsA is often used in lower doses in combination with other immunosuppressive agents. *FK 506* is a macrolide antibiotic recently isolated from a soil fungus that has been shown to have immunosuppressive activity similar to that of CsA. Although structurally unrelated to CsA, both FK 506 and CsA act to inhibit early events in the T-cell activation signal transduction pathway. FK 506 was recently approved for clinical use by the FDA based on preliminary trials in hepatic transplantation demonstrating the promise of this agent. Other agents, including the antibiotic rapamycin, mycophenolic acid, and mizorbin, are currently undergoing experimental and clinical trials.

Antibodies specific for receptors on the surface of T cells have proven to be effective in the prevention and treatment of allograft rejection. Originally, polyclonal antisera were obtained after immunization of horses, goats, or rabbits with human lymphocytes. Subsequently, monoclonal antibodies with better-defined specificities were isolated. *OKT3* is the only monoclonal antibody currently in routine clinical use. OKT3 is a murine monoclonal antibody specific for the CD3 receptor on the surface of human T cells. The administration of OKT3 results in a rapid decrease in the number of circulating lymphocytes. OKT3 binds to circulating T cells, leading to their opsonization by the reticuloendothelial system. In addition, CD3 expression is reduced in surviving T cells, making them nonfunctional. One limitation of OKT3 is that T cells can be activated by this agent prior to elimination. Thus, activated T cells secrete cytokines, leading to fever, chills, and pulmonary edema. In addition, the use of OKT3 may be limited by the formation of antimurine antibodies in patients treated. Such an immune response reduces the effectiveness of these agents. Currently, murine monoclonal antibodies specific for CD4, the α subunit of the IL-2 receptor, the T-cell adhesion molecule LFA-1, and its endothelial ligand ICAM-1 are being investigated.

Although modern nonspecific immunosuppression has dramatically increased allograft survival, it is associated with severe complications that limit its usefulness. First, immunosuppression predisposes transplant patients to conventional and opportunistic infections. Because transplant patients are immunosuppressed, infections can rapidly become life-threatening. To decrease the risk of infection, potential organ donors and recipients are carefully screened for the presence of infection. Transplant recipients are particularly susceptible to infection from 1 to 6 months after transplantation, a result of the high doses of immunosuppression administered during induction therapy and during the treatment of rejection episodes. After 6 months, when the dose of immunosuppression is generally reduced to maintenance levels, the risk of life-threatening infection also decreases. Transplant patients are susceptible to viral, bacterial, fungal, and protozoal infections. Viral infections, particularly cytomegalovirus (CMV) infections, are common and can be life-threatening. CMV infection can occur either as a primary infection or as a reactivation of a latent infection. The Epstein–Barr virus, herpes simplex virus, and hepatitis A, B, and C viruses are also commonly seen in transplant patients. Other frequently occurring infections in transplant recipients include *Pneumocystis carinii* pneumonia, bacterial urinary tract infections, and oral fungal infections. To reduce the incidence of infection, transplant patients routinely receive prophylactic medications to prevent the most common bacterial, viral, fungal, and protozoal infections.

Another major complication of immunosuppression is that immunosuppressed patients are predisposed to the development of cancers. The most frequent neoplasms in transplant patients are nonmelanotic cancers of the skin and lips, lymphoproliferative diseases, including malignant B-cell lymphoma, and cervical cancer. The incidence of malignancy increases with time in immunosuppressed patients. Transplant patients do not appear to be predisposed to the malignancies that are most common in the general population, including cancers of the breast, prostate, colon, and lung. This is consistent with the importance of immune surveillance in the prevention of virally derived neoplasms. Viruses are considered to be important elements in the pathogenesis of nonmelanotic cancers of the skin and lips, malignant lymphomas, and cervical cancers.

D. Regulation of Graft Rejection: Tolerance Induction and Antigen-Specific Immunosuppression

Specific transplantation tolerance can be defined as the long-term acceptance of grafted tissue in the absence of continuous immunosuppression, and it has been achieved in a number of different animal models. In addition, the experience with donor-specific blood transfusions demonstrates that specific manipulation of the immune system can improve allograft survival in humans. However, induction of transplantation tolerance in humans has proven elusive. Despite this difficulty, one of the goals of transplantation research is to induce tolerance in humans so that the complications associated with nonspecific immunosuppression can be avoided.

During the 1960s, patients with end-stage renal disease received frequent blood transfusions as a treatment for their anemia. It was subsequently established that these blood transfusions sensitized patients to HLA antigens, resulting in anti-HLA antibodies. Subsequently, these transfusions were generally avoided. In retrospective analyses, however, it was noted that patients receiving blood transfusions had significantly improved renal allograft survival. Therefore, renal failure patients in many centers were routinely transfused prior to transplantation. In addition, donor-specific blood transfusion was used prior to living related donor transplantation. Both random donor blood transfusions and donor-specific blood transfusions were clearly documented to improve renal allograft survival. However, the mechanisms responsible for this improvement in allograft survival have not been completely elucidated. The introduction of CsA resulted in a dramatic improvement in 1-year cadaveric renal allograft survival rates. Presently, 1-year cadaveric renal allograft survival rates approach 90%, and the potential benefit of these random donor or donor-specific blood transfusions is difficult to measure. In addition, the infectious risks of blood transfusion, including the risk of hepatitis B and human immunodeficiency virus transmission, are more completely understood. As a result, these interventions are rarely used today.

Microchimerism has been the focus of considerable interest recently as a potential mechanism of specific tolerance induction. In particular, research has focused on whether the long-term survival of allografts is associated with a state of microchimerism and tolerance. Sensitive immunohistochemical and polymerase chain reaction techniques have been used to show that donor cells are present in the lymph nodes and spleen of patients with long-term surviving renal and hepatic allografts. The significance of these findings remains to be determined. Current research protocols involve the infusion of donor bone marrow progenitors with solid organ allografts in an effort to produce microchimerism and tolerance.

Experimental studies in animal models have shown that allogeneic *bone marrow transplantation* can be successfully used to induce tolerance to solid organ allografts. Animals successfully reconstituted with allogeneic bone marrow will accept allografts long term without the need for immunosuppression. In these model systems, successful bone marrow transplantation requires that two major obstacles be overcome. First, recipient animals must undergo a preparative regimen to ensure engraftment of the donor bone marrow. Second, mature T cells in the donor bone marrow inoculum must be eliminated to prevent the development of graft-versus-host disease. Preparative regimens typically involve ablation of the recipient immune system by whole-body irradiation. Animals that are successfully engrafted have a mature immune system that is derived from the donor and that is tolerant to both donor and recipient histocompatibility antigens. Experimental evidence suggests that tolerance induction in bone marrow chimeras is dependent on repopulation of the recipient thymus with donor-derived bone marrow cells. Tolerance to donor antigens is achieved as developing cells are educated in the thymus. Mixed bone marrow chimeras are produced after the reconstitution of recipients with both allogeneic and syngeneic bone marrow. Reconstituted animals are mixed chimeras; their mature immune system is derived from both the donor and the recipient. Mixed bone marrow chimeras remain tolerant to both donor and recipient histocompatibility antigens and appear to have more functional immune systems. Despite this success in animal models, allogeneic bone marrow transplantation in humans remains experimental as a mechanism for inducing transplantation tolerance. Humans and other large animals do not tolerate whole-body irradiation in doses sufficient to permit marrow engraftment. Other preparative regimens in humans are extremely toxic, precluding their use as a mechanism of transplantation tolerance induction. Therefore, recent animal studies have focused on the use of less toxic, nonmyeloablative conditioning regimens and allogeneic bone marrow transplantation to induce mixed bone marrow chimeras. It is hoped that these regimens will be better tolerated by humans.

The study of T-cell activation events has revealed a potential strategy for the induction of specific tolerance. It has recently been shown that T-cell activation requires both antigen receptor stimulation and a costimulatory signal. Antigen receptor stimulation in the absence of a costimulatory signal results in a down-regulation of T-cell responsiveness, or *anergy* induction. Anergic T cells are refractory to subsequent stimulation. Costimulation can be provided by the interaction of accessory molecules or by soluble factors such as cytokines. Candidates for accessory molecule interaction include LFA-1/ICAM-1, CD2/LFA-3, and CD28/B7. CTLA4 is an alternate B7 ligand, expressed by the T cell. Soluble CTLA4-Ig, a recombinant fusion protein, can be used to effectively block the interaction between CD28 and B7. CTLA4-Ig has been used to inhibit alloresponses both *in vitro* and *in vivo*. Clinical trials with this agent are in progress.

The role of CD4 subsets in the induction of transplantation tolerance has been the subject of considerable interest recently. The importance of these subsets in host defense in parasite infection has been clearly demonstrated, however, the importance in transplantation is less clear. *Th-1 $CD4^+$ T cells* secrete IL-2, IFN-γ, and tumor necrosis factor-β (TNF-β) and can mediate delayed type hypersensitivity (DTH) responses. *Th-2 $CD4^+$ T cells* secrete IL-4, -5, -6, and -10 and are necessary for maturation of antibody responses and for IgE production. The two subsets are capable of cross-regulation through cytokine secretion. IL-4 and IL-10 down-regulate Th-1 responses, whereas IFN-γ can down-regulate Th-2 responses. It has been hypothesized that a Th-2 response to alloantigen may suppress graft rejection by Th-1 cells. Molecular techniques have been used to quantitate cytokine secretion in allografts undergoing rejection, and these studies appear to confirm the importance of the Th-1 cytokines IL-2 and IFN-γ in allograft rejection. However, considerable research remains to define whether Th-1/Th-2 subset interaction is an important element in transplantation tolerance, and whether manipulation of this subset interaction can be beneficial.

IV. XENOGENEIC TRANSPLANTATION

The success of clinical transplantation and the limited pool of organ donors have resulted in an acute shortage of cadaveric donor organs. It is estimated that in the United States over 40,000 patients are currently on waiting lists for donor organs. Unfortunately, approximately one in four of these patients die waiting for transplant organs to become available. Despite education programs and legislation aimed at increasing organ donation, it is generally believed that the supply of cadaveric organs is limited and will be increasingly inadequate given the increasing demand for donor organs. This shortage of organs has renewed interest in xenogeneic organ transplants. Early clinical experience with nonhuman primate xenotransplants had limited success. Although transplanted organs functioned physiologically, the intensity of the rejection response required extremely high doses of immunosuppression. Consequently, sepsis frequently resulted. Despite these setbacks, researchers continue to view xenotransplants as a potential solution to the shortage of cadaveric organs.

Interest has focused on nonhuman primates and miniature swine as potential sources of organs. Nonhuman primates are phylogenetically closer to humans and may present less of an immunologic barrier to successful transplantation. However, use of nonhuman primates as a source of transplant organs is limited by their status as endangered species in the wild. In addition, nonhuman primates reproduce poorly in captivity and are potentially the source of infectious diseases. Miniature swine are phylogenetically more distant to humans than nonhuman primates, and their tissues are less compatible with the human immune system. The use of miniature swine has many advantages, however, for they are domesticated and are in plentiful supply. Miniature swine have a relatively short gestation period and relatively large litters, allowing the establishment of defined partially inbred strains that have facilitated research on these animals. These breeding characteristics also make genetic engineering and the breeding of transgenic animals possible. Thus, researchers hope that these animals can be genetically altered to reduce the tissue incompatibility between miniature swine and humans.

As noted earlier, the main limitation to successful xenogeneic transplantation is graft rejection. The intensity of graft rejection is dependent on the donor and recipient species involved. Depending on the species combination, xenogeneic transplantation can result in hyperacute rejection or in a more delayed rejection reaction. Hyperacute rejection is the result of natural xenoreactive antibodies and is more likely to occur in more distantly related species combinations. Thus, species combinations characterized by the presence of natural xenoreactive antibodies and associated with hyperacute rejection are referred to as *discordant.* Species combinations characterized by the absence of xenoreactive antibodies and associated with a more delayed rejection response are termed *concordant.* Hyperacute rejection in discordant species combinations results in acute vascular occlusion and graft rejection within minutes to hours. This is similar to hyperacute rejection seen in presensitized allograft recipients. Natural xenoreactive antibodies are responsible for hyperacute rejection. Xenoreactive antibodies bind to endothelial cells and lead to endothelial cell activation, release of vasoactive mediators, activation of complement by the classical pathway, and ultimately cell lysis and graft thrombosis.

The use of organs from nonhuman primates such as the chimpanzee and the baboon has had limited success in clinical transplantation. These species combinations are concordant and are not susceptible to hyperacute rejection. However, the use of tissues from miniature swine is associated with hyperacute rejection. Absorption of natural xenoreactive antibodies from recipient serum has been used to prevent hyperacute rejection. Absorption is performed by perfusing a second donor organ with recipient serum prior to transplantation. However, even after the removal of all the natural xenoreactive antibody, complement deposition can still occur on the xenotransplant, leading to ultimate rejection (over a period of days rather than minutes). To overcome complement-mediated rejection of these xenografts, transgenic animals have been created that express human proteins that inhibit human complement activation and deposition. Preliminary evidence suggests that animals expressing one or a combination of the human proteins decay accelerating factor (DAF), membrane cofactor protein (MCP), and CD 59 are relatively resistant to complement-mediated tissue destruction.

In concordant species combinations, or in discordant species combinations where hyperacute rejection has been prevented, vigorous cellular rejection can occur despite high doses of immunosuppressive agents. The cellular response to xenografts appears to be focused on MHC antigens. However, direct xenorecognition may be limited by defective cell-surface molecule interaction between recipient T cells and xenogeneic antigen-presenting cells. Therefore, indirect antigen presentation may be relatively more important in the response to xenografts. Because of the intensity of the immune response to xenografts, efforts to induce xenograft-specific tolerance are considered critical to the success of xenotransplantation. The use of mixed chimerism, through mixed bone marrow transplantation, has been used to induce specific transplantation tolerance in rodent xenogeneic model systems. It remains to be seen whether these preliminary successes will be applicable in larger animals.

V. FUTURE PERSPECTIVES

Transplantation biology is a unique interface between basic science and clinical medicine. The desire to understand the complex process of graft rejection has led to major advances in basic immunology, including the discovery of the major histocompatibility complex and the characterization of the structure and function of MHC class I and II antigens. From these investigations we now know that antigen presentation by MHC molecules to T cells determines the specificity of the immune system in its discrimination between self and non-self. Thus, therapeutic intervention at the level of the molecular interaction between the MHC molecule and the T-cell receptor remains a very attractive method of preventing allograft rejection. Conceptually, intervention at the level of the MHC molecule would specifically block immune recognition of the transplant without compromising the immune system of the recipient. Several strategies to specifically block MHC antigen presentation are currently being investigated, but have yet to come to fruition. Another challenge in transplantation is overcoming the acute shortage of donor organs. One possible resolution is a better understanding of xenotransplantation and the ability to induce specific tolerance. In any event, the future interaction between basic and clinical investigators promises great advances in the area of transplantation.

BIBLIOGRAPHY

Auchincloss, H., Jr., and Sachs, D. H. (1993). Transplantation and graft rejection. *In* "Fundamental Immunology" (W. E. Paul, ed.), 3rd Ed., Chap. 31, pp. 1099–1141. Raven, New York.

Charlton, B., Auchincloss, H., Jr., and Fathman, C. G. (1994). Mechanisms of transplantation tolerance. *Annu. Rev. Immunol.* **12**, 707–734.

Hansen, T. H., Carreno, B. M., and Sachs, D. H. (1993). The major histocompatibility complex. *In* "Fundamental Immunology" (W. E. Paul, ed.), 3rd Ed., Chap. 16, pp. 577–628. Raven, New York.

Kaufman, C. L., Gaines, B. A., and Ildstad, S. T. (1995). Xenotransplantation. *Annu. Rev. Immunol.* **13**, 339–367.

Immunoconjugates

CARL-WILHELM VOGEL
REINHARD BREDEHORST
University of Hamburg

GLOSSARY

Heterobifunctional cross-linking reagent Chemical reagent with two different reactive groups, designed to react with two protein molecules for cross-linking

Monoclonal antibody Antibody with defined specificity derived from a single B-cell clone

Nude mouse Strain of hairless mice without the thymus. These animals are immunocompromised and cannot reject foreign tissue

IMMUNOCONJUGATES CONSIST OF TWO DIFFERent biomolecules linked together to create a hybrid molecule exhibiting the biological activities of the two parent compounds. The vast majority of such immunoconjugates have been constructed by chemical procedures. More recently, however, several immunoconjugates consisting of two protein moieties have been synthesized by recombinant DNA methodology. The classic immunoconjugate examples are antibody conjugates in which one component is a monoclonal antibody with specificity for a cell-surface antigen of a tumor cell and the other component is a toxic effector molecule (e.g., a protein toxin or a toxic drug). In this example, the binding function of the antibody molecule is combined with the cytotoxic function of the toxic effector molecule to create a selective cytotoxic agent for antigen-bearing tumor cells. Since the binding moiety serves to target the conjugate to a defined molecular structure, it is therefore also referred to as targeting moiety. The binding moiety of most conjugates prepared so far is an antibody, and hence the term immunoconjugates. Figure 1 shows the structure of a model immunoconjugate. Although, strictly speaking, conjugates with non-antibody-binding moieties are not immunoconjugates, these conjugates conceptually belong in the immunoconjugate category and are included in this review article.

I. EFFECTOR MOIETIES OF IMMUNOCONJUGATES

Several categories of effector molecules have been linked to antibodies and other targeting moieties to construct immunoconjugates (Table I). These effector molecules include protein toxins, low-molecular-weight drugs, biological response modifiers (BRMs), radionuclides, enzymes, and several other substances.

A. Antibody Conjugates with Protein Toxins (Immunotoxins)

Frequently coupled protein toxins of bacterial or plant origin include ricin, diphtheria toxin, and *Pseudomonas* exotoxin. Additional protein toxins used for the construction of immunoconjugates are gelonin, pokeweed antiviral protein, abrin, and several others. Immunoconjugates constructed with these protein toxins are frequently referred to as immunotoxins.

Several of these toxins (e.g., ricin, abrin) consist of two polypeptide chains attached through a disulfide bond. One polypeptide chain is responsible for the

ENCYCLOPEDIA OF HUMAN BIOLOGY, Second Edition, VOLUME 4.

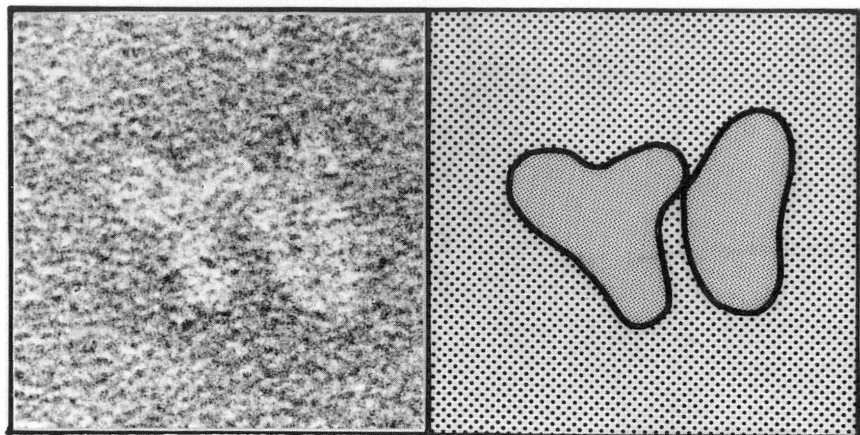

FIGURE 1 Structure of an immunoconjugate. Shown is an electron micrograph of an antibody conjugate with cobra venom factor (left) and its graphic reproduction (right). The Y-shaped IgG antibody molecule and the irregular oval cobra venom factor molecule can be easily distinguished. The cobra venom factor molecule is linked to one of the Fab arms of the antibody molecule.

toxic activity of inhibiting cellular protein synthesis and is referred to as the *toxic A chain.* The toxic A chain is an enzyme that cleaves the *N*-glycosidic bond of a single adenosine residue of 28*S* ribosomal RNA (rRNA), thereby destroying the ability of ribosomes to synthesize protein. Because of the enzymatic nature of their toxic activity, these toxins are very potent once they have entered the cell. As a matter of fact, a single toxin molecule can be sufficient to cause cell death. The biological function of the nontoxic B chain is to allow the toxins to reach the cytosol, where the components of protein synthesis are located. Upon binding of the B chain to galactose residues present on many different surface glycoproteins and glycolipids, the toxin is taken into the endocytic compartment and processed into an active fragment that translocates into the cytosol. Another group of plant toxins, which includes, for example, the pokeweed antiviral

TABLE I
Effectors Used in Immunoconjugates

Effector	Immunoconjugate	Examples of effectors
Toxin	Immunotoxin	Plant: Ricin, saporin Microbial: diphtheria toxin, *Pseudomonas* exotoxin
Low-molecular-weight drug	Chemoimmunoconjugate	Antimetabolites: 5-fluorodeoxyuridine, methotrexate Alkylating agents: chlorambucil, melphalan Antimitotic agents: vinca alkaloids DNA intercalators: doxorubicin
Biological response modifier (BRM)	BRM immunoconjugate	Interferon, cobra venom factor, interleukin 2, tumor necrosis factor, bacterial superantigens
Radionuclide	Radioimmunoconjugate	α-Emitters: ^{211}-astatine, ^{212}-bismuth β-Emitters: ^{90}-yttrium γ-Emitters: ^{131}-iodine, ^{111}-indium ^{99m}-technetium
Enzyme	Enzymoimmunoconjugate	Carboxypeptidase G, alkaline phosphatase, cytidine deaminase, β-lactamase, human β-glucuronidase
Others		Radiosensitizers, magnetic microspheres

protein and gelonin, is referred to as hemitoxins. They consist only of a toxic A chain, which is structurally similar to the A chain of ricin and other holotoxins. Lacking a B chain, hemitoxins enter the cell with much greater difficulty.

The two bacterial toxins diphtheria toxin and *Pseudomonas* exotoxin differ structurally from each other and from the aforementioned plant toxins, but functionally resemble them. They are synthesized as a single polypeptide chain and contain three different domains responsible for cell binding, membrane translocation, and toxic activity. The toxic activity of diphtheria toxin and *Pseudomonas* exotoxin is the ADP-ribosylation of elongation factor 2, which inhibits protein synthesis.

Because of their high toxicity, toxins such as ricin have been coupled to many monoclonal antibodies with specificity for tumor antigens. These immunotoxins showed good cytotoxic activity for antigen-bearing tumor cells under *in vitro* conditions. However, the nonspecific binding function of the B chain, which in the case of ricin recognizes terminal galactose residues on cells, would severely limit the selectivity of the cytotoxic activity of the immunotoxins. To overcome the unwanted toxic activity of immunotoxins for bystander cells caused by the binding domain of the toxin, several approaches have been employed. For *in vitro* applications, the binding site could be blocked by the addition of lactose to the incubation medium. The lactose competes for the B-chain binding site with cell-surface galactose so that the binding specificity of the antibody determines the selectivity of the immunotoxin. Successful *in vitro* applications using immunotoxins with holotoxins (A chain + B chain) include the purging of T lymphocytes from bone marrow for allogeneic bone marrow transplantation to reduce the incidence and severity of graft-versus-host disease.

Another approach to eliminate nonspecific immunotoxin binding was the use of hemitoxins, which lack a binding domain, resulting in selective immunotoxins for antigen-positive target cells, in which the immunotoxin binding is mediated only through the antibody portion. More frequently, however, investigators coupled toxin fragments lacking the binding domain to antibodies. The most commonly employed procedure is the separation of the toxic A chain from the binding domain-bearing B chain of ricin and coupling of the ricin A chain to antibodies. In the case of the bacterial toxins diphtheria toxin and *Pseudomonas* exotoxin, toxin fragments lacking the binding domain were used for immunotoxin generation. These toxic fragments were obtained from naturally occurring mutants or by genetic engineering, deleting the binding domain from the cloned gene.

Many *in vitro* systems have been described showing selective cytotoxic activity of immunotoxins containing toxic A chains or truncated toxins for their target cells (mostly tumor cells). In most systems, the toxic activity for antigen-bearing target cells was several orders of magnitude greater than that for antigen-negative control cells. In the case of the A-chain immunotoxins, however, it was found that their potency was significantly lower than the potency of corresponding immunotoxins made with holotoxins containing both the A and B chains. This is a consequence of the fact that the B chain mediates toxin entry into the cell. Accordingly, a smaller fraction of the A-chain immunotoxins reaches the cytosol, a prerequisite for their cytotoxic activity.

Several attempts to overcome the disadvantage of decreased internalization of A-chain immunotoxins have been reported. One such approach is the use of holotoxins with "blocked" B chains (i.e., B chains where the binding site is not functional). The inactivation of the binding site was achieved by sterical hindrance caused by the proximity of the coupled antibody or by specific chemical inactivation. Another approach was to generate A-chain immunotoxins and B-chain immunotoxins separately, which act synergistically when bound to the same target cell. Another concept to enhance the cytotoxic activity of A-chain immunotoxins was the simultaneous administration of pharmacological agents such as lysosomotropic amines or carboxylic ionophores. Lysosomotropic amines, such as ammonium chloride, methylamine, and chloroquine, and carboxylic ionophores, such as monensin, increase lysosomal pH, thereby inhibiting the degradation of A-chain immunotoxins, which enter secondary lysosomes rather rapidly. This approach enhanced the cytotoxic activity of A-chain immunotoxins *in vitro*. However, *in vivo* application of these enhancer substances proved rather difficult as sufficiently high concentrations cannot be reached because of toxicity (e.g., monensin). A prolonged plasma half-life of monensin could be achieved by coupling it to serum albumin, resulting in better *in vivo* antitumor activity of A-chain immunotoxins.

Animal studies using immunotoxins as anticancer agents have been performed by many investigators. Model animal systems include guinea pigs, mice, and, predominantly, nude mice with human tumors. It was generally found that the immunotoxins were well tolerated and that localization of immunotoxins into the tumor occurred. However, significant amounts of the immunotoxins were taken up by nontarget organs, and an antibody response to both the antibody moiety and the toxin moiety was usually observed. Nevertheless, antitumor activities were clearly observed. A few immunotoxins have been able to cause substantial or complete tumor regression. For example, complete regression of B-cell lymphomas in mice has been achieved with ricin A chain coupled to antibodies recognizing B-cell-specific antigens. Significant regression of human ovarian, colon, and breast cancers growing in immunodeficient mice has also been achieved by treatment with monoclonal antibodies coupled either to *Pseudomonas* exotoxin itself or to *Pseudomonas* exotoxin 40, a mutant form of *Pseudomonas* exotoxin in which the cell-binding domain has been deleted.

Several immunotoxins have been approved for two different kinds of human trials. The first involves the *ex vivo* addition of immunotoxins to harvested bone marrow to eliminate contaminating tumor cells before reinfusion in patients undergoing autologous bone marrow transplantation. Treatment with immunotoxins containing ricin or ricin A chain resulted in selective elimination of leukemia cells in autologous grafts of about 5 logs. The second kind of trial involves the parenteral administration of immunotoxins to patients with hematologic or solid malignancy tumors. Therapeutic effects in patients with solid tumors have thus far yielded minimal clinical responses. Only occasional transient tumor reductions after treatment with ricin A-chain-containing immunotoxins have been reported in patients with melanoma, breast cancer, and colon cancer. Slightly better responses have been noted in patients with hematologic malignancies, for example, in patients with malignant B-cell lymphomas after treatment with an immunotoxin consisting of an anti-CD22 antibody fragment coupled to chemically deglycosylated ricin A chain. The majority of responses, however, were observed in patients with minimal disease. For example, 11 out of 12 patients with malignant B-cell lymphomas remained in complete remission for up to 2 years after autologous bone marrow transplantation and subsequent adjuvant therapy with an anti-CD19 antibody conjugated to ricin with the galactose binding sites blocked. These data suggest that immunotoxins may be most efficacious in the setting of residual or debulked disease.

B. Genetically Engineered Immunoconjugates (Recombinant Toxins)

Recently, several immunoconjugates consisting of two protein moieties have been synthesized by recombinant DNA methodology. The cell-targeting moiety can be a single-chain, antigen-binding protein, an interleukin, or a growth factor. The toxic moiety is a portion of a bacterial or plant toxin, and hence the term *recombinant toxins*. Both *Pseudomonas* exotoxin lacking the native binding domain and diphtheria toxin have been used to generate recombinant toxins in *Escherichia coli*. Ricin-based recombinant toxins are more difficult to engineer since the A chain of the plant toxin must be attached to the cell-targeting moiety by a disulfide bond, which is difficult to produce in bacteria.

Recombinant toxins have been made by combining mutant forms of *Pseudomonas* exotoxin with various interleukins (e.g., IL-2, IL-4, IL-6) and growth factors, including transforming growth factor alpha and acidic fibroblast growth factor. Each was cytotoxic to cell lines containing the appropriate receptors and, for some of these recombinant toxins, antitumor activity in animals bearing human cancers has been demonstrated. Another approach takes advantage of the finding that the variable regions of the light and heavy chains of antibodies can be combined into a single chain form that retains high-affinity binding to antigen. Accordingly, the complementary DNA from several antibodies was used to construct Fv fragments that were in turn fused to truncated forms of *Pseudomonas* exotoxin lacking the cell-binding domain. The efficacy of these single-chain immunotoxins has been demonstrated in several animal studies. For example, a single-chain antibody–*Pseudomonas* exotoxin fusion protein, which specifically binds to the human erbB-2 receptor, efficiently inhibited the growth of erbB-2-overexpressing human ovarian carcinoma cells in athymic nude mice. Complete regression of human carcinomas in mice has been observed after intravenous administration of a single-chain antibody–*Pseudomonas* exotoxin fusion protein with specificity for a carbohydrate antigen expressed on the surface of many carcinomas. Similar observations have been made with recombinant toxins constructed by the combination of diphtheria toxin with various targeting moieties, including IL-2, melanocyte-stimulating hormone, and single-chain antibodies. One of these recombinant toxins, an IL-2-diphtheria toxin fusion protein, proved to be a potent cytotoxic agent against fresh leukemic cells from patients with adult

T-cell leukemia (ATL) and has been evaluated in clinical trials in individuals with various lymphoid malignancies. The results did not fulfill the hopes engendered by the *in vitro* and animal model studies, although a few complete responses (~6%) and some partial responses (~15%) have been observed despite the refractory disease in the patients treated on the clinical protocol. The same recombinant toxin, however, proved to be quite efficacious in patients with psoriasis, a hyperproliferative and inflammatory skin disorder associated with skin infiltration by IL-2 receptor-positive ($CD25^+$) T lymphocytes. Reduction of intradermal T lymphocytes by systemically administered IL-2–diphtheria toxin fusion protein resulted in striking clinical improvement in 40% and moderate improvement in another 40% of all patients.

C. Antibody Conjugates with Cytotoxic Drugs (Chemoimmunoconjugates)

Many studies have been reported in which low-molecular-weight toxins or chemotherapeutic drugs have been coupled to monoclonal antibodies. Among the substances coupled are many of the chemotherapeutic agents currently used in free form for cancer chemotherapy, such as the anthracyclines doxorubicin and daunorubicin, the antimetabolite methotrexate, the alkylating agent melphalan, and many other agents including bleomycin, mitomycin, and vinca alkaloids. Low-molecular-weight mycotoxins such as α-amanitin or trichothecenes, which are not used as therapeutic agents, have also been coupled to antibodies.

Antibody-mediated delivery of toxic drugs is primarily intended to reduce uptake of the drug by non-target tissues, thereby reducing general toxicity and widening the therapeutic window. The lower general toxicity might allow delivery of more of the cytotoxic drug to the tumor site. An alternative concept for the antibody-mediated delivery of drugs is to circumvent drug resistance when resistance is due to decreased drug uptake. Antibody-bound drugs might enter the cell by a different route and may reach different intracellular locations and may therefore circumvent or overcome drug resistance.

In contrast to antibody conjugates with protein toxins, which are enzymes and therefore require only a small number of toxin molecules to be internalized to become effective, the low-molecular-weight drugs are not enzymes and require a sufficiently high concentration in the tumor cell for their cytotoxic activity. Accordingly, it is desirable to deliver as many drug molecules in antibody-bound form to the cell as possible. The most severe limitation in derivatizing antibody molecules with numerous drug molecules is the inactivation of the binding function of the antibody. By and large, the derivatization of an IgG antibody with more than 10 drug molecules usually impairs the antigen-binding function of the antibody to a significant degree. One approach to increase the number of drug molecules delivered to a cell without interfering with the binding function of the antibody is the use of intermediate drug carriers. These are substances like serum albumin or dextran that can be derivatized with multiple drug molecules. The derivatized drug carrier can be coupled to the antibody with comparatively minimal inactivation of its binding function, similar to the coupling of a protein toxin.

Similar to the case with immunotoxins, many *in vitro* studies have shown specific cytotoxic activity of chemoimmunoconjugates for antigen-bearing target cells. Many *in vivo* studies have also been performed, mainly in nude mice with human tumor transplants, with varying degrees of success. In a recent study, however, chemoimmunoconjugates prepared from a chimeric (mouse–human) monoclonal antibody and the anticancer drug doxorubicin have been shown to induce complete regression and cures of xenografted human lung, breast, and colon carcinomas growing subcutaneously in athymic mice. The same conjugates also cured 70% of mice bearing extensive metastases of a human lung carcinoma. The extraordinary efficacy and potency of these conjugates have been attributed to several factors. First, the monoclonal antibody binds with high affinity and specificity to an antigen related to Lewis Y that is abundantly expressed at the surface of human carcinoma lines. Second, bound antibody molecules are rapidly internalized and, in the acidic environment of lysosomes and endosomes, active drug molecules are liberated efficiently from internalized conjugates via cleavage of acid-labile hydrazone bonds. These data demonstrate that the limited efficacy of chemoimmunoconjugates can be increased significantly by an appropriate combination of monoclonal antibody, anticancer drug, and linker chemistry that allows the conjugates to be stable while being delivered and to release drug selectively at the target site.

Only a few phase I clinical trials with chemoimmunoconjugates have been performed to gather information on efficacy and toxicity. No significant antitumor activity has so far been observed in these patients. For example, administration of methotrexate immunoconjugates to patients with metastatic non-small-cell lung cancer yielded only minimal clinical responses. Thus, although *in vitro* and xenograft studies have demonstrated superior tumor killing by antibody drug conjugates compared to free drug, effective clinical application of this strategy has lagged.

D. Antibody Conjugates with Biological Response Modifiers

The term BRM is used to describe a rather diverse group of biomolecules, including synthetic analogues and derivatives that modify, direct, or augment the body's defense systems. This biological response is mediated by interaction with defined receptor molecules for the BRMs. Several well-defined BRMs have been coupled to monoclonal antibodies in an attempt to generate agents with antitumor activity. The common property of these antibody–BRM conjugates is that they exert their cytotoxic activity through activation or stimulation of one or several of the host's defense mechanisms.

Cobra venom factor (CVF) is the complement-activating protein in cobra venom. CVF activates the complement system analogously to the activated form of the third component of complement (C3b). However, the complement enzyme formed with CVF in the process of activation is not subject to the stringent control mechanisms that regulate the enzyme formed with C3b. Accordingly, CVF leads to exhaustive complement activation. By coupling CVF to monoclonal antibodies with specificity for tumor antigens, this complement-activating activity of CVF is targeted to the surface of the tumor cell, where it will induce tumor cell killing by the host's complement system. Antibody conjugates with CVF have been shown to induce selective complement-dependent killing of several tumors, including melanoma, leukemia, and neuroblastoma. Although no extensive animal studies have been reported yet, CVF has two important properties that make it a promising therapeutic agent: first, CVF has no direct cytotoxic activities and, second, because CVF is a structural analogue of mammalian C3b, it offers the theoretical possibility of developing a C3b derivative of significantly reduced or even absent immunogenicity.

IL-2 has been used in humans to activate lymphokine-activated killer cells for the treatment of cancer but it is also known to produce a generalized vascular permeability. The cytokine function of IL-2 coupled to antitumor monoclonal antibodies was destroyed, but the IL-2 immunoconjugates produced a localized increased vascular permeability when injected i.v. into nude mice bearing human tumors. In a similar approach, antibody–CVF conjugates have been used recently to enhance the vascular permeability in tumor tissues by generation of the complement-derived vasoactive anaphylatoxins C3a and C5a. The study demonstrates that the uptake of a ^{99m}Tc–anti-CEA antibody could be increased significantly by pretargeted CVF.

Another BRM that has been coupled to monoclonal antibodies is the chemotactic tripeptide formyl-menthionyl-leucyl-phenylalanine (f-Met-Leu-Phe). This tripeptide is chemotactic for inflammatory cells such as macrophages and polymorphonuclear leukocytes. The f-Met-Leu-Phe tripeptide has been coupled to antitumor antibodies and been shown to increase the number of macrophages in tumor-bearing animals.

Antitumor antibodies coupled with α-interferon have been shown to augment the killing of tumor cells in the presence of human peripheral blood mononuclear cells. This increased cytotoxic activity is believed to be caused by the binding of natural killer cells to their target tumor cells by the antibody conjugates and activation of the natural killer cells by the interferon. [*See* Interferon; Natural Killer and Other Effector Cells.]

Very recently, promising results have been achieved with monoclonal antibody–superantigen fusion proteins. Superantigens are a family of bacterial and viral proteins that bind with high affinity to MHC class II molecules and subsequently interact with T cells expressing particular sequences in the variable region of the T-cell receptor β chain (TCR Vβ). Thereby bacterial superantigens activate a high frequency of T cells to cytokine release and cell-mediated cytotoxicity, which has been shown in an animal model to result in efficient suppression of tumor growth in the absence of overt systemic side effects.

E. Antibody Conjugates with Radionuclides (Radioimmunoconjugates)

Radioactive nuclides have been coupled to antibodies for two purposes: imaging and radiotherapy.

1. Radioimmunoimaging

Radioimmunoimaging or immunoscintigraphy is a new imaging modality for clinical diagnosis. A radiolabeled antitumor antibody binds to antigen-bearing tumor cells and thereby accumulates in the tumor. With the aid of a γ-camera the patient will be scanned, and a tumor nodule can be imaged because of the γ radiation of the radiolabel. This method has been found useful to detect primary, recurrent, metastatic, and occult tumor nodules in cancer patients.

Many studies were performed with iodine isotopes, most frequently ^{131}I, because of the well-established methods to radiolabel antibodies with iodine. However, ^{131}I has several drawbacks as an isotope for imaging purposes. Its high-energy γ radiation is not ideal for γ-camera imaging, its relatively long (8 days) radiation half-life and its β emission cause toxic effects, the oxidizing conditions necessary for its incorporation into the antibody damage the antigen-binding activity, and, most importantly, it exhibits low *in vivo* stability caused by deiodination of the iodine-labeled antibodies. ^{123}I has a shorter (13 hr) radiation half-life and its photon energy allows more efficient γ-camera detection, but it still carries the other limitations of iodine.

For several metals, such as technetium, indium, and gallium, radioactive isotopes exist that offer suitable γ-emission energies and radiation half-lives (^{99m}Tc, 6 hr; ^{111}In, 2.8 days; ^{67}Ga, 3.3 days). However, their use as radiolabels was only possible after methods for their stable binding to antibodies had been developed. These radioactive metals can now be bound to antibodies that have been derivatized with chelating agents. Using this chelation methodology, it has been found that these radioactive metal isotopes can be stably bound to antibodies, and radioimmunoimaging studies have been performed in experimental animals and tumor patients, particularly with ^{111}In-labeled antitumor antibodies. Tumors can be imaged with good specificity and sensitivity by radioimmunoimaging, and this procedure is a widely used experimental imaging method. However, the smallest tumor nodules imaged by radioimmunoimaging are approximately 1 cm^3 in size, and this is the reason why radioimmunoimaging has not replaced standard clinical imaging procedures [e.g., computed axial X-ray tomography (CAT scanning) or magnetic resonance imaging (MRI)]. The major problems are background radioactivity caused by the blood pool, nonspecific uptake by nontarget organs, and insufficient accumulation of the radiolabeled antibody in the tumor caused by limiting extravasation and tumor penetration. However, the intrinsic sensitivity of radioimmunoimaging is not a limited factor, and it may well become a superior imaging tool for cancer patients if the pharmacological limitations can be overcome.

Various strategies have been proposed to overcome the background problem, including computed background subtraction, the use of a second antibody to increase the clearance of the first antibody, and local application of radiolabeled antibodies into body cavities or the lymphatics. A recent approach to overcoming this problem has been the development of bifunctional monoclonal antibodies. Such antibodies, which can function as carriers of radionuclides, have a dual specificity, with one binding site for a tumor-associated antigen and one binding site for a hapten. This dual specificity allows the development of a two-phase targeting procedure for radionuclides consisting of a long-term nontoxic targeting phase with the antitumor bifunctional antibody, which takes advantage of the longer residence time of antibody in the tumor, followed by a short-term binding phase of the toxic radionuclide-derivatized hapten. Increased uptake ratios and faster localization of the radionuclide can be expected, since the radioactivity is attached to the low-molecular-weight structure of the radionuclide-derivatized hapten capable of fast distribution through the body tissues and rapid clearance through the kidney. As documented by clinical studies, this approach provides for better tumor-to-normal tissue ratios leading to improved detection of tumors. The bifunctional antibody approach, however, suffers from the fact that the antibody molecule is composed of two monovalent antibody fragments, whose affinity is orders of magnitude lower than that of a bivalent antibody molecule. As a result, dissociation of bound bifunctional antibody molecules from the target site is likely to occur during the period required for efficient clearance of nonbound bifunctional antibody from circulation before injection of the radionuclide-derivatized hapten.

Based on the avidin (streptavidin)–biotin system, alternative procedures have been developed. Two-step protocols consisting of either avidinylated (or streptavidinylated) antibody and radiolabeled biotin or biotinylated antibody and radiolabeled avidin (or streptavidin) and three-step protocols conceptually derived from the two-step protocols have been reported. These systems provide two advantages over the bifunctional antibody approach: (1) antibody binding is bivalent, thus taking advantage of the complete antibody affinity, and (2) binding of radiolabeled biotin to pretargeted to antibody–avidin conjugates is, for practical purposes, irreversible. Clinical studies have confirmed that the avidin–biotin system offers several advantages over the administration of directly labeled monoclonal antibodies. One serious limitation, however, is the immunogenicity of avidin and streptavidin. It is likely that antiavidin (streptavidin) antibodies will adversely affect a second or third course of administration.

Recently, a two-step system has been reported that is capable of circumventing the problems associated with other two- or three-step procedures. First, a nontoxic bivalent monoclonal antibody conjugated to an enzyme is targeted to the tumor cells, followed by the administration of a radionuclide-derivatized enzyme inhibitor specific for the antibody-conjugated enzyme. The recombinant human enzyme dihydrofolate reductase and its high-affinity competitive inhibitor methotrexate have been selected as a model system for this approach. The promising results of *in vitro* studies suggest that this alternative approach may allow the development of an immunoscinitigraphic technique superior to the current protocols using bifunctional antibodies or the avidin (streptavidin)–biotin system.

2. Radioimmunotherapy

Antitumor antibodies labeled with radioisotopes emitting high-energy β radiation or α radiation offer several advantages over antibodies conjugated with drugs or toxins. Radiolabeled antibodies do not need to be internalized by their target cells to become active. More importantly, they can exert cytotoxic activity for tumor cells with no or reduced surface antigen expression or with poor antibody access because the radiation has a tissue range of 100 to 1000 cells in the case of β emitters and several cells in the case of α emitters. However, as already observed in radioimaging studies, the use of radiolabeled antibody conjugates for radioimmunotherapeutic approaches poses several serious limitations, among which is their nonspecific uptake in normal tissues such as liver, bone marrow, and kidney, which can lead to toxic effects. Therefore, current strategies to improve tumor-to-normal tissue ratios are identical for radioimaging and radioimmunotherapeutic approaches. In contrast to diagnostic studies, however, for which several radionuclides with suitable properties are available, in the case of patient therapeutic approaches only a few radionuclides satisfy the more stringent therapeutic requirements.

The first β emitter used for radiotherapy was ^{131}I. This iodine isotope, however, carries all the disadvantages outlined earlier. In addition, its γ emission, which is the basis for its application in radioimmunoimaging, adds to the nonspecific toxicity, which is dose-limiting because of its effect on the bone marrow. Nevertheless, therapeutic effects with ^{131}I-labeled antibodies have been described in various animal models and human trials. As with radiolabeled antibodies for imaging purposes, better results have been observed with local, intracavitary injections into the pleura, pericardium, and peritoneum.

The β-emitting isotopes of yttrium and rhenium have been identified as more favorable isotopes for radioimmunotherapy. ^{186}Re has a physical half-life of 3.7 days and little γ emission, whereas ^{90}Y has a physical half-life of 2.7 days with no γ emission. The physical half-life of these two β-emitting isotopes is long enough for tumor localization but short enough to minimize nonspecific toxicity. Both radioisotopes decay to stable products, and methods for their stable coupling to antibodies have been developed. Using ^{90}Y-labeled antibodies, many animal studies and human trials have been reported with promising therapeutic results. These studies have shown that higher dose rates and radiation doses may be delivered to tumor tissue with antibodies labeled with ^{90}Y compared with ^{131}I, resulting in improved clinical responses. For example, treatment of refractory Hodgkin's disease with ^{91}Y-labeled antiferritin antibodies yielded complete remissions in 30% of patients, which is a higher rate than previously observed with ^{131}I-labeled antiferritin antibodies.

The α particles have a much shorter path length of only several cell diameters but exhibit an extremely powerful cytotoxic action. This is due to the high linear energy transfer and the limited ability of cells to repair damage done to DNA by α particles. Two α-emitting isotopes of bismuth and astatine have been identified for the labeling of monoclonal antibodies for therapeutic purposes. ^{212}Bi has a physical half-life of 7.2 hr. Its coupling to monoclonal antibodies has been more problematic but was recently achieved by using *N*-succinimidyl-astatobenzoate. ^{212}Bi-labeled antibodies have been shown to exhibit excellent *in vitro* toxicity as well as antitumor effects in animal studies, particularly after intraperitoneal application. Another possibility to induce α-particle emission is the use of boron-labeled antibodies. ^{10}B absorbs thermal neutrons that release an α particle. Monoclonal antibodies have been labeled with ^{10}B. Because tissues not containing ^{10}B have a low neutron capture ability, selective tumor cytotoxicity may be achieved with ^{10}B-labeled antibodies bound to the tumor using subsequent irradiation with thermal neutrons.

F. Antibody Conjugates with Enzymes (Enzymoimmunoconjugates)

Several enzymes with rather different catalytic activities have been coupled to monoclonal antibodies to achieve either cytotoxicity or other consequences of enzyme action. Strictly speaking, the ribosome-inactivating toxins such as ricin are enzymes and would also belong in this category. Other targeted enzymes to induce cytotoxicity include phospholipase C, glucose oxidase, and lactoperoxidase. Phospholipase C cleaves the phospholipids of plasma membranes and thereby induces cell death. Glucose oxidase and lactoperoxidase generate toxic peroxide and oxidizing iodine species. Urokinase and tissue plasminogen activator have been coupled to antibodies to activate plasminogen. Both enzymes generate active plasmin, the major fibrinolytic enzyme. By coupling either of the two plasminogen-activating enzymes to an antibody directed to fibrin, a significant enhancement of fibrinolysis could be achieved to dissolve blood clots.

A novel concept developed by several laboratories involves a two-step procedure in which an antibody–enzyme fusion protein or conjugate is first bound to noninternalizing cell-surface antigens, after which a prodrug of significantly lower cytotoxicity is administered. The antibody-bound enzyme converts the prodrug into an active drug extracellularly, where it can then diffuse into the tumor cells. This antibody-dependent enzyme-mediated prodrug therapy approach (ADEPT) offers several advantages over the coupling of a toxic drug directly to an antibody. First, many anticancer drugs are not potent enough to achieve a sufficiently high concentration at the tumor site by antibody-mediated delivery. In contrast, the prodrug approach allows enzymatical liberation of active drug at the time site, thereby increasing the local concentration. Second, the prodrug is released at the tumor site in free form. Consequently, it does not have to be released from the macromolecular antibody to become active. Third, the free drug liberated at the tumor site by the targeted enzyme will also be able to reach tumor cells that do not express many antigens or that cannot be reached by the macromolecular antibody–drug conjugate.

Several model systems of antibody–enzyme fusion proteins or conjugates and corresponding prodrugs have been reported. These include carboxypeptidase G2, which releases active benzoic acid mustard from its glutamic acid-conjugated prodrug form; alkaline phosphatase, which releases mitomycin or etoposide from its phosphorylated prodrug form; penicillin V amidase, which liberates doxorubicin from its phenoxyacetamide prodrug form; cytosine deaminase, which converts 5-fluorocytosine to 5-fluorouracil; and β-lactamase, which releases doxorubicin from a cephalosporin–doxorubicin prodrug. One serious limitation of these xenogeneic enzyme components is that their immunogenicity may interfere with their repeated clinical use. This limitation has long been recognized and the use of catalytic antibodies has been proposed as a possible solution. Human enzymes also offer an alternative, albeit with potential drawbacks of unwanted activation of prodrug by endogenous enzyme, interference from endogenous substrates or inhibitors, and potential immunogenicity if intracellular enzymes are selected. A human enzyme that has been successfully applied for a fusion protein-mediated prodrug activation (FMPA) approach is human β-glucuronidase. The prodrug is a nontoxic glucuronide-spacer derivative of doxorubicin that decomposes to doxorubicin by enzymatic deglucuronidation.

Although in several preclinical studies the ADEPT concept proved to be a promising approach, only few clinical trials have been initiated with enzyme–antibody conjugates. In one study, patients with colon cancer were treated with a carboxypeptidase G2 antibody conjugate. Approximately 50% of the patients showed a growth delay of 5 months. It remains to be seen how the other enzyme fusion proteins or conjugates compare with currently used therapeutic approaches in clinical studies.

G. Antibody Conjugates with Other Effectors

Several other effector molecules have been coupled to monoclonal antibodies. One example is miconidazole, a radiosensitizer that can enhance the lethality of ionizing radiation during radiotherapy of tumors. Another example of an effector molecule is hematoporphyrin, which is a photochemical that can be activated by visible light. Its cytotoxic effect is mediated through the production of an oxygen radical.

In a very recent approach, the tyrosine kinase inhibitor genistein has been coupled to a monoclonal antibody with specificity for the B-cell-specific receptor CD19. The conjugate selectively inhibited CD19-associated tyrosine kinases and triggered rapid apoptotic cell death. When administered to immunodeficient mice with human B-cell precursor (BCP) leukemia, 99.999% of human BCP leukemia cells were killed, resulting in a 100% long-term event-free survival from an otherwise invariably fatal leukemia.

A different type of effector is polystyrene microspheres containing magnetite, which can be coupled to monoclonal antibodies directed against cell-surface epitopes. The use of a flow system with a permanent magnet allows the removal of the magnetically tagged cells from bone marrow for autologous transplantation.

Liposomes are macromolecular phospholipid structures that have been used to encapsulate drugs such as anticancer agents. Such liposomes can also be targeted to specific tumor cells by covalent coupling to antitumor antibodies. For that purpose, a phospholipid derivative has been incorporated into the liposome structure, which allows the covalent binding of antibodies or their fragments to the surface of the liposome.

II. TARGETS FOR IMMUNOCONJUGATES

A few antibody conjugates have been constructed for noncellular targets such as the aforementioned antifibrin antibodies coupled to a fibrinolytic agent, but the vast majority of immunoconjugates have been made to target an effector molecule to a defined cell population characterized by the expression of a specific surface structure (Table II). As is evident from the preceding discussions, most cell-directed immunoconjugates are directed to tumor cells. Obviously, tumor cells are an interesting target for immunoconjugates, which represent a novel modality of cancer treatment.

Vessels of tumor tissues have also been the target for immunoconjugates. Interleukin 2 or CVF conjugates have been applied to enhance the vascular permeability and, thereby, the access of immunoconjugates to tumors. Immunotoxins directed against tumor vasculature have been used to attack the endothelial cells lining the blood vessels of tumor tissues rather than the tumor cells themselves. This approach provides several advantages. The vascular endothelial cells are directly accessible to circulating therapeutic agents and, since a large number of tumor cells are reliant on each capillary for oxygen and nutrients, even limited damage to the tumor vasculature results in extensive killing of tumor cells. The therapeutic potential of this approach has been demonstrated in a murine model using neuroblastoma cells transfected with the murine interferon γ (INF-γ) gene. As a result of INF-γ secretion in such neuroblastoma tumors, capillary and venular endothelial cells within the tumor mass were activated to express class II antigens of the major histocompatibility complex, which served as target for anti-class II antibody–ricin A-chain conjugates. After injection of these conjugates, dramatic regressions of large solid tumors occurred as a result of widespread thrombosis of the tumor vasculature and massive secondary necrosis of the tumor parenchyma. Although this model system cannot be directly transferred to the clinic, several differences between tumor blood vessels and normal ones have been documented, suggesting that the approach could be applicable in humans.

However, the selective killing of defined cell populations using immunoconjugates is certainly not restricted to tumor cells or tumor tissues. Another group of cells that has been the target for immunoconjugates is lymphocyte subsets. These include T cells, which are responsible for graft rejection and graft-versus-host disease after organ transplantation. For example, antibodies directed against the CD5 antigen present on most peripheral blood T lymphocytes have been coupled to the ricin A chain. This immunotoxin has been shown to produce complete clinical responses in severe graft-versus-host disease after bone marrow transplantation. These rather promising results may point out that immunoconjugates directed against normal cells rather than tumor cells may be better models to study immunoconjugate therapy, because additional difficulties inherent to tumor models such as tumor cell heterogeneity and accessibility are absent.

Another group of immunoconjugates has been constructed to kill selectively B-cell clones producing antibodies against defined antigens. By coupling of the antigen to a cytotoxic agent such as the ricin A chain

TABLE II
Targets for Immunoconjugates

Targets	Examples of applications
Tumor cells	Tumor imaging and therapy
Tumor blood vessels	Increase of vascular permeability for enhanced uptake of immunoconjugates; killing of vascular endothelial cells in tumor tissues
T lymphocytes	Graft rejection, graft-versus-host disease, cancer therapy
B lymphocytes	Autoimmune diseases, cancer therapy
Parasites	Anti-infection therapy
Blood clots	Solubilization of clots (myocardial infarction), diagnosis of deep venous thrombosis, pulmonary emboli

or a cytotoxic drug, selective immunorepressive agents have been generated. Such immunorepressive immunoconjugates have been produced with low-molecular-weight model haptens coupled to an ovalbumin carrier or model protein antigens such as tetanus toxoid. Antigen–toxin conjugates have also been generated with antigens involved in the pathogenesis of autoimmune diseases. For example, thyroglobulin has been conjugated to the ricin A chain. This thyroglobulin–ricin A-chain conjugate specifically suppressed the thyroglobulin autoantibody response of lymphocytes from patients with Hashimoto's thyroiditis. In another model system, the acetylcholine receptor had been coupled to ricin. This antigen–ricin conjugate caused reduced synthesis of antiacetylcholine receptor antibodies involved in the pathogenesis of myasthenia gravis.

A further potential class of target cells is infectious agents. However, surprisingly little work has been done in immunotoxin model systems using infectious agents as targets. Immunotoxins have been constructed with diphtheria toxin against an amoeba, *Acanthamoeba castellani,* and with the A chains of ricin and abrin against *Trypanosoma cruzi,* the causative agents of Chagas' disease.

III. TARGETING MOIETIES USED IN IMMUNOCONJUGATES

The vast majority of immunoconjugates have been constructed with antibodies or antibody fragments directed against surface structures of target cells (Table III). In the case of tumor cells, most of these antibodies were directed against tumor-associated antigens. The function and chemical nature of these antigens were usually not known; as a matter of fact, in many cases the antibodies were used to identify and chemically characterize the recognized surface antigens. In addition to monoclonal antibodies against newly discovered surface antigens of tumor cells, monoclonal antibodies to known surface receptors such as epidermal growth factor, hormones, and transferrin have been used to target cytotoxic agents to tumor cells.

Most antibody conjugates were prepared with intact antibodies of the IgG subclasses. A few antibody conjugates were constructed with IgM, mainly using human monoclonal antibodies, which frequently are of the IgM class. In addition to intact antibodies, immunoconjugates were also constructed with antibody fragments such as $F(ab')_2$ and Fab. An even smaller antibody-derived binding domain used for the construction of immunoconjugates has been the Fv fragment by combining the light- and heavy-chain variable domains with a peptide linker using recombinant DNA technology. These single chain Fv (scFv) molecules have been shown to bind antigen with high affinity and provide several advantages over intact monoclonal antibodies, which are limited in their therapeutic potential by their large size, subunit structure, and immunogenicity.

Ligands to known receptors on target cells have also been used as targeting moieties for immunoconjugates. In the case of hormone receptors, thyrotropin-releasing hormone (TRH) has been coupled to the toxic fragment of diphtheria toxin and shown to cause inhibition of protein synthesis in pituitary cells. Similarly, melanotropin (MSH) has been used as a targeting moiety for daunomycin or diphtheria toxin for the selective elimination of MSH receptor-binding melanoma cells. Other molecules for which receptors exist on target cells have also been used as targeting moieties and include transferrin, α_2-macroglobulin, epidermal growth factor, and interleukins such as IL-2, IL-4, and IL-6. [*See* Interleukin-2 and the IL-2 Receptor.]

Another molecule that has served as the targeting moiety for ricin A chain or *Pseudomonas* exotoxin is a soluble derivative of CD4, a surface protein found on T cells. CD4 has been identified as the receptor for the human immunodeficiency virus (HIV), the virus responsible for AIDS. CD4 binds to the gp120 glycoprotein that is expressed on the surface of the HIV virus. Soluble CD4 derivatives conjugated with a toxin have been shown to kill HIV-infected cells.

Finally, a rather diverse group of targeting moieties has been used in antigen–toxin conjugates to eliminate selected antibody-secreting B-cell clones. As dis-

TABLE III
Targeting Moieties of Immunoconjugates

Targeting moiety	Target site
Antibodies	Various antigenic structures
Antibody fragments	Various antigenic structures
Antigens	B-cell surface immunoglobulins
Hormones	Specific receptors
Growth factors, interleukins	Specific receptors
Transferrin, α_2-macroglobulin	Specific receptors
CD4	gp120 glycoprotein

cussed earlier, thyroglobulin, the acetylcholine receptor, and several other model antigens have been used as targeting moieties in immunosuppressive antigen–toxin conjugates.

IV. SYNTHESIS OF IMMUNOCONJUGATES

Immunoconjugates can be synthesized by two different procedures: chemical methods or recombinant DNA technology.

A. Chemical Synthesis of Immunoconjugates

The majority of immunoconjugates have been synthesized by chemical methods. In proteins, available reactive groups most frequently used for chemical coupling are amino groups (ε-amino groups of lysine residues and amino-terminal amino groups) and free sulfhydryl groups. In the case of glycoproteins, *cis*-diol groups in the carbohydrate portion have also been employed after oxidation to aldehyde functions. If no free sulfhydryl groups are available for coupling, they have been introduced either by the reduction of existing disulfide bonds or by chemical derivatization of amino groups with agents such as *N*-succinimidyl-3-(2-pyridydithio) propionate (SPDP), which generates a free sulfhydryl group after reduction of its pyridyldithio moiety.

Low-molecular-weight drugs have usually been linked to antibodies using condensing reagents such as carbodiimides or after chemical activation of a group in the drug that allowed direct reaction with an antibody molecule. Several problems have been encountered with the synthesis of antibody–drug conjugates. Because the derivatization is random, the binding of the drugs usually impaired the binding function of the antibody, with a greater extent of inactivation occurring with an increasing average number of drug molecules bound per antibody molecule. However, the more drug molecules can be targeted by antibody-mediated delivery, the greater is the cytotoxic potential. One approach to overcome this dilemma is the use of intermediate drug carriers such as dextran and polyamino acids or proteins such as albumin, which can be derivatized with multiple drug molecules and subsequently coupled to the antibody.

Another problem observed with antibody–drug conjugates is the requirement for the drug to be tightly bound to the antibody to prevent drug release before reaching the target site. However, in many instances it has been shown that release of the drug from the antibody is a prerequisite to exert its cytotoxic function in the target cell. Two approaches have been reported aiming at the release of the drug once it has reached the target cell. One such approach is the linkage of the drug to the antibody by an acid-labile linker, which releases the drug in the acidic environment of the lysosome of the target cells, an intracellular compartment that internalized antibody conjugates frequently reach. Another approach is the introduction of a short oligopeptide between the drug and the antibody, which is also cleaved by intralysosomal proteases in the target cell. As demonstrated in a recent study, optimal tumor suppression was achieved with antibody–drug conjugates made with a combination of both a peptide spacer and an acid-labile linker.

The synthesis of immunoconjugates consisting of two proteins, such as an antibody and a protein toxin, by chemical methods requires the use of a cross-linking reagent. With the exception of some early studies using rather nonspecific cross-linking reagents such as glutaraldehyde or homobifunctional reagents containing two identical reactive groups, the synthesis is performed with heterobifunctional cross-linking reagents, consisting of two differently reactive groups. One reactive group reacts with amino groups and invariably consists of an *N*-hydroxysuccinimide ester. The other reactive group usually reacts with a sulfhydryl group and can be either a pyridyldithio group, a maleimide group, or an aliphatic halogen such as iodine. The first step in the synthesis of an immunoconjugate consists of the derivatization of the first protein (e.g., an antibody) with a heterobifunctional reagent such as the above-mentioned SPDP. A second protein containing a free sulfhydryl group (e.g., ricin A chain) can then be coupled to the derivatized antibody, yielding the desired antibody–ricin A-chain conjugate. If the second protein to be coupled does not contain a free sulfhydryl group (e.g., CVF), free sulfhydryl groups can be introduced with SPDP as outlined earlier. Once free sulfhydryl groups have been incorporated, the derivatized protein can be coupled to the derivatized antibody.

The availability of heterobifunctional reagents has significantly contributed to the progress of immunoconjugate research. The major advantage of these reagents is the fact that the conjugates generated by this procedure are always heteroconjugates, consisting of at least one molecule each of the two reactants. Because of the reaction sequence of heterobifunctional cross-linking reagents with proteins, the formation of homoconjugates (polymers consisting of only one of the two protein species) is excluded. Nevertheless, immunoconjugate synthesis with heterobifunctional reagents has many inherent drawbacks. The derivatization of the protein to be coupled with the cross-linking reagent is random, which implies that individual protein molecules in a reaction mixture will be derivatized with different numbers of cross-linking molecules. In addition, a protein of the size of an IgG antibody has approximately 70 amino groups. Although they certainly differ in their individual reactivity, many amino groups will be able to react with the cross-linker. This implies that different antibody molecules derivatized with a small number of cross-linker molecules will be derivatized at different amino groups. This multiple and different derivatization of antibody molecules has important consequences. One consequence is that chemical derivatization of amino groups leads to functional inactivation of the protein similar to derivatization of proteins with drugs, and increasingly with an increased average degree of derivatization. Another consequence is the heterogeneity of the derivatized protein molecules caused by the derivatization at different amino groups. Furthermore, multiple derivatization of the conjugate components yields immunoconjugates of heterogeneous composition. For example, each sulfhydryl-reactive group of introduced cross-linking reagents can react with a free sulfhydryl group of the other reaction partner resulting in compositional heterogeneity. If the second protein contains multiple free sulfhydryl groups, the extent of compositional heterogeneity is even greater because the second reaction partner can also react with multiple molecules of the first reaction partner. Obviously, chemical inactivation and compositional (size) heterogeneity have significant effects on the biological activity of immunoconjugates.

Another disadvantage of the current heterobifunctional cross-linking reagents is that the proteins in the resulting immunoconjugates are rather close to each other so that they become functionally inactive because of steric hindrance by the other reaction partner. Further work will have to optimize the chemical procedures for immunoconjugate synthesis. An ideal process would result in the derivatization of only one defined reactive group in each protein, resulting in an immunoconjugate of predetermined composition (such as 1:1), with the linker being long enough not to interfere with the function of either protein. Several conceptual improvements in that direction have been published. Heterobifunctional cross-linking reagents have been developed that allow the coupling to carbohydrate moieties. In the case of antibodies, the use of these reagents allows coupling of effector molecules away from the antigen-binding site of the Fab arms of the antibody molecule. Although it has been clearly shown that this approach will not interfere with the binding function of the antibody, the term *site-specific* may not be justified. Although the derivatization is confined to the carbohydrate site chains, it is certainly not site-specific. Furthermore, derivatization of the proteins at multiple oligosaccharide moieties occurs. To avoid the steric hindrance of coupled proteins, several attempts to overcome this disadvantage by introducing a longer spacer between the two proteins have been published. In one example, a polypeptide spacer derived from the insulin B chain between the antibody and ricin resulted in an increased potency and efficacy of the immunoconjugate.

B. Genetic Engineering of Immunoconjugates

Recently, several bacterial and plant toxins have been cloned and fused by recombinant DNA techniques with a targeting protein such as growth factors, interleukins, hormones, or single-chain antibodies. In order to direct diphtheria toxin to cells, ligands such as MSH and single-chain antibodies have been placed at the carboxyl end of the toxin to replace the cell-binding domain. In constructing recombinant *Pseudomonas* exotoxin fusion proteins, the specific binding domain I of the toxin molecule (amino acids 1–252) was deleted and its new amino end was fused directly to the COOH-terminal amino acid of a targeting protein. Occasionally, a few additional amino acids have been added as a link between the truncated toxin and the targeting protein. It is important that the targeting moiety is placed at the amino end of *Pseudomonas* exotoxin. Additions at the carboxyl end of domain III containing the ADP-ribosylation activity place the important Arg-Glu-Asp-Leu-Lys sequence within the protein where it cannot function properly. This sequence mediates retention of the toxin molecule in the endoplasmic reticulum and is essential for its cytotoxicity. Recombinant ricin-based molecules have been difficult to produce, probably because the A chain of the plant toxin must be attached to the cell recognition domain by a disulfide bond, which is difficult to generate in bacteria. Recently, a protease cleavage site has been introduced that may allow the production of active ricin-based gene fusion proteins.

A particular advantage of such genetically engineered immunoconjugates is that they consist of the two components in a predetermined 1 : 1 ratio. These recombinant immunoconjugates are therefore truly linked by a site-specific linkage. Another advantage is that recombinant toxins can be mass-produced cheaply in bacteria as homogeneous proteins. For example, chimeric toxins made in *Escherichia coli* accumulate in large amounts within the cell in insoluble aggregates (inclusion bodies) that can contain up to 90% recombinant protein. After cell disruption, inclusion bodies are easily isolated. The protein is then dissolved in a strong denaturant such as 7 *M* guanidine-HCl, renatured by rapid dilution, and subsequently purified to near homogeneity by successive anion and size exclusion chromatography.

V. PROBLEMS AND PROSPECTS

The most important shortcomings of current immunoconjugates include their lack of sufficiently specific accumulation in tumor tissue *in vivo* and their immunogenicity. The major obstacles in achieving sufficient tumor uptake are the blood supply of the tumor, the extravasation of immunoconjugates in the tumor, and their penetration into the tumor tissue. Many factors will affect these parameters, including the vasculature of the tumor, the size of the immunoconjugate, and the affinity of the antibody portion for its antigen. The molecular weight of the immunoconjugate obviously will affect its extravasation at the tumor site and the penetration through the intercellular space. For example, large conventional immunotoxins made by chemical methods penetrate slowly into solid tumors, whereas the much smaller recombinant toxins made from growth factors or single-chain antibodies penetrate tissues more rapidly. These molecules, however, suffer from the fact that they have very short plasma half-lives ranging from 10 to 30 minutes. Their small size permits filtration by the kidney and the exposure of proteolytically sensitive sites may cause rapid degradation by circulating or cell-bound proteases. The binding affinity of the antibody for its surface antigen will also affect the distribution of the immunoconjugate throughout the tumor. High antibody affinity will saturate antigens on tumor cells close to the capillaries, thereby decreasing the antibody penetration into the tumor and causing a more heterogeneous distribution.

To improve the efficacy of immunoconjugates, good accumulation of immunoconjugates at tumor sites must be achieved. Therefore, methods that will cause an easier penetration of immunoconjugates into the tumor will have to be developed. Based on the observation that the vessels in tumors respond poorly to vasoactive agents, vasoconstrictive drugs such as propanolol, pindolol, and oxprenolol have been administered that act mainly on nontumor vessels. As a consequence of vasoconstriction in nontumor tissues, the tumor blood flow increased and resulted in a greater antitumor effect of conjugates than in the absence of such agents. In a second approach, tumor necrosis factor (TNF) was used for its ability to damage tumor vessels and increase vascular permeability. By using TNF simultaneously with or shortly after injection of the immunoconjugates, a severalfold increase in the tumor uptake of conjugates was observed. Another successful approach to enhance the tumor uptake of macromolecules has been the pretreatment with vasoactive immunoconjugates. IL-2 and CVF coupled to appropriate monoclonal antibodies proved to be useful for increasing access of immunoconjugates to tumor tissues.

The second major limitation of immunoconjugates is their immunogenicity. To date, the vast majority of immunoconjugates have been made with monoclonal antibodies of murine origin. As a protein from a different species, murine immunoglobulin elicits an immune response in humans. The generation of the antimurine antibody response restricts the use of monoclonal antibodies or immunoconjugates to a single course of therapy. Several attempts to reduce this immune response are being pursued. One obvious solution would be the use of human monoclonal antibodies. They have been generated, but for unknown reasons they are mainly of the IgM class and usually directed against intracellular antigens. A more promising approach was to "humanize" murine monoclonal antibodies. This has been done by recombinant DNA technology by fusing the variable domains of a murine monoclonal antibody to the constant domains of a human antibody. Such chimeric antibodies have indeed been shown to elicit much less of an immune response and to exhibit longer plasma half-lives. In furthering this approach, humanized antibodies have been created by inserting only the hypervariable regions of the mouse antibody with the desired binding specificity. In another approach, single-chain antibody molecules composed of two variable regions of the light and heavy chains have been fused to toxins.

It will be more difficult, however, to find a way to overcome the immunogenicity of the effector moieties of immunoconjugates. Protein effectors, such as the ricin A chain, have invariably been found to elicit an immune response that severely affects the biodistribution and efficacy of ricin A-chain-containing immunoconjugates. Furthermore, immunosuppressive agents such as cyclosporin A were unable to prevent even a single patient from developing significant titers of antiricin antibodies. For the most part, it appears unlikely to develop toxin analogues of lower or absent immunogenicity because there are no mammalian analogues to the ribosome-inactivating toxins. Possible exceptions include the use of human enzymes such as human β-glucuronidase for the prodrug activation approach and the biological response modifier CVF. Since CVF is a structura! analogue of mammalian C3, it may be possible to generate a human C3 derivative with CVF-like functions but of significantly reduced or even absent immunogenicity.

BIBLIOGRAPHY

Bosslet, K., Czech, J., and Hoffmann, D. (1994). Tumor-selective prodrug activation by fusion protein-mediated catalysis. *Cancer Res.* **54**, 2151–2159.

Burrows, F. J., and Thorpe, P. E. (1994). Vascular targeting: A new approach to the therapy of solid tumors. *Pharmac. Ther.* **64**, 155–174.

Dohlstein, M., Abrahmsén, L., Björk, P., Lando, P. A., Hedlund, G., Forsberg, G., Brodin, T., Gascoigne, N. R. J., Förberg, C., Lind, P., and Kalland, T. (1994). Monoclonal antibody-superantigen fusion proteins: Tumor-specific agents for T-cell-based tumor therapy. *Proc. Natl. Acad. Sci. USA* **91**, 8945–8949.

Gottlieb, S. L., Gilleaudeau, P., Johnson, R., Estes, L., Woodworth, T. G., Gottlieb, A. B., and Krueger, J. G. (1995). Response of psoriasis to a lymphocyte-selective toxin (DAB_{389}-IL-2) suggests a primary immune, but not keratinocyte, pathogenic basis. *Nature Med.* **1**, 442–447.

Grossbard, M. L., and Nadler, L. M. (1992). Immunotoxin therapy of malignancy. *In* "Important Advances in Oncology" (V. T. DeVita, S. Hellman, and S. A. Rosenberg, eds.), pp. 111–135. J.B. Lippincott Co., Philadelphia.

Hawkins, G. A., McCabe, R. P., Kim, C-H., Subramanian, R., Bredehorst, R., McCullers, G. A., Vogel, C-W., Hanna, M. G., Jr., and Pomato, N. (1993). Delivery of radionuclides to pretargeted monoclonal antibodies using dihydrofolate reductase and methotrexate in an affinity system. *Cancer Res.* **53**, 2368–2373.

Kuzel, T. M., and Rosen, S. T. (1994). Antibodies in the treatment of human cancer. *Curr. Opin. Oncol.* **6**, 622–626.

LeBerthon, B., Khawli, L. A., Alauddin, M., Miller, G. K., Charak, B. S., Mazumder, A., and Epstein, A. L. (1991). Enhanced tumor uptake of macromolecules induced by a novel vasoactive interleukin 2 immunoconjugate. *Cancer Res.* **51**, 2694–2698.

LoBuglio, A. F., and Saleh, M. N. (1992). Advances in monoclonal antibody therapy of cancer. *Am. J. Med. Sci.* **304**, 214–224.

Pastan, I., and FitzGerald, D. (1991). Recombinant toxins for cancer treatment. *Science* **254**, 1173–1177.

Pieterstz, G. A., and McKenzie, I. F. C. (1992). Antibody conjugates for the treatment of cancer. *Immunol. Rev.* **192**, 57–80.

Rodrigues, M. L., Presta, L. G., Kotts, C. E., Wirth, C., Mordenti, J., Osaka, G., Wong, W. L. T., Nuijens, A., Blackburn, B., and Carter, P. (1995). Development of a humanized disulfide-stabilized anti-p185^{HER2} Fv–β-lactamase fusion protein for activation of a cephalosporin doxorubicin prodrug. *Cancer Res.* **55**, 63–70.

Senter, P. D., Saulnier, M. G., Schreiber, G. J., Hirschberg, D. L., Brown, J. P., Hellstrom, I., and Hellstrom, K. E. (1988). Antitumor effects of antibody–alkaline phosphatase conjugates in combination with etopside phosphate. *Proc. Natl. Acad. Sci. USA* **85**, 4842–4846.

Stickney, D. R., Anderson, L. D., Slater, J. B., Ahlem, C. N., Kirk, G. A., Schweighardt, S. A., and Frincke, J. M. (1991). Bifunctional antibody: A binary radiopharmaceutical delivery system for imaging colorectal carcinoma. *Cancer Res.* **51**, 6650–6655.

Trail, P. A., Willner, D., Lasch, S. J., Henderson, A. J., Hofstead, S., Casazza, A. M., Firestone, R. A., Hellström, I., and Helltröm, K. E. (1993). Cure of xenografted human carcinomas by BR96–doxorubicin immunoconjugates. *Science* **261**, 212–215.

Uckun, F. M., Evans, W. E., Forsyth, C. J., Waddick, K. G., Ahlgren, L. T., Chelstrom, L. M., Burkhardt, A., Bolen, J., and Myers, D. E. (1995). Biotherapy of B-cell precursor leukemia by targeting genistein to CD19-associated tyrosine kinases. *Science* **267**, 886–891.

Vitetta, E. S. (1990). Immunotoxins: New therapeutic reagents for autoimmunity, cancer, and AIDS. *J. Clin. Immunol.* **10**, 15–18.

Vogel, C.-W. (ed.) (1987). "Immunoconjugates. Antibody Conjugates in Radioimaging and Therapy of Cancer." Oxford Univ. Press, New York.

Vogel, C.-W., and Müller-Eberhard, H. J. (1981). Induction of immune cytolysis: Tumor-cell killing by complement is initiated by covalent complex of monoclonal antibody and stable C3/C5 convertase. *Proc. Natl. Acad. Sci. USA* **78**, 7707–7711.

Waldman, T. A. (1991). Monoclonal antibodies in diagnosis and therapy. *Science* **252**, 1657–1662.

Immunodeficiencies, Genetic

RAMSAY FULEIHAN
Yale University School of Medicine

GLOSSARY

Antibody Serum proteins, immunoglobulins, that recognize (bind) a specific antigen

Antigen Molecules that are recognized by antibody molecules

Autosomal inheritance Inheritance of genes on all chromosomes except the X and Y chromosomes

Cytokines Proteins made by cells; they bind to specific receptors on other cells and affect their behavior

Lymphoproliferative disease Growth of lymphocytes in an uncontrollable manner but which are not malignant

Opportunistic organism Microorganism that cause disease only in immunodeficient patients such as in AIDS

Signal transduction Process by which events on the cell surface are translated into changes in the nucleus of a cell; the process usually involves activation of a series of molecules inside the cell, one after the other, until activated molecules move into the nucleus and either turn on or turn off expression from one or more genes

Thymus An organ located in the chest, just behind the breastbone, where T lymphocytes develop from precursor cells that leave the bone marrow and enter the thymus

INHERITED IMMUNE DEFICIENCY DISEASES RESULT from genetic defects that affect the development or function of one or more components of the immune system, leading to an enhanced susceptibility to infection and, in some cases, to cancer. The main function of the immune system is protection against invading microorganisms (bacteria, viruses, fungi, and parasites) that cause disease (pathogenic microorganisms or pathogens). In addition, it must destroy any of the body's own cells that develop into cancer. The genetic defects that alter any of the just-described components of the immune system result in an enhanced susceptibility to infection. These immunodeficiency diseases demonstrate the importance of each of the many components of the immune system. This article describes some of the known genetic immunodeficiency diseases, each of which illustrates the role of different molecules and cell populations in immunity.

I. IMMUNITY TO INVADING PATHOGENS

The first line of defense against pathogens involves mechanical barriers such as the skin and mucous membranes of the respiratory and gastrointestinal tracts. Once a pathogen has broken through these barriers, a variety of proteins and phagocytic cells attack the invading pathogen and destroy it. Phagocytic cells are white blood cells (WBC) that can ingest pathogens and play an important role in the early response to an infection. Phagocytes must be able to reach the site of infection and must be able to ingest and kill the pathogen. [*See* Phagocytes; Neutrophils.]

Some pathogens have a cell wall structure that prevents their ingestion by phagocytes. Such pathogens are cleared by antibodies, which are immunoglobulin molecules that specifically recognize molecules on the surface of the pathogen (antigen). Antibody molecules recognize the three-dimensional structure of antigen and have at least three basic mechanisms of action: (1) they can directly neutralize toxins, (2) they can kill pathogens by binding to and activating proteins of the complement system, and (3) they can facilitate the uptake of pathogens by phagocytic cells through surface receptors for immunoglobulin. Immunoglobulin molecules have two identical variable regions, antigen-binding sites and a constant region that determines the isotype and function of the immunoglobulin molecule. There are five major isotypes of immunoglobulin molecules (IgM, IgD, IgG, IgA, and IgE), each with characteristic functions. Mature B lymphocytes express IgM and IgD on their surface. The main function of antibodies is to clear the extracellular space of pathogens. IgM is secreted as a pentamer of five molecules attached to each other. Because of their large size, IgM molecules remain within the blood vessels. IgG antibodies are very efficient, are found in blood vessels and outside blood vessels, and cross the placenta to protect the newborn. IgA antibodies are secreted at mucous membranes and in breast milk; they help prevent pathogens from entering through the mucous membranes. IgE antibodies may play a role in fighting parasitic diseases, but also cause allergic diseases. When B lymphocytes are first activated by an antigen, they synthesize and secrete IgM antibodies. Immunoglobulin isotype switching, from IgM/IgD to IgG, IgA, or IgE, is the mechanism by which a B lymphocyte can synthesize an antibody with the same antigen specificity (same variable region) but with different function (different constant region, isotype). Isotype switching in human B lymphocytes requires a contact-dependent signal from T lymphocytes. [*See* Antibody Diversity (Clonal Selection); Complement System.]

Some pathogens, such as viruses, invade cells directly and take over the cell's machinery. These pathogens remain inside the cell and out of the reach of antibody molecules. T lymphocytes play a key role in clearing such intracellular pathogens by recognizing infected cells and killing them. T lymphocytes recognize antigens by antigen receptors on their surface. T lymphocyte antigen receptors have a variable region similar to that of immunoglobulin molecules, but are different in that they do not recognize the whole antigen in its three-dimensional structure but recognize only a fragment in association with (presented by) antigen-presenting molecules on the surface of infected cells. Antigen-presenting molecules are called major histocompatibility complex (MHC) molecules because they determine whether an individual tolerates a tissue graft from another individual. There are two classes of MHC molecules: MHC class I and class II molecules. MHC class I molecules are present on all cells, present fragments from proteins that are produced by the cell, and present antigen to the CD8 subpopulation of T lymphocytes. When a virus infects a cell, it takes over the cell's machinery and produces viral proteins within the cells. Fragments from these proteins bind to MHC class I molecules, while they are being formed, and travel with the MHC class I molecule to the surface of the cell, where they interact with a CD8 T lymphocyte that specifically recognizes the fragment in the MHC class I molecule. New proteins synthesized by cancer cells are also presented by MHC class I molecules. Therefore, the main function of CD8 T lymphocytes is to destroy virus-infected or cancer cells. MHC class II molecules are found on the surface of professional antigen-presenting cells, B lymphocytes, monocytes, and dendritic cells; can be induced on the surface of other cells; present fragments of proteins from pathogens that are taken up and degraded by antigen-presenting cells; and present antigen to the CD4 subpopulation of T lymphocytes. The main functions of CD4 T lymphocytes, therefore, are to help B lymphocytes make antibodies against the invading pathogen and to help monocytes kill ingested pathogens. [*See* CD8 and CD4: Structure, Function, and Molecular Biology; Major Histocompatibility Complex (MHC); T-Cell Receptors.]

II. DEFECTS OF THE PHAGOCYTIC SYSTEM: DEFECTS OF NEUTROPHIL FUNCTION

The phagocytic system plays a very important role in killing pathogens early in an infection. Phagocytes include neutrophils, eosinophils, and basophils. Neutrophils are the most abundant of the phagocytic cells. Their main function is to phagocytose (ingest) and kill pathogens. Defects in the number or function of neutrophils result in a susceptibility to recurrent bacterial infections. Patients suffer from soft tissue inflammation and abscesses. They have an enhanced susceptibility to infection by some bacteria (*Staphylococcus aureas, Pseudomonas,* and *Serratia*) and by some fungi (*Aspergillus* and *Candida*). Defects in neutrophil numbers are mostly secondary; however, congenital neutropenia does occur, leaving affected individuals susceptible to recurrent infections. Defects in neutrophil function result from a variety of genetic defects:

A. Leukocyte Adhesion Defect

Leukocyte adhesion deficiency results from defects in molecules found on the surface of neutrophils (integrins), which allow neutrophils to bind (adhere) to the cells lining the blood vessels (vascular endothelium). Adherence to vascular endothelium is the first step in the process of neutrophil migration to the site of infection. As a result of this deficiency, neutrophils remain in the blood vessels where they are found in higher numbers than normal. Leukocyte adhesion-deficient neutrophils cannot reach the sites of infection, resulting in the absence of pus. Complete absence of the adhesion molecules results in a susceptibility to severe infections, whereas the presence of reduced adhesion molecules is associated with less severe infections.

B. Chronic Granulomatous Disease

Chronic granulomatous disease (CGD) results from an inability to kill ingested bacteria. Killing of bacteria by neutrophils involves a burst of respiratory activity to generate superoxide. Defects in any of the enzymes involved in the respiratory burst can lead to a defect in intracellular killing by neutrophils. Four enzyme defects have been described to date and are inherited as autosomal recessive disorders (p22-phox, p47-phox, and p67-phox), except for the most common gene defect in CGD (gp91-phox), which is located on the X chromosome. The diagnosis of CGD is made by a test known as the nitroblue tetrazolium dye (NBT) test. The superoxide, generated by the respiratory burst in normal cells, reduces the soluble yellow NBT dye to a deep blue insoluble pigment which, using a microscope, can be visualized in the cell. Female carriers of the X-linked form of this disease will demonstrate NBT dye reduction in only half of their neutrophils, the half carrying the normal gene. Although no specific therapy exists, patients have been found to benefit from the regular administration of interferon-γ (IFN-γ), as well as from the administration of prophylactic antibiotics.

C. Other Phagocytic Deficiencies

Other deficiencies result from defects in other enzymes such as severe deficiency of (G6PD) glucose-6-phosphate dehydrogenase or deficiency in myeloperoxidase.

III. X-LINKED AGAMMAGLOBULINEMIA: THE PROTOTYPE ANTIBODY DEFICIENCY

X-linked agammaglobulinemia (XLA) was the first immunodeficiency to be described. In 1952, Ogden C. Bruton described a young boy with recurrent bacterial infections who had no immunoglobulins in his serum. He also demonstrated that administration of immunoglobulins derived from normal individuals resulted in a reduction of the frequency of the patient's bacterial infections. Plasma cells, which secrete immunoglobulins, were found to be lacking in patients with XLA, and it was later shown that the cellular defect in XLA was the absence of mature B lymphocytes. Affected males develop recurrent bacterial and viral infections early in infancy or childhood; usually after 4–6 months of age, when the protective maternal antibodies that crossed the placenta during the third trimester of pregnancy have decreased. The main finding on physical examination is the absence of tonsils and lymph nodes. Levels of all immunoglobulin types in the serum are markedly decreased if not absent, and, despite repeated immunization, the patients cannot synthesize specific antibodies. Few if any mature B lymphocytes are found in the blood. Although *in vitro* studies of T lymphocyte number and function are normal, patients with XLA are prone to develop chronic infections with enteroviruses. Indeed, these patients can contract vaccine-associated paralytic poliomyelitis, for which reason attenuated live viral vaccines should be avoided in patients with XLA. The use of intravenous immunoglobulin (IVIG) replacement therapy every 3 weeks has allowed patients to lead practically normal lives with a marked reduction in the number and severity of infections. However, IVIG treatment does not always protect against debilitating and often fatal chronic enteroviral infections. IVIG is isolated from a large pool of screened "healthy donors" and prepared such that there are few if any side effects and no viral contamination from known viruses such as human immunodeficiency virus, hepatitis B virus, and, more recently, hepatitis C virus.

Immunoglobulins are synthesized by B lymphocytes, which develop in the bone marrow from stem cells that also give rise to other cells found in the blood. B lymphocyte precursors undergo several stages of development before they become mature B lymphocytes, characterized by the presence of surface immunoglobulin molecules capable of recognizing foreign pathogens. The defect in XLA appears to affect the development of B lymphocytes, resulting in fewer and fewer B lymphocytes present at each successive stage of development. The gene responsible for XLA was identified in 1993 by two independent groups. Its protein product was found to be a tyrosine kinase, which has been named Btk (for Bruton's tyrosine kinase). Tyrosine kinases activate or inhibit the function of proteins by adding phosphate groups to tyrosine amino acids, thereby participating in the transfer of a signal from the cell surface to the nucleus (signal transduction). Although the exact function of Btk remains under investigation, it appears to have an important role in the development of B lymphocytes. Because of the absence of mature B lymphocytes, patients with XLA cannot synthesize immunoglobulin and thus lack the protective role of antibodies. The absence of antibodies in XLA results in a susceptibility to bacterial and viral infections, some of which can be debilitating or fatal and demonstrate the importance of antibodies in immunity.

XLA is also a prototype X chromosome-linked recessive disease. It predominantly affects males because they have only one X chromosome, whereas females are unlikely to have two affected X chromosomes. Females normally inactivate one of the X chromosomes in a random fashion. A particular cell, therefore, will express the genes found on the active X chromosome. B lymphocyte precursors in female carriers will express either X chromosome in a random manner (approximately 50:50). However, only the B lymphocyte precursors that have the normal Btk gene will develop into mature B lymphocytes; therefore, all mature B lymphocytes in female carriers will exhibit the X chromosome bearing the normal gene. B lymphocyte precursors and all other WBC such as T lymphocytes and granulocytes, which are not affected by the defective gene, will express either X chromosome in a random manner. The exclusive use of the normal X chromosome in B lymphocytes in carriers demonstrated that the defect in XLA was intrinsic to B lymphocytes.

IV. DEFICIENCY OF COMPLEMENT PROTEINS

The complement system involves a large number of serum proteins (at least 20) that participate in the killing of pathogens. Nine of these proteins are numbered C1–C9. Activation of the complement system results in the generation of a complex formed by the proteins C5–C9 which can destroy bacteria. Two mechanisms activate the complement system: the classical pathway via antibodies which utilizes the early components of the C1–C9 proteins, C2–C4, and an alternative pathway which activates complement proteins directly by pathogens. Therefore, complement proteins can directly kill some bacteria or can help antibodies to kill bacteria. Deficiency of any of the early components leads to a decreased efficiency of antibody function and, therefore, to an enhanced susceptibility to infections. The early complement proteins appear to play a role in the clearing of antibody/antigen complexes. Therefore, patients deficient in one of the early complement proteins are also susceptible to diseases generated by immune complexes. Deficiency of any of the late components, C5–C8 proteins, results in an enhanced susceptibility to *Neisseria meningitidis* infections. Deficiency of C9 does not result in an immunodeficiency because the C5–C8 complex has some antibacterial activity.

V. X-LINKED HYPER-IgM: A DEFECT IN COMMUNICATION BETWEEN CELLS

X-linked immunodeficiency with normal or elevated IgM (X-linked hyper IgM, X-HIM) is a rare genetic disorder characterized by markedly decreased or absent serum levels of IgG, IgA, and IgE immunoglobulins with normal or elevated levels of IgM. The syndrome was first described in 1961 and since then over 100 cases have been reported worldwide. Affected males are susceptible to recurrent bacterial infections, opportunistic infections with *Pneumocystis carinii* and with *Cryptosporidium,* autoimmune diseases, and lymphoproliferative disease. X-HIM can be differentiated from XLA by the presence of normal numbers of mature B lymphocytes and by the ability to make specific antibodies in response to immunization. However, these antibodies are restricted to the IgM isotype, with a failure to switch to the production of other isotypes (IgG, IgA, and IgE). In addition, there is a lack of a memory response, which is characterized by a rapid production of larger quantities of antibodies upon reexposure to a pathogen or vaccine. The lymph nodes, while present, have an abnormal architecture. Administration of intravenous immunoglobulin every 3 weeks allows patients to lead practically normal lives, although they remain susceptible to lymphoproliferative disease as well as to cancer.

The underlying defect in X-HIM appears to be an inability to undergo immunoglobulin isotype switching from IgM/IgD secretion to the production of other immunoglobulin (Ig) isotypes, IgG, IgA, or IgE. In 1993, five independent groups identified the genetic defects in X-HIM to be in the gene encoding for a protein that is expressed on the surface of activated T lymphocytes and provides the signal B lymphocytes require for isotype switching by binding to CD40 on the surface of B lymphocytes. This molecule was therefore named CD40 ligand. Numerous mutations that disrupt the ability of CD40 ligand to bind to CD40 have been found in X-HIM. There are a few patients who had no defect in CD40 ligand but were found to have a defect in the B lymphocyte response to CD40 ligand. This is not surprising because the basic defect is the lack of interaction between CD40 and its ligand. Indeed, disruption of either the CD40 or the CD40 ligand gene in mice results in a similar immune deficiency, which resembles the human disease, X-HIM.

The defect in X-HIM is not limited to the lack of T lymphocyte help to B lymphocytes because CD40 has been found on the surface of many other cell types. CD40 ligand on T lymphocytes also provides help to monocytes via CD40 to kill intracellular pathogens, which may be important in clearing infections by opportunistic organisms. Furthermore, interaction of CD40 and its ligand has been found to play many important roles in the immune response to infections and in the development of immune-mediated diseases.

As in XLA, X-HIM is an X-linked disease that predominantly affects males. Unlike XLA, the defect is not lethal to T lymphocytes. Carriers of X-HIM, therefore, have random inactivation of the X-chromosome in their T lymphocytes. However, carrier detection and prenatal diagnosis can be performed by direct gene analysis or by classical DNA techniques.

VI. SEVERE COMBINED IMMUNODEFICIENCY DISEASE: A DEFECT IN T AND B LYMPHOCYTES

Severe combined immunodeficiency disease (SCID) results from the absence or lack of function of T and B lymphocytes. There are a number of known genetic defects that result in SCID as well as some still undetermined genetic defects. Because there are autosomal as well as X-linked genes involved, both males and females are affected. All of the described genetic defects are recessive and therefore there is a much higher incidence of autosomally inherited cases of SCID in the children of related parents. Patients with SCID lack antibody-mediated immunity as well as cellular immunity provided by T lymphocytes. In cases in which B lymphocytes are present, the B lymphocytes may be dysfunctional or they may lack T lymphocyte help as in X-HIM. The result is a severe immunodeficiency that presents early in life with recurrent bacterial, viral, fungal, and opportunistic infections, chronic diarrhea, and failure to thrive. Patients are prone to develop disease by attenuated live viral vaccines and therefore should not receive any of these vaccines. Patients are also susceptible to a fatal reaction from a blood transfusion known as graft vs host (GVH) disease. This results from the presence of lymphocytes in the blood which recognize the patient's tissues as foreign tissue and initiate a rejection reaction that is usually fatal if not treated. The reaction starts as a rash but can extend to involve the liver, lung, and/or intestinal tract. Normally, the immune system of the recipient of a blood transfusion will destroy any lymphocyte or other WBC in the donor blood. However, in SCID, the patient has no functioning immune system to destroy the foreign lymphocytes and is thus a target for GVH disease. It is therefore very important to irradiate all blood products given to SCID patients, which inhibits the capacity of lymphocytes to mount a GVH reaction.

There are a number of known genetic defects that result in SCID and some that have not been determined yet. Some of the known defects include the following.

A. Adenosine Deaminase and Purine Nucleotide Phosphorylase Deficiency

Deficiency in enzymes that degrade nucleic acid metabolites results in the buildup of nucleic acid metabolites that are toxic to T and B lymphocytes but not to other cells in the body. Patients with these enzyme deficiencies have a virtual absence of lymphocytes with very few if any mature T and B lymphocytes and therefore suffer the effects of lack of antibodies as well as the lack of T lymphocyte function. The defective genes are not located on the X chromosome and therefore both males and females are affected equally.

B. X-Linked SCID

X-linked SCID results from defects in the common γ chain, a protein component of multiple cytokine receptors. Some of these cytokine receptors appear to play an important role in the development of T lymphocytes, whereas others are important in both T and B lymphocyte function. Because the gene for the common γ chain is located on the X chromosome, the disease is primarily restricted to males. These patients have no mature T lymphocytes but have normal numbers of B lymphocytes. However, B lymphocyte function is defective due both to the lack of T lymphocyte help and to the importance of the common γ chain for normal B lymphocyte function. Therefore, despite the presence of B lymphocytes, both T lymphocyte and antibody immunity are deficient. [*See* IL-2 and IL-2 Receptor.]

C. Jak 3 Deficiency

Jak 3 deficiency is virtually identical to X-linked SCID but affects males and females equally. It is caused by a defect in Jak 3, a protein that is associated with the common γ chain and is required by the common γ chain for signal transduction. The Jak 3 gene is not located on the X chromosome.

D. Zap 70 Deficiency

Zap 70 is a tyrosine kinase that is important for CD8 T lymphocyte development and for CD4 T lymphocyte function. Defects in Zap 70 result in an autosomal recessive form of SCID in which there is an absence of mature CD8 T lymphocytes, normal numbers of CD4 T lymphocytes that do not function, and normal numbers of B lymphocytes that probably function normally but lack T lymphocyte help. These patients, therefore, have SCID from the lack of T and B lymphocyte function.

E. Interleukin-2 (IL-2) Deficiency

IL-2 is a cytokine produced by T lymphocytes that is important for T lymphocyte function. It allows T lymphocytes to divide and to multiply. Deficiency of IL-2 results in SCID with normal numbers of T and B lymphocytes but with no T lymphocyte function. Addition of IL-2 to *in vitro* tests can correct many of the test results, and administration of IL-2 to patients improves T lymphocyte function and reduces the frequency and severity of infections. [*See* Interleukin-2 and the IL-2 Receptor.]

F. Others

There are numerous other cases of autosomally recessive SCID affecting T lymphocytes with or without affecting B lymphocytes in which the genetic defect is known or has not been identified yet. There are many proteins which if disrupted can affect the development or function of T or B lymphocytes and could potentially be involved in SCID.

SCID is a fatal disease unless a bone marrow transplant is performed. If successful, bone marrow transplantation is usually curative. In many cases, no donor B lymphocytes survive, leaving only the B lymphocytes originating from the host. As a result, some patients, especially X-linked SCID patients, will continue to have a B lymphocyte defect and will require intravenous immunoglobulin therapy. SCID due to adenosine deaminase deficiency can be treated by the administration of adenosine deaminase, which improves the number and function of T and B lymphocytes. [*See* Bone Marrow Transplantation.]

With an improved understanding in the genetic basis of immunodeficiency diseases and with the recent progress in molecular biology, it has become possible to consider using gene therapy to treat genetic immunodeficiency disease. Indeed, adenosine deaminase-deficient SCID was the first disease to undergo clinical trials in gene therapy. The advantage of this disease is that lymphocytes are critically dependent on the presence of adenosine deaminase. Even if gene therapy is not very efficient, the few cells that have been corrected will have a strong survival advantage. Also, strict control of adenosine deaminase production is not necessary because it is not toxic and a wide range of its production has no effect on the individual. Initial results from the early gene therapy attempts have been published, but are difficult to interpret because the patients continued to receive adenosine deaminase therapy. [*See* Gene Therapy.]

VII. BARE LYMPHOCYTE SYNDROME: A DEFECT IN ANTIGEN-PRESENTING MOLECULES

The bare lymphocyte syndrome is a form of SCID which results from the absence of MHC molecules on the surface of cells. T lymphocyte development in the thymus is affected and results in a few T lymphocytes in the periphery. During T lymphocytes development, the thymic tissues select for the T cells which will interact with the MHC molecules present in the thymus and select against the T lymphocytes which will react to self-antigens. This process occurs by interaction between the T lymphocyte antigen receptors and MHC molecules in the thymus. MHC class II molecules allow CD4 T lymphocytes to develop and MHC class I molecules allow CD8 T lymphocytes to develop. In the bare lymphocyte syndrome, all MHC class II molecules are absent because of a defect in one of several proteins that are necessary for expression of all MHC class II molecules. The patients have decreased numbers of CD4 T lymphocytes that do not function well because antigen-presenting cells do not have MHC molecules to present antigen to the CD4 T lymphocytes. Some patients also do not have any MHC class I proteins present and will have decreased numbers of CD8 T lymphocytes as well. In experimental mice, lack of MHC class II or class I molecules results in a virtual absence of CD4 or CD8 T lymphocytes, respectively. However, in humans, there are substantial numbers of CD4 or CD8 T lymphocytes present either because the deficiency in MHC molecules is not absolute or because the development of CD4 and CD8 is not totally dependent on MHC molecules. In any case, the result is a severe immune deficiency with failure to thrive, chronic diarrhea, and recurrent infections. Unlike SCID, the bare lymphocyte syndrome cannot be corrected by bone marrow transplantation because there is no defect in the T lymphocytes themselves but in the presence of MHC molecules in the thymus as well as on the surface of antigen-presenting cells. The prognosis for patients is therefore very grim.

VIII. DiGEORGE ANOMALY: A DEFECT SECONDARY TO THE ABSENCE OF THE THYMUS

The DiGeorge anomaly is a genetic disorder that results in a developmental anomaly affecting several tissues, including the thymus, parathyroid glands, heart, and some structures of the face. Patients usually have a congenital defect of the heart, low calcium levels in the blood secondary to low or absent parathyroid hormone, and low numbers of T lymphocytes because of a small or absent thymus. Congenital heart disease is the major life-threatening defect in this disease. Low calcium results in tetany (cramp-like involuntary spasms). The immunodeficiency varies from mild to severe and is related to the number and function of mature T lymphocytes. If patients survive the neonatal period, they become susceptible to recurrent infections, including fungal and opportunistic infections, diarrhea, and failure to thrive. Patients, especially with milder immunodeficiency, tend to develop improved immune function with age. The anomaly is associated with defects in chromosome 22, although no single gene defect has been identified yet. [*See* Thymus.]

IX. CONCLUDING REMARKS

Primary immunodeficiency diseases result from genetic defects that alter the development or function of cells involved in the immune response, leaving affected individuals with an enhanced susceptibility to severe infections, including opportunistic infections. This article described only a few of the known genetic immune deficiency diseases that illustrate how a single gene defect can have such a profound effect on the immune system. Studying patients with these diseases leads to important findings on how the immune system works. As we learn more about the immune system, we can better understand the diseases that are mediated by immune processes. For example, mechanisms similar to those that are utilized by the immune system to fight infection or to fight cancer can lead to allergic diseases or to autoimmune diseases. Allergy is a result of an immune response to innocuous agents such as foods, plant pollen, or other respiratory allergens, whereas autoimmune diseases result from an immune response to self-antigens. There is much more to the immune system and its diseases that has yet to be discovered.

BIBLIOGRAPHY

Conley, M. E., Parolini, O., Rohrer, J., and Campana, D. (1994). X-linked agammaglobulinemia: New approaches to old questions based on the identification of the defective gene. *Immunol. Rev.* **138,** 5–21.

Fischer, A. (1993). Primary T-cell immunodeficiencies. *Curr. Opin. Immunol.* **5,** 569–578.

Fuleihan, R., Ramesh, N., and Geha, R. S. (1995). X-linked agammaglobulinemia and immunoglobulin deficiency with normal or elevated IgM: Immunodeficiencies of B cell development and differentiation. *Adv. Immunol.* **60,** 37–56.

Janeway, C. A., Jr., and Travers, P. (1996). "Immunobiology: The Immune System in Health and Disease," 2nd Ed. Garland, New York.

Leonard, W. J., Noguchi, M., Russell, S. M., and McBridee, O. W. (1994). The molecular basis of X-linked severe combined immunodeficiency: The role of the interleukin-2 receptor gamma chain as a common gamma chain, gamma c. *Immunol. Rev.* **138,** 61–86.

Mach, B., Steimle, V., and Reith, W. (1994). MHC class II-deficient combined immunodeficiency: A disease of gene regulation. *Immunol. Rev.* **138,** 207–221.

Rawlings, D. J., and Witte, O. N. (1994). Bruton's tyrosine kinase is a key regulator in B-cell development. *Immunol. Rev.* **138,** 105–119.

Report of a WHO Scientific Group (1995). Primary immunodeficiency diseases. *Clin. Exp. Immunol.* **99**(Suppl. 1), 1–24.

Roos, D. (1994). The genetic basis of chronic granulomatous disease. *Immunol. Rev.* **138,** 121–157.

Immunological Tolerance

N. A. MITCHISON
Deutsches RheumaForschungsZentrum Berlin

GLOSSARY

Adjuvant Substance, often not itself an antigen, that can enhance the response to an accompanying antigen

CD4, CD8 Accessory molecules that strengthen the binding of a T-cell receptor to a major histocompatibility molecule and that themselves transmit additional signals into the T cell

Clonal Of or pertaining to a clone, which is a group of cells related by descent from a single cell; in this article the important point is that T cells belonging to a single clone share the same T-cell receptor in common

Epitope Antigenic peptide bound by a major histocompatibility complex molecule, presented to and recognized by a T-cell receptor. The term is also applied to peptides recognized by antibodies

Histocompatibility Histocompatibility antigens are responsible for transplant rejection. The major histocompatibility complex (MHC) is a group of molecules of this type, the most important of which bind immunogenic peptides and present them to the T-cell receptors of T cells (see TCR)

Idiotype Structure on a T-cell receptor or immunoglobulin recognized by other T cells or antibodies. A series of idiotypes and the anti-idiotypic T and B cells that recognize them form an immunological "network"

Immunogen Antigen, with respect to its capacity to induce an immune response

Superantigen Bacterial protein able to bind to one or more families of T-cell receptors and able to stimulate the T cell that bears the T-cell receptor to respond positively or negatively

TCR T-cell receptor; it is able to recognize an epitope in the context of a particular MHC molecule; analogous to the immunoglobulin receptor on a B cell

Transgene Gene artificially implanted into the genome of a mouse (or other animal)

IMMUNOLOGICAL TOLERANCE MAY BE DEFINED AS a condition of diminished specific immunological reactivity acquired through exposure of the immune system to either a foreign or self macromolecule. The part of the immune system most affected are T cells, tolerance of self is acquired mainly in the thymus, and (for T but not B cells) the only relevant macromolecules are proteins. However, tolerance can also be acquired under certain circumstances by B cells and, less commonly, by T and/or B cells in the peripheral (mature) immune system. The major mechanism by which tolerance is acquired is by deletion (negative clonal selection) of T or B cells reactive with the inducing molecule, although other mechanisms can also play a part in peripheral tolerance. The immune system naturally becomes tolerant of only a proportion, perhaps a small one, of all self-proteins and their constituent parts.

Immunological tolerance has a colorful history, in which studies on tolerance have played an important part in shaping ideas about the working of the immune system. The subject is of special importance in connection with the autoimmune and allergic diseases, which can up to a point be regarded as resulting from failure of the tolerance mechanism. This account starts with a brief history of the discovery of tolerance and then treats the main aspects of the subject serially, in nonchronological order.

I. DISCOVERY

At the beginning of the twentieth century the German scientist Paul Ehrlich, a founder of the science of immunology, coined the term *horror autotoxicus* to describe the importance of the body's defense systems not attacking the body itself. The immune system needs to discriminate between what is foreign and what is self. When foreign matter enters the body, it provokes an immune response and the body develops immunity to it. On no account should any part of the body itself provoke a response, as that would cause damage. This is just one of the excellent ideas that occurred to Ehrlich: it fits into the general scheme that he proposed for the immune system, in which the binding of an antibody to an antigenic structure on foreign material would cause more of that antibody to be made. To account for *horror autotoxicus,* he proposed that the immune system does not have any antibodies that can bind to structures within the body to start with (for reasons which Ehrlich could not explain), and that is why no further antibodies of this type are produced. Thus Ehrlich had a clear idea of how the immune system could make an immune response, one that is broadly in agreement with modern ideas on the subject, and he realized that one of the facts that his idea would need to explain is the distinction that the system makes between foreign and self.

In the ensuing decades, little was done to follow up this idea. Ehrlich himself was partly to blame, for although he could see the importance of these biological problems, both he and the generation of immunologists who followed him, led by Karl Landsteiner and Michael Heidelberger (both of whom migrated to the United States), believed that the way ahead lay through chemistry. Throughout the 1920s and 1930s, the goals of immunology lay in discovering the chemical structure of antigens and antibodies, not in answering questions about the cells of the immune system. Only in the 1940s did interest in immunologically active cells revive, when Macfarlane Burnet and Frank Fenner in Australia published their book on the immune response. There they predicted that some mechanism must operate that enables the immune system to distinguish between foreign and self, and they suggested that the system might learn to make this distinction during fetal life.

At the end of the Second World War the same simple idea occurred independently to three scientists in different parts of the world: that the immune system may learn to recognize as self those antigens to which it is exposed during early life, before foreign matter has had a chance to enter the body. One of these was Ray Owen, who was studying the inheritance of blood groups in cattle in the dairy country around Madison, Wisconsin. He noticed something odd about nonidentical cattle twins: when they shared a placenta in common, as sometimes happens, they often also shared blood groups in a way that could be explained only by assuming that blood cells passed from one twin to the other. That in itself was not surprising, but Owen became intrigued by the phenomenon and went on to follow up the twins as they grew older. He found that sometimes the sharing of blood group persisted into adult life, for a period far longer than could be accounted for by the persistence of short-lived red blood cells. Evidently the twins had exchanged not only their red blood cells but also the stem cells that give rise to them. These are normal tissue cells that should be rejected as foreign, yet they persisted. The rejection of foreign stem cells is nowadays well known as a barrier to bone marrow transplantation in humans; at the time, Owen already realized that the persistence that he had discovered broke the general rules of graft rejection. He drew the correct conclusion, namely, that foreign cells present during fetal life had altered the immune system of his cattle twins so that they could no longer be recognized as foreign.

At the same time, Milan Hasek in Prague studied chicken eggs that he had joined together by cutting windows in the shell and allowing the underlying vascularized respiratory membranes to fuse. The "parabiotic" eggs that result establish a common circulation, through which cells are exchanged. He found that the pair of chickens derived from these eggs would often accept skin grafts from one another, and that the grafts could even grow spectacular crops of feathers. From this experiment he drew the same conclusion as Owen.

The third group carried out the simplest experiment of all. Peter Medawar and his colleagues in London injected cells taken from one inbred strain of mouse into embryos or the newborn of another inbred strain. They found that the mice that then developed would accept test skin grafts from the original donor strain, although uninjected mice of the same combination invariably rejected their grafts. Furthermore, the tolerant condition was quite specific: mice that had been rendered tolerant of tissue from one strain of mouse remained able to reject skin grafts from another "third party" strain. This group too drew the correct conclusion, and it was Burnet and Medawar who shared the Nobel Prize awarded for this work in 1960: one for his prediction of the possible mechanism, and the other for verifying it in an experimental system that led to many further discoveries about its nature.

The next few years were a period of intense activity, in which tolerance of many different types of cells was studied in many animal species, in various experimental systems all over the world. One early discovery was of the termination of tolerance by "adoptive" transfer. When mice that had been made tolerant by the Medawar procedure were injected with lymphocytes taken from untreated mice of the same inbred strain, they would usually reject their test grafts. Evidently the injected lymphocytes could settle down in the new host and, not being tolerant themselves, use it as a base from which to mount an immune response against a foreign skin graft. This experiment implies that the tolerant condition can be narrowed down: it applies not to whole individuals, but to their lymphocytes. Other possibilities could have been imagined: for instance, the tolerant animal could make immune effector cells just as usual, but they would all get mopped up by donor cells surviving from the original tolerance-inducing injection before they could reach the test skin graft. It is explanations like this that are excluded by the adoptive transfer experiment.

Yet another possibility that the adoptive transfer experiment excludes is suppression, that is, the activation of a population of suppressor cells able to prevent other cells of the immune system from mounting an antigraft response. This possibility seemed remote at the time, but later work has shown that it is in fact possible under certain circumstances to raise populations of suppressor cells able to recognize foreign cells, a phenomenon that is further discussed later. However, these populations are evidently hard to raise by neonatal injection as practiced by the Medawar group.

Another early finding was that injection of cells from one tissue (the spleen was often used) would induce tolerance of many other tissues, not only skin but also heart, kidney, and others. In general, therefore, few tissues seemed to have unique antigens that are not represented elsewhere in the body. However, the passage of time has dulled this rule, and differentiation antigens of limited tissue distribution are now known that were overlooked by the early experiments.

A third important early line of investigation concerned reversibility of the tolerant condition. In the first experiments, the tolerant condition lasted lifelong, so the question was whether this reflected an irreversible alteration in the immune system or merely the persistence of donor cells from the original tolerance-inducing injection. An answer to this question came from an experiment with chicken red blood cells. These cells can be labeled *in vitro* with a radioactive isotope, and their fate then followed after they have been returned into the bloodstream. Within the bird from which they came, they survive for about 1 month, but in an unrelated bird they are rapidly eliminated in about 4 days. If an unrelated bird is injected at hatching, and then again and again so that it always has the foreign cells in its bloodstream, it will remain tolerant indefinitely and not show rapid elimination. In this system it was thus possible to ask the question about reversibility by leaving tolerant birds without the foreign red blood cells for varying lengths of time. This had to be done with irradiated blood, as otherwise stem cells present in the blood colonized the tolerant birds, which then behaved like Hasek's chickens and remained tolerant indefinitely. Using irradiated blood, it was found that the birds left without the foreign red cells did lose their tolerance and acquired the ability to eliminate the foreign cells rapidly. Furthermore, the older the bird at the time the supply of foreign cells was withdrawn, the longer it took to recover reactivity. From this and similar findings the conclusion was drawn that a population of mature lymphocytes cannot recover from tolerance but can be replaced by a fresh stream of cells from the thymus, a stream that diminishes with age. This conclusion was reinforced when the role of the thymus came to be investigated more fully. It was found that recovery from tolerance is dependent on the presence of the thymus and can be prevented by surgical removal of the thymus. Thus tolerance susceptibility could no longer be regarded as something unique to the embryo, but rather as a property of T cells undergoing development, a process that continues throughout life.

Medawar and the other pioneers were well aware of the most important implication of their work: the likelihood that their experiments provided a valid model of the natural mechanism of self-tolerance. This was after all the question that Burnet had raised. Yet although they expressed their hope that self-tolerance in the normal immune system would operate in the way that they had demonstrated, it was to be several years before direct evidence could be obtained. Final proof that tolerance of self-proteins is acquired came from studies of the self-proteins 4-hydroxyphenylpyruvate dioxygenase (HPPD, a liver enzyme), C5 (a protein of the complement cascade), and hemoglobin. Each of these proteins is fully tolerated in normal individuals, but those individuals who either lack one of them (as a result of congenital deficiency) or have an altered (polymorphic variant) form can recognize the normal form as foreign and react immunologically against it.

Thus far the words "antigen" and "antibody" have hardly been mentioned. Nowadays the division of lymphocytes into T cells and B cells is well understood (so much so that the very word "lymphocyte" is falling into disuse), and the fundamental similarity of the two types of cells is generally accepted, in spite of their numerous differences. It took a while to reach this consensus, and at the time when the basic concept of tolerance was being worked out it was far from clear how it would fit into the established framework of antigens and the antibody response that they evoke. A summary of the immunology of the day written in 1967 by another Nobel laureate, Nils Jerne's "Waiting for the End," illustrates this delay. He was confident about the "trans-immunology" of antigens and antibodies and uncertain about the "cis-immunology" of the cells of the immune system. All that is now water under the bridge, but it is worthwhile pausing to ask why it took so long for T-cell tolerance to be accepted as a normal function of the immune system, every bit as important as the production of antibodies.

It is not entirely by chance that our understanding of tolerance has its roots in the study of T-cell response to cellular antigens. Medawar had spent the 1939–1945 war years studying the reaction to skin grafts, a subject of great importance for the treatment of airmen suffering from burns. He had shown that a foreign graft serves as an antigen and is rejected by an immune response. This discovery made a great impression, and it was only ultracautious scientists such as George Snell (yet another Nobel laureate) who continued for a while to refer to "histocompatibility factors." If one was going to investigate tolerance, and especially tolerance of self-cells, it made sense to start with the immune response to foreign cells.

It was a stroke of luck that the discovery of tolerance was made with a T-cell response such as graft rejection, because T cells rather than B cells are the right population in which to study tolerance. This became clear later, when T cells were found not only to reject skin grafts but also to have regulatory functions, particularly in the production of antibodies by B cells. When immunologists turned to tolerance in the context of the antibody response, by good fortune they were in fact studying tolerance in the compartment of the tolerance-sensitive T cells, rather than in the relatively insensitive B-cell compartment.

To summarize, by the end of the classic period in immunology research, immunological tolerance was known to obey the following rules: (1) It is immunologically specific, in the sense that a particular antigen induces tolerance only of itself and not of other unrelated antigens. (2) Once induced, it becomes a permanent property of the population of immune cells treated, although this population can be replaced via the thymus if the supply of inducing antigen is withdrawn. (3) It does not result from suppression. (4) So far as was known, these properties apply only to T cells. (5) So far as was known, they apply to natural self-tolerance as well as to tolerance obtained by manipulation of the immune system. (6) The induction of tolerance has many of the properties of a conventional positive immune response, making it likely that immunology would remain a unified science, with all the benefits that entailed.

II. CLONAL SELECTION

From here on this account does not proceed in chronological order, but instead follows up a number of different aspects of tolerance in succession.

We start with an account of the cell mechanisms involved. In 1957 Burnet announced his new clonal theory of immunity and tolerance, and similar ideas were formulated by David Talmage and by Jerne. The central idea was that each lymphocyte has its own receptor encoded by its own gene, different from that on most other lymphocytes, and that these receptors antedate any encounter with antigen. Upon encounter with its antigen, a lymphocyte responds in one of two ways. In early life, most lymphocytes respond by dying (induction of tolerance), whereas in later life they do so by multiplying and differentiating into effector cells (induction of memory and of a normal positive response). Burnet compared the immune response to a man buying a suit: previous theories likened the immune response to a traditional tailor making one made-to-measure, whereas according to the clonal selection theory the antigen picked one ready-to-wear.

Other immunologists were at first reluctant to accept this view, because of the enormous waste of lymphocytes it seemed to imply, and because they could not imagine how so much diversity among the receptors could be generated. But gradually it came to prevail, mainly because of the growing evidence that one cell could indeed make just one kind of antibody (or, as was shown later, one T-cell receptor). What is important for the present purpose is that it gave an elegant explanation of the rules of tolerance outlined earlier, particularly its irreversibility at the cellular level and the symmetry between positive and negative (tolerance) immune responses. Eventually a position was reached in which it was widely accepted that immunological tolerance is obtained by clonal deletion, and particularly that self-tolerance results from the clonal deletion, mainly within the thymus, of many, many self-reactive T cells. All that was missing was a direct demonstration of this process.

With the introduction of T-cell receptor (TCR)-transgenic mice, that gap could be filled. To understand these experiments, we must detour into a brief explanation of allelic exclusion. For most genes, both alleles are usually switched on at the same time. The immune system cannot let this happen in the case of immunoglobulins and TCRs, because that would mean single cells having more than one receptor, which would not be compatible with clonal selection. Accordingly, lymphocytes implement allelic exclusion, that is, they allow only one receptor allele to be expressed by tightly regulating the process of DNA rearrangement that leads to expression of these receptors. As soon as one chromosome has rearranged, enhancers and silencers come into operation and prevent the other chromosome from rearranging (with certain exceptions, notably in the case of the TCR α chain). Transgenic mice are made by implanting rearranged genes encoding the α and β chains of one particular TCR into a mouse embryo, and then selecting progeny mice that carry the genes in their germ line. A line produced in this way has a high proportion of T cells expressing the transgene, because the interloper gene switches on the allelic exclusion mechanism and so largely prevents the developing T cells from rearranging their own TCR genes.

Harald von Böhmer and his colleagues made mice in this way that were transgenic for a TCR that recognized the male (H-Y) antigen. Female mice of this type, provided they have the other genes that this receptor requires [the major histocompatibility complex (MHC) gene whose product is recognized by the TCR, and the CD8 and β_2-microglobulin genes], have many transgene-expressing T cells. In males, however, these cells are missing. The mice thus show both positive clonal selection (indicated by the expanded number of T cells conditional on females having the right MHC gene) and negative clonal selection (indicated by the cell deletion in males). These two forms of selection are illustrated in Fig. 1. As a rule, the adjective "clonal" is now dropped from the terminology, and one refers simply to positive and negative selection. The experiment can be further refined in various ways, for instance, by putting the transgenes onto a SCID (severe combined immunodeficiency) genetic background and so totally preventing rearrangement of endogenous TCR genes.

Several questions remain to be answered. The extent of overlap of the two processes of positive and negative selection, their exact location, and the relationship to expression of CD4 and CD8 molecules are not fully understood. The underlying problem is that we know very little about the signals that T cells receive from one another and from their local environment during development, and even less about how these signals are translated into gene activation. One can be confident that these gaps will soon be filled. [*See* T-Cell Activation; T-Cell Receptors.]

During the processes of positive and negative selection, many T cells die: a current estimate is that 96% of the T cells within the thymus are fated to die within a few days. As one might expect, T cells taken from the thymus soon die under normal conditions of tissue culture. When they do so, they show clear signs of undergoing apoptosis (programmed cell death). It seems likely, therefore, that the cells that die during positive and negative selection in the thymus do so in this way. Yet at present this conclusion is in doubt. The problem is that genetic manipulations that would be expected to interfere with apoptosis have little or no detectable effect on selection in the thymus. On the one hand, mice transgenic for the *bcl-2* gene that protects against apoptosis and, on the other, mice with the *fas* and *gld* gene inactivated by mutation seem to behave normally with respect to self-tolerance (binding of the *fas*-encoded protein to its *gld*-encoded receptor is a major pathway to apoptosis).

III. TOLERANCE OF SOLUBLE, MONOMERIC ANTIGENS

Once tolerance of cellular antigens had been placed on a firm footing, interest naturally turned to soluble purified proteins, which offered better scope for quantitation. As expected, proteins such as foreign serum albumin or immunoglobulin injected at birth into experimental animals did induce tolerance, as judged by reduced antibody titers made in response to challenge with the same protein in adjuvant; the challenge had

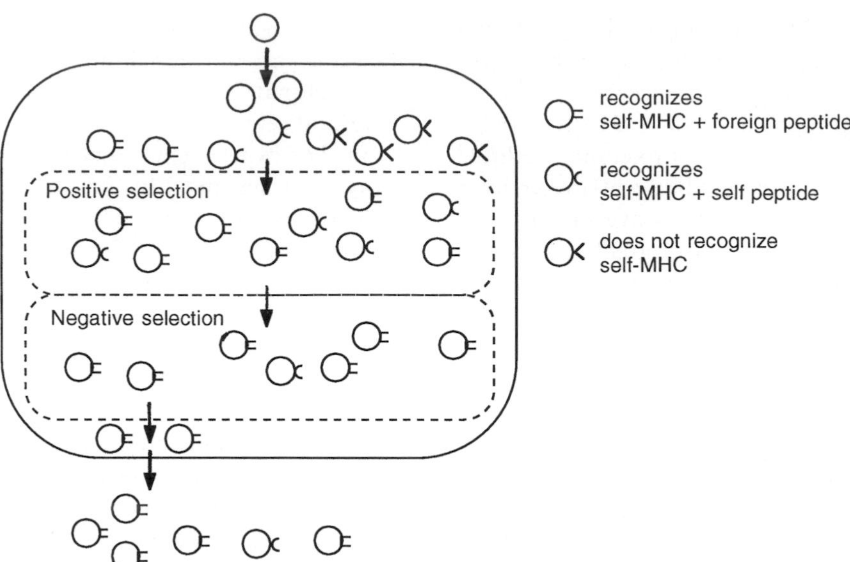

FIGURE 1 T-cell selection during development in the thymus. Precursor cells enter the thymus in small numbers, devoid of a TCR (T-cell receptor). They then express a TCR, with a wide range of antigen-binding sites. Each site belongs to one of the three possible categories shown. The cells then move into the cortex of the thymus, where those cells that have a receptor that cannot recognize self-MHC molecules die, whereas the other two categories undergo positive selection and expand in numbers. The survivors move again, into the thymic medulla and corticomedullary junction, where negative selection takes place: most of the cells that recognize self-MHC plus a self-peptide die, leaving only those cells that recognize self-MHC plus a foreign peptide. The survivors then emigrate from the thymus into the peripheral immune system. The two types of selection are not in fact as well separated as is shown here, as there is overlap of the locations in which the two processes take place. Note that some of the cells that recognize self-MHC plus a self-peptide survive and emigrate to the periphery, where they provide a pool of cells that is potentially self-reactive but normally contained by peripheral mechanisms. The importance of this pool is discussed in Section IX.

to be given with adjuvant, because these proteins are poor immunogens when injected on their own. Also as expected, animals treated simply with one neonatal injection start to recover reactivity after a few weeks, following clearance of the protein from the circulation.

Two further unexpected findings were made. One was that tolerance shows a form of nonspecificity in which it could be terminated by immunization with a cross-reactive antigen. For instance, a rabbit made tolerant by treatment with bovine serum albumin responds to challenge with sheep serum albumin, as expected. But at the same time it also makes appreciable amount of antibody to bovine serum albumin; this antibody is quite specific and could not be removed by absorption with excess sheep serum albumin. This curious phenomenon found an explanation only when the difference between tolerance in T and B cells came to be understood, as described in the following.

The second unexpected finding was that adult mice became tolerant when injected with low dose of soluble protein without adjuvant, a phenomenon that came to be termed low-zone tolerance. This is of special interest for several reasons. The phenomenon is widespread, with many proteins able to induce it, including foreign immunoglobulin and serum albumin and foreign proteins such as lysozyme and ribonuclease. Probably any monomeric protein can do so, provided that it is rendered minimally immunogenic by treatments such as ultracentrifugation to remove aggregates. The protein has to be monomeric, because polymeric proteins such as flagellin are highly immunogenic for B cells even on their own without adjuvant (flagellin is the main component of the bacterial flagellum). Monomeric self-proteins are also known to induce low-zone tolerance naturally, the best-studied example being the aforementioned HPPD, a protein that leaks from liver cells into the circulation in con-

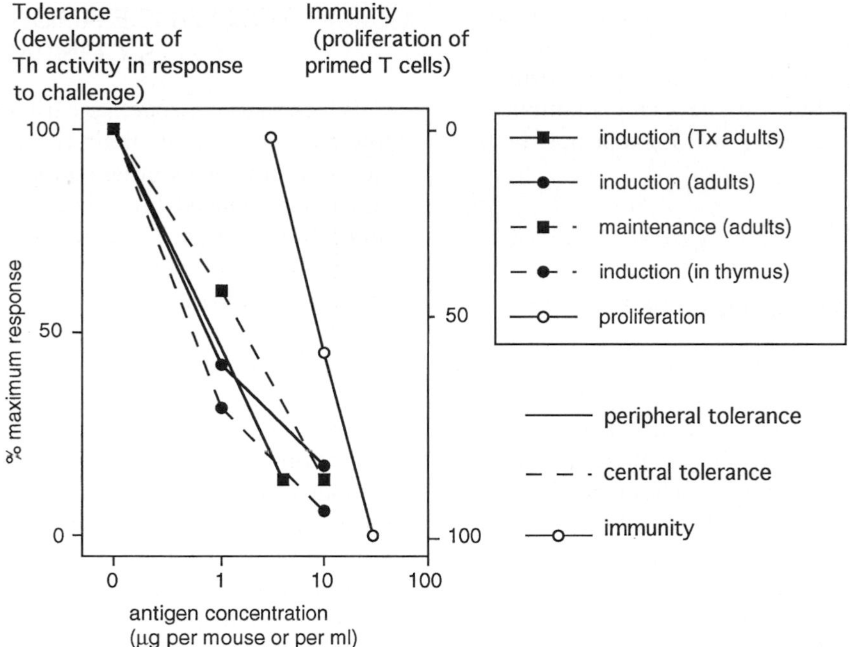

FIGURE 2 Tolerance and immunity thresholds. Low-zone tolerance in the thymus and the periphery occurs at the same concentrations of antigen, which are only a little lower than the concentrations required to stimulate a positive response in T cells. Tolerance induced in thymectomized (Tx) adults, or induced in adults, affects mainly or exclusively peripheral T cells. Tolerance maintained in an already tolerant adult, or induced in thymus organ cultures, affects cells in the thymus. The degree of tolerance is measured by the helper T-cell (Th) response developing after challenge (immunization with antigen plus adjuvant). The positive response (immunity) is measured by the proliferation *in vitro* of primed T cells. The data come from the author's group's experimental work in mice with the antigens bovine serum albumin and 4-hydroxyphenylpyruvate dioxygenase; similar data have been published for several other soluble monomeric protein antigens by this group and by that of William Weigle.

centrations just high enough to render T cells self-tolerant.

The concentration of antigen needed to induce low-zone tolerance is remarkably constant, as shown in Fig. 2. Low-zone tolerance can be induced with exactly the same dose–response relationship within the thymus and outside it, although it takes longer to do so in the periphery. There is little reason to suppose that a different cell mechanism is involve in low-zone tolerance in the thymus and the periphery, as discussed further in Section VI.

In summary, the rule is that T cells at any stage of development can respond to antigen in one of two ways: they may die (negative selection), or they may proliferate and differentiate, that is, make a positive response. The difference does not seem to be one of developmental stage, but rather the extent to which the T cell is activated by secondary signals ("costimulation"), such as are conveyed by adjuvants or aggregation, which favor a positive response. In the absence of such secondary signals, T cells become consistently tolerant. In contrast, the extent to which a particular antigen or a particular anatomical location generates activation signals is highly variable, and in addition the T cell itself may be more (or less) sensitive to these signals at particular developmental stages. One should not suppose that all the T cells within an individual behave uniformly; rather, some cells will as a rule be deleted and others will respond positively, with the net outcome depending on the balance between the two processes.

With this concept in mind, it is worth returning briefly to the cellular antigens. It has been found that foreign cells, which are normally highly immunogenic in an adult mouse, can induce a condition of tolerance if injected into a mouse that has been manipulated by treatment with sublethal irradiation or immunosuppressive drugs. Treatment with antilymphocyte antibodies, such as anti-CD4 monoclonal antibody, can have the same effect. This fits well with the rule that T cells respond relatively uniformly in becoming tolerant, but can be easily manipulated in the extent to which they will make a positive response.

IV. TOLERANCE IN CD8 T CELLS

The soluble protein antigens just described are recognized only by CD4 T cells, and to do so they have to be processed by "professional" antigen-presenting cells (dendritic cells, macrophages, B cells, activated T cells, and to some extent other cells in which expression of class II MHC molecules has been induced). CD4 T cells are largely immunoregulatory in functioning via cytokines that act on neighboring cells. In a sense, they constitute the brains of the immune system. Another, smaller set of T cells exists that bears the alternative marker CD8. These cells recognize antigen bound to class I MHC molecules, which are expressed on almost all cells of the body. Instead of loading up with peptides derived from external antigens, as do class II MHC molecules, the class I molecules load up with peptides derived from antigens located in the interior of the presenting cell. This enables CD8 T cells to attack cells of the body that harbor viruses within their interior.

On the whole, CD8 T cells obey the same rules of tolerance as do CD4 cells. They undergo negative selection in the thymus, are subject to peripheral tolerance induction, and show the same sort of selective recognition of a limited subset of peptides within a tolerated protein (as described in Section VII). Indeed, much of what is known about positive and negative selection within the thymus has been learned from study of CD8 T cells recognizing H-Y, the male-specific antigen. Relatively little is known about their induction thresholds for either tolerance or immunity, as there is no obvious way of varying the concentration of the antigens that they recognize. The main difference is in the distribution of the self-proteins toward which they are tolerant. For CD8 T cells, these are almost exclusively cell-internal proteins, many of which cannot reach the thymus or lymph nodes. Presumably, therefore, much of their tolerance-of-self is induced in tissues outside the immune system. [*See* CD8 and CD4: Structure, Function, and Molecular Biology.]

V. TOLERANCE AMONG T AND B CELLS COMPARED

This section draws together earlier threads concerning tolerance in these two main types of lymphocyte. The subject is not new: studies contrasting thymectomized and bursectomized birds laid its foundations in the mid-1960s (just as T cells use the thymus as a site of development, so in birds B cells likewise use the bursa of Fabricius). Within a few years it became clear that T cells have not only important effector functions in antiviral cytotoxicity but also regulatory ones: helper activity for B cells and other T cells, and a role in control of the Th1/Th2 balance (the latter is discussed in Section VI).

Once it has become clear that B cells cannot make most types of antibody without help from T cells, much of the earlier work could be understood. As mentioned earlier, studies on low-zone tolerance of soluble proteins in which the antibody response had been used as an end point could be reinterpreted in terms of tolerance induced in helper T cells. The phenomenon of cross-reactive termination of tolerance at last found a satisfactory explanation: T but not B cells become tolerant of soluble antigens, so when T cells later respond to a cross-reactive antigen, they find nontolerant B cells able to make antibodies reactive with the original tolerance-inducing antigen.

This explanation of cross-reactive termination of tolerance helped establish the point that B cells, unlike T cells, seldom become tolerant of soluble monomeric proteins. In other experiments, T cells primed against one antigen and B cells primed against another could be transferred together into irradiated mice, where the B cells proved able to respond to their own antigen only if it was physically linked to the antigen against which the T cells had been primed, an example of the phenomenon of "linked help." In this system it was easy to explore the possibility of making one cell population or the other tolerant, and it too helped establish that tolerance could often be induced only among T cells. Subsequent experiments with transgene-encoded antigens have fully confirmed the rule: low-level expression, in the now familiar 10^{-8} *M* range makes tolerant only T cells, although B cells can be made tolerant by increasing the level of expression through several orders of magnitude.

Nevertheless, B cells become tolerant surprisingly easily under certain conditions. One critical variable is their stage of development. Recently matured B cells, which express surface immunoglobulin M but not yet immunoglobulin D, are particularly sensitive to tolerance induction. Yet the acquisition of immunoglobulin D receptors does not itself cause the loss of this susceptibility, but only marks the end of this stage of development. Recently it has been shown that a "second window" of susceptibility to tolerance induction opens after B cells undergo hypermutation in germinal centers, and it has been suggested that this may be important for purging cells that have acquired self-reactivity through mutation (hypermutation is a phenomenon known only in germinal centers, whereby B cells mutate the combining sites of their immunoglobulin receptors at the extraordinarily high rate of about one mutation per gene per cell generation). A second critical variable is the extent of cross-linking of the antigen. It has long been known that highly cross-linked antigens such as the blood group sugars on erythrocytes or the major histocompatibility antigens must induce self-tolerance, for otherwise specific antibodies against them would be hard to produce.

The resistance of B cells to the induction of tolerance has important implications for tolerance-of-self. Mention has already been made of the prototype self-protein HPPD, which induces low-zone self-tolerance. In contrast, it does not induce tolerance in B cells. The same applies to another well-studied self-protein, the C5 component of complement. Generally, the immune system seems to rely on its T-cell component for maintaining tolerance of self, leaving tolerance in B cells as an optional extra, not essential because of the dependence of B cells on help from T cells. It is no doubt this specialization that has permitted B cells to evolve their unique mechanism of hypermutation, something that is of immense value for producing high-affinity antibodies but that is also a luxury in which cells bearing responsibility for self-tolerance could not safely indulge. Note how well adapted the immune system is: the mirror image would not work, in which B cells but not T cells were responsible for self-tolerance, because T cells with their matrix of receptors would have little use for high affinity.

This view of the dominant role of T cells is in accordance with much of what is known about the immunoregulatory diseases: inflammatory arthritis, insulin-dependent diabetes, allergy, and so on. There are good grounds for regarding these diseases as primarily T-cell-mediated: the human leukocyte antigen (HLA) associations, the role of T cells in animal models, and the success of T-cell-directed therapy with cyclosporin and with anti-T-cell antibodies (cyclosporin is a cyclic peptide produced by a fungus; it is one of the most powerful immunosuppressive drugs known). Thus it makes sense to press on with the development of novel therapies targeted at T cells, whether or not the induction of tolerance as discussed in the next section finds a place among them.

VI. PERIPHERAL TOLERANCE

A good case can be made that the immune system needs to have tolerance inducible in the periphery (i.e., within mature populations of T cells that have completed their development within the thymus and migrated out into lymph nodes and spleen). The need arises in part from the existence of self-proteins that are neither made in the thymus nor reach the thymus from the circulation. The number of proteins in this category is unknown: the common housekeeping proteins of cells are presumably excluded, as are the cell-surface molecules present on lymphocytes, dendritic cells, supporting cells, and other cells present in the thymus. So also are proteins such as HPPD and C5, which are produced elsewhere but are known to reach the thymus. Nevertheless, one guesses that a substantial number of proteins exist that a T cell will encounter only after it has emigrated from the thymus. A particularly striking group of these proteins are those of the menstrual cycle and pregnancy, such as the hormone chorionic gonadotrophin.

The need for peripheral tolerance also stems from the proximity of the thresholds for tolerance induction and immunization, as shown in Fig. 2. This is a fundamental design feature of the immune system, which has presumably evolved to balance the conservation of as much as possible of the T-cell repertoire against the need to avoid autoimmunity. One can imagine a supersafe immune system that would have a wide gap between the tolerance and immunity thresholds. But it would not be much use for defense against infection, because parasites would find it easy to evolve antigens that would escape detection. The immune system needs a way to contain potentially self-reactive T cells that escape negative selection in the thymus.

Self-reactive T cells have indeed been found to escape into the periphery. For instance, rat T cells were selected in tissue culture for reactivity to myelin basic protein, and then returned into adoptive hosts where they promptly attacked the myelinated nerves. Potentially disease-inducing T cells can be detected at low frequency in normal humans, such as those reactive with acetylcholine receptor, the target antigen in the autoimmune disease myasthenia gravis.

A variety of mechanisms of peripheral tolerance have been demonstrated in experimental systems, although their relative importance in meeting the need described here is not yet clear. In the following account, these mechanisms are presented in a speculative order of importance, starting with (1) negative selection. Mention has already been made, in connection with Fig. 2, of the constancy between the thymus and the periphery of the antigen concentration required for tolerance induction. Further evidence of negative selection comes from (i) treatment of normal adult mice over a prolonged period with low doses of bacterial superantigen and (ii) treatment of TCR-transgenic mice with the corresponding antigen (or corresponding peptide recognized by the TCR). Treatment of normal adult mice with a superantigen results in detection of the TCR family to which it binds, although a period of expansion may intervene. A mechanism proposed for negative selection in the periphery is shown in Fig. 3.

Next comes (2) majority rule. To understand this mechanism, one needs to assume that T cells function mainly within clusters arranged around antigen-presenting dendritic cells. Within these clusters it has been shown that T cells communicate with one another by means of cytokines that act in a paracrine manner (i.e., act on neighboring cells). The assumption is then made that following initial activation by contact with the central antigen-presenting cell, a T cell will need further activation signals in the form of excitatory cytokines from its neighbors within the cluster in order to complete its response. If T cells recognizing a particular antigen are very rare, the response cannot be completed. In line with this expectation, it has been found (in one instance) in the author's laboratory) that artificially increasing the frequency of a population of self-reactive T cells, whose reactivity is normally fully contained, leads to breaking of self-tolerance. To be accepted, this attractive hypothesis needs further support.

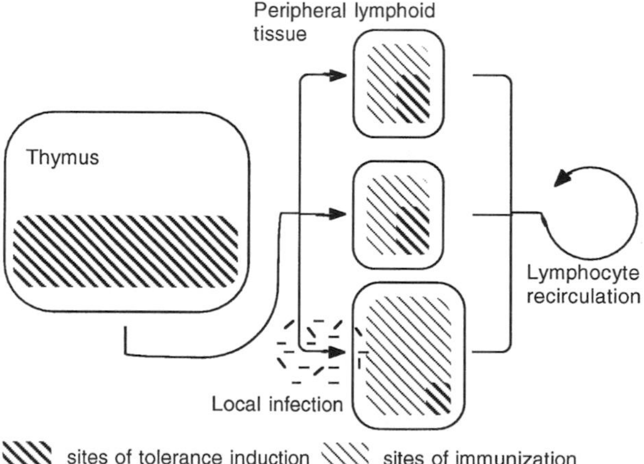

FIGURE 3 Tolerance induction and immunization sites in the thymus and the periphery T-cell traffic follows the arrow out of the thymus and then repeatedly through the peripheral lymphoid organs (recirculation through lymph nodes and spleen). The "sites" in which tolerance and immunization occur are shown in highly diagrammatic form, as they represent particular types of interaction with antigen-presenting cells (involving different cell-surface ligands and cytokines) rather than anatomical compartments. A site of local infection is also shown, where inflammatory products cause the local node to swell and the ratio of tolerance to immunization sites to shrink.

There are numerous examples of (3) cytokine balance reducing expression of an immune response. This area is dominated by the concept of a balance between Th1 and Th2 cells, which produce, respectively, gamma-interferon and other pro-inflammatory cytokines, and IL-4 and other B-cell-stimulatory cytokines. The two sets of cytokines are mutually inhibitory, and both types of T cell participate in most responses. However, even a transitory excess of one type of activity can have a long-lasting inhibitory effect on the other, and gross imbalance can under certain circumstances have an inhibitory effect on the entire response. Furthermore, imbalance is associated with autoimmunity (i.e., reactions against components of self) in animal models, and it seems likely that the same will hold true in humans. [*See* Cytokines and the Immune Response.]

Next comes (4) anergy. It has been suggested that exposure to antigen can induce a condition of unresponsiveness in an entire population of peripheral T cells. Conditions in which this is thought to occur include a variety of tissue culture models and TCR-transgenic mice in which the corresponding antigen is produced by another transgene. In the latter model the outcome varies. A drastic result is down-regulation of the transgene-encoded TCR on the entire population of T cells; a less drastic one is diminished reactivity to the TCR-encoded antigen. These are interesting and reproducible effects, but there is some doubt about their relevance to the normal immune system. For instance, there is as yet no evidence that the fraction of "normal" T cells that express low levels of TCR are enriched for self-reactive T cells; and it is possible that the diminished reactivity observed in the doubletransgenics may reflect deletion of an activated population of T cells. However, there is no doubt about the interest of anergy. Peptides, as well as whole proteins, have proved highly effective in inducing this condition. For instance, a commercial peptide vaccine, based on the concept of anergy, provides a promising treatment for cat allergy.

A mechanism that is clearly important in animal models of autoimmunity is (5) network control. This refers to the ability of T cells that recognize peptides derived from the TCRs of other T cells to regulate the latter's activity. Although clearly important in recovery from disease, the role of network control in preventing it is doubtful. An intriguing possibility is Irun Cohen's concept of the immunological homunculus, which posits that the normal immune system is highly selected for reactivity to self-antigens, a few of which dominate the T-cell repertoire just as the lips and fingers dominate the sensory nervous system (shown in the strangely disproportionate picture of a distorted homunculus that neurologists draw). Cohen proposes that this high level of self-reactivity is normally contained by an anti-idiotypic network, and that breakdown in this control causes autoimmune disease.

Oral tolerance (6) is a relatively specialized phenomenon in which ingested antigens induce tolerance. More than one mechanism is probably involved: portal filtering is one of them (proteins that pass from the gut into the circulation traverse the liver before they reach the immune system, and in that organ may have their more immunogenic aggregates removed by liver macrophages) and bystander suppression is another (peptides that pass from the gut directly into local lymph nodes may preferentially activate T cells to make suppressive cytokines). The latter possibility, recently proposed by Howard Weiner and colleagues, is illustrated in Fig. 4. The special interest of bystander suppression is that it is so obviously applicable to the therapy of immunological diseases. Indeed, several clinical trials of oral tolerance in diseases such as rheumatoid arthritis, multiple sclerosis, and sympathetic uveitis (an immunological disease of the eye) are currently under way.

Finally, mention should be made of (7) suppression. The concept of suppressive T cells has had its ups and

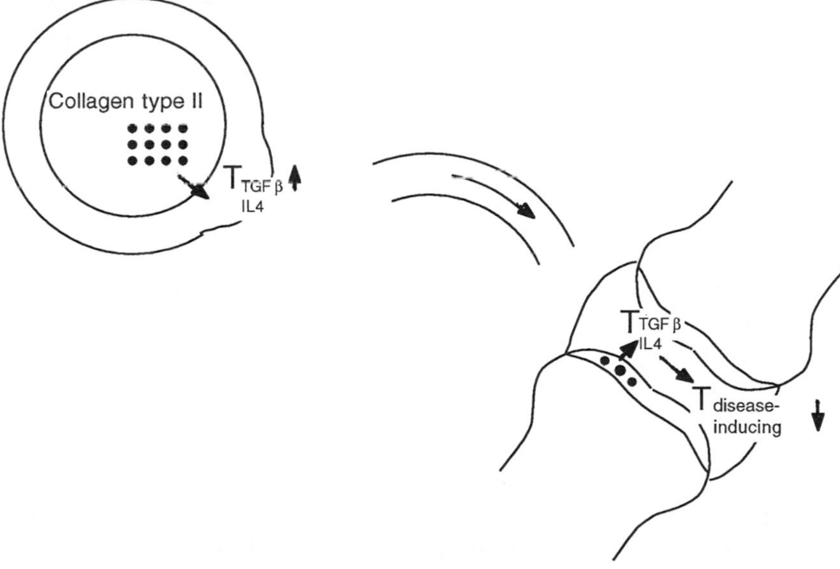

FIGURE 4 The mechanism of oral tolerance. A foreign protein exemplified here by collagen type II is ingested. It (or peptides derived from it) passes into a local lymph node in the wall of the gut, and there activates T cells that produce the cytokines transforming growth factor beta (TGFβ) and interleukin 4 (IL4). These cells then migrate into the circulation and thence into sites where self-collagen type II is available to them. This protein is largely confined to cartilage and is here depicted within an inflamed arthritic joint. The self-collagen reactivates the T cells, which again produce their characteristic cytokines, which then inhibit the activation of neighboring T cells, especially the disease-inducing ones ("bystander suppression"). Note that this hypothesis of Weiner's is supported by experiments, but that clinical trials have been less successful.

downs, and there is no doubt that at the time of writing expert opinion believes that most of the known instances of suppression can be accommodated within the concept of Th1/Th2 imbalance. Nevertheless, a few maverick immunologists continue to argue in favor of suppressor T cells as a unique cell type.

VII. ANTIGEN PRESENTATION: WHAT ARE WE TOLERANT OF?

Considering the $\sim 10^5$ different proteins encoded in the mammalian genome, it might be thought surprising that any T cells can survive negative selection. Fortunately the developing T cell is exposed to far fewer self-antigenic determinants ("epitopes") than this figure would suggest. Self-epitopes are hidden from the immune system in three principal ways. First, they may be expressed only in locations where T cells do not visit. These are the so-called "immunologically privileged sites," such as the cornea, the central nervous system, and the pigmented cells of skin (melanocytes), which were originally defined by the ability of foreign cells implanted into them to survive over the long term.

Second, we return to the activation signals needed by T cells, particularly cell-surface ligands and cytokines. In sites lacking appropriate molecules of these types, foreign antigens can survive indefinitely without provoking an immune response. This phenomenon has received names such as "immunological silence," "immunological neglect" and "immunological latency." Striking examples are the low immunogenic activity of the cell-surface molecule allo-Thy-1 (even though it is expressed on T cells themselves) and transgene-encoded viral glycoproteins expressed on the surface of pancreatic islet cells. In both cases, presentation of these antigens along with an appropriate activation stimulus (e.g., a more powerful immunogen) can evoke a brisk immune response. Presumably self-proteins expressed in this way induce negative selection in T cells minimally or not at all.

Third, a distinction can be made between cell-surface molecules, with their relatively good access to the immune system, and cell-internal ones, which are relatively sequestered. Figure 5 shows an example in which an intracellular receptor protein is ignored by T cells, whereas a similar protein that is expressed on the cell surface appears able to induce self-tolerance.

These are some of the rules of engagement for entire proteins. There are also such rules for the peptide

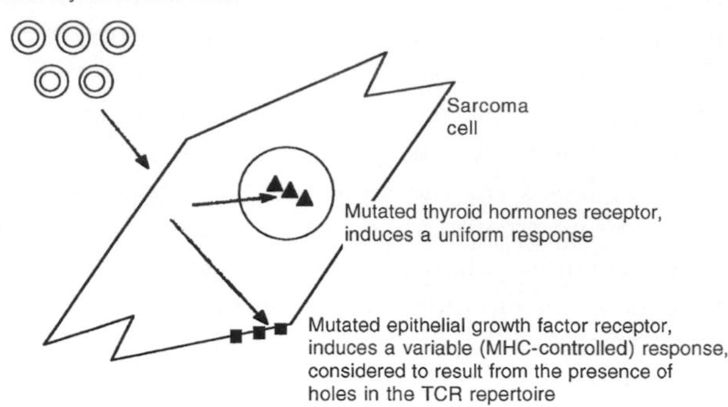

FIGURE 5 T cells are indifferent to cell-internal proteins. In the experiment shown here, tumor cells transformed by avian myeloblastosis virus grew in rats. The virus-encoded protein expressed on the cell surface induced a variable (MHC-controlled) response, considered to result from the corresponding self-protein having punched holes in the T-cell receptor repertoire during negative selection. In contrast, the nuclear protein induced a nonvariable response. This work is drawn from the author's group.

components of individual proteins, which reflect the detailed biochemistry of antigen processing within antigen-presenting cells. Eli Sercarz and his colleagues have demonstrated that proteins that contain many potential epitopes ("dominant epitopes") in fact elicit T-cell responses to only a small subset. Some of the remaining epitopes can nevertheless elicit a response if administered as peptides, rather than within the intact protein ("subdominant epitope"). Evidently there are competitive and inhibitory interactions between peptides during the processes of cleavage within endosomes, binding to MHC molecules, and further cleavage (endosomes are small intracellular vesicles budded off from the cell membrane, within which cleavage of antigens into peptides for presentation by MHC molecules takes place). This is an important principle, as it implies that a range of epitopes exist that are normally cryptic, but that can be accessed by appropriate cleavage of an antigen prior to administration. As applied to self-proteins, it indicates that only a limited portion of any one protein is likely to be presented to T cells during the induction of tolerance, and that there will be other regions of the protein that if suitably presented could break through the tolerance barrier. The Sercarz group has demonstrated the validity of this concept by immunizing with peptides derived from self-MHC molecules, some of which turn out to be immunogenic. The concept suggests that abnormal antigen processing might contribute to the etiology of autoimmune disease. Although this is a logical extension of the experimental work, its validity has yet to be verified in a human disease or animal model.

VIII. EVOLUTION

Immunological tolerance can be regarded as a highly evolved component of the immune system's functioning, likely to be better understood if more was known of its evolution. Unfortunately, little is known about tolerance in primitive immune systems. No vertebrate has yet been discovered with any substantial part of the immune system missing, and no invertebrate is known to manifest specific immune functions comparable to those of the vertebrates. That is not to deny that many of the innate defense mechanisms of invertebrates are related to those found in vertebrates. But as tolerance is clearly a specific, acquired function, we are unlikely to find a true invertebrate homologue. Nevertheless, the hunt for ancestral stages will certainly continue, in primitive vertebrates as well as in invertebrates, and is likely to intensify as new molecular tools such as the polymerase chain reaction using consensus primers become available.

IX. IMPLICATIONS FOR MEDICINE

Ever since its discovery, immunological tolerance as a subject of research has been closely allied to medicine. The Medawar group foresaw the main application of their work in organ transplantation. That hope has been partially fulfilled. Careful investigation has revealed that long-surviving transplants do establish a measure of tolerance in their hosts, but to a disappointingly small extent, so that the need for continuing administration of immunosuppressive drugs remains. Intervention designed to establish tolerance has never been tested on any substantial scale, mainly because the drugs are already so effective. Cancer immunology draws part of its inspiration from studies on tolerance. An example that is attracting much attention in the mid-1990s is the array of antigens expressed by melanoma cells, each of which could potentially be used for vaccination against this form of cancer. Several of these antigens are tissue specific, i.e., they are expressed on melanoma cells and also on normal melanocytes (mentioned in Section VII). Although the immune system does not normally react against these antigens because they occur in a privileged site, it becomes able to do so as expected once the cancer spreads.

The newer understanding of tolerance has its main impact on the diseases in which the immune system itself causes damage: the autoimmunities, allergies, and chronic tropical infections such as Chagas' disease in which there is a major component of damage to the body mediated by the immune system. Mention has already been made of the likelihood that these diseases are driven by T cells, and the hope of restoring tolerance in the periphery is propelling new approaches to their therapy. Everything in the preceding discussions about peripheral tolerance and antigen presentation is relevant to these diseases. There are at present two overall ideas about autoimmune disease, one of which is that what matters most is cytokine imbalance. Local imbalance probably resulting from infection is thought to trigger autoimmunity, in which case the logical aim of therapy would be to restore the balance by means of inhibitory cytokines (for instance, by using oral tolerance), anticytokine antibodies or constructs, or implanted cytokine genes. The other idea is that what matters is some form of disease-inducing antigen presentation, in which case one might hope to block the TCRs involved. Central to this latter idea is the concept of a "mimetope," an antigenic peptide derived from a microorganism that mimics a self-molecule and thus induces an autoimmune response; as yet no mimetope has been identified with any certainty. It is very likely that both of these ideas will be found to be relevant, although the approach via cytokines seems at present more immediately applicable.

BIBLIOGRAPHY

Arnold, B., Schönrich, G., and Hämmerling, G. J. (1993). Multiple levels of peripheral tolerance. *Immunol. Today* **14**, 12–17.

Matzinger, P. (1994). Tolerance, danger and the extended family. *Annu. Rev. Immunol.* **12**, 991–1046.

Mitchison, N. A. (1994). Immunological tolerance and its implications for autoimmunity. *In* "Immunopharmacology of Joints and Connective Tissue" (M. E. Davies and J. T. Dingle, eds.), pp. 35–52. Academic Press, San Diego.

Nossal, G. J. V. (1995). Choices following antigen entry: Antibody formation or immunological tolerance? *Annu. Rev. Immunol.* **13**, 1–28.

Sercarz, E. E., Lehmann, P. V., Ametani, A., Benichou, G., Miller, A., and Moudgil, K. (1993). Dominance and crypticity of T cell antigenic determinants. *Annu. Rev. Immunol.* **11**, 729–766.

Immunology of Parasitism

G. V. BROWN
G. F. MITCHELL
The Walter and Eliza Hall Institute of Medical Research, Australia

GLOSSARY

Antigen presentation Process by which antigens are taken up, processed, and presented to the immune system—in particular, T cells—by a variety of cells, only some of which have specificity for antigen, namely, macrophages, dendritic cells, and B cells

Cytokines Macromolecular products of cells; generally used in an immunological context, although, unlike antibodies, they do not have specificity for antigen. They affect the activities of other cell types. Lymphokines are from lymphocytes (e.g., interleukin-2) and monokines are from monocytes, although it is becoming clear that many cell types can produce cytokines of any particular type

Hypersensitivity reaction Visible reaction to injection, ingestion, or inhalation of antigen that can manifest rapidly as wheezing, tissue swelling, and the like (i.e., immediate hypersensitivity) or after some hours (i.e., delayed-type hypersensitivity). Classically, the former is dependent on immunoglobulin E antibodies bound to mast cells (as in allergy to grass pollens), whereas the latter is initiated by inflammatory T cells

Immunuglobulin E antibody Type of antibody that binds to mast cells in tissues. Other antibody types (known as immunoglobulin isotypes) include IgA, IgG, and IgM. Upon contact with antigen, the mast cell releases a variety of pharmacologically active molecules (e.g., histamine) that lead to the signs and symptoms of immediate hypersensitivity

Monoclonal antibodies Antibody products of a perpetual cell line (called a hybridoma) derived from the fusion of a modified myeloma cell and an antibody-secreting cell

T-cell subpopulations CD4$^+$ and CD8$^+$ Division of the lymphocyte population derived from the thymus (i.e., T cells) into two broad categories according to surface molecules. In general, CD8$^+$ T cells participate in cytotoxic reactions (i.e., the killing of virus-infected cells), whereas CD4$^+$ T cells, as a result of the production of cytokines (actually lymphokines), help B cells in antibody production or engage in inflammatory responses (e.g., delayed-type hypersensitivity)

T-cell subpopulations Th$_1$ and Th$_2$ Further division of the CD4$^+$ T cells according to the type of lymphokine produced; for example, in general a Th$_1$ response results in γ-interferon production, whereas a Th$_2$ response leads to interleukin-4 production

AS A CLASS OF ORGANISMS, PARASITES ARE NOTED for their ability to produce chronic infection, although they sometimes cause acute illness. These organisms have a global distribution, but are generally concentrated among the rural poor in the tropical, less industrially developed parts of the world, where the insidious effects of parasitism by one or many species often go undetected in the vicious circle of poverty, malnutrition, and disease.

Because of the dependency of the parasite on the host and its propensity to cause chronic, even lifelong, infection, a successful parasite must not kill large numbers of individuals of the host species, at least in prereproductive life. A favorable balance must be established that ensures long-term survival and perpetuation of the parasite without endangering the host. To maintain this balance, partially effective host immune effector mechanisms against the parasite are

ENCYCLOPEDIA OF HUMAN BIOLOGY, Second Edition, VOLUME 4.

balanced by partially effective parasite immune evasion mechanisms. Interactions of a single parasite with genetically diverse hosts, or of genetically diverse parasites with a single host, produce a spectrum from total resistance to life-threatening disease. Overreaction by the host to the offending organism can lead to immunopathological consequences that may be as threatening to life as the pathogen itself.

One important method for preventing life-threatening disease is for the parasite to induce immune responses that prevent successive infections and further "colonization" by the same species, a phenomenon called concomitant immunity, but also known as premunition or nonsterilizing immunity. It is likely that sterilizing immunity is very rare, a situation that may also pertain to many viral infections. Resident parasites develop immune evasion mechanisms allowing their own survival, yet at the same time they induce or maintain responses that eliminate incoming parasites. Very efficient means for eliminating some protozoa can be thwarted if the parasite changes molecules expressed at the surface (antigenic variation). Age–prevalence and age–intensity curves obtained from epidemiological data in endemic areas strongly suggest that the development of substantial immunity to reinfection occurs in some human parasitic diseases in the face of continuing exposure. Even when apparent age-related resistance is not demonstrable, concomitant immunity can be operating very efficiently.

Another distinguishing feature of parasites, as a class of potentially pathogenic organisms in humans, is that none can be prevented currently by vaccination. There are, however, a few attenuated organisms and defined antigens that are used as vaccines in veterinary medicine. The task of developing prophylactic or therapeutic vaccines against these complex organisms is enormous: parasite species are genetically diverse, and the genetic diversity of human hosts is, of course, extreme.

Topics discussed in this article are those of the most biological interest, namely, the nature of host-protective immune responses and the mechanisms of counteraction by immune evasion. Brief mention is also made of two other aspects of parasite immunology: immunodiagnosis and immunopathology.

I. HOST-PROTECTIVE IMMUNITY

It is not surprising that practically all known immune effector mechanisms have been documented or implicated, often at every vulnerable stage of the life cycle, to limit parasite entry, proliferation, or continued residency. These mechanisms include complement-fixing antibodies for larval cestodes; opsonizing antibodies for many intracellular and extracellular protozoa; eosinophil-, macrophage-, and neutrophil-mediated antibody-dependent cellular cytoxicity for schistosomules, microfilariae, and infective larvae; intracellular destruction following cytokine production by lymphocytes and/or macrophages for red blood cell-, macrophage-, and lymphocyte-dwelling protozoa; immunoglobulin A (IgA)-mediated effects (including maternal antibody effects) in *Giardia* infections; and, recently, cytotoxic T-cell activities (or at least $CD8^+$ T-cell effects) for exoerythrocytic stages of plasmodia. [*See* Cytokines and the Immune Response; Lymphocytes; Macrophages; Neutrophils.]

Many proposed mechanisms of immunity are based on observations from *in vitro* or animal model systems that are of varying relevance to human parasitism. After identifying what can happen *in vitro* or in model systems, the laboratory scientist must proceed to the more difficult task of demonstrating that the proposed mechanism actually takes place in humans. The multiplicity of interactions between a mammalian host and a rapidly proliferating antigenically variable parasite (with a genome size of approximately 2×10^7 base pairs) that completes several life cycle stages in one host is several orders of magnitude more complicated than the interaction of a host and a relatively invariant virus (with a genome size of approximately 2×10^5 base pairs or less). Despite the striking advances of the past decade, there are great deficiencies in our understanding of the relative importance in *natural* host–parasite interactions of immunological mechanisms clearly demonstrated *in vitro,* and it is likely that the host mounts several different types of immune attack against every different stage of the parasite. One powerful new approach that should facilitate understanding of the host response is provided by the technology to disable specific genes in mice (gene knockout) and then study the response to infection.

Most of the mechanisms described here reflect an interplay between various subpopulations of T cells reactive to specific antigens or B cells with the capacity to produce various kinds of antibodies and the activities of inflammatory cells (i.e., macrophages, eosinophils, neutrophils, mast cells, platelets, or fibroblasts) with no innate antigenic specificity. Of specific relevance to immunoparasitology is the method by which the host is exposed to the antigen, ranging from intravenous inoculation (e.g., hemoprotozoa) to direct introduction into lymph nodes (macrofilaria) or presentation by parasites dwelling in antigen-presenting macrophages (e.g., *Leishmania*). The route of antigen presentation can have profound effects on the immune response. For example, intravenously administered antigen is associated with low immunogenicity or tolerance, and for malaria it is clear that the immune response elicited in this way by repeated infections and drug cures might be different from the response to subcutaneous vaccination in the presence of adjuvant. [*See* Malaria.]

Characteristically, initial parasitic infection occurs in a young host, often in the presence of maternally acquired antibody. A "trickle infection" with multiple challenge at low doses in a semi-immune host might allow efficient antigen presentation and induction of effective immunity. A different result might occur with massive challenge in a nonimmune adult.

Parasites are such complex organisms that literally thousands of antigens may be released from living or dying parasites. Consequently, the immune system may be diverted toward thousands of foreign "housekeeping" or other intracellular molecules that induce responses irrelevant to host protection and many of the responses demonstrable *in vitro* may fall into this category.

As in other branches of the immunology of infectious disease, it is hoped that cytokine research will clarify the relevant signals derived from T cells and other cells that not only orchestrate the immune responses induced by pathogens, but may even have direct effects on the parasites themselves. In general, the *in vitro* tests of immune function can view only part of a phenomenal cascade of events initiated by the contact of host and parasites.

Medical textbooks state correctly that the hallmarks of helminth parasitic diseases, as for immediate hypersensitivity reactions, are eosinophilia and IgE production. Precise definition of the functions of these host responses has been difficult. Model systems do not provide good examples in which IgE antibodies are indispensible for host resistance through, for example, elicitation of an immediate hypersensitivity response that creates a hostile local environment. Eosinophils participate in responses against schistosomules and eggs in *Schistosoma mansoni* infections but the range of positive and negative effects that such cells may have on host resistance to helminths is not known. Monoclonal antibodies have been described that cause antibody-dependent cellular cytotoxicity *in vitro*, and other monoclonal antibodies of a different class can block this cytotoxic reaction.

II. IMMUNE EVASION

A particularly interesting aspect of the immunology of parasitism is the study of the mechanisms used by parasites to evade expression or inhibit induction of host-protective immune responses (i.e., immune evasion). Two broad categories of mechanism are (1) an alteration of antigen display or immunogenicity of the parasite and (2) an alteration of immune responses by the host. Clearly, certain mechanisms within these two categories (see the following sections) are closely related, and it is important to emphasize that it is highly unlikely that a successful parasite will have evolved only one mechanism of immune evasion. As indicated earlier, gene knockout approaches provide powerful means to identify host-protective immune responses; the same applies to gene knockout parasites in identification of relevant immune evasion mechanisms. Such studies are only now becoming possible.

Within the category of altered antigen display by the parasite are mechanisms such as antigenic variation (see Section II,A) and population heterogeneity with regard to antigen expression; sequestration in cells, cysts, and other immunologically privileged sites; masking and modulation of antigens; molecular mimicry; and reduced expression of associative recognition molecules (major histocompatibility complex-encoded molecules and costimulatory and adhesion molecules) for T cells.

A special situation applies to intramacrophage protozoa that thwart the hostility of their intracellular environment by inhibiting the fusion of phagosomes with lysosomes, escaping from the phagolysosome to lie in the cytoplasm, or inhibiting the action of digestive enzymes while resident in a phagolysosome within the macrophage. Some parasites at the time of entry may also alter the oxidative burst of phagocytes, which is a potent antiparasite mechanism.

Within the category of altered immune responses are specific and nonspecific immunosuppression (the latter through a multitude of possible mechanisms), immune deviation (see Section II,B), elaboration of cellular toxins, anti-inflammatory and anti-complementary factors, and digestion of bound antibodies. IgA, the immunoglobulin active at mucosal surfaces, can be cleaved by proteases secreted by some bacteria.

Two mechanisms are discussed in more detail: antigenic variation in trypanosomes and plasmodia and immune deviation in cutaneous leishmaniasis.

A. Antigenic Variation

African trypanosomes (the organisms responsible for sleeping sickness) have developed a sophisticated method for evading the host immune response. Parasites swimming freely in the blood are covered by a tough glycoprotein coat that envelops the entire plasma membrane of the organism and protects it against nonspecific host defenses. Furthermore, in the face of specific immune challenge, the parasite is able to change its surface coat, thereby rendering it invisible to the immune memory response. One hundred variant antigenic types have been documented to arise from a single cloned organism by the serial expression of a family of genes.

It is now clear that antigenic variation also occurs during chronic malaria infection. Once again, it is the erythrocytic form of the parasite that can express a wide range of variant (var) genes, but the variant antigens, rather than being expressed on the parasite, are found on the surface of the malaria-infected erythrocytes. This mechanism of immune evasion has been demonstrated in rodent and simian malaria and is generally believed to play a role during chronic human infection with *Plasmodium falciparum*. The mechanism for serial expression of antigenic variants is unknown, but these variants of cloned populations multiply the heterogeneity that already exists in antigenically diverse parasite populations.

B. Immune Deviation

Some parasites have the capacity to direct host immune responses along pathways that are unproductive for host resistance and the elimination of parasites. One of the best examples that has emerged from recent studies in the mouse concerns the causative organism of human cutaneous (Old World, zoonotic) leishmaniasis, the sandfly-transmitted protozoan *Leishmania major*. It is now clear that T cells of one broad type, namely, $CD4^+$ T cells, can promote either host susceptibility in mice or, alternatively, host resistance, by activating macrophages to kill their intracellular parasites. Activation of the $CD4^+$ T-cell type that secretes γ-interferon can aid in resistance (Th_1 cells), whereas activation of the $CD4^+$ T-cell type that secretes interleukin-4 (Th_2 cells) is largely counterproductive. If the parasite can divert the T-cell response along the interleukin-4-secreting pathway rather than the γ-interferon-secreting type (i.e., Th_1 $CD4^+$ T cells), then the parasite will not be eliminated.

Consistent with this hypothesis, the administration of antibody against γ-interferon exacerbates disease, whereas antibody to interleukin-4 is therapeutic. Moreover, injection of IL-12, which promotes γ-interferon production, is protective. Relative dominance of the two types of $CD4^+$ T cell appears to determine the outcome in hosts of different genotypes. How the different subsets are induced is not known, but a difference in the modes of antigen presentation is the favored hypothesis.

Another mechanism of immune deviation may operate in malaria and Chagas' disease. These parasites all express protein antigens with an unusual structure of blocks of tandemly repeated amino acid sequences. Immunodominant epitopes in these tandem repeat antigens may direct B cells into a network of irrelevant antibody production when host protection demands a response to a less dominant, but biologically more important, epitope. The response to a smokescreen of irrelevant antigens induces the gross splenomegaly (i.e., spleen enlargement) and hypergammaglobulinemia (i.e., high antibody content in the blood) typically associated with these infections, but the host remains unprotected.

Immune evasion mechanisms are of interest not only for the immunobiologist but also for those attempting to develop vaccines. If the molecular bases of some of the immune evasion mechanisms mentioned in this section were known, opportunities for neutralizing these mechanisms immunologically (i.e., with a therapeutic vaccine) or with drugs could emerge.

III. IMMUNODIAGNOSIS

Detecting parasites in human hosts is often difficult, impractical, or tedious. For these reasons indirect methods are often used, the time-honored method being the detection of antiparasite antibody in the serum (e.g., immunodiffusion, agglutination, or direct reaction with parasites, such as circumsporozoite precipitation for antisporozoite immunity or the circumoval precipitin test for schistosomiasis involving precipitation around eggs). The most common assays are the enzyme-linked immunosorbent assay or the related radioimmunoassay. In these tests, antigen anchored to a solid support (e.g., a plastic dish) is reacted with test serum, which could contain immunoglobulin specific for the antigen, and after washing, the complex of antigen and immunoglobulin is detected with anti-immunoglobulin labeled with enzyme or radioactivity. High radioactivity, or development of a color change after addition of the enzyme substrate, is indicative of the presence of antibody. [*See* Radioimmunoassays.]

Visualizing parasite eggs in feces or parasitized erythrocytes in a blood film provides direct evidence of current infection, but, in contrast, the presence of serum antibody is simply an indicator of past or present exposure to the parasite. Previously infected and/or highly resistant individuals in an endemic area might be positive for antibody, whereas individuals recently infected or responding weakly may be negative in an insensitive test. Skin tests for immediate (i.e., IgE-dependent) or delayed-type (i.e., T-cell-dependent) hypersensitivity reactions also present problems in interpretation akin to those of serological tests. Moreover, cross-reactions with other parasites could reduce the specificity, although this problem can be reduced by the availability of non-cross-reacting epitopes synthesized chemically or produced using recombinant DNA techniques. [*See* Recombinant DNA Technology in Disease Diagnosis.]

High specificity and sensitivity are required to increase the predictive value of a diagnostic test, and many immunodiagnostic tests for parasitic infection fall far short of the ideal. Much recent effort has been devoted to the development of immunoassays for the detection of parasite antigens or metabolites, generally in serum, but also in urine and in feces. The advent of monoclonal antibodies has provided the necessary tools to achieve this. One monoclonal antibody anchored to a solid support is reacted with serum containing circulating parasite molecules, and bound antigen is detected with a second labeled monoclonal antibody. Theoretically, the best parasite product for detection in serum is one that is poorly immunogenic so that in the infected host it is not masked or cleared by induced antibody. It is hoped that quantitative antigen detection methods [and gene amplification techniques such as the polymerase chain reaction (PCR)] will provide some quantitative information on the *number* of parasites present in individuals, a critical aspect of many epidemiological studies in which the severity of pathology appears to correlate with the intensity of infection.

Monoclonal antibodies are being used for detection of parasites in vectors (e.g., malaria sporozoites in mosquitoes or *Leishmania* promastigotes in sandflies) and provide powerful tools for species differentiation (e.g., leishmaniasis) and for the detection of antigenic variants of parasite species. Monoclonal antibodies, of course, provide highly specific tools for the analysis of parasite epitopes recognized by B cells; we still lack such convenient tools for the study of T-cell-stimulating epitopes [*See* Monoclonal Antibody Technology.]

IV. IMMUNOPATHOLOGY

Some antiparasite immune responses are beneficial for the host, but, as indicated in Section II, some are beneficial for the parasite. An exaggerated immune response to a parasite can be deleterious to the host, as shown by chronic schistosomiasis, which is the classic example of immunopathology in parasitic disease. In this snail-transmitted helminth disease, eggs laid by the female worm become impacted in organs and tissues. T cells respond to antigens emanating from eggs and produce lymphokines that result in granuloma formation and fibrosis around the eggs.

Some walling off of the eggs is desirable for the host, in that adjoining cells (e.g., hepatocytes) are protected, but large granulomas and fibrosis, reflecting an exaggerated cellular response in organs such as the liver, ultimately impede blood flow, leading to the serious complication of raised pressure in the portal vein (i.e., portal hypertension). Other pathological consequences follow from this. Interestingly, exaggerated responsiveness decreases as the disease progresses, and the host benefits by producing a small controlled granuloma that is less likely to cause obstruction. The immunological basis of this phenomenon of granuloma modulation is of great interest, for in theory the same mechanism induced by vaccination could prevent severe disease in chronic schistosomiasis by a mechanism akin to desensitization. Alternatively, the disease-preventing vaccine could act to inhibit egg production or viability. A vaccine that prevents infection would obviously be more desirable.

Onchocerciasis and Bancroft's filariasis (i.e., elephantiasis) are other human diseases in which exaggerated immune responses to antigens of adult worms or microfilariae are believed to be the major components of pathology. Immunopathology can be exacerbated by chemotherapy, which causes massive killing of microfilariae in patients with onchocerciasis. The peculiar tropical splenomegaly syndrome (i.e., hyperreactive malarious splenomegaly) seen in malaria-endemic areas is another example of an immunopathological consequence of infection in genetically predisposed individuals. It is clear that many of the manifestations of severe malaria, including respiratory distress, defective red blood cell production, hepatic dysfunction, fever, and hypoglycemia, could be produced by tumor necrosis factor, acting in concert with other cytokines, and acute manifestations of cerebral malaria may be a direct result of downstream effects of cytokine release such as production of nitric oxide or tumor necrosis factor. An accompaniment of concomitant immunity is the development of tolerance to the pathological effects of these cytokines, perhaps through masking of critical parasite epitopes by antibody or by "tolerance" with a diminution of symptoms during later infections. Other immunological responses that may have pathological consequences are hypergammaglobulinemia, the formation of complexes of antibodies with antigen in the blood and tissues, allergy, and autoantibody production.

Clearly, the analysis of immunopathology is a critical aspect of vaccine development, it being necessary to ensure that the vaccine sensitizes only for host protection in humans, but not for immunopathology. This could be particularly so in a vulnerable subset of vaccinees who might be genetically predisposed to respond differently from the majority of the population. [*See* Immunopathology.]

V. VACCINE DESIGN

The first approach to vaccine design is to induce the response that occurs naturally in the proportion of individuals who develop high levels of resistance in response to repeated infection. The target antigens are called natural antigens. There are few examples of vaccines that induce resistance better than natural infection (tetanus toxoid perhaps being one). An alternative approach is to focus an immune response on molecules of the organism that are not recognized well during natural infection and that serve a critical function for the parasite, such as a receptor or an enzyme. These are often referred to as novel antigens. An attenuated organism, that is, one altered in such a way that it loses virulence yet remains immunogenic, could provide the best vehicle for presenting natural antigens, whereas a molecularly defined product could theoretically be better as a novel antigen vaccine. Information relevant to the first approach has been gained through the study of naturally induced immunity by looking for responses correlating with immunity (e.g., antibody specificities or isotypes in serum of protected hosts that are absent in exposed but nonimmune individuals). Novel antigens are more likely to be discovered by finding specific receptors or enzymes involved in host–parasite interactions.

Vaccination to stimulate host resistance may be therapeutic for the treatment of current infection, or prophylactic to prevent initial infection. Successful immunotherapy has been reported in Venezuela against leishmaniasis using a mixture of killed organisms with bacille Calmette–Guérin (BCG) adjuvant, similar to the treatment of lepromatous leprosy using armadillo-derived *Mycobacterium leprae* with BCG. Immunotherapy is said to be as effective as chemotherapy in limited human trials performed to date. Molecular vaccines against human parasites that are at early stages of development include (1) vaccines against *P. falciparum* malaria, based on portions of the circumsporozoite protein of sporozoites and several merozoite antigens of blood stages of *P. falciparum* (the most advanced vaccine, a synthetic peptide produced by Dr. Manuel Patarroyo and colleagues, has been tested in field trials in South America, Tanzania, and The Gambia), (2) vaccines against schistosomes, based on integral membrane and other surface proteins and functional enzymes of the worms, and (3) both glycolipid and protein vaccines in leishmaniasis. Trials of a defined antigen vaccine against taeniasis have demonstrated very high level protection of susceptible sheep.

BIBLIOGRAPHY

Behnke, J. M. (1990). "Parasites: Immunity and Pathology. The Consequences of Parasitic Infection in Mammals." Taylor & Francis, New York.

Bloom, B. R. (1986). Learning from leprosy: A perspective on immunology and the Third World (Presidential address). *J. Immunol.* **137**, i–x.

Burnet, F. M., and White, D. O. (1972). "The Natural History of Infectious Diseases." Cambridge Univ. Press, Cambridge, England.

Cohen, S., and Warren, K. S. (1982). "Immunology of Parasitic Infection," 2nd Ed. Blackwell, Oxford, England.

Mims, C. A. (1995). "The Pathogenesis of Infectious Disease." 4th Ed. Academic Press, London.

Mitchell, G. F. (1986). Cellular and molecular aspects of host–parasite relationships. *Prog. Immunol.* **6**, 798–808.

Wakelin, D. (1984). "Immunity to Parasites: How Animals Control Parasitic Infection." Arnold, London.

Wyler, D. J. (1990). "Modern Parasite Biology. Cellular, Immunological and Molecular Aspects." Freeman, New York.

Immunology, Regional

J. WAYNE STREILEIN
Harvard Medical School

GLOSSARY

Immune deviation Unique spectrum of immune effectors elicited by antigens under special immunizing circumstances; typically, one or another effector modality (e.g., delayed hypersensitivity, complement-fixing antibodies) is selectively deleted from the systemic immune response to a particular antigen

Immune effectors Antibodies and T lymphocytes that possess effector functions associated with immune responses

Immunologic privilege Partial to complete failure of the immune system to respond to and eliminate antigenic tissues or grafts that are placed in special body sites (e.g., anterior chamber of the eye, brain, testis, cheek pouch of Syrian hamster)

Lymphocyte traffic Capacity of lymphocytes to travel from one site of the body to other sites, using the bloodstream and/or lymphatic channels as the medium for migration

Lymphoid organs and tissues Sites in the body wherein lymphocytes are normally generated and accumulate; these sites include lymph nodes, spleen, thymus, Peyer's patches, tonsils, appendix, and lymphocyte collections along mucosal surfaces

Mucosal immunity Specialized form of immune response to antigens encountered first through mucosal surfaces; in general, the most important feature of mucosal immunity is that the predominant immunoglobulin produced is IgA

Parenchymal cells Differentiated cell type that provides the unique function of an organ or tissue (e.g., keratinocytes in skin, hepatocytes in liver)

Regional specialization Modification of the generic systemic immune response to suit the special physiologic needs of various regions (organs and tissues) of the body

Selective immune deficiency Characteristic of individuals with immune deviation; one (or more) immune effector modality is not generated in response to an antigenic exposure

IN CARRYING OUT ITS PRIMARY FUNCTION OF PROviding protection against life-threatening pathogens, the immune system has developed an extraordinarily diverse array of antigen-specific effector modalities. Effector cells and molecules differ with respect to their ability to eliminate or eradicate various pathogens. Moreover, certain effectors are more appropriate for pathogens in one as opposed to other types of organs and tissues. As a consequence of these two considerations, the immune system has devised unique strategies for providing protection at individual tissue sites—protection that is appropriate for the pathogen and commensurate with preservation of the unique function of the individual tissue. The study of specialization in immune responsiveness according to different regions of the body is called regional immunology.

I. INTRODUCTION

Enormous advances have been made in the past few years in describing the cellular and molecular mechanisms responsible for the proper functioning of the immune system. As our knowledge of this extraordinary system has grown, we have begun to appreciate its diversity. The immune system possesses a broad range of strategies for dealing in a highly specific manner with a wide array of different antigens and pathogens. In this regard, not all immune effector modalities have comparable potential to neutralize, eliminate, or otherwise nullify specific pathogenic agents. In a sense, such a wide array of effector modalities addresses the evolutionary need to cope with the great diversity of pathogenic agents in our universe. [*See* Immune System.]

Beyond this type of diversity, there is another level of complexity that is not necessarily implied by these considerations. Experimental evidence indicates that the quality and quantity of immune responses to antigenic challenges in various organs and tissues of the body are not necessarily equivalent. One of the earliest reported examples of this phenomenon was the description of immunologic privilege in the anterior chamber of the eye. More than a century ago, investigators discovered that tumor tissues implanted into the anterior chamber of rabbit eyes often survived and even grew in that site, whereas similar tumor tissues implanted in the subcutaneous space did not grow or survive. This phenomenon was called immunologic privilege. Over the years, immunologic privilege has been described for several other unique sites of the body: brain, testis, ovary, cartilage, and cheek pouch of Syrian hamsters. Until very recently, immunologic privilege was regarded simply as a laboratory curiosity rather than as an expression of an important, if enigmatic, aspect of systemic immune responsiveness.

The most clear-cut example of regional specialization in immune responsiveness is the presence of large amounts of IgA antibodies in external body secretions. Exposure to exogenous antigens via the gastrointestinal tract results in the production of IgA antibodies, which appear in secretions at mucosal surfaces but are poorly represented in the serum or in extravascular tissue compartments.

In this article, the conceptual framework upon which the idea of regional immunology is based is developed, specific examples of regional spheres of immunologic influence are described, and examples are presented of pathogenic consequences that may result from tissue-directed specializations in the immune response.

II. CONCEPTUAL FRAMEWORK OF REGIONAL IMMUNOLOGY

A. Existence of Uniquely Differentiated Nonlymphoid Tissues and Organs

The body is composed of different organs and tissues that in the aggregate work toward the common goal of sustaining and propagating life. While processes that are responsible for creating these different tissues from the zygote are beyond the scope of this article, certain features of this ontogenetic sequence are of particular interest to immunology.

In the eventual emergence of a fully differentiated organ or tissue (such as liver or skin), specificity is created via complex genetic paradigms. Unique genes are activated in cells that will become the specific (parenchymal) cells of a particular organ or tissue, and unique protein products are produced, proteins that are responsible in part for the fully differentiated state. Development and maintenance of the functional specificity of individual tissues and organs are also influenced by stromal elements. The skin affords a good example: the diverse properties of epidermis found on the sole of the foot, on the scalp, around the eyes, and covering the tongue appear to be directed by inducing factors released from underlying dermis. This example emphasizes that parenchymal cells of organs and tissues remain susceptible to modulation by external elements, even in the fully differentiated state—a point of importance in regional immunology.

In adult organisms, specific organs and tissues maintain their physiologic form and function by several different mechanisms: (1) endogenous genes encode specific proteins, which in turn endow individual cells with specific functions; (2) cells influence each other via cell-to-cell communication networks, such as cytokines and receptors, cell-adhesion molecules, and electrical coupling (gap junctions); and (3) stromal connective tissue cells, which also are capable of elaborating cytokines, undoubtedly serve as a further stabilizing influence.

Thus, fully differentiated, nonlymphoid somatic tissues carry out their unique physiologic functions and assume their unique microanatomical forms as a consequence of sustained activation of endogenous genes of parenchymal cells and genes acting via cells of the supporting connective tissue. The dynamic nature of these cellular communication devices is a crucial consideration for the concept of regional immunology.

B. Existence of Differentiated Lymphoid Organs and Tissues

The creation and development of the various organs and tissues of the lymphoid apparatus (lymph nodes, spleen, etc.) follow similar rules of ontogeny. The major themes of intercellular communication mediated by cytokines and their receptors, and of interacting cell-adhesion molecules, including specific receptors for ligands on the surfaces of other cells, are typical for both lymphoid and nonlymphoid tissues. However, throughout the life of the organism, the immune system (as well as the entire hematopoietic system) retains the capacity to generate fully differentiated cells from the most primitive pluripotent mesenchymal stem cells. Moreover, these ontogenetic processes are conducted in an almost ad hoc fashion in diverse regions of the body. The participating cells frequently migrate, via the blood, from site to site before completing their differentiation process. Perhaps as a consequence of this need to assemble and disassemble interacting stable multicellular units, gap junctions and electrical coupling are not typically utilized by lymphoreticular cells. Therefore, interacting lymphohematopoietic cells rely primarily on cytokines and receptors and cell-adhesion molecules to exchange information and to influence each other. An extensive (if still incomplete) array of growth- and differentiation-promoting cytokines and receptors have now been described for the developing and intercommunicating cells of the immune and hematopoietic cell lineages. Our knowledge of the diversity of cell-adhesion molecules that also promote these processes is less advanced, although progress in this field of research is now moving at a rapid pace. It is important to bear in mind that not all intercellular communication is designed to promote or up-regulate an effect; some cytokines can differentially down-regulate (decrease) certain functional properties of target cells. [*See* Hemopoietic System.]

Over the past decade, it has been learned that cytokines used by nonlymphoid cells to exchange information are similar (often identical) to the cytokines used by interacting lymphoreticular cells. In fact, the receptors for these cytokines are often identical. Thus, the potential is great for signals derived from one type of cell to influence the functional activities of cells of a different tissue lineage and this influence probably has physiological significance. [*See* Cytokines in the Immune Response.]

C. The Perspective of the Immune System

Numerous immunological issues are raised when the immune system is called upon to respond to antigens in physiologically diverse regions of the body. Some of the most important issues are described in the following sections.

1. Unique Tissue-Specific Molecules (Antigens)

If differentiated tissues express one or more molecules (typically proteins) that are unique for that tissue alone, are these molecules incorporated into the immunologic definition of "self"? Autologous proteins that come to be regarded as self by the immune system do so (largely, if not completely) because lymphocytes are exposed to these autologous molecules before their developmental program is complete. For T cells, this ontogenetic process is thought to occur within the thymus. If certain proteins are first expressed in a specific organ (e.g., surface molecules expressed on mature sperm) rather than in the thymus, and if this expression occurs after the full complement of mature T cells has disseminated from the thymus to peripheral tissue, then these proteins may not be regarded by the immune system as self. Other proteins may escape designation as self by the immune system because they are synthesized and subsequently remain "sequestered" only behind blood–tissue barriers (e.g., the S antigen of the retina). Molecules, which for these (and other) reasons never become incorporated into an immunologic definition of self, may serve as autoantigens. Therefore, tissue-restricted molecules can become the targets of autoimmune attack that can destroy the organs and tissues on which they are expressed.

2. Detection of Exogenous Antigens by the Immune System

Given that organs and tissues possess physicochemical barriers that limit the penetration of antigenic and pathogenic agents into the body, how do/can cells of the immune system become aware of these exogenous agents? The epidermis is a good example of an important barrier. The barrier itself includes the stratum corneum and the basement membrane that separates the dermis from the epidermis. Langerhans cells, located within the epidermal compartment, appear to be specifically designed to pick up and transport foreign molecules to the draining lymph nodes, where T cells normally reside. This mechanism maximizes the likelihood that pathogens will be detected by the immune system. Alternatively, in the case of the eye and the brain, poor if any effective lymphatic drainage pathways exist. This circumstance severely limits the possibility that T cells in regional lymph nodes could become aware of antigens in these sheltered organs.

3. Access of Immune Effectors to Regionally Located Antigens

To rid a local tissue of a specific antigen or pathogen, how do immune effector modalities gain access to that tissue? Effector modalities, such as specifically sensitized T cells and antibodies, are not chemotactically drawn to tissue sites by antigen. Instead, accumulation of immune effectors at tissue sites occurs by nonspecific means, relying first on random diffusion and migration of cells and molecules from peripheral blood into (virtually all) accessible sites in the body, and second on non-antigen-specific chemotactic gradients and by cell-adhesion molecules that are established by random encounters between effectors and antigens at tissue sites. Endothelial cells of the microvasculature are critical to these processes; for example, organs and tissues served by capillaries with endothelial cells united by tight junctions (such as the brain) represent formidable barriers to the infiltration of immune cells and molecules, whereas fenestrated capillaries, such as those found in skin, promote the accumulation of blood-borne cells and molecules. [*See* Natural Killer and Other Effector Cells.]

4. The Problem of Immune Effector Response Amplification

In virtually all immune responses, successful eradication of antigens and pathogens relies on recruitment of nonspecific (nonadaptive) inflammatory defense mechanisms, which include macrophages, granulocytes, complement components, and so on. The remarkable amplification potential of inflammation depends on recruitment of inflammatory cells, which is mediated via cytokines secreted by immune effector cells and is further amplified by cytokines released from the recruited cells themselves. Local factors, already present within specific organs and tissues (presumably subserving other physiologic purposes), may promote or interfere with the recruitment and amplification process. A good example of the latter is the presence of transforming growth factor-β (TGFβ) in normal aqueous humor of the anterior chamber of the eye. The relative inability of the anterior chamber to display and sustain delayed hypersensitivity responses can be ascribed in part to the profoundly inhibitory properties of TGFβ on immunogenic inflammation. [*See* Inflammation.]

D. The Perspective of Specific Organs and Tissues

Important issues are raised for parenchymal cells of specific organs and tissues by the possibility of expression of immunity within their midst.

1. Avoidance of Autoimmunity

The existence of microanatomical specializations that have the effect of isolating certain tissue-specific molecules and cells from the general circulation is probably not an accident. In the testis, numerous tight junctions encircle and bind together Sertoli cells. As a consequence, developing sperm, which are imbedded in Sertoli cells, are beyond the reach of potentially hazardous molecules and cells of the blood and of the interstitial spaces of the seminiferous tubules. In an analogous fashion, the continuous capillaries of the brain and retina form a barrier, ensuring that active mechanisms alone can permit entry of molecules and cells from the blood. Arguably, part of the reason for these anatomical specializations is the need to prevent the immune system from recognizing unique molecules (antigens) expressed on cells beyond these barriers (see earlier discussion). Although simple mechanical barriers alone are not considered sufficient to account for the avoidance of autoimmunity in these tissues, barriers of this type certainly contribute.

2. Maintenance of Differentiated Tissue Function

In carrying out their unique physiologic functions, each organ and tissue must face the possibilities that (1) it may be destroyed by an intruding pathogen and (2) the pathogen may evoke tissue-localized immune responses that are themselves deleterious. In reference to the latter, immune effector cells secrete cytokines into an environment when they mediate protection. Therein lies a dilemma. Parenchymal cells often express, or can be induced to express, receptors that can bind these very same cytokines, and this may produce deleterious consequences, especially if the lymphocyte-macrophage-derived factors alter the physiologic properties of the parenchymal cells. In addition, recruitment of nonspecific host defense mechanisms may create amplified inflammation, which proves to be detrimental within a confined organ or tissue space. For example, accumulation of blood-borne cells and fluids within a restricted bodily compartment (such as the brain, confined as it is by the bony skull) may compress parenchymal cells and produce tissue ischemia and death. Finally, the release of inflammatory mediators during an ongoing immune response may cause "innocent bystander" injury to parenchymal cells. These considerations reveal that expression of immunity within an organ or tissue may be deleterious to that tissue, even though the intent is one of protection.

III. THE THESIS OF REGIONAL IMMUNOLOGY

On the basis of these considerations, the thesis is advanced that each region of the body (organ or tissue) has the potential to modify the quality of immune responses to antigens and pathogens in such a manner as to achieve a unique spectrum of immune effector modalities that are (1) specific for the antigen(s) in question, (2) sufficient to effect elimination or eradication of the pathogenic agent, and (3) unlikely to interfere with the proper physiologic function of that organ or tissue.

The thesis states that the most important force dictating the quality and quantity of the tissue-tailored immune response is the parenchymal cell, which provides the tissue with its differentiated function. Thus, in skin, the epidermal keratinocyte is thought to assume the primary responsibility for setting the "tone" of local immune reactivity, whereas in the liver, the hepatocyte would be expected to play a similar role.

According to this thesis, the magnitude and type of immune protection afforded within each region of the body results from a compromise that must be struck between parenchymal cells and lymphocytes (chiefly T cells). This compromise takes into account factors such as (1) "ideal" immune effector modality for the pathogenic agent in question and (2) preservation of the functional properties of the differentiated organ or tissue. The compromise is mediated by intercommunications between immune cells (chiefly T cells and macrophages—dendritic cells) and parenchymal cells through cytokines, cell-adhesion molecules, and the extraordinary mobility and recirculation potential of lymphoreticular cells.

IV. COMPONENTS OF REGIONAL SPHERES OF IMMUNOLOGIC INFLUENCE

To accomplish regional specialization in immune responses, certain features appear to be essential ingredients in the process.

A. Tissue-Seeking Lymphocytes

Recirculating lymphocytes, especially T cells, possess the special capacity to migrate preferentially to one or another region of the body. Because migration of this type is *not* mediated by antigen, other mechanisms operate (see earlier). Tissue-tropism displayed by subpopulations of lymphocytes is mediated by different sets of tissue-specific homing receptors and their complementary ligands. As a consequence, some T cells migrate preferentially to lymphoid tissues associated with the gastrointestinal tract, whereas other T cells demonstrate a proclivity for trafficking to lymphoid tissues that drain the skin. Differential expression of homing receptors on T cells and of tissue-restricted ligands on high endothelial cells of postcapillary venules (called high endothelial venules) in these disparate lymphoid compartments accounts for the existence of different migratory patterns. Within the past few years, our knowledge of the heterogeneity of lymphocyte migratory patterns has been expanded to include (1) homing receptors that direct certain T cells to the dermis of skin inflamed with active psoriasis or infiltrated with tumor cells and (2) homing receptors on other T cells that cause these cells to accumulate in the joints of patients with rheumatoid arthritis. Additional unique homing patterns (such as migrations to the brain, endocrine organs, and kidney) will probably be identified in the future. [*See* Lymphocytes.]

The mechanisms responsible for creating and maintaining diverse migratory pathways are essentially unknown at present. Homing receptors on lymphocytes do belong to classes of cell-surface molecules that function in cell adhesion. Beyond this, little is known about the factors that lead to the imprinting of recirculating lymphocytes in a manner that permits some to seek the skin while others seek the gastrointestinal tract.

B. Resident Antigen-Presenting Cells

The diversity of mechanisms by which antigens and pathogens gain access to different regions of the body implies that distinct strategies exist to promote recognition of antigens by lymphocytes. In some tissues, such as skin, specialized antigen-presenting cells (Langerhans cells) reside within the epidermis; by virtue of their functional properties, these cells confer upon this tissue site the specific ability to take up, process, and present antigens to lymphocytes. In addition, Langerhans cells that have captured antigen from the epidermal compartment can loosen their attachment to the epithelium, migrate across the dermal–epidermal junction, and flow with the afferent lymph to the draining lymph node. In other tissues, such as the gastrointestinal tract, parenchymal epithelial cells themselves are specialized (M cells) for the purpose of ensuring that antigenic materials in the lumen of the gut are transported across the epithelial surface into Peyer's patches. In these examples, evolution appears to have provided the organs and tissues with the capacity to maximize the likelihood that an antigen intruding into its environment will be detected by the immune system.

Alternatively, the eye and brain appear to have evolved toward a state that minimizes the likelihood that local processing and presentation of antigen can proceed in a *conventional* fashion. Although certain portions of the eye (cornea) are devoid of bone marrow-derived cells capable of antigen presentation, other portions (retina, ciliary body, iris, tubercular meshwork) contain significant densities of class II major histocompatibility complex (MHC)-bearing dendritic cells and macrophages. When tested *in vivo,* these cells are essentially incapable of activating conventionally primed or alloreactive T cells. Moreover, dendritic cells of the eye constitutively display the capacity to take up antigen, which is processed and presented in a fashion that activates the unique set of T + B lymphocytes associated with eye-derived immune deviation. By altering local antigen-presenting cells, the eye (and the brain) may be able to avoid evoking those types of immune effectors that are particularly deleterious to these organs.

C. Uniquely Differentiated Parenchymal Cells

Each distinct organ and tissue contains parenchymal cells that carry out the differentiated functions unique to that organ or tissue. The keratinocytes of skin, the absorptive epithelial cells of the small and large intestine, the hepatocytes of the liver, and the follicular epithelial cells of the thyroid gland are but four examples. These cells are directly responsible for creating a regional microenvironment that is appropriate for their physiologic function. When the cells of the immune system enter these diverse microenvironments, they may be diverted into new patterns of behavior by the factors and mediators that provide the microenvironment with its distinctive features. Because even the immune system conducts its affairs in this way, the microenvironment of a lymph node is unique and dictated largely by the secretory products of resident lymphocytes, chiefly T cells and macrophages.

The idea that each organ and tissue comprises a unique regional sphere of influence emanating from its definitive parenchymal cell is at the heart of the concept of regional immunology. According to this view, migrating lymphocytes regularly come under the temporary influence of diverse microenvironments and, as a consequence, their functions may be temporarily altered. Similarly, Langerhans cells (which are also migratory) may have their properties dictated by keratinocytes when they occupy the epidermal compartment. Alternatively, Langerhans cells that migrate to the draining lymph node may become functionally altered as they come under the influence of T cells in that site.

D. Specialized Endothelium of Postcapillary Venules

Certain T and B lymphocytes adopt at least two different migratory patterns: (1) to the lymphoid tissues associated with the gut and (2) to the lymphoid tissues draining the skin. It has already been mentioned that the pathway followed by any particular lymphocyte is governed by its own distinctive homing receptors and by complementary ligands expressed on specialized endothelial cells of postcapillary venules. The ability of endothelial cells to direct lymphocytes from the blood across the vessel wall into lymphoid organs depends, in part, on as yet unidentified factors elaborated by mature T cells. In addition, inductive influences also emanate from the parenchymal cells of tissues that drain into particular lymphoid organs, and these inductive factors help to modify the expression of appropriate ligands on postcapillary venules in these organs. Thus, specialization of stromal endothelial cells is an important component of a regionally distinct sphere of immune influence.

E. Draining Lymphoid Organs and Tissues

During primary immune responses to new antigens, the critical first step—antigen recognition by naive T cells—usually takes place within an organized peripheral lymphoid organ (lymph node, spleen) rather than in nonlymphoid tissues. The reasons for this requirement are beyond the scope of this article. Suffice it to say that the initial T-cell-inductive event initiated by antigen must take place in the protected microenvironment of a peripheral (secondary) lymphoid organ. We presume that antigenic signals that arise from nonlymphoid tissues are delivered to secondary lymphoid organs, where they can be transduced into appropriate effector modalities, but we know very little about this relationship. For example, excision of a lymph node prior to injection of antigen into skin can abort the expected immune response. A more dramatic example of a link between a tissue and its draining lymphoid tissue occurs with the eye. Antigens inoculated into the anterior chamber of the eye elicit a deviant systemic form of immunity, which selectively excludes delayed hypersensitivity and complement-fixing antibodies, whereas other types of immune effectors are retained. If splenectomy precedes the anterior chamber injection of antigen, a conventional immune response develops, replete with both delayed hypersensitivity and complement-fixing antibodies. Thus, the spleen serves as the draining lymphoid organ for the eye, and this fact emphasizes that a meaningful relationship exists between a draining lymphoid organ and the specific region it serves.

In summary, a functional regional sphere of immunologic influence comprises five elements: tissue-tropic lymphocytes, specialized antigen-presenting cells, parenchymal cells (and the microenvironment they create), specialized endothelial cells of postcapillary venules, and a draining lymphoid organ. When antigens arise from an organ or tissue, these elements interact to create and mold immune responses that are protective yet appropriate; that is, they do not interfere substantially with the physiologic function of the region in question.

V. IMMUNE CONSEQUENCES OF REGIONAL SPHERES OF IMMUNOLOGIC INFLUENCE

The existence of specialized immune responses that are regionally determined has important ramifications for the systemic immune response as well as for the well-being of the entire organism.

A. Unique Spectra of Immune Effectors

The immune system can marshal a wide variety of functionally distinct, yet antigenically specific, effector cells and molecules. On the cellular side, we recognize cytotoxic T cells, delayed hypersensitivity T cells, and helper T cells—of more than one type. On the humoral side, the human immune system generates four different isotypes of IgG and two isotypes of IgA, as well as IgM and IgE antibodies. Each of these isotypes has a special set of functional properties, typically related to molecular differences found within the Fc portion of each immunoglobulin molecule. As a further consideration, both cellular and humoral immune effectors can recruit macrophages, natural killer cells, granulocytes, complement components, clotting factors, and others to assist in mediating their effects. In response to antigens first confronted in a distinct region of the body, a unique spectrum of immune effectors is selected, presumably because of the compromise described earlier. The spectra of elicited effectors may vary from tissue to tissue and organ to organ of initial antigenic confrontation. Because these effectors are generated centrally and are then disseminated systematically via the blood, the spectrum of immune effectors elicited becomes identical with the systemic immune response itself. Thus, when antigens are first presented to mucosal surfaces, the selective local production of IgA antibodies rather than IgG also characterizes the systemic immune response.

B. Selection of Immune Effectors Creates Selective Immune Deficiencies

If a regional immune response comprises some, *but not all,* possible effector modalities, apparently the ignored effectors will eventually comprise a selective immune deficiency, at least with regard to the antigen in question. This is an important consideration because elimination of pathogens sometimes depends solely on a single immune effector modality—and all other modalities are superfluous. As an example, antibodies to antigens associated with the tubercle bacillus confer essentially no protective immunity; only T-cell-mediated immunity is able to eradicate this organism. Therefore, if a selective cell-mediated immune deficiency is created by regional exposure to the tubercle bacillus, the host would be deprived of a protective immune response, even though specific antibodies are produced. Thus, this immune deficiency would constitute a significant hazard to the host's survival.

The important conclusion from this set of statements is that although regional immune responses are created with the special needs of a specific tissue or organ in mind, the immune response that is generated is a systemic one. More specifically, a selective immune deficiency that is elicited by antigen in one region of the body will turn out to be a deficiency felt in all other regions of the body. This is the idea that brings home the importance of considering regional immune responses as "compromises."

C. When Regional and Systemic Immune Concerns Collide

Inevitably, the unique needs of a specific tissue are sometimes at odds with the needs of the entire organism. Not surprisingly, this conflict has been observed in experimental animal model systems in which regional immunity has been studied. For example, studies of experimental intraocular tumors have revealed that when tumor-associated antigens are relatively weak, the deviant systemic immune response that is generated protects the tumor and renders the host susceptible to death. By contrast, when intraocular tumors express strong antigens, a conventional immune response is generated, the host is spared, and the tumor is rejected, although frequently the eye itself is destroyed. Which factors determine whether regional concerns will dominate over systemic concerns or vice versa is not completely clear. Certainly, antigenic strength is an important variable, but much remains to be learned about other factors that influence this decision.

VI. EXAMPLES OF REGIONAL SPHERES OF IMMUNOLOGIC INFLUENCE

Several regions of the body are known to be subserved by specialized immune reactivities. These include the mucosal surfaces, the lower respiratory tract, the skin, the eye, the brain, certain endocrine organs, perhaps the liver, and the maternal–fetal interface. None is completely described or understood, but some are better characterized than others. As examples, three regional spheres of immunologic influence are described in Tables I and II.

In Table I, the five elements typically thought to contribute to regionally distinct responses have been indicated for the following regions: skin, mucosal surfaces, and eye. The parenchymal cells that dominate skin and mucosal surfaces are the respective epithelial cells. The dominant parenchymal forces within the eye are compartmentalized, with the retinal pigment epithelium and neuronal retina providing the microenvironment of the posterior portion of the eye, and the iris, ciliary body, and cornea creating the anterior microenvironment.

Local strategies for antigen presentation differ markedly for these three regions. Epidermal Langerhans cells provide the cutaneous surface with a pervasive network of dendritic processes that serve to trap antigens and pathogens crossing the stratum corneum and arising within the epidermis. By contrast, antigen uptake across the intestine and certain other mucosal surfaces is mediated by specialization of surface epithelial cells (M cells), which are strategically located on the lumenal surfaces overlying lymphocyte collections in the lamina propria (Peyer's patches). Antigens that pass through M cells are picked up by dendritic cells of the subjacent lymphoid compartment, which then prepare the antigens for presentation to lymphocytes. Thus, the capacity for antigen presentation in the skin seems to be distributed evenly across the integument, whereas this function is promoted across mucosal surfaces at discrete sites that may be separated from each other by considerable distances. Antigen presentation takes on a completely different form in the eye. On the one hand, bone marrow-derived dendritic cells and macrophages of the normal iris, ciliary body, and retina are incapable of performing conventional antigen processing and presentation. This appears to result from a local microenvironment, created by ocular parenchymal cells, which (1) modifies antigen-presenting cell function and (2) interferes with activation of T cells. Although TGFβ is a critical endogenous immunosuppressive factor in normal ocular fluids, other regulatory factors also contribute, such as α-melanocyte-stimulating hormone, vasoactive intestinal peptide, calcitonin gene-related peptide.

Tissue-seeking lymphocytes have been demonstrated for the skin and for the mucosal surfaces. Whether these populations represent completely different subsets of recirculating cells or are practically overlapping is unclear. To date, the existence of lymphocytes that display tropic properties for the eye have yet to be identified, perhaps because such cells have not been looked for.

TABLE I
Comparison of Components of Selected Regional Spheres of Immunologic Interest

Component	Skin	Mucosa	Eye
Parenchymal cells	Keratinocytes	Absorptive epithelium	Retina and/or epithelia of iris/ciliary body
Antigen presentation	Promoted by Langerhans cells	Promoted by M cells of epithelia and dendritic cells of Peyer's patches	Suppressed by Muller cells and cells of iris/ cilliary body
Tissue-seeking lymphocytes	Have been described	Have been described	Not defined
Specialized endothelial cells	HEV[a] of peripheral lymph nodes	HEV of Peyer's patches and mesenteric lymph nodes	Not defined
Draining lymphoid organ	Lymph nodes	Lymph nodes	Spleen

[a]High endothelial venules.

As might be expected, the existence of tissue-seeking lymphocytes implies the existence of tissue-specific ligands that are expressed uniquely on endothelial cells of the microvasculature supplying different regions. Ligands have been identified on high endothelial cells of mesenteric lymph nodes, and these ligands differ from similar molecules found on the high endothelial cells of lymph nodes that drain skin. The existence of lymphocytes that bind selectively to dermal microvessels indicates that a third set of molecules serves this function in the dermis. Ligands that direct lymphocytes selectively into the eye have yet to be defined.

Lymph nodes are situated along lymphatic vessels that drain skin. These organs trap antigen-bearing Langerhans cells escaping the skin and, therefore, serve as sites for antigen presentation to lymphocytes, the initiating event that leads to proliferation, differentiation, and the generation of immune effector modalities. Similarly, lymph nodes draining Peyer's patches and other mucosa-associated lymphoid collections function as regional sites for lymphocyte activation and differentiation. However, the internal compartments of the eye are not served by lymphatic vessels. As a consequence, molecules and cells that leave this organ are delivered directly into the bloodstream. Therefore, the spleen serves as the draining lymphoid organ for the eye. Considerable evidence indicates that the spleen transduces antigenic signals into effectors and regulators of the immune response that may differ considerably from those generated within lymph nodes. This may account for why immune responses of ocular antigens differ so remarkably from responses to antigens from the skin or the gastrointestinal tract.

The types of immune effector cells and molecules that are generated when antigens are introduced into the skin, the mucosa, and the eye are categorized in Table II. This table is incomplete, because several modalities are not even mentioned (IgM, IgE). The

TABLE II
Spectrum of Immune Effectors Generated in Response to Diverse Regional Exposures to Antigens[a]

	T lymphocytes		Antibodies		
Region	Delayed hypersensitivity	Cytotoxic	Complement-fixing IgG	Non-complement-fixing IgG	IgA
Skin	+++	+++	+++	+++	−
Mucosa	+++	?	+	+	+++
Eye	−	+++	−	+++	+

[a]+, immune effector is present; +++, effector is prominent; −, absence of an immune effector; ?, unknown.

data are presented primarily to indicate that the spectra of immune effectors elicited by antigens at these three regions are *not* identical.

Immune responses to cutaneous antigens are dominated by delayed hypersensitivity and complement-fixing antibodies, both of which depend on amplification through nonspecific inflammatory processes for successful elimination of the antigen or pathogen to which they are directed. The mucosal surfaces also generate delayed hypersensitivity responses, but the major immunoglobulin produced is IgA. Ocular immune responses are composed of cytotoxic T cells and non-complement-fixing antibodies, each of which can effect their specific function with minimal participation from the nonspecific mediators of inflammation.

Even this incomplete tabulation of effector responses is informative as to the teleology of regional specializations in immunity. By generating immune effectors that are unable to enlist nonspecific host defense mechanisms, the eye avoids the local accumulation of nonspecific inflammatory mediators, which have the potential to destroy vision by their tendency to disrupt the delicate microanatomy of the visual axis. The strategy of generating IgA antibodies ensures that mucosal surfaces will be bathed with antibodies that can withstand attack by proteolytic enzymes present in mucosal secretions, because protective antibodies must bind to their target antigen or pathogen in this environment. The intense inflammatory responses that attend delayed hypersensitivity and complement fixation can bring maximal defense mechanisms to bear within the skin. The logic of these vigorous responses apparently derives from the virulence of pathogens that gain access across the cutaneous barrier and from the skin's ability to withstand and recover its physiologic function after experiencing violent and destructive inflammatory responses that are necessary to rid the tissue of pathogens.

These few examples may serve to indicate the complexity and diversity that still awaits description in these and other regional spheres of immunologic interest. Only time and additional experiments will reveal the extent to which other regions of the body are similarly served. In addition, much remains to be learned about the mechanisms that operate to imprint regional patterns of reactivity on a "generic" systemic immune apparatus and response.

BIBLIOGRAPHY

Brandtzag, P. (1989). Overview of the mucosal immune system. *Curr. Topics Microbiol. Immunol.* **146,** 13–28.

Picker, L. J., and Butcher, E. C. (1993). Physiological and molecular mechanisms of lymphocyte homing. *Annu. Rev. Immunol.* **10,** 561–592.

Streilein, J. W. (1990). Skin associated lymphoid tissues (SALT): The next generation. *In* "The Skin Immune System (SIS)" (J. Bos, ed.), pp. 226–248. CRC Press, Boca Raton, Fla.

Streilein, J. W. (1993). Immune privilege as the result of local tissue barriers and immunosuppressive microenvironments. *Curr. Opin. Immunol.* **5,** 428–432.

Streilein, J. W. (1994). Ocular regulation of systemic immunity. *Reg. Immunol.* **6,** 143–150.

Immunopathology

STEWART SELL
University of Texas Medical School at Houston

GLOSSARY

Allergen Antigen that elicits an atopic or anaphylactic-type reaction

Anaphylaxis Immediate hypersensitivity reaction (hives, shock) caused by release of mediators from mast cells as a result of reaction of antigen (allergen) with mast cell-bound IgE antibody (Reagin)

Arachidonic acid 5,8,11,14-Eicosatetraenoic acid; an unsaturated fatty acid found in cell membranes that is a precursor in the synthesis of leukotrienes, prostaglandins, and thromboxanes, which are mediators of inflammation

Basement membrane Extracellular meshwork of collagen and reticular fibers in a proteoglycan matrix that supports epithelial cells and separates vessels from other tissues

Complement System of at least 13 serum proteins that are activated by enzymatic cleavages and aggregations to produce fragments or aggregates with biological (inflammatory) activity. One method of activation is by reaction of the first component of complement with antibody–antigen complexes containing aggregated Fc regions of antibody

Epitope Antigenic determinant; the smallest structural area on an antigen that can be recognized by an antibody

Granzyme Serine proteases present in granules of T-cytotoxic cells that act in synergy with perforin to produce death of target cells

HLA—human leukocyte antigens Cell-surface proteins encoded by the major histocompatibility complex (MHC) that were originally recognized on white blood cells using serum from multiparous women and graft recipients. HLA belong to two major groups: those encoded by class I MHC regions (HLA-A, -B, and -C) and those encoded by class II MHC regions (HLA-DQ, -DR, and -DP).

Interleukin Substance produced by one white cell that can act on other white cells

Isoagglutinins Antibodies that act with antigens on the cells of different individuals of the same species. Anti-blood group A is an isohemagglutinin that reacts with type A red blood cells

Lymphokine Soluble substance, produced and secreted by lymphocytes, that acts on other cells

Major histocompatibility complex (MHC) Group of genes coding for proteins expressed on the surface of cells that are recognized by immune T cells, effecting tissue grant rejection. MHC molecules are now known to serve as antigen-presenting structures. Class I MHC molecules are responsible for endogenous presentation of antigens to CD8+ T-cytotoxic cell precursors. Class II MHC molecules present exogenous antigens processed by antigen-presenting dendritic cells during induction of antibody formation. Between the class I and class II regions of the MHC is the class III region, which codes for complement components C12, factors B and C4, tumor necrosis factors α and β, cytochrome p450, and heat-shock protein

Opsonin Antibody that enhances phagocytosis

Phagocytosis Engulfment of particles by cells

Prausnitz–Kustner reaction Acute wheal and flare reaction elicited by injection of allergen into a skin site previously injected with antiserum containing IgE antibodies; in other words, passive cutaneous anaphylaxis

Perforin Cytolytic mediator present in cytotoxic T cells with homology to the membrane attack complex of complement that acts with granzymes to cause death of target cells upon reaction with T-cytotoxic cells

Prostaglandins Aliphatic acids derived from arachidonic acid that have a variety of biological activities, including increasing vascular permeability, causing smooth muscle contraction, and decreasing the threshold for pain. Originally found in prostatic fluid

Reagin Originally used for the complement-fixing antibodies detected in syphilitic patients by the Wassermann reaction with cardiolipin; now used for skin-fixing antibodies of anaphylactic reactions

Reticuloendothelial system Complex of phagocytic cells of the body, primarily the sinusoidal cells of the liver (Kupffer cells), spleen, and lymph nodes

Rheumatoid factor Autoantibody against slightly denatured immunoglobulin (usually IgM anti-IgG) found in patients with rheumatoid arthritis

Toxoid Altered form of a toxin that is immunogenic but not toxic

IMMUNOPATHOLOGY IS A HYBRID WORD, INCORporating immunity and pathology. Immunity comes from the Latin word *immunitas* and means "protection from"; pathology from the Greek words *pathos,* meaning "suffering or disease," and *-logy* indicates the "study of." Thus immunopathology is an oxymoron; a self-contradictory term meaning the study of how protective mechanisms cause disease. The statement that someone is "immune to measles," in common terms, means that the person has had measles once and will not get measles again. However, immunity is in reality a double-edged sword, cutting down our enemies with one side and causing disease with the other side. *Immunology* is the study of immunity or the immune response.

From the time of conception the human body must maintain its integrity in the face of a changing and often threatening environment. Physiologic mechanisms allow us to adjust to changes in temperature, nutrition, and so on; immune mechanisms protect us against infectious agents. Immune mechanisms may be divided into two major groups: innate and adaptive (Table I). Innate mechanisms are present in all normal individuals and operate on different agents in the same manner regardless of whether the individual has been exposed to the infectious agent previously. The adaptive immune response requires previous exposure or immunization to become active, and the products of this response react specifically with the agent that stimulates the immunization and not with other unrelated infectious agents.

TABLE I
Comparison of Innate and Adaptive Immunity

Characteristic	Innate resistance	Adaptive resistance
Specificity	Nonspecific, indiscriminate	Specific, discriminating
Mechanical	Epithelium	Immune induced reactive fibrosis (granuloma)
Humoral	pH, lysozyme, serum proteins	Antibody
Cellular	White blood cells	Specifically sensitized lymphocytes
Induction	Does not require immunization; constitutive	Requires immunization

I. IMMUNIZATION

Immunization (or vaccination) is the process of stimulating adaptive resistance by induction of a specific immune response. The immune response has three phases: afferent, central, and efferent. The afferent limb consists of the delivery of the foreign material to the reactive cellular components of the immune system. These cellular components are organized in specialized tissues known as lymphoid organs, such as lymph nodes and spleen. The central limb is composed of the interaction of the cells of the immune system within the lymphoid organs and the eventual production of specific products: antibody or sensitized cells (lymphocytes) that have acquired the capacity to react specifically with the immunizing agent. The efferent phase of the immune response consists of the delivery of the immune products to the site of antigen deposition or infection in the tissue and activation of immune effector mechanisms. Immunopathology deals with the effects of the immune effector mechanisms on tissues.

The state of immune reactivity is determined by cells of the immune system and their products. The cells that take part in the immune response are listed in Table II. Some of these cells have specific molecules (receptors) that recognize foreign material, and some do not. The molecules that are recognized as foreign are termed *antigens.* A complete antigen is able to both induce an immune response and react with the products of the immune response. The parts of the molecules that are recognized by antibody or sensitized lymphocytes are called *epitopes.* Specific antibody molecules or reactive T cells have a reactive site (paratope) that can recognize and combine with the epitope. An antigen may have more than one epitope. It is essential for the understanding of immunopathology that the role of both specifically reactive and nonspecific (accessory) cells in the inductive and efferent

TABLE II
Features of Cells of Immunity and Inflammation

Cell type	Location	Distinguishing feature	Function
Polymorphonuclear neutrophil	Blood	Segmented nucleus, lysosomal granules	Nonspecific effector cell of acute inflammation
Lymphocyte			
T-effector cell[a]	Blood	Small mononuclear, specific receptor for antigen, T-cell markers (CD4+ or 8+)[b]	Specific effector cell for delayed hypersensitivity
T helper	Tissue	CD4+	Aids macrophages in sensitization of B cells
T suppressor	Tissue	CD8+	Controls sensitization process
B cell	Tissue	Surface Ig+	Precursor for antibody-producing plasma cells
Macrophage			
Monocyte	Blood	Large mononuclear, lysosomal granules	Late nonspecific effector cell in inflammation
Dendritic	Tissue	Follicles of lymphoid organs	Process antigen for antibody response (B cells)
Interdigitating	Tissue	Diffuse cortex of lymphoid organs	Process antigen for T cells
Mast cells	Tissue	Metachromatic granules	Contain mediators of immediate vascular + smooth muscular reactions
Fibroblasts	Tissue	Elongated connective tissue cells	Proliferate and fill in zones of tissue destruction (scarring)

[a]For further classification of T-effector cells, see Table IV.
[b]CD, cluster of differentiation, an identifying site (epitope) detected by monoclonal antibodies.

phases of the immune response be detailed. Foreign materials that are able to get past epithelial barriers and enter the body pass from the tissue spaces into lymphoid organs. Two responding cellular areas in the lymphoid organs are the B-cell zone and the T-cell zone. T cells and B cells are the major populations of lymphocytes and are the two arms of immunity.

The B-cell arm of the immune response results in production of specific antibody. Antibody is a soluble protein molecule secreted by differentiated B cells. In the lymphoid organ, antigen is processed in B-cell zones by tissue-fixed macrophages (dendritic histocytes) and B cells themselves, in the presence of small numbers of T-helper (Th) cells. The T helper cells produce factors (interleukins, IL) that serve to stimulate proliferation and differentiation of B cells that have reacted with antigen on the surface of the antigen-processing accessory cell. Two types of T-helper cells control the direction of B-cell differentiation. B cells differentiate into plasma cells that synthesize and secrete specific antibody. The secreted antibody is released into the efferent lymphatics and delivered to the systemic circulation. Antibodies in the blood are then able to reach tissue sites containing antigen. Th1 cells secrete interleukins-1 and -12 and interferon-γ (IFN-γ), which direct B cells to produce IgM and IgG antibodies. Th2 cells secrete IL-4, IL-5, and IL-6, which direct B cells to produce IgA and IgE antibodies. The class of antibody produced by plasma cells that derive from B-cell differentiation will determine which immune effector mechanism will be activated (see the following).

The T-cell arm of the effector response comprises two major subpopulations of T cells: CD8+ T-cytotoxic cells (T_{CTL}) and CD4+ delayed hypersensitivity cells (T_{DTH}). T_{CTL} recognize antigens produced in any nucleated cell in association with class I MHC molecules on the surface of the target cell. Viral-infected cells are the best-documented example of this process (endogenous antigen processing). T_{DTH} recognize antigens that are processed by a subpopulation of macrophages in the T-cell zones and presented to T cells in the context of class I MHC molecules (exogenous processing). T_{CTL} effector cells specific for the antigen proliferate and are then delivered to the lymphatics, which drain the lymphoid organs into the blood (efferent lymphatic). The specifically sensitized T-effector cells then circulate and localize in tissue sites containing the antigen.

II. INFLAMMATION

Inflammation is the process of delivery and activation of blood proteins and cells to tissues as a response to injury or infection. Cells in the blood take part in the inflammation induced by immune reactivity or may also be activated nonimmunologically by bacterial products or substances present in dying (necrotic) tissue. The hallmark cell of acute inflammation is the polymorphonuclear neutrophil (PMN). PMNs move into sites of acute inflammation, phagocytose dead tissue or bacteria, and may release enzymes from their cytoplasmic granules that cause further damage. Monocytes (blood macrophages) and lymphocytes are the hallmark cells of chronic inflammation. Lymphocytes reacting with antigen, or otherwise activated, release products (lymphokines) that attract and activate macrophages. Activated monocytes phagocytose and "clean up" tissue debris resulting from the preceding chronic inflammation. [*See* Inflammation.]

A. Antibody-Mediated Inflammation

The characteristics of the effector mechanisms mediated by antibody are largely determined by the nature of the antibody molecule. Antibody molecules are made up of units of four polypeptide chains joined together by disulfide bonds. The prototype antibody molecule, IgG, has a Y-shaped appearance. The stem of the Y contains the structures that determine the biological activity (i.e., placental transfer, ability to active complement). The arms of the Y contain the antigen binding sites (also called paratopes). Binding of antigen (epitope) to the paratopes causes structural alterations in the tertiary structure of the stem of the antibody molecule that expresses the complement binding site.

Activation of the complement system plays a critical role in antibody-mediated cell lysis and inflammation. Binding complement by antigen–antibody complexes produces activation of enzymes that cleave other components of the complement system, resulting in biologically active fragments of complement molecules. Some of these fragments attach to cell-surface membranes (membrane attack complex) and cause destruction (lysis) of the cell. Other products of complement activation act to cause contraction of endothelial cells lining small blood vessels (anaphylatoxin) or attract PMNs (chemotaxis). Activation of complement in tissues results in dilation of small blood vessels, as well as contraction of endothelial cells, so that components in the blood can enter the tissues. Activated complement fragments also attract activated polymorphonuclear leukocytes, the cells of acute inflammation.

The sequence of complement activation is shown in Fig. 1. Complement consists of a series of 11 serum proteins that interact after activation. C1 through C9 are part of the activation sequence, whereas other components serve to inactivate the system. The first component of complement is activated by binding to aggregated antibody molecules in an immune complex of antibody and antigen. Activated C1 is a proteolytic enzyme and splits C4 and C2 into active products. On the cell surface, C4b and C2a join together and split C3 into C3a and C3b. C3b binds to the cell surface and makes the cell more susceptible to phagocytosis (opsonizes) and also activates C5. C5 splits into C5a and C5b. C5b joins to the cell surface and activates C6–C9 to form a complex of molecules that intercalates into the cell membrane and creates a channel in the cell that allows the cytoplasm to leak out (cell lysis). The activated C4a, C3a, and C5a products are released into the fluid phase. Together they are called anaphylatoxin and stimulate increased blood flow and increased vascular permeability. C5a is also chemotactic for PMNs. Thus the cell membrane-bound components of complement are the killer molecules of cytolytic reactions; the activated soluble split products are critical for the inflammation of immune complex (Arthus) reactions (see the following). When antibody reacts with antigen on a target cell surface, cell lysis can result; if antibody reacts with soluble antigens to form an immune complex or with basement membrane antigens, complement-mediated inflammation will occur. Complement activation can directly inactivate viruses or bacteria via the alternate complement pathway (antibody independent) and can enhance phagocytosis of infectious agents, or contribute to an inflammatory response.

Antibodies belong to a family of serum proteins termed *immunoglobulins*. The properties of the five major classes of immunoglobulins are listed in Table III. The most important from an immunopathologic standpoint are IgG, IgM, and IgE. IgG is the most prevalent class of immunoglobulin, and its reaction with soluble antigen or tissue antigens is most frequently the causative antibody in inactivation and immune complex reactions. IgM is composed of five

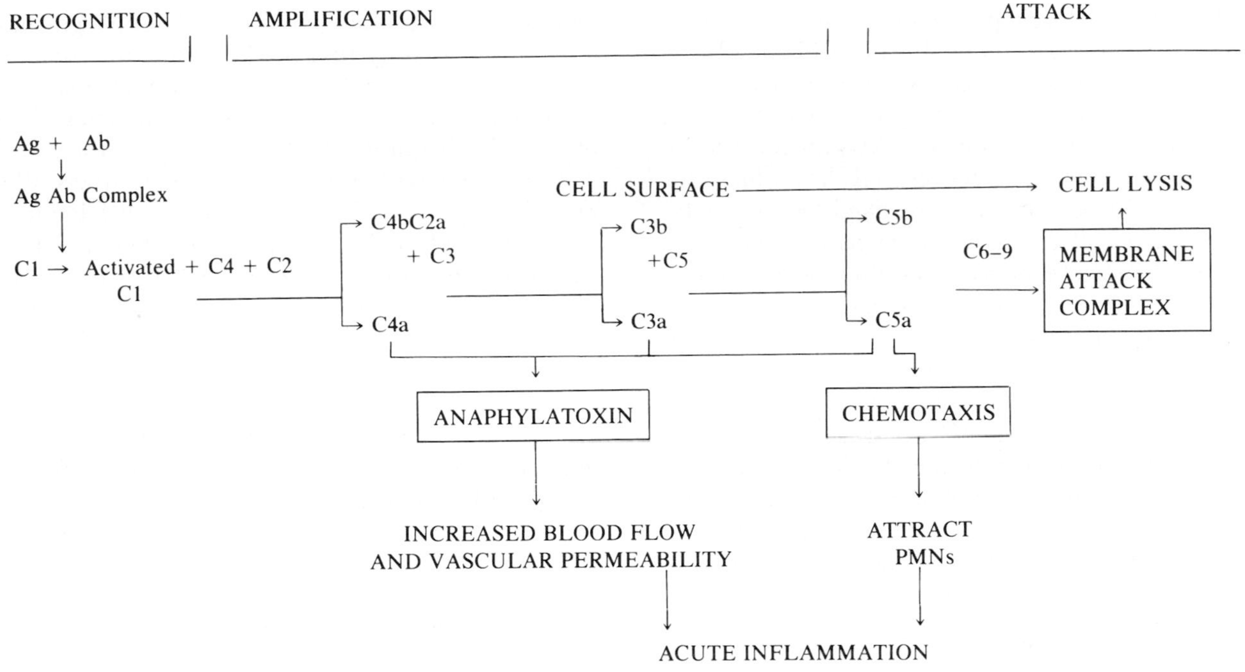

FIGURE 1 Role of complement activation of cytolytic and immune complex reactions.

TABLE III
Some Properties of Human Immunoglobulins[a]

Property	Immunoglobulin class				
	IgG	IgA	IgM	IgD	IgE
Serum concentration (g/100 ml)	1.2	0.4	0.12	0.003	<0.0005
Molecular weight	140,000	160,000[b]	900,000	180,000	200,000
Electrophoretic mobility	γ	Slow β	Between γ and β	Between γ and β	Slow β
H chains	γ	α	μ	δ	ε
L chains	λ or κ	λ or κ	λ or κ	λ or κ	λ or κ
Complement fixation	Yes	No	Yes	No	No
Placental transfer	Yes	No	No	No	No
Percent intravascular	40	40	70	—	—
Half-life (days)	23	6	5	3	2.5
Percent carbohydrate	3	10	10	13	10
Antibody activity	Most Ab to infections; major part of secondary response; Rh isoagglutinins; LE factor	Present in external secretions	First Ab formed; ABO isoagglutinins; rheumatoid factor	Antibody activity rarely demonstrated, found on lymphocyte surface	Reagin sensitizes mast cells for anaphylaxis

[a]Modified from J. L. Fahey (1966). *J.A.M.A.* **194,** 183.
[b]Serum IgA, 160,000 MW; secretory IgA, 350,000 MW; may activate alternate pathway.

antibody units joined together and is the first antibody formed after immunization. One molecule of IgM reacting on the surface of a cell is able to activate complement, whereas two or more IgG molecules reacting in close proximity are required. IgM is the most active antibody for cytolytic reactions. IgE has affinity for mast cells. Mast cells are tissue-fixed leukocytes located near small blood vessels in tissue. They have cytoplastic membrane-limited granules containing pharmacologically active agents that act on smooth muscle and endothelial cells. The resulting effect depends on the nature of the smooth muscle involved because of different types of receptors for the most active ingredient, histamine. Contraction of smooth muscle in the bronchi of the lung leads to constriction of the breathing tubes and symptoms of asthma. Dilation of the smooth muscles of small blood vessels, combined with contraction of the cells lining these vessels, results in increased blood flow and increased vascular permeability, the first manifestations of acute inflammation (see Section II,C). The operation of these IgE-mediated mechanisms (anaphylaxis) will be described in the next section. [*See* Smooth Muscle.]

B. Cell-Mediated Inflammation

Cell-mediated inflammation is effected by one of a number of cellular mechanisms. Six general cell-mediated cytotoxic effector mechanisms have been recognized *in vitro* (Table IV). In addition, other cell types (e.g., polymorphonuclear leukocytes or macrophages) may function as killer cells via antibody that binds to cells (cytophilic antibody). Macrophages may also be activated by nonspecific stimulators such as phorbol esters or polynucleotides. Immune-specific cell-mediated immunity reactions are initiated by T_{CTL} or T_{DTH} [*See* Natural Killer and Other Effector Cells.]

An infection with a pathogenic organism sets off a race between the infectious agent and the immune system of an infected individual, the outcome of which will determine whether the infected person lives or dies. For this reason, the specific immune response must be rapid and effective but also be able to be deactivated when the infection has been cleared or neutralized, as continued production of immune products could lead to continued destruction of normal tissue or unrestricted proliferation of lymphoid cells (lymphoma or leukemia) (Table V). Rapid delivery of antigens is effected by afferent lymphatics, which deliver antigens to the lymph nodes. The lymph nodes contain the immune reactive cells, act as filters for the lymphatics, and provide a site where the immune products may be rapidly manufactured and delivered to the bloodstream through efferent lymphatics, which drain into the systemic circulation. To be effective, the immune products must be able to reach the sites of infection in tissues. This is accomplished by inflammation.

C. Acute Inflammation

Acute inflammation consists of the passage of fluid, proteins, and cells from the blood into the tissues. The cells taking part in inflammation are listed in Table II. During an inflammatory response, components of both the innate and the adaptive (specific) immune systems may be utilized. Initiation of an inflammatory response begins with increasing blood

TABLE IV

Mechanisms of Lymphoid Cell-Mediated Immunity

T_{CTL}	T-cytotoxic cells. Kill target cells directly; have specific receptors for target cells.
T_{DTH}	Delayed hypersensitivity T cells. Indirect killing; release lymphokines that activate macrophages; have receptors for antigens.
Null cells + specific antibody	Antibody-dependent cell-mediated cytotoxicity (ADCC)
NK cells	Natural killer cells. Kill target cells nonspecifically.
LAK cells	Lymphokine-activated killer cells. Kill tumor cells.
Macrophages	Activated by lymphokines. Kill intracellular parasites as well as phagocytose and digest extracellular organisms.

TABLE V

Functional Abilities of the Adaptive Immune System

1. *Recognition* of many different foreign invaders specifically
2. *Rapid synthesis* of immune products on contact with invaders
3. *Rapid delivery* of the immune products to the site of infection
4. *Diversity* of effector defensive mechanisms to combat infectious agents with different properties
5. *Direction* of the defensive mechanisms specifically to foreign invaders rather than one's own tissue
6. *Deactivation* mechanisms to turn off the system when the invader has been cleared

flow to infected tissues and contraction of the endothelial cells lining the blood vessels. The cardinal signs of the earliest stage of acute inflammation were first described by the famous Roman physician Cornelius Celsus in his book "De Medicina" about 25 BC (Table VI). Increased blood flow causes redness (rubor) and increased temperature (calor). Passage of fluid into the tissue is seen as edema and swelling (tumor), and release of inflammatory mediators (e.g., prostaglandins) produces pain (dolor). In later stages of acute inflammation, infiltration of tissues with white blood cells (the cells of inflammation) produces a white color; leakage of red blood cells into tissue produces a red color. If the inflammation is severe, activation and release of digestive enzymes from the inflammatory cells may produce death of the involved tissue, which is recognized as pus. Pus containing mainly white blood cells is white; if red cells are present the pus will be yellow or red, depending on the proportion of red cells.

D. Chronic Inflammation

Chronic inflammation features an infiltration of mononuclear cells (lymphocytes and monocytes) in contrast to the polymorphonuclear cells characteristic of acute inflammation. Activated lymphocytes release lymphokines, which attract and activate monocytes. Activated monocytes (macrophages) clean up necrotic tissue and bacteria by engulfing and digesting the engulfed material in cytoplasmic vessels (phagocytosis). Macrophage means "big eater." If the tissue damage is not severe, the site of inflammation will be restored to normal (resolution); if damage is more extensive, tissue fibroblasts will proliferate and fill in the necrotic area with fibrous tissue (scarring).

E. Immune Effector Mechanisms

Immune effector mechanisms are variations on the theme of inflammation. The specificity of antibody and specifically sensitized lymphocytes provides a means of applying the defensive action of the inflammation directly to the infectious agent. The expression of the specific immune response *in vivo* may be classified into seven major mechanisms, complicated by the fact that more than one of these mechanisms may be active at the same time. A comparison of the seven immune effector mechanisms to the original classification of Philip Gell and Robin Coombs is given in Table VII. These mechanisms have both protective and destructive effects (Table VIII). For the remainder of this chapter, the destructive or disease-causing effects of these immune mechanisms will be stressed. The first four mechanisms are mediated by immunoglobulin antibodies. The characteristics of the reactions are determined by the properties of the immunoglobulin molecules, the nature and tissue location of the antigen, and on the accessory inflammatory systems that are activated. The fifth and sixth mechanisms are mediated by sensitized lymphocytes (T cells), whereas the seventh mechanism may be initiated by reaction of either antibody or effector lymphocytes with antigen. The nature of the lesions observed (granulomas) is determined by the fact that the antigen is poorly degradable.

TABLE VI
The Four Cardinal Signs of Acute Inflammation, According to Celsus (25 BC)[a]

Rubor	Redness
Tumor	Swelling
Calor	Heat
Dolor	Pain

[a]The fifth classic sign of acute inflammation, functio laesa (loss of function), was added by Rudolf Virchow (1821–1902).

TABLE VII
Classification of Immune Mechanisms

Gell and Coombs (1963)	Present classification
—	Inactivation or activation
Type II	Cytotoxicity or cytolytic
Type III	Toxic complex (Arthus)
Type I	Atopic or anaphylactic
—	T-cell cytoxicity
Type IV	Delayed hypersensitivity
—	Granulomatous

III. AUTOIMMUNITY

Tissue damage is caused by immune mechanisms when reaction to an infection is severe or inappropriate but may also occur when the immune response is directed not against foreign antigens but against self-antigens (autoimmunity). Autoimmunity occurs when an individual develops an immune response to his or her own tissue antigens. Normally during development, a person eliminates or controls self-reactive cells and develops a condition that prevents him or her from recognizing his or her own tissue as foreign.

TABLE VIII
The "Double-Edged Sword" of Immune Reactions[a]

Immune effector mechanism	Protective function: "immunity"	Destructive reaction: "allergy"
Neutralization	Diphtheria, tetanus, cholera, endotoxin neutralization, blockage of virus receptors	Insulin resistance, pernicious anemia, myasthenia gravis, hyperthyroidism
Cytotoxic	Bacteriolysis, opsonization, bacterial infections	Hemolysis, leukopenia, thrombocytopenia
Immune complex	Acute inflammation, polymorphonuclear leukocyte activation, bacterial infections	Vasculitis, glomerulonephritis, serum sickness, rheumatoid diseases
Anaphylactic	Focal inflammation, increased vascular permeability, expulsion of intestinal parasites	Asthma, urticaria, anaphylactic shock, hay fever
T-cell cytoxicity	Destruction of cells and virus-infected cells, immune surveillance of cancer	Contact dermatitis, autoallergies, viral exanthems
Delayed hypersensitivity	Macrophage activation, tuberculosis, syphilis, fungi, etc.	Postvaccinial encephalomyelitis, toxic shock, multiple sclerosis
Granulomatous[b]	Walling-off of organisms, leprosy, tuberculosis, helminths, fungi	Berylliosis, sarcoidosis, tuberculosis, filariasis, schistosomiasis

[a]Modified from S. Sell (1978). Introduction to symposium on immunopathology: Immune mechanisms in disease. *Hum. Pathol* 9, 24.

[b]Granulomatous reactions, like other inflammatory lesions, may result from nonimmune stimuli as well as from an immune reaction activated by antibody or by sensitized cells. The frequent association of granulomatous reactions with delayed hypersensitivity reactions often results in the inclusion of granulomatous reactions as a subset of delayed hypersensitivity.

This condition is known as tolerance. There are a number of possible mechanisms whereby tolerance is lost and autoimmunity may occur (Table IX). The most important part of the process of developing tolerance is the elimination of cells that recognize self-antigens in the thymus during development of T cells. Once autoimmunity develops, diverse disease manifestations may occur depending on the location of self-antigen and on which of the seven immunopathologic mechanisms are activated. [*See* Autoimmune Disease.]

IV. IMMUNOPATHOLOGIC MECHANISMS

A. Inactivation or Activation

Inactivation or activation reactions occur when antibody reacts with an antigen that performs a vital function and inhibits that function. Inactivation may occur by reaction of antibody to soluble molecules (e.g., bacterial toxins) or by reaction of antibody with

TABLE IX
Mechanisms of Autoimmunity

Immunizing event

1. Release of sequestered antigen: Tissue-specific antigens (e.g., myelin of central nervous system) normally not exposed to immune system are released into the circulation in amounts that stimulate an immune response.
2. Partial denaturation of antigen: Alteration of part of a molecule may lead to structures recognized as foreign by immune system. In responding to the denatured structures, an immune response to the nondenatured component may also take place.
3. Cross-reacting antigen: Molecules containing some "foreign" structures and some "self" structures are able to induce an immune response to both foreign and self epitopes.
4. Polyclonal activation: If the immune system is "turned on" by a strong stimulus, some of the products (antibodies or cells) may react with self-antigens.

Change in host

1. Regeneration of deleted clones by somatic mutation: Self-reactive lymphocytes that have been eliminated during the development may reappear in the adult by mutation of previously unreactive lymphocytes.
2. Recovery from anergy: Self-reactive cells may be blocked from reacting to self-antigens by a failure to express receptors or to respond to antigen stimulation. A change in the status of the individual may remove this block.
3. Loss of suppressor T cells: T cells that hold self-reactive T or B cells in check may decline in numbers as a result of a failure to maintain their production, allowing the self-reactive cells to respond.
4. Loss of controlling antibody: Feedback control of antibody production by anti-idiotype may become inactive if the level of the controlling antibody falls.

cell-surface receptors (e.g., virus receptors). Reactions with soluble molecules may produce changes in the tertiary structure of the biologically active molecule so that it no longer performs its biological function or is cleared from the circulation by the reticuloendothelial system as an immune complex. Reaction of antibody with cell-surface receptors may cause endocytosis (modulation) of the receptors so that they disappear from the cell surface. In some instances, antireceptor antibodies may mimic the action of the activating ligand for that receptor. Inactivation reactions to toxic agents (e.g., diphtheria or tetanus toxins) are beneficial, and antibodies to virus may prevent cellular infection. In fact, this is the goal of immunization to toxoids or viruses. However, when antibody reacts with something vital for normal function, the same reactions induce disease.

A list of some major diseases associated with inactivation effects of autoantibodies is given in Table X. Some examples include diabetes, myasthenia gravis, and hyperthyroidism. Although diabetes is actually a group of different diseases, the basic problem is a lack of insulin or an inability of cells of the body to respond to insulin. Diabetes may be mediated by at least three different antibodies as well as by nonimmune mechanisms. The antibodies include (1) antibodies to insulin, which block the effect of insulin, (2) antibodies to insulin receptors, which prevent the binding of insulin, and (3) antibodies to the cells that make insulin, resulting in a failure to produce insulin. Myasthenia gravis is a muscular unresponsiveness to neurogenic stimulation caused by autoantibodies that react with acetylcholine receptors and block the neuromuscular transmission effected by release by acetylcholine at the motor end plate. Antireceptor activation is exemplified by some forms of hyperthyroidism caused by antibodies that react with receptors on thyroid cells that normally bind to thyroid-stimulating hormone. The antibody to the receptor mimics the activity of thyroid-stimulating hormone.

TABLE X
Some Diseases Caused by Antibodies to Biologically Active Molecules

Disease	Target antigen
Diabetes	Insulin Insulin receptors
Juvenile diabetes	Insulin
Clotting deficiencies	Clotting factors
Pernicious anemia	Parietal cells Intrinsic factor
Myasthenia gravis	Acetylcholine receptor
Hyperthyroidism	Antithyroid-stimulating hormone receptor (LATS)[a]
Hypothyroidism	Antithyroid hormone
Asthma	Beta-adrenergic receptor

[a]LATS, long-acting thyroid stimulator.

B. Cytotoxic or Cytolytic Reactions

Reaction of antibodies to cell-surface antigens in the presence of *complement* will sometimes cause lysis or destruction of the cells. Complement components can realign membrane structures and disrupt cell membrane integrity, or the cell can be coated with complement components that render the cell susceptible to phagocytosis by the reticuloendothelial system. Complement is activated by alterations in the tertiary structure of some antibodies when they react with antigen. One molecule of IgM antibody reacting with a cell-surface antigen is capable of activating complement and of lysing a cell, whereas at least two molecules of IgG antibody must react in close proximity to one another to activate complement and accomplish lysis of a cell. For this reason, lytic reactions are much more effectively mediated by IgM than by IgG antibody. Other immunoglobulin classes do not fix complement in this way and thus do not lyse cells.

Cytotoxic or cytolytic reactions are often directed against cells in the blood. The disease caused by these antibodies reflects the loss of the function of the cell. Antibodies to red cells cause hemolytic anemia; to white blood cells, agranulocytosis and increased susceptibility to infection; to platelets, thrombocytopenia and bleeding disorders. Transfusion of red blood cells bearing blood group antigens to an individual who does not have the same antigens may result in a massive hemolysis (transfusion reaction) if the recipient has preformed antibody to the transfused cells. Erythroblastosis occurs during pregnancy when the mother has antibody to fetal red cell antigens contributed by the father and the antibody is able to cross the placenta. The maternal antibody destroys the red cells in the fetus. The fetus responds by increasing red cell production (erythroblastosis). Erythroblastosis fetalis usually does not occur with the first incompatible pregnancy because immunization leading to antibodies to the fetal red cells occurs during delivery of an incompatible baby. This can be prevented in most instances by treating the mother with the appropriate antibodies at the time of delivery, thus preventing immunization and production of antibodies that might cause erythroblastosis during the next incompatible pregnancy.

C. Immune Complex (Arthus) Reactions

Immune complex reactions are due to the formation of complexes of antibody with soluble antigens in the circulation or by reaction of circulating antibody with basement membrane components. Activation of complement by antigen–antibody complexes (usually IgG antibody) in tissues is responsible for an inflammatory reaction mediated by activated components of complement. Activation of complement on the surface of the cell results in some components that are split off of larger molecules that bind to the cell surface, with additional components joining until a membrane lesion is produced and the cell is lysed (cytolytic reaction). In addition, biologically active split fragments of complement that do not bind to the cell surface are produced. The important activated fragments for inflammation are C3a, C4a, and C5a. Activation of these products produces dilation of small arterioles as well as constriction of endothelial cells (anaphylatoxin), causing increased blood flow and egress of blood components to pass into the tissue between separated endothelial cells. C5a is also highly chemotactic (attractive) for polymorphonuclear neutrophilic leukocytes. The result is the attraction and activation of these cells to tissues when antibody has reacted with antigen. PMNs are the hallmark of acute inflammation. They contain large numbers of lysosomes that are filled with digestive enzymes. Tissue damage is caused by digestion of tissue by enzymes released from the PMNs at the inflammatory site. Injection of antigen into the skin of an animal with circulating IgG antibody to the antigen produces an acute inflammatory reaction (Arthus reaction) that peaks at 6 hours and fades by 24 hours. It is characterized histologically by perivascular necrosis and PMN infiltration of arterioles and venules. Using immune labeling techniques (immunofluorescence), antigen–antibody complexes and complement components can be identified in the walls of the affected vessels. Some illustrative examples of immune complex reactions are serum sickness and some forms of glomerulonephritis. [*See* Antibody–Antigen Complexes, Biological Consequences.]

Serum sickness is a systemic inflammatory reaction following injection of foreign serum manifested by involvement of vessels throughout the body. This reaction was first noted at the turn of the century when horse antitoxin was used to treat individuals with tetanus. Ten days to 14 weeks after the injection of horse serum, inflammation of joints (arthritis) and kidneys (glomerulonephritis) and systemic inflammation of blood vessels (vasculitis) were observed. This reaction was due to the formation of soluble immune complexes in the circulation. The complexes deposit in vessel walls, renal glomerular basement membrane, and joints. On deposition, the accumulation of these complexes leads to activation of complement and subsequent acute inflammation. If not fatal, the disease terminates when the immune complexes are cleared from the circulation and tissues. Such clearance occurs when more than one IgG antibody molecule reacts with a soluble antigen so that at least two antibody molecules are located close enough together to form a structure that is recognized by phagocytic cells or that activates complement. Thus early after an antibody response to foreign serum, antigen–antibody complexes will be found wherein antigen is in excess over antibody so that the complexes will essentially consist of one or two antigen molecules bound to one antibody molecule. These complexes are not cleared by phagocytosis and can deposit in tissues. When two or more such complexes deposit closely together in tissue, the antigen-bound antibody molecules can form the aggregates of IgG molecules required to activate complement and thus initiate inflammation in the tissue. Later, as the levels of serum antibody continue to rise, immune complexes at equivalence or in antibody excess will be cleared through Fc or complement receptors on cells of the reticuloendothelial system through binding of the aggregated Fc regions or C3b on the complexes.

Glomerulonephritis is caused by deposition of soluble immune complexes in the basement membrane of the renal glomerulus or by reaction of antibodies specific for basement membrane antigens. Acute glomerulonephritis classically occurs after infection with certain strains of streptococcal bacilli. During resolution of the infection, the bacilli release cationic antigens that have an affinity for the anionic basement membrane. When the infected individual produces antibody to this protein, reaction with the antigen deposited in the membrane results in glomerulonephritis. A second, chronic phase of the glomerular inflammation results when, as a result of the release of basement membrane components, an autoantibody to the basement membrane is formed. This may result in progressive renal failure.

A variety of other human diseases are believed to be caused by the immune complex mechanisms. These include lupus erythematosus (DNA–anti-DNA), rheumatoid arthritis (IgM–anti-IgG, so-called "rheumatoid factor"), and polyarteritis nodosa (circulating immune complexes, e.g., hepatitis B antigen–anti-HBsAg).

The protective function of immune complex reactions is to mobilize inflammatory cells at the site of infection. The activation of the chemotactic components of complement provides this signal, and the anaphylatoxic action of activated complement opens the vessel walls to permit the PMNs to pass into the tissue. The released lysosomal enzymes for the PMNs act on infectious organisms or their products and destroy or inactivate them.

D. Anaphylactic and Atopic Reactions

Anaphylactic reactions include hives (wheal and flare), systemic vascular shock, gastrointestinal allergy, and acute asthmatic attacks. Atopic reactions include hay fever and chronic asthma. The difference is most likely due to the amount of antigen (allergen) to which the genetically susceptible individual is exposed in a given period. A large exposure produces anaphylatic symptoms; repeated small exposures during a longer period produce atopic symptoms. The term *allergy* is commonly used for these reactions, and the antigen is termed *allergen*. [*See* Allergy.]

Anaphylactic or atopic reactions are initiated by pharmacologically active agents (histamine and serotonin) that are released from mast cells after antibody bound to mast cells reacts with antigen. In addition, arachidonic acid metabolites produced by macrophages from cell membrane phospholipids released from mast cell granules contribute to later inflammatory phases of the reaction. Injection of antigen into the skin of an anaphylactically sensitive individual results in an acute reaction composed of swelling (edema) surrounded by redness (erythema) that peaks within 15 to 30 minutes and fades in 2 or 3 hours (cutaneous anaphylaxis, characterized by wheal and flare). A later cellular inflammatory phase of the reaction may occur after 20 to 20 hours. This later phase is more evident in the biphasic bronchospastic response of the lung, where production of arachidonic metabolites by lung tissue may be more pronounced. The antibody responsible is almost always of the IgE class, which has the capacity to bind to receptors on mast cells through the nonantigen binding part of the antibody molecule. Mast cells are located near smooth muscle (precapillary arteriole, bronchial smooth muscle, etc.) and contain granules filled with pharmacological agents that act directly on smooth muscle and endothelial cells. When antigen reacts with mast cell-bound IgE antibody, the cell releases the contents of its granules by a nonlytic fusion mechanism (degranulation). The released pharmacologically active agents, primarily histamine and serotonin, act on end-organ target cells (e.g., vascular and bronchial smooth muscle) to cause immediate symptoms. In addition, the membrane phospholipids are metabolized by oxidative pathways to produce a series of longer-acting biologically active compounds, leukotrienes and prostaglandins, that are responsible for later effects.

Anaphylactic or atopic reactions are elicited by reaction of antigens with IgE antibody on mast cells. The ability of IgE antibody to "fix to skin" is determined by its binding to mast cells in the skin (cytophilic antibody). The reaction can be passively transferred by injection of serum containing IgE antibody into the skin (passive cutaneous anaphylaxis). Antigen can be injected up to 45 days after injection of serum containing IgE antibody, as the antibody will remain "fixed" to mast cells in the skin. This is also referred to as the Prausnitz–Kustner reaction, in honor of the authors who first described the typical reaction. Passive transfer of an Arthus reaction is demonstrable only if the antigen is injected within 24 to 48 hours after antibody, as the IgG antibody will diffuse away (it does not "fix" to mast cells). Skin-fixing antibody is called reaginic antibody.

The protective function of anaphylactic reactions is not as well understood as in other immune effector mechanisms, but there are several possible functions. Histamine and serotonin are primary mediators of increased blood flow to sites of reaction. Anaphylactic reactions serve to open vascular endothelium, thus permitting bloodborne proteins and cells to enter inflammatory sites where they are needed. Anaphylactic reactivity to intestinal parasites often occurs, and increased intestinal motility and secretion may purge the gastrointestinal tract of them. The acute sneezing and coughing that are part of the acute asthmatic attack may help eliminate agents in the tracheobronchial tree. If nothing else, the acute reactivity certainly serves as a warning to avoid contact with exciting agents.

E. T-Cell Cytotoxicity

T-cytotoxic cell (CD8+ T_{CTL}) lymphocytes attack target cells directly. T_{CTL} recognize antigens on target cells through specific receptors. After reacting with a target cell, two cytolytic mechanisms are activated:

Fas and granular proteins (perforin and granzymes). Fas ligand on the T_{CTL} binds to and activates Fas (the product of the "death gene") on the target cell. Perforin reacts with the target cell membrane to produce "holes" in the membrane that permit granzymes to enter the cell. Granzymes and active Fas convert intracellular interleukin 1β (IL-1β) converting enzyme (ICE) to an active form. The activation of ICE by Fas, but not that of granzymes, is inhibited by the product of the bcl-2 gene; thus the action of granzymes is downstream from the bcl-2 inhibition point. Activated ICE or intracellular granzymes appear to act on an as yet unidentified substrate essential for survival of the cell. This results in activation of $p34^{cdc2}$ kinase, entry of DNase into the nucleus, nuclear collapse and eventual cell death. Active ICE-mediated conversion of pro-IL-1β to IL-1β results in the release of IL-1β. IL-1β signals adjacent cells of the impending death (apoptosis) of the target cell and enhances phagocytosis. There are many steps in this process that are incompletely understood.

T-cell cytotoxicity is induced by antigens bound to class I MHC-coded molecules (HLA-A, -B, or -C) on nucleated cells, such as virus-infected cells, and presented to CD8+ lymphocytes. T_{CTL} mediate contact dermatitis (i.e., poison ivy), tissue graft rejection, some autoimmune lesions (thyroiditis), and viral exanthems (rashes), such as measles. The characteristic lesion of T_{CTL} is infiltration of an epithelial cell layer with lymphocytes. In contact dermatitis and viral exanthems, T_{CTL} pass from venules in the dermis into the overlying epithelium, where they initiate destruction of the epithelial cells. In poison ivy, for example, T_{CTL} are attracted to poison ivy antigens present in the epidermal layer of the skin. Upon reaction with the antigen, the T_{CTL} are activated to kill the epidermal cells through direct contact between the T_{CTL} and antigens on the epidermal cells.

T-cell cytotoxicity is responsible for protective active immunity against viral infections such as smallpox, measles, and chicken pox. The characteristic positive skin reaction to vaccination against smallpox (the "take") is an epidermal eschar caused by destruction of epidermal cells infected with the vaccinia virus. In measles, multiple disseminated skin lesions are produced as a result of T_{CTL} to virus-infected skin cells. T_{CTL} are particularly important in eliminating intracellular virus infections when the viral antigens are expressed on the cell surface. It is unfortunate if the infected cell is one that performs a unique vital function (e.g., a neuron), because during elimination of the infected cells, irreversible tissue damage may result.

F. Delayed Hypersensitivity

Delayed hypersensitivity reactions are mediated by a population of specifically sensitized T lymphocytes that bear receptors for antigens. The lymphocytes that effect delayed hypersensitivity reactions belong to a subpopulation of the T-lymphocyte class: CD4+ T_{DTH} cells. They are induced by processing of antigen by class II MHC interdigitating macrophages in the paracortex of the lymph node. [*See* Lymphocytes.]

The characteristic lesion of delayed hypersensitivity reactions is a perivascular mononuclear infiltrate. T_{DTH} lymphocytes reacting with antigen in tissues release lymphokines, which attract and activate macrophages. Most of the tissue damage is caused by macrophages, which have been attracted to the site of inflammation by lymphokines. The term *delayed* is used because injection of antigen into the skin of an individual expressing delayed hypersensitivity induces an inflammatory reaction that peaks at 24 to 48 hours, in contrast to immune complex reactions, which peak at 6 hours (Arthus reactions), and cutaneous anaphylactic reactions, which peak at 15 to 30 minutes. Some examples of delayed hypersensitivity reactions are the tuberculin skin test and viral-associated demyelinating diseases of the nervous system (e.g., postinfectious encephalomyelitis, Guillain-Barré, and multiple sclerosis). [*See* Macrophages.]

Delayed skin reactions become manifest when sensitized lymphocytes react with antigen deposited into the dermis of the skin. In normal skin, lymphocytes pass from venules through the dermis to lymphatics, which return these cells to the circulation. Recognition of antigen processed by dermal dendritic macrophages (presenting antigen in association with class II MHC molecules) by sensitized lymphocytes (T_{DTH} cells) results in immobilization of lymphocytes at the site and production and release of lymphocyte mediators. The mediators attract and activate macrophages with eventual destruction of antigen and resolution of the reaction. This results in an accumulation of cells seen at 24 to 48 hours after antigen injection. When the antigen is destroyed, the reactive cells return via the lymphatics to the bloodstream or draining lymph nodes. In this way, specifically sensitized lymphocytes may be distributed throughout the lymphoid system after local stimulation with antigen.

The protective function of delayed hypersensitivity is mainly directed to intracellular agents such as fungal and mycobacterial infections, in particular those organisms that directly infect macrophages. These reactions are effected by attraction and activation of infected macrophages by mediators (IL-2, IFN-γ, and

TABLE XI
Characteristics of the Seven Types of Hypersensitivity Reactions

	Immunopathologic mechanism						
Characteristic	Inactivation or activation	Cytotoxic or cytolytic	Toxic complex (Arthus)	Anaphylactic or atopic	T-cell cytotoxic	Delayed hypersensitivity	Granulomatous
Immune reactant	IgG antibody	IgM > IgG	IgG antibody	IgE antibody	T_{CTL}	T_{DTH}	T_D cells
Accessory component		Complement, macrophages (RES)	Complement, polymorphonuclear leukocytes, coagulation	Mast cells, chemical mediators	Perforin, granzymes	Lymphokines	Macrophages (epithelioid and giant cells)
Skin reaction		Pemphigus, pemphigoid	Arthus: peaks 6 hr; fades by 24 hr	Wheal and flare: peaks 15–30 min; fades by 2–3 hr	Exanthem, contact dermatitis	Delayed (tuberculin skin test)	Granuloma; takes weeks to develop and months to fade
Protective function	Inactivate toxins	Kill bacteria	Mobilize polys to site of infection	Open vessels, delivery of blood components to sites of inflammation	Kills virus-infected cells	Attracts and activates macrophages	Isolate infectious agents
Examples of protection	Tetanus, diphtheria, streptococci	Bacterial infections	Bacterial and fungal infections	Parasitic infections	Smallpox, measles, chicken pox	Tuberculosis, fungal infections, syphilis	Leprosy, tuberculosis
Pathogenic mechanism	Inactivation of biologically active agents or cell-surface receptors	Cell lysis or phagocytosis (opsonization)	Polymorphonuclear leukocyte infiltration, release of lysosomal	Bronchoconstriction edema, shock	Destruction of cells	Activation of macrophages	Replacement of tissues by granulomas
Representative disease states	Insulin-resistant diabetes, myasthenia gravis, hyperthyroidism	Hemolytic anemias, vascular purpura, transfusion reactions, erythroblastosis fetalis	Glomerulonephritis vasculitis, arthritis, rheumatoid diseases	Anaphylactic shock, hives, asthma, hay fever, insect bites	Viral exanthems, thyroiditis, graft rejection	Toxic shock, postvaccinial encephalomyelitis, multiple sclerosis	Sarcoidoses, berylliosis, tuberculosis

[a]Modified from S. Sell (1987). "Immunology, Immunopathology and Immunity," 4th Ed. Elsevier, New York, Reproduced by permission.

IL-12) released from CD4+ T cells (type 1 cytokines). Although delayed hypersensitivity reactions to infectious organisms may result in extensive tissue damage, the damage is a secondary effect of the immune attack on an infecting agent. It is incorrect to think, as was done in the past, that delayed hypersensitivity is a deleterious type of reaction. However, toxic shock is a manifestation of activation of T_{DTH} cells and macrophages by reaction of toxins (e.g., staphylococcal enterotoxin) with class II MHC on macrophages and the β chain of the T_{DTH} receptor and activation of release of IL-1 from the macrophage and IL-2 and tumor necrosis factor from the T cell.

G. Granulomatous Reactions

A variant of delayed hypersensitivity or immune complex reactions is the granulomatous reaction, which occurs when the antigen is poorly broken down and remains as a chronic irritant. Indeed, the early lesions often resemble those of delayed hypersensitivity reactions or are associated with necrotizing vasculitis. However, instead of macrophages clearing away the antigen, there is a prolonged accumulation of macrophages and lymphocytes, which organize into granulomas. Typical granulomas consist of a ball-like mass of macrophages resembling epithelial cells (epithelioid cells) and multinucleated giant cells admixed with lymphocytes. Frequently the center of larger lesions is necrotic. Grossly this form of necrosis looks like cottage cheese and is described as caseous necrosis. The major destructive mechanism is simply the occupation of organ space, although the lesions may become so extensive that the normal function of the organ is impaired. Granulomas serve to isolate the infectious agent by walling off infected areas of tissue. This operates when the delayed hypersensitivity mechanism is unable to eliminate the agent. Infectious diseases in which granulomatous reactions play a major role are tuberculosis, leprosy, parasite and fungus infections, and syphilis. Granulomas occur in children who have a deficiency in macrophage digestive function (granulomatous disease of children). In these individuals, granulomas form where macrophages ingest material that cannot be degraded and large numbers of macrophages collect in the tissue.

V. SUMMARY

Specific antibody and sensitized T-effector lymphocytes provide mechanisms for directing protective defense mechanisms to foreign invaders. These directed immune effector mechanisms serve to focus nonimmune specific inflammatory accessory systems on the specific infectious agent. Seven mechanisms are recognized: inactivation, cytotoxic or cytolytic, immune complex (Arthus), anaphylactic (atopic), T-cell cytotoxic delayed hypersensitivity, and granulomatous reactions. A listing of the characteristics of each of these reactions is given in Table XI. Each of these mechanisms has important protective functions but may also cause tissue damage. Immunopathology is the study of how immune mechanisms cause disease.

BIBLIOGRAPHY

Abbas, A. K., Lichtman, A. H., and Pober, J. S. (1991). "Cellular and Molecular Immunology." Saunders, Philadelphia.

Gell, P. G. H., and Coombs, R. R. A. (1963). "Clinical Aspects of Immunology." F. A. Davis Co., Philadelphia.

Janeway, C. A., Jr., and Travers, P. (1994). "Immunology: The Immune System in Health and Disease." Garland, New York.

Kuby, J. (1992). "Immunology." Freeman, New York.

Lackmann, P. J., Peters, D. K., Rosen, F. S., and Wallport, M. J. (eds.) (1993). "Clinical Aspects of Immunology," 5th Ed. Blackwell, Oxford, England.

Paul, W. E. (ed.) (1984). "Fundamental Immunology." Raven, New York.

Roitt, I. M., and Delves, P. J. (eds.) (1992). "Encyclopedia of Immunology." Academic Press, London.

Roitt, I. M., Brostoff, J., and Male, D. K. (1993). "Immunology," 3rd Ed. Mosby, St. Louis.

Sell, S. (1996). "Immunology, Immunopathology and Immunity," 5th Ed. Lange Med. Pub., Norwalk, Connecticut.

Immunotoxins

JON ROBERTUS
University of Texas, Austin

ARTHUR FRANKEL
Florida Hospital Cancer and Leukemia Research Center, Orlando

GLOSSARY

Allogeneic Pertaining to tissue with similar but not identical genetic makeup

CD5, CD19 Cell-surface proteins that serve as markers for T and B cells, respectively

Monokines Soluble factors released from activated monocytes or macrophages, which can alter the behavior of neighboring cells

Neuropathy Disorder of nerve cells

Palliation Alleviation of symptoms without a genuine cure

Tumor burden Amount of tumor cells present in an individual, not precisely defined; in humans, tumor masses of 1 g, 1 kg, and 10 kg are referred to as small, moderate, and large, respectively

AN IMMUNOTOXIN IS A CONJUGATE OF A cell-specific antibody and a toxic enzyme. The toxic enzymes are isolated from plant or bacterial sources and generally attack eukaryotic ribosomes or ancillary proteins. In theory, the antibody can guide the toxin to a tumor or other cell type and bind to it. After internalization, the toxin inhibits the protein-synthesis apparatus of the intoxicated cell, leading to cell death. Immunotoxins have been tested against cultured tumor cells and against tumors in animal systems to develop experience with their use. Recently, clinical trials have shown mixed results. Although some cancers have been successfully treated with the conjugates, a number of troublesome side effects are usually manifest.

I. THEORY OF CELL-SELECTIVE AGENTS

Medical science has long been fascinated by the concept of a "magic bullet," which could be administered to a patient and specifically seek out infected or malignant cells and either destroy them or deliver a therapeutic agent to them.

In theory, a wide range of cell-specific molecules could be used to make magic bullets. In fact, protein hormones, such as human gonadotropin, have been used in this way. Certain cell types have receptors on their surfaces for gonadotropin. Conjugates made with the hormone and a given toxin are able to pick out those cells with gonadotropin receptors and kill them with reasonable selectivity. This type of targeting vehicle, however, is not general and can be used only in special cases. Immunotoxins (ITs) are another class of magic bullets, although currently they are not as selective as desired. The name "immunotoxin" describes the nature of this particular class of magic bullets: "immuno" refers to an antibody or immunoglobulin, which is used to select a particular cell type; "toxin" refers to a protein molecule, which, when delivered to and taken up by a target cell, is able to intoxicate or kill that cell. When these two proteins are chemically or genetically linked together by medical scientists, they form an IT. When introduced into the body, the IT typically circulates through the blood, making contact with a variety of cells. A chance collision with a cell of the type recognized by the antibody allows the IT to bind to it, and processes are initiated, which can intoxicate that cell. Although the IT is a magic bullet in the sense that it is selective, it is not a guided missile that homes in unerringly on the target.

II. IMMUNOTOXIN STRUCTURE AND FUNCTION

A. Antibodies

In principle, antibodies can be used as targeting vehicles to a wide range of tissues. Antibodies are protein molecules produced by the B cells of all mammals. Antibodies of the G class are dimeric proteins shaped like the letter "Y." On each arm of the Y is a uniquely shaped pocket, which is complimentary in shape to some portion of a foreign molecule, called the antigen. By a complex selection mechanism, humans eliminate from circulation those B cells that produce antibodies capable of interacting with their own macromolecules, particularly their own proteins. Millions of antibody-producing cells remain, however, to recognize foreign proteins such as those on the surfaces of invading bacteria or virus particles. [*See* Antibody–Antigen Complexes: Biological Consequences.]

A common practice in biological science is to raise antibodies against a protein of interest. For example, one could isolate the coat protein of a common virus, which we can call "X." One could inject X into a rabbit and wait for the rabbit to produce a high yield of antibody that binds specifically to X and is called the "anti-X antibody." A large molecule (e.g., a protein) usually has many antigenic sites (i.e., capable of eliciting formation of an antibody) on its surface; in the foregoing experiment, different antibody types would be raised against each such site. Methods now exist to raise antibodies against just a single antigenic site—these are called "monoclonal antibodies." [*See* Monoclonal Antibody Technology.]

The ideal situation for constructing an IT would be to identify an antigenic marker that is unique to the cell type to be destroyed and is not present on healthy tissue. For example, if tumor cells had on their surfaces a protein or other molecular marker that existed on no other cell type, it would be possible to raise antibodies that bind only to the tumor cells bearing that marker. In such an ideal case, it would be possible to conjugate a toxin to the antitumor antibody and make very specific magic bullets. Unfortunately, this ideal does not actually occur. Tumor cells arise from normal progenitor cells and survive only by hiding from surveillance by the host antibody system. If tumor cells bore different markers than normal cells, they would be attacked by our own immune systems and would not, in general, cause more trouble than any other invader. As a result, attempts to raise truly specific antitumor antibodies have met with considerable difficulty. In general, the best that can be hoped for is to raise an antibody that takes advantage of slight chance differences between a given tumor cell and normal tissue and that has a slightly higher affinity for the former than the latter.

In addition to cell specificity, another desirable property of an IT antibody is its ability to be taken into the target cell, where the toxin can act. The cell membrane surrounding normal or tumor cells is designed to keep extraneous matter outside the cell. Only certain molecules are allowed to enter, usually through specific channels. This uptake process is complex and is treated separately in this encyclopedia. Molecules taken inside a cell generally bind to specific cell-surface receptors, which are then taken up by endocytosis. An example is the uptake of iron by the transferrin system. Antibodies against the transferrin receptor will be readily taken into the cell by the natural endocytosis meant to take up iron; ITs targeted against the transferrin receptor are, thus, quite effective. These ITs are of little therapeutic value, however, because they are not tumor-specific and healthy tissue is targeted as well as cancer cells. Almost any antibody that binds a cell-surface marker will eventually be taken inside the cell by nonspecific endocytotic processes, but clearly there is a design advantage in identifying and using antibodies that are taken up by the target cell at a reasonable rate. [*See* Cell Membrane Transport.]

B. Toxin Proteins

In principle, any agent toxic to the target cell could be used in the construction of ITs. Radioisotopes and cytotoxic chemicals have been carried to target cells by antibodies, but in general they lack the specificity and killing power of catalytic protein toxins. Protein toxins can be isolated from plants, fungi, and bacteria. Essentially all of these toxins act by disrupting protein synthesis in the target cell, although the modes of toxicity differ in detail.

1. Plant Toxins

Plant toxins fall into two broad classes: heterodimeric cytotoxins and single-chain ribosome-inhibiting proteins (RIPs) (Fig. 1). Heterodimeric toxins, such as ricin, abrin, and modeccin, possess two chains. One is the A chain, which enzymatically inhibits ribosomes by removing one specific adenine base from the ribosomal RNA. In the toxin, the A chain is linked by a disulfide bond to the B chain, which is a protein (lectin) capable of binding to certain galactose-containing carbohydrates located on cell-surface proteins. Binding of the B chain to cell surfaces triggers endocytosis and brings the enzymatic A chain into the cellular cytoplasm where it inhibits protein synthesis and kills the cell (Fig. 2). Ribosome-inhibiting proteins, such as pokeweed antiviral protein (PAP), trichosanthin, and gelonin, have only a single enzymatic chain, which is evolutionarily related to the A chain of the heterodimeric toxins and which catalyzes the same reaction. Because RIPs lack a surface recognition factor, they are only very weakly cytotoxic by themselves. If they are delivered to the cytoplasm, however, they can be as lethal as the cytotoxic A chains.

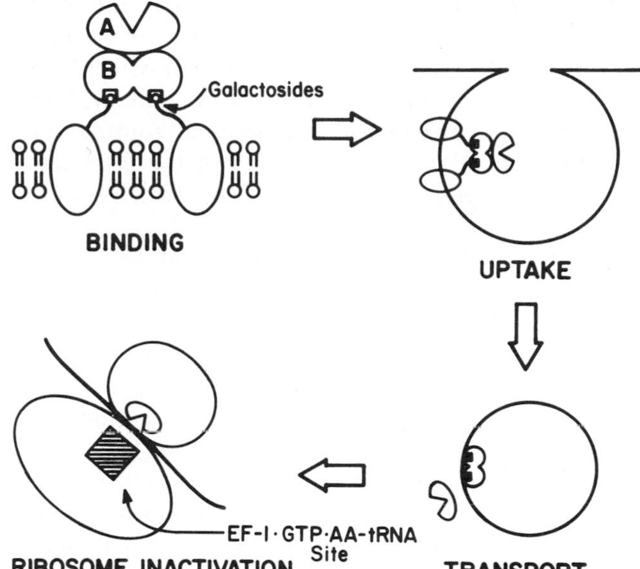

FIGURE 2 The action of cytotoxins. Ricin is the best-studied example of a true cytotoxin. It binds to cell surfaces through interaction of its lectin B chain with cell-surface galactose residues. This triggers endocytosis into a vesicle. By an unknown mechanism, the enzymatic A chain is transported into the cytoplasm, where it attacks ribosomes. As a result of the N-glycosidation of a specific adenine base, the ribosomes are no longer able to utilize incoming aminoacyl tRNAs, and protein synthesis is disrupted.

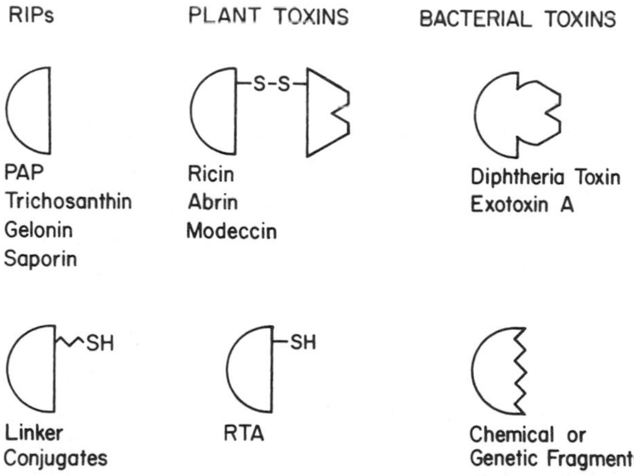

FIGURE 1 Three principal types of enzymatic toxins. Ribosome-inhibiting proteins (RIPs), also called hemitoxins, are generally isolated from higher plants and act as *N*-glycosidases on rRNA. A free thiol can be generated on the molecular surface by chemical modification or genetic engineering. True plant cytotoxins have an A chain similar to that of the RIPs and a lectin B chain. Chemical reduction of the toxin leaves a free-surface thiol on the A-chain enzyme. Bacterial toxins often have an enzymatic domain, shown as a hemisphere, and a cell recognition domain linked as a single-polypeptide chain. This chain can be cleaved into separate components by proteolytic digestion or genetic engineering.

The most thoroughly studied and commonly used plant cytotoxin is ricin, isolated from the seeds of *Ricinus communis,* the castor plant. The enzymatic A subunit and the lectin B subunits of ricin are called RTA and RTB, respectively. RTB is a glycoprotein of 262 amino acid residues. It binds the galactose moieties of cell-surface proteins (or perhaps glycolipids) and is carried into the cell by endocytosis. *In vitro* experiments show that free galactosides in the media can retard ricin binding to cell surfaces by competing with surface-bound galactose. Cellular uptake may involve transfer to coated pits, but this is not required. The exact target or targets for ricin binding are unknown, but as many as 3×10^7 potential binding sites on a cloned human HeLa cell exist. The half-time for transport of ricin through the plasma membrane depends on cell type and conditions but is generally less than 3 hr. Once ricin is endocytosed, it travels through the Golgi and other portions of the secretory apparatus by a number of pathways. The majority of ricin molecules eventually recycle to the cell surface and are released; as little as 1% of the ricin initially endocytosed translocates into the cytoplasm. The mechanism of this translocation is unknown. RTB may aid RTA in its release to the cytoplasm, perhaps by forming some sort of permeable channel through which RTA moves. Alternatively, and probably more likely, RTB may act as a marker/flag for internal routing and sorting and may direct a fraction of the RTA into a relatively permeable vesicle from which escape is likely.

RTA is an *N*-glycosidase consisting of 267 amino acid residues. Initially, it is linked to RTB by a disulfide bond, but this bond is reduced inside the cell, allowing the chains to separate. RTA can efficiently recognize ribosomes at their intracellular concentration (roughly 1–10 μM). Under these conditions, RTA can inactivate ribosomes at a rate of 1500/min. [*See* Ribosomes.]

Intact ricin is able to kill a wide variety of animal cells. A goal of IT design is to channel this toxicity to attack only those cells selected by the targeting antibody. RTA alone is roughly 10^3–10^5 times less toxic than ricin, and for this reason attempts have been made to conjugate only the enzyme chain with antibodies. By the same rationale, single-chain RIPs, such as PAP and gelonin, have been used in IT design to reduce the nonspecific toxicity caused by RTB's binding of cell surfaces.

Using recombinant DNA technology to alter natural toxins to make them more efficacious in cell-specific toxicity has prompted considerable interest.The three-dimensional structure of ricin is known, allowing a molecular description of the way in which RTB binds sugars and the way in which rRNA is recognized by RTA. This model allows functional roles to be assigned to certain amino acids. It is now possible to redesign the toxin and alter its action slightly. Experiments are under way to design toxins that will have reduced binding to normal cells but will be taken up rapidly as IT complexes by targeted cells.

2. Bacterial Toxins

In addition to plant toxins, bacterial toxins have also been used in the construction of ITs. Diphtheria toxin (DT), from *Corynebacterium diphtheriae,* is synthesized as a proenzyme of 535 amino acid residues. It is then cleaved to form an amino-terminal A chain of 193 residues linked to the B chain by a disulfide bond. The A chain is an enzyme that cleaves nicotinamide adenine dinucleotide (NAD) and inserts the adenosine diphosphate (ADP)–ribose moiety into the active site of elongation factor 2 (EF2). This covalent modification of EF2 inhibits protein synthesis. The B chain binds to cell surfaces, probably by recognizing an ion-transport protein, and thereby appears to trigger endocytosis. The B chain also assists in the translocation of the enzyme A chain across the endosomal membrane into the cytoplasm, perhaps by inserting into the membrane to form an escape pore. The transport is facilitated by low pH inside the endosome. Although DT is a potent and well-understood toxin, its utility in the design of ITs is limited because many people are specifically immunized against DT by vaccination during childhood. As a result, conjugates injected into the body are met by a fully developed immune response. A second bacterial toxin, which holds some promise in IT construction, is exotoxin A, isolated from *Pseudomonas aeruginosa.* This 613-residue protein has three domains thought to function in cell-surface recognition, membrane transport, and enzyme action, which, like DT, causes inhibition of EF2 by ADP-ribosylation.

C. Immunotoxin Linkage

As described earlier, antibodies and toxic proteins for ITs have generally been isolated separately and must then be joined to form an active agent. This is commonly accomplished by chemically attaching a reactive group, or "linker," to the antibody and allowing it to react with a toxin to form the derivatized antibody. We saw that RTA and RTB are joined by a reducible disulfide linkage, which allows the chains to separate after endocytosis. This reducible linkage is often mimicked in IT construction by using reagents, which form a disulfide bond between antibody and toxin. The most commonly used linker is *N*-succinimidyl-3-(2-pyridyldithio) propionate (SPDP), shown in Fig. 3. In this case, the reagent is allowed to interact with antibody under conditions designed to attach one or two linkers. Amine groups from lysine side chains act as nucleophiles, displacing the *N*-hydroxy succinimidyl moiety and forming a covalent bond to the linker. This process is random however, and the derivatized antibody population is heterogeneous. The 2-thiopyridine is now displaced in a disulfide exchange reaction by a free thiol on the toxin. If the toxin is isolated RTA, the thiol normally joined to RTB will carry out the displacement and form a cleavable covalent bond to the antibody. If intact ricin is to be used, or some single-chain toxin, then a thiol must be attached to that protein. SPDP could be attached randomly to lysine side chains on the toxin. A reducing agent, such as dithiothreitol, would then

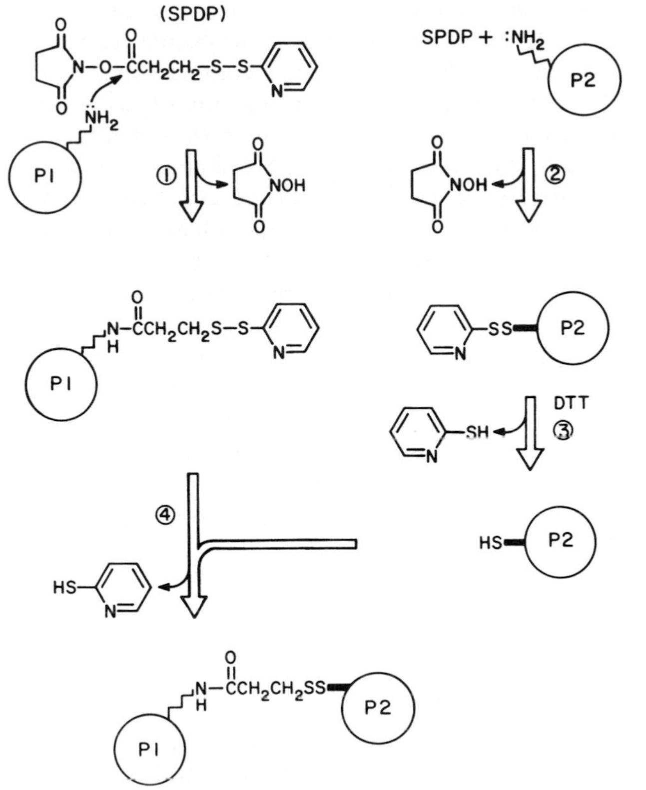

FIGURE 3 Chemical coupling of antibodies to toxins. The most commonly used reagent is *N*-succinimidyl-3-(2-pyridyldithio) propionate (SPDP). In step 1, a lysine side chain on the antibody (P1) attacks SPDP, releasing *N*-hydroxy succinimide. This creates a reactive intermediate, which can exchange with a thiol on the toxin (P2). RTA has free thiol, as shown in Fig. 1, and can react directly at step 4. This exchange reaction releases 2-thiopyridine and links P1 and P2. If the toxin of choice is an RIP, a thiol group must be added by reaction with SPDP (step 2). In step 3, an exchange reaction with dithiothreitol (DTT) creates a free thiol. The heavy bond on P2 is meant to suggest that the thiol-containing linker may substitute for the natural thiol of a cysteine side chain.

be used to remove 2-thiopyridine and create free thiols (Fig. 3).

In addition to these cleavable linkers, a wide range of noncleavable linkages can be synthesized, which irreversibly tether toxin and antibody in the IT construct. In general, isolated RTA and single-chain plant toxins must be linked to antibodies by a reducible bond, whereas whole toxins can be linked by either a reducible or nonreducible linker. In all cases, the exact choice of reagent and regimen of chemical linkage have strong effects on the efficacy of the resulting conjugate. Some reagents may react preferentially with the antigen binding site of the antibody, reducing its cell-surface recognition ability; in other cases, conjugates might form in which the toxin physically blocks the antigen binding site. Alternatively, the reagent might selectively attack a group in the active site of the toxin, reducing its potency. Several agents are generally tested for each antibody–enzyme conjugate to optimize IT action.

Finally, it should be noted that it is now possible to use recombinant DNA technology to design ITs in which the cell recognition protein and the toxin are fused into a single unit. In practice, light- and heavy-chain gene fragments coding for the antigen binding site are linked by genetic engineering methods; this unit is then fused to the gene or gene fragment coding for a toxic protein. Such a construct has several possible advantages: it is homogeneous and well defined, and it eliminates from the construct the antibody Fc fragment as well as the chemical linker, reducing the antigenicity of the IT and increasing its long-term effectiveness. Initial studies on human cell lines using this new technology suggest that in certain cases the recombinant ITs are at least as effective as those produced by chemical linkage.

D. Immunoconjugates

Once synthesized, the IT is purified to remove any free toxin or antibody and to isolate conjugates of a given size. For example, in analyzing IT action, complexes of one antibody and one toxin molecule must be separated from those that might contain two or more toxins to one antibody. Size-exclusion chromatography and antibody or toxin affinity chromatography are used. The purity of the material is assessed by nonreducing sodium dodecyl sulfate polyacrylamide gel electrophoresis followed by dye staining and densitometric scanning. The pure IT should appear as a single molecular species on such a gel.

The derivatization and conjugation may alter the antibody affinity or the toxin enzymatic activity, and this must be assessed. The IT and unmodified antibody are compared in their abilities to block the binding of radiolabeled antibody to target cells. Alternatively, immunoperoxidase staining of tissue sections using ITs and peroxidase-conjugated antitoxin can be used to reveal changes in tissue specificity of the conjugated antibody. Toxin enzymatic potency is measured by testing the inhibition of rabbit reticulocyte protein synthesis (for RIP) or by testing the ADP-ribosylation of EF2 (for bacterial toxins). A large variety of ITs have been prepared with intact antibody and

TABLE I
Immunotoxins Lethal to Human Tumors

Tumor type	Toxins used[a]
Breast adenocarcinoma	RTA
Melanoma	RTA, DTA, PAP
Colorectal carcinoma	RTA, DTA
Hepatoma	Ricin
Ovarian carcinoma	RTA, PE
Graft-versus-host disease	RTA
Prostate carcinoma	RTA
B-lymphoma	RTA, PAP
Myeloma	RTA
Bladder carcinoma	RTA
Acute lymphoblastic leukemia	RTA

[a]Abbreviations used: RTA, ricin A chain; DTA, diphtheria toxin A chain; PAP, pokeweed antiviral protein; PE, *Pseudomonas* exotoxin.

toxin functions and tested for selective cytotoxicity (Table I). [*See* Immunoconjugates.]

III. IMMUNOTOXIN ACTIVITY

A. *In Vitro* Characterization

Immunotoxins incubated separately with cells expressing the target antigen and with cells without the antigen should kill only the antigen-bearing cells. Varying concentrations of ITs are incubated with target cells for 12–36 hr, and the concentration required to kill 50% of the tumor cells (ID_{50}) and control cells is measured. In general, protein synthesis incorporation is measured. This assay measures the potency of the IT and should closely reflect both the antibody affinity and the rate of antigen internalization to intracellular compartments from which the toxin can escape to the cytosol. A number of ITs have shown potent selective cytotoxicity. ID_{50}'s (the dose that kills 50% of the cells) on sensitive cells range in concentration between 1 and 1000 p*M*, with 100- to 10,000-fold higher ID_{50}'s on control antigen negative cells. Immunotoxins made with high-affinity antibodies to cell-surface antigens present in high copy number, which are readily endocytosed, make active cytotoxins. Unfortunately, this assay does not measure better than 90% of cell kill. To accurately compare different active ITs, several tests of efficacy have been developed and appear to yield similar predictions. Saturating concentrations of IT are incubated with sensitive target cells. After prolonged incubation, the reduction in surviving cells, able to form a colony, is measured. Efficacious ITs can reduce the fraction of clonogenic cells to 1 in 1000, or even 1 in 10,000. Kinetics of cell kill can be measured by incubating saturating concentrations of IT with target cells for varying lengths of time and measuring the cell kill, usually by a protein synthesis assay. Cell kill with a survival of 10–30%/hr correlates with more prolonged assays yielding 1 in 1000 or 1 in 10,000 reduction in colony formation. Immunotoxins prepared with antibodies to different antigenic sites on the same cell-surface antigen have shown differing efficacy. Particular antigen–antibody complexes may internalize better, or the geometry of the complex may improve access of the toxin to membranes. When whole toxin and toxin fragments have been linked to the same antibody and compared, usually the whole-toxin conjugates have shown greater efficacy. The augmented killing has been attributed to enhancement functions not present in the toxin fragment conjugates. Certain low-molecular-weight agents, such as ammonium chloride, chloroquine, and carboxylic acid ionophores, have been found to enhance both the potency and efficacy of ITs. The basis for this enhancement is unknown but may be related to weakening of intracellular vesicle membranes or prolonging transit time in certain intracellular vesicle compartments.

B. *In Vivo* Characterization

Immunotoxins have been administered to rodents, rabbits, and primates. After intravenous administration, the IT circulates in the bloodstream with a half-life of minutes to hours. This half-life is significantly shorter than the half-life of unmodified antibody ($t_{1/2}$ of days) and is due to either clearance by the reticuloendothelial system or liver cells or metabolism with reduction of the disulfide bond between the toxin and antibody. Mannose-, fucose-, and xylose-terminated oligosaccharides on plant toxins are bound by specific receptors on Kupffer cells of the liver. The use of deglycosylated ricin A chain in the IT prolongs half-life twofold in mice and reduces isolated liver sinusoidal cell binding by 50%. Unglycosylated toxins, however, are bound by undefined receptors on liver cells. New disulfide linkers have been constructed in which a methyl group and a benzene ring have been attached to the carbon atom adjacent to the disulfide bond to prevent attack by thiolate anions. Abrin A-chain ITs prepared with the hindered disulfide linkers have a 10-fold longer half-life in mice. Penetration of IT into extravascular tissues, including tumors, has not been extensively studied. Intravenous administration of anti-T11–gelonin led to saturation of T cells in the lymph nodes and spleen of rhesus monkeys but required five times as much material/animal as antibody alone to achieve equal extravascular deposition. The large size of the ITs ($M_r \approx 180{,}000$) may limit penetration of capillaries.

Toxicity studies of ITs in rodents (mice and rats) and monkeys (rhesus and cynomologous) suggests toxin- and antibody-specific organ damage as well as species-specific pathology. LD_{50}'s range from 0.05–0.1 mg/kg for whole-toxin conjugates to 5–10 mg/kg for subunit ITs. The higher toxicity of whole-toxin conjugates may be due to persistent, normal tissue binding sites on the toxin. Whereas most toxin conjugates show liver toxicity, *Pseudomonas* exotoxin and saporin conjugates show more dramatic liver damage. Antitransferrin receptor antibody conjugates have much more toxicity than control antibody conjugates; presumably this is due to widespread presence of transferrin receptor in many normal cells.

Antitumor studies in mice have shown that local exposure of tumor cells to IT by regional therapies appears more efficacious than systemic treatment. The ineffectiveness of systemic therapy may be due to clearance or metabolism of the IT, or difficulty in capillary penetration by the large molecules. Tumor burden is important, because delaying the time between tumor implantation and the start of IT therapy decreases the chance of successful therapy. Greater tumor burden at the time of therapy may be associated with the presence of more resistant clones, and larger tumors may have more altered vascular beds, preventing adequate penetration of IT. Finally, several studies suggest that different antibody–toxin conjugates directed at the same tumor model can yield varying results. Many of the foregoing pharmacologic factors may be responsible, or additional factors unique to tumor cells *in vivo* may play a role.

C. Clinical Applications

The first successful application of ITs has been the prevention and palliation of graft-versus-host disease in patients undergoing allogeneic bone marrow transplantation. Patients are treated with chemoradiotherapy to destroy malignant or abnormal hematopoietic stem cells. The treatment also destroys their normal stem cells so that a bone marrow graft is required from an identical twin or donor with a matched human leukocyte-associated antigen. Unfortunately, although the donor marrow often survives in the recipient, the donor T cells recognize the recipient's tissues as foreign and attack the patient's tissues, producing graft-versus-host disease. The liver, gastrointestinal tract, immune system, and skin show the most damage. By depleting the donor graft of T cells using an anti-T-cell IT, the patient suffers fewer incidences of graft-versus-host disease. Recently, patients with active graft-versus-host disease have been treated systemically with anti-T-cell IT, and the severity and duration of the disease have been reduced. Some of the activated donor T cells in the patient are killed by the IT. Immunotoxin therapy may lead to successful human leukocyte antigen-mismatched allogeneic bone marrow transplantation. [*See* Bone Marrow Transplantation.]

The same anti-CD5–ricin A-chain IT has been used for the systemic treatment of chronic lymphocytic leukemia. No clinical responses were seen. The IT was much less cytotoxic to these chronic leukemia cells than to activated T cells. Clinical trials have begun using this IT for treatment of cutaneous T-cell lymphoma. Another, with an antibody to the IL-2 receptor conjugated to pseudomonas exotoxin, has been used on four patients with T-acute lymphoblastic leukemia (which displays IL-2 receptors). Again, no clinical benefit was observed, but does were low due to liver toxicity, and the antigen density was low on patient cells. A trial has recently begun using anti-CD19-conjugated whole ricin for B-cell lymphoma and leukemia. Solid tumors that have been treated with ITs include metastatic melanoma, colorectal carcinoma, breast carcinoma, and ovarian carcinoma, but very few clinical responses have been seen. In each case, tumor penetration by IT has been difficult to document, and immune responses to the ITs have been observed. Toxicities have included fluid retention, edema, and hypoalbuminemia in melanoma and breast cancer trials and neurologic toxicities including diffuse encephalopathy in colorectal and ovarian cancer trials and peripheral neuropathy in the breast cancer trial. Reversible renal toxicity was seen in the colorectal cancer trial. A "capillary leak" syndrome was observed in all RTA conjugate trials but in no other conjugate trials. The neurotoxicities are unexplained, except in the breast cancer trial, where antibody binding to Schwann cells was demonstrated. [*See* Leukemia; Lymphoma.]

Future trials in leukemia and solid tumors will have to address the poor penetration and strong humoral immune response to IT. Current research has focused on the use of genetically engineered ITs, which should be smaller and yet retain the enhancing functions of the whole-toxin conjugates. Several of these recombinant molecules will soon begin clinical study.

BIBLIOGRAPHY

Blakey, D. C., Wawrzynczak, E. J., Wallace, P. M., and Thorpe, P. E. (1988). Antibody toxin conjugates: A perspective. *In* "Monoclonal Antibody Therapy. Prog. Allergy" (H. Waldmann, ed.), pp. 50–90. S. Karger, Basel.

Frankel, A. E. (ed.) (1988). "Immunotoxins." Kluwer Academic Publishers, Boston.

Lord, J. M., Spooner, R. A., and Roberts, L. M. (1989). Immunotoxins: Monoclonal antibody–toxin conjugates—A new approach to cancer therapy. *In* "Monoclonal Antibodies: Production and Application," pp. 193–211. Alan R. Liss, New York.

Pastan, I., Willingham, M. C., and FitzGerald, D. J. P. (1986). Immunotoxins. *Cell* **47,** 641–648.

Vitetta, E. S., Fullton, R. J., May, R. D., Till, M., and Uhr, J. W. (1987). Redesigning nature's poisons to create anti-tumor reagents. *Science* **238,** 1098.

Vitetta, E. S., Thorpe, P. E., and Uhr, J. W. (1993). Immunotoxins: Magic bullets or misguided missiles? *Immunol. Today* **14,** 252–259.

Implantation, Embryology

ALLEN C. ENDERS
University of California, Davis, School of Medicine

GLOSSARY

Blastocyst Hollow spherical structure consisting of a shell of cells, the trophoblast (i.e., trophectoderm), and an inner cell mass located at one pole

Conceptus All of the derivatives of the blastocyst after implantation, including the embryo and the extraembryonic membranes

Endoderm Layer of cells that develops in the inner surface of the inner cell mass of the blastocyst

Endometrium Mucosal lining of the uterus, consisting of connective tissue stroma, uterine surface epithelium, and tubular glands

Eutherian mammal "True" placental mammals (i.e., mammals with efficient nutritive placentas); includes all mammals except monotremes and marsupials

Exocoelom Extraembryonic coelom; the cavity within the extraembryonic mesothelium

Inner cell mass Portion of the blastocyst giving rise to the embryo, as well as contributing to the extraembryonic membranes

Lacunae Spaces within the syncytial trophoblast, which fill with maternal blood and become intervillous spaces of the placenta

Placenta Specialized portion of the extraembryonic membranes and its associated maternal constituents, which form the major exchange area between the fetus and the maternal organism during most of pregnancy

Trophoblast Initially, the outer layer of the blastocyst; later, the outer (i.e., ectodermal) layer of the placenta

IMPLANTATION IN ANY EUTHERIAN MAMMAL begins with the assumption of a relatively fixed position of the blastocyst within the uterus and initiates a stage during which the trophoblast of the blastocyst and the uterine endometrium become progressively more intimate. When we consider that the blastocyst is initially free to move about within the uterine cavity and that the placenta formed after implantation of the blastocyst not only is firmly embedded in the endometrium, but also has its surface, the trophoblast, bathed directly in maternal blood, a sequence of necessary events of implantation can be deduced. Implantation thus includes adhesion of the blastocyst to the uterine surface epithelium, penetration through this epithelium and its basement membrane, expansion into the underlying connective tissue stroma, tapping of the maternal blood vessels within the stroma, and the formation of blood-filled spaces within the enlarging trophoblast. These spaces are then formed around the entire circumference of the blastocyst.

I. SEQUENCE OF EVENTS IN IMPLANTATION

A. Time Sequence of Human Implantation

Much of the basic information concering implantation in humans has been derived from carefully preserved specimens collected around the world, especially from the meticulously prepared specimens in the Carnegie collections. The classification of the stages of human development has recently been refined, placing

TABLE I
Human Implantation Stages

Carnegie stage	Age (days)	Developmental event	Changing condition
3a	5	Blastocyst matures	
3b	5	Blastocyst loses zona pellucida	Develops the syncytial trophoblast that adheres to the uterine surface
4a	—[a]	Blastocyst attaches	Trophoblast intrudes between the uterine epithelial cells
4b	—[a]	Epithelial penetration	Trophoblast spreads along the basal lamina
5a	8–9	Trophoblastic plate formation	Trophoblast penetrates the basal lamina, invades the stroma, and taps the maternal vessels
5b	9–11	Lacuna formation	Trophoblast differentiation forms microvillus-lined clefts and lacunae
5c	12–13	Primary villus formation	Focal cytotrophoblast proliferations extend from the chorionic plate into the syncytium
6	13–15	Secondary placental villi, secondary yolk sac formation	Mesenchymal cells proliferate on the fetal side of the primary villi
7	16–18	Branching and anchoring villi formation	Cytotrophoblast extends through the syncytial trophoblast and spreads along the basal plate
8–9	18–22	Tertiary villi formation	Vessels differentiate *in situ* and are filled with blood when the heart beat links the yolk sac, the embryo, and the chorioallantoic placenta

[a]No adequate specimens were reported.

the free blastocyst in stage 3, initial attachment of the blastocyst to the uterus in stage 4, and most of the events of implantation in stage 5 (Table I). Stage 4, the attaching blastocyst, has never been described in humans, but it has been deduced that attachment of the blastocyst probably occurs 6 days after ovulation. The events of stage 5 take place 7–14 days after ovulation, when placental villus formation has begun (i.e., the start of stage 6). By the end of the third postovulatory week, numerous branched placental villi surround the circumference of the conceptus, which is now called a chorionic vesicle. At approximately Day 21 postovulation, the heart within the developing embryo begins to beat and red blood cells appear in the placental villi. The intra- and extraembryonic circulatory system, however, is not well developed for several more days.

B. Transition Stage from Free to Attached Blastocyst

A few human blastocyst stages have been examined by light microscopy, and blastocysts resulting from *in vitro* fertilization have been examined by electron microscopy, but neither the stage of adhesion of the blastocyst to the uterine surface nor the stage of penetration of the uterine epithelium by the implanting blastocyst has been described in humans. Consequently, the nature of these crucial first events of implantation must be either extrapolated from other known events in humans or deduced from information gathered from other primates. Fortunately, both light- and electron-microscopic studies of preimplantation and early implantation stages of the rhesus monkey and the baboon are available to help interpret aspects of orientation and adhesion of the blastocyst and epithelial penetration stages. [*See* Electron Microscopy.]

The extracellular layer (i.e., the zona pellucida) that surrounds the blastocyst is shed from the blastocyst 1–2 days prior to implantation, permitting direct apposition of the trophoblast to the uterine surface. Although the blastocyst is "free" in the uterine cavity prior to implantation, there is little fluid within the cavity, and the blastocyst is probably in contact with the surface epithelium. After loss of the zona pellucida, the trophoblast cells near the inner cell mass of the blastocyst fuse to form the syncytial trophoblast and develop surface projections. It is only in this periembryonic region that the blastocyst adheres to the uterine surface.

C. Epithelial Penetration Stage

Following adhesion of the blastocyst to the uterine surface epithelium (i.e., the luminal epithelium), processes of the syncytial trophoblast intrude between cells of the epithelium, to the level of its basement membrane, which separates the epithelium from the underlying stroma (Fig. 1). The trophoblast adheres to lateral surfaces of apparently healthy uterine epithelial cells, especially in the region of the junctional complexes, which connect the epithelial cells to each other. Consequently, trophoblast keeps the epithelium sealed from the lumen at the same time that it is invading. Some of the uterine epithelial cells are surrounded by the trophoblast as it grows in the plane of the uterine epithelium and extends along the epithelial surface of the basement membrane and into adjacent glands. Although epithelial cells are phagocytized by the trophoblast, there is little evidence of necrosis of the uterine epithelium until after phagocytosis of the cells has been completed.

When the trophoblast has expanded on the uterine surface to form a broad region of attachment, the implantation site is said to be in the trophoblastic plate stage. During this stage, which in the rhesus monkey is reached 1 day after implantation, the trophoblast penetrates the basement membrane into the underlying stroma. The earliest human implantation sites that have been examined (stage 5a) are in a late trophoblastic plate stage.

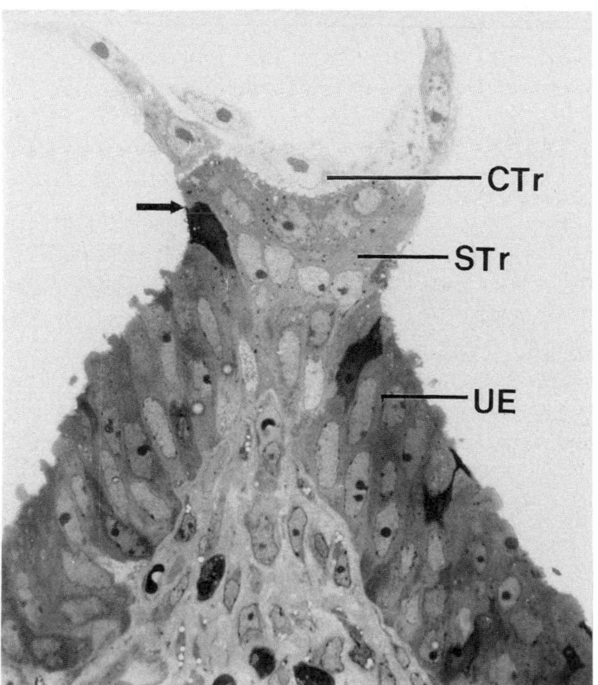

FIGURE 1 High-magnification light micrograph of a region where the trophoblast from a rhesus monkey blastocyst (above) is penetrating the uterine epithelium (below). A mass of syncytial trophoblast (STr) forms a wedge of cytoplasm extending between the uterine epithelial cells (UE). The arrow indicates the junction between the trophoblast and a darker uterine epithelial cell. This region of attachment was just to one side of the inner cell mass (not shown). CTr, cytotrophoblast. ×600.

D. Human Previllous Stages

At early (i.e., stage 5a) human implantation sites the trophoblast near the inner cell mass has expanded into masses of cellular trophoblast (i.e., cytotrophoblast) and syncytial trophoblast (i.e., multinucleate trophoblast), whereas the rest of the trophoblast toward the uterine cavity remains as a simple epithelium of flattened cells (Fig. 2). Within the trophoblast many of the nuclei in both cytotrophoblast and the syncytium are large and are thought to be polyploid. At the same time the trophoblast adjacent to the endometrium of the uterus and that following the basement membrane into glands contain nuclei that are much smaller. The tissue of origin of these small nuclei is not yet clear: it may be trophoblast, or trophoblast fused with uterine epithelium. At early human implantation sites the inner cell mass is oriented toward the endometrial surface, indicating that the human blastocyst, like that of other primates, initially attaches through the trophoblast overlying or adjacent to the inner cell mass.

Since no studies of early human implantation stages have been performed using electron microscopy, it is not clear whether the trophoblast initially pauses at the basement membrane of the uterine surface epithelium, as it does in the rhesus monkey, nor is it clear how the trophoblast penetrates through this layer and into the uterine vessels. At the earliest implantation sites (i.e., stage 5a) the maternal vessels seem to be partially surrounded by the trophoblast (Fig. 3), but later (i.e., stages 5b and 5c) trophoblast shares part of the walls of the vessels, and maternal blood flows into spaces within the trophoblast, called lacunae (Figs. 4 and 5).

A single human implantation site in late stage 5c has been examined by electron microscopy. In this specimen the trophoblast intruding into the maternal vessels is adherent to the endothelial cells lining the vessels and shares with it endothelial cell junctional complexes.

While the trophoblast expands into the endometrium of the uterus, more areas of the wall of the blastocyst become differentiated into the syncytial trophoblast and cytotrophoblast masses. By stage 5c, almost all of the wall of the blastocyst has been converted from flattened cells into syncytial trophoblast and cytotrophoblast, and the expanding conceptus is beneath the surface of the endometrium.

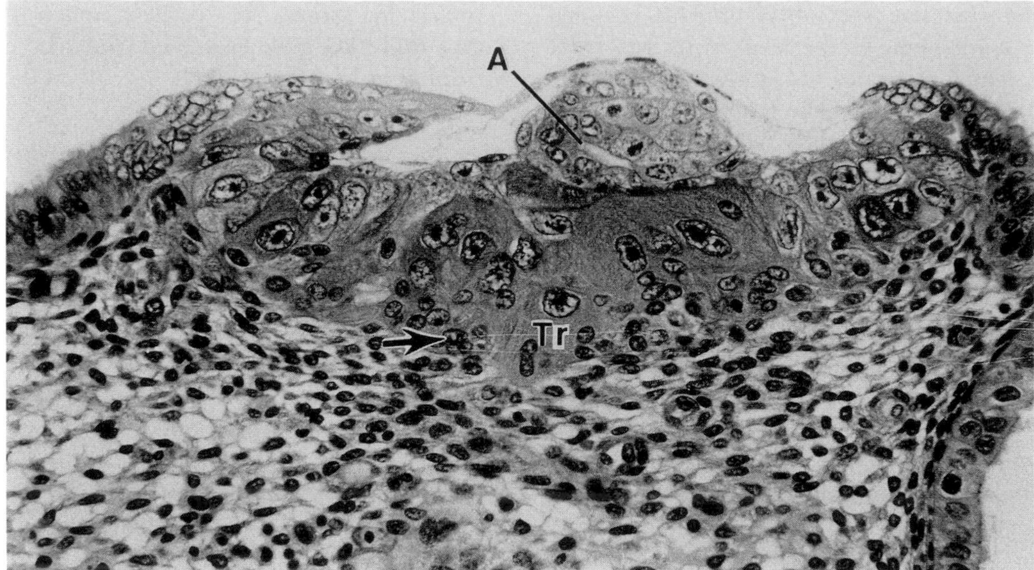

FIGURE 2 Stage 5a of human implantation. The inner cell mass already has a small amniotic cavity (A). The broad band of trophoblast (Tr) invading the endometrium has formed a trophoblastic plate, consisting of a mixture of pale cytotrophoblast cells and darker syncytial trophoblast, some of which contains large polyploid nuclei. Note that small nuclei (arrow) are present at the endometrial border of the trophoblast. Carnegie specimen No. 8020. ×350.

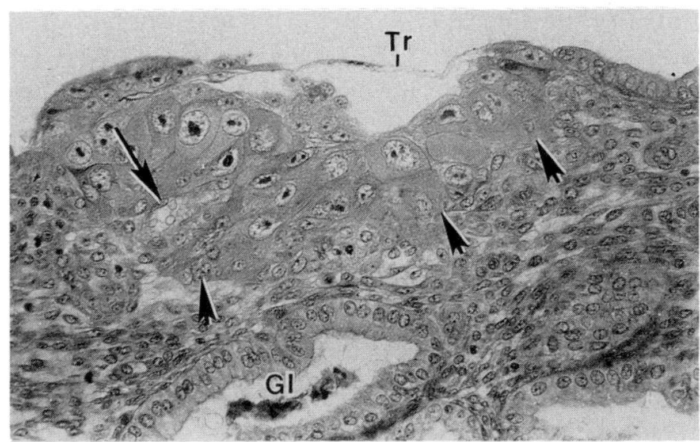

FIGURE 3 Margin of a human trophoblast plate stage (stage 5a). The arrow points to a nucleus in the wall of a maternal blood vessel that is largely surrounded by the trophoblast. Only a few cells of the inner cell mass are present, and the thin trophoblast (Tr) from the side of the blastocyst away from the implantation site has not yet begun to form the cytotrophoblast and the syncytial trophoblast. Arrowheads indicate the border of the trophoblastic plate with the endometrium. Gl, uterine gland. Carnegie specimen No. 8225. ×210.

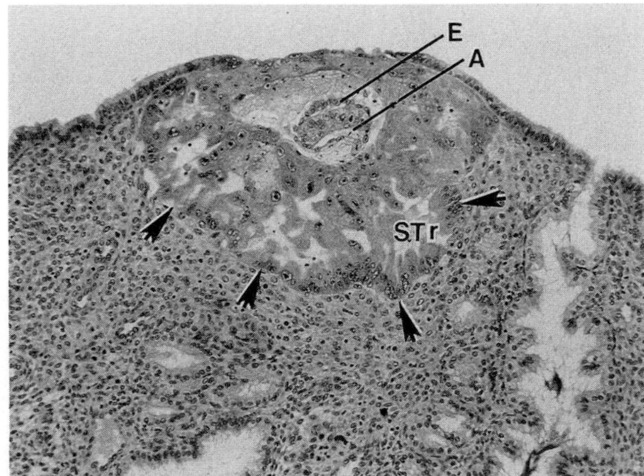

FIGURE 4 At stage 5b the endometrium has begun to undergo decidualization, which is seen as a richly cellular appearance of the endometrial stroma in this low-magnification micrograph. Many lacunae have developed within the syncytial trophoblast (STr). The endodermal layer (E) and the amnion (A) can be seen on either side of the embryonic disk. Arrowheads indicate the border between the trophoblast and the endometrium. Carnegie specimen No. 8004. ×108.

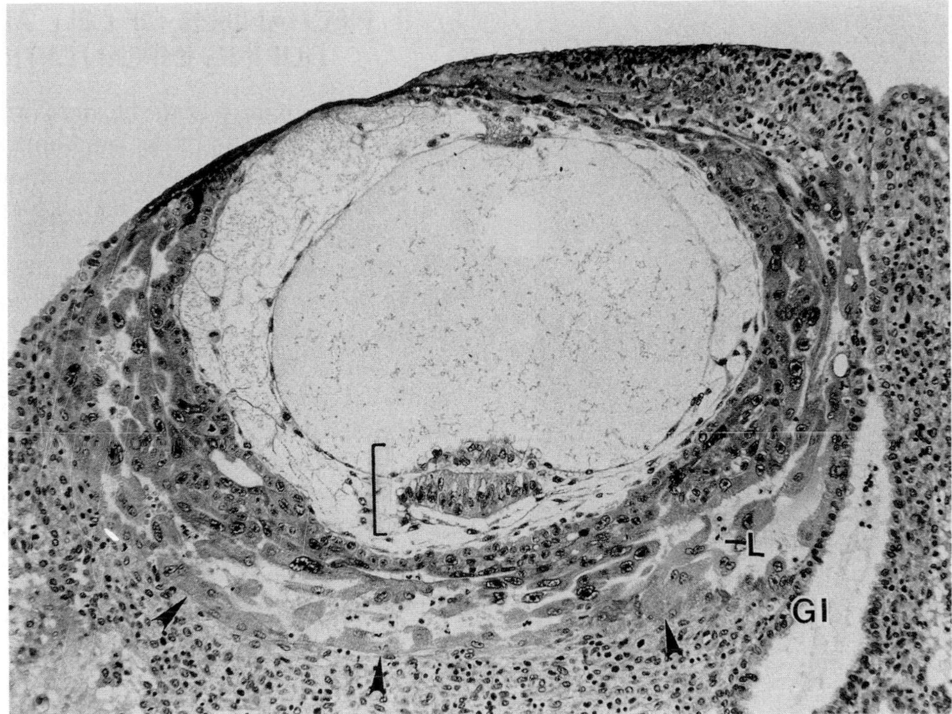

FIGURE 5 By stage 5c the regions of the cytotrophoblast and the syncytial trophoblast have spread around most of the circumference of the blastocyst. The border between the endometrium and the trophoblast is indicated by arrowheads. Note that there are now blood cells in the lacunae (L) within the trophoblast. The forming embryo (bracket) is in a disk stage. A delicate reticulum, the extraembryonic mesoderm, surrounds the embryonic disk and its amnion. Gl, gland. Carnegie specimen No. 7700. ×150.

E. Trophoblast Differentiation

Shortly after trophoblast penetration of the maternal vessels in stage 5c, it can be seen that much of the trophoblast has changed its characteristics. Rather than the syncytium and the cytotrophoblast being intermixed, most of the cytotrophoblast cells now line the inner surface of the implantation site, facing the blastocyst cavity. Furthermore, the number of polyploid nuclei is reduced, and most of the nuclei within the syncytium are uniform in size and tend to be distributed singly along the syncytial trophoblast lining the lacunae. Examination of the implantation sites in rhesus monkeys and baboons in the same period of development shows conversion of most of the syncytial trophoblast into a new type of syncytium that forms a thin polarized layer, with numerous microvilli lining the forming lacunae. This syncytium has a greatly increased surface area, does not cause clotting of the maternal blood, and is apparently less invasive than the previous multinucleate syncytial mass. The syncytial trophoblast, in its position lining lacunae filled with maternal blood, constitutes a layer without intercellular spaces, but with a large surface area that facilitates the transport of substances to the conceptus.

Since implantation is thought to begin approximately 7–8 days after fertilization, and lacunae are well formed by 11–12 days, it is clear that the stage in which the trophoblast is highly invasive lasts for only a few days. At the end of this time, the conceptus is only about 1 mm in diameter. It is farther away from the muscle layers of the uterus than it was at the start of implantation owing to the increased thickness of the endometrium and the superficial position of the conceptus (Fig. 6).

F. Site of Implantation

Since the most common site of the placenta in successful pregnancies is the dorsal wall of the uterus (with the ventral wall second), it can be deduced that the blastocyst normally implants toward the midline of the dorsal or ventral wall. This is the position in which uterine muscular contraction would be expected to place the blastocyst. The abundance of other sites,

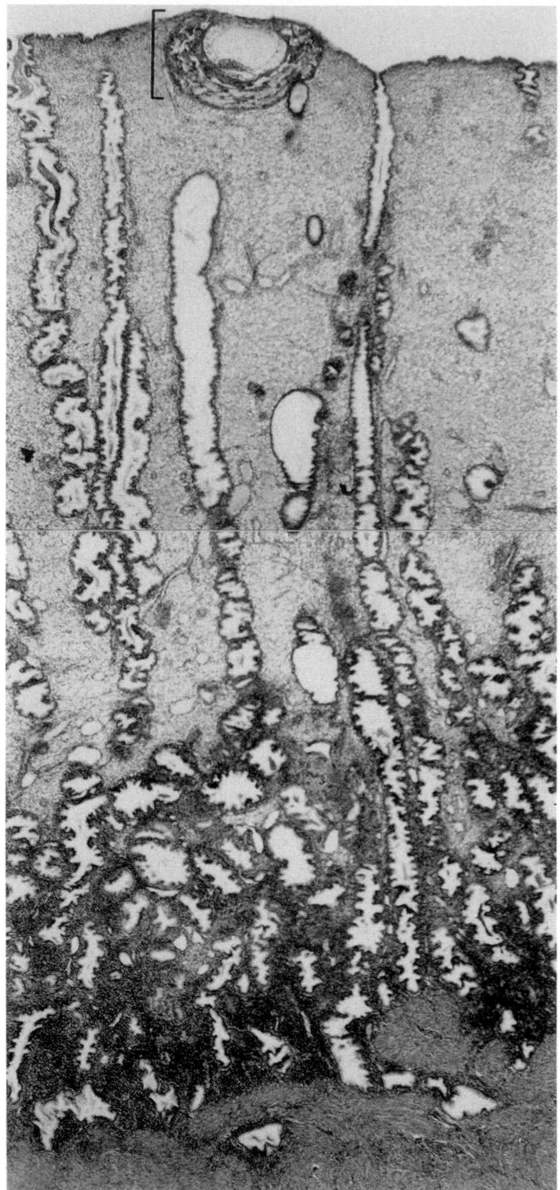

FIGURE 6 At a low magnification a conceptus from stage 5c is seen (bracket) near the surface of the endometrium. The coiled irregular uterine glands extend from the surface to the edge of the smooth muscle of the uterus (at the bottom of the micrograph). Note the small size and superficial position of the conceptus in relation to the total thickness of the endometrium. Carnegie specimen No. 7950. ×30.

including ectopic sites, indicates that the blastocyst attempts to implant at the place it occupies when development has advanced appropriately to the implantation stage.

II. MECHANISMS OF CELL ADHESION DURING IMPLANTATION

Although the nature of the changes in the cell surface that permit adhesion of the trophoblast to the uterus are of great interest, studies with laboratory animals have yet to determine the exact mechanisms by which this adhesion occurs in any species. It has been shown that there are at least three methods of penetration of the uterine surface epithelium by the blastocyst in mammals: fusion of the trophoblast with uterine cells, intrusion of the trophoblast cell processes between epithelial cells, and cell death (i.e., apoptosis) of uterine epithelial cells, followed by phagocytosis of these cells by the trophoblast. Such critical aspects as to how the uterine basement membrane is breached are not fully understood for any species. Studies of these crucial events in implantation in nonhuman primates and in laboratory mammals could lead not only to increased understanding of the events of implantation, but also to the ability to control these events.

III. DEVELOPMENT OF THE EMBRYO DURING IMPLANTATION

At the time that the blastocyst first attaches to the uterus, the inner cell mass has differentiated into a layer of flattened cells adjacent to the cavity of the blastocyst, the primitive endoderm (i.e., the hypoblast), and a compact group of cells between the endoderm and the trophoblast, the ectoderm (i.e., the epiblast). In the youngest human implantation stage described, there is already a small cavity within the epiblast that constitutes the presumptive amniotic cavity (Fig. 2). The cells enclosing this space on the side toward the trophoblast are thus early amniotic epithelial cells, and the remainder of the epiblast gives rise to the embryonic disk. In rhesus monkeys and baboons it has been shown that formation of the amniotic cavity begins with a stage in which the cells of the epiblast develop apical junctions toward the center of the inner cell mass; the potential space thus formed becomes the amniotic cavity. It is consequently clear that in humans and similar primates the amnion forms by cavitation within the epiblast. A peculiarity of these species is that, during growth of the amniotic cavity, amniotic epithelial cells are adjacent to the trophoblast for a considerable period, and as the amniotic cavity enlarges, a diverticulum of the amnion remains adherent to the trophoblast.

The primitive endoderm becomes irregular near the embryonic disk and also extends beyond it as a single layer, the parietal endoderm, around the cavity of the implanting blastocyst. The cells under the embryonic disk reorganize to form an irregular columnar layer, the visceral endoderm, and a group of thin cells separate from the rest of the endoderm to form a small secondary yolk sac adjacent to the embryonic disk (Fig. 7).

The fate of the parietal endoderm is particularly interesting and was, until recently, controversial. It has been shown that, in the rhesus monkey, the cells of this layer develop the cytological characteristics of mesenchymal cells and begin to form extracellular matrix constituents. Thus, parietal endoderm can differentiate into extraembryonic mesoderm and contributes the first cells to this layer and thus to the exocoelom. Since human endoderm forms a more extensive meshwork of cells (i.e., the endodermal reticulum) in the position of the forming extraembryonic mesoderm than does that of the rhesus monkey (Fig. 4), it is probable that parietal endoderm is also the source of the first extraembryonic mesoderm in humans. Recently, this peculiar source of extraembryonic mesoderm has helped explain some of the divergent results obtained with chorionic villus sampling. [*See* Extracellular Matrix.]

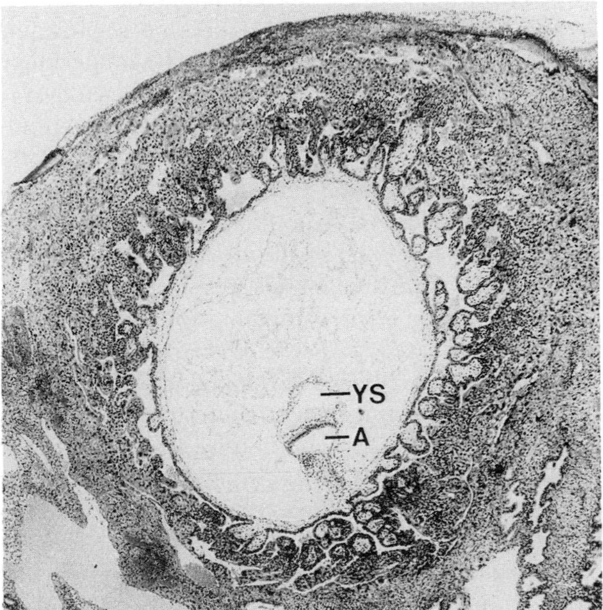

FIGURE 7 At stage 6 placental villi surround the conceptus, forming the chorionic vesicle. A secondary yolk sac (YS) has been formed, and the amnion (A) is beginning to enlarge. Although not apparent at this magnification, the embryo is in the primitive streak stage of development. Carnegie specimen No. 7802. ×32.

The primitive streak initially develops as an aggregation of cells at one end of the embryonic disk, thus establishing the caudal end of the disk. The primitive streak provides a source of cells that form the intraembryonic mesoderm and contribute to the extraembryonic mesoderm; the latter extends to the preexisting extraembryonic mesodermal layer by way of the forming embryonic stalk. The initial thickening of the embryonic disk is subtle and may be difficult to discern, but a clearly defined primitive streak and groove are present by stage 6. Whether or not the intraembryonic endoderm is also derived from the primitive streak in primates has not been established.

IV. PREPARATION OF ENDOMETRIUM FOR THE IMPLANTING BLASTOCYST

Under normal circumstances the blastocyst loses its zona pellucida prior to implantation, at a time when the lining of the uterus, the endometrium, is in an early progestational stage. That is, the endometrium shows the effects of progesterone deriving from the corpus luteum, including extensive glandular development, with coiling of the endometrial glands and good vascularity. The stroma of the endometrium tends to show fluid retention, or edema. The enlargement of the stromal fibroblasts that indicates the beginning of the decidual response does not begin until stage 5b or 5c.

As the implantation site grows, the maternal blood vessels enlarge and some of the uterine glands are blocked and become cystic; thus, at stage 6 there is often leakage of blood into the glands and into the uterine lumen (which can produce the "spotting" that is occasionally confused with menstruation). In the normal menstrual cycle, progesterone levels begin to diminish by 11–12 days after ovulation, as the corpus luteum declines. If implantation occurs, the corpus luteum continues to function under the stimulation of human chorionic gonadotropin (hCG) produced by the trophoblast. This hormone substitutes for the pituitary gonadotropins in maintaining the corpus luteum. Elevated levels of the hormone can normally be detected by 12 days after fertilization. Although undoubtedly some hCG is produced earlier, its detection is not practical before this time, and even the most sensitive methods cannot detect hCG consistently before Day 8.

Other factors have been reported to be present early in pregnancy. One of these, the early pregnancy factor, could provide evidence of the presence of a viable embryo prior to implantation. Use of these factors to detect early pregnancy is not yet practical at the clinical level.

V. IMPLANTATION RATES AND FECUNDITY

Fecundity, the chances of producing a full-term infant in a given menstrual cycle during which intercourse occurs, is surprisingly low in women. Estimates based on newlyweds in France at a time when the government was giving a bonus to increase the population and on newlywed Hutterites in the United States suggest a rate varying from 23 to 28%. This low rate includes failures of ovulation, fertilization, cleavage, implantation, and subsequent fetal development.

With the advent of *in vitro* fertilization, it has been possible to introduce into the uterus ova in which successful fertilization has already been judged to have occurred. Despite the wide variety of methods that have been developed, the success for a single cycle of *in vitro* fertilization and reintroduction remains less than that of normal occurrence just noted. However, repeated cycles have resulted in a considerably higher rate of live birth, especially in women who had previously conceived using *in vitro* fertilization. Although study of the cleavage stages indicates that some failures occur early, it is thought that much of the failure occurs during implantation. Preimplantation stages show high levels of chromosomal imbalance, but studies of normal *in vivo* pregnancies have shown that only a few, such as trisomy 21, which produces Down syndrome, are viable. [*See* Down Syndrome, Molecular Genetics.]

One of the problems of determining the extent of pregnancy failure during implantation is the inability to determine how many blastocysts begin to implant. Studies of hormonal levels on Days 12–18 after intrauterine introduction of an *in vitro*-fertilized cleavage stage suggest that a number of pregnancies are initiated, indicated by elevated hCG on Day 12, but soon fail, with menstrual flow occurring no later than Day 18. [*See* In Vitro Fertilization.]

Whereas chromosome counts and chromosome banding techniques have revealed that a number of early abortions involve genetic abnormalities, a great many abortions are missed and never provide material for genetic studies. Although embryo transfers with transient elevated hCG can be used to indicate the extent of implantation failure, no material is available from such failures that could be used to determine the cause of implantation failure. Consequently, the extent to which the high failure rate during implantation is due to inappropriate endometrial conditions, poorly or improperly developed blastocytes, or genetic factors influencing the blastocysts has yet to be determined. [*See* Abortion, Spontaneous.]

VI. MODELS OF IMPLANTATION *IN VITRO*

A number of attempts have been made during *in vitro* fertilization to use fertilized ova to study aspects of implantation in an *in vitro* situation. In one study four human blastocysts were placed on a layer of flattened endometrial cells grown *in vitro*. Three of the four showed some outgrowth of trophoblast cells, which contacted the culture dish beneath the endometrial cells. However, most of the inner cell mass cells were lost, and the flattened cell layer was dissimilar from the normal lining of the human uterus. Provided an appropriate polarized and endocrine-sensitive endometrial layer can be established, such a model might be used to study adhesion of the human blastocyst to the epithelium and penetration between junctional complexes.

Another experimental procedure involved the use of blastocysts derived from *in vitro* fertilization and donor uteri maintained by the infusion of uterine vasculature *in vitro*. Although this study reported success in establishing an *in vitro* implantation, the implantation site illustrated was several days in advance of any stage that could have developed in the short time (just over 2 days) for which the uterus was perfused, indicating that the implantation had occurred in the donor uterus prior to surgical removal. This conclusion is reinforced by the observation that, unlike studies involving the ectopic placing of mammalian blastocysts, the implantation site was well defined and well organized.

BIBLIOGRAPHY

Bianchi, D. W., Wilkins-Haug, L. E., Enders, A. C., and Hay, E. D. (1993). The origin of extraembryonic mesoderm in experimental animals: Relevance to chorionic mosaicism in humans. *Am. J. Med. Gen.* **46,** 542–550.

Enders, A. C. (1993). Overview of the morphology of implantation in primates. *In* "*In Vitro* Fertilization and Embryo Transfer in Primates" (D. P. Wolf, R. L. Stouffer, and R. M. Brenner, eds.), pp. 145–157. Springer-Verlag, New York.

Enders, A. C., and King, B. F. (1988). Formation and differentiation of extraembryonic mesoderm in the rhesus monkey. *Am. J. Anat.* **181,** 327–340.

Enders, A. C., and King, B. F. (1993). Development of the human yolk sac. *In* "The Human Yolk Sac and Yolk Sac Tumors" (F. F. Nogales, ed.), pp. 33–47. Springer-Verlag, Berlin.

Enders, A. C., Schlafke, S., and Hendrickx, A. G. (1986). Differentiation of embryonic disc, amnion and yolk sac in the rhesus monkey. *Am. J. Anat.* **177,** 161–185.

Hertig, A. T., Rock, J., and Adams, E. C. (1956). A description of 34 human ova within the first 17 days of development. *Am. J. Anat.* **98,** 435–494.

Lenton, E. A., and Woodward, A. J. (1988). The endocrinology of conception and implantation in women. *J. Reprod. Fertil., Suppl.* **36,** 1–15.

Lindenberg, S., Pedersen, B., and Sci, B. (1993). Comparative studies on human and animal implantation models. *In* "Implantation in Mammals" (L. Gianaroli, A. Campana, and A. O. Trounson, eds.), pp. 113–122. Raven, New York.

O'Rahilly, R., and Müller, F. (1987). Developmental stages in human embryos. *Carnegie Inst. Washington Publ.* **637.**

Tan, S. L., Doyle, P., Maconochie, N., Edwards, R. G., Balen, A., Bekir, J., Brinsden, P., and Campbell, S. (1994). Pregnancy and birth rates of live infants after *in vitro* fertilization in women with and without previous *in vitro* fertilization pregnancies. *Am. J. Obstet. Gynecol.* **170,** 34–40.

Van Blerkom, J. (1994). Intrinsic factors affecting the outcome of laboratory-assisted conception in the human. *In* "The Biological Basis of Early Human Reproductive Failure" (J. Van Blerkom, ed.), pp. 3–27. Oxford Univ. Press, New York.

ISBN 0-12-226974-8